Microbiology

Microbiology

Microbiology

second edition

Daniel Lim

University of South Florida

WCB
McGraw-Hill

Boston, Massachusetts   Burr Ridge, Illinois   Dubuque, Iowa
Madison, Wisconsin   New York, New York   San Francisco, California   St. Louis, Missouri

WCB/McGraw-Hill

A Division of The McGraw·Hill Companies

MICROBIOLOGY

1 2 3 4 5 6 7 8 9 0 QPD/QPD 3 2 7 0 9 8 7 6

ISBN 0-697-26186-7

Project Team
Editor *Elizabeth Sievers*
Developmental Editor *Terrance Stanton*
Production Editor *Carla D. Kipper*
Marketing Manager *Patrick E. Reidy*
Designer *Kaye Farmer*
Art Editor *Jodi K. Banowetz*
Photo Editor *Lori Hancock*
Permissions Editor *Karen L. Storlie*

Typeface *10/12 Minion*
Printer *Quebecor Printing*
Compositor *Publication Services, Inc.*

Copyedited by *Julie Wilde*

Illustrations by Art and Sciences, Inc.

Opener background photo: Courtesy of H. Farzadegan and I.L. Roth, University of Georgia

Cover and interior design by Jeff Storm

Freelance Permissions Editor *Karen Dorman*

The credits section for this book begins on page 685 and is considered an extension of the copyright page.

Cover: Background (microscope): Northwind Picture Archives; Top left (E. Coli): © Biophoto Associates/Photo Researchers, Inc.; top right (Aspergillus): © David Phillips/Visuals Unlimited; bottom left (Adenovirus): © Biophoto Associates/Photo Researchers, Inc.; bottom right (Dermatophyte Fungus): © David Scharf/Peter Arnold, Inc.

Library of Congress Catalog Card Number: 96-83887

http://www.mhcollege.com

To my parents, Don and Lucy Lim, for their guidance and support;
my wife, Carol, for her patience, understanding, and love;
and my students, who inspired me to write this book.

BRIEF CONTENTS

part one
Introduction to Microbiology

[1] Foundations of Modern Microbiology 1

[2] Cell Morphology and Microscopy 17

[3] Composition and Structure of Procaryotic and Eucaryotic Cells 41

part two
Microbial Growth and Metabolism

[4] Nutrition and Environmental Influence 79

[5] Growth and Control of Growth 105

[6] Energy and Metabolism 139

[7] Photosynthesis and Other Metabolic Pathways 177

part three
Microbial Genetics

[8] Molecular Genetics 211

[9] Microbial Genetics 245

[10] Recombinant DNA Technology 271

part four
Survey of Microorganisms

[11] Procaryotes: The Bacteria and the Archaea 305

[12] The Eucaryotic Microorganisms 349

[13] The Viruses 383

part five
Microbial Interactions with Other Organisms

[14] Symbiosis 417

[15] Host-Parasite Relationships 439

[16] Immunology 467

[17] Infectious Diseases 497

[18] Public Health and Epidemiology 549

part six
Environmental and Economic Impact of Microorganisms

[19] Microbial Ecology 569

[20] Food and Industrial Microbiology 607

CONTENTS

Preface xix
Learning Aids xxi

part one
Introduction to Microbiology

[chapter 1]

FOUNDATIONS OF MODERN MICROBIOLOGY   1

Historical Perspectives 2
Microorganisms Were Once Thought to Arise Spontaneously 2

The Germ Theory of Disease States That Infectious Diseases Are Caused by Microorganisms 5

Immunology Is the Study of Resistance of Organisms to Infection 7

Viruses Are Submicroscopic Filterable Agents Consisting of Nucleic Acid Surrounded by a Protein Coat 8

Modern Microbiology 9
Recombinant DNA Technology Involves the Manipulation of Genes in Organisms 9

Microbes in Space 10

The Polymerase Chain Reaction Permits Rapid Amplification of Genes Outside the Cell 11

Microorganisms Have Important Roles in Medicine, the Environment, Agriculture, and Industry 11

Perspective 13

Summary 14

Evolution and Biodiversity 14

[chapter 2]

CELL MORPHOLOGY AND MICROSCOPY   17

Size, Shape, and Arrangement of Procaryotic Cells 18
Size and Scale: Procaryotes Are Small 18

Procaryotes Have Different Shapes 19

Procaryotes Have Different Cell Arrangements 19

Microscopy 20
The Compound (Bright-Field) Light Microscope Magnifies Objects Too Small to Be Seen by the Human Eye 21

Stains Increase the Contrast between a Specimen and Its Background 25

The Gram Stain 26

The Dark-Field Microscope Provides Contrast in Specimens That Normally Lack Sufficient Contrast to Be Seen with the Bright-Field Microscope 28

The Phase-Contrast Microscope Amplifies Small Differences in Refractive Indices to Enhance Specimen Contrast 28

The Fluorescence Microscope Detects Fluorescent Objects That Are Illuminated by Ultraviolet or Near-Ultraviolet Light 31

The Electron Microscope Magnifies Objects Too Small to Be Seen by Light Microscopy 31

Seeing Atoms Through a New Generation of Microscopes 34

Perspective 36

Summary 38

Evolution and Biodiversity 38

[chapter 3]

COMPOSITION AND STRUCTURE OF PROCARYOTIC AND EUCARYOTIC CELLS 41

Macromolecules: Building Blocks of the Cell 42

Proteins Are Composed of Amino Acids 42

Nucleic Acids Are Composed of Nucleotides 44

Carbohydrates and Lipids Are Organic Macromolecules 46

The Procaryotic Cell 48

The Cell Envelope Is the Outer Covering of a Procaryote Consisting of the Plasma Membrane, the Cell Wall, and, in Gram-Negative Procaryotes, an Outer Membrane 48

The Plasma Membrane Is a Selectively Permeable Barrier 49

The Cell Wall Maintains Cell Shape and Conformation 52

Peptidoglycan Synthesis 56

The Outer Membrane Is the Exterior Portion of the Cell Envelope in Gram-Negative Procaryotes 57

Bacteria Without Walls 59

The Periplasm Separates the Plasma Membrane from the Outer Membrane in Gram-Negative Procaryotes 60

A Variety of Specialized Structures Are Located Outside the Cell Envelope 61

Chemotaxis Is the Movement of Procaryotes Toward or Away from Chemical Substances 63

Procaryotes Do Not Have Membrane-Enclosed Organelles, but Do Have Specialized Internal Cell Structures 65

Endospores Enable the Cell to Survive Adverse Environmental Conditions 68

The Eucaryotic Cell 70

The Eucaryotic Cell Wall and Plasma Membrane Are Chemically Different from, but Function Similarly as, Their Procaryotic Counterparts 70

Eucaryotic Cells Contain Membrane-Enclosed Organelles 72

Eucaryotic Cells Use Flagella, Cilia, and Cytoplasmic Streaming for Locomotion 73

Eucaryotes Reproduce Asexually by Fission, Fragmentation, or Spore or Bud Formation, or Sexually by Union of Haploid Gametes 73

Perspective 75

Summary 76

Evolution and Biodiversity 77

part two

Microbial Growth and Metabolism

[chapter 4]

NUTRITION AND ENVIRONMENTAL INFLUENCE 79

Nutritional Requirements 80

All Organisms Require a Source of Carbon and Energy for Growth and Reproduction 80

Nitrogen Is Required for Proteins, Nucleic Acids, Coenzymes, Cell Walls, and Other Cellular Constituents 81

Phosphorus Is Required for Nucleic Acids, Membrane Phospholipids, and Coenzymes; Sulfur Is Required for Certain Amino Acids, tRNA, and Coenzymes 83

Other Chemical Elements Are Also Required by Microorganisms 83

Some Microorganisms Require Certain Organic Nutrients as Growth Factors 84

Nutrient Transport 85

Passive Diffusion Is the Movement of Small Molecules Across the Membrane down a Concentration Gradient 86

Facilitated Diffusion Is the Carrier-Mediated Movement of Molecules Across the Membrane down a Concentration Gradient 86

Active Transport Is the Energy-Requiring, Carrier-Mediated Movement of Molecules Across the Membrane Against a Concentration Gradient 86

Siderophores and Iron Transport 88

Group Translocation Is the Chemical Modification of a Molecule During Its Movement Across the Membrane 88

Physical and Gaseous Requirements 89

Microorganisms Have Upper and Lower Ranges of Temperature for Growth 89

Storage of Bacterial Cultures 91

Microorganisms Have Upper and Lower Ranges of pH for Growth   91

Water Is Required for Microbial Growth   92

Bacteria Are Divided into Four Groups on the Basis of Their Oxygen Sensitivity and Requirement   93

Growth Media   94
Growth Media Are of Two Types: Complex and Synthetic   94

Specialized Growth Media Are Used to Isolate and Enrich for Specific Types of Microorganisms   95

Pure Culture Techniques   96
The Streak Plate Technique Separates Bacteria in a Mixed Population   96

Agar as a Solidifying Agent   98

Aseptic Technique Is Used to Transfer Bacteria Without Contamination   98

Perspective   100

Summary   102

Evolution and Biodiversity   102

[chapter 5]

GROWTH AND CONTROL OF GROWTH   105

Bacterial Growth   106
The Typical Bacterium Reproduces by Binary Fission 106

Bacterial Populations Increase in Cell Mass and Numbers over Time 107

Quantitative Relationships of Population Growth Can Be Expressed Mathematically 107

Measurement of Growth   107
Bacterial Growth Can Be Followed by Increases in Cell Numbers 108

Turbidimetric Measurements Detect Bacteria in a Solution by Light Scattering 113

Cell Mass Can Be Determined by Weighing Cells 114

Cell Growth Can Be Measured by Changes in Cell Activity 114

Growth Curve   115
There Is No Increase in Cell Numbers During the Lag Phase of Growth 115

The Population Grows at a Constant Rate During the Exponential Phase of Growth 116

Growth Slows and a Steady State in Cell Numbers Is Reached in the Stationary Phase of Growth   116

There Is a Net Decrease in Viable Cell Numbers in the Death Phase of Growth 117

Special Techniques for Culture   117
Exponential Growth Can Be Prolonged Through Continuous Culture   118

Synchronous Cultures Are Cell Populations That Are All in the Same Stage of Growth   118

Control of Microbial Growth   119
High Temperature Can Be Used to Remove All or Most Microorganisms   119

Other Methods Are Effective in Killing or Removing Microorganisms   120

Measurement of Heat-Killing Efficiency   121

Antimicrobial Agents   122
Disinfectants Are Used on Inanimate Objects, Whereas Antiseptics Are Used on External Body Surfaces   122

Antibiotics and Synthetic Drugs Are Used in Chemotherapy to Selectively Inhibit or Kill Microorganisms Without Harming Host Tissue   125

The War Against Antibiotic-Resistant Microbes   130

Antimicrobial Susceptibility Tests Are Used to Determine the Type and Quantity of Antimicrobial Agents Used in Chemotherapy   132

Automated Methods for Antimicrobial Susceptibility Testing   134

Perspective   135

Summary   136

Evolution and Biodiversity   136

[chapter 6]

ENERGY AND METABOLISM   139

Concepts of Energy   140
Thermodynamics Is the Study of Energy Transformation   140

Energy Can Be Neither Created nor Destroyed   141

All Natural Processes Proceed in Such a Manner That There Is an Increase in Entropy   141

Reversible Chemical Reactions Proceed to an Equilibrium Point   142

Enzymes Accelerate Reaction Rates Without Themselves Being Changed 142

Coupled Rections 145
Cells Obtain Energy by the Oxidation of Molecules 145

Substances Differ in Their Oxidation-Reduction Potentials 145

Electron Carriers Transfer Electrons from Electron Donors to Electron Acceptors 146

High Energy Compounds 148

Concepts of Metabolism 149

Mechanisms of ATP Synthesis 149

Pathways Involving Substrate-Level Phosphorylation 150
Chemoheterotrophic Microorganisms Obtain Energy by Fermentation and/or Respiration 150

Glucose Is Oxidized to Pyruvate in the Embden-Meyerhof Pathway 150

Microorganisms Have Other Glycolytic Pathways for the Oxidation of Glucose 152

Pyruvate Can Be Further Degraded to a Variety of Products 154

Homofermentative and Heterofermentative Microorganisms 158

Oxidative Phosphorylation 161
The Tricarboxylic Acid Cycle Generates Intermediates for Other Pathways and Cellular Activities and Is Central to Cell Metabolism 162

Energy That Is Liberated While Electrons Are Transported in an Electron Transport Chain Can Be Coupled to the Formation of ATP 162

Chemiosmosis Explains How Electron Transport Is Used to Generate ATP 165

Oxidase Test 166

Aerobic Respiration Is More Efficient Than Fermentation in Coupling Liberated Energy to ATP Formation 168

Anaerobic Respiration Is a Process Unique to Procaryotes 168

Oxidative Phosphorylation in Eucaryotes 169

ATP Yield from Glucose Oxidation Is Not the Same in All Microorganisms 171

Chemolithotrophs 171
Hydrogen Sulfide Is Oxidized to Elemental Sulfur and Eventually to Sulfate 171

Ferrous Iron Is Oxidized to Ferric Iron 172

Ammonia Is Oxidized to Nitrite, Which Then Can Be Further Oxidized to Nitrate 172

Oxygen Is Reduced to Water During the Oxidation of Hydrogen Gas 172

Perspective 173

Summary 174

Evolution and Biodiversity 174

[chapter 7]

PHOTOSYNTHESIS AND OTHER METABOLIC PATHWAYS 177

Photosynthesis 178
Chlorophylls Are the Principal Photosynthetic Pigments in Phototrophic Eucaryotes and Cyanobacteria 178

Bacteriochlorophylls Are the Principal Photosynthetic Pigments in Bacteria 179

Accessory Pigments Harvest Light from Other Portions of the Spectrum for Photosynthesis 179

Chlorophylls Are Contained Within Chloroplasts in Eucaryotes 180

The Photosynthetic Apparatus of Procaryotes Is Contained Within Specialized Membranes or Chlorobium Vesicles 182

The Nature of Photosynthesis: Photophosphorylation 182
Photophosphorylation Occurs in Assemblages of Electron Carriers Called Photosystems 184

The Mechanism of Photophosphorylation Is Similar in Plants, Algae, and Cyanobacteria 184

Procaryotic (Other Than Cyanobacterial) Photophosphorylation Is Cyclic 186

Chemiosmosis Explains the Mechanism of Photophosphorylation 187

The Nature of Photosynthesis: Carbon Dioxide Fixation 188
Carbon Dioxide Fixation Is Catalyzed by Ribulose Diphosphate Carboxylase 188

Carbohydrate Metabolism 189
Polysaccharides Are Degraded into Smaller Molecules Before They Enter the Cell 190

Carbohydrates Are Synthesized by Anabolic Reactions Using Readily Available Substrates 192

Protein Metabolism 194
Proteins Can Be Used as Sources of Carbon and Energy 194

Stickland Reaction 196

Amino Acids Are Synthesized from Intermediates in Central Metabolic Pathways   197

Lipid Metabolism   197
Decarboxylases   198

Microorganisms Degrade Fatty Acids by β-Oxidation   200

The Glyoxylate Cycle   201

Fatty Acids Are Synthesized by the Stepwise Addition of Acetyl Groups to a Growing Chain   202

Purine and Pyrimidine Metabolism   202
Nucleases Hydrolyze Nucleic Acids into Individual Mononucleotides, Which Can Then Be Catabolized   202

Purines and Pyrimidines Are Synthesized from a Variety of Molecules   202

Perspective   204

Summary   206

Evolution and Biodiversity   206

part three

Microbial Genetics

[chapter 8]

MOLECULAR GENETICS   211

The Genetic Code   213

Deoxyribonucleic Acid   215
DNA Is the Genetic Material of the Cell   215

DNA Consists of Two Complementary Polynucleotide Chains Held Together by Hydrogen Bonds   216

Nucleotide Base Ratios   217

DNA Is Replicated Semiconservatively   219

Some Circular DNAs Are Copied via a Rolling Circle Mechanism of Replication   223

DNA Replication in Procaryotes and Eucaryotes Is Basically Similar   223

Ribonucleic Acid   224
RNA is Divided into Three Classes   224

RNA is Synthesized from DNA   226

Eucaryotes Have Several Different RNA Polymerases   228

Protein   229
Protein Synthesis Begins with the Formation of an Initiation Complex   229

The Wobble Hypothesis   230

Amino Acids Are Added to the Growing Peptide During Elongation   232

Protein Synthesis Terminates When the Ribosome Reaches a Nonsense Codon on the Messenger RNA   232

Protein Synthesis Is More Complex in Eucaryotes Than in Procaryotes   233

Regulation   233
An Enzyme Is Synthesized in the Presence of an External Substance (Inducer) in Induction   233

Enzyme Synthesis Is Inhibited in the Presence of an External Substance (Corepressor) in Repression   235

Gene Expression Is Controlled by the Termination of Transcription in Attenuation   236

Catabolites Can Repress Transcription of Genes   236

Eucaryotes Have Mechanisms for Control of Gene Expression Not Found in Procaryotes   238

Perspective   240

Summary   240

Evolution and Biodiversity   242

[chapter 9]

MICROBIAL GENETICS   245

Mutation   246
Bases Can Be Substituted by Other Bases   246

Bases Can Be Deleted or Inserted   246

The Effects of Some Mutations Are Reversible   246

The Rate of Mutations Can Be Increased by Mutagenic Agents   246

Bacteria Have Different DNA Repair Mechanisms   248

Replica Plating Differentiates Mutants from Wild Types by Their Growth Differences   249

Transfer of Genetic Material   250
The Ames Test for Carcinogenesis   251

DNA Is the Transforming Principle   252

The Discovery of Transduction   254

DNA Is Transferred by a Bacteriophage from One Bacterium to Another During Transduction   255

Plasmids Are Extrachromosomal Circular Pieces of DNA   258

Transposable Genetic Elements   260

Cell-to-Cell Contact Is Required for DNA Transfer During Conjugation   263

Perspective   266

Summary   268

Evolution and Biodiversity   268

[chapter 10]

RECOMBINANT DNA TECHNOLOGY   271

Historical Perspectives   272

Gene Cloning   274
The Target Gene Can Be Obtained in Different Ways   274

Southern Blotting   276

The Target Gene Is Incorporated into a Cloning Vector   279

The Cloning Vector Is Inserted into a Host Where the Target Gene Is Amplified   285

Various Methods Are Available to Detect Host Cells with the Cloned Gene   287

The Polymerase Chain Reaction   288

Applications of Recombinant DNA Technology   291
Applications of Recombinant DNA Technology in Medicine Have Important Ramifications for Human Health Care   291

DNA Fingerprinting   292

Recombinant DNA Technology Has Improved Agriculture   297

Public and Scientific Concerns About the Regulation of Recombinant DNA Technology   297

Perspective   300

Summary   302

Evolution and Biodiversity   302

part four

Survey of Microorganisms

[chapter 11]

PROCARYOTES: THE BACTERIA AND THE ARCHAEA   305

Classification of Microorganisms   306
Classification Schemes for Microorganisms Have Evolved over the Years   306

Three Major Approaches Have Been Used to Identify and Place Microorganisms in Classification Systems   307

Ribosomal RNA and Phylogeny   311
Sequencing of Ribosomal RNA is Relatively Simple   311

Phylogenetic Trees Can Be Developed from Ribosomal RNA Sequences   311

Classification of Procaryotes   314

Gram-Negative Bacteria of General, Medical, or Industrial Importance   316
The Spirochetes   316

Aerobic/Microaerophilic, Motile, Helical/Vibrioid Gram-Negative Bacteria   318

Nonmotile (or Rarely Motile), Gram-Negative, Curved Bacteria   319

Gram-Negative Aerobic Rods and Cocci   319

Facultatively Anaerobic Gram-Negative Rods   322

Using Flowcharts to Identify Bacteria   324

Anaerobic, Gram-Negative Straight, Curved, and Helical Rods   325

Dissimilatory Sulfate- or Sulfur-Reducing Bacteria   326

Anaerobic, Gram-Negative Cocci   326

The Rickettsias and Chlamydias   326

Mycoplasmas   327

Endosymbionts   328

Gram-Positive Bacteria Other Than the Actinomycetes   328
Gram-Positive Cocci   328

Endospore-Forming Gram-Positive Rods and Cocci   328

Revival of 25-Million-Year-Old Procaryotes?   330

Regular, Nonsporing Gram-Positive Rods   331

Irregular, Nonsporing Gram-Positive Rods   331

The Mycobacteria   331

**The Archaea, Cyanobacteria, and Remaining
Gram-Negative Bacteria   332**

Anoxygenic Phototrophic Bacteria   332

Oxygenic Photosynthetic Bacteria   333

Aerobic Chemolithotrophic Bacteria and Associated
Organisms   336

Budding and/or Appendaged Bacteria   337

Sheathed Bacteria   337

Nonphotosynthetic, Nonfruiting Gliding Bacteria   338

Fruiting Gliding Bacteria: The Myxobacteria   338

Archaea   339

The Actinomycetes   341

Perspective   344

Summary   346

Evolution and Biodiversity   347

[chapter 12]
THE EUCARYOTIC
MICROORGANISMS   349

The Algae   350

There Is Considerable Variability in the Morphology and
Structure of Algae   350

Algae Exhibit a Variety of Reproductive Processes   352

Algae Are Classified into Six Divisions Based on
Structural, Chemical, and Reproductive
Characteristics   352

The Economic Importance of Algae   355

The Fungi   359

Fungi Have Two Morphological Growth Forms: Molds
and Yeasts   361

Fungi Reproduce Asexually and Sexually   362

Fungi Are Phylogenetically Grouped into Two Kingdoms
and Four Additional Phyla   365

Aspergillus and Aflatoxin   368

The Protozoa   373

Protozoa Lack Cell Walls, but Contain Membrane-Bound
Organelles   373

Protozoa Reproduce Asexually and Sexually   373

Protozoa Are Divided into Seven Phyla   375

AIDS and Eucaryotic Microbe Infections   376

Perspective   378

Summary   380

Evolution and Biodiversity   380

[chapter 13]
THE VIRUSES   383

Properties of Viruses   384

Viruses Are Extremely Small   384

Viruses Consist of DNA or RNA Surrounded by a
Protein Coat   384

Viruses Have Three Basic Forms: Polyhedral, Helical,
and Binal   386

Classification of Viruses   387

Bacteriophages Are Classified by Their Morphology and
Nucleic Acid Content   387

*Viroids and Prions—Agents Smaller Than
Viruses   388*

Viruses Infect Other Microorganisms Besides
Bacteria   390

Plant Viruses Are Classified into 23 Virus Groups and
Two Families   390

Animal Viruses Are Classified into 18 Families   393

Propagation and Assay of Viruses   395

Viruses Require a Host for Propagation   395

Viruses Can Be Assayed by Various Methods   397

Replication of Viruses   398

Virus Replication Begins with Infection of the Host   398

*Evidence That Bacteriophage Nucleic Acid, Not
Protein, Enters the Host Cell During Infection   400*

DNA Viruses Have Various Methods for Replication of
DNA   402

RNA Viruses Have Various Methods for Replication of
RNA   403

Complete Virions Are Formed During Assembly   403

One-Step Growth Curve   404

Intact Virions Are Released from the Host After
Assembly   405

Viral Infections in Animal and Plant Cells Can Result in
Cytopathic Effects   405

Are Viral Diseases Treatable?   406

Other Consequences of Viral Infections 408

Viruses Can Be Continuously Released from a Host Cell 408

There Can Be a Significant Reduction or Elimination of Virus Production in the Host 408

Virus Infection Can Lead to Cell Transformation 408

A Bacteriophage Can Be Incorporated into the Host Chromosome in Lysogeny 410

Perspective 412

Summary 414

Evolution and Biodiversity 414

part five

Microbial Interactions with Other Organisms

[chapter 14]

SYMBIOSIS 417

Mutualism 418

Rhiziobium, Frankia, Anabaena, and Other Bacteria Fix Nitrogen Symbiotically 418

A Lichen Is a Mutualistic Association Between a Fungus and an Alga 423

Flashlight Fishes Have a Symbiotic Relationship with Luminescent Bacteria 425

Some Bacteria Are Symbionts of Protozoa 425

Some Bacteria Are Symbionts of Insects 427

Microbial Symbiosis Assists Ruminants in Digestion 427

Parasitism 429

Bdellovibrio Parasitizes Gram-Negative Bacteria 429

Chlamydiae Are Animal Parasites with Unusual Developmental Cycles 430

Germ-Free Animals 432

Commensalism 433

Mutualism, Commensalism, Parasitism—Dynamic Relationships 433

Perspective 434

Summary 434

Evolution and Biodiversity 436

[chapter 15]

HOST-PARASITE RELATIONSHIPS 439

The Normal Human Microflora 440

Microorganisms Can Compete Antagonistically 440

Microorganisms Can Grow Synergistically 441

The Story of Typhoid Mary 442

General Concepts of Host-Parasite Relationships 442

Infectious Microorganisms Persist in Reservoirs 442

Diseases Can Be Transmitted Directly or Indirectly 443

Microorganisms Gain Entry to a Host Through a Portal 445

Infectious Disease Follows a Sequence from Infection to Disease Resolution 446

Measurement of Virulence 447

Parasites Must Overcome Host Defenses to Initiate Infection and Disease 448

Microbial Factors of Virulence 448

Microorganisms Have Many Mechanisms to Assist in Invasion 448

Microbes Produce Two Types of Toxins: Exotoxins and Endotoxins 450

Detection of Endotoxin 452

Host Resistance and Tuberculosis 453

Innate (Nonspecific) Host Resistance 454

The General Health and Physiological Condition of the Host Can Influence the Course of a Disease 454

Physical Barriers Protect the Human Body Against Infectious Microorganisms 455

Human Blood Contains Protective Factors 456

Inflammation Is the Body's Response to Injury, Irritation, or Infection 458

Phagocytes Digest Bacteria, Viruses, and Other Foreign Materials 460

Perspective 462

Summary 464

Evolution and Biodiversity 464

[chapter 16]

IMMUNOLOGY 467

Humoral Immunity 468

Lymphoid Tissue Is Important in the Immune Response   468

Antigens Induce and React with Antibodies   468

Antibodies Are Glycoproteins That Bind Specifically with the Antigen That Stimulated Their Production   468

Antibody Diversity Is Made Possible by Gene Rearrangement   471

   Superantigens   *473*

Different T-Cell Subpopulations Perform Different Functions   473

The Major Histocompatibility Complex Plays a Role in the Immune Response   473

   Monoclonal Antibodies   *475*

Activated B Lymphocytes Develop into Antibody-Secreting Plasma Cells   475

Humoral Immunity Is an Important Part of Host Defense Against Microbial Invasion   475

Complement Is a Group of Proteins That Augment the Action of Antibodies   481

Cell-Mediated Immunity   482

Cytokines Activate Macrophages and Increase Their Phagocytic Activity in Cell-Mediated Immunity   483

Delayed Hypersensitive Responses Are Caused by the Release of Cytokines from Sensitized T Cells   483

Cell-Mediated Immunity May Be Manifested in Other Ways   484

Assays to Measure Cell-Mediated Immunity Depend upon Delayed Hypersensitive Skin Reactions or Detection of Sensitized Lymphocytes   485

In Vitro Antibody-Antigen Reactions   485

Soluble Antigens Mixed with Multivalent Antibodies Form Large, Precipitable Aggregates   485

Antigens Can Be Separated by Immunoelectrophoresis   486

Insoluble Antigens Mixed with Multivalent Antibodies Form Aggregates Detectable by Agglutination   486

Neutralization Tests Are Used to Detect Toxins and Viruses or Their Antibodies   487

Complement Can Be Used to Detect Antigen-Antibody Reactions   487

Fluorescent Dyes Are Used in Immunofluorescence to Detect Antigens or Antibodies   489

Radioactivity Is Used in Radioimmunoassay to Detect Small Amounts of Antigen   490

Enzymes Are Used in Enzyme-Linked Immunosorbent Assays to Detect Antigens or Antibodies   490

Perspective   492

Summary   494

Evolution and Biodiversity   494

[**chapter 17**]

INFECTIOUS DISEASES   497

The Respiratory Tract   498

Many Different Types of Microorganisms Are Normally Found in the Upper Respiratory Tract   498

Respiratory Tract Pathogens Are Highly Contagious and Transmissible   498

   Classification of Streptococci   *502*

   Group B Streptococcal Neonatal Disease   *506*

   Discovery of the Diphtheria Exotoxin   *509*

Respiratory Tract Diseases Can Be Caused by Other Microorganisms   512

The Oral Cavity and Digestive System   514

Many Different Types of Microorganisms Are Normally Found in the Oral Cavity   514

Dental Diseases Are Caused by an Accumulation of Microorganisms on the Surface of Teeth   514

Oral Diseases Can Be Caused by Other Microorganisms   515

The Digestive System Is Complex and Contains a Wide Variety of Microorganisms   515

Bacterial Diseases of the Digestive System Arise from Food Infection or Food Intoxication   516

   Poultry and Salmonella *Infections*   *520*

Escherichia coli Is Responsible for Many Different Gastrointestinal Diseases   521

   An Infectious Cause of Ulcers   *522*

Gastrointestinal Diseases Can Be Caused by Other Microorganisms   523

Diseases of the Skin and the Genitourinary System   524

Few Species of Microorganisms Normally Inhabit the Skin   525

Microbial Skin Pathogens Must Overcome Microbial Antagonism and the Physical Barrier of the Skin to Cause Infection   525

The Male and Female Genitourinary Systems Differ in Their Normal Microbial Flora   532

Acquired Immune Deficiency Syndrome (AIDS) Results in a Deficiency of CD4+ Cells   536

Emerging Infectious Diseases   542

Perspective   544

Summary   544

Evolution and Biodiversity   546

[chapter 18]

PUBLIC HEALTH AND EPIDEMIOLOGY   549

Some General Concepts of Epidemiology   550
An Epidemic Is a Markedly Increased Occurrence of a Disease in a Population   552

Morbidity and Mortality Rates Are Used by Epidemiologists to Determine the Trend of Disease   553

Epidemiological Analysis of Diseases   555
Disease Rates May Change with Time   555

Disease Rates May Be Influenced by Geography   556

Disease Rates May Be Affected by the Characteristics of a Population   556

Collection of Epidemiological Data   558
Nosocomial Infections   559

There Are Potential Biases in Interview Surveys   560

Screening and Diagnostic Tests Are More Objective Than Interview Surveys   560

Observational and Experimental Studies   561
Observational Studies Are of Two Types: Retrospective and Prospective   561

The Investigator Specifies or Manipulates Conditions in Experimental Studies   563

Perspective   564

Summary   566

Evolution and Biodiversity   566

part six

Environmental and Economic Impact of Microorganisms

[chapter 19]

MICROBIAL ECOLOGY   569

Biogeochemical Cycles   570

The Carbon Cycle   570
Primary Producers Fix Atmospheric Carbon Dioxide into Organic Compounds   571

The Greenhouse Effect   572

Herbivores and Carnivores Contribute to the Carbon Cycle by Consuming Organic Carbon and Returning It to the Atmosphere as Carbon Dioxide   575

Microorganisms Contribute to the Carbon Cycle as Decomposers   575

The Nitrogen Cycle   576
Only Certain Procaryotes Can Fix Nitrogen   576

Ammonia Is Formed During Decomposition of Organic Nitrogenous Compounds   577

The Phosphate Cycle   579

The Sulfur Cycle   579

Microbes and Soil   580
Soil Varies in Its Composition and Physical Properties   581

There Are Considerable Numbers and Varieties of Microorganisms in Soil   581

Bacteria Are Important in the Leaching of Metals from Low-Grade Ores   582

Hydrocarbon-Degrading Microorganisms Are Potentially Useful in Oil Spill Cleanups   583

Microorganisms Are Useful as Biological Pesticides   584

Microbes and Water   586
A Wide Variety of Microorganisms Are Found in Marine Habitats   586

Many Different Types of Microorganisms Are Also Found in Freshwater Habitats   587

Chemical and Biological Contaminants Affect the Quality of Water   588

Chesapeake Bay: Example of a Revitalized Estuary 590

Cryptosporidiosis 594

Sewage Treatment Removes Microorganisms and Chemicals from Sewage 594

Methane Production from Wastes 598

Microbes and the Air 601

Perspective 602

Summary 604

Evolution and Biodiversity 604

[chapter 20]
FOOD AND INDUSTRIAL MICROBIOLOGY 607

Milk and Dairy Products 608

Raw Milk Contains Microorganisms Introduced from the Udder and During Human Handling 608

Bacteria Are Used to Produce Fermented Milks and Milk Products 609

Salmonellosis Outbreak from a Dairy Plant 610

Microorganisms Are Important in Cheese Production 611

An Outbreak of Listeriosis Caused by Contaminated Milk 613

Microorganisms in Meat and Meat Products 614

Microorganisms Are Responsible for Meat Spoilage 614

Many Methods Can Be Used to Preserve Meats and Reduce Spoilage 614

Poultry and Seafood 615

Canned Foods 616

Fermented Foods 616

Alcoholic Beverages 617

Yeasts Are Used in Beer Production to Ferment Grains 617

Frost-Proof Fruits and Vegetables 618

Yeasts Are Used in Wine Production to Ferment Fruit Sugars 619

Yeasts Are Used in the Production of Distilled Beverages 619

Industrial Processes 619

A Desirable Culture Medium Is Inexpensive, yet Contains All Essential Ingredients 620

Microorganisms for Industrial Fermentation Are Carefully Selected 620

Culture Conditions Are Carefully Controlled in Industrial Fermentors 621

Different Methods Are Used to Culture Microorganisms Industrially 621

Enzymes 623

Microbial Amylases Are Used in Bread Making, Beer Production, Manufacture of Sugar Syrups, and Textile Manufacture 623

Microbial Proteases Are Used in the Clothing Industry, in Photography, and in Stain Removal 623

Other Microbial Enzymes Are of Industrial Importance 624

Metabolic Products 625

Primary Metabolites Are Metabolic Compounds Involved in Cell Growth or Function 625

Secondary Metabolites Are Products of Pathways That Apparently Are Not Associated with Primary Cellular Processes 629

Recombinant DNA Techniques for the Production of Antibiotics 630

Single-Cell Protein 631

Perspective 632

Summary 633

Evolution and Biodiversity 633

Appendix A: Classification of Procaryotes According to *Bergey's Manual of Systematic Bacteriology* 635
Appendix B: The Mathematics of Bacterial Growth 642
Appendix C: Microbiology Internet Resources 645
Appendix D: Chemistry fo the Microbiologist 647
Appendix E: Answers to the End-of-Chapter Questions 653
Glossary 668
Credits 685
Index 689

PREFACE

Microbiology is designed as an introductory textbook for the college student who has a fundamental background in biology and some knowledge of chemistry. It is intended to provide the student with a general background knowledge of microorganisms and the relationships of these microorganisms to one another and to other organisms. The book is comprehensive and contains a considerable amount of material, but the material is written in a concise manner to emphasize basic principles and the applications of these principles in modern science.

The second edition of *Microbiology* has been significantly revised. In addition to a new full-color illustration program and new color photographs, chapters have been extensively revised and updated to reflect the rapid developments in microbiology. A new chapter on recombinant DNA technology has been added to accommodate the considerable advances in this area. There is expanded coverage of phylogenic relationships in the taxonomy section, and the immunology chapter has been significantly updated to include current concepts. Material has been added to describe new and emerging pathogens and the expanding problem of drug-resistant microbes. New underlying themes in evolution and biodiversity have been incorporated throughout the text to emphasize the importance of the common thread that weaves its way through all living organisms. As with the first edition, the second edition was carefully written to provide a proper balance among the various aspects of microbiology and to maintain the clarity of explanation.

Organization

The twenty chapters of *Microbiology* are divided into six parts.

Part I. Introduction to Microbiology provides the student with an overview of the history of microbiology, with emphasis on how this history has contributed to contemporary applications. The observation, morphology and composition of microorganisms are introduced early in the book to provide the student with an understanding of microscopy and the characteristics of microorganisms necessary for an accompanying laboratory course.

Part II. Microbial Growth and Metabolism discusses the growth and nutrition of microorganisms. There is a comprehensive discussion of antimicrobial agents and the emergence of drug-resistant microbes. The generation of ATP by chemotrophs and phototrophs is included in this section.

Part III. Microbial Genetics discusses the foundations of modern molecular biology. The student is provided with the basic principles of macromolecular synthesis, including regulation. This is followed by a discussion of the transfer of genetic information and procedures used in recombinant DNA technology.

Part IV. Survey of Microorganisms is a concise, yet comprehensive review of the five major groups of microorganisms: procaryotes, algae, fungi, protozoa, and viruses. The section begins with an introduction to phylogeny and the domains Bacteria, Archaea, and Eucarya. Each group of microorganisms is discussed separately, with the information organized by contemporary classification of the microbes.

Part V. Microbial Interactions with Other Organisms focuses on symbiosis, host-parasite interactions, and the consequences of these interactions. The different types of symbioses are described. Host response to infection, including innate and acquired host resistance, is discussed. The student is introduced to representative microbial diseases, by site of entry of the microbe. The section concludes with a discussion of public health and epidemiology and their importance in disease detection and prevention.

Part VI. Environmental and Economic Impact of Microorganisms emphasizes the role of microorganisms in biogeochemical cycles and the environment. Applications of microbiology to agriculture, the food and beverage industry, and production of commercially important products are presented. Attention is given to why microorganisms are important in our existence and how these microorganisms can be used to improve our quality of life.

Supplements

For the Instructor

An *Instructor's Manual with Test Item File*, prepared by the author, contains outlines of the chapters to assist in lecture preparation, along with more than 500 test questions that can be used to generate exams.

A *Computerized Test Bank*, entitled *Microtest*, is also offered to adopters of this text. It is a software version of the test bank found in the Instructor's Manual. It requires no programming experience and is available in IBM and Mac versions: IBM Windows 3.5; and Mac 3.5.

A set of 100 full-color acetate *transparencies* provides users of the text with illustrations for classroom use.

For the Student

A *Student Study Guide*, prepared by Donald McGarey of Jacksonville State University, offers students a wealth of concept explanations, summaries, outlines, and sample questions—all designed to enhance students' understanding, enjoyment, and success in the course. (ISBN 26188)

The *Microbes in Motion* CD-ROM is a state-of-the-art tutorial product that explains and demonstrates microbiology with drawings, photos, animations, and video segments. This textbook is specifically tied to *Microbes* in two ways: with a Preview in the beginning of each chapter that points to specific modules in the CD-ROM for the material to be covered in the chapter; and with specific references in the text, marked by ⊙, that link to segments in the CD-ROM. (ISBN 24596)

Acknowledgments

A book of this magnitude and breadth could not have been developed without the assistance of many people. I would like to express my appreciation and gratitude to the following reviewers for their time, efforts, and valuable suggestions:

Allan T. Andrew
Indiana University of Pennsylvania

Lydia Arciszewski
Bergen Community College

Glenn A. Bauer
Saint Michael's College

Robert J. Boylan
New York University College of Dentistry

John R. Burdick
Midwestern University

David Carlberg
California State University

Richard Cunningham
SUNY at Albany

Jim Daly
Purchase College–SUNY

Larry Elliott
Western Kentucky University

David Essar
Winona State University

Samuel Fan
Bradley University

Frank A. Fekete
Colby College

John Ferguson
Bard College

Harvey P. Friedman
University of Missouri–St. Louis

S. Marvin Friedman
Hunter College

Heather Hall
Charles County Community College

Barbara B. Hemmingsen
San Diego State University

Brian Hoffman
Park College

Timothy Hoover
University of Georgia

Thomas R. Jewell
University of Wisconsin–Eau Claire

David Kafkewitz
Rutgers University

Richard Karp
University of Cincinnati

Robert J. Kearns
University of Dayton

Beverly Kirk
Northeast Mississippi Community College

Tom Klubertanz
Peru State College

Arthur L. Koch
Indiana University

Gerald Krieder
Albright College

Susan Landon
Northern State University

Sandra L. Landuyt
Penn Valley Community College

Lee H. Lee
Montclair State University

Andy Lloyd
Delaware State University

Jeff Lodge
Rochester Institute of Technology

Bonnie Lustigman
Montclair State University

Richard L. Lyers
Southwest Missouri State University

Charlotte McCarthy
New Mexico State University

Robert J. C. McLean
Southwest Texas State University

Al Mikell
University of Mississippi

Robert Mitchell
Community College of Philadelphia

Fernando Monroy
Indiana State University

Eric L. Mustain
East Tennessee State University

Gerard A. O'Donovan
University of North Texas

Gary W. Pettibone
SUNY College at Buffalo

Marcella Piasecki
Lynn University

Sidhartha D. Ray
Long Island University

Kelynne Reed
Austin College

Quentin Reuer
University of Alaska–Anchorage

Grace Spatafora
Middlebury College

Donald P. Stahly
University of Iowa

Susan J. Stamler
College of DuPage

John Sternick
Mansfield University

Molly Thomas
HCCS–Southeast College

Monica Tischler
Benedictine University

Leslie Uhlig
Lake Superior State University

James E. Urban
Kansas State University

Suzanne M. Walsh
Westfield State College

Diane Swender Watkins
Neosha Community College

The production of *Microbiology*, Second Edition, was made possible through the diligent efforts and teamwork of many people associated with Wm. C. Brown Publishers. I would like to thank the following members of the editorial and production staff for their guidance and attention to detail: Liz Sievers, acquisitions editor; Terry Stanton, developmental editor; Carla Kipper, production editor; Jodi Banowetz, art editor; Kaye Farmer, designer; Donna Slade, production supervisor of art and design; Lori Hancock, photo researcher; Karen Storlie, permissions coordinator.

I also wish to acknowledge the expert contributions of Julie Wilde (copyeditor), Samuel Collins (artist), and my many colleagues who graciously provided photographs or other materials for *Microbiology*. Their names are listed with their contributions in the credits section of the book.

Finally, I want thank my wife, Carol, and my students for their patience and understanding during the revision of this book. Their moral and intellectual support is especially appreciated.

LEARNING AIDS

"Nature abounds with little round things," the noted biologist and writer Lewis Thomas has observed. These little round things, which were first described in detail over three centuries ago by a Dutchman named Anton van Leeuwenhoek, have since become the most studied living organisms. Today, microorganisms are intimately involved in our daily lives. Not all are round, nor are all friendly. Some cause the most serious and devastating diseases known to humans. Yet we could not live without these microscopic forms.

Microbiology is a relatively young discipline compared with the other, more classical sciences. It is a dynamic science, with new information constantly displacing old theories and assumptions. This is what makes microbiology an exciting science, one that explores new frontiers and new ideas. Within the short history of microbiology, some of the more important scientific discoveries have been made. From Louis Pasteur's simple, but ingenious, experiments in the 19th century to disprove the doctrine of spontaneous generation to contemporary discoveries in recombinant DNA technology, microbiologists have been in the forefront of science. The principles and applications of microbiology are an integral part of such areas as medicine, genetics, biochemistry, molecular biology, ecology, agriculture, and biotechnology. The goal of this book is to impart some of this excitement and enthusiasm about microbiology to you.

Several learning aids have been incorporated into this book to assist you in understanding and retaining the material.

Chapter outlines provide you with a summary of the contents of each chapter.

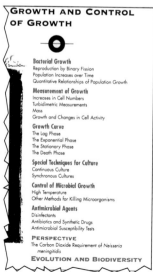

Descriptive, narrative chapter openings introduce the chapter subject matter by previewing a chapter idea or thought.

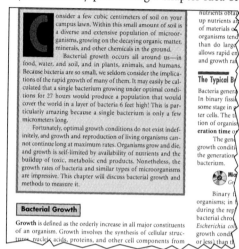

Key terms in boldface type assist you in recognizing important terms in the text.

Focus boxes augment the text and provide a closer examination of practical or interesting areas of microbiology.

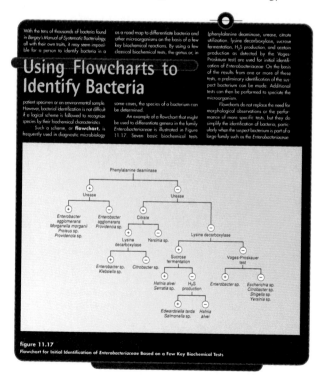

Figure and table references in the text immediately direct attention to the appropriate figure or table. In addition, chapter cross-references in the text guide you to other text locations where the topic may be more fully discussed.

CD-ROM icons flag references to Wm. C. Brown's *Microbes in Motion* CD-ROM microbiology tutorial, by Delisle and Tomalty. In addition, most chapters begin with a CD-ROM Preview Link, referring to segments of the CD that will help students focus on key concepts.

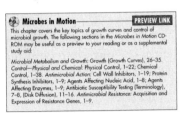

Microbes in Motion — PREVIEW LINK

This chapter covers the key topics of growth curves and control of microbial growth. The following sections in the *Microbes in Motion* CD-ROM may be useful as a preview to your reading or as a supplemental study aid:

Microbial Metabolism and Growth: Growth (Growth Curves), 26–35. *Control—Physical and Chemical:* Physical Control, 1–22; Chemical Control, 1–38. *Antimicrobial Action:* Cell Wall Inhibitors, 1–19; Protein Synthesis Inhibitors, 1–9; Agents Affecting Nucleic Acid, 1–8; Agents Affecting Enzymes, 1–9; Antibiotic Susceptibility Testing (Terminology), 7–8, (Disk Diffusion), 11–16. *Antimicrobial Resistance:* Acquisition and Expression of Resistance Genes, 1–9.

Microbial Metabolism and Growth
Growth: Growth Curves • p. 27

Evolution and biodiversity boxes at the end of each chapter link the chapter contents to the common theme of evolution and diversity of organisms.

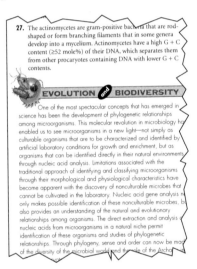

27. The actinomycetes are gram-positive bacteria that are rod-shaped or form branching filaments that in some genera develop into a mycelium. Actinomycetes have a high G + C content (≥52 mole%) of their DNA, which separates them from other procaryotes containing DNA with lower G + C contents.

EVOLUTION and BIODIVERSITY

One of the most spectacular concepts that has emerged in science has been the development of phylogenetic relationships among microorganisms. This molecular revolution in microbiology has enabled us to see microorganisms in a new light—not simply as culturable organisms that are to be characterized and identified by artificial laboratory conditions for growth and enrichment, but as organisms that can be identified directly in their natural environments through nucleic acid analysis. Limitations associated with the traditional approach of identifying and classifying microorganisms through their morphological and physiological characteristics have become apparent with the discovery of nonculturable microbes that cannot be cultivated in the laboratory. Nucleic acid gene analysis not only makes possible identification of these nonculturable microbes, but also provides an understanding of the natural and evolutionary relationships among organisms. The direct extraction and analysis of nucleic acids from microorganisms in a natural niche permit identification of these organisms and studies of phylogenetic relationships. Through phylogeny, sense and order can now be made of the diversity of the microbial world and the role of the Archaea.

Perspective boxes conclude the chapter and examine interesting examples or phenomena that illuminate chapter themes.

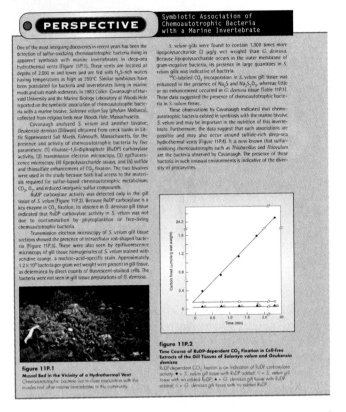

PERSPECTIVE — Symbiotic Association of Chemoautotrophic Bacteria with a Marine Invertebrate

One of the most intriguing discoveries in recent years has been the detection of sulfur-oxidizing chemoautotrophic bacteria living in apparent symbiosis with marine invertebrates in deep-sea hydrothermal vents (Figure 11P.1). These vents are located at depths of 2,500 m and lower and are fed with H_2S-rich waters having temperatures as high as 350°C. Similar symbioses have been postulated for bacteria and invertebrates living in marine muds and salt marsh sediments. In 1983 Colleen Cavanaugh of Harvard University and the Marine Biology Laboratory at Woods Hole reported on the symbiotic association of chemoautotrophic bacteria with a marine bivalve, *Solemya velum* Say (phylum Mollusca), collected from eelgrass beds near Woods Hole, Massachusetts.

Cavanaugh analyzed *S. velum* and another bivalve, *Geukensia demissa* (Dillwyn) obtained from creek banks in Little Sippewissett Salt Marsh, Falmouth, Massachusetts, for the presence and activity of chemoautotrophic bacteria by five parameters: (1) ribulose-1,5-diphosphate (RuDP) carboxylase activity, (2) transmission electron microscopy, (3) epifluorescence microscopy, (4) lipopolysaccharide assays, and (5) sulfide and thiosulfate enhancement of CO_2 fixation. The two bivalves were used in the study because both had access to the materials required for sulfur-based chemoautotrophic metabolism: CO_2, O_2, and reduced inorganic sulfur compounds.

RuDP carboxylase activity was detected only in the gill tissue of *S. velum* (Figure 11P.2). Because RuDP carboxylase is a key enzyme in CO_2 fixation, its absence in *G. demissa* gill tissue indicated that RuDP carboxylase activity in *S. velum* was not due to contamination by phytoplankton or free-living chemoautotrophic bacteria.

Transmission electron microscopy of *S. velum* gill tissue sections showed the presence of intracellular rod-shaped bacteria (Figure 11P.3). These were also seen by epifluorescence microscopy of gill tissue homogenates of *S. velum* stained with acridine orange, a nucleic-acid-specific stain. Approximately 1.2×10^9 bacteria per gram wet weight were present in gill tissue, as determined by direct counts of fluorescent-stained cells. The bacteria were not seen in gill tissue preparations of *G. demissa*.

S. velum gills were found to contain 1,000 times more lipopolysaccharide (2 µg/g wet weight) than *G. demissa*. Because lipopolysaccharide occurs in the outer membrane of gram-negative bacteria, its presence in large quantities in *S. velum* gills was indicative of bacteria.

^{14}C-labeled CO_2 incorporation in *S. velum* gill tissue was enhanced in the presence of Na_2S and $Na_2S_2O_3$, whereas little or no enhancement occurred in *G. demissa* tissue (Table 11P.1). These data suggested the presence of chemoautotrophic bacteria in *S. velum* tissue.

These observations by Cavanaugh indicated that chemoautotrophic bacteria existed in symbiosis with the marine bivalve *S. velum* and may be important in the nutrition of this invertebrate. Furthermore, the data suggest that such associations are possible and may also occur around sulfide-rich deep-sea hydrothermal vents (Figure 11P.4). It is now known that sulfur-oxidizing chemoautotrophs such as *Thiobacillus* and *Thiovulum* are the bacteria observed by Cavanaugh. The presence of these bacteria in such unusual environments is indicative of the diversity of procaryotes.

figure 11P.1
Mussel Bed in the Vicinity of a Hydrothermal Vent
Chemoautotrophic bacteria live in close association with the mussels and other marine invertebrates in this community.

figure 11P.2
Time Course of RuDP-dependent CO_2 Fixation in Cell-free Extracts of the Gill Tissues of *Solemya velum* and *Geukensia demissa*
RuDP-dependent CO_2 fixation is an indication of RuDP carboxylase activity. ● = *S. velum* gill tissue with RuDP added; ○ = *S. velum* gill tissue with no added RuDP; ▲ = *G. demissa* gill tissue with RuDP added; △ = *G. demissa* gill tissue with no added RuDP.

Chapter summaries provide a concise summary and overview of the chapter.

Supplementary readings augment the text with a list of readings for additional information on key chapter concepts. A short summary is included for each listed reference.

End-of-chapter questions are intended to pique student interest in the chapter contents.

An comprehensive **Glossary** contains definitions for the important terms used in the text. The Glossary specifically directs you to text locations where the term is more extensively defined or used to provide you with a better understanding of the application of the term.

The **Appendixes** contains the complete Table of Contents for *Bergey's Manual of Systematic Bacteriology*, answers to chapter questions, a detailed discussion of the mathematics of microbial growth, an illustrated guide to chemistry, and a guide to Internet web sites and to a site specific to this text.

chapter one

FOUNDATIONS OF MODERN MICROBIOLOGY

Historical Perspectives
Microorganisms and Spontaneous Generation
The Germ Theory of Disease
Immunology
Viruses as Submicroscopic Filterable Agents

Modern Microbiology
Recombinant DNA Technology
The Polymerase Chain Reaction
Important Roles of Microorganisms

PERSPECTIVE
Disproving the Doctrine of Spontaneous Generation

EVOLUTION AND BIODIVERSITY

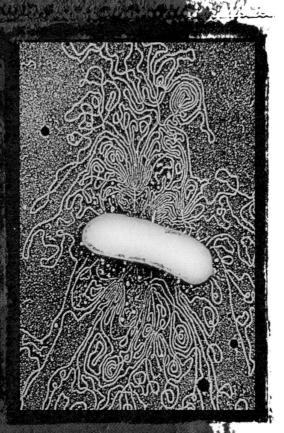

Electron micrograph of DNA released from *Escherichia coli* (colorized, ×56,000).

Imagine a world in which life-threatening diseases are diagnosed in minutes using a genetic road map to locate and identify the microscopic pathogen. Picture a society in which highly specific antibodies are used as magic bullets to target drugs to cancerous cells and where recombinant DNA in viruses is used as live vaccines, for cancer immunotherapy, and to treat cystic fibrosis through gene therapy. In this world, "superbugs" devour toxic pollutants, clean up oil spills, and eliminate agricultural pests without harm to animals, humans, and plants. Agricultural crops are protected from frost damage, saving farmers millions of dollars in potential economic loss and providing the public with year-round produce and fruits; genetically engineered bacteria fertilize crops and significantly increase agricultural production; and plants are genetically modified to protect them from insects, drought, and disease.

The world described in this scenario is not in the future—it exists today and was made possible through a series of technological advances. At the forefront of these exciting discoveries is the traditional discipline of **microbiology.** Microbiology is the study of agents too small to be seen with the unaided eye, specifically bacteria (bacteriology), viruses (virology), fungi (mycology), protozoa (protozoology), and algae (phycology) (Figure 1.1). Through the techniques developed in microbiology, it is now possible, at least in principle, to use microorganisms as living factories capable of producing any biochemical a person desires. These minute organisms also provide scientists with the capability of understanding in greater detail the genetic composition of all forms of life.

Microbes play indispensible roles in our daily lives. They are important not only in molecular biology and biotechnology, but also in medicine, ecology, agriculture, and industry. This knowledge of microorganisms did not emerge overnight, but resulted from about a century of careful and deliberate scientific advances in the field of microbiology.

Historical Perspectives

Microbiology as a science began with the development of the microscope. No one knows who invented the microscope. In A.D. 60, Seneca, a Roman philosopher and statesman, noted in his writings that "letters, however small and indistinct, are seen enlarged and more clearly through a globe of glass filled with water." Early observations of this kind led to the use of crude lenses made of certain clear minerals. Eventually glass was set in frames to help older people see better.

The invention of the modern microscope is often attributed to Zacharias Janssen (1580–c. 1638), a Dutch maker of spectacles who, as a young boy, combined two lenses to form a crude compound microscope in 1590. The microscope was subsequently improved by the German mathematician and astronomer Johann Kepler (1571–1630), the noted Italian astronomer Galileo (1564–1642), and the English physicist and mathematician Robert Hooke (1635–1703). Hooke experimented extensively with the microscope and illustrated his written observations with detailed drawings of plant tissues. It was Hooke who first described the small cavities separated by walls as "cells" in *Micrographia,* a book commissioned in 1665 by King Charles II of England because of his interest in Hooke's microscopic observations.

However, it was not until the late 1600s that microscopes became popular as instruments to observe objects normally not visible to the naked eye. A Dutch linen draper and town civil servant named Anton van Leeuwenhoek (1632–1723) became interested in the microscope and made hundreds of them in his lifetime (Figure 1.2). Leeuwenhoek's microscopes were noted for their lenses, which were meticulously ground and highly polished. These early single-lens instruments gave magnifications of only 50 to 300 diameters, or approximately one-third the magnification of modern light microscopes (Figure 1.3). Nonetheless, these microscopes provided glimpses into an entirely new world consisting of microbes—small organisms that Leeuwenhoek called "animalcules" in a letter written to the Royal Society of London in 1674:

> I found floating (in lake water) therin divers earthly particles, some green streaks, spirally wound serpent-wise, and orderly arranged, after the manner of the copper or tin worms... The whole circumstance of each of these streaks was about the thickness of a hair on one's head... These animalcules had divers coulors, some being whitish and transparent; others with green and very glittering little scales; others again were green in the middle, and before and behind white.

In this and other communications to the scientific community, Leeuwenhoek described in detail many of the different types of microbial shapes that we recognize today (Figure 1.4). Because of his accomplishments, Leeuwenhoek is considered to be the "father of microbiology."

Microorganisms Were Once Thought to Arise Spontaneously

The invention of the microscope paved the way for other significant discoveries in microbiology. One of these discoveries was initiated because of the theory of **spontaneous generation,** which proposed that life could arise spontaneously from nonliving matter. This was a longheld belief, dating back hundreds of years. People commonly believed that mice, toads, flies, maggots, and other forms of life could appear spontaneously from piles of litter, dirt, and manure. A recipe for generating mice from soiled undergarments, written by the prominent physician, chemist, and physicist J. B. van Helmont (1577–1644), echoed these sentiments:

> If a dirty undergarment is squeezed into the mouth of a vessel containing wheat, within a few days (say 21) a ferment drained from the garments and transformed by the smell of the grain, encrusts the wheat itself with its own skin and turns it into mice. And what is more remarkable, the mice from the grain and undergarments are neither weanlings or sucklings nor premature but they jump out fully formed.

figure 1.1

Representative Types of Microorganisms
a. Scanning electron micrograph of *Escherichia coli*, a bacterium (colorized, ×12,000).
b. Electron micrograph of poliovirus, a virus (colorized, ×34,000). c. Scanning electron micrograph of *Aspergillus flavus*, a fungus (colorized, ×160). d. Scanning electron micrograph of *Tetrahymena pyriformis*, a protozoan (colorized, ×8,000). e. Scanning electron micrograph of *Cyclotella meneghiniana*, an alga (colorized, ×750).

figure 1.2
Anton van Leeuwenhoek (1632–1723)
Leeuwenhoek opened the doors to a new world of microorganisms with his early microscopes.

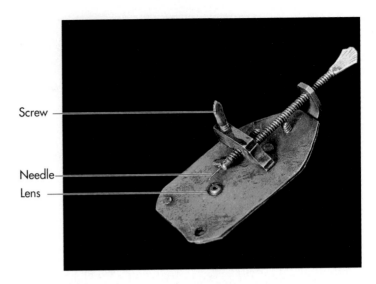

figure 1.3
Photograph of an Early Leeuwenhoek Microscope
The sample is placed at the tip of the needle and brought into focus by turning the screw. The sample is then viewed through the lens. Illumination is by light from a candle or sunlight.

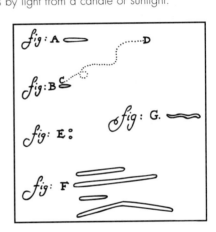

figure 1.4
Examples of Bacteria Drawn by Leeuwenhoek
Leeuwenhoek's drawings included rod-shaped bacteria (A, B, F, and G) and spherical, or coccus-shaped, bacteria (E).

Francesco Redi (1626–1698), an Italian physician, was one of the first persons to challenge the theory of spontaneous generation. He showed that meat placed in glass-covered or gauze-covered containers remained free of maggots, whereas meat in uncovered containers eventually became infested with maggots from flies laying their eggs on the meat. This and similar experiments by others dispelled the idea that organisms such as insects and mice could arise spontaneously. Despite these findings disproving spontaneous generation of macroorganisms, people still believed that microorganisms could arise this way.

With the discovery of microscopic organisms by Leeuwenhoek, the controversy surrounding spontaneous generation was renewed. A common argument used by proponents of spontaneous generation was that fresh food contained few if any microorganisms. However, if such food were allowed to sit at room temperature and spoil, large numbers of microscopic organisms would appear in the putrefied food. It was therefore only logical to assume that these spoilage organisms arose spontaneously within the food. John Needham (1713–1781), an English priest, fueled this argument by showing in 1749 that heated nutrient solutions poured into flasks that were subsequently stoppered soon became cloudy with microorganisms. The Italian priest and scientist Lazzaro Spallanzani (1729–1799) attempted to counter these arguments in 1776 by showing that no growth occurred in a wide variety of liquid nutrients that were boiled in flasks *after* sealing. However, supporters of spontaneous generation contended either that heating the liquids destroyed their ability to support life or sealing the flasks excluded oxygen that was required for life.

The controversy surrounding the theory of spontaneous generation remained unresolved until 1861 when Louis Pasteur, a French chemist (1822–1895), showed that microorganisms found in spoiled food were similar to those commonly found in the air. (Figure 1.5). He passed large amounts of air through guncotton (nitrocellulose) filters and observed with a microscope entrapped particles of inorganic matter and also microbes that were indistinguishable from those seen in putrefying foods. Pasteur furthermore showed that nutrient solutions boiled in swan-necked flasks that were exposed to the air remained sterile and untouched by microorganisms present in the environment. However, when the necks of these flasks were broken, the liquids inside quickly came in contact with microbes in the air and became contaminated. Pasteur concluded that these microorganisms found commonly in the air were responsible for the decomposition of foods and did not arise spontaneously. These simple but fundamental experiments by Pasteur were instrumental in disproving the theory of spontaneous generation.

figure 1.5
Louis Pasteur (1822–1895)
Pasteur was responsible for many accomplishments in microbiology, including disproving the theory of spontaneous generation, developing vaccines for anthrax and rabies, and developing a process (pasteurization) to destroy spoilage organisms in wine.

figure 1.6
Robert Koch (1843–1910)
Koch established the microbial cause for many infectious diseases through his postulates.

The Germ Theory of Disease States That Infectious Diseases Are Caused by Microorganisms

After Pasteur had shown that microorganisms existed in the environment and did not arise spontaneously, some other scientists focused their attention on the role of these organisms in disease. The belief that microorganisms might cause disease was known as the **germ theory of disease.**

During the latter part of the nineteenth century a German country physician named Robert Koch (1843–1910) investigated the nature of disease transmission from person to person (Figure 1.6). Koch conducted a series of experiments on animals he had infected with the spores of anthrax bacilli. Anthrax was a lethal, highly contagious bacterial disease of livestock that killed large numbers of animals and caused great economic losses to European farmers during Koch's time. In his meticulous studies reported in 1876, Koch discovered that anthrax could be transmitted from ani-mal to animal by injecting blood from a diseased animal into a healthy animal. This process could be repeated as many as 20 times, with similar symptoms appearing in each newly infected animal (Figure 1.7).

Furthermore, Koch showed that blood from infected animals contained large numbers of bacteria that were directly responsible for the disease. He transformed his observations and experiments into a series of criteria necessary for the logical association of specific microbes with specific diseases; these criteria are now known as **Koch's postulates.** Koch's postulates state that the following are needed to prove an organism causes a specific disease:

1. The specific microorganism should be shown to be present in all cases of the disease.

2. The specific microorganism should be isolated from the diseased host and grown in pure culture.

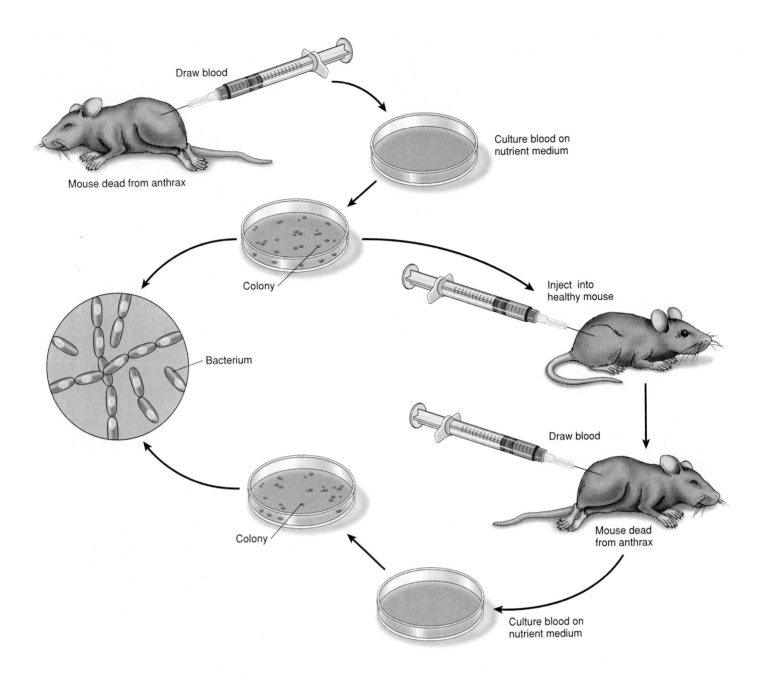

Draw blood

Mouse dead from anthrax

Culture blood on nutrient medium

Colony

Bacterium

Inject into healthy mouse

Draw blood

Mouse dead from anthrax

Culture blood on nutrient medium

Colony

figure 1.7

Koch's Postulates

Koch's postulates describe the steps required for logical proof that a microorganism causes a disease. The suspected microbe is isolated in pure culture from a diseased animal and then injected into a healthy animal. The infected animal should develop the same disease symptoms, and identical microorganisms should be isolated from this inoculated animal.

table 1.1

Dates During the Golden Age of Microbiology (1876–1906) When Infectious Diseases Were Associated with Specific Microorganisms

Disease	Year	Infectious Agent[a]	Discoverer(s)
Anthrax	1876	*Bacillus anthracis*	Koch
Pear fire blight	1877	*Erwinia amylovora*	Burrill
Gonorrhea	1879	*Neisseria gonorrhoeae*	Neisser
Malaria	1880	*Plasmodium malariae*	Laverans
Wound infections	1881	*Staphylococcus aureus*	Ogston
Tuberculosis	1882	*Mycobacterium tuberculosis*	Koch
Erysipelas	1882	*Streptococcus pyogenes*	Fehleisen
Diphtheria	1883	*Corynebacterium diphtheriae*	Klebs and Loeffler
Cholera	1883	*Vibrio cholerae*	Koch
Typhoid fever	1884	*Salmonella typhi*	Eberth and Gaffky
Bladder infections	1885	*Escherichia coli*	Escherich
Salmonellosis	1888	*Salmonella enteritidis*	Gaertner
Tetanus	1889	*Clostridium tetani*	Kitasato
Gas gangrene	1892	*Clostridium perfringens*	Welch and Nuttall
Plague	1894	*Yersinia pestis*	Yersin and Kitasato
Botulism	1897	*Clostridium botulinum*	Van Ermengem
Shigellosis	1898	*Shigella dysenteriae*	Shiga
Syphilis	1905	*Treponema pallidum*	Schaudinn and Hoffmann
Whooping cough	1906	*Bordetella pertussis*	Bordet and Gengou

[a]The names for the infectious agents are those in current use rather than those given at the time of their discovery.

3. This freshly isolated microorganism, when inoculated into a healthy, susceptible host, should cause the same disease seen in the original host.

4. The microorganism should be reisolated in pure culture from the experimental infection.

Koch's postulates initiated the **Golden Age of Microbiology** (1876–1906), an era during which the causes of most bacterial diseases were discovered. With Koch's criteria scientists were able to identify major microbial pathogens. It should be recognized that whereas Koch's postulates are invaluable for identifying specific microbial agents of diseases, they are not applicable to all microorganisms; there are technical limitations in using these criteria for all infectious diseases. Some transmissible microorganisms cannot readily be cultured on artificial media at this time because they are obligate intracellular parasites (for example, viruses; *Rickettsia rickettsii,* the causative agent of Rocky Mountain spotted fever; and *Chlamydia trachomatis,* the causative agent of trachoma and one of the causes of the sexually transmitted disease nongonococcal urethritis, or NGU). Other transmissible microorganisms cannot be grown in vitro (outside a living organism) and thus cannot be artificially cultured (for example, *Treponema pallidum,* the causative agent of syphilis). Clinical signs and symptoms of some infectious diseases are manifested differently in experimental laboratory animals and in humans (for example, *Neisseria gonorrhoeae,* which causes gonorrhea in humans but not in animals). It should also be noted that not all diseases are infectious (for example, cardiovascular diseases, diabetes, black lung disease, and certain types of cancer), and knowing which are infectious can be very difficult (for example, duodenal and gastric ulcers, previously thought to be induced by stress, emotional disturbance, and other noninfectious factors, are now known to be caused by the bacterium *Helicobacter pylori*).

Despite these limitations, Koch's postulates have been instrumental in showing the relationships between microorganisms and the diseases they cause. The importance of Koch's work as a foundation of modern microbiology is evident in the numerous diseases that were proven to be caused by specific microbes in the few years immediately after his historical studies of anthrax (Table 1.1).

Immunology Is the Study of Resistance of Organisms to Infection

While Koch was demonstrating the infectious nature of many diseases, other scientists were making discoveries that would significantly reduce the incidence of these diseases in the world. In 1884 the Russian zoologist Élie Metchnikoff (1845–1916) discovered that certain cells in the blood could engulf foreign particles such as

figure 1.8
Paul Ehrlich (1854–1915)
Ehrlich proposed that proteins (antibodies) in serum could combine with and destroy foreign substances (antigens). For his work in immunology he shared with Élie Metchnikoff the 1908 Nobel Prize in physiology and medicine.

figure 1.9
Edward Jenner (1749–1823)
Jenner's experiment with vaccination proved that cowpox conferred immunity against smallpox, and laid the foundation of modern immunology as a science.

bacterial cells and protect the body against infectious diseases. Metchnikoff called these cells *phagocytes*. This discovery formed the basis for the field of **immunology,** the study of resistance of organisms to infection.

Other scientists believed that noncellular components of the blood could also be protective (humoral immunity). Two of these scientists, Paul Ehrlich (1854–1915) and Emil von Behring (1854–1917), worked together to show that specific antitoxins (antibodies, or protein molecules in the blood of an immunized individual) effectively neutralized potent diphtheria toxin (antigen). In the 1890s Ehrlich conducted further experiments to demonstrate that immunity to toxins could be transferred from one animal to another by injection of immune serum (serum containing antibodies). These and other observations led Ehrlich to hypothesize that a toxin or other foreign substance entering the body induced the formation of protein molecules (antitoxins or antibodies) that combined specifically with the foreign substance inducing their formation (Figure 1.8).

The field of immunology actually began with an experiment performed in 1798 by Edward Jenner (1749–1823). Jenner was an English country doctor who showed that immunity could be conferred by vaccination (Figure 1.9). During the eighteenth century, smallpox frequently occurred in epidemic proportions in continental Europe and England. It was observed at the time that individuals who at one point had cowpox (a similar, less serious disease than smallpox that was contracted from cows) seldom contracted the disease agent of smallpox. In a classical experiment, Jenner inoculated an eight-year-old boy with pus taken from the lesion of a dairymaid with cowpox. The boy was later inoculated with material from a smallpox patient but did not develop smallpox. Jenner repeated this experiment several times with different people, and the results of these experiments led Jenner to propose the procedure that came to be known as **vaccination** [Latin *vacca,* cow].

Viruses Are Submicroscopic Filterable Agents Consisting of Nucleic Acid Surrounded by a Protein Coat

The word **virus** [Latin *virus,* poison] was originally used by the ancient Romans to describe any type of poisonous material. Later, when microbiology began to emerge as a discipline, the word *virus* implied any infectious microbe. It was not until the late 1800s that viruses were described as distinct submicroscopic filterable entities capable of replication and infection.

The first evidence that organisms smaller than bacteria might exist and be associated with disease was proposed in 1892 by a Russian scientist, Dmitri Ivanowsky (1864–1920). In his studies of mosaic disease of tobacco plants, Ivanowsky observed that the agent responsible for the disease easily passed through filters that stopped all known bacteria. Ivanowsky surmised that the infectious agent was a filterable microorganism that could not be seen with a light microscope. These observations established filterability as an early criterion for distinguishing viruses from bacteria and other microorganisms.

Martinus Willem Beijerinck (1851–1931), a Dutch soil microbiologist, studied the infectious agent of tobacco mosaic disease in the 1890s and discovered that not only was it filterable, but, unlike bacteria, it could also easily diffuse through solid growth

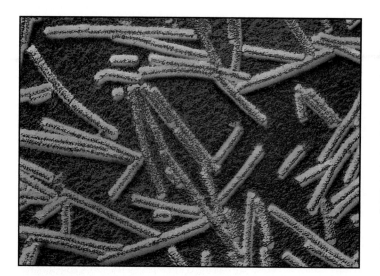

figure 1.10
Electron Micrograph of Tobacco Mosaic Viruses
The virus consists of a single long RNA molecule enclosed in a cylindrical protein coat (colorized, ×220,000).

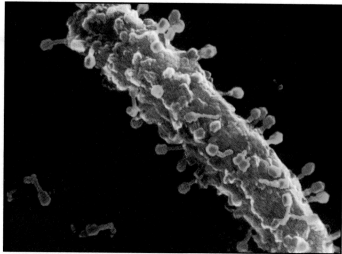

figure 1.11
Scanning Electron Micrograph of Bacteriophages Attached to a Bacterium
Bacteriophages (blue) are shown attached to an *Escherichia coli* cell (colorized, ×36,000).

media. He used the term *contagium vivum fluidum* (fluid infectious principle), or **virus,** to describe the agent. A major contribution to the understanding of viruses was made by chemist Wendell Stanley (1904–1971) of the University of California, Berkeley, who crystallized the tobacco mosaic virus (TMV) in 1935 (Figure 1.10). Crystallization of the TMV was a first step in elucidating the chemical structure and composition of viruses. Within a few years of Stanley's work, viruses were found to be composed mainly of nucleic acid contained within a protein coat.

The observations of Ivanowsky, Beijerinck, and Stanley were all made on the TMV, a plant virus. The first animal virus characterized was the virus responsible for bovine foot-and-mouth disease. This virus was isolated in 1898 from infected cattle by Friedrich Loeffler (1852–1915) and Paul Frosch (1860–1928). Like Ivanowsky, Loeffler and Frosch found that the virus causing foot-and-mouth disease was filterable. They concluded that the responsible agent was not a bacterium, but a filterable particle.

Bacterial viruses, more commonly called **bacteriophages,** or **phages** [Greek *phagein,* to devour], were discovered by Frederick W. Twort (1877–1950), a British scientist, and Felix d'Hérelle (1873–1949), a French scientist, in the early 1900s. Twort, in 1915, and d'Hérelle, in 1917, both reported observing a filterable agent that destroyed bacteria growing on solid media. D'Hérelle considered this filterable agent to be a parasite of bacteria and named it *bacteriophage* (Figure 1.11). Despite d'Hérelle's extensive work on the growth and infectious nature of bacteriophages, few scientists at the time accepted his findings. It was not until later in the 1930s, when the German biochemist Martin Schlesinger purified and characterized bacteriophages, that these viruses established their own unique place in the microbial world.

Modern Microbiology

Through the early work of these and many other scientists, microbiology began to emerge as a distinct discipline. A new classification of scientists, **microbiologists,** and new terminology appeared in the scientific world. Microbiology began to play an important and vital role in many areas. This role has expanded as microbiologists continue to make significant contributions to science. In recent years, with the discovery of new techniques in molecular biology and the application of these techniques to microorganisms, microbiology has entered a new golden age.

Recombinant DNA Technology Involves the Manipulation of Genes in Organisms

Several significant discoveries led to this new golden age of microbiology. In the early 1960s Werner Arber, a Swiss microbiologist, discovered that certain bacteria produced enzymes that cleaved viral DNA at specific sites to protect the bacteria from viral infection. These enzymes, called **restriction endonucleases,** were later purified and characterized by Hamilton Smith and Daniel Nathans. The discovery led to the development of laboratory techniques to manipulate genes in living organisms, called **recombinant DNA technology,** and genetic engineering. Using these enzymes, it now was possible to isolate and move genes from one organism to another. Arber, Smith, and Nathans were awarded the Nobel Prize in physiology or medicine in 1978 for their discovery and studies of restriction endonucleases.

In 1972 Paul Berg of Stanford University created the first recombinant DNA molecule by successfully splicing DNA from

Microbes in Space

An exciting recent development in microbiology has been the study of the presence and spread of microorganisms in the closed microgravity environment of spacecrafts. On earth, gravity is an important physical force in reducing aerosols and thus the spread of infectious microorganisms. However, under the microgravity conditions of outer space, droplets of microorganisms generated by coughing, sneezing, and talking may remain suspended for hours in the air and be a source of infection. Consequently, microorganisms that normally would not be considered a problem on earth could potentially be harmful to astronauts in the closed environment of space vehicles and the proposed international space station. Evidence of this potential health hazard was seen in the early Apollo missions in which crew members experienced upper respiratory illnesses (influenza, rhinitis, and pharyngitis), gastroenteritis, and mild dermatological problems. On the Apollo 13 mission, an in-flight malfunction in the service module resulted in a cold, moist environment, which led to a severe Pseudomonas aeruginosa urinary tract infection in one of the crew members. In-flight antibiotic therapy was ineffective until he could be returned to earth, where bacterial isolation was possible.

Because of these problems and to minimize future health hazards due to microorganisms, the National Aeronautics and Space Administration (NASA) initiated extensive preflight quarantine measures for astronauts, starting with the Apollo 14 mission. During the quarantine period, astronauts are restricted to limited-access areas to reduce contact with contaminated objects. Only individuals screened for the presence of potential pathogens may come in contact with the crew members. These quarantine procedures have been successful, and no major illnesses have been reported during the Apollo 14 through Apollo 17 or space shuttle missions.

These early space missions have been short, typically lasting only a few days or weeks. In comparison, the space station will be a permanent platform, with confinements as long as 90 days planned for each team of astronauts. To prevent gross contamination of the space station environment by pathogenic microorganisms, NASA plans to require a minimum 14-day isolation period (the incubation time for most pathogenic microorganisms) for astronauts prior to launch. Family members of the crew will be monitored for infectious diseases during the 30 to 60 day preflight period, and all crew members will be immunized against infectious disease agents. The environment of the space station will be monitored continuously with air, water, and surface samples. Although it will be impractical to have a sterile, noncontaminated environment on the space station, these measures by NASA will ensure an environment that is safe for human habitation (Figure 1.12).

figure 1.12

An Astronaut Demonstrates the Shirtsleeve Environment of Life Aboard the Space Station

The astronaut is taking notes at the NASA health maintenance facility, which has capabilities for diagnostic and environmental microbiology tests.

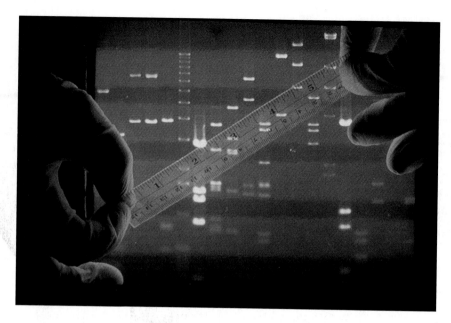

figure 1.13

Examining DNA Fragments on a Gel

DNA fragments can be separated by electric current into distinct patterns on a gel. These patterns are unique to each organism.

a virus into a bacterial chromosome. Berg was honored as a co-recipient of the 1980 Nobel Prize in chemistry for this work. Berg's studies were followed in 1973 by elegant experiments conducted by Herbert Boyer of the University of California in San Francisco and Stanley Cohen of Stanford University in which foreign genes were inserted into a plasmid, or extrachromosomal piece of DNA, which could then be transferred into a bacterium (Figure 1.13). With these techniques and genetic tools, it was now possible to isolate a gene from one organism, splice it into a plasmid, and insert the plasmid with the foreign gene into a bacterium. The process of **gene cloning** was thus born.

The Polymerase Chain Reaction Permits Rapid Amplification of Genes Outside the Cell

The most recent technique developed in the area of recombinant DNA technology is the **polymerase chain reaction (PCR),** developed by Kary Mullis in 1985. This innovative laboratory procedure, which was conceived by Mullis two years earlier during an evening drive through the mountains of northern California, amplifies a gene a billionfold in as little as one hour without the need for a living cell. The PCR technique has wide applications in microbiology, including rapid and accurate diagnosis of infectious diseases using genetic probes to identify amplified genes from microbial pathogens, and amplification of microbial genes for further study and characterization. Kary Mullis received the 1993 Nobel Prize in chemistry for his discovery of PCR.

Microorganisms Have Important Roles in Medicine, the Environment, Agriculture, and Industry

As we enter the next century, microbiology continues to play important roles in our world. In *medical microbiology,* new and emerging infections such as the Ebola virus and drug-resistant bacteria continue to plague and perplex scientists. Although smallpox and a few other diseases have been effectively eliminated, there has been a resurgence of tuberculosis, polio, and other diseases that once were thought conquered. Infectious diseases remain the leading cause of death in the world. Diseases such as Acquired Immune Deficiency Syndrome (AIDS), caused by the human immunodeficiency virus (HIV), destroy our body's immune system and affect millions of people worldwide.

Microbes are indispensable in the recycling of chemical elements, interaction with other organisms in ecosystems, and sewage treatment—a field called *microbial ecology.* In recent years microorganisms have especially been useful in the biodegradation of chemical pesticides, pollutants, and petroleum products. Such bioremediation benefits the environment because it not only removes the toxic chemical, but also does it in such a manner as to minimize additional environmental damage. The 1989 oil spill by the supertanker *Exxon Valdez,* which contaminated Prince William Sound, Alaska, with 11 million gallons of oil, provided scientists with an opportunity to use microorganisms for cleanup of the spill. Nutrients added to shorelines contaminated by oil stimulated growth of indigenous oil-degrading bacteria, resulting in

a.

b.

figure 1.14

Contamination Caused by the *Exxon Valdez* Oil Spill
a. Example of a contaminated beach on Knight Island and Prince William Sound, Alaska, after the oil spill. b. Cleanup operation on Knight Island and Prince William Sound.

significantly increased bioremediation rates (Figure 1.14). These results suggested that naturally occurring microbes may be adequate for the bioremediation of oil spills, if these microbes are provided with sufficient nutrients for growth and metabolism.

Microbiology is important to *agriculture* and the *food industry*. Certain bacteria (for example, *Rhizobium* and *Bradyrhizobium*) live in symbiotic associations with plants called legumes (soybeans, alfalfa, clover, peas, and so forth). In these associations, the bacterium converts (fixes) atmospheric nitrogen into nitrogen-containing chemical compounds, which can then be used by the plant as a source of nitrogen. Crop rotation between legumes and nonlegumes is one method by which farmers replenish the soil with nitrogen without using expensive chemical fertilizers that may be environmentally damaging. Cheeses, yogurt, buttermilk, sauerkraut, pickles, and beer are examples of

foods and beverages prepared using microorganisms. Yeasts, a rich source of protein, are being considered as a possible major source of food for human consumption in a world where traditional food supplies may become inadequate for a rapidly growing population.

Microorganisms are important in *industry* as producers of antibiotics, vitamins, and chemicals. Monosodium glutamate, a flavor enhancer in foods, is commercially produced using the bacterium *Corynebacterium glutamicum*. Enzymes such as amylases (starch digestion), used in desizing agents and detergents, and proteases (protein digestion), used for spot removal, meat tenderizing, and wound cleansing, are commercially produced by bacteria and fungi. The multibillion-dollar pharmaceutical industry is based on microbial production of antibiotics such as bacitracin, chloramphenicol, neomycin, streptomycin, and tetracycline.

In the eighteenth and nineteenth centuries, many of the world's leading scientists believed that microorganisms arose spontaneously from nonliving matter. This doctrine of spontaneous generation originated from ancient beliefs that maggots that appeared on decaying meat and flies found on manure arose spontaneously. Several scientists, including Francesco Redi (1626–1697), Lazzaro Spallanzani (1729–1799), and Theodor Schwann (1810–1882), challenged the idea of spontaneous generation. Redi, an Italian physician, conducted a simple experiment in which meat was placed in three containers. One was covered with paper, a second covered with a fine gauze that excluded flies from entering, and a third was left uncovered. Flies laid their eggs on the meat in the uncovered container, and maggots, the larvae of flies, subsequently developed from the eggs. Meat in the other two containers did not generate maggots spontaneously, although flies attracted to meat in the gauze-covered container laid eggs on the gauze and the eggs subsequently produced maggots on the gauze. Spallanzani, an Italian naturalist, and Schwann, a German physiologist, provided further evidence that microorganisms do not arise spontaneously. Spallanzani showed that heating animal and vegetable juices, called infusions, in hermetically sealed flasks prevented the appearance of microorganisms (which he called *animalcula*). Breaking the seals exposed the flasks' contents to the air and resulted in microbial growth. Schwann repeated Spallanzani's experiments, but he heated air entering the flask through a bent tube to kill any microorganisms in the air (Figure 1P.1). Although the experiments of these three scientists showed that contaminating microorganisms originated from air, the results could not be reproduced by other scientists, and the controversy concerning spontaneous generation continued.

In 1861 Louis Pasteur convincingly showed that microorganisms could not arise spontaneously. Pasteur designed an ingenious series of experiments demonstrating that microorganisms from the air were responsible for contaminating heated infusions. Pasteur first filtered large volumes of air through guncotton to collect any microorganisms that might be in the air. The guncotton was dissolved in a mixture of alcohol and ether, and the entrapped microorganisms were examined under a microscope. Pasteur observed that microorganisms in the air had various sizes and shapes, and some contained starch granules, which could be stained with iodine.

To show that these airborne microbes were capable of contaminating infusions, Pasteur designed long-necked flasks that allowed the contents to be effectively heated, while at the same time let air in. Infusions such as yeast water, sugared yeast water, sugar beet juice, or pepper water were placed into the flasks and boiled for several minutes to kill any microorganisms already present. The flasks were allowed to cool. Although the infusions were exposed to air, no growth appeared in the flasks. Pasteur concluded that the curved necks of the flasks slowed the movement of airborne microorganisms and prevented their contamination of the infusions. Pasteur furthermore observed that when the necks of the flasks were broken, the infusions became contaminated (Figure 1P.2). Several of Pasteur's original swan-necked flasks have been preserved at the Pasteur Institute in Paris and still contain their original nutrient solutions—untouched by microorganisms and uncontaminated.

Pasteur's experiments were logical, simple, and easily reproducible. As a result, they left little doubt that the doctrine of spontaneous generation was untenable.

References

Pasteur, L. 1861. Mémoire sur les corpuscles organisés qui existent dans l'atmosphère. Examen de la doctrine des générations spontanées. *Annales des sciences naturelles*, 4th series 16:5–98.

Schwann, T. 1837. Vorläufige Mittheilung betreffend Versuche über die Weingahrung und Fäulnis. *Annalen der Physik und Chemie* 41:184–193.

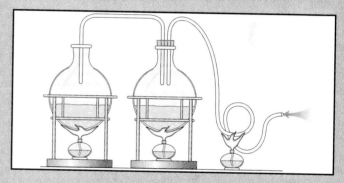

figure 1P.1
Schwann's Experimental System
The stoppered center flask containing infusion is heated to boiling. Air entering the center flask from the right is heated through a bent tube to kill any microorganisms in the air. Air leaving the center flask to the left is sent through a second heated infusion to prevent contamination of the center flask from backflow.

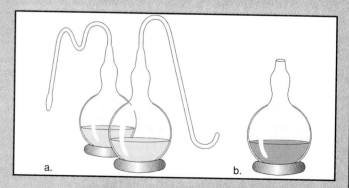

figure 1P.2
Pasteur's Experimental System
a. Infusions in swan-necked flasks are heated and then cooled. No growth appears in the infusions, even though they are exposed to the environment. b. Growth occurs if the flask necks are broken, allowing the infusions to become contaminated.

1. Microbiology is the study of agents too small to be seen with the unaided eye, specifically bacteria (bacteriology), viruses (virology), fungi (mycology), protozoa (protozoology), and algae (phycology).

2. Zacharias Janssen combined two lenses to form the first crude compound microscope. Robert Hooke made detailed drawings of plant tissues from his observations with the microscope. Anton van Leeuwenhoek is considered to be the father of microbiology because he constructed hundreds of microscopes and described in detail many of the microbial shapes he saw with his microscopes.

3. The theory of spontaneous generation was disproved by the work of scientists such as Lazzaro Spallanzani and Louis Pasteur, who showed that microorganisms could not arise spontaneously and grow in properly sterilized nutrient solutions.

4. Robert Koch established a series of criteria necessary for association of specific microorganisms with specific diseases in 1876. These criteria, known as Koch's postulates, initiated the Golden Age of Microbiology (1876-1906), a period during which the causes of most bacterial diseases were discovered.

5. Élie Metchnikoff discovered in 1884 that cells in the blood could engulf foreign particles and protect the body against infectious diseases. Earlier, in 1798, Edward Jenner showed that immunity could be conferred by vaccination. These two discoveries were instrumental in the development of the field of immunology.

6. Virology developed as a field of study with Dmitri Ivanowsky's discovery in 1892 that filterable agents were responsible for tobacco mosaic disease. The term virus was later coined by Martinus Willem Beijerinck to describe this filterable infectious agent. Bacterial viruses, or bacteriophages, were discovered by Frederick W. Twort and Felix d'Hérelle in the early 1900s in their observations of filterable agents that were parasites of bacteria.

7. A new golden age of microbiology has emerged in recent years with the discovery of new techniques in molecular biology and the applications of these techniques to microorganisms. The first recombinant DNA molecule was created by Paul Berg in 1972 when he successfully spliced DNA virus into a bacterial chromosome. In 1985, Kary Mullis developed the polymerase chain reaction, a process that amplifies a gene a billionfold in as little as one hour.

8. Today microbiology plays an important role in medicine, ecology, agriculture, food production, and industry. As we enter the twenty-first century, microorganisms will continue to be important in solving environmental problems and the intricate mysteries of the living cell.

Microbiology is a relatively young discipline that has evolved from the minuscule microbes observed through Leeuwenhoek's single-lens microscopes three centuries ago, to the discovery during the Golden Age of Microbiology that specific microorganisms cause infectious diseases, to the genetic engineering of microbes to benefit humankind and society in a new golden age of microbiology. As microbiology has evolved, it also has diversified into a discipline that significantly impacts many other fields. The foundations of molecular biology are based on detailed investigations of DNA, RNA, and proteins in bacteria and viruses. Many of the early, definitive studies in genetics were conducted using the bread mold *Neurospora crassa*. Microbes are indispensable in the recycling of chemical elements and have been instrumental in recent advances in the bioremediation of chemical pollutants and toxic wastes. The modern multibillion-dollar pharmaceutical industry developed after World War II, based on the earlier discoveries of the antibiotic penicillin, produced by the mold *Penicillium notatum*, in 1928 by the Scottish physician and bacteriologist Alexander Fleming (1881–1955) and of the antibiotics gramicidin and tyrocidine, produced by the soil bacterium *Bacillus brevis*, in 1939 by the French microbiologist René Dubos. In the same manner as microbiology is interwoven in the fabric of biology, the underlying theme of this textbook will be the roles of evolution and diversity in the microbial world. This theme of evolution and biodiversity in the microbial world will be reinforced in a special text box at the end of each chapter.

Questions

Short Answer

1. Identify 10 subdisciplines or fields of microbiology.

2. Who was the first person to describe cells? When?

3. Who was the first person to describe microbes? When?

4. What is the theory of spontaneous generation? When was it developed?

5. Describe Francesco Redi's experiment. What did he conclude?

6. What did critics say about Spallanzani's experiment?

7. What innovative strategy did Pasteur use in his experiments relating to spontaneous generation?

8. What is the germ theory? When was it developed?

9. State Koch's postulates.

10. When is/was the golden age of microbiology? What is special about that time period?

11. What were the first immune cells to be discovered? Who discovered them? When?

12. Who developed the process known as vaccination? When?

13. What two materials were found to constitute viruses?

14. What are restriction endonucleases? Why are they important?

15. Name several diseases that are infectious.

16. Name several diseases that are not infectious.

17. What type of disease is the leading cause of death in the world today?

Critical Thinking

1. Pasteur is credited with disproving the theory of spontaneous generation. Explain why previous experiments by Redi and Spallanzani were not conclusive.

2. Explain the significance of Koch's postulates. Give examples of exceptions to Koch's postulates.

3. Explain the significance of the germ theory of disease.

4. Many early microbiologists (for example, Hooke, Leeuwenhoek, Spallanzani, and Needham) were not scientists. Explain how this was possible. Does this make their work any less significant?

 Supplementary Readings

Aharonowitz, Y., and G. Cohen. 1981. The microbiological production of pharmaceuticals. *Scientific American* 245:140–152. (A discussion of the role of microorganisms in synthesis of antibiotics, vitamins, antiviral drugs, and hormones.)

Beardsley, T. 1994. Big-time biology. *Scientific American* 271:90–97. (A discussion of how a new industry has been generated from the new technology of genetic engineering.)

Brill, W. 1980. Biochemical genetics of nitrogen fixation. *Microbiological Reviews* 44:449–467. (A review of the biochemistry and genetics of nitrogen fixation by bacteria.)

Brock, T., ed. 1961. *Milestones in microbiology.* Englewood Cliffs, N.J.: Prentice-Hall. (A compilation of historical papers written by scientists making significant contributions to the field of microbiology.)

Capecchi, M.R. 1994. Targeted gene replacement. *Scientific American* 270:52–59. (A discussion of how genes cloned and propagated in bacteria can be introduced into living cells to replace defective genes.)

Gilbert, W., and L. Villa-Komaroff. 1980. Useful proteins from recombinant bacteria. *Scientific American* 242:74–94. (A discussion of how recombinant DNA methods can be used to produce insulin, interferon, and other industrially useful compounds.)

Schlessinger, D., ed. 1981. *Genetics and molecular biology of industrial organisms. Microbiology—1981.* Washington, D.C.: American Society for Microbiology. (A series of short research articles describing the use of microorganisms for production of important chemical compounds.)

chapter two 2

CELL MORPHOLOGY AND MICROSCOPY

Size, Shape, and Arrangement of Procaryotic Cells
Size and Scale
Shapes
Cell Arrangements

Microscopy
The Microscope
Stains
The Dark-Field Microscope
The Phase-Contrast Microscope
The Fluorescence Microscope
The Electron Microscope

PERSPECTIVE
Streptococci and Cell Wall Growth

EVOLUTION AND BIODIVERSITY

Nomarski differential interference contrast micrograph of diatoms (×500).

 Microbes in Motion ⸻⸻⸻ `PREVIEW LINK`

This chapter covers the key topic of classification of bacteria into groups based on shape and Gram reaction. The following sections of the *Microbes in Motion* CD-ROM may be useful as a preview to your reading or as a supplemental study aid:

Miscellaneous Bacteria: Mycoplasma, 1–6. *Bacterial Structure and Function:* Cell Wall (Bacterial Shapes) 6–9; Bacterial Groups, 1–4.

Most life-forms that are familiar to us are organisms easily seen with the naked eye. These **macroorganisms** come in all sizes, shapes, and forms. It has been conservatively estimated that over 1 million animal species and 350,000 plant species are known to humans. Beyond the realm of visibility there exists an extensive and diverse group of organisms so small that they normally cannot be seen without the aid of a microscope. These **microorganisms** are ubiquitous and include such widely different types as bacteria, algae, fungi, and protozoa. Most microbes are unicellular and measure, at most, only several micrometers in diameter. A few, such as certain algae and fungi, can form large multicellular structures that rival the sizes of some macroorganisms. Examples include seaweed, which can have lengths approaching 10 m, and mushrooms, with their stalks and prominent umbrella-shaped or cone-shaped caps. Simply because an organism is large, visible, and multicellular does not necessarily exclude its constituent cells from the classification of microorganisms.

There are two kinds of cells: **procaryotic**, from Latin *pro* (before), Greek *karyon* (nucleus), and **eucaryotic**, from Latin *eu* (true). The most notable difference between procaryotic and eucaryotic cells is that procaryotic cells do not have a membrane-enclosed nucleus. Instead of a distinct nucleus, procaryotic cells have a **nucleoid,** a region that contains a single, circular DNA molecule. Eucaryotic cells have, in addition to a well-defined membrane-enclosed nucleus, a variety of specialized membrane-enclosed organelles, including mitochondria and chloroplasts, as well as a high level of internal structural organization not found in procaryotic cells.

Among the five groups of microorganisms, the bacteria are procaryotes,[a] and protozoa, algae, and fungi are eucaryotes. In the past, all procaryotes were considered to be bacteria. It is now recognized that three major groups of organisms (Bacteria, Archaea, and Eucarya) evolved from a universal ancestor. The unique ribosomal RNA structure, plasma membrane lipid composition, cell-wall components, RNA polymerase, and mechanism of protein synthesis in Archaea make them distinct from Bacteria (see Archaea, page 307). In fact, from a phylogenetic perspective, Archaea are more closely related to Eucarya than to Bacteria. Evidence from ribosomal RNA sequencing suggests that the universal phylogenetic tree initially evolved in two directions, Bacteria and Archaea/Eucarya. Archaea, which includes microbes that inhabit extreme environments (high temperature, high salinity, low pH, and so on), are among the most primitive organisms known and may very well represent examples of earth's earliest life-forms.

Viruses are not cellular and therefore are not considered either procaryotes or eucaryotes. A virus is an obligate intracellular parasite composed of nucleic acid (DNA or RNA) within a host cell.

Much of our knowledge of microorganisms as well as macroorganisms comes from studies of procaryotes. Procaryotes are the best-understood microbes. Although procaryotes are diverse in their physical and metabolic properties and include a wide variety of cell forms, they nonetheless have been extensively studied and are well characterized.

Size, Shape, and Arrangement of Procaryotic Cells

Procaryotes come in a variety of sizes, shapes, and arrangements. These characteristics, which frequently are unique to certain groups of procaryotes, help in the identification and classification of these microorganisms. The form and structure, or morphology, of a bacterium usually provides the first clue to its identity.

Size and Scale: Procaryotes Are Small

Procaryotes are the smallest free-living organisms known. They typically cannot be seen without the aid of a microscope; some are so small that specialized microscopes must be used to see them. Even among these minute organisms, however, there is extensive variation in size.

The smallest procaryotes—and probably the smallest organisms capable of autonomous growth—are members of the genus *Mycoplasma*. Mycoplasmas occur as normal inhabitants of the respiratory and genitourinary tracts of humans and other animals. They may cause such diseases as urethritis and primary atypical pneumonia (PAP)—a mild form of pneumonia. Mycoplasmas have diameters of only 125 to 250 nm[b]—5 million of these organisms could be laid side-by-side on the head of a straight pin. These microbes lack the distinctive cell wall that is found in other bacteria and therefore have no specific shape; rather, they have a variety of shapes—a trait known as **pleomorphism** [Greek *pleon*, more, and *morphe*, form]. As a consequence of pleomorphism, *Mycoplasma* cells are capable of passing through membrane filters that would retain larger and more rigid bacteria. The size of a

[a]Procaryotes consist of the domain **Bacteria** (from Greek *bakterion* [a small rod]) and the domain **Archaea** (from Greek *archaios* [ancient]). Eucaryotes are taxonomically placed in the domain **Eucarya.** The phylogenetic distinctions within these three domains are explained in Chapter 11. In this chapter and throughout this book, the word *Bacteria* (uppercase "B") refers to organisms in the domain Bacteria, as contrasted phylogenetically to organisms in the domains Archaea and Eucarya. The word *bacteria* (lowercase "b") refers to procaryotes in general, without reference to phylogeny.

[b]Units of linear measurement commonly used in microbiology are:
1 millimeter (mm) = 10^{-3} meter
1 micrometer (μm) = 10^{-6} meter
1 nanometer (nm) = 10^{-9} meter
1 angstrom (Å) = 10^{-10} meter

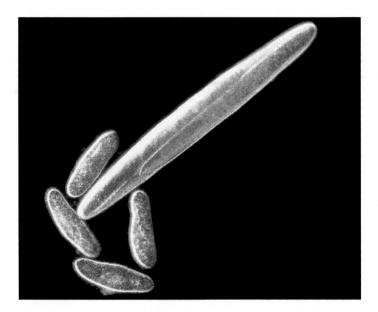

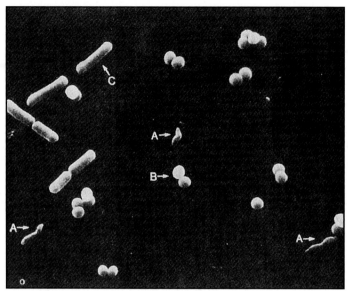

figure 2.1
Photomicrograph of *Epulopiscium fishelsoni*
Epulopiscium, the largest known procaryote, is compared in size with four eucaryotic paramecia (colorized, ×200).

figure 2.2
Scanning Electron Micrograph Comparing Shapes of Procaryotic Cells
Shown are *Campylobacter fetus* (A), spirilla; *Staphylococcus aureus* (B), cocci; and *Bacillus subtilis* (C), rods (×4,200).

Mycoplasma cell approaches the minimum required for a living, independent cell that can still contain the macromolecules and machinery required for self-reproduction.

 **Miscellaneous Bacteria**
Mycoplasma • pp. 1–6

At the other extreme of size among procaryotes are *Spirillum* (1.7 μm × 28 μm) and *Epulopiscium fishelsoni* [Latin *epulatus*, to feast, *piscis*, fish] (80 μm × 600 μm) (Figure 2.1). Members of the genus *Spirillum* are sometimes large enough to be seen without the aid of a microscope. It is believed that some of the first bacteria seen by Leeuwenhoek with his crude microscopes may have been *Spirillum*. In 1993 researchers at Indiana University isolated *E. fishelsoni* from the intestinal tract of the surgeonfish off the Great Barrier Reef in Australia; it is so large that it initially was thought to be a eucaryote. However, analysis of its DNA and cellular features confirms that it is a procaryote.

Procaryotes Have Different Shapes

Procaryotes exist in two basic forms: the **coccus** [Greek *kokkos*, berry], a spherical cell; and the **rod,** or **bacillus** [Latin *baculus*, stick], a cylindrical cell. There are many variations of these two fundamental shapes (Figure 2.2). Some bacteria have shapes that are not exclusively spherical or cylindrical, but a combination of these two forms. These organisms are known as **coccobacilli.** A few bacteria form spirals and are called **spirilla,** [Latin *spira*, coil, twist], or **spirochetes** [Greek *speira*, coil, *chaite*, bristle], depending on their mechanism of motility. A spirillum has a rigid form and moves by flagella (specialized appendages), whereas a spirochete flexes its entire body and moves by axial filaments in an undulating or rotat-

ing motion. Filamentous bacteria form long threads of cellular mass that can branch in several directions up to several hundred micrometers in length. Pleomorphism is a trait of certain types of bacteria that have no definitive shape or have a variety of shapes either because they lack cell walls (*Mycoplasma*), or because their cells divide in unusual manners to produce different forms (*Corynebacterium*) or reproductive buds or appendages (*Hyphomicrobium*).

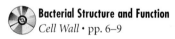 **Bacterial Structure and Function**
Cell Wall • pp. 6–9

Certain groups of bacteria change forms during the course of their life cycle. For example, *Chlamydia*, a bacterium that causes trachoma and other diseases, alternates between two shapes: (1) a small resistant cell form called an elementary body (0.3 μm in diameter) that is geared for survival in the environment, but is unable to replicate and (2) a much larger reticulate body (1 μm in diameter) that is involved in cellular multiplication within a host. Thus *Chlamydia* can assume two completely different forms, one suited for cellular survival in a harsh environment, and one for intracellular replication and proliferation. *Rhizobium*, a bacterium that converts gaseous nitrogen to ammonium in a process called nitrogen fixation, undergoes a similar type of transformation, changing from rods in a free-living phase to swollen, misshapen forms called bacteroids during infection of plant roots. These bacteroids, incapable of division, carry out nitrogen fixation (see nitrogen fixation, page 418).

Procaryotes Have Different Cell Arrangements

Unlike *Chlamydia* and *Rhizobium*, most bacteria maintain constant, representative shapes that are useful aids in identification. Some bacteria also have cells that remain attached after division,

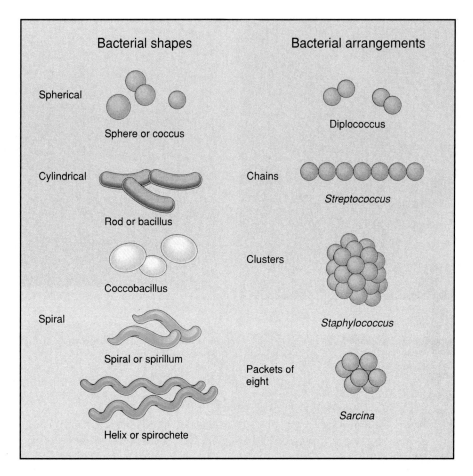

figure 2.3
Representative Cell Shapes and Arrangements of Procaryotes

clumped in distinctive cell arrangements (Figures 2.3 and 2.4). Bacterial names frequently reflect both cell morphologies and cell arrangements, and can be descriptive aids in identification. Thus the genus name *Streptococcus* [Greek *streptos,* winding, twisted] appropriately describes a bacterium that is often found as chains of cocci due to incomplete separation among dividing cells and division in a single plane. Bacteria of the genus *Staphylococcus* [Greek *staphyle,* bunch of grapes] are also cocci but are attached in clusters, which form as the bacteria divide in random planes rather than in a single plane. Other arrangements of bacteria include cubical packets of cells, as in the genus *Sarcina* [Latin *sarcina,* a package, bundle]; angular arrangements of cells, as in *Corynebacterium* [Greek *koryne,* club]; and filaments, as in *Leptothrix* and *Clonothrix* [Greek *trich,* hair].

Cell arrangements may change in response to environmental conditions. The chaining of streptococci is an example (Figure 2.5). Streptococci form chains of cells when they are exposed to nutrient-limiting conditions. Chaining results from incomplete cell division in which new cells remain attached to the original parent cells because of uncleaved cell wall material between adjoining cells (see Perspective at the end of this chapter). Chains also have been found to form in the presence of specific antibodies that inhibit enzymes responsible for cell separation. Streptococci are generally found in chains when recovered from isolated lesions or grown on laboratory media. When the same bacteria are isolated from an actively spreading lesion in the human body, the cells generally occur as single cocci and diplococci (pairs of cocci).

Microscopy

The light microscope is the workhorse of microbiology. Through the microscope we are not only able to see microorganisms, but also can determine their morphology, arrangements, and structural features. The various morphologies of the bacteria were not appreciated until the development of the microscope, and each new refinement of the microscope has brought enticing new views of the microbial world.

Many different types of microscopes have been developed over the past two centuries. Each has its own characteristics and features that provide it with a specific value in microscopy. Microscopes basically are two kinds: light and electron. The **light microscope** uses light as its source of illumination and may permit four types of microscopy: **bright-field, dark-field, fluorescence,** and **phase-contrast.** The **electron microscope** uses a beam of electrons

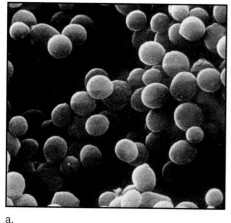

a.

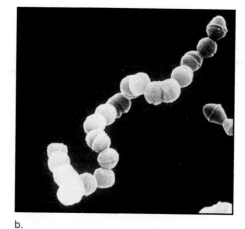

b.

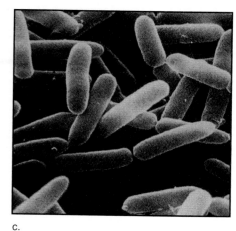

c.

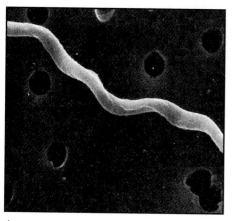

d.

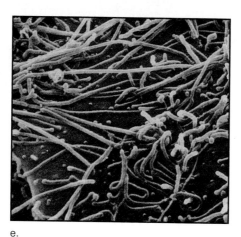

e.

figure 2.4

Scanning Electron Micrographs of Different Bacteria

a. *Staphylococcus aureus,* clusters of cocci (colorized, ×34,000). b. *Streptococcus pyogenes,* chains of cocci (colorized, ×16,000). c. *Bacillus subtilis,* rods (colorized, ×11,000). d. *Cristispira pectinis,* a spirochete (×23,000). e. *Mycoplasma pneumoniae,* pleomorphic cells (colorized, ×18,000).

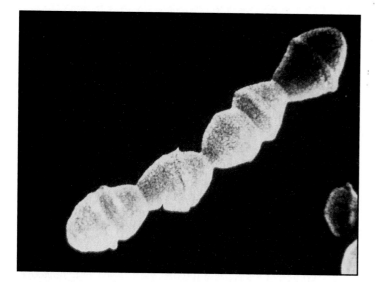

figure 2.5

Chaining of *Streptococcus pyogenes* from Incomplete Cell Division

Under nutrient-limiting conditions, streptococci form chains of cells from incomplete cell division (colorized, ×40,000).

instead of light waves to produce an image. There are two types of electron microscopes: **transmission electron microscopes (TEMs)** and **scanning electron microscopes (SEMs)**.

The Compound (Bright-Field) Light Microscope Magnifies Objects Too Small to Be Seen by the Human Eye

The standard instrument used in the laboratory to observe microorganisms is the compound (bright-field) light microscope (Figure 2.6). This instrument is called **compound** because it contains two or more sets of lenses, as compared to a single-lens system such as a magnifying glass or the microscopes of Leeuwenhoek.

A modern compound microscope has a **condenser lens,** which focuses light on the specimen; **objective** lenses that are close to the specimen and magnify it; and **ocular lenses** that are near the eye and further magnify the image. Light rays pass through the specimen in a bright-field microscope. Since objects are seen against a light background, staining of the specimen enhances its contrast against the background. Specimens that are stained and viewed with the bright-field light microscope appear dark against a brightly lit background.

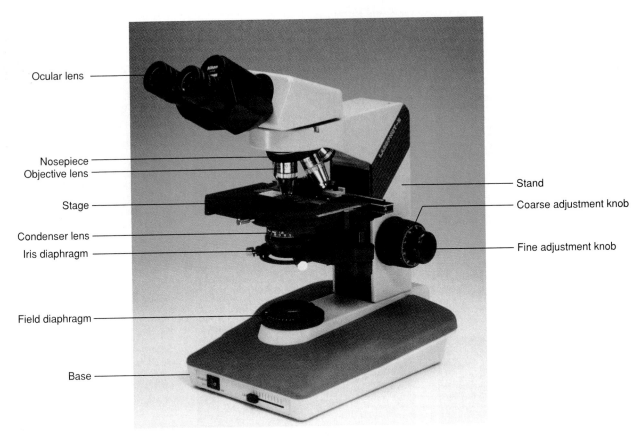

figure 2.6
Basic Elements of the Compound Bright-Field Light Microscope

Labels: Ocular lens, Nosepiece, Objective lens, Stage, Condenser lens, Iris diaphragm, Field diaphragm, Base, Stand, Coarse adjustment knob, Fine adjustment knob

---

Objects Are Magnified by the Lenses of a Light Microscope

Most microscopes have several objective lenses, each with a different magnifying power and optical property, located on a revolving nosepiece. The nosepiece is attached to the lower end of a hollow body tube.

Images of a specimen, illuminated by a lamp or by light reflected by a mirror, pass through the objective lens and body tube to the second set of magnifying lenses: the ocular lenses. The ocular lens system, also known as the eyepiece, is located at the upper end of the body tube and is placed next to the observer's eyes. This lens system serves to magnify the primary image formed by the objective lens and renders the image visible.

The total magnification of a specimen seen through a compound microscope is determined by multiplying the magnifications of both the ocular and objective sets of lenses. For example, an objective lens with a magnification of 40× used in combination with an ocular lens having a magnification of 10× would produce a total magnification of 400× (10× times 40×). Most ocular lenses have magnification of 10×, whereas objective lenses have magnifications normally ranging from 4 to 100×.

The objective lenses of a microscope should be free of spherical, chromatic, and other types of aberrations (Figure 2.7). A **spherical aberration** arises when light rays from an object passing through different portions of a curved lens do not meet at a single point for sharp focus. Much like the human eye focuses on an object by adjusting its lens to the path of light rays entering the curved cornea, a microscope lens must minimize spherical aberrations. Light rays must cross at the focal point of the specimen; otherwise the result is a fuzzy and distorted image. Although such aberrations cannot be completely eliminated, they can be minimized by the use of different lens shapes and combinations.

A **chromatic aberration** occurs because white light, not monochromatic light, is typically used for visual observation. White light is composed of a continuum of wavelengths, from 400 to 800 nm. As white light passes through a lens, the light is split into the different colors of the spectrum, ranging from violet to red, because violet has a shorter wavelength and is refracted more than red. As a result, even with a lens corrected for spherical aberration, the image produced by white light is not sharp and is surrounded by colored fringes.

Chromatic aberration can be reduced by using compound lenses of different materials and types. Most widely used objectives are **achromatic,** compensating for red and blue aberrations by combining two types of lenses, a convex and concave, to provide the same focal length for the two colors. There are still minor aberrations with an achromatic lens, but such lenses are popular for student microscopes because of their low cost. The effect of chro-

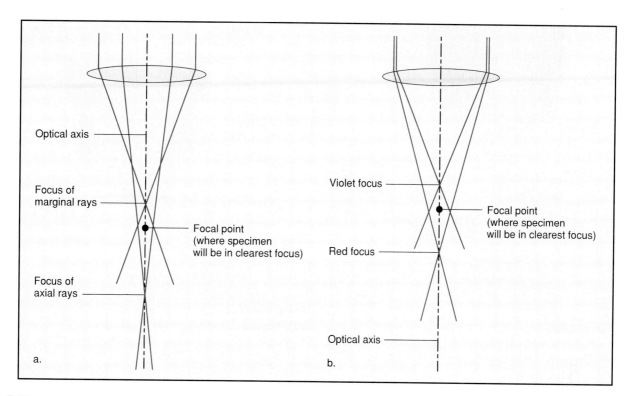

figure 2.7

Aberrations of Lenses

a. Spherical lens aberration. Light rays from the center of a spherical lens (axial rays) focus farther from the lens than the light rays from the margin of the lens, resulting in a distorted image. b. Chromatic lens aberration. Red light has longer wavelengths than violet light; therefore it focuses farther from the lens than violet light.

matic aberrations can also be minimized by using filters to eliminate wavelengths of light for which the lens is uncorrected.

For the detection of faint color or high sensitivity in color photomicrography, **apochromatic** objectives are preferred. Apochromatic objectives combine optical glass with the clear mineral fluorite (crystalline CaF_2) to focus light at three different wavelengths (red, blue, and green) at the same point. Such lenses usually require the use of a compensating eyepiece to make the images of each color the same size in all parts of the field. The images produced by apochromatic lenses are sharp and provide high contrast. Apochromatic systems are expensive and are used primarily for research applications.

A conventional light microscope can produce useful magnifications of approximately 1,000×. Although magnification can be increased beyond 1,000× in a light microscope, this increase is **empty magnification,** because the resolution of the specimen is not increased.

The Light Microscope Can Resolve Objects 0.2 μm Apart

Resolution, or **resolving power,** is defined as the ability of a microscope to distinguish two closely spaced objects as separate and distinct entities. The conventional light microscope can be used to observe specimens that are at least 0.2 μm apart, and therefore is said to have a resolving power of 0.2 μm.

The resolving power of a microscope is determined by three factors: (1) the wavelength of light (λ) used for illumination of the specimen, (2) the numerical aperture of the condenser, and (3) the numerical aperture of the objective lens used. The wavelength of light is a fixed characteristic in most light microscopes, although shorter wavelengths of light (for example, ultraviolet light) can be used in some instruments to achieve greater resolution. The resolving power of a microscope, therefore, depends primarily on the numerical apertures of the objective and condenser lenses. The resolving power of a microscope is mathematically expressed by the formula:

$$\text{Resolving power} = \frac{(0.61)\lambda}{\text{NA objective lens} + \text{NA condenser lens}}$$

The numerical aperture, generally abbreviated N.A., is the light-gathering capability of a lens. Ernst Abbe (1840–1905), a German mathematician and physicist, established a mathematical relationship that correlates the light-gathering ability of a lens with its aperture. This relationship for an objective lens is:

$$\text{N.A.} = n \sin \theta$$

where n is the refractive index of the medium between the objective lens and the specimen, and θ is one-half the angle of light entering the objective lens through the stage (Figure 2.8).

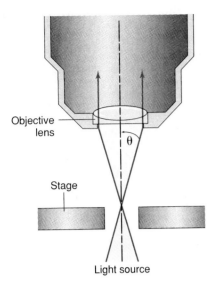

figure 2.8
Determination of Angle θ
The angle θ, which is used to calculate the numerical aperture of a lens, is one-half the cone of light entering the objective lens from the object, at focus.

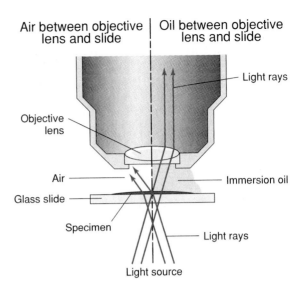

figure 2.9
The Refractive Indices of Oil Versus Air
Comparison of light rays sent through immersion oil and through air. The refractive index of immersion oil is identical to that of the microscope slide (1.52). Consequently, light rays are not bent as much when they pass through the slide and the oil as when they pass through air (refractive index = 1.00). More light from the specimen reaches the objective lens when oil is used.

The refractive index varies with the medium used between the lens and the specimen (Figure 2.9). Air is generally used as the standard surrounding medium. The refractive index of air is assumed to be 1.00. Since the value of sin θ remains constant for any given lens, the numerical aperture of an objective lens can be increased only by inserting a medium with a refractive index higher than 1.00 between the specimen and the lens.

Water has a refractive index of 1.33, whereas immersion oil has a refractive index of 1.52. By changing the medium from air to oil, one thus is able to increase the numerical aperture of the objective lens, thereby improving the resolving power of the microscope. Oil-immersion lenses are objective lenses with high magnifications and high numerical apertures that require oil as an immersion medium to increase resolution. These lenses are used to view bacteria in detail. Dry objective lenses, which use air to occupy the space between the objective and the specimen, are employed for lower magnifications where detail is not as critical. The values of the numerical aperture and magnifying power are generally stamped on the side of objective lenses.

Proper Illumination of a Specimen Is Important in Achieving Optimal Resolution with a Light Microscope

The resolving power of a microscope also is dependent upon the numerical aperture of the **condenser.** The condenser of a light microscope, located below the stage on a focusable mount, gathers light from the light source and concentrates it onto the specimen. Condensers improve resolution by eliminating stray light rays that otherwise would cause glare around the specimen. The uncor-

rected Abbe condenser is the most common type of condenser used in student microscopes. This condenser consists of two or more lenses that are not corrected for spherical or chromatic aberration. Such a condenser is relatively inexpensive and is generally adequate for most routine applications where color correction or precise images of the lamp source are not essential. Other types of condensers (some with as many as six elements) correct for aberrations and provide near-perfect images of the light source.

An adjustable multileaf **iris diaphragm** is located in the condenser. The diaphragm controls the diameter of light leaving the condenser and striking the specimen and therefore functions much like the diaphragm of the eye or in a camera. By adjusting the opening of the diaphragm, one is able to reduce glare and increase clarity of the image. A diaphragm opened too fully results in excessive illumination and glare. One that is too closed decreases resolution of the specimen by effectively reducing the numerical aperture of the objective.

Light intensity in microscopy is not controlled by the condenser or the iris diaphragm. It is determined by adjusting the voltage of the light source or by using filters to block out certain wavelengths of light.

Proper illumination of a specimen is important in achieving optimal resolution with a light microscope. Two basic methods are used to achieve correct illumination in a bright-field light microscope: **critical illumination** and **Köhler illumination** (Figure 2.10). In critical illumination, an image of the light source filament is focused on the specimen with the substage condenser. This method gives bright, but uneven, illumination and is most often used for high-power visual work.

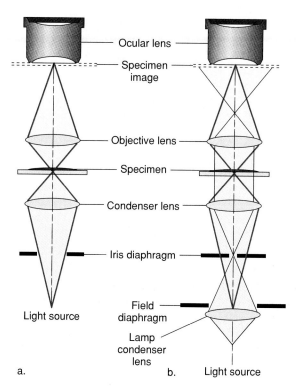

figure 2.10

Critical and Köhler Illuminations

a. In critical illumination, an image of the light source filament is focused on the specimen with the substage condenser to produce bright, but uneven, illumination of the specimen. b. In Köhler illumination, the filament of the light source is focused onto the iris diaphragm by a condenser attached to the microscope lamp. The position of the substage condenser lens is adjusted so as to form an image of the field diaphragm in the plane of the specimen. The field diaphragm is then opened or closed to reduce any spurious light, resulting in even illumination of the specimen.

The Köhler method of illumination uses a separate condenser attached to the microscope lamp to focus an image of the lamp filament onto the iris diaphragm of the substage condenser. The substage condenser is then slowly racked up to focus the image of the lamp condenser in the plane of the specimen. In this procedure the lamp condenser actually becomes the source of illumination. The Köhler method results in a uniform illumination of the specimen without unwanted background glare. Köhler illumination is typically used for general visual work and photomicrography.

Stains Increase the Contrast between a Specimen and Its Background

Bacteria are generally difficult to see, even with the aid of a microscope, because of their transparency. For this reason bacterial cells are often stained with organic dyes to increase the contrast between the specimen and the background. Organic dyes may be separated on the basis of their affinity for specific groups of compounds.

figure 2.11

Example of a Positive Stain

A Gram stain of *Bacillus anthracis*, the cause of anthrax.

Basic dyes are positively charged and combine with negatively charged cell constituents such as nucleic acids and acidic polysaccharides. Examples of basic dyes are crystal violet, methylene blue, and safranin. Acidic dyes have an affinity for positively charged materials such as basic proteins, and include Congo red, eosin, nigrosin, and basic fuchsin.

Bacterial stains are generally one of three types: **simple, differential,** or **special.** Simple stains consist of a single dye that is used to enhance the contrast of the specimen against the background. Such stains are useful in examining cell size, morphology, or arrangement, since the bacterial cell appears as a single, distinct color against an unstained background.

Differential stains use two dyes (the primary dye stains the structure in question and a counterstain is for contrast). They involve several staining and destaining solutions and are used to differentiate cells on the basis of their staining characteristics. The **Gram stain** is a differential stain that is widely used in bacteriology. Most bacteria can be differentiated into two groups based on their Gram stain reaction: **gram positive** (stained violet) and **gram negative** (stained pink or red) (see basis of Gram stain, page 60).

Special stains are used to enhance the features of specific cell constituents such as endospores, nuclear regions, and granules. Dyes used in special stains have an affinity for specific cell structures and highlight these structures (Figure 2.11). All of these staining techniques result in a stained bacterial cell or stained components against an unstained background, and are called **positive stains.**

Thousands of bacterial species are known. How can these organisms be most easily and quickly identified? A Danish physician, Hans Christian Gram

The Gram Stain

(1853–1938), answered this question in part during the period he worked at the morgue of the City Hospital of Berlin. Gram was interested in distinguishing between difficult-to-see bacteria that caused pneumonia and mammalian cell nuclei in infected tissue. Gram was disappointed in the stain he developed, which did not stain all bacteria equally, and this disappointment was evident in his statement in a published paper that he hoped "the method would be useful to other workers." Unknown to Gram at that time, the technique he developed in 1884 would eventually become the most widely used stain in microbiology. The **Gram stain** is used to separate bacteria into two major groups: **gram positive** and **gram negative** (see cell structure basis for differentiation by the Gram stain, page 60). Although this differential stain does not provide definitive identification of bacteria, it is routinely used in clinical and industrial laboratories as one of the first steps in bacterial identification. When the Gram stain is combined with information on cellular morphology and arrangement, as well as biochemical characteristics, a conclusive identification can usually be made.

The Gram stain procedure involves the suspension of freshly grown bacterial cells in a drop of water on a glass slide (Table 2.1). This suspension is gently heated over a low flame to heat-fix the smear onto the slide. The smear is then stained with a primary dye, crystal violet. The smear is washed with water and then covered with iodine. The iodine acts as a mordant (fixative) to form an insoluble complex with crystal violet inside the cell. The smear is then decolorized with alco-hol or acetone-alcohol and counterstained with safranin. After a final wash with water, the material is examined under the microscope. **Gram-positive** bacteria (those that retain the original crystal violet stain even after decolorization) stain a deep violet. **Gram-negative** bacteria (those that are decolorized and are stained with the counterstain safranin) stain pink or red.

Most bacteria are classified as either gram positive or gram negative. Some organisms, however, do not give an obvious Gram stain reaction because of the thickness or composition of their cell walls. These organisms, termed **gram variable,** appear as both gram-positive and gram-negative cells when viewed with a microscope. Other bacteria, which might normally stain gram positive in young cultures, lose their abilities to stain gram positive in older cultures. These older cultures frequently have dying cells with decomposing cell walls that are unable to retain the crystal violet dye during decolorization. Another problem associated with the Gram stain and other similar staining techniques is the distortion of cellular features during heat fixation. This distortion is of minimum consequence with the Gram stain, since one is interested in only the staining characteristics and gross morphology of the cell. Distortions may present problems, however, in other types of stains.

Despite these limitations, the Gram stain is used universally for the preliminary identification of bacteria. It is a simple and rapid technique that can be performed in any laboratory equipped with a microscope and the necessary staining reagents.

 Bacterial Structure and Function
Bacterial Groups • pp. 1–4

table 2.1

Gram Stain Procedure

Reagents	Time Applied	Reactions	Appearance
Unstained smear			Cells are colorless and difficult to see.
Crystal violet	1 minute, then rinse with water	Basic dye attaches to negatively charged groups in the cell wall, membrane, and cytoplasm.	Both gram-negative and gram-positive cells are deep violet.
Gram's Iodine (mordant)	1 minute, then rinse with water	Iodine strengthens the attachment of crystal violet to the negatively charged groups.	Both gram-negative and gram-positive cells remain deep violet.
Alcohol or acetone-alcohol mix (decolorizer)	10 to 15 seconds, then rinse with water	Decolorizer leaches the crystal violet and iodine from the cells. The color diffuses out of gram-positive cells more slowly than out of gram-negative cells because of the chemical composition and thickness of the gram-positive cell walls.	Gram-positive cells remain deep violet, but gram-negative cells become colorless and difficult to see.
Safranin (counterstain)	1 minute; then rinse thoroughly, blot dry, and observe under oil immersion	Basic dye attaches to negatively charged groups in both cell types. Few negative groups are free of crystal violet in gram-positive cell, whereas most negative groups are free in gram-negative bacteria. Consequently, gram-positive bacteria remain deep violet, whereas gram-negative bacteria become pink or red.	Gram-positive cells remain deep violet, whereas gram-negative cells are stained pink or red.

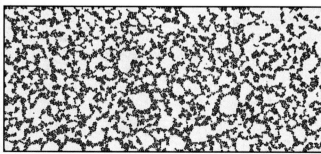

Gram stain of *Staphylococcus aureus*, a gram-positive coccus (×252).

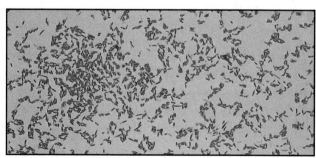

Gram stain of *Escherichia coli*, a gram-negative rod (×252).

Negative stains make cells or other cell materials appear light against a dark background. A negative stain is usually prepared by mixing cells with a substance such as India ink or nigrosin to provide the contrasting, dark background. The mixture is then spread in a thin film across a glass slide and allowed to dry. Negative stains are used to enhance cellular features that are difficult to observe with positive stains, and to do size determinations. An example of a negative stain is the capsule stain. Capsules are coverings surrounding a bacterial cell, composed of polysaccharides or proteins. In the capsule stain, only the background is stained. The stain itself does not penetrate or stain the cell capsule (Figure 2.12). All of these staining techniques provide the microbiologist with tools to enhance certain structures in bacterial cells or to see the cell better.

The Dark-Field Microscope Provides Contrast in Specimens That Normally Lack Sufficient Contrast to Be Seen with the Bright-Field Microscope

Bright-field illumination is used for observation of specimens commonly encountered in the laboratory. Dark-field illumination is useful in situations where high contrast is important. The dark-field method permits minute particles—particularly small and thin line structures such as bacterial flagella, which normally lack sufficient resolution to be seen with bright-field microscopy—to become visible to the eye. In dark-field microscopy, light is focused on the specimen at an oblique angle through a specially constructed dark-field condenser, resulting in the specimen (stained or unstained) appearing light against a dark background (Figure 2.13).

Dark-field illumination is generally achieved by placing an opaque disk, known as a dark-field stop, in the center of the substage condenser. This dark-field stop blocks direct light in the center of the condenser and does not permit it to reach the objective, but allows a hollow cone of light to illuminate the specimen. Thus while no direct light enters the objective, the specimen is seen by scattered or reflected light against a dark background. The effect achieved through dark-field illumination is comparable to seeing dust particles in a darkened room because of the reflection of rays of sunlight coming through a window.

The resolution of specimens by dark-field microscopy is quite good. For this reason, dark-field illumination is frequently used to observe thin or small objects. One use is the examination of *Treponema pallidum,* the very thin (≤0.2 μm in diameter), spiral-shaped bacterium that causes syphilis, in scrapings from syphilitic lesions. An intense, nondiffused light source is essential for good dark-field illumination. Since dirt and other foreign matter on a dirty slide can also scatter light, it is important that clean slides be used in dark-field microscopy.

The Phase-Contrast Microscope Amplifies Small Differences in Refractive Indices to Enhance Specimen Contrast

Unstained microorganisms are transparent and normally do not reflect or refract sufficient light to distinguish them from the background. Specimen contrast can be enhanced with the phase-contrast microscope, which detects small differences in refractive

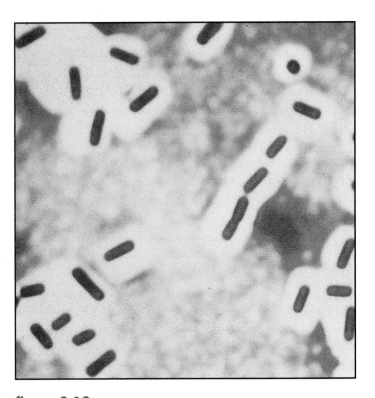

figure 2.12
Example of a Negative Stain
India ink stain of *Clostridium perfringens,* the cause of gas gangrene, showing a thick capsule around each cell (colorized).

indices of the specimen and its surrounding medium. The phase-contrast microscope was first devised by Fritz Zernicke (1888–1966) of the Netherlands in 1932. Zernicke received the Nobel Prize in physics in 1953 for his achievement.

Structures may differ in their refraction of light passing through them. In phase-contrast microscopy, these differences are amplified and converted into variations in light intensity. Phase-contrast microscopy is achieved by mounting an annular aperture diaphragm below the substage condenser so that the specimen is illuminated with a ring of light. Different annular diaphragms are used with objectives of different magnifications to achieve optimum optical results. Light from the lamp source passes around the diaphragm in the substage condenser, illuminating the specimen in a ring of light. Since the specimen and the surrounding medium have different indices of refraction, light waves passing through the specimen and entering the objective lens are somewhat refracted and out of phase compared with those coming from the medium. These differences in refractive indices are converted into variations in light intensity based on the interference of light waves passing through the specimen and the surrounding medium. Light waves that are in phase result in **constructive interference** and increased light intensity. **Destructive interference** occurs when waves are out of phase, resulting in a decrease in light intensity. The naked eye cannot detect these phase differences in light waves. For this reason, the objective lenses of a phase-contrast microscope contain a phase-shifting element (phase ring) that alters light waves passing through the objective by 90°,

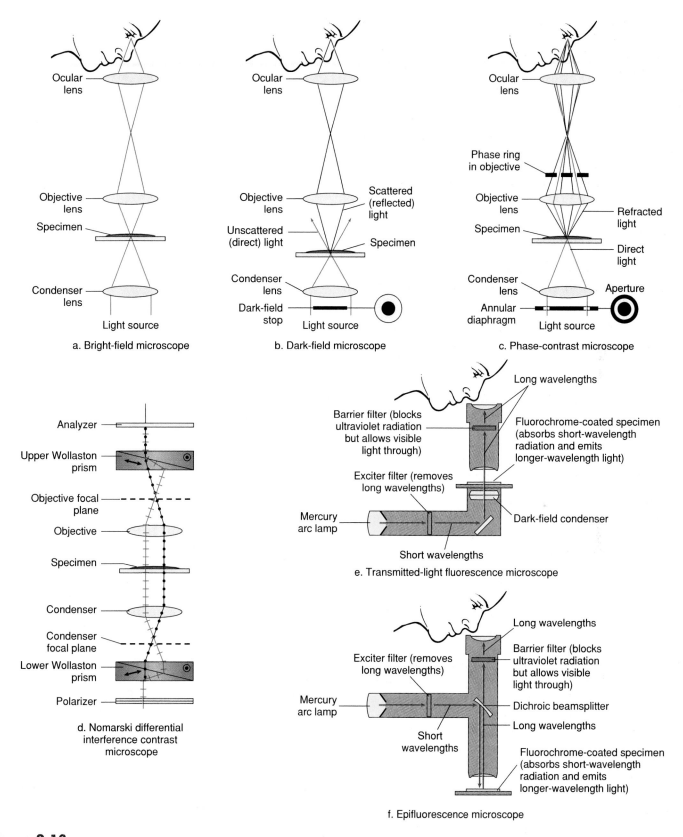

figure 2.13

Schematic Diagrams of Bright-Field, Dark-Field, Phase-Contrast, Nomarski Differential Interference Contrast, Transmitted-Light Flourescence, and Epifluorescence Microscopes

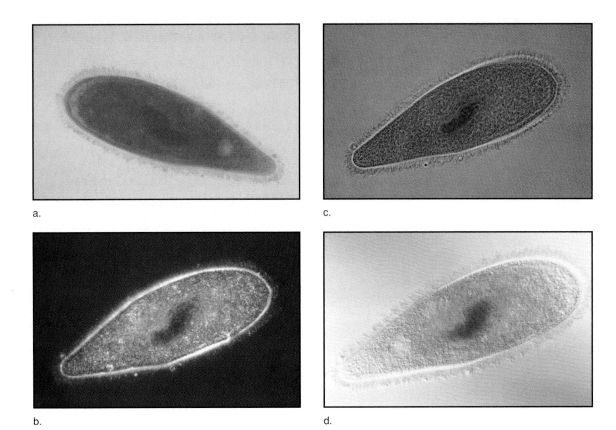

figure 2.14

The Protozoan *Paramecium* as Seen by Different Types of Light Microscopy

a. Bright-field illumination shows the outlines and general features of the specimen. b. Dark-field illumination outlines the edges of the specimen. c. Phase-contrast microscopy details the internal structures of a living specimen. d. Nomarski differential interference contrast microscopy results in a three-dimensional image of the specimen. Magnification is ×240 in all examples (colorized).

or one-quarter of a wavelength. This shift causes an increase in amplitude of the light waves reaching the eyes, resulting in increased brightness and, consequently, contrast of the specimen. Since this contrast depends on proper focusing of light entering the phase-shifting element from the annular diaphragm, precise alignment of the optics in a phase-contrast microscope is critical to achieve optimum results. A special centering telescope, placed on the eyepiece of the microscope, is used to align the adjustable annular diaphragm in the condenser with the phase-shifting element in the objective lens.

The advantage of phase-contrast microscopy is that living microorganisms in their natural state can be studied, since the specimen does not have to be fixed or stained. The specimen appears either in positive contrast (dark against a light background) or in negative contrast (bright against a dark background), depending on the type of wave retardation that occurs. Negative contrast is more commonly used, because it generally gives better resolution and contrast of specimens. The major limitation of phase-contrast microscopy is that it produces a halo, or bright ring, around dark details in a specimen. In spite of this shortcoming, phase-contrast microscopy is valuable in visualizing living cells' minute structures and details that would not be seen by other forms of microscopy.

A variation of the phase-contrast microscope is the **interference contrast microscope.** There are several types of interference contrast microscopes, but one of the more common involves double-beam interference. This method uses a beam of light that is split into two separate beams. The object beam passes through the specimen and is thus retarded in relation to the reference beam, which does not contact the specimen. When two beams reconverge with the assistance of prisms, the phase difference between the two beams provides relieflike, high-contrast images of the specimen.

A different type of double-beam interference contrast is **Nomarski differential interference contrast** microscopy. In the Nomarski microscope, two beams are produced from a single beam of polarized light, and each beam passes through a different part of the specimen. Differences in the refractive index of parts of the specimen alter the phase relationship between the two beams of light so that an interference pattern is produced when the beams are recombined. Interference microscopy does not produce the halos around specimens that are a characteristic drawback of phase-contrast microscopy. Because contrast in interference microscopy depends on differences in refractive indices across a specimen, this type of microscopy gives a relieflike three-dimensional image (Figure 2.14). Structures such as cell walls are sharply defined by interference contrast microscopy.

table 2.2

Summary of the Different Types of Microscopy and Their Applications in Microbiology

Type of Microscopy	Magnification	Specimen Appearance	Applications
Bright field	1,000 to 2,000	Stained or unstained; bacteria are generally stained to enhance contrast against background.	For gross morphological examination of bacteria, yeasts, molds, algae, and protozoa.
Dark field	1,000 to 2,000	Generally unstained; appears bright or "lighted" against a dark background.	For situations where high contrast, particularly of microbial structures, is required.
Fluorescence	1,000 to 2,000	Fluorescent	For diagnostic techniques where the fluorescent dye (antibody) fixed to the organism identifies it.
Phase contrast	1,000 to 2,000	Contrasting degrees of darkness (positive contrast) or brightness (negative contrast).	For examination of minute cellular structures and details in living cells.
Electron	100,000 and higher	Viewed on a screen	For examination of atomic and cell structures and viruses not observable with the light microscope.

The Fluorescence Microscope Detects Fluorescent Objects That Are Illuminated by Ultraviolet or Near-Ultraviolet Light

Many materials naturally fluoresce when examined under ultraviolet or near-ultraviolet radiation. Fluorescence occurs as these materials absorb the short wavelengths of ultraviolet light and then emit, in less than one-millionth of a second, light of longer wavelengths that can be seen by the eye. Fluorescence microscopy was initially developed to detect natural fluorescence of plant and animal structures, but is now extensively used for the observation of specimens that do not fluoresce unless a fluorescent dye (fluorochrome) is added.

Fluorescent dyes such as fluorescein isothiocyanate (FITC), which absorbs blue light and emits green light, are frequently conjugated with antibodies that bind only to specific antigens on a specimen (see immunofluorescence, page 489). The FITC-conjugated antibodies are added to a glass slide with an unknown specimen, and excess conjugate is removed by washing. The specimen can then be observed for fluorescence of the specifically attached dye using a specially constructed microscope. This procedure, known as **immunofluorescence,** is very useful in clinical microbiology for the diagnosis of diseases such as syphilis and rabies and in environmental microbiology for studies of natural microbial populations.

Early systems for observing fluorescence used transmitted light with dark-field illumination. In **transmitted-light fluorescence,** light passes through an excitation filter before it strikes the specimen. The excitation filter permits the passage of only short-wavelength radiation to the specimen. The fluorochrome in the specimen absorbs this radiation and emits longer-wavelength light, which passes through the objective and a barrier filter before it reaches the eye. The barrier filter blocks out ultraviolet radiation from the specimen and permits only visible light to reach the eye. The dark-field condenser removes excessive light from the objective to provide a contrasting black background. Poor image brightness, however, makes transmitted-light fluorescence less than optimal.

Modern systems use a different arrangement in which light is transmitted through the objective lens to the specimen. This method, called **incident-light excitation,** or **epifluorescence,** directs filtered short-wavelength light from the lamp to a chromatic beam splitter. The beam splitter reflects short-wavelength radiation to the specimen and additionally transmits emitted longer-wavelength light from the specimen through the barrier filter to the eye. Epifluorescence both increases the sensitivity of fluorescence, and permits the specimen to be illuminated and visualized with normal substage optics. Figure 2.13 compares transmitted-light fluorescence and epifluorescence.

Objective lenses used in either type of fluorescence microscopy must be made of glass that does not itself fluoresce. The lamps must emit strongly in the excitation range of the fluorochrome. In the case of dyes such as FITC, a mercury arc lamp or a tungsten halogen lamp is used. Shutters are used to limit the ultraviolet exposure times of fluorochromes, which minimizes fading of fluorescence over a period of time. The objective lenses should have the highest possible numerical aperture to collect the faintest fluorescence. As a result of these requirements, objectives used in fluorescence microscopy are generally more expensive than those used in other forms of microscopy. Despite these limitations, fluorescence microscopy is a valuable tool, particularly for fluorochrome-tagged specimens that might not be noticed with ordinary light microscopy.

The Electron Microscope Magnifies Objects Too Small to Be Seen by Light Microscopy

The light microscope can be used to observe specimens with diameters as small as 0.2 μm. For even smaller specimens, observations can be made with the electron microscope, which has a resolution of about 0.001 μm and is capable of magnifying objects 100,000× or more (Table 2.2).

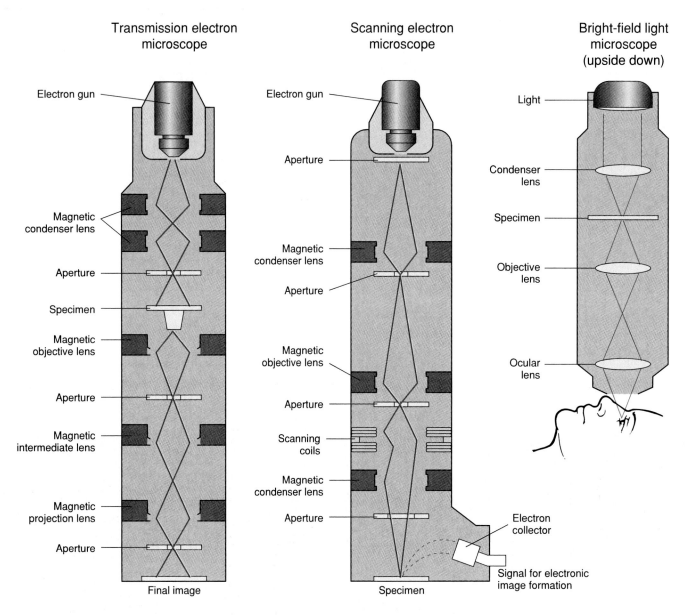

Transmission electron microscope

- Electron gun
- Magnetic condenser lens
- Aperture
- Specimen
- Magnetic objective lens
- Aperture
- Magnetic intermediate lens
- Magnetic projection lens
- Aperture

Final image

Scanning electron microscope

- Electron gun
- Aperture
- Magnetic condenser lens
- Aperture
- Magnetic objective lens
- Aperture
- Scanning coils
- Magnetic condenser lens
- Aperture
- Electron collector
- Signal for electronic image formation

Specimen

Bright-field light microscope (upside down)

- Light
- Condenser lens
- Specimen
- Objective lens
- Ocular lens

figure 2.15

Transmission and Scanning Electron Microscopes Compared with the Bright-Field Light Microscope

There are two basic types of electron microscopes: the transmission electron microscope (TEM) and the scanning electron microscope (SEM) (Figures 2.15 and 2.16). Both instruments use beams of electrons having wavelengths 100,000 times smaller than the wavelengths of visible light to react with the atomic nuclei of the specimen. The electrons thus substitute for the light in a conventional light microscope and, because of the electrons' extremely short wavelengths, there is increased resolving power.

The Transmission Electron Microscope Focuses a Fine Beam of Electrons Through the Specimen

The beam of electrons in a transmission electron microscope is generated and accelerated by an electron gun. The electrons are emitted in a vacuum and focused by an electromagnetic lens to a small spot on the specimen. Electrons pass through the specimen; the amount of scattering depends on the electron density of the specimen. Areas that are electron dense—that have large numbers of atoms of high atomic densities within a small region—are not penetrated by electrons and therefore cause their extensive scattering. Since electrons cannot be seen by the human eye, this contrast in electron scattering is visualized by projection of the electrons onto a phosphorus-coated screen that fluoresces when struck with electrons. Electron images can be permanently recorded on a photographic plate.

Biological specimens typically have low electron densities and must generally be treated by negative staining or by shadowing to increase contrast (Figure 2.17). In **negative staining,** electron-dense stains such as phosphotungstic acid are applied to the specimen to provide greater contrast. When the preparation is viewed,

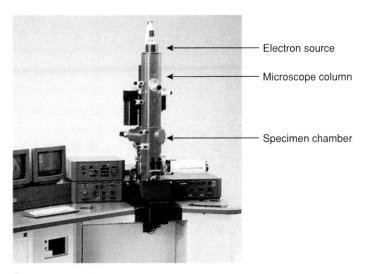

Electron source

Microscope column

Specimen chamber

a.

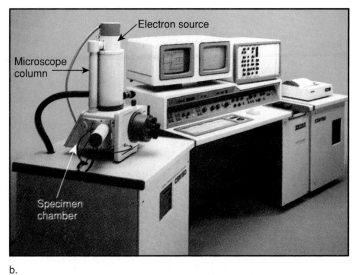

Electron source

Microscope column

Specimen chamber

b.

figure 2.16

Transmission Electron Microscope and Scanning Electron Microscope
a. Transmission electron microscope. b. Scanning electron microscope.

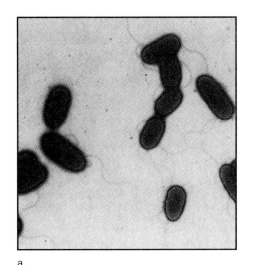

a.

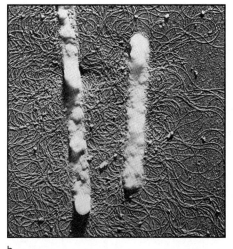

b.

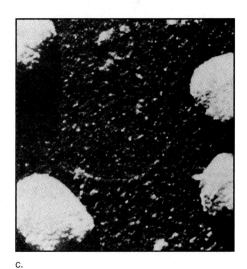

c.

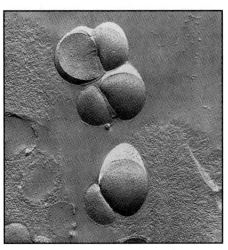

d.

e.

figure 2.17

Microorganisms as Seen by Electron Microscopy

a. *Pseudomonas aeruginosa*, stained with phosphotungstic acid (×6,450).
b. *Clostridium tetani*, metal shadowed, showing flagella (×4,400). c. Vaccinia virus, metal shadowed (×60,000).
d. *Neisseria gonorrhoeae*, freeze-etched (×12,300). e. *Physarum polycephalum*, scanning electron micrograph (×30).

Seeing Atoms Through a New Generation of Microscopes

Just as Leeuwenhoek's single-lens microscopes once opened a hitherto unseen world of microorganisms to human observation, a new generation of powerful microscopes developed in recent years has brought images to the atomic scale. These precision instruments—the scanning tunneling microscope (STM), the atomic force microscope (AFM), and the laser-scanning confocal scanning microscope—bring a new and exciting dimension to microscopy, enabling scientists to explore the world of atoms, nucleic acids, and proteins.

The STM, developed only 15 years ago by IBM's Gerd Binnig and Heinrich Rohrer, who shared the 1986 Nobel Prize in physics for their invention, uses a minute probe with a tip that ideally terminates in a single atom to trace the contours of a surface. Unlike a conventional scanning electron microscope, which scans an external beam of electrons across the surface of a specimen, the STM measures the shuttling of electrons between the probe and the specimen's surface, a phenomenon known as the tunneling current. This current is directly related to the distance between the probe and specimen. Changes in probe-to-specimen distance of as little as 0.01 nm can result in exponential tunneling current changes. Consequently, the STM has a resolution well within the subatomic range.

The STM can magnify objects up to one hundred millionfold and has already been used to image DNA, protein, and other biological materials (Figure 2.18).

The AFM is a more recent invention derived from the STM. It is similar to the STM except that its probe touches the specimen's surface and traces the outline of atoms on the surface in a manner similar to the way a person reads Braille. The result is a topographical image. The probe in an AFM is mounted on a spring in the form of a cantilever. Minute deflections of the cantilever are recorded as changes in the specimen's topography. The AFM has been used to successfully image viruses, red blood cells, and amino acids (Figure 2.19).

The laser-scanning confocal scanning microscope focuses an illuminating cone of laser light through a pinhole aperture to a specific point within a specimen. As the beam of light is scanned across this area, imaging light from the specimen focuses through a second (confocal) pinhole to a detector. This second pinhole screens out light from all but the focal point, resulting in a crisp and precise two-dimensional image called an optical section. Multiple optical sections can be combined to form three-dimensional images of chromosome arrangements and other

figure 2.18
Scanning Tunneling Microscope Image of the DNA Double Helix (colorized)

cellular structures. The confocal microscope is especially useful in viewing thick specimens, when light otherwise might be diffused by conventional microscopy.

These new developments in microscopy have enabled scientists to visualize what they have theorized for many years. With these sophisticated instruments, scientists will now be able to understand the complexity of atoms and molecules.

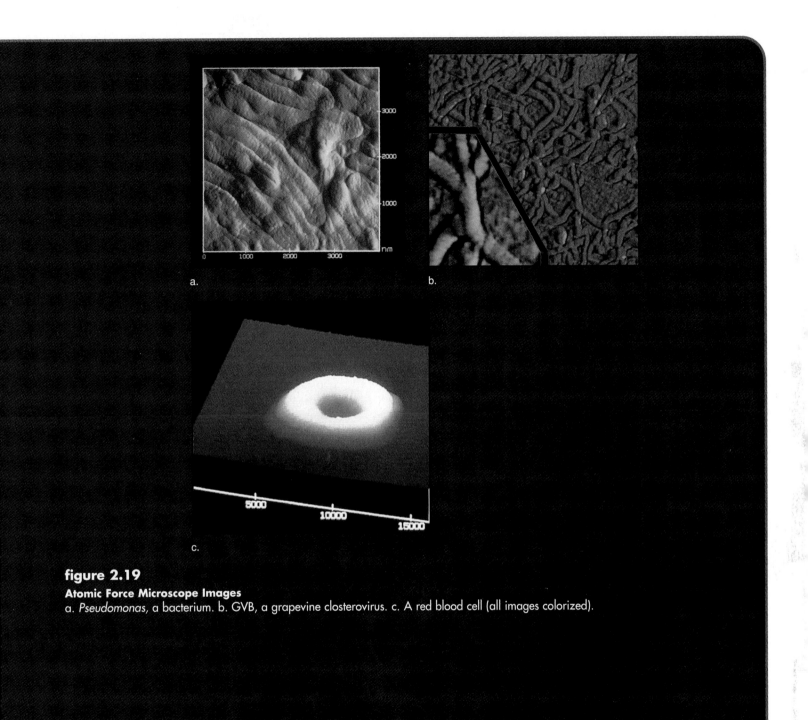

figure 2.19
Atomic Force Microscope Images
a. *Pseudomonas,* a bacterium. b. GVB, a grapevine closterovirus. c. A red blood cell (all images colorized).

regions of the specimen not penetrated by the stain are electron-transparent (light) against an electron-dense (dark) background. In **shadowing,** the specimen is coated at an angle with a thin layer of a heavy metal such as gold, platinum, or chromium. The shadow cast from such a preparation can then be used to visualize the specimen. Addition of an electron-dense stain can result in distortion of the specimen and an artifactual image.

Specimens for examination on transmission electron microscopes are routinely fixed and embedded in a supporting plastic resin to permit slicing of very thin sections of material for observation. Thin sections are made by slicing the embedded specimen with a glass knife or diamond in an instrument called an ultramicrotome. Thin sections are necessary because of the low penetration power of electrons and the large depth of field of the electron microscope. Slicing limits the quantity of material that can be seen, and can result in distortion of the specimen. **Freeze etching** is a process

used to observe structures of cellular organelles and membranes without distortion. In freeze etching, superficial layers are fractured from a frozen specimen, and carbon replicas of the exposed specimen surfaces are made for observation. This method permits examination of the detailed surface or internal structures of cells.

The Scanning Electron Microscope Provides a Three-Dimensional Image

The scanning electron microscope is a recent development in electron microscopy that permits the observations of the surface structure of a specimen. In the scanning electron microscope, a finely focused beam of electrons strikes the specimen. As these electrons are scanned across the surface of the specimen, secondary electrons, as well as other particles such as backscattered electrons and X rays, are released at various angles to produce a

PERSPECTIVE   Streptococci and Cell Wall Growth

Streptococci are gram-positive bacteria that form chains of cocci under certain growth conditions. Chaining of streptococci is useful in microscopic identification. Although chain formation has been observed in streptococci and other bacteria for many years, the structural events in cell wall growth that accompany chaining were not known until 1962, when Roger M. Cole and Jerome J. Hahn of the National Institute of Allergy and Infectious Diseases in Bethesda, Maryland, reported on the use of immunofluorescence to demonstrate the process of cell wall growth in *Streptococcus pyogenes.*

Twenty-four-hour cultures of *S. pyogenes* were mixed with fluorescein isothiocyanate-labeled antibodies directed specifically toward bacterial cell wall antigens. The cultures were incubated for growth and, at 15-minute intervals, samples were removed and fluorescent antibody attachment to the cell wall was stopped by one of two methods: (1) acid extraction of the streptococci or their cell wall protein antigens or (2) addition of a nonfluorescent-labeled cell wall antibody. Either technique inhibited further fluorescent labeling of the wall but did not affect the fluorescent label already present on the existing cell wall. The cells then were examined by fluorescence, phase-contrast, and dark-field microscopy.

From these studies it was observed that new segments of cell wall pushed away older segments of the wall (Figures 2P.1 and 2P.2). With cells collected at successively longer incubation periods, the original fluorescent cell wall segment moved farther away from the center of the parent cell. To verify that cell wall formation occurred equatorially along the circumference of the cell and that new wall material was not diffusely intercalated with the old wall, Cole and Hahn conducted a different experi-

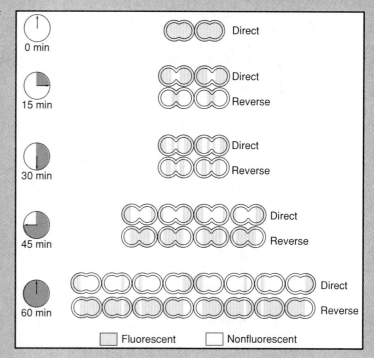

figure 2P.1
Diagrammatic Representation of Stages Shown in Figure 2P.2
In the direct method, the fluorescent label is moved farther from the center of the cell with longer incubation. In the reverse method, older cell wall segments are unlabeled and are seen pushed farther from the center of the cell with increased incubation.

three-dimensional image of the specimen. This image, which is composed of scanner (raster) lines similar to the line patterns seen on a television set, is transferred to a cathode-ray tube (CRT) screen. Images produced by scanning electron microscopy are striking and provide excellent resolution and distinct images of surface structures.

Specimen preparation for scanning electron microscopy differs from preparation for transmission electron microscopy. Since the scanning electron microscope focuses on the surface structure of a specimen, the specimen should be free of surface artifacts such as wrinkles or distortions. Drying and fixation procedures routinely used for transmission electron microscopy are not suitable because they stress the specimen and can result in shrinking or tearing. More delicate methods are used for scanning electron microscopy.

One common procedure used to minimize surface changes during specimen preparation is critical-point drying. This technique is based on the concept that there is a critical temperature and pressure at which conversion of a liquid to its gaseous phase reaches an equilibrium. When this critical point is reached, no forces are required to convert a liquid to a gas. In critical-point drying, the specimen is first dehydrated in ethanol or acetone to remove water. The specimen is then placed in a critical-point drying chamber with a transitional fluid such as carbon dioxide or Freon 13. As the pressure and temperature of the chamber are raised to the critical point, any remaining liquid in the specimen is converted to its gaseous phase without surface distortion. The chamber is vented, and the dried specimen is transferred to a desiccator to protect it from rehydration.

The dehydrated specimen is next coated with a very thin layer of metal, because the scanning electron microscope uses an electron-emitting surface to produce an image. A common method for metal layering requires the specimen to be coated with carbon to provide an electrically conducting base. A heavy metal such as gold or a gold/palladium alloy, which serves as a source of secondary electrons, is attached to the carbon. The coated specimen can then be mounted and viewed.

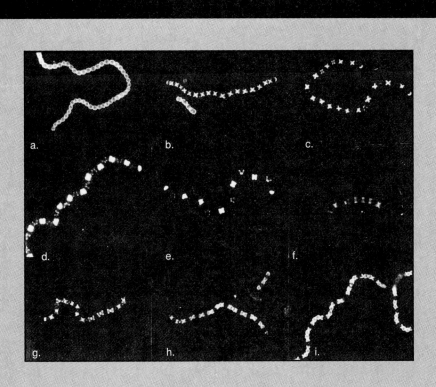

figure 2P.2

Ultraviolet Micrograph of _Streptococcus pyogenes_ Showing Cell Wall Growth

a–e. _S. pyogenes_ grown in fluorescein-labeled homologous antibody and examined at intervals after the addition of unlabeled homologous antibody (direct method). a. 0 minutes; b. 15 minutes; c. 30 minutes; d. 45 minutes; e. 60 minutes. f–i. _S. pyogenes_ grown in unlabeled homologous antibody, removed at intervals after precipitation of the antibody, and stained with fluorescein isothiocyanate-labeled homologous antibody (reverse method). f. 15 minutes; g. 30 minutes; h. 45 minutes; i. 60 minutes.

ment involving **reverse fluorescent labeling.** Streptococci were first incubated in the presence of unlabeled antibody and then at intervals incubated with fluorescein-labeled antibody. As a result of this reverse technique, older cell wall segments were unlabeled and were seen to be pushed away equatorially along the cell circumference by newly fluorescein-labeled cell wall segments.

The results of these experiments were instrumental in showing that cell wall formation in streptococci occurs by extension of the wall from the cell equator, not by diffuse intercalation of new material within the existing cell wall. The data provided significant insight into the mechanism of cell wall growth and chaining in streptococci.

Source:

Cole, R.M., and J.J. Hahn. 1962. Cell wall replication in _Streptococcus pyogenes. Science_ 135:722–724.

Summary

1. There are two kinds of cells: procaryotes and eucaryotes. Procaryotic cells do not have a membrane-enclosed nucleus.

2. There are five groups of microorganisms. Bacteria are procaryotes, whereas protozoa, algae, and fungi are eucaryotes. Viruses are not cellular and therefore are neither procaryotes nor eucaryotes. All living organisms are phylogenetically divided into three domains: Bacteria, Archaea, and Eucarya.

3. Procaryotes come in a variety of sizes, shapes, and arrangements. These morphological characteristics help in the identification and classification of these microorganisms.

4. Microorganisms can be visualized with the aid of a microscope. The light microscope uses light as a source of illumination. There are four types of light microscopy: bright-field, dark-field, fluorescence, and phase-contrast.

5. The conventional microscope is called a compound microscope because it contains two or more sets of lenses. The condenser lens focuses light on the specimen; the objective lenses are close to the specimen and magnify it; and the ocular lenses are near the eye and further magnify the image.

6. Resolution, or resolving power, is the ability of a microscope to distinguish two closely spaced objects as separate and distinct entities. The resolving power of a microscope depends on the wavelength of light used for illumination of the specimen, the numerical aperture of the condenser, and the numerical aperture of the objective lens. The light microscope has a resolving power of 0.2 μm.

7. Stains increase the contrast between a specimen and its background. The Gram stain is an example of a differential stain that separates bacteria into two groups: gram positive (stained violet) and gram negative (stained pink or red).

8. The bright-field light microscope is used for the general observation of microorganisms. The specimen appears against a light background.

9. The dark-field light microscope is used to provide high contrast to specimens. The specimen appears light against a dark background.

10. The phase-contrast light microscope amplifies small differences in refractive indices to enhance specimen contrast. An advantage of phase-contrast microscopy is that the specimen does not have to be fixed or stained and therefore can be observed in its natural state.

11. The fluorescence microscope detects objects that are illuminated by ultraviolet or near-ultraviolet light.

12. The electron microscope magnifies specimens 100,000× or more by substituting electrons with extremely short wavelengths for the light in a conventional light microscope. Greater magnifications can be obtained by use of a new generation of microscopes: the scanning tunneling microscope, the atomic force microscope, and the laser-scanning confocal scanning microscope. These newer instruments can be used to observe objects as small as atoms.

Fluorescence microscopy is used to observe fluorochrome-tagged specimens that might not be noticed with ordinary light microscopy.

EVOLUTION and BIODIVERSITY

The single cell organization of procaryotes necessitates that these cells function as independent units. Unlike multicellular eucaryotic organisms, which can apportion form and function among specialized cell types like muscle, heart, or nerve cells, a procaryote is autonomous and must perform all functions required for its nutrition, survival, and division. Because they are self-sufficient, procaryotes are considered the ancestral forms of life that existed on earth billions of years ago. An intermediate stage in the evolution of single-celled procaryotes into multicellular eucaryotes may have been the associations that developed in bacterial cells that formed chains or other cell arrangements because of incomplete cell wall formation and, therefore, remained attached after cell division. Today the wide range of size, shape, and arrangements of procaryotes is further evidence of the diversity of these microorganisms and their ability to adapt to changing environments.

Questions

Short Answer

1. Give several examples of macroorganisms. Give several examples of microorganisms.

2. Identify several traits which may help in the identification of bacteria.

3. Identify the two basic shapes of bacteria. What are some variations of these shapes?

4. Identify six types of microscopes.

5. How does a bright-field microscope differ from early microscopes (for example, Leeuwenhoek's)?

6. Identify the major parts of a bright-field microscope.

7. What is the total magnification for a microscope with 10× ocular lenses and a 95× objective lens?

8. When is immersion oil used? Why?

9. What is the smallest size specimen visible with a light microscope? With an electron microscope?

10. Identify several types of staining procedures and their importance.

11. What is the most commonly used bacterial staining procedure?

12. When is dark-field microscopy more beneficial than bright-field?

13. When is phase-contrast microscopy more beneficial than bright-field?

14. When is fluorescence microscopy more beneficial than bright-field?

15. In addition to increased magnification and resolution, what benefits(s) do SEM and TEM offer? Why aren't they routinely used?

16. Why are some bacteria found in chains, while others are clustered or randomly arranged?

Multiple Choice

1. Which of the following does *not* affect the resolving power of a microscope?
   a. the wavelength of the light source
   b. the numerical aperture of the condenser
   c. the numerical aperture of the objective lens
   d. the magnification

2. Positive stains include:
   a. Gram stain
   b. differential stain
   c. simple stain
   d. all of the above

3. Various staining methods can be used to increase the contrast between a specimen and its background; however, staining procedures can kill a specimen. How can a live specimen be viewed through a microscope under high contrast?
   a. Use a bright-field microscope.
   b. Use a phase-contrast microscope.
   c. Use a scanning electron microscope.
   d. Use a transmission electron microscope.

4. The resolving power of an electron microscope is greater than that of a light microscope because:
   a. the magnification can be increased beyond 1,000×.
   b. the wavelength of an electron is 100,000 times smaller than the wavelength of visible light.
   c. there is higher contrast between the specimen and its background.
   d. all of the above.

Critical Thinking

1. Discuss the role of morphology in bacterial identification and explain its limitations.

2. Explain the relationship between magnification and resolution. How can each be improved?

3. Describe how improvements in microscopy led to significant advances in microbiology.

4. Discuss the need for, and benefits of, the various staining procedures described in this chapter.

 Supplementary Readings

Aldrich, H.C., and W.J. Todd, eds. 1986. *Ultrastructure techniques for microorganisms.* New York: Plenum Publishing Corporation. (Individually written chapters dealing with various types of microorganisms and how to prepare them for transmission and scanning electron microscopy. A handbook of techniques.)

Dawes, C.J. 1988. *Introduction to biological electron microscopy: Theory and techniques.* Burlington, Vt.: Ladd Publishing Company. (A general preparation book for both scanning and transmission electron microscopy, with procedures for microorganisms as well as plant and animal tissues. The text includes defined procedures as well as the theory behind each preparation step.)

Gerhardt, P., R.G.E. Murray, W.A. Wood, and N.R. Krieg. 1994. *Methods for general and molecular bacteriology,* pp. 5–103. Washington, D.C.: American Society for Microbiology. (A discussion of various techniques in microbiology, including sections on the principles of light and electron microscopy.)

Hayat, M.A. 1986. *Basic techniques for transmission microscopy.* New York: Academic Press, Inc. (A general preparation handbook that gives fixation and handling procedures for a wide variety of biological samples, including microorganisms.)

Lichtman, J.W. 1994. Confocal microscopy. *Scientific American* 271:40–45. (A discussion of the operation of the confocal scanning microscope and examples of images taken with this type of microscopy.)

Perlman, P. 1971. *Basic microscope techniques.* New York: Chemical Publishing Company, Inc. (A primer on the use of light microscopes.)

Rochow, T. G., and E. G. Rochow. 1978. *An introduction to microscopy by means of light, electrons, X rays, or ultrasound.* New York: Plenum Press. (A detailed survey of the types of microscopy.)

Spencer, M. 1982. *Fundamentals of light microscopy.* London, England: Cambridge University Press. (A highly technical description of the principles of light microscopy.)

Wickramasinghe, H.K. 1989. Scanned-probe microscopes. *Scientific American* 261:98–105. (A discussion of the principles of the scanning tunneling microscope.)

chapter three 3

COMPOSITION AND STRUCTURE OF PROCARYOTIC AND EUCARYOTIC CELLS

Macromolecules: Building Blocks of the Cell
Proteins
Nucleic Acids
Carbohydrates and Lipids

The Procaryotic Cell
The Cell Envelope
The Plasma Membrane
The Cell Wall
The Outer Membrane
The Periplasm
Specialized Structures Outside the Cell Envelope
Chemotaxis
Procaryotes
Endospores

The Eucaryotic Cell
The Cell Wall and Plasma Membrane
Membrane-Enclosed Organelles
Locomotion
Asexual and Sexual Reproduction

PERSPECTIVE
Calcium Dipicolinate and Heat Resistance of Endospores

EVOLUTION AND BIODIVERSITY

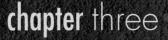

Transmission electron micrograph of the rod-shaped, gram-negative bacterium *Escherichia coli* (colorized, ×38,000).

 Microbes in Motion ─────── **PREVIEW LINK**

This chapter covers the key topics of cellular structures for both bacteria (procaryotic cells) and fungi. The following sections of the *Microbes in Motion* CD-ROM may be useful as a preview to your reading or as a supplemental study aid:

Bacteria Structure and Function: Cell Membrane, 1–10; Cell Wall, 1–5 (Antibiotic Sites), 13–15; Gram-Positive Cell, 1–5; Gram-Negative Cell 1–9; External Structures, 1–24; Internal Structures, 1–4 (Protein Synthesis), 12–13. *Fungal Structure and Function:* General Eucaryotic Structures, 1–11; Metabolism and Growth (Reproduction), 8–23.

illions of years ago, as the earth began to form a hospitable environment for biological molecules, the first forms of life appeared. Although there is no written record of this event, it is believed that the first living organisms were microorganisms, probably Archaea. Archaea, which include organisms that live in extreme environments of high temperature, low pH, and no oxygen—conditions similar to those in the prebiotic soup—have many of the physiological and metabolic characteristics that would be expected of early microbial life. These early organisms were simple in form and structure, but as time passed they evolved structures and characteristics that enabled them to adapt to their changing environments. Today's diverse organisms can trace their origins to these early cells.

The cell is the basic unit of life and is the smallest fundamental unit of all living organisms; it has all of the chemical and physical components necessary for its own maintenance and reproduction. These components of a cell function together to form a dynamic, interactive system capable of carrying on the essential life processes.

Although eucaryotes may generally be larger (see *Epulopiscium*, page 19) and more complex than procaryotes, the macromolecules and structural components of a procaryotic cell are assembled in such a manner as to comprise a functional, living organism. This chapter will examine the chemical composition of living cells and compare procaryotic and eucaryotic microorganisms.

Macromolecules: Building Blocks of the Cell

The chemical composition of a microorganism can easily be determined by analytical chemical techniques. Such analyses reveal that a large portion of the total weight of an average bacterial cell consists of **water.** In a rapidly growing *Escherichia coli* cell, water constitutes 70% or more of the total cell weight. Proteins, nucleic acids, polysaccharides, lipids, an assortment of other organic compounds such as vitamins and metabolic intermediates, and inorganic compounds constitute the remainder.

Proteins Are Composed of Amino Acids

Protein can make up as much as 55% of the dry weight (the cell weight minus its water content) of a cell. Proteins have a variety of important functions in the cell (Table 3.1). One of their most important functions is to act as enzymes. Enzymes catalyze biochemical reactions necessary for cell growth, reproduction, and metabolism. Although not all proteins are enzymes, most enzymes are proteins (catalytic RNAs, called **ribozymes,** have been discovered in procaryotes and eucaryotes and are involved in important biological reactions). Proteins are also found as structural components, transport molecules, and toxins in a cell.

Proteins are polymers of **amino acids.** Twenty different amino acids initially are used to make proteins (Table 3.2). With the exception of proline, all amino acids have an amino group (NH_3^+) and a carboxyl group (COO^-) attached to the same carbon (the α-carbon) and are thus called α-**amino acids.** Proline is similar in structure to the α-amino acids, with the exception that its α-amino nitrogen is in a five-membered ring. Because of this feature, proline is called an α-imino acid. The ring in the α-imino acid structure, which also includes the α-carbon, has a significant effect on protein structure, since it restricts the extent of protein folding.

Also attached to the α-carbon in amino acids are a hydrogen atom and a side chain, or **R group.** The composition of the R group differs in each of the 20 amino acids and is responsible for the structural and charge differences that exist among amino acids.

table 3.1

Some Cellular Functions of Proteins

Protein Function	Example	
	Procaryotes	**Eucaryotes**
Catalysis	DNA polymerase—replication and repair of DNA	DNA polymerase—replication and repair of DNA
Transport	Carrier molecules—active transport	Hemoglobin—transport of oxygen in blood
Motion	Flagellin—component of flagella	Tubulin—component of microtubules
Structure	Wall subunits—cell walls of certain Archaea	Collagen—fibrous connective tissue
Toxins	Diphtheria toxin—inactivation of elongation factor 2 in eucaryotic cells	Snake venoms—enzymes that hydrolyze phosphoglycerides
Chemical messengers/receptors	Binding proteins—binding of specific chemicals in chemotaxis	Insulin receptor—specific binding of insulin to modulate glucose metabolism
Ion transport/binding	Bacteriorhodopsin—energy transformation	Transferrin—iron-binding protein in blood

table 3.2

Structural Formulas of the 20 Amino Acids Commonly Occurring in Proteins

Amino Acids with Neutral (uncharged) Polar Side Chains

Glycine (Gly)
Molecular weight 75

$$H-\overset{\overset{\displaystyle H}{|}}{\underset{\underset{\displaystyle +}{\overset{\displaystyle |}{NH_3}}}{C}}-COO^-$$

Serine (Ser)
Molecular weight 105

$$HO-CH_2-\overset{\overset{\displaystyle H}{|}}{\underset{\underset{\displaystyle +}{\overset{\displaystyle |}{NH_3}}}{C}}-COO^-$$

Threonine (Thr)
Molecular weight 119

$$CH_3-\overset{\overset{\displaystyle OH}{|}}{\underset{\underset{\displaystyle H}{|}}{C}}-\overset{\overset{\displaystyle H}{|}}{\underset{\underset{\displaystyle +}{\overset{\displaystyle |}{NH_3}}}{C}}-COO^-$$

Cysteine (Cys)
Molecular weight 121

$$HS-CH_2-\overset{\overset{\displaystyle H}{|}}{\underset{\underset{\displaystyle +}{\overset{\displaystyle |}{NH_3}}}{C}}-COO^-$$

Tyrosine (Tyr)
Molecular weight 181

$$HO-\langle \bigcirc \rangle-CH_2-\overset{\overset{\displaystyle H}{|}}{\underset{\underset{\displaystyle +}{\overset{\displaystyle |}{NH_3}}}{C}}-COO^-$$

Asparagine (Asn)
Molecular weight 132

$$\overset{\overset{\displaystyle NH_2}{|}}{\underset{\underset{\displaystyle O}{\|}}{C}}-CH_2-\overset{\overset{\displaystyle H}{|}}{\underset{\underset{\displaystyle +}{\overset{\displaystyle |}{NH_3}}}{C}}-COO^-$$

Glutamine (Gln)
Molecular weight 146

$$\overset{\overset{\displaystyle NH_2}{|}}{\underset{\underset{\displaystyle O}{\|}}{C}}-CH_2-CH_2-\overset{\overset{\displaystyle H}{|}}{\underset{\underset{\displaystyle +}{\overset{\displaystyle |}{NH_3}}}{C}}-COO^-$$

Amino Acids with Acidic (negatively charged) Side Chains

Aspartic Acid (Asp)
Molecular weight 133

$$\overset{\overset{\displaystyle ^-O}{|}}{\underset{\underset{\displaystyle O}{\|}}{C}}-CH_2-\overset{\overset{\displaystyle H}{|}}{\underset{\underset{\displaystyle +}{\overset{\displaystyle |}{NH_3}}}{C}}-COO^-$$

Glutamic acid (Glu)
Molecular weight 147

$$\overset{\overset{\displaystyle ^-O}{|}}{\underset{\underset{\displaystyle O}{\|}}{C}}-CH_2-CH_2-\overset{\overset{\displaystyle H}{|}}{\underset{\underset{\displaystyle +}{\overset{\displaystyle |}{NH_3}}}{C}}-COO^-$$

Amino Acids with Basic (positively charged) Side Chains

Lysine (Lys)
Molecular weight 146

$$H_3\overset{+}{N}-CH_2-CH_2-CH_2-CH_2-\overset{\overset{\displaystyle H}{|}}{\underset{\underset{\displaystyle +}{\overset{\displaystyle |}{NH_3}}}{C}}-COO^-$$

Amino Acids with Basic (positively charged) Side Chains (cont.)

Arginine (Arg)
Molecular weight 174

$$H_2N-\overset{\overset{\displaystyle }{}}{\underset{\underset{\underset{\displaystyle +}{\displaystyle }}{\overset{\displaystyle \|}{NH_2}}}{C}}-NH-CH_2-CH_2-CH_2-\overset{\overset{\displaystyle H}{|}}{\underset{\underset{\displaystyle +}{\overset{\displaystyle |}{NH_3}}}{C}}-COO^-$$

Histidine (at pH 6.0) (His)
Molecular weight 155

$$\underset{\displaystyle HN}{HC}=\overset{\displaystyle }{\underset{\displaystyle NH}{C}}\diagdown C-CH_2-\overset{\overset{\displaystyle H}{|}}{\underset{\underset{\displaystyle +}{\overset{\displaystyle |}{NH_3}}}{C}}-COO^-$$

Amino Acids with Nonpolar (hydrophobic) Side Chains

Alanine (Ala)
Molecular weight 89

$$CH_3-\overset{\overset{\displaystyle H}{|}}{\underset{\underset{\displaystyle +}{\overset{\displaystyle |}{NH_3}}}{C}}-COO^-$$

Valine (Val)
Molecular weight 117

$$\overset{\displaystyle CH_3}{\underset{\displaystyle CH_3}{\diagup}}CH-\overset{\overset{\displaystyle H}{|}}{\underset{\underset{\displaystyle +}{\overset{\displaystyle |}{NH_3}}}{C}}-COO^-$$

Leucine (Leu)
Molecular weight 131

$$\overset{\displaystyle CH_3}{\underset{\displaystyle CH_3}{\diagup}}CH-CH_2-\overset{\overset{\displaystyle H}{|}}{\underset{\underset{\displaystyle +}{\overset{\displaystyle |}{NH_3}}}{C}}-COO^-$$

Isoleucine (Ile)
Molecular weight 131

$$CH_3-CH_2-\overset{\overset{\displaystyle }{}}{\underset{\underset{\displaystyle CH_3}{|}}{CH}}-\overset{\overset{\displaystyle H}{|}}{\underset{\underset{\displaystyle +}{\overset{\displaystyle |}{NH_3}}}{C}}-COO^-$$

Proline (Pro)
Molecular weight 115

$$\begin{matrix} CH_2 & & \\ \diagup & \diagdown & \\ CH_2 & & \overset{\overset{\displaystyle H}{|}}{C}-COO^- \\ \diagdown & \diagup & \underset{\underset{\displaystyle +}{\overset{\displaystyle |}{NH_2}}}{} \\ & CH_2 & \end{matrix}$$

Phenylalanine (Phe)
Molecular weight 165

$$\langle \bigcirc \rangle-CH_2-\overset{\overset{\displaystyle H}{|}}{\underset{\underset{\displaystyle +}{\overset{\displaystyle |}{NH_3}}}{C}}-COO^-$$

Tryptophan (Trp)
Molecular weight 204

$$\begin{matrix} \bigcirc\bigcirc & C-CH_2-\overset{\overset{\displaystyle H}{|}}{\underset{\underset{\displaystyle +}{\overset{\displaystyle |}{NH_3}}}{C}}-COO^- \\ \underset{\underset{\displaystyle H}{|}}{N} & CH \end{matrix}$$

Methionine (Met)
Molecular weight 149

$$CH_3-S-CH_2-CH_2-\overset{\overset{\displaystyle H}{|}}{\underset{\underset{\displaystyle +}{\overset{\displaystyle |}{NH_3}}}{C}}-COO^-$$

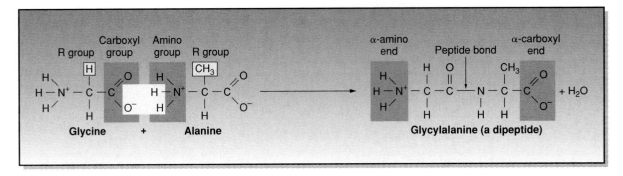

figure 3.1

Formation of a Dipeptide

Two amino acids, glycine and alanine, are joined to form the dipeptide glycylalanine. The free α-amino and α-carboxyl ends are shown.

Amino acids are joined together in proteins by covalent bonds called **peptide bonds** (Figure 3.1). The sequence of amino acids in a protein determines the biological properties of the protein molecule. Peptide bonds are formed by the linking of the α-amino group on one amino acid to the α-carboxyl group on an adjacent amino acid. When two amino acids are joined in this manner, a dipeptide is formed. A tripeptide consists of three amino acids linked by peptide bonds. Short chains of amino acids are called **polypeptides,** and longer chains or combinations of polypeptides are called **proteins.** Since peptides consist of individual amino acids joined by peptide bonds, one end of a peptide (the N-terminal) may have a free, unattached α-amino group, while the opposite end (the C-terminal) may have a free α-carboxyl group. This orderly arrangement of amino acids gives orientation to a protein—a property that is important in biological systems and can be useful in chemical extraction and identification of proteins.

A hierarchy of protein structure has been established by chemists (Figure 3.2). This hierarchy begins with the **primary structure** of a protein, defined as the sequence of amino acids in the protein. Proteins ordinarily do not lie flat or in random coils, but exhibit regular, repeated configurations caused by hydrogen bonding between the atoms within the polypeptide chain. Protein configurations that result from such bonds constitute the **secondary structure** of proteins. There are two regularly occurring types of secondary protein structures. The β-**sheet,** which occurs in silk, consists of extended polypeptide chains folding back and forth while being held together by hydrogen bonds. Wool and collagen have secondary protein structures arranged in an α-**helical** configuration. In such structures the polypeptide backbone is twisted in a helix and held in that manner by hydrogen bonds. The term **tertiary structure** refers to the three-dimensional conformation of a protein molecule. These three-dimensional structures, which can be determined by X-ray diffraction patterns, generally are folded in a particular fashion and represent active forms of proteins. Many proteins, such as DNA-dependent RNA polymerase (the enzyme responsible for synthesis of RNA from DNA) and hemoglobin,

consist of several polypeptide chains that are linked together as a unit. These proteins are said to possess a **quaternary structure.** Quaternary structures of proteins are common. In many cases, each of the polypeptide chains in the quaternary structure has a unique function in the overall activity of the protein.

Environmental effects such as high temperatures or extreme pH can cause tertiary and quaternary protein structures to become disrupted, or **denatured.** When this happens, proteins (especially enzymes) may become biologically inactive. This intolerance of many proteins of environmental extremes is used in microbiology to kill microorganisms by exposure to heat or acid pH.

Nucleic Acids Are Composed of Nucleotides

Nucleic acids are macromolecules composed of individual organic units called **nucleotides** (Figure 3.3). There are two types of nucleic acids: **deoxyribonucleic acid (DNA)** and **ribonucleic acid (RNA).** DNA is the genetic, or hereditary, material in all cells. This genetic information is used for the synthesis of proteins.

Nucleotides are molecules that consist of (1) a pentose (5-carbon) sugar attached to (2) a nitrogenous organic compound called a **base** and (3) one or more phosphates. A **nucleoside** consists of the pentose and the base, without any phosphates. The pentose sugar in DNA is deoxyribose (with a hydrogen atom instead of a hydroxyl at the second carbon position of the sugar), whereas in RNA the sugar is ribose. The bases pyrimidines and purines commonly occur in nucleic acids. Pyrimidines have a single-ring structure, whereas purines consist of two rings. DNA contains the pyrimidines cytosine and thymine and the purines adenine and guanine. RNA contains the same bases, with the exception that thymine is replaced by the pyrimidine uracil. Some RNAs (for example, transfer RNA, which is involved in protein synthesis) have unusual bases such as pseudouridine and dihydrouridine, which contribute to the activity and structure of the RNA molecule.

Most DNA molecules (the exceptions are those in some viruses) consist of two polydeoxynucleotide chains arranged in a

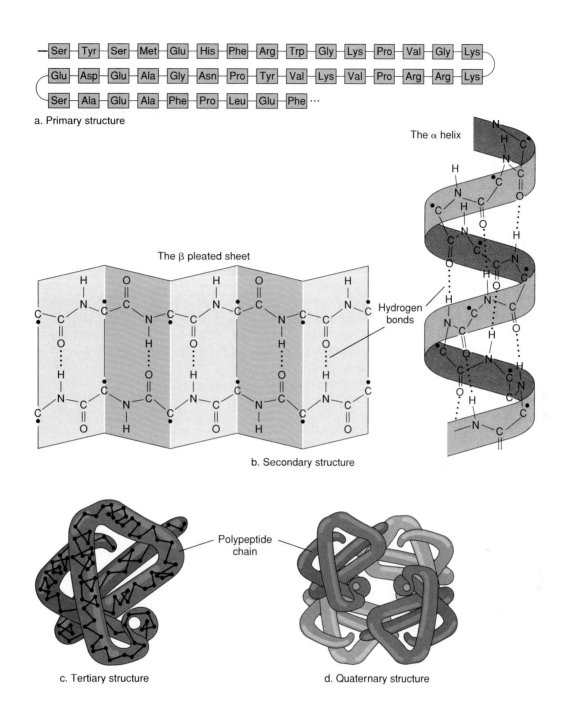

a. Primary structure

The α helix

The β pleated sheet

Hydrogen bonds

b. Secondary structure

Polypeptide chain

c. Tertiary structure

d. Quaternary structure

figure 3.2

Hierarchy of Protein Structure

a. Primary structure, with amino acids connected by peptide bonds.
b. Secondary structure, formed by hydrogen bonds. Dots (•) represent α-carbons of amino acids. c. Tertiary structure, often formed by disulfide bonds. Dots (•) represent α-carbons of amino acids.
d. Quaternary structure, consisting of several polypeptides linked to one another.

double-stranded helix (Figure 3.4). The nucleotides are linked together within each chain by phosphodiester bonds that connect the OH group on the 3′ carbon atom of one deoxyribose to the phosphate on the 5′ carbon atom of the adjoining deoxyribose to give a sugar phosphate backbone. The two strands in the DNA helix are held in place by hydrogen bonds between nucleotide base pairs. This base pairing is not random, but occurs in such a manner that adenine is always paired with thymine by two hydrogen bonds and guanine is always paired with cytosine by three hydrogen bonds.

In contrast to the double-stranded DNA molecule, RNA molecules are usually single-stranded (some viruses have double-stranded RNA molecules). Cells have three types of RNA molecules: messenger RNA (mRNA), ribosomal RNA (rRNA), and transfer RNA (tRNA). All three types of RNA molecules are polyribonucleotide chains consisting of nucleotides linked by 3′, 5′-phosphodiester bonds, giving a sugar phosphate backbone. In cells, RNA is associated with protein synthesis. Messenger RNA carries the genetic information from DNA to the ribosomes, where this information is translated into proteins with the assistance of rRNA and tRNA. The role of DNA and RNA in storage of genetic information and transfer of this information to proteins is discussed in further detail in Chapter 8 (see central dogma, page 212).

Carbohydrates and Lipids Are Organic Macromolecules

Carbohydrates and **lipids** are important classes of organic macromolecules that are found universally in plants, animals, and microorganisms. These molecules are frequently used as carbon and energy sources by these organisms and are also important as major constituents of structures such as cell walls and membranes. For example, two carbohydrate derivatives, *N*-acetylglucosamine and *N*-acetylmuramic acid, form the backbone of the cell wall in the domain Bacteria. Phospholipids (lipids containing one or more phosphate groups) are the principal structural components of biological membranes, which are semipermeable because of the solubility properties of phospholipids.

Carbohydrates are desirable carbon and energy sources for microorganisms because they are often readily available in the environment and can be directly incorporated into metabolic pathways. These organic macromolecules consist of simple sugar units that either exist independently or are joined together as larger polymers (Figure 3.5). Several different levels of classification are used by chemists for carbohydrates: **monosaccharides,** or **simple sugars,** which have the general formula $(CH_2O)_n$, where **n** has a value of three or greater; **disaccharides,** which consist of two similar or dissimilar monosaccharides joined by a glycosidic bond between the aldehyde group of one monosaccharide and an alcohol group of the other monosaccharide; **oligosaccharides** (carbohydrates having up to ten monosaccharide units); and **polysaccharides** (very large carbohydrates, sometimes containing several thousand monosaccharide units). Carbohydrates may also be covalently linked to proteins or lipids to form **glycoproteins** and **lipopolysaccharides.**

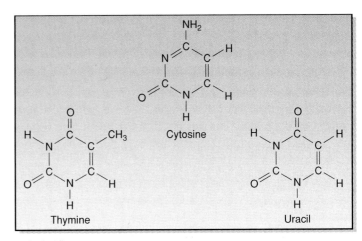

a. Pyrimidines

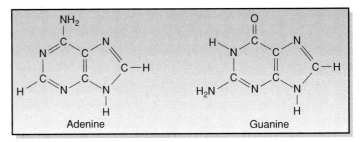

b. Purines

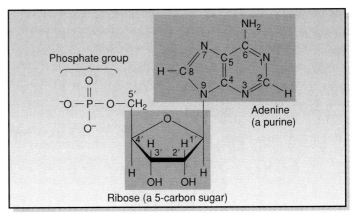

c. Nucleotide

figure 3.3
Building Blocks of Nucleic Acids
a. The three pyrimidine bases commonly found in nucleotides. b. The two purine bases commonly found in nucleotides. c. A nucleotide (adenosine monophosphate, or AMP).

The 6-carbon (hexose) simple sugar glucose is representative of many monosaccharides and is an important intermediate in carbohydrate metabolism. D-glucose, also known as dextrose, is found in the sap of most plants, in fruit juices, and as a component of animal blood. It is also a constituent of higher carbohydrates such as lactose, sucrose, maltose, starch, and glycogen, and is a major source of energy for both procaryotes and eucaryotes.

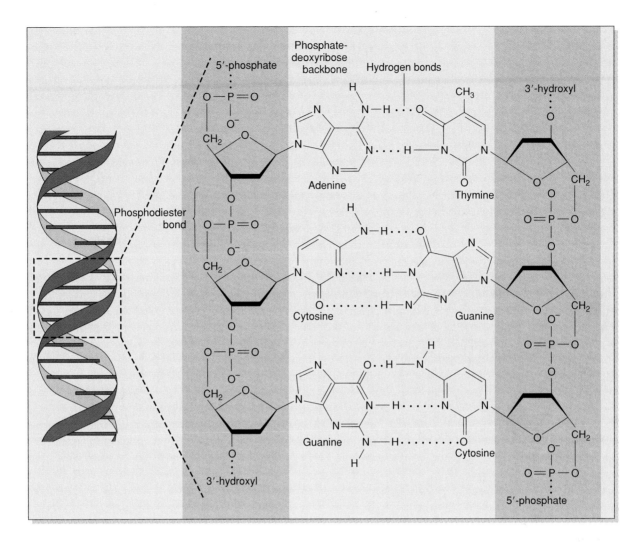

figure 3.4
Structure of a DNA Molecule

Most carbohydrates in nature exist in the form of polysaccharides. The most common polysaccharide is cellulose, the major structural component of plant cell walls and also a component of the cell walls of algae and some fungi. It has been estimated that more than 50% of all organic carbon is cellulose. Cellulose is a straight-chain polymer of glucose units joined by β(1,4) glycosidic bonds. Starch is also a polymer of glucose, but with the glucose units linked by α(1,4) bonds with α(1,6) branching, resulting in a nonlinear molecule. Animals store glucose as polymers, but in the form of glycogen. Starch and glycogen have similar structures, except that glycogen is more highly branched. Microorganisms that use polysaccharides as carbon and energy sources must first hydrolyze these large molecules to smaller monosaccharides and disaccharides, which can then be transported across the plasma membrane into the cell. These microorganisms synthesize enzymes such as cellulases (cellulose) and amylases (starch and glycogen), which hydrolyze the polysaccharides (see polysaccharide hydrolysis, page 190).

The lipids are a class of organic molecules characterized by insolubility in water and solubility in nonpolar solvents such as benzene, chloroform, and acetone. Biologically important lipids can be divided into three major classes: **neutral fats, phospholipids,** and **steroids** (Figure 3.6).

Neutral lipids, so called because they have no charged groups at physiological pH, generally consist of the alcohol glycerol attached by ester linkages to long-chain fatty acids. Such lipids are called mono-, di-, or triglycerides, depending upon whether one, two, or three fatty acids, respectively, are attached to glycerol. Fatty acids are saturated or unsaturated compounds, usually having from 14 to 22 carbon atoms, with a carboxyl group at one end. Saturated fatty acids have only single bonds between carbon atoms, whereas unsaturated fatty acids contain one or more double bonds within the molecule. Fats constitute a major food reserve in animals and, to some extent, in plants. These molecules, because of their bonding characteristics, generally contain more energy than carbohydrates with equivalent numbers of carbon atoms (see lipid metabolism, page 197).

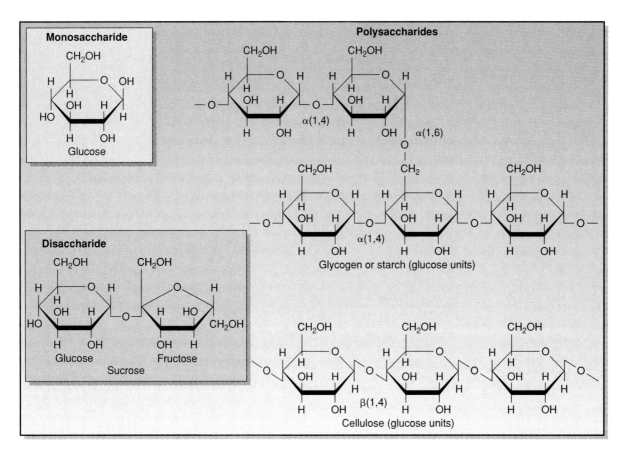

figure 3.5
Structures of Some Common Carbohydrates

Phospholipids, or phosphoglycerides, are fats in which one of the fatty acid chains has been replaced by a phosphate group, frequently with a nitrogen-containing compound attached to it. The resultant molecule is amphipathic—the fatty acid chains are hydrophobic, whereas the ionic phosphate group is hydrophilic.

Steroids, complex organic molecules composed of four interlocking ring structures, do not resemble other lipids. Steroids and the related sterols, which have hydroxyl (OH) groups attached to certain carbons in the structure, are classified as lipids because of their insolubility in water. Steroids and sterols have important biological roles. Cholesterol, a common sterol, is a structural component in eucaryotic plasma membranes and the primary constituent of gallstones, and also forms fatty deposits on the interior walls of arteries. Some vitamins and hormones, including several forms of vitamin D, testosterone, estrogen, progesterone, and cortisone, are steroids. Sterols rarely are found in bacteria except in the membranes of mycoplasmas. Mycoplasmas do not have cell walls, and the sterols appear to be responsible for the increased stability of the plasma membrane necessary to resist osmotic lysis.

The Procaryotic Cell

The electron microscope has made the study of the fine structure of procaryotes possible. Examination of a bacterial cell reveals a variety of structures (Figure 3.7). All bacteria have a plasma membrane, DNA, RNA, and ribosomes. Most bacteria have a cell wall. Other structures such as capsules, flagella, pili, endospores, and inclusion bodies are found only in certain bacteria.

The Cell Envelope Is the Outer Covering of a Procaryote Consisting of the Plasma Membrane, the Cell Wall, and, in Gram-Negative Procaryotes, an Outer Membrane

The cytoplasmic contents of a typical procaryotic cell are encased within an outer covering called the **cell envelope.** This envelope's specific content may vary in different types of procaryotes, but in most procaryotes it consists of a semipermeable **plasma membrane** surrounded by a rigid **cell wall.** In addition

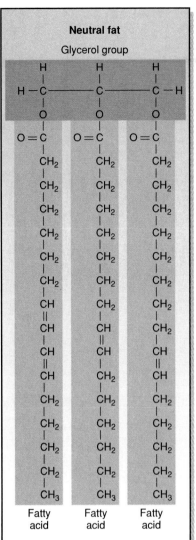
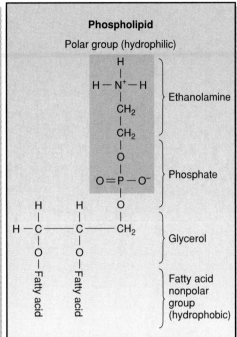
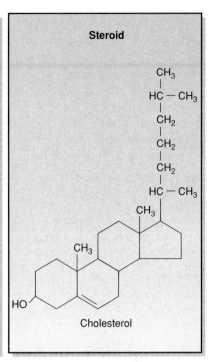

figure 3.6
Some Biologically Important Lipids

to these components, the envelope in gram-negative procaryotes has an extra outer layer known as the **outer membrane** that is part of the cell wall.

The Plasma Membrane Is a Selectively Permeable Barrier

The plasma membrane, also called the cell membrane or cytoplasmic membrane, is a semipermeable barrier that regulates the passage of molecules and ions (Figure 3.8). This discriminant permeability of the membrane is an important property of the cell. If it didn't have this selectivity, the cell would not retain internal (endogenous) pools of metabolites required for its growth and nutrition, nor would the cell prevent entry of undesirable substances from a hostile environment. The plasma membrane also provides structural integrity to bacteria that lack cell walls (for example, *Mycoplasma*); these microorganisms must rely on this sole outer barrier to withstand environmental osmotic stress. Some microorganisms can exist without a cell wall, but none can survive without a plasma membrane.

The plasma membrane in procaryotes is similar in composition to the basic unit membrane structure found in higher organisms. This procaryotic plasma membrane, as seen by high-resolution electron microscopy, consists of a continuous bimolecular sandwich structure having two thin, dense bands separated by a lighter, less dense middle band. Chemical extraction of membrane fragments isolated from cells reveals that such fragments consist of 60% to 70% protein, and 30% to 40% lipid. The main lipids in the plasma membrane are phospholipids. These

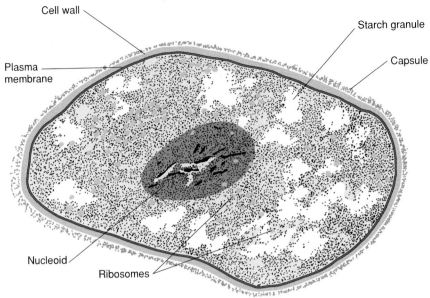

Cell wall

Starch granule

Capsule

Plasma membrane

Nucleoid

Ribosomes

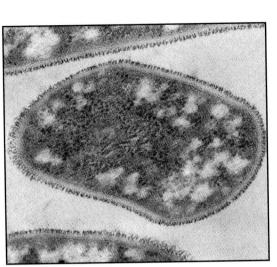

a.

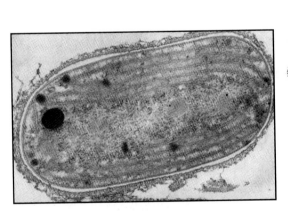

b.

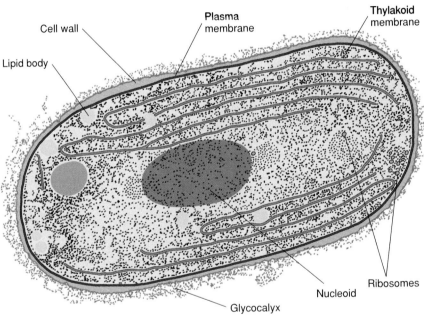

Cell wall

Lipid body

Plasma membrane

Thylakoid membrane

Glycocalyx

Nucleoid

Ribosomes

figure 3.7
Structural Organization of a Typical Procaryote
a. Electron micrograph of *Arthrobacter,* a nonphotosynthetic gram-positive bacterium, showing cell structures (×195,000).
b. Electron micrograph of *Agmenellum,* a photosynthetic cyanobacterium, showing cell structures. Not all morphological features shown are necessarily present in a single cell type (×45,000).

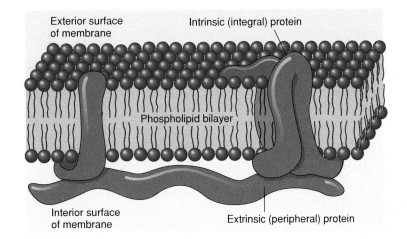

Exterior surface of membrane

Intrinsic (integral) protein

Phospholipid bilayer

Interior surface of membrane

Extrinsic (peripheral) protein

figure 3.8

The Plasma Membrane

The bacterial plasma membrane consists of a lipid bilayer and proteins (intrinsic and extrinsic) that are associated with the bilayer.

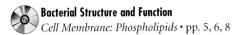

Bacterial Structure and Function
Cell Membrane: Phospholipids • pp. 5, 6, 8

figure 3.9

Differences Between Membrane Lipids of Bacteria and Archaea

a. Membrane lipid of Bacteria with ester linkage between glycerol and fatty acids. b. Membrane lipid of Archaea with ether linkage between glycerol and repeating branched aliphatic side chains. c. Structure of isoprene, the side chain in Archaea membrane lipids.

are amphipathic compounds that have hydrophilic polar groups attached to hydrophobic nonpolar chains. Membrane lipids of Bacteria have *ester* linkages between glycerol and the fatty acids, whereas lipids of Archaea have *ether* linkages between glycerol and repeating branched aliphatic chains of isoprene (Figure 3.9). The ether linkage and the substitution of isoprene for fatty acids in membrane lipids are characteristics that help distinguish Archaea from Bacteria. The domain Archaea includes the thermoacidophilic (high-temperature and acid-loving), halophilic (sodium chloride-loving), and methanogenic (methane-producing) procaryotes.

The phospholipids in the procaryotic plasma membrane are oriented with their polar groups extended outward and the nonpolar chains internalized within the membrane. Although the membrane exists as a bilayer, it is not symmetrical. The two sides of the membrane are dissimilar and may contain different proteins and phospholipids. X-ray diffraction patterns and microscopic observations show that membrane proteins and lipids are not fixed, but instead are mobile and freely diffuse within the phospholipid bilayer structure (fluid mosaic model). Fluidity is apparently an important characteristic of the plasma membrane, since microbes living in extreme temperature environments change the fatty acid composition of their membranes to maintain their fluidity. The precise role of the mobile lipids within the membrane is not known, whereas the membrane proteins are involved in molecule transport across the membrane, enzymatic reactions associated with synthesis of membrane and cell wall components, respiration, and energy generation.

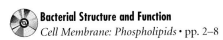

Bacterial Structure and Function
Cell Membrane: Phospholipids • pp. 2–8

The proteins that are found in the membrane exist primarily in globular form and are dispersed individually and in aggregates throughout the membrane, giving a mosaic appearance to this structure. Two classes of proteins have been identified, based on their relative solubility and location in membranes:

1. **Intrinsic (integral) proteins,** embedded in the interior of the lipid bilayer, make up approximately 70% to 80% of all membrane proteins. These intrinsic proteins are bound to the membrane lipids by nonpolar interactions and can be removed only by the use of detergents or nonpolar solvents.

2. **Extrinsic (peripheral) proteins** are found on the lipid bilayer and are removable by relatively mild treatments such as pH adjustments or changes in the ionic strength of the medium surrounding the membrane. The extrinsic proteins are believed to bind to the intrinsic proteins at the surface of the membrane, and possibly to some of the phospholipids.

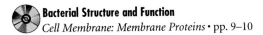
Bacterial Structure and Function
Cell Membrane: Membrane Proteins • pp. 9–10

Because of the lipid nature of the membrane, lipid-soluble molecules (for example, alcohols) are readily able to enter this barrier. However, such molecules encounter problems when they attempt to exit the plasma membrane and enter into the aqueous cytoplasm; the lipid-soluble molecules must first break their bonds with membrane lipids before they can move into the cytoplasm. Molecule passage through the plasma membrane depends not only on lipid solubility, but also on other parameters. Since the membrane possesses a net negative charge, molecules with similar charges (acidic proteins or acidic polysaccharides) are routinely repelled from entering or leaving the cell. Compounds that are too large to cross the membrane must first be degraded into smaller constituents by extracellular enzymes (amylases, proteases, and lipases) before entry (see polysaccharide, protein, and lipid hydrolysis, pages 190, 194, and 197). Bacteria that synthesize such enzymes often have a selective growth advantage in environments containing these large compounds, which serve as sources of nutrients.

Some procaryotes possess internal membranes formed from invaginations of the plasma membrane. The photosynthetic apparatus of photosynthetic bacteria is found in distinct membranous sacs continuous with the plasma membrane (Figure 3.10). These membranous sacs contain bacteriochlorophylls and accessory light-gathering pigments that are associated with the transfer of light energy in photosynthesis.

Bacteria that oxidize substances such as ammonia, nitrite, and methane have extensive internal membrane systems. The membranes in these bacteria are involved in the formation of hydrogen ion gradients used in the synthesis of ATP.

In certain gram-positive bacteria and a few gram-negative species, the plasma membrane appears to be folded into large convoluted invaginations, which have been called **mesosomes.** Mesosomes were first observed by electron microscopy and were thought to provide additional cell surface area for nutrient absorption, DNA attachment, and septum formation (the cross-

wall that is formed in a bacterium during cell division). It is now believed that mesosomes are artifacts of preparations for electron microscopy.

The Cell Wall Maintains Cell Shape and Conformation

The unique shapes of most procaryotes are preserved by a complex, inflexible cell wall, a rigid layer of material external to the plasma membrane. The wall maintains conformation and also enables the cell to survive in hypotonic environments, where the osmotic pressure is higher inside than outside the cell. Under such conditions cells would swell and eventually burst were it not for the rigidity of the wall. Bacteria that lack a cell wall, either naturally or through laboratory manipulations, must in some way compensate for the absence of this structure either with sterol-containing plasma membranes (for example, *Mycoplasma*) or restricted growth in isosmotic habitats.

Cell walls are not unique to procaryotes. Other cells, including those of fungi, algae, and plants, also possess walls. The cell walls of procaryotic and eucaryotic microorganisms have basically similar architecture and function. Cell walls consist of polysaccharides linked by chemical bridges to form a rigid structure. The cell wall of procaryotes, however, is unique in its composition. Most of what is known about procaryotic cell walls has been obtained through electron microscopy, antibiotic studies, and chemical analysis of bacterial walls.

The substance of the bacterial cell wall is **peptidoglycan** (Figure 3.11), which is found only in Bacteria. It is arranged in one to several layers on the cell surface and provides the strength that is necessary for wall integrity. This unique chemical substance consists of repeating units of two carbohydrate derivatives, *N*-acetylglucosamine (NAG) and *N*-acetylmuramic acid (NAM), linked to each other by β(1,4) bonds. NAM is found only in peptidoglycan and nowhere else in nature. Four amino acids make up what is known as a tetrapeptide attached to the NAM. The sequence of these amino acids in gram-negative bacteria and certain gram-positive bacteria is generally L-alanine, D-glutamate, *meso*-diaminopimelate, and D-alanine. Most gram-positive cocci have a similar sequence, with the exception that diaminopimelate is replaced by another diamino acid, L-lysine. The amino acid sequence of the tetrapeptide may be further modified in other bacteria. NAM and diaminopimelate are not found in eucaryotic cell walls. NAM is also not found in Archaea. The two amino acids with the D configuration (D-glutamate and D-alanine) are novel to peptidoglycan, since optically active amino acids in proteins are always in the L configuration. The alternating D and L configurations of the amino acids in the tetrapeptide, the β(1,4) linkages between NAM and NAG in the repeating carbohydrate structure, and the layered sheet configuration of peptidoglycan contribute to the mechanical strength of this structure in the cell wall.

The major strength of peptidoglycan comes from periodic cross-links between the third and fourth amino acids of adjacent

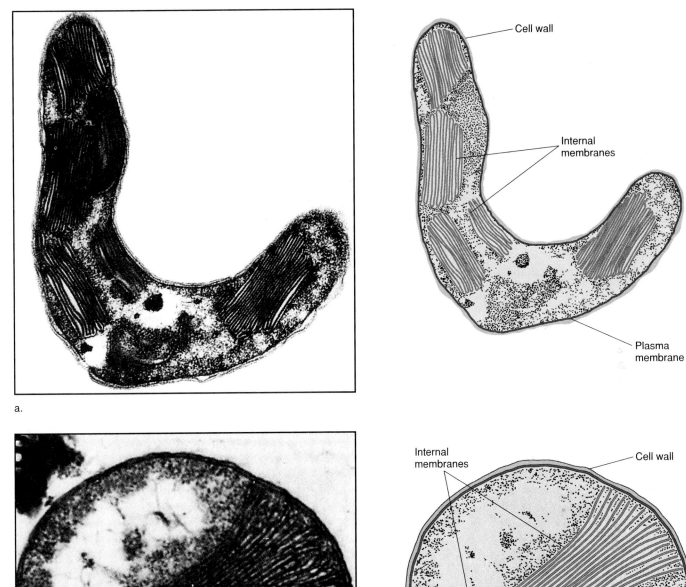

a.

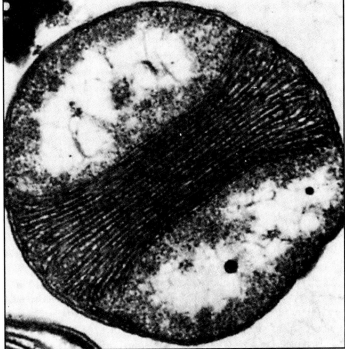

b.

figure 3.10

Internal Procaryotic Membranes

Internal membranes in a. *Ectothiorhodospira mobilis*, a photosynthetic bacterium (×53,000); b. *Nitrococcus oceanus*, a nitrifying (ammonia-oxidizing) bacterium (×22,500).

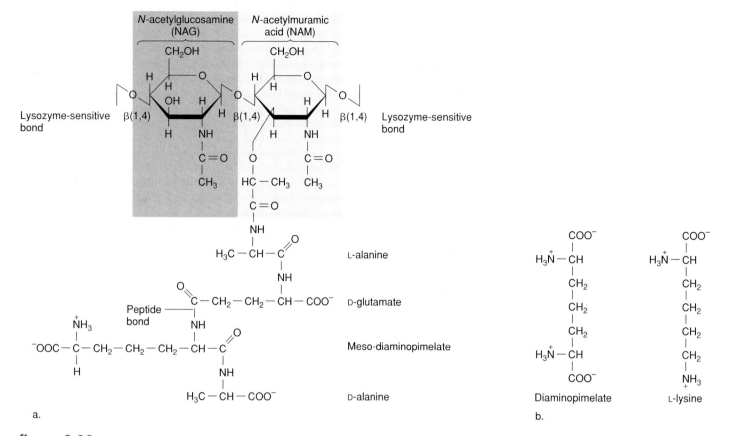

a.

b.

figure 3.11

Chemical Structure of a Repeating Unit in Peptidoglycan

a. A repeating unit of N-acetylglucosamine (NAG) and
N-acetylmuramic acid (NAM) as found in *Escherichia coli* and other
gram-negative bacteria. b. *Staphylococcus aureus* and other gram-
positive cocci replace diaminopimelate with L-lysine in the tetrapeptide.

Antimicrobial Action
Cell Wall Inhibitors: Mechanisms • p. 7

tetrapeptides. The frequency, composition, and length of these cross-links vary among different bacteria, thus influencing the rigidity and strength of the wall in these organisms. Cross-linkage may occur either directly between amino acids in neighboring tetrapeptides, as occurs in most gram-negative cells, or alternatively, as in many gram-positive cells, through a peptide bridge (Figure 3.12).

The cell wall of *Staphylococcus aureus*, a gram-positive coccus, has been extensively studied; it is known that in this bacterium, tetrapeptides are linked by a pentaglycine bridge consisting of five molecules of the amino acid glycine. Although not all tetrapeptides may be involved in cross-linking bridges, the frequency of cross-linking in *S. aureus* is considerable and approaches 75%. In contrast, in *Escherichia coli*, like other gram negatives, modifications of the peptidoglycan layer cause a significant

portion of the tetrapeptide chains to cleave from the glycan chains. As a result of these losses, *E. coli*, which has peptidoglycan cross-links consisting of a direct bridge between D-alanine and *meso*-diaminopimelate in adjacent tetrapeptides, has a cross-linking frequency of only 25%. Cross-linking bridges in bacterial cell walls not only result in a joining of linear polysaccharide chains to form large sheetlike layers of peptidoglycan, but also are responsible for the attachment of these sheets to one another to produce a saclike structure—a single molecule—encircling the cell (Figure 3.13).

Not all procaryotes have peptidoglycan in their cell walls. The Archaea do not have NAM as a constituent of their cell walls, but their walls may be composed of proteins, glycoproteins, and polysaccharides. Another group of procaryotes, the mycoplasmas, lack cell walls and therefore are not inhibited by antibiotics that act on the wall (for example, penicillin).

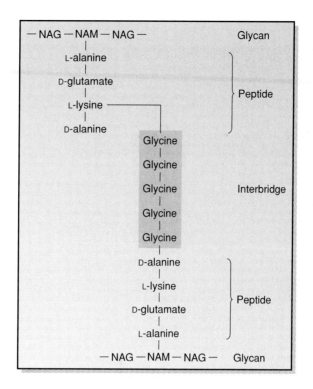

figure 3.12

Glycine Interbridge in the Peptidoglycan of *Staphylococcus aureus*

Antimicrobial Action
Cell Wall Inhibitors: Mechanisms • p. 7

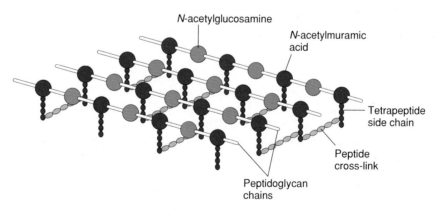

figure 3.13

Structure of Peptidoglycan Layer in Gram-Positive Cells

Antimicrobial Action
Cell Wall Inhibitors: Mechanisms • p. 7

Peptidoglycan Synthesis

The peptidoglycan of the bacterial cell wall is a complex chemical structure. Building blocks for the construction of peptidoglycan form in the cell cytoplasm and then transfer through the plasma membrane to the cell wall, where they link with existing peptidoglycan. The synthesis of these polymers and their incorporation into existing wall material occurs in four distinct stages (Figure 3.14):

1. *N*-acetylmuramic acid is synthesized in the cytoplasm of the cell from precursor molecules. During its synthesis, NAM attaches to uridine diphosphate (UDP), a carbohydrate carrier, by a high-energy phosphate bond to form uridine diphospho-*N*-acetylmuramic acid (UDP-NAM). The tetrapeptide portion of peptidoglycan, as well as a fifth amino acid (D-alanine) located at the terminus of the tetrapeptide, is added to the UDP-NAM by the step-wise addition of amino acids requiring energy input from the hydrolysis of ATP. Unlike normal protein synthesis, this peptide synthesis occurs without ribosomes, mRNA, or tRNA.

2. This complex moves to the plasma membrane, where it attaches to a second carrier, bactoprenol, by phosphate ester bonds. The energy for this reaction is derived from cleavage of the high-energy phosphate bond between UDP and NAM. Bactoprenol's role in cell wall synthesis is to make NAM hydrophobic enough for transport across the primarily phospholipid membrane.

3. After attachment to bactoprenol, the NAM-lipid complex is linked to (UDP)-*N*-acetylglucosamine. The energy requirements for this linkage are met by cleaving the bond between UDP and NAG. The cross-linking peptide bridge (pentaglycine in *S. aureus*) is added to the pentapeptide at this time. The polymer may be further modified, depending on the bacterium. This entire complex (NAG-NAM-pentapeptide-bactoprenol) now moves across the plasma membrane to the site for incorporation into the existing cell wall.

4. The newly synthesized section of peptidoglycan attaches to the existing wall by transglycosylation. During transglycosylation, the bactoprenol lipid carrier is released from the NAG-NAM disaccharide and moves back across the plasma membrane to the cytoplasm, where it is regenerated and reused. The NAG-NAM is linked by β(1,4) bond to the existing cell wall peptidoglycan backbone. The terminal D-alanine on NAM is cleaved at this time. The energy released from the cleavage of this peptide bond is used to form the peptide bond that attaches the cross-linking peptide to the existing wall (transpeptidation). Since cross-links are not always formed between adjacent tetrapeptides, in those instances where cross-links are absent, the terminal D-alanine is cleaved by the enzyme D-alanine carboxypeptidase.

Bacterial Structure and Function
Cell Wall • pp. 1–5

The antibiotic penicillin specifically inhibits the transpeptidation step in cell wall synthesis. This explains why penicillin is ineffective in nongrowing cells that are not synthesizing new wall material. In the presence of penicillin, growing cells are unable to synthesize an intact cell wall. As a result, the cells lyse. Cells that do not have cell walls or walls containing peptidoglycan, such as mycoplasmas, animal cells, and fungi, are not affected by penicillin. Gram-negative bacteria are relatively resistant to penicillin because the antibiotic must pass through the outer membrane to reach the peptidoglycan of the cell wall, and it cannot easily do so.

Bacterial Structure and Function
Cell Wall: Antibiotic Sites • pp. 13–15

The peptidoglycan layer is frequently associated with other components in the cell wall. In gram-positive bacteria, peptidoglycan constitutes approximately 90% of the cell wall and can be up to 40 layers thick (Figure 3.15). The remainder of the wall in these organisms consists of acidic polysaccharides. These polysaccharides, the **teichoic acids,** are repeating units of glycerol or ribitol joined by phosphodiester linkages, and are located throughout the wall. In some gram-positive organisms (most notably *Bacillus* and *Staphylococcus*), up to 50% of the peptidoglycan can be complexed through the 6-OH group of NAM to teichoic acids. Teichoic acids, which can be as long as 30 units, are also found in the plasma membrane. Because they are negatively charged, teichoic acids are believed to bind and regulate cation levels at the cell surface.

Bacterial Structure and Function
Gram-Positive Cell • pp. 1–5

Gram-negative bacteria have a considerably thinner peptidoglycan layer that constitutes only about 5% to 20% of the total cell envelope. Chemical analyses of the wall in gram-negative bacteria such as *E. coli* suggest that there may only be one layer of peptidoglycan in these organisms.

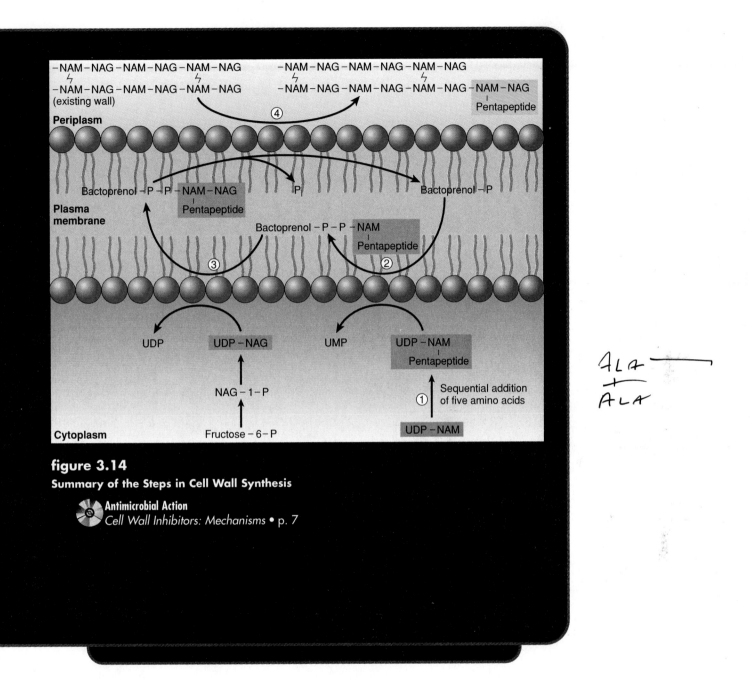

figure 3.14
Summary of the Steps in Cell Wall Synthesis

Antimicrobial Action
Cell Wall Inhibitors: Mechanisms • p. 7

The Outer Membrane Is the Exterior Portion of the Cell Envelope in Gram-Negative Procaryotes

An outer membrane is part of the cell wall in gram-negative procaryotes. The outer membrane, when examined in transverse sections by electron microscopy, resembles the trilaminar fine structure characteristic of most other biological membranes. This outer covering consists of a phospholipid bilayer interspersed with proteins and lipoproteins. The proteins and lipoproteins are firmly embedded in the matrix of the outer membrane, with the lipoproteins anchoring the outer membrane to the underlying

peptidoglycan layer. The matrix proteins may also be responsible for the permeability of the outer membrane to small molecules having molecular weights of up to 800 daltons. The pores responsible for this permeability are formed through specific arrangements of membrane proteins that form part of the lipid bilayer. Rearrangements of these proteins after gene mutation and subsequent protein modification can cause changes in membrane permeability.

Lipopolysaccharide (LPS) molecules are arranged within the external phospholipid layer of the outer membrane and account for approximately 40% of the total cell surface in gram-negative bacteria

Gram-Positive Cell Envelope

Cell wall consists of thick peptidoglycan layers

Tetrapeptide side chain

Plasma membrane

Phospholipid

Protein

a.

Gram-Negative Cell Envelope

Protein

Pore

O antigen

Lipid A

Lipopolysaccharide (LPS)

Outer membrane

Lipoprotein

Periplasm

Peptidoglycan

Plasma membrane

d.

b.

e.

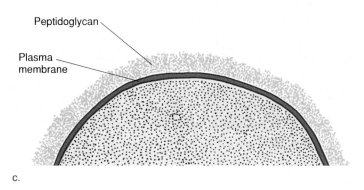

Peptidoglycan

Plasma membrane

c.

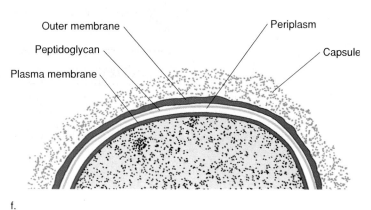

Outer membrane

Peptidoglycan

Plasma membrane

Periplasm

Capsule

f.

figure 3.15

Cell Envelopes of Gram-Positive and Gram-Negative Bacteria

a and c. Schematic diagrams of a gram-positive cell envelope.
b. Transmission electron micrograph of the cell envelope of the gram-positive bacterium *Bacillus fastidiosus* (×90,000). d and f. Schematic diagrams of a gram-negative cell envelope. e. Transmission electron micrograph of the cell envelope of the gram-negative bacterium *Pseudomonas aeruginosa* (×120,000).

Bacteria Without Walls

Although most bacteria possess distinct cell walls, some exist without this outer covering (Figure 3.16). As might be expected, such bacteria are shapeless and have greater exposure to the surrounding environment. The cell walls of certain bacteria can be removed in the laboratory by treatment with the enzyme lysozyme. Lysozyme is commonly found in body secretions such as tears and saliva and is an important first line of defense for the body in repelling bacterial invasion. This enzyme hydrolyzes peptidogly-can β(1,4) linkages in which the number 1 carbon of N-acetylmuramic acid is joined to the number 4 carbon of N-acetylglucosamine, resulting in lysis (cell disruption) and release of cellular contents into the environment. Cell lysis can be prevented by incubation of lysozyme-treated cells in an isotonic environment. Under such conditions the intact cell (without the cell wall) is called a **protoplast.** Cells that retain portions of the wall on their surfaces even after such treatment are called **spheroplasts.**

Bacteria exposed to penicillin often, but not always, develop into forms without walls, called **L-phase variants,** or **L-forms** (the *L* stands for the Lister Institute in London, where these unique organisms were first observed). L-forms also sometimes arise spontaneously when bacteria are grown in hypertonic media. The role of these unusual bacteria in human disease is not well-defined. They may be associated with persistent diseases (for example, pulmonary infections, nephritis, and endocarditis).

Bacteria of the genus *Mycoplasma*, which causes primary atypical pneumonia, are naturally occurring microbes that do not have cell walls. Since they lack walls that provide shape and rigidity, mycoplasmas have no definitive forms and are pleomorphic. To compensate for the lack of cell walls, mycoplasmas have sterol-strengthened plasma membranes. As one might expect, mycoplasmas are not affected by penicillin or other antibiotics that act solely on the cell wall, but they are sensitive to antibiotics that have targets other than the wall.

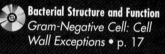

Bacterial Structure and Function
Gram-Negative Cell: Cell Wall Exceptions • p. 17

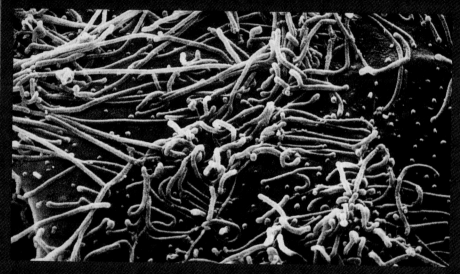

figure 3.16
Scanning Electron Micrograph of *Mycoplasma pneumoniae*
The irregular morphology is due to the lack of a cell wall (colorized, ×18,000).

(Figure 3.17). The lipopolysaccharide molecule consists of three components (listed in order from the outside to the inside of the cell):

1. An O-specific side chain (O antigen) composed of polysaccharides of varying types and proportions, depending on the bacterium

2. A core polysaccharide of relatively constant composition

3. A Lipid A portion, composed of glycophospholipid, which is associated with toxic activity in gram-negative bacteria

The LPS is more than a simple structural component of the cell wall. The O-specific side chain is responsible for surface antigenicity, which induces antibody formation in vertebrates. This region of the LPS consists of repeating oligosaccharide sequences containing up to 40 monosaccharides. The types and arrangement of the monosaccharides in the O-specific side chain determine the antigenicity of the LPS.

The Lipid A portion of the LPS is responsible for toxicity. Fever, shock, and other general physiological symptoms associated with diseases caused by *Salmonella, Escherichia,* and other gram-negative bacteria are caused by the endotoxic activity of the LPS (see endotoxins, page 450). Lipid A is usually attached to the core polysaccharide of LPS by three units of ketose-linked ketodeoxyoctonate (KDO), a molecule that is unique to bacteria.

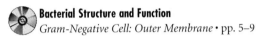

Bacterial Structure and Function
Gram-Negative Cell: Outer Membrane • pp. 5–9

The differences in wall composition between gram-positive and gram-negative bacteria explain in part why these bacteria stain differently in the Gram stain procedure. During decolorization with alcohol, the thick peptidoglycan layer in gram-positive bacteria becomes dehydrated. This dehydration causes the crystal violet-iodine complex, formed during the initial steps of the Gram stain, to remain trapped within the cell. The result is a violet-stained cell. Gram-negative bacteria, in contrast, have a thinner peptidoglycan layer and do not retain the crystal violet-iodine complex during destaining as effectively. The large proportion of lipids in the cell wall of gram-negative bacteria may also explain the response of these organisms to the Gram stain. It has been suggested that these lipids are extracted during the decolorization step, resulting in increased wall permeability. In either case, the gram-negative cell, which has not retained the primary stain, is subsequently counterstained with the red safranin. The wall composition of bacteria, however, is only partially responsible for the Gram stain reaction. Cyanobacteria, which have a wall almost identical in composition to the wall of gram-negative bacteria, stain gram positive. The reason for this difference is not known, but it suggests that more than simple wall composition is involved in the Gram stain reaction.

The Periplasm Separates the Plasma Membrane from the Outer Membrane in Gram-Negative Procaryotes

The plasma membrane is separated from the outer membrane in gram-negative procaryotes by a gap known as the **periplasm.** Within this periplasm are the peptidoglycan and approximately 50 different kinds of proteins, many of which are hydrolytic

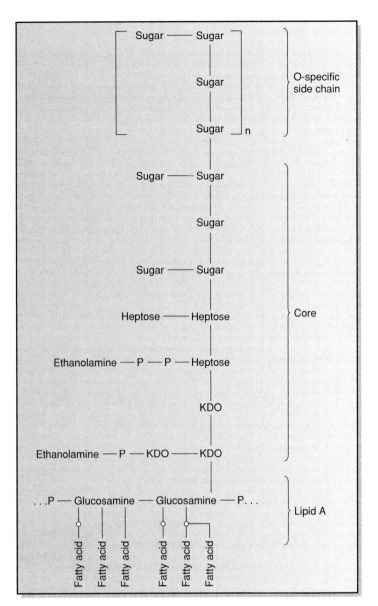

figure 3.17
The Lipopolysaccharide of the Gram-Negative Cell Envelope

enzymes and binding proteins. Examples of enzymes found in the periplasm are those that bind sugar molecules, amino acids, and inorganic ions. Specialized mechanisms are postulated for the translocation of many of these periplasmic proteins from the cell cytoplasm (where they are initially synthesized) across the plasma membrane and into the periplasm. One of the main functions of the periplasm may be to sequester enzymes in this space and prevent them from leaving the cell. Gram-positive procaryotes, which have no outer membrane, apparently do not have a periplasm.

Bacterial Structure and Function
Gram-Negative Cell: Periplasm • pp. 2–4

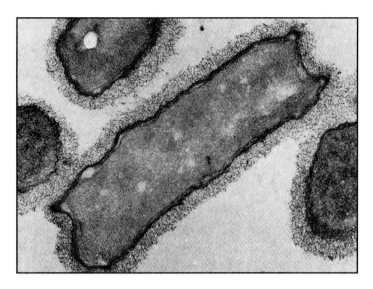

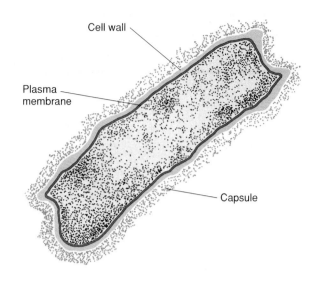

Cell wall

Plasma membrane

Capsule

figure 3.18
Electron Micrograph Showing the Thick Capsule of *Klebsiella pneumoniae* (×22,536)

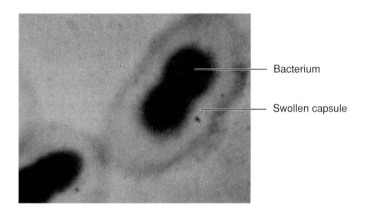

Bacterium

Swollen capsule

figure 3.19
Light Micrograph Showing the Quellung Reaction
Cells of *Streptococcus pneumoniae* are surrounded by apparently swollen capsules from the quellung reaction.

A Variety of Specialized Structures Are Located Outside the Cell Envelope

Some bacteria have specialized structures external to the cell envelope. These structures are involved in the protection of the bacteria from phagocytes (host cells that ingest particulate matter), adhesion of the bacteria to solid surfaces, and motility.

Capsules Protect the Cell from Phagocytosis, Desiccation, and Nutrient Loss

Certain bacteria have an additional layer of material, called a capsule, that is external to the cell wall (Figure 3.18). Capsules are found on many pathogenic bacteria and protect the bacteria from ingestion by phagocytes (see phagocytosis, page 460). Removal of the capsule from bacteria by chemical treatment or genetic manip-

ulation results in an unprotected cell that is easily phagocytized and is far less virulent for the host organism.

External layers also protect microbial cells from desiccation and nutrient loss. Carboxyl groups found in these layers bind cations and may function in cation absorption. In some instances, capsular polysaccharides act as receptors for bacterial viruses, enabling the virus to attach to the cell.

Bacterial capsules are composed of polysaccharide, protein, or a combination of the two (glycoprotein). The thickness of the capsule is not fixed, and varies depending on the growth medium and growth conditions. Bacteria with particularly thick capsules often form mucoid colonies on solid media. Capsules on bacterial cells can be demonstrated by a simple capsule stain, in which cells are suspended in India ink and examined under a microscope. The colloidal particles of the India ink are displaced by the capsule, which appears as an unstained halo around the cell. An alternative procedure for visualizing capsules is the **quellung** reaction [German *quellen*, to swell]. In the quellung reaction, encapsulated cells are mixed with antibodies directed specifically against capsular components. The antibodies increase the refractivity (light deflection) of the capsule, giving the appearance of a swollen capsule (Figure 3.19). This can easily be seen with a light microscope using the oil-immersion lens and reduced illumination. Some bacteria have surface macromolecules that promote adherence to solid surfaces. The **glycocalyx** is a fibrous matrix of polysaccharides that binds cells together in an aggregate mass and attaches the cell to solid surfaces. In the oral cavity, bacteria of the genus *Streptococcus,* responsible for dental plaque, produce an exopolysaccharide called dextran that binds these organisms to the tooth surface. Dextran is a glucose polymer synthesized from the hydrolysis products of sucrose, a sugar that is well-known for its role in cariogenicity. The dental plaque that forms on tooth surfaces is a matrix composed of bacteria intermixed with epithelial cells, leukocytes (white blood cells), dextran, and other polymers. As the

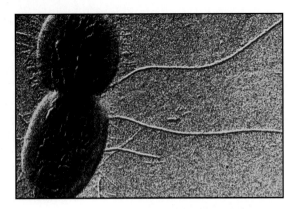

figure 3.20

Electron Micrograph of Pili (Short Appendages) Extending from *Proteus vulgaris*
The longer appendages are flagella (colorized, ×52,000).

bacteria grow and metabolize, they secrete lactic acid and other organic acids that are believed to be responsible for the demineralization of tooth enamel, leading to dental caries.

Bacterial Structure and Function
External Structures • pp. 1–11

Pili Are Involved in Conjugation and Attachment

Pili (singular, **pilus**) [Latin *pilus*, hair] are short, thin, straight appendages (10 nm × 300 to 1,000 nm) found in large quantity on the surfaces of some bacteria, notably Gram-negative cells (Figure 3.20). There are 10 to 250 pili, also called **fimbriae**, per cell. Pili known as **type I pili** are involved in attachment of bacteria to cell surfaces, particularly those of eucaryotic cells. Some pathogenic bacteria, such as *Neisseria gonorrhoeae*, the etiologic agent of gonorrhea, use pili for attachment to epithelial cells during infection. One specialized type of pilus (the **F**, or **sex pilus**) is associated with the transfer of genetic material between bacterial cells (a process called *conjugation*) (see conjugation, page 263). Although the exact function of the sex pilus in genetic transfer is not known, it apparently is involved in stabilizing the association between two mating bacterial cells while DNA is transferred.

Pili originate from the plasma membrane and are composed almost entirely of a protein called **pilin.** Pili are too small to be seen under a light microscope and can only be observed with an electron microscope.

Bacterial Structure and Function
External Structures: Pili • pp. 15–17

Flagella Are Used for Cellular Locomotion

Flagella (singular, **flagellum**) [Latin *flagellum*, whip] are appendages on bacteria that are involved in locomotion. These appendages are long, thin (average 10 to 20 nm × 1 to 70 μm) helical filaments.

The bacterial flagellum is composed of three parts: a **basal body** attached to the plasma membrane, a **hook,** and a **filament** (Figure 3.21). The basal body is a thin structure composed of protein and

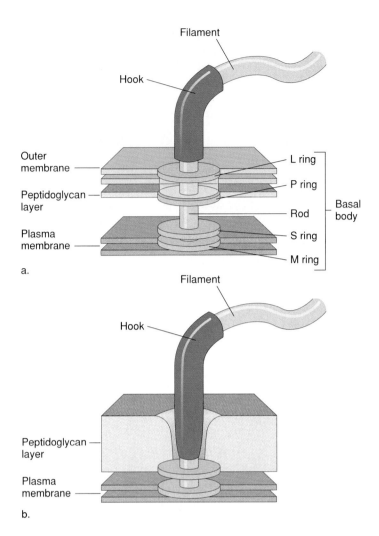

figure 3.21

Structure of the Bacterial Flagellum
The flagellum of bacteria consists of three parts: a basal body attached to the plasma membrane, a hook, and a filament. a. Gram-negative bacterial flagellum. b. Gram-positive bacterial flagellum.

constitutes a relatively small proportion (approximately 1%) of the total mass of the flagellum. In gram-negative bacteria the basal body is surrounded by four collars, or rings. The S and M rings are attached to the plasma membrane. The P ring is anchored to the peptidoglycan; the fourth ring, the L ring, is associated with the outer membrane. Basal bodies of gram-positive bacteria have only two rings. One ring is attached to the plasma membrane, and the other is associated with the cell wall. The rings around the basal body probably aid in anchoring the flagellum to the membrane and wall.

The filament is linked to the basal body by a thick, curved hook. Filaments are composed of protein subunits called **flagellin.** Flagellin has a unique amino acid composition (some flagellins contain the unusual amino acid ε-*N*-methyllysine) and is different from the other proteins that make up the basal body and hook.

Flagella on a bacterium are less numerous than pili and may be arranged in different patterns on the cell surface (Figure 3.22). Bacteria that possess only one flagellum are termed **monotrichous** [Greek *trichos*, hair], whereas those with more than one flagellum

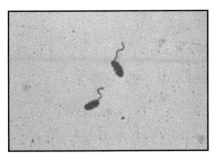

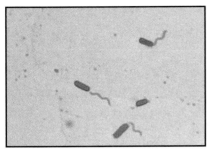

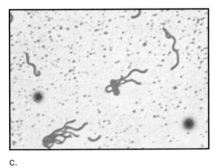

figure 3.22

Flagellation Patterns in Bacteria

a. Light micrograph of *Vibrio parahaemolyticus*, showing a single polar flagellum. b. Light micrograph of *Pseudomonas aeruginosa*, showing a single polar flagellum. c. Light micrograph of *Bacillus megaterium*, showing peritrichous flagellation.

are **multitrichous. Polar** flagellated organisms have flagella on one or both ends of the cell. The term **lophotrichous** is used to describe bacteria with a tuft of flagella. **Peritrichous** flagellation refers to the presence of flagella throughout the cell surface. Flagellar arrangement can sometimes be used to differentiate between different types of bacteria. For example, *Escherichia coli* and *Pseudomonas aeruginosa* are both gram-negative rods. However, *E. coli* has peritrichous flagellation, whereas *P. aeruginosa* is polarly flagellated.

The cell moves by the rotation of its flagella. For many years it was thought that flagella propelled bacteria by a whiplike motion. It is now known that flagella rotate around their bases in a counterclockwise direction with the expenditure of energy. This rotational movement was demonstrated in 1974 by Michael Silverman and Melvin Simon of the University of California at San Diego, who anchored monotrichously flagellated bacteria by their flagella to a slide. Antibodies on the slide attached to the flagella,

and the tethered cells were observed to rotate around the immobilized flagella.

Flagellar movement propels the cell at velocities of approximately 20 to 200 µm/sec. This speed may not appear to be very fast, but it is 10 to 100 cell lengths per second. A dog with a length of 1 meter would have to move at a speed of 30 to 300 km/hr to reach an equivalent velocity.

The type of flagellation also influences the way bacteria move through an environment. Polar flagellated bacteria dart around rapidly, whereas those that are peritrichously flagellated move in a directional manner, followed by a tumble and roll.

Bacterial Structure and Function
External Structures: Flagella • pp. 18–24

Chemotaxis Is the Movement of Procaryotes Toward or Away from Chemical Substances

Some bacteria have the ability to move toward or away from chemical substances. This movement, called **chemotaxis,** occurs in response to chemicals (for example, carbohydrates or amino acids) that either attract them (attractants) or repel them (repellents).

The movement of a bacterium can be followed by a tracking microscope, a special microscope with a moving stage that follows the movement of an individual cell. Using such an instrument, it has been observed that in the absence of a chemical gradient *E. coli,* a peritrichously flagellated bacterium, moves randomly. The bacterium travels in one direction with steady counterclockwise rotation of its flagella in a motion called a *run,* followed by an abrupt reversal of flagellar rotation, resulting in a *tumble* (Figure 3.23). When the bacterium is exposed to a chemical gradient, the runs are longer and the general movement is toward the higher concentration of an attractant or away from a repellent.

The responses of microbes to specific chemical substances are mediated by chemoreceptors located in the plasma membrane or in the periplasm. These receptors consist of binding proteins that tightly and specifically bind the chemical and signal the bacterium to move. The binding of chemoreceptors to specific chemicals activates a series of events, including the methylation or demethylation of certain bacterial proteins called **methyl-accepting chemotaxis proteins (MCPs),** or **transducers,** that signals flagellar movement.

MCPs sense the presence of and directly bind attractants and repellents or indirectly interact with these chemicals after they have bound to the chemoreceptors. If the concentration of the attractant remains constant or decreases, the enzyme CheB, a methylesterase, demethylates the MCP. The result is a tumble by the cell. As the concentration of an attractant increases, MCP is methylated by the enzyme CheR, a methyltransferase, with the methyl group coming from the donor S-adenosylmethionine. This methylation of the MCP results in a straight run by the cell.

A cascading series of events occurs that transfers the signal from the MCP to the bacterial flagellum (Figure 3.24). Demethylation of the MCP (under constant or decreased concentrations of attractant) causes a cytoplasmic protein called CheW to couple the receptor-generated signal to autophosphorylation of the protein kinase CheA, which in turn transfers its phosphate to CheB and

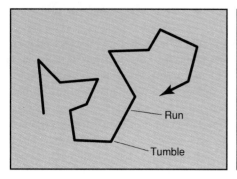

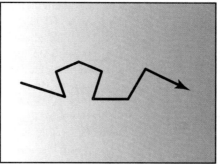

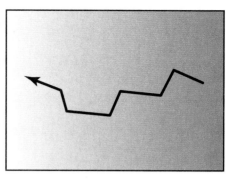

a. No chemical gradient

b. Gradient of chemical attractant

c. Gradient of chemical repellent

figure 3.23

Movement of Bacteria as Seen by a Tracking Microscope
a. Random movement of a bacterium in an area without a chemical gradient. Straight runs are followed by tumbles. b. Directed movement in a gradient of a chemical attractant. Runs are longer and the general movement is toward the higher concentration of the attractant. c. Directed movement in a gradient of a chemical repellent. The general movement is away from the higher concentration of the repellent.

Bacterial Structure and Function
External Structures: Flagella • p. 24

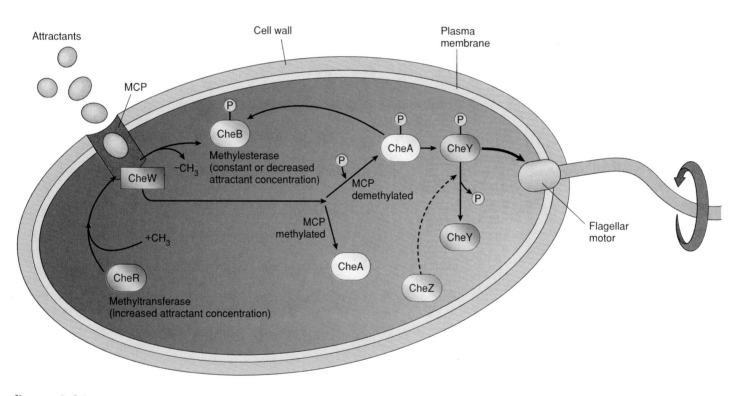

figure 3.24

Role of Methyl-Accepting Chemotaxis Proteins (MCPs) in Chemotaxis

Demethylation of MCPs when attractant concentration is constant or decreases results in a cascade phosphorylation of CheA and CheY. The CheY interacts with the bacterial flagellum, causing it to rotate clockwise and resulting in a tumble by the cell. When attractant concentration increases, methylation of MCP results in dephosphorylation of CheY by CheZ. Under these conditions, the flagellum rotates counterclockwise and the cell makes a straight run.

CheY, a protein that interacts with the flagellum. ATP is the initial source of the phosphate for phosphorylation of CheA. Phosphorylation stimulates CheB methylesterase activity and contributes to adaptation by the cell to attractant concentrations. When CheY is phosphorylated, the flagellum rotates clockwise, resulting in a tumble. When MCP is methylated (under high concentrations of attractant), CheY is dephosphorylated by the phosphatase CheZ and the flagellum switches to a counterclockwise rotation, resulting in a straight run by the cell.

Bacterial cells may have several different kinds of chemoreceptors. For example, *Escherichia coli* has approximately 20 attractant receptors and 10 repellent receptors. Some of these receptors, such as those for galactose, ribose, and maltose, are present in substantial concentrations (10,000 or more receptors per cell). With such receptors, the cell is able to migrate toward areas containing higher concentrations of these nutrients.

Chemotaxis is not unique to bacteria. It also occurs in other types of organisms, such as the amoeba *Dictyostelium* and the protozoan *Paramecium*. The phenomenon, however, has been most extensively studied in bacteria. Responses of microbes to attractants or repellents depend on the species of microorganism involved, the type of chemical, and the concentration of the substance. Chemotaxis is of obvious advantage to microorganisms that live in environments containing harmful chemicals or limited supplies of nutrients, and is an example of microbial adaptation to changes in the environment.

Bacteria are also able to respond to other environmental stimuli. Photosynthetic bacteria move toward areas of high light intensity (**phototaxis**). Movements of bacteria can be affected by oxygen concentrations (**aerotaxis**). Bacteria requiring oxygen will move toward oxygen, whereas those inhibited by oxygen will be repelled by it.

One of the more unique responses is to the earth's magnetic fields (**magnetotaxis**). Certain aquatic bacteria such as *Aquaspirillum magnetotacticum* have intracellular crystal particles of the iron oxide magnetite (Fe_3O_4) called **magnetosomes** that act as tiny magnets (Figure 3.25). The magnetosomes help the bacteria orient their movements along magnetic fields and move primarily in northward and southward directions, as well as downward toward nutrient-rich sediments. Magnetosomes are also found in eucaryotes such as some algae.

Procaryotes Do Not Have Membrane-Enclosed Organelles, but Do Have Specialized Internal Cell Structures

Although procaryotic cells generally do not have internal membrane-enclosed organelles like those found in eucaryotic cells, they do have certain specialized structures. These structures are involved in the storage of genetic information, protein synthesis, and other cell activities.

The Nucleoid Is the Chromosome-Containing Region of the Cell

The **nucleoid** is the nuclear region of the bacterial cell that contains the bacterial chromosome (Figure 3.26). Most of the genetic information of the cell is contained within the chromosome; some may be in extrachromosomal structures called **plasmids** (see plasmid, page 258). Plasmids are small, circular pieces of double-stranded DNA that replicate autonomously and may be transferred between closely related bacteria. Plasmids contain genes that are generally not crucial for cell growth and reproduction; not all bacteria have them. Plasmids can carry a variety of genes, including those for resistance to antibiotics, tolerance to toxic metals, and pili-mediated DNA transfer.

Bacteria do not possess the membrane-enclosed nucleus that is characteristic of eucaryotic cells. The DNA of bacteria, however, can be visualized by electron microscopy as thin, fibrillar material confined within the nucleoid. Bacterial DNA occurs as a double-stranded helix arranged in a covalently closed circle near or associated with the plasma membrane. In *E. coli*, the DNA has a molecular weight of approximately 2.5×10^9 daltons and constitutes only about 2% to 3% of the dry weight of the cell. Yet, when stretched linearly, the DNA measures 1,100 μm long, or 550 times the length of the *E. coli* cell. The length of DNA is such that it must be tightly folded in a twisted supercoil to fit within the cell. Negative charges of DNA, resulting from the sugar phosphate backbone, are neutralized with divalent cations such as Mg^{2+} or Ca^{2+} that prevent electrostatic repulsions between the DNA strands. Recently a histonelike protein (labeled HU) was discovered in *E. coli* that may neutralize the negative charges of bacterial DNA in the same manner as histones do in eucaryotic cells.

Bacterial Structure and Function
Internal Structures • pp. 1–4

Ribosomes Are RNA-Rich Particles That Are the Sites of Protein Synthesis

Ribosomes are the sites of protein synthesis. It is not unusual to find 30,000 or more in a rapidly growing bacterium. Ribosomes have characteristic sizes that are usually described by the rate of movement during ultracentrifugation. The Svedberg (S) is a unit of measure for the sedimentation rate of molecules in a gravitational field. Bacterial ribosomes have a sedimentation constant of 70S and consist of two subunits, 30S and 50S. (Since sedimentation behavior of molecules is affected by shape as well as mass, the individual subunit values add up to a higher value than that of the intact ribosome.) Eucaryotic ribosomes are larger and have a sedimentation constant of 80S, with subunits of 40S and 60S.

Bacterial ribosomes, composed of 40% protein and 60% rRNA, are often found on mRNA as polyribosomes. The 30S subunit consists of about 21 proteins and a 16S rRNA molecule. Two rRNA molecules (a 23S rRNA and a 5S rRNA) and about 34 proteins make up the 50S ribosome subunit. Both subunits come together at one end of a mRNA to form an intact 70S ribosome during protein synthesis.

Bacterial Structure and Function
Internal Structures • pp. 12–13

Inclusion Bodies Are Storage Depots for Lipids, Polysaccharides, and Inorganic Compounds

Bacteria frequently are found in adverse environments where nutrients are scarce. They can survive in these environments by drawing on reserve materials that have accumulated in **inclusion**, or **storage**,

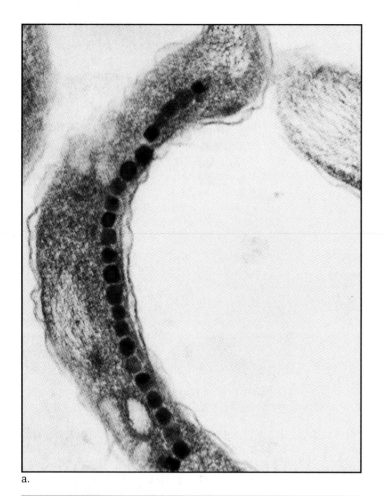

a.

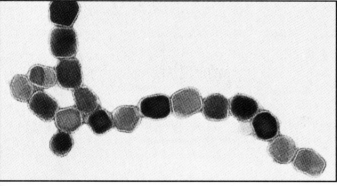

b.

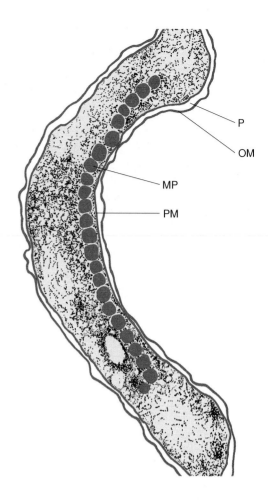

P
OM
MP
PM

figure 3.25

Magnetotactic Bacteria

a. Transmission electron micrograph of the magnetotactic bacterium, *Aquaspirillum magnetotacticum,* showing intracellular crystal particles of magnetite (MP). PM, plasma membrane; OM, outer membrane; P, periplasm (×123,000). b. Isolated magnetosomes (×140,000).

bodies. These electron-dense structures range from 20 to 100 nm in diameter. Among the substances that are stored in inclusion bodies are lipids, polysaccharides, and inorganic compounds. A bacterial cell usually produces only one kind of inclusion body. The storage compounds found in these bodies often are negatively charged and can be preferentially stained with basic dyes. For example, *Corynebacterium* stores the inorganic compound polymetaphosphate in inclusion bodies. The inclusion bodies, containing the anionic phosphates, are readily stained with the basic dye methylene blue. The stained bodies are referred to as **metachromatic granules** (because of their staining characteristics) or **Babès-Ernst bodies** (named after their discoverers). The staining technique is called the **granule stain** and is useful in identifying metachromatic granules not only in *Corynebacterium,*

but also in other bacteria (Figure 3.27). Poly-β-hydroxybutyrate, a lipidlike storage molecule for carbon and energy, is commonly found in *Bacillus* and *Pseudomonas* and is preferentially stained with a fat-soluble dye such as Sudan black. Poly-β-hydroxybutyrates form long polymers that have plasticlike properties. These microbial plastics are becoming increasingly popular as a source of raw materials for the manufacture of biodegradable plastic products.

Gas Vacuoles Retain Gases and Provide Buoyancy to the Cell

Some procaryotes (photosynthetic bacteria and cyanobacteria) control their buoyancy in water by using hollow protein cylinders called **gas vacuoles** that retain gases (Figure 3.28). A vacuole measures

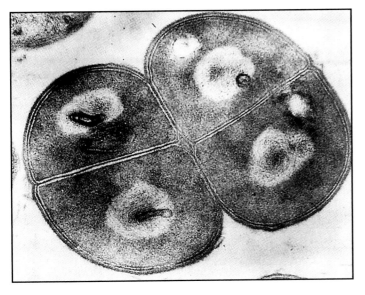

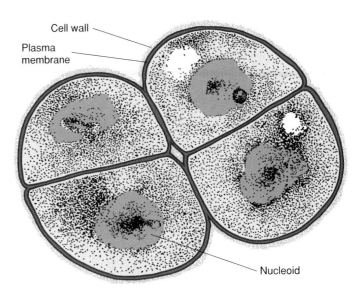

figure 3.26
Nucleoids Within Dividing *Sporosarcina ureae*
The nucleoids are the light areas within each of the four cells.

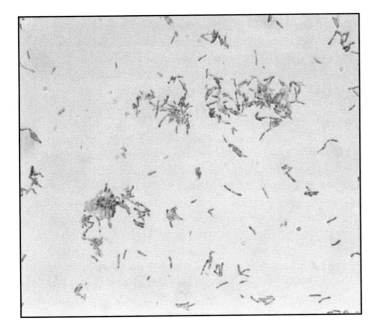

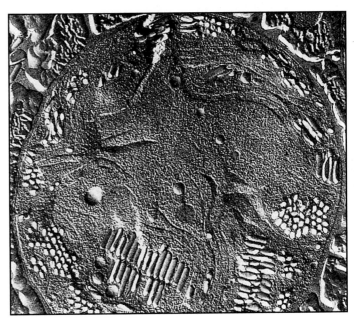

figure 3.27
Granules Inside Cells of *Corynebacterium diphtheriae*
Dark blue metachromatic granules are shown inside *C. diphtheriae* cells.

figure 3.28
Gas Vacuoles Within a Freeze-Etched *Microcystis aeruginosa* Cell
The gas vacuoles are the cylinders inside the cell (×21,000).

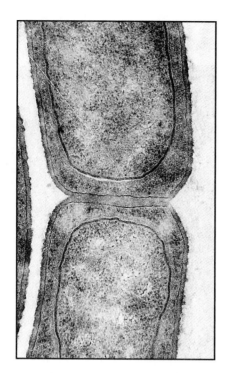

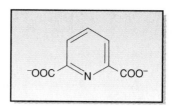

figure 3.30
Chemical Structure of Dipicolinate

75 nm in diameter and 200 to 1,000 nm in length. It is bounded by a rigid layer of a proteinaceous membrane 2 nm thick that is impermeable to water and solutes, but is freely permeable to gases. The buoyancy of the cell can be controlled through the uptake of gases into, or release of gases from, the vacuoles. The outer protein layer provides the strength and rigidity necessary for the vacuole to withstand the pressure extremes encountered in aquatic environments. By controlling their buoyancy in water, bacteria with gas vacuoles are able to maintain a depth that is optimal for nutrient levels, oxygen concentration, and light intensity.

Endospores Enable the Cell to Survive Adverse Environmental Conditions

The bacterial cells that have been described are referred to as **vegetative cells.** Vegetative cells of bacteria are associated with growth, metabolism, and reproduction. These cells are not exceptionally resistant to environmental stress and therefore are susceptible to lysis and destruction when such conditions prevail. Under nutrient-limiting conditions, certain types of bacteria (mainly in the genera *Bacillus* and *Clostridium*) have the unusual ability to form a resistant, dormant cell type called an **endospore** (so called because the spore is formed within the cell) (Figure 3.29).

The bacterial endospore is a highly refractile multilayered cell that can survive adverse environmental conditions. Bacterial endospores contain only 15% water, compared with the 70% water found in vegetative cells. Because of this low water content, endospores are dormant and do not metabolize or reproduce. Bacterial endospores can withstand drying and are up to 10,000 times more resistant to heat and 100 times more resistant to ultraviolet radiation than vegetative cells. Spores have been known to survive for hundreds of years under dormant conditions. Although many theories have been postulated to explain the unusual resistance of spores, none has proved completely satisfactory. The most attractive theory proposes that high levels in spores of the chemical compound dipicolinate, in combination with calcium ions, may be responsible for their heat resistance (Figure 3.30). Calcium dipicolinate forms a gel-like polymer that extrudes water from the spore; hence structures and macromolecules are not subject to denaturation or breakdown by water molecules upon exposure to high temperatures. Calcium dipicolinate comprises as much as 15% of the dry weight of a spore. Another theory suggests that expansion of the peptidoglycan in the spore cortex in the absence of divalent cations may play a role in heat resistance. Whatever the mechanism involved in spore resistance, spores clearly pro-

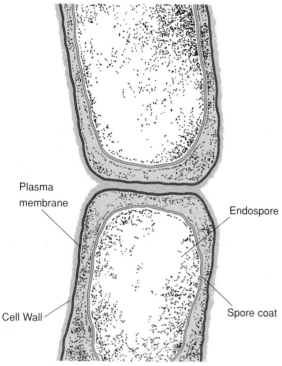

Plasma membrane

Endospore

Cell Wall

Spore coat

figure 3.29

Developing Endospores in *Bacillus*

Transmission electron micrograph showing endospores in *Bacillus cereus* (×49,500).

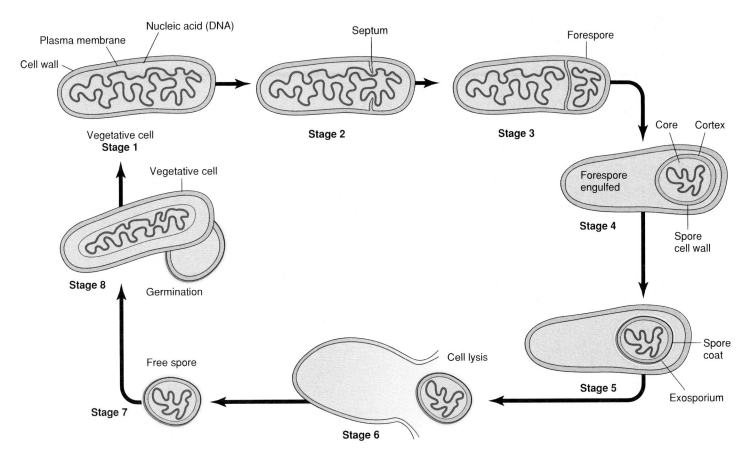

figure 3.31

Stages in Endospore Formation and Germination
Stage 1. Replication of nucleic acid in vegetative cell to form two complete sets of DNA. **Stage 2.** Invagination of the plasma membrane to form a spore septum separating the two sets of DNA. **Stage 3.** Development of a forespore. **Stage 4.** Forespore is engulfed and the core is surrounded by the spore cell wall and the cortex. **Stage 5.** Completion of endospore formation with the formation of the spore coat and, in some cases, an exosporium around the spore. **Stage 6.** Release of the endospore through lysis of the sporangium. **Stage 7.** Free spore. **Stage 8.** Germination.

vide bacteria with a specialized structure to withstand environmental stress and nutrient deprivation.

Bacterial endospore formation is a complex process that involves several distinct morphological stages (Figure 3.31):

Stage 1. In the first stage the vegetative cell replicates its nucleic acid to form two complete copies of DNA. The nucleic acid extends as a continuous filament along the entire length of the cell.

Stage 2. The second stage of endospore differentiation begins with an invagination of the plasma membrane to form a spore septum. The two copies of DNA separate and compartmentalize in two membrane-enclosed areas of the cell. Only one of these two areas will eventually differentiate into the endospore.

Stage 3. During the next stage of spore formation, a recognizable **forespore** develops from one of the compartments.

Stage 4. Next, the forespore is engulfed by the mother cell and a thick cortical layer (the **cortex**), composed of peptidoglycan, is formed. A highly cross-linked peptidoglycan layer, the **spore cell wall,** is found between the cortex and the **core,** or interior of the spore (plasma membrane, protoplasm, DNA, ribosomes, and so forth). At this stage the forespore becomes increasingly refractile. Calcium dipicolinate is synthesized and apparently incorporated into the core.

Stage 5. The fifth stage of spore formation begins with the formation of a **spore coat** around the spore. The proteinaceous spore coat is exceptionally rich in the amino acid cysteine. The proteins of the coat are different from those synthesized by vegetative cells and probably contribute to the coat's hydrophobic nature. The completed endospore consists of the spore coat, cortex, spore wall, and core in association with calcium dipicolinate. In some endospore-forming bacteria, such as *Bacillus cereus,* a loose outer envelope surrounds the spore. This envelope, called the **exosporium,** is composed mainly of protein, carbohydrate, and lipid.

Following maturation, the intact spore is released from the cell through lysis of the sporangium. The sporangium is the cell that contains the endospore. This **free spore** is a highly resistant structure, able to withstand desiccation and environmental stress (Figure 3.32). The entire spore differentiation process is commonly called **sporogenesis.**

Certain characteristics of sporogenesis have recently been defined. It is known that spore formation in bacteria is triggered by nutrient deprivation, environmental stress, or both. Under such conditions, genes that are normally repressed in vegetative cells are activated. These genes code for enzymes that catalyze the synthesis of proteins that are incorporated into the surrounding layers of the developing spore.

Spores convert back to vegetative cells again by a process called **germination.** Although some spores will germinate spontaneously in a favorable growth medium, most must first be activated. Activation occurs when spores are exposed for short periods of time to a nutrient-rich environment, to certain types of traumatic conditions such as heat or low pH, or to reducing agents like mercaptoethanol. The activated spore goes through a series of germination steps in which up to 30% of the dry weight of the spore (primarily its cortical layer and calcium dipicolinate) is excreted into the surrounding medium. Degradation of the protective spore layers is followed by a period of **outgrowth,** during which DNA, RNA, and proteins are actively synthesized. Outgrowth continues until the vegetative cell matures and divides. The complete process of germination usually takes less than 90 minutes.

Endospores are impermeable to dyes used in normal staining procedures. In Gram-stained preparations, the refractile endospores appear as unstained structures against the remaining stained cell material. A special technique called the **spore stain** is used to selectively stain endospores. In the spore stain, a dye such as carbolfuchsin is driven into the spore by steaming the smear. The cells are stained with a secondary dye and spores are seen as red. Endospores can be found in different areas of the cell (Figure 3.33). Some are located at the end of the cell (**terminal spores**), whereas others are found in the middle (**central spores**). Still others are situated between the end and the middle portion of the cell (**subterminal spores**). The location of the endospore is characteristic for some species of bacteria and can be used in identification.

Endospore-forming bacteria represent a unique group of organisms with the ability to differentiate and develop into a specialized form adapted for survival in a hostile environment. The mechanisms involved in sporogenesis and germination have only recently begun to be understood and should provide scientists with an insight into the differentiation process not only of spores, but also of other cell types.

The Eucaryotic Cell

Many of the structures in procaryotic cells are also found in eucaryotic cells. However, eucaryotic cells possess certain structures that are different from, or do not occur, in procaryotic cells.

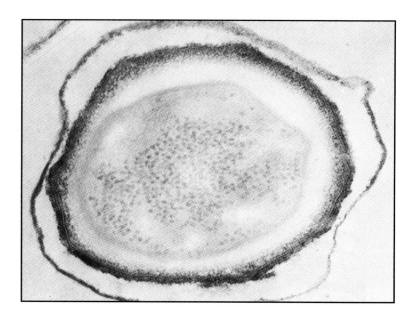

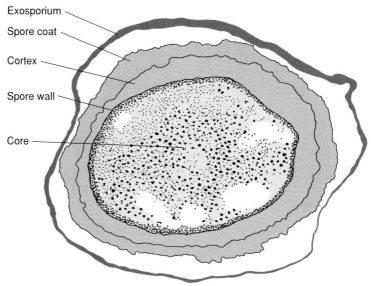

Exosporium
Spore coat
Cortex
Spore wall
Core

figure 3.32
Endospore Structure
Bacillus anthracis endospore, showing the core, surrounded by the spore wall, cortex, the spore coat, and the exosporium (×151,000).

The Eucaryotic Cell Wall and Plasma Membrane Are Chemically Different from, but Function Similarly as, Their Procaryotic Counterparts

Eucaryotic cells, like procaryotic cells, are surrounded by a plasma membrane and, in some cases, a cell wall (Figure 3.34). The cell wall in eucaryotes is found only in algae, plants, and fungi, and is one of the principal distinctions between these organisms and animals. The eucaryotic cell wall functions much like its counterpart in the procaryotic cell. It maintains shape and rigidity, and protects the cell from osmotic stress. Eucaryotic walls, however, do not possess peptidoglycan. Instead they typically contain cellulose and 6-carbon sugars known as hexuronic acids. In some instances, as in certain fungi, the cellulose in

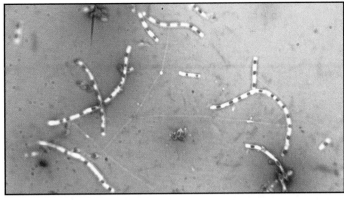

a.

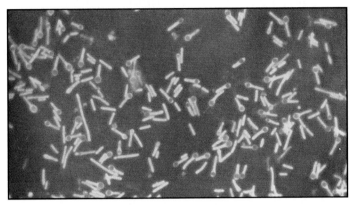

b.

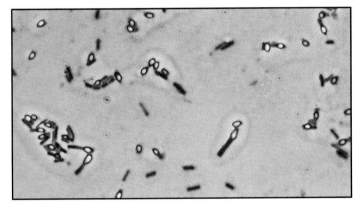

c.

figure 3.33
Endospore-Forming Bacteria
a. *Bacillus subtilis* endospore stain showing red endospores (×500).
b. *Clostridium tetani*, terminal endospores with characteristic "drumstick" appearance (×500). c. *Clostridium botulinum*, subterminal endospores (phase contrast, ×600).

figure 3.34
Organization of a Procaryotic Cell Compared with a Eucaryotic Cell
a. A typical procaryotic cell. b. A typical eucaryotic plant cell. c. A typical eucaryotic animal cell. Illustrated structures are not found in all cells.

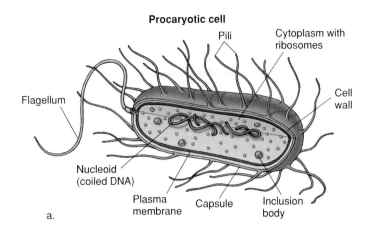

Procaryotic cell

a.

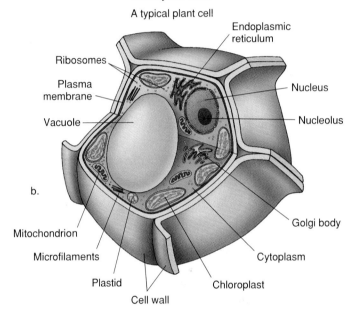

Eucaryotic cells

A typical plant cell

b.

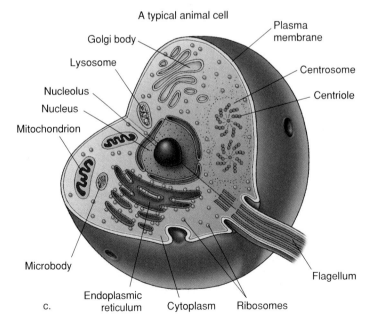

A typical animal cell

c.

the wall is replaced in part by glucans, chitin (the same tough, resistant polysaccharide that is found in the exoskeletons of arthropods), and other substances. Structural polysaccharides in the wall are interwoven into strands called microfibrils that are cross-linked with a matrix of other sugars. The wall is located external to the plasma membrane and, in some mature plant cells, a nonflexible secondary wall may form underneath the primary wall. This secondary wall contains cellulose and the polyphenol lignin, molecules that do not occur in the primary wall and that impart strength to older, nongrowing cells. Cells that are adjacent to one another often have an additional layer between them called the middle lamella, composed of pectin (the same organic substance that forms the basis of jellies), which holds these cells together.

 Fungal Structure and Function
General Eucaryotic Structures: Cell Wall • pp. 2–3

Eucaryotic plasma membranes are similar in structure to procaryotic membranes, with the exception that sterols are present. The sterols in the eucaryotic membrane impart strength to it, important because many eucaryotic cells lack a rigid wall. Both eucaryotic and procaryotic membranes perform similar functions. They act as semipermeable barriers for the movement of substances into and out of the cell.

 Fungal Structure and Function
General Eucaryotic Structures: Cell Membrane • pp. 4–5

Eucaryotic Cells Contain Membrane-Enclosed Organelles

A fundamental difference between eucaryotes and procaryotes is the presence of certain membrane-enclosed organelles in the eucaryotic cell cytoplasm. These organelles include mitochondria, chloroplasts, the nucleus, and other specialized structures.

Eucaryotic DNA is contained within a membrane-enclosed **nucleus.** Inside the nucleus is the nucleolus, a small body that is the site of ribosomal RNA synthesis. The nucleus is surrounded by a nuclear envelope consisting of two separate unit membranes. Each membrane is approximately 75 Å thick; the two membranes are separated by a space of 400 to 700 Å. The outer membrane is connected at different points to a structure called the **endoplasmic reticulum,** an extensive network of internal membranes that is often associated with ribosomes and is involved in protein synthesis. Small pores (30 to 100 nm in diameter) in the nuclear envelope serve to connect the nucleoplasm with the cell cytoplasm.

The nucleus contains the genetic information of the cell, which is complexed with **histones** (positive-charged proteins) and **nonhistone proteins** (Figure 3.35). The DNA and its associated proteins form structures called chromosomes. These chromosomes are linear, longer, and more numerous than the circular chromosomes found in bacteria. Eucaryotic cells are **diploid** (have two sets of chromosomes), in comparison with procaryotes, which usually are **haploid** (have one set of chromosomes). However, at some point in their life cycles, eucaryotic cells that sexually reproduce form haploid sex cells called **gametes.** When gametes fuse during mating, the diploid state is reestablished. Although multicellular

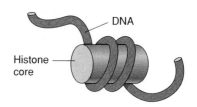

figure 3.35
Association of Histones with DNA
Eucaryotic DNA is complexed with positive-charged proteins called histones.

eucaryotes usually exist in the diploid phase, the haploid growth phase generally predominates in eucaryotic microorganisms.

Located near the nucleus in many eucaryotic cells is a dense area of cytoplasm called the **centrosome.** Within the centrosome are two small (about 0.2 μm in diameter) cylindrical structures, the **centrioles,** which are involved in chromosomal movements during cell division. Each centriole consists of nine groups of triplet microtubules (hollow protein cylinders, each approximately 18 to 25 nm in diameter) that are spaced around the circumference of the centriole.

Two other membrane-enclosed organelles found in eucaryotic cells are the **mitochondrion** (plural, **mitochondria**) and the **chloroplast** (Figure 3.36). Both structures, like the nucleus, have two sets of membranes. The mitochondrion, the site of cellular respiration and ATP production, is considered the energy factory of the cell. Cells with high energy requirements contain many more mitochondria than cells that have more modest energy requirements. The two membranes found in mitochondria are the external membrane, which lacks sterols and is not as rigid as the plasma membrane, and the inner membrane. The inner membrane is folded into convoluted structures called **cristae** that form internal compartments. The **matrix,** a semifluid material that contains many of the enzymes involved in respiration and energy production, is found within these compartments.

Chloroplasts are part of a larger group of membrane-enclosed organelles called **plastids,** which are found only in algae and plants. Plastids are surrounded by double membranes and are of three different types: **leucoplasts** (nonpigmented organelles found in root and underground stem cells, and associated with the synthesis and storage of starch, proteins, and oils), **chromoplasts** (pigmented plastids responsible for the bright colors seen in flowers, fruits, and fall leaves), and **chloroplasts.**

Chloroplasts are the sites of photosynthesis in algae and plants. The disk-shaped organelles have an extensive internal structure, with parallel saclike internal membranes known as thylakoids or lamellae that are stacked neatly in layers called grana (singular, granum). The interior compartment of the chloroplast, the stroma, contains the enzymes involved in the dark reactions of photosynthesis and carbon dioxide fixation (see chloroplast, page 180). Chlorophylls and carotenoids are pigments associated with photosynthesis and the thylakoid membranes. Both chloroplasts and mitochondria contain DNA and RNA, numerous proteins, and 70S ribosomes like those of bacteria. Because of their similarity to bacteria, chloroplasts and mitochondria are believed by some scientists

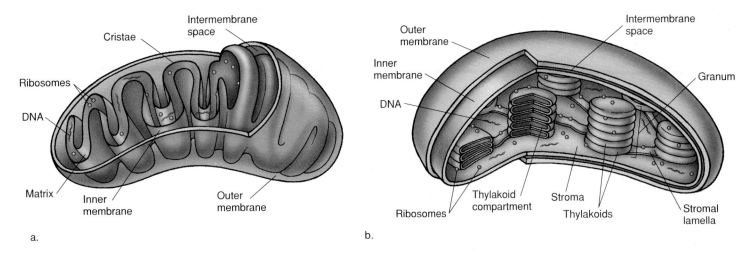

figure 3.36
Structures of the Mitochondrion and the Chloroplast
a. Mitochondrion. b. Chloroplast.

to have been procaryotes that developed into eucaryotic organelles (see endosymbiotic hypothesis, page 206).

Other types of membrane-enclosed organelles are found in some eucaryotic cells. **Microbodies** are spherical organelles, bounded by a single membrane, that are depositories for enzymes involved in cellular metabolism. **Vacuoles** are organelles filled with fluid and found more often in plant cells than animal cells. Plant vacuoles contain a fluid mixture of organic compounds known as the cell sap. These vacuoles occupy most of the volume in mature cells, exert pressure on the cell wall, and maintain the turgidity typical of plant cells. Animal cells often have organelles called **lysosomes,** which store enzymes that are involved in intracellular digestion of macromolecules and particles taken into the cell. The segregation of these enzymes within the lysosomes prevents them from attacking the cell's own macromolecules. Lysosomes fuse with vacuoles containing foreign particles and food materials and release hydrolytic enzymes into the vacuoles for the digestion of these materials. Lysosomes arise from membrane-enclosed structures called the **Golgi complex,** or **Golgi apparatus,** located near the nuclear envelope. The Golgi complex, first described by the Italian cytologist Camillo Golgi (1844–1926) in 1899, consists of flattened sacs called **cisternae.** The complex is involved in secretion of chemical substances from the interior of the cell and through the plasma membrane to the exterior; it also may be responsible for inserting proteins into the plasma membrane.

Fungal Structure and Function
General Eucaryotic Structures: Internal Structures • pp. 6–11

Eucaryotic Cells Use Flagella, Cilia, and Cytoplasmic Streaming for Locomotion

Locomotion is achieved in eucaryotic cells by the use of specialized structures (flagella and cilia) or by cytoplasmic streaming. Eucaryotic flagella are more complex than procaryotic flagella. The eucaryotic flagellum is organized with nine pairs of microtubules surrounding two separate microtubules in the center of the flagellum (Figure 3.37a). This "nine-plus-two" arrangement is surrounded by an extension of the plasma membrane. The flagellum is connected to the cell surface by a basal body (an extension of the flagellum containing nine triplets of microtubules but no central microtubules). Flagellar movement occurs by a sliding motion of the microtubules in relationship to one another, driven by ATP hydrolysis and resulting in flagellar bending. Cilia are similar in structure to flagella but are more numerous and shorter (Figure 3.37b).

Some eucaryotic cells move across surfaces by a process known as **cytoplasmic streaming.** This type of locomotion is most evident in amoebae, which form **pseudopodia** (extensions of the cell cytoplasm) in the direction of movement. An analogous type of motility is seen in plasmodial slime molds. These organisms are thin streaming protoplasmic masses that creep along surfaces. Composed of the protein actin, microfilaments within the cytoskeleton of eucaryotic cells are associated with cytoplasmic streaming.

Eucaryotes Reproduce Asexually by Fission, Fragmentation, or Spore or Bud Formation, or Sexually by Union of Haploid Gametes

Eucaryotic cells reproduce either by asexual methods similar to those seen in procaryotes (fission, fragmentation, spore or bud formation) or by sexual exchange of genetic material and division. Sexual reproduction occurs through the union of haploid gametes (which are either similar or dissimilar in morphology and composition) to form a diploid zygote. Both the individual haploid gametes and their nuclei fuse during this process. The result is a doubling of the chromosome number (diploidy) in the nucleus of the newly formed zygote. If chromosomal doubling were to occur repeatedly in subsequent generations, the nucleus would eventually be filled with genetic material. This is prevented

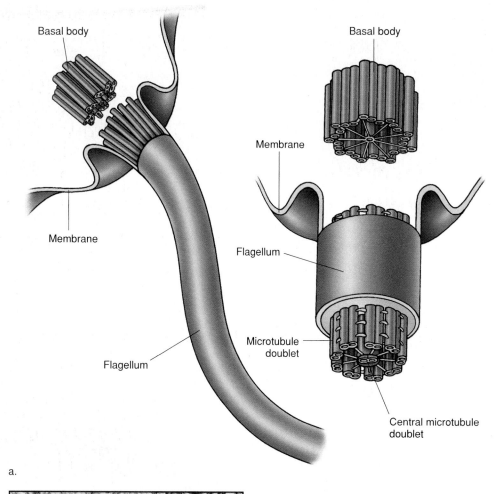

Basal body

Basal body

Membrane

Membrane

Flagellum

Flagellum

Microtubule
doublet

Central microtubule
doublet

a.

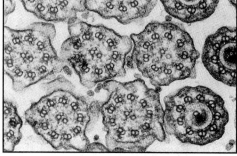

b.

figure 3.37

The Eucaryotic Flagellum and Cilium

a. Diagram of a flagellum, showing its microtubules and underlying basal body. b. Electron micrograph of the cilia of the protozoan *Tetrahymena*, showing its structure in cross section. The distinctive "nine-plus-two" arrangement of the microtubules can be seen (×66,200).

by a halving or reduction of the chromosome number at some stage in the life cycle. This halving of chromosome number, called **meiosis** or reduction division, typically accompanies sexual reproduction. The process of meiosis and its location in the life cycle may vary from organism to organism, but the result is always the same: the formation of haploid cells.

As eucaryotic cells grow, they eventually reach a stage and size at which they divide into daughter cells. Eucaryotic cell division, termed **mitosis,** consists of two distinct steps: **karyokinesis** (nuclear division) and **cytokinesis** (cytoplasmic division). Chro-

mosomes divide and migrate to opposite sides of the cell during karyokinesis. Cytokinesis takes place during the later stages of karyokinesis. The result is the formation of two indistinguishable daughter cells, each identical to the original parent cell. In certain eucaryotes, such as some algae and fungi, cytokinesis does not necessarily follow karyokinesis; a multinucleated (coenocytic) organism is then produced.

Fungal Structure and Function

Metabolism and Growth: Fungal Reproduction • pp. 8–23

PERSPECTIVE

Calcium Dipicolinate and Heat Resistance of Endospores

The heat resistance of bacterial endospores is an important characteristic to the survival of *Bacillus* and *Clostridium* in harsh, unfavorable environments. It currently is believed that large quantities of calcium dipicolinate probably confer heat resistance to endospores. Evidence for this association came from a series of studies performed by Joan F. Powell and R. E. Strange in 1953.

Powell and Strange examined products excreted by *Bacillus subtilis* and *Bacillus megaterium* during spore germination. Cultures of each bacterium were grown and then harvested by centrifugation. It was observed that centrifuged bacterial spores washed three to five times with water were deprived of glucose and did not germinate, even when exposed to heat shock (60°C). However, unwashed or incompletely washed spores apparently retained endogenous (internal) pools of glucose and, when exposed to heat shock, germinated spontaneously. These germinated spores lost their heat resistance because they were not vegetative cells, but remained viable.

Using this principle, Powell and Strange measured decreases in dry weight of *Bacillus* spores during germination and found a loss of 27 to 30 mg per 100 mg cell dry weight. At the same time, the surrounding culture medium increased by approximately an equivalent amount, indicating that material lost from the spore during germination had been excreted to the surrounding medium (Table 3P.1).

The material released by germinating spores was chemically analyzed and found to consist primarily of amino acids, peptides, glucosamine, and a substance having an absorption spectrum in the ultraviolet range, with maxima at 263, 270, and 278 nm (Figure 3P.1). This ultraviolet-absorbing substance was also recovered from resting spores of *Bacillus* that were mechanically disrupted in a tissue disintegrator. The substance occurred as a sparingly soluble calcium salt and upon crystallization was found to contain 36.3% carbon, 2.7% hydrogen, 5.7% nitrogen, and 15.8% calcium. Further chemical analyses revealed this substance to be calcium dipicolinate. Calcium

dipicolinate was found to constitute 50% of the solids excreted by germinating spores of *B. megaterium*, and to make up approximately 15% of the spore dry weight. Powell and Strange were the first to report on the isolation of calcium dipicolinate and its association with endospores.

Sources:

Powell, J.F. 1953. Isolation of dipicolinic acid (pyridine-2:6-dicarboxylic acid) from spores of *Bacillus megaterium*. *Biochemical Journal* 54:210–211.

Powell, J.F., and R.E. Strange. 1953. Biochemical changes occurring during the germination of bacterial endospores. *Biochemical Journal* 54:205–209.

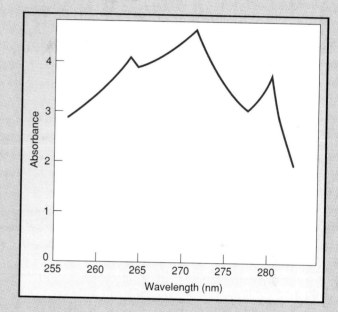

figure 3P.1

Ultraviolet Absorption of Calcium Dipicolinate Recovered from Germinating Spores of *Bacillus subtilis* and *Bacillus megaterium*

table 3P.1

Decrease in Dry Weight of *Bacillus megaterium* Spores During Germination

Medium	Number of Spores	Spore Dry Weight (mg)	Germination (%)	Change in Dry Weight for Complete Germination (mg/100mg)	
				Cells	Surrounding Medium
Nutrient tryptic digest broth	1.08×10^{11}	105	90	−30	+33
Nutrient tryptic digest broth	1.15×10^{11}	108	73	−27	+36
5 mM Glucose, 30 mM phosphate, pH 7.3	1.08×10^{11}	105	69	−29	+33

⊖ Summary

1. Macromolecules are the building blocks of the cell and consist of proteins, nucleic acids, carbohydrates, and lipids.

2. The cell envelope is the outer covering of a bacterium, and consists of the plasma membrane, cell wall, and in gram-negative bacteria, an outer membrane.

3. The plasma membrane is a semipermeable barrier that regulates the passage of molecules and ions. The procaryotic plasma membrane is similar in composition to the basic unit membrane structure found in eucaryotes. Membrane lipids of Bacteria have ester linkages between glycerol and the fatty acids, whereas membrane lipids of Archaea have ether linkages between glycerol and isoprene.

4. The cell wall is a rigid layer comprised of peptidoglycan in most Bacteria and of proteins, glycoproteins, and polysaccharides in the Archaea. The cell wall maintains conformation and also enables the cell to survive in hypotonic environments.

5. Gram-negative bacteria have an outer membrane containing lipopolysaccharide molecules. The periplasm, which contains the peptidoglycan, enzymes, and binding proteins, separates the outer membrane from the plasma membrane.

6. Some bacteria have capsules, which protect the cell from phagocytosis, desiccation, and nutrient loss; pili, which are involved in attachment and the transfer of genetic information between bacterial cells; and/or flagella, which are involved in cell motility.

7. Various structures can be found inside the bacterial cell, including inclusion bodies, which are used as storage depots for lipids, polysaccharides, and inorganic compounds; and gas vacuoles, which retain gases and provide buoyancy to the cell.

8. *Bacillus* and *Clostridium* are examples of bacteria that form endospores. Endospores are resistant, dormant cell types that protect the cell from heat, desiccation, and other adverse environmental conditions. The unusual heat resistance of endospores is believed to be associated with high levels of calcium dipicolinate.

9. Eucaryotic cells have membrane-bound organelles such as mitochondria, chloroplasts, and nuclei that are not found in procaryotes. A summary of the principal differences between procaryotes and eucaryotes is shown in Table 3.3.

table 3.3

Summary of Differences Between Procaryotes and Eucaryotes

Characteristic	Procaryote	Eucaryote
Genetic material	Circular DNA molecule in nucleoid; no discrete nucleus; plasmids	Linear DNA arranged in chromosomes within a membrane-enclosed nucleus; chromosomes complexed with histones
Membrane-enclosed organelles	Absent	Present; examples include lysosome, Golgi complex, endoplasmic reticulum, mitochondrion, and chloroplast
Ribosomes	Small size (70S)	Large size (80S), except for small size (70S) in mitochondrion and chloroplast
Vacuoles	Rare	Common
Plasma membrane	Sterols generally lacking	Sterols present
Cell wall	Present in most procaryotes and usually contains peptidoglycan in Bacteria and proteins, glycoproteins, and polysaccharides in Archaea	Present in algae, plants, and fungi; usually contains polysaccharide
Capsule	Present in some procaryotes	Absent
Locomotion	Submicroscopic flagella of simple composition; some move by gliding motility	Microscopic flagella and cilia of complex composition; cytoplasmic streaming; gliding motility
Cell division	Usually by binary fission	By mitosis (usually accompanied by meiosis)

EVOLUTION and BIODIVERSITY

Although there are significant structural differences between procaryotes and eucaryotes, many similarities exist between the two groups of organisms. Both types of organisms have nucleic acids, proteins, carbohydrates, and lipids as the building blocks for their cell structures and constituents. Procaryotes and eucaryotes have closely similar plasma membranes consisting of a basic unit membrane structure. Genetic information is carried on DNA and proteins are synthesized from an mRNA template in association with ribosomes. Differences between these two groups of organisms have slowly evolved over the years and have led to the diversity that exists today between procaryotes and eucaryotes. Even with this divergence, there remains evidence of the common ancestry of these organisms. Eucaryotic mitochondria and chloroplasts contain DNA and RNA, numerous proteins, and 70S ribosomes like those of procaryotes. Both organelles function somewhat independently of the cell in the generation of ATP by processes (aerobic respiration and photosynthesis) found in procaryotes. Mitochondria are similar to procaryotes in size. Although procaryotes do not have membrane-bound organelles like those found in eucaryotic cells, photosynthetic bacteria have invaginations of the plasma membrane that function in photosynthesis, and some bacteria have extensive specialized internal membranes that are associated with ammonia, nitrite, and methane oxidation. The plasma membrane of the cell wall-less *Mycoplasma* is strengthened by sterol, a lipid that is found in the membranes of all eucaryotes. It is evident that over time procaryotes and eucaryotes have evolved distinctive structures and characteristics, but yet retain many similar properties that suggest a common thread that is woven through all living organisms.

Questions

Short Answer

1. Which microorganisms are procaryotic cells? Which microorganisms are eucaryotic cells?

2. Why are viruses considered to be neither eucaryotic nor procaryotic?

3. What is the major difference between eucaryotic and procaryotic cells?

4. Identify four major groups of macromolecules which are components of all cells and identify their building blocks.

5. Other than the four groups referred to in question 4, what compounds are major constituents of cells?

6. Identify the structural components found in all procaryotic cells. What structures are not found in procaryotes?

7. Identify at least six organelles.

8. Explain some of the basic differences between Archaea and Bacteria.

9. How do the membranes of the Bacteria, Archaea, and eucaryotes differ?

10. How do procaryotic cell walls differ from those of eucaryotic cells?

11. How do the cell walls of gram-positive and gram-negative bacteria differ?

12. How can you distinguish gram-positive from gram-negative bacteria?

13. What is a capsule? What is (are) its function(s)?

14. What is a pilus? What is its function?

15. Describe the bacterial flagellum. How does it differ in structure from eucaryotic flagella or cilia?

16. What is a nucleoid? How does it differ from a nucleus?

17. What is a plasmid? How does it differ from a chromosome?

18. What is a ribosome? How do procaryotic ribosomes differ from eucaryotic ribosomes?

Multiple Choice

1. The group of microorganisms classified as Archaea:
   a. are among the most primitive organisms.
   b. can inhabit extreme environments.
   c. do not have a membrane-enclosed nucleus.
   d. all of the above.

2. Which of the following statements about proteins is always true?
   a. All proteins are enzymes.
   b. All proteins have a quaternary structure.
   c. All proteins have a primary structure.
   d. All of the above statements are true.

3. The outer membrane is present in:
   a. gram-positive cells.
   b. gram-negative cells.
   c. eucaryotic cells.
   d. all of the above.

4. The arrangement of flagella on the cell surface can sometimes help in the identification of an organism. For example, *E. coli* has flagella throughout the cell surface that is referred to as:
   a. lophotrichous flagellation.
   b. monotrichous flagellation.
   c. peritrichous flagellation.
   d. none of the above.

5. The movement of an organism toward or away from a chemical substance in its environment is called:
   a. tracking.
   b. chemotaxis.
   c. tumbling.
   d. none of the above.

Critical Thinking

1. Compare and contrast procaryotic and eucaryotic cells.

2. Compare and contrast gram-positive and gram-negative cells.

3. Compare and contrast the phospholipid membrane and the LPS outer membrane.

4. Explain how to fight bacterial infections targeting the bacterial cell wall and membrane structures.

5. Present evidence of procaryotic and organelle structures that supports the endosymbiont hypothesis of eucaryotic evolution.

 ## Supplementary Readings

Alberts, B., D. Bray, J. Lewis, M. Raff, K. Roberts, and J.D. Watson. 1994. *Molecular biology of the cell*, 3d ed., New York: Garland Publishing, Inc. (An in-depth textbook on cell structure and function.)

Armitage, J.P., and J.M. Lackie, eds. 1990. *Biology of the chemotactic response*. London, England: Cambridge University Press. (A series of papers on the structure of bacterial flagellum and the mechanisms involved in chemotaxis.)

Berg, H.C. 1975. How bacteria swim. *Scientific American* 234:40–47. (A discussion of the role of flagella in bacterial motility.)

Beveridge, T.J., and L.L. Graham. 1991. Surface layers of bacteria. *Microbiological Reviews* 55:684–705. (A review of the surface layers of bacteria, as seen by electron microscopy.)

Errington, J. 1993. *Bacillus subtilis* sporulation: regulation of gene expression and control of morphogenesis. *Microbiological Reviews* 57:1–33. (A review of the genetics of sporogenesis in *B. subtilis*.)

Hoch, J.A., and P. Setlow. 1985. *Molecular biology of microbial differentiation*. Washington, DC: American Society for Microbiology. (Symposium presentations on molecular biology of sporulation in *Bacillus subtilis* and differentiation in other procaryotes.)

Lake, J. 1981. The ribosome. *Scientific American* 245:84–97. (A discussion of the structure and function of the ribosome.)

Rogers, H.J. 1978. Biogenesis of the wall in bacterial morphogenesis. *Advances in Microbial Physiology* 19:1–62. (A review of bacterial cell wall synthesis.)

Rogers, H.J., H.R. Perkins, and J.B. Ward. 1981. *Microbial cell walls and membranes*. New York: Chapman & Hall/Mehuen. (A review of cell wall and membrane structure and function.)

Rose, A.H. 1976. *Chemical microbiology: An introduction to microbial physiology*. New York: Plenum Publishing Corporation. (A detailed discussion of bacterial structure and metabolism.)

Salton, M.R.J., and P. Owen. 1976. Bacterial membrane structure. *Annual Review of Microbiology* 30:451–482. (A review of membrane structure in bacteria.)

Walsby, A.E. 1994. Gas vesicles. *Microbiological Reviews* 58:94–144. (A review of the structure and function of gas vesicles.)

Young, M., and J. Mandelstam. 1979. Early events during bacterial endospore formation. *Advances in Microbial Physiology* 20:103–162. (A discussion of the biochemical events that occur during endospore formation in bacteria.)

chapter four

NUTRITION AND ENVIRONMENTAL INFLUENCE

Various types of growth media are used to cultivate microorganisms.

Nutritional Requirements
Carbon and Energy
Nitrogen
Phosphorus and Sulfur
Other Chemical Elements
Organic Nutrients as Growth Factors

Nutrient Transport
Passive Diffusion
Facilitated Diffusion
Active Transport
Group Translocation

Physical and Gaseous Requirements
Upper and Lower Ranges of Temperature for Growth
Upper and Lower Ranges of pH for Growth
Water Requirements
Oxygen Sensitivity and Requirement: The Four Groups

Growth Media
Complex and Synthetic Media
Specialized Growth Media

Pure Culture Techniques
The Streak Plate Technique
Aseptic Technique

PERSPECTIVE
Role of Iron in Virulence of *Neisseria gonorrhoeae*

EVOLUTION AND BIODIVERSITY

Microbes in Motion ———————— PREVIEW LINK

This chapter covers the key topics of growth requirements for microorganisms. The following sections of the *Microbes in Motion* CD-ROM may be useful as a preview to your reading or as a supplemental study aid: *Microbial Metabolism and Growth*: Growth, 1–24, (Media) 36–43.

early 2,500 meters below the Pacific Ocean's surface off the coast of the Galápagos Islands near Ecuador, the research submersible vessel *Alvin* probes the ocean's floor in search of life. **Hydrothermal vents,** which form when near-freezing seawater sinks deep into fissures in the ocean floor and mixes with lava, are abundant in this and other areas of the ocean. These natural vents spew black geysers of superheated (up to 350°C) water rich in hydrogen sulfide and other reduced inorganic compounds. It would seem that life would be nonexistent under such harsh conditions in depths where pressure can exceed thousands of pounds per square inch and light is absent. Yet scientists on the *Alvin* unexpectedly find flourishing invertebrate communities of giant clams, mussels, and large tube worms over two meters in length. The existence of these vent communities is dependent on huge, thriving populations of bacteria (Figure 4.1). How do these bacteria that serve as the primary source of nutrients and energy for the marine communities living around the vents survive and grow in this unusual environment? They apparently grow not at the high temperatures of the superheated water emerging from the vents, but in water that rapidly cools as it comes in contact with the surrounding cold seawater. These bacteria derive all of the energy they need from the oxidation of hydrogen sulfide, and their carbon from available carbon dioxide; they serve as the beginning of the food chain for this ecosystem. The discovery of these vent communities and their associated bacteria was revolutionary, but not entirely unexpected. Procaryotes are known to exist in a variety of environments, ranging from high temperature (110°C), to low pH (pH 1), to absence of oxygen (anaerobic conditions). This chapter will examine the nutritional, physical, and gaseous requirements of procaryotes.

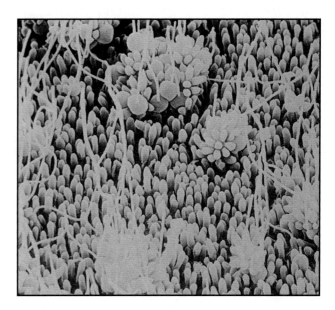

figure 4.1

Microbial Mats of Hydrothermal Bacteria

A variety of different forms are shown, including cocci, rods, and filaments, in this scanning electron micrograph of a mixed population of chemoautotrophic bacteria from the Galápagos Rift vents (colorized, ×6,000).

Nutritional Requirements

All living organisms require nutrients for growth and reproduction. These chemical substances may be organic or inorganic and are found in the environment in various forms. Microorganisms utilize nutrients by removing them from the surrounding environment and transporting them through the plasma membrane into the cell. There, some nutrients are converted into chemical building blocks for cell growth, whereas others are used to generate energy for cellular processes (Table 4.1). The chemical form of these nutrients and the methods by which different types of bacteria obtain and use them may vary, but the organism's need for them is absolute.

All Organisms Require a Source of Carbon and Energy for Growth and Reproduction

Microorganisms have developed several methods of using nutrients for energy and carbon. Nutritional patterns among microorganisms can be distinguished by two criteria: energy source and carbon source (Table 4.2).

Microorganisms Obtain Energy from Light or Chemical Compounds

All organisms require energy to stay alive and grow. Microorganisms are divided into two major categories based on the nature of their energy source: **phototrophs** and **chemotrophs.** Phototrophs capture the radiant energy of sunlight and transform it into chemical energy that is stored in the bonds of carbohydrates and other molecules (see photosynthesis, page 178). Green plants, algae, cyanobacteria, and photosynthetic bacteria other than cyanobacteria (*Rhodospirillaceae, Chromatiaceae, Chlorobiaceae,* and *Chloroflexaceae*) are examples of phototrophs.

Most organisms, however, are unable to use radiant forms of energy and must instead rely on the oxidation of chemical compounds as a source of energy. These organisms, called chemotrophs, are divided into two groups on the basis of the types of chemical compounds used: **(chemo)organotrophs** and **(chemo)lithotrophs.** Chemoorganotrophs obtain their energy from the oxidation of organic compounds. Most bacteria are chemoorganotrophic and use organic compounds such as carbohydrates, organic acids, and proteins for energy sources.

Chemolithotrophs obtain their energy from the oxidation of inorganic compounds such as hydrogen sulfide (H_2S), hydrogen gas (H_2), nitrite (NO_2^-), ammonia (NH_3), and ferrous iron (Fe^{2+}) (see chemolithotrophs, page 171). The concept of chemolithotrophy was first proposed in the late 1800s by Sergei Winogradsky (1856–1953), a Russian microbiologist. Winogradsky showed that *Beggiatoa,* a filamentous gliding bacterium found primarily in sulfide-rich habitats, could oxidize H_2S in these environments first to elemental sulfur (S^0) and then to sulfate (SO_4^{2-}). Members

table 4.1
Chemical Composition of a Typical Procaryotic Cell

table 4.1
Chemical Composition of a Typical Procaryotic Cell

Component	Percentage of Total Cell Weight	Number of Types of Each Molecule
Water	70	1
Inorganic ions	1	20
Sugars and precursors	3	200
Amino acids and precursors	0.4	100
Nucleotides and precursors	0.4	200
Lipids and precursors	2	50
Other small molecules	0.2	~200
Macromolecules (proteins, nucleic acids, and polysaccharides)	22	~5,000

table 4.2
Terminology of Procaryotes Based on Carbon and Energy Sources

Classification	Carbon Source	Energy Source
Autotroph	CO_2	—
Heterotroph	Organic compounds	—
Chemotroph	—	Chemical compounds
Chemolithotroph	—	Inorganic compounds
Chemoorganotroph	—	Organic compounds
Phototroph	—	Light
Chemoautotroph	CO_2	Chemical compounds
Photoautotroph	CO_2	Light
Chemoheterotroph	Organic compounds	Chemical compounds
Photoheterotroph	Organic compounds	Light
Methylotroph	1-carbon compounds	1-carbon compounds

Microbial Metabolism and Growth
Growth: Metabolic Nutrition • p. 4

of this group of bacteria often contain granules filled with elemental sulfur; the granules accumulate in the cell during the oxidation process. Chemolithotrophy is restricted to bacteria and is not found in higher life-forms. Chemolithotrophic bacteria are widely distributed in soil and water, taking advantage of the inorganic energy sources available in these habitats.

Microorganisms Obtain Carbon from Carbon Dioxide or Organic Compounds

Carbon is required by living organisms as an important constituent of cellular structure and metabolic compounds. This element exists in many forms in the environment. Carbon may appear in a simple form as gaseous carbon dioxide or as more complex organic compounds. Microorganisms are remarkably diverse in their carbon requirements and are divided into two groups—**autotrophs** and **heterotrophs**—based on their source of carbon.

Organisms that use carbon dioxide as a sole source of carbon are known as autotrophs. The term *autotroph* literally means "self-nourishing," an appropriate description of these organisms. Autotrophic microorganisms synthesize organic substances from carbon dioxide through a process known as carbon dioxide fixation (see carbon dioxide fixation, page 188). Autotrophs are important in nature because through carbon dioxide fixation, they provide the organic substrates that form the basis of the food chain for other organisms. The autotrophic bacteria include those that obtain their energy from light (**photoautotrophs**) and those that obtain their energy from oxidation of chemical compounds (**chemoautotrophs**).

Most microorganisms are unable to use carbon dioxide as their principal source of carbon and instead require preformed organic compounds for carbon. These microorganisms are called **heterotrophs. Chemoheterotrophs** are organisms that oxidize chemical compounds for their energy and require organic forms of carbon. Heterotrophic bacteria vary considerably in their require-

ments for organic substrates. Some, like the methane-oxidizing bacteria of the family *Methylomonadaceae*, use methane and a few other 1-carbon compounds as their sole source of energy and carbon. The term **methylotroph** is frequently used to describe organisms that grow on 1-carbon compounds. Other bacteria use a wide variety of organic substrates as energy and carbon sources. Some versatile strains of *Pseudomonas* are capable of using more than 100 different types of organic compounds. Because of this capability, pseudomonads are widely distributed in water and soil.

Although distinctions can be made among groups of bacteria with respect to their energy and carbon sources, some bacteria can vary in their requirements. For example, although photosynthetic bacteria are photoautotrophic in the presence of light, these same microbes function chemoautotrophically in darkness, using compounds such as hydrogen sulfide (H_2S), thiosulfate ($S_2O_3^{2-}$), and molecular hydrogen (H_2) as energy sources. Certain photosynthetic bacteria can use organic compounds as carbon sources and are thus **photoheterotrophic.**

The types of substances a microbe uses for its energy and/or carbon sources depend on many factors, among them the availability of these substances, environmental conditions, and the presence of functional biochemical pathways in the organism for the metabolism of the compound. The ability to adapt to changing conditions has made it possible for procaryotes to survive and grow in a variety of environments that are uninhabitable for other forms of life.

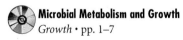

Microbial Metabolism and Growth
Growth • pp. 1–7

Nitrogen Is Required for Proteins, Nucleic Acids, Coenzymes, Cell Walls, and Other Cellular Constituents

Nitrogen is found in many important chemical compounds and, in the form of nitrogen gas (N_2), makes up a significant portion of the earth's atmosphere. In living organisms, nitrogen is a component of proteins, nucleic acids, coenzymes, cell walls, and other cellular constituents. As such, it is indispensable for microbial growth.

Bacteria obtain their nitrogen from either inorganic or organic sources. Major inorganic sources of nitrogen are nitrate (NO_3^-), ammonia (NH_3), and nitrogen gas (N_2). Organic sources of nitrogen include amino acids, purines, and pyrimidines. Organic nitrogenous compounds frequently are used not only as nitrogen sources by bacteria, but also as sources of energy.

Nitrogen from Nitrate Can Be Assimilated into Protein and Other Cellular Molecules

Nitrate is one of the most common forms of inorganic nitrogen that is available for microbial use. The assimilation of nitrogen from nitrate into protein and other cellular molecules is a multistep process called **assimilatory nitrate reduction.** In the initial reaction, nitrate (NO_3^-) is reduced (that is, gains electrons) to nitrite (NO_2^-) by the molybdenum-containing enzyme nitrate reductase. Nitrite is then further reduced by a second enzyme, nitrite reductase, and eventually is reduced to ammonia. The reduction of nitrate to ammonia is thus a complex process involving the addition of electrons in a series of reactions. The electrons for assimilatory nitrate reduction are donated by the reduced pyrimidine nucleotide, nicotinamide adenine dinucleotide (NADH + H$^+$). Assimilatory nitrate reduction is widespread among microorganisms and plants and is a major process by which nitrogen is incorporated into cellular material.

Ammonia Can Be Assimilated into Organic Compounds

Ammonia is assimilated by bacteria into a number of different organic compounds. A principal pathway of ammonia assimilation in bacteria is its incorporation into the amino acid glutamate by the following reaction:

$$
\begin{array}{l}
\text{COO}^- \\
|\\
\text{C=O} \\
|\\
\text{CH}_2 \quad + \text{ NH}_3 \\
|\\
\text{CH}_2 \\
|\\
\text{COO}^- \\
\alpha\text{-ketoglutarate}
\end{array}
\xrightarrow[\text{NADPH + H}^+ \quad \text{NADP}^+]{\text{Glutamate dehydrogenase}}
\begin{array}{l}
\text{COO}^- \\
|\\
\text{H}_3\overset{+}{\text{N}}-\text{C}-\text{H} \quad + \text{ H}_2\text{O}\\
|\\
\text{CH}_2 \\
|\\
\text{CH}_2 \\
|\\
\text{COO}^- \\
\text{glutamate}
\end{array}
$$

This reaction is reversible and requires NADPH + H$^+$ (see coenzymes, page 145). It is mediated by glutamate dehydrogenase, an enzyme that has a low affinity for ammonia. As a result of this low affinity, when ammonia concentrations fall below a certain level the pathway ceases to function.

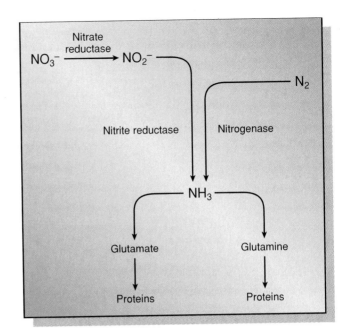

figure 4.2
Bacterial Nitrogen Assimilation
Microbes assimilate nitrogen by several methods. The assimilation of atmospheric nitrogen requires nitrogenase. Nitrate is assimilated to nitrite by its reduction by the enzyme nitrate reductase. The resultant nitrite is further reduced to ammonia by nitrite reductase. Ammonia is then incorporated into amino acids and other organic compounds.

Under these conditions, a second enzyme is activated within the cell. This enzyme, glutamine synthetase, has a high affinity for ammonia and catalyzes the reaction:

$$
\begin{array}{l}
\text{COO}^- \\
|\\
\text{H}_3\overset{+}{\text{N}}-\text{C}-\text{H} \quad + \text{ NH}_3\\
|\\
\text{CH}_2 \\
|\\
\text{CH}_2 \\
|\\
\text{COO}^- \\
\text{glutamate}
\end{array}
\xrightarrow[\text{ATP} \quad \text{ADP + P}_i]{\text{Glutamine synthase}}
\begin{array}{l}
\text{COO}^- \\
|\\
\text{H}_3\overset{+}{\text{N}}-\text{C}-\text{H}\\
|\\
\text{CH}_2 \\
|\\
\text{CH}_2 \\
|\\
\text{CONH}_2 \\
\text{glutamine}
\end{array}
$$

This reaction allows the assimilation of ammonia at low concentrations, but it has the distinct disadvantage of using up energy in the form of one molecule of ATP for each molecule of glutamine produced. Bacteria thus have two pathways for ammonia assimilation—one pathway that is inefficient at low concentrations of ammonia but requires no energy expenditure, and a second, more efficient pathway that is energetically more expensive (Figure 4.2).

Atmospheric Nitrogen Can Be Converted to Nitrogen in Organic Compounds by Nitrogen Fixation

An abundant source of nitrogen is nitrogen gas (N_2), more correctly termed **dinitrogen** because it consists of two atoms of nitrogen. Dinitrogen makes up 78% of air by weight and is thus the most abundant gas in the atmosphere. Dinitrogen is also the most stable form of nitrogen and cannot be assimilated directly by microorganisms. Conversion of atmospheric dinitrogen to combined forms of nitrogen in chemical compounds occurs by **nitrogen fixation,** a process that is unique to certain bacteria (see nitrogen fixation, page 422). It occurs either nonsymbiotically in free-living bacteria (for example, *Azotobacter, Klebsiella,* and *Clostridium*) or symbiotically in associations between bacteria (for example, *Rhizobium*) and certain plants such as the legumes. As a consequence of nitrogen fixation, atmospheric dinitrogen (which cannot be used directly by plants and animals) is converted into a usable chemical form. It is estimated that 175 million tons of atmospheric dinitrogen are fixed by bacteria each year. Compare this biological fixation of nitrogen with the 100 million tons fixed by abiotic means—lightning discharges, ultraviolet radiation, artificial combustion, and industrial production of fertilizers.

Biological nitrogen fixation is a reductive process, one that requires a large amount of energy, in which dinitrogen is reduced to two molecules of ammonia by the enzyme nitrogenase. The ammonia can then be converted to other fixed chemical forms of nitrogen for metabolic use.

Phosphorus Is Required for Nucleic Acids, Membrane Phospholipids, and Coenzymes; Sulfur Is Required for Certain Amino Acids, tRNA, and Coenzymes

Phosphorus is an element found in nucleic acids, membrane phospholipids, coenzymes, and in many intermediate compounds associated with metabolism and energy storage. Sulfur is an important constituent of the amino acids cysteine and methionine, tRNA, and some coenzymes. Both elements are obtained by bacteria from inorganic or organic sources. Inorganic phosphate is utilized directly by bacteria. Organic phosphate compounds are first hydrolyzed at their ester linkages by enzymes called phosphatases before the released phosphate is incorporated into other substances.

Most microorganisms are able to use inorganic sulfate (SO_4^{2-}) as their source of sulfur. The sulfate is relatively stable and cannot be used by the microbe without activation. Sulfate is reduced to H_2S by **assimilatory sulfate reduction** (Figure 4.3). This pathway begins with the activation of sulfate with two molecules of ATP to form adenosine-5'-phosphosulfate (APS) and subsequently adenosine-3'-phosphate-5'-phosphosulfate (AP-phosphosulfate). AP-phosphosulfate is then reduced to adenosine-3',5'-diphosphate and sulfite (SO_3^{2-}). The sulfite is further reduced to H_2S through action of the enzyme sulfite reductase. The H_2S formed is incorporated into amino acids by reaction with O-acetyl-serine (serine that has been activated by acetyl-CoA) to form cysteine:

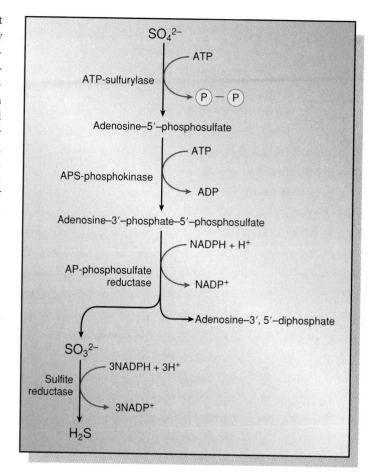

figure 4.3
Assimilatory Reduction of Sulfate to Produce H₂S

The sulfur from cysteine can then be transferred to other compounds in the cell. Sulfur can also be supplied to the cell directly in organic form through the addition of the amino acids cysteine and/or methionine to growth media.

Other Chemical Elements Are Also Required by Microorganisms

Besides oxygen and hydrogen (both of which are often supplied by organic compounds used for carbon and energy) and these four elements (C, N, P, and S), numerous other elements are required in smaller amounts by bacteria (Table 4.3). These minor elements include potassium (a principal cation and a cofactor for

table 4.3

Chemical Element Composition of a Typical Procaryotic Cell

Element	Percentage of Cell Dry Weight	Source of Element During Growth of Cell	Function in the Cell
Carbon	50	Organic compounds and CO_2	Primary constituent of cellular material
Oxygen	20	H_2O, organic compounds, CO_2, and O_2	Primary constituent of cellular water and material; as O_2, the electron acceptor in aerobic respiration
Nitrogen	15	NO_3^-, NH_3, N_2, and organic compounds	Constituent of amino acids, purines, pyrimidines, and coenzymes
Hydrogen	10	H_2O, H_2, and organic compounds	Constituent of cellular water and material
Phosphorus	3	Organic and inorganic compounds	Constituent of nucleic acids, phospholipids, coenzymes, and metabolic intermediates
Sulfur	1	SO_4^{2-}, H_2S, and organic compounds	Constituent of cysteine, methionine, tRNA, and coenzymes
Potassium	<1	Inorganic salts	Principal inorganic cation and cofactor for some enzymes
Magnesium	<1	Inorganic salts	Stability of ribosomes, membranes, and nucleic acids; component of chlorophyll; cofactor for enzymes
Calcium	<1	Inorganic salts	Endospore heat resistance; cofactor for some enzymes
Iron	<1	Inorganic salts	Constituent of cytochromes; cofactor for some enzymes

some enzymes), magnesium (important in stability of ribosomes, membranes, and nucleic acids; a component of chlorophyll; and a cofactor for many enzymes), calcium (associated with the heat resistance of bacterial endospores and a cofactor for some enzymes), and iron (a constituent of electron carriers called cytochromes and a cofactor for some enzymes). Other elements, such as silicon (found in the walls of certain algae and protozoa) and sodium (required by many marine organisms), are required only by selected microorganisms.

Some Microorganisms Require Certain Organic Nutrients as Growth Factors

Growth factors are organic nutrients required by microorganisms for growth, but not necessarily synthesized by them. The most common growth factors are amino acids, purines and pyrimidines, and vitamins.

Amino acids, the constituents of proteins, can also be precursors of intermediates of metabolic pathways. They must either be synthesized by the bacterium or provided as exogenous nutrients. Bacteria unable to synthesize certain amino acids generally lack one or more enzymes in the specific biosynthetic pathways for those amino acids. Required amino acids can be provided as free amino acids or as small peptides that are degraded by bacterial proteases before or after entry into the cell.

Purines and pyrimidines are used for the synthesis of nucleic acids and coenzymes. Like amino acids, they are either formed by the bacterial cell or must be provided. Some microorganisms are incapable of forming nucleosides and therefore must be provided with an exogenous supply. *Lactobacillus* is an example of a nucleoside-requiring bacterium.

A number of different vitamins and similar compounds (for example, hemin) are important as constituents of enzymes and coenzymes (Table 4.4). For example, vitamin B_1 (thiamine)

table 4.4

Common Vitamins and Their Functions in the Cell

Vitamin	Coenzyme Form	Function
p-Aminobenzoic acid	Tetrahydrofolic acid	Transfer of one-carbon units
Biotin	Biotin	Carboxylation
Cyanocobalamin (B_{12})	5'-deoxyadenosylcobalamin and methylcobalamin	Transfer of hydrogen and methyl groups
Folic acid	Tetrahydrofolic acid	Transfer of one-carbon units
Niacin	Nicotinamide adenine dinucleotide (NAD) and nicotinamide adenine dinucleotide phosphate (NADP)	Electron transport and dehydrogenation
Pantothenic acid	Coenzyme A	Activation of acetyl groups
Pyridoxine (B_6)	Pyridoxal phosphate	Transamination, decarboxylation, and racemization of amino acids
Riboflavin (B_2)	Flavin adenine dinucleotide (FAD) and flavin mononucleotide (FMN)	Electron transport and dehydrogenation
Thiamine (B_1)	Thiamine pyrophosphate	Group transfers; oxidation and decarboxylation of keto acids
Vitamin K	Quinones	Electron transport

is part of the structure of the coenzyme thiamine pyrophosphate, which is important in group transfer reactions and in the oxidation and decarboxylation of keto acids. Flavin adenine dinucleotide (FAD) and flavin mononucleotide (FMN), coenzymes important in electron transport and dehydrogenation reactions, are derivatives of vitamin B_2 (riboflavin). Nicotinamide adenine dinucleotide (NAD) and nicotinamide adenine dinucleotide phosphate (NADP), also important in electron transport and dehydrogenations, are formed from the vitamin niacin. The vitamin pantothenic acid forms part of the structure of coenzyme A, which is involved in the activation of acetyl groups during the catabolism of pyruvate and fatty acids. Vitamin requirements vary among bacteria. Some microbes have extensive requirements, whereas others are able to make their own vitamins. Lactic acid bacteria (*Lactobacillus* and *Streptococcus*) are examples of bacteria that require as many as five or six vitamins for growth.

Some growth factors used by bacteria are required in only very small amounts. For this reason microorganisms serve as useful tools for the sensitive and specific assay of vitamins and other similar growth factors. Such assays are performed by placing a microbe in a culture medium containing all but one growth factor. The missing growth factor is then added in graded amounts to the culture medium. Since extent of growth of the bacterium is directly proportional to the available concentration of the added growth factor, this technique provides an assay of the quantity of the added growth factor. Unlike chemical assays, microbiological assays differentiate between biologically active and inactive forms of a chemical compound and thus are potentially of more practical use in studies of living systems.

Nutrient Transport

Many of the nutrients required for microbial growth are either too large or too highly charged to pass freely through the plasma membrane. In addition, these compounds must often move against a concentration gradient to enter the cell. Without a system to transport these molecules, microbes would be unable to exist in most environments.

Microbial mechanisms for movement of molecules and ions across the plasma membrane are varied, but they can be divided into four basic categories: **passive diffusion, facilitated diffusion, active transport,** and **group translocation.** The first two types of mechanisms move molecules down a concentration gradient. Movement of molecules against a concentration gradient is possible with active transport and group translocation.

Passive Diffusion Is the Movement of Small Molecules Across the Membrane down a Concentration Gradient

Small molecules often are able to move down a concentration gradient through the plasma membrane without any input of energy or carrier mechanism. This process, called passive diffusion, depends on the size of the molecule and its solubility in lipids (Figure 4.4). Since the membrane is composed primarily of phospholipids in a bilayer, molecules that dissolve easily in lipids enter the membrane readily. One exception to this rule is water (membrane lipids have limited solubility in aqueous solution), which easily diffuses through the membrane because it is a small, uncharged molecule present at very high concentrations.

The rate of movement by passive diffusion is slow in comparison to other transport mechanisms and depends on the magnitude of the concentration gradient across the membrane. Molecular movement continues as a result of this gradient until a state of equilibrium is reached. Water and small chemical molecules generally enter or leave the cell by passive diffusion, but the rate of diffusion of chemical molecules is too slow to be of use to the cell.

Facilitated Diffusion Is the Carrier-Mediated Movement of Molecules Across the Membrane down a Concentration Gradient

Passive diffusion is nonselective in substrate specificity though it is a method for movement of water and small molecules. Facilitated diffusion and active transport are selective processes, because each uses protein carriers that recognize and bind to specific substrates (Figure 4.5a, b). Since substrate specificity is inherent to carrier-mediated movement, substances that resemble the designated substrate can compete for carrier binding sites.

In facilitated diffusion, the membrane-bound carriers, called permeases (more commonly known as transport proteins), attach to substrates to which they have an affinity. These substrates are carried down a concentration gradient to the other side of the membrane. Energy is not required for this movement. The molecular rate of movement in facilitated diffusion is more rapid than passive diffusion and is limited by the number of available carriers. Mutation or alteration of carrier molecules will also affect carrier-mediated transport. Facilitated diffusion mechanisms are found more commonly in eucaryotic cells than in procaryotic cells, because procaryotes usually use other transport mechanisms (active transport or group translocation) to transport sugars and other molecules into the cell.

Glycerol is one of the few molecules known to be transported by facilitated diffusion in bacteria. After glycerol is transported into the cell, it is phosphorylated by a cytoplasmic ATP-dependent kinase. The phosphorylated glycerol is now trapped

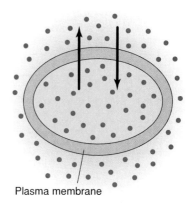

figure 4.4
Passive Diffusion
No energy is expended by the cell in passive diffusion. The rate of movement of small molecules through the plasma membrane is directly proportional to the difference in concentration between the inside and outside of the cell.

within the cell and is unable to move back across the membrane. The phosphorylation of glycerol explains how glycerol remains within the cell following transport.

Active Transport Is the Energy-Requiring, Carrier-Mediated Movement of Molecules Across the Membrane Against a Concentration Gradient

Active transport is also carrier mediated, but molecules are moved against a concentration gradient. As a consequence, the concentration of an actively transported molecule can be several hundred-fold or thousandfold greater inside the cell than outside. This concentration of substrate in active transport requires the expenditure of energy.

There are two possible explanations for energy requirements in active transport. The energy might be used to reversibly modify the conformation of the protein carrier, which spans the membrane. When this occurs, the carrier accepts the substrate, opens a channel through the membrane, and extrudes the substrate on the other side of the membrane.

Alternatively, active transport, in at least some cases, is linked to an ion gradient established across the membrane as a result of cellular metabolism, ATP hydrolysis, or both. In aerobic respiration, protons are pumped across the bacterial plasma membrane to form a proton gradient, resulting in higher proton concentrations on the outside of the membrane (see proton motive force, page 165). A proton gradient also may be established across the membrane by coupling the pumping of protons with ATP hydrolysis. Carrier proteins have specific binding sites for substrates and for protons. As the protons move down their concentration gradient across the membrane and reenter the cell, they carry the protein carriers and substrates along with them.

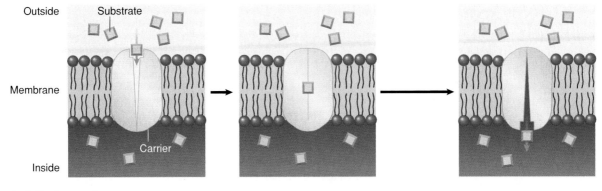

a. Facilitated diffusion

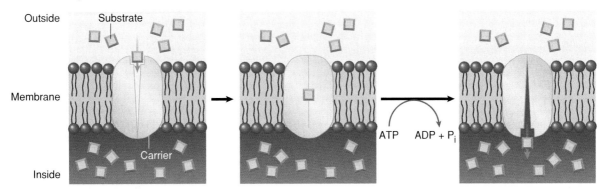

b. Active transport

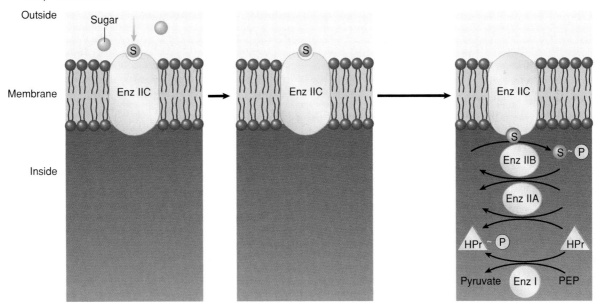

c. Group translocation

figure 4.5

Models of Carrier-Mediated Transport

a. In facilitated diffusion, the substrate is transported by a protein carrier down a concentration gradient across the membrane. No energy is expended. b. In active transport, the substrate is transported by a protein carrier against a concentration gradient. Concentration of the substrate is possible through coupling of transport with energy expenditure. c. In group translocation, the substrate is chemically modified during transport.

Siderophores and Iron Transport

Iron transport in microorganisms represents a special problem, because ferric iron (Fe^{3+}) is generally found in nature in highly insoluble hydroxides, carbonates, and phosphates. Since iron is required by microbes for cytochromes, enzymes, and other molecules important in metabolism, microorganisms have evolved special methods for iron uptake.

Bacteria living in environments with low iron concentrations or where iron exists in insoluble forms synthesize and excrete special compounds that complex specifically with iron. These iron chelators, called **siderophores,** bind and solubilize iron and transport it into the cell, where it is deposited in the cytoplasm.

There are two chemical classes of siderophores: the **hydroxamates** and the **phenolates.** Ferrichrome is a hydroxamate synthesized by many fungi.

Enterobactin is a phenolate produced by *E. coli* and other enteric bacteria.

Iron-binding proteins are found in higher organisms. Animals have **transferrins** in their blood that bind ferric ions and transport them to different parts of the body. The transferrins (and hemoglobin) lower the concentration of available iron in blood, making blood a poor habitat for microbial growth. Evidence indicates that the withholding of iron from invading microorganisms, on the one hand, and the ability of pathogenic microorganisms to successfully compete for available iron in the host, on the other hand, are important factors in determining the outcome of microbial infections.

Although any of a number of different ions (Na$^+$, K$^+$, or H$^+$) may serve to couple carrier-mediated active transport, most bacteria use H$^+$ as the main coupling ion.

Sugars, amino acids, and other organic molecules are transported via active transport. Inhibitors of energy generation also affect active transport and consequently the movement of these molecules across the plasma membrane.

The most extensively studied example of active transport is the ion gradient-mediated uptake of lactose by *Escherichia coli*. Lactose is transported across the plasma membrane by a membrane-associated permease. This permease is coded for by a structural gene of the *lac* operon (see lactose operon, page 233). The influx of lactose is linked to the entry of H$^+$ down the proton concentration gradient. Elimination of the proton gradient across the membrane (for example, by depriving the cell of energy) results in reversion of the active transport system to a facilitated diffusion system.

Many active transport systems in gram-negative bacteria are associated with specific soluble **binding proteins** located in the periplasm. These proteins are not carriers, but are capable of binding such molecules as galactose, maltose, glutamine, sulfate, and a number of amino acids. Binding proteins are released from cells by cold osmotic shock treatment, which ruptures the outer membrane. After cold osmotic shock, the uptake of certain substrates is impaired. Mutations that eliminate these binding proteins also have the same effect. Binding proteins may function in active transport by binding to substrates in the initial stages of transport and presenting the substrates to the actual membrane-bound carrier proteins.

Osmotic shock-sensitive transport systems associated with binding proteins are distinguished from other, shock-insensitive transport systems (for example, the lactose uptake system). The latter require coupling to hydrogen ions for transport by a proton-dependent gradient. Shock-sensitive systems draw their energy for transport from hydrolysis of ATP or a similar high-energy compound.

Microbial Metabolism and Growth
Growth: Nutrient Transport • pp. 8–13

Group Translocation Is the Chemical Modification of a Molecule During Its Movement Across the Membrane

Group translocation transports certain sugars, fatty acids, bases, and other molecules against a concentration gradient but uses a mechanism different than active transport. In group translocation, the substrate is chemically modified during passage through the membrane, resulting in a form to which the membrane is impermeable (Figure 4.5c). The altered substrate now is unable to cross through the membrane to the cell exterior. As a consequence, high concentrations of the modified substrate can accumulate within the cell.

The most extensively studied group translocation system is the phosphotransferase system in *E. coli* and *Salmonella* by which sugars are phosphorylated as they enter the cell. This phosphorylation occurs in two steps. The first step is the activation of a heat-stable carrier protein (HPr) by phosphoenolpyruvate (PEP):

$$\text{HPr} + \text{PEP} \xrightarrow{\text{Enzyme I}} \text{pyruvate} + \text{P—HPr}$$

Next, the phosphate group is transferred from the carrier protein to the sugar to be transported:

$$\text{P—HPr} + \text{sugar (outside)} \xrightarrow{\text{Enzymes IIA, IIB, and IIC}} \text{sugar—P (inside)} + \text{HPr}$$

The sugar is changed to the sugar-phosphate form during its movement across the membrane. Inside the cell, the sugar-P is metabolized into other compounds. The HPr is recycled.

The cytoplasmic enzyme I associated with the first reaction is relatively nonspecific for transport of different sugars. Cytoplasmic enzyme IIA and membrane-associated enzymes IIB and IIC are sugar specific and are involved in sequential phosphate transfer from HPr to the sugar. Enzyme IIC, a hydrophobic protein embedded in the membrane, actually transports and phosphorylates the sugar.

Group translocation, like active transport, requires energy. In the phosphotransferase system, this is the high-energy phosphate bond in PEP that is involved in the first reaction step of transport (see high-energy compounds, page 148).

Group translocation is a common transport system for many different types of sugars (glucose, fructose, and mannose) as well as purines, pyrimidines, and fatty acids. Group translocation transport systems conserve energy better than active transport systems because in translocation transport, the uptake of an external sugar and its phosphorylation occur in a single step. In active transport, energy is expended in transport of the sugar across the membrane and is expended again if the sugar is phosphorylated. Group translocation systems therefore are more widespread among anaerobic and facultatively anaerobic organisms (organisms that can live aerobically or anaerobically) than in obligate aerobes. This relates to the fact that ATP yields are lower in anaerobic fermentation pathways as compared to aerobic respiratory pathways (see respiration, page 168). Hence less total ATP is available to anaerobic bacteria when compared to aerobic bacteria. Bacteria known to possess mechanisms of group translocation include the anaerobes *Clostridium* and *Fusobacterium* and the facultative anaerobes *Escherichia, Salmonella, Vibrio, Bacillus, Photobacterium,* and *Staphylococcus.*

The different types of transport mechanisms enable bacteria to take up a wide variety of substrates for growth and metabolism. Different organisms, however, may transport the same substrate by different mechanisms. For example, lactose uptake occurs by a phosphotransferase system in *Staphylococcus aureus,* but by carrier-mediated active transport in *E. coli.*

Physical and Gaseous Requirements

A bacterium provided with an ample supply of all the requisite nutrients for growth still might not grow in an inhospitable environment. Procaryotes, like other living organisms, grow only within certain ranges of temperature, pH, osmotic pressure, and availability of certain gases. Although bacteria are more versatile and ubiquitous than other organisms, each species has a specific environmental range for growth.

Microorganisms Have Upper and Lower Ranges of Temperature for Growth

Microorganisms are unique in their ability to exist and grow at temperatures ranging from as low as −10°C to about 110°C. Temperature restrictions are the result of limitations in cell metabolism. As temperature increases, molecules collide more frequently and the rates of chemical reactions increase. Consequently, enzyme activity and growth rate also increase. However, at a certain point, the increase in temperature becomes detrimental to cells and results in disruption of their chemical bonds and subsequent enzyme inactivation. At low temperatures, enzyme activity slows and ultimately cellular metabolism ceases.

These restrictions are used to define the **maximum temperature** (temperature above which growth ceases) and the **minimum temperature** (temperature below which growth ceases) for microbial growth. Each microorganism also has an **optimum temperature** at which its growth rate is the fastest. These three temperatures define the range of growth for each type of microorganism and are collectively called the **cardinal temperatures.**

An example of how temperature affects growth of a specific bacterium is illustrated in a plot of the growth rate of *E. coli* B/r in a rich complex medium at different incubation temperatures (Figure 4.6). A plot of the logarithm of the growth rate versus the inverse of the growth temperature is commonly called an **Arrhenius plot** after the Swedish chemist Svante Arrhenius (1859–1927), who discovered the relationship of temperature and the velocity of chemical reactions. It is seen in this plot that *E. coli* has an optimum growth temperature of 39°C, with a growth range extending from 13.5°C (minimum temperature) to 48°C (maximum temperature). The growth rate declines on either side of 39°C, although the decline is more rapid at temperatures above 39°C than at temperatures below 39°C. The abrupt drop in growth rate at high temperatures is probably caused by denaturation of enzymes and thermal disruption of cell structures such as membranes.

Bacteria are divided into four groups on the basis of their cardinal temperatures for growth: **psychrophiles,** −10°C to 25°C (optimum temperature: 10°C to 20°C); **mesophiles,** 10°C to 45°C (optimum temperature: 20°C to 40°C); **thermophiles,** 30°C to 80°C (optimum temperature: 40°C to 70°C); and

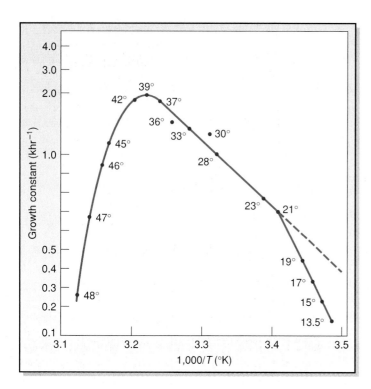

figure 4.6

Arrhenius Plot of the Growth Rate of *Escherichia coli* B/r in a Rich Complex Medium

The growth constant (khr^{-1}) is plotted on a logarithmic scale against the inverse of the absolute temperature (°K). Individual data points are marked with the corresponding degrees centigrade.

extreme thermophiles, 80°C and above (Figure 4.7). These are general guidelines for separation of bacteria; the actual ranges of temperature for growth will vary among bacteria of each group.

Psychrophiles Grow Optimally at Low Temperatures

Psychrophiles are frequently found in refrigerated foods and in areas of the world with relatively low temperatures. Psychrophilic microbes are able to grow at low temperatures because their enzymes function at these temperatures. Psychrophiles also have plasma membranes containing large amounts of unsaturated fatty acids, which allows the membranes to remain fluid at lower temperatures. Algae are examples of microorganisms found in polar regions, where temperatures often drop below 0°C. Polar oceanic waters are liquid below 0°C because of their sodium chloride content. The presence of these algae in and under ice sheets often results in hues of color in the ice. Many psychrophilic bacteria grow at refrigeration temperatures (4°C or lower) and cause spoilage of such foods as meat, milk, vegetables, and fruits. Both *Pseudomonas* and *Staphylococcus* are associated with meat spoilage. Refrigeration slows but does not always stop growth of these microbes. As a consequence, although refrigeration prolongs the shelf life of certain kinds of foods, it does not prevent spoilage and damage. Strict microbiological standards at the time of processing can significantly extend the shelf life of most food products (see food spoilage, page 614).

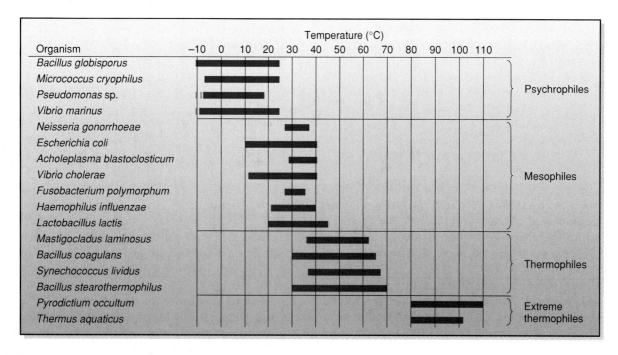

figure 4.7

Temperature Ranges for Psychrophiles, Mesophiles, Thermophiles, and Extreme Thermophiles

Storage of Bacterial Cultures

Bacteria are stored for extended periods of time as frozen cell suspensions or as freeze-dried (**lyophilized**) cultures. Under these conditions, all metabolic activity ceases and the bacteria are maintained in a state of suspended animation. Short-term (3 to 5 years) storage of bacteria is achieved by suspension of bacteria in a nutrient broth containing 15% to 25% glycerol and freezing at temperatures of −70°C or lower. The glycerol (a water-miscible liquid) in the suspension reduces ice crystal formation that otherwise would increase solute concentration and cause subsequent cell damage through dehydration and disruption of biological structures. For long-term preservation (more than five years), heavy suspensions of bacteria are dried rapidly at ultralow temperatures in a vacuum. This freeze-drying is accomplished by drawing off moisture from the frozen bacterial suspension by use of a high-vacuum pump. Such lyophilized cultures have been known to remain viable for 20 years or more.

Mesophiles Grow Optimally at Body Temperatures

Mesophiles are bacteria that have optimal growth temperatures of 20°C to 40°C. Most pathogenic bacteria (for example, *Neisseria*, *Salmonella*, and *Streptococcus*) fall within this group. Also included in this group are many of the bacteria normally found in the human body (for example, *Escherichia* and *Proteus*). Mesophiles typically grow poorly at very low or very high temperatures.

Thermophiles and Extreme Thermophiles Grow Optimally at High Temperatures

Thermophilic and extreme thermophilic procaryotes are found in such environments as hot springs, hot industrial process water, and hot decomposing material, where the temperature usually exceeds 50°C. These organisms apparently are able to survive and grow at high temperatures because their enzymes and cellular structures are heat stable. Photosynthetic organisms (whether procaryotic or eucaryotic) and eucaryotes seldom live in high-temperature environments because their internal membranes are heat labile. Endospore-forming bacteria (*Bacillus* and *Clostridium*), although not very temperature resistant as vegetative cells, can resist relatively high temperatures as spores. *Thermus aquaticus,* a bacterium that has been isolated from hot springs at Yellowstone National Park in Wyoming, has an optimum growth temperature of 80°C to 85°C. Organisms resembling *Thermus* have been found in hot water heaters. The heat stability of the DNA polymerase of *T. aquaticus* has made this enzyme, commonly called *Taq polymerase,* ideally suited for use in the polymerase chain reaction, which employs high temperatures (see polymerase chain reaction, page 289).

Methanobacterium thermoautotrophicum, a methane-producing procaryote (Archaea) widely distributed in decomposing material, and *Sulfolobus acidocaldarius,* a chemolithotrophic procaryote (Archaea) found at high temperatures and low pHs, are examples of extreme thermophilic microorganisms. The Archaea *Pyrodictium occultum* and *Pyrococcus woesei* are among the most thermophilic of all known organisms and can grow at 110°C and 104.8°C, respectively. The ability of extreme thermophiles to survive and grow at high temperatures may be due in part to the unique membrane architecture of Archaea. Instead of the conventional fatty acids bonded by ester linkages to glycerol in the plasma membranes of Bacteria, the membrane lipids of Archaea have ether linkages between glycerol and repeating branched aliphatic chains of isoprene (see membrane lipids of Bacteria and Archaea, page 51).

Microbial Metabolism and Growth
Growth: Metabolic Environment • pp. 15–19

Microorganisms Have Upper and Lower Ranges of pH for Growth

The concentration of hydrogen ions (H^+) in an environment (pH) is critical to microbial growth because it can affect enzyme activity. An extremely high or low pH can denature and inactivate enzymes and disrupt cell processes.

Most procaryotes grow best at a pH between 6.5 and 7.5, whereas some (for example, *Thiobacillus thiooxidans* and *Sulfolobus acidocaldarius*) thrive in acidic environments as low as pH 1, where they directly contribute to the acidity through the formation of such metabolic end products as sulfuric acid (Table 4.5). Such acid-loving procaryotes are called **acidophiles.** Most fungi are able to grow well at the relatively low pH of 5. Although these microbes live in acidic environments, most maintain a pH of close to neutral

table 4.5

Examples of pH Tolerance and Optimum Ranges for Certain Types of Procaryotes

Organism	pH Range for Growth	pH Optimum for Growth
Thiobacillus thiooxidans	1.0–6.0	2.0–2.8
Lactobacillus acidophilus	4.0–6.8	5.8–6.6
Proteus vulgaris	4.4–8.4	6.0–7.0
Escherichia coli	4.4–9.0	6.0–7.0
Enterobacter aerogenes	4.4–9.0	6.0–7.0
Clostridium sporogenes	5.0–9.0	6.0–7.6
Nitrobacter sp.	6.6–10.0	7.6–8.6
Nitrosomonas sp.	7.0–9.4	8.0–8.8

table 4.6

Approximate Limiting Water Activities for Growth of Microorganisms

Water Activity (a_w)	Procaryotes	Fungi	Algae
1.00	Caulobacter Spirillum		
0.90	Bacillus Lactobacillus	Fusarium Mucor	Chlamydomonas
0.85	Staphylococcus	Debaryomyces	
0.80		Penicillium	
0.75	Halobacterium Halococcus	Aspergillus Chrysosporium	Dunaliella
0.60		Saccharomyces	

Source: Data from A. D. Brown, Microbial water stress, *Bacteriological Reviews* 40:803–846 (1976).

within their cells by either preventing entry of, or actively removing, H⁺ from the cell. *Helicobacter pylori,* which is found in the stomach and causes peptic ulcers, is not acid-tolerant but protects itself from stomach acid by growing in protective mucus layers of the stomach and breaking down urea in the stomach into carbon dioxide and ammonia. The ammonia produced by this reaction is alkaline, which helps neutralize stomach acid in the vicinity of the bacteria.

Certain bacteria such as *Streptococcus* and *Lactobacillus* produce acidic end products during metabolism, which lower the environmental pH. The growth of these microorganisms is eventually inhibited due to the lowering of the pH by the acids. Extreme pH changes in culture media can be minimized by the addition of buffering agents such as phosphates or carbonates. Buffers are salts of weak acids or bases that combine with or release hydrogen ions as the hydrogen ion concentration changes in the medium, thereby slowing change in the hydrogen ion concentration. Although bacteriological media often contain nutritional ingredients such as peptones and amino acids that have some buffering capacity, most incorporate phosphate salts as buffering agents. Phosphate salts are good buffers because they provide a source of phosphorus for growth, are relatively nontoxic to bacteria, and are most effective in resisting radical changes in hydrogen ion concentrations at the pH growth range of most bacteria.

Alkalophiles, bacteria that can live and grow between pH 8.5 and 11.5, maintain a neutral pH within their cells by exchanging internal Na⁺ for external H⁺. The alkalophilic procary-

otes, which include *Bacillus alcalophilus* and *Microcystis aeruginosa,* are usually found in high pH environments such as alkaline soils and soda lakes.

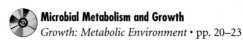 **Microbial Metabolism and Growth**
Growth: Metabolic Environment • pp. 20–23

Water Is Required for Microbial Growth

Water is required for all active cell processes. The availability of water in an environment varies, and depends not only on the actual amount of water, but also on the concentration of solutes in the water. Environments with a high concentration of solutes have less water available for cellular use, since the solutes tend to bind any available water. **Water activity (a_w)** is a measure of the water that is available for use by an organism. The water activity of a substance or solution is determined by measuring the relative humidity of the air space over the material. Water activity may be lowered by adding salt or sugar to a solution, thereby increasing the solute concentration of the solution and the osmotic pressure. A saturated solution of sodium chloride (30 g NaCl/100 ml water) has an a_w of 0.80, whereas seawater (3.5 g NaCl/100 ml water) has an a_w of 0.98. The water activity of maple syrup (140 g sucrose/100 ml water) is 0.90.

Most microorganisms require a water activity of 0.90 or greater for metabolic activity and growth (Table 4.6). Some microorganisms, particularly marine species, are able to grow in environments of high osmolarity. These organisms are called **osmotolerant** (if they can tolerate such conditions) or **osmophilic** (if they require such conditions for optimal growth). The Archaea *Halobacterium, Halococcus, Natronobacterium,* and *Natronococcus* are found in environments of high osmolarity, but have specific requirements for sodium chloride and so are called **extreme halophiles** [Greek *hals,* salt, *philein,* to love]. *Natronobacterium* and *Natronococcus,* which are found in high saline soda lakes where the pH is 9 to 11, are also extreme alkalophiles. The natronobacteria contain unusual diether lipids not found in other extreme halophiles, a characteristic that can be used to separate them phylogenetically from the halobacteria.

Because osmotic pressure and water availability influence bacterial growth, foods are often preserved by adding solutes (salts or sugars) or by removing water by dehydration. Increases in solute concentration are achieved by salting (for example, bacon and salt pork) and by increasing the sugar content of foods (for example, jams, jellies, and preserves). Salting is used in combination with low pH in the process known as pickling. Dehydration has become a popular and widely used method for food preservation. Bacteria are not necessarily killed by dehydration, but their growth is retarded under such conditions because only soluble nutrients can permeate the plasma membrane.

Bacteria Are Divided into Four Groups on the Basis of Their Oxygen Sensitivity and Requirement

The earth's atmosphere contains approximately 20% oxygen. Most types of organisms living today require molecular oxygen for growth. Many types of bacteria and fungi have the ability to grow in the absence of oxygen. Bacteria are divided into four major groups on the basis of their responses to molecular oxygen.

1. **Aerobic** bacteria grow in the presence of oxygen.

2. **Microaerophilic** bacteria grow best at oxygen concentrations lower than those of air.

3. **Facultatively anaerobic** bacteria grow in the presence or absence of oxygen (they respire in the presence of oxygen and ferment in the absence of oxygen).

4. **Anaerobic** bacteria grow best in the absence of oxygen.

Within each of these categories there are ranges of oxygen sensitivity and requirement. For example, some anaerobic microorganisms are **obligate anaerobes**—any trace of oxygen is toxic for these organisms. Examples of obligate anaerobes are *Clostridium* (found in soil, lake sediments, intestinal tracts, and canned foods, where oxygen is absent), *Methanobacterium* (useful in sewage degradation), and *Ruminococcus* (inhabitant of rumens). Other anaerobic microbes are **aerotolerant;** they survive and grow in the presence of small amounts of oxygen, although they do not use the oxygen metabolically.

Why is oxygen toxic to some organisms and not to others? To answer this question, we must first understand what happens to oxygen within the cell. Bacteria have enzymes that convert oxygen into different chemical compounds, including the toxic substances superoxide free radical (O_2^-); peroxide (O_2^{2-}), which usually appears in the form of hydrogen peroxide (H_2O_2); and hydroxyl free radical (OH·). These compounds cause oxidation of cell components and are lethal to the cell. Most aerobic, microaerophilic, and facultatively anaerobic bacteria possess enzymes that convert these toxic substances to harmless compounds. The superoxide free radical is converted to oxygen and hydrogen peroxide by the enzyme superoxide dismutase:

$$2\,O_2^- + 2\,H^+ \xrightarrow{\text{superoxide dismutase}} O_2 + H_2O_2$$

Hydrogen peroxide is subsequently decomposed to oxygen and water by catalase, or is reduced to water by peroxidase:

$$2\,H_2O_2 \xrightarrow{\text{catalase}} 2\,H_2O + O_2$$

$$H_2O_2 + NADH + H^+ \xrightarrow{\text{peroxidase}} 2\,H_2O + NAD^+$$

Aerotolerant anaerobes have superoxide dismutase, but lack catalase to degrade hydrogen peroxide. All obligate anaerobes studied to date lack both superoxide dismutase and catalase; therefore obligate anaerobes are unable to degrade hydrogen peroxide.

Oxygen-free environments can be obtained for laboratory cultures in one of several ways. Oxygen in a sealable container can be displaced with another gas, such as nitrogen or hydrogen. Alternatively, a GasPak anaerobic system can be used (Figure 4.8). Here, hydrogen is generated by the addition of water to an envelope containing sodium borohydride and sodium bicarbonate. The hydrogen reacts with available oxygen in the presence of a catalyst, palladium, to form water:

$$2\,H_2 + O_2 \xrightarrow{\text{palladium}} 2\,H_2O$$

The result is an anaerobic atmosphere. The carbon dioxide aids growth of anaerobes. The oxygen content of liquid media can be decreased or counteracted by the addition of chemical compounds called reducing agents (for example, sodium thioglycollate, cysteine, or ferrous sulfide). These remove any available oxygen from the environment by reducing it to H_2O.

Certain pathogenic bacteria (for example, *Mycobacterium* and *Neisseria*) are aerobic, but grow best in the presence of elevated carbon dioxide. These organisms are grown in a special incubator containing 5% carbon dioxide or are grown in a candle jar—a large glass container that is sealed after cultures and a lighted candle are placed inside (Figure 4.9). As the candle goes out, a 3% to 5% carbon dioxide atmosphere is established within the jar.

Microbial Metabolism and Growth
Growth: Metabolic Environment • p. 24

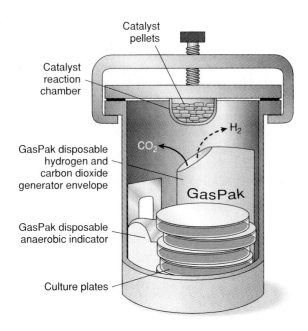

figure 4.8

GasPak Anaerobic System

The GasPak anaerobic system consists of an airtight container, a hydrogen-carbon dioxide generator envelope (GasPak), and a palladium catalyst. Water added to the envelope generates hydrogen (which combines with any free oxygen to form water) and carbon dioxide (which stimulates growth of some anaerobes). An anaerobic indicator strip containing methylene blue changes from blue to colorless as oxygen is removed.

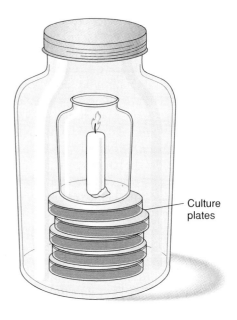

figure 4.9

Candle Jar

A 3% to 5% carbon dioxide atmosphere is produced as a lighted candle burns out in the candle jar.

Growth Media

Nutrients and physical factors are combined to provide a suitable growth environment for cultivation of microorganisms. Growth media vary in their composition, depending on the type of bacterium cultivated and the environment from which the bacterium has been obtained.

Growth Media Are of Two Types: Complex and Synthetic

Two basic types of growth media are used by microbiologists for the cultivation of bacteria: **complex** (or **undefined**) media and **synthetic** (or **chemically defined**) media (Table 4.7). A complex medium's exact composition is not known. Complex media have undefined nutrients as their ingredients, such as animal extracts (for example, beef or yeast extract) that are rich in vitamins, amino acids, minerals, and other constituents, and peptones (digested proteins). These media are usually supplied in dehydrated form and are reconstituted with distilled water to form a nutrient broth for bacterial growth. Complex media are easy to prepare and relatively inexpensive.

Microbiologists sometimes require a synthetic, chemically defined medium to study the growth requirements and characteristics of bacteria. A defined medium's exact composition is known. Such a medium is more difficult to prepare than a complex medium, but the exact quantity and composition of ingredients permit precise laboratory studies of bacterial growth. An example of the usefulness of a chemically defined medium might be an experiment to determine the types of biosynthetic pathways for amino acids in a particular bacterium. The experiment would be performed by monitoring growth of the bacterium in a defined medium under conditions in which specific amino acids were removed from the growth medium. The absence of growth in a medium missing a certain amino acid would suggest the absence of the biosynthetic pathway for that particular amino acid in the bacterium. Also, this particular amino acid would be a growth factor for this bacterium.

Both complex and synthetic media can be prepared in two forms: **broth** or **agar.** Broths are liquid media prepared in test tubes, Erlenmeyer flasks, or other containers. The ratio of broth to total container volume is an important factor to consider in broth cultures, since too much broth reduces the surface area for

table 4.7

Examples of Complex and Synthetic Media

Complex Media for Cultivation of *Escherichia coli*

Nutrient Broth:

Beef extract	3 g
Peptone	5 g
Distilled water	1,000 ml

Trypticase Soy Broth:

Pancreatic digest of casein	17.0 g
Papaic digest of soybean meal	3.0 g
Sodium chloride	5.0 g
K_2HPO_4	2.5 g
Glucose	2.5 g
Distilled water	1,000 ml

Synthetic Medium for Cultivation of *E. coli*

$NH_4H_2PO_4$	1 g
Glucose	5 g
Sodium chloride	5 g
$MgSO_4 \times 7H_2O$	0.2 g
K_2HPO_4	1 g
Distilled water	1,000 ml

oxygen exchange. Broth cultures are grown either statically (with no agitation) or with shaking. Shaking increases aeration of broths and is usually used for the cultivation of aerobic microorganisms.

Agar media are broths solidified by the addition of agar, a derivative of red algae. Solid media are prepared in test tubes as slants (media solidified at an angle) or deeps (media solidified with no angle) and in special flat containers called Petri dishes, named after one of Robert Koch's assistants, Richard J. Petri, who introduced these containers for growth of microorganisms. Agar media are useful for the isolation of bacteria, since mixed populations (two or more types of microorganisms) often can be separated by streaking them onto an agar surface.

Microbial Metabolism and Growth
Growth: Media • pp. 36–39

Specialized Growth Media Are Used to Isolate and Enrich for Specific Types of Microorganisms

Special types of growth media are often used in microbiology to isolate and distinguish specific types of microorganisms. Such specialized media are especially useful for isolation of nutritionally fastidious organisms that may not grow as rapidly as other organisms.

An **enrichment culture** enhances growth of specific types of microorganisms while often inhibiting growth of other organisms. Enrichment cultures are especially helpful in isolating small numbers of a desired microbe in a mixed population such as in soil or fecal samples. An enrichment broth typically consists of nutrients (for example, a specific carbon or energy source) or environmental factors (for example, pH, temperature, or presence or absence of oxygen) that promote growth of the desired microbe, but are not necessarily conducive for growth of other organisms, and inhibitors that prevent growth of unwanted organisms. A sample containing the desired microorganism is inoculated into the enrichment broth and incubated under conditions favorable for the microbe. The desired microbe outgrows other organisms during the enrichment procedure and can then be streaked onto an agar medium for isolation and identification. Selenite broth is an enrichment broth frequently used to selectively enrich for growth of *Salmonella* from samples of feces, urine, or sewage that have heavy concentrations of other types of bacteria. *Salmonella* grows on the nutrients in selenite broth, but *E. coli* and other gram-negative bacteria are inhibited by the sodium selenite in the broth. Another example of the use of enrichment cultures is the isolation of lipid-degrading bacteria from nature. Soil samples containing these bacteria in a mixed population are inoculated into an enrichment broth containing only lipids as a source of carbon and energy. Because only bacteria producing lipid-degrading enzymes called lipases are able to metabolize the lipids and grow, the broth selectively enriches for these organisms. Lipase-producing bacteria isolated in this manner have been especially useful in the reduction of grease that accumulates in restaurant grease traps. Unlike detergents, acids, and other chemicals that solubilize the grease for drainage into the sewage, lipase-producing bacteria *degrade* the grease. Furthermore, these living organisms continue to grow and reproduce harmlessly while metabolizing the grease, thereby reducing the need for constant maintenance and replenishment.

Differential media help distinguish different types of microorganisms by their metabolism of media components and subsequent colony appearance. Blood agar, which consists of 5% (volume/volume) sheep blood and trypticase soy agar, is an example of a medium that differentiates bacteria by their ability to completely lyse, partially lyse, or not lyse sheep red blood cells (see classification of streptococci and hemolysis, page 502). Hemolysis patterns on blood agar can help distinguish bacteria such as *Streptococcus pyogenes* (cause of streptococcal pharyngitis, or strep throat), which completely lyses red blood cells and produces clear zones around colonies on blood agar (β hemolysis); *Streptococcus pneumoniae* (one of the causes of bacterial pneumonia), which partially lyses red blood cells (α hemolysis); and *Streptococcus mutans* (cause of dental caries), which is nonhemolytic (γ hemolysis) (Figure 4.10).

Microbial Metabolism and Growth
Growth: Media • pp. 40–43

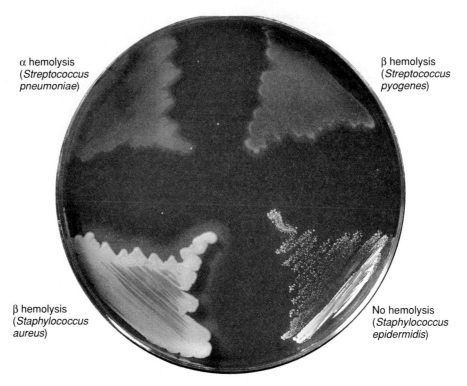

α hemolysis
(*Streptococcus pneumoniae*)

β hemolysis
(*Streptococcus pyogenes*)

β hemolysis
(*Staphylococcus aureus*)

No hemolysis
(*Staphylococcus epidermidis*)

figure 4.10
Hemolytic Patterns of Bacteria on Blood Agar
Blood agar plate showing β-hemolytic, α-hemolytic, and nonhemolytic colonies.

Selective media contain inhibitors that prevent the growth of unwanted microorganisms without affecting growth of the desired microbe. Levine's eosin methylene blue (EMB) agar and M-Endo medium are examples of differential selective media used to distinguish between lactose-fermenting and non-lactose-fermenting organisms (see coliform testing, page 590). Eosin methylene blue agar contains the dyes eosin and methylene blue, which inhibit most gram-positive bacteria. Bacteria that strongly ferment lactose produce colonies with a characteristic green metallic sheen, caused by accumulation of eosin and methylene blue in the colonies at acid pHs (Figure 4.11). Weak lactose fermenters produce purple colonies. Bacteria that do not ferment lactose but use other nutrients such as peptone in EMB agar form colorless colonies. M-Endo medium contains sodium sulfite, sodium lauryl sulfate, sodium desoxycholate, and basic fuchsin, which inhibit growth of gram-positive bacteria. *E. coli* and other lactose-fermenting bacteria produce colonies with a metallic sheen on M-Endo medium as a result of acid and acetaldehyde production from lactose (Figure 4.11b). Non-lactose-fermenting bacteria that use peptone in M-Endo medium produce colorless or faint colonies.

Pure Culture Techniques

Microorganisms generally are found in nature as mixed populations and must be isolated in **pure culture** if they are to be characterized and identified. A pure culture consists of a population of cells that all arise from one cell. Because microbes are too small to be seen and separated by the naked eye, pure cultures are necessary in microbiology for the isolation from a mixed population of a culture that consists of an individual species.

The Streak Plate Technique Separates Bacteria in a Mixed Population

A method that is often used to obtain pure cultures is the **streak plate technique** (Figure 4.12). A mixed culture sample is deposited in one area of an agar plate. An inoculating loop that has been heated and sterilized is streaked several times through the inoculated area and in uninoculated agar areas. The procedure is repeated two to three times to separate the inoculated culture further. On a properly streaked plate, individual bacterial cells multiply and grow

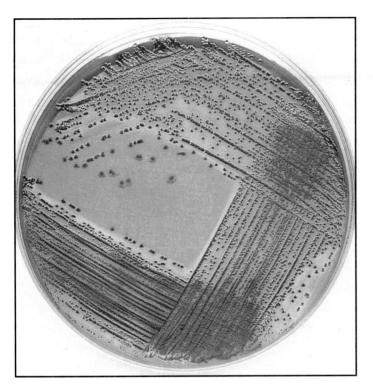

a.

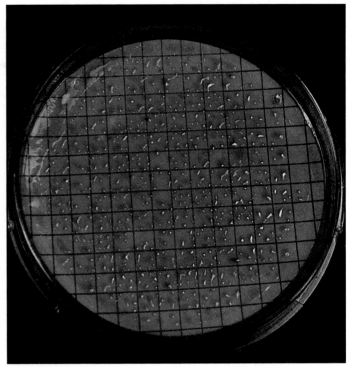

b.

figure 4.11
Lactose-Fermenting *Escherichia coli* Colonies on EMB Agar and M-Endo Medium.
a. *E. coli* produces colonies with a characteristic green metallic sheen on EMB agar. b. A membrane filter with *E. coli* colonies on M-Endo medium. *E. coli* produces colonies with a metallic sheen on M-Endo medium.

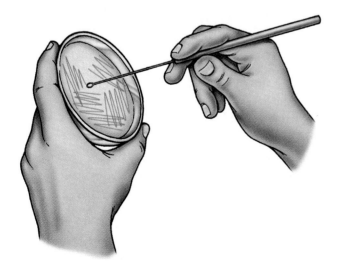

figure 4.12
Streak Plate Technique for Isolation of Bacteria
A sterile inoculating loop is used to streak bacteria in straight lines from an initial inoculum on an agar surface. The loop is sterilized and the procedure is repeated three more times. In this manner, the concentration of bacteria is diluted and isolated colonies are obtained.

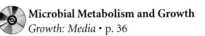 **Microbial Metabolism and Growth**
Growth: Media • p. 36

into distinct populations called colonies. Colonies formed by different types of bacteria often have different shapes, sizes, and pigmentations (Figure 4.13). These differences in colony morphology aid in the identification of bacteria, particularly from a mixed population. Other methods that can be used to obtain pure cultures are the spread plate technique and the pour plate technique (see spread plate and pour plate, page 110).

Microbial Metabolism and Growth
Growth: Growth Curves • p. 29

Aseptic Technique Is Used to Transfer Bacteria Without Contamination

Manipulation of cultures is necessary in the microbiology laboratory, but can introduce airborne or surface contaminants to a culture. The term **aseptic technique** is used to describe the procedure for handling cultures in such a manner as to eliminate contamination by unwanted microorganisms.

One example of aseptic technique is the transfer of a pure bacterial culture from one test tube to another (Figure 4.14). Contamination is reduced by carefully heating the lips of the test tubes and the inoculating loop before the transfer of the culture. Airborne contaminants are common in a microbiology laboratory where many different cultures are processed and can enter the test tubes during the handling of pure cultures. Therefore it is important to work rapidly and keep the test tubes at an angle as the culture is transferred to prevent airborne contaminants from entering the test tubes. Aseptic technique also can be used to transfer a culture from a test tube to an agar plate, except that the lid of the plastic Petri dish is not heated.

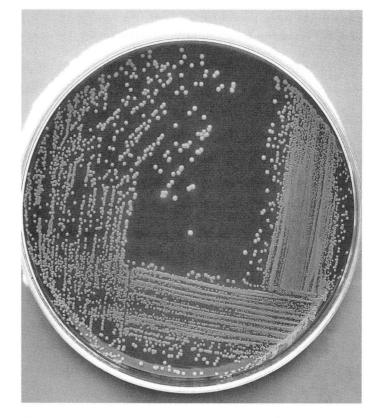

figure 4.13
Isolated Colonies of Bacteria on an Agar Plate
Individual colonies are shown on a streak plate.

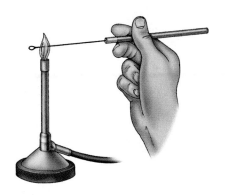

1. Sterilize the inoculating loop.

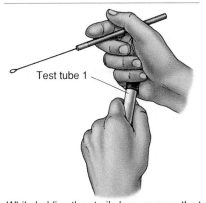

2. While holding the sterile loop, remove the test tube cap.

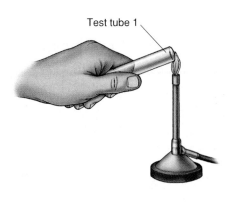

3. Briefly flame the lip of the test tube.

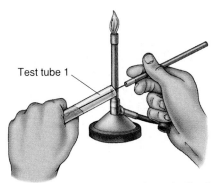

4. Remove a loopful of broth, reflame the lip of the tube, and recap.

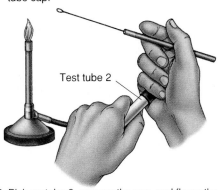

5. Pick up tube 2, remove the cap, and flame the tube lip.

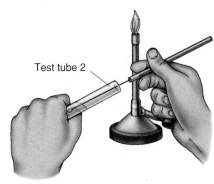

6. Insert the loopful of broth from tube 1 into tube 2, shake the loop to dislodge the broth, and remove the loop from the tube. Flame the loop and tube lip and recap the tube.

figure 4.14
Aseptic Transfer of a Bacterial Culture
Individual steps are shown for the aseptic transfer of a bacterial culture.

Iron is a nutrient that is required only in microgram quantities by microorganisms, yet its availability significantly influences their growth. It is an essential component of cytochromes and other heme and nonheme proteins, and is a cofactor for many enzymes. Given the abundance of iron in nature (iron is the fourth most abundant element in the earth's crust), it would seem that this mineral would be readily accessible for microbial assimilation. This is not the case, however. At neutral pH, iron in its ferric form (Fe^{3+}) occurs as highly insoluble hydroxides, carbonates, and phosphates, which are not easily taken up by the cell. Furthermore, microbes must compete with other organisms for this nutrient. This competition for available iron determines the growth and survival of a bacterium to a large extent.

In the case of microbial pathogens, the availability of iron is frequently a determinant of virulence. Infected vertebrate hosts often restrain microbial growth by withholding iron through such mechanisms as: (1) decreased intestinal absorption of exogenous iron, (2) shifting of iron from plasma to storage compartments, and (3) increased synthesis and positioning of host iron-binding proteins such as transferrin and lactoferrin. In vivo and in vitro studies have shown that added iron dramatically increases the virulence of many pathogenic microorganisms.

In one of these studies, Shelley M. Payne and Richard A. Finkelstein examined the effects of added iron on virulence of *Neisseria gonorrhoeae* (the cause of gonorrhea). Eleven-day-old chicken embryos were inoculated intravenously with 0.1 ml of *N. gonorrhoeae* at different serial dilutions, along with an iron supplement or a control containing no iron. Among the added iron compounds were ferric ammonium citrate, Imferon (an iron dextran complex), and Blutal (a solution of chondroitin sulfate colloidal iron).

The embryos were monitored, and the number of bacteria required to kill 50% of the embryos (the lethal dose 50% end point or LD_{50}) was determined. The results indicated that 1 to 2.5 log fewer bacteria were needed to establish a lethal infection in the presence of added iron (Table 4P.1). In a parallel experiment, Payne and Finkelstein also examined the effect of iron supplement (ferric ammonium citrate) on the level of gonococcemia (*N. gonorrhoeae* in the blood) in embryos. Their data showed that ten hours after inoculation, the level of gonococcemia was significantly higher ($p<0.001$) in embryos inoculated with bacteria and supplemented iron than embryos inoculated with only bacteria (Table 4P.2).

These experiments provide supportive evidence that the virulence of microbial pathogens is enhanced by the presence of iron. They suggest that host iron-binding compounds may effectively defend against microbial infection by removing this essential nutrient from the microbe's environment.

Source

Payne, S.M., and R.A. Finkelstein. 1975. Pathogenesis and immunology of experimental gonococcal infection: Role of iron in virulence. *Infection and Immunity* 12:1313–1318.

References

Weinberg, E.D. 1971. Roles of iron in host-parasite interactions. *The Journal of Infectious Diseases* 124:401–410.

Weinberg, E.D. 1978. Iron and infection. *Microbiological Reviews* 42:45–66.

table 4P.1

Effect of Iron Supplements on Virulence of *Neisseria gonorrhoeae* for Chicken Embryos

Experiment Number	Strain	Iron Supplement	LD_{50} (Colony-Forming Units)	
			Control	With Added Iron
1	F62	FAC[a] (62.5)[b]	4.9×10^5	1.2×10^3
2	F62	FAC (62.5)	4.3×10^5	2.4×10^3
3	2686	FAC (62.5)	1.2×10^5	7.0×10^2
4	2686	Imferon (125)	1.2×10^5	1.0×10^4
5	F62	Blutal (400)	3.7×10^5	9.0×10^3
6	F62	Imferon (250)	6.0×10^5	2.0×10^4
7–10	F62	Imferon (500)	1.0×10^{6} [c]	1.8×10^4 [c]
11	F62	Imferon (500)	9.9×10^5	2.4×10^4

[a]Ferric ammonium citrate.
[b]Numbers in parentheses indicate micrograms of iron.
[c]Geometric mean of four determinations.

table 4P.2

Effect of Iron Supplements on the Level of Gonococcemia in Chicken Embryos[a]

Iron[b]	Time	
	4 Hr	10 Hr
−	2.7×10^3 [c]	4.0×10^2 [d]
	$(7.5 \times 10^2 - 1.2 \times 10^4)$	$(<10^1 - 6.2 \times 10^3)$
+	5.9×10^3	2.4×10^4 [d]
	$(7.5 \times 10^2 - 1.9 \times 10^4)$	$(7.5 \times 10^2 - 3.0 \times 10^5)$

[a]Inoculum was 10^4 colony-forming units of strain F62.
[b]FAC (62.5 µg Fe).
[c]Viable count (colony-forming units per milliliter of blood) at indicated time; geometric mean of six determinations; numbers in parentheses give range.
[d]$p<0.001$ (Student's t-test).

Summary

1. Microorganisms can be divided into two major groups based on the nature of their energy source. Phototrophs obtain their energy from light, whereas chemotrophs obtain their energy from the oxidation of chemical compounds.

2. Microorganisms can also be divided into two groups based on their sources of carbon. Autotrophs obtain their carbon from carbon dioxide through a process called carbon dioxide fixation. Heterotrophs use organic compounds as a source of carbon.

3. A major source of nitrogen for living organisms is nitrogen gas, or dinitrogen. Certain bacteria are capable of converting nitrogen to combined forms of nitrogen in chemical compounds by a process called nitrogen fixation.

4. Growth factors are organic nutrients that are required by microorganisms for growth, but are not necessarily synthesized by them. Amino acids, purines and pyrimidines, and vitamins are examples of growth factors.

5. Microorganisms have four mechanisms for nutrient and ion transport across the plasma membrane: passive diffusion, facilitated diffusion, active transport, and group translocation. Passive diffusion is the movement of small molecules down a concentration gradient. Facilitated diffusion is the carrier-mediated movement of molecules down a concentration gradient. Active transport is an energy-requiring, carrier-mediated movement of molecules against a concentration gradient. Group translocation is the chemical modification of a molecule during movement across the membrane.

6. Microorganisms can be divided into four groups on the basis of their growth temperature ranges: psychrophiles (–10°C to 25°C), mesophiles (10°C to 45°C), thermophiles (30°C to 80°C), and extreme thermophiles (80°C and above).

7. Most bacteria grow best at a pH between 6.5 and 7.5. Acidophiles can grow at pHs as low as pH 1, whereas alkalophiles can grow at pHs as high as pH 11.5. Both groups of bacteria can live at such pH extremes because they control the movement of H^+ across the plasma membrane.

8. Water is required for microbial growth. Some microorganisms are osmotolerant or osmophilic and can survive in environments of high osmolarity. Halophiles have a specific requirement for high concentrations of sodium chloride.

9. Bacteria can be divided into four groups based on their oxygen sensitivity and requirement: aerobic bacteria, which grow in the presence of oxygen; microaerophilic bacteria, which grow best at low oxygen concentrations; facultative anaerobes, which grow in the presence or absence of oxygen; and anaerobes, which grow best in the absence of oxygen.

10. Growth media are used to grow microorganisms. Specialized growth media are often used to isolate and distinguish specific types of microorganisms. An enrichment culture enhances growth of certain microbes, while inhibiting growth of other organisms. Selective media have inhibitors that prevent growth of unwanted microorganisms without affecting growth of the desired microbe. Differential media help distinguish different types of microorganisms by their metabolism of media components and subsequent colony appearance.

11. A pure culture is one that consists of a population of cells that all arise from one cell. Pure cultures can be obtained by the streak plate technique in which a mixed culture is streaked and bacteria are separated from each other on an agar plate. Aseptic techniques are used when handling microbial cultures to minimize contamination.

 ## EVOLUTION *and* BIODIVERSITY

Microorganisms are ubiquitous, found in every environment where there are other forms of life as well as in environments where no other forms of life can exist. This widespread occurrence of microbes is possible because of their diverse structures, physiology, and metabolism. Many Archaea live in extreme environments because they have adapted to the conditions in these environments. For example, *Halobacterium*, a procaryote that is found in salt lakes, marine salterns, and other areas with high levels of sodium chloride, does not have peptidoglycan in its cell wall (a characteristic common among Archaea). Instead its cell wall is composed of glycoprotein containing an excess of acidic (negatively charged) amino acids such as aspartic acid and glutamic acid that help balance the high concentration of sodium ions in its surroundings. In high-sodium environments (20% to 25% sodium chloride), the sodium ions bind to the cell wall and maintain wall integrity. In low-sodium environments where there are insufficient amounts of sodium ions, the negative charges of the wall repel each other, resulting in cell lysis. Thus *Halobacterium* not only can survive under high-sodium conditions, but must live in such an environment because of the composition of its cell wall. *Thermoplasma*, another procaryote (Archaea), lives in low pH (pH 2), high-temperature (55°C) environments such as self-heating coal refuse piles. Like *Mycoplasma*, *Thermoplasma* does not have a cell wall but is able to survive extreme conditions because of its unique plasma membrane. This membrane, which consists primarily of lipopolysaccharide containing a tetraether lipid linked to glucose and mannose units, is able to withstand hot acid environments. *Halobacterium* and *Thermoplasma* are two of the many examples of microorganisms that have adapted to their environments by evolving unique cell structures.

Short Answer

1. Identify the requirements for growth of bacteria.

2. Most bacteria oxidize chemicals for energy and require organic carbon sources. These bacteria are referred to as _____.

3. If the organism in question 2 oxidizes inorganic chemicals for energy, it is referred to as _____.

4. Microbes and plants are responsible for converting nitrates into cellular components for the rest of the food chain. What reactions may they use to achieve this?

5. Identify several "growth factors" and explain why they are necessary.

6. Identify several methods of membrane transport.

7. What is the name of the process for gas exchange across the cellular membrane?

8. Which method of membrane transport is responsible for maintaining the cell's osmotic concentration?

9. Compare and contrast facilitated diffusion and active transport.

10. How does group translocation differ from active transport?

11. Identify several physical factors that must be met for growth.

12. Explain how bacteria have adapted to grow at various temperatures.

13. How do bacteria affect the pH in their environment?

14. What is meant by "water activity" and what determines the a_w for a culture?

15. Explain why oxygen is toxic to some bacteria.

16. How do complex media differ from synthetic media?

17. Identify three categories of specialized media and explain their functions.

18. Although few organisms can utilize agar as a nutrient, agar is a part of many types of media. Explain why.

19. Explain the importance of aseptic technique.

Multiple Choice

1. Which of the following is *not* required for growth by all bacteria?
   a. carbon
   b. nitrogen
   c. oxygen
   d. sulfur

2. Which of the following requires carrier proteins, but does not require energy?
   a. active transport
   b. facilitated diffusion
   c. group translocation
   d. passive diffusion

3. The optimum growth temperature for _____ is 40°C to 70°C.
   a. extreme thermophiles
   b. mesophiles
   c. psychrophiles
   d. thermophiles

4. Which of the following in *not* likely to kill bacteria?
   a. temperatures above the optimum
   b. pH below the optimum
   c. oxygen level below the optimum
   d. dehydration

5. _____ bacteria live with or without oxygen.
   a. Aerobic
   b. Anaerobic
   c. Facultatively anaerobic
   d. Microaerophilic

Critical Thinking

1. Explain how bacteria can live and grow in the absence of oxygen.

2. Thermophilic bacteria are generally considered to be harmless; however, some thermophiles may cause disease. Explain.

3. Why is most food safe after it has been heated to boiling for a few minutes? Explain why there are exceptions.

4. You have been growing organism X on a synthetic medium in the laboratory, but you are out of an essential ingredient. Would you expect organism X to grow better or worse on complex media?

5. Describe the media and incubation conditions you would provide for an aerobic, halophilic, mesophilic, acidophilic bacterium.

● Supplementary Readings

Brock, T.D. 1978. *Thermophilic microorganisms and life at high temperatures.* New York: Springer-Verlag. (A discussion of thermophiles.)

Fridovich, I. 1977. Oxygen is toxic! *BioScience* 27:462–466. (A review of toxic products formed in the presence of oxygen and how these products are destroyed.)

Gerhardt, P., R.G.E. Murray, W.A. Wood, and N.R. Krieg. 1994. *Methods for general and molecular bacteriology,* pp. 135–292. Washington, D.C.: American Society for Microbiology. (A discussion of various techniques in microbiology, including sections on microbial nutrition and the effects of temperature, oxygen, pH, and other parameters on growth.)

Gottschal, J.C., and R.A. Prins. 1991. Thermophiles: A life at elevated temperatures. *Trends in Ecology and Evolution* 6:157–162. (A discussion of thermophiles and their habitats.)

Gould, G.W., and J.E. Corry. 1980. *Microbial growth and survival in extremes of environment.* New York: Academic Press. (An in-depth discussion of environmental effects on microbial growth.)

Hobbs, A.S., and R.W. Albers. 1980. The structure of proteins involved in active membrane transport. *Annual Review of Biophysics and Bioengineering* 9:259–291. (A review of transport mechanisms in procaryotes and eucaryotes.)

Jannasch, H.W. 1984. Microbes in the oceanic environment. In *The microbe 1984,* D.P. Kelly, and N.G. Carr, eds. London, England: Cambridge University Press. (A review of microorganisms found in deep-sea environments of extreme pressures and temperatures.)

Krulwich, T.A., and A.A. Guffanti. 1989. Alkalophilic bacteria. *Annual Review of Microbiology* 43:435–563. (A review of alkalophilic bacteria and their physiological characteristics.)

Kushner, D.J., ed. 1978. *Microbial life in extreme environments.* London: Academic Press. (A series of papers on growth and survival of microorganisms in extreme environments.)

Lowe, S.E., M.K. Jain, and J.G. Zeikus. 1993. Biology, ecology, and biotechnological applications of anaerobic bacteria adapted to environmental stress in temperature, pH, salinity, or substrates. *Microbiological Reviews* 57:451–509. (A review of how anaerobic bacteria have evolved special mechanisms for physiological adaptation to environmental extremes.)

Marr, A.G. 1991. Growth rate of *Escherichia coli. Microbiological Reviews* 55:316–333. (A review of carbon source and other requirements on ATP yield and growth rate of *E. coli.*)

Morita, R.Y. 1976. Psychrophilic bacteria. *Bacteriological Reviews* 39:144–167. (A review article on physiology and metabolism of psychrophiles.)

Ollivier, B., P. Caumette, J-L. Garcia, and R.A. Mah. 1994. Anaerobic bacteria from hypersaline environments. *Microbiological Reviews* 58:27–38. (A review of strictly anaerobic bacteria that live in hypersaline habitats.)

chapter five 5

GROWTH AND CONTROL OF GROWTH

Bacterial Growth
Reproduction by Binary Fission
Population Increases over Time
Quantitative Relationships of Population Growth

Measurement of Growth
Increases in Cell Numbers
Turbidimetric Measurements
Mass
Growth and Changes in Cell Activity

Growth Curve
The Lag Phase
The Exponential Phase
The Stationary Phase
The Death Phase

Special Techniques for Culture
Continuous Culture
Synchronous Cultures

Control of Microbial Growth
High Temperature
Other Methods for Killing Microorganisms

Antimicrobial Agents
Disinfectants and Antiseptics
Antibiotics and Synthetic Drugs
Antimicrobial Susceptibility Tests

PERSPECTIVE
The Carbon Dioxide Requirement of *Neisseria meningitidis*

EVOLUTION AND BIODIVERSITY

Transmission electron micrograph of *Bacillus subtilis*, showing an individual, rod-shaped, gram-positive bacterium beginning to divide by binary fission. The densely colored cell wall (red) appears pinched at the point of division (colorized).

Microbes in Motion — PREVIEW LINK

This chapter covers the key topics of growth curves and control of microbial growth. The following sections in the *Microbes in Motion* CD-ROM may be useful as a preview to your reading or as a supplemental study aid:

Microbial Metabolism and Growth: Growth (Growth Curves), 26–35. *Control—Physical and Chemical:* Physical Control, 1–22; Chemical Control, 1–38. *Antimicrobial Action:* Cell Wall Inhibitors, 1–19; Protein Synthesis Inhibitors, 1–9; Agents Affecting Nucleic Acid, 1–8; Agents Affecting Enzymes, 1–9; Antibiotic Susceptibility Testing (Terminology), 7–8, (Disk Diffusion), 11–16. *Antimicrobial Resistance:* Acquisition and Expression of Resistance Genes, 1–9.

Consider a few cubic centimeters of soil on your campus lawn. Within this small amount of soil is a diverse and extensive population of microorganisms, growing on the decaying organic matter, minerals, and other chemicals in the ground.

Bacterial growth occurs all around us—in food, water, and soil, and in plants, animals, and humans. Because bacteria are so small, we seldom consider the implications of the rapid growth of many of them. It may easily be calculated that a single bacterium growing under optimal conditions for 27 hours would produce a population that would cover the world in a layer of bacteria 6 feet high! This is particularly amazing because a single bacterium is only a few micrometers long.

Fortunately, optimal growth conditions do not exist indefinitely, and growth and reproduction of living organisms cannot continue long at maximum rates. Organisms grow and die, and growth is self-limited by availability of nutrients and the buildup of toxic, metabolic end products. Nonetheless, the growth rates of bacteria and similar types of microorganisms are impressive. This chapter will discuss bacterial growth and methods to measure it.

Bacterial Growth

Growth is defined as the orderly increase in all major constituents of an organism. Growth involves the synthesis of cellular structures, nucleic acids, proteins, and other cell components from nutrients obtained outside the cell. All living organisms must take up nutrients and be able to excrete waste products. This exchange of materials occurs via the outer surface of the organism. Smaller organisms tend to have larger ratios of surface to internal volume than do larger organisms. This large surface-to-volume ratio allows rapid entry of nutrients and permits more rapid metabolic and growth rates.

The Typical Bacterium Reproduces by Binary Fission

Bacteria generally reproduce by a process known as **binary fission.** In binary fission, a single cell grows progressively larger until at some stage in growth it divides into two separate and equal daughter cells. The time required for the cell to divide (or for a population of organisms to double in number) is referred to as the **generation time** or the **doubling time.**

The generation time depends on the bacterial species and its growth conditions (Table 5.1). Under specified growth conditions, the generation time is constant and characteristic for a particular bacterium.

Microbial Metabolism and Growth
Growth: Growth Curves • p. 27

Binary fission in bacteria differs from mitosis in higher organisms; in binary fission there is continuous DNA synthesis during the replication cycle of bacteria. Replication of the entire bacterial chromosome takes approximately 40 minutes at 37°C in *Escherichia coli* cells. The doubling time for *E. coli* under ideal growth conditions, however, is much shorter (about 20 minutes or less) than the DNA replication time. In order to maintain con-

table 5.1

Generation Times of Selected Bacteria

Bacterium	Growth Medium	Temperature (°C)	Generation Time (min)
Bacillus subtilis	Complex medium	36	35
Clostridium botulinum	Glucose broth	37	35
Escherichia coli	Broth	37	17
Lactobacillus acidophilus	Milk	37	66
Mycobacterium tuberculosis	Synthetic medium	37	792
Pseudomonas aeruginosa	Glucose broth	37	31
Pseudomonas aeruginosa	Lactose broth	37	34
Pseudomonas aeruginosa	Tryptic meat broth	35	32
Salmonella typhimurium	Trypticase soy broth	37	24
Shigella dysenteriae	Milk	37	23
Staphylococcus aureus	Glucose broth	37	32
Streptococcus lactis	Lactose broth	30	48
Streptococcus lactis	Glucose milk	37	26
Streptococcus lactis	Peptone milk	37	37
Streptococcus pneumoniae	Glucose broth	37	30
Xanthomonas campestris	Glucose broth	25	74

From Philip L. Altman and David S. Dittmer, *Biology Data Book*, Volume 1. Copyright © 1972 Federation of American Societies for Experimental Biology. Reprinted by permisiion.

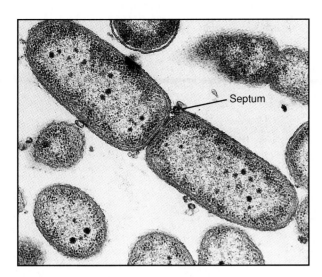

figure 5.1

Electron Micrograph of Dividing _Pseudomonas aeruginosa_, a Gram-Negative Bacterium

Two rod-shaped cells are formed from binary fission of the parent cell. A septum is shown at the point of cell division between the two cells (×34,200).

Microbial Metabolism and Growth
Growth: Growth Curves • p. 26

tinuity between cell division and DNA replication, the bacterial cell initiates a new round of DNA replication approximately every 20 minutes. By doing this, the bacterium has completed copies of its chromosome ready for each daughter cell at the time of division.

Not all aspects of bacterial DNA replication are known. During replication, DNA appears to migrate and attach to a portion of the plasma membrane. As DNA replicates in this area of the membrane, there is concomitant synthesis of new membrane material. The formation and partition of new membrane probably aids in separation of the DNA to opposite ends of the dividing cell.

After the DNA replication and separation, the cell forms a transverse septum of membranous material (Figure 5.1). New cell wall is synthesized, and two separate and equal daughter cells form. Incomplete cleavage of the septum sometimes can result in formation of such cellular arrangements as chains, clusters, or filaments. Some bacterial cells have multiple copies of DNA; such a condition is encouraged, particularly if the cells are grown in a nutritionally rich environment in which cell division lags behind DNA synthesis.

Bacterial Populations Increase in Cell Mass and Numbers over Time

Growth is typically considered as an increase in cell mass, but it may also be defined as an increase in cell numbers within a population. Both growth parameters are interrelated, since increases in cell mass eventually result in cell division and an increase in cell numbers. Under optimal conditions, the growth of bacteria

is **balanced**— there is an orderly increase in the DNA, RNA, and protein of the cell population. Growth of bacteria and other microorganisms is usually expressed as population growth because of the small size of microbes. This expression is valid since bacterial populations consist of millions of cells that are identical products of binary fission.

Quantitative Relationships of Population Growth Can Be Expressed Mathematically

Bacterial reproduction by binary fission results in a doubling of the population for each generation. Consequently, since the population is in balanced growth, there is also a doubling of the DNA, RNA, and protein of the cell population. The generation time of a bacterial population can be expressed mathematically as follows:

$$g = t - t_0/(\log_2 N - \log_2 N_0)$$

where g is the generation time, and N and N_0 represent cell concentrations at times t and t_0, respectively.

In microbiology, populations are usually plotted on graph paper as the logarithm to the base 10. Conversion of the equation to logarithms to the base 10 results in the following equation:

$$g = 0.301 \ (t - t_0)/(\log_{10} N - \log_{10} N_0)$$

Using this equation, the generation time of a culture can be determined if the cell concentration of the culture is known at two different incubation times. For example, the generation time of a culture containing 10^3 cells/ml at time t_0 and 10^7 cells/ml 7 hours later would be calculated as follows:

$$g = 0.301 \ (t - t_0)/(\log_{10} N - \log_{10} N_0)$$

$$g = 0.301 \ (7 - 0)/(\log_{10} 10^7 - \log_{10} 10^3)$$

$$g = 0.301 \ (7)/(7 - 3)$$

$$g = 2.107/(7 - 3)$$

$$g = 2.107/4$$

$$g = 0.527 \ \text{hour, or } 31.6 \ \text{minutes}$$

Because bacteria grow exponentially (at least for a portion of their growth), bacterial growth curves are plotted on a semilogarithmic scale (logarithm of number or mass of bacteria versus time) instead of an arithmetic (linear) scale (Figure 5.2). A semilogarithmic graph provides a linear relationship of the logarithm of cell concentration against time and can be used directly to determine population doubling time from the slope of the line.

Measurement of Growth

A zoologist can easily enumerate elephants in a herd and record increases in their numbers, although such monitoring may take decades. An ornithologist monitors changes in bird populations through field studies. Even ant behavior and development can be observed in the field or in the laboratory under controlled conditions. But how does a bacteriologist follow growth of billions of microorganisms that cannot be seen without the aid of a microscope? Are there limitations to the study of bacterial population growth?

Bacterial populations are easy to monitor. Bacteria grow rapidly and may produce several generations of identical progeny within a few hours. Their propagation does not require acres of pasture. A bacterial population with 10 billion members can easily be grown in a few milliliters of broth in a small Erlenmeyer flask.

Because bacteria grow in so little space in such a short time, it is simple to alter growth parameters to determine the effects on growth rate and cellular metabolism. Temperature, atmospheric conditions, rate of aeration, and composition of the medium can be changed to customize an experiment.

Bacterial growth can be measured in a number of different ways. Each type of measurement has advantages and disadvantages in terms of ease, accuracy, and sensitivity. The primary methods used to monitor bacterial growth involve following changes in cell numbers, cell mass, or cell activity.

Bacterial Growth Can Be Followed by Increases in Cell Numbers

The size of a bacterial population can be determined by counting the individual cells in a measured volume. These cells can be counted directly by using a microscope or indirectly by counting colonies on a plate of agar.

The Total Number of Cells in a Population Can Be Counted Using a Microscope

The most direct way to measure population growth is to count the number of cells in a suspension under a microscope (**direct microscopic count**). Since the field of vision in a slide preparation is limited, particularly at high magnifications, only a small volume of cell suspension can be enumerated by microscopy at any one time. Counting by microscopy is performed with special chambers (hemocytometer chambers or Petroff-Hausser counting chambers) that are divided into squares of known areas. The Petroff-Hausser counting chamber, which is designed specifically for bacterial cell enumeration, consists of a whole grid containing 25 large squares for a total area of 1 mm^2 and a total volume of 0.02 mm^3 (Figure 5.3). A small volume of cell suspension is added to the counting chamber. A special coverslip is then placed over the chamber so that each square is filled with suspension to a depth of 0.02 mm, and thus has a constant volume. By counting the number of cells in designated squares, one can extrapolate and estimate the total cell count in the original cell suspension.

Microbial Metabolism and Growth
Growth: Growth Curves • p. 30

Although the direct microscopic count method is rapid and performed with just a counting chamber and a light microscope, the technique is not very sensitive. A dense culture (10^6 or more per ml of suspension) is required to make the count statistically

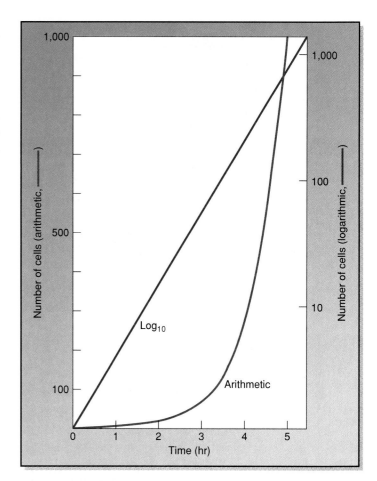

figure 5.2
Bacterial Growth Plotted on Arithmetic and Semilogarithmic Scales

accurate. Furthermore, the count obtained by microscopic observation includes both viable cells and nonviable cells; visual inspection cannot differentiate between living and dead cells.

The Number of Viable Cells in a Population Can Be Determined by Growing the Cells on Plates of Agar Media

Direct microscopic counts provide the bacteriologist with the total number of cells (viable and nonviable) in a population. In most instances, however, it is desirable to know the number of viable bacteria in a suspension. This is particularly true if one is interested in the infectious capability of live organisms. In these cases, the **viable count** is an alternative to a microscopic cell count (Figure 5.4).

The viable count technique involves growth of bacteria from a suspension on plates of agar media. When a single bacterium grows and divides on an agar medium, it forms a colony of cells, which is usually easily seen with the naked eye. Colonies that appear on the medium are enumerated to determine the original number of cells in the suspension. Problems are often encountered with bacteria that grow in chains or filaments, such as many of the bacilli. When these cells fail to separate on plating, the formation of fewer colonies than the actual number of cells plated results. Because of this effect,

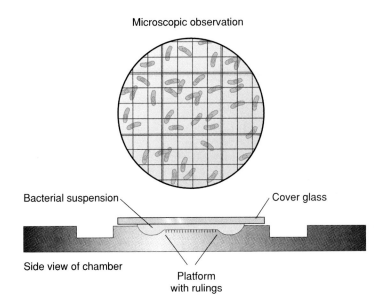

Microscopic observation

Bacterial suspension

Cover glass

Side view of chamber

Platform
with rulings

figure 5.3

Direct Microscopic Count with the Petroff-Hausser Counting Chamber

The Petroff-Hausser counting chamber consists of 400 small squares (0.0025 mm² each) and 25 large squares (0.04 mm² each). The number of cells is counted in several large squares and averaged. In the example given, 12 cells are counted in the large square. If this count is consistent with the counts obtained from several other large squares, then the number of cells per milliliter in the original cell suspension is determined in the following manner:

$$12 \text{ cells} \times 25 \text{ large squares } (0.04 \text{ mm}^2 \text{ each}) = 300 \text{ cells/mm}^2$$

$$300 \text{ cells/mm}^2 \times 50/\text{mm (each square has a depth}$$
$$\text{of } 0.02 \text{ mm}; 0.02 \times 50 = 1 \text{ mm}) = 1.5 \times 10^4 \text{ cells/mm}^3$$

$$1.5 \times 10^4 \text{ cells/mm}^3 \times 1,000 \ (1,000 \text{ mm}^3 = 1 \text{ cm}^3) = 1.5 \times 10^7$$
$$\text{cells/cm}^3, \text{ or } 1.5 \times 10^7 \text{ cells/ml}$$

figure 5.4

Dilution of Bacteria for a Viable Count

In the example shown, bacteria from a cell suspension are diluted in 9-ml water blanks, and 0.1 ml of each dilution is plated onto a nutrient agar plate. After incubation, only the agar plate prepared from dilution tube 4 has between 30 and 300 colonies. The 35 colonies on the agar plate represent the number of bacteria in 0.1 ml from this dilution tube. Thus tube 4 contains 350 colony-forming units (CFU)/ml (35 CFU/0.1 ml × 10 = 350 CFU/ml). Tube 3 contains ten times this quantity, (3,500 CFU/ml), tube 2 contains 35,000 CFU/ml, and tube 1 contains 350,000 CFU/ml (3.5 × 10⁵ CFU/ml). Since tube 1 has 10 ml and contains 1 ml of the original suspension, the original suspension has 3.5 × 10⁶ CFU/ml.

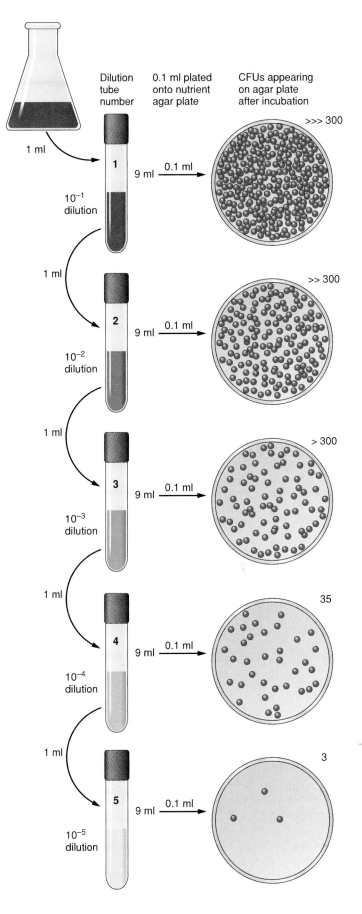

Dilution tube number | 0.1 ml plated onto nutrient agar plate | CFUs appearing on agar plate after incubation

1 ml

1   9 ml   0.1 ml   >>> 300
10^{-1} dilution

1 ml

2   9 ml   0.1 ml   >> 300
10^{-2} dilution

1 ml

3   9 ml   0.1 ml   > 300
10^{-3} dilution

1 ml

4   9 ml   0.1 ml   35
10^{-4} dilution

1 ml

5   9 ml   0.1 ml   3
10^{-5} dilution

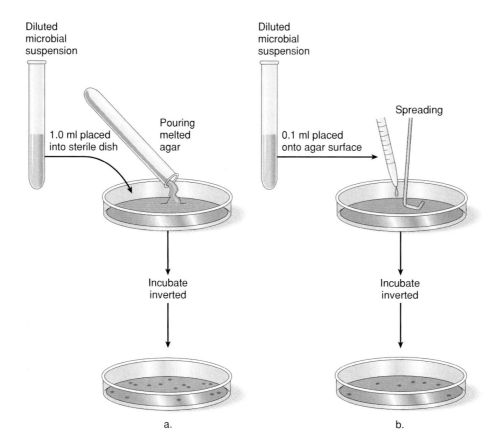

a.   b.

figure 5.5

Pour and Spread Plate Techniques for Enumerating Microorganisms

a. In the pour plate technique, a known volume of a diluted microbial suspension is placed into a sterile dish and melted agar is poured into the dish. After incubation, the number of colonies in the agar is counted, and the concentration of microorganisms in the original suspension is calculated. b. In the spread plate technique, a known volume of diluted microbial suspension is placed onto a nutrient agar medium. The suspension is spread evenly on the agar surface with a

bent glass rod that has been sterilized by dipping in alcohol and flaming. The agar medium is incubated, and the number of colonies that develop on the agar is counted to determine the concentration of microorganisms in the original suspension.

Microbial Metabolism and Growth
Growth: Growth Curves • p. 31

viable count results are usually expressed as colony-forming units (cfu) per milliliter of suspension rather than as cells per milliliter.

Since the number of bacteria in a heavy suspension easily approaches 10^9 cells/ml, dilution of the original cell suspension prior to placing the bacteria on agar media is often necessary to reduce sufficiently the number of cells per milliliter in order that growth on the plate will not be too sparse or too crowded. Ideally, plated dilutions of the culture result in 30 to 300 colonies on an agar plate. The growth of fewer than 30 colonies on a plate results in a statistically invalid count, and more than 300 colonies on a plate can cause interference among cells and colonies during growth. The dilution of bacteria from a suspension is performed aseptically, using diluents of water, saline, or buffered saline (dilution blanks). Buffered saline is

sometimes required as a diluent because some bacteria (for example, *Neisseria gonorrhoeae*) lyse in the absence of a buffered solution. By counting the number of colonies appearing on the agar plate after a period of incubation, the researcher can extrapolate and determine the number of living cells in the original suspension.

There are two versions of the viable count: the **spread plate** and the **pour plate** (Figure 5.5). A spread plate is prepared by spreading a small volume of diluted cell suspension onto the surface of an agar plate with a glass rod bent into the shape of an L. The plate is then inverted (to prevent condensation on the agar surface, which would result in confluent growth) and incubated, after which colonies are counted. A pour plate is prepared by mixing diluted bacteria into a sterile Petri dish. Melted agar (at a tem-

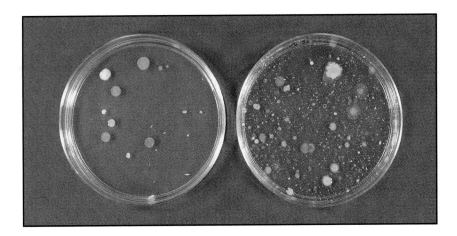

figure 5.6
Examples of Pour Plates, Showing Distinct Colony Morphologies

perature of 45°C to 50°C) is added and the mixture is allowed to solidify at room temperature. The plate is inverted and incubated for 24 to 48 hours; then colonies are counted.

There are advantages and disadvantages to each of these techniques for viable counts. Pour plates can be used to count larger volumes of bacteria (1 milliliter or more), particularly those in viscous samples such as food suspensions or milk. However, some microorganisms (for example, psychrophiles and many mesophiles) are killed at the temperature of melted agar. The pour plate is also more time-consuming than the spread plate, since tubes of melted agar must be prepared and maintained at a constant temperature of 45°C to 50°C, usually in a water bath, before being poured into Petri dishes. Spread plates have the disadvantage that bacteria that are unable to tolerate large amounts of oxygen grow poorly on their surfaces. Although spread plates could be incubated anaerobically, this is not practical.

The viable count technique is accurate because each colony theoretically is derived from one viable cell in the suspension. Nonviable cells do not form colonies on the agar medium and are therefore not included in the count. The technique is more sensitive than direct microscopic counts, since suspensions containing small quantities of bacteria can be counted. Furthermore, because each colony on the agar plate is derived from a single cell, mixed populations of bacteria in the original suspension are often discernible through differences in colony morphology and pigmentation (Figure 5.6). Single isolated colonies from these plates can also be used to obtain a pure culture of a bacterium.

 **Microbial Metabolism and Growth**
Growth: Growth Curves • pp. 29, 31

Are there any disadvantages to the viable count procedure? The technique is more time-consuming than direct microscopic counts. Agar plates and dilution blanks must be prepared prior to examining a suspension for bacteria. Since bacteria growth is a necessary prerequisite for colony formation, the viable count enumerates only those microorganisms capable of reproduction in the growth medium and under the environmental incubation conditions. Some bacteria such as *Mycobacterium* grow slowly on agar

media, even under optimal conditions. Other bacteria such as *Treponema* and *Chlamydia* cannot be propagated on artificial media. Despite these limitations, the viable count remains as one of the most important techniques for the enumeration of bacteria.

The Number of Viable Cells in a Population Can Be Determined by Filtration

The viable count technique is not feasible for the enumeration of bacteria in samples that contain very low concentrations of cells (for example, water or air samples). In these situations, bacteria are concentrated by filtration onto a gridded nitrocellulose filter (Figure 5.7). The filter, with minute pores (0.22 or 0.45 µm in diameter), traps the bacteria and is then placed onto agar in a Petri dish and incubated. Colonies appearing on the filter are counted (see filtration of water samples, page 591).

The Most Probable Number Method Can Be Used to Estimate the Number of Viable Cells in a Population

Some microorganisms do not grow well on solid media or are more easily identified by growth in liquid media. The **most probable number (MPN)** method for liquid media provides an alternative to the viable count for enumeration of microbes (Figure 5.8). The MPN technique is based on the statistical distribution of organisms in a suspension. When samples of fixed volume are taken from a cell suspension, some may have more cells than others. Repeated samples from the suspension, however, will yield an average number of organisms, termed the *most probable number*.

For example, consider a 100-ml suspension with a total of 100 cells. Ten 10-ml aliquots from such a suspension will yield an average of ten organisms per sample (some aliquots will contain more organisms and others will have fewer). If each of these 10-ml aliquots is inoculated into broth, all of the broth tubes would be expected to show growth after incubation. In a similar manner, each 1-ml sample from the same suspension is expected to average one organism per sample. Many of these samples will contain one organism or more and will yield growth when inoculated into broth.

figure 5.7

Membrane Filtration Technique for Enumerating Microorganisms

a. A gridded membrane filter is placed on a filter holder and clamped in position. A water sample is poured into the funnel and passed through the filter with the aid of a vacuum pump. b. The filter is placed onto agar or a nutrient medium-impregnated absorbant pad in a Petri dish and incubated. c. Colonies appearing on the filter are counted to determine the concentration of microorganisms in the water sample.

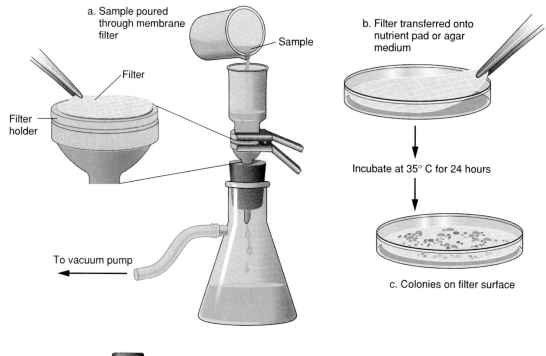

a. Sample poured through membrane filter

Sample

Filter

Filter holder

To vacuum pump

b. Filter transferred onto nutrient pad or agar medium

Incubate at 35° C for 24 hours

c. Colonies on filter surface

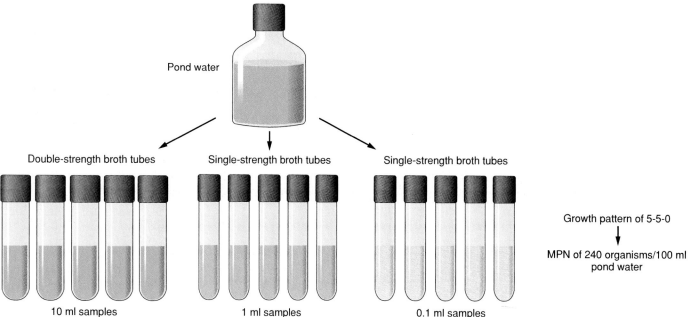

Pond water

Double-strength broth tubes

Single-strength broth tubes

Single-strength broth tubes

Growth pattern of 5-5-0

MPN of 240 organisms/100 ml pond water

10 ml samples

1 ml samples

0.1 ml samples

figure 5.8

Most Probable Number Technique for Estimating Numbers of Microorganisms

A sample of pond water is examined for bacterial density by the most probable number (MPN) technique. Five tubes of double-strength broth (10 ml per tube) are inoculated with 10-ml samples of the water. Five tubes of single-strength broth (10 ml per tube) are inoculated with 1-ml samples of the water, and five more tubes of single-strength broth (10 ml per tube) are inoculated with 0.1-ml samples. The broth tubes are incubated and examined for growth. The growth pattern is compared with a standard MPN table, and the "most probable number" of bacteria in the original pond water sample is determined. As an example, if the pattern growth were 5-5-0 (all of the broth tubes inoculated with 10 ml and 1 ml of pond water were turbid and indicated growth, and none of the broth tubes inoculated with 0.1 ml

of the pond water were turbid), then the MPN for the sample would be 240 organisms/100 ml pond water.

Other volumes of sample can be used for the MPN determination. For example, if the pond water were believed to contain very few bacteria, samples of 100, 10, and 1 ml would be used for inoculation of broth tubes or bottles. In this case, the MPN table would be used to estimate cell numbers in the original pond water, but all MPN index values would be multiplied by a factor of 0.1 (to compensate for the difference in sample volumes).

The table shown on page 113 is based on the inoculation of five tubes of broth with each dilution of sample, but other MPN tables are available for three-tube series of inoculations.

Number of Tubes Giving Positive Reaction			MPN Index per 100 ml	Number of Tubes Giving Positive Reaction			MPN Index per 100 ml
a	b	c	100 ml	a	b	c	100 ml
0	0	0	<2	4	2	1	26
0	0	1	2	4	3	0	27
0	1	0	2	4	3	1	33
0	2	0	4	4	4	0	34
1	0	0	2	5	0	0	23
1	0	1	4	5	0	1	30
1	1	0	4	5	0	2	40
1	1	1	6	5	1	0	30
1	2	0	6	5	1	1	50
				5	1	2	60
2	0	0	4				
2	0	1	7	5	2	0	50
2	1	0	7	5	2	1	70
2	1	1	9	5	2	2	90
2	2	0	9	5	3	0	80
2	3	0	12	5	3	1	110
				5	3	2	140
3	0	0	8				
3	0	1	11	5	3	3	170
3	1	0	11	5	4	0	130
3	1	1	14	5	4	1	170
3	2	0	14	5	4	2	220
3	2	1	17	5	4	3	280
				5	4	4	350
4	0	0	13	5	5	0	240
4	0	1	17	5	5	1	300
4	1	0	17	5	5	2	500
4	1	1	21	5	5	3	900
4	1	2	26	5	5	4	1,600
4	2	0	22	5	5	5	≥1,600

Note: a = Number of broth tubes showing growth after inoculation with 10-ml samples.

b = Number of broth tubes showing growth after inoculation with 1-ml samples.

c = Number of broth tubes showing growth after inoculation with 0.1-ml samples.

From *Standard Methods for the Examination of Water and Wastewater,* 19th edition. Copyright 1995 by the American Public Health Association, the American Water Works Association, and the Water Environment Federation. Reprinted with permission.

figure 5.8 (*continued*)
Most Probable Number Technique for Estimating Numbers of Microorganisms

However, a few samples will have no organisms and, when inoculated into broth, will yield no growth. Now if several 0.1-ml samples are removed from the original 100-ml suspension, there will be an average of one organism per every ten samples (the majority of the samples will contain no organisms and will result in no growth when inoculated into broth). On the basis of this principle, the number of organisms in a cell suspension can be estimated by the pattern of growth in broth culture of several different dilutions of the sample.

The MPN technique, based on successive dilutions of the original cell suspension until there are no bacteria in the final dilution used to inoculate into broth, is not as accurate as the viable count method. It is used most often to estimate the number of coliform bacteria in water samples (see processing of water samples, page 590). Coliform bacteria are gram-negative aerobic or facultatively anaerobic non-spore-forming rods that ferment lactose with gas production within 48 hours at 35°C. The technique allows not only bacterial enumeration, but also presumptive identification of the bacteria by their growth in liquid media. Only certain types of bacteria will grow in the media used for the MPN technique, which depends on the biochemical activities of the bacteria. In addition, the MPN technique is used in the enumeration of algae, which do not form distinct colonies on solid media.

Turbidimetric Measurements Detect Bacteria in a Solution by Light Scattering

Although measurements of bacterial numbers in populations provide useful information, these measurements are often time-consuming. Turbidimetric measurements are based on the principle that particles such as bacteria in a solution scatter light when a beam of light is transmitted through the solution. The number of bacteria in the solution is directly proportional to the amount of light scattered (and inversely proportional to the amount of unscattered light transmitted through the solution).

Turbidity is measured by absorbance—or, as it is often called, optical density (OD)—using a spectrophotometer or photometer (Figure 5.9). A spectrophotometer measures the amount of light absorbed by a specimen. Absorbance (A) is the logarithm of the ratio of the light intensity striking the suspension (I_0) to that passing through the specimen (I):

$$A = \log(I_0 / I)$$

where absorbance is directly proportional to the concentration of the suspension.

Measurements of turbidity are rapid and easy to perform, and provide an estimate of cell numbers and cell mass. If several similar aliquots are removed from a culture at the same time, then turbidimetric measurements, total viable cell counts, and cell mass can be determined simultaneously. By plotting the values obtained for these growth parameters on a semilogarithmic graph, a standard curve can be generated. This standard curve, which remains the same under identical conditions of growth, can be used to derive cell counts or cell mass indirectly from turbidimetric data (Figure 5.10).

There are limitations to turbidity measurements. Particles other than cells in the suspension can scatter or absorb light and

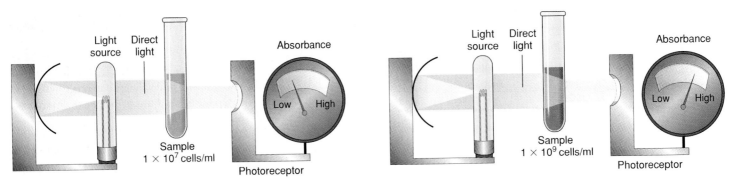

figure 5.9
Turbidity Measurements with a Spectrophotometer
The turbidity of a cell suspension is measured by light transmission through it. Light transmission is detected by a photoreceptor and is read on a meter as absorbance.

may therefore mask bacterial turbidity. Turbidimetric measurements do not differentiate between viable cells and nonviable cells. A large number (greater than 10^6 cells/ml) of bacteria must be present in the sample to provide enough turbidity for measurement. When interpreted with a standard curve, however, turbidimetric measurements provide a rapid and simple method to estimate bacterial population densities.

 **Microbial Metabolism and Growth**
Growth: Growth Curves • p. 32

Cell Mass Can Be Determined by Weighing Cells

Cell mass can be measured directly by determining the dry weight of a cell suspension. Moisture content varies in any given cell, and dry weight (the weight of cells minus any moisture) is preferred over wet weight in measurements of cell mass. Dry weight is determined with a known volume of cell suspension that is washed to free the cells of extraneous materials (for example, media components), dried in an oven, and then weighed. If the number of cells in the original volume is known (by total cell count), the average mass of an individual cell can be calculated.

Cell Growth Can Be Measured by Changes in Cell Activity

Sophisticated methods have been developed in recent years to measure bacterial growth through changes in cell activity. All living cells must synthesize macromolecules (DNA, RNA, and proteins) during growth. Increases in amounts of these molecules as a function of growth can be monitored. Changes in nucleic acid and protein quantities are determined by measuring the actual amounts of these macromolecules through chemical analyses. Alternatively, the incorporation of radiolabeled precursors into newly formed nucleic acids or proteins can be used to follow cellular growth (Figure 5.11).

Most bacteria take up oxygen and release carbon dioxide during growth. Changes in the levels of these gases can be measured with an oxygen meter and electrode to measure cellular growth (Figure 5.12). However, these measurements are subject to changes in environmental barometric pressures and furthermore cannot be performed with anaerobic organisms, which do not require oxygen for growth.

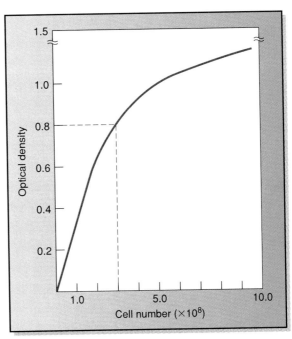

figure 5.10
Derivation of Cell Count from Turbidity Measurements
Optical density readings can be converted to cell numbers by using a standard curve in which optical density units are plotted against the viable cell count. In the example shown, an optical density reading of 0.8 corresponds to a cell count of 3×10^8.

Which of these many different procedures is the procedure of choice to measure bacterial population growth? Each method has advantages and disadvantages (Table 5.2). Measurements of environmental samples normally involve viable cells (viable counts, membrane filter counts, and MPN), since the primary interest is in the number of viable bacteria present in the environment. In many industrial applications, where optimal product development is of primary concern, cell mass or synthesis of certain products is used to monitor growth. Laboratory research experiments often use macromolecular synthesis as a measurement of growth, particularly when there is interest in the effects of chemical or inhibitory agents on growth and metabolism.

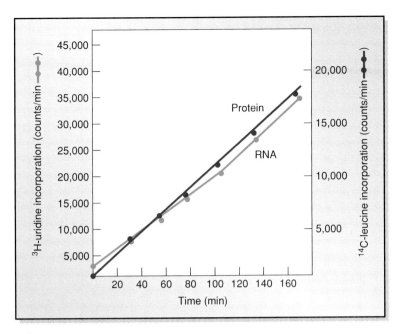

figure 5.11
RNA and Protein Synthesis During Microbial Growth
The growth of microorganisms can be measured by RNA synthesis (determined by the incorporation of radiolabeled uridine into newly synthesized RNA) or by protein synthesis (determined by the incorporation of a radiolabeled amino acid, such as leucine, into newly synthesized protein). As growth occurs, increased amounts of radioactive uridine and leucine appear in the newly synthesized RNA and protein.

Growth Curve

Bacterial population growth occurs in a series of distinct steps referred to as a **population growth curve.** A bacterial population growth curve is used to characterize changes in a cell population during growth. There are four distinct phases in a closed system, or **batch culture:** lag phase, exponential (sometimes called logarithmic) phase, stationary phase, and death (or decline) phase (Figure 5.13). The growth curve is plotted with cell mass, numbers, or content against time. Under identical conditions of incubation, media, and inoculum, a specific bacterium will always generate the same growth curve. Although not all cells in a culture divide at the same time, the rapid growth rate of most bacterial populations in combination with the vast number of cells in the population result in a common growth curve for the entire population.

There Is No Increase in Cell Numbers During the Lag Phase of Growth

Growth usually does not commence immediately when bacteria are inoculated into a new growth medium. In many cases, the bacteria must first synthesize enzymes required for growth under the new conditions. This period is called the **lag phase** of growth. During this phase, there is no increase in cell numbers. However, this is not

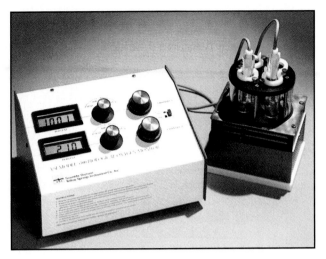

a.

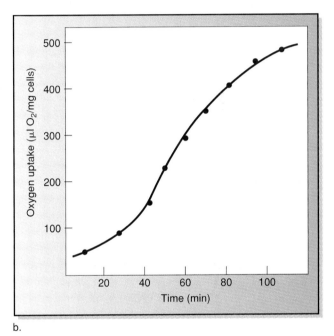

b.

figure 5.12

Determination of Dissolved Oxygen Concentration in a Solution

The concentration of oxygen in a solution can be determined with an oxygen electrode coated with an oxygen-permeable membrane. As oxygen diffuses across the membrane, it alters the conductivity. Changes in oxygen concentration are indicated on the monitor, which reads from 0% to 100% oxygen saturation. The quantity of oxygen consumed (or evolved) by microorganisms in the solution can be determined by standard conversions and is expressed as $\mu l\ O_2$/hour or $\mu l\ O_2$/mg cells. a. Oxygen monitor and electrode. b. Oxygen uptake curve.

table 5.2

Summary of Methods for Measurement of Microbial Growth

Method	Use	Limitations
Direct microscopic count	Rapid laboratory enumeration of cell suspension	Requires large number (≥10^6 cells/ml) of cells for accuracy
Viable count	Enumeration of viable cells in water, milk, and other products	Time-consuming, requires proper medium for growth
Membrane filtration	Enumeration of bacteria from water, milk, and other products, especially when numbers are low	Time-consuming, requires proper medium for growth
Most probable number (MPN)	Enumeration of bacteria from water, milk, and other products	Time-consuming, requires proper medium for growth, provides indirect estimate of numbers
Turbidimetric measurement	Rapid estimation of cell density in a suspension	Does not differentiate between viable and nonviable cells, provides only estimate of cell density, requires >10^6 cells/ml
Dry weight determination	Determination of cell mass for industrial or laboratory applications	Time-consuming, does not differentiate between viable and nonviable cells
Cell activity measurement	Research applications to follow cell metabolism	Involves extensive preparatory time

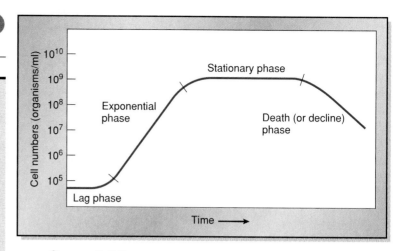

figure 5.13

Population Growth Curve

A population growth curve consists of four distinct growth phases: lag phase, exponential phase, stationary phase, and death, or decline, phase. Each growth phase reflects changes in the environment and metabolism of the cells.

Microbial Metabolism and Growth
Growth: Growth Curves • pp. 33–34

increase the synthesis of certain enzymes to produce the precursors needed for biosynthesis.

Not all bacterial populations exhibit a lag phase. This phase is absent when an exponentially growing culture is transferred to a medium of the same chemical composition and incubation conditions. Under such conditions, the transferred cells have all of the requisite enzymes and metabolic pathways to begin exponential growth in the fresh medium immediately.

The Population Grows at a Constant Rate During the Exponential Phase of Growth

The **exponential,** or **logarithmic, phase** of the growth curve is that period in which the population is actively growing at a constant rate. This growth results in a doubling of the population per generation time. A semilogarithmic plot, against time, of growth during the exponential phase yields a straight line from which the generation time can easily be derived. The exponential phase of growth cannot continue indefinitely.

Growth Slows and a Steady State in Cell Numbers Is Reached in the Stationary Phase of Growth

As nutrient supplies are depleted and toxic waste products accumulate, growth begins to slow, and soon thereafter the bacteria enter the **stationary phase.** Here, although growth and cell division may still occur, a steady state in cell numbers is reached—there is neither an increase nor a decrease in cell numbers. The net result is a plateau of the growth curve that represents the maximum population dividing under the defined growth conditions for a bacterial species.

a dormant period; the bacteria are actively synthesizing enzymes in preparation for the next, exponential phase of growth (Figure 5.14).

The length of the lag phase depends on many factors, including environmental conditions (temperature, pH, and presence or absence of oxygen) and growth medium constituents. For example, a shift of cells from a nutrient-rich growth medium to one that is deficient in a number of nutrients may result in an extended lag phase as the bacteria adapt to the new environment and synthesize enzymes necessary for metabolic pathways. In environments that lack nutrients present in the former environment, the cell must

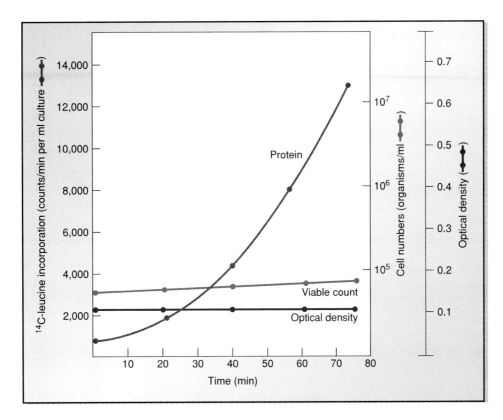

figure 5.14

Protein Synthesis and Viable Count During the Lag Phase of Growth

There is a continuous increase in protein synthesis (measured by the incorporation of radiolabeled leucine into newly synthesized protein) but no increase in cell numbers or optical density during the lag phase of growth. The example here is *Escherichia coli.*

There Is a Net Decrease in Viable Cell Numbers in the Death Phase of Growth

In many populations the steady state is eventually replaced by a **death phase,** or **decline phase.** Here, more cells die than are replaced by new cells, and there is a net decrease in certain parameters, especially viable cell numbers, used to measure cell growth.

There are several explanations for the death phase of the growth curve. Some bacteria (for example, *Streptococcus* and *Lactobacillus*) produce acidic compounds as by-products of cellular metabolism. These acids lower the pH of the medium, resulting in cell death. Growth of such bacteria in media may be extended by buffering the medium to maintain a neutral pH. Some bacteria synthesize autolytic enzymes as they age. These enzymes cause cell lysis. *Streptococcus pneumoniae*, for example, produces an autolytic enzyme, *N*-acetylmuramyl-L-alanine amidase, during the latter stages of growth. The enzyme cleaves the linkage between *N*-acetylmuramic acid and L-alanine in the cell wall peptidoglycan, resulting in lysis of the cell. *N*-acetylmuramyl-L-alanine amidase is also activated by surface-active agents (for example, bile salts and sodium deoxycholate). This characteristic is used in a test, the bile solubility test, commonly used to identify *S. pneumoniae*. In the bile solubility test, addition of sodium deoxycholate to a growing culture of *S. pneumoniae* results in a decrease in the turbidity of the culture with time.

Regardless of the manner of cell death, the population growth curve is directly affected by environmental conditions and composition of the medium. Although conditions in the microenvironment of a bacterium may mimic a batch culture, nature is not a closed system. Growth of bacteria in nature is limited by nutrient availability and accumulation of toxic waste products.

 **Microbial Metabolism and Growth**
Growth: Growth Curves • pp. 33–35

Special Techniques for Culture

Up to this point we have talked about batch cultures. In batch cultures, exponential growth continues for relatively few generations until nutrients are exhausted or toxic products accumulate. However, it is desirable in some instances to maintain exponential, or balanced, growth for a longer period of time. This is especially true if we wish to examine the effects of environmental factors on bacterial growth.

Exponential Growth Can Be Prolonged Through Continuous Culture

Bacteria can be maintained in the exponential phase in a growth curve through the use of **continuous cultures.** In continuous culture, the culture volume and the growth rate are maintained at a constant level by the addition of fresh medium and the removal of an equal amount of used medium and old cells. In a continuous culture system, fresh medium is supplied to a fixed-volume culture vessel, and spent medium is removed through an overflow tube. The culture is said to be in **steady state** when the system reaches an equilibrium point where cell numbers and nutrient levels in the culture vessel become constant.

There are two types of continuous culture systems that are commonly used in microbiology. A **chemostat** is a system in which growth is controlled by the flow rate of the system and the concentration of a limiting nutrient (Figure 5.15). In such a system, the growth rate is controlled by the rate at which nutrients are added to the growth vessel, and the population density is controlled by nutrient limitation. The investigator adjusts both parameters to meet the requirements for the system. A **turbidostat,** in contrast, contains a light-sensing device that measures the turbidity of the culture. The flow rate of fresh medium is automatically adjusted to maintain the desired turbidity level.

The chemostat and turbidostat have similar operating principles. As fresh medium is added to the culture vessel and spent medium and excess cells are removed through the overflow tube, the bacteria in the culture vessel are maintained at a constant growth rate. This is because bacteria in the culture vessel grow at a rate just fast enough to replace the cells that are removed through the overflow tube. If the flow rate is increased, there is a concomitant increase in removal of spent medium and cells. As a consequence, the population grows at a fast rate (up until a point at which it cannot keep up with the flow rate and the culture is "washed out" of the vessel). If the flow rate is decreased, the growth rate of the culture also decreases. In instances where the limiting nutrient is the energy source for the culture, there is a minimum amount of energy (maintenance energy) that must exist in the culture vessel for cells to carry out nongrowth functions such as regulation of ion balance, motility, and accumulation of substrates against a concentration gradient. The cell density in a chemostat is controlled by regulating the concentration of a limiting nutrient in the culture medium. The fresh growth medium used in a chemostat contains an excess of all but one nutrient. As this nutrient is depleted by the bacteria and is replenished at a constant rate, cell density stabilizes.

Continuous culture systems have widespread applications in microbiology. They are used to examine the effects of substrates and environmental factors (the limiting nutrient in the system) on population growth. These systems also are used in industry, providing information on optimal growth conditions for microorganisms in the synthesis of commercial products. Geneticists use continuous cultures to select mutant strains of bacteria. It is possible to select for a population of cells that can outgrow all other cell

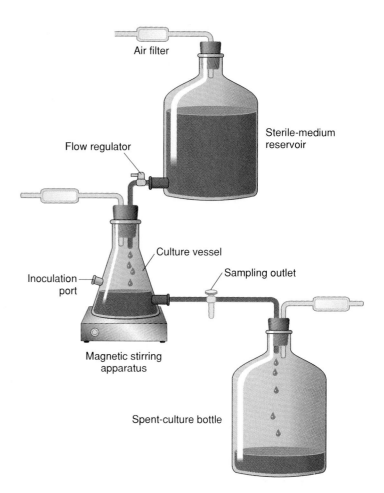

figure 5.15

A Chemostat Used for Continuous Culture of Bacteria
A chemostat is used to maintain microbial populations in continuous culture. It consists of a reservoir of sterile culture medium, a flow regulator to control the rate at which fresh medium is added to the culture vessel, the culture vessel, and a spent-culture container to collect the used culture medium.

types under certain conditions simply by varying the growth conditions in a chemostat. Continuous culture has proven to be a valuable tool in the study of microbes.

Synchronous Cultures Are Cell Populations That Are All in the Same Stage of Growth

The typical population growth curve, with its distinct phases, represents the combined growth pattern of a large number of individual cells, each one of which is growing and dividing without regard to the others. Some experimental work requires information on the growth behavior of individual cells. This information can be obtained through **synchronous cultures.** These cultures consist of populations of cells that are all in the same stage of growth—virtually every cell in the population grows and divides at the same time. This synchrony results in a steplike growth curve that plateaus each time the population divides (Figure 5.16). Synchronous populations rapidly lose their synchrony, because cells of

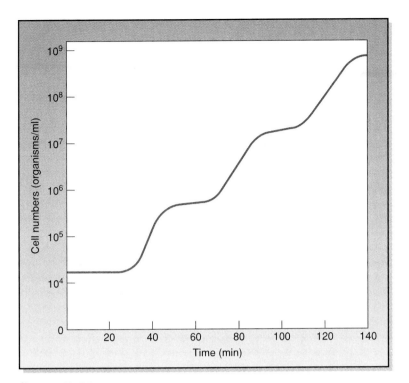

figure 5.16

Synchronous Growth Curve

Synchrony can be maintained for only a few generations. Eventually cells divide at different times, and synchrony is lost.

a population do not all age at the same rate, eventually resulting in different division times.

Synchronous growth may be achieved by altering the environment (temperature or nutrient supply) so that all cells in the population complete cellular growth and begin division simultaneously. For example, cells can be shifted to a low temperature at which they grow to full size, but do not divide. Upon a return to their optimal growth temperature, the cells will undergo a synchronized division. Synchronous growth may also be obtained by the filtration of a cell population so that organisms of the same size, and thus of the same stage of growth, are collected on the filter. These bacteria then can be inoculated into a medium to obtain a synchronized culture.

The value of synchronous culture is obvious. Bacteria are too small to study as individual cells, so a synchronous population of bacteria that are all in the same phase of growth, differentiation, or cellular organization can serve as a model of a single cell.

Control of Microbial Growth

Growth of microorganisms is influenced by environmental factors such as temperature, pH, water availability, and the presence or absence of oxygen (see physical and gaseous requirements of microorganisms, page 89). These factors not only affect growth of microbes in their natural environment, but also can be used to control or eliminate growth in the laboratory. For example, removal of oxygen in an environment can control growth of an aerobic bacterium such as *Mycobacterium tuberculosis.* Prior to the

discovery of antibiotics, treatment of tuberculosis included artificial collapse of the lung to limit the amount of oxygen available to the bacterium.

Sterilization is the killing or removal of all living organisms, a necessary prerequisite for culture media. Unsterilized media are not suitable for cultivation of microorganisms, because the media may contain unwanted organisms that mask the growth of the desired microbe.

High Temperature Can Be Used to Remove All or Most Microorganisms

Different methods are used for sterilization. The most common is **autoclaving.** An autoclave is, in essence, a large pressure cooker (Figure 5.17). Unlike boiling, which kills vegetative cells but not endospores, an autoclave kills all living organisms by a combination of high temperature and moist heat, or steam. A sterilization temperature of 121°C is obtained by increasing the pressure within the instrument to 15 lb/sq inch at sea level (higher pressures are required at higher altitudes, where barometric pressures are lower, to reach the temperature of 121°C). The air within the autoclave chamber must be replaced with saturated, flowing steam during the sterilization process to achieve a temperature of 121°C. This temperature reached during autoclaving is sufficient to kill even bacterial endospores, given an adequate time of exposure. The time required for sterilization by autoclaving depends on the volume of media and materials being processed. Large volumes require longer sterilization periods to allow heat penetration. Small volumes of media (test tube volumes and small flask volumes) are autoclaved routinely for 15 minutes.

An alternative to the autoclave is fractional sterilization **(tyndallization),** developed by the British physicist John Tyndall (1820–1893) in 1877. In tyndallization, the solution to be sterilized is heated to 100°C for 30 minutes in the presence of flowing steam on three successive days. After each cycle of heating, the material is incubated, usually at 37°C, until the next day. In the first heating cycle, all vegetative cells are destroyed. Bacterial endospores are not killed, but they germinate into vegetative cells as a result of the heat shock. These cells are subsequently killed during the second day of heating. The material is heated a third time as a precautionary measure. Tyndallization is a more tedious process than autoclaving but uses less expensive equipment and can be used for liquids that are degraded at the higher autoclave temperature. It is rarely used today, but was used in the past as an alternative to the autoclave.

Pasteurization is the process of treating liquids with heat below the boiling point to kill most pathogenic and spoilage microorganisms. Pasteurization was devised by Louis Pasteur (1822–1895) as a method to control microbial contamination of wine. It is commonly used to kill pathogenic bacteria such as *Mycobacterium tuberculosis, Brucella abortus, Salmonella,* and *Streptococcus* in milk and other beverages. Pasteurization also reduces the total bacterial population in these products and extends shelf life. The dairy industry uses two methods to pasteurize milk: the low-temperature holding (LTH) method, in which

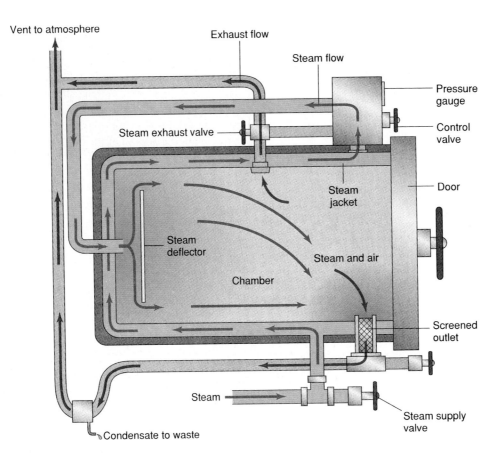

Vent to atmosphere

Exhaust flow

Steam flow

Pressure gauge

Control valve

Steam exhaust valve

Door

Steam jacket

Steam deflector

Steam and air

Chamber

Screened outlet

Steam

Steam supply valve

Condensate to waste

figure 5.17

An Autoclave
During autoclaving, air is forced out of the chamber by incoming steam. The steam pressure increases to 15 lb/sq in, and the temperature rises to 121°C. When sterilization is completed, the steam exhaust valve opens and steam flows out of the chamber. Liquids are autoclaved with slow steam exhaust to prevent their evaporation and boiling over.

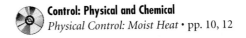

Control: Physical and Chemical
Physical Control: Moist Heat • pp. 10, 12

milk is heated to 62.8°C for 30 minutes, and the high-temperature short-time (HTST), or flash pasteurization, method which exposes milk to a temperature of 71.7°C for 15 seconds (see pasteurization of milk, page 608) .

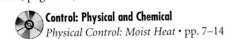

Control: Physical and Chemical
Physical Control: Moist Heat • pp. 7–14

Other Methods Are Effective in Killing or Removing Microorganisms

Some chemical compounds (for example, amino acids, vitamins, and enzymes) are sensitive even to moderate levels of heat and cannot be sterilized by high temperatures. These substances are sterilized by **filtration** through membrane filters with pore sizes of 0.22 or 0.45 μm. The pores in the filters are small enough to retain most bacteria (bacteria without cell walls, such as *Mycoplasma*, are pliable and may squeeze through these filters), but they do not retain the smaller viruses. Because viruses are unable to grow without suitable

hosts, they seldom present problems in sterilization of bacteriological media by filtration. However, these viruses can be a major problem in filter-sterilized media used for tissue culture (culture of human, animal, or plant cells for the propagation of viruses).

Air is commonly filtered to remove microorganisms. **High-efficiency particulate air (HEPA) filters** capable of removing 99.97% of particles 0.3 μm and larger are placed in laminar flow biological safety cabinets to provide a curtain of sterile air within the cabinet. This prevents contamination of the work space from outside the cabinet and also protects the laboratory worker from any escape of dangerous microorganisms inside the cabinet. Laminar flow biological safety cabinets are used when one is working with infectious agents such as tumor viruses (see tumor viruses, page 408), *Mycobacterium tuberculosis* (see tuberculosis, page 510), and *Legionella pneumophila* (see legionellosis, page 505), and recombinant DNA (see recombinant DNA, page 272).

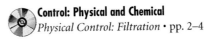

Control: Physical and Chemical
Physical Control: Filtration • pp. 2–4

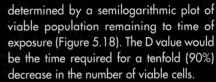

Heat is a useful sterilizing procedure, but its effectiveness can vary depending on the procedure used. One method for measuring the effectiveness of heat sterilization is the **thermal death time (TDT),** the shortest time required to kill all of the microorganisms in a suspension at a specific temperature. The TDT is determined by heating aliquots of a microbial suspension for different periods of time and mixing the heated suspensions with a culture medium to determine the absence or presence of growth from any surviving cells.

Another method for measuring heat inactivation is to determine the **decimal reduction time (DRT),** or **D value,** the time required to kill 90% of the microorganisms in a suspension at a specific temperature. The DRT of a bacterium can be determined by a semilogarithmic plot of viable population remaining to time of exposure (Figure 5.18). The D value would be the time required for a tenfold (90%) decrease in the number of viable cells.

Measurement of Heat-Killing Efficiency

The effectiveness of heat sterilization depends on several factors, including the types and initial concentrations of microorganisms in the population, the nature of material to be treated, pH, and the presence of organic matter. Endospore-forming bacteria such as *Clostridium botulinum* are more resistant to heat sterilization than vegetative cells of *Staphylococcus aureus* and *Salmonella enteritidis.* Sterilization of a powder may require a different temperature and time of exposure than sterilization of a liquid. High-acid foods (for example, sauerkraut, pickles, and strawberry preserves) can be heat sterilized at lower temperatures than low-acid foods (for example, carrots, peas, and beans), partially because most microorganisms do not survive in high-acid environments and those that do have lower thermal resistance. Materials heavily covered with organic matter may be partially protected from the effects of heat sterilization.

The D value is used extensively by the canning industry to determine the shortest time required for effective heat sterilization in canning operations. Because excessive exposure to heat can alter the texture and flavor of foods, it is important to know the shortest time and lowest temperature necessary to kill food-borne pathogens.

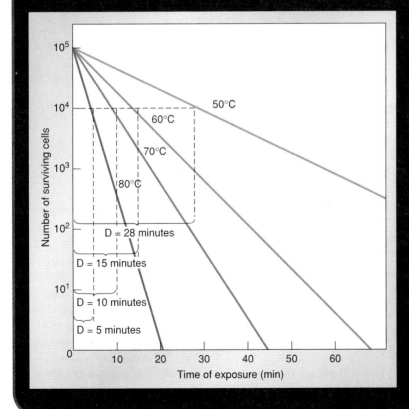

figure 5.18
Effect of Temperature on the D Values for a Bacterium
The D values are shown for a bacterium at four different temperatures. Note that the D values decrease as the temperature increases.

Dry heat is frequently used for the sterilization of glassware and laboratory equipment. Microbial cells are apparently killed by oxidation of their constituents and protein denaturation in dry heat sterilization. The process requires a longer period of time than autoclaving, since flowing steam is not used. Incineration is one form of dry heat sterilization. Inoculating loops are sterilized by incineration in the flame of a Bunsen burner or in an electric heating coil. Animal carcasses, hospital and laboratory infectious wastes, and disposable gowns, bedding, bandages, syringes, and needles are incinerated. Glassware and other similar objects can be sterilized by placing them in an oven at 160°C to 170°C for two to three hours. Although dry heat sterilization is a slower process than autoclaving, it has the advantages of not corroding metal instruments and not introducing moisture into powders and oils. A disadvantage of dry heat sterilization is that it cannot be used to sterilize objects or materials that would be affected by high temperatures.

Control: Physical and Chemical
Physical Control: Dry Heat • pp. 5–6

Ionizing radiation with gamma rays is used for the sterilization of pharmaceuticals and disposable medical supplies such as plastic syringes and surgical gloves. Gamma rays from cobalt 60 have high penetration for solids and liquids and destroy microorganisms by the formation of hydroxyl free radicals and peroxides from water and other substances (see hydroxyl free radicals and peroxides, page 93). Ionizing irradiation has been used to reduce bacterial populations in foods such as strawberries, apples, poultry, and fish, thereby increasing their shelf lives.

Nonionizing radiation, or **ultraviolet radiation,** is not as penetrating as ionizing radiation and is only effective on microorganisms that are fully exposed such as those on a surface or in a thin layer of water or other liquid. This type of radiation does not penetrate glass, paper, or solids. At a wavelength of 260 nm, ultraviolet radiation damages DNA by the formation of pyrimidine dimers that distort the structure of DNA and interfere with replication and transcription (see pyrimidine dimers, page 246). Ultraviolet lamps are sometimes used in laboratories, hospital rooms, operating rooms, and biological safety cabinets to sterilize the air and any exposed surfaces, and in aquariums to reduce microbial populations in circulating water. Because prolonged exposure to ultraviolet radiation can damage the eyes and burn the skin, its use is carefully monitored.

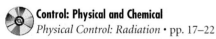

Control: Physical and Chemical
Physical Control: Radiation • pp. 17–22

Other methods for control of microbial growth include refrigeration, deep freezing, dehydration, osmotic pressure, and the use of chemicals such as sodium benzoate, calcium propionate, boric acid, and sorbic acid for food preservation. These methods used to preserve foods and reduce spoilage are further discussed in Chapter 20 (see food preservation, page 614).

Antimicrobial Agents

Antimicrobial agents are chemicals that inhibit or destroy microbial growth. Some of these agents are **cidal** (**algicidal, bactericidal,** and **fungicidal**) and kill the microbe (but not necessarily their

table 5.3

Common Disinfectants and Antiseptics

Class	Examples	Mode of Action
Phenol and phenolic compounds	Carbolic acid, hexachlorophene, cresol, phenol, orthophenylphenol	Disruption of plasma membrane through protein denaturation
Alcohols	Ethanol, isopropanol (50%–70% aqueous solution)	Lipid solvents and protein denaturants
Surfactants	Quaternary ammonuim compounds	Disruption of plasma membrane through charge interactions with phospholipids
	Soaps, detergents	Disruption of plasma membrane through charge interactions with lipoproteins
Halogens	Iodine (tincture of iodine, povidone-iodine)	Reaction with tyrosine to inactivate proteins
	Chlorine	Oxidizing agent
Alkylating agents	Formaldehyde, glutaraldehyde, ethylene oxide	Denaturation of proteins and nucleic acids through attachment of methyl or ethyl groups to these molecules
Heavy metals	Mercury (merbromin, nitromersol, thimerosal), silver (1% silver nitrate), copper (copper sulfate)	Protein denaturant

spores). The term **germicide** often is used to collectively describe cidal agents. Other antimicrobial agents are **static** (**algistatic, bacteriostatic,** and **fungistatic**) and reversibly inhibit growth. A bacteriostatic agent is effective only as long as it is in the immediate vicinity of the susceptible bacterium. If the bacteriostatic agent is removed, the bacterium may resume growth. Some bacteriocidal agents are also **bacteriolytic** and kill cells by cell lysis. Agents that inhibit cell wall synthesis (for example, penicillin) or damage the plasma membrane (for example, polymyxin) are bacteriolytic.

Disinfectants Are Used on Inanimate Objects, Whereas Antiseptics Are Used on External Body Surfaces

Antimicrobial agents may be divided into four distinct categories: **disinfectants, antiseptics, antibiotics,** and **synthetic drugs.** Disinfectants are chemical compounds that destroy disease-causing

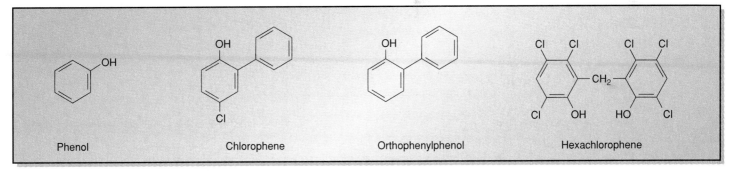

figure 5.19
Structures of Some Phenolic Disinfectants

microorganisms and their products on inanimate objects; they are distinguished from antiseptics, which are used on external body surfaces (Table 5.3). Antiseptics are less toxic than disinfectants and usually inhibit, but do not always kill, microorganisms.

A convenient way to measure the germicidal action of a chemical is to compare its killing action to a known standard. The **phenol coefficient** of a compound is the ratio of its effectiveness to that of phenol (the standard) against a test organism. For example, if a standard suspension of *Staphylococcus aureus* is killed by a 1:2,500 dilution of hexachlorophene but requires a 1:20 dilution of phenol to cause the same results, the phenol coefficient of hexachlorophene is 2,500/20 or 125.0. This phenol coefficient means that hexochlorophene is 125 times more effective than phenol as a germicide in killing *S. aureus* in vitro. The phenol coefficient provides a reasonable index of germicidal activity, but it is more accurate for phenolic compounds (which have chemical properties and modes of action similar to phenol) than for other agents.

Phenolic compounds, alcohols, surface-active agents (**surfactants**), halogens, alkylating agents, and heavy metals are examples of disinfectants and antiseptics. Some of the more important chemicals are described in the following sections.

Phenol Causes Membrane Damage and Leakage of Cell Cytoplasmic Contents

Phenols are compounds that have a hydroxyl group attached directly to an aromatic ring (Figure 5.19). Phenols have long been recognized as effective disinfectants—since 1865, when Joseph Lister (1827–1912) used phenol to disinfect surgical instruments and dressings. Lister discovered that phenol applications dramatically reduced the incidence of postsurgical infections. A 5% phenol solution effectively kills all vegetative bacteria and many spores by causing membrane damage and leakage of cell cytoplasmic contents. Cresol, a methylphenol that is a distillation product of coal tar, is used as a preservative and in some soaps and detergents. In the past, one of the most widely used phenolic derivatives was hexachlorophene, which consists of two phenol groups. Hexachlorophene is usually mixed with soaps and is particularly effective against gram-positive bacteria such as *S. aureus*, a common source of infections in newborn infants. It has been discovered, however, that hexachlorophene penetrates

the skin and can cause neurological damage; therefore its use is more restricted today.

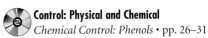

Control: Physical and Chemical
Chemical Control: Phenols • pp. 26–31

Alcohols Are Lipid Solvents that Disrupt Plasma Membrane Lipids and Denature Proteins

Alcohols are lipid solvents that kill cells by disrupting the lipids in the plasma membranes and also by denaturing proteins. Ethanol and isopropanol are widely used disinfectants that are most effective at a concentration of approximately 70%. Isopropanol alcohol is slightly more effective as a germicidal agent than ethanol, is less expensive, and is not subject to legal regulations. For these reasons, isopropanol is used for such purposes as the disinfection of thermometers.

Control: Physical and Chemical
Chemical Control: Alcohols • pp. 17–21

Surfactants Disrupt the Plasma Membrane by Reacting with Membrane Lipids

Surface-active substances (surfactants) are compounds that have both hydrophilic (water-attracting) and hydrophobic (water-repelling) groups that reduce surface tension and the miscibility of molecules. Surfactants have widespread applications as emulsifiers, soaps, detergents, and cleansing agents. The most important surface-active agents are quaternary ammonium compounds, cationic (positively charged) compounds with positively charged groups that react with the negatively charged phosphate groups of the membrane phospholipids to disrupt the plasma membrane (Figure 5.20). Cationic detergents are frequently used to disinfect food and beverage utensils in restaurants, equipment in food-processing plants, and floors and walls in hospitals and nursing homes. Anionic (negatively charged) organic molecules affect the lipoprotein portion of the plasma membrane, are milder than cationic agents (because they are repelled by the net negative charge of the bacterial surface), and therefore are often used as soaps and detergents. The cleansing action of soaps and detergents in everyday use, particularly by restaurant workers and food handlers, plays an important role in reducing potential food-poisoning outbreaks.

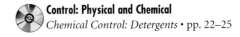

Control: Physical and Chemical
Chemical Control: Detergents • pp. 22–25

Benzethonium chloride

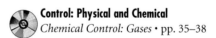

Benzalkonium chloride

Cetylpyridinium chloride

figure 5.20
Structures of Some Surfactants

Halogens Are Strong Oxidizing Agents

The halogens (chlorine, fluorine, bromine, and iodine) are strong oxidizing agents that react with enzymes, proteins, fats, or other cellular molecules to disrupt cell function. Iodine and chlorine are among the most widely used disinfectants. A 2% solution of iodine in sodium iodide and dilute alcohol (tincture of iodine) is commonly used as an antiseptic for treating minor cuts, abrasions, and wounds. Povidone-iodine (Betadine) is an iodine-detergent mixture (iodophore) that is frequently used as an alternative to tincture of iodine, especially for children because it does not irritate the skin or sting as much when applied to open wounds. Iodine inactivates many proteins by altering their structure through iodination of the amino acid tyrosine.

Chlorine compounds are potent disinfectants for swimming pools, water supplies, and hot tubs. When elemental chlorine or a chlorine compound is added to water, the chlorine reacts with the water to form hypochlorous acid, a powerful oxidizing agent. Because chlorine's activity is reduced by the presence of organic matter, which combines with chlorine, it is customary to add sufficient amounts of chlorine to water so that the free chlorine concentration is 0.5 to 1.0 parts per million (ppm); this ensures effective disinfection.

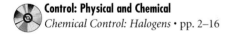

Control: Physical and Chemical
Chemical Control: Halogens • pp. 2–16

Alkylating Agents Generally Attach Methyl or Ethyl Groups to Cellular Molecules Such as Proteins and DNA, Making Them Nonfunctional

Formaldehyde, glutaraldehyde, ethylene oxide, and β-propiolactone are alkylating agents that often are used as preservatives and disinfectants. Alkylating agents generally attach methyl or ethyl groups to proteins and DNA, rendering these molecules nonfunctional and causing death of the microorganism. Formaldehyde is used primarily as a preservative in embalming. Formalin, a 37% solution of formaldehyde gas, is used for the preservation and fixation of tissue. It is also used in the preparation of vaccines, because it destroys microorganisms but does not affect their anti-

genic properties. Glutaraldehyde is more effective and less toxic than formaldehyde as a disinfectant and has been used increasingly for sterilizing surgical instruments.

Control: Physical and Chemical
Chemical Control: Aldehydes • pp. 32–34

Ethylene oxide (EtO) readily penetrates paper, cloth, and certain types of plastics and is used to sterilize disposable plastic Petri dishes, syringes, catheters, and sutures. Ethylene oxide has wide applications in the food industry as a preservative for spices, dried fruit, and nuts. It also has been used by hospitals to sterilize instruments and surgical equipment and by NASA to decontaminate spacecraft components. The gas is released into a tightly sealed chamber containing the items to be sterilized for three to four hours in a temperature-controlled and humidity-controlled environment. Because EtO is toxic if inhaled, and explosive, it is normally mixed with an inert gas (Freon or carbon dioxide) to reduce the danger of its use. After sterilization, the chamber is extensively flushed with an inert gas to remove any residual EtO.

Control: Physical and Chemical
Chemical Control: Gases • pp. 35–38

β-propiolactone (BPL) is occasionally used as an alternative to EtO. BPL is less explosive and destroys microorganisms more readily than EtO, but has poor penetration capabilities and may be carcinogenic. Therefore it is not used as a sterilizing agent as often as EtO.

Heavy Metals Bind to the Sulfhydryl Groups of Proteins, Resulting in Protein Denaturation

Heavy metal ions of mercury, silver, and copper are toxic to microorganisms because they bind to the sulfhydryl groups of proteins and denature them. Organic mercuric compounds such as merbromin (Mercurochrome), nitromersol (Metaphen), and thimerosal (Merthiolate) are often used for sterilization of instruments, skin antisepsis, and irrigation of the urethra. An eyedrop solution containing 1% silver nitrate ($AgNO_3$) is sometimes used to prevent an eye disease called ophthalmia neonatorum, which is often caused by *Neisseria gonorrhoeae* and occurs in infants deliv-

table 5.4

Major Categories of Chemotherapeutic Agents

Mode of Action	Examples
Inhibition of cell wall synthesis	Penicillins, cephalosporins, vancomycin, bacitracin, oxacillin, nafcillin
Damage to plasma membrane	Polymyxin, nystatin, amphotericin B
Inhibition of protein synthesis	Streptomycin, kanamycin, gentamicin, amikacin, neomycin, chloramphenicol, erythromycin, tetracyclines
Inhibition of nucleic acid synthesis	Rifamycins, actinomycin D, nalidixic acid, ciprofloxacin, norfloxacin
Structure analogs	Sulfonamides

ered through an infected birth canal. Copper sulfate ($CuSO_4$) is useful as an algicide. A 1-ppm concentration of copper sulfate is usually sufficient to control the obnoxious and odoriferous algal blooms that often occur in lakes and reservoirs.

Antibiotics and Synthetic Drugs Are Used in Chemotherapy to Selectively Inhibit or Kill Microorganisms Without Harming Host Tissue

Unlike disinfectants and antiseptics, which are applied on nonliving materials or on external tissue, antibiotics and synthetic drugs can be taken internally. The term *antibiotic* was originally used by Selman Waksman (1888–1973) to describe any antimicrobial chemical substance produced by microorganisms. Waksman, who received the Nobel Prize in physiology and medicine in 1952 for his discovery of streptomycin, was a microbiologist who searched for antimicrobial agents in bacteria. Today many antimicrobial agents are produced synthetically or semisynthetically. These synthetic chemicals are called synthetic drugs to distinguish them from the naturally produced antibiotics.

To be useful, antimicrobial agents must have **selective toxicity.** The agent must inhibit or kill the microorganism without damaging the cells of the animal host. Although many natural and synthetic chemicals kill microorganisms, only a much smaller group has selective toxicity and can be used as **chemotherapeutic agents.** Chemotherapeutic agents vary in their range of activity. **Narrow-spectrum** agents are effective against only specific microorganisms or groups of microorganisms. Bacitracin and gentamicin are examples of narrow-spectrum drugs that are targeted specifically at gram-positive bacteria and gram-negative bacteria, respectively. **Broad-spectrum** agents such as chloramphenicol, streptomycin, and tetracycline are effective against a relatively wide range of microorganisms. Physicians sometimes prescribe

broad-spectrum drugs when time is of the essence in initiating chemotherapy for a potentially lethal infectious disease in which the etiologic agent has not yet been identified. With the emergence of drug-resistant bacteria in recent years, concern has been raised about the overuse of antimicrobial agents and the initiation of treatment before diagnosis of the disease. It is now recommended that cultures be performed and antimicrobial susceptibilities be determined before the use of antimicrobial agents.

Chemotherapeutic agents may be classified by their mechanism of action. These mechanisms include: (1) the inhibition of cell wall synthesis, (2) damage to the plasma membrane, (3) the inhibition of protein and nucleic acid synthesis, and (4) the inhibition of cell metabolism (Table 5.4). Some examples of chemotherapeutic agents in each category are given in the following sections.

Penicillins Inhibit Biosynthesis of the Cell Wall Peptidoglycan

Some of the more common antibiotics used today are those that inhibit the biosynthesis of the cell wall peptidoglycan structure. The peptidoglycan gives strength and rigidity to the cell wall, and antibiotics such as the penicillins, cephalosporins, vancomycin, bacitracin, oxacillin, and nafcillin interfere with the synthesis of this important structure (Figure 5.21).

The penicillins are among the most widely used class of antibiotics in the world. They inhibit the transpeptidation step of peptidoglycan synthesis (see peptidoglycan synthesis, page 56) and are effective only on microorganisms actively synthesizing peptidoglycan. All of the penicillins have a common basic structure: a thiazolidine ring joined to β-lactam ring. Different forms of penicillin have different side groups attached to the basic structure. One of the original natural penicillins isolated from microorganisms is benzylpenicillin (penicillin G). Penicillin G is effective against gram-positive organisms and gram-negative cocci, but can cause hypersensitive reactions. Penicillin G is usually administered intramuscularly because it is readily inactivated by gastric acids. Phenoxymethyl penicillin (penicillin V) is more resistant to acids and therefore, unlike penicillin G, can be given orally.

Many of the natural and synthetic penicillins are sensitive to the action of β-**lactamase (penicillinase),** a plasmid-coded enzyme produced by many clinically significant bacteria, including *Staphylococcus aureus* and *Neisseria gonorrhoeae* (see β-lactamase, pages 527 and 534). β-lactamase hydrolyzes the β-lactam ring of penicillin and renders the antibiotic ineffective. Because penicillinase is an extracellular enzyme, its production by certain bacteria in an area of the body may destroy penicillin that is designated for use against other, non-penicillinase-producing bacteria in the general vicinity. For this reason, penicillinase-producing microorganisms can represent significant problems in the clinical management of a disease. Some of the semisynthetic penicillins (for example, oxicillin or nafcillin) that have bulky side groups next to the β-lactam ring are resistant to the action of penicillinase and are effective against penicillinase-producing bacteria.

Antimicrobial Action
Cell Wall Inhibitors • pp. 1–19

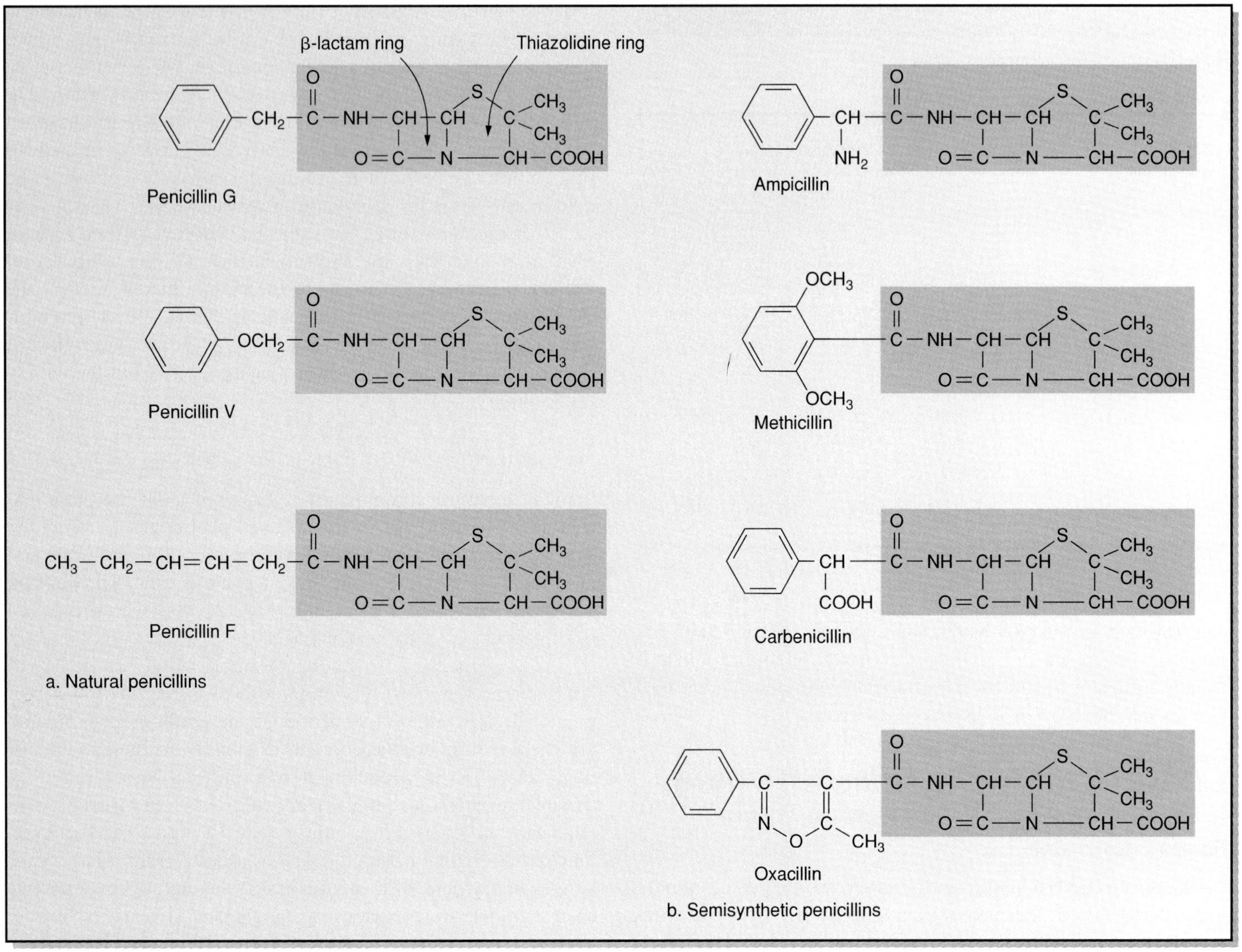

figure 5.21

Structures of Some Common Penicillins

The shaded areas represent the common nucleus; the unshaded areas are the side chains that are specific for each type of penicillin.

Damage to the Plasma Membrane Results in Leakage of Metabolites and Other Cytoplasmic Constituents

The plasma membrane is an important microbial structure that regulates the transport of substances into and out of the cell and is also the site of respiratory electron transport and other cellular activities. Antibiotics that damage or impair the function of the plasma membrane are toxic for microorganisms. Polymyxin, nystatin, and amphotericin B are examples of these antibiotics (Figure 5.22).

Polymyxin is a natural polypeptide synthesized by *Bacillus polymyxa*. It binds to the outer surface of plasma membranes to disrupt membrane structure, causing leakage of metabolites. However, because polymyxin is toxic to tissue, its use is reserved for serious infections caused by *Pseudomonas aeruginosa,* a bacterium

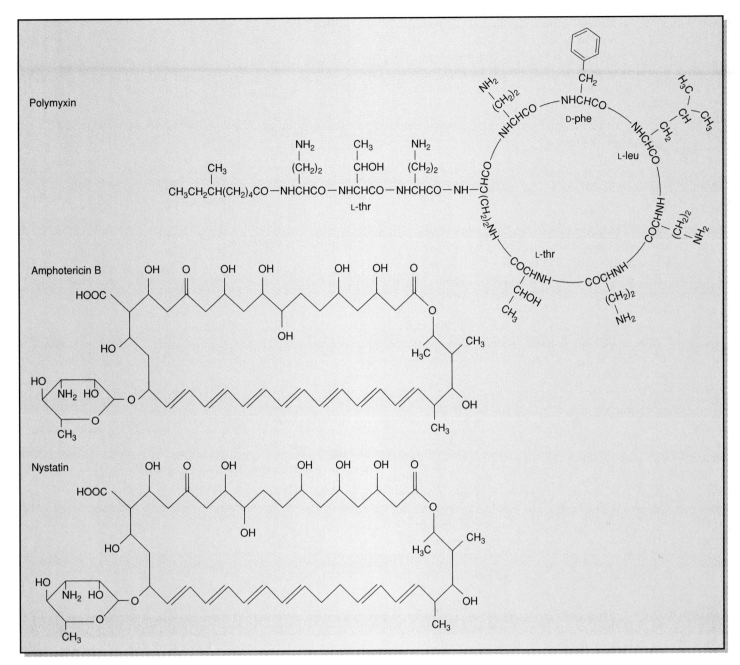

figure 5.22
Polymyxin, Amphotericin B, and Nystatin

that is resistant to most other antibiotics, or for topical application to prevent minor skin wound infections.

Nystatin and amphotericin B are polyene antibiotics that have large ring structures with many double bonds. The polyenes selectively inhibit organisms containing sterols in their membranes and therefore are used for the treatment of fungal infections.

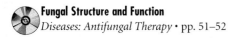

Fungal Structure and Function
Diseases: Antifungal Therapy • pp. 51–52

Inhibitors of Protein and Nucleic Acid Synthesis Disrupt Cell Function

Antibiotics that inhibit protein and nucleic acid synthesis seriously impair cell function. The aminoglycosides (streptomycin, kanamycin, gentamicin, tobramycin, amikacin, and neomycin) are a large class of antibiotics that have an aminocyclitol ring and one or more amino sugars (Figure 5.23). The aminoglycosides impair procaryotic ribosome function by interacting with the 30S ribosomal subunit and causing, in many cases, misreading of the messenger RNA (Table 5.5). These antibiotics are bactericidal and have a broad spectrum, but they are not effective against obligate anaerobes. Apparently, anaerobiosis affects assimilation of aminoglycoside antibiotics into the cell.

Antimicrobial Action
Protein Synthesis Inhibitors: Mechanisms • pp. 4–5

The tetracyclines are antibiotics that inhibit bacterial growth by binding to the small (30S) ribosomal subunit and interfering with the attachment of aminoacyl-transfer RNA to the ribosome. Unlike the aminoglycosides, tetracyclines bind to procaryotic and eucaryotic ribosomes, but they are more selective for procaryotes because eucaryotic membranes are less permeable to the antibiotic.

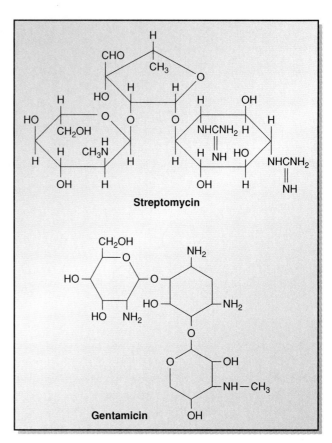

figure 5.23
Structures of Representative Aminoglycosides

Antimicrobial Action
Protein Synthesis Inhibitors: Protein Inhibition Mechanisms • pp. 4–5

---

table 5.5

Inhibitors of Protein Synthesis

Antimicrobial Agent	Ribosome Subunit Affected		Protein Function Inhibited			
	30S	50S	Initiation	Codon Recognition	Peptide Bond Formation	Translocation
Streptomycin	+		+	+		
Kanamycin	+					+
Gentamicin	+					+
Amikacin	+					+
Neomycin	+			+		
Chloramphenicol		+			+	
Erythromycin		+			+	+
Tetracycline	+			+		

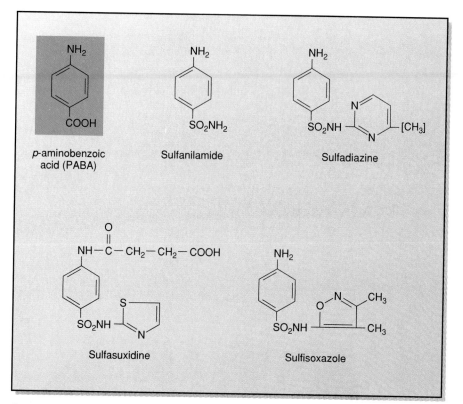

figure 5.24
Sulfonamides
The sulfonamides are structurally similar to *p*-aminobenzoic acid (PABA), a precursor of folic acid.

Chloramphenicol is an antibiotic that binds to the 23S rRNA on the 50S ribosomal subunit and inhibits peptidyl transferase, thereby preventing peptide bond formation. Because chloramphenicol can be neurotoxic, it is used sparingly and generally in situations when no other antibiotic is effective. Both chloramphenicol and the tetracyclines are broad-spectrum bacteriostatic agents that are effective against many gram-positive and gram-negative bacteria.

Erythromycin, an antibiotic that binds to the 50S ribosomal subunit, blocks peptide bond formation and translocation. Erythromycin is generally bacteriostatic and is effective against gram-positive bacteria and some gram-negative bacteria. It is frequently used as an alternative to penicillin in patients who are allergic to penicillin or who are infected with penicillin-resistant organisms.

Antimicrobial Action
Protein Synthesis Inhibitors • pp. 1–9

Erythromycin belongs to a class of antibiotics called macrolides, in which a lactone ring is attached to one or more sugars. Erythromycin is produced by a strain of the bacterium *Streptomyces erythraeus* that was originally isolated from soil in the Philippines. It is the preferred drug for treating legionellosis, mycoplasma pneumonia, and whooping cough. Although erythromycin is the most widely used macrolide, newer semisynthetic macrolides such as clarithromycin and azithromycin have been developed.

The rifamycins are antibiotics that inhibit bacterial DNA-dependent RNA polymerase of bacteria. Rifampicin, a synthetic derivative of the naturally occurring rifamycin B, binds to the β subunit of RNA polymerase and inhibits the initiation of RNA synthesis. Rifampicin is a broad-spectrum drug that is especially effective against mycobacterial infections.

Actinomycin D, an antibiotic produced by *Streptomyces antibioticus*, binds between guanine-cytosine base pairs on DNA and blocks RNA chain elongation in both procaryotes and eucaryotes. Nalidixic acid, ciprofloxacin, and norfloxacin are synthetic drugs called quinolones, which act by binding to DNA gyrase (topoisomerase), the enzyme involved in the unwinding of DNA. Gyrase activity is affected by the binding, and DNA replication is inhibited.

Antimicrobial Action
Protein Synthesis Inhibitors • p. 8

Sulfonamides Are Structure Analogs That Structurally Resemble and Compete with Metabolites in Cellular Enzymatic Reactions

Some of the first antimicrobial agents used in chemotherapy were the sulfonamides. There are numerous sulfonamides, but all are derivatives of *p*-aminobenzenesulfonamide (Figure 5.24). Sulfonamides are **structure analogs**—chemical compounds that structurally resemble cellular metabolites and compete with these metabolites in cellular enzymatic reactions. Sulfonamides are structurally similar to *p*-aminobenzoic acid (PABA), a precursor of folic acid. Folic acid is a vitamin that is synthesized by most bacteria. Sulfonamide competes with folic acid synthesis, thereby blocking formation of this vitamin, which is involved in purine and pyrimidine synthesis. Sulfonamides are selectively toxic for certain bacteria for two reasons: (1) Some bacteria are impermeable to folic acid, and therefore exogenous folic acid cannot enter and (2) mammals cannot synthesize folic acid and must be provided with preformed folic acid as a vitamin in their diets. Therefore mammalian cells can survive sulfonamide because they receive preformed folic acid into their cells, whereas folic acid-impermeable bacteria, in the presence of sulfonamide, cannot make the folic acid they need.

Antifungal and Antiprotozoan Chemotherapy Represents Special Challenges

Treatment of human fungal and protozoan infections with antimicrobial agents represents a special challenge because the infecting eucaryotic pathogens share structural, metabolic, and molecular characteristics with human cells. Fewer drugs are available to treat fungal and protozoan diseases than are available for procaryotic diseases, and some of the drugs that are used are potentially toxic to human tissue.

Amphotericin B and nystatin, produced by *Streptomyces*, are polyene antibiotics that are among the most effective antifungal

The War Against Antibiotic-Resistant Microbes

The discovery of penicillin and other antibiotics in the early 1900s established a new era of miracle drugs that appeared to give humans a marked edge over bacteria in their seemingly endless host-parasite war. Infectious diseases such as pneumonia, tuberculosis, and gonorrhea that once were the scourge of humanity and affected millions of people could now be effectively controlled by magic bullets bearing names like penicillin, tetracycline, and rifamycin. By the 1970s, it appeared that complete victory over many once deadly diseases was just a matter of time.

However, in the early 1970s, this scenario began to dramatically change. In 1976, penicillin-resistant *Neisseria gonorrhoeae* harboring β-lactamase genes appeared in the United States. Since then, the percentage of these resistant strains has increased in some years to more than 10% of all gonorrhea cases reported, making treatment more difficult and expensive (Figure 5.25). Tetracycline-resistant strains of *N. gonorrhoeae* also began to appear in 1986. Other bacterial pathogens have become more resistant to antibiotics as well (Figure 5.26). *Enterococcus*, which causes bladder infections, endocarditis, and septicemia, has developed resistance to many antibiotics including vancomycin. Most strains of *Staphylococcus aureus*, which causes a wide variety of diseases ranging from pyoderma to toxic shock syndrome, are now penicillin resistant; an increasing number are also methicillin and gentamicin resistant. In fact, methicillin-resistant *S. aureus* (MRSA) have become a major problem in nosocomial, or hospital-acquired, infections and have become resistant to most other antibiotics. The potential nightmare, which may become reality in a few years, is a supermicrobe resistant to all known antibiotics.

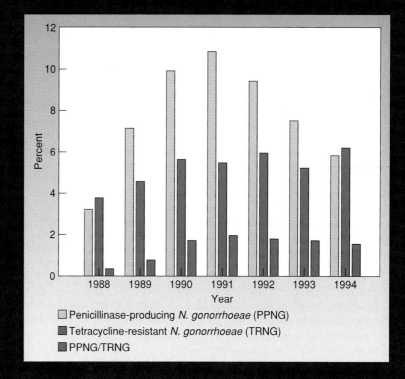

figure 5.25

Percentage of *Neisseria gonorrhoeae* Strains Resistant to Penicillin and/or Tetracycline

What has caused this recent emergence of antibiotic-resistant strains? The indiscriminate use of antibiotics in the 1950s and 1960s to treat upper respiratory tract infections before diagnosis (most upper respiratory tract infections are caused by viruses, not by bacteria) has contributed to selection of drug-resistant strains. Patients who do not complete their entire regimen of prescribed antibiotics add to the problem. When antibiotic therapy is ended early, drug-resistant strains can develop. The widespread use of antibiotics in animal feed to increase weight gain in cattle, pigs, and chickens is a contributing factor. In recent years, it has been discovered that some bacteria can transfer plasmid-borne antibiotic-resistant genes to other bacteria. There is evidence of the transfer of the β-lactamase gene between *S. aureus* and other bacteria.

Fortunately, new drugs developed by pharmaceutical companies help shift the balance of scales in this seesaw biological battle on the side of humanity. However, as bacteria continue to evolve new mechanisms to evade drugs, it becomes increasingly important for physicians and scientists to maintain their vigilance against antibiotic-resistant microbes.

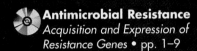 **Antimicrobial Resistance**
Acquisition and Expression of Resistance Genes • pp. 1–9

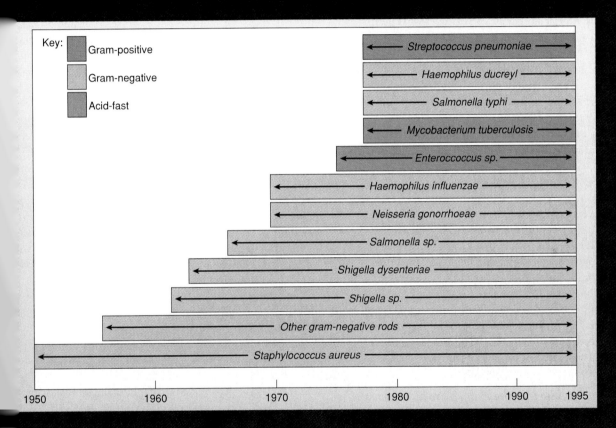

Figure 5.26

Emergence of Antibiotic Resistance in Bacterial Pathogens Since the Beginning of Antibiotic Therapy

medicines. Both drugs bind to sterols in fungal membranes, causing alterations in membrane permeability and leakage of cell constituents. Amphotericin is given intravenously for the treatment of cryptococcosis, histoplasmosis, coccidioidomycosis, and candidiasis (see fungal respiratory tract diseases, page 512; and fungal oral diseases, page 515), but frequently causes toxic side effects including kidney damage and anemia, and therefore its use is closely monitored. Nystatin is applied as a topical ointment or suppository and is useful in the treatment of candidiasis.

Griseofulvin, produced by *Penicillium,* is an antimicrobial agent that is used for ringworm, athlete's foot, and other fungal infections of the skin, nails, and hair (see fungal skin diseases, page 530). Griseofulvin is taken orally and is absorbed from the intestinal tract into the bloodstream, where it is carried to the deep skin layers and accumulates in keratinous tissue (epidermis, nails, and hair). The antibiotic is thought to disrupt the mitotic spindle and to inhibit fungal cell division.

Flucytosine is a synthetic derivative of cytosine that is useful in the treatment of cryptococcosis and candidiasis. Fungi deaminate the compound to yield fluorouracil, a highly toxic substance that prevents normal nucleic acid synthesis.

Many protozoa have several stages in their life cycles (see protozoan life cycle, page 377) and, therefore, treatment often requires different antibiotics at different stages. For example, treatment of malaria, which is caused by *Plasmodium,* involves the use of primaquine during the liver phase of infection and chloroquine during infection of red blood cells. Primaquine and chloroquine are synthetic derivates of quinine, a drug obtained from the bark of the chinchona tree. Quinine was used for many years to treat malaria until it was replaced by these less toxic derivatives. Because of the development of chloroquine-resistant strains of *Plasmodium,* quinine and a newer drug, mefloquine, have been used in recent years for the treatment of drug-resistant malarial infections.

The synthetic drug metronidazole is used to treat several protozoan diseases, including amoebic dysentery, caused by *Entamoeba histolytica;* diarrhea caused by *Giardia lamblia* (see protozoan gastrointestinal diseases, page 523); and vaginitis caused by *Trichomonas vaginalis.* Pentamidine isethionate, another antiprotozoan drug, is used to treat African sleeping sickness, caused by *Trypanosoma.*

Antiviral Drugs Take Advantage of the Unique Properties of Viruses

Viruses are unable to reproduce independently and replicate only within a living host cell (see virus replication, page 395). Since viruses use the host metabolic machinery during replication, treatment of viral diseases is difficult without affecting the host cell. There are few antiviral drugs and those that have been approved in the United States take advantage of the unique properties of viruses (Table 5.6).

Most antiviral agents are synthetic nucleosides that mimic the structures of purines and pyrmidines. These analogs block viral DNA or RNA synthesis either by inhibiting the viral polymerase or by becoming incorporated into the growing viral DNA or RNA molecule, resulting in a nonfunctional viral nucleic acid. Acyclovir (Zovirax), a structure analog of guanosine, is used in the

table 5.6

Some Examples of Antiviral Drugs

Pathogen	Disease	Drugs of Choice
Herpesvirus	Genital herpes, shingles, chickenpox keratoconjunctivitis	Acyclovir (Zovirax), idoxuridine (IudR), trifluridine, vidarabine (ara-A)
Human immuno-deficiency virus	AIDS	Azidothymidine (AZT), didanosine (ddI), dideoxycytosine (ddC)
Influenza A virus	Influenza	Amantadine

treatment of genital herpes, shingles, chickenpox, and other infections caused by herpesviruses. The drug, which is administered as an ointment, orally, or by injection, is phosphorylated by herpesvirus kinase and inhibits viral DNA polymerase, thus blocking viral DNA replication. Because acyclovir is activated specifically by herpesvirus kinase, the drug has selective antiviral activity, making it useful in chemotherapy. Idoxuridine (IUdR) and trifluridine, analogs of thymidine, and vidarabine (ara-A), an analog of adenine, are also used in the treatment of herpesvirus infections.

Azidothymidine (AZT), also called zidovudine, is an analog of thymidine that is used for treatment of acquired immune deficiency syndrome (AIDS) patients. AIDS is caused by human immunodeficiency virus (HIV), a retrovirus that synthesizes DNA from RNA by the enzyme reverse transcriptase (see AIDS, page 536). AZT inhibits reverse transcriptase, thereby blocking further DNA synthesis and viral replication. Although AZT does not cure AIDS, it does slow the course of the disease in many instances. Didanosine (ddI) and dideoxycytosine (ddC) are other nucleoside analogs that are used to inhibit reverse transcriptase.

Amantadine is a drug that specifically inhibits the replication of influenza A viruses. Because the drug blocks the penetration and uncoating of virus particles, it is usually given prophylactically to prevent disease. Use of this drug can reduce the incidence of influenza by 50% to 80%. Amantadine has some side effects, including dizziness, confusion, and insomnia.

Interferons are glycoproteins that inhibit the translation of viral mRNA. Interferons are naturally produced by mammalian cells during viral infection and may be useful in the treament of hepatitis, genital herpes, influenza, and the common cold (see interferon, page 406).

Antimicrobial Susceptibility Tests Are Used to Determine the Type and Quantity of Antimicrobial Agents Used in Chemotherapy

One of the most important functions of a hospital laboratory is to determine the antimicrobial susceptibilities of pathogens causing disease in patients. This information is used by the physician to determine the type and quantity of antimicrobial agent to use in chemotherapy. There are two methods that can be used to determine the susceptibility of a bacterium to antimicrobial agents: the **tube dilution method** and the **disk diffusion method.**

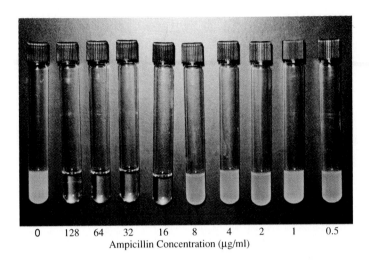

figure 5.27

Tube Dilution Method for Determining Antimicrobial Susceptibility
This series shows the growth of *Staphylococcus aureus* in tubes containing up to 8 µg/ml of the antibiotic ampicillin. There is no growth in tubes containing 16 µg/ml or more of ampicillin. The minimal inhibitory concentration (MIC) of ampicillin for *S. aureus* is thus 16 µg/ml.

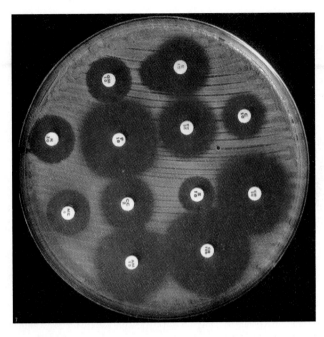

figure 5.28

Disk Diffusion Method for Determining Antimicrobial Susceptibility
Disks containing different antibiotics are placed on an agar plate inoculated with *Staphylococcus aureus*. Clear zones appear around disks with antibiotics that are inhibitory for the bacterium.

 **Antimicrobial Action**
Antibiotic Susceptibility Testing: Terminology • pp. 7–8

The Tube Dilution Method Is Used to Determine Susceptibility to Precise Quantities of an Antimicrobial Agent

In the tube dilution method, the test bacterium is inoculated into broth tubes containing serial dilutions of the antimicrobial agent (Figure 5.27). The inoculated cultures are incubated for a suitable period of time (usually 24 hours), and the presence or absence of growth is determined by the turbidity in each tube. Tubes containing moderate to high concentrations of the antimicrobial agent would normally be expected to have no growth. The **minimal inhibitory concentration (MIC)** of the antimicrobial agent for the test bacterium can be determined by the lowest concentration of the agent that produces no observable growth.

At the MIC end point, bacterial growth is inhibited, but there may still be viable bacteria in the tube. To check for viable cells, an aliquot of tubes showing no growth can be inoculated onto agar plates. The plates are incubated and then observed for growth. The **minimal bactericidal concentration (MBC)** is the concentration of antimicrobial agent in the tube that has produced no growth.

The tube dilution method is considered accurate for determining susceptibility to precise quantities of an antimicrobial agent. However, the method is time-consuming, expensive, and not practical for use in most hospital laboratories for routine susceptibility testing.

 **Antimicrobial Action**
Antibiotic Susceptibility Testing: Broth Dilution • pp. 14–16

The Disk Diffusion Method Uses Filter Paper Disks Containing Known Concentrations of Antimicrobial Agents to Determine Antimicrobial Susceptibility

A more common procedure used for susceptibility testing in hospitals is the disk diffusion method. Filter paper disks containing known concentrations of antimicrobial agents are placed onto the surface of an agar plate inoculated with the test bacterium. The plate is incubated for 16 to 18 hours, and the zones of inhibition are read around each paper disk (Figure 5.28).

During the incubation period, the antimicrobial agent diffuses through the agar, and a concentration gradient of the agent is established. At some point in this gradient, growth of the susceptible bacteria is suppressed, and no growth is observed within a circular zone around the disk. The size of the zone of inhibition is determined by the type of agar medium used, the incubation conditions, the type of antimicrobial agent used, and the susceptibility of the test organism to the diffusing agent. In 1966, William Kirby and Alfred Bauer first introduced the principle of measuring zones of inhibition around disks to determine antimicrobial agent susceptibilities, so this test is often called the **Kirby-Bauer test.** The Kirby-Bauer test is a rapid, convenient method to determine the susceptibilities of microorganisms to antimicrobial agents.

Antimicrobial Action
Susceptibility Testing: Disk Diffusion • pp. 11–13

Automated Methods for Antimicrobial Susceptibility Testing

Antimicrobial susceptibility tests are important for the clinical management of infectious diseases. Without rapid methods to determine the susceptibility of microbial pathogens to antibiotics, disease can run rampant in a host and possibly cause death.

In the past, susceptibility tests using the tube dilution or disk diffusion method required several days to perform. Today automation has decreased the processing time of antimicrobial susceptibility tests and

has significantly impacted the microbiology laboratory. Most automated instruments for susceptibility testing use one of two basic principles: (1) microbial growth is measured by turbidity or (2) microbial growth is measured by the detection of metabolites.

Instruments such as the MS-2 of Abbott Laboratories, the AutoMicrobic System (AMS) of Vitek Systems, and the Autobac IDX of General Diagnostics use turbidity

to detect microbial growth (Figure 5.29). A microorganism in pure culture is inoculated into several broth tubes containing different antimicrobial agents. Growth is then monitored turbidimetrically over a short time period (usually three to eight hours). Antimicrobial susceptibility is determined by the absence of growth in broth tubes containing the effective antimicrobial agents.

An alternative to turbidity measurements is the detection of carbon dioxide released during microbial growth and metabolism in a closed system. The BACTEC System (Becton Dickinson Diagnostic Instrument Systems) measures carbon dioxide as a product of microbial metabolic activity by infrared spectrophotometry. Any level of carbon dioxide above a preset baseline indicates microbial growth (Figure 5.30).

Both the turbidimetric and metabolic activity systems have been automated. Following inoculation of the microorganism into broths, the systems automatically read and analyze all results and provide a printout of the results within a few hours. These systems have significantly improved the ability of the microbiology laboratory to provide antimicrobial susceptibility test results to the physician within a short time.

figure 5.29
The AutoMicrobic System

Disposable test cards containing broths with different antimicrobial agents are inoculated with the microorganism. The test cards are loaded into the AutoMicrobic System Reader/Incubator (second chamber from the left). After incubation, the test cards are read and analyzed automatically, and a computer printout is produced.

figure 5.30
The BACTEC System

Vials containing metabolic substrates (for example, glucose), antimicrobial agents, and the microorganism are monitored for CO_2 production by infrared spectrophotometry.

Many bacteria, fungi, and protozoa require carbon dioxide (CO_2) for growth, but for many years it was not known why. One of the first studies to show the role of CO_2 in cell growth and metabolism was performed by Dorothy M. Tuttle and Henry W. Scherp in 1952. They examined the effects of different nutrients on the CO_2 requirement for the bacterium *Neisseria meningitidis*.

Cells of *N. meningitidis* were serially diluted and inoculated into parallel series of defined media containing different growth factors (yeast extract, purines and pyrimidines, casein hydrolysate, and vitamins) and incubated in the presence or absence of added CO_2. The incubation of bacteria under added CO_2 was achieved by placing cultures in sealed jars in an atmosphere of 3% CO_2 added from a commercial tank. Cultures were incubated at 37.5°C and then examined for the presence or absence of turbidity after 24 and 48 hours.

The results showed that larger inocula of bacteria were required to obtain growth without CO_2 than with 3% CO_2 (Table 5P.1). Furthermore, it was observed that 0.01% yeast extract or a combination of 0.03% guanine/uracil/cytosine + 0.5% casein hydrolysate + B vitamins could effectively substitute for the CO_2 requirement. Other combinations of growth factors were not as effective in promoting bacterial growth with no added CO_2.

Previous studies by G.P. Gladstone, P. Fildes, and G.M. Richardson had suggested that the CO_2 requirement of bacteria was related to cell growth; CO_2 reduced the length of the lag phase of the growth curve. Based on this information, Tuttle and Scherp surmised that the function of the growth factors in their experiments was to stimulate the production of CO_2, thereby reducing the length of the lag phase for growth of *N. meningitidis* and stimulating the growth of these bacteria. Subsequent experiments by other investigators have shown that growth factors such as those used by Tuttle and Scherp probably supply CO_2 that is necessary for CO_2-fixing reactions in the cell. The experiments of Tuttle and Scherp were instrumental in first showing that these growth factors could replace atmospheric CO_2 for the growth of bacteria.

Source:

Tuttle, D.M., and H.W. Scherp. 1952. Studies on the carbon dioxide requirement of *Neisseria meningitidis*. *Journal of Bacteriology* 64:171–181.

References

Gladstone, G.P., P. Fildes, and G.M. Richardson. 1935. Carbon dioxide as an essential factor in the growth of bacteria. *British Journal of Experimental Pathology* 16:335–348.

Griffin, P.J., and S.V. Rieder. 1957. A study on the growth requirements of *Neisseria gonorrhoeae* and its clinical application. *Yale Journal of Biology and Medicine* 29:613–621.

Wimpenny, J.W.T. 1969. Oxygen and carbon dioxide as regulators of microbial growth and metabolism. In *19th Symposium of the Society for General Microbiology, volume 19, Microbial growth,* pp. 161–197. Cambridge: Cambridge University Press.

table 5P.1

Substitutes for Supplemental CO_2 in the Growth of *Neisseria meningitidis*

Supplement	Number of Organisms Inoculated per 5 ml ($\times 2.0$)[a]							
	10^7	10^6	10^5	10^4	10^3	10^2	10^1	10^0
None	+	+	–	–	–	–	–	–
3% CO_2	+	+	+	+	+	+	+	+
0.01% yeast extract	+	+	+	+	+	+	+	+
0.03% GUC (total)[b]	+	+	+	+	c	–	–	–
0.5% casein hydrolysate	+	+	+	+	c	–	–	–
B vitamins[d]	+	+	–	–	–	–	–	–
0.03% GUC + 0.5% casein hydrolysate	+	+	+	+	c	c	c	–
0.03% GUC + B vitamins	+	+	+	+	c	c	–	–
0.5% casein hydrolysate + B vitamins	+	+	+	+	+	+	c	–
0.03% GUC + 0.5% casein hydrolysate + B vitamins	+	+	+	+	+	+	+	+

[a]+ = growth in 24 hr; – = nongrowth in 48 hr.

[b]GUC = 0.004% guanine, 0.013% uracil, and 0.013% cytosine.

[c]growth in 48 hr, but not in 24 hr.

[d]B vitamins = thiamine, pantothenic acid, pyridoxine, and nicotinic acid, 1 µg/ml; riboflavin, 0.5 µg/ml; biotin, 0.001 µg/ml; and folic acid concentrate, 0.01 µg/ml.

⬤— Summary

1. Growth is the orderly increase in all major constituents of an organism. Growth of microorganisms is usually expressed as population growth because of the small size of microbes.

2. Bacteria generally divide by binary fission. The generation time, or the doubling time, is the time required for a population of bacteria to double in number.

3. Bacterial population growth can be followed by increases in cell numbers and cell mass, turbidimetric measurements, and changes in cell activity.

4. A common procedure to monitor bacterial population growth is the viable count. The viable count technique involves diluting bacteria from the sample population and placing the diluted suspensions on an agar medium by the spread plate or pour plate procedure.

5. The most probable number (MPN) procedure estimates the number of viable cells in a sample population by successive dilutions of the sample until no bacteria are left.

6. A bacterial population growth curve in a closed system, or batch culture, consists of four phases: lag, exponential, stationary, and death, or decline. Bacteria synthesize enzymes for growth during the lag phase, but cell numbers do not increase. Bacteria actively grow at a constant rate during the exponential phase, and then enter the stationary phase as nutrient supplies are depleted and toxic waste products accumulate. The death phase occurs when more cells die than are replaced by new cells.

7. Continuous cultures are established when a culture reaches an equilibrium point wherein cell numbers and nutrient levels in the culture vessel become constant. A chemostat is an example of a continuous culture system in which growth is controlled by the flow rate of the system and the concentration of a limiting nutrient. Synchronous cultures consist of populations of cells that are all in the same stage of growth.

8. Sterilization is the killing or removing of all living organisms. An autoclave sterilizes materials by a combination of high temperature (121°C) and moist heat (steam). Pasteurization is the treatment of liquids with heat at 62.8°C for 30 minutes (low-temperature holding method) or 71.7°C for 15 seconds (high-temperature short-time method) to kill most pathogenic and spoilage microorganisms.

9. Antimicrobial agents are chemicals that inhibit or destroy microbial growth. These agents can be divided into four categories: disinfectants, antiseptics, antibiotics, and synthetic drugs. Disinfectants are used on inanimate objects, whereas antiseptics are less toxic and are used on external body surfaces. Antibiotics are naturally produced antimicrobial agents, whereas synthetic drugs are synthetic chemicals developed by the pharmaceutical industry.

10. Chemotherapeutic agents are antimicrobial agents that are selectively toxic for microorganisms and do not damage the animal host. Narrow-spectrum agents are effective against only specific microorganisms or groups of microorganisms. Broad-spectrum agents are effective against a wide range of microorganisms. Chemotherapeutic agents can inhibit peptidoglycan synthesis, protein synthesis, and nucleic acid synthesis; damage the plasma membrane; or act as structure analogs to compete with cellular metabolites in enzymatic reactions.

11. Antimicrobial susceptibility tests are useful in determining the proper antimicrobial agent to use in chemotherapy. Use of these tests on a routine basis prior to treatment can help reduce the development of antibiotic-resistant microorganisms.

EVOLUTION *and* BIODIVERSITY

The emergence of antimicrobial drug resistance in recent years has brought renewed concerns that microbial pathogens that were once thought to be nearly eradicated by successful chemotherapy may now be reappearing as more highly resistant forms. Increased microbial drug resistance has resulted in higher death rates, increased health care costs, and often more toxic drugs or drug combinations used in chemotherapy. When the first antibiotics were introduced for general use in the 1940s, it was thought that these "miracle drugs" might eradicate infectious disease. Yet today, drugs that at one time were used successfully to treat tuberculosis, gonorrhea, pneumonia, and hospital-acquired staphylococcal infections have lost their effectiveness. What has happened to cause this increased resistance? As antimicrobial agents became widely used, they were overprescribed in many instances. Physicians would administer antibiotics before identifying the disease agent, which in the case of respiratory tract infections was often a virus not affected by antibiotics. Patients would take only part of the prescribed dosage of antibiotics, not realizing that this partial treatment would only help microbes develop resistance. Antibiotics were incorporated into livestock feeds to prevent disease and promote growth of animals. As a consequence of these practices, microorganisms adapted to increasingly higher levels of antibiotics and became more resistant to these drugs. This evolution of microbes has resulted in the highly resistant strains that now present so many problems in chemotherapy and are an indication of the remarkably rapid ability of microorganisms to adapt to changes in their environments.

Short Answer

1. How do bacterial cells divide?

2. What is meant by bacterial "growth"?

3. Given g = 30 minutes, how long would it take for a cell to yield a population of 1,024?

4. List several methods for counting bacteria.

5. Which method of viable count determination is best for cultures that are anaerobic? Which is best for heat-sensitive cultures?

6. What method of enumeration would you use if you wanted:
   a. to determine the utilization of various carbon sources?
   b. to count bacteria in a relatively clean water supply?
   c. a quick answer?

7. Identify, in sequence, the phases of a bacterial growth curve. During which phase(s) is (are) there no cell division?

8. How do we know bacteria are not dormant during the lag phase?

9. Why doesn't the constant rate of growth observed during the exponential phase continue indefinitely?

10. Could exponential growth be extended artificially? Explain.

11. Which method of microbial control would you use to sterilize:
    a. surgical instruments or glassware?
    b. surfaces (for example, operating or transfer room)?
    c. culture media?

12. List several antiseptics. List several disinfectants.

13. Identify the mechanisms of action for chemotherapeutic antimicrobial agents.

14. Identify several factors which have led to the development of drug-resistant bacteria.

15. The term "antibiotic" is generally used for any drug given to control microbial growth. Why is this a misnomer?

16. Describe the Kirby-Bauer test. Explain its significance.

Multiple Choice

1. Which of the following methods of control does *not* sterilize?
   a. autoclave
   b. ionizing radiation
   c. pasteurization
   d. tyndallization
   e. All of the above sterilize.

2. Which of the following bind sulfhydryl groups to denature proteins?
   a. phenols, alcohols
   b. surfactants
   c. halogens (for example, iodine and chlorine)
   d. formalin, glutaraldehyde, ethylene oxide
   e. heavy metals (for example, copper and mercury)

3. Which of the following inhibit cellular metabolism?
   a. penicillins
   b. polymyxin, nystatin, amphotericin B
   c. aminoglycosides
   d. sulfonamides

4. Which of the following is (are) narrow-spectrum agents?
   a. chloramphenicol
   b. gentamicin
   c. streptomycin
   d. tetracycline

Critical Thinking

1. Explain why we would employ methods (for example, pasteurization and disinfection) that do *not* kill 100 percent of the organisms.

2. Compare and contrast antiseptics and disinfectants.

3. Describe the characteristics of bacteria that are exploited by antimicrobial therapy.

4. Discuss (pros and cons) the use of antimicrobial agents.

Supplementary Readings

Baron, E.J., L.R. Peterson, and S.M. Finegold. 1994. *Bailey & Scott's diagnostic microbiology*, 9th ed. St. Louis: Mosby. (A textbook of diagnostic microbiology with chapters on antimicrobial susceptibility testing.)

Davies, J. 1994. Inactivation of antibiotics and the dissemination of resistance genes. *Science* 264:375–382. (A research article describing common mechanisms of antibiotic resistance in bacteria.)

Garrod, L.P., H.P. Lambert, and F. O'Grady, eds. 1981. *Antibiotics and chemotherapy*, 5th ed. Edinburgh: Churchill Livingstone. (Detailed discussion of modes of action of chemotherapeutic agents and their uses in treatment of diseases.)

Mandelstam, J., K. McQuillen, and I.W. Dawes. 1973. *Biochemistry of bacterial growth*, 2d ed. New York: Blackwell Scientific Publications, Ltd. (A discussion of physiological and chemical changes during bacterial growth.)

Murray, P.R., E.J. Baron, M.A. Pfaller, F.C. Tenover, and R.H. Yolken, eds. 1995. *Manual of clinical microbiology*, 6th ed. Washington, D.C.: American Society for Microbiology. (An extensive clinical laboratory manual with chapters on antimicrobial susceptibility testing.)

Neidhardt, F.C., J.L. Ingraham, and M. Schaechter. 1990. *Physiology of the bacterial cell: A molecular approach*. Sunderland, Mass.: Sinauer Associates. (A well-written textbook of bacterial physiology and genetics.)

Norris, J. A., and D.W. Robbins, eds. 1969. *Methods in microbiology, volume 1*. New York: Academic Press. (A description of methods for culture of microorganisms and the various techniques used for measurement of microbial growth.)

Norris, J. A., and D.W. Robbins, eds. 1970. *Methods in microbiology, volume 2*. New York: Academic Press. (A description of continuous culture techniques, methods to optimize microbial growth, and the effects of temperature and pH on growth.)

Pirt, S. J. 1975. *Principles of microbe and cell cultivation*. New York: John Wiley & Sons. (An in-depth discussion of the effects of environmental and chemical parameters on growth and the chemistry of microbial growth.)

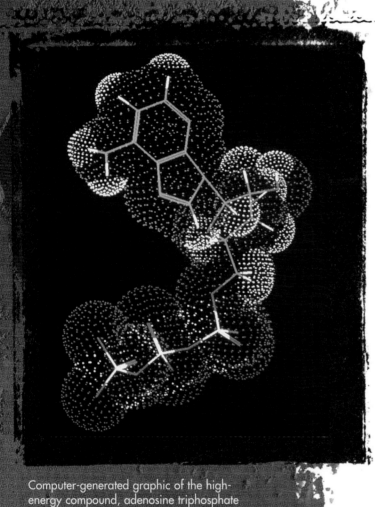

ENERGY AND METABOLISM

chapter six

Concepts of Energy
Thermodynamics
Energy: Neither Created Nor Destroyed
Entropy
Reversible Chemical Reactions
Enzymes

Coupled Reactions
Energy and Oxidation
Oxidation-Reduction Potentials
Electron Carriers

High-Energy Compounds

Concepts of Metabolism

Mechanisms of ATP Synthesis

Pathways Involving Substrate-Level Phosphorylation
Fermentation, Respiration in Chemoheterotrophs
The Embden-Meyerhof Pathway
Other Glycolytic Pathways
Pyruvate Degradation

Oxidative Phosphorylation
The Tricarboxylic Acid Cycle
The Electron Transport Chain and ATP
Chemiosmosis
Aerobic Respiration vs. Fermentation
Anaerobic Respiration: A Unique Process
ATP Yield from Glucose Oxidation

Chemolithotrophs
Hydrogen Sulfide Oxidation
Ferrous Iron Oxidation
Ammonia Oxidation
Hydrogen Gas Oxidation

PERSPECTIVE
Bacterial Luminescence

EVOLUTION AND BIODIVERSITY

Computer-generated graphic of the high-energy compound, adenosine triphosphate (ATP). The positions of the component atoms are represented by different colors: carbon (green), nitrogen (blue), oxygen (red), phosphorus (white), and hydrogen (white).

 Microbes in Motion ———— PREVIEW LINK

This chapter covers the key topics of microbial energy mechanisms—aerobic and anaerobic. The following sections in the *Microbes in Motion* CD-ROM may be useful as a preview to your reading or as a supplemental study aid:

Bacterial Structure and Function: Cell Membrane (Energy Production), 12–21. *Microbial Metabolism and Growth:* Metabolism, 1–20.

One of the most fascinating aspects of life is the ability of a living cell to oxidize simple chemical substances and release **energy** that can be used to drive all of life's processes. Even more amazing is the fact that these minute, microscopic cells can perform this transformation more rapidly and efficiently than any machine or artificial device. Like a steam engine or electric generator, the cell performs its job continuously. But unlike these manufactured devices, a cell can use energy to reproduce itself, generate other similar cells, and convert disordered chemicals from the environment into highly organized cellular constituents.

However, to do this, cells must have a constant source of energy. Energy, defined as the capacity to do work, is rarely available to the cell in the proper form to perform chemical work. Living organisms have devised a variety of ways to extract energy from their surroundings and convert it to useful forms.

Concepts of Energy

Microorganisms are dynamic, self-propagating entities that perform different types of work: mechanical, electrical, and chemical. Mechanical work is physical, such as the movement of flagella or cilia or the streaming cytoplasm of amoebae. The movement of chemical compounds or ions against a concentration gradient, such as the active transport of glucose by a bacterium, is also considered work. When the transported substance is a charged molecule, the concentration gradient is also a potential gradient, and electrical work is involved. Organisms are constantly synthesizing complex organic molecules that function as the chemical building blocks of cells; this biosynthesis is an example of chemical work.

To perform work, microbes and other living organisms must have a source of energy. Energy exists in different forms (mechanical, thermal or heat, electrical, and chemical) and can be converted, though not with perfect efficiency, from one form to another. Several units may be used to measure energy; the most frequently used unit in biological systems is the **kilocalorie (kcal)**. A kilocalorie is the quantity of heat energy required to raise the temperature of 1 kg of water 1°C.

Other than phototrophs, which derive energy from light, all other organisms obtain their energy ultimately from chemical substances. Chemical energy resides in the bonds of chemical molecules, but not all molecules have the same number of bonds or the same amount of energy.

For example, the equation

$$C_6H_{12}O_6 + 6\,O_2 \rightarrow 6\,CO_2 + 6\,H_2O$$

is balanced and has the same number and types of atoms on both sides of the equation, but the number and types of bonds differ. Consequently, the energy contained in the molecules differs. When molecules are rearranged in a chemical reaction, energy is either liberated or required. Energy-liberating processes (**exergonic reactions**) provide the cell with the energy required to perform work, which is an energy-requiring process (**endergonic reaction**). For example, ATP hydrolysis (an exergonic reaction) can be coupled with phosphorylation of glucose (an endergonic reaction):

$$ATP + H_2O \rightarrow ADP + PO_4 \qquad \Delta G = -7.3 \text{ kcal/mole}$$

$$Glucose + PO_4 \rightarrow glucose\text{-}6\text{-}PO_4 + H_2O \quad \Delta G = +3 \text{ kcal/mole}$$

This **coupling** of exergonic and endergonic reactions is fundamentally important to all cellular processes.

Energy, in addition to its different forms, can also assume two different states: **potential energy** and **kinetic energy.** Potential energy is defined as stored energy or energy of position. Kinetic energy is energy of work or motion. The energy contained within any given substance or object is initially all in the state of potential energy. This potential energy, however, is gradually converted to kinetic energy as the condition or position of the substance or object changes. An example of such change is the conversion of potential energy (energy of position) in a stationary rock at the top of a hill to kinetic energy (energy of motion) as the rock moves and rolls down the hill. Chemical molecules represent storehouses of potential energy. As these molecules are oxidized by living organisms, the potential energy stored within the chemical bonds is liberated and made available for performing work.

Thermodynamics Is the Study of Energy Transformations

Energy not only exists in different forms, but also has the ability to change from one form to another. During fuel combustion, chemical energy is converted into heat (thermal energy) and light (electromagnetic energy). Electromagnetic energy from the sun is transformed by phototrophs into chemical energy in the form of organic compounds through the process of photosynthesis. The chemical energy that is released when molecules are broken down into simpler compounds is used by living organisms for movement, metabolism, and growth. Each of these examples represents transformation of energy.

The study of energy transformations forms a branch of science known as **thermodynamics.** In thermodynamics, energy exchanges are considered within the framework of a defined system and its surroundings. A thermodynamic system is defined by such variables as temperature, volume, pressure, and chemical content.

Bioenergetics refers specifically to the study of energy transformations in living systems (Figure 6.1). Energy exchanges are especially important to living organisms, since energy is usually available to these organisms only in the form of chemical or electromagnetic energy and must first be converted to perform work. Because energy is not always readily available in a usable form, organisms must be able to store energy in forms such as chemical energy for later use.

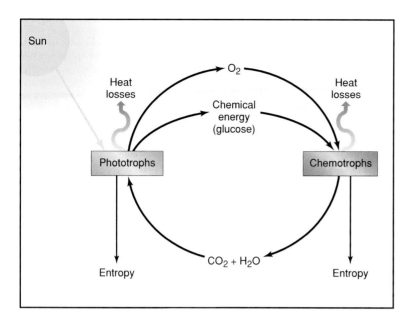

figure 6.1

Energy Transformations in the Biosphere

Energy from sunlight is transformed to chemical energy (for example, glucose) by phototrophs. These chemical compounds (for example, glucose) are then used by chemotrophs during respiration or fermentation to produce ATP for cellular activity. Carbon dioxide and water are produced during respiration and fermentation, completing the cycle.

Energy Can Be Neither Created Nor Destroyed

Although energy is convertible from one form to another, it can be neither created nor destroyed. This principle of energy conservation, known as the **first law of thermodynamics,** states that in any transformation of energy, whether the transformation is a chemical reaction in a laboratory beaker or photosynthesis by a living cell, energy cannot be created or destroyed; it is conserved. The total amount of energy present after the reaction must equal the initial amount of energy.

The energy contained within the products is less than the energy of the reactants in spontaneous reactions. The difference in energy content between the initial and final states of a reaction (the total amount of energy released during the reaction) is called **enthalpy** and is designated by ΔH. Not all energy released during a reaction, however, is available for doing work. That portion of released energy that can potentially be used for work is known as **free energy.** Changes in free energy of a reaction are designated by ΔG ($\Delta G°$ if determined under standard conditions of one atmosphere pressure and one molar concentration; $\Delta G°'$ if determined under standard conditions at pH 7.0, which approximates biological conditions). Free energy changes are expressed in units of kilocalories per mole. Reactions that occur spontaneously have a decrease in free energy level from reactants to products (negative $\Delta G°'$) and are exergonic. Endergonic reactions have a positive $\Delta G°'$ and require the input of energy to proceed.

The term *free energy* includes not only energy that is actually used for work, but also energy that has the potential for work. Heat or thermal energy, although seldom used for work in biological systems, is thus included as part of free energy. In most reactions, however, heat does not perform work and is simply lost to the surroundings. When heat is lost to the surroundings during energy transformations, as in the combustion or oxidation of glucose, the reaction is termed **exothermic. Endothermic** reactions are those in which heat is absorbed from the surroundings. An example of an endothermic reaction is the addition of ammonium sulfate to water. As the salt dissolves in water, heat is absorbed, resulting in cooling of the solution.

All Natural Processes Proceed in Such a Manner That There Is an Increase in Entropy

Free energy constitutes the portion of liberated energy that can be used for work. The other portion of energy released during a reaction cannot under any condition be used for work and is considered to be unrecoverable energy. This unrecoverable energy is termed **entropy** and is designated by the letter **S.** Entropy is a measurement of the disorder of a system. Spontaneous reactions always proceed with an increase in entropy, a characteristic embodied in the **second law of thermodynamics.**

The second law of thermodynamics states that all processes in the universe occur in such a manner that there is an increase in total entropy. As a consequence of this law, the universe constantly moves toward maximum disorder, or entropic doom. Fortunately there is enough usable energy in the universe to make entropic doom a distant and remote possibility. The increase in entropy under the second law explains why a rock that has rolled down a hill cannot return up the hill on its own; additional energy from some other source must be made available to push it. The additional energy is necessary because a part of the original energy in the rock has been lost to entropy and cannot be recovered.

Chemical reactions also occur in such a manner that there is an increase in entropy. These reactions move in the direction of the products, have a negative $\Delta G°'$, and cannot be reversed without an input of energy. The oxidation of glucose to carbon dioxide and water is an example; it is accompanied by a $\Delta G°'$ of −673 kcal/mole and increased entropy:

$$C_6H_{12}O_6 + 6\,O_2 \rightarrow 6\,CO_2 + 6\,H_2O \qquad \Delta G°' = -673 \text{ kcal/mole}$$

The concept of increased randomness in chemical reactions can be compared to differences in order observed with an assembled automobile engine and a disassembled engine. A completely assembled engine is more ordered than one that is disassembled.

The possibilities for disorder are greater with several hundred parts than with one large entity. In a similar sense, a large chemical molecule is considered more ordered than a rearrangement of the molecule and its bonds into several smaller molecules. Glucose is a more ordered molecule than the products of its oxidation, carbon dioxide and water. Thus when glucose is oxidized to its products, there is an increase in entropy.

The relationships among the terms free energy, enthalpy, and entropy are brought together in the equation

$$\Delta G^{o'} = \Delta H - T\Delta S$$

where $\Delta G^{o'}$ represents the free energy change, ΔH is the total energy change, T is the absolute temperature (degrees Kelvin) of the system, and ΔS represents the change in entropy. This equation not only provides information on the quantity of energy that is available for work ($\Delta G^{o'}$) as a result of a reaction; it also indicates whether a reaction is exergonic ($\Delta G^{o'} < 0$) or endergonic ($\Delta G^{o'} > 0$). The equation also takes into consideration entropy changes during a reaction.

Reactions that release free energy and have a negative $\Delta G^{o'}$ move in the direction of the products. In contrast, reactions that have a positive $\Delta G^{o'}$ value cannot proceed without an input of energy.

Reversible Chemical Reactions Proceed to an Equilibrium Point

Reversible chemical reactions continue until they reach an **equilibrium point,** at which there is a balance between reactants and products. At the equilibrium point, the amounts of reactants associating into products equal the amounts of products dissociating into reactants. For reactions that occur at given temperatures, the ratio of the concentrations of reactants and products at the equilibrium point takes on a specific value called the **equilibrium constant (K_{eq}).** In the reaction $A + B \rightleftarrows C + D$, the equilibrium constant is represented by the equation

$$K_{eq} = \frac{[C][D]}{[A][B]}$$

where the square brackets denote concentrations in moles per liter.

The free energy of the reaction will be at a minimum at equilibrium, with no further change in free energy (that is, $\Delta G^{o'} = 0$). It is therefore possible to calculate the standard free energy change of a reaction from its equilibrium constant using the equation

$$\Delta G^{o'} = -RT \ln K_{eq}$$

where R is the gas constant (1.99 cal/mole/degree), T is the absolute temperature, and ln is the natural logarithm (2.3 times the logarithm to the base 10) of the equilibrium constant.

If the K_{eq} of a reaction is known, we can calculate the $\Delta G^{o'}$ and the direction of the reaction under standard conditions and 1 molar concentrations of substrate and products. The standard free energy change value is useful because it provides a way to compare different reactions and the quantity of energy released or required by the reactions.

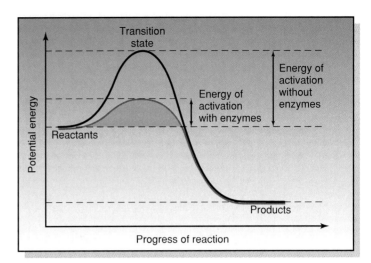

figure 6.2

Energy Requirements of a Chemical Reaction

Activation energy is required for all chemical reactions. Enzymes lower the amount of activation energy required.

Enzymes Accelerate Reaction Rates Without Themselves Being Changed

Not all spontaneous (exergonic) reactions occur at fast rates or velocities. The velocity or speed of a reaction at a given temperature and concentration is determined by the **activation energy (E_a),** the difference between the free energy of the reactants and the highest free energy state of the reactants during their transition to product (Figure 6.2).

The activation energy barrier can be significant, and depends on the types of molecules. The rate at which chemical reactions occur can be increased by increasing the rate at which reactants move and collide. Increases in temperature or reactant concentrations result in increased reaction rates. An example of this is the mixing of hydrogen and oxygen gases. The reaction $H_2 + 1/2\ O_2 \rightarrow H_2O$ has a $\Delta G^{o'}$ of -57 kcal/mole; yet when the two gases are mixed at room temperature, no discernible reaction occurs. If, however, the mixture is ignited with an electrical spark, water forms immediately. The spark—or rise in temperature—provides the E_a required for the reaction to proceed. After the initial spark, the ΔG continues the reaction.

Reaction rates may also be increased by the action of **catalysts,** chemical agents that increase reaction rates without themselves being changed. In the reaction of H_2 and O_2, instead of using a spark, we could have used a catalyst; a palladium catalyst lowers the E_a sufficiently to allow the reaction to occur at room temperature. This principle is illustrated by the GasPak system, in which an enclosed chamber is used to grow anaerobic bacteria (see GasPak system, page 93). Oxygen, which is toxic to anaerobic bacteria, is removed from the chamber's atmosphere by the palladium-catalyzed reaction of oxygen with hydrogen to form water.

Enzymes are protein (or, in the case of ribozymes, RNA) catalysts that occur in living organisms. The molecule (or molecules)

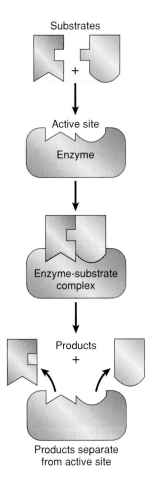

figure 6.3

Enzyme-Substrate Interaction

The lock-and-key mechanism is one model for enzyme action. The substrate fits into the active site, much as a key fits a lock, to form an enzyme-substrate complex. The enzyme catalyzes the reaction and releases the product. Note that the enzyme is not changed by the reaction and is now free to interact with another substrate molecule.

on which an enzyme acts is called its **substrate.** An average bacterial cell has thousands of different enzymes, each catalyzing specific chemical reactions. The kinds of enzymes present or absent in an organism determine the types of chemical pathways active in that particular organism. Enzymes do not change the equilibrium and the ΔG of the chemical reaction, but they do lower the activation energy level and permit the reaction to proceed at sufficient speed at lower temperatures to meet the cell's needs. As a consequence of this reduced activation energy, cellular reactions proceed at a faster rate.

Enzyme-substrate interactions are very specific. This specificity allows for fine control of enzymatic reactions and prevents unnecessary and energy-wasting reactions from operating continuously in the cell. Substrates bind with the enzyme at a site called the **active site** to form temporary enzyme-substrate complexes resulting in lower activation energy requirements for the reaction (Figure 6.3). After the activated transition state is reached, the enzyme-substrate complex disso-

ciates into reaction products and enzyme. The enzyme has not changed during the reaction and is now free to react with another substrate. The entire enzyme-catalyzed reaction is summarized as

$$E + S \rightleftarrows ES \rightleftarrows E + P$$

where E is enzyme, S is substrate, and P is product.

Enzymes not only are specific in their ability to react with substrates; their activity is also influenced by environmental factors such as pH and temperature. The pH of a reaction mixture affects the charge characteristics of amino acids comprising the structure of the enzyme and its active site or sites. Temperature affects enzyme activity in the same manner that it affects other chemical reactions. As temperature decreases, molecules move at slower rates and the reaction proceeds more slowly. However, since enzymes are proteins, an increase in temperature can lead to denaturation of the enzyme and a dramatic decrease in reaction rate. At higher temperatures the reaction rate increases until the point at which the enzyme becomes denatured and inactivated.

Certain chemical substances called **inhibitors** prevent or slow down enzyme reactions by binding to the enzyme and preventing substrate attachment. The inhibition of enzyme activity by such inhibitors includes two categories: reversible and irreversible. Reversible inhibition can be either competitive or noncompetitive. Competitive inhibitors structurally resemble the normal substrate of the enzyme and compete with the substrate for binding at the enzyme active site. Competitive inhibition can thus be reversed by increasing substrate concentration. An example of a competitive inhibitor is malonate, which has a structure that closely resembles the structure of succinate (Figure 6.4). Malonate competes with succinate for the active site on succinate dehydrogenase, an enzyme involved in cellular metabolism. By increasing the concentration of the substrate succinate, the inhibitory effects of the competitive inhibitor can be decreased.

Noncompetitive inhibitors are not structurally related to the enzyme substrate. Increasing the proportion of substrate to inhibitor therefore does not reverse the effects of noncompetitive inhibition. Substances such as cyanide, mercury, and arsenic are noncompetitive inhibitors that bind to enzymes at sites other than the substrate-binding site and thus reduce the catalytic activity of the enzyme. The effects of such inhibition depend on the relative concentrations of enzyme and inhibitor.

Allosteric inhibition is a special type of noncompetitive inhibition important in regulation of metabolic pathways. **Allosteric enzymes** are enzymes that have two types of specific binding sites: an active site, where substrate binds, and an allosteric site, where small molecules called **effectors** specifically bind. The binding of an effector at the allosteric site changes the conformation of the enzyme at the active site and results in a corresponding inhibition (or activation, in the case of allosteric activation) of enzyme activity (Figure 6.5a). **Feedback inhibition** is a mechanism regulated by allostery. In feedback inhibition, the effector is an end product of a biosynthetic pathway; it interacts with and inhibits the activity of a key enzyme (usually the first) early in the pathway (Figure 6.5b). End-product accumulation

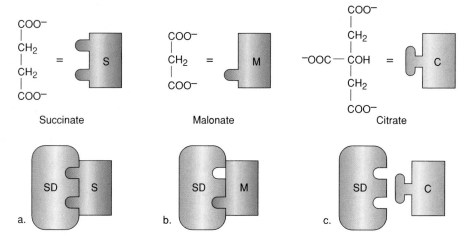

figure 6.4

Competitive Inhibition

Malonate competes with succinate for the active site on succinate dehydrogenase. a. Proper fit between succinate (S) and succinate dehydrogenase (SD). b. Competitive inhibition by malonate (M). Malonate binds, but not as well, to succinate dehydrogenase and prevents the binding of succinate. c. Citrate (C), which is not recognized by the succinate dehydrogenase active site, does not bind to the enzyme.

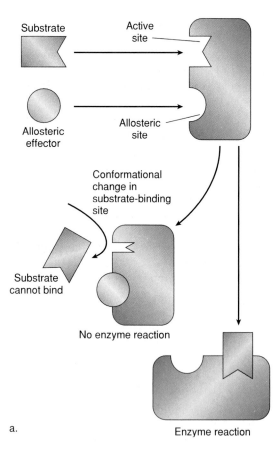

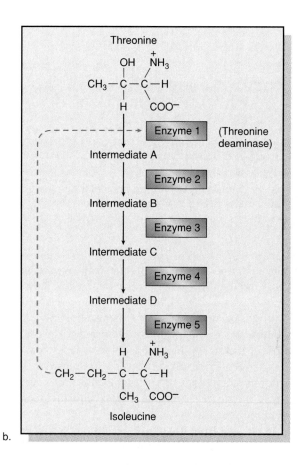

figure 6.5

Allosteric Enzymes and Feedback Inhibition

a. The mechanism of enzyme inhibition by an allosteric effector. The effector combines with the allosteric site, changing the conformation of the enzyme so that the substrate can no longer bind to the active site.

b. Feedback inhibition in a biosynthetic pathway. Five enzymes catalyze the biosynthesis of isoleucine from threonine. If the product of the pathway (isoleucine) accumulates, it allosterically inhibits threonine deaminase.

table 6.1

Some Important Coenzymes and Their Functions in the Cell

Coenzyme	Vitamin Source	Function
Biotin	Biotin	Carboxylation
Coenzyme A	Pantothenic acid	Activation of acetyl groups during pyruvate and fatty acid catabolism
Flavin adenine dinucleotide (FAD)	Riboflavin (B_2)	Electron transport and dehydrogenations
Flavin mononucleotide (FMN)	Riboflavin (B_2)	Electron transport and dehydrogenations
Nicotinamide adenine dinucleotide (NAD)	Niacin	Electron transport and dehydrogenations
Nicotinamide adenine dinucleotide phosphate (NADP)	Niacin	Electron transport and dehydrogenations
Pyridoxal phosphate	Pyridoxine (B_6)	Transamination, decarboxylation, and racemization of amino acids
Tetrahydrofolic acid	Folic acid	Transfer of one-carbon units
Thiamine pyrophosphate	Thiamine (B_1)	Group transfers; oxidation and decarboxylation of keto acids

results in feedback inhibition of enzyme activity. If the end product is used up, enzyme activity resumes. Feedback inhibition is an important mechanism by which microorganisms and other cells regulate the synthesis of amino acids and other low molecular weight compounds without accumulation of unwanted metabolic intermediates. Synthesis of glycogen, a cellular storage product, is regulated by allosteric activation. When intermediates of pathways such as glycolysis (see the Embden-Meyerhof pathway, page 150) accumulate, allosteric enzymes associated with glycogen synthesis are activated.

Enzymes can also be inhibited noncompetitively in a permanent, or irreversible, manner. Irreversible inhibition may be complete or partial and generally involves the denaturation or alteration of enzyme structure. This type of inhibition is caused by chemical or physical treatment of enzymes.

A special class of compounds called **cofactors** often binds tightly to enzyme surfaces and is required for enzymatic activity. Cofactors are nonprotein and can be either organic or inorganic. The inorganic cofactors are often metallic ions (for example, Mg^{2+} or Na^+), whereas the organic cofactors are vitamin derivatives. Organic cofactors, generally called **coenzymes,** are involved in numerous important cellular reactions (Table 6.1). Coenzymes frequently are carriers of electrons, chemical groups, or atoms and can move these substances from one substrate to another during chemical reactions.

Coupled Reactions

Biosynthetic reactions in living systems are endergonic reactions that cannot proceed without an input of energy. Endergonic reactions generally obtain this requisite energy by linking, or **coupling,** with exergonic reactions. For coupling reactions to be successful, the free energy released by the exergonic portion of the reaction (ΔG) must be greater than the energy required by the endergonic portion (that is, the overall ΔG of the coupled reaction must be negative).

Cells Obtain Energy by the Oxidation of Molecules

The coupled reactions most common in biological systems are those involving oxidation and reduction (**oxidation-reduction,** or **redox,** reactions). These reactions involve the transfer of electrons from one molecule to another.

Oxidation describes any reaction in which electrons are lost from a substance. The substance losing the electrons is called the electron donor (or reducing agent) and is said to be oxidized upon loss of the electrons. Electrons do not remain unassociated; those that are released from a substance must be accepted by another compound or element in a reaction known as **reduction.** The substance receiving electrons is called an electron acceptor (or oxidizing agent) and is reduced upon accepting these electrons. The two substances involved in an oxidation-reduction reaction are referred to as the **redox pair.** Redox pairs are written with the oxidized form listed first, such as SO_4^{2-}/H_2S, Fe^{3+}/Fe^{2+}, and NO_3^-/NO_2^-.

Oxidations in biological systems frequently involve dehydrogenation reactions, the removal of hydrogen ions (protons) along with electrons from a substance. Addition of hydrogen ($H^+ + e^-$) to substances during reduction occurs in reactions called hydrogenations. Such reductions accompany oxidations, and the oxidation and reduction reactions are therefore said to be coupled. Substances that become oxidized upon loss of electrons are subsequently able to act as electron acceptors, and thus become reduced again. In a similar manner, substances that gain electrons through reduction are subsequently able to serve as electron donors.

Substances Differ in Their Oxidation-Reduction Potentials

Substances differ in their abilities to donate or accept electrons. Those that serve as electron acceptors must have greater affinities for electrons than the electron donors. The ability of a substance to donate or accept electrons is described by a measurable physical parameter called the **oxidation-reduction potential,** or **redox potential** (E_0). The redox potential for a redox pair is electrically measured relative to the standard hydrogen electrode: $1/2\ H_2 \rightleftarrows H^+ + e^-$. At a pH equal to 0 and standard conditions, the redox potential of the hydrogen electrode is 0.00 V. At pH 7.0, which is more meaningful for biological systems, the redox potential is expressed as E'_0 and for the hydrogen electrode is -0.42 V. The potential of all other redox pairs can be determined electrically and placed on a scale relative to the redox potential of the $2\ H^+/H_2$ redox pair (Figure 6.6).

From this redox potential gradient, the relative affinity of different substances for electrons may be obtained. Two general rules should be remembered with respect to oxidation-reduction reactions. The first rule is that the reduced substance of the redox pair, which has the more negative E'_0, donates electrons to the oxidized substance of the pair, which has the more positive redox value. For example, the oxidation reaction

$$Fe^{2+} \rightleftarrows Fe^{3+} + e^- \ (+0.77 \text{ V})$$

can be coupled with any reduction reaction having a less negative (more positive) redox potential, including the reaction

$$1/2 \ O_2 + 2 \ H^+ + 2 \ e^- \rightleftarrows H_2O \ (+0.82 \text{ V})$$

The oxidation of ferrous ion (Fe^{2+}), however, cannot be coupled with the reaction

$$Fumarate + 2 \ H^+ + 2 \ e^- \rightleftarrows succinate \ (+0.03 \text{ V})$$

which has a more negative (less positive) redox potential than the oxidation half of the coupled reaction.

The second rule is that the greater the difference (ΔE) in redox potential between the redox pair serving as the electron donor and the pair serving as an electron acceptor, the greater the energy available via the oxidation-reduction reaction. The standard free energy that is made available through electron transfer in oxidation-reduction reactions is expressed by the equation

$$G = -nF\Delta E$$

where n is the number of electrons and F is a constant (Faraday's constant) equal to 23 kcal/volt/mole.

Compounds or elements that have high redox potentials and are very reduced (for example, glucose) generally liberate large amounts of energy as their electrons are released and donated to electron acceptors. This occurs only if the electron acceptor available to the cell has a sufficiently more positive redox potential. Other substances lower on the redox potential gradient are less reduced and consequently do not release as much energy during their oxidation. Some compounds are completely oxidized (for example, CO_2); these cannot serve as sources of energy, although they can accept electrons.

Oxidation-reduction reactions, like any other type of chemical reaction, are accompanied by changes in free energy. Part of the energy that is liberated during the exergonic oxidative portions of these reactions is conserved in the bonds of the reduced products when the released electrons are accepted in endergonic reduction reactions. The remaining portion of the energy, which is not conserved, is used for work or lost as heat or to entropy.

Electron Carriers Transfer Electrons from Electron Donors to Electron Acceptors

Electrons released by a chemical compound during a biological oxidation often are not directly accepted by the ultimate electron acceptor. Instead, these electrons are transferred through a series of **electron carriers** before reaching the final electron acceptor. Electron carriers act as relays to transfer electrons from the initial donor to the final acceptor. As these transfers occur, the electron

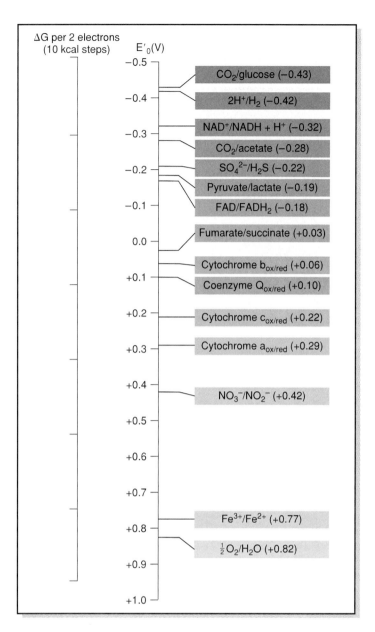

figure 6.6

Gradient of Redox Potentials

The susceptibility of a substance to oxidation or reduction is measured quantitatively as its oxidation-reduction (redox) potential relative to the voltage required to remove or add an electron to H_2. Thus the standard hydrogen electrode is assigned an arbitrary redox potential (E'_0) of 0.00 V under standard conditions (all reactants and products at 1 molar or 1 atm). At pH 7.0 (biological conditions), the redox potential (E'_0) for the hydrogen electrode is -0.42 V. The more reduced components have more negative voltages.

carriers themselves are oxidized and reduced. These oxidation-reduction reactions are accompanied by free energy changes and a liberation of energy. During electron transport, electrical potential may be converted to chemical energy that is stored in the high-energy phosphate bonds of ATP.

Among the more important electron carriers are those that transfer not only electrons but also hydrogen atoms (Figure 6.7).

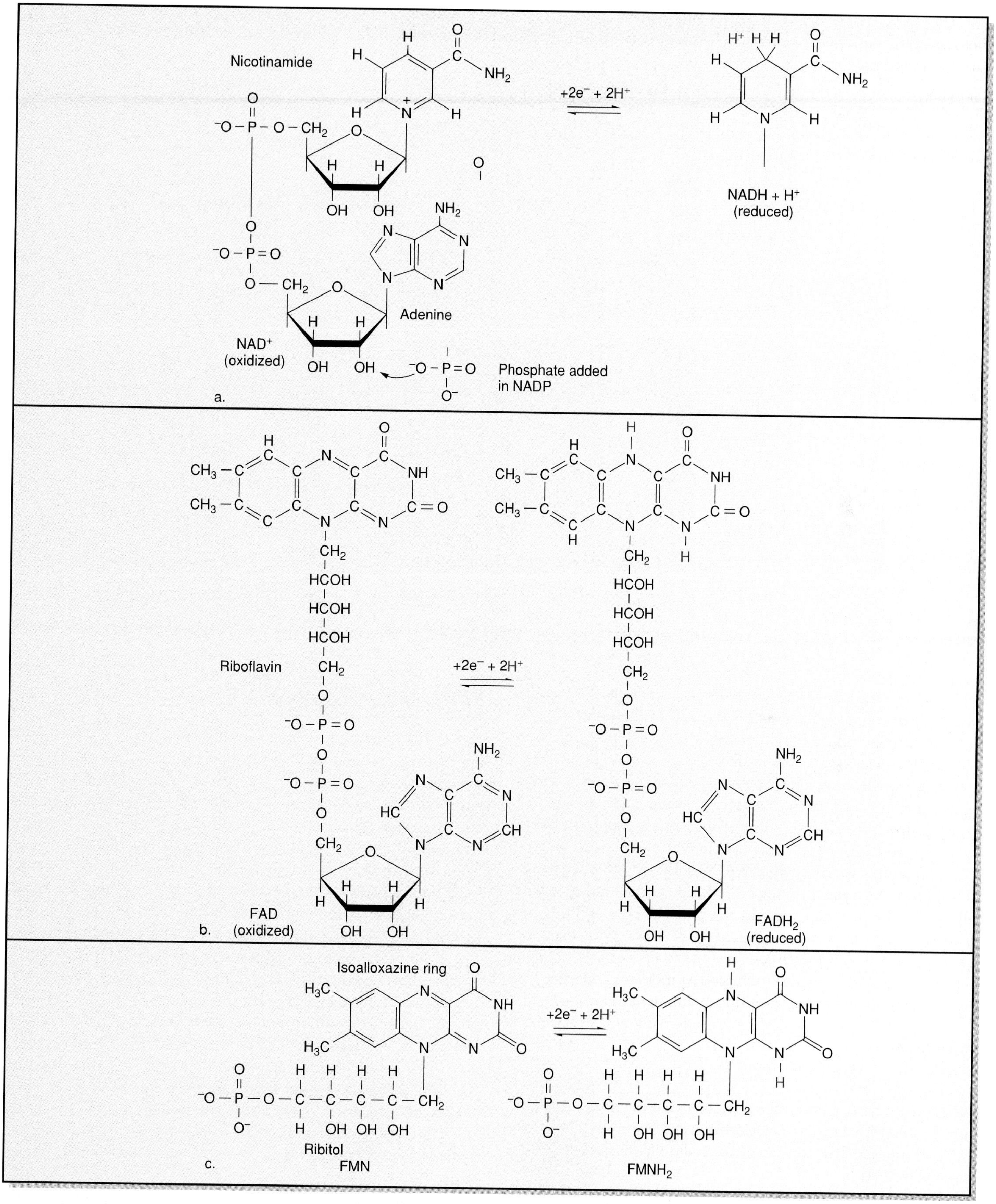

figure 6.7

Examples of Electron Carriers

a. Nicotinamide adenine dinucleotide (NAD reduced to NADH + H$^+$). b. Flavin adenine dinucleotide (FAD reduced to FADH$_2$). c. Flavin mononucleotide (FMN reduced to FMNH$_2$).

table 6.2

Examples of High-Energy Compounds

Compound	Reaction (hydrolysis)	ΔG (kcal/mole)
Adenosine triphosphate	adenosine—P~P~P $\rightleftarrows$ adenosine—P~P + P_i	−7.3
Adenosine diphosphate	adenosine—P~P $\rightleftarrows$ adenosine—P + P_i	−7.3
Creatine phosphate		−8.0
Acetyl-CoA		−8.0
1,3-diphosphoglycerate		−11.8
Phosphoenolpyruvate		−14.8

These carriers include (1) the pyridine nucleotides NAD (nicotin-amide adenine dinucleotide) and NADP (NAD phosphate) and (2) the flavoproteins FAD (flavin adenine dinucleotide) and FMN (flavin mononucleotide). As a result of the association of these carriers with specific biochemical pathways in the cell, released energy is conserved in the form of high-energy compounds such as ATP. ATP can then be used by the cell at its discretion to supply the energy needs for work.

Electron carriers such as NAD⁺ undergo reversible oxidations and reductions. In the oxidized state, NAD⁺ has a positive charge on the nitrogen atom in the pyridine ring and one less hydrogen atom than in the reduced form. Upon reduction, two electrons and two hydrogen ions are picked up by NAD⁺. One of the hydrogen atoms is added to the carrier molecule while the proton from the second hydrogen atom remains in solution as a hydrogen ion, H⁺. The reversible oxidation-reduction reaction is expressed as

$$NAD^+ + 2\,e^- + 2\,H^+ \rightleftarrows NADH + H^+$$

NAD⁺ and other similar compounds serve only as intermediate proton and electron carriers. The electrons and hydrogen atoms they carry are transferred to other acceptors in the cell, which subsequently become reduced.

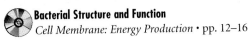

Bacterial Structure and Function
Cell Membrane: Energy Production • pp. 12–16

High-Energy Compounds

Energy obtained from chemical compounds can be converted to high-energy compounds that can be used by the cell to perform work (Table 6.2). Molecules are typically considered to be high-energy if their hydrolysis is characterized by a large negative free-energy change (greater than −7 kcal/mole). Without high-energy compounds, energy that is released during normal oxidations would be wasted as heat. High-energy compounds are like organic batteries that provide the organism with stored energy for work and other cellular functions. Like batteries, these energy store-houses can be recharged after they have discharged their energy.

Many high-energy compounds have one or more high-energy phosphate bonds. These phosphate bonds occur in such organic substances as ATP, PEP (phosphoenolpyruvate), acetyl phosphate, and 1,3-diphosphoglycerate. Energy stored and later liberated by bond hydrolysis can be used for cellular work.

The most common of these high-energy compounds is ATP (Figure 6.8). Only the outer two phosphate linkages (γ- and β-linkages, generally called the pyrophosphate) of ATP are considered to be high-energy bonds. Cleavage of the bonds is accompanied by energy release. Hydrolysis of each of the two phosphate bonds results in a ΔG°' of −7.3 kcal/mole.

The high negative ΔG°' that results from hydrolysis of the high-energy bonds of ATP is a consequence of the charge relationships in

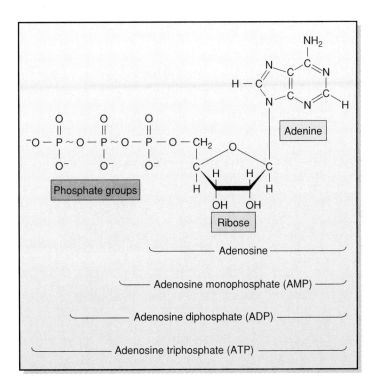

figure 6.8

Structure of ATP Showing Location of High-Energy Phosphate Bonds

High-energy bonds are designated by a ~.

the pyrophosphate linkages. The P—O bonds in pyrophosphate are actually small dipoles, with the phosphorus atoms having partial positive charges and the oxygen atoms surrounding each phosphorus having partial negative charges. As a result, there are repulsive positive charges within the anhydride linkages and repulsive negative charges among the peripheral oxygen atoms. Hydrolysis of the anhydride linkages not only dissipates these charge repulsions, but also results in a more stable structure and a negative free-energy change. The innermost ester linkage (α-phosphate) between phosphorus and carbon (P—O—C) lacks such charge repulsions. As a consequence its cleavage does not result in as large a free-energy change (−3 kcal/mole), and this bond is not considered to be a high-energy linkage.

Concepts of Metabolism

Most microorganisms obtain their energy by the oxidation of chemical compounds. This breakdown of chemical compounds, or **catabolism,** is one of the two general classes of reactions that occur within a cell. **Anabolism,** the processes associated with the biosynthesis of chemical compounds, is the other class of cell reactions. **Metabolism** is the total of the reactions (catabolic and anabolic) carried out by the cell.

Catabolic reactions serve two purposes in living cells. They are an important source of chemical energy for the cell, because such reactions are accompanied by the release (exergonic reaction) of free energy (the net ΔG is negative). Part of

this released energy is conserved in the form of ATP and other high-energy compounds. Catabolism is also the process by which the cell is able to degrade low molecular weight compounds such as glucose, as well as large polymeric substances, into smaller, simpler constituents. These smaller molecules are the carbon skeletons used as precursors for synthesis of the subunits needed for growth.

In comparison to the exergonic catabolic reactions, anabolic reactions often are endergonic and require an input of energy (the net ΔG is positive). The necessary energy is made available through catabolic pathways. Both catabolic and anabolic reactions constitute cellular metabolism and are coupled through energy exchanges usually involving ATP and other high-energy compounds. Certain biochemical pathways have dual roles, functioning in both catabolism and anabolism. Key metabolites in these **amphibolic pathways** serve as substrates for catabolism and as precursors for anabolism. The tricarboxylic acid cycle, which is discussed in this chapter, is an example of an amphibolic pathway.

Mechanisms of ATP Synthesis

Microorganisms use three mechanisms to generate ATP: **oxidative phosphorylation**, **photophosphorylation,** and **substrate-level phosphorylation.** In oxidative phosphorylation and photophosphorylation, ATP is synthesized during the transfer of electrons through an electron transport chain to some final electron acceptor. Electron transport chains are sequences of oxidation-reduction reactions. As electrons are transferred through a series of carriers having sequentially lower redox potentials, a portion of the free energy that is liberated is conserved in the form of ATP. Oxidative phosphorylation and photophosphorylation are similar to each other; they differ primarily in the original source of electrons. Electrons are supplied by reduced chemicals in oxidative phosphorylation and by reduced chlorophyll molecules, which have recently absorbed light energy, in photophosphorylation.

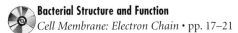
Bacterial Structure and Function
Cell Membrane: Electron Chain • pp. 17–21

In substrate-level phosphorylation, phosphate is added to an organic compound. ATP is synthesized when a phosphorylated metabolic intermediate transfers its phosphate to ADP. The energy used to drive substrate-level phosphorylation is provided by the oxidation of a reduced chemical compound. An example of substrate-level phosphorylation is the oxidation of glyceraldehyde-3-phosphate to 1,3-diphosphoglycerate (Figure 6.9). Some of the energy from this oxidation reaction is used to phosphorylate the glyceraldehyde-3-phosphate in the first carbon position of the molecule. The phosphate bond formed is a high-energy bond. The 1,3-diphosphoglycerate may transfer this high-energy phosphate bond to adenosine diphosphate (ADP) to form the energy-rich storage compound ATP and 3-phosphoglycerate.

This chapter discusses some of the major pathways by which microorganisms generate ATP via substrate-level phosphorylation and oxidative phosphorylation. Photophosphorylation will be discussed in Chapter 7.

Carbohydrates are the most common energy sources for microorganisms. However, proteins, lipids, and even nucleic acids also can be used as sources of energy. Carbohydrates are useful energy sources because they are compounds with large quantities of electrons to donate during oxidation. These organic substances are universally available to plants, animals, and microorganisms and are central in the metabolism of these organisms.

Chemoheterotrophic Microorganisms Obtain Energy by Fermentation and/or Respiration

Chemoheterotrophic microorganisms obtain energy from carbohydrates by two basic processes: **fermentation** and **respiration.** In fermentation, an organic substrate serves as the electron donor; an oxidized intermediate of the substrate acts as the final electron acceptor and subsequently becomes reduced. Because fermentation does not require the presence of oxygen, microorganisms that ferment carbohydrates may do so in the absence of oxygen.

In contrast to fermentation, respiration requires an external electron acceptor for substrate oxidation. When molecular oxygen is the terminal electron acceptor, it is reduced to H_2O; this process is called **aerobic respiration.** In **anaerobic respiration,** another inorganic molecule such as nitrate (NO_3^-) or sulfate (SO_4^{2-}) serves as the terminal external electron acceptor and becomes reduced to nitrite (NO_2^-), nitrous oxide (N_2O), and nitrogen (N_2), or to hydrogen sulfide (H_2S), respectively. Although different in their mechanisms for electron transfer, both fermentation and respiration use substrate oxidation to channel energy from chemical compounds into energy-rich molecules such as ATP.

Glucose Is Oxidized to Pyruvate in the Embden-Meyerhof Pathway

The oxidation of glucose is an important example in nature of carbohydrate oxidation by chemoheterotrophic microorganisms. Glucose may be oxidized by several different pathways, but in many organisms glucose oxidation begins by the **Embden-Meyerhof pathway.** This pathway, named after two of its discoverers, employs substrate-level phosphorylation to generate ATP. The Embden-Meyerhof scheme is an example of a **glycolytic pathway**—a catabolic pathway that literally lyses (splits) sugars. In the Embden-Meyerhof pathway, a molecule of glucose is oxidized to two molecules of pyruvate with the net production of two molecules of ATP and two molecules of NADH + H^+.

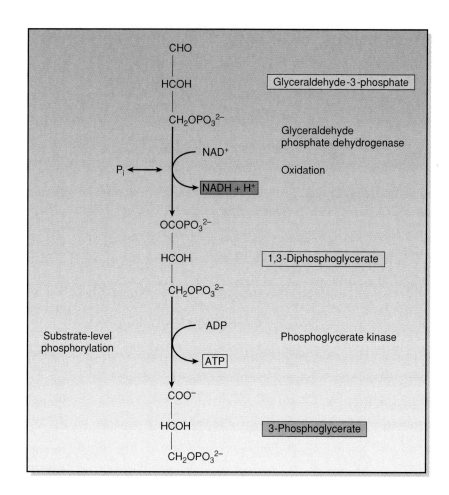

figure 6.9

Substrate-Level Phosphorylation
Glyceraldehyde-3-phosphate is phosphorylated and oxidized to 1,3-diphosphoglycerate. This high-energy compound donates its C-1 phosphate to ADP to produce ATP, an example of substrate-level phosphorylation.

The Embden-Meyerhof pathway consists of ten distinct reactions, each of which is catalyzed by a different enzyme. Figure 6.10 provides a schematic overview of the Embden-Meyerhof pathway. A detailed discussion of the pathway follows.

The first reaction of the Embden-Meyerhof pathway is the phosphorylation of glucose to glucose-6-phosphate by hexokinase (EM-1). Glucose-6-phosphate is important not only as a metabolic intermediate in this pathway, but also as an intermediate in other pathways of carbohydrate metabolism. It therefore is an important link between the Embden-Meyerhof scheme and other metabolic pathways.

The second reaction of the Embden-Meyerhof pathway is the isomerization of glucose-6-phosphate to fructose-6-phosphate by the enzyme phosphoglucoisomerase (EM-2). Third, fructose-6-phosphate is phosphorylated by phosphofructokinase (EM-3), which transfers a phosphate group from ATP to fructose-6-phosphate to form fructose-1,6-diphosphate. Phosphofructokinase is a key

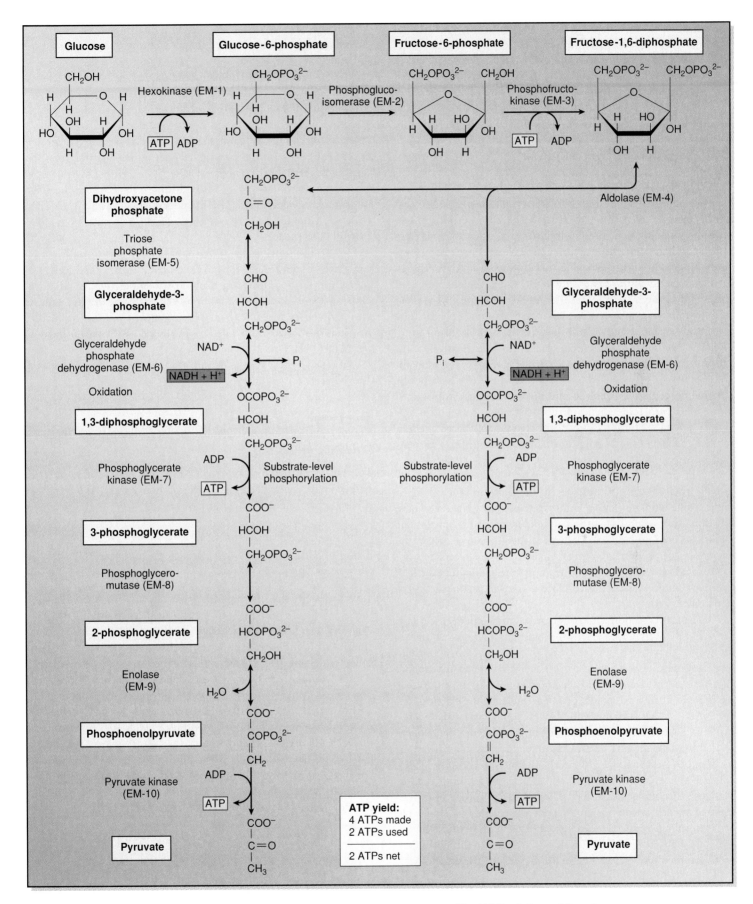

figure 6.10
The Embden-Meyerhof Pathway for Glucose Metabolism

Microbial Metabolism and Growth
Metabolism: Catabolism • pp. 2–4

regulatory enzyme that is activated by high concentrations of ADP and AMP and conversely inhibited by high concentrations of ATP. This regulation, linked to the ATP supply, allows the cell to determine if additional quantities of ATP are required for cell processes. If ATP supply is low—resulting in high cellular levels of ADP and AMP—phosphofructokinase is activated. Under such conditions of low ATP levels in the cell, the Embden-Meyerhof pathway is "turned on" for synthesis of additional molecules of ATP. As ATP concentrations increase in the cell, the pathway is "turned off."

This inhibition of phosphofructokinase by ATP is an example of allosteric inhibition. ATP is an allosteric effector that has a negative effect on phosphofructokinase activity. When phosphofructokinase is present in large quantities, ATP binds to it and stabilizes the enzyme in an inactive state, thereby effectively blocking the further oxidation of glucose in glycolysis. In his experiments with fermenting yeast, Louis Pasteur observed that yeasts grown in the presence of oxygen use glucose less rapidly than those grown in the absence of oxygen. This decrease in the rate of glucose consumption during aerobic respiration, known as the **Pasteur effect,** is explained by the allosteric inhibition of phosphofructokinase by ATP. When oxygen is available, more ATP is produced through the complete oxidation of a glucose molecule via glycolysis and respiration than is produced per glucose molecule metabolized by fermentation. The ATP that is generated inhibits phosphofructokinase, thereby reducing the quantity of substrate that is sent through the Embden-Meyerhof pathway.

The fourth step in the Embden-Meyerhof pathway is cleavage of fructose-1,6-diphosphate by the enzyme fructose-1, 6-diphosphate aldolase (EM-4) into two triose phosphate molecules: dihydroxyacetone phosphate and glyceraldehyde-3-phosphate. The two trioses are isomers of each other and are interconvertible by the enzyme triose phosphate isomerase (EM-5). Since glyceraldehyde-3-phosphate and dihydroxyacetone phosphate are interconvertible, two triose molecules from each glucose actually participate in the remaining reactions.

The sixth reaction is the only oxidation step of the Embden-Meyerhof pathway. Glyceraldehyde-3-phosphate is oxidized to 1,3-diphosphoglycerate by the enzyme glyceraldehyde-3-phosphate dehydrogenase (EM-6) in the presence of inorganic phosphate (P_i). The pair of electrons and accompanying hydrogen atoms that are released during this oxidation are accepted by the electron carrier NAD^+, which is reduced to $NADH + H^+$. Part of the energy made available from the oxidation of glyceraldehyde-3-phosphate is used to make the high-energy bond of the phosphate group on the number 1 carbon of 1,3-diphosphoglycerate. The energy contained within each 1,3-diphosphoglycerate is subsequently used to synthesize two molecules of ATP. The first molecule of ATP is synthesized as each 1,3-diphosphoglycerate is converted to a molecule of 3-phosphoglycerate. The energy released by the bond cleavage ($\Delta G^{o'} = -11.8$ kcal/mole) is sufficient to transfer the phosphate group on the number 1 carbon of 1,3-diphosphoglycerate to ADP

to generate ATP ($\Delta G^{o'} = +7.3$ kcal/mole) by substrate-level phosphorylation. This reaction is catalyzed by the enzyme phosphoglycerate kinase (EM-7).

The second ATP molecule generated by substrate-level phosphorylation is synthesized when the remaining phosphate on 3-phosphoglycerate is eventually transferred to ADP. 3-phosphoglycerate is first converted to 2-phosphoglycerate by phosphoglyceromutase (EM-8) and then to 2-phosphoenolpyruvate by enolase (EM-9). Phosphoenolpyruvate, unlike its precursor 2-phosphoglycerate, is an energy-rich substance with a high-energy phosphate bond ($\Delta G^{o'} = -14.8$ kcal/mole). ATP is produced as phosphoenolpyruvate transfers its phosphate group to ADP. This reaction, catalyzed by pyruvate kinase (EM-10), results in the formation of pyruvate.

The ultimate result of the Embden-Meyerhof pathway is the gross formation of four ATP molecules by substrate-level phosphorylation. However, because two molecules of ATP are expended in the initial reactions of the pathway, the net gain is two molecules of ATP from the oxidation of one molecule of glucose to two molecules of pyruvate and two molecules of $NADH + H^+$.

Microbial Metabolism and Growth
Metabolism: Catabolism • pp. 2–5

Microorganisms Have Other Glycolytic Pathways for the Oxidation of Glucose

The Embden-Meyerhof scheme is only one of several glycolytic pathways by which glucose is converted to pyruvate. At least three other glycolytic pathways occur in bacteria.

6-phosphogluconate Dehydratase and 2-keto-3-deoxy-6-phosphogluconate Aldolase Are Key Enzymes in the Entner-Doudoroff Pathway

One alternative glycolytic pathway was reported in *Pseudomonas saccharophila* by Nathan Entner and Michael Doudoroff in 1952 (Figure 6.11). This pathway, the **Entner-Doudoroff pathway,** has since been found in other species of *Pseudomonas* and in certain other gram-negative procaryotes, but not in eucaryotes. In the Entner-Doudoroff pathway, glucose is phosphorylated to glucose-6-phosphate, which is then oxidized to 6-phosphogluconate. The 6-phosphogluconate is dehydrated to yield 2-keto-3-deoxy-6-phosphogluconate, which is cleaved by an aldolase to produce one molecule of pyruvate and one molecule of glyceraldehyde-3-phosphate. The glyceraldehyde-3-phosphate is transformed to pyruvate via reactions identical to reactions 6 through 10 in the Embden-Meyerhof pathway. The overall reaction for the Entner-Doudoroff pathway follows:

$$Glucose + NADP^+ + NAD^+ + ADP + P_i \rightarrow$$

$$2\ pyruvate + NADPH + H^+ + NADH + H^+ + ATP$$

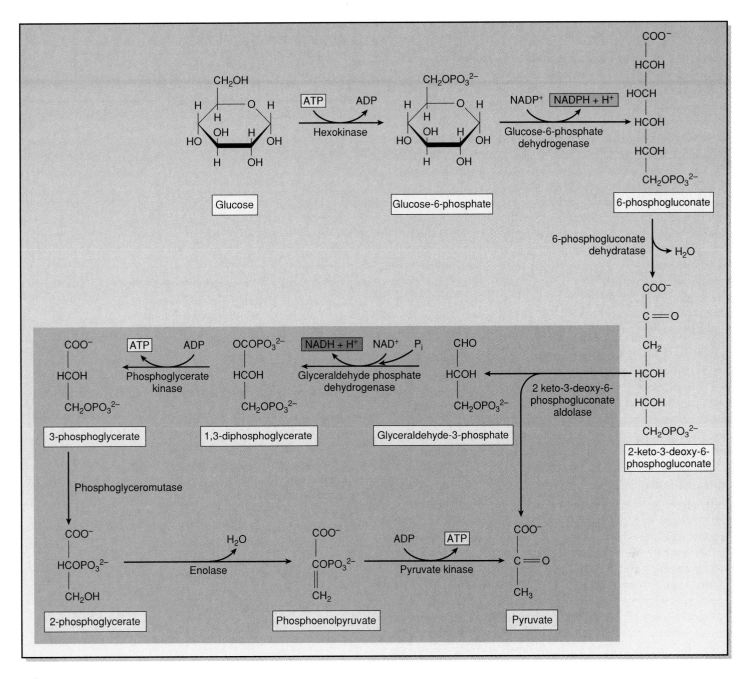

figure 6.11
The Entner-Doudoroff Pathway for Glucose Metabolism
The shaded area is identical to the last portion of the Embden-Meyerhof pathway.

The key enzymes of the Entner-Doudoroff pathway are 6-phospho-gluconate dehydratase and 2-keto-3-deoxy-6-phosphogluconate aldolase. The Entner-Doudoroff pathway can be distinguished from the Embden-Meyerhof pathway by assaying for these unique enzymes.

Both pathways result in formation of NADH + H⁺ during the oxidation of glyceraldehyde-3-phosphate. One net molecule of ATP is produced by substrate-level phosphorylation for each molecule of glucose oxidized in the Entner-Doudoroff pathway (compared with

two net ATP molecules in the Embden-Meyerhof pathway). Despite the lower yield of ATP, the Entner-Doudoroff pathway is useful because it generates NADPH + H⁺ from the oxidation of glucose-6-phosphate to 6-phosphogluconate. The generation of NADPH + H⁺ is very important, since this reduced coenzyme is required as the immediate electron donor in many biosynthetic reactions. The Entner-Doudoroff pathway is used more frequently by aerobes than anaerobes, which can obtain more energy from the Embden-Meyerhof pathway.

The Pentose Phosphate Pathway Is a Shunt of Glycolysis

Another pathway of glucose utilization is the **pentose phosphate pathway** (Figure 6.12). This pathway is also known as the **Warburg-Dickens pathway** (named after Otto Warburg and Frank Dickens), the **phosphogluconate pathway,** and the **hexose monophosphate shunt.** The pathway involves some reactions of the glycolytic pathway and therefore is frequently called a shunt of glycolysis. Although energy can be generated from the pentose phosphate pathway by the transfer of liberated electrons to the respiratory electron transport chain, the pathway generally is not considered to be a major energy-yielding pathway in microorganisms. Certain bacteria use the pentose phosphate pathway as yet another way to produce biosynthetic reducing power in the form of NADPH + H^+. The pathway is also a principal source of pentoses used in nucleotide synthesis and is important in the interconversion of hexoses and pentoses.

Phosphoketolase Is a Key Enzyme in the Phosphoketolase Pathway

The enzyme phosphoketolase, which cleaves acetyl phosphate from phosphorylated pentoses and hexoses, is found in some bacteria, including heterofermentative lactic acid bacteria such as *Leuconostoc mesenteroides* and *Bifidobacterium bifidus;* it forms the basis for pathways that are branches of the pentose phosphate pathway. The **phosphoketolase pathway** was originally discovered in *L. mesenteroides* and is summarized by the following overall reaction:

$$\text{Glucose} + \text{ADP} + \text{P}_i \rightarrow \text{pyruvate} + \text{ethanol} + CO_2 + \text{ATP}$$

Although the products of this reaction give the appearance of glucose oxidation via the Embden-Meyerhof pathway, a different set of enzymes and intermediates is used (Figure 6.13). Phosphoketolase, the key enzyme in the pathway, catalyzes the phosphorylation and breakdown of xylulose-5-phosphate to glyceraldehyde-3-phosphate and acetyl phosphate. The glyceraldehyde-3-phosphate is converted to pyruvate via the normal glycolytic pathway, with the synthesis of two ATP molecules. Since one ATP molecule was used in the initial phosphorylation of glucose to glucose-6-phosphate, the net yield of the pathway is one molecule of ATP produced by substrate-level phosphorylation per molecule of glucose. The pathway also generates NADPH + H^+, which is used in biosynthetic reactions, and the key intermediate xylulose-5-phosphate, which serves as an entry point for the metabolism of pentoses.

These pathways of glucose metabolism are not universally found in all organisms (Table 6.3). The Embden-Meyerhof and pentose phosphate pathways occur in both procaryotes and eucaryotes. The Entner-Doudoroff and phosphoketolase pathways are found only in certain procaryotes; they have not been found in other organisms. In many cases, alternative pathways for glucose oxidation are less energy efficient than a primary pathway such as the Embden-Meyerhof pathway. Nonetheless, these alternative pathways make it possible for microorganisms to utilize a wide variety of substrates (an evolutionary consideration) and also to produce NADPH + H^+ and metabolic intermediates that may be important precursors for synthesis of other compounds involved in cell function.

table 6.3

Distribution of Major Pathways of Glucose Metabolism among Procaryotes

Organism	Embden-Meyerhof	Entner-Doudoroff
Alcaligenes eutrophus	−	+
Arthrobacter sp.	+	−
Azotobacter chroococcum	+	−
Bacillus sp.	+	−
Escherichia coli and other enteric bacteria	+	−
Pseudomonas sp.	−	+
Rhizobium sp.	−	+
Thiobacillus sp.	−	+
Xanthomonas sp.	−	+

From G. Gottschalk, *Bacterial Metabolism,* 2nd edition. Copyright © 1986 Springer-Verlag, New York. Reprinted by permission.

Pyruvate Can Be Further Degraded to a Variety of Products

In the formation of pyruvate by the Embden-Meyerhof pathway and other glycolytic pathways, NAD^+ is reduced to NADH + H^+. Because the cell has a limited supply of oxidized NAD (NAD^+), there must be a mechanism to regenerate NAD^+ from NADH + H^+ if these glycolytic pathways are to continue functioning. In fermentation, electrons and hydrogens from NADH + H^+ are passed to organic compounds during the processing of pyruvate. This transfer of electrons to an internal electron acceptor (that is, to a product of the pathway) is a requisite of fermentation. Some of the resulting metabolic end products formed during fermentations are discussed in this section (Figure 6.14). The end products that accumulate depend on the particular enzymes of the fermenting organism and the organic intermediates used as electron acceptors.

Lactic Acid Bacteria Reduce Pyruvate to Lactate

A number of bacteria regenerate NAD^+ by reducing pyruvate to lactate (lactic acid) using the enzyme lactate dehydrogenase. Many, though not all, of the bacteria that carry out this reaction are called **lactic acid bacteria** and include members of the genera *Lactococcus, Lactobacillus, Leuconostoc,* and *Pediococcus.* Lactic acid bacteria are important in the dairy industry, where they are used in the preparation of fermented milk products including yogurt, kefir, acidophilus milk, and Bulgarian milk (see milk and dairy products, page 609). The acids synthesized during growth of lactic acid bacteria reduce the pH of the surrounding environment and eventually inhibit bacterial growth. Synthetic media used for growth of lactic-acid-producing microorganisms are often buffered to neutralize these acids and prolong growth.

Microbial Metabolism and Growth
Metabolism: Catabolism • pp. 6–8

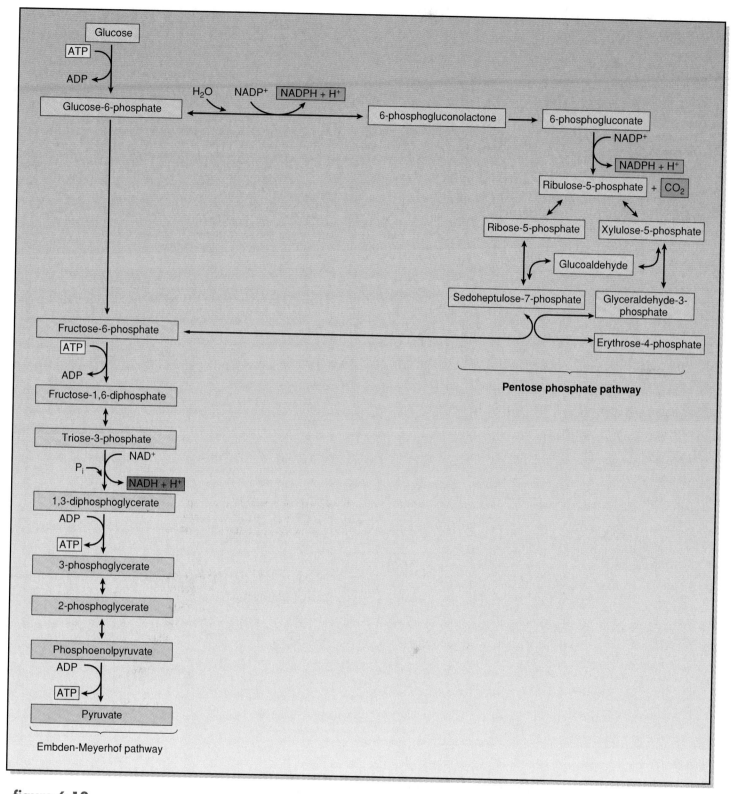

figure 6.12

The Pentose Phosphate Pathway in Relationship to the Embden-Meyerhof Pathway

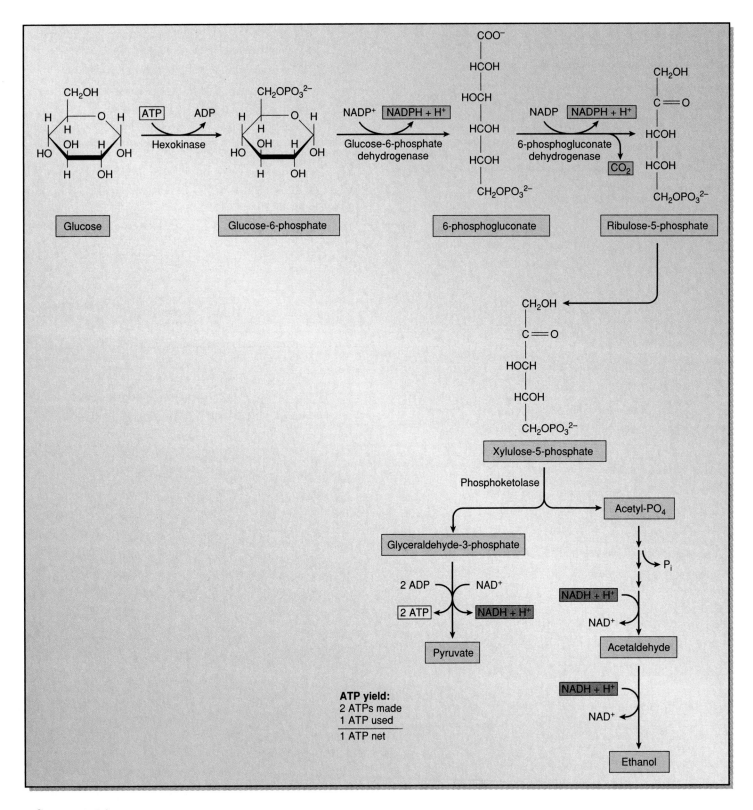

figure 6.13
The Phosphoketolase Pathway

Yeasts Produce Ethanol and Carbon Dioxide from Pyruvate

Fermenting yeasts decarboxylate pyruvate to acetaldehyde. The CO_2 released during decarboxylation causes bread dough to rise (leavening); the carbohydrates in the dough are fermented by the yeast. Acetaldehyde is subsequently reduced to ethanol, which is driven off during baking. The electrons for this reduction are provided by NADH + H+, which is thus reoxidized to NAD+. Ethanol is an important product of fermentation in the chemical, beverage, and baking industries. Alcohol is formed during the fermentation of the carbohydrates present in grains by yeasts, particularly those of the genus *Saccharomyces*.

Propionic Acid Bacteria Form Propionate, Acetate, and Carbon Dioxide from Lactate

Propionic acid bacteria (genus *Propionibacterium*) ferment lactate (the end product of fermentation of many bacteria) to propionate (propionic acid), acetate (acetic acid), and CO_2. These bacteria are thus able to extract some of the bond energy from lactate. Pyruvate, derived from oxidation of lactate, is carboxylated to oxaloacetate; this is subsequently reduced to succinate, resulting in the reoxidation of NADH + H+ to NAD+. The succinate is then decarboxylated to yield propionate as an end product. The conversion of pyruvate to propionate involves a cyclic series of reactions that can be summarized as follows:

Pyruvate + methylmalonyl-CoA → oxaloacetate + propionyl-CoA

Oxaloacetate + NADH + H+ → succinate + NAD+

Succinate + propionyl-CoA → succinyl-CoA + propionate

Succinyl-CoA → methylmalonyl-CoA

Such bacteria are used in the production of Swiss cheese (see cheese production, page 611). The propionate is responsible for the flavor of the cheese, and the CO_2 released during the decarboxylation of succinate contributes to the natural characteristic holes (eyes) of the cheese.

Butyrate, Butanol, Acetone, Isopropanol, and Carbon Dioxide Are Produced in Butyrate Fermentation

Organisms in the genus *Clostridium* actively ferment a variety of compounds (including amino acids and carbohydrates) to obtain energy. Many clostridia metabolize carbohydrates to pyruvate, which is then converted to a number of different products, including

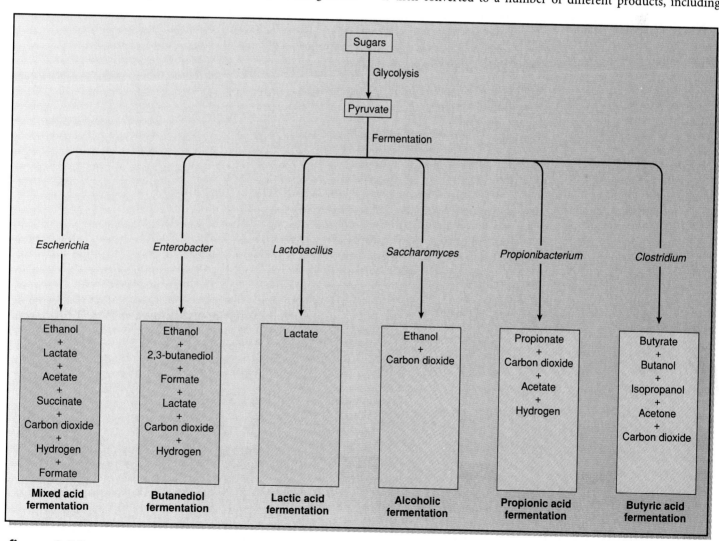

figure 6.14

Major Fermentation Pathways and Fate of Pyruvate in Bacteria

Homofermentative and Heterofermentative Microorganisms

Lactic-acid-producing microorganisms fall into two categories: **homofermentative** bacteria and **heterofermentative** bacteria. These two groups are distinguished by the types of products synthesized from carbohydrate fermentation (Figure 6.15).

Homofermentative bacteria synthesize a single product, lactate (lactic acid). These organisms include *Lactococcus, Pediococcus,* and some species of *Lactobacillus.*

Heterofermenters, in contrast, produce equal quantities of lactate, ethanol, and carbon dioxide as well as minor quantities of other products (acetate, formate, and so forth) from fermentation of glucose. Examples of heterofermentative organisms are *Leuconostoc* and certain species of *Lactobacillus.*

The difference in the fermentation products formed by these two groups of organisms is explained by the presence or absence of the enzyme **fructose-1,6-diphosphate aldolase,** which cleaves fructose-1,6-diphosphate to dihydroxyacetone phosphate and glyceraldehyde-3-phosphate in the Embden-Meyerhof pathway. This enzyme is found in homofermentative bacteria but not in heterofermentative bacteria. As a result, heterofermentative bacteria cannot ferment glucose by the Embden-Meyerhof pathway; they use the phosphoketolase pathway. Heterofermentative organisms phosphorylate glucose to glucose-6-phosphate, and convert it to 6-phosphogluconate. The 6-phosphogluconate is decarboxylated to pentose phosphate and is cleaved to form glyceraldehyde-3-phosphate and acetyl phosphate by the enzyme phosphoketolase. The glyceraldehyde-3-phosphate goes through an oxidation step and a reduction step, and eventually forms lactate. The acetyl phosphate is converted to ethanol through two consecutive reductions. Because of the initial oxidations and the formation of one instead of two triose phosphates in this pathway, only one net ATP is gained by substrate-level phosphorylation from each glucose fermented in heterofermentative organisms. Two net ATP molecules are generated from glycolysis in homofermenters. This difference in ATP yield is evident in the greater cell yield of homofermenters as compared to heterofermenters that are provided the same quantity of glucose. The two groups of organisms can be distinguished either by assaying fructose-1,6-diphosphate aldolase (which is absent in heterofermenters) or by inspecting for the production of CO_2 (produced by heterofermenters, but not homofermenters).

figure 6.15

Pathways for Glucose Fermentation in Homofermentative and Heterofermentative Bacteria

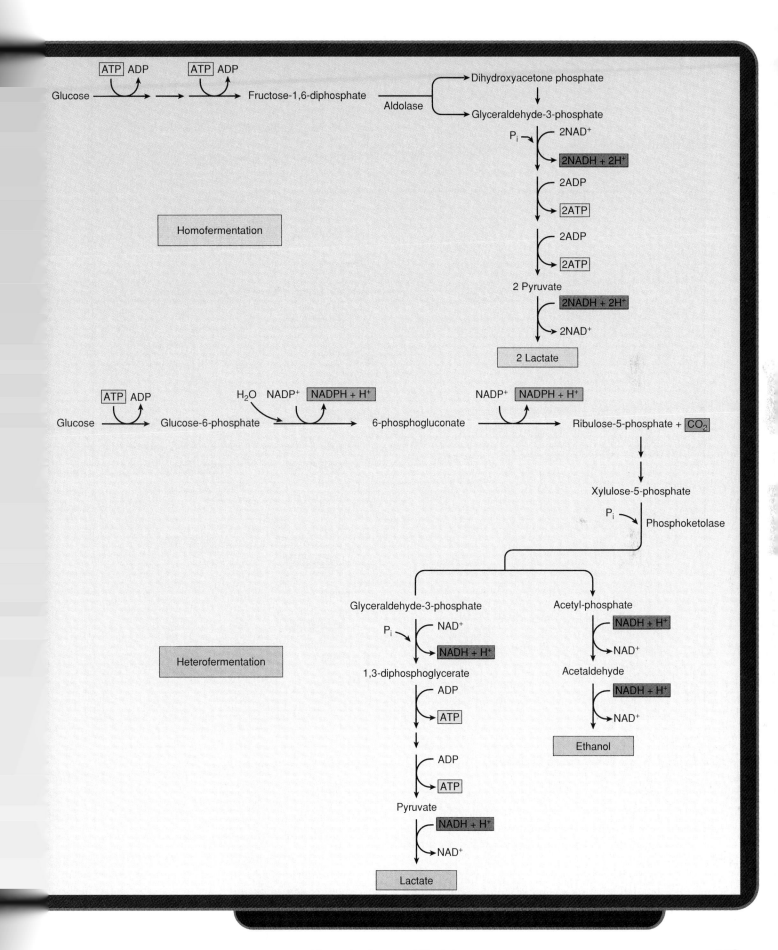

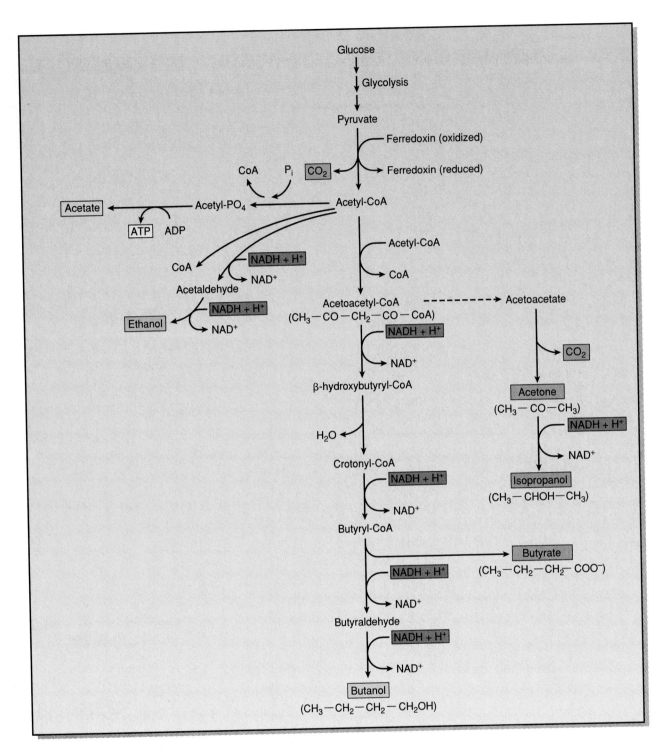

figure 6.16
Pathway for Butyrate Formation in *Clostridia*

butyrate (butyric acid), butanol, acetone, isopropanol, and CO_2. The key reaction in butyrate fermentation is the formation of acetoacetyl-CoA by the condensation of two molecules of acetyl-CoA derived from pyruvate or acetate (Figure 6.16). The acetoacetyl-CoA subsequently is reduced by two molecules of NADH + H^+, and butyrate is formed. In addition to *Clostridium*, bacteria of the genera *Butyrivibrio*, *Eubacterium*, and *Fusobacterium* also produce butyrate.

Mixed Acid Fermentation and Butanediol Fermentation Are Distinguished by the Quantity of Acid Produced

Pyruvate that is produced from glucose metabolism in gram-negative enteric bacilli such as *Escherichia*, *Salmonella*, and *Enterobacter* can be further degraded under anaerobic conditions into different products, depending on the species and their

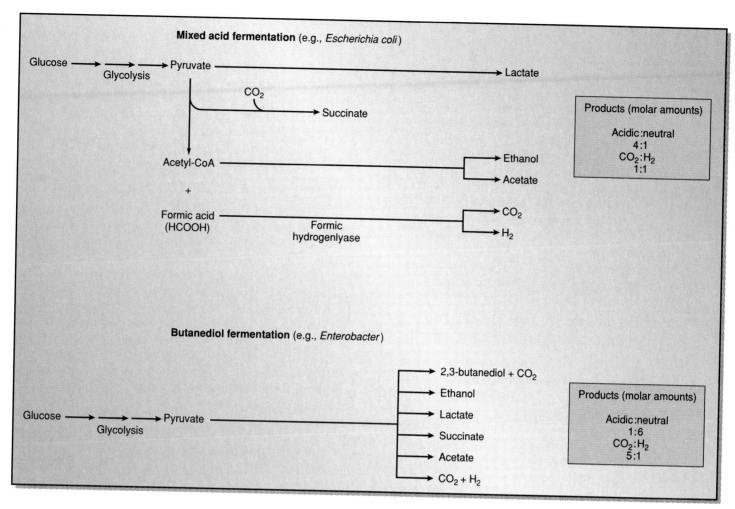

figure 6.17
Mixed Acid versus Butanediol Fermentation

enzyme complement. This degradation of pyruvate occurs by one of two distinct fermentative pathways: **mixed acid fermentation** and **butanediol fermentation** (Figure 6.17). Mixed acid fermentation produces (1) three major acids: lactate (lactic acid), succinate (succinic acid), and acetate (acetic acid), (2) ethanol, (3) and either formate or CO_2 and H_2, depending on the absence or presence of the enzyme formic hydrogenlyase. Butanediol fermentation also produces lactate, succinate, and acetate, but in smaller quantities. Other products of butanediol fermentation are butanediol, ethanol, CO_2, and H_2.

The two pathways may be easily distinguished by the quantity of acid produced. More acid is produced during mixed acid fermentation than in butanediol fermentation. The ratio of CO_2 to H_2 produced in the two pathways is also different: it is equimolar (1:1) in mixed acid fermentation and 5 (CO_2):1 (H_2) in butanediol fermentation. The difference in acid production in these two types of fermentations can be detected by the **methyl red test.** This test is performed by growing bacteria in glucose broth and quantitatively measuring the amount of acid produced after 48 hours of growth, using the pH indicator methyl red. When added to the growth medium, the indicator remains red at low pHs (<5) and turns yellow-

orange at higher pHs (>5.5). A red color is indicative of a positive test for mixed acid fermentation, whereas a yellow-orange color indicates a negative test. The 48-hour incubation period prior to the methyl red test is necessary, since all enterics will initially produce some acid from glucose fermentation. With prolonged incubation, mixed acid fermenters continue to excrete large amounts of acid in comparison to butanediol fermenters. However, the 48-hour incubation period for the methyl red test can be reduced to 18 to 24 hours if smaller substrate volumes and larger inocula are used.

Another test, the **Voges-Proskauer test,** may be used to identify the butanediol fermenters such as *Enterobacter.* The reagents used in the test detect acetylmethylcarbinol (acetoin), an intermediate in butanediol production. This intermediate is not produced by *E. coli, Salmonella,* and the other enteric mixed acid producers.

Oxidative Phosphorylation

One or two net ATP molecules are synthesized by substrate-level phosphorylation through the metabolism of glucose to two molecules of pyruvate. This ATP yield can be increased to as many as 38 ATP molecules in bacteria if these pyruvates are subsequently channeled into the tricarboxylic acid (TCA) cycle, with accompanying electron transport and oxidative phosphorylation. This increased ATP yield is possible because the complete oxidation of pyruvate is realized through the TCA cycle. The electrons released from these oxidations and the earlier oxidations in glycolysis are sent through an electron transport chain to a final acceptor, resulting in the generation of additional molecules of ATP. Oxygen or inorganic molecules such as ions of nitrate (NO_3^-) or sulfate (SO_4^{2-}) may be the terminal electron acceptor of these electrons. The energy available during the transport of electrons to the final acceptor (an acceptor with a highly positive redox potential) is conserved in ATP by a process known as oxidative phosphorylation. Up to three ATP molecules are formed by the energy change realized from the difference in the reduction potential of the $NAD^+/NADH + H^+$ pair ($E'_0 = -0.32$ V) and the $1/2$ O_2/H_2O pair ($E'_0 = +0.82$ V) when oxygen is the final electron acceptor.

The Tricarboxylic Acid Cycle Generates Intermediates for Other Pathways and Cellular Activities and Is Central to Cell Metabolism

The **TCA cycle** (also known as the **Krebs cycle** or the **citric acid cycle**) was elucidated by Sir Hans Krebs, an English biochemist who received the Nobel Prize in physiology and medicine in 1953 jointly with Fritz A. Lipman for their studies on intermediary metabolism. This major pathway generates intermediates required for other pathways and cellular activities and is the center of cell metabolism. Activation of the TCA cycle can therefore lead to increased activity of other metabolic pathways. Inhibition of the TCA cycle has deleterious effects on overall cellular metabolism. Figure 6.18 provides a schematic overview of the TCA cycle. A detailed discussion of the cycle follows.

The pyruvate generated by glycolysis or other pathways can feed into the TCA cycle by first being oxidatively decarboxylated in a reaction requiring coenzyme A (CoA) to form acetyl-CoA, with the release of CO_2 and the reduction of NAD^+. The acetyl group of acetyl-CoA then enters the TCA cycle by combining with oxaloacetate to form citrate. The enzyme responsible for this reaction, citrate synthase, is a key enzyme in the TCA cycle. Citrate synthase is an allosteric enzyme with activity regulated by $NADH + H^+$, α-ketoglutarate, and/or ATP concentrations, depending on the type of bacterium. Citrate synthase activity in gram-negative bacteria is inhibited by high concentrations of $NADH + H^+$ and/or α-ketoglutarate. The activity of the same enzyme in gram-positive bacteria is controlled by ATP levels.

In the next series of reactions following formation of citrate, citrate is converted first to isocitrate (*cis*-aconitate is an unstable intermediate of this reaction) and then to α-ketoglutarate. The conversion of isocitrate to α-ketoglutarate is an oxidative decarboxylation reaction catalyzed by the enzyme isocitrate dehydrogenase. The immediate product of the oxidation is oxalosuccinate, but this product is rapidly decarboxylated to yield α-ketoglutarate. NAD^+ is reduced to $NADH + H^+$, and a molecule of CO_2 is released during these reactions.

α-ketoglutarate becomes the substrate for a second oxidative decarboxylation reaction, which produces succinyl-CoA. This reaction is catalyzed by α-ketoglutarate dehydrogenase and is accompanied by the release of one molecule of CO_2 and the reduction of NAD^+ to $NADH + H^+$.

The next reaction in the TCA cycle is the conversion of succinyl-CoA to succinate. This is the only exergonic reaction in the TCA cycle directly coupled with the generation of a high-energy compound—either ATP, as in bacteria and higher plants, or guanosine triphosphate (GTP), as in mammals—via substrate-level phosphorylation. GTP can be used directly by the cell or converted to ATP by the following reaction:

$$GTP + ADP \xrightarrow{\text{nucleoside diphosphate kinase}} ATP + GDP$$

The TCA cycle continues with the oxidation of succinate to fumarate by the enzyme succinate dehydrogenase. The electrons and hydrogen atoms released during this oxidation are accepted by flavin adenine dinucleotide (FAD), not NAD. FAD is a coenzyme similar to NAD in its action. In this particular instance FAD serves as the specific coenzyme and is covalently bound to succinate dehydrogenase.

Fumarate is converted to malate. Malate is subsequently oxidized to oxaloacetate, with NAD accepting the electrons and protons of this oxidation. This final oxidation completes one turn of the TCA cycle. The cycle can now repeat with the condensation of oxaloacetate with a new molecule of acetyl-CoA entering the pathway from pyruvate.

The result of one turn of the TCA cycle, starting from acetyl-CoA, is the production of two CO_2 molecules, one ATP (GTP) molecule, three molecules of $NADH + H^+$, and one molecule of $FADH + H^+$. An additional CO_2 molecule and $NADH + H^+$ are generated in the preliminary step to the cycle, in which pyruvate is oxidized to acetyl-CoA and CO_2. Since two molecules of pyruvate are formed from each molecule of glucose sent through the Embden-Meyerhof pathway, it takes two complete turns of the TCA cycle to process these pyruvates.

Most of the energy that is liberated from the oxidation of pyruvate through the TCA cycle is conserved in reduced cofactors and then in ATP molecules synthesized by oxidative phosphorylation. These ATP molecules are produced as electrons released during oxidation, transferred by such carriers as $NADH + H^+$ and $FADH_2$ to an electron transport chain, and accepted ultimately by the terminal acceptor, oxygen, or some other molecule.

Energy That Is Liberated While Electrons Are Transported in an Electron Transport Chain Can Be Coupled to the Formation of ATP

The **electron transport chain** is an alternating oxidation-reduction chain of electron carriers located in the plasma membrane in procaryotes and in the inner mitochondrial membrane in eucaryotes. Electrons from $NADH + H^+$ or $FADH_2$ are transported in the chain through a set of electron carriers. The number and types of electron carriers vary from bacterial species to species. The most common carriers in bacterial electron transport chains are pyridine nucleotides, flavoproteins, quinones, iron-sulfur proteins, and cytochromes. Pyridine nucleotides, flavoproteins, and quinones are carriers of protons and electrons; iron-sulfur proteins and cytochromes are electron carriers.

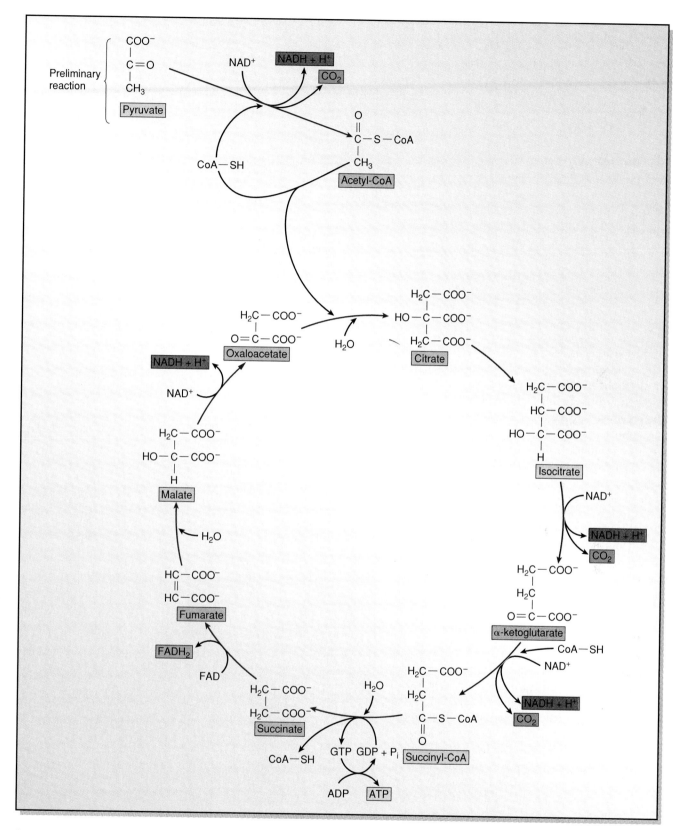

figure 6.18

The Tricarboxylic Acid Cycle

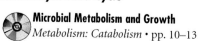
Microbial Metabolism and Growth
Metabolism: Catabolism • pp. 10–13

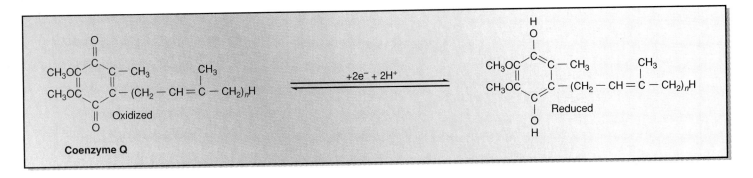

figure 6.19

Coenzyme Q

Electrons usually enter the electron transport chain through the pyridine nucleotide NADH + H⁺. Occasionally flavoproteins accept electrons in the oxidation reactions of metabolism. Flavoproteins are proteins containing a prosthetic group derived from riboflavin (see Figure 6.7). The prosthetic group, which may be either FAD or flavin mononucleotide (FMN), is reduced as it accepts electrons (and protons) or oxidized as the electrons (and protons) are passed on. The flavoprotein FMN has a more positive redox potential than NAD and therefore serves as an intermediate acceptor for electrons passed from NADH + H⁺ in electron transport.

Quinones are lipid-soluble substances of low molecular weight. One of the more common quinones, coenzyme Q, or ubiquinone (so named because it is ubiquitous and is found in all organisms), is a dual electron-proton carrier (Figure 6.19).

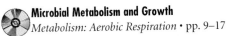

Microbial Metabolism and Growth
Metabolism: Aerobic Respiration • pp. 9–17

Iron-sulfur proteins are proteins containing iron and sulfur atoms complexed with four cysteine residues of the protein. The iron atoms in these proteins are the actual electron carriers and can be in either the oxidized (Fe^{3+}) or the reduced (Fe^{2+}) state. Ferredoxin is an iron-sulfur protein present in the electron transport chain of some bacteria.

Cytochromes are heme proteins, with an iron-containing porphyrin ring attached to proteins. The iron ligand of the cytochrome serves as the electron carrier by becoming alternately reduced and oxidized from the ferrous (Fe^{2+}) form to the ferric (Fe^{3+}) form. Cytochromes carry only electrons, not protons. Each cytochrome can carry only one electron at a time. The cytochromes were first discovered in 1886 by C.A. McMunn. They were divided into three major classes by D. Keilin in 1930 on the basis of their absorption spectra. Cytochrome a absorbs at

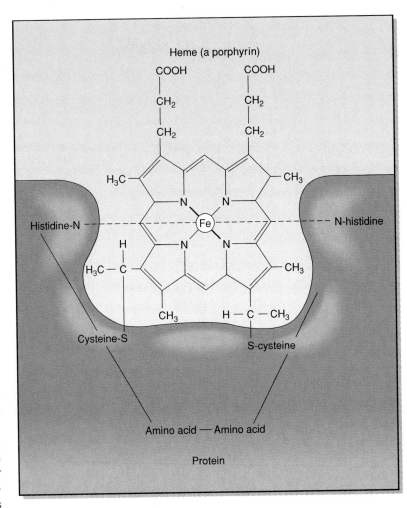

figure 6.20

Cytochrome c

The iron-porphyrin group of cytochrome c, showing the manner in which the porphyrin is attached to the protein.

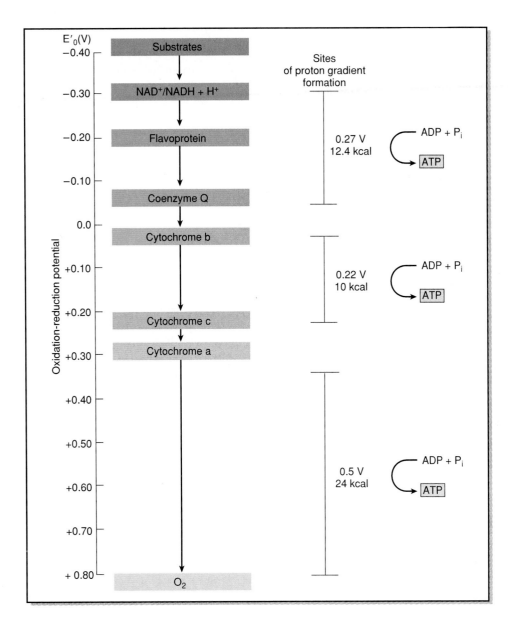

figure 6.21

The Electron Transport Chain

Electrons during glucose oxidation enter the electron transport chain through NADH + H$^+$ or FADH$_2$. The electrons are transported via a series of carriers with progressively more positive redox potentials, and eventually reach oxygen. The energy liberated during this process is partially conserved in the synthesis of ATP + P$_i$ at discrete coupling sites. There are three coupling sites for ATP generation between NADH + H$^+$ and oxygen.

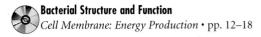

Bacterial Structure and Function
Cell Membrane: Electron Transport Chain • pp. 18–19

the longest wavelength; cytochrome b, at an intermediate wavelength; and cytochrome c, at the shortest wavelength (Figure 6.20). Additional cytochromes have since been discovered and are designated by subclasses (a$_1$, a$_2$, a$_3$, and so forth).

Each of the different electron carriers has a different redox potential and is arranged in the electron transport chain in a specific location (Figure 6.21). Electrons are passed from carriers having more negative redox potentials to carriers having more positive redox potentials; eventually electrons reach the terminal electron acceptor. The energy liberated while electrons move through this gradient is partially conserved in the synthesis of ATP from ADP + P$_i$. Although enough free energy is liberated during electron transport to form approximately seven ATP molecules from ADP and P$_i$, at most three ATP molecules are actually made because there are only three discrete coupling sites of ATP generation between NADH + H$^+$ and the terminal electron acceptor. ATP generation is associated with the three main enzyme

complexes of electron transport: NADH + H$^+$ dehydrogenase (site I), cytochrome c reductase (site II), and cytochrome c oxidase (site III). FADH$_2$, which has a more positive redox potential than NADH + H$^+$, feeds electrons into the transport chain via coenzyme Q, bypassing one of the three ATP coupling sites. As a consequence, only two ATP molecules at most are formed for each electron pair entering the electron transport chain through FADH$_2$ instead of NADH + H$^+$.

Bacterial Structure and Function
Cell Membrane: Energy Production • pp. 12–18

Chemiosmosis Explains How Electron Transport Is Used to Generate ATP

Chemiosmosis, or **proton motive force,** first proposed in 1961 by Peter Mitchell, explains how ATP is synthesized during electron transport. Mitchell, who received a Nobel Prize in chemistry in

Oxidase Test

The types of cytochromes and associated enzymes located in the electron transport chain vary from organism to organism. Some bacteria, specifically facultatively anaerobic organisms such as the enterics, lack cytochrome c oxidase. The absence or presence of cytochrome c oxidase can be detected in a diagnostically useful test, the **oxidase test,** which is used to distinguish enteric bacteria from bacteria that have cytochrome c oxidase (for example, *Pseudomonas* and *Neisseria*).

The oxidase test is performed by mixing a drop of *N,N*-dimethyl-*p*-phenylenediamine on a piece of filter paper with a young culture of bacteria, or by adding a drop of the reagent to bacteria on an agar plate. The dye serves as an electron donor and is oxidized to a purple-blue product (indophenol) by organisms that have cytochrome c oxidase (Figure 6.22). Cytochrome c oxidase oxidizes cytochrome c, which in turn oxidizes *N,N*-dimethyl-*p*-phenylenediamine to indophenol. The dye is not oxidized and remains colorless when reacted with oxidase-negative organisms.

The oxidase test provides a simple, rapid technique to distinguish oxidase-positive bacteria, particularly pathogens such as *Neisseria gonorrhoeae* and *Neisseria meningitidis*, from those that are oxidase negative. It is commonly used in clinical laboratories as a diagnostic aid in the preliminary identification of these bacteria.

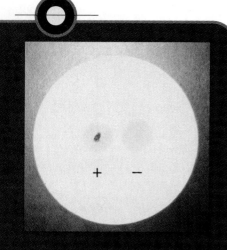

figure 6.22
Oxidase Test for the Presence of Cytochrome c Oxidase
The oxidase test is performed by mixing a drop of *N,N*-dimethyl-*p*-phenylenediamine on filter paper with a young culture of bacteria. Oxidase-positive bacteria produce cytochrome c oxidase, which indirectly oxidizes the colorless *N,N*-dimethyl-*p*-phenylenediamine to a purple-blue product (indophenol).

1978 in recognition of his work, hypothesized that electron transport is accompanied by the translocation of protons from the inside to the outside of the plasma membrane to establish an electrochemical gradient that is the driving force (proton motive force) for ATP synthesis (Figure 6.23). This proton gradient represents an electrical potential, similar to the potential in a battery between the positive and negative poles. As the gradient dissipates, its energy may be transferred in part to the synthesis of ATP.

The proton gradient is established during electron transport, a process that occurs in the membrane (the plasma membrane in procaryotes and the mitochondrial membrane in eucaryotes). The electron carriers are oriented in the membrane in such a way that there is a separation of electrons and hydrogen ions (protons) during transport. Between NADH + H$^+$ (or FADH$_2$) and the cytochromes, carriers that transport both electrons and protons (flavoproteins and quinones) alternate with carriers that transport only electrons (the iron-sulfur proteins and cytochromes); consequently protons are translocated across and to the outside of the membrane during electron transport.

As each pair of electrons is sent through the transport chain, protons are released to the outside of the membrane. This translocation results in an excess of protons on the exterior side of the membrane and the establishment of a proton gradient. The gradient does not automatically dissipate, since the membrane is impermeable to protons. However, protons are able to traverse the membrane via enzyme complexes (ATPases) spanning the membrane at certain sites. These enzyme complexes act as proton channels across the membrane. The inward movement of protons across the membrane is coupled with ATP synthesis.

This proton pump mechanism for ATP synthesis is reversible. When the pump operates in the reverse reaction, ATP hydrolysis drives protons across the membrane to the cell exterior, resulting in an energized membrane. This reverse reaction is important in active transport (see transport mechanisms, page 86).

The establishment of a proton gradient, not electron transport, is responsible for ATP synthesis. Evidence for this role of the proton motive force came from studies with bacteriorhodopsin, a protein in the purple plasma membrane fraction of *Halobacterium* (domain Archaea). Bacteriorhodopsin, which is responsible for the patches of purple membrane on the cell surface of *Halobacterium,* has a chemical structure similar to the sensory pigment rhodopsin of the vertebrate retina. Covalently bound to bacteriorhodopsin is the carotenoid derivative retinal, a chromophore that normally exists in the *trans* configuration. Bacteriorhodopsin strongly absorbs light at 570 nm and functions as a light-driven proton pump. When the pigment is illuminated, the retinal chromophore absorbs the light and temporarily changes to a *cis* configuration. This transformation results in deprotonation and the transfer of protons across the plasma membrane to the periplasm. The retinal molecule then slowly takes up a proton from the cytoplasm and isomerizes back to the *trans* form. The proton gradient generated across the plasma membrane is used to

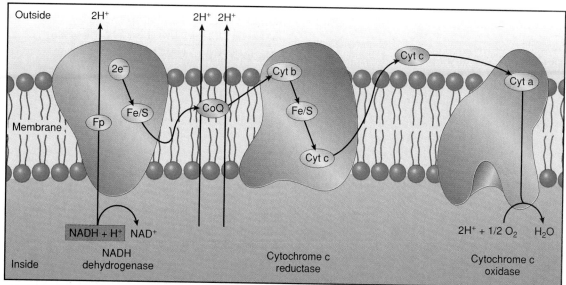

figure 6.23

The Chemiosmotic Theory of ATP Synthesis and Proton Translocation

a. Electron carriers in the electron transport chain pump protons (H^+) to the outside of the cell, establishing a proton gradient across the membrane.
b. As the protons reenter the cell via ATPases, the gradient dissipates, and the released energy is coupled to the synthesis of ATP. Fp, flavoprotein; Fe/S, iron-sulfur protein; CoQ, coenzyme Q; Cyt, cytochrome.

Bacterial Structure and Function
Cell Membrane: Electron Transport Chain • p. 20

drive ATP synthesis via an ATPase complex, as in oxidative phosphorylation, but in the absence of electron transport. Halobacteria benefit from such an oxygen-independent proton pump mechanism, since they are commonly found in salt-saturated environments where oxygen solubility is low and there is ample light. In such environments, these bacteria have an alternative light-mediated system for ATP production.

Bacterial Structure and Function
Cell Membrane: Energy Production • pp. 19–21

Uncouplers Inhibit ATP Synthesis Without Inhibiting Electron Transport

Some of the strongest evidence for ATP synthesis by a proton pump comes from the use of chemical agents that specifically inhibit electron transport, oxidative phosphorylation, or both (Table 6.4). One group of chemical agents, called uncouplers, causes membranes to be leaky to protons. These uncouplers are lipid-soluble substances that combine with protons on the exterior

table 6.4

Inhibitors of Electron Transport

Inhibitor	Result
Uncouplers:	
2,4-dinitrophenol	Uncoupling of oxidative
Dicumarol	phosphorylation from
Salicylanilide	electron transport
Inhibitors of electron transport:	
Amobarbital (Amytal)	Blockage of electron transport between FAD and quinone
Antimycin A	Blockage of electron transport between cytochrome b and cytochrome c
Cyanide	Blockage of electron transport at cytochrome a
Inhibitors of phosphorylation:	
Oligomycin	Inhibition of ATPase
Atractyloside	Inhibition of membrane carrier for adenine nucleotides in plants

table 6.5

Examples of Electron Acceptors in Anaerobic Respiration

Electron Acceptor	Reduced Product	Genus
NO_3^-	NO_2^-, N_2O, N_2	Alcaligenes Bacillus
NO_2^-	N_2O, N_2	Paracoccus Pseudomonas Spirillum
SO_4^{2-}	H_2S	Desulfococcus Desulfomonas Desulfovibrio
Fumarate	Succinate	Desulfovibrio Vibrio

of the cell and carry them across the membrane to the cell interior, bypassing the ATPase in the membrane. Although electron transport still occurs in the presence of uncouplers, no proton gradient is established. The result is an uncoupling of oxidative phosphorylation from electron transport—energy released during electron transport is lost as heat and not conserved in the form of ATP. Examples of uncouplers are 2,4-dinitrophenol (DNP), dicumarol, and salicylanilide. For example, when *E. coli* cells are incubated in the presence of 2,4-dinitrophenol, glucose is oxidized, but ATP is not formed during electron transport. Uncouplers provide evidence that although electron transport and oxidative phosphorylation usually are coupled, each is able to operate separately.

Other chemical agents directly affect electron transport by combining with carriers of the chain. Cyanide, an irreversibly bound noncompetitive inhibitor, binds with cytochrome a to block electron transport at that step. Antimycin A inhibits electron transport between cytochrome b and cytochrome c. These chemical agents have proven useful in establishing the precise locations of carriers in the electron transport chain.

Aerobic Respiration Is More Efficient Than Fermentation in Coupling Liberated Energy to ATP Formation

The majority of the ATP molecules formed during respiration is provided by oxidative phosphorylation. The NADH $+H^+$ and $FADH_2$ formed in the TCA cycle, as well as the NADH $+H^+$ produced during glycolysis, can be shunted into the electron transport chain when oxygen is available. As many as 3 ATP molecules are synthesized for each NADH $+H^+$, and 2 ATP molecules may be synthesized for each $FADH_2$ sent through the chain. Thus in addition to the 2 net ATP molecules formed during glycolysis and the 2 ATP molecules formed from GTP in the TCA cycle as a result of glucose oxidation, 34 additional ATP molecules (30 ATP molecules

from ten molecules of NADH $+ H^+$ and 4 ATP molecules from two molecules of $FADH_2$) are made available through oxidative phosphorylation under optimal conditions.

Most microorganisms obtain energy from chemical compounds by fermentation and aerobic respiration. The total amount of free energy available from the oxidation of glucose to carbon dioxide and water is -688 kcal/mole. Theoretically as many as 38 net ATP molecules are generated by bacteria in aerobic respiration from the oxidation of one molecule of glucose. If it is assumed that each ATP formed has -7.3 kcal/mole of free energy available in its terminal phosphate bond, approximately 40% of the available energy released as a result of glucose oxidation (38 ATP molecules × -7.3 kcal/mole) is conserved in the form of ATP.

The fermentation of glucose to ethanol and carbon dioxide, in comparison, releases 57 kcal/mole of free energy, of which only 14.6 kcal/mole (26%) is conserved in the 2 ATP molecules that are formed. Aerobic respiration is thus more efficient than fermentation in the conservation of released energy. The greater yield of ATP and the higher efficiency of aerobic respiration is evident when the growth of facultatively anaerobic organisms is compared under anaerobic and aerobic conditions (that is, the Pasteur effect) (see Pasteur effect, page 152). When grown aerobically with the same quantities of carbohydrates, the same microorganism reaches a greater cell mass (fourfold greater) than when grown under anaerobic conditions.

Anaerobic Respiration Is a Process Unique to Procaryotes

Fermentation and aerobic respiration occur in procaryotes and eucaryotes. Additionally, some procaryotes have a variation of aerobic respiration called anaerobic respiration, by which they synthesize ATP. This process, which is unique to procaryotes, is similar to aerobic respiration; the major exception is that the terminal electron acceptor in the electron transport chain is a chemical compound other than molecular oxygen. A wide variety of substances can serve as alternate electron acceptors to oxygen (Table 6.5).

Nitrate (NO_3^-) is an example of a terminal electron acceptor used by some microbes in anaerobic respiration. Organisms such as *Pseudomonas* and *Bacillus* that can respire aerobically generally uti-

Oxidative phosphorylation is similar in eucaryotes and procaryotes, with certain exceptions. Glycolysis in eucaryotes takes place in the cytoplasm, whereas the TCA cycle

example of such a system is the transfer of electrons from NADH + H⁺ to glycerol-3-phosphate, which is synthesized from dihydroxyacetone phosphate by the cyto-

acetone phosphate exits the mitochondrion to be reduced at the expense of NADH + H⁺ again to continue the cycle. Since the electrons transported into the mitochondrion by

Oxidative Phosphorylation in Eucaryotes

and oxidative phosphorylation operate in the mitochondrion. The proton gradient in oxidative phosphorylation is established across the inner mitochondrion membrane. The outer mitochondrion membrane, however, is impermeable to NADH + H⁺. Consequently the two molecules of NADH + H⁺ formed from the glycolytic pathway are unable to enter the mitochondrion through the membrane.

The electrons of NADH + H⁺ enter the mitochondrion via shuttle systems. One

plasmic enzyme glycerol phosphate dehydrogenase (dihydroxyacetone phosphate/glycerol-3-phosphate shuttle); (Figure 6.24). Glycerol-3-phosphate is able to penetrate the mitochondrion membrane. Once inside the mitochondrion, glycerol-3-phosphate is oxidized back to dihydroxyacetone phosphate by an FAD-linked glycerol phosphate dehydrogenase, which passes the electrons to coenzyme Q in the electron transport chain. The dihydroxy-

the dihydroxyacetone phosphate/glycerol-3-phosphate shuttle are transferred to coenzyme Q, they bypass the first ATP coupling site of the transport chain and yield only two ATP molecules per electron pair. As a consequence, since glycolysis contributes two molecules of cytoplasm-derived NADH + H⁺ for each molecule of glucose oxidized, only 36 ATP molecules are theoretically produced from complete oxidation of glucose.

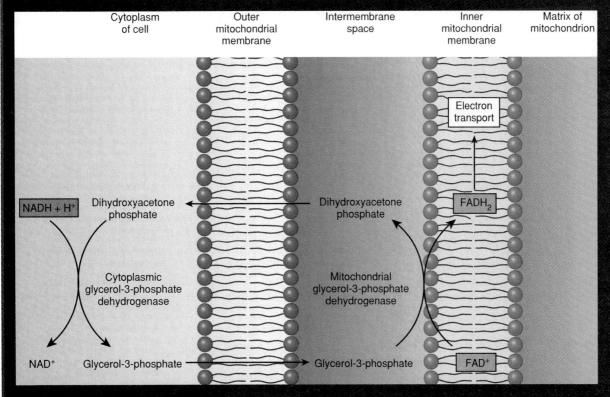

figure 6.24
The Dihydroxyacetone Phosphate/Glycerol-3-Phosphate Shuttle
This shuttle transports electrons from cytoplasmic NADH + H⁺ into the electron transport chain of the inner mitochondrial membrane. Electrons move inward via glycerol-3-phosphate, which can be oxidized to dihydroxyacetone phosphate by an FAD-linked dehydrogenase in the inner membrane.

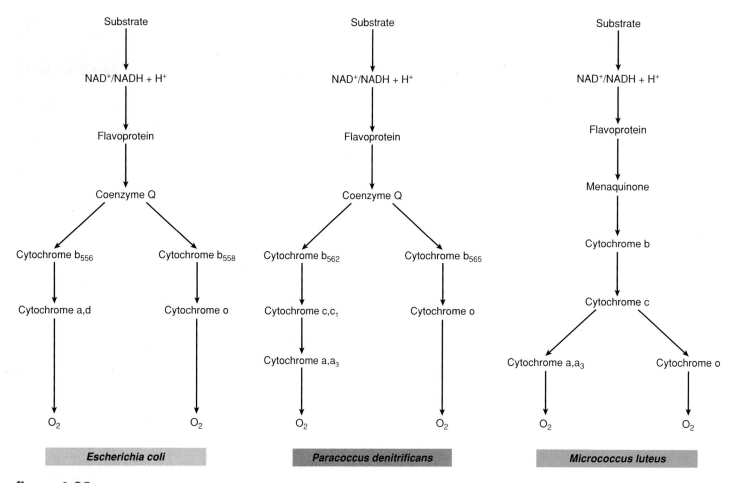

figure 6.25

Electron Transport Chains of Three Types of Bacteria

Escherichia coli, Paracoccus denitrificans, and *Micrococcus luteus* have electron transport chains that branch in two directions before reaching the terminal electron acceptor. Branch points occur at the coenzyme Q level and at the cytochrome c level.

lize oxygen as an electron acceptor. However, if oxygen is absent, some species of these genera may shift to anaerobic respiration. They reduce nitrate to nitrite (NO_2^-) by the enzyme nitrate reductase. The reduction potential for the NO_3^-/NO_2^- pair is only +0.42 V, compared with +0.82 V for the $1/2\ O_2/H_2O$ pair used in aerobic respiration. Consequently, when nitrate is used as a terminal electron acceptor in anaerobic respiration, less energy is released during electron transport, an ATP coupling site is lost, and at most only two instead of three molecules of ATP are generated. Since the nitrite that is formed during anaerobic respiration is toxic to many procaryotes, it frequently is further reduced to nontoxic nitrous oxide (N_2O) or nitrogen gas (N_2). The process of nitrate reduction to gaseous nitrogen is called **denitrification,** which is considered a **dissimilatory** process. In comparison, **assimilatory nitrate reduction** involves the uptake of nitrate as a source of cell nitrogen and its reduction to organic nitrogen. Most organisms that carry out denitrification are soil inhabitants and use the nitrate in soil in a dissimilatory fashion. Reduction of nitrate to gaseous products is detrimental to plants, since it results in loss of fixed nitrogen from the soil. The antithesis of this process is nitrogen fixation by cyanobacteria and certain other procaryotes, symbiotically or nonsymbiotically.

Other procaryotes (for example, *Desulfovibrio*) use sulfate (SO_4^{2-}) as the electron transport system terminal external electron acceptor, and reduce sulfate to hydrogen sulfide (H_2S) by **dissimilatory sulfate reduction.** Sulfate-reducing bacteria typically are obligate anaerobes that exist in aquatic environments and anaerobic soils. They are economically important in corrosion of metals such as iron and copper, both of which react with sulfides to form insoluble sulfide precipitates. Some organisms accumulate sulfate and reduce it intracellularly for incorporation into organic sulfur in a process called **assimilatory sulfate reduction.**

A number of other chemical compounds, including fumarate ($^-OOC—CH=CH—COO^-$) and carbon dioxide, are used as electron acceptors in anaerobic respiration. The yield of ATP generated by anaerobic respiration depends on the difference in reduction potential of the electron donor and acceptor pairs, but it is typically less in anaerobic respiration than in aerobic respiration. Nonetheless, this alternative for electron transport gives procaryotes the flexibility that enables them to grow and survive in a wide range of environments.

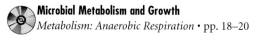 **Microbial Metabolism and Growth**
Metabolism: Anaerobic Respiration • pp. 18–20

Examples of Chemolithotrophs

Group		Oxidizable Substrate	Oxidized Product	Genus
Sulfur oxidizers		H_2S, S^0 $S_2O_3^{2-}$	SO_4^{2-}	Sulfolobus Thiobacillus
Iron bacteria		Fe^{2+}	Fe^{3+}	Gallionella Sphaerotilus Thiobacillus
Nitrifying bacteria	Ammonia oxidizers	NH_3	NO_2^-	Nitrosococcus Nitrosolobus Nitrosomonas Nitrosospira
	Nitrite oxidizers	NO_2^-	NO_3^-	Nitrobacter Nitrococcus Nitrospina
Hydrogen bacteria		H_2	H_2O	Alcaligenes Nocardia Pseudomonas

ATP Yield from Glucose Oxidation Is Not the Same in All Microorganisms

The possession of an electron transport chain undoubtedly increases the ATP yield from oxidation of chemical compounds. The examples given in this chapter indicate a theoretical maximum yield of 38 molecules of ATP from the complete aerobic oxidation of one molecule of glucose. This ATP yield assumes the synthesis of three ATP molecules from the transport of electrons from NADH + H^+ to oxygen in the electron transport chain.

However, this ATP yield is only theoretical and may be much lower among bacterial species, depending on the nature of their electron transport chains and the reduction potential of the terminal electron acceptor in these chains. For example, the electron transport chains of *Escherichia coli*, *Paracoccus denitrificans*, and *Micrococcus luteus* differ in their components (Figure 6.25). The chain of *E. coli* lacks cytochrome c, but has a cytochrome d that is not found in *P. denitrificans* and *M. luteus*. *M. luteus* has menaquinone instead of coenzyme Q as an intermediate electron carrier in its chain. These differences in electron carriers affect ATP generation during electron transport in these chains. Experimental evidence indicates that the electron transport chain of *E. coli* has two ATP coupling sites, but the chains of *P. denitrificans* and *M. luteus* (at least the branch via cytochrome a, a_3) have three ATP coupling sites. These differences would suggest that the ATP yield from complete aerobic oxidation of one mole of glucose in *E. coli* is significantly less than in the other two types of bacteria with different electron transport chains.

A similar situation occurs in bacteria that generate ATP by anaerobic respiration. The basic mechanism of electron transport is the same as in aerobic respiration. The terminal electron acceptors in the transport chain of anaerobically respiring bacteria, however, frequently have less-positive reduction potentials than

the $1/2 \ O_2/H_2O$ pair. For example, *P. denitrificans* forms three molecules of ATP per atom of oxygen when grown aerobically. Under anaerobic conditions, this same bacterium uses nitrate as the terminal external electron acceptor and generates at most two molecules of ATP per molecule of nitrate because of the loss of the third ATP coupling site in electron transport.

As may be seen from these examples, there are variations in electron transport chains in different bacteria. One must consider these differences when estimating the efficiency of ATP yield from oxidative phosphorylation.

Chemolithotrophs

Chemolithotrophs are procaryotes that oxidize reduced inorganic compounds for energy (Table 6.6). Most of these organisms can also obtain all of their carbon from carbon dioxide and therefore are autotrophs. Chemolithotrophs generate ATP by the oxidation of inorganic compounds and electron transport phosphorylation. Inorganic compounds commonly used as energy sources by chemolithotrophs include reduced sulfur compounds, ferrous compounds, reduced nitrogenous compounds, and H_2.

Hydrogen Sulfide Is Oxidized to Elemental Sulfur and Eventually to Sulfate

Microbes that oxidize reduced sulfur compounds are found in such sulfide-rich environments as sulfur springs, acid mine waters, sewage, and marine mud. The procaryotes *Thiobacillus* (domain Bacteria) and *Sulfolobus* (domain Archaea) are among those that oxidize hydrogen sulfide (H_2S) to elemental sulfur and eventually to sulfate (SO_4^{2-}). The elemental sulfur formed in the first oxidation step is highly insoluble and is deposited in or outside the procaryotic cell as sulfur granules. Production of sulfate in a subsequent oxidation step results in acidic conditions. Many of the sulfur-oxidizing procaryotes are able to tolerate low pH environments and, in fact, maintain the acidity of these environments through their metabolic products.

Ferrous Iron Is Oxidized to Ferric Iron

Ferrous iron (Fe^{2+}) is aerobically oxidized to ferric iron (Fe^{3+}) by bacteria that are typically encrusted with coats of iron oxide and inhabit aquatic environments containing large quantities of reduced iron salts. Among these organisms are *Sphaerotilus*, *Gallionella*, and *Thiobacillus ferrooxidans*. Many of these iron-oxidizing bacteria also oxidize sulfur to sulfuric acid and are therefore acidophiles. The acidic environment generated by these bacteria results in a proton gradient across the bacterial plasma membrane, which is used to drive ATP synthesis by a classical chemiosmotic ATPase reaction.

T. ferrooxidans is able to oxidize ferrous or sulfur compounds and is often found in drainage water from mines. This organism oxidizes ferrous iron in the following manner:

$$4\ FeSO_4 + O_2 + 2\ H_2SO_4 \rightarrow 2\ Fe_2(SO_4)_3 + 2\ H_2O$$

When the ferric sulfate that is produced from this oxidation comes in contact with water, it is hydrated:

$$2\ Fe_2(SO_4)_3 + 12\ H_2O \rightarrow 4\ Fe(OH)_3 + 6\ H_2SO_4$$

As a result of these reactions, mine waters are very acidic and are often uninhabitable for other forms of life (Figure 6.26) (see leaching of metals from ores, page 582).

Ammonia Is Oxidized to Nitrite, Which Then Can Be Further Oxidized to Nitrate

The oxidation of reduced nitrogen compounds is a two-step process. The first step of the series—the oxidation of ammonia (NH_3) to nitrite (NO_2^-)—is carried out by *Nitrosomonas*, *Nitrosospira*, *Nitrosococcus*, and *Nitrosolobus*. Oxidation of nitrite to nitrate (NO_3^-) is accomplished by *Nitrobacter*, *Nitrospina*, and *Nitrococcus*. The complete oxidation of ammonia to nitrate is called **nitrification**, and the bacteria that carry out this process in sequence are known as nitrifying bacteria. These bacteria are typically found in the soil where they play an important role in the nitrogen cycle (see nitrogen cycle, page 576).

figure 6.26
Acid Mine Drainage from a Coal Strip Mine
Iron-oxidizing bacteria oxidize ferrous compounds associated with the coal, resulting in the formation of insoluble ferric hydroxide deposits. Since the pH of these acidic mine waters is very low, damage to the environment is common.

Oxygen Is Reduced to Water During the Oxidation of Hydrogen Gas

Bacteria using hydrogen gas as an energy source oxidize it by this simple reaction:

$$H_2 \xrightarrow{\text{hydrogenase}} 2\ H^+ + 2\ e^-$$

The electrons released from the oxidation of H_2 typically go through an electron transport chain, resulting in ATP synthesis and reduction of oxygen to water. Most of the hydrogen-oxidizing bacteria were formerly classified in the genus *Hydrogenomonas*. This genus has been eliminated, and the hydrogen-oxidizing bacteria are now placed in several genera, including *Pseudomonas*, *Alcaligenes*, and *Nocardia*. In contrast to other groups of chemolithotrophs, hydrogen-oxidizing bacteria are **facultative** chemolithotrophs; most can utilize a wide variety of organic substrates as carbon and energy sources.

PERSPECTIVE Bacterial Luminescence

One of the most remarkable characteristics of some bacteria is their ability to emit light. These luminescent bacteria, which include *Photobacterium* and *Lucibacterium*, frequently live as symbionts of fish, and their light emission may serve to assist their hosts in predation, in avoiding predators, and in communication with other members of the host species. The bacterial enzyme responsible for the light-emitting reaction is **luciferase.**

Bacterial luciferase catalyzes the bioluminescent oxidation of $FMNH_2$ and a long-chained aliphatic aldehyde (RCHO) by molecular oxygen:

$$FMNH_2 + RCHO + O_2 \xrightarrow{\text{luciferase}}$$
$$0.1 \; hv \; + FMN + H_2O + RCOOH$$

Oxidized FMN, water, acid (RCOOH), and light are produced as a result of this reaction. The bioluminescent pathway is hypothesized to be a branch of electron transport in luminescent bacteria, where electrons may either be sent through a conventional electron transport chain to the terminal electron acceptor oxygen or be shunted to oxygen via the luciferase system (Figure 6P.1). Because oxygen is required for bioluminescence, luminescent bacteria provide a sensitive assay for oxygen.

The chemiluminescent quantum yield of bacterial luciferase is quite significant; it was measured in a series of experiments reported by J.W. Hastings, W.H. Riley, and J. Massa in 1965. In these experiments, the luminescent bacterium *Photobacterium fischeri* was grown in a 50-gallon fermenter. Cells were harvested at the time of maximum light emission, and luciferase was extracted from the cell paste. Luciferase activity was then measured by incubating purified enzyme with aldehyde, $FMNH_2$, and other chemicals (Table 6P.1). Light intensity was measured with a photomultiplier photometer and calcu-

lated as units of quanta per second, using [3]H- or ^{14}C-labeled hexadecane as a liquid light standard. It was determined that pure luciferase gave a chemiluminescent quantum yield of 0.27 per enzyme molecule.

Hastings and his colleagues not only were able to calculate the quantum yield of luciferase from *P. fischeri*, but also observed that there were large amounts of luciferase (approximately 5% of the total cell protein) in the cell and that luminescence accounted for as much as 20% of the total oxygen consumption of the cell. These data suggest that luciferase synthesis represents a significant proportion of the cell's biosynthetic activities.

Sources

Hastings, J.W., and K.H. Nealson. 1977. Bacterial bioluminescence. *Annual Reviews of Microbiology* 31:549–595.

Hastings, J.W., W.H. Riley, and J. Massa. 1965. The purification, properties, and chemiluminescent quantum yield of bacterial luciferase. *The Journal of Biological Chemistry* 240:1473–1481.

table 6P.1

Reaction Mixture for Assay of Luciferase

Reaction Mixture	Volume (ml)
2-mercaptoethanol (5×10^{-3} M)	0.95
Phosphate buffer (1.25 M, pH 7.0)	0.20
Bovine serum albumin (1%)	0.20
Enzyme preparation	0.05
n-decyl aldehyde (10^{-2} M)	0.10
$FMNH_2$ (5×10^{-5} M), to initiate reaction	1.00

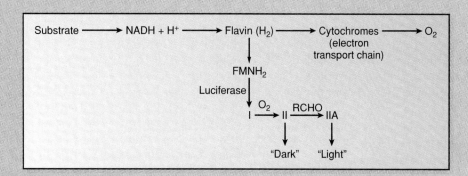

figure 6P.1
Possible Pathways for Electron Transport in Luminescent Bacteria

⬤ Summary

1. All living organisms must have a source of energy to perform work. Most organisms obtain their energy ultimately from chemical substances (phototrophs obtain their energy from light). Energy-liberating processes (exergonic reactions) provide the cell with the energy to carry out energy-requiring processes (endergonic reactions). The coupling of exergonic and endergonic reactions is fundamentally important to all cellular processes.

2. Thermodynamics is the study of energy transformations; bioenergetics is the study of energy transformations in living systems. The first law of thermodynamics states that energy can be neither created nor destroyed during its transformation. The second law of thermodynamics states that spontaneous reactions always proceed with an increase in entropy.

3. Enzymes are protein catalysts that occur in living cells and accelerate reaction rates without themselves being changed. Inhibitors are chemical substances that prevent or slow down enzyme reactions by binding to the enzyme. Allosteric inhibition occurs when a small molecule called an effector binds to an allosteric site on the enzyme, resulting in a change in conformation of the active site and inhibition of enzyme activity.

4. Oxidation-reduction, or redox, reactions involve the transfer of electrons from one molecule to another. Oxidation describes any reaction in which electrons are lost from a substance, whereas reduction describes a reaction in which electrons are accepted by a substance.

5. The ability of a substance to donate or accept electrons is determined by its oxidation-reduction, or redox, potential. The reduced substance of a redox pair can donate electrons to the oxidized substance of the pair, which has the more positive redox potential. The greater the difference in redox potential between the redox pair serving as the electron donor and the pair serving as an electron acceptor, the greater the energy available as a result of the oxidation-reduction reaction.

6. Electron carriers are oxidized and reduced as they transfer electrons from a donor to an acceptor molecule. Some electron carriers such as NAD and FAD transfer hydrogen atoms as well as electrons.

7. Energy liberated as a result of oxidation-reduction reactions can be stored in high-energy compounds. ATP, phosphoenolpyruvate, acetyl phosphate, and 1,3-diphosphoglycerate are examples of high-energy compounds.

8. Chemical compounds are broken down by exergonic catabolic pathways and synthesized by endergonic anabolic pathways. Both types of pathways constitute the metabolic pathways of a cell (Figure 6.27).

9. Microorganisms use three mechanisms to generate ATP: oxidative phosphorylation, photophosphorylation, and substrate-level phosphorylation. ATP is synthesized during the transfer of electrons via an electron transport chain to a final electron acceptor in oxidative phosphorylation and photophosphorylation. In substrate-level phosphorylation, ATP is synthesized when a phosphorylated metabolic intermediate transfers its phosphate to ADP.

10. An organic substrate serves as the electron donor, and an oxidized intermediate of the substrate acts as the final electron acceptor and subsequently becomes reduced in fermentation. An external terminal electron acceptor receives electrons and is reduced in respiration. Molecular oxygen is the terminal electron acceptor in aerobic respiration, whereas some other molecule such as nitrate or sulfate is the electron acceptor in anaerobic respiration.

11. The Embden-Meyerhof pathway, an example of glycolysis, consists of ten distinct reactions in which glucose is oxidized to pyruvate. ATP is synthesized in the Embden-Meyerhof pathway by substrate-level phosphorylation. The pyruvate that is formed in glycolysis is further degraded to a variety of products during fermentation.

12. The tricarboxylic acid (TCA) cycle is a central pathway of cellular metabolism. Electrons from NADH + H$^+$ and FADH$_2$ generated by the TCA cycle are sent through an electron transport chain to generate ATP by chemiosmosis.

13. Chemolithotrophs are procaryotes that oxidize reduced inorganic compounds such as hydrogen sulfide, ferrous iron, ammonia, and hydrogen for energy. Many chemolithotrophs also obtain their carbon from carbon dioxide and therefore are autotrophs. Chemolithotrophs generate ATP by the oxidation of inorganic compounds and electron transport phosphorylation.

EVOLUTION *and* BIODIVERSITY

Although procaryotes are the smallest of all living cells, as a group they show the most metabolic diversity. Procaryotes can use sunlight, organic compounds, and inorganic compounds as sources of energy. They use a variety of substrates, and some—the autotrophs—can even produce their own organic compounds from carbon dioxide. They are able to live in anaerobic environments and in locales that would be inhospitable to other forms of life. This wide diversity of procaryotes is made possible by their myriad of enzymes and metabolic pathways. Some of the first living organisms were probably heterotrophic procaryotes capable of fermenting organic substances on a primitive earth lacking atmospheric oxygen. As organic compounds were used, heterotrophs may have evolved into autotrophs, which replenished the supply of organic molecules. With the evolution of cyanobacteria, algae, and plants came photosynthesis and oxygen. Despite the diversity of metabolism among organisms today, common metabolic threads run through the various forms of life. ATP is synthesized during electron transport; fundamental pathways such as the TCA cycle are found in a wide variety of organisms; and certain molecules such as ATP, NAD, and cytochromes are remarkably similar in all organisms. This biochemical unity among even the most diverse of organisms suggests that all organisms evolved from a common ancestral stock.

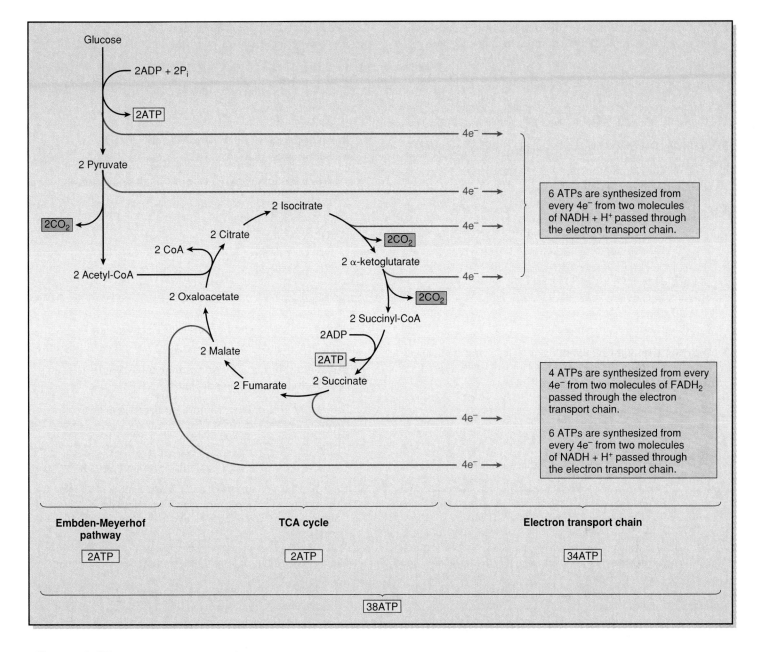

figure 6.27 Summary of Pathways of Energy Generation in Chemotrophic Bacteria

 **Questions**

Short Answer

1. State the first and second laws of thermodynamics.

2. Explain what enzymes are and why they are important.

3. Identify several factors that will decrease the rate of enzymatic reactions.

4. Compare and contrast competitive and noncompetitive inhibitors.

5. Explain feedback inhibition and its importance.

6. Compare and contrast oxidation and reduction reactions.

7. Identify several common electron carriers. Explain their importance.

8. Identify several high-energy compounds.

9. Compare and contrast anabolism and catabolism.

10. Identify three methods of ATP generation.

11. Compare and contrast fermentation and respiration.

12. Identify several pathways for glycolysis. Which is the major pathway for bacteria? Why does more than one pathway exist?

13. Are all fermentation products acidic?

14. Briefly describe by equations or sentences the fate of glucose when it is completely catabolized by aerobic respiration.

15. Compare and contrast aerobic and anaerobic respiration.

16. Explain how chemolithotrophs differ from chemoorganotrophs.

Multiple Choice

1. Which of the following reactions represents an endergonic reaction?

   a. $C_6H_{12}O_6 + 6 O_2 \rightarrow 6 CO_2 + 6 H_2O$
   b. Glucose-6-PO_4 + H_2O → Glucose + PO_4
   c. ADP + PO_4 → ATP + H_2O
   d. All of the above are endergonic.

2. Which of the following methods of ATP generation involves the transfer of electrons through an electron transport chain and requires sunlight?

   a. oxidative phosphorylation
   b. photophosphorylation
   c. substrate-level phosphorylation
   d. None of the above.

3. Pyruvate may be used to make:

   a. lactate, proprionate, butyrate
   b. ethanol and CO_2
   c. glucose
   d. All of the above can be formed from pyruvate.
   e. Both a and b, but not c.

4. Which of the following represents a final electron acceptor reaction?

   a. $H_2S \rightarrow SO_4^{2-}$
   b. $NO_3^- \rightarrow NO_2^-$
   c. $FeSO_4 \rightarrow Fe_2(SO_4)_3$
   d. $NH_3 \rightarrow NO_2^-$

Critical Thinking

1. Explain how, and why, protons (not electrons) are responsible for ATP synthesis by oxidative phosphorylation.

2. Explain how bacteria could continue to grow if their carbohydrate supply were depleted.

3. Discuss the role of fermentation for bacteria in an aerobic environment.

⬤— Supplementary Readings

Caldwell, D.R. 1995. *Microbial physiology & metabolism*. Dubuque, Iowa: Wm. C. Brown Publishers. (A microbial physiology textbook with discussions of metabolism, fermentation, and genetics.)

Gerhardt, P., R.G.E. Murray, W.A. Wood, and N.R. Krieg. 1994. *Methods for general and molecular bacteriology*, pp. 463–599. Washington, D.C.: American Society for Microbiology. (A discussion of various techniques in microbiology, including sections on microbial metabolism and assays for metabolic and enzymatic activities.)

Haddock, B. A., and C. W. Jones. 1977. Bacterial respiration. *Bacteriological Reviews* 41:47–99. (A comprehensive discussion of bacterial electron transport chains and ATPase complexes.)

Hinckle, P.C., and R.E. McCarthy. 1978. How cells make ATP. *Scientific American* 238:104–123. (A review article on ATP synthesis.)

Hobbs, A.S., and R.W. Albers. 1980. The structure of proteins involved in active membrane transport. *Annual Reviews of Biophysics and Bioengineering* 9:259–291. (A review of transport mechanisms in procaryotes and eucaryotes.)

Ingledew, W.J., and R.K. Poole. 1984. The respiratory chains of *Escherichia coli. Microbiological Reviews* 48:222–271. (An in-depth discussion of electron transport and respiration in *E. coli.*)

Neidhardt, F.C., J.L. Ingraham, and M. Schaechter. 1990. *Physiology of the bacterial cell: A molecular approach*. Sunderland, Mass.: Sinauer Associates. (A textbook of bacterial physiology and genetics.)

Nichols, D.G. 1982. *Bioenergetics: An introduction to the chemiosmotic theory*. New York: Academic Press. (An excellent discussion of chemiosmosis.)

Stryer, L. 1988. *Biochemistry*, 3d ed. San Francisco: W. H. Freeman. (A detailed review of biochemistry, with discussions of chemical reactions and structures.)

Wood, W.B., J.H. Wilson, R.B. Benbow, and L.E. Hood. 1981. *Biochemistry: A problems approach*, 2d ed. Menlo Park, Calif.: Benjamin/Cummings Publishing Company. (A concise review of major topics in biochemistry, with extensive and thought-provoking problems.)

Zubay, G. 1993. *Biochemistry*, 3d ed. Dubuque, Iowa: Wm. C. Brown Publishers. (A general textbook of biochemistry.)

chapter seven

PHOTOSYNTHESIS AND OTHER METABOLIC PATHWAYS

Tapering filaments of the heterocystous cyanobacterium *Gloeotrichia echinulata* (**×63**).

Photosynthesis
Chlorophylls
Bacteriochlorophylls
Accessory Pigments
Chlorophylls in Chloroplasts
Procaryotes

The Nature of Photosynthesis: Photophosphorylation
Photosystems
Plants, Algae, and Cyanobacteria
Procaryotic Photophosphorylation
ATP Generation

The Nature of Photosynthesis: Carbon Dioxide Fixation
Ribulose Diphosphate Carboxylase

Carbohydrate Metabolism
Polysaccharides
Carbohydrates

Protein Metabolism
Proteins as Sources of Carbon and Energy
Amino Acid Synthesis

Lipid Metabolism
β-Oxidation
Stepwise Addition of Acetyl Groups

Purine and Pyrimidine Metabolism
Nucleases
Purines and Pyrimidines

PERSPECTIVE
Prodigiosin—A Secondary Metabolite

EVOLUTION AND BIODIVERITY

ike highways of life, a cell's metabolic pathways provide the essential ingredients to sustain life and support growth. These diverse, interconnected chemical arteries take nutrients found in the surroundings and transform them into the energy and building blocks necessary to construct the myriad of chemicals and structures needed for a functional organism. Energy released through the catabolism of nutrients is used by the cell to drive anabolic reactions for the synthesis of cellular constituents.

Among metabolic pathways, none are more important to life than those that participate in photosynthesis. Without photosynthesis, the cycling of carbon into glucose (the basic energy source and fundamental building block of carbohydrates for all organisms) would be incomplete. Oxygen, which is important in aerobic respiration, would not be available. Photosynthesis is considered the ultimate source of life for all plants and animals because it provides the fundamental necessary ingredients.

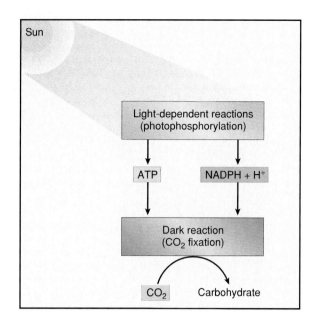

figure 7.1
Schematic Diagram of Photosynthesis

Photosynthesis

Photosynthesis, a process known for many years, occurs in green plants and algae. It was not until the 1880s—when Sergei Winogradsky found that purple and green bacteria oxidized H_2S to SO_4^{2-} with transient accumulation of sulfur granules in their cells—that these photosynthetic microorganisms were studied in greater detail. Winogradsky, however, thought that these microbes were ordinary autotrophic sulfur bacteria and not photosynthetic. At about the same time, Theodor Wilhelm Engelmann (1843–1909), a German microbiologist, discovered that these bacteria were phototactic and grew better when exposed to certain wavelengths of the light spectrum. Although he was unable to show that these organisms produced oxygen in the presence of light, Engelmann nonetheless proposed that they might be photosynthetic. Unfortunately, it was widely believed at the time that oxygen production was an indispensable part of the photosynthetic process, and Engelmann's proposal received little support. Not until the 1930s was it discovered that not all photosynthetic organisms formed oxygen.

Photosynthesis is now known to occur in algae, plants, and several groups of procaryotes. Photosynthesis consists of two major sets of reactions: **photophosphorylation** and **carbon dioxide fixation** (Figure 7.1). Photophosphorylation, or the **light-dependent reactions** of photosynthesis, occurs only in the presence of light and generates ATP and reducing power in the form of $NADPH + H^+$. The ATP and reducing power are used to drive carbon dioxide fixation, or the **dark (light-independent) reactions** of photosynthesis, which can occur without light. Although photophosphorylation and carbon dioxide fixation are both part of photosynthesis, they are autonomous sets of reactions able to function independently.

Chlorophylls Are the Principal Photosynthetic Pigments in Phototrophic Eucaryotes and Cyanobacteria

Photosynthesis is carried out in specialized membranes found only in phototrophs. These structures contain pigments that trap light (electromagnetic) energy and transform it to chemical energy in the form of ATP. The principal light-trapping pigment molecule in phototrophic plants, algae, and cyanobacteria is **chlorophyll.** Chlorophylls and cytochromes both contain porphyrins, but they differ in two ways. First, chlorophylls have magnesium, not iron, as the central atom in the porphyrin ring. Second, a long-chain hydrophobic alcohol (phytol) is attached to one of the four pyrrole groups in the chlorophyll porphyrin ring structure. The solubility of phytol in lipids determines the orientation of the chlorophyll molecules in the internal membranes of chloroplasts.

At least three major types of chlorophylls are found in phototrophic plants, algae, and cyanobacteria: a, b, and c. These three types of chlorophylls are distinguished by slight differences in their chemical structures and absorption spectra. The absorption spectrum is a plot of the absorption of a substance over a continuous spectrum of wavelength of light. The absorption spectra of chlorophylls depends not only on the chemical structure of the pigment molecules, but also on the chlorophylls' interactions with binding proteins in photosynthetic membranes or, in the case of purified chlorophylls, the solution in which they are suspended.

Chlorophyll a is the most common and extensively studied chlorophyll (Figure 7.2). When suspended in acetone, chlorophyll a shows strong absorption of red light and blue light at wavelengths of 675 nm and 420 nm, respectively. Chlorophyll b, which occurs in green algae and higher plants, also absorbs red and blue light but at different wavelengths—645 nm and 470 nm. Chlorophyll c replaces chlorophyll b in brown algae, diatoms, and dinoflagellates.

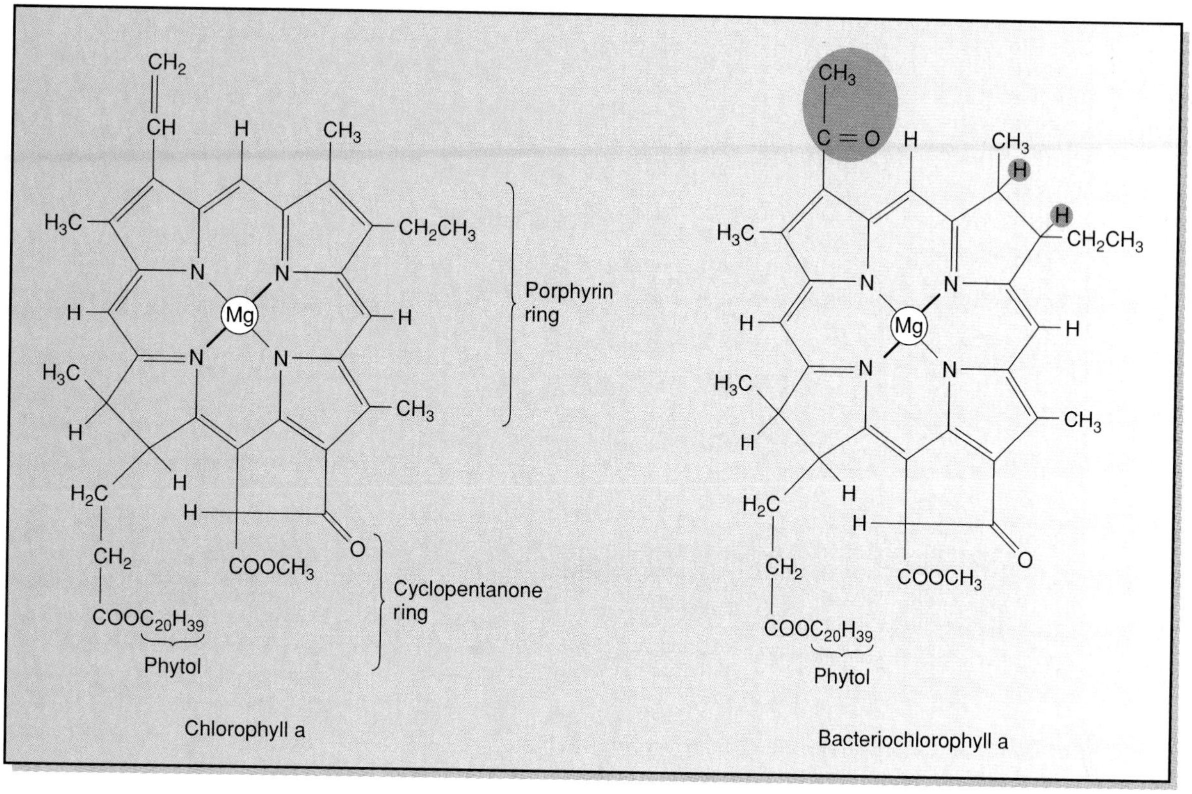

figure 7.2

Structures of Chlorophyll a and Bacteriochlorophyll a
The two structures are identical except for the atoms circled and shaded on the bacteriochlorophyll a structure.

Bacteriochlorophylls Are the Principal Photosynthetic Pigments in Bacteria

The photosynthetic pigments of phototrophic procaryotes other than cyanobacteria are similar in structure to chlorophylls, with slight differences in certain side groups. They are called **bacteriochlorophylls** because they are found in bacteria (Figure 7.2). There are five major types of bacteriochlorophylls: a, b, c, d, and e. The absorption peaks of bacteriochlorophylls c, d, and e approximate those of chlorophylls a, b, and c, with wavelengths ranging from approximately 400 to 650 nm (Figure 7.3).

Bacteriochlorophylls a and b have an extra double bond in one of the pyrrole groups and are able to absorb light further toward the infrared portion of the spectrum (near 750 nm). The purple phototrophic bacteria (*Chromatiaceae* and *Rhodospirillaceae*) have either bacteriochlorophyll a or b. In comparison, the green phototrophic bacteria (*Chlorobiaceae* and *Chloroflexaceae*) contain large quantities of one major bacteriochlorophyll (bacteriochlorophyll c, d, or e) and smaller amounts of bacteriochlorophyll a as a minor pigment.

Although chlorophylls and bacteriochlorophylls are found in phototrophic organisms, what proof do we have that these pigments are directly responsible for photosynthesis? Several major lines of evidence indicate a clear association with photosynthesis.

1. Only those organisms that possess these pigments photosynthesize.

2. Light provided at wavelengths that match the absorption spectra of chlorophylls or bacteriochlorophylls in an organism is most effective for photosynthesis.

3. Organisms do not photosynthesize if cultured under conditions in which chlorophylls or bacteriochlorophylls are not produced.

4. Mutant organisms that no longer synthesize chlorophylls or bacteriochlorophylls lose their ability to photosynthesize.

Accessory Pigments Harvest Light from Other Portions of the Spectrum for Photosynthesis

Chlorophylls and bacteriochlorophylls absorb light from only a small portion of the spectrum. Were it not for a special group of pigments called **accessory pigments,** which harvest light from other portions of the spectrum, much of the sun's light would be lost as heat. The accessory pigments fall into two classes: **carotenoids** and **biliproteins** (or **phycobilins**).

Carotenoids are lipid-soluble, long-chain, conjugated (having alternating single and double bonds) hydrocarbons found in photosynthetic eucaryotes and procaryotes. As a rule, carotenoids

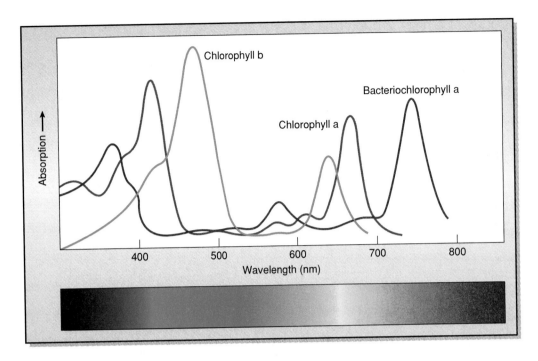

figure 7.3

Absorption Spectra of Common Chlorophylls and Bacteriochlorophylls

The absorption spectra of chlorophylls a and b and bacteriochlorophyll a in ether are shown. Note that chlorophyll a has different absorption peaks than bacteriochlorophyll a.

are yellow or orange and absorb light in the blue range of the spectrum (400 to 550 nm). These accessory pigments usually are associated with chlorophylls, to which they transfer absorbed light energy (Figure 7.4). Two major types of carotenoids occur in photosynthetic organisms. The carotenes (for example, β-carotene) are pure hydrocarbons. Carotenols (for example, xanthophyll) have hydroxyl (–OH) groups on both ends of the hydrocarbon chain.

The second class of accessory pigments, the biliproteins, occurs only in red algae and cyanobacteria. Biliproteins are water-soluble, linear tetrapyrroles coupled to proteins. Phycoerythrin (a red pigment that absorbs light at a wavelength of 550 nm) and phycocyanin (a blue pigment that absorbs light at 620 to 640 nm) are examples of biliproteins.

The accessory pigments not only assist chlorophylls in capturing light energy; they also protect the photosynthetic apparatus from a destructive process known as **photooxidation.** Phototrophs are especially sensitive to bright light because chlorophylls and bacteriochlorophylls act as photosensitizers. As these pigment molecules are exposed to light, they become oxidized and the excited molecule combines with oxygen. This light-dependent oxidation is termed photooxidation. Carotenoids protect chlorophylls and bacteriochlorophylls from photooxidation with their more readily oxidizable double bonds. These accessory pigments absorb most of the harmful light and thus act as shields for the light-sensitive chlorophylls. The role of carotenoids as photoprotective agents is evident, since bacteriochlorophylls in photosyn-

thetic bacterial mutants that no longer synthesize carotenoids are sensitive to photooxidation.

Chlorophylls Are Contained Within Chloroplasts in Eucaryotes

The pigments that harvest light energy in eucaryotic photosynthesis occur in specialized organelles called **chloroplasts** (Figure 7.5). Chloroplasts vary in number from a few in algal cells to several hundred in cells of higher plants.

In chloroplasts, chlorophylls and carotenoids are located in flattened disklike membranous sacs known as **thylakoids,** which form part of the elaborate internal membranes. Biliproteins are found in granules called **phycobilisomes,** which are located on the surfaces of thylakoids. The thylakoids in most algal and plant cells are arranged in stacks called **grana** [singular, granum]. Thylakoids of adjacent grana are periodically connected by stromal lamellae, which are extensions of the thylakoid membranes. Most chloroplasts contain from 40 to 60 grana, each consisting of anywhere from a few to a hundred or more thylakoids. The structures found within the chloroplast are separated from the cell cytoplasm by two sets of membranes, an inner and outer membrane. These components are embedded in a chloroplastic ground substance called the **stroma.** All three sets of chloroplast membranes (thylakoids, inner membrane, and outer membrane) consist of lipids and proteins. Eighty percent of the membrane lipids are glycolipids, whereas most of the proteins are involved in electron transport and photophosphorylation.

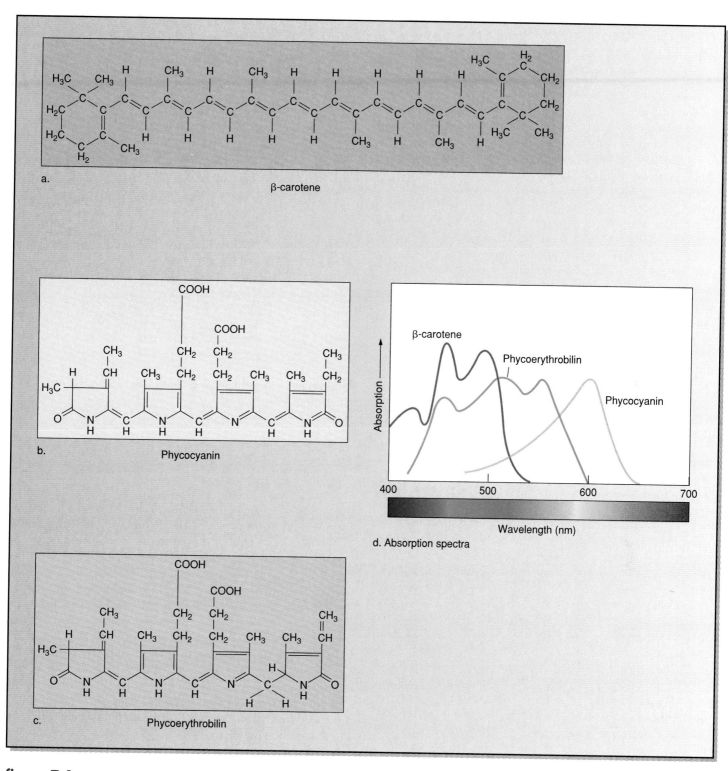

figure 7.4

Accessory Pigments and Their Absorption Spectra

The structures of β-carotene, phycocyanin, and phycoerythrobilin (the chromophore of phycoerythrin) and their absorption spectra in aqueous solution are shown. The accessory pigments and the chlorophylls or bacteriochlorophylls together absorb light at wavelengths spanning the spectrum. a. β-carotene. b. Phycocyanin. c. Phycoerythrobilin. d. Absorption spectra.

The Photosynthetic Apparatus of Procaryotes Is Contained Within Specialized Membranes or Chlorobium Vesicles

Unlike eucaryotic phototrophs, photosynthetic procaryotes do not carry out photosynthesis in chloroplasts. The photosynthetic apparatus of cyanobacteria is found in thylakoids, but these are not contained within chloroplasts. Cyanobacterial thylakoids are found in elaborate invaginated membranes dispersed throughout the cell.

Phototrophic bacteria have specialized membranes or vesicles to carry out photosynthesis (Figure 7.6). The photosynthetic membranes of *Rhodospirillaceae* and *Chromatiaceae* are infolded portions of the plasma membrane. The extent of these membranes depends on the quantity of pigment and degree of photosynthesis in the organism. Bacteria that are actively photosynthetic have large quantities of photosynthetic pigments and more extensive membrane systems than those that are not as active in photosynthesis. Phototrophic bacteria belonging to the families *Chlorobiaceae* and *Chloroflexaceae* have specialized structures called **chlorosomes** or **chlorobium vesicles,** which are adjacent to the plasma membrane but remain as distinct structures. These vesicles have dimensions of approximately 50 nm by 100 to 150 nm and are enclosed by a nonunit membrane. Photosynthetic pigments are present within the chlorobium vesicles.

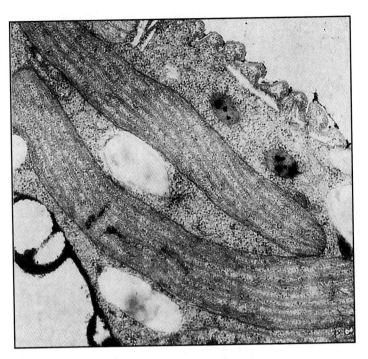

figure 7.5

Electron Micrograph Showing Chloroplasts from the Alga *Euglena gracilis*
Note the parallel thylakoid membranes within each chloroplast (×15,000).

The Nature of Photosynthesis: Photophosphorylation

Although photophosphorylation and carbon dioxide fixation are separate sets of reactions in photosynthesis, they are coupled because carbon dioxide fixation requires ATP and NADPH + H$^+$, which are provided by photophosphorylation. Photophosphorylation depends upon electromagnetic energy, which occurs in discrete packets (quanta) called **photons** or **Einsteins.** A photon of light contains an amount of energy equal to the frequency of the light (velocity in centimeters per second/wavelength in nanometers) times Planck's constant (1.58×10^{-34} cal-sec). Since Planck's constant and the velocity of light are constant values, the energy of a photon increases as the wavelength of light decreases. Thus the energy content of a quantum of red light at a wavelength of 650 nm (43 kcal/Einstein) is considerably less than that of a quantum of blue light at a wavelength of 490 nm (58 kcal/Einstein).

When light interacts with matter, photons are annihilated. According to the first law of thermodynamics, energy can be neither created nor destroyed (see first law of thermodynamics, page 141). However, it can be transformed. Several things can happen to the energy that is contained within photons striking a molecule (Figure 7.7).

1. The energy can be dissipated as heat.

2. The energy can be reemitted as a new photon of light at a longer wavelength, with a portion of the initial energy lost as heat. This change commonly occurs in fluorescence.

3. The energy contained within the photons can cause a chemical change in the compound that absorbs them.

This last possibility occurs in photosynthesis when a molecule of chlorophyll is struck by a photon of light. When the quantum of energy in the photon is absorbed by an electron in the chlorophyll, the electron moves to an orbital having a higher energy level. The electron has gone from an unexcited state known as the ground state to an excited state. The excited electron in its new orbital is extremely unstable and, if located adjacent to an acceptor molecule, is transferred to that acceptor. Electron carriers associated with photosynthesis accept these ejected electrons. The chlorophyll molecules become stronger reducing agents (that is, have lower redox potentials) after absorbing light energy. The electrons are shuttled through carriers that have more positive redox potentials. In plants, algae, and cyanobacteria, these electrons eventually reduce NADP$^+$ to NADPH + H$^+$. During this electron transport, energy is liberated and used to synthesize ATP through the process known as photophosphorylation, similar to oxidative phosphorylation.

The overall sequence of events that occurs in the light-dependent reactions of photosynthesis (that is, photophosphorylation) is summarized by the following equation:

$$12 \text{ H}_2\text{O} + 12 \text{ NADP}^+ + 18 \text{ ADP} + 18 \text{ P}_i \xrightarrow{\text{light}}$$

$$6 \text{ O}_2 + 12 \text{ NADPH} + 12 \text{ H}^+ + 18 \text{ ATP}$$

The ATP and NADPH + H$^+$ produced by photophosphorylation can subsequently be used in the dark reactions of photosynthesis

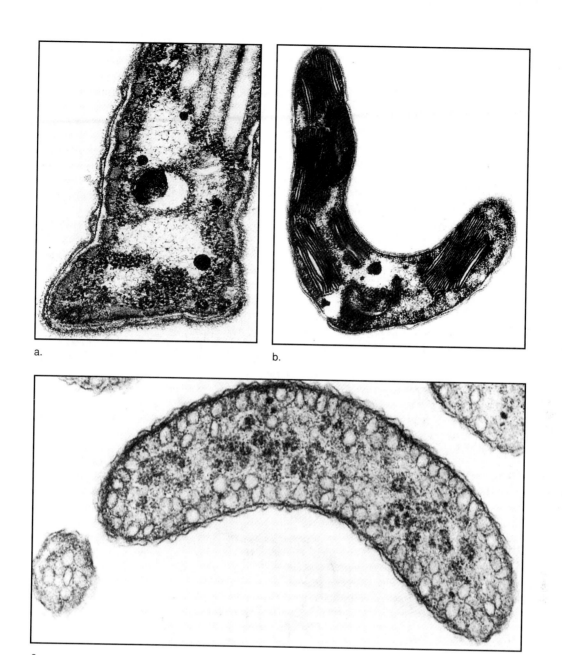

figure 7.6

Electron Micrographs of Bacterial Photosynthetic Vesicles and Membranes

a. Chlorobium vesicles in *Pelodictyon clathratiforme*. Chlorobium vesicles (dark gray) are shown underlying and attached to the plasma membrane. Gas vacuoles (light gray with pointed ends) are in the upper portion of the cell (×72,000). b. Intracytoplasmic photosynthetic membranes in *Ectothiorhodospira mobilis* (×30,000). c. Intracytoplasmic photosynthetic membranes (vesicles) in *Rhodospirillum rubrum* (×51,000).

to fix carbon dioxide into organic compounds. Carbon dioxide fixation is summarized by this equation:

$$12\ NADPH + 12\ H^+ + 18\ ATP + 6\ CO_2 \longrightarrow$$

$$C_6H_{12}O_6 + 12\ NADP^+ + 18\ ADP + 18\ P_i + 6\ H_2O$$

These two equations can be combined to obtain an overall equation for photosynthesis:

$$6\ CO_2 + 6\ H_2O \xrightarrow{\text{light}} C_6H_{12}O_6 + 6\ O_2$$

In this equation, six molecules of carbon dioxide are fixed into one molecule of hexose ($C_6H_{12}O_6$). Water serves as the electron and hydrogen donor in this series of reactions. Oxygen is released as the water is lysed to replace electrons ejected from chlorophylls during light-dependent electron transport and photophosphorylation.

Although this formula may be representative of photophosphorylation in plants, algae, and cyanobacteria, which carry out **oxygenic** photosynthesis, it does not reflect photophosphorylation in phototrophic bacteria, in which photosynthesis is **anoxygenic** (without production of oxygen). Phototrophic bacteria carry out photophosphorylation using electron and hydrogen donors other than water. Cornelius B. van Niel, who studied photosynthesis in bacteria, realized this; in 1935 he proposed a generalized equation that applies to photosynthesis in both eucaryotes and procaryotes:

$$6\ CO_2 + 6\ H_2A \xrightarrow{\text{light}} C_6H_{12}O_6 + H_2O + 6\ A$$

In this equation the electron and hydrogen donor is represented by H_2A rather than by H_2O. Whereas H_2O is the electron and hydrogen donor in plants, algae, and cyanobacteria, other reduced compounds such as H_2S and even H_2 are used in the same capacity by photosynthetic bacteria.

Photophosphorylation Occurs in Assemblages of Electron Carriers Called Photosystems

Photophosphorylation in all phototrophs (plants, algae, and procaryotes) takes place in **photosystems,** assemblages of chlorophylls and accessory pigments in groups of 250 to 400 molecules. As individual pigment molecules are struck by photons, they become excited. Through electron ejection, these light-absorbing molecules (chlorophylls and accessory pigments) transfer a portion of the absorbed energy to molecules having longer wavelength absorption maxima. Energy transfer continues until the light energy eventually reaches a chlorophyll molecule in the photosystem having the longest wavelength absorption maximum. This special chlorophyll acts as an "energy sink," absorbs light at a wavelength of approximately 700 nm (870 nm in phototrophic bacteria), and is called the **reaction center** of the photosystem. The chlorophylls and accessory pigments that attract and funnel light energy to this reaction center are called **antenna pigments.** These function as the first step in light energy transfer to the reaction center.

The Mechanism of Photophosphorylation Is Similar in Plants, Algae, and Cyanobacteria

Plants, algae, and cyanobacteria have similar mechanisms for the generation of ATP by photophosphorylation. ATP is generated from the flow of electrons through a series of electron carriers.

ATP and NADPH + H⁺ Are Generated during Noncyclic Photophosphorylation

Present evidence indicates that there are two separate photosystems that operate in plants, algae, and cyanobacteria. The reaction center in both photosystems is a chlorophyll a molecule. In photosystem I the chlorophyll a that is the reaction center is called **P700**—a *P*igment that absorbs light at a wavelength of *700* nm. The reaction center in photosystem II is chlorophyll **P680,** which absorbs light at *680* nm.

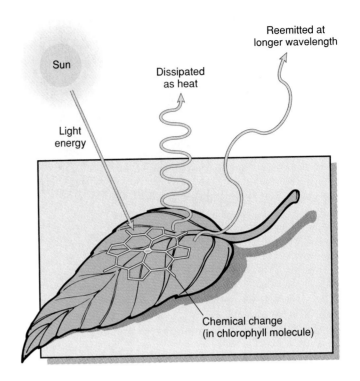

figure 7.7
Fate of Light Striking an Object
The light can be (1) dissipated as heat, (2) reemitted as a new photon at a longer wavelength, or (3) used to cause a chemical change in the object absorbing it.

The two photosystems are coupled by an electron transport system. As carriers transport electrons from one photosystem to the next, energy is released and conserved in the form of ATP. The electrons also function to reduce $NADP^+$ to $NADPH + H^+$.

The sequence of events in photophosphorylation was first outlined by Robin Hill and Fay Bendall in 1960. The Hill-Bendall scheme, often called the Z-pathway because of its resemblance to the letter Z, postulates that one electron is raised from the ground state to an activated state and released from chlorophyll P680 in photosystem II for every quantum of light energy reaching that reaction center (Figure 7.8). The electron is replaced in the reaction center chlorophyll by electrons released from the splitting of a water molecule. This light-dependent splitting of water is termed **photolysis.** Since two electrons are made available from the splitting of one water molecule, electrons are believed to move in pairs in the Hill-Bendall model for the light-dependent reactions. The oxygen evolved during photosynthesis also is derived from photolysis, whereas the hydrogen made available through the splitting of water is used for the eventual reduction of $NADP^+$ to $NADPH + H^+$.

High-energy electrons released from chlorophyll P680 are passed to an unidentified primary electron acceptor. Electrons pass from the primary electron acceptor through a series of electron carriers, which have sequentially more positive redox potentials, until they arrive at reaction center chlorophyll P700 in photosystem I. The electron carriers in this transport chain between chloro-

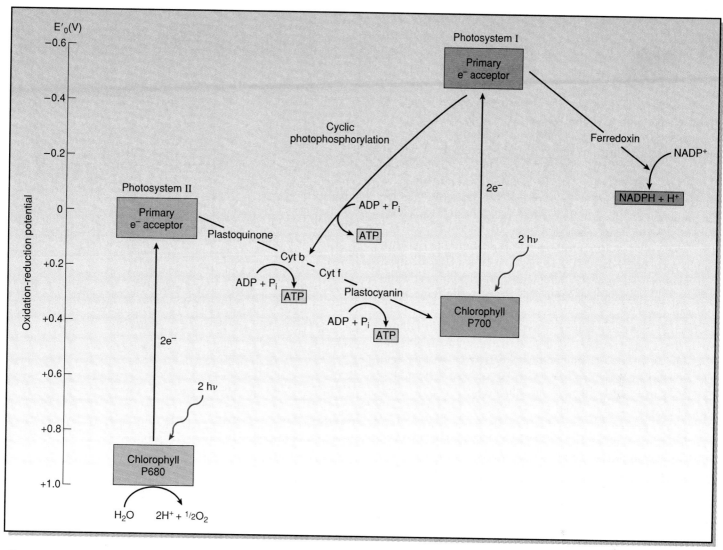

figure 7.8

The Hill-Bendall Scheme for Photophosphorylation

Two photosystems (I and II) are involved in photophosphorylation in plants, algae, and cyanobacteria. Electrons released from chlorophyll P680 are sent to a primary electron acceptor in photosystem II. The electrons are then passed through a series of electron carriers to chlorophyll P700 in photosystem I. Electrons from chlorophyll P700 are sent to an electron acceptor and eventually through ferredoxin to NADP$^+$, which becomes reduced to NADPH + H$^+$. ATP is synthesized during the movement of electrons from photosystem II to photosystem I. The electrons lost from chlorophyll P680 are replaced by the splitting of H$_2$O (photolysis).

phylls P680 and P700 include plastoquinone, cytochrome b, cytochrome f, and plastocyanin. As each pair of electrons is passed to lower energy level acceptors (that is, electron carriers with more-positive redox potentials), the released energy is conserved in the formation of one—possibly two—molecules of ATP from ADP + P$_i$. This light-dependent synthesis of ATP is known as **photophosphorylation.**

The electrons reaching chlorophyll P700 are energized once more as the reaction center chlorophyll absorbs light energy. The energized electrons are passed from chlorophyll P700 to an iron-sulfur protein (FeS) and eventually through ferredoxin and flavoprotein to NADP$^+$. The NADP$^+$ is reduced to NADPH + H$^+$.

As a result of electrons passing through the Z-pathway, NADP$^+$ is reduced and ATP is synthesized. One molecule of NADPH + H$^+$ and one to two molecules of ATP are formed for every four quanta of light. Two quanta of light strike chlorophyll P680 in photosystem II, releasing a pair of electrons. These electrons reach chlorophyll P700 in photosystem I, where two additional quanta of light are required for the ejection of a pair of electrons from this reaction center chlorophyll. The electrons lost from chlorophyll P680 are replaced by the splitting of a molecule of H$_2$O, resulting in liberation of 1/2 O$_2$. This process, in which electrons ejected from chlorophyll P680 move through the Z-pathway and eventually are accepted by NADP$^+$, is known as **noncyclic photophosphorylation.**

Electrons Are Cycled Continuously in Cyclic Photophosphorylation

Electrons are not always passed through the entire Z-pathway. They alternatively can be cycled continuously through a small portion of the pathway. This recycling of electrons is accomplished by shunting electrons reaching the FeS protein in photosystem I through cytochrome b in the electron transport chain. From cytochrome b, the electrons continue in the Z-pathway to chlorophyll P700 and then back to the FeS protein in a continuous cycle. As each pair of electrons is passed from the FeS protein to cytochrome b, which has a more-positive redox potential than the FeS protein, there is a drop in energy level. The result is the synthesis of at least one molecule of ATP.

This **cyclic photophosphorylation** provides the cell with ATP in the absence of $NADP^+$ reduction and photolysis-induced O_2 evolution. Cells probably carry out cyclic photophosphorylation when there is insufficient total light or light at the wavelengths needed for photosystem I (only reaction center chlorophyll P700 is activated in a cyclic scheme). Alternatively, if there is enough $NADPH + H^+$ in the cell, but additional ATP is required, there may be a shift from noncyclic photophosphorylation to cyclic photophosphorylation. The synthesis of ATP by cyclic photophosphorylation is believed to be more primitive than noncyclic photophosphorylation and in many ways resembles the process of photophosphorylation in bacteria.

There are several pieces of evidence that support the Hill-Bendall scheme of two photosystems in the light reactions of oxygenic photosynthesis. Monochromatic (single wavelength) light activates only one of these two photosystems in plants. Far-red light at a wavelength of 700 nm is inefficient in bringing about oxygen evolution. However, if this light is combined with light of a shorter wavelength (680 nm), oxygen is evolved—evidence that noncyclic photophosphorylation with both photosystems I and II is required for photolysis and the release of oxygen.

Further evidence for the existence of two distinct photosystems comes from the effect of the plant herbicide 3-(3,4-dichlorophenyl)-1,1-dimethylurea (DCMU) on photosynthesis. DCMU specifically inhibits photosystem II by blocking electron flow to plastoquinone. As a result of this inhibition, only cyclic electron flow occurs. This specific inhibition of only one photosystem by DCMU while a second photosystem is still operative indicates that two separate photosystems exist in phototrophic eucaryotes. DCMU has no effect on bacterial photosynthesis, since only one photosystem is present in these microbes.

Procaryotic (Other Than Cyanobacterial) Photophosphorylation Is Cyclic

Photosynthetic bacteria are anoxygenic; that is, unlike plants, algae, and cyanobacteria, they do not produce O_2 during photosynthesis. In fact, photosynthetic bacteria can carry out the light reactions only under anaerobic conditions, although they may grow aerobically without photosynthesis. In bacterial photosynthesis, ATP is synthesized by cyclic photophosphorylation, only

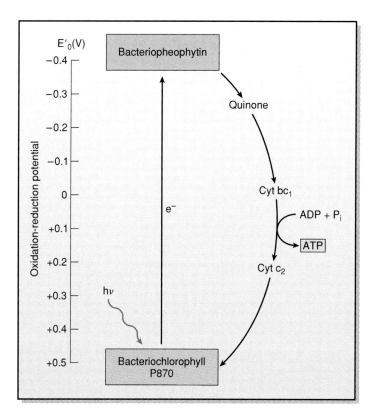

figure 7.9
Bacterial Cyclic Photophosphorylation
Photosystem I is involved in bacterial cyclic photophosphorylation. Electrons released from bacteriochlorophyll P870 are sent to bacteriopheophytin. The electrons are then cycled back through electron carriers to bacteriochlorophyll P870, resulting in the synthesis of ATP.

using a single photosystem containing bacteriochlorophyll P870 as the reaction center (Figure 7.9). As light strikes bacteriochlorophyll P870, an electron is passed to bacteriopheophytin within the reaction center. The electron is then cycled through a quinone molecule, a cytochrome bc_1 complex, cytochrome c_2, and back to the reaction center. One or more molecules of ATP are synthesized as electrons are transferred down to electron carriers of more-positive redox potentials in this cyclic pathway.

In ATP synthesis via cyclic photophosphorylation in bacteria, electrons ejected from bacteriochlorophyll P870 eventually cycle back to it. In bacterial photosynthesis, photolysis of water and the evolution of oxygen do not occur because oxidized bacteriochlorophyll P870 is not sufficiently oxidized to receive electrons from water. The synthesis of ATP fulfills one of the requirements for carbon dioxide fixation. But where does the bacterium obtain the reducing power also required for carbon dioxide fixation?

This reducing power comes from reduced substances in the environment. Photosynthetic bacteria are typically found in such anaerobic environments as sulfur springs or at the bottom of stratified lakes and ponds. Reduced compounds in these environments transfer their reducing power directly or indirectly to $NADP^+$ (Figure 7.10). Molecules such as hydrogen gas (H_2), which has a more nega-

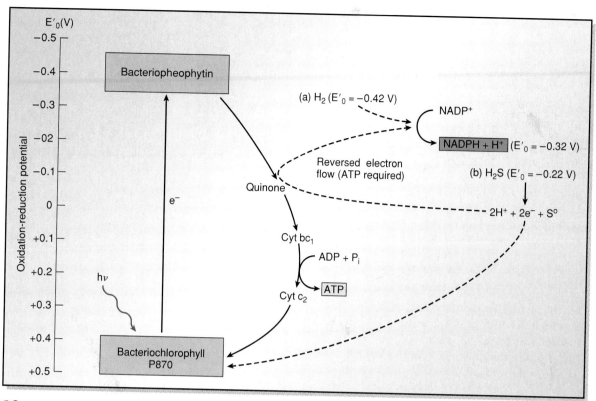

figure 7.10

Generation of Reduced NADP (NADPH + H⁺) by Photosynthetic Bacteria

Photosynthetic bacteria generate NADPH + H⁺ by one of two possible mechanisms: a. Directly, by the transfer of reducing power from reducing compounds having a more negative redox potential than NADP⁺. b. Indirectly, by the transfer of reducing power to NADP⁺ via an ATP-requiring reversal of the electron transport chain used in oxidative phosphorylation. Reversed electron flow occurs when reduced compunds have a more positive redox potential than NADP⁺ and cannot transfer their reducing power directly to NADP⁺.

tive redox potential ($E'_0 = -0.42$ V) than NADP⁺ ($E'_0 = -0.32$ V), transfer their electrons directly to NADP⁺. Other reduced compounds, such as hydrogen sulfide (H_2S), which has a more positive redox potential ($E'_0 = -0.22$ V) than NADP⁺, cannot directly transfer their reducing power to NADP⁺ but are still able to be used as reductants by photosynthetic bacteria. It is postulated that the reducing power of these compounds is transferred to NADP⁺, probably by an ATP-requiring reversal of the electron transport chain used in oxidative phosphorylation. Reversed electron flow has been demonstrated in chemolithotrophic procaryotes, and therefore it is reasonable to assume that such a mechanism may also exist in other procaryotes.

As H_2S is oxidized by photosynthetic bacteria, elemental sulfur accumulates either inside the cell, as in many purple sulfur bacteria, or outside the cell, as in green sulfur bacteria (Figure 7.11). These sulfur granules, although not unique to photosynthetic bacteria (they are also produced by sulfur-oxidizing bacteria such as *Thiobacillus*), are useful morphological traits in the identification of these organisms. Under certain conditions, some cyanobacteria and algae carry out anoxygenic photosynthesis using reduced substances as electron donors. When this occurs, the cyanobacteria can accumulate sulfur granules outside their cells if H_2S is used as a reducing agent.

Chemiosmosis Explains the Mechanism of Photophosphorylation

The mechanism for ATP generation during photosynthesis is explained by chemiosmosis (see chemiosmosis, page 165). Components of the electron transport chain in the light reaction are oriented across the thylakoid or photosynthetic membranes in such a manner that protons are extruded to one side of the membrane during photosynthesis. In plants, algae, and cyanobacteria, the protons move from the stroma to the interior of the thylakoid during electron transport. In photosynthetic bacteria, electrons are believed to move to the exterior of membrane vesicles, or infoldings. This movement in both cases results in the establishment of a pH or proton gradient across the membrane. Although the membrane is impermeable to protons, protons are able to reenter it by linking with ATPases spanning the membrane. The energy released from dissipation of the proton gradient as these protons are brought back across the membrane by the ATPases is conserved in the formation of ATP from ADP + P_i.

Chemiosmosis for photophosphorylation is thus essentially the same as it is for oxidative phosphorylation. Such a model's validity is supported by numerous experimental observations. Intact impermeable photosynthetic membranes are necessary for

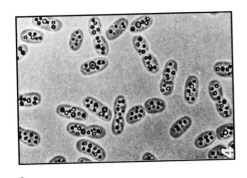
a.

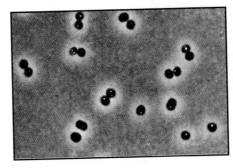
b.

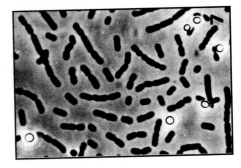
c.

figure 7.11

Sulfur Granules Produced by Photosynthetic Bacteria
a. *Chromatium vinosum*, a purple sulfur bacterium, bright field. Note the intracellular sulfur granules. b. *Thiocapsa roseopersicina*, a purple sulfur bacterium, phase contrast. Note the intracellular sulfur granules.

c. *Chlorobium limicola*, a green sulfur bacterium, phase contrast. Note the extracellular sulfur granules.

the establishment of a proton gradient and ATP synthesis. The formation of an artificial proton gradient across photosynthetic membranes results in a burst of ATP synthesis. It has been repeatedly observed that electron transport results in an accumulation of protons on one side of these membranes. The system can also be made to run in reverse by placing photosynthetic membranes in a solution of excess ATP. Under such conditions, the ATPase is forced to run in reverse, leading to the formation of a proton gradient of opposite orientation across the membrane.

The Nature of Photosynthesis: Carbon Dioxide Fixation

The light-dependent reactions of photosynthesis are used to capture light energy in the form of ATP and—in plants, algae, and cyanobacteria—also to produce $NADPH + H^+$. The term *photosynthesis*, however, is most often associated with the synthesis of glucose from carbon dioxide and water. Because the fixation of carbon dioxide into organic compounds can occur in the absence of light—as long as sufficient quantities of ATP and $NADPH + H^+$ are available—it is known as the dark reactions of photosynthesis. Organisms that fix carbon dioxide are called autotrophs and may be either phototrophic (generating ATP from light as an energy source) or chemotrophic (generating ATP from the oxidation of chemical compounds)(see carbon and energy requirements of procaryotes, page 80).

Carbon Dioxide Fixation Is Catalyzed by Ribulose Diphosphate Carboxylase

The incorporation of carbon dioxide into organic compounds was first observed by chemist Melvin Calvin of the University of California at Berkeley during the 1940s and 1950s. Calvin, who received the Nobel Prize in chemistry in 1961 for his work on carbon dioxide fixation, followed the uptake of radioactively labeled carbon dioxide ($^{14}CO_2$) in the eucaryotic alga *Chlorella*. He noted from algal extracts that the ^{14}C label first appeared in the 3-carbon compound 3-phosphoglycerate. By limiting the quantity of carbon

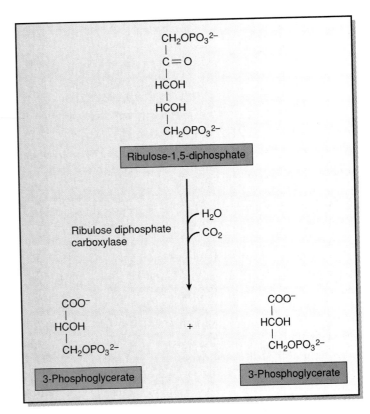

figure 7.12

Carboxylation of Ribulose-1,5-Diphosphate to Two Molecules of 3-Phosphoglycerate

dioxide made available to the algae, Calvin further found that a nonradioactive 5-carbon sugar, ribulose-1,5-diphosphate (RuDP) accumulated in the chloroplasts as 3-phosphoglycerate quantities decreased. This observation suggested that RuDP is the carbon dioxide acceptor in carboxylation and must be regenerated for carbon dioxide fixation to continue (Figure 7.12).

The regeneration of RuDP is fundamental to carbon dioxide fixation. The pathway responsible for RuDP regeneration after

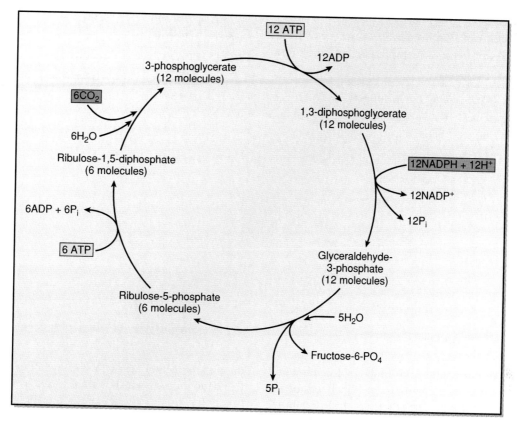

figure 7.13

Fundamental Scheme for Carbon Dioxide Fixation

Carbon dioxide enters the Calvin cycle through the carboxylation of ribulose-1,5-diphosphate to two molecules of 3-phosphoglycerate.

Six turns of the cycle, accompanied by the entry of six molecules of carbon dioxide, produce one fructose-6-phosphate molecule.

carbon dioxide fixation is called the **Calvin cycle** or the **Calvin–Benson–Bassham cycle** after the three individuals who worked out the reactions in the cycle. The Calvin cycle is sometimes known as the **C_3 cycle** because a key intermediate is the 3-carbon compound 3-phosphoglycerate, or it is known as the **reductive pentose cycle** because the fixed carbon dioxide is reduced, and the acceptor molecule is a pentose.

In the Calvin cycle, RuDP is carboxylated, and the resultant 6-carbon molecule is split to form two molecules of 3-phosphoglycerate (PG). These molecules are phosphorylated to two molecules of 1,3-diphosphoglycerate (diPG) at the expense of hydrolysis of two ATP molecules to two ADP molecules. The diPGAs are next reduced to form two molecules of glyceraldehyde-3-phosphate (GAP), with the electrons and hydrogens for this reduction coming from two molecules of NADPH+ H^+. Through a series of subsequent reactions, two molecules of GAP eventually produce one molecule of ribulose-5-phosphate, with the sixth carbon of the two GAPs used as a building block for hexose. The cycle is completed with the ATP-driven phosphorylation of ribulose-5-phosphate to RuDP (Figure 7.13).

The energy requirement for the conversion of one molecule of carbon dioxide into one-sixth of a hexose molecule in the Calvin cycle is three molecules of ATP and two molecules of NADPH + H^+. Six turns of the cycle are required to form a complete hexose molecule. The ATPs and NADPH + H^+ molecules required for the Calvin cycle are provided by the light reactions of photosynthesis. The fructose-6-phosphate generated as a result of carbon dioxide fixation is used to form more complex carbohydrates or other organic compounds.

Most of the enzymes and intermediates in the Calvin cycle also occur in other metabolic pathways. For example, 3-phosphoglycerate, glyceraldehyde-3-phosphate, 1,3-diphosphoglycerate, and fructose-6-phosphate are also intermediates in the Embden-Meyerhof pathway (see Figure 6.10). Unlike the Embden-Meyerhof pathway, which is catabolic and produces ATP, the Calvin cycle is an anabolic pathway that requires ATP.

Ribulose-1,5-diphosphate carboxylase is a key enzyme that is unique to the Calvin cycle. RuDP carboxylase, an enzyme found in aggregates bound loosely to membrane surfaces, is found in large quantities in photosynthetic organisms (up to 50% of total chloroplast protein). Because RuDP carboxylase has a low affinity for carbon dioxide and has a slow turnover time in the reaction it catalyzes, large quantities of the enzyme are required for carbon dioxide fixation.

Carbohydrate Metabolism

Photosynthesis is an important process that occurs in phototrophic bacteria; however, nonphototrophic bacteria can derive energy and carbon from a wide variety of chemical compounds,

table 7.1

Common Natural Disaccharides and Polysaccharides and Enzymes That Degrade Them

Carbohydrate	Composition	Source	Degradative Enzyme
Lactose	Galactose-glucose disaccharide, $\beta(1,4)$ linkage	Milk	β-galactosidase
Maltose	Glucose-glucose disaccharide, $\alpha(1,4)$ linkage	Plants (grain)	Maltase
Sucrose	Glucose-fructose disaccharide, $\alpha(1,2)$ linkage	Plants (sugar beet, sugarcane)	Sucrase
Agar	Galactose and galacturonic acid polymer	Algae	Agarase
Cellulose	Glucose polymer, $\beta(1,4)$ linkage	Plants (cell walls)	Cellulase
Chitin	N-acetyl-D-glucosamine polymer, $\beta(1,4)$ linkage	Arthropods (exoskeleton)	Chitinase
Glycogen	Glucose polymer, $\alpha(1,4)$ and $\alpha(1,6)$ linkages	Animals (liver, muscle)	Amylase, phosphorylase
Lignin	Coniferyl alcohol polymer	Plants (woody cell walls)	Ligninase
Pectin	Galacturonic acid polymer	Plants (fruits, vegetables)	Pectinase
Starch	Glucose polymer, $\alpha(1,4)$ and $\alpha(1,6)$ linkages	Plants (tubers, roots, seeds, fruits)	Amylase

including carbohydrates, proteins, and lipids. In microbes and other organisms, these metabolic pathways are vitally important and determine the distinguishing characteristics of the organism.

Carbohydrates represent major carbon and energy sources for microorganisms. Polysaccharides are the most common forms of carbohydrates and constitute the bulk of organic matter on the earth. These organic compounds are used extensively for food reserves and for cellular and organismal structure. However, most polysaccharides are complex, insoluble molecules that are too large to pass through plasma membranes. These polysaccharides must first be hydrolyzed by extracellular enzymes into smaller, transportable molecules that can then be used in cellular metabolism (Table 7.1).

Polysaccharides Are Degraded into Smaller Molecules Before They Enter the Cell

Cellulose is an example of a large, insoluble polysaccharide that must be digested into smaller molecules before it can enter the microbial cell. Cellulose, composed of repeating D-glucose subunits linked by $\beta(1,4)$ bonds, is the chief structural component of plant cell walls (see Figure 3.5, page 48). It makes up about 10% of the dry weight of leaves, 50% of the woody structures of plants, and over 90% of cotton fiber; overall, cellulose accounts for 50% or more of all the carbon in plants. Many fungi are able to digest cellulose and are thus important in the decomposition of decaying plant material. Cellulose digestion is not as common among bacteria, but does occur among certain bacteria residing in soil or in the intestinal tracts of animals that feed on grass and plant material (see ruminants, page 427).

In most microorganisms, cellulose hydrolysis is a two-step process. The cellulose is first cleaved to the disaccharide cellobiose by cellulase. Cellobiose is then either transported directly across the plasma membrane or further hydrolyzed by the enzyme cellobiase to glucose, which is transported across the plasma membrane. Inside the cell, glucose is metabolized via the Embden-Meyerhof pathway or other metabolic pathways. Starch and glycogen (sometimes known as animal starch) are major storage forms of carbohydrate in plants and animals. Both polymers contain D-glucose subunits joined by $\alpha(1,4)$ glycosidic linkages and have similar structures, although glycogen is a more highly branched molecule (see Figure 3.5). Starch is found abundantly in tubers, roots, seeds, and fruits, and is digestible by amylases produced by many fungi, bacteria, and animals. Microbial amylases are used extensively in the food and textile industry for breadmaking, beer production, the manufacture of sugar syrups, and textile manufacture (see industrial microorganisms, page 623).

Glycogen is produced and stored in liver and muscles and is additionally found in large quantities in the human vagina after puberty. Vaginal bacteria such as *Lactobacillus* degrade this polymer into acidic end products that contribute to the acidic pH characteristic of the vagina.

Glycogen storage granules found in bacteria such as *Escherichia coli* serve as food reserves for these organisms. Bacteria utilize endogenous glycogen as a carbon and energy source by the phosphorylysis of glycogen to glucose-6-phosphate in a two-step process:

$$\text{glycogen} + n\,P_i \xrightarrow{\text{phosphorylase}} n\ \text{glucose-1-phosphate}$$
$$\xrightarrow{\text{phosphoglucomutase}} n\ \text{glucose-6-phosphate}$$

The glucose-6-phosphate can then enter the Embden-Meyerhof pathway for further catabolism. Unlike the conventional phosphorylation of glucose to glucose-6-phosphate in the Embden-Meyerhof pathway, the entry of glycogen into the pathway via glucose-6-phosphate does not require the expenditure of energy through ATP hydrolysis.

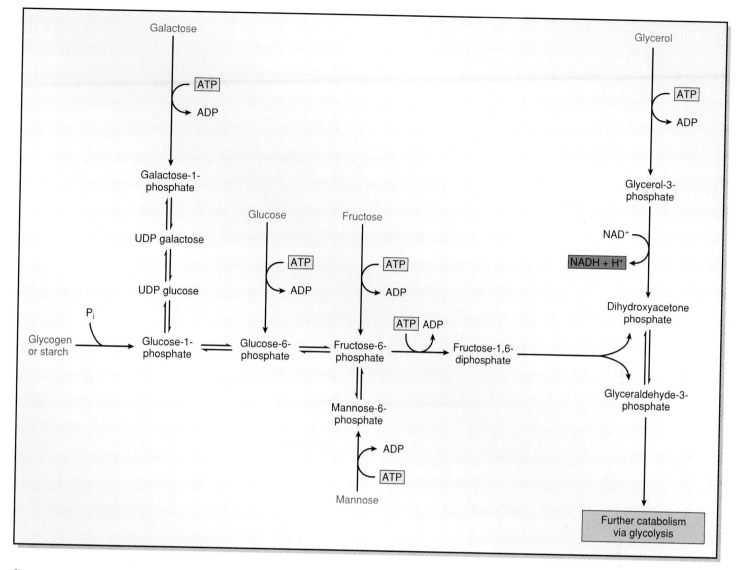

figure 7.14

Hexose Catabolism

Hexoses other than glucose can be catabolized via glucose and its derivatives in the Embden-Meyerhof pathway.

Other polysaccharides degradable by microbial enzymes include chitin (the main constituent of the exoskeletons of arthropods), agar (produced by many algae), lignin (a polymer common in woody plants), and pectin (a carbohydrate occurring in ripe fruits and certain vegetables). Soil fungi and bacteria contribute significantly to the enzymatic degradation of these natural polymers.

Many microorganisms use disaccharides for growth. Disaccharides are first hydrolyzed to their respective monosaccharides, which are then catabolized. Lactose (milk sugar) is present in mammalian milk at a concentration of approximately 5% and is a significant source of nutrients for microbes that are commonly found in milk. The souring of milk occurs when *Streptococcus lactis,* the primary milk-souring organism, ferments lactose to lactic acid and other end products. The glycosidic bond in lactose is cleaved by β-galactosidase, an enzyme that has been important in our understanding of regulatory mechanisms in the cell (see lactose operon,

page 233). Other common disaccharides such as sucrose (cane sugar) and maltose (malt sugar) are hydrolyzed by sucrase (invertase) and maltase, respectively. Sucrose and maltose are abundantly found and are important food sources for bacteria and yeast.

Hexoses generated by enzymatic hydrolysis of disaccharides and polysaccharides are convertible to glucose or isomers of glucose. Glucose occupies an important central position in cellular metabolism and is the entry point into metabolic pathways for many carbohydrates.

Hexose catabolism often occurs via the Embden-Meyerhof pathway (Figure 7.14), but this is not always the case. Chapter 6 discusses other pathways for glucose metabolism. One of these pathways, the phosphoketolase pathway, leads to the formation of xylulose-5-phosphate as a key intermediate of pentose metabolism. In lactobacilli and many enteric bacteria, pentoses such as ribose, xylose, and arabinose, which are found in many plant

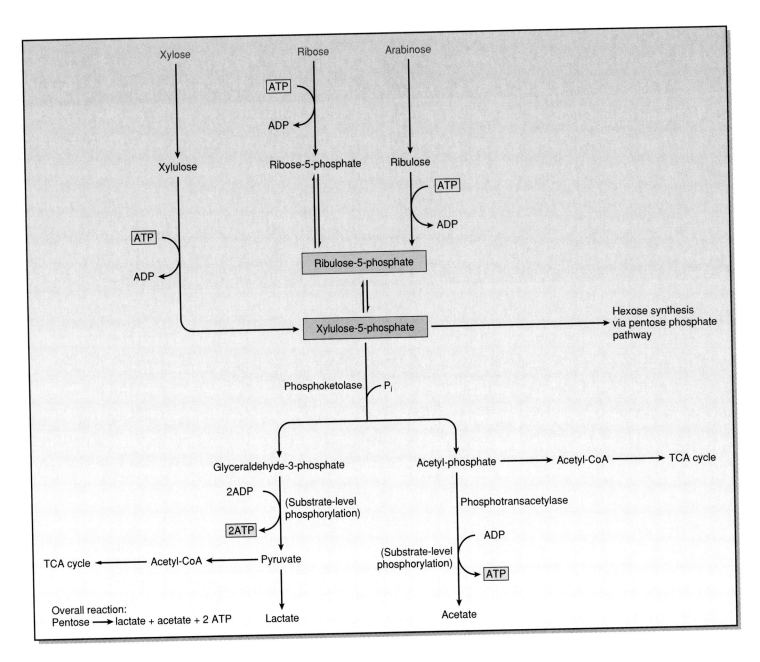

figure 7.15

Pentose Catabolism

Xylulose-5-phosphate is a key intermediate in the catabolism of pentoses in the phosphoketolase pathway.

polysaccharides, are phosphorylated and converted to xylulose-5-phosphate. The xylulose-5-phosphate is subsequently oxidized to lactate and acetate, with a net yield of two molecules of ATP per molecule of pentose (Figure 7.15).

Carbohydrates Are Synthesized by Anabolic Reactions Using Readily Available Substrates

Microorganisms require a variety of carbohydrates that serve as important constituents of cellular macromolecules. Some microorganisms (autotrophs) synthesize organic compounds by fixing carbon dioxide. These organisms were discussed earlier in this

chapter. Other microorganisms (heterotrophs) generate their organic molecules, including carbohydrates, through anabolic reactions using readily available substrates.

Hexoses Are Produced by Gluconeogenesis

Hexoses are synthesized in microbial cells through reversal of the Embden-Meyerhof pathway by a process called **gluconeogenesis** (Figure 7.16). Gluconeogenesis occurs when the cell must make hexoses for structural purposes or when the cell has sufficient energy, with excess energy stored in the form of reduced chemical compounds. Not all steps in the Embden-Meyerhof pathway

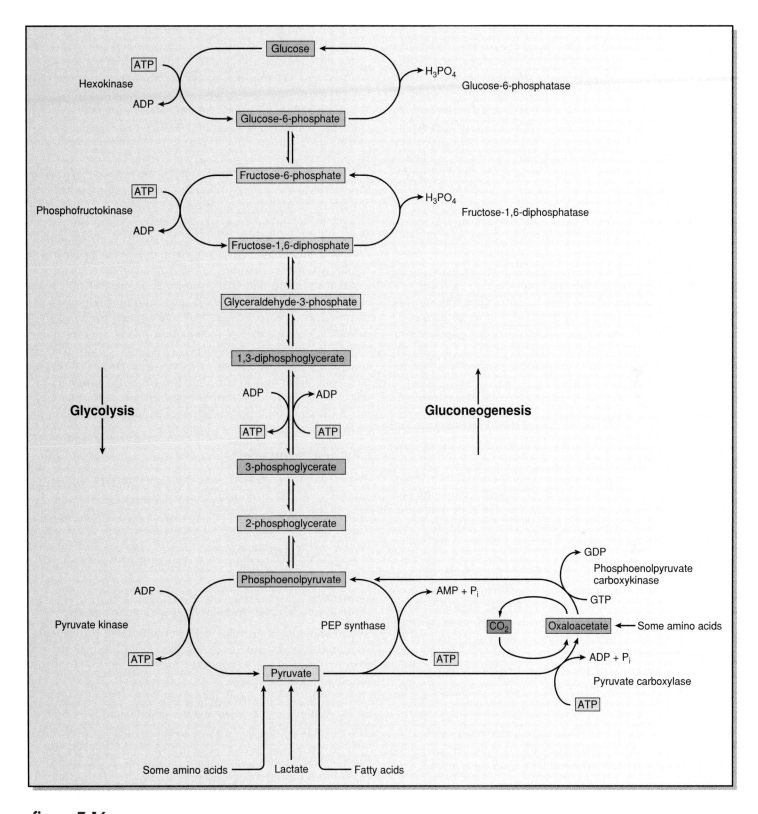

figure 7.16

Gluconeogenesis

The Embden-Meyerhof and gluconeogenic pathways are compared. Many steps in the Embden-Meyerhof pathway are reversible, but some are not and require unique enzymes: glucose-6-phosphatase, fructose-1,6-diphosphatase, phosphoenolpyruvate synthase, pyruvate carboxylase, and phosphoenolpyruvate carboxykinase.

are directly reversible. Three steps are thermodynamically irreversible but are bypassed in gluconeogenesis by using alternative enzymes and pathways. Two of the "irreversible reactions" in the Embden-Meyerhof pathway involve ATP: the phosphorylation of glucose to glucose-6-phosphate and the phosphorylation of fructose-6-phosphate to fructose-1,6-diphosphate. The reversal of these steps would involve the removal of the phosphates by either glucose-6-phosphatase or fructose-1,6-diphosphatase. The formation of phosphoenolpyruvate (PEP) from pyruvate in gluconeogenesis occurs in one of two ways. In one scheme, pyruvate is converted to PEP by PEP synthase, with the consumption of two high-energy phosphate bonds. In the other scheme, pyruvate is first carboxylated to oxaloacetate, which then is decarboxylated to produce PEP. The enzymes participating in this second scheme are pyruvate carboxylase and phosphoenolpyruvate carboxykinase.

Direct reversal of the phosphorylysis of glycogen to glucose-1-phosphate is thermodynamically unfavorable. Glycogen is synthesized by an alternative pathway in which glucose-1-phosphate is first converted to ADP-glucose through the action of the enzyme ADP-glucose pyrophosphorylase. The glucose monomers are then joined to an existing glycogen molecule by glycogen synthetase.

The Embden-Meyerhof pathway is an example of an **amphibolic pathway,** a metabolic pathway that functions in dual roles of catabolism and anabolism. Many important pathways in bacterial cells, including the Embden-Meyerhof pathway and the tricarboxylic acid (TCA) cycle, are amphibolic. For example, glyceraldehyde-3-phosphate in the Embden-Meyerhof pathway can be catabolized to form pyruvate. This same compound can also be used as a precursor for amino acids such as serine and glycine. In the TCA cycle, the intermediates α-ketoglutarate and oxaloacetate are used as precursors for biosynthesis of amino acids as well.

Metabolic intermediates that are used in anabolic pathways must be replenished; otherwise anabolic pathways could not continue. These intermediate compounds are replenished by **anaplerotic** ("filling up") reactions, which are necessary to keep primary pathways going. Although the same intermediates in anabolic pathways may be involved in catabolic and anabolic reactions, the direction of their flow is frequently controlled by different enzymes. For instance, the two coenzymes NAD^+ and $NADP^+$ are involved in oxidation-reduction reactions, but NAD^+ is used in catabolic reactions, whereas $NADP^+$ is used in anabolic reactions. NAD^+ and $NADP^+$ have similar structures, but $NADP^+$ has an extra phosphate group that allows it to be recognized by enzymes involved in biosynthesis. This type of control is one way in which the cell regulates rates of catabolism and anabolism.

Pentoses Are Produced by the Pentose Phosphate Pathway

Pentoses are synthesized from hexoses via the **pentose phosphate pathway** (Figure 7.17). Pentoses are an important part of cellular metabolism, because these 5-carbon monosaccharides form the backbone of nucleic acids, are precursors for the aromatic amino acids (phenylalanine, tryptophan, and tyrosine), and participate in the dark reactions of photosynthesis. The interconversion of pentoses and hexoses allows microbial cells to merge major metabolic pathways with nucleic acid synthesis and photosynthesis.

The pentose phosphate pathway begins with glucose, which is phosphorylated to glucose-6-phosphate. Glucose-6-phosphate is oxidized to 6-phosphogluconate, with 6-phosphogluconolactone as an intermediate. The 6-phosphogluconate is then decarboxylated and oxidized to ribulose-5-phosphate, a pentose. The electron carrier participating in these oxidations is NADP. Reduced NADP is required for anabolic reactions. The ribulose-5-phosphate formed as a consequence of the second oxidation step is interconvertible with two other pentoses, ribose-5-phosphate and xylulose-5-phosphate. These pentoses not only are precursors of nucleotides, but are also used for the synthesis of other intermediates in the pentose phosphate pathway. These other intermediates include glyceraldehyde-3-phosphate, erythrose-4-phosphate, fructose-6-phosphate, and sedoheptulose-7-phosphate. The presence of 3-, 4-, 5-, 6-, and 7-carbon monosaccharides in the pentose phosphate pathway provides metabolic versatility. Many of these compounds are found in other metabolic cycles. Erythrose-4-phosphate, for example, is a precursor for the aromatic amino acids. Glyceraldehyde-3-phosphate and fructose-6-phosphate are intermediates in the Embden-Meyerhof pathway.

Protein Metabolism

Proteins constitute the largest class of macromolecules in a rapidly growing bacterial cell. In the exponential phase of growth, approximately 60% of the dry weight of a typical *E. coli* cell is protein. This weight represents 2,000 to 3,000 different types of proteins. Protein diversity is important to the cell, since proteins not only make up cell enzymes, but also form a significant portion of many cell structures.

Proteins Can Be Used as Sources of Carbon and Energy

Amino acids and proteins, as major constituents of living matter, are also used as nutrients by living organisms. The composition of these organic molecules makes them potential sources of carbon, nitrogen, sulfur, and energy.

Before proteins enter catabolic cycles, they are degraded by microorganisms into smaller peptides or amino acids by proteolytic enzymes. These enzymes are produced by many microbes and often contribute to tissue damage during bacterial infection. For example, *Pseudomonas aeruginosa* produces several proteases (including an elastase that is capable of digesting elastin in connective tissue) that may be responsible for the hemorrhagic skin lesions observed in some infections. Proteolytic tissue damage is especially a problem in anaerobic wound infections, because the large quantities of proteases produced by clostridia cause toxic degradation products to accumulate in the tissue. Microbial proteases have a number of industrial applications. Proteases are used

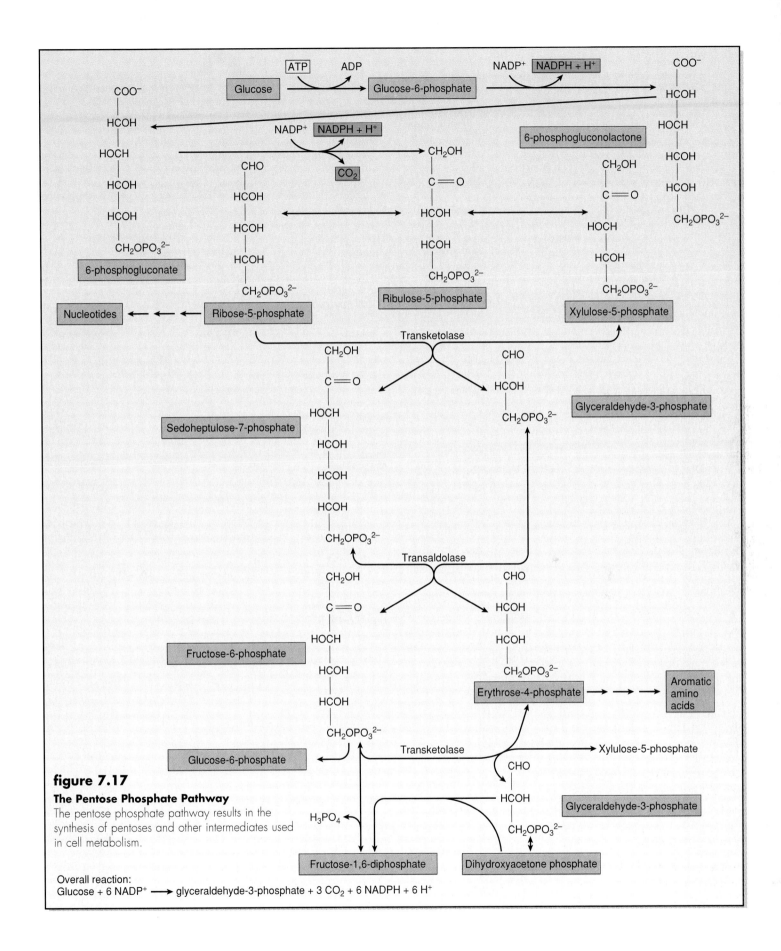

figure 7.17

The Pentose Phosphate Pathway

The pentose phosphate pathway results in the synthesis of pentoses and other intermediates used in cell metabolism.

Overall reaction:
Glucose + 6 NADP$^+$ $\longrightarrow$ glyceraldehyde-3-phosphate + 3 CO$_2$ + 6 NADPH + 6 H$^+$

Some species of *Clostridium* are able to ferment pairs of amino acids that cannot be fermented singly. This process, discovered by Leonard Hubert Stickland and H. Albert Barker and commonly called the **Stickland reaction,** uses one amino acid as an electron donor and another amino acid as an electron acceptor (Figure 7.18). The fermentations of both amino acids are thus coupled through the transfer of electrons.

An example of such a coupled system is the dual fermentation of alanine and glycine. Alanine serves as the electron donor in this system, transferring its electrons to glycine. In the process of this dual fermentation, the alanine is oxidatively deaminated to pyruvate, which then is cleaved to form acetyl-CoA and CO_2. ATP is synthesized from the conversion of acetyl-CoA to acetate. The Stickland reaction is especially useful to clostridia growing in protein-rich environments where amino acids serve as electron donors and electron acceptors are plentiful.

Stickland Reaction

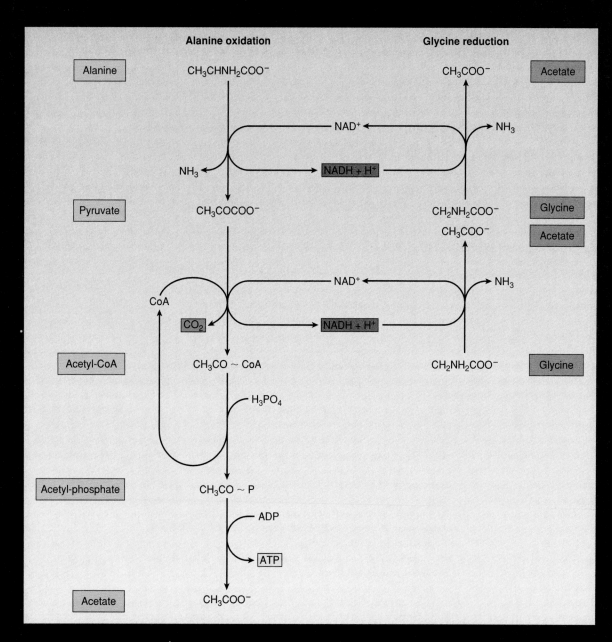

figure 7.18

The Stickland Reaction in *Clostridium*

The fermentations of alanine and glycine are coupled through the transfer of electrons in the Stickland reaction. Alanine donates electrons, which are accepted by glycine.

Origin of the 20 Amino Acids Commonly Occurring in Proteins

Family	Precursor		Amino Acid

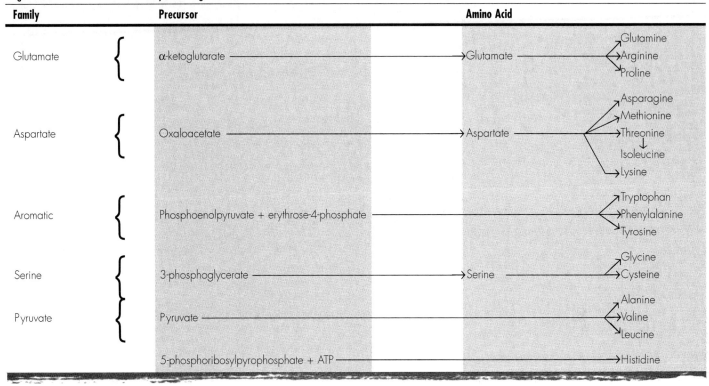

to soften and prepare hides for tanning; as an addition to washing detergents to dissolve mucus, blood, and other proteinaceous stains; and as meat tenderizers.

When proteins are used as sources of carbon and energy for microbial growth, the amino acids formed by proteolysis are transported across the plasma membrane. Inside the cell, the amino acids may first be deaminated to produce organic acids, which then enter the TCA cycle. The NH_3 resulting from deamination can serve as a source of nitrogen for biosynthesis.

Amino Acids Are Synthesized from Intermediates in Central Metabolic Pathways

Amino acids are required for protein synthesis. Some microbes are able to synthesize the 20 amino acids commonly found in proteins. These have the necessary enzymes to form amino acids from common intermediates in central metabolic pathways—the Embden-Meyerhof pathway, the pentose phosphate pathway, and the TCA cycle. Amino acids can be conveniently divided into several different groups or families, based on shared pathways of biosynthesis and a common use of metabolic precursors (Table 7.2).

Many of the amino acids are formed by the transfer of an amino group from an existing amino acid to an α-keto acid, or **transamination.** For example, aspartate is made by a transamination reaction in which the amino group from the amino acid glutamate is transferred to the α-keto acid oxaloacetate:

$$\text{Glutamate + oxaloacetate} \xrightarrow{\text{pyridoxal phosphate}}$$
$$\text{α-ketoglutarate + aspartate}$$

The other amino acids in the aspartate family (asparagine, methionine, threonine, isoleucine, and lysine) are synthesized through pathways leading from aspartate or one of its products.

Lipid Metabolism

Lipids are biological molecules that are insoluble in water but soluble in organic solvents. This large heterogeneous group includes fats and fatty acids, waxes, and steroids. Lipids are important in eucaryotes as protective coverings for plants, skin, and fur and as hormones that regulate organismal activity. In procaryotes, lipids are used as structural components of the cell or, in some instances,

Certain bacteria, particularly those of the family *Enterobacteriaceae*, produce enzymes that decarboxylate specific amino acids to amines or diamines. These **decarboxylases** are not synthesized by all species or genera of this family, so decarboxylase reactions provide a useful marker for bacterial identification.

ornithine. L-lysine is decarboxylated to CO_2 and cadaverine, a foul-smelling diamine associated with the putrefaction of flesh.

Decarboxylases

no acids to amines or diamines. These **decarboxylases** are not synthesized by all species or genera of this family, so decarboxylase reactions provide a useful marker for bacterial identification.

Tests for decarboxylases for three amino acids—arginine, lysine, and ornithine (the last is not one of the 20 amino acids in proteins)—are used for bacterial identification in clinical laboratories (Figure 7.19). The amino acid L-arginine is catabolized in one of two manners, using a decarboxylase or a dihydrolase. In the decarboxylase system, arginine is decarboxylated to agmatine and CO_2. Agmatine is then split by agmatine ureohydrolase to form putrescine, CO_2, and urea. In the dihydrolase system, arginine is eventually catabolized to the same products, but by a three-step reaction. In the first step, L-arginine is deaminated by arginine dihydrolase to form citrulline and NH_3. Citrulline is then split to form ornithine, CO_2, and NH_3. The ornithine is subsequently decarboxylated to putrescine and CO_2 by ornithine decarboxylase. The ornithine decarboxylase used in the arginine dihydrolase system is also used by bacteria for the direct decarboxylation of

The presence or absence of specific amino acid decarboxylases is easily determined by pH changes in a test medium. Bacteria are inoculated into a base medium containing low amounts of glucose and a 1% solution of one of the three amino acids (arginine, lysine, or ornithine). The medium also contains a pH indicator, bromcresol purple, which is yellow at acidic pHs and purple at alkaline pHs. The base medium is purple at the time of inoculation. After incubation, the medium will be either yellow or purple. Bacteria that do not decarboxylate the amino acid in the medium will catabolize glucose, forming acidic end products that drop the pH of the medium, and resulting in a yellow color. Bacteria that do decarboxylate the amino acid will initially produce acid from glucose, but as the amino acid is decarboxylated to putrescine or cadaverine and CO_2, the pH of the medium will rise. The result is a purple color that indicates a positive decarboxylase test. Since diamines are unstable in the presence of oxygen, the test must be performed under anaerobic conditions by overlaying the surface of the medium with paraffin or mineral oil after inoculation.

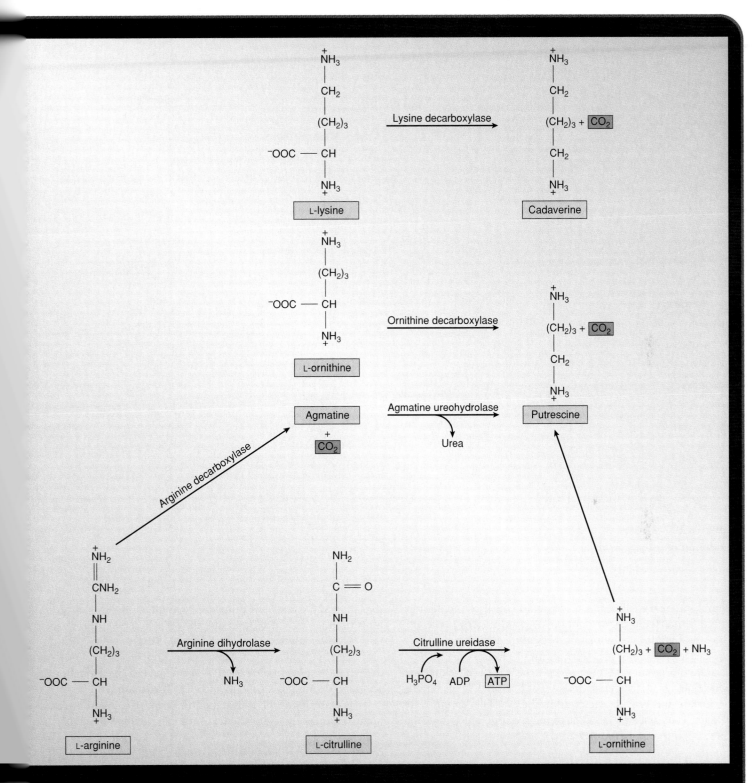

Figure 7.19
Decarboxylation of L-lysine, L-ornithine, and L-arginine

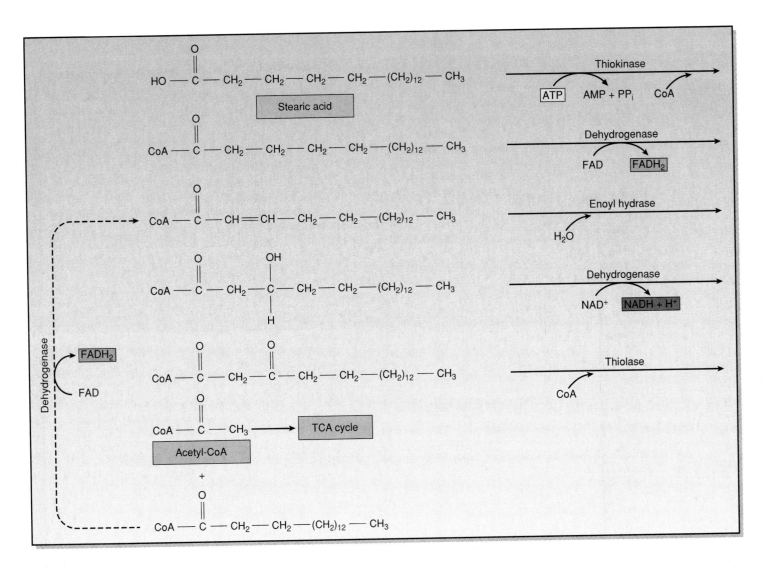

figure 7.20
β-**Oxidation of a Fatty Acid**

as sources of energy. Procaryotic structures such as the plasma membrane and the outer membrane of gram-negative cells are composed of lipids in combination with other organic molecules. The cell envelopes of mycobacteria, agents of tuberculosis and leprosy, contain large amounts of lipid (up to 60% of the envelope dry weight). Some bacteria secrete lipases that hydrolyze lipids on the skin's surface and provide energy from the available glycerol and fatty acids. Fatty acids represent large reservoirs of energy that are released upon their oxidation.

Microorganisms Degrade Fatty Acids by β-Oxidation

Fats are highly reduced organic compounds and therefore can be sources of energy for living organisms. Microbes degrade fats extracellularly by first hydrolyzing the ester bonds between the

glycerol and the fatty acid groups. The resultant glycerol and fatty acids can then be transported into the cell. Glycerol enters glycolysis via phosphorylation by glycerol kinase to glycerol-3-phosphate and subsequent oxidation by glycerophosphate dehydrogenase to dihydroxyacetone phosphate. The individual fatty acids are oxidized by a process known as β-**oxidation.**

β-oxidation involves the repetitive cleavage of 2-carbon fragments from fatty acids (Figure 7.20). The oxidation sequence begins with activation of the fatty acid with coenzyme A. The resulting fatty acyl CoA is then oxidized, and a double bond is formed between the α-carbon and β-carbon atoms of the molecule. The electrons and protons released during this reduction are passed to FAD. After the addition of a water molecule at this double bond, a second oxidation takes place, and the electrons and protons are transferred to NAD. The final step in the repetitive

Many of the intermediates of the TCA cycle are used in pathways for the biosynthesis of amino acids and other compounds. As a result of this dual nature of TCA cycle intermediates, oxaloacetate (an intermediate of the TCA cycle) quickly becomes depleted; its replacement is necessary for the continued function of the TCA cycle. An anaplerotic pathway that replenishes TCA cycle intermediates during growth on acetate or fatty acids is the **glyoxylate cycle,** a pathway found in microorganisms and plants (Figure 7.21).

The Glyoxylate Cycle

The glyoxylate cycle, so named because it includes glyoxylate as an intermediate, channels acetate and acetyl-CoA into the TCA cycle intermediates citrate, isocitrate, malate, and oxaloacetate. Vertebrates lack the two enzymes that are unique to the glyoxylate cycle, isocitrate lyase and malate synthase, and are unable to convert acetyl-CoA into TCA cycle intermediates. With insufficient oxaloacetate, the acetyl-CoA produced by fatty acid β-oxidation cannot enter the TCA cycle in these animals, and the excess acetyl-CoA is converted into ketone bodies. Most ketone bodies are not metabolized and are excreted. The consequence is ketosis, a condition characterized by the enhanced production of ketone bodies, as in diabetes mellitus. In bacteria, the glyoxylate cycle allows the TCA cycle to be bypassed during the catabolism of acetate or fatty acids.

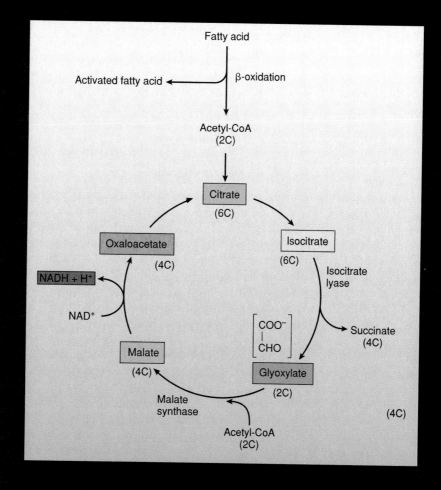

figure 7.21
The Glyoxylate Cycle
The glyoxylate cycle contains two unique enzymes: isocitrate lyase, which cleaves isocitrate to glyoxylate and succinate, and malate synthase, which catalyzes the synthesis of malate from acetyl-CoA and glyoxylate.

cycle is cleavage of acetyl-CoA from the fatty acid molecule. When this occurs, the remaining fatty acid chain is recharged with another coenzyme A, and the cycle repeats. The acetyl-CoA that is cleaved at each turn of the cycle is sent through the TCA cycle and oxidative phosphorylation to yield 12 molecules of ATP. The reduced $FADH_2$ and NADH + H^+ formed as a result of the two oxidation steps in the cycle yield 5 molecules of ATP through oxidative phosphorylation. Thus β-oxidation of a fatty acid results in 17 molecules of ATP formed for each 2-carbon fragment cleaved from the fatty acid. For an 18-carbon fatty acid such as stearic acid, the total ATP yield is 146. This total is obtained from the energy released by β-oxidation of the first eight 2-carbon fragments (8 × 17 = 136 ATP molecules) and the channeling of the final 2-carbon acetyl-CoA fragment through the TCA cycle (12 ATP molecules), minus 2 molecules of ATP that are required to activate the fatty acid initially. β-oxidation of unsaturated fatty acids and odd-numbered fatty acids yields slightly different quantities of ATP. Although fatty acids represent a potentially rich energy source for microorganisms, their oxidation is an aerobic process. Consequently, microbial degradation of fats and lipids does not occur readily under anaerobic conditions, such as in sewage treatment (see sewage treatment, page 594).

Fatty Acids Are Synthesized by the Stepwise Addition of Acetyl Groups to a Growing Chain

Fatty acids are generally synthesized through stepwise addition of 2-carbon units to a growing chain. These 2-carbon units are initially part of the 3-carbon molecule malonyl-CoA, which is formed by the carboxylation of acetyl-CoA. The enzyme acetyl-CoA carboxylase, which catalyzes this reaction, has the vitamin biotin as a cofactor. The biotin apparently binds CO_2 and transfers it to acetyl-CoA to form malonyl-CoA. The malonyl group is transferred from coenzyme A to a small carrier protein called the acyl carrier protein (ACP). The initial portion of the fatty acid chain is synthesized from acetyl-ACP combining with malonyl-ACP to form the 4-carbon acetoacetyl-ACP, with release of CO_2 and ACP. The acetoacetyl-ACP is reduced with two molecules of NADPH + H^+ to form butyryl-ACP. Additional 2-carbon units are added to this structure, with each addition of an acetyl group followed by reduction with two molecules of NADPH + H^+. The fatty acid thus grows in units of two carbon atoms at a time. Unsaturated and branched fatty acids follow similar pathways, but with minor variations.

Fats are formed from fatty acids via esterification of glycerol phosphate by fatty acyl-CoA or fatty acid-ACP. The glycerol phosphate is derived either from the phosphorylation of glycerol or from the reduction of dihydroxyacetone phosphate by NADH + H^+. The phosphate on the glycerol moiety is cleaved after two fatty acids are added, and is followed by the addition of the third fatty acid.

Purine and pyrimidine nucleotides are found in nucleic acids, many coenzymes, and certain antibiotics. They are also important in the activation of other compounds in anabolic pathways.

Nucleases Hydrolyze Nucleic Acids into Individual Mononucleotides, Which Can Then Be Catabolized

Nucleic acids and their constituents, pentoses, purines, and pyrimidines, can be used as sources of carbon, nitrogen, and energy. Bacteria that utilize nucleic acids in this manner produce nucleases that hydrolyze nucleic acids into individual mononucleotides, which are then broken down into such end products as NH_3 and CO_2. Some pathogenic bacteria such as *Staphylococcus aureus* and *Serratia marcescens* produce large quantities of extracellular DNases, which are potential aids in identification. DNase production is detected by plating bacteria onto a DNA-containing agar medium. After an appropriate incubation period, the medium is flooded with 1N HCl. Colonies that produce DNase hydrolyze the DNA in the surrounding medium, resulting in zones of clearing (Figure 7.22). The DNases synthesized by these bacteria can degrade viscous DNA released from damaged host cells and enable the bacteria to penetrate and invade deeper into tissue.

Purines and Pyrimidines Are Synthesized from a Variety of Molecules

The biosynthesis of purines and pyrimidines is complex and involves contributions by several different molecules (Figure 7.23). Carbon and nitrogen atoms of the purine structure are derived from carbon dioxide, formyl-tetrahydrofolate, methyl-tetrahydrofolate, and the amino acids aspartate, glutamine, and glycine. The atoms of the purine structure are synthesized onto a ribose-5-phosphate. Nucleoside diphosphates and triphosphates are formed from the monophosphates by successive phosphorylations.

In contrast, the pyrimidine ring is constructed first before addition of the ribose-5-phosphate. The pyrimidine ring is formed from carbamoyl phosphate and aspartate. The pentose sugar is then added to the base.

Deoxyribonucleotides are formed directly from the ribonucleotides by reduction of the number 2 carbon on the ribose sugar. This reduction occurs with either the ribonucleoside diphosphate or the triphosphate, depending on the species of the bacterium, and results in formation of the corresponding deoxyribonucleotide derivative.

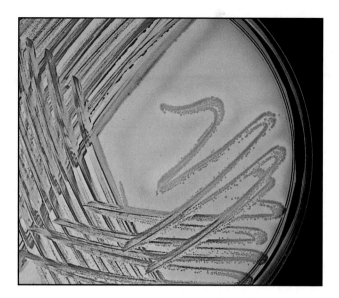

figure 7.22
Testing for DNase Production

Organisms producing extracellular DNase can be detected by adding dilute hydrochloric acid to the surface of the DNA-containing agar medium. In this example, clear zones appear around the areas of growth of the DNase-producing bacterium *Serratia marcescens*.

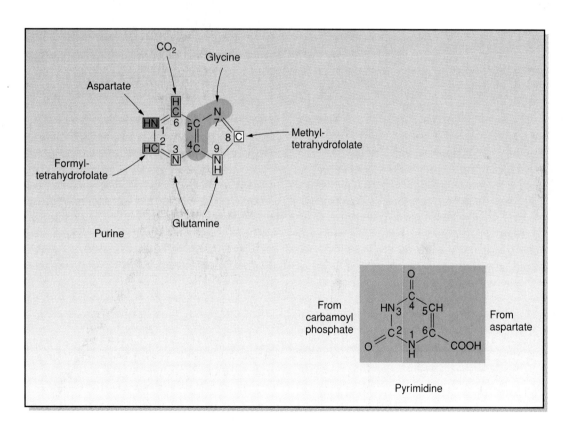

figure 7.23
Origin of Atoms in Purine and Pyrimidine Bases

Biochemical substances (precursors, intermediates, or products) that play a role in cellular metabolism are classified as either **primary metabolites** or **secondary metabolites**. Metabolites constitute a diverse group of chemical compounds and include such substances as vitamins, amino acids, sugars, carbon dioxide, urea, and antibiotics. Primary metabolites are involved in cell growth or function. In comparison, secondary metabolites apparently are not necessary for primary cellular processes.

Secondary metabolites are interesting biochemicals and include the antibiotics, a clinically important group of antimicrobial metabolites. Antibiotic production is a multimillion dollar pharmaceutical industry; studies on secondary metabolism have focused on optimizing the microbial synthesis of these compounds. However, very little is known about the biochemical pathways or regulatory mechanisms that are associated with secondary metabolism. Why does a cell expend a significant part of its energy to produce compounds that seemingly are of little or no value to it?

A model of secondary metabolism that may provide answers is the pathway that leads to the production of the microbial pigment prodigiosin (Figure 7P.1). Prodigiosin, a red compound, is a secondary metabolite synthesized by the gram-negative rod *Serratia marcescens* during the late logarithmic or stationary phase of growth. This tripyrrole pigment is formed only at temperatures below 39°C—*S. marcescens* cells grown at 39°C or above are nonpigmented.

Several years ago an intriguing method was introduced to determine the precursors for prodigiosin biosynthesis. S. M. Hussain Qadri and Robert P. Williams developed a nonproliferating cell system to follow prodigiosin synthesis without interference from the primary metabolism of cellular growth and multiplication. This system involved washing cells of *S. marcescens* previously grown in broth to remove endogenous metabolites. These washed cells were then resuspended in

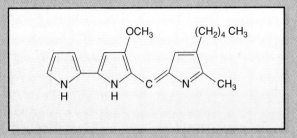

figure 7P.1

Prodigiosin (2-methyl-3-pentyl-6-methoxyprodigiosene)

0.85% (w/v) NaCl and examined for the effects of added nutrients on prodigiosin biosynthesis. Without endogenous pools of metabolites, these nonproliferating cells could not grow or multiply.

Using this nonproliferating system, Qadri and Williams added ^{14}C-labeled amino acids to the cells to follow the incorporation of these amino acids into prodigiosin. After different periods of incubation, the bacteria were centrifuged and prodigiosin was extracted from the cells. It was discovered from such experiments that the prodigiosin molecule was constructed from several amino acids, including proline (which is incorporated intact), histidine (which is indirectly incorporated), methionine (which contributes its methyl group), and alanine (which contributes all but its carboxyl group) (Figure 7P.2).

These data provided some of the first evidence that amino acid intermediates in primary metabolism contribute to the structure of a secondary metabolite (prodigiosin). Although the importance of prodigiosin to *S. marcescens* is unknown, it has been suggested that one function of prodigiosin may be the removal of excess accumulations of metabolites such as amino acids. The data of Qadri and Williams appear to lend support to such a theory.

Sources

Lim, D.V., S.M.H. Qadri, C. Nichols, and R.P. Williams. 1977. Biosynthesis of prodigiosin by non-proliferating wild-type *Serratia marcescens* mutants deficient in catabolism of alanine, histidine, and proline. *Journal of Bacteriology* 129:124–130.

Qadri, S.M.H., and R.P. Williams. 1974. Incorporation of amino acid carbon into prodigiosin synthesized by nonproliferating cells of *Serratia marcescens. Canadian Journal of Microbiology* 20:461–468.

Williams, R.P. 1973. Biosynthesis of prodigiosin, a secondary metabolite of *Serratia marcescens. Applied Microbiology* 25:396–402.

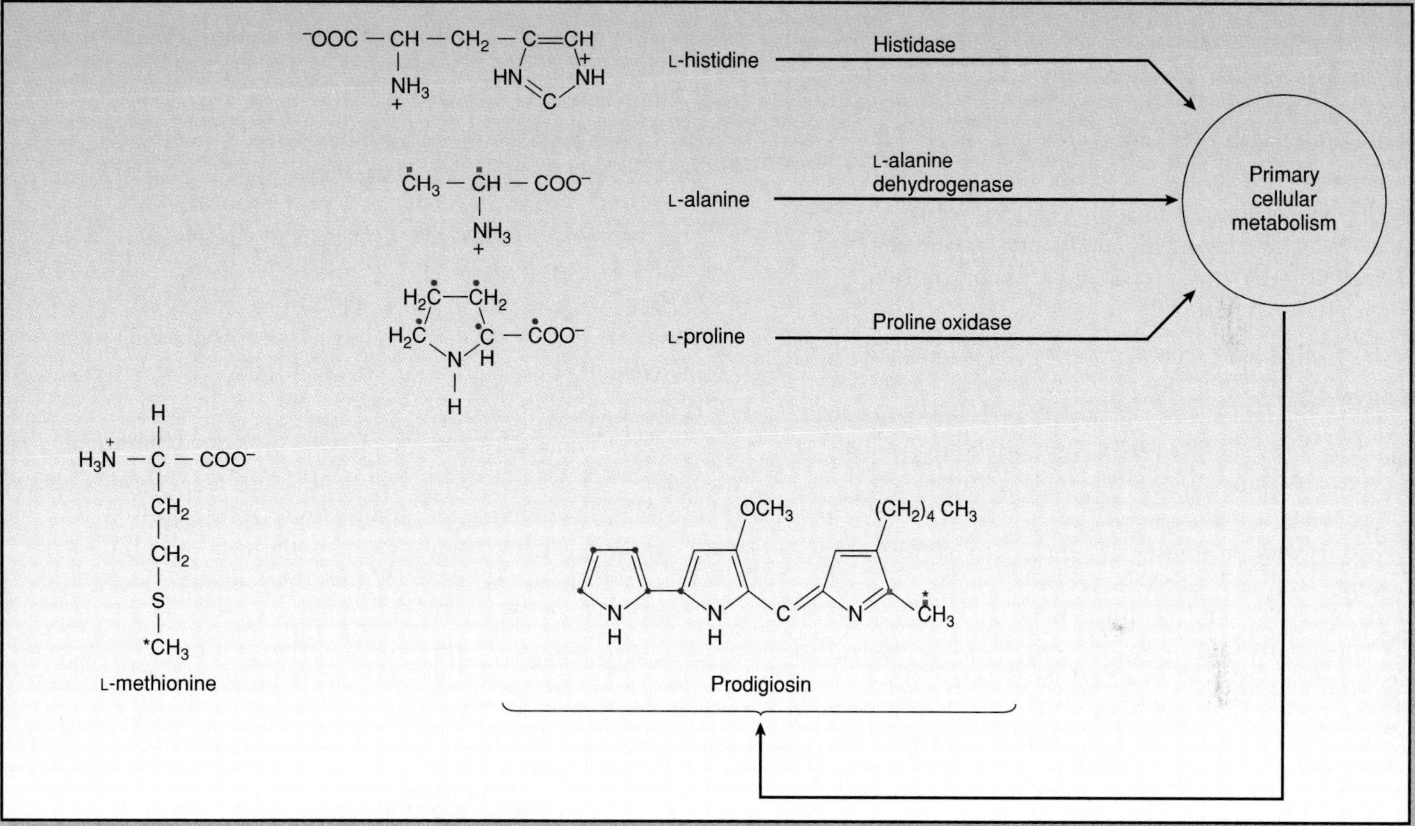

figure 7P.2

Proposed Scheme for Biosynthesis of Prodigiosin from L-alanine, L-histidine, and L-proline

The three amino acids L-alanine, L-histidine, and L-proline contribute to the structure of prodigiosin. The methyl group of the third pyrrole may alternatively be derived from L-methionine. Symbols (●, ■, and ★) represent carbon atoms of amino acids detected in prodigiosin by radiolabeling studies.

1. Photosynthesis consists of two major sets of reactions: photophosphorylation, or the light-dependent reactions, and carbon dioxide fixation, or the dark reactions.

2. The principal light-trapping pigment molecule in phototrophic plants, algae, and cyanobacteria is chlorophyll. Bacteriochlorophyll is the photosynthetic pigment in phototrophic bacteria.

3. Accessory pigments harvest light from other portions of the light spectrum in photosynthesis and also protect the photosynthetic apparatus from photooxidation. Carotenoids are lipid-soluble, long-chain, conjugated hydrocarbons founds in photosynthetic eucaryotes and procaryotes. Biliproteins, water-soluble, linear tetrapyrroles coupled to proteins, occur only in red algae and cyanobacteria.

4. The photosynthetic apparatus is found within chloroplasts in eucaryotes, thylakoids in elaborated membranes in cyanobacteria, and specialized membranes or chlorobium vesicles in phototrophic bacteria.

5. The Hill-Bendall scheme describes the sequence of events occurring in eucaryotic and cyanobacterial photo-phosphorylation. ATP and NADPH + H$^+$ are generated during noncyclic photophosphorylation, whereas only ATP is generated during cyclic photophosphorylation.

6. Bacterial photophosphorylation is cyclic. NADPH + H$^+$ is produced by the transfer of reducing power from reduced compounds in the bacterium's environment.

7. Bacterial photosynthesis is anoxygenic, whereas eucaryotic photosynthesis is oxygenic. ATP generation in both eucaryotic and procaryotic photophosphorylation is explained by chemiosmosis.

8. Autotrophs fix carbon dioxide into organic compounds via the Calvin cycle. Carbon dioxide fixation requires sufficient quantities of ATP and NADPH + H$^+$, which can be generated in phototrophs by the light-dependent reactions of photo-synthesis or in chemotrophs by other chemical reactions.

9. Microorganisms have a wide variety of metabolic pathways for metabolism of carbohydrates, proteins, lipids, and nucleic acids (see Figure 7.25).

10. Polysaccharides are degraded into smaller molecules before they enter the cell. Gluconeogenesis, which is reversal of several steps in the Embden-Meyerhof pathway, is an example of a pathway for hexose synthesis. Pentoses are synthesized from hexoses via the pentose phosphate pathway.

11. Microorganisms can produce proteases, which are enzymes that hydrolyze proteins. Amino acids are synthesized from common intermediates in central metabolic pathways.

12. Fatty acids can be degraded by some microorganisms via β-oxidation. Fatty acids are generally synthesized through the stepwise addition of acetyl groups to a growing fatty acid chain.

13. Some microorganisms produce nucleases that hydrolyze nucleic acids into individual mononucleotides. Purines and pyrimidines are synthesized from a variety of molecules.

EVOLUTION and BIODIVERSITY

Chloroplasts and mitochondria show many similarities to procaryotes: they are approximately the same size, reproduce by dividing, contain RNA and circular DNA, have 70S ribosomes, and have double membranes similar to the plasma and outer membranes of gram-negative bacteria. These similarities suggest that these organelles may have arisen from endosymbiotic procaryotes (procaryotes that live within another organism). According to the **endosymbiont hypothesis,** eucaryotic cells originally existed without mitochondria or chloroplasts. With time, these primitive eucaryotic cells established an endosymbiotic relation with procaryotes. Chloroplasts probably evolved from cyanobacteria that provided ATP through photosynthesis for their host cells for growth and synthesis of specialized structures in exchange for the protected and nutrient-rich environment of the host. Such a theory is plausible since present-day eucaryotic cells exist that contain endosymbiotic cyanobacteria (Figure 7.24). If the endosymbiont theory is correct, then it would also be evident that mitochondria and chloroplasts have changed over the years. These organelles have very small quantities of DNA compared to procaryotes and have lost certain procaryotic structures such as the cell wall.

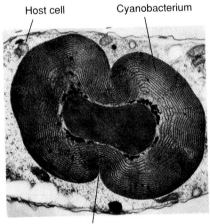

Host cell   Cyanobacterium

Cleavage furrow

figure 7.24

A Present-Day Cyanobacterium Living in Permanent Symbiosis Within a Eucaryotic Cell

The two organisms are jointly called *Cyanophora paradoxa.* Note that the cyanobacterium is undergoing cleavage.

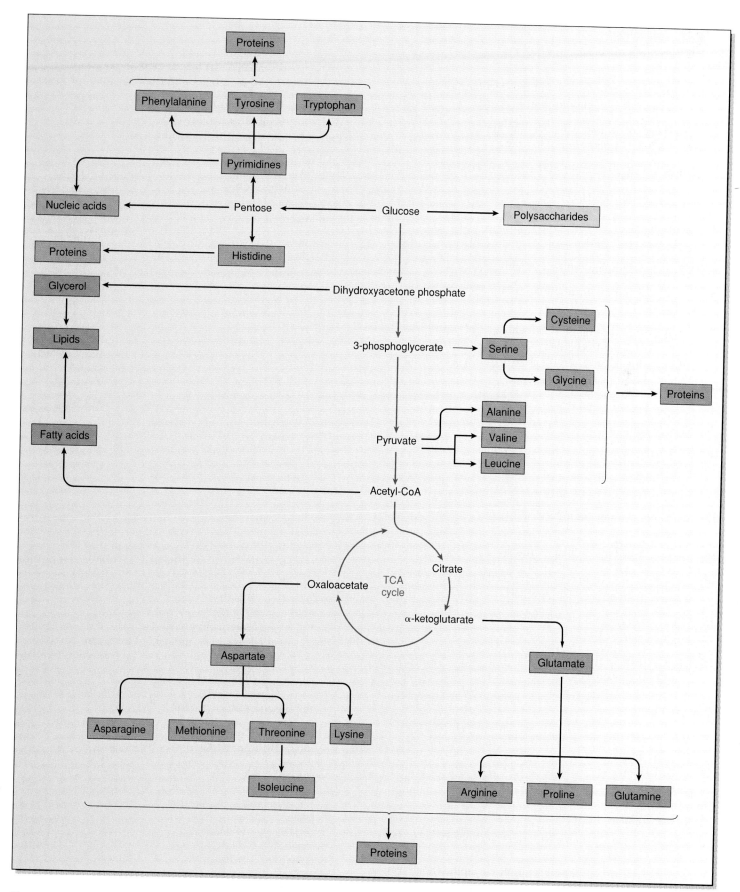

figure 7.25

Short Answer

1. Compare and contrast photosynthesis in plants, algae, and cyanobacteria and in phototrophic bacteria.

2. What classes of accessory pigments are found in cyanobacteria? in phototrophic bacteria?

3. Identify the electron donor in the following reaction:

$$6\,CO_2 + 6\,H_2O \rightarrow C_6H_{12}O_6 + 6\,O_2$$

4. Are the light and dark reactions of photosynthesis always coupled? Explain your answer.

5. Define photolysis and explain its significance.

6. Why is magnesium an important component of many fertilizers?

7. Would a reduced compound "X" with a redox potential of -0.41 V be able to donate electrons directly to $NADP^+$ without going through reverse electron transport in bacterial photosynthesis? Explain your answer.

8. Which is the reduction step in the Calvin cycle?

9. Where might cellulose-digesting bacteria be found?

10. What is the significance of glycogen storage granules in bacteria?

11. Briefly describe gluconeogenesis.

12. Give some examples of industrial uses of microbial proteases.

13. Briefly describe the Stickland reaction.

14. Explain the basis of tests to determine the presence or absence of amino acid decarboxylases.

15. What two enzymes do vertebrates lack that are required for the glyoxylate cycle?

16. Can CO_2 ever be used as an energy source? If yes, explain how. If no, explain why not.

Multiple Choice

1. Which of the following are not found in cyanobacteria?
   a. chloroplasts
   b. thylakoids
   c. biliproteins
   d. carotenoids
   e. photosystems I and II

2. CO_2 enters the Calvin cycle in the step involving formation of
   a. glyceraldehyde-3-phosphate from 1,3-diphosphoglycerate
   b. ribulose-1,5-diphosphate from (2) glyceraldehyde-3 phosphate
   c. (2) 3-phosphoglycerate from ribulose-1,5-diphosphate
   d. 1,3-diphosphoglycerate from 3-phosphoglycerate
   e. ribulose-1,5-diphosphate from ribulose-5-phosphate

3. Organism A is a bacterium that can grow aerobically in the dark on a medium containing NO_2^-, CO_2, KH_2PO_4, minerals, and water. This bacterium is a
   a. photoheterotroph
   b. photoautotroph
   c. chemoheterotroph
   d. chemoautotroph
   e. phototroph

4. A metabolic pathway that functions in the dual roles of catabolism and anabolism is said to be
   a. anaplerotic
   b. synergistic
   c. amphibolic
   d. plerotic
   e. anabolic

5. Aspartate is formed by a transamination reaction in which the amino group from glutamate is transferred to
   a. pyruvate
   b. serine
   c. oxaloacetate
   d. alanine
   e. succinate

Critical Thinking

1. The Hill-Bendall scheme for photophosphorylation generates ATP and $NADPH + H^+$ required for CO_2 fixation. Assume that a pair of electrons is ejected from a reaction center chlorophyll in the Hill-Bendall scheme each time the chlorophyll is struck with two quanta of light. Also assume that two ATPs are synthesized by noncyclic photophosphorylation for every pair of electrons ejected from reaction center chlorophyll P680 and sent through the Hlll-Bendall scheme to eventually reduce $NADP^+$ to $NADPH + H^+$. How many quanta of light would be required to produce enough $NADPH + H^+$ for the synthesis of two hexoses in CO_2 fixation, assuming only noncyclic photophosphorylation occurs?

2. What would be the effect of blocking light below a wavelength of 700 nm to cyanobacteria? to phototrophic bacteria?

3. Cyanobacteria have characteristics of both phototrophic bacteria and plants. Comment on the evolutionary significance of this observation.

4. β-oxidation is a process used by some microorganisms to degrade fatty acids. Of what industrial significance is degradation of fatty acids by microorganisms?

5. What common role do O_2, NO_3^-, and CO_2 have in bacterial metabolism?

Caldwell, D.R. 1995. *Microbial physiology & metabolism.* Dubuque, Iowa: Wm. C. Brown Publishers. (A microbial physiology textbook with discussions of metabolism, fermentation, and genetics.)

Doelle, H. W. 1975. *Bacterial metabolism,* 2d ed. New York: Academic Press. (A comprehensive and detailed description of the biochemistry of bacterial metabolism.)

Drews, G. 1985. Structure and functional organization of light-harvesting complexes and photochemical centers in membranes of phototrophic bacteria. *Microbiological Reviews* 49:59–70. (An extensive review of photosynthetic apparatus in bacteria.)

Gerhardt, P., R.G.E. Murray, W.A. Wood, and N.R. Krieg. 1994. *Methods for general and molecular bacteriology,* pp. 463–599. Washington, D.C.: American Society for Microbiology. (A discussion of various techniques in microbiology, including sections on microbial metabolism and assays for metabolic and enzymatic activities.)

Gottschalk, G. 1986. *Bacterial metabolism,* 2d ed. New York: Springer-Verlag. (An extensive discussion of metabolic pathways in bacteria.)

Mandelstam, J., K. McQuillen, and I.W. Dawes. 1982. *Biochemistry of bacterial growth.* Oxford, England: Blackwell Scientific Publications. (A detailed discussion of bacterial growth under different conditions.)

Neidhardt, F.C., J.L. Ingraham, and M. Schaechter. 1990. *Physiology of the bacterial cell: A molecular approach.* Sunderland, Mass.: Sinauer Associates. (A textbook of bacterial physiology and genetics.)

Stryer, L. 1988. *Biochemistry,* 3d ed. San Francisco, Calif.: W. H. Freeman. (A detailed review of biochemistry, with discussions of chemical reactions and structures.)

Wood, W.B., J.H. Wilson, R.B. Benbow, and L.E. Hood. 1981. *Biochemistry: A problems approach,* 2d ed. Menlo Park, Calif.: The Benjamin/Cummings Publishing Company. (A concise review of major topics in biochemistry, with extensive and thought-provoking problems.)

Zubay, G. 1993. *Biochemistry,* 3d ed. Dubuque, Iowa: Wm. C. Brown Publishers. (A general textbook of biochemistry.)

chapter *eight*

MOLECULAR GENETICS

The Genetic Code

Deoxyribonucleic Acid
DNA: The Genetic Material
Structure
Replication
The Rolling Circle Mechanism
Replication in Procaryotes and Eucaryotes

Ribonucleic Acid
mRNA, rRNA, and tRNA
Synthesis from DNA
RNA Polymerases in Eucaryotes

Protein
The Initiation Complex
Elongation
Termination
Protein Synthesis in Eucaryotes

Regulation
Induction
Repression
Attenuation
Catabolite Repression
Unique Mechanisms in Eucaryotes

PERSPECTIVE
Discontinuous Synthesis of DNA During Replication

EVOLUTION AND BIODIVERSITY

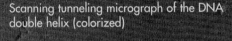
Scanning tunneling micrograph of the DNA double helix (colorized)

 Microbes in Motion ———————— **PREVIEW LINK**

This chapter covers the key topics of protein synthesis and replication of bacteria and viruses. The following sections in the *Microbes in Motion* CD-ROM may be useful as a preview to your reading or as a supplemental study aid:

Bacterial Structure and Function: Internal Structures, 1–21. *Microbial Metabolism and Growth*: Genetics (Recombination), 15–18. *Viral Structure and Function*: Viral Replication; DNA Replication, 8–11.

eoxyribonucleic acid (DNA) has often been called the blueprint of life, because the characteristics of all life-forms ultimately depend upon the information specified by DNA. If DNA is the blueprint of life, then proteins are among the major building blocks. The information in DNA is translated into proteins via ribonucleic acid (RNA). DNA, RNA, and protein are the principal macromolecules of a cell and are responsible for the remarkable ability of organisms to pass their characteristics from one generation to the next with relatively few mistakes.

Genetics is a field of biology that deals with mechanisms responsible for the transfer of traits from one generation to another. Procaryotes and eucaryotes store their genetic information in DNA (viruses store genetic information in either DNA or RNA, depending on the type of nucleic acid in the virus). Each organism has its own distinct set of DNA found in the nucleoid, as the DNA-containing nuclear region in procaryotes is often called (DNA is contained within the nucleus in eucaryotes).

The chromosome of an organism can be divided into functional regions called **genes.** A gene is an information-containing region of the chromosome that is transcribed into an RNA message (or a single structural RNA such as an rRNA or tRNA molecule). The typical *Escherichia coli* cell contains about 3,000 to 4,000 different genes. The total complement of genes found in a cell or a virus is considered to be its **genome.** The term **genotype** indicates the genetic composition of an organism. As a result of its genotype, an organism expresses certain observable characteristics, known as its **phenotype.**

Each gene codes for a functional protein product that has an amino acid sequence determined by the sequence of bases within that gene. Early theories hypothesized that proteins were formed directly from a DNA template, but this is incorrect; a more complex flow of genetic information from DNA to proteins actually occurs. The genetic information contained within DNA is first copied into messenger RNA (mRNA) in a process known as **transcription.** The mRNA, acting as a messenger, relays information to synthesize protein molecules. The synthesis of proteins from an RNA messenger is called **translation.** The sequence of events that describes the flow of genetic information in the cell from DNA to RNA and eventually to proteins is known as the **central dogma** of molecular biology (Figure 8.1).

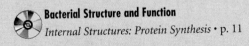

Bacterial Structure and Function
Internal Structures: Protein Synthesis • p. 11

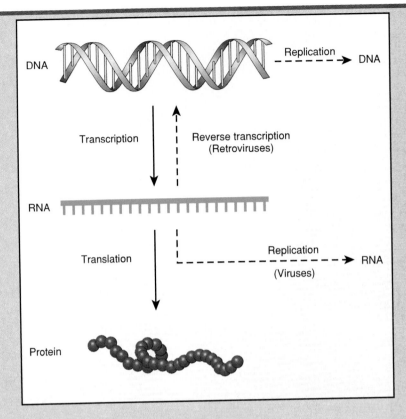

figure 8.1
The Central Dogma
The central dogma describes the flow of genetic information in the cell.

The central dogma is universal and applies to both procaryotes and eucaryotes. However, procaryotic mRNA is often **polycistronic** (more than one cistron, or gene), with more than one coding region (Figure 8.2a). Because one gene codes for a single protein, a polycistronic mRNA encodes several proteins separately translated from the same messenger. In comparison, eucaryotic mRNA is monocistronic, with only one protein translated per mRNA molecule (Figure 8.2b). Eucaryotic mRNA also frequently contains noncoding regions (**introns**) that are excised from the messenger molecule before the coding regions (**exons**) are translated. This chapter will focus on DNA, RNA, and protein synthesis in procaryotes.

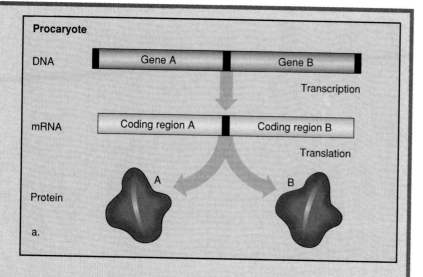

Procaryote

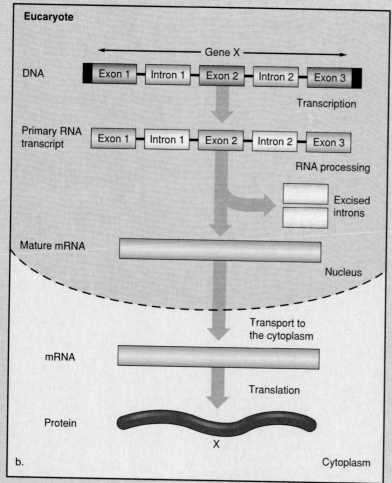

Eucaryote

figure 8.2

Comparison of Transcription and Translation in Procaryotes and Eucaryotes

a. Procaryotes have polycistronic mRNA that can encode more than one protein. b. Eucaryotes' mRNA is monocistronic and frequently must be processed to remove noncoding regions (introns) before translation.

The Genetic Code

The genetic information of a cell—or, in the case of procaryotes, of an entire organism—is organized like the writing in a book. This living book's DNA is written in a universal language that is transcribed into RNA, a macromolecule consisting of the four bases—adenine, uracil, guanine, and cytosine (Figure 8.3). These bases, linked to ribose molecules to form nucleosides and then linked to phosphate to form nucleotides, are organized into three-letter words known as **codons.** Since there are four different bases arranged in these codon triplets, there is a possibility of 4^3, or 64, different codons that can be formed from various combinations of the four bases. These 64 codons constitute what is known as the vocabulary, or **genetic code,** of living cells (Table 8.1).

Although there are 64 possible codons, only 61 of these code for one of the 20 common amino acids. The remaining three codons (UAA, UAG, and UGA) do not code for a specific amino acid and are therefore called **nonsense codons** or **termination codons.** These codons are important in signaling the termination of protein synthesis. Because many of the amino acids are represented by more than one codon, the genetic code is said to be degenerate. This redundancy means that in many cases, several codons may specify the insertion of the same amino acid into a protein.

The third position of a codon is the most degenerate of the three positions. An example of this degeneracy is the coding of the amino acid alanine. Alanine is coded by four different triplet base sequences: GCU, GCC, GCA, and GCG. The first two bases in each of these codon triplets are identical. The base in the third position of the codon can be uracil, cytosine, adenine, or guanine, yet the same amino acid (alanine) is specified by any of these four different codons.

The information within the codons in mRNA is translated into amino acids during protein synthesis. These amino acids are strung together into the specific sequences of proteins, some of which serve as catalysts (enzymes) for chemical reactions in the cell and others of which are structural proteins. Although the same bases and codons are universally found in all cells, the variations in DNA and protein compositions in different types of organisms make genetic diversity possible among these organisms.

The genetic information in any given cell is more or less constant, and this DNA is replicated and passed on to its progeny. Although errors occasionally occur in DNA replication, the relative constancy of the genetic

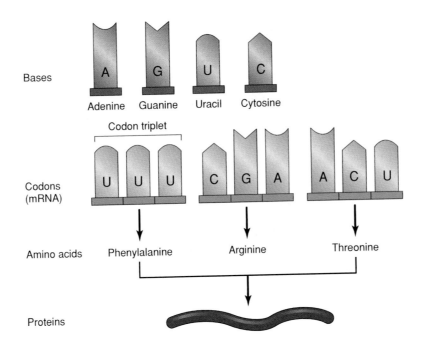

figure 8.3
The Genetic "Alphabet" of Living Cells
The genetic alphabet consists of four bases (adenine, uracil, guanine, and cytosine), which are arranged into codon triplets in mRNA. These codons make up the genetic code of the cell.

table 8.1

The Genetic Code

Codons are given as they would appear in mRNA

UUU	Phenylalanine	CUU	Leucine	GUU	Valine	AUU	Isoleucine
UUC	Phenylalanine	CUC	Leucine	GUC	Valine	AUC	Isoleucine
UUG	Leucine	CUG	Leucine	GUG(start)[b]	Valine	AUG (start)[b]	Methionine
UUA	Leucine	CUA	Leucine	GUA	Valine	AUA	Isoleucine
UCU	Serine	CCU	Proline	GCU	Alanine	ACU	Threonine
UCC	Serine	CCC	Proline	GCC	Alanine	ACC	Threonine
UCG	Serine	CCG	Proline	GCG	Alanine	ACG	Threonine
UCA	Serine	CCA	Proline	GCA	Alanine	ACA	Threonine
UGU	Cysteine	CGU	Arginine	GGU	Glycine	AGU	Serine
UGC	Cysteine	CGC	Arginine	GGC	Glycine	AGC	Serine
UGG	Tryptophan	CGG	Arginine	GGG	Glycine	AGG	Arginine
UGA	Termination[a]	CGA	Arginine	GGA	Glycine	AGA	Arginine
UAU	Tyrosine	CAU	Histidine	GAU	Aspartic acid	AAU	Asparagine
UAC	Tyrosine	CAC	Histidine	GAC	Aspartic acid	AAC	Asparagine
UAG	Termination[a]	CAG	Glutamine	GAG	Glutamic acid	AAG	Lysine
UAA	Termination[a]	CAA	Glutamine	GAA	Glutamic acid	AAA	Lysine

[a]Termination codon.
[b]GUG and AUG, at the beginning of the mRNA, code for N-formylmethionine in Bacteria.

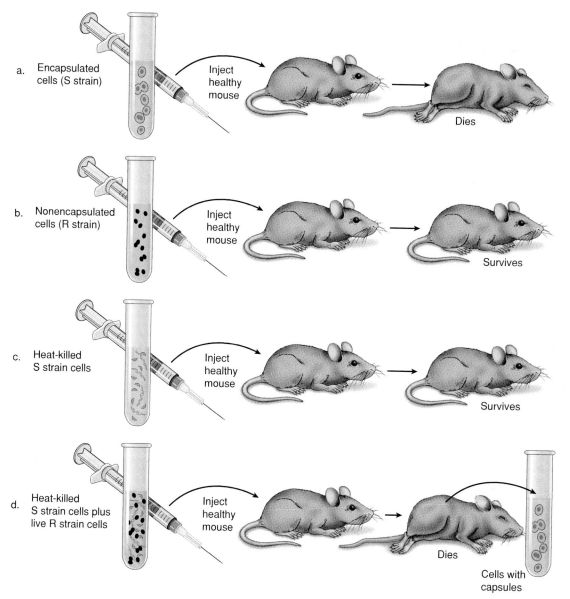

figure 8.4

Bacterial Transformation

a. Encapsulated (smooth, or S, strain) cells of *Streptococcus pneumoniae* produce a fatal infection in mice. b. Nonencapsulated (rough, or R, strain) cells do not. c. Heat-killed cells are avirulent.

d. A mixture of heat-killed S strain cells and live R strain cells of *S. pneumoniae* kills mice. DNA from the S strain cells entered the R strain cells and has transformed them into virulent S strain bacteria.

information in a cell makes it possible for generations of cells to remain identical. Without this stability, living organisms would be in a state of constant genetic change.

Deoxyribonucleic Acid

Proteins composed of 20 different amino acids can have enormous sequence differences and can assume many different physical conformations. Because of this property, scientists for many years believed that genes were made of protein. It was not until the mid-twentieth century that DNA was identified as the genetic material. Several landmark experiments were instrumental in this discovery.

DNA Is the Genetic Material of the Cell

The first of these landmark experiments was conducted in 1928 by the British scientist Fred Griffith. Griffith was interested in the epidemiology of pneumonia caused by *Streptococcus pneumoniae*, a bacterium surrounded by a thick polysaccharide capsule. Griffith observed in experiments that encapsulated strains (called smooth, or S, strains) of *S. pneumoniae* were virulent (capable of causing disease) in mice, whereas heat-killed S-strain cells and nonencapsulated strains (called rough, or R, strains) of the bacterium were avirulent (not capable of causing disease) when injected into mice (Figure 8.4a, b, c). However, when heat-killed S strains and live R strains of *S. pneumoniae* were

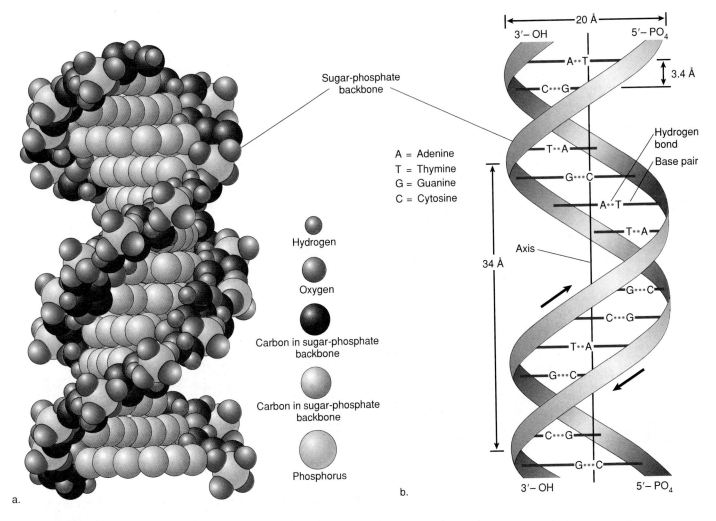

figure 8.5

Structure of DNA
a. A space-filling model of the DNA double helix showing arrangement of atoms within the DNA. b. A diagrammatic model of DNA showing its dimensions and pairing of the bases.

injected simultaneously into mice, the R cells converted to encapsulated, virulent S cells lethal to the mice (Figure 8.4d). *S. pneumoniae* has different types of polysaccharide capsules, and Griffith discovered that R cells were always transformed into the capsule type of the co-injected, heat-killed S cells.

 **Microbial Metabolism and Growth**
Genetics: Recombination • pp. 15–18

The substance responsible for the transformation was identified in 1944 by Oswald T. Avery, Colin M. MacLeod, and Maclyn McCarty of the Rockefeller Institute. This material, resistant to proteases and RNases but sensitive to DNases, had all the characteristics of DNA. Avery and his colleagues also discovered that DNA isolated from cells of one strain of bacteria and taken up and inserted into cells of another strain could successfully transform the recipient cells. Convincing evidence that DNA is the genetic material was provided by the Hershey and Chase T2 bacteriophage isotope-labeling experiment, in which it was shown that the phage DNA, not the protein, carried the genetic information required for phage replication (see Hershey and Chase experiment, page 400).

DNA Consists of Two Complementary Polynucleotide Chains Held Together by Hydrogen Bonds

The structure of DNA was elucidated by James D. Watson and Francis H. C. Crick in 1953 on the basis of X-ray diffraction data supplied by Maurice H. F. Wilkins and Rosalind Franklin. Watson, Crick, and Wilkins shared the 1962 Nobel Prize in physiology and medicine for their work on DNA structure.

These scientists discovered that DNA consists of two polynucleotide chains arranged in a long, slender, double-stranded helix (Figure 8.5). The helix has a diameter of 20 Å (2 nm) and makes a complete turn every 34 Å (3.4 nm). Because the distance between adjacent nucleotides is 3.4 Å, there are ten base pairs per turn of the helix. The chains are composed of four different bases: two pyrimidines (cytosine and thymine) and two purines (adenine and guanine). The numbers and order of these bases are variable and unique for each species. This uniqueness provides organisms with their distinct hereditary characteristics. The nucleotides are joined to each other in a chain by covalent phosphodiester bonds that link the number

The distinctive complementary pairing of bases in DNA is reflected in a constant 1:1 ratio for adenine:thymine and guanine: cytosine. However, the proportion of G + C to

turation of double-stranded DNA occurs over a narrow temperature range. As the strands separate, there is a sharp increase in their ultraviolet light absorption

melting temperature, or **T_m** (Figure 8.6). The T_m of double-stranded DNA increases as the mole percent of G + C increases, thereby providing an estimate of the G + C content in the DNA.

Nucleotide Base Ratios

A + T varies among different species and can be used for classification and identification. This ratio, conveniently expressed as the mole percentage of G + C in DNA, ranges from 22% to 74% in bacteria. Since guanine-cytosine base pairs have three hydrogen bonds, in comparison with the two hydrogen bonds of adenine-thymine base pairs, DNA containing a high mole percent of G + C tends to be more stable to unwinding or denaturation into single strands.

This stability can be used to estimate the content of G + C in DNA. The dena-

at 260 nm. The heterocyclic rings of nucleotides constituting DNA absorb ultraviolet light strongly at 260 nm. This absorption, however, is approximately 40% greater when the two complementary strands of DNA are separated than when they are in a doubled-stranded structure. This change in absorption of denatured DNA, called **hyperchromicity,** can be used to follow the dissociation of the double helix into two single strands. The temperature corresponding to the midpoint of this increase in absorption is called the

Denatured strands of DNA can be reannealed by slow cooling during which heat-denatured, single-stranded DNA re-forms double-stranded DNA. The similarity of two organisms can be determined by denaturing their DNA and then reassociating the separate strands to form hybrids. The greater the degree of hybridization, the closer the genetic similarity of the organisms. This technique of DNA molecular analysis provides one of the most precise methods of classification (see classification of bacteria, page 309).

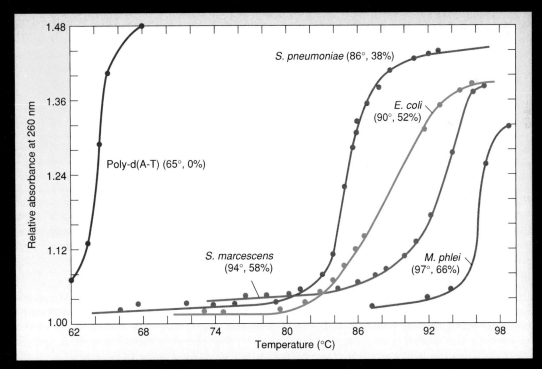

figure 8.6

Melting Curves of Double-stranded DNA Extracted from Four Different Bacteria (*Streptococcus pneumoniae, Escherichia coli, Serratia marcescens,* and *Mycobacterium phlei*) and of an Artificially Synthesized Poly-d(A-T) Nucleic Acid

The numbers in parentheses represent the melting temperature (T_m) and the G + C content, respectively, of each DNA sample.

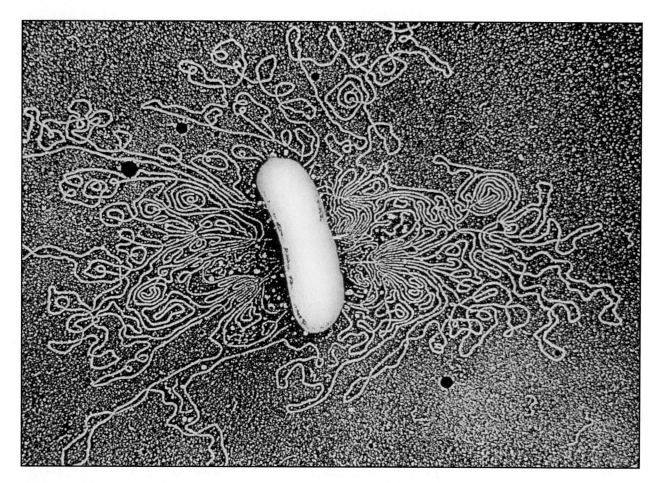

figure 8.7

Chromosomal DNA from a Bacterial Cell

DNA released from an *Escherichia coli* cell (colorized, ×56,000).

3 carbon (3′ carbon) of one deoxyribose to the number 5 carbon (5′ carbon) of the adjoining deoxyribose (see purine and pyrimidine structures, page 46).

The two chains in the DNA helix are held in place by hydrogen bonds between base pairs. The base pairing is such that adenine is always paired with thymine by two hydrogen bonds, and guanine is always paired with cytosine by three hydrogen bonds. This pyrimidine-purine pairing is necessary, since pyrimidine-pyrimidine pairing would occupy too little space and purine-purine pairing would take up too much space to allow a regular helix of constant diameter. As a consequence of this base pairing, the two polynucleotide chains in the DNA helix are complementary—that is, a sequence of GCATAGC on one chain is complemented by a sequence of CGTATCG on the opposite chain. The polynucleotide chains are also arranged in such a manner that they run in opposite directions and therefore are antiparallel. This arrangement results in one chain having a polarity of 5′ to 3′—with a nonphosphorylated 3′-deoxyribose carbon (3′—OH) at one end and a phosphorylated 5′-deoxyribose carbon (5′—PO_4) at the other end—and the other chain having a reverse polarity of 3′ to 5′:

$$5'—PO_4 \quad\quad —G—C—A—T—A—G—C— \quad\quad 3'—OH$$
$$3'—OH \quad\quad —C—G—T—A—T—C—G— \quad\quad 5'—PO_4$$

The presence of phosphodiester bonds between nucleotides in each DNA chain imparts a net negative charge to the nucleic acid. In eucaryotes this negative charge is neutralized by small proteins called histones that have high proportions of positively charged amino acids (lysine and arginine). Because of this negative charge of DNA, basic dyes such as methyl green and pyronine can be used to preferentially stain the nucleic acid. Other techniques to identify DNA include colorimetric assays by reaction of diphenylamine with the deoxyribose sugar moieties of DNA and susceptibility of DNA to DNases.

The chromosome of *Escherichia coli* contains 4.7 million base pairs, with a molecular weight of 2.5×10^9 daltons. This size corresponds to 1.1 mm, or approximately 550 times the length of an *E. coli* cell. Although this amount of DNA is only 1/1,000 the DNA quantity found in an animal cell, it is a lot to fit into a bacterium. This large amount of nucleic acid fits into a bacterial cell because the DNA is twisted into a highly supercoiled, circular structure (Figure 8.7). DNA supercoiling is comparable to taking

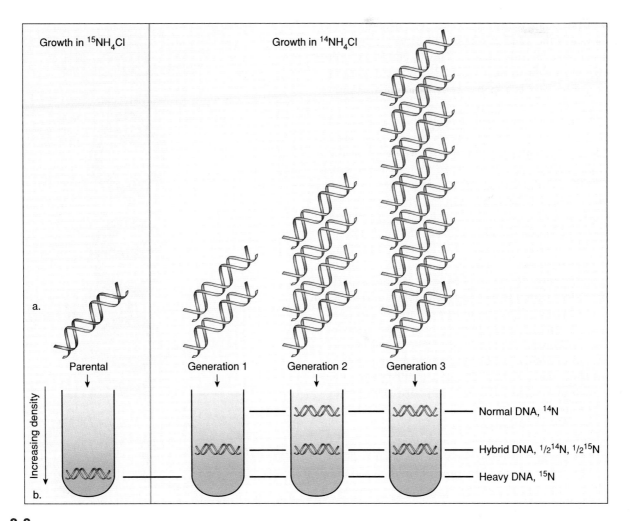

figure 8.8

The Meselson and Stahl Experiment

Meselson and Stahl grew *Escherichia coli* in the presence of ^{15}N-labeled NH_4Cl and then shifted growth to $^{14}NH_4Cl$. DNA extracted from bacteria before and after the shift showed heavy DNA (labeled with ^{15}N), hybrid DNA (labeled with ^{15}N and ^{14}N), or a mixture of normal (labeled with ^{14}N) and hybrid DNA. This pattern of labeling verified the existence of semiconservative replication of DNA. a. DNA labeling patterns. b. DNA in cesium chloride gradients.

a double-stranded rope and twisting it into a tightly constricted, compact, coiled structure. Bacterial DNA contains as many as 50 topological domains held in place by RNA and protein.

Bacterial Structure and Function
Internal Structures: Chromosome • pp. 2–10

DNA Is Replicated Semiconservatively

The complementary nature and precise base pairing of double-stranded DNA immediately suggest a mechanism for DNA replication in which both strands are copied concurrently. In fact, DNA replication does occur in this manner, with each strand serving as a template (that is, an existing pattern) for synthesis of a new daughter strand. This suggests a **semiconservative** mode of replication, which was confirmed by the work of Matthew Meselson and Franklin W. Stahl in 1958.

These microbiologists ingeniously grew *E. coli* in a medium containing ^{15}N-labeled (density-labeled) NH_4Cl as the sole source of nitrogen (Figure 8.8). The ^{15}N (considered a heavy isotope) was incorporated into newly synthesized DNA, thereby increasing its density. After several generations of growth, the bacteria were transferred into a medium containing normal ^{14}N-labeled NH_4Cl. DNA was extracted from the bacteria at different time intervals, purified, and centrifuged in a solution of cesium chloride. Under such conditions (known as density-gradient centrifugation), substances equilibrate at the point at which their density equals the density of the cesium chloride solution. Heavy substances, with higher densities than light substances, band near the bottom of the gradient, where there is a higher concentration of cesium chloride.

Meselson and Stahl observed that DNA extracted from bacteria grown for several generations in the $^{15}NH_4Cl$ contained only heavy ^{15}N-labeled DNA and settled lower in the cesium chloride gradients. However, when these bacteria were shifted to a medium containing $^{14}NH_4Cl$, the DNA isolated from cells was a mixed hybrid of heavy ^{15}N and normal ^{14}N after one generation of growth. Denaturation of this hybrid DNA yielded strands containing either ^{15}N or ^{14}N, but not both. When the bacteria were incubated in the normal ^{14}N medium for additional generations, extracted DNA consisted of either ^{15}N-^{14}N hybrids or pure ^{14}N forms. The results of this experiment were consistent with a semiconservative mode of replication. By this mechanism DNA strands are separated and each strand serves as a template for the synthesis of daughter strands. If replication were **conservative,** the two DNA strands would not separate, and synthesis of an entire new daughter duplex would occur from the parental double-stranded template. Under such conditions, either ^{14}N- or ^{15}N-labeled DNA would be observed, but not the ^{15}N-^{14}N hybrids seen by Meselson and Stahl.

DNA replication is also sequential and bidirectional. Replication begins at an **origin of replication** on the circular DNA molecule of procaryotes and proceeds, usually in both directions, to a terminus. Evidence for this divergent mode of replication came from autoradiography. An autoradiograph is a photographic film that picks up small amounts of radioactivity from a sample on, for example, a cell or a gel. Cells are briefly exposed to a pulse of a specific radioactive compound. The cells are then lysed gently, and the location of the radioactivity in newly synthesized cell material is determined by exposing the cell preparation on photographic film.

John Cairns, an Australian scientist, showed that the bacterial chromosome was circular by autoradiographs of DNA extracted from *E. coli* after one generation of growth in ^3H-thymidine (tritium-labeled thymidine). In his experiments Cairns grew the *E. coli* in a medium containing radioactive thymidine (^3H-thymidine). The ^3H-thymidine was incorporated into newly synthesized DNA. After two generations the DNA was extracted from lysozyme-treated cells and placed on photographic film. As the ^3H decayed, the emulsion was exposed, revealing the shape and replication pattern of the DNA molecule by the lines of dark-silver grains that form over the radioactive DNA. DNA was seen as a circular molecule with two replication forks (so called because of their Y-shaped structures) moving in opposite directions from the point of strand separation and replication. This branched circular structure resembled the Greek letter theta (θ), so this type of replication is sometimes called **theta replication** (Figure 8.9).

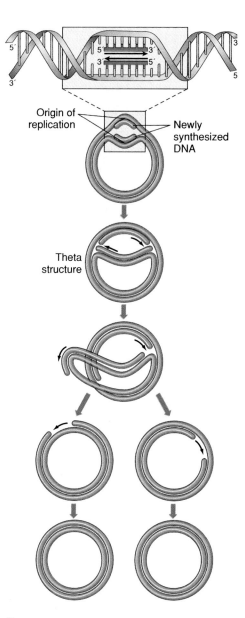

figure 8.9

The Theta (θ) Mode of DNA Replication
DNA replication is bidirectional from the origin, resulting in a structure resembling the Greek letter theta (θ).

Replication Begins with the Binding of Initiator Proteins to a Specific DNA Sequence at the Origin of Replication

DNA replication is a complex process that involves several proteins. The process of DNA replication is started by initiator proteins that recognize and bind to a specific DNA sequence of about 300 bases at the origin of replication. In *Escherichia coli* this protein complex then binds to a **DNA helicase** that uses energy from ATP to unwind short sections of the helix in advance of the repli-

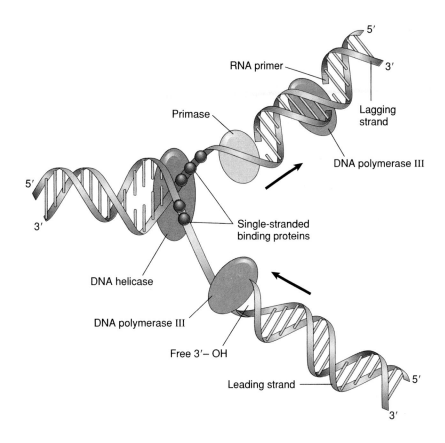

RNA primer

5′
3′

Primase

Lagging strand

DNA polymerase III

Single-stranded binding proteins

5′
3′

DNA helicase

DNA polymerase III

Free 3′– OH

Leading strand

5′
3′

figure 8.10
DNA Replication
DNA replication is a complex process, starting with the unwinding of the DNA helix.

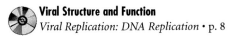

Viral Structure and Function
Viral Replication: DNA Replication • p. 8

cation fork (Figure 8.10). This unwinding (and subsequent rejoining of the DNA helix) is assisted by enzymes called **DNA topoisomerases,** plus helix-destabilizing **single-stranded DNA-binding proteins** that bind preferentially to single-stranded DNA and keep the denatured strands apart. The DNA topoisomerases nick one of the DNA strands and furthermore relax the compact supercoiled DNA duplex. The topoisomerases relieve torsional stress due to the unwinding of the DNA helix in front of the replication fork. The new daughter strands are always formed in a $5'$—PO_4 to $3'$—OH direction, with bases added to the $3'$ ends of the growing strands. The parental double-stranded DNA is antiparallel and complementary, and the newly synthesized daughter strands are also antiparallel and complementary.

This antiparallel arrangement presents a problem in replication, because for the two daughter strands to be synthesized simultaneously from both parental templates, the final direction of chain growth would have to be $5'$—PO_4 to $3'$—OH for one daughter strand and $3'$—OH to $5'$—PO_4 for the other. However, enzymes that catalyze DNA synthesis only add nucleotides to the $3'$—OH end of nascent DNA, extending the new DNA from the $5'$ to $3'$ direction. How then is DNA replicated from both parental strands in a $5'$ to $3'$ direction?

One of the Newly Replicated DNA Strands Is Synthesized Discontinuously in Short Segments

This paradoxical question was answered by Reiji Okazaki (1930–1975), a Japanese scientist who showed in 1964 that one of the DNA daughter strands was synthesized discontinuously in short segments of 1,000 to 2,000 nucleotides, known as Okazaki fragments, in a $5'$ to $3'$ direction of chain growth (away from the replication fork). Okazaki pulse labeled *E. coli* with ^{3}H-thymidine for very short periods of time (5 to 30 seconds) and found through sedimentation analysis on alkaline sucrose gradients that radioactively labeled DNA extracted from these bacteria occurred in short pieces. When the same bacteria were pulse labeled for longer periods of time (one minute or longer), the radioactive label appeared in larger continuous pieces of DNA (see the Perspective at the end

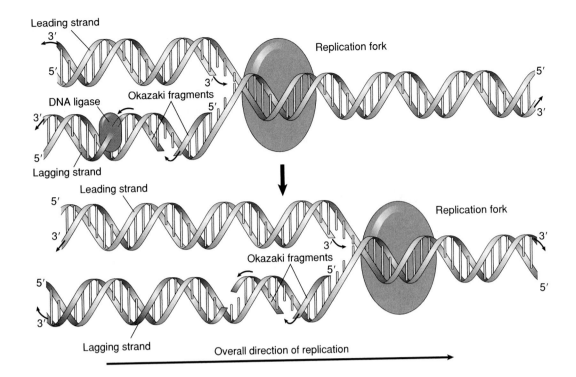

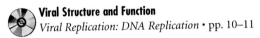

Overall direction of replication

figure 8.11

Okazaki Model for DNA Synthesis

Short fragments of DNA (Okazaki fragments) are synthesized for one of the two daughter DNA strands. The discontinuous daughter DNA strand is called the lagging strand in replication. The fragments are eventually joined by a DNA ligase to form a continuous strand. The other daughter strand (the leading strand) is formed continuously.

Viral Structure and Function
Viral Replication: DNA Replication • pp. 10–11

of this chapter). It was later shown that Okazaki fragments are joined by a DNA ligase to form a continuous daughter strand.

Okazaki fragments are formed for only one of the two DNA daughter strands; the complementary daughter strand is probably synthesized continuously in the 5′ to 3′ direction (toward the replication fork) because new bases logically can be added to the 3′—OH end (Figure 8.11). Because the daughter strand that is synthesized discontinuously has transient gaps that must be filled in, its synthesis lags behind the synthesis of the continuous strand. For this reason the discontinuous daughter DNA strand is frequently termed the **lagging strand** in replication, whereas the continuous daughter strand is called the **leading strand.**

Viral Structure and Function
Viral Replication: DNA Virus Replication • pp. 7–11

E. coli *Has Three DNA Polymerases*

Three different **DNA polymerases, I, II,** and **III,** have at one time or another been implicated in DNA replication in *E. coli* (Table 8.2). All catalyze the addition of mononucleotides to the 3′—OH end of a growing polynucleotide. The enzymes also are capable of cleaving mononucleotides in a 3′ to 5′ direction. This exonuclease (a nuclease that cleaves terminal nucleotides) activity probably corrects mistakes during DNA synthesis.

However, only one of the three enzymes, DNA polymerase III, is the true replication enzyme. This complex protein consists of a catalytic core of three polypeptide subunits. Seven additional subunits make up the complete enzyme. DNA polymerase III is inactivated at 45°C in temperature-sensitive mutants of *E. coli* that are able to synthesize DNA at 30°C but not at a temperature of 45°C. Furthermore, DNA polymerase III has a polymerization rate of 30,000 nucleotides polymerized per minute per molecule of enzyme at 37°C. This rate is many times greater than the polymerization rates of the other two DNA polymerases.

DNA polymerase I—also known as the Kornberg enzyme after Arthur Kornberg, a biochemist from the United States who discovered it—is involved in one of the mechanisms for the repair of chromosome damage resulting from ultraviolet radiation (see DNA repair mechanisms, page 248). This repair occurs by an excision-repair mechanism in which the altered bases are excised

table 8.2

Properties of *Escherichia coli* DNA polymerases

Property	DNA Polymerase I	DNA Polymerase II	DNA Polymerase III
Molecular weight	103,000	90,000	167,500[a]
Number of molecules per cell	400	?	10–20
Nucleotides polymerized per minute per molecule of enzyme	~600	~50	~30,000
$5' \rightarrow 3'$ polymerization	+	+	+
$5' \rightarrow 3'$ exonuclease	+	−	−
$3' \rightarrow 5'$ exonuclease	+	+	+
Probable function	Repair of ultraviolet damage; DNA synthesis in Okazaki fragment gaps	?	Replication of chromosomes

[a]Molecular weight of catalytic core, consisting of α (alpha, MW 130,000 daltons) ε (epsilon, MW 27,500 daltons), and θ (theta, MW 10,000 daltons).

and the resultant gap filled by DNA polymerase I. Evidence for this function of DNA polymerase I comes from observations that *E. coli* mutants containing almost no DNA polymerase I are particularly sensitive to ultraviolet radiation. DNA polymerase I probably also is involved in filling the gaps of Okazaki fragments in semiconservative replication. DNA polymerase I has a 5' to 3' exonuclease activity and may use this activity to remove an RNA primer present on Okazaki fragments while simultaneously replacing the primer with newly synthesized DNA. Nothing is known about the in vivo function of DNA polymerase II.

A Short RNA Primer Is Required for DNA Synthesis

All three polymerases have a common requirement for a preexisting polynucleotide chain on which to add mononucleotides on the 3'—OH end. Since none of the DNA polymerases can initiate DNA synthesis de novo, the requirement for a preexisting polynucleotide is satisfied by a short RNA primer. This RNA primer (approximately 10 nucleotides long) is synthesized by an RNA-polymerizing enzyme, **primase,** which is able to initiate RNA synthesis without a primer-dependent preexisting chain. In *E. coli,* the primase is associated with other replication proteins, including DNA helicase, to form a complex called the **primosome** that moves along the replication fork to form RNA primers as needed. The primer is later excised from the DNA chain, probably by DNA polymerase I, which has 5' to 3' exonuclease activity, in combination with an RNA-degrading enzyme, RNase H, that may be involved in removing the last RNA nucleotide (the one at the RNA-DNA junction).

The mononucleotides added to growing DNA chains during replication are obtained from deoxynucleoside triphosphates. When these precursors are added to the growing chain, the two terminal phosphates are cleaved from the nucleotide. The energy liberated after this cleavage and the bond rupture within pyrophosphate is sufficient to form a phosphodiester bond that links the new mononucleotide to the nucleotide on the 3' end of the chain.

Some Circular DNAs Are Copied via a Rolling Circle Mechanism of Replication

Circular DNA is copied in some instances via a **rolling circle** mechanism of replication (see virus replication, page 402). In the rolling circle model, one of the DNA strands is nicked at a specific site by an endonuclease and then the free 3'—OH end generated by the nick is extended by DNA polymerase, using the intact parental strand as a template. The newly synthesized strand displaces the nicked parental strand. A second new strand may also be synthesized, using the nicked parental strand that has "rolled off" the DNA circle. In microorganisms, the rolling circle mode of replication is used in bacteriophages that have circular DNA, and in DNA transfer during bacterial conjugation.

DNA Replication in Procaryotes and Eucaryotes Is Basically Similar

The structure and replication of DNA in procaryotes and eucaryotes are basically the same, although there are certain important differences. The eucaryotic genome is contained within a defined nucleus and consists of a nucleoprotein complex known as **chromatin.** This chromatin is organized into **nucleosomes** (basic structural subunits of the chromatin, each containing approximately 200 base pairs and histone proteins) and becomes tightly packed at cell division to form individual mitotic chromosomes. Eucaryotic DNA is highly repetitive and contains interrupted gene sequences divided into coding (exons) and noncoding (introns) segments. Both coding and noncoding segments are transcribed into RNA, but the noncoding segments are removed from the mature RNA before translation. Procaryotic DNA is not as highly repetitive and is not divided into exons and introns.

Although the details of eucaryotic DNA replication are not as well-defined as those of procaryotic DNA replication, it appears that there are more similarities than differences in the two modes of replication. Like procaryotic DNA, eucaryotic DNA is replicated semiconservatively at divergent replication forks and involves an RNA primer. Okazaki fragments are formed on one strand of the DNA, but these DNA fragments are smaller (100 to 150 nucleotides) than those produced in procaryotes. Furthermore, whereas procaryotic DNA generally replicates from a single origin, eucaryotic DNA always has multiple replication origins. On the average, about 100 replication forks form on each eucaryotic chromosome. Although the total quantity of DNA in a typical eucaryotic chromosome is greater than in the procaryotic nucleoid, the stretch of DNA synthesized from each origin (known as a **replicon**) is much shorter in eucaryotes (10 to 100 μm) than in procaryotes (1,500 μm). Unlike procaryotes, which synthesize DNA throughout their cell cycles, eucaryotes synthesize DNA only at a specific stage of their cell cycles. This stage is called the **S phase** (synthesis phase), and is characterized by DNA synthesis and chromosome duplication.

Ribonucleic Acid

Ribonucleic acid (RNA) has a structure similar to DNA, with the exceptions that the base uracil replaces thymine and that the sugar ribose replaces deoxyribose. RNA also commonly exists in single-stranded form.

RNA Is Divided into Three Classes

There are three major classes of RNA: messenger RNA (mRNA), ribosomal RNA (rRNA), and transfer RNA (tRNA). All three forms of RNA are transcribed from the DNA template by an enzyme called **DNA-dependent RNA polymerase.**

RNA polymerase of Bacteria is a multisubunit enzyme consisting of four polypeptides (α, β, β′, and σ) arranged in the ratio 2:1:1:1. The complete enzyme (the holoenzyme) can be reversibly separated by phosphocellulose column chromatography into a core enzyme (α₂ββ′) and the σ factor (sigma factor). The core enzyme retains its ability to bind to DNA and initiate transcription, but without the σ factor it is unable to bind tightly and selectively at specific sites on the DNA (promoter regions) for initiation of RNA synthesis. Because of its role in transcription, the σ factor is often referred to as the initiation factor of RNA polymerase. Unlike the single RNA polymerase of Bacteria, Archaeal RNA polymerases are of several types, structurally more complex, and resemble eucaryotic RNA polymerases (see Archaea, page 312).

Messenger RNA Carries Information for Protein Synthesis

Messenger RNA is the most varied of the three forms of RNA, yet it comprises only about 5% of the total RNA in the cell because most mRNAs have a life span of only a few minutes in bacterial cells. This short half-life eliminates wasteful synthesis of excessive proteins. Messenger RNA carries the genetic message from the DNA template to ribosomes, whereupon it is translated into proteins. This message is transcribed from only one strand of the DNA double helix at a time.

Ribosomal RNA Is Associated with Protein Synthesis

The genetic message on mRNA is carried to ribosomes, where it is translated into proteins. Ribosomes contain approximately one-third protein and two-thirds rRNA. Three different types of rRNA molecules are found in procaryotic ribosomes: a 5S rRNA of 125 bases (molecular weight, 4.0×10^4 daltons) and a 23S rRNA of about 2,900 bases (molecular weight, 1.1×10^6 daltons), both found in the larger 50S ribosomal subunit, and a 16S rRNA of 1,540 bases (molecular weight, 6.0×10^5 daltons), found in the smaller 30S ribosomal subunit. The eucaryotic ribosome contains four types of ribosomal RNAs: a 5S rRNA, 18S and 28S rRNAs (corresponding to the 16S and 23S procaryotic rRNAs), and a 5.8S rRNA. The rRNA molecules not only contribute to the ribosome structure, but also have functional roles in protein synthesis. In bacteria, the 16S rRNA of the small ribosomal subunit base pairs with an initiation sequence site on mRNA and positions the mRNA on the ribosome. The 23S rRNA associated with the large ribosomal subunit is believed to be the peptidyl transferase associated with peptide bond formation during protein synthesis.

Transfer RNA Matches Amino Acids to Nucleotides in Messenger RNA

The smallest of the three RNAs are the tRNAs. These are 73 to 93 bases in length (molecular weight of about 25,000 daltons) and are folded back upon themselves in a three-dimensional configuration to produce cloverleaf shapes (Figure 8.12). This unique structure is maintained by the internal complementary base pairing that results in three, and in some cases four loops, or closed arms, and one open arm. Transfer RNA contains modified bases, including methylated forms of the usual purines and pyrimidines found in RNA. Methylguanosine, dimethylguanosine, pseudouridine (with an exchange of a carbon and a nitrogen atom), dihydrouridine (with two hydrogens added to uridine), and ribothymidine (thymidine attached to a ribose) are some of the modified bases found in tRNA.

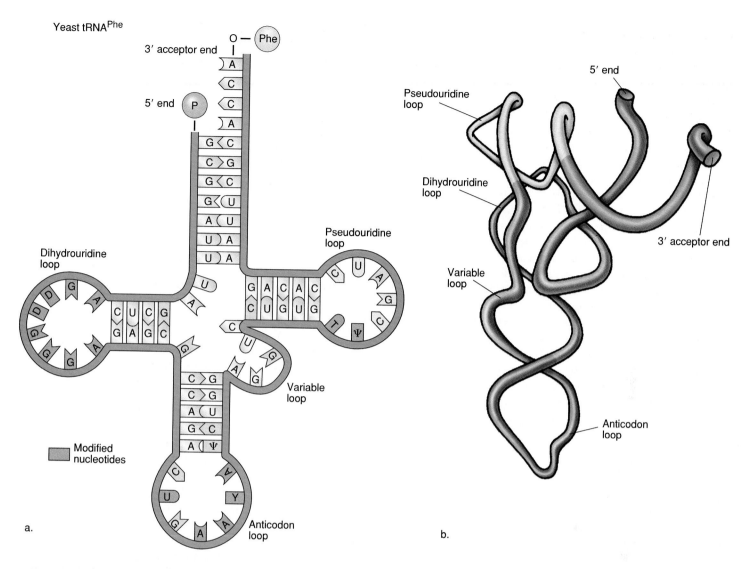

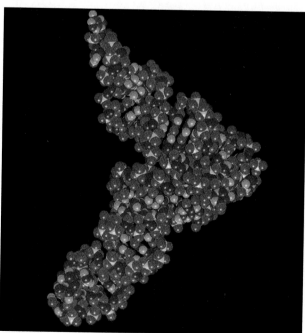

figure 8.12

The Cloverleaf Structure of Yeast Phenylalanine tRNA

a. The complete nucleotide structure of yeast phenylalanine tRNA. The amino acid is attached to the 3' acceptor end of the tRNA. Abbreviations: A, adenine; C, cytosine; G, guanine; U, uracil; D, dihydrouradine; ψ, pseudouradine; Y, a modified purine. b. A schematic drawing of the actual three-dimensional shape of the molecule based on X-ray diffraction analysis. c. A computer-drawn simulation of the space-filling model of the molecule. Colors: carbon, dark blue; hydrogen, white; nitrogen, blue; oxygen, red; phosphorus, yellow.

Transfer RNAs are involved in the translation of the genetic message. The tRNA molecule is an adapter of amino acids to mRNA codons at the ribosome and ultimately for incorporation into growing proteins. Each type of tRNA carries a specific amino acid. The activation of tRNA with an amino acid occurs via a specific aminoacyl-tRNA synthetase that is linked with the appropriate amino acid and then transfers the amino acid to the 3′ end of the tRNA. ATP is hydrolyzed to AMP and pyrophosphate during this event and provides the energy necessary for attachment of the amino acid to the tRNA. The aminoacyl-tRNA can then move to a ribosome for insertion of the amino acid into a protein. After insertion, the tRNA moves away from the ribosome and is activated again with an appropriate amino acid.

There is at least one tRNA for each of the 20 different amino acids; probably many more types of tRNA actually exist because of the degeneracy of the genetic code. How does the cell determine which of these different types of activated tRNA is needed at a ribosome? To avoid errors in protein synthesis, this recognition must be highly specific and is accomplished through one of the three loops in the tRNA structure. This loop, the anticodon loop, contains three bases that are in a sequence unique for each tRNA. These bases are complementary to a three-base codon sequence in the mRNA that has carried the genetic message from DNA to the ribosome. The other two loops in the tRNA structure, the dihydrouridine (DHU) loop and the pseudouridine (ψ) loop, are thought to be involved in binding of the tRNA to the aminoacyl-tRNA synthetase and the ribosome, respectively. Both rRNA and tRNA are relatively stable forms of RNA when compared with the less stable mRNA.

RNA Is Synthesized from DNA

All three forms of RNA—mRNA, rRNA, and tRNA—are synthesized from specific genes on the chromosome. Although rRNA and tRNA make up the majority of the RNA found in the cell, less than 1% of the chromosomal DNA codes for their synthesis. Both of these RNA species are initially formed as larger precursors, which are then enzymatically cleaved into their smaller final products. In the case of procaryotic rRNA, all three rRNA types (5S, 16S, and 23S) are believed to arise from a single larger precursor.

RNA Synthesis Begins with Binding of RNA Polymerase to a Promoter on DNA

RNA is synthesized in a 5′ to 3′ direction of chain growth from the DNA template. Theoretically, two different mRNA molecules could be transcribed from the two complementary DNA strands. In actuality, only one of the two DNA strands is transcribed in any given DNA region (Figure 8.13). A promoter DNA sequence (41 to 44 bases in length) at the beginning of each transcription unit on the DNA template not only specifies which of the two DNA strands is to be transcribed, but also is a recognition site for the binding of RNA polymerase to a promoter and for the initiation of transcription.

The DNA sequences that serve as promoters for E. coli RNA polymerase have been well-characterized and consist of 2 DNA sequences 6 bases long that are separated by approximately 25 unrecognized bases on the DNA. One of these sequences is called the **Pribnow box** and lies 10 bases before the point of initiation of transcription. The second sequence is appropriately called the **–35 sequence**, because it lies 35 bases from the start of RNA synthesis. Both of these sequences are highly conserved regions of the DNA, with few variations in their base sequences among different promoters. By comparing different promoter sequences in E. coli, one is able to determine a **consensus sequence** for the Pribnow box (**TATAAT**) and the –35 sequence (**TTGACA**) that represents bases occurring most often in these regions.

The initiation of RNA synthesis begins with binding of the RNA polymerase to the promoter, causing the strands of DNA to separate. The σ subunit recognizes specific binding sites and ensures that the RNA polymerase forms stable associations with DNA only at promoters, not at other sites. Although RNA synthesis can occur in vitro without the σ subunit, this type of synthesis is nonspecific and occurs at random sites on the DNA. Most transcription in E. coli occurs with a specific σ (called σ^{70} because its molecular weight is 70,000). However, other σ factors have been discovered in E. coli, including σ^{32} (with a molecular weight of 32,000), which is associated with heat shock regulation. The σ^{32} recognizes a specific class of promoters (the heat shock promoters) that are involved in the expression of heat shock genes when there is a temperature increase. The σ^{32} of E. coli normally is unstable and degraded shortly after synthesis, but this degradation is inhibited when the temperature is increased from 30°C to 42°C. Under these conditions, the levels of σ^{32} increase in the cell and more RNA polymerase molecules bind to heat shock promoters. The heat shock proteins that are synthesized protect the cell against the higher temperature.

Multiple σ factors with different specificities have also been detected in Bacillus subtilis. At least four different σ factors (σ^{29}, σ^{32}, σ^{37}, and σ^{55}) with molecular weights of 29,000, 32,000, 37,000, and 55,000, respectively, occur in B. subtilis. Each time the σ subunit in RNA polymerase is changed, the RNA polymerase is able to recognize a different promoter. The different σ proteins in B. subtilis are believed to be associated with the sporulation and germination cycles of the bacterium.

Ribonucleotides Are Added Sequentially in Chain Elongation

The binding of RNA polymerase causes approximately one turn of the DNA helix to be unwound at a time. This unwinding exposes a short segment of single-stranded DNA that is used as a template for transcription. As transcription occurs, an RNA chain is formed by the successive binding of ribonucleotides complementary to the DNA template. RNA polymerase catalyzes the formation of phosphodiester bonds between ribonucleotides accompanied by pyrophosphate release and hydrolysis. After approximately eight to

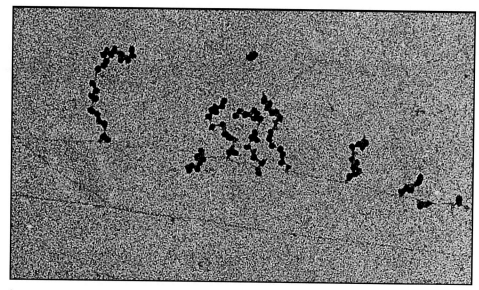

a.

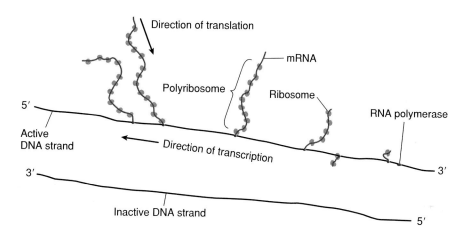

b.

figure 8.13

Transcription Coupled with Translation

a. Electron micrograph of two strands of DNA in *Escherichia coli*. Polyribosomes are attached to mRNA actively transcribed by RNA polymerase from one of the two DNA strands (×66,700).

b. Diagrammatic illustration of pertinent details shown in the electron micrograph.

ten ribonucleotides are added to the growing RNA chain, the σ factor dissociates from the holoenzyme to yield the core enzyme, which mediates the elongation. As the core polymerase moves along the DNA, ribonucleotides are paired with the complementary DNA bases in the open region of the DNA helix. This open region of the DNA involves only a few bases at a time. The area immediately behind the core polymerase reanneals, and the newly synthesized RNA is dissociated from the DNA template as the helix re-forms.

Termination of RNA Synthesis Occurs at Specific Sites on the DNA

Transcription continues until the RNA polymerase encounters a specific termination site on the DNA. Several different termination sequences have thus far been identified. In *E. coli* the termination sequence codes for a double-stranded RNA stem-and-loop structure followed by a polyU sequence (rho-independent termination)

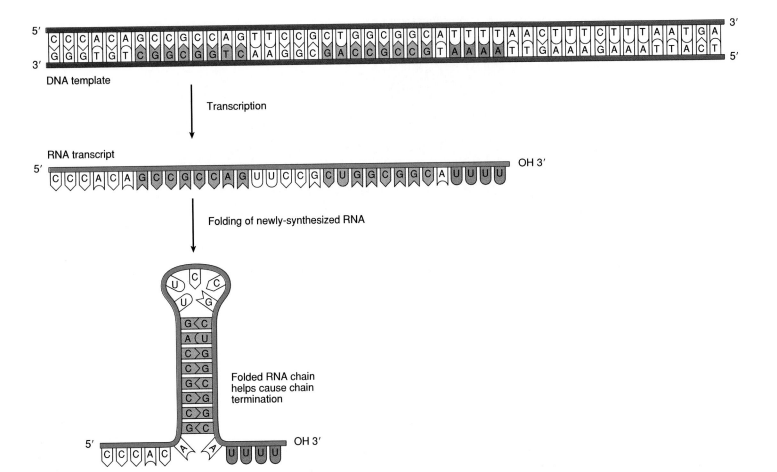

figure 8.14
Rho-Independent Termination of Transcription
Transcription terminates when RNA polymerase synthesizes a polyU sequence that is located immediately after a self-complementary base sequence. The self-complementary sequence forms a hairpin in the newly synthesized RNA chain, resulting in chain termination.

(Figure 8.14). A second type of termination in *E. coli,* which occurs in the absence of a polyU sequence after the stem-and-loop structure, involves a protein factor called ρ (rho) that is a hexamer of 46,000-dalton subunits. In rho-dependent termination, rho is thought to bind to RNA and move along the RNA until it catches up with the RNA polymerase at a terminator, where the enzyme unwinds the RNA-DNA hybrid. Hydrolysis of ATP provides the energy for this reaction.

Eucaryotes Have Several Different RNA Polymerases

Procaryotes (other than Archaea) have a single DNA-dependent RNA polymerase that synthesizes mRNA, rRNA, and tRNA, whereas eucaryotes have at least four different enzymes that catalyze the RNA synthesis. The eucaryotic RNA polymerases are more complex in structure than the RNA polymerase of Bacteria and have from eight to ten subunits each. Three of the eucaryotic RNA polymerases are found in the nucleus and are responsible for the synthesis of different classes of RNA. These RNA polymerases can furthermore be distinguished by their sensitivity to the antibiotic α-amanitin, their activity under different ionic conditions, and minor differences in their subunit structures.

RNA polymerase I is located in the nucleolus and is used to synthesize a large precursor to rRNA. RNA polymerase II is located in the nucleoplasm and catalyzes the synthesis of precursors that are eventually processed to form cytoplasmic mRNAs. RNA polymerase III, also located in the nucleoplasm, is responsible for the synthesis of tRNA precursors and 5S rRNA. In addition to nuclear RNA polymerases, eucaryotic cells also have mitochondrial and/or chloroplast RNA polymerases, which transcribe DNA in these organelles.

Transcription and translation occur simultaneously in procaryotes. In eucaryotes, however, where the chromosome is located in a different cell compartment (nucleus) than the protein-

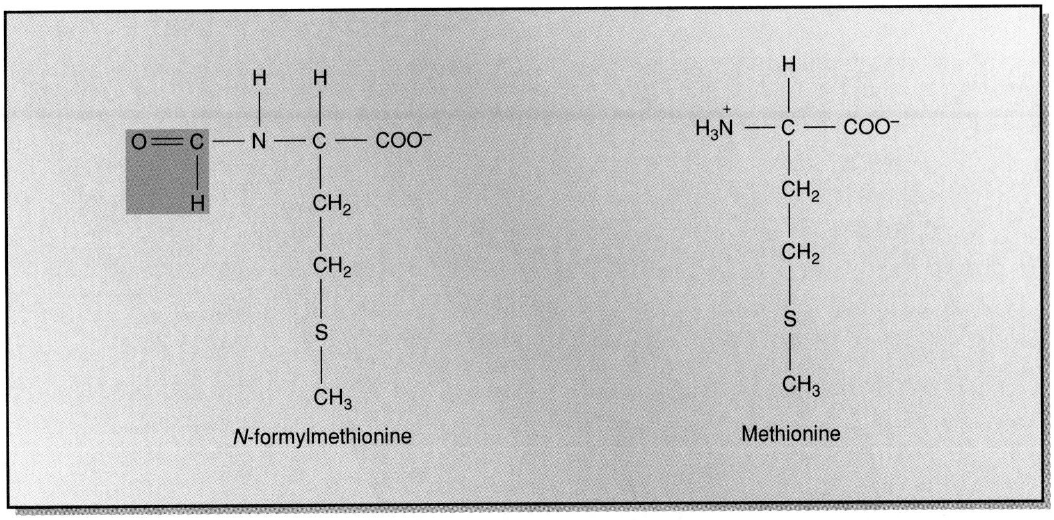

figure 8.15
Structures of *N*-formylmethionine and Methionine
The formyl group is shaded blue.

synthesizing machinery (cytoplasm), a different situation exists. A large mRNA precursor, usually five to ten times greater in size than the final transcript, is formed in the nucleus and is **capped,** or **blocked,** at the 5′ terminus by 7-methylguanosine. Capping occurs shortly after the initiation of RNA synthesis and appears to assist in the recognition of the mRNA by the small ribosomal subunit. Several hundred adenine nucleotides are also added to the 3′ terminus of the mRNA. This polyA tail is characteristic of eucaryotic mRNAs and is believed to increase the stability of the mRNA and protect it from exonucleases.

Following the addition of these nucleotides, the mRNA precursor is then processed to a final transcript before translation. During mRNA processing, noncoding portions (introns) of the RNA are excised, and the remaining translatable fragments (exons) are rejoined. The processed RNA molecule is then transported to the cytoplasm, where it is translated.

This **RNA splicing** to form a final mRNA product, a phenomenon that has only recently been discovered, generally does not occur in procaryotes. RNA splicing is not restricted to mRNA; it also occurs in the processing of rRNA and some tRNAs. The biological significance of RNA processing is unclear, but it is important in the maturity of RNA molecules.

Until recently it was believed that all enzymes were proteins. In 1987 Thomas R. Cech at the University of Colorado discovered RNA molecules with catalytic properties. These RNA enzymes, or **ribozymes,** are now known to occur in procaryotes and eucaryotes. Most ribozymes are RNA self-splicing enzymes that excise themselves from an RNA molecule. For example, in the rRNA of the protozoan *Tetrahymena thermophila,* a 413-nucleotide intron sequence acts as a ribozyme to splice itself out

of the rRNA and join the adjacent exons. The discovery of ribozymes has caused scientists to consider that RNA having both genetic and catalytic properties may have been the ancestral universal macromolecule.

Protein

The genetic message that is transcribed from DNA to mRNA is eventually translated into specific proteins. Amino acids specified by the codons on the mRNA are assembled into polypeptide chains. This polymerization of amino acids is only part of the process of protein synthesis. The polypeptide chains are modified and folded into active configurations during and after the assembly steps. The translation of mRNA codons into specific amino acids, however, begins protein synthesis, which is a multistep process that takes place at the ribosomes. The distinct steps in protein synthesis are **initiation, elongation,** and **termination.**

Protein Synthesis Begins with the Formation of an Initiation Complex

The first step in protein synthesis is initiation. In the early 1960s, it was observed that 45% of the proteins in *E. coli* began with the amino acid methionine (Met), yet this amino acid constitutes only about 4% of the internal amino acids in these proteins. This observation led to the discovery that protein synthesis begins with methionine. In Bacteria and in eucaryotic mitochondria and chloroplasts, the initiating amino acid is a formylated version of methionine: *N*-formylmethionine, or fMet (Figure 8.15).

The Wobble Hypothesis

The codon-anticodon pairing between mRNA and tRNA during protein synthesis can sometimes result in a mismatch at the third base position of the codon.

This wobble was used by Francis Crick in 1966 to explain the degeneracy of the genetic code. Crick proposed that as a result of this mismatched base pairing at the

nine in the third codon position binds to uracil in the complementary anticodon position. When adenine, uracil, or cytosine is the third codon base, the complementary anticodon base can be inosine, a base that is derived from the deamination of adenine. This flexibility in base pairing at the third codon position explains why the third position of the codon is the most degenerate. It also explains why fewer than 61 aminoacyl-tRNAs are required for protein synthesis—many of the aminoacyl-tRNAs, as a result of the wobble, are able to bind to more than one codon.

Whereas the first two bases of the codon faithfully pair with their complementary bases in the anticodon loop of tRNA, the third codon base (at the 3' end of the codon) can tolerate a mismatch, or **wobble.**

third codon position, a number of different types of hydrogen bondings could occur (Table 8.3). For example, either uracil or cytosine in the third codon position is able to bind to guanine in the complementary anticodon position. Either adenine or gua-

table 8.3

Possible Base Pairing Combinations with the Wobble Hypothesis	
Third Base in Anticodon (5' end)	Third Base in Codon (3' end)
U	A or G
C	G
A	U
G	U or C
I[a]	U, C, or A

[a]Inosine.

The methionine that initiates protein synthesis is specified by the codon AUG or, occasionally, GUG. These two codons normally specify methionine and valine, respectively, as internal amino acids in proteins. However, in the presence of an **initiation complex,** AUG and GUG code specifically for a modified aminoacyl-tRNA, fMet-tRNA. In contrast, Archaea have an initiator tRNA that carries *unmodified* methionine (see Archaea, page 312).

This initiation complex in protein synthesis is formed from the binding of the 30S ribosomal subunit to the initiation codon AUG of the mRNA—or alternatively, GUG (Figure 8.16). The initiation codon establishes the **reading frame** for the mRNA. Because the genetic code is read in consecutive blocks of three bases at a time, mRNA theoretically could be translated in any one of three different *reading frames* depending on which base is read as the beginning of the RNA message. Since a single base shift in the reading of the RNA transcript would result in synthesis of a completely different and probably nonfunctional protein, a correct reading frame is important in translation.

An initiator tRNA, carrying fMet, pairs with the initiation codon. The formyl group is attached to the terminal amino group of methionine after methionine has attached to the initiator tRNA. Since this aminoacyl-tRNA contains a modified methionine, it can be used only for initiation.

A short purine-rich region, consisting of three to nine bases (for example, AGGA for the *lacZ* mRNA in *E. coli*), precedes the 5'

a.

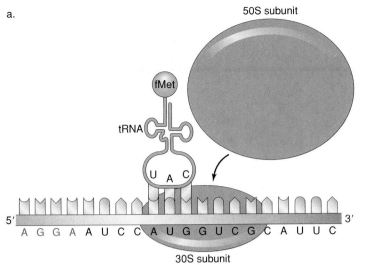

Initiation: An initiator tRNA, carrying fMet, pairs with the initiation codon AUG in the presence of the 30S ribosome subunit and following the Shine-Dalgarno sequence.

b.

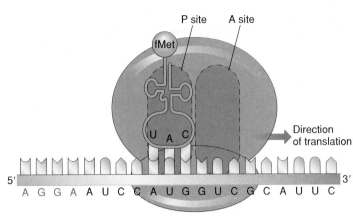

The 50S ribosome subunit joins with the 30S subunit to form the intact 70S ribosome. This completes the initiation complex.

c.

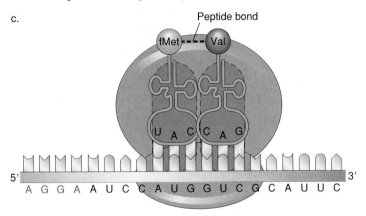

Elongation: An aminoacyl tRNA (carrying valine in this example) binds to the complementary codon (GUC in this example) at the A site of the ribosome. A peptide bond is formed between fMet and Val, and the ester bond between the tRNA in the P site and its amino acid (fMet) is cleaved.

d.

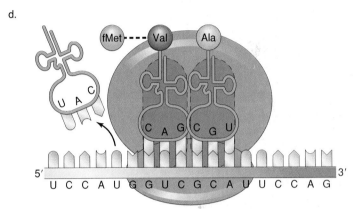

In translocation, the peptidyl tRNA on the A site is transferred to the P site on the ribosome and displaces the free tRNA in the P site. The A site is now free to receive a new aminoacyl-tRNA and the cycle of chain elongation is repeated.

e.

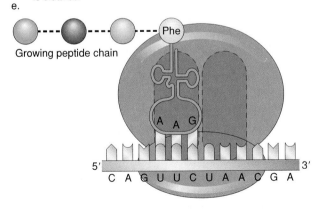

Termination: When the ribosome reaches a nonsense codon (UAA in this example) on the mRNA, protein synthesis is terminated.

f.

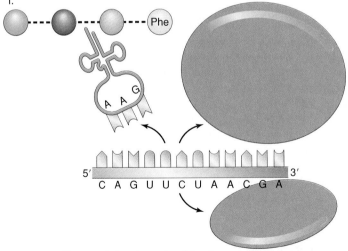

The protein is released and the ribosome is dissociated.

figure 8.16

Steps in Protein Synthesis

A site = aminoacyl or acceptor site. P site = peptidyl or donor site.

end of the initiating AUG codon in procaryotes and specifies the ribosome binding site on the mRNA. This **Shine-Dalgarno sequence,** named after the Australian biochemists John Shine and Larry Dalgarno who first described its existence in 1974, is complementary to a pyrimidine-rich sequence near the 3′ terminus of the 16S rRNA of the 30S ribosomal subunit and serves to thread the mRNA into the ribosome. Mutations in the Shine-Dalgarno sequence that decrease complementary base pairing adversely affect formation of the mRNA-ribosome complex, thereby illustrating the importance of this sequence.

Initiation requires not only formation of an initiation complex with a 30S ribosomal subunit, mRNA, and fMet-tRNA, but also three protein initiation factors (IF1, IF2, and IF3) and GTP. IF3 catalyzes the dissociation of the 70S ribosome to form the individual subunits used in initiation. It also is associated with the binding of the 30S subunit to the appropriate recognition site on mRNA. IF2 and GTP appear to be involved in the positioning of fMet-tRNA to the correct codon on the 30S subunit-mRNA complex. IF1 is believed to assist IF3 in the binding of 30S subunits to mRNA. IF2 is also involved in the IF2-mediated fMet-tRNA binding reaction. The initiation complex is completed when the 50S subunit joins with the 30S subunit to form the intact 70S ribosome. The formation of the 70S ribosome is accompanied by hydrolysis of GTP to GDP and P_i, and release of the three initiation factors from the ribosome-mRNA complex. The complete 70S ribosome has two sites for tRNA-codon binding: the A (aminoacyl, or acceptor) site, which accepts incoming tRNAs, and the P (peptidyl, or donor) site, which contains the fMet-tRNA-initiation codon complex.

Amino Acids Are Added to the Growing Peptide During Elongation

The ribosomal complex that forms as a result of initiation is now ready to accept additional aminoacyl-tRNAs. Elongation begins with the binding of an aminoacyl-tRNA to the complementary codon on the mRNA at the A site of the ribosome. The codon determines the type of amino acid inserted into the growing protein. The recognition step requires two protein elongation factors, EF-Ts (a heat-stable factor) and EF-Tu (a heat-unstable factor), and hydrolysis of GTP to GDP + P_i.

Elongation factors Ts and Tu promote the binding of incoming aminoacyl-tRNA molecules to the ribosome. A Tu-GTP complex is apparently formed by action of the factor Ts. The Tu-GTP complex binds to aminoacyl-tRNA and transfers the tRNA to the ribosome. When the aminoacyl-tRNA attaches to the A site on the ribosome, GTP is cleaved to yield a Tu-GDP complex and inorganic phosphate (P_i). The Ts factor then regenerates Tu-GTP via an exchange of GDP and GTP.

A peptide bond is formed between the α-carboxyl group of the amino acid in the P site of the ribosome and the α-amino group of the amino acid in the A site. When N-formylmethionine is the amino acid in the P site, only the carboxyl group of the amino acid is free to participate in this peptide bond. During linkage, the ester bond between the tRNA in the P site and its amino acid is cleaved. The breakage of this ester bond liberates the energy required for peptide bond formation. The aminoacyl-tRNA on the A site is thus transferred to the peptide at the P site, which results in a growing peptide that is now one residue longer. This peptidyl transfer step is catalyzed by the enzyme, peptidyl transferase, which forms a part of the 50S ribosomal subunit.

The peptidyl-tRNA on the A site is next transferred to the P site on the ribosome, where it displaces the free tRNA. This transfer is accompanied by movement of the ribosome along the mRNA by one codon length in the 5′ to 3′ direction. As a result of this translocation—which requires the elongation factor G (or EF-G, often called the translocase), as well as the hydrolysis of one molecule of GTP—the A site on the ribosome is now free to accept a new aminoacyl-tRNA, and the cycle of chain elongation is repeated.

Protein Synthesis Terminates When the Ribosome Reaches a Nonsense Codon on the Messenger RNA

Protein synthesis continues until the ribosome reaches one of the three nonsense codons (UAA, UGA, or UAG) on the mRNA. At this point a protein release factor (RF) binds to the ribosome. *E. coli* has three release factors: RF1 recognizes UAA and UAG; RF2 recognizes UAA and UGA; and RF3 stimulates the termination process. The release factor (RF1 or RF2 in concert with RF3) hydrolyzes the ester bond that holds the protein to the last tRNA that entered the ribosome. The protein is released and a dissociation factor causes the ribosome to separate into its individual subunits.

The protein, released from the ribosome, still contains a formyl group on the initial amino acid methionine. The formyl group can be removed by the enzyme deformylase, leaving methionine as the amino acid on the N-terminal end of the protein. Alternatively the entire *N*-formylmethionine, as well as additional amino acids on the N-terminus of the protein, can be removed by the action of a peptidase.

The mRNA is read in a 5′ to 3′ direction during protein synthesis. This direction of reading, similar to the direction of RNA synthesis from the DNA template, allows translation to occur simultaneously with transcription in procaryotes. Ribosomes most likely attach to mRNAs as the message is simultaneously transcribed from the DNA template. A single mRNA, furthermore, can have several ribosomes (polyribosomes) attached to it, each of which is a functional site of protein synthesis. Polyribosomes result in several identical protein chains being synthesized concurrently from the RNA transcript. At any particular time, however, the peptides nearer the 3′ end of the mRNA are longer than those at the front of the nucleic acid. The ability of mRNA to interact with ribosomes and form polyribosomes explains in part why little mRNA is required by cells for protein synthesis. For any given message, several copies of the protein (as many as 20 copies, in the case of *E. coli*) can be made simultaneously on the mRNA.

Translation is efficient and rapid. It takes only 10 to 20 seconds for a rapidly growing bacterium to synthesize a protein of 300 to 400 amino acids. The protein begins to fold into its three-dimensional shape as it is being synthesized and before its release from the ribosome. After termination and release from the ribosome, the protein can undergo further chemical and structural modifications. These **posttranslational modifications** result in altered protein configurations that are essential to protein function and activity. Examples of such modifications include the acetylation and amidation of amino acids.

Bacterial Structure and Function
Internal Structures: Protein Synthesis • pp. 11–21

Protein Synthesis Is More Complex in Eucaryotes Than in Procaryotes

Protein synthesis is more complex in eucaryotes than in procaryotes. Messenger RNA molecules in procaryotes are frequently polycistronic (encode more than one protein), whereas eucaryotic mRNA molecules appear to specify only single proteins. Eucaryotic mRNA is processed to remove introns and moves from the nucleus into the cytoplasm before translation can occur. In the cytoplasm, the processed mRNA is attached to both free 80S ribosomes and 80S ribosomes on the endoplasmic reticulum, and is translated. Eucaryotic initiation involves as many as ten separate factors. There also is a difference in the way the initiation complex forms. In eucaryotes, the 40S subunit first recognizes the 5′ capped end of the mRNA and then moves to an initiation site, where it is joined by the 60S subunit. In comparison, procaryotic mRNAs do not have a 5′ cap. Instead, ribosomes bind at the Shine-Dalgarno sequence on the mRNA. Despite these differences, the fundamental mechanism of protein synthesis is similar in procaryotes and eucaryotes.

Regulation

Microorganisms are able to synthesize hundreds of different types of proteins. Not all of these proteins, however, are synthesized at the same time or in equal amounts. Control mechanisms regulate the types and quantities of proteins formed by microbes. Without regulatory mechanisms, cells would needlessly waste their efforts and energy by continuously synthesizing unnecessary proteins.

Protein synthesis in microorganisms may be controlled by **induction** or by **repression.** Unlike feedback inhibition and feedback activation (see feedback inhibition, page 143), in which already produced substances regulate the activity of an allosteric enzyme, induction and repression are regulatory phenomena that involve control at the level of transcription or translation.

An Enzyme Is Synthesized in the Presence of an External Substance (Inducer) in Induction

One of the first models proposed for transcriptional control of protein synthesis was the operon model formulated in 1961 by two French biologists, François Jacob and Jacques Monod. Jacob and Monod's operon model is based on the enzymes associated with lactose utilization in *E. coli* (Figure 8.17).

Three different enzymes are involved in lactose uptake and utilization in this bacterium: (1) β-galactosidase, which hydrolyzes β-galactosides such as lactose into their individual monosaccharides (glucose and galactose, in the case of lactose hydrolysis); (2) β-galactoside permease, which transports the sugar into the cell; and (3) thiogalactoside transacetylase, which has an unknown cellular function. When *E. coli* cells are grown in the absence of a β-galactoside, they each contain an average of 0.5 to 5.0 molecules of β-galactosidase. However, when a suitable substrate such as lactose is added, there is a rapid increase in synthesis of β-galactosidase (and the other two enzymes of the lactose operon). Within two to three minutes, enzyme begins to appear, and up to 5,000 molecules of β-galactosidase may be found per bacterium. If the substrate is removed, the synthesis of enzyme stops as rapidly as it started.

The lactose operon model for induction is based on the concept that chromosomes are divided into distinct units of transcription called **operons.** An operon consists of one or more structural genes that are under the control of a common **operator.** In the lactose operon, the structural genes code for the three different enzymes (β-galactosidase, β-galactoside permease, and thiogalactoside transacetylase) involved in lactose utilization. The operator is a region on the DNA that interacts with a **repressor protein** coded by a **regulator gene,** also known as the *i* **gene.** Adjacent to the operator site, within the operon, is a **promoter** to which RNA polymerase binds. There is a certain amount of overlap between promoter and operator regions, so that an occupied operator site prevents binding of RNA polymerase to the promoter. The operator has a key role in transcription control.

The repressor protein attaches to the operator site and prevents binding of RNA polymerase to the promoter in the lactose operon. The repressor protein is inactivated in the presence of an **inducer** (allolactose, which is a derivative of lactose, in the case of the lactose operon). The inducer attaches to the repressor protein, changing its configuration so it is unable to bind to the operator. Under such conditions, RNA polymerase is able to bind to the promoter region of the DNA and transcribe the structural genes

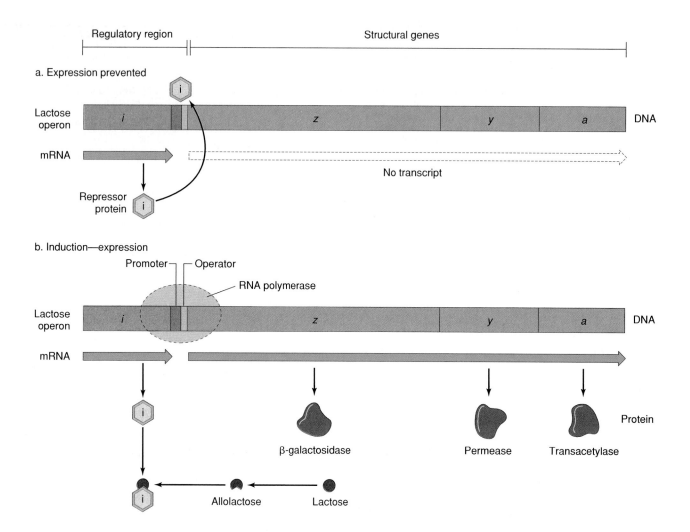

figure 8.17

The Lactose Operon and Induction

a. The repressor protein, coded by the *i* gene, binds to the operator and prevents expression of the structural genes (*z*, *y*, and *a*). b. The inducer (allolactose, which is a derivative of lactose) binds to the repressor and inactivates it, allowing gene expression.

within the operon. This operon model explains why the genes for lactose utilization are expressed only in the presence of lactose. Lactose *induces* enzyme synthesis only when such enzymes are necessary.

Not all enzymes are inducible or repressible; some are produced continuously, or **constitutively,** in constant amounts by the cell. In fact, inducible or repressible enzymes may become constitutive through mutation. For example, mutation of the regulatory gene so that a repressor protein is not produced (i^c mutants) results in constitutive synthesis of β-galactosidase and other enzymes of the lactose operon. Alternatively, the operator site can be modified so that the repressor protein no longer binds to it (o^c mutants). Consequently, RNA polymerase will always bind to the promoter, and the operon is always turned on. Constitutive mutants have proven useful in industrial applications for continuous synthesis of important enzymes or proteins.

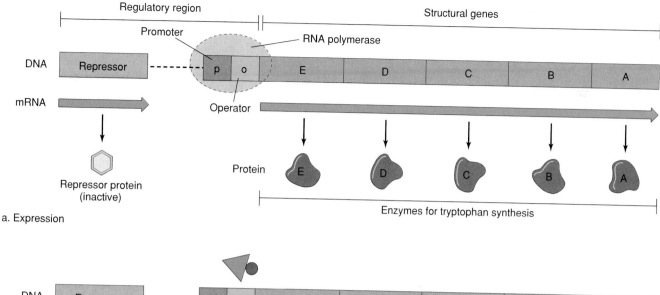

a. Expression

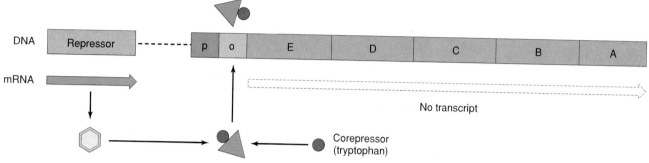

b. Repression — expression prevented

figure 8.18

The Tryptophan Operon and Repression

In repression, the repressor does not bind to the operator unless a corepressor (tryptophan, in this case) is also present. a. In the absence of the corepressor, RNA polymerase is able to bind to the promoter region; the structural genes, coding for enzymes in tryptophan synthesis, are expressed. b. Binding of the repressor protein–corepressor complex to the operator blocks the binding of RNA polymerase to the promoter region.

Enzyme Synthesis Is Inhibited in the Presence of an External Substance (Corepressor) in Repression

In repression, a system similar to induction, the repressor protein is normally inactive and unable to bind to the operator. The repressor protein is activated only in the presence of a **corepressor**—usually the end product of a biosynthetic pathway.

Tryptophan biosynthesis is an example of a pathway controlled by repression (Figure 8.18). The tryptophan operon in *E. coli* consists of five structural genes (*trpE, trpD, trpC, trpB,* and *trpA*), which code for the three enzymes that convert chorismic acid to tryptophan. A promoter region (*trpP*), an operator region (*trpO*), and a gene (*trpR*) that codes for a repressor protein are also part of the operon and regulate the expression of the structural genes.

When tryptophan is limiting or absent, the repressor protein is inactive and unable to bind at the operator region. Consequently, RNA polymerase is able to bind to the promoter region and initiate transcription. When tryptophan is abundant, it acts as a **corepressor,** and binds to the repressor protein and activates it. The repressor protein–corepressor complex binds at the operator region and blocks the binding of RNA polymerase to the promoter region. The operon is turned off, and the transcription of genes for enzymes involved in tryptophan biosynthesis is repressed.

Induction and repression are similar regulatory phenomena, with one major difference. In repression, the presence of a corepressor (tryptophan, in the case of the tryptophan operon) increases the affinity of the repressor protein for the operator. In induction, the presence of an inducer (allolactose, in the case of the lactose operon) decreases the affinity of the repressor protein for the operator.

Gene Expression Is Controlled by the Termination of Transcription in Attenuation

In addition to *trpP*, *trpO*, and *trpR*, there is another region (*trpL*, or *trp* leader) on the tryptophan operon that controls the expression of genes for tryptophan biosynthesis. The *trpL* region, located between the *trpO* and *trpE* of the operon, contains within it a stop signal called the **attenuator.** Transcription of the leader usually stops at this attenuator and a leader mRNA is produced (Figure 8.19). When tryptophan is abundant, the first part of the leader mRNA sequence is translated into a short (14 amino acids) **leader peptide.** The leader peptide contains two tryptophan residues in immediate succession and is only synthesized when there is a plentiful supply of tryptophan and therefore also tryptophanyl-tRNA. Synthesis of the leader peptide results in termination of transcription of the tryptophan structural genes. When tryptophan is in short supply, the leader peptide is not synthesized, and the tryptophan structural genes are transcribed. This regulation of transcription termination is called **attenuation.**

How does translation of the leader sequence regulate transcription of structural genes? The tryptophan leader mRNA sequence can base-pair in several alternative hairpin loop structures (Figure 8.19). The type of hairpin loop that forms is apparently determined by the position of the ribosome during translation of the leader sequence. When tryptophan is present, the leader peptide is synthesized and the ribosome moves to the stop codon (UGA) at the end of the transcript. The remainder of the leader sequence then forms a terminator hairpin loop (Figure 8.19a), which physically prevents RNA polymerase from transcribing the structural genes. When tryptophan is limited and there is a low concentration of tryptophanyl-tRNA, the ribosome

stalls at the *trp* codons and a different hairpin loop forms (Figure 8.19b). This alternative hairpin loop is not a terminator, and RNA polymerase is able to continue transcription past the attenuator.

Evidence for regulation by attenuation in the tryptophan operon is shown by studies with mutants unable to make a repressor protein (R^- mutants). Normally, if repression were the only regulatory mechanism, R^- mutants would be constitutive and produce continuous, constant amounts of enzymes associated with tryptophan biosynthesis. However, in these derepressed R^- mutants, synthesis of enzymes can be further stimulated by the removal of tryptophan, as well as by deletion of the operon region between the operator and the first structural gene. These findings indicate the presence of another regulatory phenomenon, which is attenuation.

Attenuation is a major regulatory phenomenon that reduces transcription about eightfold to tenfold in the presence of tryptophan. Attenuation is used in at least five other operons that control the synthesis of amino acids (histidine, phenylalanine, leucine, threonine, and isoleucine operons). The combination of attenuation and repression allows a fine control of gene expression for the biosynthesis of these amino acids.

Catabolites Can Repress Transcription of Genes

Both induction and repression are examples of **negative control systems;** in the absence of the controlling factor (activated repressor protein) there is enzyme synthesis. Negative control systems are common in bacteria and are well-studied. Not as much is known about **positive control systems,** in which the presence of the controlling factor turns on enzyme synthesis.

One of the better understood mechanisms of positive control is **catabolite repression,** in which enzyme synthesis is repressed when organisms are grown in the presence of glucose or some other catabolite. An example of such repression occurs with glucose on the lactose operon. If a culture of *E. coli* is grown in the presence of lactose and glucose, the cells will first utilize glucose because the lactose operon is repressed by the presence of glucose. The cells use lactose only after the supply of glucose is depleted. At this point the lactose operon is no longer under catabolite repression and the operon genes are transcribed. Catabolite repression leads to **diauxic growth,** a growth pattern in which there are two exponential phases of the population growth curve as the cells use one substrate and then the other. In Figure 8.20, the first substrate was glucose and the second substrate was lactose.

How does the presence of a catabolite such as glucose repress the transcription of the lactose operon genes? Catabolite repression involves a small effector molecule, cyclic AMP (cAMP), which interacts with a catabolite activator protein (CAP) to form

Excess tryptophan

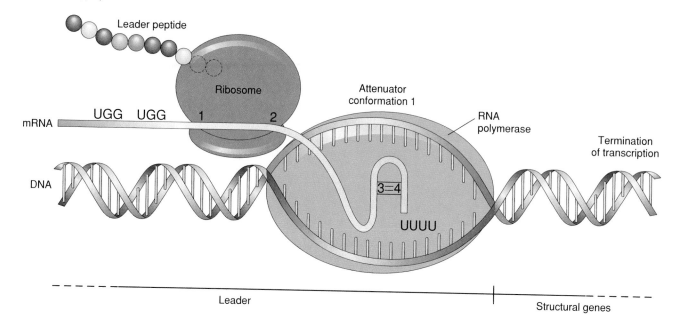

b. Limited tryptophan

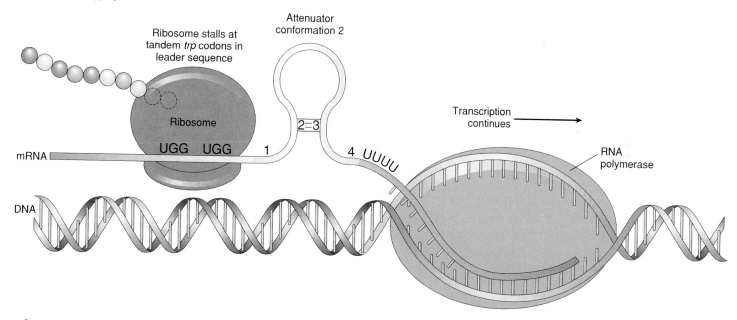

figure 8.19

Attenuation of the Tryptophan Operon in *Escherichia coli*

The leader sequence in the tryptophan operon contains an attenuator region that codes for a tryptophan-rich peptide. a. Under conditions of excess tryptophan, the ribosome proceeds through region 1 of the RNA transcript, which contains tandem tryptophan codons (UGG), and translates the leader peptide. The ribosome is situated on the RNA in such a manner that it prevents region 2 from pairing with region 3. Regions 3 and 4, which are complementary, pair to form a hairpin loop, terminating transcription at the end of the leader RNA. b. When tryptophan is in short supply (and little activated trp-tRNA is available), the ribosome stalls at the tandem tryptophan codons in region 1, leaving region 2 free to pair with region 3 and preventing formation of the 3,4 termination loop. RNA polymerase proceeds through the attenuator into the operon, allowing expression of the *trp* structural genes.

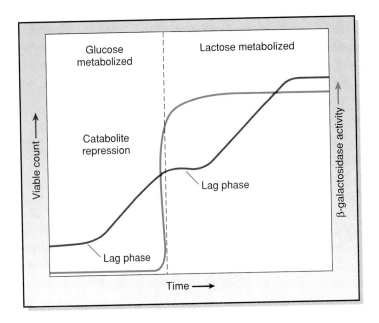

figure 8.20

Diauxic Growth

In an environment containing both glucose and lactose, *Escherichia coli* will metabolize glucose before lactose. Preferential metabolism of glucose results from catabolite repression of the lactose operon.

a CAP-cAMP complex (Figure 8.21). The CAP-cAMP complex binds to the promoter region of the operon and prepares the promoter for binding by RNA polymerase. The binding of the CAP-cAMP complex apparently causes local unwinding of the DNA, changes the configuration of the promoter, and opens it to RNA polymerase binding. CAP by itself does not bind to the promoter, and therefore the cell must have sufficient levels of cAMP to form CAP-cAMP complexes.

In catabolite repression, cell utilization of glucose (the catabolite) results in decreased cAMP levels within the cell. Decreases in cAMP levels may be due either to inhibition of adenylate cyclase, the enzyme that synthesizes cAMP from ATP, or due to increased degradation of cAMP. When the supply of glucose is depleted, levels of cAMP are restored, CAP-cAMP complexes are formed, and may bind to the promoter and prepare it for RNA polymerase binding.

The lactose operon, as well as many other operons, is under both positive and negative control. As a result of this dual control, cells synthesize enzymes only when they are needed. In the presence of glucose and lactose, the lactose operon is turned off as the cell utilizes the more readily metabolized glucose. When glucose levels are depleted, the lactose operon becomes functional, and the cell now synthesizes the enzymes required for lactose utilization.

Eucaryotes Have Mechanisms for Control of Gene Expression Not Found in Procaryotes

Eucaryotes have regulatory mechanisms that differ in several respects from those of procaryotes. Eucaryotes do not appear to have a mechanism equivalent to the classical operon system of procaryotes. An operon system enables procaryotes to rapidly adapt to changes in nutrient concentrations in the environment. The resultant adjustments are short term and usually do not involve permanent alterations in the organism's genotype. In contrast, eucaryotes are not generally exposed to continuously changing environments and therefore do not need to respond as rapidly to changes in their surroundings.

However, there are several mechanisms by which eucaryotes are able to control genetic expression. These include gene amplification (duplication of gene copies), gene loss, gene rearrangement, posttranscriptional modification of RNA, and posttranslational modification of protein.

For example, certain yeasts that form small colonies when grown on media with low levels of glucose are called **cytoplasmic petite mutants.** Petite mutants have mitochondrial DNA containing large deletions that abolish the mutants' ability to make functional mitochondria. Although the amount of mitochondrial DNA

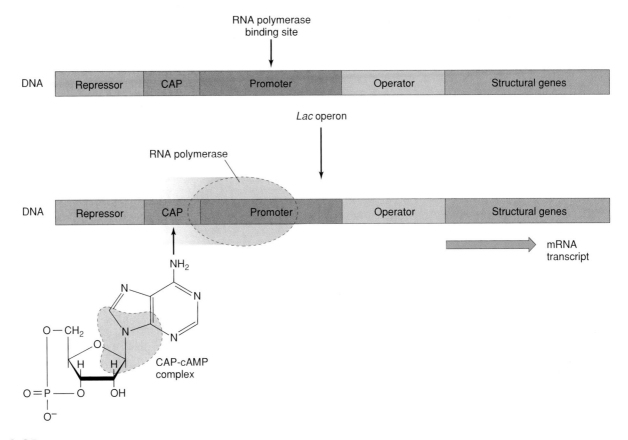

figure 8.21

Catabolite Repression in the Lactose Operon

RNA polymerase binds to the promoter only after binding of the CAP-cAMP complex to the CAP site on the operon.

deleted may range from 20% to 99.9%, the total amount of mitochondrial DNA in petite mutants is almost always the same as in normal yeasts that have all of their mitochondrial DNA. This similarity is due to an amplification of the remaining mitochondrial DNA in the petite mutants. A petite mutant that has 25% of its mitochondrial DNA would amplify this DNA fourfold, whereas one that has only 10% of the mitochondrial DNA would amplify it tenfold. Because of this characteristic, petite mutants are useful for obtaining large amounts of a product from a mitochondrial gene that has been amplified.

The discovery of discontinuous DNA synthesis in the late 1960s by Reiji Okazaki and colleagues at Nagoya University, Nagoya, Japan, was important to understanding the mechanism for synthesis of DNA in the 5' to 3' direction of chain growth (away from the replication fork). Okazaki and colleagues showed discontinuous DNA synthesis by following the uptake of radioactively labeled precursors into newly formed nucleic acid over a short period of time.

Growing *Escherichia coli* cells were exposed to ^{3}H-labeled thymidine for various periods (pulse labeling). After allowing the cells to incorporate the ^{3}H-thymidine for the desired time, the culture was poured onto crushed ice and KCN to stop growth and thymidine incorporation. The cells then were centrifuged, and the DNA was extracted by NaOH-EDTA. The resultant denatured DNA was layered onto a 5% to 20% linear sucrose gradient and centrifuged. After centrifugation, fractions were collected from the gradient and examined for radioactivity using a scintillation counter. Using this pulse-labeling method, newly synthesized DNA could be labeled for various periods of time.

It was discovered that most of the radioactivity incorporated into DNA during a 5-second pulse was recovered in a distinct, slow-moving component of the sucrose gradient—that is, a component containing small pieces of DNA (Figure 8P.1). This component corresponded to a sedimentation rate of 11S. Some radioactivity was located in a second component having a faster sedimentation rate—that is, a component containing larger pieces of DNA. Increasing the pulse time to 10 or 30 seconds increased the amount of radioactivity in both components of the gradient. Increasing the pulse time to 60 seconds or longer resulted in large increases of radioactivity in the fast-sedimenting DNA, but no increases in the 11S DNA. The sedi-

mentation rate of the fast-sedimenting DNA increased with longer pulse times and was 50S after a 10-minute pulse. The 11S component was confirmed as DNA by its susceptibility to degradation by pancreatic DNase, *E. coli* exonuclease I, and *Bacillus subtilis* nuclease, as well as by its resistance to alkali, pancreatic RNase, and bacterial α-amylase.

From these results, Okazaki and colleagues theorized that newly formed DNA was initially synthesized in short fragments and later joined into a longer strand. This conclusion was reached from the observation that during pulse labeling, ^{3}H-thymidine appeared first in a slow-moving component of the sucrose gradient. In a sucrose gradient, small molecules move slower than large molecules. As the pulse-labeling time increased, the radioactivity shifted to the fast-moving component, indicating that the short fragments had joined into a longer piece of DNA and eventually into a long strand with a sedimentation rate of 50S. The sedimentation rate (11S) of the short DNA fragments suggested that the length of a fragment was about 1,000 to 2,000 nucleotides.

Although Okazaki and colleagues initially believed that discontinuous DNA synthesis occurred in both directions (toward and away from the replication fork), it later was shown that discontinuous synthesis occurred in only one direction (away from the replication fork). In honor of Okazaki's discovery, the fragments synthesized during discontinuous DNA synthesis were named **Okazaki fragments**.

Source

Okazaki, R., T. Okazaki, K. Sakabe, K. Sugimoto, and A. Sugino. 1968. Mechanism of DNA chain growth: I. Possible discontinuity and unusual secondary structure of newly synthesized chains. *Proceedings of the National Academy of Science.* 59:598–605.

Summary

1. Genetics is the field of biology that deals with mechanisms responsible for the transfer of traits from one generation to another.

2. The chromosome of an organism can be divided into information-containing regions called genes. The total complement of genes found in a cell or a virus is considered to be its genome. The term genotype indicates the genetic composition of an organism; phenotype indicates the observable characteristics of an organism.

3. The central dogma of molecular biology describes the flow of genetic information in the cell: DNA is transcribed to mRNA; the mRNA is then translated to proteins.

4. Procaryotic mRNA is often polycistronic, with more than one coding region. Eucaryotic mRNA is monocistronic, with only one protein translated per mRNA molecule. Eucaryotic mRNA also frequently contains noncoding regions (introns) that are excised from the mRNA before the coding regions (exons) are translated.

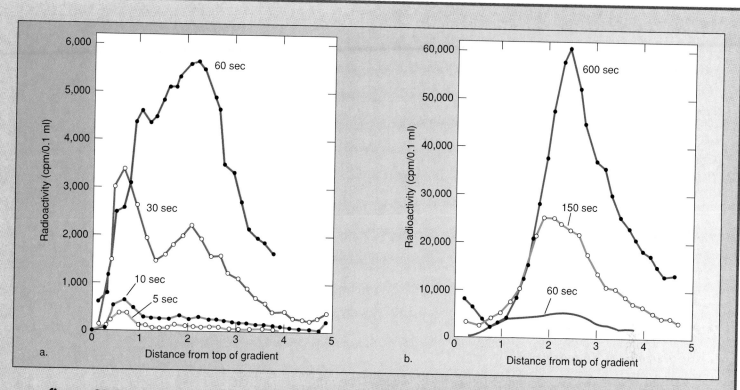

figure 8P.1

Alkaline Sucrose Gradient Sedimentation of Pulse-Labeled DNA from *Escherichia coli*

Ten-milliliter cultures of *E. coli* B, grown to a concentration of 5×10^8 cells per milliliter, were pulse-labeled with 10^{-7} M, ^{3}H-thymidine for the indicated times: a. 5 seconds, 10 seconds, 30 seconds, and 60 seconds; and b. 60 seconds, 150 seconds, and 600 seconds. DNA was extracted by NaOH-EDTA treatment and sedimented on an alkaline sucrose gradient by centrifugation. Peaks of radioactivity are shown. The distance from the top of the gradient is relative to that of infective DNA from bacteriophage δA (19S, reference).

5. The genetic code consists of 64 codons, of which three (UAA, UAG, and UGA) are nonsense codons, or termination codons, that do not code for a specific amino acid. The third position of a codon is the most degenerate of the three positions. This degeneracy is explained by the wobble hypothesis, which states that there is flexibility in base pairing in the third codon position.

6. DNA is replicated semiconservatively by DNA polymerase III in a 5′ to 3′ direction of chain growth starting at an origin of replication. Replication begins with unwinding and relaxation of the double helix by several enzymes including DNA helicase and DNA topoisomerase as well as initiator proteins and helix-stabilizing, single-stranded, DNA-binding proteins. A short RNA primer, required for DNA synthesis, is synthesized by a primase. Short DNA segments, or Okazaki fragments, are synthesized on one of the two daughter strands and later joined by a DNA ligase.

7. There are three major classes of RNA: mRNA, which carries the genetic message from DNA; rRNA, which contributes to ribosome structure and has functional roles in protein synthesis; and tRNA, which carries amino acids to the ribosome during translation. RNA synthesis begins with the

binding of RNA polymerase to a promoter on the DNA template. After chain elongation in a 5′ to 3′ direction of chain growth, transcription terminates when the RNA polymerase either reaches a region where a double-stranded RNA stem-and-loop structure is formed followed by a polyU sequence, or it terminates when it encounters a protein factor called rho that causes unwinding of the RNA-DNA complex.

8. Protein is synthesized in a 5′ to 3′ direction from the mRNA. In procaryotes, a short purine-rich region, the Shine-Dalgarno sequence, precedes the 5′ end of an initiating codon (AUG or GUG) and specifies the ribosome binding site on the mRNA. An initiation complex is formed, with fMet-tRNA binding to the initiating codon in Bacteria (Met-tRNA binds to the initiating codon in Archaea). During elongation, aminoacyl-tRNAs bind to complementary codons on the ribosome-mRNA complex and peptide bonds are formed between adjacent amino acids. Protein synthesis terminates when the ribosome reaches a nonsense codon on the mRNA and release factors bind to the ribosome, resulting in dissociation of the ribosome and release of the newly formed protein.

9. Protein synthesis in microorganisms may be controlled by induction or by repression. In induction (for example, the lactose operon), an enzyme is synthesized in the presence of an external substance (the inducer). In repression (for example, the tryptophan operon), enzyme synthesis is inhibited in the presence of an external substance (the corepressor).

10. Attenuation regulates transcription of structural genes by the formation of alternative hairpin loops on the mRNA that either physically inhibit RNA polymerase or permit RNA polymerase to transcribe. Catabolite repression is the repression of the transcription of structural genes when a CAP–cAMP complex does not form in the presence of a catabolite such as glucose.

EVOLUTION and BIODIVERSITY

As the genetic fiber that determines the composition and characteristics of an organism, DNA (or the equivalent RNA, in the case of some viruses) is unique to life. It distinguishes living organisms from nonliving objects such as rocks, metal, and dirt, which do not have DNA. Yet even with its presence in life-forms ranging from the smallest bacteria to the enormous blue whale to the intricate and complex human body, DNA is surprisingly constant in its fundamental composition and structure. It is this constancy that makes DNA the molecular foundation for the unity of life across species and genus lines. At the same time, DNA is also responsible for the diversity of life. Differences in DNA sequences result in differences in proteins, the molecules that are the essence of a cell and define its physiological and metabolic characteristics. It is this dichotomy that makes this relatively simple molecule so remarkably extraordinary and important to life. Over the billions of years that life has existed and evolved, DNA has provided the genetic information that has played such an important role in both similarities and differences among organisms.

Questions

Short Answer

1. Rank the following from smallest to largest: chromosome, codon, gene, genome, nucleotide.

2. Compare and contrast transcription and translation.

3. Compare and contrast an organism's genotype and phenotype.

4. How did researchers prove that DNA is the hereditary information?

5. Describe the arrangement of nucleotides forming DNA.

6. What are histone proteins and what is their function?

7. How did researchers prove that DNA is replicated semiconservatively?

8. Explain how DNA replication begins and in what direction synthesis occurs.

9. What are Okazaki fragments and why are they significant?

10. Since DNA polymerases cannot initiate DNA synthesis, how does it begin?

11. How does eucaryotic RNA differ from procaryotic RNA?

12. Identify three types of RNA and describe the role of each.

13. Explain how the correct strand of DNA is chosen for transcription and how RNA synthesis begins.

14. Why is procaryotic RNA referred to as "polycistronic"?

15. Briefly describe how protein synthesis occurs.

16. Briefly describe the operon model of regulation.

17. Briefly describe regulation by attenuation.

Multiple Choice

1. Which of the following is *not* found in DNA?
   a. adenine
   b. cytosine
   c. guanine
   d. thymine
   e. uracil

2. Anticodons are found in the:
   a. gene
   b. mRNA
   c. rRNA
   d. tRNA

3. Which of the following is *not* the result of transcription?
   a. mRNA
   b. rRNA
   c. tRNA
   d. DNA
   e. All of the above except a.

4. Which of these errors in transcription would have a significant impact?
   a. UAU rather than UAC
   b. CCG rather than CCA
   c. CGG rather than AGG
   d. UAG rather than UAC

5. Which of the following is the least efficient method of regulation?
   a. attenuation
   b. feedback inhibition
   c. operon repression and induction

Critical Thinking

1. The central dogma states that genetic information flows from DNA to RNA to protein. Discuss the need for such a pattern in evolution. Provide examples that contradict this pattern.

2. Discuss the possible applications of DNA hybridization (as mentioned on page 217). Which of these are now common practice?

3. The rolling circle mechanism of DNA replication used by some viruses, and during bacterial conjugation, would not work for replication of procaryotic or eucaryotic genomes. Explain.

4. It has been said that evolution would not be possible without mutation. Explain.

 Supplementary Readings

Alberts, B., D. Bray, J. Lewis, M. Raff, K. Roberts, and J.D. Watson. 1994. *Molecular biology of the cell,* 3d ed. New York: Garland Publishing, Inc. (An in-depth textbook on cell structure and function.)

Lewin, B. 1994. *Genes V.* Oxford, England: Oxford University Press. (A thorough coverage of DNA replication, RNA synthesis, protein synthesis, and regulation of gene expression.)

Lodish, H., D. Baltimore, A. Berk, S.L. Zupursky, P. Matsudaira, and J. Darnell. 1995. *Molecular cell biology,* 3d ed. New York: Scientific American Books, Inc. (An in-depth textbook on molecular biology.)

Russell, P.J. 1996. *Genetics,* 4th ed. New York: HarperCollins. (A fundamental textbook of genetics, with sections on DNA replication, transcription, translation, genetic exchange in bacteria, transposable elements, and recombinant DNA technology.)

Watson, J.D., N.H. Hopkins, J.W. Roberts, J.A. Seitz, and A.M. Weiner. 1987. *Molecular biology of the gene,* 4th ed. Menlo Park, Calif.: The Benjamin/Cummings Publishing Company. (An in-depth textbook on macromolecular synthesis, gene function, and molecular genetics.)

Wood, W.B., J.H. Wilson, R.M. Benbow, and L.E. Hood. 1981. *Biochemistry: A problems approach,* 2d ed. Menlo Park, Calif.: The Benjamin/Cummings Publishing Company. (A review of macromolecular structures and functions, including a problems section.)

Zubay, G. 1993. *Biochemistry,* 3d ed. Dubuque, Iowa: Wm. C. Brown Publishers. (A well-written textbook of biochemistry.)

chapter nine

MICROBIAL GENETICS

Mutation
Base Substitution
Deletion and Insertion
Reversible Effects
Mutagenic Agents
DNA Repair Mechanisms in Bacteria
Replica Plating

Transfer of Genetic Material
Transformation
Transduction
Plasmids
Conjugation

PERSPECTIVE
Lysogenic Conversion of *Corynebacterium diphtheriae*

EVOLUTION AND BIODIVERSITY

Electron micrograph of *Escherichia coli*
infected with bacteriophage T4 (colorized).

💿 Microbes in Motion   PREVIEW LINK

This chapter covers the key topics of transfer of genetic material between
microorganisms and the effect this has on development of antibiotic
resistance. The following sections in the *Microbes in Motion* CD-ROM may
be useful as a preview to your reading or as a supplemental study aid:

Microbial Metabolism and Growth: Genetics (Mutation), 4–8,
(Recombination) 11–24. *Antimicrobial Resistance:* Transfer and Spread of
Resistance (Conjugation), 4–5.

The most impressive feature of DNA is its accuracy in replication. The precision of DNA replication is such that errors occur in approximately 1 out of every 10^9 bases copied. This faithful copying of DNA is important; otherwise many genes would be changed during replication, causing extensive alterations in genotype and phenotype with each cell division. Nonetheless, there is a strong advantage to small and limited modifications in the genome of a microorganism. These genetic changes enable a microbial species to adapt to constantly changing physical and biological environments; therefore they are important in microbial evolution.

DNA in microorganisms may be altered by two methods: mutation, and transfer of genetic material.

Mutation

A **mutation** is defined as an inheritable change in the sequence of DNA. Organisms that have undergone mutations are termed **mutants.** The result of a mutation is usually a phenotypic change in an organism. This new phenotype can be used to differentiate the mutant from its original **wild-type** form.

Mutations may be either **spontaneous** or **induced.** Spontaneous mutations occur naturally, usually as a result of errors in DNA replication. Induced mutations arise as a result of chemical or radiation effects on genes. Mutagenic agents increase the mutation rate (the average number of mutations per cell per generation) in comparison with that of spontaneous mutations and also can be used to cause more specific classes of mutations in the genome.

Bases Can Be Substituted by Other Bases

The simplest type of mutation is one in which there is a **substitution** of one or more bases in the DNA (Figure 9.1). Most substitutions are **point mutations** involving a single base change. There are different consequences of such a mutation. If the change occurs in the third position of the codon, which is degenerate, the event is known as a **silent mutation** because there is no effect on the amino acid coded by the affected area of the DNA. For example, any base can be inserted in the third position of the codon for serine (UC_) with no alteration in the reading of the codon. Substitutions in either the first or second bases of a codon, however, generally have a more serious effect because they will code for an entirely different amino acid. Such changes are known as **missense mutations.** If the amino acid occupies an indispensable position in the protein, there will be a significant alteration in the protein, possibly resulting in a change in protein structure or—in the case of enzymes—inactivation of protein function. Sometimes there may be no change in function. A serious consequence of a substitution occurs in **nonsense mutation,** where the mutation results in the formation of a nonsense codon. With such an event, premature termination of protein synthesis often results in a nonfunctional protein or a lack of expression of downstream genes (genes further along the 3' end of the ribonucleic acid) in polycistronic messages.

Bases Can Be Deleted or Inserted

Mutations in which segments of the genome are removed are termed **deletions.** When one or more bases are added to the DNA, they are called **insertions.** These changes can involve one or many bases. Single-base deletions and insertions can cause a **reading frame shift,** which completely changes the amino acid sequence of the protein starting from the point of the mutation. The effects of such mutations may be minimized if another frame shift mutation occurs downstream of the affected region. A single-base deletion may offset the effects of a single-base insertion if it occurs near the site of the first mutation. Larger deletions and insertions, however, generally have irreversible effects and typically produce unusable proteins.

The Effects of Some Mutations Are Reversible

Mutation effects can sometimes be reversed. The consequences of a second deletion or insertion in reversing the effects of a reading frame shift have already been mentioned. The simplest type of reversion is a **back mutation** in which, for example, a mutated base is changed back to its original form. In other cases biosynthetic pathways turned off by a mutation that affects synthesis of a particular enzyme can be made functional again by an alternative pathway that bypasses the affected enzyme.

Suppressor mutations overcome or suppress the effects of the initial mutation without any alteration in the original gene. For example, a suppressor mutation might alter a tRNA anticodon. Suppose a single-base substitution causes the codon GCU to be changed to ACU. The new codon codes for threonine instead of alanine. If, however, the anticodon on the alanine-tRNA is also changed by a suppressor mutation so that it now recognizes ACU instead of GCU, the effects of the original mutation will have been suppressed. Because there are multiple genes for alanine-tRNA, the cell will now have alanine-tRNA that recognizes the codon GCU as well as alanine-tRNA modified by the suppressor mutation that recognizes the codon ACU.

The Rate of Mutations Can Be Increased by Mutagenic Agents

Different mutagenic agents can be used to increase the rate of mutations (Table 9.1). Unfortunately, mutagenic agents are found in the environment and undoubtedly affect the mutation rates not only of microorganisms, but also of higher organisms, including humans. Since many mutagens are also carcinogens, increased exposure to these mutagens may lead to higher incidences of cancer in humans.

Ultraviolet radiation causes formation of pyrimidine dimers in the chromosomes. Examples of such dimers, formed between adjacent pyrimidines on a DNA strand, are thymine-thymine, thymine-cytosine, and cytosine-cytosine. The dimers, which consist of adjacent pyrimidines joined by covalent linkages on the same DNA strand, distort the DNA structure and interfere with replication and transcription.

Base analogs are chemical compounds that resemble DNA bases but cause faulty base pairing. Examples of base analogs are 5-bromouracil and 2-aminopurine. The analog 5-bromouracil substitutes for thymine, but unlike thymine, it occasionally pairs with

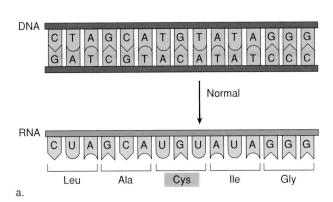

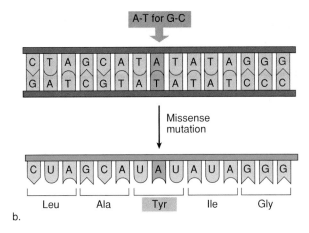

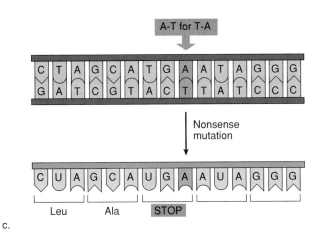

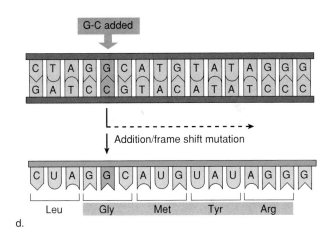

figure 9.1

Different Types of Mutations

a. Normal gene segment. b. Missense mutation: The substitution of adenine for guanine in the DNA leads to an A-T instead of a G-C base pair and the substitution of tyrosine for cysteine in the peptide. c. Nonsense mutation: The substitution of adenine for thymine leads to an A-T instead of a T-A base pair. The result is coding for a nonsense codon and termination of protein synthesis. d. Addition/frame shift mutation: The addition of an extra guanine (G-C base pair) in the DNA leads to a shift in the reading frame. The result is a completely different peptide starting with the second codon.

guanine instead of adenine. This incorrect pairing leads to errors in replication, with a G-C base pair replacing the original A-T base pair. Instead of thymine, which normally would pair with adenine to form a final A-T base pair in the replicated DNA, the analog 5-bromouracil pairs with guanine to form a final G-C base pair after DNA replication. A similar result occurs with 2-aminopurine, which substitutes for adenine but pairs with cytosine instead of thymine.

Microbial Metabolism and Growth
Genetics: Mutation • p. 7

Other mutagens cause specific chemical changes in bases, resulting in transitions (replacement of a purine by a different purine or replacement of a pyrimidine by a different pyrimidine) or transversions (replacement of a purine by a pyrimidine or replacement of a pyrimidine by a purine). Nitrous acid deaminates adenine and cytosine, forming hypoxanthine (which pairs with cytosine) and uracil (which pairs with adenine), respectively. Hydroxylamine converts cytosine to a compound that pairs with adenine instead of guanine. Monofunctional alkylating agents (ethyl methanesulfonate and methyl methanesulfonate) alkylate guanine

table 9.1

Common Mutagens and Their Modes of Action

Mutagen	Basis of Action	Mutagenic Effect
Chemical mutagens		
Ethyl methanesulfonate, methyl methanesulfonate	Alkylation of purines followed by depurination	G-C ⇌ A-T transitions; transversions
Hydroxylamine	Conversion of cytosine to hydroxylaminocytosine	G-C ⇌ A-T transitions
N-methyl-*N'*-nitro-*N*-nitrosoguanidine	Alkylation of guanine	G-C ⇌ A-T transitions
Nitrous acid	Deamination; conversion of cytosine to uracil and of adenine to hypoxanthine	G-C ⇌ A-T transitions
Base analogs		
2-aminopurine	Adenine analog; causes mispairing with cytosine	G-C ⇌ A-T transitions
5-bromouracil	Thymine analog; causes mispairing with guanine	G-C ⇌ A-T transitions
Others		
Ultraviolet radiation	Formation of pyrimidine-pyrimidine dimers	G-C ⇌ A-T transitions, frame shifts
Acridine, ethidium bromide	Intercalation	Frame shifts

so that it pairs with thymine and not cytosine. *N*-methyl-*N'*-nitro-*N*-nitrosoguanidine, a bifunctional alkylating agent, acts specifically at the DNA replication fork and causes crosslinking of the DNA strands and errors in replication. Other agents such as acridine and ethidium bromide are mutagenic because they are intercalated directly into DNA between successive bases and distort the structure of the double helix. The result of their action is the addition or removal of a single base, causing a shift in the reading frame.

Microbial Metabolism and Growth
Genetics: Mutation • pp. 4–8

Bacteria Have Different DNA Repair Mechanisms

Damage to DNA can have serious consequences for a bacterium. Bacteria have developed several different mechanisms for repair and replacement of damaged DNA.

Excision repair corrects damage that causes distortion in the DNA helix such as what occurs with pyrimidine dimers that are formed as a result of exposure to ultraviolet radiation (Figure 9.2). A UvrABC endonuclease (Uvr means *Uv repair*) encoded by three genes (*uvrA*, *uvrB*, and *uvrC*) makes cuts on both sides of the damaged area. This excision results in release of a 12-nucleotide single-stranded DNA segment containing the damaged bases. The single-stranded gap left in the DNA is filled by DNA polymerase I, and DNA ligase joins the fragments.

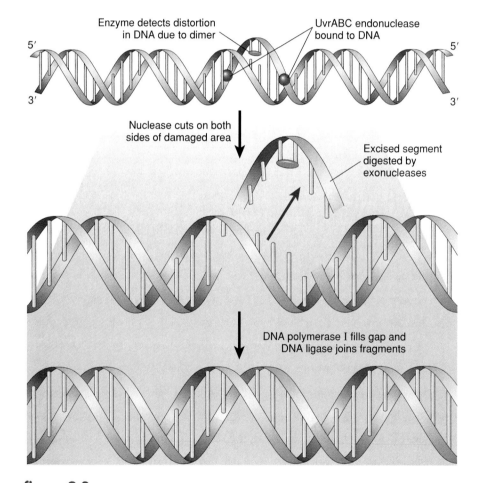

figure 9.2
Excision Repair of Distortion-Related Damage to DNA
A UvrABC endonuclease excises the damaged DNA segment. DNA polymerase I fills the resultant gap and DNA ligase joins the fragments.

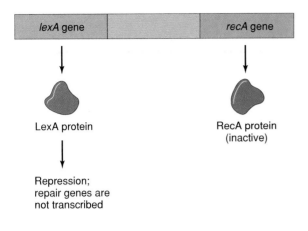

lexA gene   recA gene

LexA protein

RecA protein
(inactive)

Repression;
repair genes are
not transcribed

b. Induced state

lexA gene   recA gene

LexA protein

Active RecA
stimulates LexA
to cleave itself

RecA protein
is activated by
DNA damage

Inactivated
LexA protein

No repression;
repair genes
are transcribed

figure 9.3

The SOS Response

a. The SOS system is normally repressed by a *lexA*-encoded protein, LexA. b. When there is DNA damage, a *recA*-encoded protein, RecA, stimulates LexA to cleave itself and become inactivated. As a result of LexA inactivation, the SOS system is derepressed.

Correction of pyrimidine dimers can occur also by **photoreactivation.** In photoreactivation, a light-activated enzyme, photolyase, cleaves the covalent linkages between the damaged pyrimidines. This type of repair does not require the removal and replacement of bases as in excision repair.

Damaged bases can be removed by enzymes called **glycosylases** that detect and remove an unnatural base. The resultant hole is called an AP site (for *apyrimidinic* or *apurinic,* depending on what type of base has been removed). DNA polymerase I removes a few additional bases in this region and fills in the gap. DNA ligase then joins the fragments.

In instances where DNA damage is so serious that it affects DNA replication, such as leaving gaps in the helix, a distress signal is sent to the cell to induce a complex repair system called the **SOS response** (Figure 9.3). In *E. coli,* the SOS system consists of about 17 genes involved in the excision and repair of various types of DNA damage. Expression of these genes is controlled by a *lexA* gene and a *recA* gene. The SOS system is normally repressed by a *lexA*-encoded protein, LexA. When there is DNA damage, a *recA*-encoded protein, RecA (which also is involved in homologous recombination, as described on p. 252), stimulates LexA to cleave itself and become inactivated. As a result, the DNA repair genes in the SOS system are derepressed. The repair genes are transcribed and the DNA is repaired.

Replica Plating Differentiates Mutants from Wild Types by Their Growth Differences

Mutagenic agents are typically used in quantities that will kill most of the cells in a population. The small number of surviving cells have a greater chance of having undergone a mutagenic event. Even under such optimal conditions, the mutation rate is quite low, usually 10^{-3} to 10^{-5} mutation per bacterium per generation (that is, one bacterium in 10^3 to 10^5 is likely to undergo a mutational event each generation). This rate is much higher than the spontaneous mutation rate of 10^{-6} to 10^{-12}. Nonetheless, the rate of induced mutations is small enough to make detection of mutants difficult unless a selection technique is used.

Mutagenesis results in a genotypic and, frequently, a phenotypic change from the **wild type** (the unaltered original genotype). These differences in characteristics can be used to separate mutants from wild types. Antibiotic-resistant mutants can be separated from wild-type susceptible strains by growing both types of bacteria on a medium containing the antibiotic; only the mutant will grow. Mutants that have undergone changes in cell or colony morphology, such as pigmentation, encapsulation, or motility, are usually distinguished by their readily observable differences in colony morphologies from wild-type organisms. **Auxotrophic** mutants (strains that require one or more growth factors because of a genetic mutation) can be distinguished from **prototrophs** (the wild-type strain that does not have this growth factor requirement) by the inability of the auxotroph to grow on a medium lacking the required growth factor(s). The change in genotype of an auxotrophic mutant is usually symbolized by a minus sign after the compound that is required for growth. Thus an auxotroph of *Escherichia coli* having growth requirements for arginine, histidine, and leucine is indicated by the notation: *E. coli* (*arg⁻, his⁻, leu⁻*). The genotypes of prototrophs are sometimes symbolized by a plus sign to emphasize their differences from auxotrophs: *E. coli* (*arg⁺, his⁺, leu⁺*).

A technique that is commonly used to identify mutants is **replica plating** (Figure 9.4). Mutated and wild-type bacteria are grown on an enriched medium. Each cell develops into an individual colony. The colonies of both types are then transferred either to several different media, each of which lacks a certain type of nutrient, or to the same medium incubated under different environmental conditions. Transfer from the original plate (called the master plate) to the new plates is performed by carefully placing a pad of sterile velvet onto the surface of the master plate. The velvet picks up some cells from each colony. If this inoculated velvet pad is removed from the master plate and placed consecutively onto one or more uninoculated plates of different test media, a replica of the master plate is made on these subsequent plates. Since theoretically only one cell is required to initiate growth and form a colony per plate of medium, several replica plates can be made from one inoculated velvet pad. The inoculated plates are incubated, and the patterns of growth are then compared to determine the nutritional or environmental requirements of the bacteria. For example, in screening for arginine-requiring auxotrophs by replica plating, colonies would grow on a replicate plate with arginine, but would not grow on a plate lacking this amino acid.

Microbial Metabolism and Growth
Genetics: Mutation • p. 8

Transfer of Genetic Material

Mutation is important in changing the genetic information in a cell, but it is not the only process that causes genetic modifications. Procaryotes and eucaryotes can produce new genotypes from the transfer of genetic material from one cell to another cell. In eucaryotes genetic transfer occurs after two haploid gametes fuse to form a diploid zygote. Procaryotes do not produce male and female gametes and do not exchange DNA by this sexual method. However, procaryotes can occasionally transfer genetic material from a **donor cell** to a **recipient cell.** The recipient becomes a partial diploid for the genetic material transferred; such partial zygotes are called **merozygotes.** The genetic material introduced into the recipient cell is called the **exogenote.** The exogenote may recombine with the recipient genome, the **endogenote;** may persist as a self-replicating plasmid; or may eventually be lost through dilution during cell division.

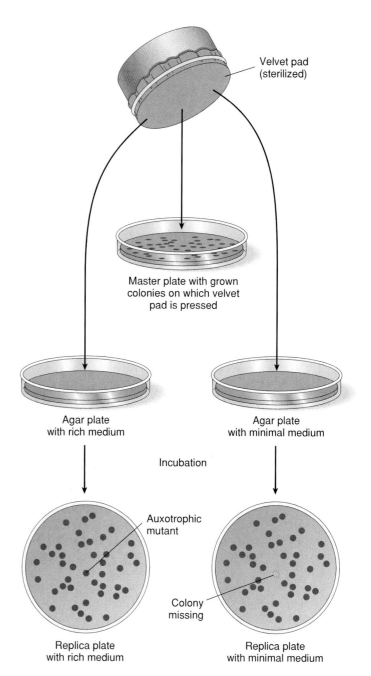

figure 9.4
Replica Plating

Colonies from a master plate are transferred to replica plates with a sterile velvet pad. The replica plates are incubated and the patterns of growth are then compared to determine the presence of mutants.

The Ames Test for Carcinogenesis

Carcinogens are substances that produce cancer. Many mutagens are also carcinogens because they cause changes in the organism that can eventually lead to cell transformation.

Several years ago Bruce Ames, a scientist at the University of California at Berkeley, developed a bacterial test to screen carcinogens. This test, called the **Ames test,** is based on the assumption that carcinogens either are or can generate mutagens. If this assumption is true, the potency of a carcinogen can be determined by its ability to cause mutations.

The Ames test measures the reversion of histidine auxotrophs of *Salmonella* *typhimurium* to nonmutant, histidine-synthesizing prototrophs in the presence and absence of the substance being tested. The test is performed with strains of *S. ty-phimurium* that rarely revert spontaneously to prototrophy, and that contain either a base substitution or a frame shift mutation. Furthermore, the test strains are devoid of DNA repair enzymes, so they are more sensitive to DNA alterations, and they are mutated so that their envelopes are more permeable to chemical agents.

The Ames test is carried out by mixing the bacteria with a rat liver extract (Figure 9.5). The rat liver extract is used because many chemicals do not become carcinogenic unless they are first converted to other metabolites; this conversion in animals usually takes place in the liver. A small amount of the mixture is spread onto a solid growth medium that lacks histidine. The test chemical is then added to the medium. If the chemical is mutagenic, some histidine mutants (*his⁻*) will revert to the wild type (*his⁺*) and produce visible colonies on the medium. Chemicals that result in large numbers of revertants are considered highly mutagenic and thus also highly carcinogenic. The Ames test can be performed inexpensively and rapidly with bacterial populations, unlike other tests for carcinogens that require laboratory animals. The test takes advantage of the short generation times and selection techniques of bacteria. The Ames test is used extensively to screen for carcinogens that are mutagens, and is comparable to animal testing in its detection abilities.

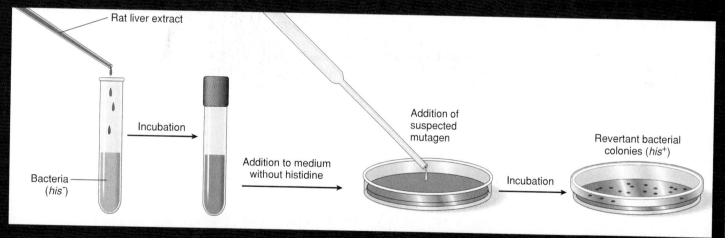

figure 9.5

The Ames Test

His⁻ mutants of *Salmonella typhimurium* are mixed with rat liver extract. This mixture is spread onto an agar plate lacking histidine, and the suspected mutagen is added to the agar. The greater the mutagenicity of the chemical, the more revertant (*his⁺*) colonies will develop on the medium.

At the molecular level, the reciprocal exchange between two homologous (identical or nearly identical sequences) DNA molecules is called **general,** or **homologous, recombination** (Figure 9.6). This recombination between homologous DNA molecules begins with the nicking of a single strand of one of the DNA molecules by an endonuclease. Next, a RecA protein binds to the single-stranded fragment and forms a complex between the nicked single-stranded region and the complementary sequence in the homologous DNA duplex. Following this strand invasion of the DNA duplex, there is cross-strand exchange and ligation. This breakage and reunion of homologous DNA molecules results in **recombinant molecules,** which consist of DNA from each of the two original DNA molecules.

There are three methods of genetic transfer in bacteria. The first method, **transformation,** involves the insertion of naked DNA from a donor cell into a recipient cell. If a bacteriophage (bacterial virus) is the intermediary vector for the genetic material, the process is termed **transduction.** The third method of bacterial gene transfer, **conjugation,** requires contact between donor and recipient cells for DNA transfer. Regardless of the DNA transfer mechanism, natural genetic exchange in procaryotes is a rare event. Nonetheless, genetic exchange is an important event that leads to genetic diversity and evolution.

DNA Is the Transforming Principle

Fred Griffith's discovery of bacterial transformation in 1928 was a historic milestone in microbiology (see transformation of *Streptococcus pneumoniae,* page 215). As a result of these early studies on transformation in *Streptococcus pneumoniae,* it is now known that DNA—the "transforming principle"—is the source of genetic information for all living cells. Natural transformation occurs in only certain bacteria, including *Haemophilus, Neisseria, Staphylococcus, Bacillus, Pseudomonas,* and *Rhizobium,* but it has been most extensively investigated in the bacterium *S. pneumoniae.* Studies of transformation in these and other bacteria have provided us with an insight into the physiological and molecular mechanisms associated with this gene transfer process.

Only Competent Cells Can Receive DNA During Transformation

Genetic transformation occurs optimally between closely related bacteria in the late logarithmic phase of growth. Even under such ideal conditions, only a small portion of the total DNA from a donor cell is transferred to the recipient cell. Not all bacteria are transformable. Even among those organisms that can be transformed, transformation appears to depend on the physiological state of the recipient cell. A bacterium that is able to receive DNA from a closely related donor cell and be transformed by it is said to be **competent.**

Competence is determined by the presence of a **competence factor** on the cell surface. This factor is found on the surfaces of competent *S. pneumoniae* cells and is released into the medium by the cells. The competence factor has been partially characterized

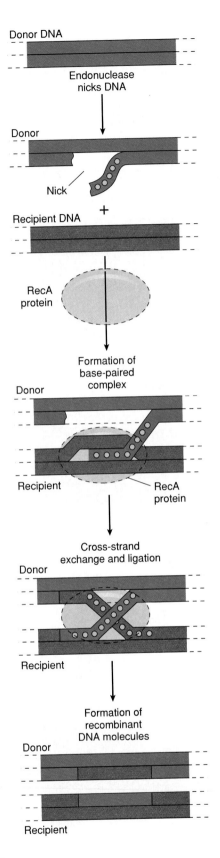

figure 9.6

Homologous Recombination

The breakage and reunion of homologous (identical or nearly identical) DNA molecules results in recombinant molecules.

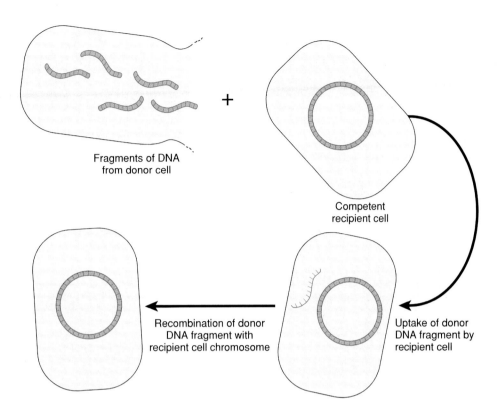

figure 9.7

Transformation

Donor DNA is taken up by a competent recipient cell and recombines with the recipient cell's chromosome in transformation.

and appears to be a protein of low molecular weight. When competence factor is added to noncompetent cells, it binds to their surfaces and changes them into competent cells. Competency depends on several conditions, including composition of the growth medium and growth stage of the culture.

Naked DNA Enters a Competent Recipient Cell During Transformation

Transformation begins with the release or removal of DNA from a donor cell (Figure 9.7). In natural transformation, DNA is released into the surrounding environment when a cell lyses. These DNA fragments may be taken up by competent cells and transformed. In artificial transformation carried out in the laboratory, DNA is extracted from donor bacteria by chemical or enzymatic lysis of the donor cells. The nucleic acid is sheared into short pieces during its extraction, purified from contaminating proteins through phenol treatment, and then is precipitated with ethanol. This relatively pure DNA preparation can now be added to a suspension of competent bacteria treated with calcium chloride to permeabilize their plasma membranes to DNA.

The naked double-stranded donor DNA binds to receptor sites on the cell surface of a competent bacterium. These receptor sites are exposed only when the bacterium is in a competent stage. There is evidence that the competence factor may play a role in unmasking receptor sites during periods of competence. The externally bound DNA remains on the cell surface only long enough for it to be cleaved by a membrane-associated endonuclease into fragments of 7,000 to 10,000 base pairs in length.

Bacteria differ in their uptake of DNA fragments. In *Streptococcus pneumoniae, Bacillus subtilis,* and other gram-positive bacteria, the double-stranded DNA bound to the cell surface is degraded to yield single-stranded DNA. These single-stranded DNA fragments are transported into the cell. In *Haemophilus influenzae,* a gram-negative bacterium, double-stranded DNA fragments are transported into the cell, but only single-stranded fragments are incorporated into the chromosome through recombination.

A single-stranded DNA fragment incorporated into the bacterial chromosome displaces a portion of one strand of the chromosome. The two strands comprising this heteroduplex region of the cell genome may not be identical, and there may be areas in which mismatched bases are not held together by hydrogen bonds. Under these circumstances, the integrity of the chromosome may be restored by replacement of the mismatched bases on one strand with complementary matched bases. This correction mechanism shows a preference for removing mismatched bases on the donor DNA. Alternatively, if replication occurs first, the two strands including the heteroduplex region are copied, and the exogenote is maintained in intact form through the progeny.

Transduction was first observed in 1952 by Joshua Lederberg and his graduate student Norton Zinder, who actually were looking for conjugation in *Salmonella typhimurium*.

U separated by a sintered glass filter impervious to bacteria (Figure 9.8). Bacteria incubated in the same culture medium in either arm of the U-tube thus are unable

exchange was not affected by the addition of DNase to the bacterial cultures, a procedure that would normally degrade naked DNA in transforming populations.

The Discovery of Transduction

In the course of their experiments, Lederberg and Zinder found that the transfer of genetic characteristics from donor to recipient cells did not require the cell-to-cell contact characteristic of bacterial conjugation. The two scientists made this discovery while utilizing a U-tube first devised by Bernard D. Davis in 1950 in his studies of conjugation.

The Davis U-tube consists of a U-shaped glass tube with the two sides of the

to come in contact with each other.

Davis originally used the U-tube to show that such contact was necessary for bacterial conjugation. Lederberg and Zinder also used the Davis U-tube in their experiments and hoped to show that conjugation was possible in *Salmonella*. However, much to their surprise, transfer of genes occurred between donor and recipient bacteria in the U-tube. The genetic

Further investigation revealed that the genes had been transferred not directly via cell-to-cell contact, but with an intermediate agent, a bacteriophage (*Salmonella* phage P22) that could pass through the glass filter in the tube. These studies provided the first evidence that bacteriophages are important in bacterial genetic exchange.

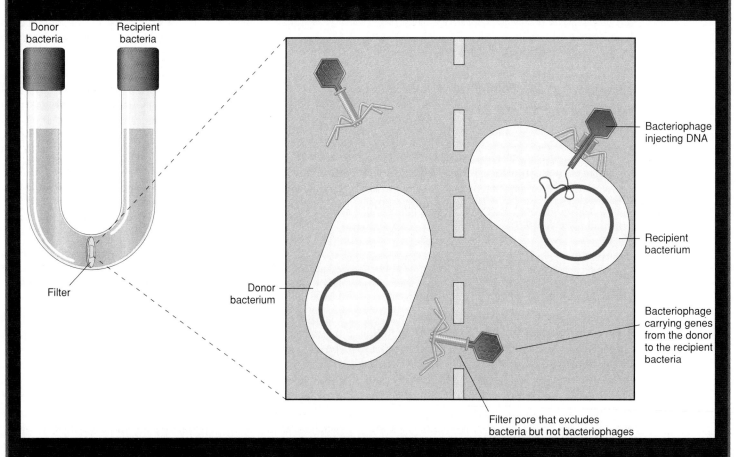

figure 9.8
The Davis U-tube
A sintered glass filter separates bacteria in the two arms of the U-tube. Bacteriophages and naked DNA can pass through the pores of the filter, whereas the larger bacteria cannot.

Although many DNA fragments may be available for uptake by the recipient cell, only a few such fragments (generally ten or less) are actually incorporated into it. The recipient is unable in principle to distinguish one fragment from another and therefore cannot selectively bind and incorporate fragments that contain specific genetic markers. As a consequence of this random uptake of DNA, the possibilities are slight that any given gene will be in the DNA fragments entering the cell. The frequency of transformation of a specific genetic marker therefore is very low (usually less than 1%).

Microbial Metabolism and Growth
Genetics: Recombination • pp. 15–18

Naked Bacteriophage DNA Enters the Competent Recipient Cell During Transfection

Transfection, which involves the introduction of bacteriophage DNA into a cell, is a special case of transformation in *Escherichia coli.* The most common technique for transfection is the treatment of bacteria with cold calcium chloride to make their membranes permeable to DNA. Following this treatment, the bacteria can be infected with phage DNA that has been isolated from a suspension of phage.

The process of transfection is a valuable tool in genetic engineering, where the phage genome can be used as a vehicle to carry foreign genetic material into a competent host bacterium. Transfection is also useful in identifying DNA isolated from an infected bacterium as phage DNA. The isolated DNA is separated into different fractions and transfected to other bacteria. The fraction containing the phage DNA causes phage production in the transfected bacteria.

DNA Is Transferred by a Bacteriophage from One Bacterium to Another During Transduction

A bacteriophage serves as a vehicle for the transfer of bacterial DNA from one bacterium to another in transduction. Transduction was first described by Norton Zinder and Joshua Lederberg in 1952 for *Salmonella* and the bacteriophage P22. This phage-mediated method of genetic transfer has since been found to occur in many other bacteria and to involve many different types of DNA-containing bacteriophages.

Transduction of host genes by viruses happens in two ways. Random fragments of DNA are transferred in **generalized transduction,** resulting in a low transduction frequency of a specific genetic marker. **Specialized,** or **restricted, transduction** occurs only with **temperate bacteriophages** (those able to incorporate their DNA into the host chromosome without cell lysis). Specific portions of the host genome are excised with the prophage during induction. The phage injects this attached host DNA into a recipient cell, where it recombines with the recipient genome. No successful infection takes place because the virus is incomplete. The result is highly efficient transduc-

tion of specific donor genes. Not all bacteriophages transduce, but the variety and number that do is significant enough to make transduction an important phenomenon for microbial genetic exchange.

Random Portions of the Bacterial Chromosome Are Transferred During Generalized Transduction

Generalized transduction can occur in any population of bacterial cells susceptible to infection by bacteriophages (Figure 9.9). Following infection of a host, bacteriophages often initiate a **lytic cycle** in which the phage nucleic acid is replicated and packaged into phage particles (see bacteriophage replication, page 398). Under normal conditions, these phage particles are released upon lysis of the host cell. Occasionally, however, the host chromosome is fragmented into small portions (fragments that contain only 1% to 2% of the total bacterial DNA) during the lytic cycle, and these DNA fragments are mistakenly packaged into some of the progeny phage particles and released upon cell lysis. Other progeny phages contain only phage DNA. The lysate produced by generalized transduction thus consists of a mixture of both phages that contain random fragments of the bacterial chromosome and those that contain only phage nucleic acid.

Transducing phage particles containing host DNA have very little additional room within their available space for phage DNA and are considered **defective.** Defective phages, although able to infect susceptible bacteria, do not have a full complement of phage genes. Such phages are unable to carry through a complete lytic cycle in the infected bacterium. However, they are able to infect the recipient cell and insert the donor bacterial DNA. This donor DNA subsequently is incorporated into the recipient chromosome in much the same manner as in transformation.

Bacterial lysates of transducing phage also have large quantities of normal, nontransducing phage. Since each transducing phage particle in the lysate contains only a small portion of the entire donor chromosome, the probability that any given donor gene is transferred to a recipient bacterium is quite low—usually only one out of every 10^5 to 10^8 recipient cells infected with such a lysate is transduced for a specific gene.

Microbial Metabolism and Growth
Genetics: Recombination • pp. 19–20

Specific Genes Are Transferred During Specialized Transduction

The probability that a given gene is transferred to a recipient cell is considerably increased in specialized transduction, which occurs only with temperate phages. In specialized transduction the phage DNA integrates at a specific attachment site in the bacterial chromosome (Figure 9.10). The phage DNA in the chromosome is called a **prophage,** and a bacterium carrying such a prophage is

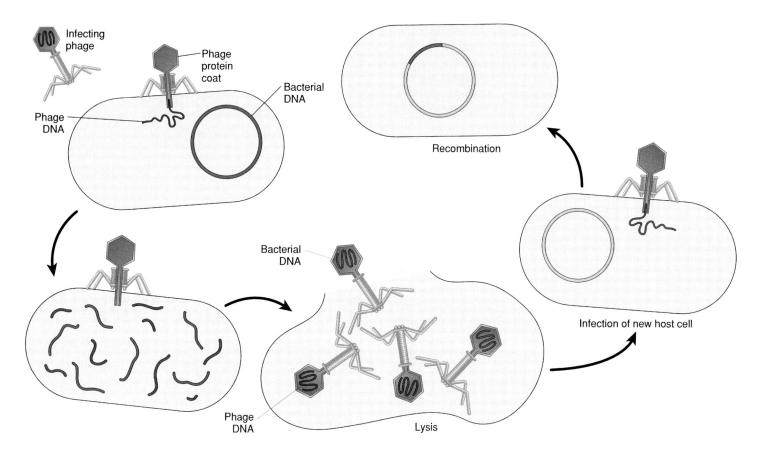

figure 9.9

Generalized Transduction

In generalized transduction, a bacteriophage infects a bacterium. A lytic cycle begins, but portions of the bacterial chromosome are incorporated inside phage particles. Upon lysis of the host cell, phages containing these host DNA fragments can infect other bacteria, resulting in the insertion of the DNA into the chromosome of the newly infected cell.

Microbial Metabolism and Growth
Genetics: Recombination • p. 19

said to be **lysogenic** (see lysogeny, page 410). A prophage—for example lambda (λ)—always inserts at the same location in the DNA of a specific host bacterium. Different prophages may be integrated at specific sites in the host DNA, and a host cell may be lysogenized for more than one prophage at the same time. Upon induction with ultraviolet radiation or other inducing agents, a portion of the bacterial genome is removed with some of the phage DNA. This material is replicated and packaged into maturing phage particles, which are released from the lysed cell. Since the size of the DNA molecule incorporated into phage particles remains relatively constant, an equal amount of phage DNA is lost for the host DNA that is inserted into the phage. These progeny

bacteriophages lack a complete phage genome and are considered defective. Defective phage particles can still inject their genetic material into recipient cells, where genetic exchange occurs between donor and recipient DNA. The efficiency of gene transfer for specific genes is very high in specialized transduction, because bacteriophages carry only those bacterial genes adjacent to the prophage at the time of excision.

The most widely used and best understood example of specialized transduction is the bacteriophage-mediated transfer of the galactose (*gal*) genes, which control galactose fermentation in *E. coli*. Lambda (λ) phage particles infecting a culture of *E. coli* have their DNA inserted specifically into the host chromosome at a site

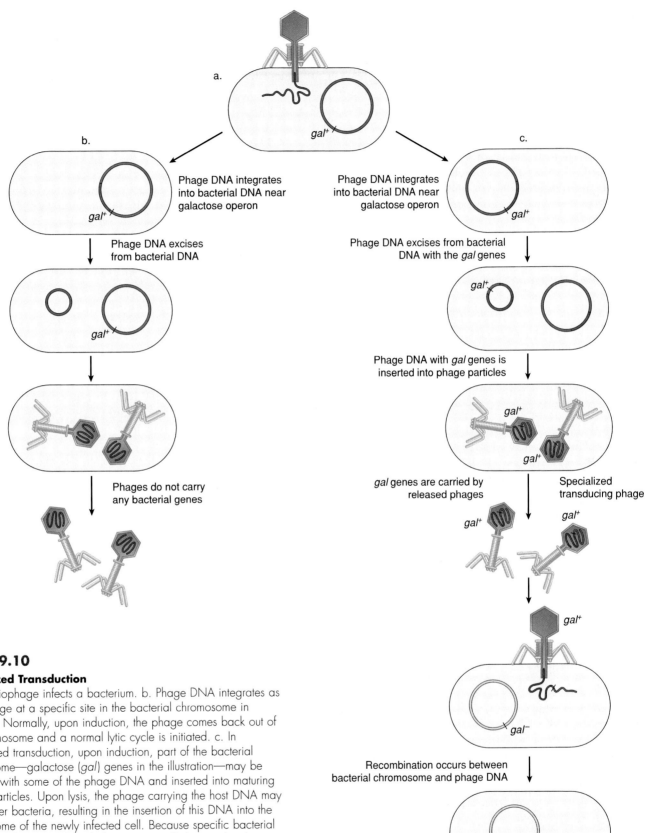

a.

b.

Phage DNA integrates
into bacterial DNA near
galactose operon

Phage DNA excises
from bacterial DNA

Phages do not carry
any bacterial genes

c.

Phage DNA integrates
into bacterial DNA near
galactose operon

Phage DNA excises from bacterial
DNA with the *gal* genes

Phage DNA with *gal* genes is
inserted into phage particles

gal genes are carried by
released phages

Specialized
transducing phage

Recombination occurs between
bacterial chromosome and phage DNA

Bacterium has been transformed
from *gal* negative to *gal* positive

figure 9.10

Specialized Transduction

a. Bacteriophage infects a bacterium. b. Phage DNA integrates as
a prophage at a specific site in the bacterial chromosome in
lysogeny. Normally, upon induction, the phage comes back out of
the chromosome and a normal lytic cycle is initiated. c. In
specialized transduction, upon induction, part of the bacterial
chromosome—galactose (*gal*) genes in the illustration—may be
removed with some of the phage DNA and inserted into maturing
phage particles. Upon lysis, the phage carrying the host DNA may
infect other bacteria, resulting in the insertion of this DNA into the
chromosome of the newly infected cell. Because specific bacterial
genes have been transferred, this event is called specialized
transduction.

between the loci for *gal* and *bio* (the *bio* locus contains genes that control the synthesis of the vitamin biotin).

One of two events occurs when the bacterial culture is induced. The prophage may come out of the bacterial chromosome with its entire complement of phage genes and initiate a normal lytic cycle. Alternatively, only a portion of the prophage may detach from the genome, carrying with it either the *gal* genes or the *bio* genes, and enter a lytic cycle. This latter possibility occurs in specialized transduction and results in progeny phage containing only a portion of their normal phage genetic complement and part of the host DNA. Since a given prophage is always inserted at the same sites in a specific bacterium, only those host genes on either side of the prophage insertion site can be transduced in specialized transduction.

As a consequence of this specificity, transducing λ phage particles are either *λdg* (defective and carrying the *gal* genes) or *λdbio* (defective and carrying the *bio* genes). These progeny phages, although defective and unable to carry out a complete lytic cycle, can still infect *E. coli* cells and insert the donor genes into the recipient chromosome. Such defective λ phage particles can replicate if normal nondefective λ phages, termed **helpers,** also infect the bacterial cell and supply the phage genes missing in the defective particle. Although specific genes are transferred in specialized transduction, the frequency of transduction in a population of bacteria is low. Because not every bacterium in a culture produces transducing phage, a typical lysate contains only about 1 *λdg* or *λdbio* phage particle per every 10^5 normal nondefective particles.

The frequency of specialized transduction can be significantly increased through a phenomenon known as **high-frequency transduction.** A lysate containing, for example, *λdg* and normal phages is used to infect *gal⁻ E. coli* at high multiplicity. A high multiplicity of infection is achieved by infecting a culture with defective transducing phage particles and a large number of normal helper phage particles. Under such conditions, bacteria infected by a *λdg* phage are also infected with a normal λ phage and are thus considered to be double lysogens. If such a culture is now induced, each infected bacterium in the population will produce both normal and *λdg* phage particles. The lysate contains a one-to-one ratio of these bacteriophages. When this lysate is used to infect a culture of *gal⁻ E. coli* cells, roughly 50% of the infected cells will be transduced by the *λdg* phages—a high frequency of transduction for the *gal* genes.

Donor DNA Is Not Incorporated into the Recipient Chromosome During Abortive Transduction

In certain instances of transduction, the transducing phage DNA is not incorporated into the host chromosome. The exogenote neither replicates nor is destroyed; it remains separate from the endogenote and is transmitted unilinearly to only one of two daughter cells formed at cell division (Figure 9.11). Such **abortive transduction** is very common and is easily recognized by expression of exogenote genes in bacteria carrying this genetic material.

For example, if a bacterium lacking the gene for proline synthesis (*pro⁻*) is infected with a transducing phage carrying the proline gene (*pro⁺*) and an abortive transduction occurs, the *pro⁺* gene is initially expressed in the infected bacterium. This bacterium synthesizes proline and therefore is able to grow on a plate medium containing no proline. As this originally infected bacterium grows and divides, the exogenote is passed to only one of the two daughter cells. The daughter cell receiving the *pro⁺* gene continues to synthesize proline and grows like the parent. The daughter cell without the exogenote, however, grows only until the endogenous pool of proline carried over from the parent is depleted. This pattern of unilinear transmission of the exogenotes continues in subsequent cell divisions. The result is the formation of slow-growing microcolonies of daughter cells lacking exogenotes that are detectable with magnification. Abortive transduction is quite common among transducing bacteriophages and, in some phage suspensions, accounts for 90% of all transductions.

Properties of a Bacterium Can Be Influenced by the Expression of Bacteriophage Genes in the Bacterial Chromosome

The properties of some bacteria depend upon the expression of phage genes in the bacterial chromosome, a phenomenon known as **lysogenic conversion** (see lysogenic conversion, page 411). Lysogenic conversion occurs when a normal (nondefective) temperate bacteriophage lysogenizes a cell and one or more of the prophage genes are expressed. For example, *Corynebacterium diphtheriae* produces toxin only when it is lysogenized with the phage β. The phage clearly contains the structural gene for toxin production, since lysogenization with a mutated β phage lacking this gene results in a non-toxin-producing strain of *C. diphtheriae*. A second example of phage conversion is the change in the lipopolysaccharide structure of *Salmonella anatum* when this bacterium is lysogenized by the phage ε[15]. In both of these examples, phage genes are expressed in the lysogenic state.

Microbial Metabolism and Growth
Genetics: Recombination • pp. 19–20

Plasmids Are Extrachromosomal Circular Pieces of DNA

Bacteria have a genetic exchange process called conjugation that involves cell-to-cell contact. It is during this association that DNA is transferred from either the bacterial chromosome or extrachromosomal genetic material (**plasmid**). The DNA entering the recipient cell from the donor cell may recombine with the recipient chromosome, altering the genotype of that bacterium. Whether or not a bacterium conjugates depends upon genes that are found on a plasmid.

Microbial Metabolism and Growth
Genetics: Recombination • pp. 21–24

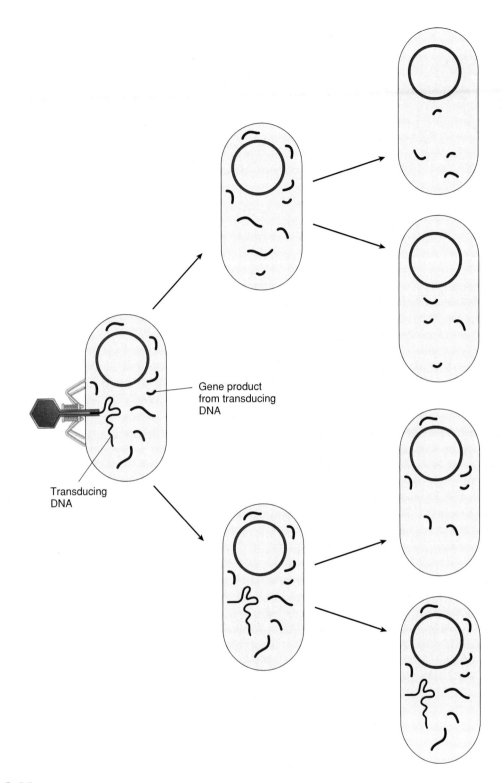

Gene product from transducing DNA

Transducing DNA

figure 9.11

Abortive Transduction

In abortive transduction, the transducing DNA is not integrated into the bacterial chromosome. The exogenote neither replicates nor is destroyed; it is transmitted unilinearly to only one of the two daughter cells formed at cell division. Gene products from the transducing DNA are transmitted to both daughter cells. However, in those cells that do not receive the exogenote, the gene products are progressively diluted as cell multiplication proceeds.

Transposable Genetic Elements

Transfer of genetic material between two organisms requires not only similarity between the organisms involved, but also a homology between the donor and host DNA. Fragments of donor DNA injected into a bacterium are incorporated into the recipient chromosome only at those sites having complementary base sequences. The low frequency of natural DNA transfer results from this specificity of genetic exchange.

Mutation and transfer of DNA were long considered to be the only mechanisms for modification and evolution of genetic sequences in living organisms. In the 1940s, however, Barbara McClintock discovered a new mechanism for chromosomal modification. McClintock, in her studies on pigmentation in the corn plant *Zea mays*, noted a variegation that was caused not by mutation, but by the movement of discrete genetic elements among chromosomes. These genetic elements appeared to control the expression of genes adjacent to their insertion sites on the chromosomes. It was not until two decades later, in the 1960s, that similar controlling genetic elements were discovered and characterized in *Escherichia coli*. Additional DNA sequences were found inserted into portions of the bacterial chromosome. Like the genetic elements first observed by McClintock, these **transposable elements,** or **transposons,** move blocks of genetic material from one location to another on the bacterial chromosome or to plasmids or bacteriophages.

Transposons are now recognized to be common constituents of bacterial chromosomes, plasmids, and bacteriophages. They contain genes coding for antibiotic resistance, carbohydrate utilization, toxin production, amino acid synthesis, and hydrocarbon degradation. Electron micrographs have revealed that the ends of transposons contain inverted repeating base sequences of 9 to 40 bases. It is believed that these inverted repeating ends recognize regions in the DNA for insertion—possibly those DNA regions that have base sequences complementary to the repeat sequence.

One type of mobile genetic element is the **insertion sequence (IS).** An insertion sequence is a short piece of DNA less than 2,000 base pairs long that can be inserted at a specific site in a chromosome or plasmid. Unlike transposons, which are longer and contain genes that are unrelated to insertion as well as insertion genes, ISs contain only insertion genes. ISs can be instrumental in regulating the expression of certain phenotypes. For example, the gram-negative flagellated bacterium *Salmonella* can synthesize two distinct types of flagellin (Figure 9.12). The genes coding for the flagellins (*H1* and *H2*) are located in operons residing at different chromosomal locations. The *H2* gene is closely linked to another gene (*RH1*) that codes for a repressor of the *H1* operon. Thus when the *H2* operon is turned on, the repressor is synthesized and represses the *H1* operon. The promoter for the *H2* operon is located within an IS. This IS contains a gene, *hin*, with a product that is a protein mediating the inversion of the entire IS and its site-specific recombination with the DNA. When the IS is in the normal orientation, the *H2* gene is transcribed and the *H1* operon is repressed. However, if the IS is inverted with the promoter facing in the opposite direction of the *H2* operon, the *H2* gene (including the *RH1* gene for the repressor) is not transcribed. The absence of repressor results in transcription of the *H1* gene. Thus the orientation of the IS determines the type of flagellin synthesized by *Salmonella*.

The discovery of transposons indicates that cells do not have to rely on only mutation and genetic exchange for chromosome modification and the regulation of gene expression. Transposons represent nature's way of genetically engineering DNA sequences in evolution.

Plasmids are small, circular pieces of DNA that exist independently of the bacterial chromosome (Figure 9.13). These extrachromosomal elements replicate autonomously in the cytoplasm and have been found to carry different types of genetic markers, including those for antibiotic resistance and production of toxins, virulence factors, and metabolic enzymes (Table 9.2). The amount of plasmid DNA may constitute as much as 1% to 2% of the total cellular DNA in a bacterium. A bacterium may contain several plasmids.

Plasmids Are Both Conjugative and Nonconjugative

Plasmids, like bacterial chromosomes, are circular, double-stranded DNA molecules. Unlike larger chromosomes, plasmids carry only a few genes and these genes are not essential for cell growth. These extrachromosomal elements are divided into two major classes: **conjugative plasmids,** which are associated with conjugation, and **nonconjugative plasmids.** Conjugative plasmids carry a sequence of genes called the *tra* (for *transfer*) **genes**

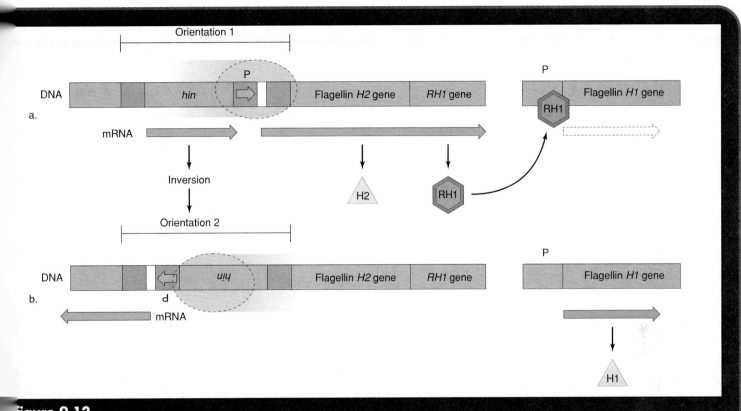

Figure 9.12

Regulation of Flagellin Synthesis in *Salmonella* by an Insertion Sequence (IS)

a. The IS, containing the *hin* gene, is oriented so that the flagellin *H2* gene and the *RH1* gene are transcribed. The *RH1* gene codes for a repressor that represses the *H1* operon. b. The IS is inverted, with the promoter (P) facing in the opposite direction of the *H2* operon. Under such conditions, the flagellin *H2* gene and the *RH1* gene are not transcribed, resulting in transcription of the flagellin *H1* gene.

that are associated with conjugative transfer. Genes in the *tra* region are associated with cell attachment and plasmid transfer between bacteria during conjugation. An example of a conjugative plasmid is the **F** (for *fertility*) **plasmid** of *E. coli*, which has been extensively studied. The F plasmid is 94.5 kilobase pairs long. Because the *tra* region, which consists of about 20 genes, occupies one-third of the plasmid genome, conjugative plasmids are much larger than nonconjugative plasmids. Nonconjugative plasmids (for example, many Col plasmids) do not have *tra*

genes, but can have *mob* (**mobilization**) **genes** that allow them to take advantage of the *tra* function of a conjugative plasmid for transfer between bacteria.

Resistance Plasmids Carry Genes for Antibiotic Resistance

Bacteria can become resistant to antibiotics by mutating existing chromosomal genes or by transferring antibiotic-resistant genes. Although transformation and transduction of antibiotic-resistant genes can occur between bacteria of the same species, transfer of resistance genes across genus and species lines is more likely to occur by conjugation. Plasmids that carry genes for antibiotic resistance are found in strains of many pathogenic bacteria. These **resistance plasmids (R plasmids)** spread by conjugative transfer through bacterial populations and confer resistance to a number of antibiotics. Classical R plasmids have two functionally distinct parts: (1) *tra* **genes,** which are associated with conjugative transfer, and (2) **resistance genes (R determinant),** which code for antibiotic resistance and may consist of as many as seven or eight genes for resistance to multiple antibiotics. R plasmids generally acquire resistance genes by transposons, which can move between plasmids. Multiple resistance plasmids were first discovered in Japan in the 1950s during a period when there was a rapid increase in multiple drug resistance in clinical isolates of *Shigella,* which causes dysentery. When cultures of *Shigella* were exposed to chloramphenicol, streptomycin, tetracycline, or sulfonamide, the bacteria exhibited resistance to more than one of the antimicrobial agents. Furthermore, this multiple resistance could be transmitted to other nonresistant strains of *Shigella* as well as to other intestinal bacteria such as *E. coli.* Since then, multiple resistance plasmids have been found in other clinically significant bacteria, including *Staphylococcus aureus, Pseudomonas aeruginosa, Klebsiella pneumoniae,* and *Enterococcus faecalis.* The emergence of these multiple-drug-resistant bacteria has presented major problems in chemotherapy as fewer and fewer antimicrobial agents are effective in treatment of microbial diseases (see antibiotic resistance, page 130).

One hypothesis for the origin of antibiotic resistance genes is that these genes may have evolved in antibiotic-producing bacteria such as *Streptomyces* as a mechanism by which the bacterium could protect itself from its own antimicrobial products. This theory is supported by the observation that genes coding for antibiotic resistance and antibiotic production often occur in the same gene cluster and that antibiotic resistance genes are found in low frequency in bacterial strains stored before the widespread use of antibiotics in chemotherapy.

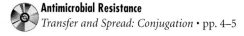
Antimicrobial Resistance
Transfer and Spread: Conjugation • pp. 4–5

Plasmids Can Carry Genes That Code for Bacteriocins, Toxins, Resistance to Heavy Metals, and Metabolic Processes

Bacteriocinogenic plasmids carry genes for the synthesis of **bacteriocins,** proteins that inhibit or kill other bacteria. Bacteriocins are distinguished from antibiotics, which have a broader antimicrobial spectrum and are less potent. One molecule of a bacteriocin is often sufficient to kill a bacterium. **Colicins,** bacteriocins produced by *E. coli* and carried by **Col plasmids,** have been the most extensively studied, and the mode of action of some colicins is known in detail. Colicin E2 causes single-stranded nicks in DNA and interferes with DNA replication. Colicin E3 inactivates 30S ribosomes by cleaving the 16S rRNA, thereby stopping protein synthesis.

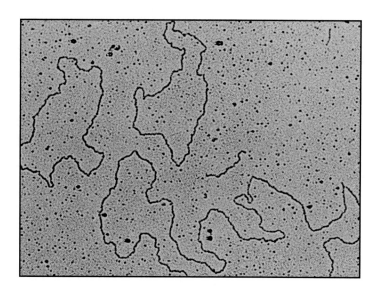

figure 9.13
Electron Micrograph Showing Two Different Plasmids
The large molecule is plasmid pBF4 (containing genes for resistance to clindamycin and erythromycin) from *Bacteriodes fragilis;* the small molecules are plasmid pSC101 (containing the gene for resistance to tetracycline) from *Escherichia coli* (colorized).

Other bacterial plasmids contain genes that code for toxins, resistance to heavy metals, and special types of metabolism. Plasmid-coded enterotoxins in enteropathogenic *E. coli* enable the bacteria to cause disease. Some bacteria, such as *Staphylococcus aureus* and species of *Pseudomonas,* have plasmids that confer resistance to certain heavy metals such as cadmium, lead, and mercury. Many species of *Pseudomonas* are able to utilize an extraordinarily large number of organic compounds as energy sources. In many instances, genes for utilization of these organic compounds are carried on plasmids.

Plasmids Can Be Integrated into the Chromosome

Most kinds of plasmids entering a cell remain separate from the chromosome, but some can be integrated into the chromosome. The term **episome** is used to describe a plasmid that is capable of existing either extrachromosomally or as part of the chromosome. Since plasmids carry genetic markers, they influence the genotype of an organism.

Although a bacterial cell may contain more than one type of plasmid, plasmids tend to prevent any other plasmids of the same or closely related type from establishing themselves in the same cell. This **plasmid incompatibility** is determined by plasmid genes that regulate their DNA replication. Plasmids can be separated, therefore, into incompatibility groups. Plasmids of the same incompatibility group are closely related and interfere with each other.

Plasmids, like prophages in lysogenic bacteria (see curing of lysogenic bacteria, page 411), can be removed from bacterial cells

table 9.2

Examples of Major Types of Plasmids

Type	Representatives	Hosts	Copy Number (Copies/ Chromosome)	Approximate Size (kb)	Phenotypic Features[a]
Fertility Factor[b]	F factor	E. coli, Salmonella, Citrobacter	1–3	95–100	Sex pilus, conjugation
Col Plasmids	ColE1	E. coli	10–30	9	Colicin E1 production
	ColE2	Shigella	10–15		Colicin E2
	CloDF13	Enterobacter cloacae			Cloacin DF13
R Plasmids	RP4	Pseudomonas and many other gram-negative bacteria	1–3	54	Sex pilus, conjugation, resistance to Ap, Km, Nm, Tc
	R1	Gram-negative bacteria	1–3	80	Resistance to Ap, Km, Su, Cm, Sm
	R6	E. coli, Proteus mirabilis	1–3	98	Su, Sm, Cm, Tc, Km, Nm
	R100	E. coli, Shigella, Salmonella, Proteus	1–3	90	Cm, Sm, Su, Tc, Hg
	pSH6	Staphylococcus aureus		21	Gm, Tm, Km
	pSJ23a	S. aureus		36	Pn, Asa, Hg, Gm, Km, Nm, Em
	pAD2	Enterococcus faecalis		25	Em, Km, Sm
Metabolic Plasmids	CAM	Pseudomonas		230	Camphor degradation
	SAL	Pseudomonas		56	Salicylate degradation
	TOL	Pseudomonas putida		75	Toluene degradation
	pJP4	Pseudomonas			2,4-dichlorophenoxyacetic acid degradation
		E. coli, Klebsiella, Salmonella			Lactose degradation
		Providencia			Urease
	sym	Rhizobium			Nitrogen fixation and symbiosis
Virulence Plasmids	Ent (P307)	E. coli		83	Enterotoxin production
	K88 plasmid	E. coli			Adherence antigens
	ColV-K30	E. coli		2	Siderophore for iron uptake; resistance to immune mechanisms
	pZA10	S. aureus		56	Enterotoxin B

[a]Abbreviations used for resistance to antibiotics and metals: Ap, ampicillin; Asa, arsenate; Cm, chloramphenicol; Em, erythromycin; Gm, gentamycin; Hg, mercury; Km, kanamycin; Nm, neomycin; Pn, penicillin; Sm, streptomycin; Su, sulfonamides; Tc, tetracycline.
[b]Many R plasmids, metabolic plasmids, and others are also conjugative.
Source: Data from Lansing M. Prescott et al., *Microbiology,* 3rd edition, Times Mirror Higher Education Group, 1996.

by **curing.** Plasmid curing either occurs spontaneously or is induced by treatment of the host cell with chemical agents such as ethidium bromide or acridine, or with ultraviolet irradiation. The curing process apparently inhibits the replication of plasmids without affecting chromosomal replication.

Plasmids share many features that are similar to viruses and, in fact, are believed by some scientists to be the ancestors of viruses. It is possible that plasmids originated from pieces of chromosome that were excised and were not necessary for cell growth and replication. With the passage of time, portions of this extrachromosomal material evolved viral coats and became modern-day viruses.

Cell-to-Cell Contact Is Required for DNA Transfer During Conjugation

Conjugation occurs between two closely related but different bacterial cells—a donor and a recipient. In *E. coli,* the donor cell contains a conjugative F plasmid. The recipient cell does not contain this plasmid. As a result of this difference, the donor cell is often called an **F⁺ cell,** and the recipient cell is termed an F⁻ **cell.** Like other plasmids, the F plasmid can be lost from the cell through curing. F⁺ cells incubated in the stationary phase of growth for prolonged periods are cured of the F plasmid and changed into F⁻ cells.

F⁺ cells contain not only the F plasmid, but also have a spe-cial, elongated pilus known as the **sex pilus,** the production of which is determined by genes in the *tra* region of the F plasmid. If F⁺ cells are mixed in culture with F⁻ cells, the two cell types come in contact with each other through the aid of the sex pilus. One strand of the F plasmid is nicked at a site called *oriT* (origin of transfer), and the 5′ end of the cut strand is transferred into the recipient cell (Figure 9.14). During transfer, the plasmid DNA is replicated via a rolling circle mechanism of replication. Inside the recipient cell, the single strand of donor plasmid is replicated and recircularized (a complete copy of the plasmid remains in the donor). Because the recipient cell now contains an F plasmid, it becomes an F⁺ cell like the donor. Genetic markers other than the sex factor located on the F plasmid are also transferred to the recipient cell. However, the chromosomal genes of the donor bac-terium are not transferred in mating between F⁺ and F⁻ cells.

Donor Genes Can Be Transferred at High Frequencies During High-Frequency Recombination

Under certain conditions the F plasmid is integrated into the donor chromosome. Although this event rarely happens—approximately once per every 10⁵ cells—when it does occur, the result is a **high-frequency recombination (Hfr)** cell. The F plas-mid in Hfr cells is integrated at one of several specific sites and orientations in the chromosome. When an Hfr cell donates genes to an F⁻ cell, the Hfr cell chromosome opens internally at the *oriT* where the F plasmid is located. Part of the F plasmid moves into the recipient cell, followed by the transfer of chromosomal genes in a sequential manner. Only if and when all of the chromosomal genes are transferred to the recipient cell is the remainder of the F plasmid transferred. Thus during conjugation between an Hfr cell and an F⁻ cell, the F⁻ cell rarely receives the entire F plasmid. Although the recipient cell remains F⁻, it does receive donor genes at high frequencies—this is the reason why such genetic exchange is termed high-frequency recombination. The distal end of the F plasmid is transferred to the recipient cell only if mating remains uninterrupted long enough for the entire donor chromosome, including the F plasmid, to be transferred. In the case of *E. coli*, this time is between 90 to 100 minutes, an extremely long period

figure 9.14

Transfer of DNA During Conjugation

In conjugation, donor and recipient cells come in contact. One strand of the F plasmid in the donor cell is cut at the *oriT* site and enters the recipient cell. During transfer, the plasmid DNA is replicated via the rolling circle model. Inside the recipient cell, the donor plasmid is recircularized. The recipient cell, which now contains an F plasmid, becomes a donor (F⁺ or F′) cell.

Microbial Metabolism and Growth
Genetics: Recombination • p. 11

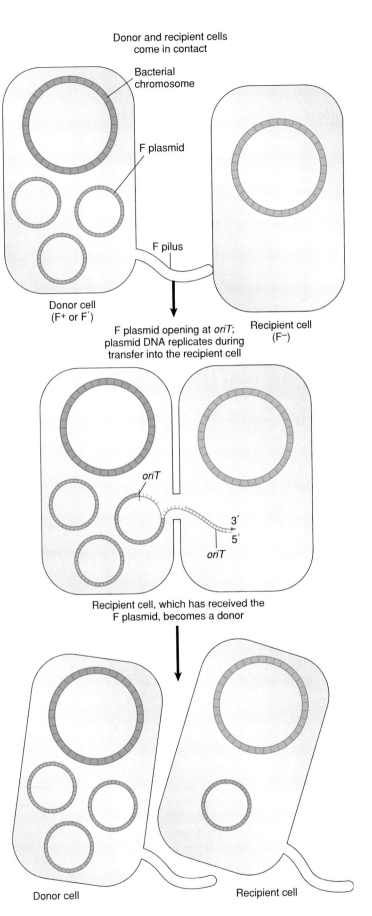

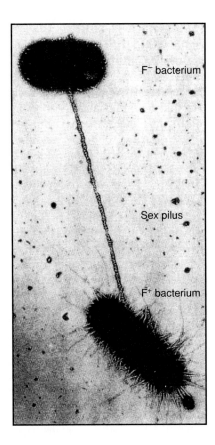

figure 9.15

Cell-to-Cell Contact During Bacterial Conjugation

Two bacteria (a donor F⁺ bacterium and a recipient F⁻ bacterium) are shown connected by the sex pilus.

Antimicrobial Resistance
Transfer and Spread of Resistance: Conjugation • pp. 4–5

of time for bacterial mating. Under precise laboratory conditions in which mating cells are undisturbed, such long mating periods are possible.

F' Cells Have F' Plasmids Carrying Chromosome Genes

F plasmids that have integrated into the donor chromosome in Hfr cells occasionally are excised from the chromosome. When such an event occurs, the excised F plasmid may carry with it a part of the bacterial chromosome. Such a plasmid is called an **F' plasmid.** If the genes now on the F' plasmid are indispensable to the bacterium, loss of the plasmid through curing will result in cell death.

Conjugation between F' cells and F⁻ cells results in the transfer of not only the F plasmid, but also the donor cell genes that are now part of the F' plasmid. Donor genes are transferred at high frequencies in matings between F' and F⁻ cells. Under these conditions, the recipient cell is changed to an F' cell and receives the donor cell genes that are part of the F' plasmid. Since the bacterial genes in the F' plasmid may also be present in the recipient cell chromosome, the recipient, in effect, can become a partial diploid as a result of such transfer.

Donor Cells Have a Sex Pilus Associated with Conjugation

The sex pilus in donor cells is a hollow structure consisting of subunits of a phosphoprotein, **pilin** (see pilin structure, page 62). Although the sex pilus is large enough to allow transfer of DNA between the donor and recipient cells, there is no evidence that this occurs. The sex pilus does appear to form a bridge between the two cells and may be involved in the recognition of the recipient cell during conjugation (Figure 9.15).

An alternative hypothesis is that DNA transfer occurs between cells as a result of wall-to-wall contact. This hypothesis is supported by the observation that the sex pilus immediately retracts upon cell-to-cell contact. This retraction may in fact promote formation of wall-to-wall contact during conjugation.

Microbial Metabolism and Growth
Genetics: Recombination • pp. 11–14

The Sequential Transfer of Chromosomal Genes During Conjugation Can Be Used to Map Genes

Conjugation between Hfr cells and F⁻ cells has been used to map bacterial chromosomes. In such mapping, Hfr cells are mixed with F⁻ cells. Mating is allowed to proceed for a given period of time and then is deliberately interrupted by disrupting the cell mixture, usually by vigorous agitation in a mixer or blender. The recipient cell is phenotypically examined for the markers it has received from the Hfr donor. Since in any specific Hfr strain, the F factor is always inserted at the same site in the chromosome, and the order of gene transfer is the same from Hfr cell to F⁻ cell, the order of genes on the chromosome can be determined by allowing mating to proceed for set periods of time before interruption. It has been discovered that the *E. coli* chromosome takes 90 to 100 minutes to completely transfer over to the recipient cell during mating. The *E. coli* chromosome is thus considered to have a 90- to 100-minute genetic map. The genes on that map are now well-known and precisely located through interrupted mating experiments and other techniques. These interrupted mating experiments also provided the first evidence for the circular nature of the bacterial chromosome.

Lysogenic Conversion of *Corynebacterium diphtheriae*

Lysogenic conversion is a characteristic known to be responsible for the toxigenic properties of some medically important bacteria, including *Corynebacterium diphtheriae* (diphtheria toxin), *Streptococcus pyogenes* (erythrogenic toxin), and some strains of *Clostridium botulinum* (botulinum toxin). The ability of a temperate bacteriophage to carry genes that determine bacterial properties was discovered by Victor Freeman in 1951.

Freeman examined lysogenic conversion in *Corynebacterium diphtheriae* by combining diphtheria bacteria known to be avirulent by guinea pig inoculation with bacteriophages from filtrates of lysed virulent cultures of *C. diphtheriae*. Five avirulent strains (numbers 411, 444, 770, 1174, and 1180) of *C. diphtheriae* and two bacteriophages (A and B) were used in Freeman's studies. Bacteria and phage were mixed in two different ways. In the first method, heart infusion agar was inoculated with 0.5 ml of an 18-hour broth suspension of bacteria, dried for half an hour, and then inoculated with 0.5 ml of filtered bacteriophage lysate. In the second method, 3 ml of heart infusion broth was inoculated with 0.15 ml of an 18-hour culture of bacteria, incubated for one hour until growth was visible, and then inoculated with 0.15 ml of filtered bacteriophage lysate. Table 9P.1 shows the susceptibility of the five *C. diphtheriae* strains to bacteriophages A and B. Only *C. diphtheriae* strain 411 was resistant to phage infection. Phage A failed to show any visible lysis of diphtheria bacteria growing in broth.

Freeman then injected suspensions of each culture's lysates into guinea pigs (0.1 ml intradermally or 1.0 ml subcutaneously) for in vivo toxigenicity tests. Suspensions of agar plate lysates were prepared by washing the surface growths of the plates with 0.85% NaCl and using the washes as material for injection. Broth lysates were injected directly into guinea pigs. A naturally virulent strain of *C. diphtheriae* was used as a positive control for virulence tests. In addition, diphtheria antitoxin was injected with lysates in a separate series of experiments to confirm the specificity of toxicity. The results of the in vivo toxigenicity tests are summarized in Table 9P.2. Phage B lysates contained a dermal necrotic factor that phage A lysates lacked. The virulence factor in the phage B lysates appeared to be diphtheria toxin, since diphtheria antitoxin prevented necrosis from occurring. Furthermore, the absence of toxicity by strain 411 provided two important controls. First, it indicated that extracellular phage was not toxic. Second, this result ruled out any possibility that carryover toxin from the original phage suspensions (those used to infect diphtheria bacteria) was responsible for the necrotic reactions. Similar results were obtained from subcutaneous injections of culture lysates into guinea pigs, except that the animals died (Table 9P.3).

These studies by Freeman were important because they were the first evidence that bacteriophage-infected strains of *C. diphtheriae* were toxigenic. We now know that a prophage can carry toxin genes not only for *C. diphtheriae*, but also for other bacteria.

Source

Freeman, V.J. 1951. Studies on the virulence of bacteriophage-infected strains of *Corynebacterium diphtheriae*. *Journal of Bacteriology* 61:675–688.

table 9P.1

Bacteriophage Susceptibility of Avirulent Strains of *Corynebacterium diphtheriae*

Strain Number	Degree of Clearing on Agar Medium		Degree of Clearing in Broth	
	Phage A[a]	*Phage B*[a]	*Phage A*	*Phage B*
444	2+[b]	4+	0	1–2+
1174	2+	4+	0	1–2+
1180	2+	4+	0	1–2+
770	2+	4+	0	1–2+
411	0	0	0	0

[a]Material spotted on plate previously inoculated with the culture indicated in the first column.
[b]Degree of clearing: 0 = none; 1+ = slight; 2+ = moderate; 3+ = marked; 4+ = complete.

Intradermal Tests of Bacteriophage Lysates in Guinea Pigs

	Avirulent Culture Grown on Agar						Avirulent Culture Grown in Broth					
Strain Number	Control[a]		Phage A Filtrate[a]		Phage B Filtrate[a]		Control		Phage B Added		Filtrate of Incubated Culture Phage B Mixture	
	T	C	T	C	T	C	T	C	T	C	T	C
444	0	0	0	0	4+[b]	0	0	0	4+	0	2+	0
1174	0	0	0	0	4+	0	0	0	4+	0	2+	0
1180	0	0	0	0	4+	0	0	0	4+	0	2+	0
770	0	0	0	0	4+	0	0	0	4+	0	2+	0
411	0	0	0	0	0	0	0	0	0	0	—	—

Note: The intradermal method used involves a second inoculation of the test substances 4 hours after the initial injections and immediately following an intraperitoneal injection of 1,000 units of diphtheria antitoxin. The test results are recorded under "T" and the control antitoxin results, under "C."
[a]Material used for the second inoculation of the plates; saline was used in the control cultures.
[b]The symbol 4+ indicates the presence of both erythema and necrosis involving an area greater than 1 cm^2, 2+ indicates a similar reaction, except for the absence of necrosis.

Subcutaneous Tests of Bacteriophage Lysates in Guinea Pigs

Strain Number	Culture plus Saline	Culture plus Phage A	Culture plus Phage B	Culture plus Phage B and Antitoxin
444	0/3[a]	0/1	4/4	0/2
1174	0/1	0/1	2/2	0/1
1180	0/1	0/1	2/2	0/1
770	0/1	0/1	2/2	0/1
411	0/1	0/1	0/1	0/1
Total	0/7	0/5	10/11	0/6

Note: All cultures and culture lysates were washed off the agar media with 0.85% saline and inoculated in 1.0-ml doses.
[a]The numerator represents the number of guinea pigs that died; the denominator, the total number tested.

1. A mutation is an inheritable change in the sequence of DNA. The rate of mutation can be increased by mutagenic agents.

2. Bacteria have several different mechanisms for repair and replacement of damaged DNA. Excision repair excises and replaces damaged DNA. Photoreactivation uses the enzyme photolyase and light to cleave the covalent linkages within pyrimidine dimers. Glycosylases are enzymes that detect and remove unnatural bases. The SOS system is normally repressed by LexA, but when there is DNA damage, RecA stimulates LexA to cleave itself and become inactivated, resulting in expression of the SOS system genes for repair of the DNA damage.

3. A technique commonly used to identify mutants is replica plating, which involves the transfer of colonies from a master plate to one or more uninoculated plates of different test media. Autotrophic mutants can be identified by their absence of growth on replicated plates that lack the required growth factor for the mutant.

4. Bacteria can transfer genetic material between cells by three methods: transformation, transduction, and conjugation.

5. Transformation involves the entry of free, naked DNA from a donor cell into a competent recipient cell and recombination into the chromosome. Transfection, a type of transformation, involves the insertion of bacteriophage DNA into a cell.

6. Transduction involves the transfer of DNA from one cell to another cell via a bacteriophage. In generalized transduction, a fragment of the host cell genome may be packaged into a phage particle during a lytic cycle. In specialized transduction, a specific portion of a host cell genome is packaged into the phage particle following lysogeny.

7. Cell-to-cell contact is required for conjugation between a donor cell containing a conjugative F plasmid and a recipient cell lacking this plasmid. A high-frequency recombination (Hfr) cell is one in which the F plasmid is integrated into a specific region of the bacterial chromosome, resulting in transfer of the chromosomal genes in a sequential manner to the recipient cell during conjugation.

8. A plasmid is a small, circular piece of DNA that exists independently of the chromosome. Conjugative plasmids carry a sequence of genes called the *tra* genes associated with conjugation and plasmid transfer. Nonconjugative plasmids do not have *tra* genes. Plasmids can carry genes for antibiotic resistance, the synthesis of bacteriocins and toxins, resistance to heavy metals, and metabolic activity.

EVOLUTION *and* BIODIVERSITY

With the thousands of genes and millions of base pairs in an average bacterial chromosome, one would think that mutations would be common during DNA replication. Surprisingly, the rate of mutation during DNA replication is extremely low. Mistakes occur in only 1 out of every 10^9 bases copied. This accuracy of DNA replication is important because it enables the stable inheritance of genetic traits from generation to generation. In those rare instances when mutations do occur, they arise not only from errors in replication but also from the action of mutagenic agents such as ultraviolet light and chemicals in the environment. Not all mutations are harmful. Some are neutral or beneficial. Occasionally mutations confer a selective advantage to a microorganism and enhance its competitiveness in nature. For example, a mutation resulting in increased resistance to an antibiotic might enable the affected microbe to survive in environments where it might normally be inhibited or killed by the antibiotic. Over evolutionary time, the cumulative effects of mutations can play an important role in the gradual selection of certain phenotypic traits. In this sense, mutations can serve as a molecular clock for measuring the divergence of species.

Questions

Short Answer

1. Explain why procaryotes are dependent on mutation or genetic exchange for evolution.

2. How do spontaneous mutations occur? What factors will increase the rate of mutation?

3. Explain how it is possible that a mutation could have no effect on the cell.

4. Explain why it is likely that the deletion or insertion of a single nucleotide will have a major impact on a cell.

5. Can the substitution, addition, or deletion of a single nucleotide be lethal to a cell? Can a mutation be helpful?

6. Identify several mutagens and explain how they affect a cell.

7. Explain why many mutations are not permanent.

8. If auxotrophic mutants cannot grow without the necessary growth factor, how do you identify them?

9. Without extensive biochemical or genetic tests, how might you identify a bacterial mutant?

10. Identify three mechanisms of genetic exchange used by bacteria.

11. Why is general or homologous recombination necessary, after genetic exchange, for the permanent acquisition of new traits?

12. How does transfection differ from transduction?

13. How does generalized transduction differ from specialized transduction?

14. What are plasmids?

15. Why are plasmids helpful to a cell?

16. What role do plasmids play in conjugation?

Multiple Choice

1. Which of the following types of mutations will result in the premature termination of protein synthesis?
   a. missense mutations
   b. nonsense mutations
   c. silent mutations

2. Which of the following requires the assistance of a virus?
   a. conjugation
   b. transformation
   c. transduction
   d. None of the above.

3. Griffith's classic experiment (discussed in chapter 8) is an example of:
   a. conjugation
   b. transformation
   c. transduction
   d. All of the above.

4. Which of the following would be associated with the highest rate of conjugation and recombination?
   a. F plasmid
   b. R plasmid

   c. Col plasmids
   d. episomes

5. After conjugation with a(n) _____, the F⁻ cell will become F⁺.
   a. F⁻
   b. F⁺
   c. Hfr

Critical Thinking

1. Many mutagens are also carcinogens. Describe how a change in a cell's hereditary information may result in carcinogenesis.

2. You have been given a culture of *Serratia marcescens* grown at room temperature and notice the colonies are pink. After transferring the culture and incubating at 37°C, your culture has produced white colonies. What hypothesis can you form? Design an experiment to test your hypothesis.

3. Genetic exchange among procaryotes is a rare occurrence. Even so, which process would you expect to be most common? Why?

4. Describe several factors (both internal and external) which prevent genetic exchanges between procaryotes. How are the mechanisms which prevent genetic exchange advantageous to procaryotes?

5. After decades of abuse, the medical profession has called for restraint in the use of antimicrobial agents. Discuss why and present guidelines for the appropriate use of antimicrobial agents.

Dale, J.W. 1989. *Molecular genetics of bacteria*. New York: John Wiley & Sons. (A textbook of microbial genetics.)

Miller, J.H. 1992. A *short course in bacterial genetics: A laboratory manual and handbook for* Escherichia coli *and related bacteria*. Cold Spring Harbor, NY: Cold Spring Harbor Laboratory Press. (An in-depth laboratory manual describing techniques used in molecular biology and microbial genetics.)

Neidhardt, F.C., J.L. Ingraham, and M. Schaechter. 1990. *Physiology of the bacterial cell: A molecular approach*. Sunderland, Mass.: Sinauer Associates, Inc. (A textbook of microbial physiology and genetics.)

Russell, P.J. 1996. *Genetics*, 4th ed. New York: HarperCollins. (A fundamental textbook of genetics, with sections on DNA replication, transcription, translation, genetic exchange in bacteria, transposable elements, and recombinant DNA technology.)

Streips, U.N., and R.E. Yasbin, eds. 1991. *Modern microbial genetics*. New York: Wiley-Liss. (An advanced textbook on the genetics of *Escherichia coli* and other bacteria.)

Watson, J.D., N.H. Hopkins, J.W. Roberts, J.A. Seitz, and A.M. Weiner. 1987. *Molecular biology of the gene,* 4th ed. Menlo Park, Calif.: The Benjamin/Cummings Publishing Company. (An in-depth textbook on macromolecular synthesis, gene function, and molecular genetics.)

chapter ten

RECOMBINANT DNA TECHNOLOGY

Plasmid pBR322, a typical cloning vector used in genetic engineering (colorized, ×100,000).

Historical Perspectives

Gene Cloning
Obtaining the Target Gene
Cloning Vectors
The Host
Detection Methods

The Polymerase Chain Reaction

Applications of Recombinant DNA Technology
Medicine
Agriculture

Public and Scientific Concerns About the Regulation of Recombinant DNA Technology

PERSPECTIVE
Expression in *Escherichia coli* of Genes for Human Insulin

EVOLUTION AND BIODIVERSITY

Microbes in Motion PREVIEW LINK

This chapter covers the key topic of the use of genetic engineering techniques to multiply nucleic acid (PCR), detect microbes (probes), and produce commercially important products (vaccines, medicines). The following sections in the *Microbes in Motion* CD-ROM may be useful as a preview to your reading or as a supplemental study aid:

Microbial Metabolism and Growth: Genetics (Genetic Engineering), 25–30. *Viral Structure and Function:* Viral Detection (Direct Detection Nucleic Acid), 11–14, 20. *Vaccines:* Vaccine Development (Vaccine Composition), 9.

As a new business day begins on Wall Street one cool fall morning in 1980, the floor of the stock exchange is suddenly transformed into a bustling hub of activity. Thousands of buyers rush to purchase stock in a new, obscure company that has decided to go public. Within minutes the price of a share of the new stock has increased more than 150% over its opening price, from $35 to $89 a share. Brokers are so swamped with orders to buy that rationing is quickly imposed. Seldom in its history has the financial district seen such an impulsive response to a new offering.

Interestingly, the company involved in this intense trading is not a large oil conglomerate that has located a new offshore oil reserve. Nor is it a well-known firm that employs thousands of workers in a large factory. The new company, which has only 140 employees the day it goes public—most of whom are recent college graduates—has no established track record or proven product to offer to the public. Its stock is considered highly speculative by most prestigious brokerage houses. Yet, the potential impact of the new field in which the company is involved is such that investors are willing to speculate on the company's future.

What is this unusual company? The name of the company is Genentech, Inc.—an acronym for **gen**etic **en**gineering **tech**nology. This venture was founded by Herbert Boyer, a 44-year-old scientist-entrepreneur from the University of California at San Francisco, and Robert Swanson, a 32-year-old venture capitalist. Genentech, Inc., located in South San Francisco, was formed in 1976 by these two enterprising individuals to apply the new technology of **genetic engineering,** or the deliberate modification of the genetic makeup of a cell or organism, to development of products for science and commerce.

Why would thousands of people rush to invest their money in such a speculative and unproven company? They saw the potential in this new technology that could lead to new and innovative products. The refinement of procedures for bacterial gene transfer has recently led to expanded use of these procedures in the laboratory to manipulate genes in living organisms, a process commonly called **recombinant DNA technology.** As a result of advances in this area, it is now possible not only to transfer eucaryotic genes into bacteria, but also to manipulate eucaryotic chromosomes directly. Such recombinant DNA technology has provided important information on eucaryotic gene regulation and expression.

Recombinant DNA technology is a relatively new field that has emerged only in the last 35 years. The concept of exchanging genetic material between different species of organisms had its beginnings in the 1960s, when Julius Marmur isolated bacteria that contained DNA from two different species, *E. coli* and *Serratia marcescens*. This work was eventually extended to other groups of bacteria. It was not until several years later, however, that researchers began to examine the possibilities of gene transfer between procaryotes and eucaryotes.

In the late 1960s, Werner Arber discovered that a certain class of enzymes, termed **restriction endonucleases,** in *E. coli* could restrict invasion by foreign DNA, such as viral DNA, by cleaving the DNA into fragments. In 1970, Hamilton Smith isolated a restriction endonuclease (*Hind*III) from the bacterium *Haemophilus influenzae*, and Daniel Nathans later used this enzyme to cut specific regions of the DNA chromosome of simian virus 40 (SV40), a virus that causes tumors in monkeys. As a result of these studies, the first restriction enzyme cleavage maps (restriction maps) of the SV40 chromosome were constructed.

Unlike other nucleases that nonspecifically split nucleic acids, restriction nucleases recognize and cleave specific sequences in double-stranded DNA that exhibit twofold symmetry around a central axis:

$$5'—PO_4 \quad —A—G—C—T— \quad 3'—OH$$
$$3'—OH \quad —T—C—G—A— \quad 5'—PO_4$$

Such sequences that read the same in opposite directions are called **palindromes.** They occur quite commonly in bacteriophage nucleic acids—one reason why some bacteria may be resistant to phage infection. These bacteria possess restriction endonucleases that degrade infecting phage nucleic acid into smaller, nonfunctional fragments, thereby restricting invasion by the foreign phage DNA. A wide variety of restriction enzymes—more than 500—have been discovered and isolated from different species of bacteria (Table 10.1). Most of these enzymes recognize different base sequences and are specific in their breakage of DNA into several restriction fragments, ranging in length from a few to several thousand nucleotide pairs. Whereas some restriction endonucleases produce straight cuts in double-stranded DNA, many cleave DNA in a staggered manner such that single-stranded overhangs, or **sticky ends,** are formed. For example, the restriction enzyme *Eco*RI from *E. coli* staggers cuts when cleaving the following DNA palindrome:

$$5'—PO_4 \quad —G\overset{\downarrow}{}—A—A—T—T—C— \quad 3'—OH$$
$$3'—OH \quad —C—T—T—A—A\underset{\uparrow}{}—G— \quad 5'—PO_4$$

$$5'—PO_4 \quad —G \quad A—A—T—T—C— \quad 3'—OH$$
$$3'—OH \quad —C—T—T—A—A \quad G— \quad 5'—PO_4$$

Examples of Restriction Endonucleases

Enzyme Name	Microbial Source	Recognition Sequence[a]
AluI	Arthrobacter luteus	5'—A—G↓C—T—3' 3'—T—C↑G—A—5'
BamHI	Bacillus amyloliquefaciens H	5'—G↓G—A—T—C—C—3' 3'—C—C—T—A—G↑G—5'
EcoRI	Escherichia coli	5'—G↓A—A—T—T—C—3' 3'—C—T—T—A—A↑G—5'
EcoRV	Escherichia coli	5'—G—A—T↓A—T—C—3' 3'—C—T—A↑T—A—G—5'
HaeIII	Haemophilus aegyptius	5'—G—G↓C—C—3' 3'—C—C↑G—G—5'
HindIII	Haemophilus influenzae	5'—A↓A—G—C—T—T—3' 3'—T—T—C—G—A↑A—5'
PstI	Providencia stuartii	5'—C—T—G—C—A↓G—3' 3'—G↑A—C—G—T—C—5'
SalI	Streptomyces albus	5'—G↓T—C—G—A—C—3' 3'—C—A—G—C—T↑G—5'

[a]The arrows indicate the cleavage sites on each strand.

Sticky ends from two different DNA fragments cleaved by the same restriction enzyme are complementary and can base-pair to form intact DNA molecules. The reannealed DNA's open ends can then be joined by the enzyme DNA ligase, which synthesizes phosphodiester bonds.

This grafting of DNA fragments forms the basis of recombinant DNA technology. During the 1970s, a number of scientists, including Stanley Cohen, Herbert Boyer, and Paul Berg, used this technology to insert DNA from different sources into bacteria via a vehicle called a vector. Target gene insertion into a vector for introduction into a cell for replication is the first step in gene cloning. Plasmids and bacteriophages are examples of vectors frequently used in genetic engineering to insert and amplify foreign genes in rapidly replicating bacteria. Gene amplification is a process by which a cell can produce large quantities of a specific gene product, provided the gene is expressed. As a consequence of such amplification, gene products previously difficult to obtain in large quantities can now be produced rapidly and in large quantities in procaryotic gene factories.

Gene Cloning

Gene cloning is a technique by which a gene of interest is inserted into a new host cell, where it is amplified and expressed. The insertion and amplification of a gene in a bacterial cell is a process that involves several steps:

1. Isolation or synthesis of the gene of interest. The gene to be cloned must first be isolated from genomic DNA or synthesized from an RNA template or from nucleotides in vitro.

2. Incorporation of the gene into a vector. The gene is inserted into a **cloning vector.** Cloning vectors are independently replicating DNA molecules that serve as vehicles to protect genes from degradation and transport them into host cells.

3. Insertion of the vector into a host. The cloning vector containing the target gene is inserted into a host cell where the gene is amplified and expressed.

4. Detection of the cloned gene. Many different genes generally are inserted into host cells; therefore an important step in recombinant DNA technology is the detection of cells containing the target gene. This process of detecting the desired gene can be difficult and time-consuming since thousands of other clones may be present in the mixture.

Microbial Metabolism and Growth
Genetics: Genetic Engineering • pp. 25–30

The Target Gene Can Be Obtained in Different Ways

The gene of interest can be obtained directly from genomic DNA, synthesized from an mRNA template by the enzyme reverse transcriptase, or constructed from nucleotides in vitro. The goal in each case is to obtain a DNA fragment with a single gene that can be incorporated into a cloning vector.

The typical approach to gene cloning involves cleavage of the donor genome by restriction endonucleases, making sure that no cleavage sites are located within the target gene. This cleavage, although specific for palindromic sequences, results in a large number of different DNA pieces derived from the original genome. If the sequence of the target gene and bordering DNA is known, it is possible to use specific restriction endonucleases to create a DNA fragment that includes this gene. DNA fragments can be isolated by **agarose gel electrophoresis,** a technique in which DNA molecules are separated based on their electrical charges (the negatively charged DNA molecules migrate toward the positive electrode, or anode) and relative sizes (smaller fragments migrate farther) (Figure 10.1). DNA fragments differing in length by as little as 30 to 50 nucleotides are separated into distinct bands that can be visualized by staining with ethidium bromide. Ethidium bromide is a dye that intercalates, or inserts, between the bases of the DNA molecules and fluoresces bright orange when exposed to ultraviolet light. The size of a DNA fragment can be determined by comparing its band position to the position of bands produced by electrophoresis of reference DNA fragments with known sizes. The band representing DNA fragments of the desired size can then be cut out of the gel and its DNA extracted for subsequent insertion into a cloning vector.

The Target Gene Can Be Synthesized from an RNA Template

Since mRNA is transcribed from genomic DNA, mRNA molecules in a donor cell can be used as a source of genetic information for the target gene. A major advantage in using mRNA as a template for isolation of genes from eucaryotic donor cells is that unlike genomic DNA, which may contain coding sequences separated by noncoding sequences, eucaryotic mRNA transcripts consist of only coding sequences corresponding to genes (see mRNA processing, page 229).

Using the enzyme **reverse transcriptase,** which synthesizes single-stranded DNA chains from RNA templates (see reverse transcriptase, page 403), a **complementary DNA (cDNA)** copy of each mRNA molecule is made from mRNA extracted from the donor cell. The cDNA is then replicated by DNA polymerase to

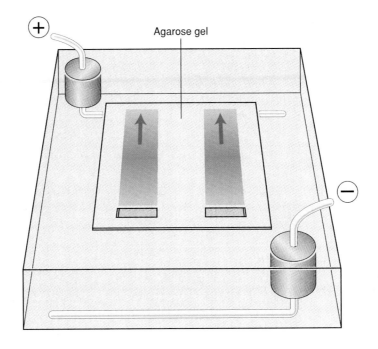

a.

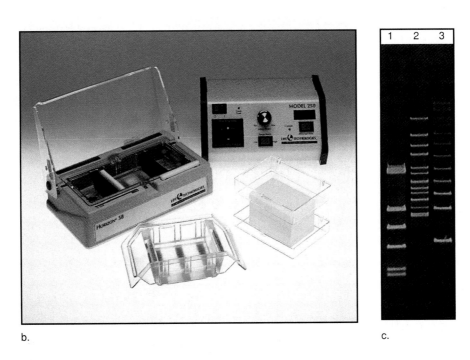

b.

c.

figure 10.1

Agarose Gel Electrophoresis of DNA

a. A diagram of a horizontal gel electrophoresis system. During electrophoresis, DNA molecules are separated by charge and size on the agarose gel. b. A photograph of a horizontal gel electrophoresis system. c. (Lane 1) electrophoretic pattern of λ bacteriophage DNA digested with the restriction enzyme HindIII; (lane 2) an 8–48 kb standard, showing DNA fragments of known size ranging from 8.3 kilobases (bottom) to 48.5 kilobases (top); and (lane 3) a 5 kb ladder, showing DNA fragments of known size ranging from 4.9 kilobases (bottom) to 120 kilobases (top) in 5-kilobase increments.

The sizes and characteristics of genes in DNA fragments can be determined by using a technique developed by Edward M. Southern at the University of Edinburgh. In this **Southern blotting** method, high molecular weight DNA is digested with a restriction endonuclease, and the resultant fragments are separated, based on their size, by electrophoresis on an agarose gel. The double-stranded DNA fragments are denatured (separated into single strands) by immersing the gel in an alkaline solution of sodium hydroxide. The single-stranded DNA fragments from the gel are transferred by blotting onto a sheet of nitrocellulose.

The DNA fragments bind tightly to the nitrocellulose to create a replica of the gel separation pattern. A radioactively labeled probe (purified RNA, cDNA, or a cloned segment of DNA) specific for the gene examined is added to the nitrocellu-lose and hybridized with complementary sequences of the bound DNA fragments. The nitrocellulose sheet is washed to remove unassociated probe, dried, and autoradi-ographed. DNA fragments that anneal to the labeled probe appear as dark bands on the autoradiograph (Figure 10.2).

Southern blotting has been especially useful in recombinant DNA technology in determining the number of times a cloned gene is represented in a genome. It is also helpful in locating restriction sites in DNA.

A similar technique, called **Northern blotting** (not named after a person, but to indicate that it is a technique similar to Southern blotting), is used to separate RNA. In Northern blotting, RNA is separated by gel electrophoresis and then transferred onto nitrocellulose. A replica of the separated RNA pattern is annealed with a radioactively labeled DNA probe and autoradiographed.

Southern Blotting

figure 10.2

Southern Blot

a. DNA fragments that have been electrophoresed and separated on an agarose gel are transferred to nitrocellulose. A radioactively labeled DNA probe is added to the nitrocellulose and allowed to hybridize with complementary sequences of the bound DNA fragments. After washing to remove unassociated probe, the nitrocellulose sheet is autoradiographed. DNA fragments annealed to the labeled probe appear as dark bands on the autoradiograph. b. Agarose gel electrophoresis of DNA fragments from bacteriophage Mu. c. Southern blot of the DNA gel shown in (b). The dark bands show those DNA fragments that have annealed to the labeled probe.

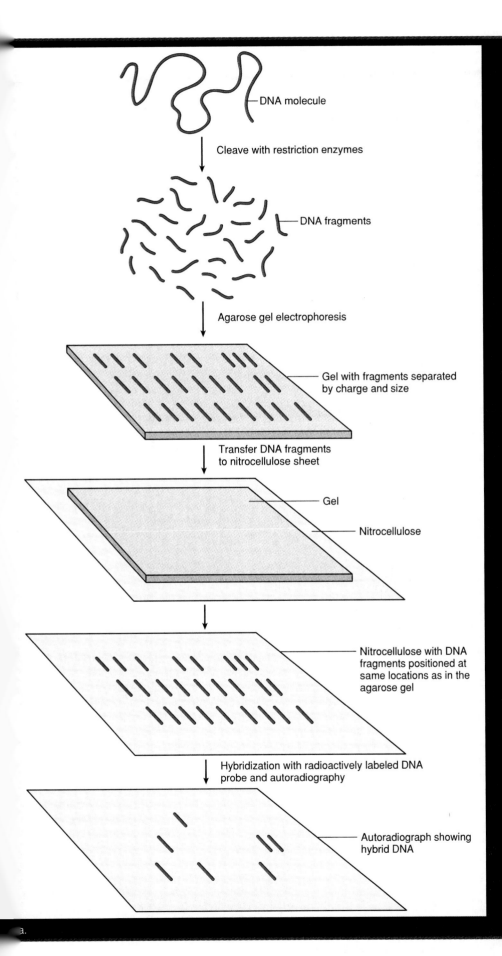

DNA molecule

Cleave with restriction enzymes

DNA fragments

Agarose gel electrophoresis

Gel with fragments separated by charge and size

Transfer DNA fragments to nitrocellulose sheet

Gel

Nitrocellulose

Nitrocellulose with DNA fragments positioned at same locations as in the agarose gel

Hybridization with radioactively labeled DNA probe and autoradiography

Autoradiograph showing hybrid DNA

a.

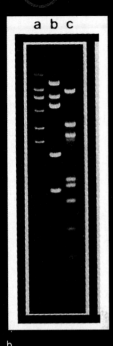

a b c

b.

8

c.

form double-stranded DNA, which can then be inserted into a cloning vector (Figure 10.3). Clones obtained in this manner are called **cDNA clones.**

Various methods can be utilized to enrich for specific mRNAs used for making a cDNA library. Antibodies against the protein translated from the target mRNA can be used to isolate polyribosomes containing the growing polypeptide chain and the associated mRNA. The mRNA is purified from this mixture and serves as the template for preparing cDNA. Alternatively, **subtractive hybridization** can be used to selectively enrich for cDNA containing the target gene. In this procedure, mRNA is isolated from two closely related cell types, one of which does not express the gene of interest. Next, cDNA molecules prepared from the mRNA of the cell type containing the target gene are hybridized with an excess of mRNA molecules from the cell type not expressing the gene. Unpaired single-stranded cDNAs that fail to find a complementary mRNA partner are likely to represent mRNA sequences found only in the cell type containing the target gene. These single-stranded DNAs can be separated from DNA:mRNA hybrids by a hydroxyapatite column, which binds double-stranded but not single-stranded nucleic acids.

The Target Gene Can Be Synthesized from Nucleotides in Vitro

If the nucleotide sequence of the target gene is known, its DNA can be synthesized. Short oligonucleotides containing 20 to 30 bases can be synthesized automatically in a few hours by a DNA synthesizer, or "gene machine." Longer chains are prepared by enzymatically joining two or more oligonucleotides by DNA ligase. The DNA is synthesized by the stepwise addition of single nucleotides to the 5' end of a growing chain attached to a solid phase support such as silica gel particles. After the oligonucleotide is completed, it is removed from the support and purified to eliminate contaminants.

Sometimes the amino acid sequence of the target protein is known. This information and knowledge of the genetic code can be used to predict the nucleotide sequence of the target gene for oligonucleotide synthesis. Synthetic oligonucleotides are used widely not only for gene cloning but also for site-directed mutagenesis, as a source of DNA primers for the polymerase chain reaction, and as probes for detection of specific DNA sequences.

In site-directed mutagenesis, a short oligonucleotide is prepared with a desired base change at a specific site (Figure 10.4). The altered oligonucleotide is annealed to a single-stranded copy of the gene of interest. The oligonucleotide is then extended using DNA polymerase to produce a new copy of the gene with the mutated sequence. The double-stranded DNA can be inserted into a host cell, where many copies of the modified gene can be produced for

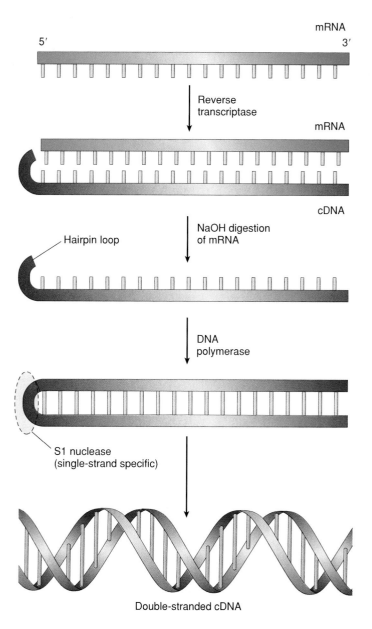

figure 10.3

Steps in the Synthesis of cDNA from an mRNA Template

A cDNA containing a hairpin loop is produced from the mRNA by the enzyme reverse transcriptase. After degradation of the mRNA by treatment with alkali, the cDNA is copied into a complementary DNA strand by DNA polymerase. The single-stranded hairpin loop connecting the two DNA strands is cleaved by S1 nuclease, which is specific for single-stranded DNA. The double-stranded cDNA can now be inserted into a cloning vector.

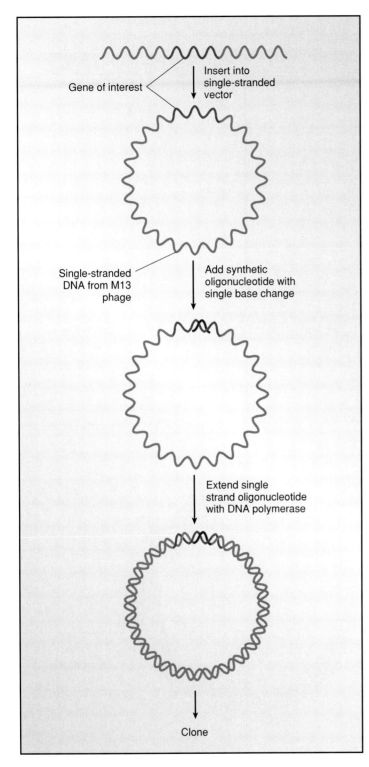

figure 10.4

Site-Directed Mutagenesis

A synthetic oligonucleotide with a single base change is bound to a specific region of the M13 phage DNA. The oligonucleotide is extended with DNA polymerase. The resultant double-stranded DNA can then be cloned.

further studies of its function. Whereas conventional mutagens cause random, unpredictable changes in DNA (see mutation, page 246), site-directed mutagenesis results in *predictable* modifications at *specific* sites on the gene. This specificity of mutation makes it possible to study the relationship between gene structure and function. Bacteriophage M13 is often used as a vector for site-directed mutagenesis because its genome can exist as single-stranded DNA (see bacteriophage M13, page 391).

The Target Gene Is Incorporated into a Cloning Vector

In order to clone a gene, it must be incorporated into a genetic vehicle for insertion into a cell. Cloning vectors for DNA fragments should have one or more restriction enzyme recognition sites that occur only *once* within the vector (that is, there should only be one *Eco*RI or one *Hin*dIII recognition site on the vector). These unique sites make it possible to cleave the vector at a specific location with an endonuclease for desired gene insertion without disrupting other areas of the vector. Using the same restriction enzyme, DNA fragments with complementary overlapping sticky ends can be generated on both the target gene and the cloning vector (Figure 10.5). The two fragments are then mixed to form a recombinant molecule.

An effective cloning vector should also have a detectable characteristic, or **selective marker,** that is conferred upon the host cell. Most cloning vectors carry genes encoding antibiotic resistance. Their presence within cells can be detected by the ability of these cells to grow on antibiotic-containing media.

Plasmids Can Serve as Cloning Vectors

There are two basic types of genetic molecules that have these desirable characteristics: plasmids and bacteriophage genomes. Plasmids are useful cloning vehicles because they (1) are small—1,000 to 300,000 base pairs in length compared to the 4.7 million base pairs in the *E. coli* genome—and therefore easily manipulated; (2) consist of circular double-stranded DNA, which makes them more stable during manipulation; (3) have an autonomous **origin of replication,** which enables them to replicate independently of the bacterial genome; (4) are maintained in multiple numbers within the cell, making amplification possible; (5) have restriction enzyme recognition sites that occur only once within the vector; and (6) have selectable markers such as antibiotic resistance, which facilitates their detection within host cells. Furthermore, plasmids can accept insertion of foreign DNA up to 5,000 base pairs in length and still maintain their integrity.

An example of an effective plasmid frequently used for cloning genes is pBR322, which replicates in *E. coli*. pBR322 has all

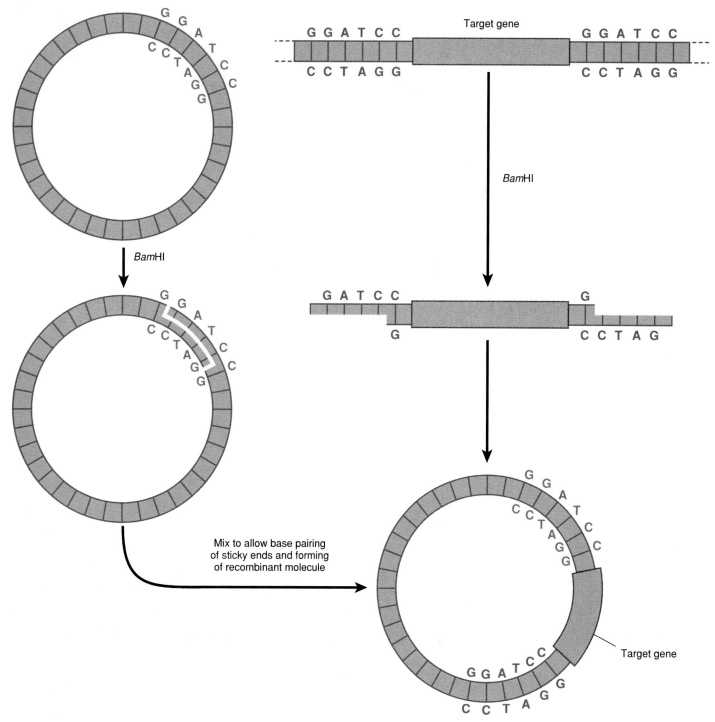

figure 10.5

Insertion of a Target Gene into a Cloning Vector

The same restriction enzyme is used to generate DNA fragments with complementary cohesive ends on both the target gene and the cloning vector. The two fragments are then ligated to form a recombinant molecule.

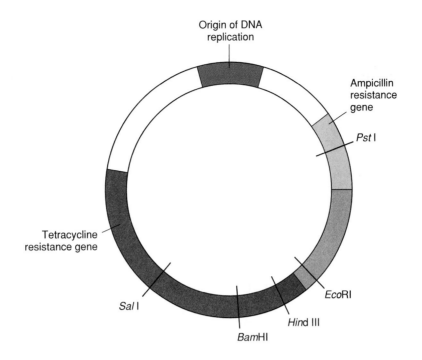

figure 10.6
The pBR322 Plasmid
The structure of plasmid pBR322, a typical cloning vector, is shown with restriction enzyme cleavage sites, resistance genes, and the origin of replication.

of the desirable characteristics of a good cloning vector. It is a relatively small (4,361 base pairs), circular, double-stranded DNA plasmid that replicates independently of the bacterial genome. About 50 copies of pBR322 exist within a typical *E. coli* cell. These plasmid molecules can be amplified to as many as 1,000 to 2,000 per cell (or about 50% of the total cellular DNA!) in the presence of chloramphenicol. Chloramphenicol inhibits protein synthesis and stops cell replication but does not affect the independent replication of pBR322. The increased ratio of plasmid DNA to chromosomal DNA makes it easier to isolate and identify the plasmid DNA.

pBR322 contains single cleavage sites for many restriction endonucleases, including *Eco*RI, *Hind*III, *Bam*HI, *Pst*I, and *Sal*I (Figure 10.6). The plasmid also carries genes for tetracycline and ampicillin resistance. The *Bam*HI restriction site is located within the tetracycline resistance gene and the *Pst*I site is located within the ampicillin resistance gene. These unique restriction sites in pBR322 allow the detection by **insertional inactivation** of DNA fragments inserted at these specific sites. If DNA is inserted at the *Bam*HI site, the tetracycline resistance gene is inactivated. Cells

that have undergone insertional inactivation of the tetracycline resistance gene are unable to grow in tetracycline-supplemented media, but are able to grow in ampicillin-supplemented media. This phenomenon is helpful in determining the successful insertion of foreign DNA into the plasmid.

The procedure for insertion of foreign DNA into plasmid pBR322 is shown in Figure 10.7. The plasmid and the foreign DNA are digested by specific restriction endonucleases. The foreign DNA and plasmid are recombined by a DNA ligase. The resulting hybrid plasmid, called a **chimera** (after the multicreature imaginary monster in Greek mythology that had a lion's head, a goat's body, and a serpent's tail), is then introduced into the host cell.

Bacteriophage Genomes Can Serve as Cloning Vectors

Bacteriophage lambda (λ), previously discussed in Chapter 9 (see bacteriophage λ and specialized transduction, page 255), is frequently used as a cloning vector because of several useful features. Approximately one-third of the genome is not essential for

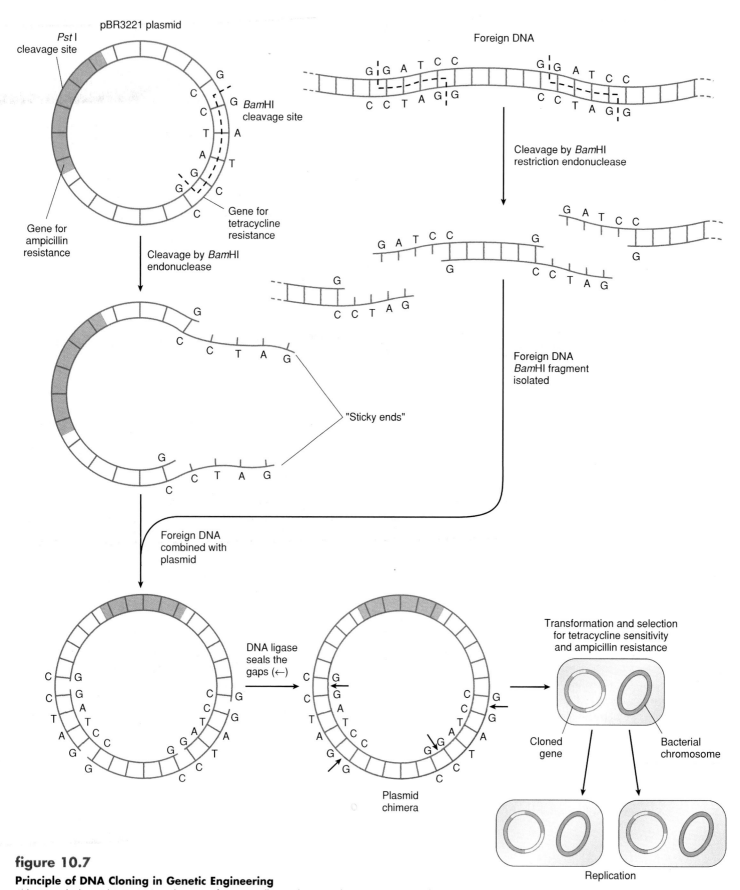

figure 10.7

Principle of DNA Cloning in Genetic Engineering

A bacterial plasmid (pBR322 with genes for ampicillin and tetracycline resistance, in this example) and the DNA to be cloned are cleaved with a restriction endonuclease (*Bam*HI, in this example) and combined to form a plasmid chimera. The hybrid plasmid is then inserted into the bacterium. Transformed bacteria containing the plasmid are selected by their sensitivity to tetracycline and resistance to ampicillin. Although the plasmid remains separate from the chromosome in these bacteria, the plasmid is replicated autonomously and expresses its genes.

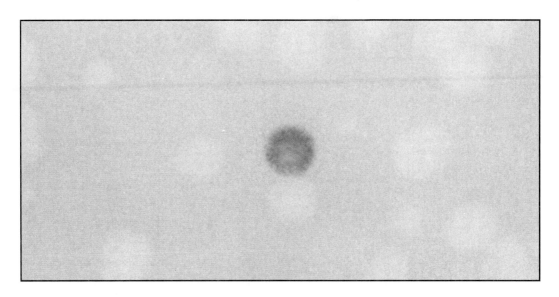

figure 10.8
Lactose-Positive (*lac*+) and Lactose-Negative (*lac*⁻) Phage Plaques
The *lac*+ phage plaque (blue) is distinguished from *lac*⁻ phage plaques (clear) by an insoluble blue dye released when X-gal is hydrolyzed by β-galactosidase.

replication of the phage and therefore can be replaced with foreign DNA. The linear λ DNA becomes circular inside the host cell, where it replicates by the theta (θ) mode of replication (see theta mode of replication, page 220).

The genome of λ is well-characterized and, in fact, has been altered in modified λ phages to accommodate gene cloning. Whereas wild-type λ has five *Eco*RI restriction sites, the modified phage λgt11 has a single *Eco*RI site and substitutes the β-galactosidase gene for a nonessential region of the genome. The reduction of *Eco*RI sites allows cleavage of this single site without disrupting the rest of the phage genome. When λgt11 phages with the β-galactosidase gene successfully replicate on a *lac*⁻ strain of *E. coli* (a strain that normally lacks the lactose operon and therefore the β-galactosidase gene), β-galactosidase is produced and *lac*+ plaques (plaques indicating the presence of the β-galactosidase gene and therefore the lactose operon) can be detected by the chemical 5-bromo-4-chloro-3-indolyl-β-D-galactopyranoside (X-gal), a color indicator (Figure 10.8). X-gal releases an insoluble blue dye when it is hydrolyzed by β-galactosidase. If foreign DNA is inserted into the β-galactosidase gene and inactivates it, the resultant *lac*⁻ plaques will be colorless.

Another class of modified λ phages, named Charon phages (after the Greek mythical boatman in Hades who ferried the dead across the Styx River), contains deletions in their genomes to allow the cloning of larger DNA fragments. Charon 4A and 40 are examples of two such modified λ phages capable of cloning very large (up to 24 kilobase pairs) DNA fragments.

The filamentous bacteriophage M13 is often used in recombinant DNA technology because it produces both double-stranded and single-stranded circular DNA during infection of a cell. Consequently, foreign DNA inserted into the M13 genome is packaged as both double-stranded and single-stranded DNA. The double-stranded M13 DNA is used as a cloning vector to amplify the target gene in infected host cells. The single-stranded DNA is packaged and secreted from host cells that do not lyse. Because this M13 DNA containing the foreign DNA is single-stranded, it is ideal for DNA sequencing that is performed on single-stranded DNA.

The general procedure for inserting foreign DNA into a bacteriophage genome is shown in Figure 10.9. The foreign DNA and phage genome are digested with the same restriction endonucleases. Following insertion of the foreign DNA into the phage genome and reassociation of the DNA with the

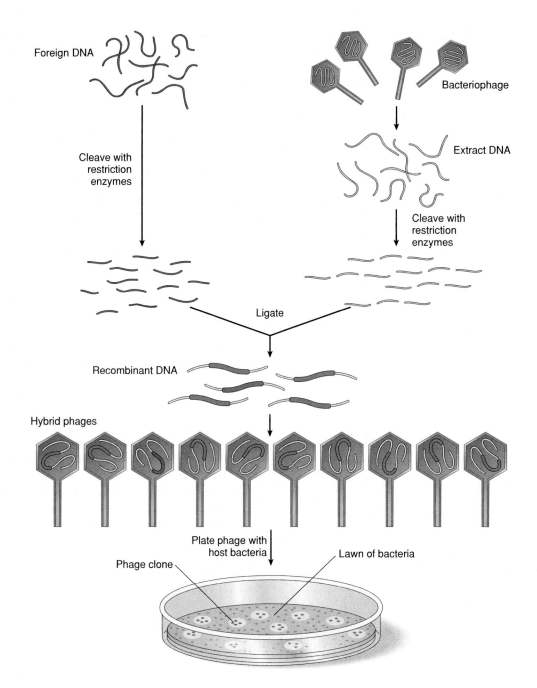

figure 10.9

The Use of a Bacteriophage as a Cloning Vector

The foreign DNA and phage genome are cleaved with restriction enzymes and ligated. The recombinant DNA is packaged into the phage head. Host bacteria are infected with the phage.

aid of DNA ligase, the recombinant phage genome is packaged into the phage head (the protein coat surrounding the genome) (see phage head, page 387). The intact phage can now be used to infect the host bacterium. Naked phage DNA can also be taken up directly by a bacterium through a process called transfection (see transfection, page 255). However, transfection is less efficient for cloning than uptake of the complete phage particle.

Other Genetic Molecules Can Serve as Cloning Vectors

Other types of cloning vectors have been developed and used in genetic engineering. **Cosmids** are plasmid vectors containing λ phage *cos* (cohesive end) sites incorporated into the plasmid DNA. *Cos* sites complement one another and can bind together to circularize the DNA molecule. Cosmids typically contain an origin of replication, several unique restriction sites, and one or two selective markers for antibiotic resistance. These hybrid vectors are capable of accepting DNA fragments as large as 45 to 50 kilobase pairs (or 10 times the amount of foreign DNA as standard plasmid cloning vectors). This ability to clone very large genes is especially useful in cloning eucaryotic genes as well as groups of genes adjacent to one another on a donor genome. Cloning with a cosmid vector is achieved by digesting the plasmid and the foreign DNA segment with a common restriction enzyme. The digested pieces are ligated and packaged into bacteriophage particles that can be used to infect susceptible host cells.

Phagemids are hybrid vectors containing DNA from a filamentous phage (for example, M13) and a plasmid. These vectors contain origins of replication for both the phage and the plasmid. Normally replication proceeds from the plasmid origin to yield double-stranded DNA molecules. However, when a cell containing a phagemid is infected with a single-stranded filamentous phage, the replication mode changes. One strand of the phagemid double-stranded DNA is nicked and new DNA is synthesized by the rolling circle mechanism of replication (see rolling circle mechanism of replication, page 402). This mode of replication results in single-stranded copies of the phagemid DNA (including the cloned gene it is carrying) that can be isolated and used for sequencing.

Yeast artificial chromosome, or **YAC,** vectors are linear plasmids composed partially of a yeast chromosome. YAC vectors have an origin of replication and unique restriction sites, but also carry a centromere (for chromosome segregation and stability during mitosis) and telomeres on each end of the chromosome to enable the vectors to function as normal eucaryotic chromosomes in a host yeast cell. YAC vectors accept foreign DNA fragments up to one million base pairs in length and are especially useful in generating gene libraries of eucaryotes such as humans.

Shuttle vectors are used to move DNA between two unrelated organisms. Such vectors can replicate in both organisms and have been especially helpful in transferring genes between bacteria and eucaryotic cells or between two different species of bacte-

ria. **Expression vectors** are vectors that contain not only the target gene but also regulatory sequences that can be used to control gene expression.

DNA Libraries Are Collections of Cloned DNA Fragments

A collection of cloned DNA fragments from the genome of an organism constitutes a **DNA library.** Whereas a library of books is housed in a building, a DNA library is generally housed in host bacterial cells of which only a few may harbor the desired gene. There are two types of DNA libraries: a **genomic library,** which contains DNA fragments from a digested genome, and a **cDNA library,** which contains cDNA constructed from mRNA. A cDNA library is particularly useful when working with eucaryotic genes because it is derived from mRNA transcripts containing only coding sequences corresponding to genes (see exons and introns, page 213) (Figure 10.10).

A DNA library generally will consist not only of clones containing the target gene (or parts of the gene sequence), but also clones with other unrelated DNA fragments. DNA libraries often are constructed from DNA fragments representing all or most of the donor cell genome. One approach to preparing a DNA library, called **shotgun cloning,** involves enzymatically cleaving an entire donor genome into small DNA fragments and inserting these fragments into cloning vectors. Although it may take examination of many DNA fragments to locate a particular gene, this approach is much easier than trying to predict the location of a specific gene in an intact chromosome. An ideal DNA library is one that has clones containing all of the target gene sequence. Unfortunately, ideal DNA libraries are seldom encountered in genetic engineering and therefore screening a library for the specific complete target gene can be tedious and time-consuming. As will be discussed later in this chapter, techniques for detection of the target gene are available that make the screening process easier.

The Cloning Vector Is Inserted into a Host Where the Target Gene Is Amplified

Following incorporation of the target gene into a cloning vector, the vector is inserted into a host cell. The ideal host for gene cloning is one that readily accepts the vector, grows rapidly, and is genetically stable in culture. The bacteria *Escherichia coli* and *Bacillus subtilis* and the yeast *Saccharomyces cerevisiae* are hosts with these characteristics, and are typically used for gene cloning.

Plasmid vectors are introduced into competent host cells by transformation, using calcium chloride to make the plasma membrane permeable to the DNA. Entry of bacteriophage cloning vectors into host cells occurs by the process of infection and is generally tenfold more efficient than transformation in the successful insertion and expression of the cloned gene.

Although mammalian cells grown in tissue culture and exposed to calcium chloride can also take up cloning vectors by

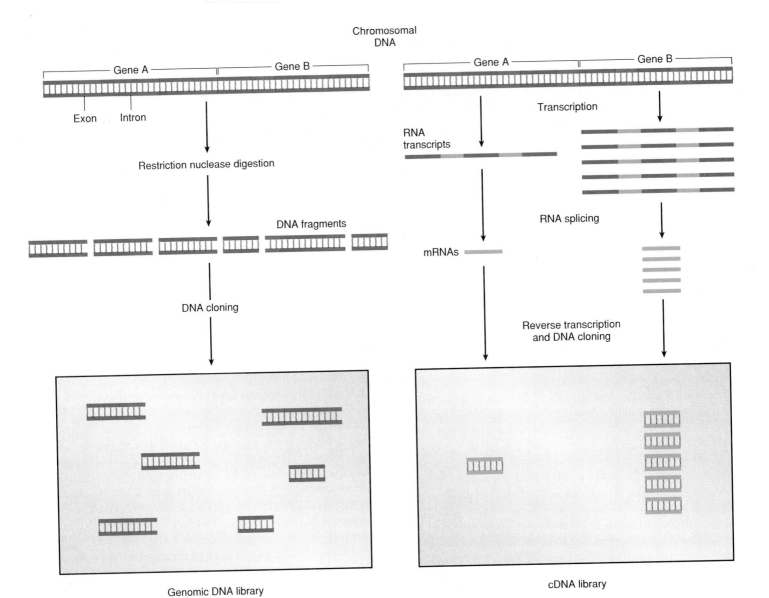

figure 10.10

Comparison of Genomic and cDNA Libraries
A genomic library prepared from eucaryotic DNA consists of exons (coding sequences of a gene) and introns (noncoding sequences). In comparison, the introns are removed by RNA splicing during the formation of the mRNA used to prepare a cDNA library.

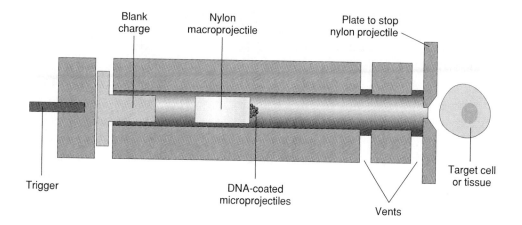

figure 10.11

The Gene Gun

DNA-coated microprojectiles on a nylon macroprojectile are propelled into the cell with the gene gun.

endocytosis, or **transfection** (a term that has a different meaning than **bacterial transfection,** which is the introduction of bacteriophage DNA into a bacterium), there are other ways to incorporate foreign DNA into eucaryotic cells. Small holes can be opened in the membranes of cells by electrical shock to permit entry of DNA. This process, called **electroporation,** can also be used on bacteria to make their plasma membranes permeable to DNA. DNA may be directly inserted by microinjection into animal cells such as fertilized eggs. Inside the cell, the DNA becomes incorporated into the host genome to create a **transgenic animal** that has new genetic information from the foreign DNA. One of the more interesting ways to transform eucaryotic cells is to shoot DNA-coated microprojectiles with a **gene gun** into the cell (Figure 10.11). The nucleic acid-coated particles penetrate the cell without killing it and the DNA can then recombine with the cell DNA. This procedure has been used successfully with yeast, algae, mammalian cells, and plant cells.

After entry into the host cell, the cloning vector is amplified and expressed. Control of gene expression by the genetic engineer is an essential part of cloning to ensure efficient and maximum production of the protein product. Vectors that have promoters for binding of RNA polymerase and control of gene expression are commonly used in gene cloning. It is especially desirable to have **strong promoters** within a vector, which lead to high levels of gene expression. The *lac* promoter is one such strong promoter. Bacteriophages λgt11, Charon 4A, and Charon 40 are examples of cloning vectors that include regulatory regions for the *lac* operon. These regulatory regions provide a convenient switch for manipulating the expression of *lac* genes as well as other downstream genes on the vector. Thus host cells can be grown to high density before inducer is added to turn on the operon.

Various Methods Are Available to Detect Host Cells with the Cloned Gene

After the cloned gene is inserted into the host cell, there must be suitable methods to identify and isolate cells with the target gene. Populations of bacteria, which typically contain billions

of individual cells, will have a mixture of clones. Some cells will carry the target gene, while others may harbor random DNA fragments inserted in vectors during cloning. Detection of cells with the desired gene is especially important when working with DNA libraries that may contain thousands of different clones.

Protein Products Can Be Used to Identify Cells Containing the Target Gene When the Gene Is Expressed

In instances where the foreign gene is expressed in the cloning host, one can screen for the protein product as an indicator of successful cloning. It is important that the host not produce the same protein. Otherwise cells with the target gene would not have a unique and detectable product.

A colorful example of this type of screening is the detection of cloned luciferase genes in *E. coli*. The Jamaican click beetle, *Pyrophorus plagiophthalamus*, has four distinct luciferase genes, each of which produces a different colored light. When these genes are inserted into *E. coli* and the bacteria are exposed to luciferin, the bacteria glow different colors depending on the specific gene received (Figure 10.12).

Host Cells Containing the Target Gene Can Be Detected by Nucleic Acid Probes

Nucleic acid probes are commonly used to screen large numbers of cells for a cloned gene. A nucleic acid probe is a molecule that recognizes a specific DNA or RNA sequence. Probes consist of single-stranded DNA or RNA with a base sequence that is complementary to the sequence of the target gene.

One approach to screening clones is to replica plate cells suspected of containing the target gene onto a nitrocellulose filter and a reference agar plate (Figure 10.13). Bacteria in colonies formed on the filter are lysed and their double-stranded DNA denatured by heating. As double-stranded DNA is slowly heated, its strands separate, or denature. A radiolabeled probe (generally a DNA probe

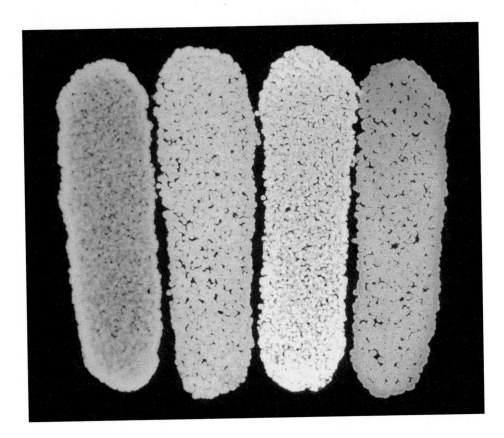

figure 10.12

Luciferase Genes Expressed in *Escherichia coli*

These four colony streaks of *E. coli* glow different colors because each contains a different cloned luciferase gene from the Jamaican click beetle, *Pyrophorus plagiophthalamus*.

radioactively labeled with ^{32}P) is added to the filter and allowed to **hybridize,** or bind specifically, with the single-stranded DNA derived from the cells. Under highly restrictive conditions of high temperature and low salt concentration of the hybridization reaction, the probe will hybridize only to perfectly complementary sequences. As these **high stringency** conditions are lowered, the probe will hybridize to partially complementary sequences. The filter is washed to remove unbound radiolabeled probe and is covered with a sheet of X-ray film for analysis by autoradiography. Spots appearing in areas of radioactivity on the film can be used to identify colonies on the reference agar plate that contain the target gene.

Cloned Genes Can Be Detected by Using Reporter Genes

The successful cloning of a gene into a host cell can be determined by the activity of a **reporter gene** located on the cloning vector. Reporter genes code for an easily detectable trait in the host cell. The *E. coli lacZ* gene, which codes for the enzyme β-galactosidase, is frequently used as a reporter gene in genetic engineering. Its expression can be monitored on agar plates containing the indicator X-gal, which turns blue when split by β-galactosidase (see X-gal, page 283). Alternatively, if the target gene is inserted into the *lacZ* region of the cloning vector, the gene is inactivated, and white colonies instead of blue form from cells containing the target gene.

The Polymerase Chain Reaction

Gene cloning is a valuable tool for amplifying DNA in vivo. In 1983, during an evening drive through the mountains of northern California, Cetus scientist Kary Mullis conceived a novel procedure for rapidly amplifying DNA in vitro. When he returned to his laboratory, Mullis performed experiments to bring his idea, now known as the **polymerase chain reaction (PCR),** to reality. The PCR technique multiplies a DNA molecule exponentially to virtually unlimited copies in a short period of time and has wide applications in biomedical research. Mullis was awarded the 1993 Nobel Prize in chemistry for his development of PCR.

The PCR technique is a relatively simple procedure that amplifies DNA by a cyclic repetition of three steps (Figure 10.14):

1. The original double-stranded DNA molecule is denatured at high temperature.

2. Oligonucleotide primers are annealed to the DNA at low temperature.

3. The primers are extended on the DNA template by a DNA polymerase.

These three incubation steps are linked in a thermal cycle. As the two strands of the target DNA are separated by heat denatura-

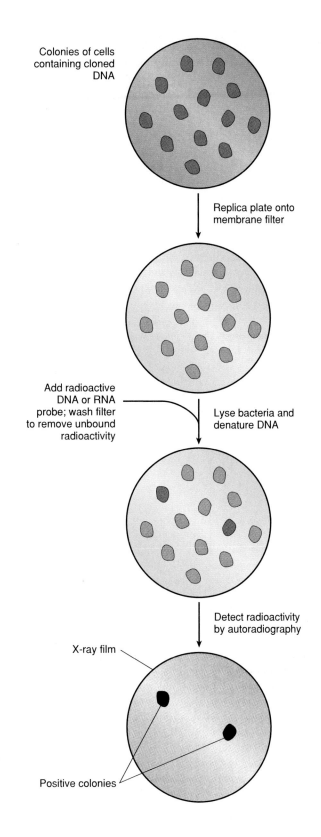

Colonies of cells containing cloned DNA

Replica plate onto membrane filter

Add radioactive DNA or RNA probe; wash filter to remove unbound radioactivity

Lyse bacteria and denature DNA

Detect radioactivity by autoradiography

X-ray film

Positive colonies

figure 10.13

Screening of Clones by Colony Hybridization with a Radioactive Nucleic Acid Probe

A radioactive DNA or RNA probe is used to screen colonies on a membrane filter for the target gene sequence.

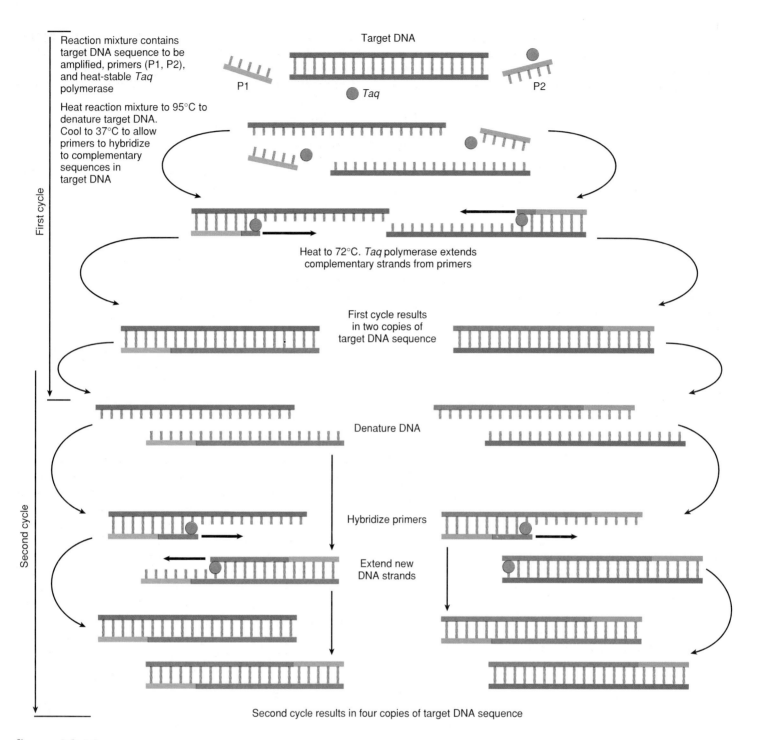

Reaction mixture contains target DNA sequence to be amplified, primers (P1, P2), and heat-stable *Taq* polymerase

Target DNA

P1

Taq

P2

Heat reaction mixture to 95°C to denature target DNA. Cool to 37°C to allow primers to hybridize to complementary sequences in target DNA

First cycle

Heat to 72°C. *Taq* polymerase extends complementary strands from primers

First cycle results in two copies of target DNA sequence

Second cycle

Denature DNA

Hybridize primers

Extend new DNA strands

Second cycle results in four copies of target DNA sequence

figure 10.14

DNA Amplification Using the Polymerase Chain Reaction

In the polymerase chain reaction, DNA is denatured by heating. Oligonucleotide primers are annealed to the DNA at a low temperature and are extended by DNA polymerase.

Viral Structure and Function
Viral Detection: Direct Detection Nucleic Acid • p. 20

tion, two synthetic oligonucleotide primers (added in excess) anneal to their respective recognition sequences on the now separated complementary strands. DNA polymerase is added, resulting in extension of the primers on the DNA single-stranded templates. After one round of this cycle, there are two copies of the original double-stranded DNA molecule. In the second cycle, oligonucleotide primers (added in excess) anneal to each of these four DNA single strands to initiate a new round of DNA synthesis. At the end of this second cycle, there are four copies of the original target DNA molecule. The number of copies of the initial DNA molecule continues to double with each subsequent cycle and can increase to a billionfold in as few as 30 cycles in 1 hour!

Many improvements have been made in the PCR procedure since its conception to increase its sensitivity and specificity. Originally *E. coli* DNA polymerase was used to extend the primers at 37°C. However, the high temperature required for denaturation of the DNA also denatured the polymerase and it had to be replenished after each cycle. Substitution of a thermostable DNA polymerase (*Taq* polymerase) isolated from the thermophilic bacterium *Thermus aquaticus* not only eliminated this problem of enzyme denaturation, but also greatly increased the specificity of the PCR reaction. With *Taq* polymerase and higher temperatures, hybridization of primers to specific DNA sequences occurs at higher stringency, resulting in a more homogenous DNA product than was possible with the original *E. coli* polymerase.

Development of a programmable thermal cycler, essentially a machine that automatically conducts successive heating and cooling cycles, eliminated the tedious task of manually heating and cooling the reaction mixture. The convenience and popularity of performing PCR has increased through the introduction of commercial kits containing PCR reagents and its cost has been significantly reduced through the cloning of *Taq* polymerase genes into *E. coli* to optimize production.

A modification of the PCR uses **nested amplification** to improve the sensitivity and specificity of the procedure. In nested amplification, the target DNA is amplified for 15 to 30 cycles with a primer pair and the product is transferred to a new reaction tube. A second primer pair that is specific for sequences within the DNA produced by the first primer pair is then added. After another 15 to 30 cycles of amplification, the product is identified by agarose gel electrophoresis or other procedures. The sensitivity of nested amplification procedures is extremely high—as little as a single copy of the target DNA molecule is required for amplification. The second amplification series also confirms the specificity of the first round of amplification. Furthermore, any inhibitors that might be present in the original sample are diluted during product transfer to a new reaction tube for the second series of amplification.

A variation of the PCR technique has important applications in the study of comparative evolutionary relationships among organisms, also known as **phylogeny.** In this modification, rRNA isolated from a cell is converted to cDNA by reverse transcriptase. The cDNA is then amplified by PCR and product sequences are compared. Organisms that are closely related through evolution have been found to have similar rRNA sequences (see phylogenetics, page 311).

Applications of Recombinant DNA Technology

The techniques associated with recombinant DNA technology have resulted in many practical applications in medicine and agriculture. Some of these applications are described in this section.

Applications of Recombinant DNA Technology in Medicine Have Important Ramifications for Human Health Care

Currently there is a concerted international scientific effort, known as the **Human Genome Project,** to clone, map, and sequence the entire human genome. There are about 3 billion base pairs and 100,000 genes in the human genome; therefore this is an immense task that is estimated to span 15 years of work from 1990 to 2005. The benefits gained through the Human Genome Project are potentially enormous. Scientists will be able to identify and treat genetic defects in humans, develop diagnostic tests for a wide variety of diseases, and understand the genetic basis for many diseases. Similar projects have sequenced or are under way to sequence the genomes of the eucaryotic microbe *Saccharomyces cerevisiae* (12 million base pairs), which is important in bread making and the production of alcoholic beverages (see *S. cerevisiae,* page 618), and procaryotes such as *E. coli* (4.7 million base pairs); *Haemophilus influenzae* (1.8 million base pairs), which causes pneumonia and meningitis in children (see *Haemophilus influenzae,* page 512); *Mycoplasma genitalium* (500,000 base pairs); *Helicobacter pylori,* which causes peptic ulcers (see *Helicobacter pylori* and peptic ulcers, page 522); and *Methanococcus jannaschii* (see *M. jannaschii,* page 312). Sequencing these genomes should help scientists identify genetic switches for microbial virulence factors and industrially important enzymes. Although mapping a gene does not necessarily define gene function or the gene product, it does provide information about the gene location on the chromosome and an understanding of biochemical changes that might occur when the gene is mutated.

Identification of Genes Can Lead to Potential Earlier Diagnosis and More Effective Therapy

One of the most significant achievements of the Human Genome Project has been identification of the gene causing cystic fibrosis (CF). Cystic fibrosis is a hereditary disease of children characterized by excess mucus production in the respiratory tract. It is one of the most prevalent genetic diseases in the United States, affecting 1 out of every 2,500 individuals. Victims often die of

DNA Fingerprinting

Recent advances in molecular biology have made new and helpful techniques possible for the genetic characterization of organisms. One of these techniques, called **DNA fingerprinting,** was devised in the mid-1980s by Alex Jeffreys of Great Britain. It is based on the identification of repetitive DNA sequences that occur in the genomes of higher organisms such as humans.

Although the human genome contains approximately 3 billion base pairs, only about 5% of this DNA codes for its estimated 100,000 genes. The remaining noncoding DNA lies between genes where it often occurs as repeated sequences linked in tandem. The number of repeats of an individual sequence can vary from one allele to another and among individuals. In DNA fingerprinting, alleles containing these repeated sequences, called **variable number of tandem repeats** or **VNTR,** are digested with a restriction enzyme that does not cut within the VNTR. Small base sequence differences in alleles, such as a single base deletion or insertion, can result in different length DNA fragments generated by restriction digestion. The variable fragment lengths caused by base-pair variations in the DNA molecule are called **restriction fragment length polymorphisms** or **RFLPs.** In DNA fingerprinting, these different length fragments generated by restriction digestion are separated by size (and number of repeat units) using gel electrophoresis, and the resulting bands can be detected by Southern blotting using a probe specific for the VNTR sequence (Figure 10.15).

PCR is not necessary for DNA fingerprinting, but can amplify and enhance the resolution of DNA bands observed by gel electrophoresis, particularly from samples containing minute amounts of blood, semen, or saliva. The sequence of the VNTR region to be analyzed must be known so that primers can be prepared that bracket and selectively amplify only that portion of the genome. The amplified product is then analyzed by gel electrophoresis.

DNA fingerprinting has been especially helpful in criminal cases where evidence such as a hair or semen left at a crime scene can be linked to the perpetrator of the crime. Like real fingerprints, DNA fingerprints are molecular markers that are unique for each person. Only identical twins would be expected to have the exact same DNA fingerprint. Many law enforcement agencies, including the FBI, are now using this new technology to establish molecular databanks and track criminals.

---

congestive lung complications before the age of 30. Through chromosome mapping, researchers have identified an abnormal gene in the chromosome responsible for CF. Although discovery of this gene will not necessarily end the problem of CF, it will make earlier diagnosis of the disease possible and enhance efforts to find a cure.

The treatment of certain genetic diseases by **gene therapy** is a revolutionary approach using recombinant DNA technology. Severe Combined Immunodeficiency (SCID) is an inherited disorder in which the immune system is seriously impaired. In about one-quarter of infants who have SCID, a gene that normally produces adenosine deaminase (ADA) is defective. ADA is an enzyme that routinely rids the bloodstream of harmful metabolic products that can destroy immune system cells called T cells (see T cells, page 473). In the first approved gene therapy experiments on humans in 1990, normal genes for ADA were inserted into T cells removed from two young girls with SCID. The genetically altered T cells were then reinjected into the girls. The patients' immune systems amazingly began to function normally again as their T cells produced natural ADA. Although the modified T cells eventually died after several months, the success of this landmark experiment provides hope that gene therapy

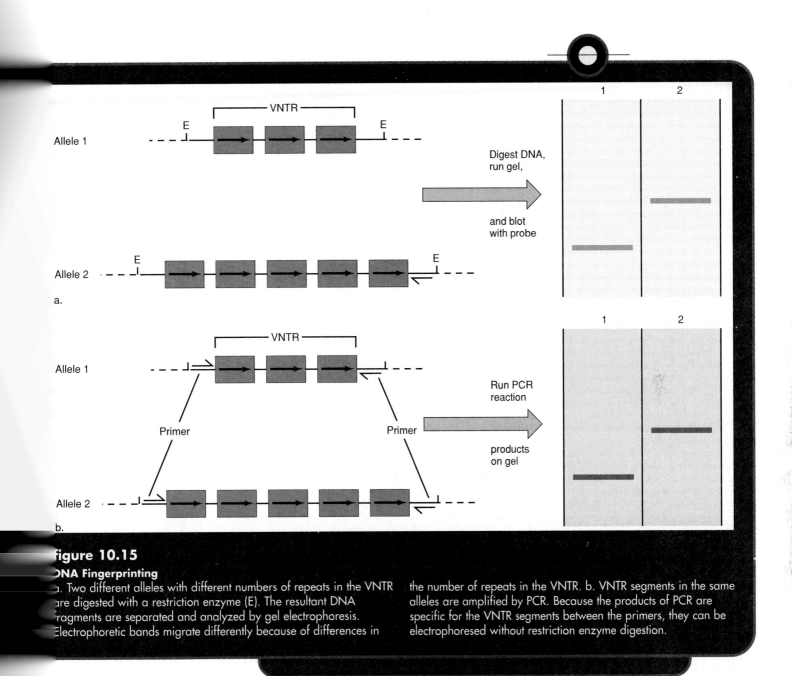

Figure 10.15

DNA Fingerprinting

a. Two different alleles with different numbers of repeats in the VNTR are digested with a restriction enzyme (E). The resultant DNA fragments are separated and analyzed by gel electrophoresis. Electrophoretic bands migrate differently because of differences in the number of repeats in the VNTR. b. VNTR segments in the same alleles are amplified by PCR. Because the products of PCR are specific for the VNTR segments between the primers, they can be electrophoresed without restriction enzyme digestion.

might provide a cure not only for ADA-associated SCID, but also for other diseases linked to defective genes such as muscular dystrophy, hemophilia, Tay-Sachs disease, sickle-cell anemia, and Huntington's disease.

Antisense RNA or DNA Can Selectively Turn Off or Modify the Activity of a Gene

One relatively recent development that is the result of recombinant DNA technology is the use of **antisense RNA or DNA** molecules to selectively turn off or modify the activity of a specific gene. Antisense molecules bind specifically with a targeted gene's RNA message (the "sense," or complementary, strand to the "antisense" RNA or DNA) and interrupt the flow of information from the gene to its protein product. The use of antisense RNA and DNA to regulate gene expression originated from a similar strategy that is used by bacteria to regulate plasmid numbers (Figure 10.16). An RNA molecule (RNA II, approximately 500 nucleotides long) in bacteria is associated with the initiation of ColE1 plasmid DNA replication. RNA II activity is inhibited by another RNA molecule (RNA I, approximately 100 nucleotides long) that binds to and inactivates RNA II. The concentration of RNA I in the cell increases in proportion to the number of

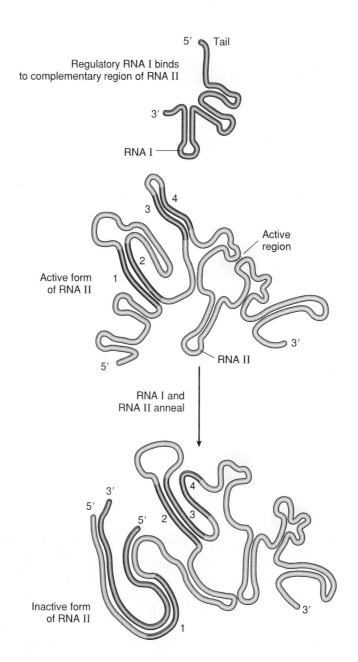

5′ Tail

Regulatory RNA I binds
to complementary region of RNA II

3′

RNA I

4

3

Active
region

2

Active form
of RNA II

1

5′

RNA II

3′

RNA I and
RNA II anneal

3′

5′

5′

2

4

3

1

Inactive form
of RNA II

3′

figure 10.16

Antisense RNA Regulation of Plasmid DNA Replication

RNA I is a regulatory RNA that has a sequence that is complementary to region 1 of RNA II, which helps initiate plasmid DNA replication. As the concentration of RNA I increases in proportion to the number of plasmid molecules in the cell, RNA I binds to RNA II. The conformation of RNA II is distorted and the RNA II is inactivated.

plasmid molecules. Consequently, the level of RNA I molecules regulates plasmid DNA replication. Such RNA-mediated reactions are especially of interest to evolutionary biologists who have hypothesized that the first cells may have lacked DNA and proteins and used RNA-catalyzed reactions for much of their metabolism. Today antisense RNA and DNA molecules provide hope for searching out and destroying aberrant gene products inside cells.

Recombinant DNA Technology Has Led to Improved Vaccines

Vaccines have proven extremely valuable in containing infectious diseases that have killed millions of people in the past. Unfortunately, problems have been encountered with impurities in vaccines that can result in severe adverse reactions. Recombinant DNA technology has enabled the elimination of these impurities through genetically

for foot-and-mouth disease, a viral disease that causes serious debilitation, weight loss, cessation of lactation, and abortion in cattle, sheep, and other ruminants. The first recombinant vaccine for humans was approved by the FDA in 1986. This vaccine, which protects against hepatitis B, is marketed by companies under the tradenames Engerix-B and Recombivax HB (Figure 10.17). Additional subunit vaccines are currently under development for AIDS and cholera.

 Vaccines
Vaccine Development: Vaccine Composition • p. 9

Nucleic Acid Probes Aid in Identification of Disease-Causing Pathogens

One of the most widely used applications of recombinant DNA technology has been in the area of nucleic acid probes. These probes are used in the clinical laboratory to recognize pathogenic microbes and viruses by their unique DNA (or RNA) sequences (Figure 10.18). DNA isolated from the pathogen is denatured and the resultant single-stranded DNA molecules are mixed with a probe labeled with a reporter such as a radioisotope (^{32}P), an enzyme (alkaline phosphatase or horseradish peroxidase), or an affinity label (biotin or digoxigenin) that can be detected after hybridization. If the pathogen contains a DNA sequence complementary to the probe, the complementary strands will hybridize. In some instances, the number of pathogens (and amount of their DNA) in a specimen is low and the DNA must be amplified prior to detection by a probe. PCR can be used to selectively amplify target sequences present in small quantities in a DNA mixture in a specimen. This ability to rapidly amplify specific DNA targets has made PCR especially effective for the diagnosis of pathogens that otherwise might be difficult to identify in a specimen containing other microorganisms. PCR is particularly useful for identifying pathogens such as the human immunodeficiency virus (HIV), which causes AIDS; the hepatitis virus; *Chlamydia trachomatis*, which causes trachoma and nongonococcal urethritis (NGU); *Borrelia burgdorferi*, which causes Lyme disease; and *Mycobacterium tuberculosis*, which causes tuberculosis (see infectious diseases, page 497). Most of these organisms cannot easily be cultivated in the laboratory. In some instances their identification by conventional laboratory tests takes weeks. PCR reduces the identification time to hours or days. Furthermore, by amplifying the target gene through PCR, a nucleic acid probe should theoretically be able to detect a single pathogenic bacterial cell in a specimen. Detection of such low numbers of organisms is useful not only in medicine, but also in the food industry where many serious diseases are caused by microbial and viral pathogens initially present as only a few organisms.

 Viral Structure and Function
Viral Detection: Detection by Nucleic Acid • pp. 11–14

figure 10.17

Hepatitis B Vaccine Produced by Recombinant DNA Technology
A technician uses a chromatography column to separate key proteins from batches of yeast cells.

engineered vaccines called **subunit vaccines** (see vaccines, page 478). These vaccines are prepared from the gene or genes responsible for eliciting the immune response during infection by the pathogen. The gene is removed from the pathogen, incorporated into a cloning vector, and expressed in a host cell. The gene's protein product is isolated, purified, and used as a subunit vaccine. The advantage of a subunit vaccine is that it contains only that portion of the pathogen's genome that encodes for the specific antigen triggering the immune response. The first subunit vaccine developed in 1981 was a vaccine

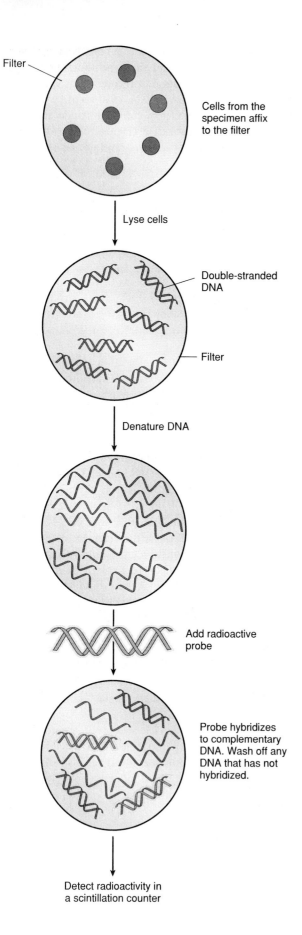

Filter

Cells from the specimen affix to the filter

Lyse cells

Double-stranded DNA

Filter

Denature DNA

Add radioactive probe

Probe hybridizes to complementary DNA. Wash off any DNA that has not hybridized.

Detect radioactivity in a scintillation counter

Recombinant Proteins Have Been Manufactured Through Recombinant DNA Technology

By using recombinant DNA techniques, it is now possible to custom manufacture recombinant proteins (Table 10.2). The first medically important product of genetic engineering was human insulin, released for commercial sale in the fall of 1982 (Figure 10.19). Other recombinant products that are now commercially available include human growth hormone (used to treat dwarfism), Factor VIII (a clotting protein lacking in many hemophiliacs), tissue plasminogen activator (an enzyme that dissolves blood clots and is used to treat heart attacks), and interferon (an inhibitor of viral replication, useful for treatment of hepatitis, genital warts, and some types of cancer). Many of these compounds previously have been expensive and difficult to obtain in large amounts. With the cloning of these genes into bacterial gene factories, it has been possible to reduce the price and increase the availability of these materials.

table 10.2

Genetically Engineered Products

Gene Cloned	Application
Factor VIII	Protein required for blood clotting, lacking in many hemophiliacs
Insulin	Treating diabetes
Interferon-α-2b	Treating hepatitis B infection
Human growth hormone	Treating dwarfism
Gamma interferon	Treating cancer
Interferon-α	Treating cancer
Tissue plasminogen activator	Dissolving blood clots
Streptokinase	Dissolving blood clots
Deoxyribonuclease	Treating cystic fibrosis
Interferon-β	Treating multiple sclerosis

 Microbial Metabolism and Growth
Genetics: Genetic Engineering • p. 30

figure 10.18

General Procedure for Using a Nucleic Acid Probe to Detect a Specific DNA Sequence in a Microorganism

A radiolabeled DNA probe hybridizes to complementary DNA in the microbe.

Recombinant DNA Technology Has Improved Agriculture

As the world's population continues to proliferate and land for farming becomes limited, farmers seek out new ways to improve crop productivity. Advances in biotechnology play an important role in many areas of agriculture, including crop nutrition and yield, plant resistance to pests and environmental extremes, and animal husbandry.

Plant cells present a special challenge to genetic engineers because there are few cloning vectors capable of successfully infecting and functioning within these cells. One of the most widely used cloning vectors for genetic manipulation of plant cells is the **Ti plasmid,** or **tumor-inducing plasmid.** The Ti plasmid, found in *Agrobacterium tumefaciens,* is responsible for the ability of this soil bacterium to form a tumor called a crown gall during infection of a plant. When *A. tumefaciens* infects a plant cell, a 30,000-base-pair segment of the plasmid known as T-DNA is incorporated into the host cell genome (Figure 10.20). This DNA segment can be used to shuttle genes into plant cells. For the Ti plasmid to be useful in genetic engineering, the following conditions are necessary: (1) the tumor-causing gene must be inactivated and (2) the target gene must be inserted into the T-DNA segment of the plasmid that will be incorporated into the host cell genome. Using *A. tumefaciens* and the Ti plasmid, it has been possible to generate transgenic crops such as soybean, alfalfa, lettuce, tomato, tobacco, squash, papaya, and potato. Genetically altered tomatoes resist spoilage and diseases, tobacco plants are more resistant to infection by tobacco mosaic virus (TMV), cotton plants are protected against the bollworm and other potentially devastating pests, and potato plants become immune to beetle damage.

One of the most exciting advances in plant biotechnology has been research to increase the quantity of nitrogen available for crops. Each year in the United States, farmers spend billions of dollars on nitrogenous fertilizers. Bacteria of the genus *Rhizobium* are capable of nitrogen fixation when grown symbiotically with certain crops called legumes (for example, peas, beans, alfalfa, clover, and so forth) (see nitrogen fixation, page 418). Scientists have identified the genes responsible for nitrogen fixation, called the *nif* genes, and have successfully transferred these genes to bacteria and yeasts, which normally do not fix nitrogen. This discovery now makes possible the potential transfer of the *nif* genes to plants, that normally do not fix nitrogen. These plants would then become self-fertilizing.

Other recent biotechnological advances in agriculture have included development of a strain (*ice⁻*) of *Pseudomonas syringae* that protects plants from frost damage because it cannot make a protein that promotes ice crystal formation; insertion of the *Bacillus thuringiensis* gene, which codes for an insecticidal protein (Bt-toxin), into plants to make the plants more resistant to insect larvae (see biological pesticides, page 584); and creation of herbicide-tolerant plants that can resist the effects of weed-killing herbicides. In addition to these areas of research, scientists also hope to engineer plants that have increased tolerances for drought, high salinity, soil pH extremes, and environmental contaminants.

In the area of animal husbandry, biotechnology has enabled farmers to increase the productivity and disease resistance of their animal stock. Genetically engineered bovine growth hormone was approved by the FDA in the 1980s to increase milk production in cows and is now routinely used to improve milk yield as much as 25%. Other recombinant molecules have produced pigs and cattle with less fat, cattle that are more resistant to shipping fever disease, and fish that are larger and more resistant to disease and toxic chemicals in polluted waterways.

Public and Scientific Concerns About the Regulation of Recombinant DNA Technology

The new science of genetic engineering is not without its problems. As gene splicing became popular in the 1970s and thousands of researchers began to experiment with the new technique in their unrestricted laboratories, the public became increasingly concerned about the inherent dangers of recombinant DNA technology. Newspapers, magazines, and television documentaries warned of the dangers of "monstrous microbes" escaping laboratories and infecting populations with rare and incurable diseases. City and town councils debated extensively the perils of recombinant DNA research within their municipalities. In 1976, the city of Cambridge, Massachusetts, went so far as to curtail the research activities of scientists at Harvard University and the Massachusetts Institute of Technology (although this curtailment was eased one year later).

This public outcry did not escape the attention of scientists. In February, 1975, a group of 134 distinguished scientists from throughout the world met at the Conference on Recombinant DNA Molecules at the Asilomar Conference Center in Pacific Grove, California. The scientists discussed the risks associated with gene splicing and the importance of biological containment in such work. The result of the conference was the establishment of a set of guidelines to be administered by the National Institutes of Health (NIH). These NIH guidelines listed not only the levels of containment to be used for recombinant DNA research, but also specified the types of experiments that should not be performed because of their potential hazards. Among these experiments are the introduction of genes from oncogenic viruses and pathogenic organisms into such common bacteria as *E. coli,* the insertion of genes for drug resistance into certain types of microbes, and the construction of recombinant DNA molecules containing such toxins as those for botulism or diphtheria. Although the NIH guidelines are enforceable only for federally funded research projects, most scientists working in industry or privately funded projects have also accepted and adhered to them.

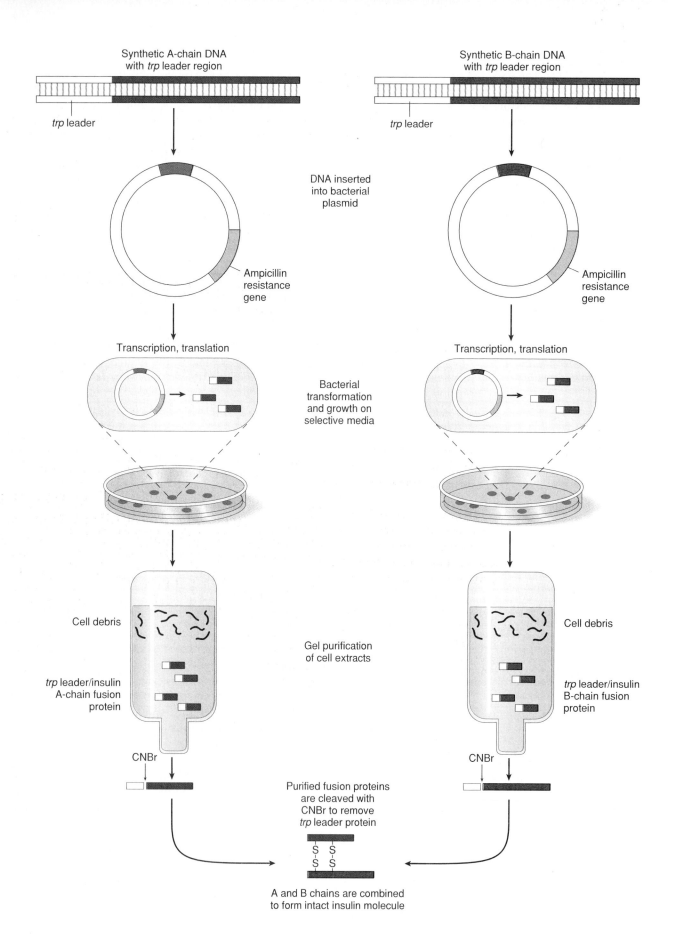

Synthetic A-chain DNA
with *trp* leader region

trp leader

Synthetic B-chain DNA
with *trp* leader region

trp leader

DNA inserted
into bacterial
plasmid

Ampicillin
resistance
gene

Ampicillin
resistance
gene

Transcription, translation

Transcription, translation

Bacterial
transformation
and growth on
selective media

Gel purification
of cell extracts

Cell debris

trp leader/insulin
A-chain fusion
protein

CNBr

Cell debris

trp leader/insulin
B-chain fusion
protein

CNBr

Purified fusion proteins
are cleaved with
CNBr to remove
trp leader protein

S—S   S—S
S—S   S—S

A and B chains are combined
to form intact insulin molecule

figure 10.19

Commercial Production of Human Insulin in *Escherichia coli* by Genetic Engineering

DNA coding for insulin A and B chains is synthetically prepared. The synthetic DNA is fused with a *trp* leader to control gene expression and is placed into bacterial plasmids. The plasmids, which contain the gene for ampicillin resistance, are inserted into recipient *E. coli*. The bacteria are grown in a medium containing ampicillin to select for cells that contain the plasmid. The *trp* leader/insulin fusion proteins are isolated from the bacteria and purified. The purified fusion proteins are cleaved with cyanogen bromide (CNBr) to remove the *trp* leader protein. Cyanogen bromide specifically cleaves polypeptide chains at methionine residues; the *trp* leader is fused to the insulin gene by a nucleotide triplet base sequence coding for methionine (insulin, which does not contain methionine, is not affected by the cyanogen bromide). The A and B peptides are connected by chemical treatment to form an intact insulin molecule.

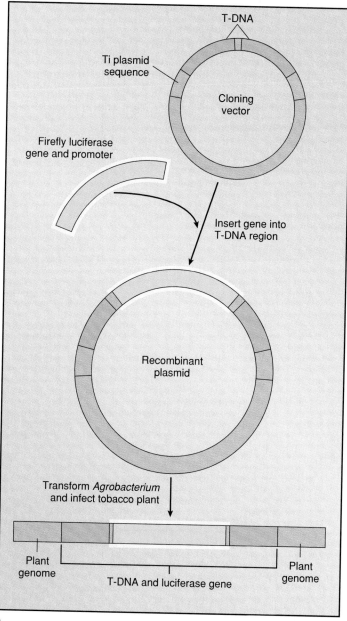

a.

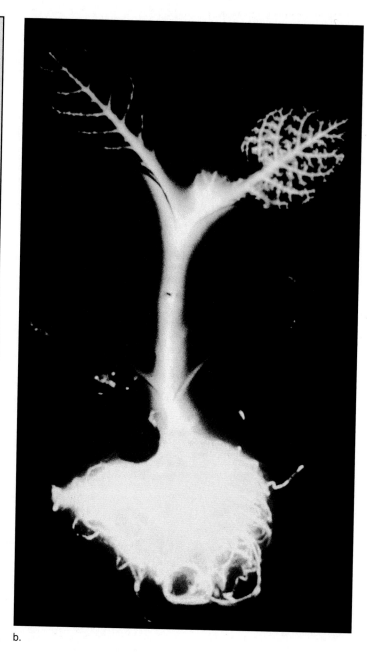

b.

figure 10.20

Use of the Ti Plasmid as a Cloning Vector

a. The firefly luciferase gene is incorporated into the T-DNA region of the Ti plasmid. The recombinant plasmid is then inserted into *Agrobacterium tumefaciens*, which is used to infect a tobacco plant.

b. Photograph of the transgenic tobacco plant, *Nicotiana tabacum*, made bioluminescent by the luciferase gene. When the plant is watered with a solution of luciferin, the substrate for luciferase, it brilliantly glows.

Diabetes mellitus is a chronic disorder usually caused by a deficient secretion of insulin. There are an estimated 14 million diabetics in the United States, and almost 40,000 die annually as a result of this disorder, making it the third most common cause of death. Many of these diabetics rely upon daily injections of insulin. In the past, insulin was extracted from the pancreatic glands of pigs and cattle slaughtered for food, but with the increasing number of diabetics, demand may exceed available supplies. In addition, many patients are allergic to nonhuman insulin. Recombinant DNA technology has made it possible to overcome these problems by the insertion of the genes for human insulin into *Escherichia coli*.

The first report of this accomplishment was in 1979 by David V. Goeddel and coworkers at Genentech, Inc., South San Francisco. Insulin is a protein consisting of two polypeptide chains, an A chain containing 21 amino acids and a B chain containing 30 amino acids, joined by two disulfide bridges. Goeddel and coworkers cloned separately the synthetic genes for human insulin A and B chains in plasmid pBR322 using the following procedure (Figure 10P.1). The right half (BB) of the B-chain gene was chemically constructed from oligonucleotides, and the ends of the BB half were made cohesive ("sticky ends") by digestion of the DNA with the restriction enzymes *Hind*III and *Bam*HI. The BB fragment was then inserted into plasmid pBR322 that had

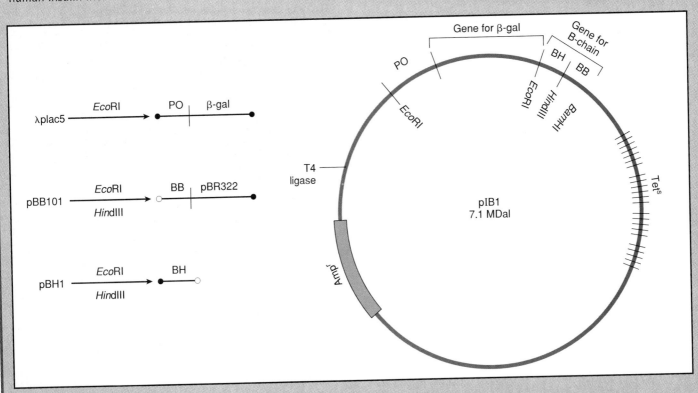

figure 10P.1

Construction of the *Lac*-Insulin Plasmid for Human Insulin B Chain

Plasmid pBB101, containing the right half (BB) of the human insulin B chain and most of plasmid pBR322, and plasmid pBH1, containing the left half (BH) of the insulin B chain, are purified by gel electrophoresis, mixed, and inserted into *Eco*RI-cleaved plasmid λplac5 (which contains a *lac* control region and most of the structural gene for β-galactosidase) to form plasmid pIB1. Plasmid pIB1 now has the gene for the insulin B chain and the *lac*

control region, which can be used as an on-off promoter switch. The plasmid also codes for tetracycline resistance, which aids in the selection of transformed *Escherichia coli* cells. A similar procedure is used to construct the *lac*-insulin A plasmid (not shown). o represents *Hind*III restriction site ends; • represents *Eco*RI restriction site ends.

been cleaved with *Hin*dIII and *Bam*HI. The left half (BH) of the B-chain gene was prepared similarly and inserted into a second cloning vehicle, pBR322, using *Eco*RI and *Hin*dIII restriction endonucleases. The A chain gene was inserted as one fragment into a third pBR322 at the *Eco*RI-*Bam*HI sites. In each instance, fragment insertion into the pBR322 plasmid was confirmed by the conversion of the plasmid from tetracycline-resistant to tetracycline-sensitive.

Plasmids pBB101 (containing fragment BB) and pBH1 (containing fragment BH) were then digested with *Eco*RI and *Hin*dIII endonucleases. The BH fragment of pBH1 and a large fragment of pBB101 (containing the BB fragment and most of pBR322) were purified by gel electrophoresis, mixed, and inserted into *Eco*RI-cleaved plasmid λplac5 to form plasmid pIB1. Plasmid λplac5 contains a *lac* control region and most of the structural gene for β-galactosidase and provides an on-off promoter switch for efficient plasmid gene transcription and translation. A second plasmid (pIA1) was constructed from *Eco*RI-treated pA11 (pBR322 containing the A chain gene) and the *lac* portion of λplac5.

The resultant plasmids—pIB1, which contained the B-chain genes, and pIA1, which contained the A-chain genes—were separately used to transduce *E. coli*. *E. coli* containing pIB1 and grown in the presence of isopropyl-β-D-thiogalactoside (IPTG), an inducer of β-galactosidase, produced β-galactosidase-insulin B-chain hybrids. *E. coli* containing pIA1 and induced with IPTG produced β-galactosidase-insulin A-chain hybrids.

The insulin peptides were cleaved from β-galactosidase and assayed for activity by radioimmunoassay. ^{35}S-labeled A and B chains were prepared by S-sulfonation (SSO$_3^-$) and mixed together or separately with porcine-derived insulin A chain or bovine-derived B chain in the presence of anti-insulin antibody. *E. coli*-prepared A and B chains showed radioimmune activity when mixed together or separately with porcine or bovine insulin peptides (Table 10P.1).

This experiment was a major accomplishment and one of the early milestones in recombinant DNA technology. The successful insertion of eucaryotic genes in a procaryote indicated that it was then possible to use microorganisms as factories for the production of medically and industrially important molecules.

Source

Goeddel, D.V., D.G. Kleid, F. Bolivar, H.L. Heyneker, D.G. Yansura, R. Crea, T. Hirose, A. Kraszewski, K. Itakura, and A.D. Riggs. 1979. Expression in *Escherichia coli* of chemically synthesized genes for human insulin. *Proceedings of the National Academy of Sciences* 76:106–110.

table 10P.1

Reconstitution of Radioimmune Human Insulin

Escherichia coli–prepared A and B chains of human insulin are mixed together or with porcine insulin A chain or bovine insulin B chain. Results are given as nanograms of radioimmune active insulin produced.

"A Chain" Sample	"B Chain" Sample	Radioimmune Active Insulin (ng)
E. coli 58	—	<0.5
—	*E. coli* DE117	<0.5
Porcine A	*E. coli* DE117	74
E. coli 58	Bovine B	45
E. coli 58	*E. coli* DE117	20

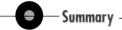
1. Genetic engineering is the deliberate modification of the genetic makeup of a cell or organism. As a result of recent advances in genetic engineering, it is now possible to manipulate genes in living organisms, a process called recombinant DNA technology.

2. Restriction endonucleases are enzymes that cleave specific sequences in double-stranded DNA that exhibit twofold symmetry around a central axis (palindromes), often resulting in single-stranded overhangs, or sticky ends, that are useful in genetic engineering.

3. In genetic engineering, the gene of interest is obtained directly from genomic DNA, synthesized from an mRNA template by the enzyme reverse transcriptase, or constructed from nucleotides in vitro.

4. The gene is inserted into a cloning vector, which is an independently replicating DNA molecule that serves as a vehicle to protect the gene from degradation and transport the gene into the host cell. Plasmids and bacteriophage genomes are common cloning vectors used in genetic engineering. A collection of cloned DNA fragments from the genome of an organism constitutes a DNA library.

5. The vector containing the gene of interest is inserted into a host cell. Plasmid vectors are introduced into host cells by transformation, whereas bacteriophage vectors enter host cells by infection.

6. The gene of interest can be detected in a host cell by various methods. Protein products of the gene can be used to screen for host cells containing the gene. Nucleic acid probes can be used to screen cells for the gene. Presence of the gene in the cell can also be determined by the activity of a reporter gene located on the cloning vector.

7. The polymerase chain reaction rapidly amplifies DNA by cyclic repetition of three steps: denaturing of the double helix, annealing of oligonucleotide primers to the DNA, and extending of the primers by a DNA polymerase.

8. Recombinant DNA technology has applications in many areas. Disease-causing genes can be identified and further characterized. Antisense RNA or DNA can be developed to selectively turn off or modify the activity of a specific gene. Subunit vaccines can be prepared from the genes responsible for eliciting an immune response during infection by a pathogen. Nucleic acid probes can be prepared for identifying target genes in pathogenic microorganisms in disease diagnosis. Recombinant proteins such as human insulin and human growth hormone can be custom manufactured at reduced expense. Agricultural crops can be genetically modified to become more resistant to pests and disease.

EVOLUTION *and* BIODIVERSITY

Recombinant DNA technology has made possible not only advances in medicine, agriculture, and industry, but also is an invaluable research tool to explore the evolution and diversity of life. Through this technology we can compare the DNA or RNA sequences of two organisms and measure the degree of divergence in their sequences as a molecular barometer of biological evolution. With nucleic acid hybridization, the extent of base pairing can be used to determine the evolutionary distance between species and genera. Gene transfer and genetic recombination have been occurring naturally for billions of years and have resulted in the diversity among organisms that we see today. With the development of recombinant DNA technology, it is now possible to manipulate DNA molecules and genetically customize procaryotes and eucaryotes. This rapidly emerging technology raises legal, ethical, and biological questions regarding its use. As microbial, animal, plant, and human genomes are sequenced and manipulated in the next few years, will this new knowledge benefit humankind or will it replace natural selection and upset the natural progression of evolution?

Questions

Short Answer

1. What discoveries made genetic engineering possible?

2. What is unique about the class of nucleases called restriction endonucleases? It is believed that bacteria make restriction endonucleases for what purpose?

3. How can "sticky ends" of just a few bases be manipulated?

4. Explain what is meant by the term *vector* and give examples.

5. How do researchers isolate the "target gene" for genetic engineering?

6. Explain how a gene can be synthesized from mRNA.

7. Describe the process of electrophoresis.

8. Compare and contrast Southern blotting and northern blotting.

9. Explain how researchers are able to introduce the vector, carrying a target gene, into a new cell.

10. Explain how researchers know if the target gene has been incorporated into the organism's genome.

11. Why are "selective markers" desirable traits for cloning vectors?

12. Briefly describe the PCR technique.

13. Briefly describe DNA fingerprinting.

14. How do transgenic organisms differ from those of selective breeding programs?

Multiple Choice

1. Which of the following can *not* be cloned?
   a. chromosomal DNA
   b. plasmid DNA
   c. mRNA
   d. proteins
   e. All of the above could be cloned.

2. Which of the following is required for genetic engineering?
   a. DNA
   b. ligase and restriction endonuclease
   c. vector
   d. host cell
   e. All of the above except a.

Critical Thinking

1. Explain why bacteria and viruses have been featured in most genetic engineering experiments. Describe the difficulties encountered when trying to genetically engineer complex organisms like mammals. Suggest strategies to overcome each difficulty.

2. You have been given a project to clone the melatonin gene from humans into *Escherichia coli*.

   a. What characteristics would you require of your vector?
   b. Would an expression vector be helpful for this project? Why or why not?
   c. How will you monitor your success? Would insertional inactivation be helpful for this project? Why or why not? What advantage is there to having two marker sequences?

3. Discuss the impact of recombinant DNA technology on society. Is this a political issue? Explain.

 Supplementary Readings

Alberts, B., D. Bray, J.Lewis, M. Raff, K. Roberts, and J.D. Watson. 1994. *Molecular biology of the cell*, 3d ed. New York: Garland Publishing, Inc. (An in-depth textbook on cell structure and function.)

Ausubel, F.M., R. Brent, R.E. Kingston, D.D. Moore, J.G. Seidman, J.A. Smith, and K. Struhl, eds. 1995. *Short protocols in molecular biology,* 3d ed. New York: John Wiley & Sons. (A detailed laboratory manual on gene cloning.)

Gasser, C.S., and R.T. Fraley. 1992. Transgenic crops. *Scientific American* 266:62–69. (A review of transgenic crops produced by genetic engineering.)

Kelly, D.P., and N.G. Carr, eds. 1984. *The microbe 1984: Part II. Procaryotes and eucaryotes.* London, England: Cambridge University Press. (A review of microorganism impact in the environment and in genetic engineering.)

Mullis, K.B. 1990. The unusual origin of the polymerase chain reaction. *Scientific American* 262:56–65. (A description by Kary Mullis of how he devised the innovative polymerase chain reaction.)

Pershing, D.H., T.F. Smith, F.C. Tenover, and T.J. White. 1993. *Diagnostic molecular microbiology: Principles and applications.* Washington, DC: American Society for Microbiology. (An in-depth manual of molecular tests for identification of pathogenic microbes, with focus on PCR-based procedures.)

Peters, P. 1993. *Biotechnology: A guide to genetic engineering.* Dubuque, Iowa: Wm. C. Brown Communications, Inc. (A review of basic techniques in gene cloning and recombinant DNA technology.)

Russell, P.J. 1996. *Genetics*, 4th ed. New York: HarperCollins. (A fundamental textbook of genetics, with sections on DNA replication, transcription, translation, genetic exchange in bacteria, transposable elements, and recombinant DNA technology.)

Watson, J.D., N.H. Hopkins, J.W. Roberts, J.A. Steitz, and A.M. Weiner. 1987. *Molecular biology of the gene*, 4th ed. Menlo Park, Calif.: The Benjamin/Cummings Publishing Company. (An excellent discussion of macromolecular synthesis, gene function, and molecular genetics.)

Watson, J.D., M. Gilman, J. Witkowski, and M. Zoller. 1992. *Recombinant DNA*, 2d ed. New York: W. H. Freeman and Company. (A review of techniques and principles of cloning.)

Zubay, G. 1993. *Biochemistry*, 3d ed. Dubuque, Iowa: Wm. C. Brown Communications, Inc. (A comprehensive textbook of biochemistry, including detailed chapters on genetic engineering and recombinant DNA technology.)

chapter eleven

PROCARYOTES: THE BACTERIA AND THE ARCHAEA

Classification of Microorganisms

Ribosomal RNA and Phylogeny

Classification of Procaryotes

Gram-Negative Bacteria of General, Medical, or Industrial Importance

Gram-Positive Bacteria Other Than the Actinomycetes

The Archaea, Cyanobacteria, and Remaining Gram-Negative Bacteria

The Actinomycetes

PERSPECTIVE
Symbiotic Association of Chemoautotrophic Bacteria with a Marine Invertebrate

EVOLUTION AND BIODIVERSITY

Colorized scanning electron micrograph of the spirochete *Leptospira interrogans*, showing a periplasmic flagellum wrapped around the flexible helical cell (×22,000).

 Microbes in Motion ⎯⎯⎯⎯⎯ PREVIEW LINK

Among the key topics in this chapter is characteristics of certain groups of bacteria: Yersinia, Vibrio, Chlamydia, and Mycoplasma. The following sections in the *Microbes in Motion* CD-ROM may be useful as a preview to your reading or as a supplemental study aid:

Miscellaneous Bacteria: Zoonoses (Yersinia), 57–67. *Gram-Negative Organisms:* Bacilli—Facultative Anaerobes (Vibrionaceae), 30–38. *Miscellaneous Bacteria:* Chlamydiae, 1–6: Mycoplasma, 1–6.

Imagine a university library, with its many thousands of volumes, without a logical method for arranging these books. Such a disordered library would be practically useless. Books could not be easily found, and once found, could not be replaced in locations where they could be found again. There would be no method for the systematic addition of new publications to the collection. Books on a similar subject would be randomly dispersed throughout the building.

Now consider a living library that contains as many as 5 million different kinds of organisms. The sheer numbers and diversity of organisms, including microorganisms, make it necessary to have an organized system for classification. Clearly the identification of microorganisms becomes a formidable, if not impossible, task without such a system. The systematic categorization of organisms into a coherent scheme is known as **taxonomy.** Taxonomy serves not only to organize plants, animals, and microbes into categories, but also can be useful in showing possible evolutionary relationships among similar types of organisms. Taxonomy provides order; organisms with common characteristics are recognized as such, and assigned to the same taxon (group or rank).

Classification of Microorganisms

Microorganisms occupy a unique position among living organisms because they are able to exist as independent, self-sufficient cells. Microbes may be either procaryotes (lacking a nucleus) or eucaryotes (possessing a true nucleus). Unlike animals or plants, which have distinctive organs, tissues, or structures for specialized functions, microbes are autonomous units of living matter. Each microbial cell is a dynamic system, able to grow, metabolize, and reproduce without the assistance of sister cells. In contrast, cells in an animal or a plant work together for the benefit of the entire organism. They act collectively to gather food and energy from the environment for the growth and reproduction of the whole organism. Even in instances where several thousand microbial cells aggregate to form a large structure, such as in the case of algae, fungi, or the fruiting bodies of myxobacteria, each cell within the structure maintains its independence and autonomy. It is this characteristic of independence and autonomy that distinguishes microbial cells from cells of other forms of life.

Microorganisms are named by the traditional binomial system first used by the Swedish physician and naturalist Carolus Linnaeus (1707–1778). This method of **nomenclature** (the systematic naming of organisms), widely used for the classification of plants and animals, assigns a species name and a genus name to each organism. Microbial taxonomists refer to a species as a group of microorganisms that have similar **phenotypic** (observable) characteristics. Organisms of the same species often have similar, although not necessarily identical, **genotypic** (genetically derived) characteristics. Similar species are placed into a higher taxon called

table 11.1

Nomenclature of the Bacterium *Treponema pallidum*

Taxonomic Rank	Example
Domain	Bacteria
Division	Gracilicutes
Class	Scotobacteria
Order	*Spirochaetales*
Family	*Spirochaetaceae*
Genus	*Treponema*
Species	*pallidum*
Strain	Nichols

a genus. Similar genera (plural of genus), in turn, are grouped into families, and families are part of orders (Table 11.1). An additional classification category used in bacteriology is the **strain.** A species can consist of several different strains; cells of a strain are all derived from the same ancestor and retain the characteristics of the ancestor. Thus a bacterium such as *Escherichia coli* (genus and species name) may consist of several thousand different strains, each of which shares the defining set of phenotypic traits.

Classification Schemes for Microorganisms Have Evolved over the Years

Many different classification schemes have been proposed for microorganisms (Table 11.2). Early classification systems established before the discovery of microbes divided all living organisms into either plants (kingdom **Plantae**) or animals (kingdom **Animalia**) on the basis of characteristics such as photosynthetic capabilities, motility, and structural features. Plants were photosynthetic, stationary, and had rigid cell walls. Animals were generally motile, nonphotosynthetic, and lacked cell walls. This two-kingdom system was used without major difficulty for many years. As microorganisms were discovered, however, many problems arose. Microbes had unusual characteristics, neither plantlike nor animal-like. Some microorganisms were photosynthetic and motile, whereas other, nonphotosynthetic microbes were stationary. Still others had the ability to change their metabolic characteristics under different environmental conditions. As a result of such traits, taxonomists began to search for classification schemes that would take into consideration these newly discovered forms of life.

An expanded three-kingdom classification system to accommodate microorganisms was proposed in 1866 by the German zoologist Ernst H. Haeckel (1834–1919). Haeckel proposed that those organisms that were basically unicellular and simple in organization be placed into a third distinctive kingdom, designated the **Protista.** The protists consisted of bacteria, algae, fungi, and protozoa. Haeckel's proposal never gained complete acceptance, but his use of the word *Protista* to denote microorganisms was incorporated into subsequent classification schemes.

table 11.2

Some Classification Systems for Living Organisms

System's Original Proposer	Classification by Kingdom or Domain and Major Groups of Organisms
Linnaeus (1753)	Animalia — Animals, protozoa Plantae — Plants, algae, bacteria, fungi
Haeckel (1866)	Animalia — Animals Plantae — Plants, multicellular algae Protista — Unicellular algae, bacteria, fungi, protozoa
Whittaker (1969)	Animalia — Animals Plantae — Plants Protista — Algae, protozoa Fungi — Fungi Monera (Procaryotae) — Procaryotes
Woese (1981)	Archaea } Bacteria } Domains Eucarya }

As additional traits of microorganisms became known through improvements in microscopy and the development of biochemical techniques, more detailed and scientific classification schemes were proposed. One of the most widely accepted systems is the five-kingdom system suggested in 1969 by Robert H. Whittaker. Whittaker's system recognizes five kingdoms of living organisms based on cell structure and modes by which organisms obtain their nutrients (*photosynthesis, absorption* of dissolved nutrients, and *ingestion* of undissolved food particles): **Monera** (also called **Procaryotae), Fungi, Protista, Plantae,** and **Animalia.** Other classification systems separate microorganisms—bacteria (including the cyanobacteria, or blue-green bacteria), algae, fungi, and protozoa—into two groups: the **procaryotes** and the **eucaryotes.** This distinction is determined by cellular structure. In such classifications, bacteria, which do not have membrane-enclosed nuclei, are defined as procaryotes and placed in the kingdom Monera.

Unlike procaryotes, eucaryotic cells undergo mitotic division and have well-defined chromosomes with associated histones (basic proteins that neutralize the negative charges of the DNA) and membrane-enclosed organelles such as mitochondria, chloro-plasts, lysosomes, and the endoplasmic reticulum. Eucaryotes are generally physically larger than procaryotes and have more complex mechanisms of locomotion. In Whittaker's classification system, the various eucaryotic microbes are assigned to the kingdoms Fungi (fungi) and Protista (protozoa and algae). Whittaker's system is based to a large extent on evolutionary divergence. In this evolutionary scheme, Monera is regarded as the most primitive kingdom. A few organisms in this kingdom are believed to have evolved into the Protista, some members of which in turn developed into the other three kingdoms. The orderly arrangement and evolution-oriented progression of Whittaker's system makes it one of the more widely accepted classification schemes today.

A different system was proposed in 1981 by Carl R. Woese. Woese argued that the division of organisms into procaryotes and eucaryotes was insufficient and that the procaryotes could be further separated based on their 16S rRNA sequences. He therefore split the procaryotes into two groups based on the nucleotide sequences of their 16S rRNA sequences: **Archaea** [Greek *archaios,* ancient, as in the Archaean era 3.9 to 2.6 billion years ago], and **Bacteria.** The two groups of procaryotes also had other distinguishing differences. The Archaea are procaryotes, because they do not have a membrane-enclosed nucleus, and morphologically resemble the Bacteria. Archaea have several different kinds of cell walls or no cell wall, but no true peptidoglycan, which is always found in the cell walls of Bacteria. Archaea, furthermore, possess plasma membrane lipids and ribosomal ribonucleic acids that are different from those found in Bacteria and eucaryotes (see Archaea and Bacteria membrane lipids, page 51). These dissimilarities are significant enough, in the opinion of Woese and other scientists, to warrant a separate taxonomic classification for the Archaea based on genetic content, specifically the sequence of nucleotides in 16S ribosomal RNA. Under this classification the Archaea include procaryotes currently divided into three main categories: the methanogens, the extreme halophiles, and the extreme thermophiles. Woese's classification system groups all living organisms into one of three domains: Bacteria, Archaea, and Eucarya.

Despite the variety and complexity of classification schemes that have been proposed, no one scheme is completely satisfactory. Living organisms, particularly microbes, which grow and reproduce at rapid rates, undergo constant modifications in structure, physiology, and genetic composition. As a consequence of these evolutionary changes, the characteristics of organisms change. Such changes, although not easily incorporated into most classification systems, are readily adaptable to schemes that take evolutionary divergence into consideration.

Three Major Approaches Have Been Used to Identify and Place Microorganisms in Classification Systems

Regardless of the system used to classify microorganisms, varied methods have been used to identify and place these microbes in their appropriate places in classification schemes. These methods are: the classical approach, numerical taxonomy, and molecular approaches.

table 11.3

Criteria Used in the Classical Approach to Procaryotic Taxonomy

Category	Examples of Criteria
Structure/morphology	Shape
	Size
	Arrangement
	Capsules
	Sheaths
	Flagella and their arrangement
	Endospores
	Inclusion bodies
	Gram stain
Biochemical/physiological	Range of carbohydrates used as carbon and energy sources
	Optimum temperature for growth
	Range of temperature for growth
	Optimum pH for growth
	Range of pH for growth
	Growth factor requirements
	Aerobic, anaerobic, facultatively anaerobic
	End products of respiration and fermentation
	Antibiotic sensitivities

The Classical Approach Orders Procaryotes on the Basis of Morphology and Physiology

The **classical approach** to procaryotic taxonomy is one in which procaryotes are grouped into **species** and **genera** primarily on the basis of their structural and morphological characteristics and, secondarily, on the basis of biochemical and physiological traits (Table 11.3). Organisms with similar cell shapes or structures are thus in the same group. Other physical attributes used in the classical approach to taxonomy include the presence and type of flagella, staining properties, and the presence of external coverings such as capsules. Examples of secondary characteristics used in taxonomy are nutritional or atmospheric growth requirements and by-products of cellular metabolism.

Numerical Taxonomy Groups Microorganisms into Phenoms Based on Their Similarities

Although the classical approach is used by many taxonomists, it has the disadvantage of an inherent bias. The taxonomist must determine which characteristics among the many in an organism are the most important. This bias led to the establishment in the eighteenth century of an alternate system for classification known as **numerical,** or **Adansonian, taxonomy,** named after its inventor, the French botanist Michael Adanson (1727–1806). In numerical taxonomy, all observable characteristics of an organism carry equal weight and therefore are considered equally in determining similarities among organisms (Figure 11.1).

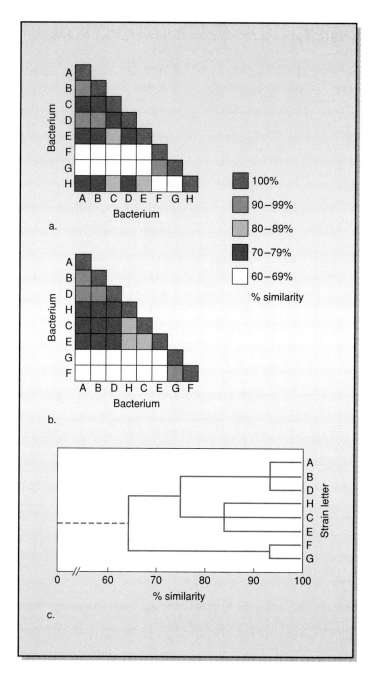

figure 11.1

Numerical Taxonomy

Eight different bacterial strains (A through H) are compared with one another by similarity coefficients. a. Matrix of strains before rearrangement. b. Matrix of strains after rearrangement by percent similarity. Strains with similar characteristics are clustered.
c. Dendrogram constructed from similarity values. Vertical lines define the percent similarity between different strains or groups of strains. For example, strains A, B, and D are 90% to 99% similar, as are strains F and G. Strains H, C, and E are 80% to 89% similar. However, when strains A, B, and D are compared to strains F and G, there is only a 60% to 70% similarity.

table 11.4

Determination of Similarity Coefficient and Matching Coefficient by Numerical Taxonomy for Two Procaryotes

Number of characteristics present in both organisms: a

Number of characteristics present in organism #1 and absent in organism #2: b

Number of characteristics absent in organism #1 and present in organism #2: c

Number of characteristics absent in both organisms: d

Matching coefficient $(S_S) = \dfrac{a + d}{a + b + c + d}$

Similarity coefficient $(S_J) = \dfrac{a}{a + b + c}$

Similarities between two or more organisms are determined using one of two methods (Table 11.4). Both methods compare the number of identical characteristics to the total number of characteristics observed in the organisms. The **matching coefficient (S_S) method** is based on the percentage of characteristics that are common to two organisms when compared (that is, characteristics present in both organisms or absent in both organisms). The **similarity coefficient (S_J) method** determines the percentage of characteristics that are present in both organisms and does not consider characteristics absent in the two organisms. Both types of comparisons cluster similar organisms into groups called **phenoms** (categories comparable to species or genera in classical taxonomy), with boundaries established by predetermining matching or similarity coefficient values. In instances where several organisms and many characteristics are compared, a diagrammatic matrix called a **dendrogram** is used to illustrate similarity levels (Figure 11.1c).

Numerical taxonomy eliminates the biases common to other taxonomic approaches and provides an objective comparison among organisms. Comparisons become statistically more significant as larger numbers of characteristics are compared. Such analyses of many organisms and their similarities clearly cannot be performed without the aid of computers.

Molecular Approaches Group Microorganisms Based on Nucleic Acid Comparisons

Development of modern techniques for genetic analysis in microorganisms has led to more refined methods for microbial classification. **Molecular approaches** to taxonomy involve comparisons of nucleic acids among organisms. Such comparisons include determinations of DNA or RNA sequences and studies of hybridization among nucleic acids from different organisms.

DNA contains four different bases (adenine, thymine, guanine, and cytosine). Their structures and bonding arrangements are such that in double-stranded DNA (the usual form of bacterial DNA), guanine and cytosine are always paired, as are adenine and thymine. The proportion of guanine-cytosine base pairs in the DNA of the genome (commonly expressed as the mole percent of guanine plus cytosine content or **mole% G + C**) will vary in different organisms. Two different species of organisms usually will not have identical mole% G + C. Among procaryotes, the mole% G + C ranges from 25 to 75. By determining the proportion of G + C in isolated chromosomal DNA, limits on the genetic relatedness of two organisms can be established. Organisms that have vastly different mole% G + C contents are assumed to be genetically dissimilar and therefore not related. However, the corollary to this statement—procaryotes with similar mole% G + C are closely related—is not always valid because the mole% G + C values do not take into consideration the sequence of bases in the DNA.

Mole% G + C values can be determined by a procedure known as the thermal denaturation method (Figure 11.2). In this method, a

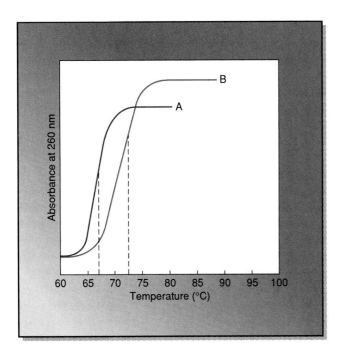

figure 11.2

DNA Melting Curve

Double-stranded DNA is heated, and the extent of hydrogen bonding between the two strands is determined by the melting temperature (T_m, the temperature at the midpoint of the absorbancy increase). In the example, the T_m of curve A is 67°C and of curve B, 72.5°C.

solution of double-stranded DNA is slowly heated, and its strand separation, or denaturation, is followed by an increase in absorbance as detected by a spectrophotometer. Single-stranded DNA has approximately 40% greater absorbance than double-stranded DNA when measured at a wavelength of 260 nm. The temperature at which strand separation occurs depends on the extent of hydrogen bonding between the two DNA strands and can be plotted on a melting curve of temperature versus absorbance. Since guanine and cytosine base pairs are joined by three hydrogen bonds, whereas adenine and thymine base pairs are joined by only two hydrogen bonds, the melting temperature (T_m, the temperature at the midpoint of this curve) indicates the extent of hydrogen bonding in double-stranded DNA. As the mole% G + C value increases, the T_m also increases.

A more precise approach to molecular taxonomy is a comparison of genetic relatedness by nucleic acid hybridization (Figure 11.3). In this procedure, nucleic acids from two different organisms are isolated and combined together as either DNA-DNA or DNA-mRNA double strands—a process known as hybridization. Many different methods have been developed to determine DNA homology between two organisms. In one method, one of the organisms is grown in the presence of radio-labeled thymine, which is incorporated into newly synthesized DNA of the organism. The second organism is grown in a medium containing nonradiolabeled thymine; its newly synthesized DNA is nonradioactive. The DNAs from both organisms are isolated and denatured with high temperature (100°C) or high pH (pH ≥ 13). The resultant single strands are allowed to hybridize on a nitrocellulose filter. After hybridization, any remaining single-stranded DNA is digested with a nuclease, such as S1 nuclease from *Aspergillus,* that digests single-stranded but not double-stranded DNA. The extent of hybridization is determined by measuring the amount of radioactivity remaining in the hybrid molecules.

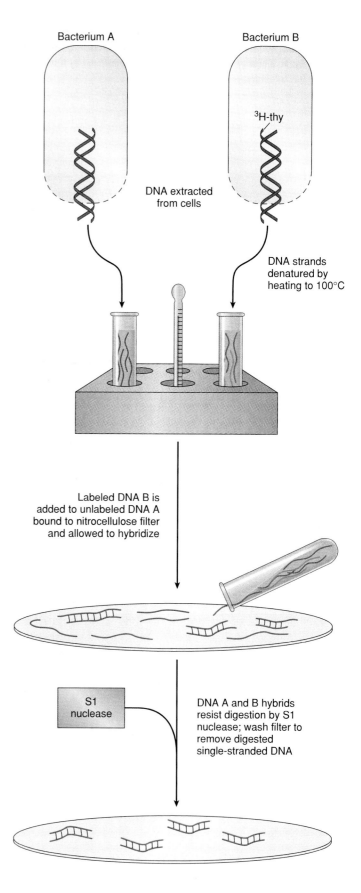

figure 11.3

Nucleic Acid Hybridization

DNA is extracted from two organisms (bacterium A and bacterium B), one of which has been grown in the presence of tritiated thymine (³H-thy) to label radioactively its DNA. Each DNA suspension is heated to 100°C to denature the double-stranded nucleic acids. The nucleic acids are allowed to hybridize by adding the labeled single-stranded DNA to unlabeled single-stranded DNA immobilized on a nitrocellulose filter. After hybridization, any remaining single-stranded DNA is digested by a nuclease such as S1 nuclease from *Aspergillus* (double-stranded DNA is not attacked by nucleases) and the filter is washed. The amount of radioactivity remaining in the double-stranded hybrid molecules is a measure of the genetic relatedness of the two organisms.

Organisms that are closely related would be expected to have a higher degree of DNA homology than organisms that are not as closely related.

Other methods used for comparisons of genetic relatedness among organisms are determinations of nucleic acid (DNA and RNA) base sequences and the extent of genetic exchange between organisms (see exchange of genetic material, page 250). A popular method recently developed for clinical diagnosis of microorganisms involves the use of DNA probes—small pieces of DNA that recognize specific genes—to identify gene sequences in an organism (see nucleic probes, page 295). This procedure, which is based on DNA:DNA hybridization, is a powerful diagnostic aid because of its high degree of specificity in rapidly identifying microorganisms. Such genetic tools provide what many believe may be the most accurate approaches to taxonomy, since relatedness is ultimately traced to the genetic composition of organisms.

Ribosomal RNA and Phylogeny

The advent of molecular techniques made possible not only genetic analysis of microorganisms, but also the study of evolutionary relationships among living organisms, or **phylogeny.** Although there are many ways to study phylogenetic relationships (for example, amino acid sequences of proteins, nucleotide sequences of nucleic acids found in organisms, or presence or absence of essential enzymes), it is now recognized that ribosomal RNA (rRNA) sequences are important indices of phylogeny, particularly among procaryotes. Ribosomal RNAs have characteristics that are important in studying evolutionary divergences among organisms. These universal characteristics have identical functions in all living organisms. This functional constancy makes rRNAs ideal molecular chronometers to measure evolutionary change over the billions of years that procaryotes have existed. Because rRNA is a small molecule that cannot tolerate much structural change and still retain its function, its sequence is moderately well **conserved,** or constant, across phylogenetic lines. Consequently, small differences in rRNA sequences can be used to determine evolutionary distances between organisms. Among the three rRNA molecules (5S, 16S, and 23S) in procaryotes, 16S rRNA is used most commonly as a phylogenetic tool. The small size (125 nucleotides) of the 5S rRNA limits the amount of information that can be generated from this molecule, whereas the large size (2,900 nucleotides) of the 23S rRNA makes this molecule more difficult to experimentally analyze than the 16S rRNA (about 1,500 nucleotides). Carl R. Woese and Norman R. Pace revolutionized microbial systematics in the 1970s and 1980s by using 16S rRNA as a molecular chronometer to follow evolutionary divergence among procaryotes.

Sequencing of Ribosomal RNA Is Relatively Simple

The sequencing of ribosomal RNA is relatively simple. After cell lysis and phenol extraction of the total RNA, the RNA is precipitated with alcohol and salt. A small DNA primer that has a complementary base sequence to a highly conserved region of the 16S rRNA is added to the mixture. Reverse transcriptase and deoxyribonucleotides are then added to generate cDNA from the 16S rRNA template. The cDNA is sequenced and this information can be used to deduce the sequence of the 16S rRNA. The polymerase chain reaction (PCR) can also be used to amplify the rRNA genes (the cDNA). PCR amplification requires far fewer cells than direct rRNA sequencing and is more rapid and convenient when processing many samples.

Phylogenetic Trees Can Be Developed from Ribosomal RNA Sequences

Because 16S rRNA has several regions containing highly conserved sequences, slight differences in the 16S rRNA sequences from different organisms can be used to determine their phylogenetic relationships. The 16S rRNA sequences from organisms are compared and phylogenetic trees showing evolutionary distances and relationships among these organisms are constructed from differences in RNA alignment. The concept behind this type of comparison is that procaryotes branching off from other procaryotes a long time ago will have had time to diversify, and their 16S rRNA sequences will be less similar than those of procaryotes that have branched off more recently.

As a result of such analyses, phylogenetic trees have been developed for different groups of procaryotes. **Signature sequences,** or short nucleotide sequences unique to certain groups of procaryotes, have been useful in defining three specific domains (Bacteria, Eucarya, and Archaea) of a **universal phylogenetic tree** (Figure 11.4). Organisms within each of these three domains are distinctly separated from each other by large sequence distances in their 16S rRNA (18S rRNA in eucaryotes). The root of the universal tree represents the common ancestor from which all living organisms evolved.

Interestingly, this universal tree shows the Archaea to be more closely related phylogenetically to Eucarya than to Bacteria. It is clear from rRNA sequence analyses that the universal ancestor branched in two directions: the Bacteria and the Archaea/Eucarya. With the passage of time, the Archaea and the Eucarya diverged from one another.

These phylogenetic relationships were further confirmed in 1996 when the 1.66 million base pair genome of the Archaeon *Methanococcus jannaschii* (named after Holger Jannasch, leader of an expedition that discovered this methane-producing microorganism in a deep-sea hydrothermal vent beneath the Pacific Ocean

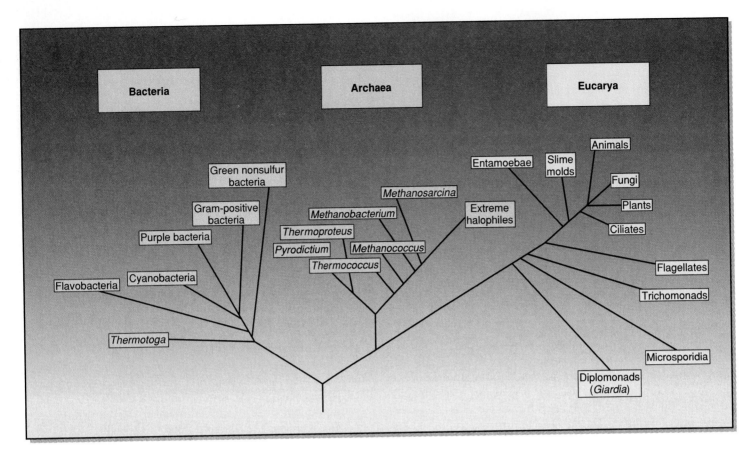

figure 11.4
The Universal Phylogenetic Tree as Determined by Ribosomal RNA Sequencing
Phylogenetic relationships among organisms in the three domains are shown.

in 1982) was sequenced. The genome of *M. jannaschii* (domain Archaea) could now be compared to the genomes of organisms from the other two branches of the universal tree: *Haemophilus influenzae* (domain Bacteria), *Mycoplasma genitalium* (domain Bacteria), and *Saccharomyces cerevisiae* (domain Eucarya). When such a comparison was made, it was discovered that 56% of the 1,738 genes identified in *M. jannaschii* were unique and had not previously been detected in Bacteria and eucaryotes. Of the remaining genes, those involved in cell division, metabolism, and energy production were similar to genes found in Bacteria, whereas most of the genes associated with transcription, translation, and replication were more similar to those found in eucaryotes.

The universal tree has Archaea branching off the root earlier than Bacteria and Eucarya, suggesting that the Archaea are the most primitive among the three domains (Figure 11.5). This placement of the Archaea near the universal ancestor is reasonable considering that many of these organisms inhabit extreme environments (for example, high temperature and low pH) that resemble conditions on a primitive planet. It appears that the ancestral procaryote was

an anaerobic, thermophilic, sulfur-metabolizing organism (domain Archaea). This ancestor eventually gave rise to two major branches that are phylogenetically and phenotypically distinct: the extreme thermophile branch and the methanogen/extreme halophile branch. Eucaryotes eventually arose from the Archaea and over time diverged into the various branches associated with this domain. Although the Archaea are procaryotes and share many common characteristics with the Bacteria, they also share characteristics with the phylogenetically related Eucarya, (Table 11.5). Unlike the single RNA polymerase of Bacteria, Archaeal and Eucaryal RNA polymerases are of several types and, in the case of the extreme thermophiles, are structurally more complex (Figure 11.6). Translation in Bacteria generally begins with an initiator tRNA carrying a modified methionine, *N*-formylmethionine (see initiation of translation, page 229), whereas the initiator tRNA in Archaea and Eucarya contains an unmodified methionine. Some antimicrobial agents such as chloramphenicol, erythromycin, and streptomycin, which affect protein synthesis in Bacteria, do not affect protein synthesis in Archaea or Eucarya.

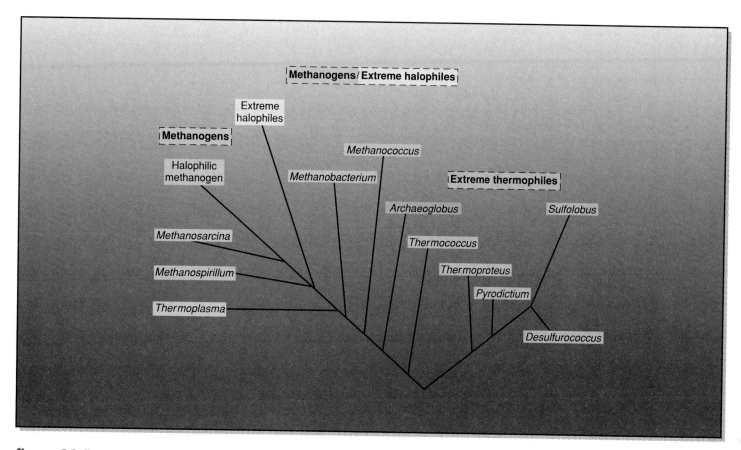

figure 11.5
The Archaea Phylogenetic Tree
The relationships among the Archaea are based on 16S rRNA comparisons.

table 11.5

Distinguishing Characteristics of Bacteria, Archaea, and Eucarya

Characteristic	Bacteria	Archaea	Eucarya
Cell wall	Contains muramic acid	Lacks muramic acid	Lacks muramic acid
Membrane lipids	Ester-linked straight hydrocarbon chains	Ether-linked branched aliphatic hydrocarbon chains	Ester-linked straight hydrocarbon chains
Membrane-bound nucleus	Absent	Absent	Present
Chromosome	Single circular chromosome	Single circular chromosome	Linear chromosomes
RNA polymerase	One type, with 6 different subunits	Several types, each with 8 to 12 different subunits	Several types, each with 12 to 14 different subunits
Ribosomes	70S	70S	80S (70S in mitochondria and chloroplasts)
Amino acid carried by initiator tRNA	N-formylmethionine	Methionine	Methionine
tRNA	Thymine and dihydrouridine usually present	Thymine absent and dihydrouridine usually absent	Thymine and dihydrouridine usually present
Cellular organelles	Absent	Absent	Present
Sensitivity to chloramphenicol erythromycin, and streptomycin	Sensitive	Resistant	Resistant

The Bacteria are divided phylogenetically into at least eleven different phyla (Figure 11.7). With the exception of the *Thermotoga* phylum, the remaining ten phyla of Bacteria originate from a common root. *Thermotoga* and *Thermosipho*, the two known genera of the *Thermotoga* phylum, are extreme thermophiles with cell walls containing peptidoglycan (a signature characteristic of Bacteria). Members of this phylum are clearly Bacteria, although they appear to represent a phylogenetic position between the Archaea (which include many extreme thermophiles) and the Bacteria.

The advent of molecular biology, along with its tools to determine phylogenetic relationships among living organisms, has brought profound changes in how we view procaryotes and eucaryotes. Previously, procaryotes were considered distinct and unique from eucaryotes. Through RNA sequencing, we now know that procaryotes can be divided into two domains, the Bacteria and the Archaea, and that the Archaea are more closely related to the Eucarya than to the Bacteria. Although the Bacteria and Archaea are both irrefutably procaryotes, they occupy different ecological niches and appear to have evolved in separate directions. As rRNAs from additional species are sequenced, it should be possible to further delineate the phylogenetic relationships among living organisms.

Classification of Procaryotes

Procaryotes represent a heterogeneous group of microorganisms with a vast range of characteristics. The most widely accepted organization of this diversity is found in **Bergey's Manual of Determinative Bacteriology,** an extensive reference manual used for bacterial classification. The concept for this manual, considered by many to be the bible of bacteriology, was first conceived by David H. Bergey with the assistance of a special committee of the Society of American Bacteriologists (organized in 1899 and now known as the American Society for Microbiology), chaired by Francis C. Harrison. The first edition of *Bergey's Manual* was published in 1923, with subsequent editions published as procaryotes were reclassified.

Prior to the first edition of *Bergey's Manual,* a number of scientists had attempted to organize bacteria systematically in concise publications. In the late 1800s, Ferdinand Cohn (1828–1898), a botanist, drew comparisons between bacteria and the cyanobacteria. K. B. Lehmann and R. E. Neumann of Germany organized procaryotes into distinct groups in their *Bakteriologische Diagnostik.* The Society of American Bacteriologists assisted Frederick D. Chester in the publication of a *Manual of Determinative Bacteriology* in 1901. By the early 1920s, *Chester's Manual* had become outdated, and it was at this point

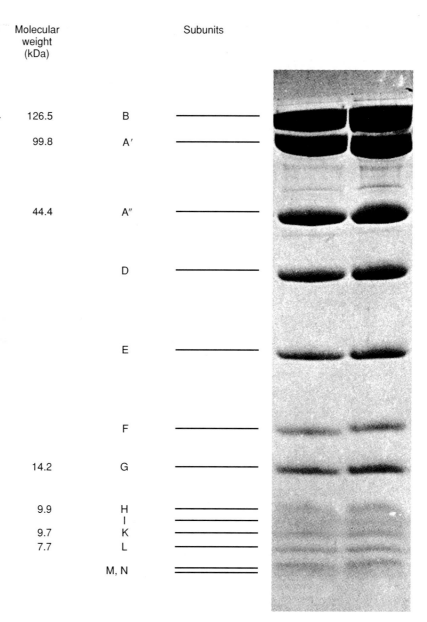

figure 11.6

Subunit Composition of RNA Polymerase of *Sulfolobus acidocaldarius*, Domain Archaea

S. acidocaldarius, an extremely thermophilic elemental sulfur metabolizer, has a structurally complex RNA polymerase consisting of 13 different subunits named A′, A″, B, D, E, F, G, H, I, K, L, M, and N. The purified RNA polymerase protein subunits are denatured and separated by polyacrylamide gel electrophoresis.

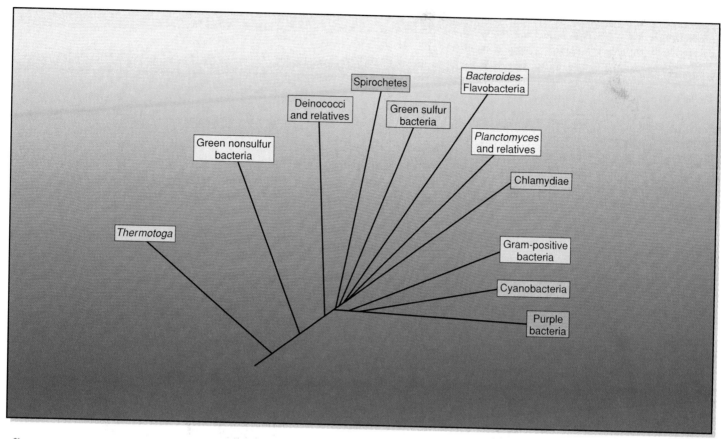

figure 11.7

The Bacterial Phylogenetic Tree

The relationships among the Bacteria are based on 16S rRNA comparisons.

that Bergey and Harrison decided to revise the publication into what has since been known as *Bergey's Manual of Determinative Bacteriology.* The ninth edition of *Bergey's Manual,* published in 1994, separated procaryotes into 35 groups. Since 1936 the *Manual* has been compiled by a group of bacteriologists under the auspices of the Bergey's Manual Trust, which was established with accumulated royalties by Bergey prior to his death in 1937.

The rapid evolution of procaryotic nomenclature combined with the constant discovery of new microorganisms, however, has made it necessary for the Bergey's Manual Trust to reexamine its approach to publication of *Bergey's Manual.* Members of the Trust began to realize in the late 1970s and early 1980s that a new approach was needed to minimize the manual's obsolescence upon publication. The Trust decided on a plan to publish *Bergey's Manual* not as a single volume, but as a series of four subvolumes. It was further decided that the new format would include a greater amount of information on the ecology, enrichment and isolation, and general characteristics of procaryotes as they relate to taxonomy. The new format, which has an expanded scope and examines relationships between organisms (systematics), is called ***Bergey's Manual of Systematic Bacteriology.***

Work on the first volume in the four-volume sequence began in 1980 and was completed in 1982, with subsequent publication in 1984. *Bergey's Manual of Systematic Bacteriology* groups procaryotes into four categories: gram-negative Bacteria of general, medical, or industrial importance (Volume 1); gram-positive Bacteria other than the actinomycetes (Volume 2); the Archaea, cyanobacteria, and remaining gram-negative Bacteria (Volume 3); and the actinomycetes (Volume 4). Members of the Bergey's Manual Trust believe that the division of procaryotes into four groups for treatment in separate volumes will permit the timely revision and publication of material.

The classical approach to classification in *Bergey's Manual of Systematic Bacteriology* will be used to organize the description of representative groups of procaryotes in the rest of this chapter. This description of procaryotes is intended not to be an exhaustive coverage of *Bergey's Manual,* but to acquaint the student with the diverse forms and metabolic types that make up the procaryotes. Although the phylogenetic approach is becoming more defined and will be used in the next edition of *Bergey's Manual of Systematic Bacteriology,* the classical approach is traditionally used by most laboratories today to identify and classify procaryotes.

table 11.6

Gram-Negative Bacteria of General, Medical, or Industrial Importance

Section	Major Characteristics
The Spirochetes	Slender, winding, or helically coiled cells; outer sheath surrounds protoplasmic cylinder and periplasmic flagella
Aerobic/microaerophilic, motile, helical/vibrioid, gram-negative bacteria	Rigid, helical to vibrioid cells with polar flagella; aerobic or microaerophilic
Nonmotile (or rarely motile), gram-negative, curved bacteria	Curved, vibrioid, or helical cells; nonmotile
Gram-negative aerobic rods and cocci	Straight or curved rods and cocci; respiratory modes of metabolism
Facultative anaerobic, gram-negative rods	Straight or curved rods; respiratory and fermentative modes of metabolism; includes many human parasites
Anaerobic, gram-negative, straight, curved, and helical rods	Obligate anaerobes; form wide variety of fermentation products
Dissimilatory sulfate- or sulfur-reducing bacteria	Obligate anaerobes; use sulfur, sulfate, or other oxidized sulfur compounds as electron acceptors in anaerobic respiration
Anaerobic, gram-negative cocci	Obligate anaerobes; nonmotile
The Rickettsias and Chlamydias	Small, obligate intracellular parasites of humans, other animals, and arthropods
The Mycoplasmas	Small free-living organisms; lack cell walls
Endosymbionts	Microorganisms that exist intracellularly as endosymbionts of fungi, protozoa, insects, and other invertebrates

Gram-Negative Bacteria of General, Medical, or Industrial Importance

The first volume of *Bergey's Manual* includes gram-negative Bacteria of general, medical, or industrial importance and divides them into 11 separate sections (Table 11.6). These Bacteria are further separated into orders, families, genera, and species.

The Spirochetes

The first section of Volume 1 of *Bergey's Manual*, on the spirochetes, contains five genera of bacteria (*Spirochaeta, Christispira, Treponema, Borrelia,* and *Leptospira*) that are *flexuous*, slender, and helically coiled. The cytoplasm and nuclear region of a spirochete are contained within a plasma membrane–cell wall complex and form what is known as a protoplasmic cylinder. Wrapped around the protoplasmic cylinder are periplasmic flagella (also called axial fibrils or axial filaments) that are attached to one end of the cell and extend along most of the cell length (Figure 11.8). The number of periplasmic flagella in a cell ranges from 2 to more than 100. The flagella and their arrangement around the protoplasmic cylinder are responsible for the unique rotating, screwlike motility of these bacteria. Spirochetes also move by flexing their bodies as well as by creeping or crawling across solid surfaces. Both the protoplasmic cylinder and the periplasmic flagella are surrounded by a trilaminar outer sheath. The entire cell, which may be up to 250 μm

long, is typically only 0.1 to 0.75 μm in diameter (some members of the genus *Christispira* have diameters up to 5 μm). It is difficult to see spirochetes by bright-field light microscopy because of their typically very small diameter. However, they can be seen by dark-field microscopy, and by fluorescence microscopy when tagged with the appropriate fluorescent antibodies.

Many types of human disease are associated with the spirochetes. Bacteria of the genus *Treponema* [Greek *trepo*, turn, *nema*, thread] cause such diseases as syphilis (*T. pallidum*)(see syphilis, page 534), yaws—a non-sexually-transmitted tropical disease characterized by painful, debilitating lesions on the soles of the feet and the palms of the hand—(*T. pertenue*), and pinta—a non-sexually-transmitted tropical disease characterized by hyperpigmentation followed by depigmentation of the skin (*T. carateum*) (Figure 11.9). *Borrelia burgdorferi* and *Leptospira interrogans* are responsible for Lyme disease and leptospirosis, respectively. Lyme disease was first discovered in 1975 during an investigation of several hundred people with a mysterious illness in the town of Old Lyme, Connecticut. Since 1975, Lyme disease has been reported in over 40 states in the United States and in other countries. *B. burgdorferi*, the etiologic agent of Lyme disease, is generally transmitted to humans by the deer tick, *Ixodes scapularis*, which is found in association with animals such as white-tailed deer and white-footed mice. The disease begins as a rash with concentric red rings resembling a "bull's eye" at the site of the tick bite, followed by fatigue, fever, headache, and arthritis. Over 7,000 cases

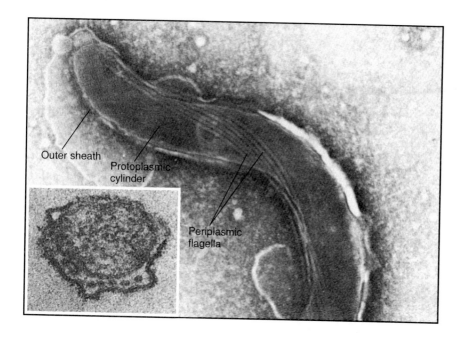

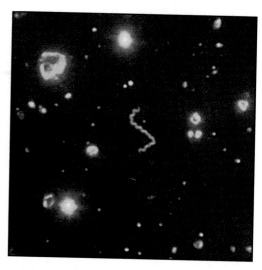

figure 11.9

Fluorescent Micrograph of the Spirochete *Treponema carateum*

Fluorescent antibodies bound to the bacterium illuminate it against the dark background.

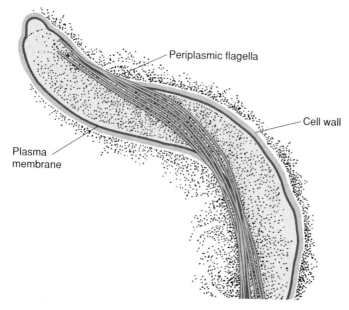

Periplasmic flagella

Cell wall

Plasma membrane

figure 11.8

Electron Micrograph and Schematic Diagram of *Treponema pallidum* Showing Periplasmic Flagella

Several periplasmic flagella can be seen extending from one end of the cell (×50,000). Cross section inset shows the outer sheath, periplasmic flagella, and central protoplasmic cylinder (×120,000).

Bacterial Structure and Function
External Structures: Flagella • p. 25

of Lyme disease are reported annually in the United States. Leptospirosis is characterized by influenza-like symptoms and jaundice. *L. interrogans*, a thin (6 to 20 μm in length by 0.1 μm in diameter) spirochete with a characteristic hook on one or both ends of the cell, is transmitted to humans by contact with water contaminated with urine from infected animals (Figure 11.10).

Other spirochetes do not cause disease, but can be found in symbioses with insects such as wood-eating cockroaches and termites. *Pillotina* (host: wood-eating cockroaches and termites), *Diplocalyx* (host: termites), *Hollandina* (host: termites), and *Clevelandina* (host: termites) are genus names that have been proposed for these bacteria that live in the anaerobic or microaerophilic hindguts of the insects but have not yet been cultivated in the laboratory.

Many of the spirochetes are difficult to identify in the laboratory, not only because of their small size, but also because many of these bacteria are anaerobic and have not been successfully cultivated on artificial media. The spirochetes are distinguished by characteristics such as size, morphology, relationship to oxygen, habitat, and pathogenicity (Table 11.7).

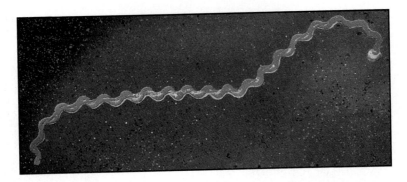

figure 11.10

Electron Micrograph of *Leptospira interrogans*

Note the hook at one end of the cell, characteristic of this bacterium (colorized).

table 11.7

Characteristics of Spirochetes (family: *Spirochaetaceae*)

Genus	Morphology	Oxygen Requirement	Diseases
Spirochaeta	Helical cells, 0.2–0.75 μm in diameter and 5–250 μm in length	Obligately anaerobic or facultatively anaerobic	None known to be pathogenic
Cristispira	Helical cells, 0.5–3.0 μm in diameter and 30–180 μm in length; generally contain 2 to 10 complete helical turns	Unknown (have not been grown in pure culture)	Widely distributed among marine and freshwater mollusks (clams, mussels, and oysters), but not believed to be pathogenic to these hosts
Treponema	Helical rods, 0.1–0.4 μm in diameter and 5–20 μm in length; cells have tight regular or irregular spirals	Obligately anaerobic or microaerophilic	Pinta, syphilis, yaws
Borrelia	Helical cells, 0.2–0.5 μm in diameter and 3–20 μm in length; composed of 3 to 10 loose coils	Microaerophilic	Tick-borne and louse-borne relapsing fever

Aerobic/Microaerophilic, Motile, Helical/Vibrioid Gram-Negative Bacteria

In the second section of Volume 1 of *Bergey's Manual*, the bacteria, unlike the spirochetes, are *rigid*, helical to vibrioid (comma-shaped) cells that move by conventional polar flagella. These bacteria, 0.2 to 1.7 μm in diameter and up to 60 μm in length, are larger than the spirochetes and thus can be seen with a bright-field light microscope. Members of the genus ***Spirillum*** are considered to be among the largest bacteria discovered (Figure 11.11).

Spirillum and bacteria of two other related genera, ***Aquaspirillum*** and ***Oceanospirillum,*** are aerobic to microaerophilic organisms frequently found in freshwater and marine habitats. It has been hypothesized that these bacteria may be important in the recycling of organic matter from decomposing plants in such habitats.

Campylobacter [Greek *campylo,* curved, *bacter,* rod], another member of this group, is found in the reproductive organs, gastrointestinal tract, and oral cavity of humans and animals (see *Campylobacter* infections, page 520). ***Campylobacter fetus,*** first isolated as *Vibrio fetus* in 1909 and given its current name in 1973, causes a sexually transmitted disease of animals that results in abortions and infertility in infected cattle and sheep. Since the bacterium is carried asymptomatically for long periods of time in the genitourinary and intestinal tracts of these animals, it can persist undetected in herds. *Campylobacter* infections can lead to serious economic losses and are of great concern to ranchers. In humans, *C. fetus* is responsible for opportunistic infections—typically fever and blood infections—primarily in people who have debilitating conditions. A sister species, ***Campylobacter jejuni,*** is an important human intestinal pathogen that causes gastroenteritis.

The bacterium ***Bdellovibrio*** [Greek *bdella*, leech] a small, curved, motile rod, has the unique ability to parasitize other bacteria (see *Bdellovibrio* parasitism, page 429). Parasitic strains of this bacterium attack gram-negative organisms, inserting themselves into the periplasm between the cell wall and membrane. Within this narrow space, the bacterium utilizes the host cell as a substrate

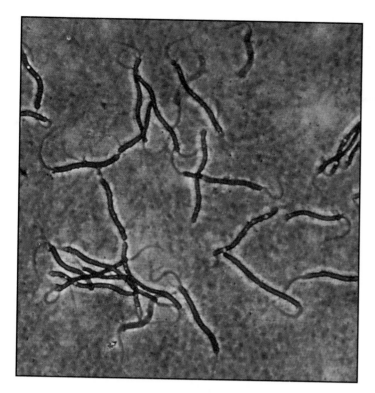

figure 11.11

Phase-Contrast Micrograph of *Spirillum volutans*

S. volutans is a helical, polar-flagellated bacterium found in aquatic environments.

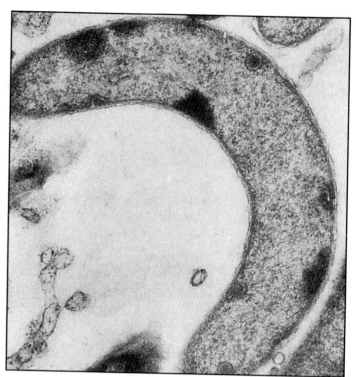

figure 11.12

Electron Micrograph of *Microcyclus aquaticus*, a Curved Rod (×31,400)

for development and replication. The growing *Bdellovibrio* elongates into a snakelike form, which eventually fragments into smaller, highly motile progeny that leave the host, thereby reinitiating the life cycle. The predilection of bdellovibrios for gram-negative prey is due to the absence of a periplasm in gram-positive bacteria. Evidence suggests that components of the lipopolysaccharide in the gram-negative cell envelope may serve as receptor sites for the predator. Bdellovibrios are found in diverse habitats, including soil, oceans, rivers, streams, estuaries, and sewage systems. High concentrations of these bacteria are found in polluted environments, an observation that has led some scientists to infer that prey densities may have an influence on the levels of bdellovibrios in such areas.

Nonmotile (or Rarely Motile), Gram-Negative, Curved Bacteria

The third section of Volume 1 of *Bergey's Manual* consists of gram-negative curved, vibrioid, or helical cells that are nonmotile. These bacteria are nonpathogenic and are generally found in soil, freshwater, and marine environments. An example of a bacterium in this section is ***Microcyclus,*** a curved rod that has a tendency to form rings after division and prior to cell separation (Figure 11.12). Some species of *Microcyclus* produce gas vacuoles, which make these bacteria buoyant in their aquatic habitat.

Gram-Negative Aerobic Rods and Cocci

A large number of bacteria with diverse properties are found in the fourth section of Volume 1 (Table 11.8). These gram-negative bacteria all possess the common characteristic of having a respiratory mechanism of metabolism, although some can use electron acceptors other than oxygen.

The ***Pseudomonadaceae*** are polar-flagellated straight or curved rods. Most of these bacteria use a wide variety of chemical compounds as energy sources and are useful in the degradation of hydrocarbons in oil spills (see *Pseudomonas putida* and hydrocarbon degradation, pages 11 and 584) and grease in restaurant grease traps (see grease degradation, page 95). Because they are so nutritionally diverse, members of this family are widespread in nature and cause many different diseases in plants, humans, and animals.

Pseudomonas aeruginosa is a common inhabitant of soil and water and is an opportunistic pathogen, causing wound, burn, and urinary tract infections in humans (see *Pseudomonas* infections, page 528). The bacterium is especially a problem in hospital environments, where it is a frequent contaminant of nonsterile wet surgical instruments and respiratory apparatus. Infections by *P. aeruginosa* are a great risk for burn victims, where these bacteria are the single greatest cause of death. Hospitals are particularly concerned about pseudomonad infections, since most strains of *Pseudomonas* are unusually resistant to antibiotic therapy.

Characteristics of the Gram-Negative Aerobic Rods and Cocci

Family	Morphology	Flagellar Arrangement	Distinctive Characteristics
Pseudomonadaceae	Rods	Polar (most)	Use a wide variety of chemical compounds as energy sources; are common inhabitants of soil and water; some are human, animal, or plant pathogens
Azotobacteraceae	Rods	Peritrichous, polar, or nonmotile	Are capable of fixing nitrogen nonsymbiotically; are common inhabitants of soil and water
Rhizobiaceae	Rods	Peritrichous or polar	Are capable of fixing nitrogen symbiotically; are common soil inhabitants
Methylococcaceae	Rods or cocci	Polar or nonmotile	Use 1-carbon compounds (methane, methanol, and formaldehyde) as sole carbon and energy sources
Halobacteriaceae	Rods or other forms (bent and swollen rods, clubs, ovoids, spheres, spindles, and other irregular forms)	Polar or nonmotile	Most require at least 2.5 M NaCl for growth
Acetobacteraceae	Rods	Peritrichous, polar, or nonmotile	Oxidize ethanol to acetic acid
Legionellaceae	Rods or filaments, 20 μm or more in length	Peritrichous, polar, or nonmotile	Are capable of causing pneumonia-like illness; are commonly found in streams, lakes, air-conditioning cooling towers, and evaporative condensers
Neisseriaceae	Cocci or Rods	Nonmotile	Cause gonorrhea (Neisseria gonorrhoeae), cerebrospinal meningitis, (Neisseria meningitidis), and other diseases of humans and animals

A few species of *Pseudomonas,* including *P. aeruginosa,* produce pigments that assist in identification. *P. aeruginosa* synthesizes two such pigments: pyoverdin and pyocyanin. Pyoverdin, a siderophore (see siderophore, page 88), is an unstable pigment that fluoresces upon excitation by ultraviolet irradiation. Pyocyanin is a blue phenazine pigment that diffuses freely into the surrounding medium. These two pigments give cultures of *P. aeruginosa* a characteristic blue-green fluorescent appearance.

Xanthomonas, another genus of the family *Pseumonadaceae,* is a common plant pathogen. The 1974 edition of *Bergey's Manual* condenses more than 100 of these plant pathogens into five species, including the species *Xanthomonas campestris.* Although this condensation may have been a rational taxonomic decision, it caused considerable confusion among plant pathologists who no longer could identify specific diseases with a specific bacterium. *Bergey's Manual of Systematic Bacteriology* corrects this problem by listing specific strains of *X. campestris* and their characteristics by the term **pathovar.** Among these is **X. campestris pathovar citri,** a highly infectious bacterium that causes citrus canker, a disease of citrus trees that is characterized by lesions on leaves and unsightly scabs on the fruit. *X. campestris* pathovar *citri* is not common in the United States, but causes endemic (habitually present within a geographical area) citrus disease in South America and the Far East. In 1984 citrus in Florida was affected by citrus bacterial leaf spot disease. This disease is similar to citrus canker, but is caused by a different pathovar of *X. campestris* (Figure 11.13).

Azotobacter and ***Rhizobium*** are gram-negative rods capable of fixing nitrogen nonsymbiotically or symbiotically, respectively. Rhizobia enter into symbiotic associations with leguminous plants (peas, beans, alfalfa, clover), causing characteristic root nodules to form during nitrogen fixation. Although azotobacters and rhizobia are aerobic organisms, their nitrogen-fixing enzymes are oxygen sensitive. These enzymes, called nitrogenases, remain active, apparently because the bacterial cells are able to reduce oxygen tensions in areas of the cytoplasm where the enzymes are located. An oxygen-binding protein called leghemoglobin is produced when rhizobia enter a symbiotic relationship with leguminous plants. This hemoglobinlike protein binds oxygen and removes it from the vicinity of nitrogenases (see nitrogen fixation, page 418). Azotobacters are large ovoid cells that probably have the highest respiration rate of any living organism. This high respiration rate undoubtedly contributes to the rapid removal of oxygen from the cytoplasm. Biological nitrogen fixation contributes significantly to the total quantity of fixed nitrogen added to the soil each year.

figure 11.13

Grapefruit Leaf Showing Distinctive Lesions of Citrus Bacterial Leaf Spot Disease Caused by *Xanthomonas campestris*

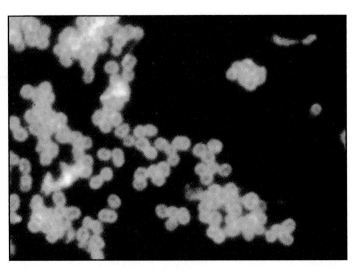

figure 11.14

Fluorescent Antibody Stain of *Neisseria gonorrhoeae*, Cause of Gonorrhea

Fluorescent antibodies bound to the gram-negative diplococci illuminate them against the dark background.

The family ***Methylomonadaceae*** consists of two genera of bacteria, ***Methylomonas*** and ***Methylococcus,*** which use 1-carbon compounds such as methane, methanol, and formaldehyde as sole carbon and energy sources. These bacteria are found in aerobic environments adjacent to areas where these compounds are located, such as natural gas deposits, coal formations, and anaerobic muds.

In the family ***Neisseriaceae*** is the bacterium ***Neisseria gonorrhoeae,*** which causes gonorrhea (Figure 11.14) (see gonorrhea, page 533). This sexually transmitted disease occurs in an estimated 1 million people in the United States annually. Gonorrhea affects all races and socioeconomic classes, male and female, and is so prevalent that it is considered endemic in the United States and many other countries. ***N. gonorrhoeae*** is a gram-negative aerobic coccus that is extremely fastidious and requires a moist environment for growth. Such environments are generally found in such areas of the human body as the conjunctiva, the nasopharynx, and the urogenital area. Thus it is not surprising that the transmission of *N. gonorrhoeae* occurs by sexual contact involving these areas. ***Neisseria meningitidis,*** another bacterium in this family, causes meningitis, particularly in children. Both

pathogens are identified in the clinical laboratory as gram-negative, kidney bean–shaped pairs of cocci (diplococci). These organisms contain the enzyme cytochrome c oxidase, which is not present in many other bacteria (see oxidase test, page 166). The bacteria require increased levels of carbon dioxide (3% to 5%) and specialized media containing hemoglobin, vitamins, amino acids, and other nutrients for their growth and subsequent isolation.

Several other genera, in addition to genera of the eight recognized families, are found in this section of *Bergey's Manual* under the category "Other Genera." These bacteria have the common characteristics of this section (they are aerobic gram-negative rods or cocci) but are not easily classified under any of the eight recognized families. Such designations of bacteria are not uncommon in *Bergey's Manual;* in fact, they are used quite often to place certain bacteria in specific sections. Among the genera in this section are ***Brucella*** (the cause of infectious abortion in cattle), ***Bordetella*** (the causative agent of whooping cough), and ***Francisella*** (responsible for tularemia, an infectious disease of rabbits that is transmissible to humans who come in close contact with diseased animals).

table 11.9

Characteristics of Facultatively Anaerobic Gram-Negative Rods

Family	Representative Genera	Diseases or Characteristics
Enterobacteriaceae	Escherichia	Gastroenteritis, urinary tract infections
	Shigella	Bacillary dysentery
	Salmonella	Typhoid fever, gastroenteritis
	Klebsiella	Pneumonia, urinary tract infections
	Serratia	Opportunistic pathogen[a]
	Proteus	Urinary tract infections, opportunistic pathogen[a]
	Yersinia	Bubonic plague, diarrhea
Vibrionaceae	Vibrio	Cholera, gastroenteritis
	Aeromonas	Wound infections, meningitis, septicemia[b]
	Plesiomonas	Diarrhea
Pasteurellaceae	Pasteurella	Septicemia[b] (mice and rabbits)
	Haemophilus	Meningitis (especially in children)

[a]An organism that normally is nonpathogenic, but is capable of causing disease in a compromised host.
[b]An infection of the blood.

Facultatively Anaerobic Gram-Negative Rods

The organisms in the fifth section of Volume 1 of *Bergey's Manual* are called facultative anaerobes because they can respire or ferment carbohydrates and therefore can exist and grow in the presence or absence of oxygen. These gram-negative rods are divided into three families, ***Enterobacteriaceae, Vibrionaceae,*** and ***Pasteurellaceae*** (Table 11.9).

The family *Enterobacteriaceae* includes a number of significant human parasites, including ***Escherichia, Salmonella, Shigella,*** and ***Yersinia. Escherichia coli,*** a common inhabitant of the gastrointestinal tract and an organism frequently found in soil and water, is the most studied microbe in the scientific world. *E. coli* is frequently chosen for study because it is easily cultivated and grows rapidly in the laboratory, and much is already known about its genetics, physiology, and structure. It is considered by many to be the darling of molecular biology.

E. coli is the predominant facultative anaerobe in the large intestine. Normally *E. coli* is not a problem in the gastrointestinal tract and actually is beneficial to the human host, since it synthesizes some important vitamins (for example, vitamin K) and prevents growth of some potentially harmful bacteria by competing with them for nutrients and oxygen. However, under certain conditions, *E. coli* can cause disease—most notably gastroenteritis and urinary tract infections (see gastrointestinal diseases by *E. coli*, page 521). Some strains of *E. coli* that are enteroinvasive (invade the gastrointestinal tract) or enterotoxigenic (produce toxins that affect the intestines) cause short-term diarrheal illnesses. Such ill-

nesses are especially common in infants in newborn nurseries and in travelers in countries having poor sanitary conditions, where contaminated drinking water or food cause what is often known as traveler's diarrhea. Enterohemorrhagic *E. coli* produces a toxin that causes hemolytic-uremic syndrome, which is characterized by lysis of erythrocytes and kidney failure. A 1993 oubreak of enterohemorrhagic *E. coli* infections affecting more than 400 people was traced to inadequately cooked hamburgers served by a fast-food restaurant chain.

E. coli is the most common cause of urinary tract infections in females (in males, the number one cause is *Proteus mirabilis*). Such infections generally affect elderly people with structural abnormalities in their urinary tract and women having gynecologic problems. Urinary tract infections persist until the obstruction is removed or the physical abnormality is corrected. *E. coli* and similar bacteria that ferment lactose are routinely used as indices of fecal contamination of water (see coliform, page 591). The presence of these bacteria in water suggests that other intestinal microbes, including pathogenic organisms, may also be present in the water.

Many other bacteria in the family *Enterobacteriaceae* are capable of causing human disease. ***Shigella dysenteriae,*** a nonmotile gram-negative rod, invades the mucous membrane, causing bacillary dysentery. The genus ***Salmonella*** consists of a large group of bacteria divided into more than 1,700 serotypes, or serovars, on the basis of their antigenic characteristics. These bacteria have undergone several changes in taxonomic classification. *Bergey's Manual of Systematic Bacteriology* divides the sal-

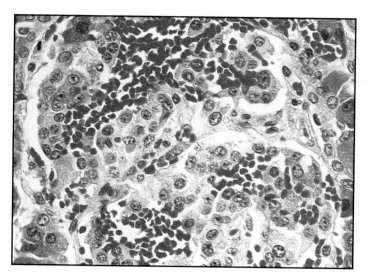

figure 11.15

Gram Stain of *Yersinia pestis* in the Pancreas of a Fatal Human Case of Plague
Gram-negative cells are seen throughout the tissue specimen.

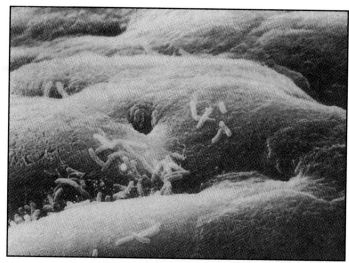

figure 11.16

Scanning Electron Micrograph of *Vibrio cholerae* Adhering to Intestinal Mucosa
Numerous curved cells of *V. cholerae* are visible on the intestinal surface (×4,260).

monellae into five subgenera, under which are listed the different serovars. Among the serovars of salmonellae are ***Salmonella typhi,*** which causes typhoid fever, and ***Salmonella typhimurium,*** the most common cause of *Salmonella*-induced gastroenteritis in humans.

Yersinia pestis, the agent responsible for the dreaded disease bubonic plague, is also a member of the family *Enterobacteriaceae* (Figure 11.15). This bacterium is named after the French bacteriologist Alexandre J.E. Yersin (1863–1943), who first isolated it from plague victims in Hong Kong in 1894. Plague is not as prevalent today as it was in the sixth century A.D., when it killed more than 100 million people during a 50-year period, and in the fourteenth century when it killed one-quarter of Europe's population and was known as the Black Death because of the many deaths and the black patchy hemorrhaging that quickly developed under the skin. Sporadic outbreaks still occur, however, especially in the Far East, and sporadic cases are reported in other areas of the world. In the western and southwestern United States, prairie dogs, rock squirrels, and wild rodents are the main reservoirs for *Y. pestis* and occasionally transmit the bacterium to a human host through bites or infected rat fleas.

 **Miscellaneous Bacteria**
Zoonoses: Yersinia • pp. 57–67

The family ***Vibrionaceae*** contains several human pathogens, most notably ***Vibrio cholerae.*** This gram-negative curved rod was first associated with cholera in 1883 by Robert Koch (1843–1910),

the German bacteriologist who isolated the bacterium from cholera patients during epidemics in Egypt and India (Figure 11.16). Unlike *Enterobacteriaceae* and many other bacteria, vibrios grow well in media of highly alkaline pH (pH 9.0). This characteristic is frequently used to isolate vibrios selectively from other microorganisms present in clinical specimens. *V. cholerae* is generally endemic in countries having poor sanitary conditions, but the bacterium has also been isolated in the United States, especially in waters along the Gulf coast. In recent years several cases have been documented of human illness and deaths in states along the Gulf coast from the ingestion of shellfish such as raw oysters contaminated with *V. cholerae* (see cholera, page 518). The extent of shellfish contamination and the public health relevance of such contamination with *V. cholerae* remains to be determined. *V. cholerae* is presently considered to be authochthonous (indigenous) to water. ***Vibrio vulnificus,*** a halophilic bacterium that is also associated with the ingestion of raw oysters, can cause septicemia (infection of the blood), wound infections, and gastroenteritis in people with compromised immune systems and liver disease. ***Vibrio parahaemolyticus,*** another halophilic bacterium, is a major cause of gastroenteritis in Japan and other countries where raw seafood is widely consumed. Other marine species of *Vibrio* are nonpathogenic and occur as normal flora in the gastrointestinal tracts of fish and other aquatic organisms.

 **Gram-Negative Organisms**
Bacilli—Facultative Anaerobes: Vibrionaceae • pp. 30–38

Using Flowcharts to Identify Bacteria

With the tens of thousands of bacteria found in *Bergey's Manual of Systematic Bacteriology*, all with their own traits, it may seem impossible for a person to identify bacteria in a patient specimen or an environmental sample. However, bacterial identification is not difficult if a logical scheme is followed to recognize species by their biochemical characteristics.

Such a scheme, or **flowchart,** is frequently used in diagnostic microbiology as a road map to differentiate bacteria and other microorganisms on the basis of a few key biochemical reactions. By using a few classical biochemical tests, the genus or, in some cases, the species of a bacterium can be determined.

An example of a flowchart that might be used to differentiate genera in the family *Enterobacteriaceae* is illustrated in Figure 11.17. Seven basic biochemical tests (phenylalanine deaminase, urease, citrate utilization, lysine decarboxylase, sucrose fermentation, H_2S production, and acetoin production as detected by the Voges-Proskauer test) are used for initial identification of *Enterobacteriaceae*. On the basis of the results from one or more of these tests, a preliminary identification of the suspect bacterium can be made. Additional tests can then be performed to speciate the microorganism.

Flowcharts do not replace the need for morphological observations or the performance of more specific tests, but they do simplify the identification of bacteria, particularly when the suspect bacterium is part of a large family such as the *Enterobacteriaceae*.

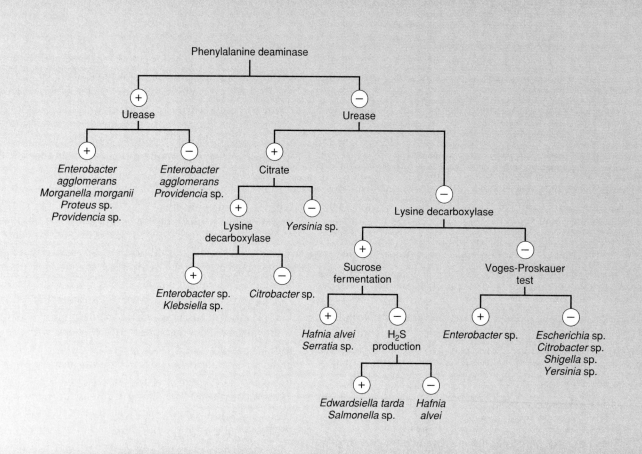

figure 11.17

Flowchart for Initial Identification of *Enterobacteriaceae* Based on a Few Key Biochemical Tests

table 11.10

Characteristics of Anaerobic, Gram-Negative Straight, Curved, and Helical Rods

Genus	Morphology	Major End Products of Fermentation
Bacteroides	Rods; nonmotile or motile by peritrichous flagella	Mixture of succinic, acetic, formic, lactic, and propionic acids
Fusobacterium	Rods; nonmotile	Butyric acid (major product); smaller amounts of acetic, propionic, formic, or lactic acid possible
Leptotrichia	Straight or slightly curved rods, with one or both ends pointed or rounded; nonmotile	Lactic acid (major product); small amounts of formic, acetic, or succinic acid possible
Butyrivibrio	Curved rods; polar or subpolar flagella	Butyric acid
Succinimonas	Short rods or coccobacilli; single polar flagellum	Succinic and acetic acids
Succinivibrio	Helical or spiral-shaped cells; single polar flagellum	Succinic and acetic acids
Anaerobiospirillum	Helical or spiral-shaped cells; bipolar tufts of flagella	Succinic and acetic acids
Wolinella	Helical, curved, or straight rods; single polar flagellum	Carbohydrates not fermented (hydrogen and formic acid used as energy sources)
Selenomonas	Crescent-shaped cells; tuft of flagella on side of cell	Propionic and acetic acids
Anaerovibrio	Slightly curved rods; single polar flagellum	Propionic and acetic acids
Pectinatus	Slightly curved rods; lateral flagella on only one side of cell	Propionic and acetic acids
Acetivibrio	Straight to slightly curved rods; single flagellum or multiple flagella on side of cell	Acetic acid
Lachnospira	Straight to slightly curved rods; single lateral to subpolar flagellum	Ethanol; carbon dioxide; hydrogen; formic, lactic, and acetic acids

Anaerobic, Gram-Negative Straight, Curved, and Helical Rods

Three sections of *Bergey's Manual* are devoted exclusively to anaerobic bacteria (bacteria that do not tolerate oxygen well and exist and grow best in its absence). The bacteria in the section on anaerobic gram-negative straight, curved, and helical rods form different fermentation products. The types of products formed and the morphology and flagellation of these organisms are characteristics used in their classification and identification (Table 11.10).

Many of these anaerobic organisms are found in the gastrointestinal tract or oral cavity of humans and other animals. How are anaerobic bacteria able to exist in body locations that are usually considered to be aerobic? Facultatively anaerobic bacteria in the gastrointestinal tract use up the available oxygen during cellular metabolism, leaving an oxygen-free environment. Anaerobic bacteria in the oral cavity are found primarily in areas of plaque

and dense pockets of bacteria along the gingiva and periodontal membrane—sites in which anaerobic conditions prevail.

Contrary to popular opinion, *E. coli* is not the most common bacterium in normal human feces. **Bacteroides,** an anaerobic bacterium, and other obligately anaerobic bacteria outnumber facultative anaerobic bacteria in the colon by a ratio of 100 to 1 (Figure 11.18a). **Fusobacterium,** also an obligate anaerobe, is a common inhabitant of the mouth, where it may constitute a large proportion of the microbial population and may be involved in periodontal disease (Figure 11.18b).

Anaerobic bacteria are important agents of disease, particularly in patients with abscesses or puncture wounds. These provide the anaerobic conditions necessary for growth and multiplication. Foul-smelling discharges, gas formation in tissues, and tissue discoloration are clinical signs frequently associated with anaerobic infections. Many hospitals now recognize the clinical significance and prevalence of anaerobic infections and have bacteriology laboratories specifically equipped for the identification of these microbes.

Dissimilatory Sulfate- or Sulfur-Reducing Bacteria

The diverse group of anaerobic bacteria in the section of *Bergey's Manual* on dissimilatory sulfate- or sulfur-reducing bacteria use sulfur, sulfate, or other oxidized sulfur compounds as electron acceptors in anaerobic respiration (see anaerobic respiration, page 168). These bacteria live in anaerobic muds and sediments of freshwater, brackish water, and marine environments, as well as in the gastrointestinal tracts of animals and humans. The best-known genus in this group is **Desulfovibrio,** which can be found in polluted waters showing blackening and sulfide production. *Desulfovibrio* is a major problem in the oil industry, because it can corrode iron pipes. This bacterium is environmentally important, however, because it neutralizes acidic sulfur compounds such as sulfur dioxide, and thus mitigates the effects of pollution and has been used to help reduce acid mine wastes (see acid mine drainage, pages 580 and 583).

Anaerobic, Gram-Negative Cocci

The section in *Bergey's Manual* on anaerobic, gram-negative cocci is small and contains only three genera. These bacteria are all anaerobic, nonmotile, and found in the alimentary tract of humans and animals. **Veillonella,** a representative genus, has been isolated from dental abscesses and urinary tract infections. However, because it is always isolated along with other bacterial pathogens, its role in disease, if any, is unknown. Another bacterium in this group, **Acidaminococcus,** is able to use amino acids as sole energy sources for growth; most strains require several amino acids. The acidaminococci are frequently found in the intestinal tract of humans and animals, but are apparently not pathogenic.

The Rickettsias and Chlamydias

The organisms described in the rickettsiae and chlamydiae section of *Bergey's Manual* were for many years thought to be viruses rather than bacteria. They were mistakenly identified as viruses because they are small (most are 0.2 to 0.5 μm in diameter) and are obligate intracellular parasites. These organisms are now known to be bacteria. Rickettsiae and chlamydiae possess both DNA and RNA (viruses have only one type of nucleic acid), have cell walls similar to those found in gram-negative bacteria (viruses have a protein coat, but no cell wall), divide by binary fission (viruses assemble within a host after infection, but do not divide by binary fission), and are susceptible to antibiotics that affect bacteria (bacterial antibiotics are ineffective against viruses).

The genus **Rickettsia** contains several species that cause such diseases as Rocky Mountain spotted fever and different forms of typhus (Table 11.11). These diseases are transmitted by arthropod vectors and are characterized by body rashes and fever (see rickettsial diseases, page 529).

A second genus in this group, **Chlamydia,** is responsible for trachoma, the leading cause of blindness in the world. *Chlamydia* also causes the sexually transmitted disease lymphogranuloma

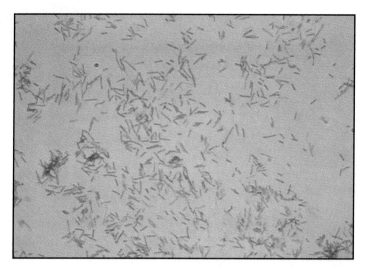

a.

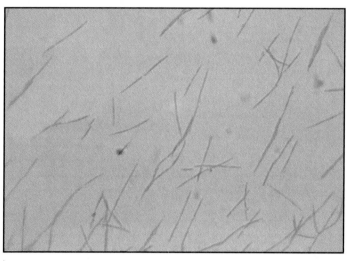

b.

figure 11.18
Gram-Negative Anaerobes
a. *Bacteroides.* b. *Fusobacterium.*

venereum (LGV) and is one of the agents responsible for nongonococcal urethritis (NGU), a generalized type of urethritis that is not associated with *Neisseria gonorrhoeae* (see nongonococcal urethritis, page 536). *Chlamydia* has a unique developmental cycle in which it alternates between a noninfectious form that reproduces by binary fission inside the cell and a smaller, dense-centered infectious form that is released upon cell death and lysis (see *Chlamydia* developmental cycle, page 430).

Diseases caused by rickettsiae and chlamydiae are difficult to diagnose in a bacteriology laboratory because these obligate intracellular parasites cannot be grown on artificial bacteriological media. Diagnosis is usually based on clinical symptoms and serological tests.

 **Miscellaneous Bacteria**
Chlamydia • pp. 1–6

Diseases Caused by Rickettsiae and Chlamydiae

Family	Organism	Diseases	Vector or Mode of Transmission
Rickettsiaceae	Rickettsia prowazekii	Epidemic typhus	Human body louse
	Rickettsia typhi	Endemic typhus	Rat flea
	Rickettsia rickettsii	Rocky Mountain spotted fever	Tick
	Rickettsia tsutsugamushi	Scrub typhus	Mite
	Rochalimaea quintana	Trench fever	Human body louse
	Coxiella burnetii	Q fever	No known arthropod vector (transmitted by dust and food)
Chlamydiaceae	Chlamydia trachomatis	Trachoma	Use of common washing utensils
		Lymphogranuloma venereum (LGV)	Sexual intercourse
		Nongonococcal urethritis (NGU)	Sexual intercourse
	Chlamydia psittaci	Psittacosis	Contact with infected birds (parrots and parakeets)

figure 11.19

Typical Fried Egg Appearance of *Mycoplasma pneumoniae* Colonies on Agar

Mycoplasmas from colonies 10 to 600 µm in diameter on solid media (×100).

The Mycoplasmas

The smallest free-living organisms known to humans are those of the genus ***Mycoplasma.*** Most rickettsiae, although smaller than *Mycoplasmas,* are obligate parasites and are not free-living. The mycoplasmas are not only small in size (125 to 250 nm in diameter), they also are the only bacteria that normally exist without a cell wall (see bacteria without cell walls, page 59). Unlike other bacteria, some mycoplasmas have sterols in their plasma membranes. These sterols provide the strength needed in the membranes to maintain cellular integrity without a wall. Mycoplasmas are unable to synthesize sterols and therefore must be provided with them in their growth media, a requirement usually met by growing mycoplasmas on media containing animal serum. On solid media, most mycoplasmas form very small colonies (10 to 600 µm in diameter) that have a fried egg appearance; an opaque, granular central area; and a flat, translucent peripheral zone (Figure 11.19).

Mycoplasma pneumoniae causes a disease known as primary atypical pneumonia (PAP), a mild form of pneumonia confined to the lower respiratory tract. Because mycoplasmas do not have cell walls, penicillin is ineffective in chemotherapy. Tetracycline, which inhibits protein synthesis, is recommended as the preferred antibiotic for treatment of PAP.

Miscellaneous Bacteria
Mycoplasma • pp. 1–6

Endosymbionts

A large number of microorganisms have been found to exist intracellularly as endosymbionts (microorganisms growing within the host cell) of other, larger organisms (see endosymbiosis, page 425). Many of these symbionts are bacteria; these have their own section of *Bergey's Manual* and include bacteria symbiotic with fungi, protozoa, insects, and other invertebrates.

Not much is known about these endosymbionts. Their growth requirements are difficult to duplicate in the laboratory, and only a few have been cultured. All of the endosymbionts seem to adapt well to the host environment, and many appear to benefit the host by providing it with necessary vitamins and other nutrients. Such symbiotic relationships lend support to the endosymbiont hypothesis, which describes the origin of mitochondria and chloroplasts in eucaryotes (see endosymbiont hypothesis, page 206). An example of an endosymbiont is *Lyticum flagellatum,* which is symbiotic for the ciliate protozoan *Paramecium tetraurelia.*

Gram-Positive Bacteria Other Than the Actinomycetes

The second volume of *Bergey's Manual of Systematic Bacteriology* describes the gram-positive Bacteria other than the actinomycetes. A diverse group of Bacteria is listed in this volume, including cocci, endospore-forming bacteria, filamentous forms, and irregularly shaped bacteria (Table 11.12).

Gram-Positive Cocci

The gram-positive cocci include bacteria that innocuously inhabit the human body as well as those that cause many different types of diseases.

Bacteria of the genus *Staphylococcus* are found in many parts of the human body. The external areas of our body are colonized with the nonpathogenic species, *Staphylococcus epidermidis.* This species does not cause any major problems to the host but is a nuisance when it contaminates clinical cultures taken from other areas of the body. *Staphylococcus aureus,* another species in the same genus, is pathogenic, causing such human diseases as impetigo, osteomyelitis, and toxic shock syndrome (see staphylococcal diseases, page 526). It is also one of the most common causes of food poisoning in the United States (see staphylococcal food intoxication, page 518). Staphylococci are gram-positive bacteria typically arranged as clusters of cocci and can be distinguished from the chain-forming streptococci (Figure 11.20). Staphylococci also produce the enzyme catalase, whereas streptococci do not.

Members of the genus *Streptococcus* are responsible for a number of different diseases, including streptococcal pharyngitis (strep throat), bacterial pneumonia, rheumatic fever, scarlet fever, endocarditis, and necrotizing fasciitis (the so-called flesh-eating disease) (see streptococcal diseases, page 498). Both staphylococcal and streptococcal diseases are frequently characterized by pus for-

table 11.12
Gram-Positive Bacteria Other Than the Actinomycetes

Section	Major Characteristics
Gram-positive cocci	Chemoorganotrophic, mesophilic, non-spore-forming cocci
Endospore-forming gram-positive rods and cocci	Mostly gram-positive motile rods; many commonly found in the soil; all form endospores
Regular, nonsporing gram-positive rods	Chemoorganotrophic, mesophilic, non-spore-forming rods
Irregular, nonsporing gram-positive rods	Bacteria that may exhibit club-shaped forms, rod/coccus cycles, filamentous forms, or other unusual cell morphologies
The Mycobacteria	Rod-shaped to filamentous bacteria with unusually large quantities of lipids in cell envelopes; stain acid fast
Nocardioforms	Aerobic bacteria that form mycelia; frequently have large amounts of lipids in cell envelopes and stain acid fast

mation, so these bacteria are often described as *pyogenic cocci.* *Neisseria* are also pyogenic cocci, causing pus-forming diseases such as gonorrhea and meningitis. Staphylococci and streptococci are distinguished from organisms of the genus *Micrococcus,* which are frequently present on normal skin. Micrococci utilize glucose by respiration, whereas staphylococci and streptococci ferment glucose. Staphylococci and streptococci produce a variety of enzymes and toxins that aid in invasion and infection of the host (see staphylococcal and streptococcal enzymes and toxins, pages 448 and 498).

Ruminococcus, another gram-positive coccus, is a normal inhabitant of animal rumens, where it is important in cellulose digestion. Without this bacterium, ruminants would not be able to use hay and grass as food (see ruminants, page 427).

Endospore-Forming Gram-Positive Rods and Cocci

The bacteria in the next section of *Bergey's Manual* all are gram-positive and form endospores (Figure 11.21). The genus *Bacillus* consists of organisms that are frequently found in the soil. *Bacillus anthracis* causes anthrax in animals and humans. Other species of *Bacillus* are producers of important antibiotics such as polymyxin and bacitracin (see antibiotics, page 126), biological pesticides (see biological pesticides, page 585), and industrial enzymes (see industrial microbial enzymes, page 623).

The genus *Clostridium* consists of anaerobic sporeformers that are responsible for a number of human diseases, including botulism (see botulism, page 517), gas gangrene, and tetanus. Clostridia are inhabitants of the intestinal tracts of animals, sewage, and soil. These bacteria ferment a wide variety of organic compounds such as proteins, purines, alcohols, and many different types of polysaccharides (see cellulose and chitin degradation, page 575).

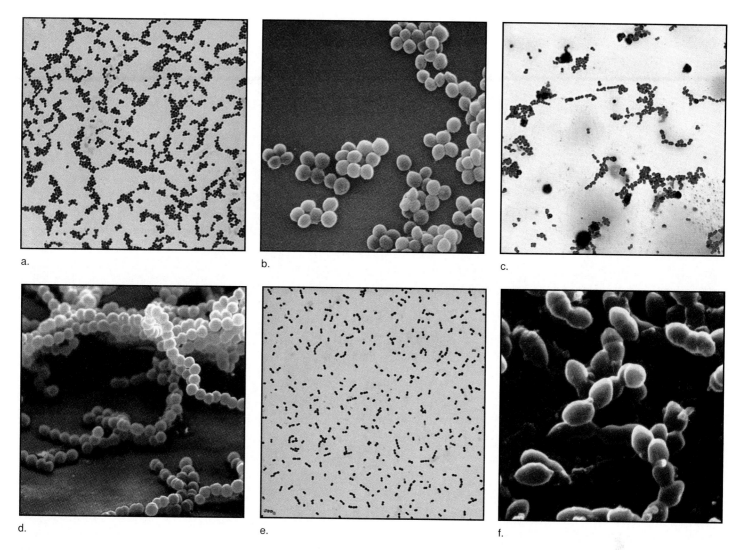

a.   b.   c.

d.   e.   f.

figure 11.20

Representative Gram-Positive Cocci

a. *Staphylococcus aureus,* Gram-stained smear (×400). b. Clusters of staphylococci, scanning electron micrograph (colorized, ×10,000). c. *Streptococcus lactis,* carbolfuchsin stain (×400). d. Chains of strep- tococci, scanning electron micrograph (colorized, ×6,500). e. *Streptococcus pneumoniae,* Gram-stained smear (×322). f. *Streptococcus pneumoniae,* scanning electron micrograph (colorized, ×17,000).

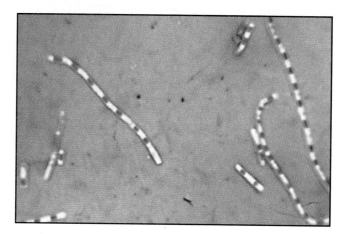

figure 11.21

Spore Stain of *Bacillus subtilis,* Showing Central Endospores Stained Red (×1,000)

Bacillus spores are hardy and have been known to survive dormant in the soil for many years. In 1995 an extraordinary discovery was reported by the microbiologist

Revival of 25-Million-Year-Old Procaryotes?

Raúl Cano and graduate student Monica Borucki at California Polytechnic State University at San Luis Obispo. These two scientists claimed that they had successfully

revived bacillus spores from the digestive tract of an amber-encased stingless Dominican bee, *Proplebeia dominicana*, that existed 25 to 40 million years ago

(Figure 11.22). Abdominal tissue from the fossil bee was aseptically removed and inoculated into trypticase soy broth. Within two weeks the bacillus spores germinated

and produced viable vegetative cells. After extensive analysis of the microbe's DNA and comparison of its ribosomal gene sequences with the sequences of genes from other procaryotes, including about 50 different bacillus species, Cano and Borucki concluded that this ancient procaryote most closely resembled the modern *Bacillus sphaericus*.

Skeptics questioned whether this isolated microorganism may actually have been a present-day bacillus introduced into the bee by laboratory contamination. Although Cano and Borucki used careful techniques to avoid contamination, including working in a clean laboratory hood and checking for contamination by trying to grow microbes from other pieces of the amber, some scientists feel that contamination is difficult to avoid.

If Cano and Borucki's finding is proven to be true, it would eclipse the previous documented record for longest spore survival—70 years for spores stored in ampules by Louis Pasteur and revived in 1956. Amber—the sticky, honey-colored, translucent tree resin that encases fossil life-forms—may be the reason for this long-term survival of spores. As amber hardens, it becomes waterproof and virtually airtight. Together with the ability of bacterial spores to survive adverse conditions (see endospores, page 68), the amber provides a hardened crypt for the entombed microbe. The recovery of viable ancient procaryotes has potential scientific and industrial value for unique drugs, natural pesticides, and microbial enzymes that are not produced by present-day microbes. Furthermore, nucleic acid analysis of such organisms may provide useful information on the rates of evolutionary change in genes.

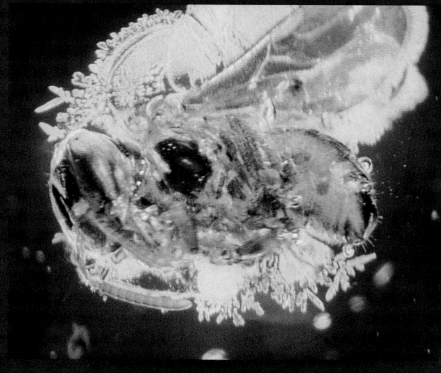

figure 11.22
Amber-Encased Stingless Dominican Bee
Viable spores resembling *Bacillus sphaericus* were revived from the digestive tract of a bee similar to the one illustrated.

Roles of *Lactobacillus* in the Food Industry

Organism	Product That Bacterium Is Used To Prepare
L. acidophilus	Acidophilus milk
L. brevis	Green olives, pickles
L. bulgaricus	Bulgarian milk, kefir, mozzarella cheese, Parmesan cheese, Swiss cheese, yogurt
L. casei	Cheddar cheese
L. helveticus	Swiss cheese
L. lactis	Swiss cheese
L. plantarum	Cheddar cheese, green olives, pickles, sauerkraut

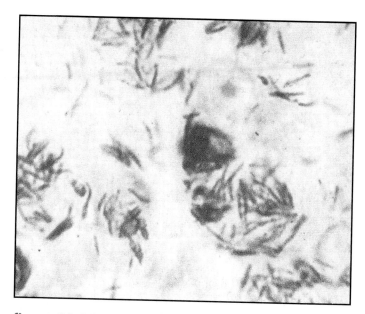

figure 11.23

Acid-Fast Stain of *Mycobacterium leprae*, Showing Retention of the Carbolfuchsin Dye by the Acid-Fast Bacteria

Regular, Nonsporing Gram-Positive Rods

Lactobacillus is the primary genus in the section of *Bergey's Manual* on non-spore-forming, gram-positive rods of regular shape. Lactobacilli are frequently associated with cattle and are found in dairy products. As their name implies, lactobacilli form lactic acid (and other products) from carbohydrate metabolism, and consequently tolerate pHs as low as 5. This resistance to acid is useful in the selective isolation of these bacteria from among other less-acid-resistant bacteria on media of low pH. *Lactobacillus* is normally part of the human vaginal flora, where it competes with and inhibits other members of the flora such as yeasts (*Candida albicans*). This competition reduces the chances for vaginal yeast infections (see microbial antagonism, page 440). Lactobacilli are used extensively in the food industry for the preparation of yogurt, cheeses, sauerkraut, and pickles (Table 11.13) (see food microbiology, page 610). In the past, streptococci, which also produce lactic acid, were included with the lactobacilli in a group called lactic acid bacteria. The streptococci and lactobacilli are now distinguished from each other on the basis of their cell morphologies.

Irregular, Nonsporing Gram-Positive Rods

The next section of *Bergey's Manual* includes non-spore-forming, gram-positive rods of irregular shape, ranging from straight rods to club-shaped forms. Included in this group are some human pathogens and industrially important bacteria.

The childhood disease diphtheria, characterized by the formation of a pseudomembrane in the throat, is caused by the bacterium *Corynebacterium diphtheriae* (see diphtheria, page 507). Corynebacteria [Greek *coryne*, club] are easily identified by their distinctive club-shaped morphology and grouping of cells to form arrangements resembling Chinese letters. This unusual arrangement is caused by a snapping cell division in which cells do not completely separate after binary fission. Cells of *Corynebacterium* also contain stainable polyphosphate granules, which can

be an aid in identification. Most corynebacteria are not pathogenic and are found as part of the normal flora in the nasopharynx and on the skin. The term **diphtheroid** (coryneform) is used to describe gram-positive pleomorphic bacteria that resemble *C. diphtheriae* in morphology but are less virulent (see diphtheroids, page 498).

The genus ***Arthrobacter,*** which is widely distributed in soils, is included in this group. *Arthrobacter* cells have an unusual life cycle in which there is a change from rod-shaped cells to coccoid cells. The coccoid cells appear during the stationary growth phase and are referred to as arthrospores.

Propionibacterium, another bacterium in this section, produces propionic acid and carbon dioxide during cellular metabolism. The acid and the gas are responsible for the unique flavor and holes seen in Swiss cheese (see cheese production, page 611). Propionibacteria have morphologies ranging from rods to club-shaped forms.

The Mycobacteria

The mycobacteria section of *Bergey's Manual* contains only one genus: *Mycobacterium.* Mycobacteria are rod-shaped to filamentous bacteria that have unusually large quantities of lipids in their cell envelopes; up to 60% of the envelope's dry weight consists of lipids as compared with only 2% normally in other bacteria. This characteristic makes mycobacteria extremely resistant to destaining by acid-alcohol and easily identifiable by this acid-fast trait (Figure 11.23). The high lipid content of the envelope is responsible for the slow growth rate of mycobacteria and their resistance to most ordinary bactericides. Mycobacteria are found in soil, water, animals, and humans. In humans they cause a wide variety of diseases, including leprosy (see leprosy, page 527), tuberculosis (see tuberculosis, page 510), and respiratory ailments.

table 11.14

The Archaea, Cyanobacteria, and Remaining Gram-Negative Bacteria

Section	Major Characteristics
Anoxygenic phototrophic bacteria	Bacteria that contain bacteriochlorophyll, do not produce oxygen during photosynthesis, have only one photosystem, and can use light as an energy source
Oxygenic photosynthetic bacteria	Bacteria that contain chlorophylls, produce oxygen during photosynthesis, have two photosystems, and can use light as an energy source
Aerobic chemolithotrophic bacteria and associated organisms	Bacteria that utilize inorganic compounds as an energy source or that oxidize metals or deposit metals on their cell surfaces
Budding and/or appendaged bacteria	Bacteria that reproduce by budding, by the production of appendages, or by a combination of the two
Sheathed bacteria	Bacteria that form an external sheath around chains of cells
Nonphotosynthetic, nonfruiting gliding bacteria	Nonfruiting gliding bacteria that may exist as rods or filaments
Fruiting gliding bacteria: the myxobacteria	Fruiting gliding bacteria that may have a complex developmental cycle in which cells aggregate to form fruiting bodies
Archaea	Procaryotes that are phylogenetically distinct from Bacteria and are distinguished from them by their unusual rRNA structure, different RNA polymerase, membrane lipid composition, mechanism of protein synthesis, and lack of muramic acid in the cell wall

The Archaea, Cyanobacteria, and Remaining Gram-Negative Bacteria

The procaryotes in Volume 3 of *Bergey's Manual* have diverse types of metabolism and unusual structures (Table 11.14). They include the Archaea, cyanobacteria, and remaining gram-negative Bacteria. Some of these procaryotes use light as a source of energy (phototrophic), whereas others obtain their energy from inorganic compounds (chemolithotrophic). There are gliding, budding, stalked, and sheathed bacteria. The Archaea include procaryotes that live in extreme environments and are phylogenetically distinct from other procaryotes.

Anoxygenic Phototrophic Bacteria

Bacteria that use light as an energy source are called phototrophs [Greek *phot*, light, *trephein*, to nourish]. Anoxygenic phototrophic bacteria differ from photosyntheic plants and the oxygenic photosynthetic bacteria (the cyanobacteria) because they: (1) have bacteriochlorophylls instead of chlorophylls as their photosynthetic pigments; (2) do not produce oxygen during photosynthesis; and (3) possess only one photosystem in comparison to the two photosystems (photosystems I and II) that are found in the cyanobacteria (see photosynthesis, page 186). Many of the phototrophic bacteria are also autotrophs (able to use CO_2 as a sole source of carbon).

The anoxygenic phototrophic bacteria stain gram negative and are divided into two major groups on the basis of their pigmentation: purple bacteria and green bacteria (Table 11.15). The green bacteria have an array of photosynthetic pigments (bacteriochlorophylls c, d, e, and some a) different from that of the purple bacteria (bacteriochlorophylls a and b). The photosynthetic apparatus of the green bacteria is located in specialized cylindrical vesicles called chlorosomes, which underlie and are attached to the plasma membrane. The photosynthetic machinery of the purple bacteria, in comparison, is contained within elaborate internal membranes found in the cell cytoplasm. In addition to their major pigments, the purple and green bacteria may be pigmented in shades of brown, orange, and yellow.

Anoxygenic Phototrophic Bacteria

Group	Family/Subgroup	Genera
Purple bacteria	Chromatiaceae	Chromatium
		Thiocystis
		Thiospirillum
		Thiocapsa
		Lamprobacter
		Lamprocystis
		Thiodictyon
		Amoebobacter
		Thiopedia
	Ectothiorhodospiraceae	Ectothiorhodospira
	Purple nonsulfur bacteria	Rhodospirillum
		Rhodopila
		Rhodobacter
		Rhodopseudomonas
		Rhodomicrobium
		Rhodocyclus
Green bacteria	Green sulfur bacteria	Chlorobium
		Prosthecochloris
		Pelodictyon
		Ancalochloris
		Chloroherpeton
	Multicellular, filamentous, green bacteria	Chloroflexus
		Heliothrix
		Oscillochloris
		Chloronema
Genera incertae sedis		Heliobacterium
		Erythrobacter

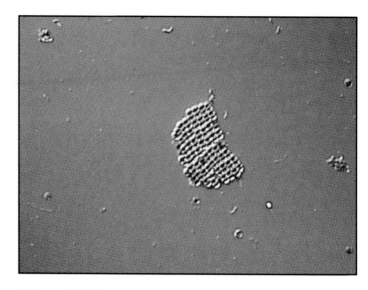

figure 11.24

Nomarski Differential Interference Contrast Micrograph of *Thiopedia rosea,* a Photosynthetic Purple Sulfur Bacterium that Grows in Sheets

sulfur bacteria can oxidize simple organic molecules for phototrophic growth, provided that a reduced sulfur compound is also available as a sulfur source. ***Chloroflexus,*** a multicellular, filamentous, green bacterium, is versatile and able to grow heterotrophically in the dark under aerobic conditions, as well as phototrophically in the light.

Phototrophic bacteria are generally found in anaerobic aquatic environments, where reduced compounds (compounds that contain extra electrons and are normally more stable in environments without oxygen) are readily available as sources of electrons for photosynthesis. These bacteria must reside in habitats close enough to the water surface to allow light penetration for photosynthesis. In contrast, the cyanobacteria and algae photosynthesize aerobically. Electrons for these photosyntheses are derived from the photolysis of water, resulting in the release of oxygen.

Oxygenic Photosynthetic Bacteria

The oxygenic photosynthetic bacteria include the **cyanobacteria** [Greek *kyanos,* blue] and the **prochlorophytes.** Members of this group became part of the chloroplasts in photosynthetic eucaryotic cells, according to the endosymbiont hypothesis. These microbes fit phylogenetically into the Bacteria tree, and molecular analysis indicates that chloroplasts, cyanobacteria, and the prochlorophytes shared a common ancestor.

Cells of cyanobacteria resemble the cells of other bacteria because they have a cytoplasm surrounded by a plasma membrane and cell wall, but no membrane-enclosed nucleus (Figure 11.25).

Both the purple and green bacteria are further subdivided on the basis of their metabolism. Purple sulfur bacteria (***Chromatiaceae*** and ***Ectothiorhodospiraceae***) use sulfur compounds such as H_2S as electron donors in photosynthesis (Figure 11.24). As H_2S is oxidized, sulfur granules accumulate as globules inside the bacterial cell (*Chromatiaceae*) or outside the cell (*Ectothiorhodospiraceae*). In the past the purple sulfur bacteria were distinguished from the purple nonsulfur bacteria (for example, ***Rhodospirillum***) because of this sulfur metabolism. Recent studies, however, have shown that the purple nonsulfur bacteria, although not able to utilize elemental sulfur, do use sulfide at low concentrations. The purple nonsulfur bacteria also oxidize organic acids, alcohols, and other organic molecules.

The green bacteria are divided into two groups: green sulfur bacteria and multicellular, filamentous, green bacteria. Most green

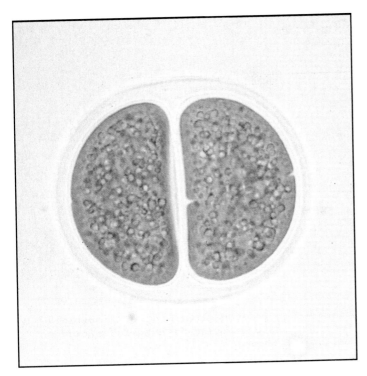

a.

figure 11.25

Cyanobacteria

a. Bright-field micrograph of *Chroococcus turgidus,* showing cells remaining together with a surrounding sheath following cell division (×300). b. Phase-contrast micrograph of *Spirulina,* showing the helical structure of this oscillatorian cyanobacterium (×300).

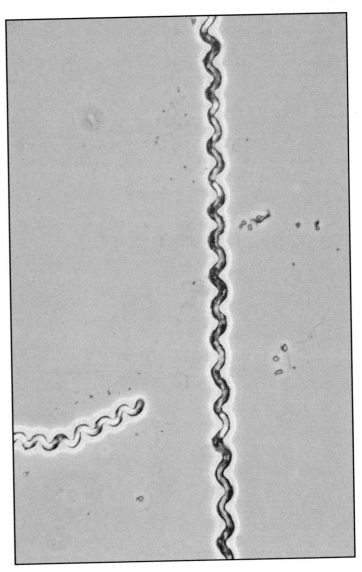

b.

The cyanobacterial cell wall is composed of murein (peptidoglycan) and several additional amino acids. External to the wall is a sheath of slimy material that the organism uses for gliding motility along solid surfaces. The cyanobacteria have photosynthetic saclike membranes (thylakoids) resembling those of eucaryotic plants and algae, but the thylakoids are not segregated from the cytoplasm by a membrane. The cyanobacteria do have chlorophylls and a light-gathering photosystem that parallel similar pigments and systems in photosynthetic eucaryotes. However, all the other characteristics of these organisms strongly indicate that they are procaryotes, not eucaryotes.

The cyanobacteria are a large heterogeneous group of microorganisms (approximately 150 genera with 1,500 species). Like other bacterial groups, they are found in many habitats. The Red Sea is so named because of the red color imparted by occasional blooms of **Oscillatoria,** a cyanobacterium. Some cyanobacteria exist in thick mats in water beneath the ice in Antarctica, and they even grow in the fur of polar bears, giving the fur a green tint. Other cyanobacteria are found in hot springs and desert soils. Thick mats of dome-shaped, layered chalk deposits called stromatolites, consisting of cyanobacteria bound to calcium carbonate, occur in a few places such as shallow pools of water in hot, dry climates (Figure 11.26). Fossilized stromatolites, dated at over 3 bil-lion years old, contain the remains of ancient cyanobacteria and indicate that these microbes played an important evolutionary role in the introduction of oxygen to the atmosphere of the earth.

Although sometimes called blue-green bacteria, the cyanobacteria display a variety of colors, including red, brown, yellow, dark purple, and even black. This color range results from the presence of different photosynthetic pigments, including chlorophyll a, carotenoids, and phycobilins. Even though cyanobacteria are phototrophic, a few are able to grow slowly in darkness by using carbohydrates as sources of energy and carbon.

Many of the cyanobacteria are able to fix nitrogen to ammonia. Consequently these microbes are often found in symbiosis with other organisms in such areas as bare rock and soil. The genus *Oscillatoria,* containing organisms that are common inhabitants of seas, contributes extensively to fixation of atmospheric nitrogen. Nitrogen fixation by cyanobacteria generally occurs within heterocysts, enlarged cells usually found along the filament. The thick-

figure 11.26

Stromatolites

Stromatolites consist of cyanobacteria bound to calcium carbonate to form domed structures like the ones illustrated.

table	11.16

Classification of Cyanobacteria

Family	Major Characteristics
Chroococcaceae	Unicellular rods or cocci; reproduction by binary fission or budding
Pleurocapsaceae	Single cells enclosed in a fibrous layer; reproduction by multiple fission
Oscillatoriaceae	Vegetative cells in trichomes[a]; reproduction by trichome fragmentation
Nostocaceae	Vegetative cells or heterocysts in nonbranching trichomes; reproduction by trichome fragmentation; can form akinetes[b]
Stigonemataceae	Vegetative cells or heterocysts in branching trichomes; reproduction by trichome fragmentation; can form akinetes

[a] a strand or chain of cells

[b] a vegetative cell that is transformed into a resistant spore

walled heterocysts appear to be resting stages of vegetative cells and lack photosystem II, which is associated with oxygen evolution. Consequently the oxygen-labile nitrogenase enzyme is stable and active, and nitrogen fixation is possible within the heterocysts.

The cyanobacteria have three different types of cellular organization: unicellular, colonial, and filamentous. Unicellular forms exist as single cells that are either free-living or attached to rocks, walls, and other organisms. Colonial forms arise when dividing cells adhere to one another. Filaments are formed when cell division occurs in only one direction (unbranched) or in several directions (branched). These filaments form large masses that sometimes exceed 1 m in length. The individual cells of cyanobacteria, however, are typically 0.5 to 60 μm in diameter. Reproduction occurs by cell division in unicellular forms and by fragmentation in colonial and filamentous forms. Some species form resistant resting cells (akinetes) for protection against environmental extremes such as low temperatures or desiccation.

Classification of the cyanobacteria is relatively simple. There are five families: **Chroococcaceae, Pleurocapsaceae, Oscillatoriaceae, Nostocaceae,** and **Stigonemataceae** (Table 11.16). The *Chroococcaceae* are unicellular rods or cocci that reproduce by binary fission or budding. The *Pleurocapsaceae* consist of single cells enclosed in a fibrous layer. Unlike the *Chroococcaceae*, these cyanobacteria reproduce by multiple fission. Numerous small coccoid daughter cells, called baeocytes, form as a result of this division. Oscillatorian cyanobacteria form long filamentous strands, called trichomes, containing vegetative cells. Reproduction is by fragmentation of the trichomes. Cyanobacteria of the families *Nostocaceae* and *Stigonemataceae* form trichomes containing vegetative cells and also heterocysts. In some instances, heterocysts can differentiate into akinetes (Figure 11.27).

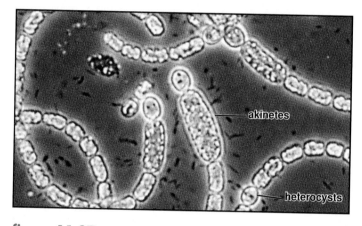

figure 11.27

Phase-Contrast Micrograph of the Cyanobacterium *Nostoc*, Showing Akinetes and Heterocysts

The prochlorophyta are similar to the cyanobacteria, except for three distinguishing characteristics: (1) The prochlorophyta contain chlorophyll b in addition to chlorophyll a, (2) they do not have phycobilins and therefore appear bright green instead of blue-green, and (3) they have thylakoids with double membranes (unlike the single membranes in cyanobacterial thylakoids). There are two recognized genera of prochlorophytes. One is **Prochloron,** a spherical, single-celled organism that lives as an extracellular symbiont of marine vertebrates; it has not yet been cultured. The other is **Prochlorothrix,** a free-living cylindrical organism that forms filaments; it has been grown in pure culture and is found in Dutch lakes. There are also prochlorophytes that

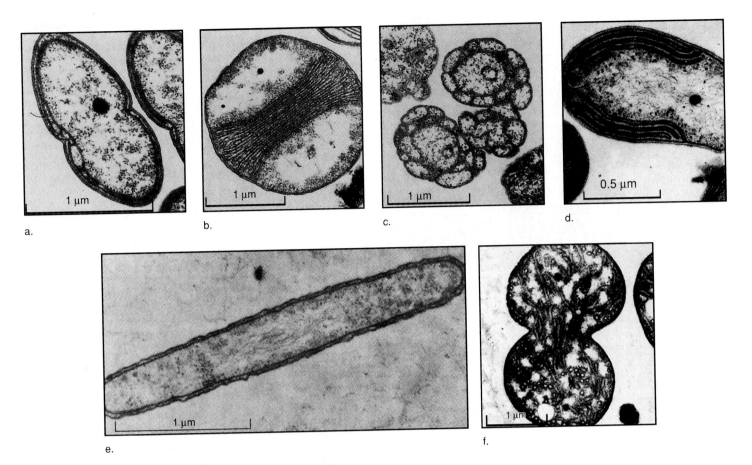

figure 11.28

Nitrifiers

a. Electron micrograph of *Nitrosomonas europaea*, showing peripheral cytomembranes (×32,000). b. Electron micrograph of *Nitrosococcus oceanus*, showing cytomembranes arranged as flattened lamellae in the center of the cell (×22,500). c. Electron micrograph of *Nitrosolobus multiformis*, showing its lobular shape and cytomembranes that partially compartmentalize the cell (×22,500). d. Electron micrograph of *Nitrobacter winogradskyi*, showing polar cap of peripheral cytomembranes (×41,000). e. Electron micrograph of *Nitrospina gracilis*, showing the absence of extensive cytomembranes found in other nitrifiers but the presence of small bleb-like intrusion of the plasma membrane (×37,500). f. Electron micrograph of *Nitrococcus mobilis*, showing tubular type of cytomembranes extending throughout the cytoplasm (×16,000).

are found free-living in open ocean waters and are referred to as picoplankton. These organisms are thought to be important ecologically as the primary producers in these waters.

Aerobic Chemolithotrophic Bacteria and Associated Organisms

Chemolithotrophic organisms utilize inorganic compounds as an energy source, an activity that is unique to bacteria (see chemolithotrophy, page 171). *Bergey's Manual* divides the chemolithotrophs into five groups: (1) nitrifiers, (2) colorless sulfur bacteria, (3) obligate hydrogen oxidizers, (4) iron and manganese oxidizing and/or depositing bacteria, and (5) magnetotactic bacteria.

The nitrifiers are bacteria that oxidize ammonia to nitrite and nitrite to nitrate in two separate stages. Nitrosofying bacteria such as **Nitrosomonas, Nitrosospira, Nitrosococcus,** and **Nitrosolobus** oxidize ammonia to nitrite. The nitrite can then be oxidized to nitrate by the true nitrifying bacteria (**Nitrobacter, Nitrospina,** and **Nitrococcus**). Both types of bacteria are required for the complete oxidation of ammonia to nitrate. Many of the nitrifiers have complex and distinctive internal membrane systems that are involved in ammonia or nitrite oxidation (Figure 11.28). Nitrifiers are found in soil, freshwater, and marine environments that are rich in ammonia, nitrite, and other inorganic salts.

Microorganisms that oxidize reduced sulfur compounds to sulfate are frequently found in acid environments, where the pH of

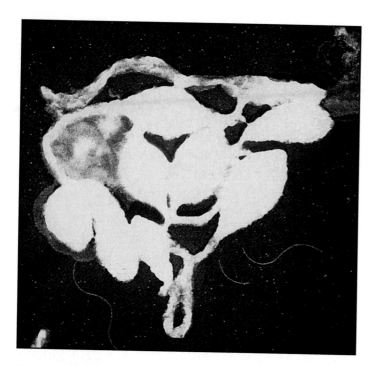

figure 11.29

Hyphomicrobium

Electron micrograph of *Hyphomicrobium*, showing hyphae and terminal buds (colorized, ×9,000).

the environment is lowered by the sulfuric acid that is produced by the oxidation. An example of a sulfur-oxidizing bacterium is ***Thiobacillus ferrooxidans,*** which is used in mining operations to recover valuable metals from sulfur-containing ores (see leaching of metals, page 582). ***Hydrogenobacter,*** an example of an obligate hydrogen oxidizer, uses H_2 as an electron donor and O_2 as an electron acceptor to produce H_2O from the reduction of O_2 with H_2.

Bacteria that oxidize metals or deposit metals on their cell surfaces use organic compounds as a source of energy and are not chemolithotrophs. Nonetheless, they are included with the chemolithotrophs because they are not easily placed in any other section of *Bergey's Manual*. This group includes such bacteria as ***Siderocapsa*** [Greek *sidero,* iron] ***Siderococcus,*** and ***Naumanniella.*** These bacteria are found in iron-bearing waters, where their cells are encrusted with iron, manganese oxides, or both.

Magnetotactic bacteria (for example, ***Aquaspirillum***) demonstrate directed movement in a magnetic field, a phenomenon called **magnetotaxis** (see magnetotaxis, page 65). These bacteria synthesize intracellular crystals of magnetite (Fe_3O_4) arranged as chains along the long axis of the cell. These chains of magnetite serve as internal magnets to orient the cell along a specific magnetic line. Although the function of magnetotaxis is unknown, it has been suggested that such behavior may direct these microaerophilic, aquatic bacteria downward along magnetic lines, away from oxygen-enriched surface waters and toward the more anaerobic sediments.

Budding and/or Appendaged Bacteria

Although most bacteria reproduce by binary fission, some do not. The bacteria in the section of *Bergey's Manual* on budding and/or appendaged bacteria reproduce by budding, by the production of appendages, or by a combination of the two. Unlike the nonliving stalks of the myxobacteria (see gliding, fruiting bacteria, page 338), the buds and appendages formed by these bacteria are usually direct cytoplasmic extrusions of the cell, called **prosthecae.** The prosthecae provide the bacterium with a greater surface area for the absorption of nutrients from the surrounding environment.

An example of a budding bacterium is ***Hyphomicrobium*** [Greek *hyphe,* thread], which forms buds at the tips of filamentous outgrowths (hyphae) of the cell (Figure 11.29). As each bud matures, it synthesizes a flagellum and breaks away from the parent cell. This daughter cell eventually loses its flagellum, forms its own hyphae, and repeats the budding process.

Bacteria of the genus ***Caulobacter*** [Latin *caulis,* stalk] reproduce in a different manner. These microbes attach themselves to a solid surface by a stalk that is an extension of the cell, with a wall, membrane, and cytoplasm. Not infrequently, several cells may adhere to each other by the bases of their stalks to form rosettes. Cell division in *Caulobacter* commences with the elongation of the cell at the end opposite the point of surface attachment. As the cell elongates, the newly formed portion develops a flagellum and the flagellated cell is released into the environment. This flagellated swarmer cell eventually loses its flagellum and replaces it with a stalk that anchors the bacterium to the solid surface. The life cycle is then repeated. *Caulobacters* are generally found in aquatic environments that have low levels of organic matter. In these environments, the stalk serves not only as an anchor, but also as additional surface area for the absorption of limited nutrients. In fact, the length of the stalk increases dramatically when nutrient supply (especially phosphorus) is limited.

Sheathed Bacteria

The sheathed bacteria are gram-negative cells, often arranged as chains within filaments surrounded by outer sheaths composed of proteins, polysaccharides, and lipids (Figure 11.30). These organisms are usually found in aquatic environments, particularly slow-running, fresh water contaminated with sewage or wastewater, where the organisms' sheaths may become encrusted with iron or manganese oxides. In some instances the sheaths have an adhesive holdfast that is used for attachment to solid surfaces.

Members of the genus ***Sphaerotilus,*** an example of the sheathed bacteria, are nutritionally versatile and widespread in nature. These bacteria use a variety of organic acids and sugars as sources of carbon and energy. They frequently are found in activated sludge (a product of sewage treatment), where they can cause a detrimental condition called bulking. In this phenomenon, tangled filaments of bacteria increase the bulk of the sludge so that it does not properly settle during wastewater treatment.

The sheathed bacteria reproduce by binary fission, with the release of motile daughter cells called **swarmer cells** from one end of the sheath. Swarmer cells migrate and eventually form their own filaments.

Nonphotosynthetic, Nonfruiting Gliding Bacteria

The nonphotosynthetic, nonfruiting gliding bacteria have no flagella, but they exhibit gliding motility. Gliding bacteria include those that form fruiting bodies (see the next section) and those that do not form fruiting bodies (this section). They move across solid surfaces either on a slime layer deposited by the cell or on small rotating protein particles, acting like ball bearings, which lie between the plasma membrane and the outer envelope of these gram-negative bacteria.

Nonfruiting gliding bacteria of the genus *Cytophaga* have the unusual ability to digest such compounds as cellulose, chitin, and agar, using extracellular enzymes synthesized by the bacteria. Some *Cytophaga* are pathogens of fish, causing such diseases as fin rot, tail rot, and bacterial gill disease.

Beggiatoa, Thiothrix, and *Leucothrix* are examples of gliding bacteria having long filaments. These organisms live in aquatic environments rich in H_2S (sulfur springs, decaying seaweed beds, and waters heavily polluted with sewage). For energy, the H_2S is oxidized to elemental sulfur, which is deposited as granules inside the filaments (Figure 11.31).

Fruiting Gliding Bacteria: The Myxobacteria

The fruiting gliding bacteria move by gliding and often produce colorful fruiting bodies. This section consists of only one order of bacteria: *Myxobacterales.*

The *Myxobacterales* have complex developmental cycles. Under certain environmental conditions that are not well-defined but generally associated with nutrient limitation, the rod-shaped vegetative cells of these bacteria aggregate to form a colorful structure visible to the naked eye, called the **fruiting body** (Figure 11.32). This represents the resistant, or resting, stage of the bacterium, and either lies on the surface of the ground or is raised above it on a stalk of slime. Contained within the fruiting body are resistant cells called **myxospores.** As the fruiting body matures, it ruptures and releases the myxospores into the environment. The mature myxospores germinate and give rise to vegetative cells.

Myxobacteria are soil organisms found in abundant numbers in environments containing rich organic matter. Their distinctive and strikingly colorful fruiting bodies are frequently seen on tree barks, decomposing plant material, and manure.

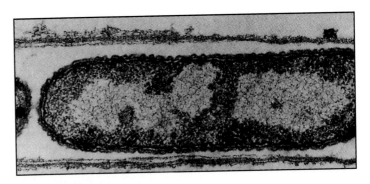

figure 11.30
Sheathed Bacteria
Electron micrograph of *Sphaerotilus natans,* showing cell enclosed by a sheath. The sheath is composed of a lipoprotein-lipopolysaccharide complex external to the cell wall (×36,000).

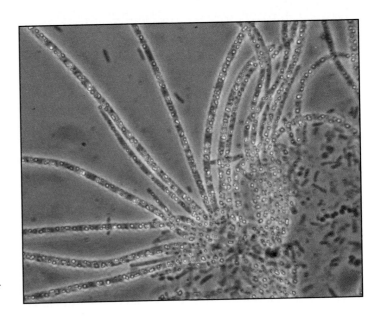

figure 11.31
Gliding Bacteria
Phase-contrast micrograph of *Thiothrix,* showing long filaments and bright sulfur granules inside the cells (×1,000).

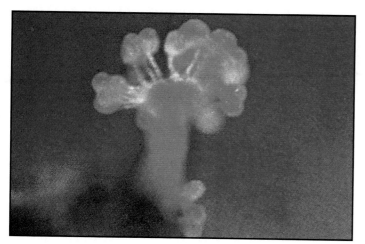

figure 11.32
Colorful Fruiting Body of the Myxobacterium *Stigmatella aurantiaca*

a.

b.

c.

Archaea

The Archaea are a group of organisms distinguished from other procaryotes by their unusual rRNA structure, different RNA polymerase, plasma membrane lipid composition, mechanism of protein synthesis, and absence of peptidoglycan in the cell wall. The phylogenetic relationship of these unique procaryotes to the Bacteria and the Eucarya was described earlier in this chapter.

The Archaea Are Diverse in Their Morphology and Physiology

The Archaea are a diverse group of procaryotes, ranging morphologically from cocci to rods to spiral-shaped cells. They consist of gram-positive and gram-negative cells, with diameters ranging from 0.1 to over 15 μm. The Archaea can be aerobic, facultatively anaerobic, or anaerobic. Some are organotrophs, while others are autotrophs. Their habitats extend from deep sea, geothermally heated vents and hot sulfur springs (the extreme thermophiles) to the rumens and intestinal systems of animals, and the anaerobic sediments of lakes, swamps, and bogs (the methanogens) to marine salterns and salt lakes like the Great Salt Lake in Utah and the Dead Sea between Israel and Jordan (the extreme halophiles) (Figure 11.33).

None of the Archaea have muramic acid or D-amino acids in their cell walls, chemical molecules that are characteristic of the peptidoglycan in the Bacteria. Consequently penicillin, which inhibits the transpeptidation step in peptidoglycan synthesis (see peptidoglycan synthesis, page 56), is ineffective against these procaryotes. A pseudopeptidoglycan is formed in some Archaea, notably *Methanobacterium*, *Methanothermus*, and *Methanobrevibacter*, but consists of alternating repeats of N-acetylglucosamine and N-acetyltalosaminuronic acids and L-amino acids (not the D-amino acids found in traditional peptidoglycan). The N-acetylglucosamine and N-acetyltalosaminuronic acids are linked by lysozyme-resistant β(1,3) bonds instead of the lysozyme-sensitive β(1,4) bonds found in peptidoglycan of the Bacteria.

figure 11.33
Habitats of Archaea
a. Hot sulfur spring, a common habitat for the thermoacidophile *Sulfolobus*. b. Boiling springs and geysers in Yellowstone National Park. c. Great Salt Lake, Utah, a habitat for *Halobacterium* and other extreme halophiles.

table 11.17

The Archaea

Group	Representative Genera	Major Characteristics
Methanogenic Archaea	Methanobacterium Methanobrevibacter Methanococcus Methanolobus Methanomicrobium Methanosarcina Methanospirillum Methanothermus	Strictly anaerobic; chemoautotrophic or chemoheterotrophic, with methane always the product of metabolism; produce coenzyme F_{420} and methanopterin
Sulfate reducers	Archaeoglobus	Strictly anaerobic; chemolithotrophic or chemoorganotrophic growth; autotrophic growth occurs with thiosulfate and H_2; produce coenzyme F_{420} and methanopterin
Extremely halophilic Archaea	Halobacterium Halococcus Haloferax Natronobacterium Natronococcus	Aerobic, although some can grow anaerobically in the presence of nitrate; chemoheterotrophic; require at least 1.5 M NaCl for growth; some members are alkalophilic, growing only at pH>8.5
Cell wall-less Archaea	Thermoplasma	Obligately thermophilic; obligately acidophilic; facultatively anaerobic; chemoorganotrophic
Extremely thermophilic S^0-metabolizers	Desulfurolobus Pyrococcus Pyrodictium Sulfolobus Thermococcus Thermoproteus	Obligately thermophilic; acidophilic or neutrophilic; chemoautotrophic or chemoheterotrophic; most are sulfur metabolizers

Halobacterium, some methanogens (for example, *Methanolobus*), and several extreme thermophiles (for example, *Sulfolobus, Pyrodictium,* and *Thermoproteus*) have cell walls made of glycoprotein, whereas certain methanogens (for example, *Methanococcus* and *Methanomicrobium*) have cell walls composed exclusively of protein.

The plasma membranes of the Archaea are chemically unique. They lack fatty acids and have the branched chain hydrocarbon isoprene attached to glycerol by ether (instead of ester) linkages. However, their membranes have polar and nonpolar orientations and form lipid bilayers similar to the membranes of the Bacteria and the Eucarya. The difference in membrane lipids of the Archaea may be associated with the extreme environments of these procaryotes. Branched chain hydrocarbons may impart greater mechanical strength and chemical resistance to the membranes of such Archaea as *Sulfolobus* (an extreme thermophile that grows in sulfur-rich hot acid springs) and *Halobacterium* (an extreme halophile).

The Archaea Are Separated into Five Groups in Bergey's Manual

The Archaea are divided into five groups in *Bergey's Manual:* the methanogenic Archaea, the sulfate reducers, the extremely halophilic Archaea, the Archaea lacking cell walls, and the extremely thermophilic elemental sulfur metabolizers (Table 11.17).

The methanogenic Archaea are strict anaerobes that oxidize compounds such as H_2 or formate for energy and use carbon dioxide as an electron acceptor to produce methane (CH_4). Methanogens are found in anaerobic environments such as swamps, marshes, and the intestinal tract of animals. ***Methanobacterium*** and ***Methanosarcina*** are examples (Figure 11.34). It is generally believed that methanogenic procaryotes were the predominant organisms on a primitive earth that contained a large amount of carbon dioxide and little or no oxygen.

Only one genus, ***Archaeoglobus,*** is currently recognized as a sulfate reducer. This organism, isolated from marine hydrothermal vents, is unique among the Archaea in its ability to use sulfate as an electron acceptor. Cells of *Archaeoglobus* grow at temperatures as high as 92°C with an optimum of 83°C. Interestingly, *Archaeoglobus* produces coenzyme F_{420} and methanopterin, two coenzymes found in methanogenic Archaea and associated with methane production. Phylogenetically, *Archaeoglobus* lies between the methanogens and the extremely thermophilic elemental sulfur metabolizers, and may be an intermediate link between these two groups.

The extremely halophilic Archaea require high concentrations of NaCl for survival. These organisms, which include ***Halobacterium*** and ***Halococcus,*** are generally found in high-salt

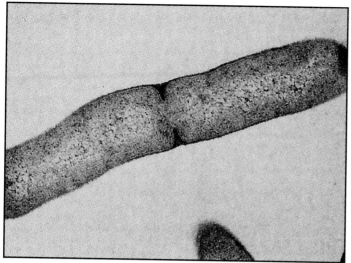

a.

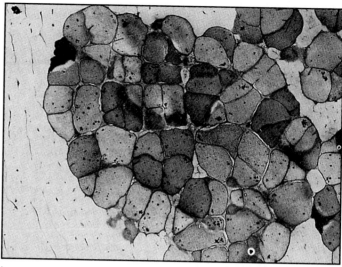

b.

figure 11.34

Methanogenic Archaea

a. Electron micrograph of *Methanobacterium thermoautotrophicum*, a thermophile, in the process of division (×30,000). b. Electron micrograph of *Methanosarcina barkeri*. Cells of the organism remain attached in sarcinoid colonies, reaching 2 to 3 mm in diameter. Dark spots are polyphosphate granules inside the cells (×10,000).

habitats such as the Dead Sea, the Great Salt Lake, and salt-drying beds formed by the evaporation of seawater in shallow ponds. Halophilic procaryotes are able to tolerate the high-salt content of these environments because their enzymes have adapted to require a high-salt concentration for activity. In fact, the ribosomes, cell wall proteins, and other cell constituents of these bacteria require high-salt environments for stability.

Thermoplasma is a cell wall-less procaryote that grows in coal refuse piles. These piles contain iron pyrite (FeS), coal fragments, and organic by-products of coal mining operations that become very hot and acidic through spontaneous combustion and leaching of the organic compounds. *Thermoplasma*, a thermoacidophile that grows optimally at 55°C and pH 2, thrives in this environment where it apparently uses the organic compounds leached from the coal. The plasma membrane of *Thermoplasma* contains a chemically unique lipopolysaccharide composed of a tetraether lipoglycan containing glucose and mannose units. Together with membrane glycoproteins, this and other membrane molecules enable *Thermoplasma*, without a cell wall, to withstand the low pH and high temperature of its environment. Phylogenetically, *Thermoplasma* more closely resembles the methanogen/extreme halophile branch of the Archaea phylogenetic tree than the extremely thermophilic sulfur metabolizer branch.

The extremely thermophilic Archaea include three orders (**Thermococcales, Thermoproteales,** and **Sulfolobales**) and at least nine genera. Many of the microorganisms in this group are acidophilic and sulfur metabolizers. **Thermococcus** and **Pyrococcus,** members of the order *Thermococcales*, are obligate anaerobes that grow near deep-sea hydrothermal vents. Both are chemoorganotrophs that grow on proteins, starch, and other organic matter and use elemental sulfur (S^0) as an electron acceptor, reducing it to H_2S during anaerobic respiration. **Thermoproteus** has similar metabolic properties as *Thermococcus* and *Pyrococcus*. It is an anaerobic chemoorganotroph that can use S^0 as an electron acceptor, but it can also grow chemolithotrophically on H_2. *Thermoproteus* grows at temperatures ranging from 60°C to 96°C and at pH values between 1.7 and 6.5, and is found in sulfur-rich, hot, aquatic environments such as hot springs. **Sulfolobus,** a member of the order *Sulfolobales*, is a thermoacidophile that grows best at temperatures of 75°C to 85°C and pH values between 2 and 3. *Sulfolobus* is an obligate aerobe capable of oxidizing S^0 or H_2S to H_2SO_4, and of fixing CO_2 to organic forms of carbon. *Sulfolobus* is also able to grow chemoorganotrophically. It is generally found in hot sulfur springs and similar habitats.

The Actinomycetes

The fourth volume of *Bergey's Manual* contains Bacteria called the actinomycetes—a category with an extremely large number of organisms, as is evident because they make up an entire volume. This volume is divided into eight sections, including a section on nocardioform actinomycetes that is repeated, with two additional genera, from Volume 2 (Table 11.18). The actinomycetes [Greek *aktis*, a ray, beam, *mykes*, fungus] are gram-positive bacteria that are rod-shaped or form branching filaments that in some genera develop into a mycelium. A common property of the actinomycetes is the high G + C content (≥52 mole%) of their DNA, which separates these organisms from other procaryotes containing DNA with lower G + C contents. For a long time, these microbes, because of their similarities to both the bacteria and the fungi, were considered

The Actinomycetes

Section	Genera	Major Characteristics
Nocardioform actinomycetes	Nocardia Rhodococcus Nocardioides Pseudonocardia Oerskovia Saccharopolyspora Faenia (Micropolyspora) Promicromonospora Intrasporangium Actinopolyspora Saccharomonospora	Aerobic bacteria that form mycelia; frequently have large amounts of lipids in their cell envelopes; stain acid fast
Actinomycetes with multilocular sporangia	Geodermatophilus Dermatophilus Frankia	Bacteria that produce branching filaments that divide by longitudinal and transverse septa, giving rise to coccoid-like elements
Actinoplanetes	Actinoplanes Ampullariella Pilimelia Dactylosporangium Micromonospora	Bacteria in which spores are produced within spore vesicles or sporangia
Streptomyces and related genera	Streptomyces Streptoverticillium Kineosporia Sporichthya	Aerobic bacteria that are highly oxidative and form extensive branching substrate and aerial mycelia
Maduromycetes	Actinomadura Microbispora Microtetraspora Planobispora Planomonospora Spirillospora Streptosporangium	Bacteria that contain the sugar madurose in their cell walls
Thermomonospora and related genera	Thermomonospora Actinosynnema Nocardiopsis Streptoalloteichus	Mesophilic and thermophilic bacteria with cell walls containing meso-diaminopimelate and no other characteristic sugars or amino acids; produce spores (not endospores) that may be borne on substrate and aerial hyphae
Thermoactinomycetes	Thermoactinomyces	Thermophilic bacteria with cell walls containing meso-diaminopimelate and no other characteristic sugars or amino acids; single spores are borne on hyphae and have the typical structure of endospores
Other genera	Glycomyces Kibdelosporangium Kitasatospora Saccharothrix Pasteuria	

to be intermediates of the two groups. They are now considered to be bacteria because of their procaryotic properties.

Actinomycetes systematics has undergone many changes in the last 20 years. The current division of the actinomycetes in *Bergey's Manual* separates these bacteria on the basis of such characteristics as cell wall structure and composition, spore production, and temperature resistance. Representative examples of actinomycetes are described here.

Nocardia is a nocardioform actinomycete that is widely distributed in water and soil, but can also cause pulmonary and tissue infections in animals and humans. The nocardioform actinomycetes are aerobic bacteria that form mycelia, or masses of hyphae (see hyphae, page 361). In most instances, the hyphae fragment to form rod-shaped or coccoid cells. Like mycobacteria, nocardioforms frequently have large amounts of lipids in their cell envelopes and therefore are also acid fast.

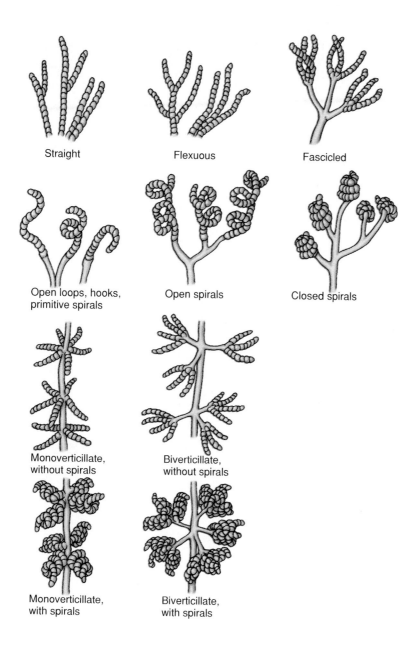

Straight

Flexuous

Fascicled

Open loops, hooks, primitive spirals

Open spirals

Closed spirals

Monoverticillate, without spirals

Biverticillate, without spirals

Monoverticillate, with spirals

Biverticillate, with spirals

figure 11.35

Conidia Arrangements of Aerial Mycelia in the Streptomyces

Dermatophilus is a pathogen that causes skin infections, usually in the hooves of cattle, sheep, and similar animals. The lesions result in a pustular dermatitis that eventually becomes crusty. Human infection is rare. Dermatophili are short, branched, and filamentous, with the filaments dividing by longitudinal and transverse septa. Filament septation leads to the formation of coccoid-like bodies that mature into zoospores.

Frankia is a nitrogen-fixing symbiont of plants such as the alder, Australian pine, bayberry, sweet fern, and autumn olive. These organisms produce nodules on the roots of these plants and fix nitrogen by a mechanism similar to that used by *Rhizobium*. Like *Dermatophilus*, *Frankia* forms branching filaments with transverse and longitudinal septation.

Actinoplanes is a bacterium that forms extensive mycelia, with spores contained within sporangia. The spores are flagellated and may be spherical, ovoid, club-shaped, or cylindrical. Actinoplanes are present in most soils and are especially abundant in soils with a neutral pH. These organisms are capable of producing secondary metabolites (see secondary metabolites, pages 204 and 629), and in recent years some species have been used in the production of new antibiotics (see antibiotics, page 125).

Streptomyces comprises a large group of bacteria that are important in antibiotic production and as soil organisms. More than 300 species of *Streptomyces* have been characterized, and many of these produce antibiotics such as streptomycin, tetracycline, chloramphenicol, and erythromycin. In the soil, streptomyces produce exoenzymes that degrade polysaccharides (starch, cellulose, pectin, and chitin), proteins, fats, and other large molecules. Metabolites (geosmins) of streptomyces are responsible for the distinctive aroma of soil. Streptomyces form aerial mycelia that give rise to asexual reproductive spores called conidia (Figure 11.35). The aerial mycelia allow airborne dispersal of these spores.

The maduromycetes are actinomycetes that contain the sugar madurose (3-O-methyl-D-galactose) in their cell walls. *Streptosporangium,* a genus in this group, forms single or clustered sporangia on aerial mycelia. Nonmotile spores are released as the sporangia rupture. Streptosporangia are a significant component of the actinomycete population in soil.

Mesophilic and thermophilic actinomycetes with cell walls containing *meso*-diaminopimelate and no other characteristic sugars or amino acids are placed into two groups: (1) thermomonospora and related genera and (2) thermoactinomycetes. Members of both groups form spores on aerial hyphae. However, spores produced by *Thermoactinomyces* contain calcium dipicolinate and have the typical structure of endospores. Most *Thermoactinomyces* species have a growth temperature range of 30°C to 60°C, with an optimum growth temperature of 50°C. *Thermoactinomyces* organisms are commonly found in natural high-temperature habitats such as leaf and compost heaps and overheated stores of hay, grain, and other plant materials.

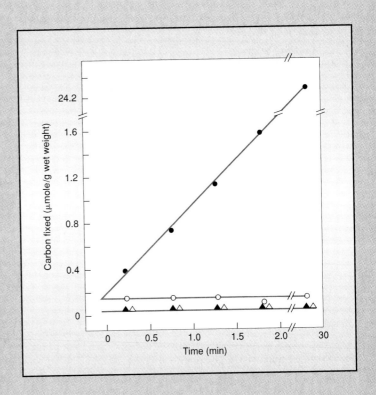
One of the most intriguing discoveries in recent years has been the detection of sulfur-oxidizing chemoautotrophic bacteria living in apparent symbiosis with marine invertebrates in deep-sea hydrothermal vents (Figure 11P.1). These vents are located at depths of 2,500 m and lower and are fed with H_2S-rich waters having temperatures as high as 350°C. Similar symbioses have been postulated for bacteria and invertebrates living in marine muds and salt marsh sediments. In 1983 Collen Cavanaugh of Harvard University and the Marine Biology Laboratory at Woods Hole reported on the symbiotic association of chemoautotrophic bacteria with a marine bivalve, *Solemya velum* Say (phylum Mollusca), collected from eelgrass beds near Woods Hole, Massachusetts.

Cavanaugh analyzed *S. velum* and another bivalve, *Geukensia demissa* (Dillwyn) obtained from creek banks in Little Sippewussett Salt Marsh, Falmouth, Massachusetts, for the presence and activity of chemoautotrophic bacteria by five parameters: (1) ribulose-1,5-diphosphate (RuDP) carboxylase activity, (2) transmission electron microscopy, (3) epifluorescence microscopy, (4) lipopolysaccharide assays, and (5) sulfide and thiosulfate enhancement of CO_2 fixation. The two bivalves were used in the study because both had access to the materials required for sulfur-based chemoautotrophic metabolism: CO_2, O_2, and reduced inorganic sulfur compounds.

RuDP carboxylase activity was detected only in the gill tissue of *S. velum* (Figure 11P.2). Because RuDP carboxylase is a key enzyme in CO_2 fixation, its absence in *G. demissa* gill tissue indicated that RuDP carboxylase activity in *S. velum* was not due to contamination by phytoplankton or free-living chemoautotrophic bacteria.

Transmission electron microscopy of *S. velum* gill tissue sections showed the presence of intracellular rod-shaped bacteria (Figure 11P.3). These were also seen by epifluorescence microscopy of gill tissue homogenates of *S. velum* stained with acridine orange, a nucleic-acid–specific stain. Approximately 1.2×10^9 bacteria per gram wet weight were present in gill tissue, as determined by direct counts of fluorescent-stained cells. The bacteria were not seen in gill tissue preparations of *G. demissa*.

S. velum gills were found to contain 1,000 times more lipopolysaccharide (2 µg/g wet weight) than *G. demissa*. Because lipopolysaccharide occurs in the outer membrane of gram-negative bacteria, its presence in large quantities in *S. velum* gills was indicative of bacteria.

[14]C-labeled CO_2 incorporation in *S. velum* gill tissue was enhanced in the presence of Na_2S and $Na_2S_2O_3$, whereas little or no enhancement occurred in *G. demissa* tissue (Table 11P.1). These data suggested the presence of chemoautotrophic bacteria in *S. velum* tissue.

These observations by Cavanaugh indicated that chemoautotrophic bacteria existed in symbiosis with the marine bivalve *S. velum* and may be important in the nutrition of this invertebrate. Furthermore, the data suggest that such associations are possible and may also occur around sulfide-rich deep-sea hydrothermal vents (Figure 11P.4). It is now known that sulfur-oxidizing chemoautotrophs such as *Thiobacillus* and *Thiovulum* are the bacteria observed by Cavanaugh. The presence of these bacteria in such unusual environments is indicative of the diversity of procaryotes.

figure 11P.1

Mussel Bed in the Vicinity of a Hydrothermal Vent
Chemoautotrophic bacteria live in close association with the mussels and other marine invertebrates in this community.

figure 11P.2

Time Course of RuDP-dependent CO_2 Fixation in Cell-free Extracts of the Gill Tissues of *Solemya velum* and *Geukensia demissa*

RuDP-dependent CO_2 fixation is an indication of RuDP carboxylase activity. ● = *S. velum* gill tissue with RuDP added; ○ = *S. velum* gill tissue with no added RuDP; ▲ = *G. demissa* gill tissue with RuDP added; △ = *G. demissa* gill tissue with no added RuDP.

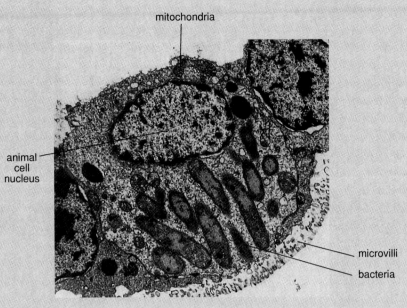

mitochondria

animal
cell
nucleus

microvilli

bacteria

a.

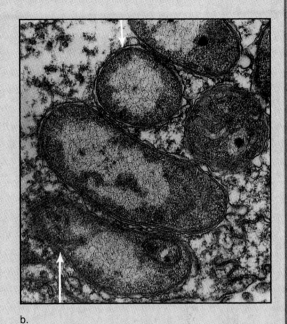

b.

figure 11P.3

Transmission Electron Micrographs Showing Bacteria in Gill Tissue Cells of _Solemya velum_

a. Transverse section of gill filament showing rod-shaped bacteria within an animal cell (×4,600). b. Higher magnification showing oblique and transverse sections of rod-shaped bacteria. Arrows point to an outer unit membrane, possibly that of the host animal cell, surrounding the bacteria (×28,500).

table 11P.1

Rates of Carbon Dioxide Fixation in Gill Tissue of _Solemya velum_ and _Geukensia demissa_

Organism	Sulfur Compound	+ Sulfur	− Sulfur
Solemya velum	0.2 mM Na_2S	4.50	0.70
		4.60	0.49
	1.0 mM $Na_2S_2O_3$	6.50	0.57
		8.30	0.43
		9.10	0.35
Geukensia demissa	0.2 mM Na_2S	0.13	0.08
		0.18	0.09
	1.0 mM $Na_2S_2O_3$	0.10	0.08
		0.10	0.09

Note: Data are expressed as micromoles of CO_2 fixed per gram wet weight per hour. Whole gills (_S. velum_) or pieces of gill (_G. demissa_) were incubated in filtered seawater containing radiolabeled $NaH^{14}CO_3$. After 3 hrs of incubation, gill tissues were solubilized and radioactivity determined. Values are given for each of several experiments.

Source

Cavanaugh, C.M. 1983. Symbiotic chemoautotrophic bacteria in marine invertebrates from sulphide-rich habitats. _Nature_ 302:58–61. Reprinted by permission from _Nature_, Vol. 302, No. 5903, pp. 58–61, ©1983 Macmillan Journals Limited.

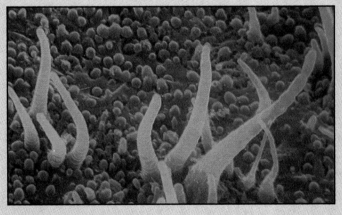

figure 11P.4

Microbial Mats of Deep-Sea Vent Bacteria

A variety of different forms are shown, including cocci, rods, and filaments, in this scanning electron micrograph of a mixed population of chemoautotrophic bacteria from the Galápagos Rift vents (colorized, ×6,000).

References

Grassle, J.F. 1985. Hydrothermal vent animals: Distribution and biology. _Science_ 229:713–725.

Ruby, E.G., and H.W. Jannasch. 1982. Physiological characteristics of _Thiomicrospira_ sp. L-12 isolated from deep-sea hydrothermal vents. _Journal of Bacteriology_ 149:161–165.

Ruby, E.G., C.O. Wirsen, and H.W. Jannasch. 1981. Chemolithotrophic sulfur-oxidizing bacteria from the Galápagos Rift hydrothermal vents. _Applied and Environmental Microbiology_ 42:317–324.

1. Phylogeny is the study of evolutionary relationships among organisms. All living organisms can be separated into one of three domains on the basis of their 16S rRNA sequence: Bacteria, Archaea, and Eucarya.

2. Three major approaches are used to identify and place microorganisms in classification systems. The classical approach is based primarily on structural and morphological characteristics and secondarily on biochemical and physiological traits. Numerical taxonomy groups microorganisms based on their similarities. Molecular approaches compare nucleic acid sequences among organisms.

3. *Bergey's Manual of Systematic Bacteriology* provides bacteriologists with a guide to the classification and identification of most bacteria. It groups procaryotes into four categories: gram-negative Bacteria of general, medical, or industrial importance (Volume 1); gram-positive Bacteria other than the actinomycetes (Volume 2); the Archaea, cyanobacteria, and remaining gram-negative Bacteria (Volume 3); and the actinomycetes (Volume 4).

4. The spirochetes are slender, winding, or helically coiled bacteria that are motile by periplasmic flagella (also called axial fibrils or axial filaments) wrapped around the cell.

5. The bacteria in the second section of Volume 1 of *Bergey's Manual* are rigid, helical to vibrioid cells that move by conventional polar flagella.

6. The third section of Volume 1 of *Bergey's Manual* consists of gram-negative curved, vibrioid, or helical cells that are nonmotile.

7. *Pseudomonas* and *Xanthomonas* are examples of aerobic, gram-negative rods that are polar flagellated.

8. *Escherichia, Salmonella, Shigella,* and *Yersinia* are facultatively anaerobic, gram-negative rods that are often found associated with the human body, and that may cause disease.

9. *Bacteroides,* an anaerobic, gram-negative bacterium, is the most common bacterium in the human intestine.

10. *Desulfovibrio* is an example of a dissimilatory sulfate-reducing or sulfur-reducing bacterium that uses oxidized sulfur compounds as electron acceptors in anaerobic respiration.

11. *Veillonella* and *Acidaminococcus* are anaerobic, gram-negative cocci found in the alimentary tract of humans and animals.

12. *Rickettsia* and *Chlamydia* are obligate intracellular parasites that are responsible for diseases such as Rocky Mountain spotted fever, typhus, and trachoma.

13. *Mycoplasma* is a cell wall-less bacterium that has sterols in its plasma membrane.

14. *Staphylococcus* and *Streptococcus* are gram-positive cocci that cause diseases such as impetigo, toxic shock syndrome, streptococcal pharyngitis (strep throat), rheumatic fever, and necrotizing fasciitis (flesh-eating disease).

15. *Bacillus* and *Clostridium* are examples of endospore-forming, gram-positive bacteria.

16. *Lactobacillus* is a non-spore-forming, gram-positive rod that is used extensively in the food industry for the preparation of yogurt, cheeses, sauerkraut, and pickles.

17. *Corynebacterium* is a club-shaped, gram-positive bacterium that has stainable polyphosphate granules and is the etiologic agent of diphtheria.

18. *Mycobacterium* is a rod-shaped to filamentous bacterium that has unusually large quantities of lipid in its cell envelope, is acid-fast, and causes diseases such as tuberculosis and leprosy.

19. Nocardioforms such as *Nocardia* are aerobic bacteria that form mycelia, or masses of hyphae.

20. The anoxygenic phototrophic bacteria have bacteriochlorophylls, do not produce oxygen during photosynthesis, have only one photosystem, and can use light as an energy source.

21. The oxygenic photosynthetic bacteria have chlorophylls, produce oxygen during photosynthesis, have two photosystems, and can use light as an energy source.

22. Chemolithotrophic bacteria use inorganic compounds as a source of energy. *Nitrosomonas* oxidizes ammonia to nitrite; *Nitrobacter* oxidizes nitrite to nitrate; *Thiobacillus* oxidizes sulfur to sulfate.

23. *Hyphomicrobium* is a budding bacterium that forms buds at the tips of filamentous outgrowths (hyphae) of the cell. *Caulobacter* attaches itself to solid surfaces by a stalk that is an extension of the cell.

24. *Sphaerotilus* is an example of a sheathed bacterium that forms an external sheath around its cells.

25. Fruiting gliding bacteria move by gliding and often produce colorful fruiting bodies that represent the resting stage.

26. The Archaea are distinguished from other procaryotes by their unusual rRNA structure, different RNA polymerase, plasma membrane lipid composition, mechanism of protein synthesis, and lack of muramic acid in the cell wall. They are divided into five groups: the methanogenic Archaea, the

sulfate reducers, the extremely halophilic Archaea, the Archaea lacking cell walls, and the extremely thermophilic elemental sulfur Archaea.

27. The actinomycetes are gram-positive bacteria that are rod-shaped or form branching filaments that in some genera develop into a mycelium. Actinomycetes have a high G + C content (≥52 mole%) of their DNA, which separates them from other procaryotes containing DNA with lower G + C contents.

EVOLUTION and BIODIVERSITY

One of the most spectacular concepts that has emerged in science has been the development of phylogenetic relationships among microorganisms. This molecular revolution in microbiology has enabled us to see microorganisms in a new light—not simply as culturable organisms that are to be characterized and identified by artificial laboratory conditions for growth and enrichment, but as organisms that can be identified directly in their natural environments through nucleic acid analysis. Limitations associated with the traditional approach of identifying and classifying microorganisms through their morphological and physiological characteristics have become apparent with the discovery of nonculturable microbes that cannot be cultivated in the laboratory. Nucleic acid analysis not only makes possible identification of these nonculturable microbes, but also provides an understanding of the natural and evolutionary relationships among organisms. The direct extraction and analysis of nucleic acids from microorganisms in a natural niche permit identification of these organisms and studies of phylogenetic relationships. Through phylogeny, sense and order can now be made of the diversity of the microbial world and the role of the Archaea in linking procaryotes and eucaryotes.

Questions

Short Answer

1. Compare and contrast nomenclature and taxonomy.

2. Identify the kingdoms of living organisms according to Whittaker's system. Identify the domains according to Woese's system.

3. Which system of classification is phylogenetic?

4. Identify three approaches to identifying and classifying microorganisms.

5. Explain how and why flowcharts are commonly used to identify bacteria. Discuss their advantages and disadvantages.

6. Compare and contrast Archaea and Bacteria phenotypically.

7. Identify the major groups of procaryotes as described in *Bergey's Manual of Systematic Bacteriology*.

8. Identify and describe the 11 sections of gram-negative Bacteria described in Volume 1 of *Bergey's Manual*.

9. Identify and describe the six sections of gram-positive Bacteria other than actinomycetes described in Volume 2 of *Bergey's Manual*.

10. Identify and describe the eight sections of Archaea, cyanobacteria, and the remaining gram-negative Bacteria described in Volume 3 of *Bergey's Manual*.

11. Identify and describe the eight sections of actinomycetes described in Volume 4 of *Bergey's Manual*.

12. Compare and contrast mycoplasmas and other bacteria.

13. Compare and contrast *Rickettsia* and *Chlamydia*. How do these organisms differ from other bacteria?

14. Compare and contrast actinomycetes and fungi.

Multiple Choice

1. Which of the following is a spirochete?
   a. *Escherichia*
   b. *Pseudomonas*
   c. *Treponema*
   d. *Streptomyces*

2. Which of the following is most similar to *Rickettsia* and *Chlamydia*?
   a. *Bdellovibrio*
   b. *Clostridium*
   c. *Mycobacterium*
   d. *Mycoplasma*

3. How could you distinguish *Pseudomonas* species from *E. coli*?
   a. Gram stain reaction
   b. morphology
   c. glucose fermentation vs. respiration
   d. All of the above.

4. How could you distinguish staphylococci from streptococci?
   a. Gram stain reaction
   b. morphology
   c. glucose fermentation vs. respiration
   d. All of the above.

5. Which of the following produce endospores?
   a. *Bacillus* and *Clostridium*
   b. *Neisseria* and *Treponema*
   c. *Rickettsia* and *Chlamydia*
   d. *Salmonella* and *Shigella*

Critical Thinking

1. Several classification schemes have been proposed over the years. Both Whittaker's and Woese's systems are widely accepted today. Discuss the advantages and disadvantages of each system. Which is correct? Explain.

2. Examine a copy of *Bergey's Manual* and, if possible, compare it with a botony or zoology taxonomy reference. How is it similar? How does it differ? Why haven't Whittaker and Woese produced references to rival *Bergey's Manual?*

3. Considering the rate of advances in phylogenetics, outline or describe a possible organization of *Bergey's Manual* for the year 2010.

4. As a clinician striving to diagnose and treat disease, which of the three approaches to identification and classification would you use? Why?

Supplementary Readings

Amann, R.I., W. Ludwig, and K-H. Schleifer. 1995. Phylogenetic identification and in situ detection of individual microbial cells without cultivation. *Microbiological Reviews* 59:143–169. (A discussion of the procedures and applications of identifying bacteria in their natural environments without cultivation.)

Buchanan, R.E., and N.E. Gibbons, eds. 1994. *Bergey's manual of determinative bacteriology.* 9th ed. Baltimore: Williams & Wilkins. (An extensive reference guide to all of the procaryotes known through 1994. Each species is characterized with respect to morphological and biochemical traits.)

Bult, C.J., et al. 1996. Complete genome sequence of the methanogenic Archaeon, *Methanococcus jannaschii. Science* 273:1058–1073. (A scientific article reporting the genome sequence of *M. jannaschii* and comparing the genes of this Archaeon with the genes of Bacteria and eucaryotes.)

Goodfellow, M., M. Mordarski, and S.T. Williams, eds. 1983. *The biology of the actinomycetes.* London: Academic Press. (A comprehensive, authoritative survey of the current knowledge of actinomycete biology. There are detailed reviews of systematics, morphology, cell wall composition, genetics, and ecology of the actinomycetes.)

Holt, J.G., editor-in-chief. *Bergey's manual of systematic bacteriology.* Vol. 1, 1982 (Krieg, N.R., ed.). Vol. 2, 1986 (Sneath, P.H.A., ed.). Vol. 3, 1988 (Staley, J.T., ed.). Vol. 4, 1988 (Williams, S.T., ed.). Baltimore: Williams & Wilkins. (The new format of procaryotic taxonomy in four volumes. This reference contains extensive, up-to-date descriptions of each major group of procaryote.)

Olsen, G.J., C.R. Woese, and R. Overbeek. 1994. The winds of (evolutionary) change: Breathing new life into microbiology. *Journal of Bacteriology* 176:1–6. (A mini-review summarizing the advances in phylogenetics and including a comprehensive figure of the procaryotic phylogenetic tree.)

Starr, M.P., H. Stolp, H.G. Truper, A. Balows, and H.G. Schlegel, eds. 1981. *The prokaryotes.* New York: Springer-Verlag. (An extensive, detailed survey of procaryotes, with discussions on laboratory methods for the growth and isolation of these microorganisms and their clinical significance.)

Woese, C.R. 1987. Bacterial evolution. *Microbiological Reviews* 51:221–271. (An extensive review of phylogenetics and the universal phylogenetic tree.)

Woese, C.R. 1994. There must be a prokaryote somewhere: Microbiology's search for itself. *Microbiological Reviews* 58:1–9. (A treatise on the scientific search for phylogenetic relationships among the Archaea, the Bacteria, and the Eucarya.)

Woese, C.R., and R.S. Wolfe, eds. 1985. *The bacteria: A treatise on structure and function.* Vol. 8, *Archaebacteria.* London: Academic Press. (A comprehensive review of the Archaea, including their distinctive biological and molecular properties.)

chapter twelve

THE EUCARYOTIC MICROORGANISMS

The Algae
Morphology and Structure
Reproductive Processes
Classification: Six Divisions

The Fungi
Molds and Yeasts
Reproduction
Phylogenetic Groups

The Protozoa
Structure
Reproduction
Classification: Seven Phyla

PERSPECTIVE
Control of Yeast-Mycelium Dimorphism by Hexoses and
    Carbon Dioxide Levels

EVOLUTION AND BIODIVERSITY

The alga *Volvox*, showing the release of a
daughter colony from a larger mother colony.

 Microbes in Motion ————————— **PREVIEW LINK**

Among the key topics in this chapter are structure, reproduction, and
classification of the fungi and protozoa. The following sections in the
Microbes in Motion CD-ROM may be useful as a preview to your
reading or as a supplemental study aid:

Fungal Structure and Function: Specific Fungal Structures, 1–12;
Metabolism and Growth (Fungal Reproduction), 8–23. *Parasitic Structure
and Function:* Protozoa Classification (Classification), 4–15.

Nestled among the dense herbaceous undergrowth and decaying organic matter on moist forest floors, glistening masses of slime spread across the ground and gradually creep onto rotting logs, weathered rocks, and leaf piles. This colorful mass of oozing living material is neither plant nor animal; rather, each mass contains millions of amoeboid cells that move slowly and feed on bacteria and organic particles. As the food supply diminishes, these cells transform into multicelled structures to form a stalked, colorful reproductive (fruiting) organ filled with hardy spores. Eventually the spores are released and carried by wind, rain, dust particles, and passing animals to new locations, where the spores give rise to individual amoeboid cells, and the life cycle is renewed.

These unusual, colorful creatures are cellular slime molds. The cellular slime molds are one type of eucaryotic microorganism (Figure 12.1). Mushrooms, seaweed, and amoebas are other examples of microorganisms that have eucaryotic features.

Eucaryotes are similar to procaryotes in many respects. Both types of organisms have the cell as their basic unit. Although cells vary in size and shape, all are dynamic entities, able to assimilate nutrients for growth, metabolism, and reproduction, as well as to excrete waste products. Eucaryotic cells, however, are generally more complex in structure and organization than the procaryotic cell. Eucaryotes have a defined nucleus, membrane-enclosed organelles, linear DNA arranged in chromosomes, and 80S ribosomes that are not found in procaryotes (see differences between procaryotes and eucaryotes, page 76). It is this structural and organizational complexity, not simply size, that differentiates eucaryotes from procaryotes.

figure 12.1
The Slime Mold *Tubifera ferruginosa* on a Piece of Wood

The Algae

The term *algae* [sing., *alga*, Latin *alga*, seaweed] describes a large and diverse group of microorganisms that contains over 22,000 species. Algae are considered part of Whittaker's Protista kingdom in his five-kingdom classification scheme (see Whittaker classification, page 307). These organisms are distinguished from fungi and protozoa primarily by their photosynthetic ability. In the past the cyanobacteria were included among the algae. However, the procaryotic structure of the cyanobacteria (and other photosynthetic procaryotes) clearly separates them from the more complex eucaryotic algae.

Algae range from single-celled organisms to large, multicellular aggregates that can weigh hundreds of pounds and extend to lengths of over 100 feet (for example, kelp, a form of seaweed). The cells that make up these large structures still maintain their independent existence, a characteristic that makes these microorganisms different from the photosynthetic macroorganism green plants.

Algae are ubiquitous, occurring in diverse environments that include fresh and marine waters, soil, rocks, trees, plants, and animals. Some thermophilic species are able to exist at high temperatures (for example, in hot springs), whereas others are psychrophilic and found in the polar regions. Algae are associated in water with microscopic animals as **plankton.** Planktons are an important link in the food chain, serving as food for larger animals in the aquatic environment. Seafood farms rely heavily on algal plankton as the beginning step of the food chain.

Some algae live within plants or animals and are an important source of oxygen and nutrients for these organisms. These symbiotic algae are called **zooxanthellae.** In the symbiotic relationship between dinoflagellates (a type of alga) and reef-forming corals, for instance, the shape and location of the corals are influenced by the light requirements of the algae. Because of this association, coral reefs are found only in shallow waters, where light can easily reach the algae. The algae contribute nutrients through photosynthesis, resulting in coral reef formation in tropical waters that generally are deficient in nutrients.

Regardless of their niche, algae constitute a significant and vital part of the living world. Through photosynthesis they provide a large portion of the oxygen and organic materials that are required by other life-forms. Algae are important sources of protein and iodine, particularly for humans living in the Far East. Without eucaryotic algae, life as we know it would have a much more restricted existence.

There Is Considerable Variability in the Morphology and Structure of Algae

Single algal cells exist in a variety of forms, including spherical, curved, and rodlike shapes (Figure 12.2). Aggregates of cells can form multicellular colonies or filaments that are either branched

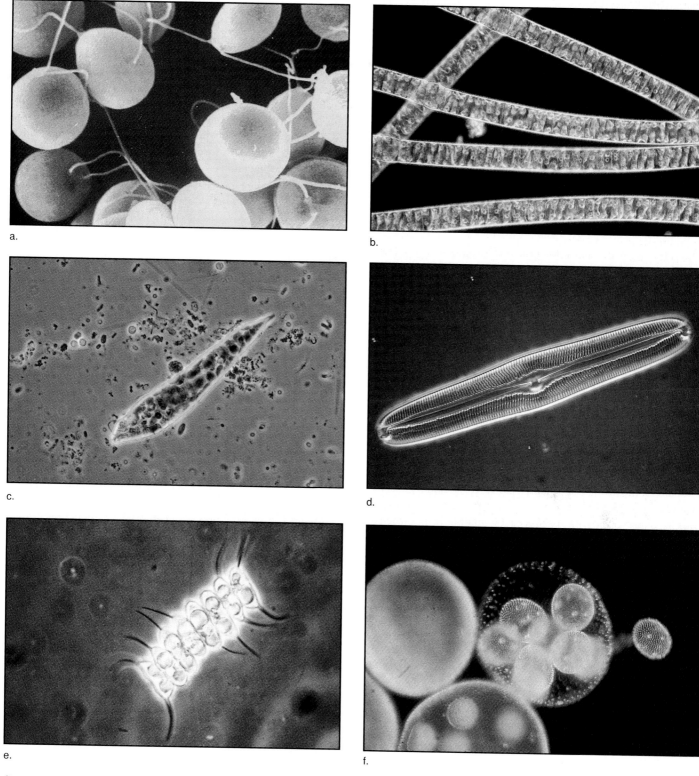

figure 12.2

Representative Examples of Algae

a. *Chlamydomonas*, scanning electron micrograph (colorized, ×1,600). b. *Spirogyra*, dark-field micrograph (colorized, ×400). c. *Euglena*, phase-contrast micrograph (colorized, ×250).

d. *Pinnularia*, interference contrast micrograph (colorized, ×280). e. *Scenedesmus*, bright-field micrograph (colorized, ×250). f. *Volvox*, dark-field micrograph (colorized, ×100).

or unbranched. In some respects these cell aggregates resemble more complex multicellular organisms; the individual cells of the aggregate act cooperatively to benefit the entire organism.

Except for certain algae that lack cell walls (euglenoids), the algal cell wall is composed of cellulose in association with other polysaccharides. Certain cell walls may also contain calcium carbonate, silica, proteins, or a combination of these. All of these components give the wall a mucilaginous consistency resembling the capsules of bacteria. Algal cell walls, like procaryotic cell walls, impart shape and rigidity to the cells.

Algae possess some cellular structures found characteristically in other eucaryotes but not in procaryotes. Among these structures are a defined nucleus, chloroplasts, and mitochondria. Algal chloroplasts are similar to those found in green plants. These photosynthetic structures contain a number of different light-gathering pigments, among them chlorophylls, carotenoids, and biliproteins. Three major types of chlorophylls (a, b, and c), two kinds of carotenoids (xanthophylls and carotenes), and two types of phycobiliproteins (phycocyanin and phycoerythrin) are found in eucaryotic algae. All three types of pigments are involved in trapping light energy for photosynthesis. Pigments are used as aids in algal classification. The carotenoids and biliproteins also protect sensitive photosynthetic structures from photooxidation (see photooxidation, page 180).

Algae Exhibit a Variety of Reproductive Processes

Algae reproduce either asexually or sexually. Most algae reproduce asexually by mitotic division, but others reproduce by fragmentation of cells from colonies or multicellular aggregates, or by spore formation. The spores formed by algae during reproduction are either flagellated and motile (zoospores) or nonmotile (aplanospores). Zoospores are commonly produced by aquatic algae, whereas aplanospores (which in some organisms may subsequently develop into motile zoospores) are usually found in terrestrial algae.

Sexual reproduction in algae is similar to that in higher eucaryotic organisms. Haploid sex cells (gametes) combine to form a diploid zygote. Unions of identical gametes (no sex differences) are called **isogamous,** whereas unions involving distinct male and female gametes are termed **heterogamous.** When the uniting gametes are quite different (for example, one gamete is small and motile and the other gamete is large and nonmotile), the combination is **oogamy.** Certain terrestrial plants and some algae, such as the multicellular green alga *Ulva* (sea lettuce), also exhibit a process called **alternation of generations.** In this process, haploid generations (the gametophytes) alternate with the diploid generations (the sporophytes) during the life cycle of the organism (Figure 12.3).

Members of the genus **Chlamydomonas** (a unicellular green alga) are examples of algae that can reproduce either asexually or sexually (Figure 12.4). During asexual reproduction, each cell reaches a certain size and its chromosomes quadruple in number. Four separate sets of cell walls develop around these sets of chromosomes. Four flagellated daughter cells (zoospores) form and are released when the wall of the parent cell ruptures.

A similar process occurs in sexual reproduction, with the exception that there is an exchange of genetic material during a phase of the alternation of generations. Haploid gametes of opposite mating strains pair through flagellar attraction. A thin protoplasmic thread links the two mating gametes. Parts of their walls dissolve, their flagella are lost, and the cells fuse (along with their respective nuclei, chloroplasts, and mitochondria) into a single diploid zygote. The zygote enters a period of dormancy, at the conclusion of which its genetic material undergoes a reduction division (meiosis). During meiosis, the chromosomes of the zygote separate into four motile, haploid daughter cells, each of which contains two flagella and a separate cell wall. These daughter cells are released following rupture or dissolution of the parent cell wall. The released haploid cells then either reproduce asexually, or sexually pair with haploid cells from another mating algal strain. Such sexual reproduction is unique to eucaryotes and is one characteristic that distinguishes algae as eucaryotes.

Algae Are Classified into Six Divisions Based on Structural, Chemical, and Reproductive Characteristics

Algal classification is complex and varied, depending on the system used. Structural, chemical, and reproductive characteristics are routinely used to classify these organisms (Table 12.1). Current classification schemes generally separate eucaryotic algae into six separate phyla, or divisions, on the basis of the following characteristics:

1. Morphological and cellular organization

2. Cell wall composition and physical properties

3. Presence or absence of flagellation

4. Type and nature of pigmentation

5. Nature of reserve food products

6. Reproductive structures and method of reproduction

Three of the recognized divisions (*Chrysophyta, Pyrrophyta,* and *Euglenophyta*) contain almost exclusively unicellular organisms. Two divisions (*Chlorophyta* and *Rhodophyta*) contain both unicellular and multicellular algae. The division *Phaeophyta* consists of multicellular algae. Some classification systems include the term *phyco* [Greek *phykos,* seaweed] in algal divisional names (for example, *Chlorophycophyta*) to distinguish algae from the green plants. *Chlorophyta* literally means "green plants" [Greek *chloros,* green, *phyton,* plant]. This discussion of algae will use the classification system described in most textbooks, in which the term *phyco* is omitted.

Chrysophyta, Pyrrophyta, and Euglenophyta Are Unicellular Algae

The cells constituting unicellular algae are sometimes found in colonies. Members of the division **Chrysophyta** [Greek *chrysos,* gold] are known as "golden algae" because many contain a yellow-brown carotenoid pigment, fucoxanthin. The cell walls of these organisms are unique because they do not contain cellulose, but silicon compounds often are found in the walls. Many species in this division are called **diatoms** [Greek *diatomos,* cut in two] (Figure 12.5). Diatoms have two thin, overlapping shells (frustules) of

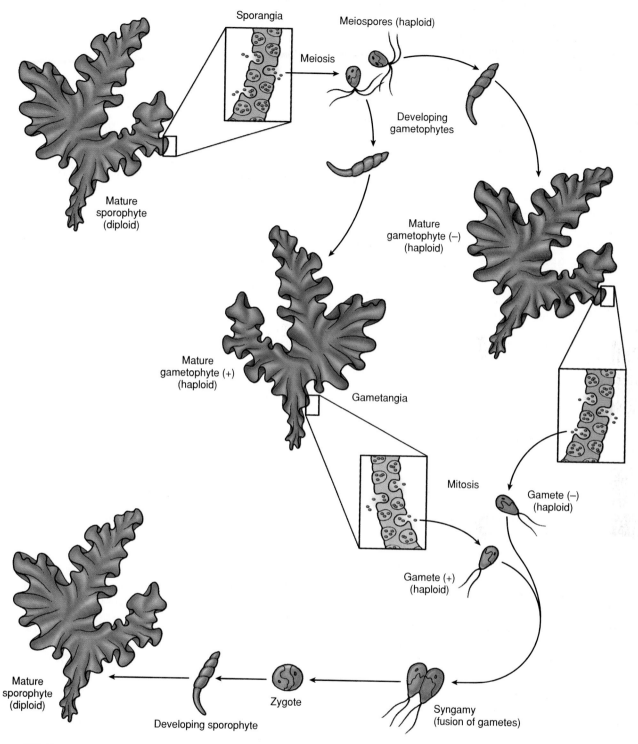

Sporangia

Meiospores (haploid)

Meiosis

Mature
sporophyte
(diploid)

Developing
gametophytes

Mature
gametophyte (−)
(haploid)

Mature
gametophyte (+)
(haploid)

Gametangia

Mitosis

Gamete (−)
(haploid)

Gamete (+)
(haploid)

Mature
sporophyte
(diploid)

Zygote

Developing sporophyte

Syngamy
(fusion of gametes)

figure 12.3

Alternation of Generations in the Sea Lettuce *Ulva*

The gametophyte (haploid generation) and sporophyte (diploid generation) are morphologically similar except for their reproductive structures.

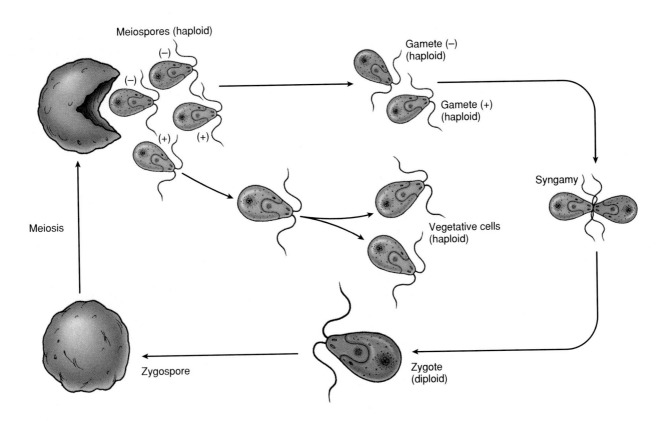

figure 12.4

The Life Cycle of the Green Alga *Chlamydomonas*

The organism is haploid during most of the life cycle. A diploid organism is produced by fusion (syngamy) of two haploid gametes. Meiotic division results in the formation of four new haploid cells, which then can either enter another sexual cycle or divide asexually by mitosis.

table 12.1

Algal Classification

Phylum	Photosynthetic Pigments	Food Reserve	Flagella	Cell Wall Components	Remarks
Chrysophyta (golden algae and diatoms)	Chlorophyll a and often c, carotenoids	Leucosin and oils	1 or 2; apical	Pectic compounds with siliceous material	Mostly marine
Pyrrophyta (dinoflagellates)	Chlorophylls a and c, carotenoids	Starch and oils	2; in grooves	Cellulose	Marine and freshwater; sexual reproduction rare
Euglenophyta (euglenoids)	Chlorophylls a and b, carotenoids	Paramylum and fats	1, 2, or 3; apical	No cell wall	All unicellular, most freshwater; sexual reproduction unknown
Chlorophyta (green algae)	Chlorophylls a and b, carotenoids	Starch	Usually 2	Cellulose	Mostly freshwater, but some marine
Phaeophyta (brown algae)	Chlorophylls a and c, carotenoids	Laminarin and fats	2; lateral, in reproductive cells only	Cellulose and algin	Almost all marine, flourish in cold ocean waters
Rhodophyta (red algae)	Chlorophyll a and sometimes d, carotenoids, phycobilins	Starch	None	Cellulose, pectin, xylans	Mostly marine, some freshwater; complex sexual cycle

The Economic Importance of Algae

Algae are economically important in many ways. For centuries red and brown algae have been used as food sources by different cultures, especially in the Far East. The red alga **Porphyra,** also known as nori, has been cultivated by many generations of Chinese and Japanese and is a popular food today in the Far East. Nori commonly is prepared in the form of sushi or as a toasted soup additive. More than 30,000 people are employed in the nori industry in Japan alone. Other red algae, such as **Acanthopeltis, Chondrus, Eucheuma,** and **Nemalion,** are also eaten, usually as vegetables or sweetened jellies. Extracts of **Carrageen** (carrageenin), a dark purple seaweed found on the coasts of northern Europe and North America, are used as suspending agents in foods, pharmaceuticals, cosmetics, and industrial liquids, and as a clarifying agent for beverages. Agar, used in preparing bacteriological culture media, in baking and canning, and in making the capsules that contain vitamins and drugs, is extracted from the cell walls of **Gelidium, Gracilaria,** and other red algae.

Although seaweeds, which contain undigestible cellulose in their cell walls, are not a rich source of carbohydrates for humans, they are valuable as diet supplements because of their vitamin and mineral contents. The brown seaweed known as kelp is a commercial source of iodine. Kelps produce a number of unusual and commercially valuable carbohydrates. One group of these carbohydrates is algin, a colloidal substance used as a gelling and thickening agent in cosmetics, toothpaste, and foods such as ice cream. The use of algin in the ice cream industry enables manufacturers to use less cream and still obtain a smooth, apparently creamy product.

Over the years, there has been increasing interest in the economic importance of algae, especially as food or food additives and commercially important products. In the future, as industry continues to exploit the economic value of algae, additional uses for these microorganisms most likely will be discovered.

figure 12.5

Nomarski Differential Interference Contrast Micrograph of Diatoms, ×500

figure 12.6
Scanning Electron Micrographs of Representative Unicellular Algae

a. *Gymnodinium*, a dinoflagellate (colorized, ×4,000). b. *Trinacria regina*, a diatom with finely detailed silica cell walls (colorized, ×3,500). c. *Campylodiscus hibernicus*, a diatom with intricately patterned silica cell walls (colorized, ×250). d. *Licmophora*, a marine diatom (colorized, × 64). e. *Euglena*, a euglenophyte (colorized, ×6,000).

polymerized silica that fit together, very much like the halves of a Petri dish (Figure 12.6). These shells are marked with elaborate patterns that actually represent small porous openings that connect the cellular protoplasm with the external environment. Shells of diatoms, stored for millions of years deep in the earth, form a fine, granular substance known as diatomaceous earth that is used as an abrasive, a filtration material, and an insulation. *Chrysophyta* cells often contain oil droplets that serve as food reserves. Fish and water associated with these algae have an oily taste. *Chrysophyta* as a group are important components of plankton and are considered to be among the major food-producing organisms in the world's oceans.

The division ***Pyrrophyta*** [Greek *pyrrhos*, flame color] is represented by the **dinoflagellates** and, along with the diatoms, is an important constituent of plankton. Outbreaks of red tide, a form of plankton that periodically appears off the shores of Pacific, New England, and Gulf Coast states, are caused by the dinoflagellate ***Gymnodinium breve*** (among other organisms). Red

figure 12.7
Red Tide Bloom
A fish kill resulting from a red tide outbreak.

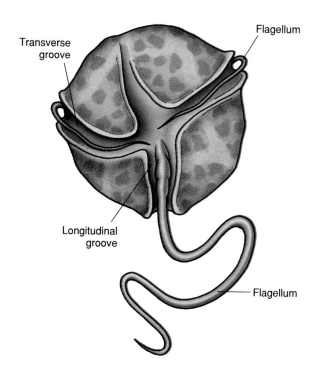

figure 12.8
Flagellation of *Pyrrophyta*
Flagella are arranged perpendicular to each other, with one flagellum in a transverse groove encircling the cell and the second flagellum in a longitudinal groove.

tide outbreaks typically occur during the spring and fall when temperature changes in the water cause mixing of warm surface water with cooler water from deeper strata, resulting in churning of nutrients from the sediments and increased activity of the organisms that cause red tide. During outbreaks the sea is colored red by the presence of enormous numbers of *Pyrrophyta* (Figure 12.7). These cells release products that are toxic for fish and other marine life; shellfish that have concentrated these toxins can cause human poisoning. Water that has been affected by red tide develops a strong stench caused by dead fish and decaying organic matter. In humans the toxins can cause neuromuscular problems such as numbness of the lips, tongue, face, and extremities. *Pyrrophyta* are motile and usually have two flagella. The flagella lie in grooves, one of which encircles the entire cell and the second of which lies in a longitudinal groove perpendicular to the first flagellum (Figure 12.8). As these flagella beat, they cause the cell to spin through the water like a top. This spinning motility gives *Pyrrophyta* the name dinoflagellates [Greek *dinein,* to whirl].

Algae belonging to the division **Euglenophyta** [Greek *eu,* true, *glene,* cavity] live in freshwater environments. *Euglenophyta* lack cell walls, but have plasma membranes containing flexible strips that form an outer covering called the pellicle. The flexible pellicle enables members of this division (particularly **Euglena**) to change their shape in a wiggling type of locomotion through mud. *Euglena,* the most common genus in this division, possess a pair of special orange-pigmented structures known as **stigmata,** or eyespots. These are located on one end of the cell at the base of the flagellum, and act as light shields. In combination with an associated photoreceptor, stigmata enable the cell to sense light and respond to it. *Euglenophyta* store food in the form of paramylon, a polysaccharide unique to this group of organisms. Although most species of this division have chloroplasts, some lack these organelles and are thus nonphotosynthetic.

Chlorophyta, Phaeophyta, and Rhodophyta Are Unicellular and/or Multicellular Algae

The green algae (division **Chlorophyta**) are the most diverse of the algae. The 7,000 species of *Chlorophyta* [Greek *chloros,* green] are found in many habitats, including fresh and marine waters, tree trunks, arctic environments, and animals (Figure 12.9). The green color of these organisms results from their possession of chlorophylls a and b, both of which absorb light in the red and blue portions of the spectrum and transmit green light. Carotenoids produced by the green algae help shield against intense light and give the algae red, orange, and rust hues (Figure 12.10). *Chlorophyta* are often considered to be the predecessors of multicellular land plants because of similarities in structure, specialized functions, and pigmentation between these two groups of organisms. The *Chlorophyta* are similar to land plants in their possession of chlorophylls a and b and β-carotene as photosynthetic pigments. Both chlorophytes and green plants have starch as a storage product and have cell walls composed of cellulose. The green algae are an ancient group. Fossils resembling unicellular green algae dating back 900

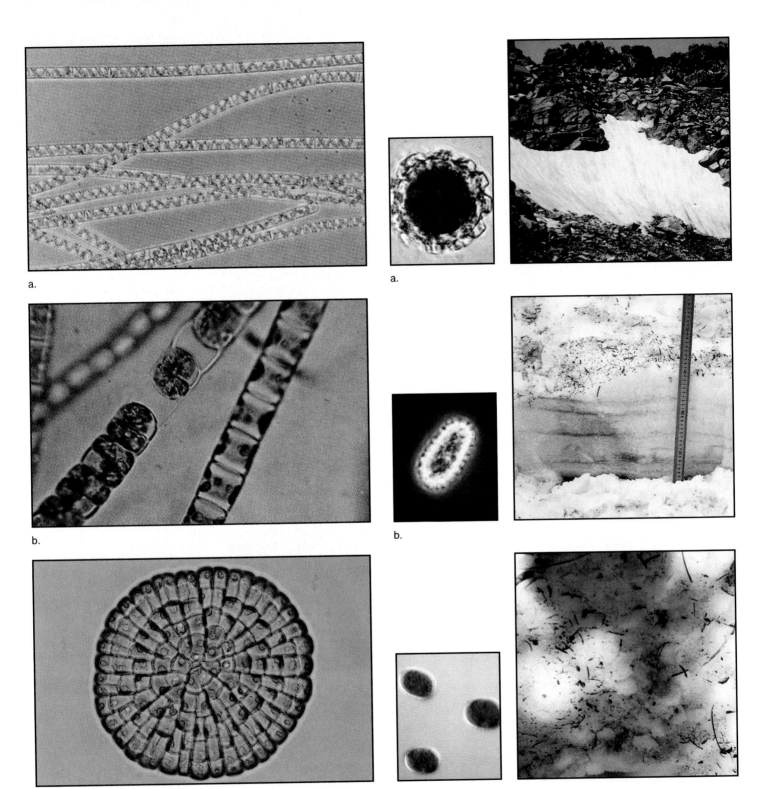

figure 12.9

Examples of Green Algae

a. *Spirogyra*, a filamentous alga that forms floating masses in bodies of fresh water. b. *Ulothrix*, a filamentous alga commonly found in streams, ponds, and lakes. Zoospores can be seen within the filaments. c. *Coleochaete*, a disk-shaped alga that grows on the surface of submerged freshwater plants.

figure 12.10

Snow Algae

a. *Chlamydomonas nivalis*, responsible for red snow. Photograph taken at Cedar Breaks Mountain National Monument in Utah.
b. *Chloromonas brevispina*, responsible for green snow. Photograph taken at Cayuse Pass, Mt. Ranier National Park in Washington.
c. *Chloromonas granulosa*, responsible for orange snow. Photograph taken at Bill Williams Mountain in Arizona.

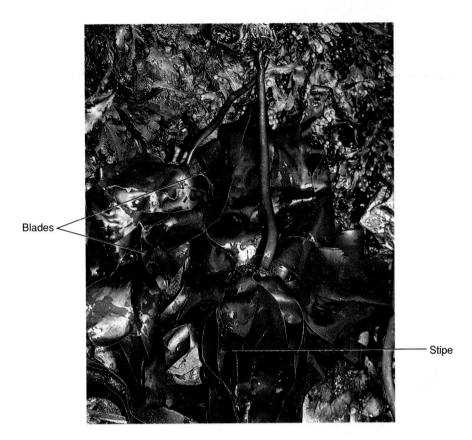

Blades

Stipe

figure 12.11
The Brown Alga *Laminaria digitata* Showing Stipe and Blades

million years have been discovered in the Bitter Springs Formation of central Australia. Multicellular green algae are believed to have evolved about 550 million years ago.

Phaeophyta [Greek *phaios*, dark], commonly called brown algae, are structurally similar to plants. The brown algae are typically found in marine environments, where they attach to seabeds by holdfasts and usually have gas-filled floats that aid in buoyancy. Other portions of the organism are a stemlike structure called the stipe, and a blade, or frond, that contains the photosynthetic organelles. These three structures (holdfast, stipe, and blade) are analogous to similar structures (root, stem, and leaf) in plants (Figure 12.11).

The brown algae and another group of algae, the red algae, or *Rhodophyta* [Greek *rhodon*, red], exist as large, multicellular structures and constitute what is commonly known as seaweed. Red algae are usually attached to other algae or rocks (few are found as floating forms) and are generally found at greater depths than any other type of algae. Phycobilins, accessory pigments that aid in the harvesting of green and blue-green light that penetrate into deep water, are responsible for their red color and make it possible for the red algae to exist at greater depths than other forms of algae. *Rhodophyta* are especially important in bacteriology as the source of agar. Some of the red algae (family: *Corallinaceae*), which have cell walls comprised of calcium carbonate, play important roles in building coral reefs. Coralline algae existed more than 700 million years ago as evidenced by fossil records.

The diversity of algae and their role as the primary organisms in the food chain for marine life make them appealing as a potential food source for humans and domestic animals. Algae, when grown under appropriate conditions, are rich in proteins, carbohydrates, fats, and vitamins. As the world's population increases and conventional methods for food production are outpaced, algae will undoubtedly become a more important part of the human diet.

The Fungi

The fungi [sing., fungus, Latin *fungus*, mushroom] are a distinct group of organisms that are neither plants nor animals. They resemble plants because they have rigid cell walls and are nonmotile, but unlike plants, the fungi lack chlorophyll and are unable to photosynthesize. The fungi also lack the multicellular complexity and organization of most animals. For these reasons, taxonomists consider fungi distinct from plants and animals.

Fungi are as diverse as algae. There are approximately 70,000 known species of fungi (and probably 10 to 20 times as many undiscovered species) found in such varied environments as tropical forests, oceans, and deserts. Various fungi adapt to pH extremes and are found in hot and cold climates, although they are not as heat or cold resistant as procaryotes. Fungi are primarily terrestrial heterotrophic organisms; their principal role is the decomposition of organic matter. Fungi, along with the bacteria, are considered to be the decomposers of the biosphere. Because they most often live on nonliving organic matter, fungi are called **saprophytes** [Greek *sapros*, rotten, *phyton*, plant]. The products released by the decaying action of fungi from their metabolism of organic matter are returned to the environment to be used once again by other organisms.

Fungi are also important in industry. Yeasts, a type of fungus, are used in the baking, wine, and brewery industries for the fermentation of sugars into carbon dioxide and ethanol. Mushrooms, another group of fungi, are a food source. Other fungi are important for cheese and antibiotic production.

Fungi can be destructive and cause great economic loss. Pathogenic fungi cause many plant diseases (wheat rust, corn smut, potato blight, root rot, and stem rot) (Figure 12.12), some human diseases (ringworm, athlete's foot, and histoplasmosis), and some animal diseases (mange in dogs and cats), as shown in Table 12.2. Fungi that flourish in moist, warm climates are responsible for mildew and the decay and destruction of wood, clothing, and food.

figure 12.12

Corn Smut, a Disease Caused by the Basidiomycete *Ustilago zeae* and Characterized by Large, Tumorlike Galls
The ears of corn are considered a delicacy in Mexico and Central America.

table 12.2

Some Diseases Caused by Fungi

Phylum	Fungus	Disease
Chytridiomycota	Physoderma sp.	Corn brown spot
	Synchytrium endoboticum	Potato wart
	Urophlyctis sp.	Alfalfa crown wart
Zygomycota	Rhizopus stolonifer	Strawberry leak
Ascomycota	Ajellomyces capsulatus (Histoplasma capsulatum)	Histoplasmosis (lung disease)
	Ajellomyces dermatitidis (Blastomyces dermatitidis)	Blastomycosis (lung, skin, and bone disease)
	Arthroderma sp.	Ringworm and athlete's foot
	Candida albicans	Thrush
	Coccidioides immitis	Coccidioidomycosis (lung disease)
	Endothia parasitica	Chestnut blight
	Microsporum sp.	Ringworm
	Monilinia fructicola	Brown rot of peaches
	Pneumocystis carinii	Pneumonia
	Taphrina deformans	Peach and almond leaf curl
	Trichophyton sp.	Ringworm and athlete's foot
Basidiomycota	Cronartium ribicola	White pine blister rust
	Cryptococcus neoformans	Cryptococcosis (lung and central nervous system disease)
	Gymnosporangium juniperi-virginianae	Apple and juniper rust
	Phragmidium disciflorum	Rose rust
	Puccinia graminis	Wheat rust
	Ustilago tritici	Loose smut of wheat
Oomycota	Phytophthora infestans	Potato blight
	Plasmopara viticola	Downy mildew of grapes
	Saprolegnia parasitica	Fish dermatitis
Plasmodiophoromycota	Plasmodiophora brassicae	Club root of cabbage
	Spongospora subterranea	Powdery scab of potatoes

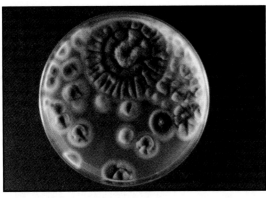

a.

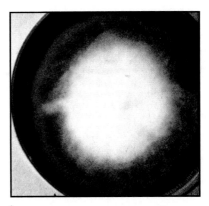

b.

c.

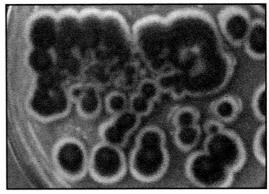

d.

figure 12.13

Examples of Fungal Colonies

a. *Penicillium notatum,* a common blue-green mold. Different species produce the antibiotic penicillin and are used in the production of cheeses such as Roquefort and Camembert. b. *Trichophyton gourvilli,* a cause of hair infections in Africa. Colony shows the typical white aerial mycelium. c. *Aspergillus niger,* widely used for the commercial production of citric acid. Conidia are visible in the lactophenol cotton blue-stained colony. d. *Penicillium chrysogenum,* a producer of penicillin. Vegetative mycelia surround the blue-green colonies.

Fungi Have Two Morphological Growth Forms: Molds and Yeasts

The fungi are eucaryotic microorganisms that can exist as two basic morphological growth forms: filamentous **molds** (hyphal types) and unicellular **yeasts**. Some fungi, such as *Ajellomyces* and *Candida,* are **dimorphic** and can alternate between both morphological forms, depending on environmental conditions of nutrition and temperature. For example, **Ajellomyces capsulatus** (*Histoplasma capsulatum*) grows at room temperature in a filamentous form on the culture medium Sabouraud glucose agar, but grows as a yeast at 37°C on blood agar (tryptic soy agar containing 5% sheep blood).

Molds typically consist of filaments called **hyphae,** [Greek *hyphe,* web]. Hyphae, which often exist as branching structures, can form a mass called a **mycelium** [Greek *mykes,* fungus] (Figure 12.13). Mycelial masses have the consistency of cotton fiber. The walls of the hyphae are composed primarily of the polysaccharide chitin, except the walls of the Oomycetes, which are of cellulose. Portions of the mycelium extend below the surface (vegetative mycelium) and function in attachment and nutrient absorption. Reproductive portions of the mycelium are called aerial mycelia because they are found above the surface. Some fungal species have rhizoids—root-like extensions that anchor the mycelium to the surface.

In some fungi, a hypha consists of individual cells divided by septa (Figure 12.14). In other fungal species, distinct septa are absent or very rare, resulting in an organism in which the cytoplasm is continuous and may contain hundreds of nuclei (coenocytic hyphae). In these organisms, there is a continuous movement of cytoplasm toward the tip of the hypha, where growth is localized. The older portions of the hypha consequently contain less cytoplasm. Cytoplasmic movement occurs through pores in the septa, even in those species with septated hyphae; consequently, even these septated hyphae are, in fact, coenocytic.

Yeasts are unicellular, oval organisms, usually 3 to 5 μm in diameter. Yeastlike growth is formed by some animal parasitic molds. Dimorphic fungi convert from molds to yeasts under conditions of high temperature and rich nutrients, such as are common in warm-blooded animal hosts. Although yeasts constitute only a small portion of fungal species, they are important

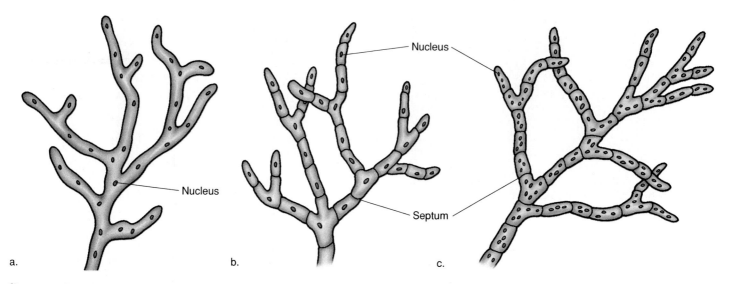

figure 12.14

Types of Hyphae Found in Fungi

a. Coenocytic (aseptate). b. Septate with uninucleate cells. c. Septate with multinucleate cells.

in medicine, food production, and molecular biology research. They are common in environments with high sugar concentrations—they ferment these sugars. Yeasts are used extensively in industry to produce ethanol and are also involved in the production of leavened breads, biscuits, cakes, wine, beer, and whiskey (Figure 12.15) (see food microbiology, page 617). In molecular biology, yeasts are excellent cloning and expression systems for eucaryotic genes and gene libraries (see yeast artificial chromosome, page 285).

 **Fungal Structure and Function**
Specific Fungal Structures • pp. 1–12

Fungi Reproduce Asexually and Sexually

Fungi exhibit both asexual and sexual modes of reproduction (Figure 12.16). Asexual reproduction in the haploid vegetative growth phase occurs by spore formation, generally at the tips of aerial

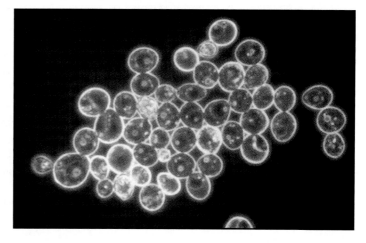

figure 12.15

Saccharomyces cerevisiae **(Brewer's Yeast), a Representative Budding Yeast, ×800**

figure 12.16

Asexual and Sexual Modes of Fungal Reproduction

a. Life cycle of *Rhizopus*. Asexual spores released by the sporangium are carried by air currents to other locations. The spores germinate, giving rise to hyphal masses. Sexual reproduction occurs when hyphae from two different mating types come together, forming gametangia, which fuse to form a zygote. The zygote undergoes meiosis and germinates, producing a new sporangium. b. Life cycle of a basidiomycete. Basidiospores released from the mushroom give rise to a primary mycelium consisting of monokaryotic (having a single nucleus) cells. Two hyphae from different mating types join together to form a dikaryotic secondary mycelium, in which each cell contains two nuclei, one from each mating type. The hyphae of the mycelium grow and eventually form a fruiting body known as the basidiocarp (the typical mushroom structure).

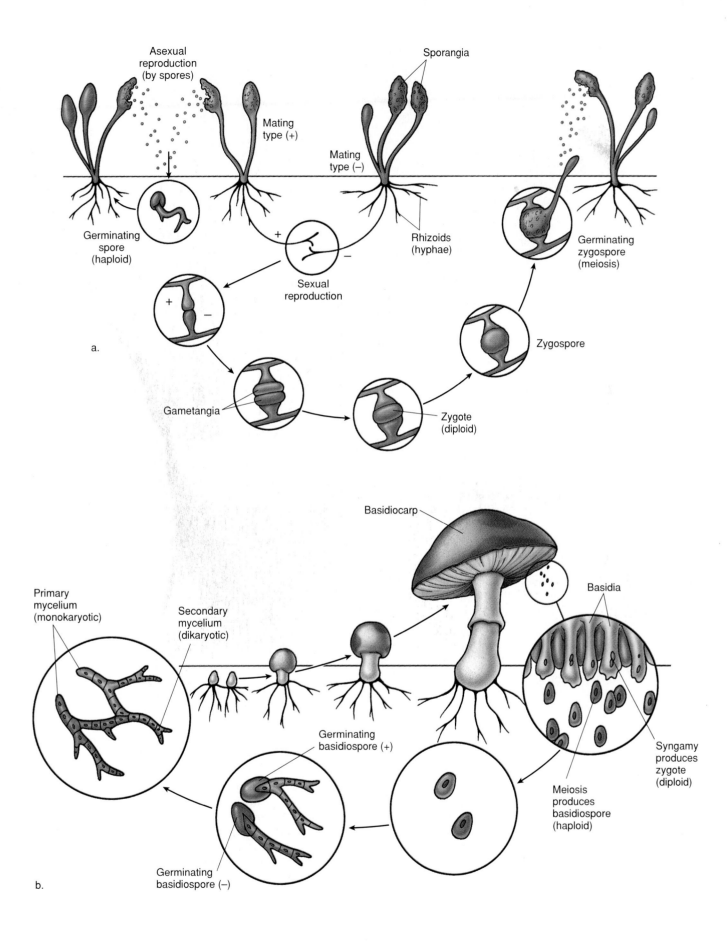

Asexual reproduction (by spores)

Sporangia

Mating type (+)

Mating type (−)

Germinating zygospore (meiosis)

Rhizoids (hyphae)

Germinating spore (haploid)

+

−

Sexual reproduction

+ −

Zygospore

Gametangia

Zygote (diploid)

a.

Basidiocarp

Primary mycelium (monokaryotic)

Secondary mycelium (dikaryotic)

Basidia

Germinating basidiospore (+)

Syngamy produces zygote (diploid)

Meiosis produces basidiospore (haploid)

Germinating basidiospore (−)

b.

a.

b.

c.

d.

figure 12.17

Scanning Electron Micrographs of Fungal Spores
a. Single-celled conidia shown on conidiophores of *Verticillium albo-atrum* (colorized, ×1,350). b. Sporangium of *Mycotypha africana*, a fungus that causes the plant disease powdery mildew (colorized, ×255). c. Conidiophores of *Penicillium caseicolum* (colorized, ×970). d. Conidia of *Aspergillus* (colorized, ×500).

hyphae (Figure 12.17). Some spores are uninucleated, and others are multinucleated. Spores are flagellated and motile (**zoospores**) in the oomycetes (aquatic lower fungi) and nonmotile in all other fungi. **Conidia** (sometimes called **conidiospores**) are nonmotile spores formed at the tips of specialized hyphae called conidiophores (Figure 12.18). Some conidiophores form single conidia, whereas others have clusters or chains of these asexual spores. Certain conidia are thick-walled and resistant to heat and dry conditions. In contrast, some fungi produce spores (**arthrospores**) that are not as resistant to these harsh conditions. Some filamentous fungi form asexual **sporangiospores** that are enclosed within a sporangium, a specialized reproductive structure raised above the surface by hyphae known as sporangiophores. As the sporangium matures, it ruptures and the sporangiospores are released into the environment for dispersal.

Sexual reproduction in vegetative cells occurs by the fusion of haploid gametes from different hyphae to form a diploid zygote. The zygote undergoes meiosis to form four haploid gametes. In some instances, fusion of gametes from different hyphae is not accompanied by nuclear fusion. The result is a **dikaryon,** a cell with paired but unfused nuclei derived from different parent hyphae. In other instances, the gamete is found within a specialized structure called the gametangium, located at the tip of the hypha and separated by a septum from the other

figure 12.18

Penicillium, **Showing Conidia at the Ends of Conidiophores that Arise from Septate Hyphae,** ×**640**

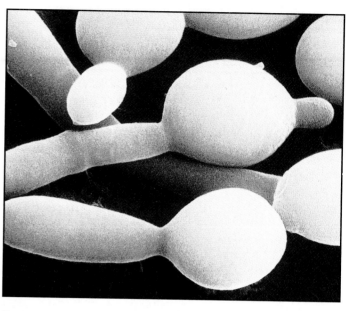

figure 12.19

Scanning Electron Micrograph of the Yeast Phase of *Candida albicans*

New daughter cells are seen as buds arising from the mother cells (colorized, ×4,300).

portions of the hypha. Gametangia from different organisms fuse to form a diploid zygote.

Yeasts have several different mechanisms for reproduction. Some yeasts in the diploid state immediately undergo meiosis, whereas others grow asexually in the diploid stage until certain conditions are encountered. Yeasts reproduce asexually by binary fission and also by budding. In binary fission, the yeast increases in size, the nucleus divides, and the cell is split into two new cells. Budding is a process in which a new daughter cell arises as a bud from the mother cell (Figure 12.19). The new bud grows until it reaches the same size as the mother cell; a cross-wall then forms and the bud separates from the original cell. In sexual reproduction, two haploid yeast cells fuse to form a single diploid cell. The resultant diploid cell buds, and the diploid nucleus moves into the bud. The bud separates from the mother cell (which disintegrates) and begins to grow vegetatively.

Fungal Structure and Growth
Metabolism and Growth: Fungal Reproduction • pp. 8–23

Fungi Are Phylogenetically Grouped into Two Kingdoms and Four Additional Phyla

In the past, fungal classification was based mainly on the nature of their life cycles, reproductive structures, and sexual spores. Because of the diversity of fungi, not all taxonomists agreed on how to clas-sify these organisms. One of the more widely accepted fungal clas-sification systems is based on evolutionary relationships. This phylogenetic classification separates the fungi into two kingdoms and four additional phyla that are considered protists (Table 12.3). Some of the major groups of fungi are discussed in the following paragraphs.

The Kingdom Fungi Includes the Phyla Chytridiomycota, Zygomycota, Ascomycota, and Basidiomycota

The **Chytridiomycota** are aquatic (marine and freshwater) and soil fungi that have a single whiplash flagellum and cell walls contain-ing chitin and glucans, but no cellulose. The whiplash flagellum is a long, relatively rigid filament with a flexible, whiplike tip. Sexual reproduction occurs by fusion of two gametes or by fusion of hyphalike filaments called rhizoids, resulting in the formation of a thick-walled resting body.

Chytrids are widely distributed in aquatic habitats and soils and are associated with vascular plants, mosses, phytoplankton, in-sects, and nematodes. Members of this phylum are responsible for such plant diseases as potato wart (**Synchytrium**), alfalfa crown wart (**Urophlyctis**), and corn brown spot (**Physoderma**).

Members of the phylum **Zygomycota**, [Greek *zygou,* yoke] are haploid organisms having few or no cross-walls in their hy-phae. Reproduction can be asexual, producing sporangiospores, or sexual. Sexual reproduction between compatible mating types,

table 12.3

Phylogenetic Classification of the Fungi

Kingdom	Phylum	Distinguishing Characteristics	Representative Groups	Examples
Fungi	Chytridiomycota	Motile cells produced at some stage of life history	Class: Chytridiomycetes	Synchytrium
	Zygomycota	Asexual or sexual reproduction, with zygospores produced during sexual reproduction	Class: Trichomycetes Class: Zygomycetes	Amoebidium Smittium Mucor Rhizopus
	Ascomycota	Ascus formed during sexual reproduction in some members; compartmentalized mycelium with distinctive walls found in some members	Order: Eurotiales Order: Saccharomycetales Order: Schizosaccharomycetales Order: Sordariales	Aspergillus Penicillium Candida Saccharomyces Pneumocystis Neurospora
	Basidiomycota	Dikaryotic mycelium and basidiospores	Order: Agaricales Order: Aphyllophorales Order: Uredinales Order: Ustilaginales	Amanita Coprinus Polyporus Puccinia Ustilago
Stramenopila	Oomycota	Reproduction by gametangial contact that results in the production of a thick-walled spore called an oospore	Order: Peronosporales Order: Saprolegniales	Phytophthora Plasmopara Pythium Dictyuchus Saprolegnia
	Hypochytriomycota	Single anterior flagellum covered with flagellar hairs	Order: Hyphochytriales	Rhizidiomyces
	Labyrinthulomycota	Presence of an ectoplasmic network of branched, wall-less filaments produced by cells	Family: Labyrinthulaceae Family: Thraustochytriaceae	Labyrinthula Thraustochytrium
Protists	Plasmodiophoromycota	Endoparasites of vascular plants and stramenopiles, producing an enlargement of host cells called hypertrophy	Class: Plasmodiophoromycetes	Plasmodiophora Polymyxa
	Dictyosteliomycota	Cells aggregate to produce a slug-shaped organism called a pseudoplasmodium		Dictyostelium Polysphondylium
	Acrasiomycota	Cylindrical amoebae having phagotrophic nutrition		Acrasis Guttulina
	Myxomycota	Life cycle includes formation of a multinucleate, amoeboid structure, the plasmodium	Class: Myxomycetes	Dictydium Physarum Tubifera

referred to as + and −, results in the formation of a zygote (2n) that eventually gives rise to a thick-walled, resistant **zygospore** (n). Common bread mold was historically caused by a member of this phylum, **Rhizopus stolonifer,** although most molding of bread is now caused by **Penicillium** (member of the phylum *Ascomycota*) and other ascomycetes. Mold appears on bread when spores germinate and form mycelia that cover the surface of the bread.

The phylum **Ascomycota** [Greek *askos,* bag] constitutes a large group of fungi that occupy a broad range of habitats. Most members of this phylum form multinucleated conidia during asexual reproduction. Sexual reproduction is more complex and involves formation of a unique saclike structure called an **ascus** (pl., **asci**), in which diploid nuclei can form. These asci are borne in open or closed containers called ascocarps. Ascus formation is generally preceded by the fusion of two hyphae from neighboring organisms. The diploid nuclei, which result from fusion in the young ascus, undergo meiosis and eventually form four haploid nuclei. These nuclei may undergo mitotic division and subsequently form up to eight haploid nuclei. Each haploid nucleus is surrounded by a thick wall to form an ascospore (Figure 12.20). The ascospores are enclosed within the original walls of the ascus (Figure 12.21). At maturity, the ascus ruptures and releases the ascospores into the environment.

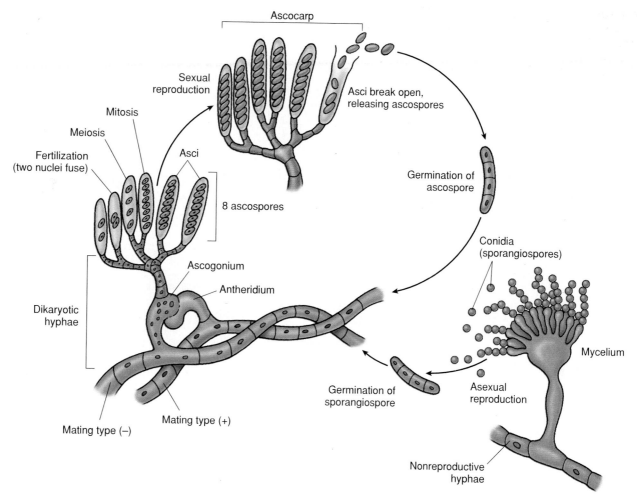

figure 12.20

Life Cycle of an Ascomycete

Asexual reproduction involves the production of conidia. Sporangiospores released from the conidia germinate to form septated hyphae. Sexual reproduction occurs when male nuclei of the antheridium pass into the female mating type, the ascogonium, resulting in fusion of the two protoplasts. Hyphal filaments grow out of the ascogonium, and cell division in the developing ascogenous hyphae occurs in such a way that the resultant cells are dikaryotic (contain two haploid nuclei). Asci form at the tips of the dikaryotic hyphae. Ascus formation is generally preceded by the fusion of two hyphae from neighboring organisms. Adjacent daughter nuclei within the ascus fuse to form a diploid nucleus (zygote), which undergoes meiosis followed by one mitotic division to produce up to eight haploid nuclei. These haploid nuclei are surrounded by a thick wall to form ascospores. At maturity the ascus becomes turgid and bursts, releasing its ascospores into the environment. In some species, the ascospores are propelled as far as 30 cm.

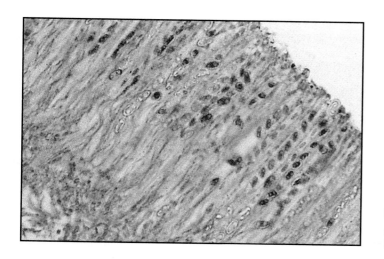

figure 12.21

An Ascomycete, Showing Tightly Packed Asci with Eight Red-Stained Ascospores in Each Ascus, ×768

Aspergillus and Aflatoxin

Aspergillus (phylum *Ascomycota*), a common blue-green mold that grows in environments with a minimum of nutrients, is responsible for the decay of many foods and materials. There are over 600 species of *Aspergillus*, but only a dozen or so of these species have been associated with infection. Certain strains of **Aspergillus flavus** (and also **Penicillium puberulum**) secrete aflatoxins (*Aspergillus flavus*

toxins), which cause severe liver damage and even liver cancer in poultry and other animals. Aflatoxins are also extremely potent mutagens; as little as 50 parts per billion (ppb) are required to cause cancer in animals.

Animals contract these mycotoxins from ingesting contaminated feed. The first association of aflatoxin with animal feed

came in 1960, when more than 100,000 turkeys died suddenly and inexplicably in England. It was later discovered that the birds had ingested Brazilian peanut meal contaminated with aflatoxin. Although there is no direct evidence to link aflatoxin with human cancer, there appears to be a correlation between the lifetime ingestion of grains containing aflatoxin and the incidence of human liver cancer in Southeast Asia. Aflatoxin levels in peanuts and peanut products are now monitored, and the Food and Drug Administration has established the safe level as 15 ppb.

 Fungal Structure and Function
Diseases: Allergies and Intoxication • p. 49

Ascus formation has been extensively studied in the genus *Neurospora,* an organism that has contributed significantly to our understanding of genetics because of its reproductive cycle. Unlike most higher organisms, *Neurospora* is haploid. Because it has only one set of chromosomes, there is no masking of a mutation by a dominant wild-type allele, and the effects of the mutation are expressed in the phenotype. The ascomycetes include blue-green, red, and brown molds that are responsible for food spoilage, many disease-causing fungi, and economically significant organisms. *Saccharomyces cerevisiae,* which is used in baking, is an ascomycete. During baking *S. cerevisiae* catabolizes glucose and other carbohydrates in the dough and produces carbon dioxide, which causes the dough to rise. Other species of *Saccharomyces* are used extensively in the production of alcoholic beverages such as beer and wine (see production of alcoholic beverages, page 617). The ascomycete *Penicillium* is important in the production of antibiotics (see penicillin production, page 629) and various types of cheeses (see cheese production, page 611).

The phylum **Basidiomycota** [Latin *basidium,* little pedestal] includes the rusts, smuts, and mushrooms (Figure 12.22). Mushrooms are commercially grown by planting the primary mycelium (spawn), formed after germination of the fungal spores, in well-fertilized soil in a cool, dark location. Growth in the dark prevents competition from green plants. The mycelium, the vegetative portion of the organism, quickly expands extensively below the surface. Under warm, moist conditions, a dikaryotic mycelium (one with two paired nuclei) appears and a characteristic fruiting body forms on the surface (Figure 12.23). The fruiting body is composed of densely packed hyphae. Sexually produced spores (basid-

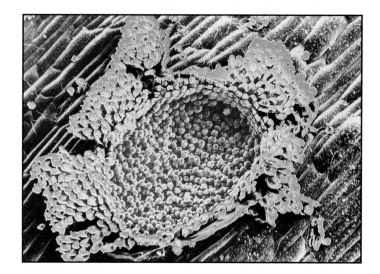

figure 12.22

Scanning Electron Micrograph of Rust on Wheat
Spores of rust, which are shown, are released and can infect other wheat plants (colorized, ×55).

iospores) are located on specialized hyphae (basidia). In common mushrooms, the basidia are arranged among radiating gills along the underside of the cap. Mature mushrooms discharge millions of basidiospores into the environment, effectively dispersing their progeny. There are hundreds of known species of mushrooms,

figure 12.23
Polyporus sulphureus, **a Basidiomycete**

figure 12.24
Amanita muscaria, **a Mushroom that Is Hallucinogenic when Eaten in Small Quantities but Poisonous in Larger Doses**

identified by spore color, shape, and markings. Some mushrooms are edible, but others are highly poisonous and their mycotoxins can affect the central nervous system. For example, ingestion of a small amount of the mushroom *Amanita virosa,* appropriately called the "destroying angel," is fatal to humans. The colorful mushroom *Amanita muscaria,* commonly called the fly mushroom because of its past use as an insecticide, is hallucinogenic if ingested and is used in orgiastic celebrations of some Siberian tribes (Figure 12.24).

The Kingdom Stramenopila Includes the Phyla Oomycota, Hypochytriomycota, and Labyrinthulomycota

The term stramenopile was introduced in 1989 to classify fungi having lineages similar to certain groups of algae possessing chlorophylls a and c. **Hypochytriomycota** and **Labyrinthulomycota,** two of the three phyla in the kingdom Stramenopila, are small groups relative to the third phylum, **Oomycota,** commonly called **oomycetes.**

The oomycetes [Greek *oion,* egg], known as the water molds because they are often found in aquatic environments, have both asexual and sexual methods of reproduction in their life cycles (Figure 12.25). The name *oomycete* refers to the method of sexual reproduction in these fungi. In sexual reproduction, some of the fungal hyphae develop enlarged tips called oogonia that contain one or more egg cells (female gametes). Other slender, branched hyphae called antheridia contain male gametes. Meiosis is thought to occur within the oogonia and the antheridia. During fertilization, haploid male gametes pass through fertilization tubes between the antheridia and the oogonia to fuse with the female gametes and form diploid zygotes. These develop into thick-walled resting oospores. Under favorable conditions, the oospores germinate, forming short coenocytic hyphae that differentiate into oogonia and antheridia, and the cycle begins again.

In asexual reproduction, the hyphae differentiate into sporangia in which zoospores develop from mitotic divisions. Flagellated zoospores (containing a whiplash flagellum and a tinsel flagellum) released from the sporangium may encyst and remain dormant. As the zoospores germinate and develop hyphae, the asexual reproduction cycle begins again.

Oomycetes are responsible for such diseases as potato blight (*Phytophthora infestans*), which caused extensive damage to the potato crop in Ireland in the mid-1800s, and downy mildew in grapes (*Plasmopara viticola*), a disease that almost destroyed the French wine industry in the late 1800s. Other oomycetes cause diseases of fish. For example, the oomycete *Saprolegnia* causes ich, a severe dermatitis of freshwater fish. *Saprolegnia* is especially a problem in overcrowded fish aquariums and hatcheries.

Slime Molds Have Characteristics Resembling Both Fungi and Protozoa

Members of the phyla *Plasmodiophoromycota, Dictyosteliomycota, Acrasiomycota,* and *Myxomycota* have at one time or another been called slime molds. The slime molds are organisms whose

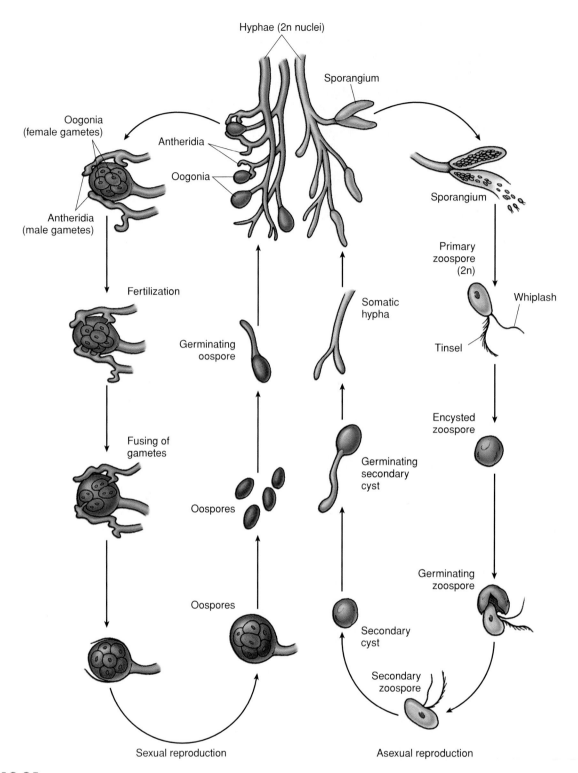

figure 12.25

Life Cycle of *Saprolegnia*, an Oomycete

Saprolegnia has a life cycle that includes both sexual and asexual reproduction. In sexual reproduction, haploid male and female gametes fuse to produce diploid zygotes. The zygotes develop into oospores, which germinate and form hyphae to start the cycle again.

In asexual reproduction, the hyphae on the oospores differentiate into sporangia. Zoospores develop from mitotic divisions in the sporangia. The liberated zoospores, which may become encysted, eventually germinate and develop hyphae, and the cycle begins again.

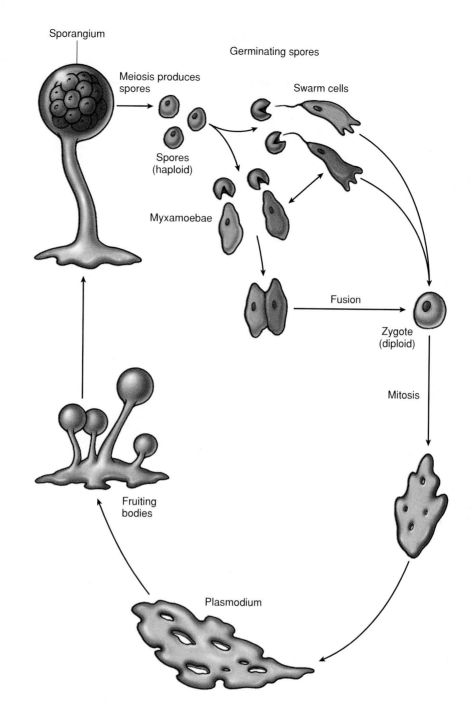

figure 12.26

Life Cycle of an Acellular Slime Mold
Under limited nutrient conditions, the plasmodium forms fruiting bodies. Spores released from these fruiting bodies become flagellated gametes (swarm cells), which unite in pairs to form zygotes. The zygotes then develop into new plasmodia.

classification has been controversial. The controversy stems from the fact that these organisms, unlike true fungi, are phagocytic, and their vegetative forms lack a cell wall (spores of slime molds, however, do possess cell walls). Some taxonomists consider the slime molds to be protozoa because their vegetative structures often resemble amoebas. However, since the slime molds produce fruiting bodies and typical fungal spores, they are also considered to be fungi. Currently these four phyla are grouped as protists and are separated from the kingdoms Fungi and Stramenopila.

The most common slime molds are the ***Myxomycota*** [Greek *myxa*, slime], the true acellular or plasmodial slime molds (Figure 12.26). These organisms are coenocytic and their vegetative structures consist of streaming protoplasmic masses called plasmodia. The plasmodia stream over surfaces and engulf

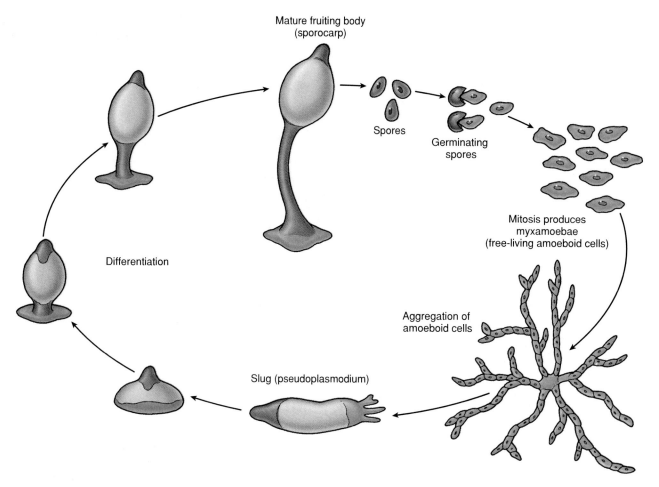

figure 12.27

Life Cycle of the Cellular Slime Mold *Dictyostelium discoideum*
Under limited nutrient conditions, cells aggregate to produce a slug (pseudoplasmodium). The slug differentiates into a mature fruiting body (sporocarp) containing spores. Spores released from the sporocarp germinate to produce myxamoebae.

bacteria, other fungi, and decaying material as sources of nutrients. The plasmodia increase in bulk to as much as several hundred grams. Under limited nutrient conditions, the plasmodium develops into a fruiting structure with spores. As the spores are released, they form flagellated gametes (swarm cells). The gametes unite in pairs to form zygotes, which in turn develop into new plasmodia. As a result of the animal-like and funguslike characteristics they display, slime molds provide interesting systems for the study of morphogenesis.

The cellular slime molds (phyla ***Plasmodiophoromycota, Dictyosteliomycota,*** and ***Acrasiomycota***) are free-living amoeboidlike organisms (myxamoebae) that feed upon soil bacteria, fungi, and organic matter. As the food supply in the immediate vicinity is depleted, these cells aggregate to produce, in some species, a slug-shaped organism called a pseudoplasmodium (Figure 12.27). This cell mass crawls along the surface and responds to such stimuli as heat and light. Eventually the pseudoplasmodium forms a colorful, stalked reproductive (fruiting) body containing hundreds of spores (sporocarp). When the spores are mature, they are released from the fruiting body and dispersed in the environment. The spores germinate to produce individual amoeboidlike cells, and the life cycle is completed.

The fungi are important inhabitants of the living world. They have important ecological roles in the decomposition of decaying organic matter and are also associated with antibiotic synthesis and food production. Fungi are responsible for many plant, animal, and human diseases. An understanding of their composition and taxonomy is therefore an important part of the discipline of microbiology.

The Protozoa

The protozoa (sing., protozoan) [Greek *protos*, first, *zoion*, animal] are heterotrophic unicellular organisms, many of which are motile and all of which are nonphotosynthetic. Most protozoa are microscopic, although some are large enough to be seen without the aid of a microscope. Protozoa are found in many environments and, like the algae, are important links in the planktonic food chain of aquatic animals. Like the fungi, they play an important part in the decomposition of organic matter through the engulfment of other organisms and particulate matter. Some protozoa are agents of disease, and some cause infections among the most serious known to humanity.

Protozoa Lack Cell Walls, but Contain Membrane-Bound Organelles

Most protozoa lack cell walls and are protected from the environment by a protective cell envelope. The envelope varies in composition in different organisms, but usually consists of an organic matrix interspersed with inorganic compounds such as silica or calcium carbonate. These inorganic compounds often result in the deposition of calcified fossil remains in ocean bottoms and terrestrial sites. The outer envelope is known by different names (shell, test, or pellicle) and is distinct from the plasma membrane, which is closely associated with, and located internal to, the envelope. Certain protozoa can form both a vegetative cell (known as a **trophozoite**) and a resting stage (known as a **cyst**). All the factors influencing cyst formation have not been defined, but cysts are known to develop under conditions of nutrient limitation or environmental stress and, in some species, during cellular reproduction. Cysts have thick external coverings that pro-

tect the cells from drying and toxic chemicals. In general, encystment is a dormant stage in the life cycle enabling environmental survival.

Parasitic Structure and Function
Protozoa: Classification • pp. 4–8

Similar to other eucaryotic cells, protozoa have membrane-bound organelles in their cytoplasm. These membranous structures include the endoplasmic reticulum, Golgi complexes, mitochondria, and nuclei. It is not unusual for a protozoan to have multiple nuclei during a portion of its life cycle. Members of the class of ciliated protozoa, for example, have two nuclei of different sizes and functions. A large macronucleus regulates metabolism, and a smaller micronucleus is associated with the sequestration of genetic material for exchange during conjugation.

An important characteristic of protozoa is their ability to ingest materials by pinocytosis and phagocytosis. During pinocytosis, liquids are surrounded by invagination of the plasma membrane, with subsequent entry into the cell as vacuoles. Larger particulate matter is ingested by a similar process called phagocytosis. Enzymes that are introduced into the vacuoles are believed to be responsible for the digestion of engulfed material in both processes.

Protozoa Reproduce Asexually and Sexually

The haploid protozoa reproduce either asexually or sexually. Asexual reproduction is relatively simple. The parent cell may divide mitotically into two equal daughter cells (binary fission) or into several daughter cells (multiple fission). Multiple fission is preceded by repeated division of the nucleus. Some protozoa reproduce asexually by budding. During budding, a smaller, motile daughter cell develops either on the cell surface (exogenous budding) or in the cell interior (endogenous budding). The budding daughter cell is eventually released into the environment.

Protozoan sexual reproduction may involve **syngamy, conjugation, autogamy,** or **cytogamy.** In syngamy, two haploid gametes unite to form a diploid zygote, which then divides meiotically to reestablish the haploid phase. Conjugation, which occurs in ciliated protozoa, involves the joining of two cells and the exchange of nuclei (Figure 12.28). Following this exchange of genetic material, the cells produce progeny through fission or budding. Autogamy is similar to conjugation but involves only one cell. The micronucleus in the cell divides into two separate nuclei,

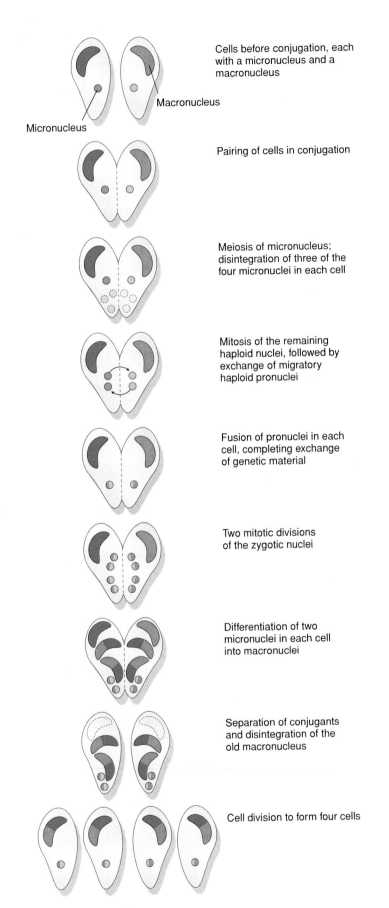

Cells before conjugation, each with a micronucleus and a macronucleus

Macronucleus

Micronucleus

Pairing of cells in conjugation

Meiosis of micronucleus; disintegration of three of the four micronuclei in each cell

Mitosis of the remaining haploid nuclei, followed by exchange of migratory haploid pronuclei

Fusion of pronuclei in each cell, completing exchange of genetic material

Two mitotic divisions of the zygotic nuclei

Differentiation of two micronuclei in each cell into macronuclei

Separation of conjugants and disintegration of the old macronucleus

Cell division to form four cells

figure 12.28

Conjugation in the Ciliate *Tetrahymena thermophila*

table 12.4

Phyla of Protozoa and Their Characteristics

Phylum	Major Characteristics	Representative Genera
Sarcomastigophora	Typically, single type of nucleus; sexual reproduction, when present, by syngamy; locomotion by flagella, pseudopodia, or both	Amoeba Entamoeba Giardia Leishmania Naegleria Trichomonas Trypanosoma
Labyrinthomorpha	Formation of ectoplasmic network with spindle-shaped or spherical, nonamoeboid cells; in some genera, movement of amoeboid cells within a network by gliding	Labyrinthula
Apicomplexa	Formation of apical complex containing structures distinguishable with an electron microscope; cilia and flagella typically absent; sexual reproduction by syngamy	Cryptosporidium Plasmodium Toxoplasma
Microspora	Unicellular spores; spore walls complete, without suture lines, pores, or other openings	Encephalitozoon Metchnikovella Nosema
Ascetospora	Multicellular spores; without polar capsules or polar filaments	Marteilia Paramyxa Urosporidium
Myxozoa	Spores of multicellular origin, with one or more polar capsules and with one, two, or three (rarely more) valves	Hexacapsula Myxidium Sphaeromyxa
Ciliophora	Cilia present in at least one stage of life cycle; two types of nuclei; sexual reproduction by conjugation, autogamy, and cytogamy; asexual reproduction by transverse binary fission; budding and multiple fission also possible	Balantidium Ichthyophthirius Nyctotherus Paramecium Tetrahymena Trichodina

and they join to form a zygote nucleus. The cell then divides into two individual daughter cells. In cytogamy, which is another variation of conjugation, the two individual cells fuse, but do not exchange nuclei.

Some protozoa may also be able to regenerate lost or damaged parts of the cell. Typically in regeneration only the portion of the cell containing the nucleus will regenerate; the portion without a nucleus is discarded.

Protozoa Are Divided into Seven Phyla

Protozoa are a subkingdom in the kingdom Protista. This subkingdom includes over 65,000 recognized species. In 1980 the Committee on Systematics and Evolution of the Society of Protozoologists classified the protozoa into seven phyla (Table 12.4).

This classification replaced earlier systems that divided the protozoa into four phyla: *Mastigophora, Sarcodina, Ciliophora,* and *Sporozoa.* Although protozoan classification constantly changes because of the vast amounts of new information regarding these organisms generated in recent years, the current traditional classification system that recognizes seven phyla is accepted by most protozoologists. Examples of protozoa from some of the major phyla are discussed in the following paragraphs.

The Sarcomastigophora Are Motile by Flagella, Pseudopodia, or Both

The phylum **Sarcomastigophora** is divided into two subphyla: **Mastigophora** and **Sarcodina**. The *Mastigophora* [Greek *mastigos,* whip] are flagellated protozoa that reproduce primarily asexually

AIDS and Eucaryotic Microbe Infections

One of the serious consequences of acquired immune deficiency syndrome (AIDS) is the accompanying, frequently fatal infections by eucaryotic microorganisms. AIDS is a disease that cripples the body's immunity to infection (see acquired immune deficiency syndrome, page 536). By overwhelming the immune system, AIDS permits a variety of infections to develop in the body. Among these diseases are three caused by eucaryotic microorganisms: pneumocystis pneumonia, cryptosporidiosis, and toxoplasmosis.

Pneumocystis pneumonia, a lung infection caused by the fungus **Pneumocystis carinii,** is one of the two diseases most often leading to death in AIDS patients (the other disease is a cancer of blood vessel walls known as Kaposi's sarcoma). Through June, 1996, 548,102 people in the United States had been diagnosed with AIDS, and 63% (343,000) are known to have died. P. carinii is commonly found in the lungs of humans, but seldom causes disease in persons with normal immunity. However, in persons with a defect in their immunity—for example, patients receiving immunosuppressive drugs and AIDS victims—P. carinii is immunologically unchecked and causes the alveoli to fill with a foamy acidophilic material. Patients are likely to be extremely weak, dyspneic (short of breath), and cyanotic (having a purplish discoloration of skin due to deficient oxygenation of the blood). Although previously classified as a protozoan because of its phenotypic characteristics, P. carinii recently has been found to be more closely related phylogenetically to the fungi. The rRNA sequences of P. carinii are more similar to those of yeasts than protozoa. Furthermore, lung infections by P. carinii and other fungi are acquired by inhalation, whereas protozoan infections usually are transmitted by person-to-person contact, insect bites, ingestion of contaminated food, or other similar means. This reclassification of P. carinii is an example of the advantages in using rRNA signature sequences in determining the phylogenetic placement of an organism (see ribosomal RNA and phylogeny, page 311).

Cryptosporidiosis and toxoplasmosis are caused by the protozoa **Cryptosporidium** and **Toxoplasma gondii,** respectively. Cryptosporidium is carried in animal intestines and can cause waterborne diseases (see Cryptosporidium waterborne disease, page 594). Cryptosporidiosis is characterized by severe, prolonged diarrhea that may last for several months or years, with particularly serious effects in immunosuppressed individuals like those with AIDS. T. gondii, a protozoan that can infect cats, dogs, cattle, sheep, and humans, is usually transmitted by the ingestion of raw or undercooked meat, blood transfusion, transplacental infection of the fetus, or inhalation of dust contaminated with cat droppings. Although most cases of toxoplasmosis in nonimmunosuppressed individuals are asymptomatic, the disease can result in massive lesions of the brain, lungs, liver, and other organs in AIDS patients.

---

by longitudinal binary fission. This group contains mostly parasitic species, several of which are human pathogens, including *Trypanosoma gambiense,* the cause of African sleeping sickness, which is transmitted by the tsetse fly; *Giardia lamblia,* a widespread intestinal pathogen usually acquired through contaminated drinking water; *Trichomonas vaginalis,* the cause of trichomoniasis, an infection of the genital tract; and *Leishmania donovani,* the etiological agent of oriental sores and other tropical skin diseases and transmitted by sandfly species (Table 12.5).

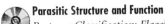

Parasitic Structure and Function
Protozoa Classification: Flagellated Protozoa • pp. 11–13

Members of the group *Sarcodina,* the other subphylum in the phylum *Sarcomastigophora* [Greek *sarkos,* flesh], move by pseudopodia and include the extensively studied amoebas. Amoebas reproduce by binary fission and are relatively simple in structure compared with other protozoa. Members of this group cause such diseases as amebic dysentery (*Entamoeba histolytica*) and meningoencephalitis (*Naegleria fowleri*). *Sarcodina* are often found as fossils. The White Cliffs of Dover are actually ocean sediments composed of calcified shells of *Forameniferida* (a marine member of the *Sarcodina*) that have been deposited over millions of years.

The Apicomplexa Form Apical Complexes Distinguishable with an Electron Microscope

The phylum **Apicomplexa** derives its name from the complex arrangement of a polar ring, microtubules, and other structures in the anterior, or apical, end of the trophozoite. The structures form-

table 12.5

Some Human Diseases Caused by Protozoa

Phylum	Protozoan	Disease
Sarcomastigophora	Entamoeba histolytica	Amebic dysentery
	Giardia lamblia	Gastroenteritis
	Leishmania donovani	Oriental sores and other tropical skin diseases
	Naegleria fowleri	Meningoencephalitis
	Trichomonas vaginalis	Trichomoniasis (vaginitis or urethritis)
	Trypanosoma gambiense	African sleeping sickness
Apicomplexa	Cryptosporidium sp.	Cryptosporidiosis (diarrhea)
	Plasmodium sp.	Malaria
	Toxoplasma gondii	Toxoplasmosis
Ciliophora	Balantidium coli	Balantidiasis (diarrhea, dysentery)

ing this apical complex are distinguishable with an electron microscope. The phylum consists of two classes: **Perkinsea** and **Sporozoea**.

Members of the class *Sporozoea* [Greek *spora*, seed] produce resistant spores (oocysts) that survive between hosts. All *Sporozoea* without exception are animal parasites, and some may cause serious diseases in infected hosts. The sporozoan **Plasmodium** causes malaria in humans and other warm-blooded animals. The disease toxoplasmosis, which may result in congenital defects, is caused by **Toxoplasma gondii**. This organism also causes intestinal and extraintestinal infections in cats and other felines. Domestic cats are important in the life cycle, and the cat's feces serve as a common source of human infections (Figure 12.29). Sporozoa have complex life cycles involving asexual and sexual stages that may or may not occur in an infected host. Asexual reproduction is by multiple fission. Sexual reproduction within the infected host occurs by fusion of haploid gametes (syngamy), resulting in the formation of a zygote, the only diploid stage in the life cycle.

Parasitic Structure and Function
Protozoa Classification: Sporozoan Protozoa • pp. 14–15

The Ciliophora Have Short, Hairlike Projections Called Cilia

The protozoan **Ciliophora** [Latin *cilium*, eyelid] uses short, hairlike projections (cilia) for locomotion and for the ingestion of food particles. Most of these protozoa are free-living, but some are parasitic. Ciliates differ from other protozoa by having two nuclei per cell: a micronucleus associated with sexual reproduction, and a macronucleus that controls cell metabolism and growth (see Figure 12.28). One common free-living ciliate, **Paramecium,** ingests other organisms into a cytopharynx through a specialized mouth pore located at one side of the cell. Hundreds of cilia on the cell direct food particles toward this cavity. Ciliated protozoa reproduce either asexually by transverse fission, budding, and multiple fission, or sexually by conjugation, autogamy, and cytogamy.

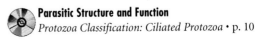

Parasitic Structure and Function
Protozoa Classification: Ciliated Protozoa • p. 10

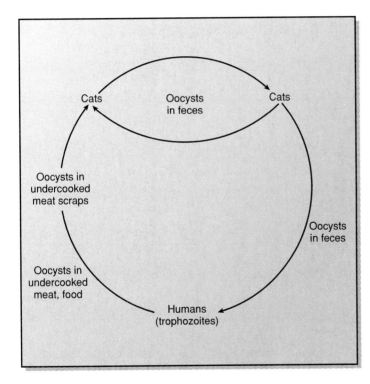

figure 12.29

Life Cycle of *Toxoplasma gondii*

Cats are the definitive host for *Toxoplasma gondii*. In the cat, *T. gondii* male and female haploid gametes fuse to form a zygote. The zygote develops into an oocyst, which is shed in the feces. The mature oocyst consists of two sporocysts, each with four sporozoites. Humans contract toxoplasmosis by ingesting the mature oocysts, usually by contact with an infected cat or by consuming undercooked, contaminated meat. Inside the human host, the sporozoites invade the intestinal lining, body tissues, and blood. Within infected host cells, the sporozoites can undergo a process of asexual fission called schizogony to form trophozoites.

PERSPECTIVE

Control of Yeast-Mycelium Dimorphism by Hexoses and Carbon Dioxide Levels

Yeast-mycelium dimorphism is a trait exhibited by many fungal species. Fungal dimorphism is of interest to researchers for two reasons: (1) Morphological conversion occurs among clinically important fungi and (2) dimorphism is a useful model of morphogenesis and cellular differentiation in a relatively simple eucaryote. Many different environmental factors, including atmospheric conditions, amino acid concentrations, vitamin levels, and carbohydrate concentrations, have been implicated as determinants in fungal morphogenesis. The influence of hexose concentration and CO_2 levels on fungal dimorphism was studied by S. Bartnicki-Garcia in 1968.

Bartnicki-Garcia grew strains of *Mucor rouxii* and other dimorphic fungi on solid or liquid media containing yeast extract, peptone, and a hexose as the carbon source. Fungal cultures were incubated at 28°C under three types of strict anaerobic conditions: (1) 100% N_2, (2) a mixture of 30% CO_2 plus 70% N_2, and (3) 100% CO_2. The morphological development of fungal cultures was followed by removing cell samples and examining them after centrifugation by light microscopy.

It was discovered that dimorphism could be controlled by glucose concentration. Fungi grown anaerobically under 30% CO_2 and 70% N_2 on media containing low concentrations of glucose (0.05% or less) developed exclusively as thin mats of mycelia (Figures 12P.1 and 12P.2). On media containing higher concentrations of glucose, the fungi changed progressively from hyphal to yeast forms (Figures 12P.1 and 12P.3). This control of the dimorphism of *M. rouxii* by glucose occurred only under anaerobic conditions. In aerobic cultures there was no evidence of yeast formation at glucose concentrations of 0% to 20%; only mycelia developed.

Furthermore, it was discovered that fungal morphogenesis was influenced by the CO_2 level as well as the glucose concentration. Dimorphic strains and species of *Mucor* produced yeast forms when cultivated in the presence of 2% glucose and 100% CO_2; however, when the CO_2 was lowered from 100% to 30%, most *Mucor* cultures yielded yeast and mycelial forms (Table 12P.1).

Bartnicki-Garcia was careful to note that other chemical and environmental factors besides hexoses and CO_2 may also serve as determinants of dimorphism. However, these observa-

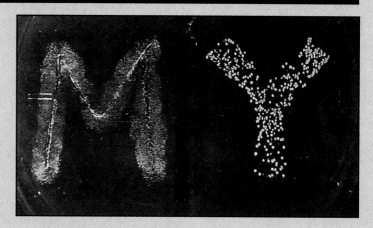

figure 12P.1

Control of Dimorphism by Glucose Concentration

Mucor rouxii on an agar plate after 30 hours of growth under 30% CO_2 and 70% N_2. Left: Mycelia on a medium containing 0.05% glucose. Spores were inoculated in a single M-shaped streak. Right: Yeast colonies on a medium containing 5% glucose. Spores were inoculated in a Y-shaped streak.

tions of Bartnicki-Garcia provided some of the first evidence that specific chemical substances and atmospheric conditions influenced fungal morphogenesis.

Source

Bartnicki-Garcia, S. 1968. Control of dimorphism in *Mucor* by hexoses: Inhibition of hyphal morphogenesis. *Journal of Bacteriology* 96:1586–1594.

References

Mardon, D., E. Balish, and A.W. Phillips. 1969. Control of dimorphism in a biochemical variant of *Candida albicans*. *Journal of Bacteriology* 100:701–707.

Odds, F.C. 1985. Morphogenesis in *Candida albicans*. *Critical Reviews in Microbiology* 12:45–93.

Orlowski, M. 1991. *Mucor* dimorphism. *Microbiological Reviews* 55:234–258.

San-Blas, G., and F. San-Blas. 1984. Molecular aspects of fungal dimorphism. *Critical Reviews in Microbiology* 11:101–127.

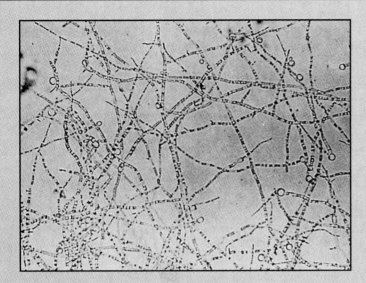

figure 12P.2

Mucor rouxii Hyphae

Micrograph of hyphae after 24 hours of growth in a liquid culture in 0.01% glucose under 30% CO_2 and 70% N_2 (×100).

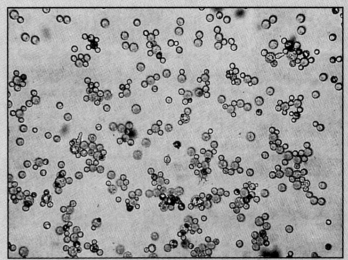

figure 12P.3

Mucor rouxii Yeast Cells

Micrograph of yeast cells after 24 hours of growth in a liquid culture in 1% glucose under 30% CO_2 and 70% N_2 (×100).

table	12.P1

Morphological Response of Diverse Species and Strains of *Mucor* to Different Levels of Glucose and CO_2 in Anaerobic Cultivation[a]

Organism	100% CO_2		30% CO_2 + 70% N_2	
	2% Glucose	0.01% Glucose	2% Glucose	0.01% Glucose
M. subtilissmis, NRRL 1909	Y	Y	Y	Y + M
M. rouxii, NRRL 1894	Y	M	Y + M	M
M. racemosus, NRRL 1427	Y	M	Y + M	M
M. rouxii, IM 80	Y	M	Y + M	M
M. racemosus, NRRL 1428	Y	No	Y + M	M

[a]On agar plates incubated for 48 hrs. Y = yeast development; M = mycelium; No = no growth.

Summary

1. Algae are eucaryotic microorganisms that are distinguished from fungi and protozoa by their photosynthetic ability. Algae have chloroplasts similar to those found in green plants and have chlorophylls, carotenoids, and biliproteins as light-gathering pigments.

2. Algae are classified into six divisions based on morphology and cellular organization, cell wall composition and physical properties, presence or absence of flagellation, type and nature of pigmentation, nature of reserve food products, and reproductive structures and nature of reproduction.

3. *Chrysophyta, Pyrrophyta,* and *Euglenophyta* are unicellular algae. *Chrysophyta* are often called golden algae because many contain a yellow-brown carotenoid pigment, fucoxanthin. *Pyrrophyta* are called dinoflagellates because of their spinning motility, and are important constituents of plankton. *Euglenophyta* lack cell walls, but have plasma membranes containing flexible strips that form an outer covering called the pellicle that enables the cells to change their shapes in a wiggling type of locomotion through mud.

4. *Chlorophyta, Phaeophyta,* and *Rhodophyta* are unicellular and/or multicellular algae. *Chlorophyta,* the green algae, are the most diverse of the algae and can be found in many habitats, including fresh and marine waters, tree trunks, arctic environments, and animals. *Phaeophyta,* the brown algae, are typically found in marine environments, where they attach to seabeds by holdfasts and usually have gas-filled floats that aid in buoyancy. *Rhodophyta,* the red algae, are important in bacteriology as the source of agar. Both *Phaeophyta* and *Rhodophyta* constitute what is commonly known as seaweed.

5. Fungi have rigid cell walls, are nonmotile, and are nonphotosynthetic. Fungi can exist as two basic morphological forms: filamentous molds and unicellular yeasts. Some fungi, such as *Ajellomyces* and *Candida,* are dimorphic and can alternate between both morphological forms, depending on environmental conditions of nutrition and temperature.

6. Molds consist of filaments called hyphae, which can form a mass called a mycelium. Yeasts are unicellular, oval organisms.

7. Fungi are phylogenetically classified into the kingdoms Fungi and Stramenopila and four protist phyla.

8. The kingdom Fungi consists of the phyla *Chytridiomycota, Zygomycota, Ascomycota,* and *Basidiomycota.* This kingdom includes *Rhizopus,* which causes bread mold; *Neurospora,* an organism that is used in genetic research because of its reproductive cycle; *Saccharomyces,* which is used in baking and in the production of alcoholic beverages; *Penicillium,* which is important in the production of antibiotics and various types of cheeses; and the basidiomycetes (rusts, smuts, and mushrooms).

9. The kingdom Stramenopila consists of the phyla *Oomycota, Hypochytriomycota,* and *Labyrinthulomycota.* This kingdom includes *Phytophthora infestans,* the cause of potato blight; *Plasmopara viticola,* the cause of downy mildew in grapes; and *Saprolegnia,* a fish pathogen.

10. Members of the phyla *Plasmodiophoromycota, Dictyosteliomycota, Acrasiomycota,* and *Myxomycota* are slime molds having characteristics resembling both fungi and protozoa. These four phyla are grouped as protists and are separated from the kingdoms Fungi and Stramenopila.

11. Protozoa are heterotrophic unicellular organisms that are nonphotosynthetic. Most protozoa are nonmotile and lack cell walls.

12. *Sarcomastigophora* are motile by flagella, pseudopodia, or both, and include *Trypanosoma,* which causes African sleeping sickness, and *Entamoeba,* which causes amebic dysentery. *Apicomplexa* form complex arrangements of a polar ring, microtubules, and other structures in the anterior, or apical, end of the trophozoite (vegetative cell). *Ciliophora* use cilia for locomotion and for the ingestion of food particles.

EVOLUTION *and* BIODIVERSITY

The first eucaryotes appeared on earth 1.5 billion years ago, approximately 2 billion years after the first procaryotes. The fossil record indicates that eucaryotic microbes preceded plants and animals by about 800 million years. Although eucaryotic microorganisms represent a large, heterogeneous group, little was known about their evolutionary relationships until recently. In the past the origins of eucaryotes and their relationships were interpreted from the fossil record and phenotypic characteristics. This information traditionally has been used to separate eucaryotes into distinct kingdoms. With the emergence of procaryote phylogeny, rRNA-based phylogeny of eucaryotes developed and scientists were able to explore phylogenetic relationships among eucaryotes. The eucaryotic phylogenetic tree, which is based on 18S rRNA instead of the 16S rRNA found in procaryotic ribosomes, suggests that diplomonads (for example, *Giardia lamblia*) are modern representatives of the earliest known eucaryotes (see universal phylogenetic tree, page 311). This early divergence of these protozoans is corroborated by similarities in the rRNA of *G. lamblia* and the 16S rRNAs of Bacteria and Archaea. *G. lamblia* has an rRNA of approximately 16S like procaryotes and is unique among eucaryotes in having a Shine-

Dalgarno mRNA binding site in its 16S-like rRNA (see Shine-Dalgarno sequence, page 232). Furthermore, *G. lamblia* lacks mitochondria, does not appear to have an endoplasmic reticulum or a Golgi complex, and lacks evident sexual life cycle stages (see eucaryotic cell structures, page 70). These procaryotic traits suggest that *G. lamblia* evolved prior to the full development of cellular structures found in other eucaryotes and before the endosymbiotic events that led to the formation of mitochondria (see endosymbiont hypothesis, page 206). Soon thereafter, microsporidia (for example, *Vairimorpha necatrix*) and trichomonads (for example, *Trichomonas vaginalis*) evolved. Protozoa in both groups are parasitic and lack mitochondria. Modern eucaryotes with their distinctive mitochondria and other internal membrane-bound organelles then evolved as a result of endosymbiotic events.

 Questions

Short Answer

1. How do procaryotes differ from other microorganisms?

2. How do algae differ from other microorganisms?

3. Why were cyanobacteria once grouped with algae? Why are they now considered to be true procaryotes?

4. Which division of algae lacks a cell wall? What is the primary component for the cell walls of other algae?

5. The commercially important kelp is a member of which division? The medically significant dinoflagellates belong to which division?

6. How do fungi differ from other microorganisms?

7. Compare and contrast molds and yeasts.

8. Identify several products we would not have without fungi.

9. How do fungal and bacterial spores differ?

10. How do slime molds differ from other fungi? Why are they classified as fungi?

11. How do protozoa differ from other microorganisms?

12. Protozoa do not form spores; however, they do enter a dormant stage known as _____. What is the active stage called?

13. Identify several medically significant protozoa. Do all protozoa cause disease?

Multiple Choice

1. Diatoms belong to which phylum?
   a. *Chlorophyta*
   b. *Chrysophyta*
   c. *Euglenophyta*
   d. *Phaeophyta*
   e. *Pyrrophyta*
   f. *Rhodophyta*

2. Which of the following is (are) dimorphic?
   a. *Aspergillus*
   b. *Candida*
   c. *Penicillium*
   d. *Rhizopus*

3. Which of the following is (are) an asexual form of reproduction found among protozoa?
   a. autogamy
   b. conjugation
   c. cytogamy
   d. fission
   e. syngamy

4. Flagellated protozoa belong to the (sub)phylum
   a. *Apicomplexa.*
   b. *Ciliophora.*
   c. *Mastigophora.*
   d. *Sarcodina.*

Critical Thinking

1. Through this chapter, you learned about several organisms that do not fit conveniently into their groups (for example, slime molds, *Euglena*, *Pneumocystis*). Given these difficulties, what criteria would you use for classification of a new organism?

2. Although algae, fungi, and protozoa are abundant in our environment, we rarely suffer from diseases caused by them. Explain why. (Be sure to consider algae, fungi, and protozoa separately, as well as collectively.)

3. Fungal and protozoan (along with helminthic) diseases are often covered in courses titled "Exotic Microbiology" or "Strange and Unusual Pathogens." Explain why. Is this sentiment valid today?

4. You won't find the organisms discussed in this chapter in *Bergey's Manual.* Explain why. Do you expect this will be true 10 or 20 years from now? Why or why not?

Supplementary Readings

Adam, R. 1991. The biology of *Giardia* spp. *Microbiological Reviews* 55:706–732. (A review of the biology of *Giardia*.)

Alexopoulos, C.J., C.W. Mims, and M. Blackwell. 1996, *Introductory mycology*, 4th ed. New York: Wiley. (A fundamental textbook in introductory mycology.)

Anderson, D.M. 1994. Red tides. *Scientific American* 71:62–68. (A review of the different types of toxic algae responsible for red tide blooms.)

Bold, H.C, C.J. Alexopoulos, and T. Delevoryas. 1980. *Morphology of plants and fungi.* New York: Harper & Row. (An in-depth, timely description of plants—including algae—and fungi.)

Bold, H.C., and M.J. Wynne. 1985. *Introduction to the algae: Structure and reproduction*, 2d ed. Englewood Cliffs, N.J.: Prentice-Hall. (A textbook in introductory phycology.)

Carmichael, W.W. 1994. The toxins of cyanobacteria. *Scientific American* 70:78–86. (A review of cyanobacterial toxins, their deadly properties, and their potential usefulness for humans.)

Garcia, L.S., and D.A. Bruckner, eds. 1991. *Diagnostic medical parasitology*, 2d ed. Washington, D.C.: American Society for Microbiology. (A description of parasites important in medicine and human disease.)

Moore-Landecker, E. 1990. *Fundamentals of the fungi*, 3d ed. Englewood Cliffs, N.J.: Prentice-Hall. (A textbook on fungi.)

Noble, E.R., and G.A. Nobel. 1982. *Parasitology: The biology of animal parasites*, 5th ed. Philadelphia: Lea and Febiger. (A discussion of the physiology, biochemistry, immunology, and pathology of major groups of parasites, including descriptions of host-parasite relationships and parasite life cycles.)

Raven, P.H., R.F. Evert, and S.E. Eichhorn. 1992. *Biology of plants*, 5th ed. New York: Worth Publishers. (A detailed survey of plants, with special chapters devoted to procaryotes, algae, and fungi. The structure, physiology, and genetics of plants are extensively covered.)

Schmidt, G.D., and L.S. Roberts. 1981. *Foundations of parasitology*. 3d ed. St. Louis: Mosby. (A comprehensive survey of protozoa, with detailed discussions of morphology and biochemistry of these organisms.)

Sogin, M.L. 1989. Evolution of eukaryotic microorganisms and their small subunit ribosomal RNAs. *American Zoologist* 29:487–499. (A review of the phylogeny of eucaryotic microorganisms.)

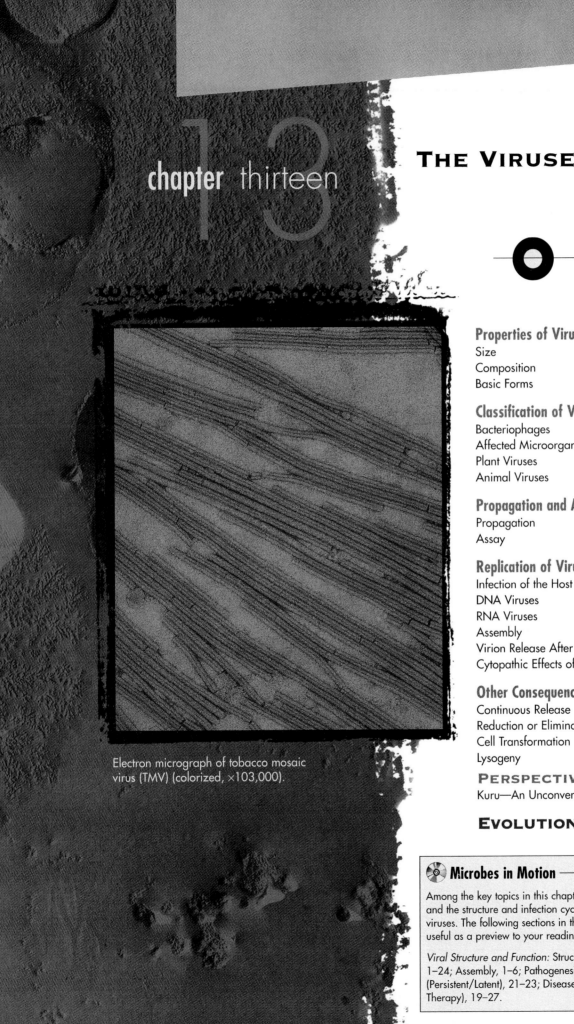

chapter thirteen

THE VIRUSES

Properties of Viruses
Size
Composition
Basic Forms

Classification of Viruses
Bacteriophages
Affected Microorganisms
Plant Viruses
Animal Viruses

Propagation and Assay of Viruses
Propagation
Assay

Replication of Viruses
Infection of the Host
DNA Viruses
RNA Viruses
Assembly
Virion Release After Assembly
Cytopathic Effects of Viral Infections

Other Consequences of Viral Infections
Continuous Release
Reduction or Elimination of Virus Production
Cell Transformation
Lysogeny

PERSPECTIVE
Kuru—An Unconventional Disease Caused by a Prion

EVOLUTION AND BIODIVERSITY

Electron micrograph of tobacco mosaic
virus (TMV) (colorized, ×103,000).

 **Microbes in Motion** — PREVIEW LINK

Among the key topics in this chapter are viral pathogenesis and disease,
and the structure and infection cycle (invasion, replication, assembly) of
viruses. The following sections in the *Microbes in Motion* CD-ROM may be
useful as a preview to your reading or as a supplemental study aid:

Viral Structure and Function: Structure, 1–9; Invasion, 1–8; Replication,
1–24; Assembly, 1–6; Pathogenesis (Host Cell Damage), 7–13,
(Persistent/Latent), 21–23; Diseases (Enveloped DNA Virus), 6–8, (Viral
Therapy), 19–27.

ach winter a serious infectious disease stalks young infants in nurseries throughout the United States. Children between the ages of three or four weeks and one year appear to be most susceptible to serious disease, whereas younger or older children are more resistant. The disease begins as a mild infection of the upper respiratory tract and then rapidly spreads into the lower respiratory tract, where it causes bronchitis (an infection of the small airways entering the lungs) and pneumonia.

Unlike most other infectious agents, which invade a body once and then are blocked in subsequent attacks by antibodies of the body's immune system, the agent responsible for this respiratory disease may return and strike the same person again and again. Furthermore, this is one of the few respiratory pathogens that consistently causes a major disease outbreak year after year. After each outbreak, the agent mysteriously disappears, only to reappear in the winter of the following year.

This "Houdini" pathogen is a virus called **respiratory syncytial virus (RSV),** named after the respiratory disease it produces and the characteristic syncytial (cell-fusing) masses it forms in infected cell cultures. RSV is a major cause of respiratory disease in young children. Infections are especially severe in infants with congenital heart disease; RSV infections in these infants can lead to mortality rates as high as 37%. Unfortunately there is no method to prevent RSV infection effectively. It appears to be transmitted in the hospital environment by unsuspecting doctors and nurses who spread the virus from one infant to another. Exposed infants do not seem able to effectively mount a defense against the viral invasion and vaccines do not help in preventing this disease.

As strange and unorthodox as RSV disease may seem, it is typical of many viral diseases. **Viruses** are obligate intracellular parasites that rely upon a host for metabolism and reproduction. Viruses that use bacteria as their host are called **bacteriophages** (or simply **phages**). Bacteria such as rickettsiae and chlamydiae are also obligate parasites (see rickettsiae and chlamydiae, page 326), but viruses (1) possess a single type of nucleic acid, either DNA or RNA (not both) and (2) possess a protein coat that surrounds the nucleic acid. Unlike most bacteria, algae, and fungi, viruses are metabolically inert and lack the metabolic machinery to generate energy or synthesize macromolecules. Although viruses contain genetic information encoded in their nucleic acid, they do not have ribosomes or any of the enzymes required to fully process this information. Consequently, viruses are no more alive outside a host than fragments of DNA. However, inside a suitable host the inert virus particle comes "alive" as it takes over the host biosynthetic machinery to synthesize the viral nucleic acids and proteins necessary for replication. Viruses therefore are unlike any other form of microorganism. These distinctive acellular forms are neither procaryotic nor eucaryotic and thus are usually considered separately from other types of microorganisms (Table 13.1).

Comparison of Viruses with Other Microorganisms

Characteristic	Viruses	Other Microorganisms
Size	Generally ≪200 nm	Generally ≥200 nm
Nucleic acid	DNA or RNA	DNA and RNA
Outer covering	Usually simple protein coat	Complex membrane, wall, or both
Reproduction	Requires host	Generally self-reproducing
Metabolism	Utilizes host metabolic machinery	Macromolecular or synthetic machinery; has own metabolic machinery
Cultivation	Cannot be cultivated on cell-free media	Usually can be cultivated on cell-free media

Properties of Viruses

Regardless of the type of host they infect, all viruses have the same general structure: genetic material in the form of either DNA or RNA surrounded by a protein coat and, in some cases, an outer membrane. The complete virus particle with its nucleic acid and its outer covering is called a **virion.**

Viruses Are Extremely Small

Most viruses are smaller than bacteria (≪200 nm) and can only be seen with an electron microscope (Figure 13.1). A few viruses, such as the poxviruses, have diameters exceeding 300 nm and are within the theoretical limit of resolution of the light microscope. All viruses, however, are so small that the quantity of genetic material they carry is limited. As a consequence, most viral genomes code for only the minimum amount of information required for structural integrity. This characteristic is reflected in the simple composition of a virus.

 **Viral Structure and Function**
Viral Structure: Virus Size • p. 2

Viruses Consist of DNA or RNA Surrounded by a Protein Coat

Viruses consist of a single type of nucleic acid, either single- or double-stranded DNA or RNA, enclosed within a protein coat (Figure 13.2). The size of the nucleic acid in viruses ranges from only a few thousand bases in the case of **picornavirus** to several hundred thousand bases for the **poxvirus.** An average gene is estimated to contain approximately 1,000 bases (1 kilobase). Most viruses are able to fit only a few genes into their limited genomes.

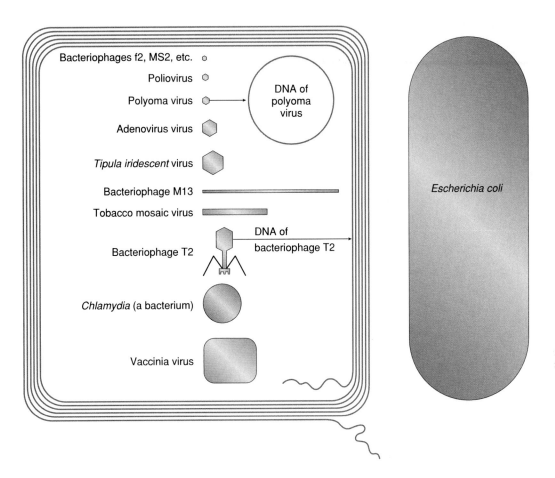

figure 13.1
Comparative Sizes of Some Common Viruses and Bacteria

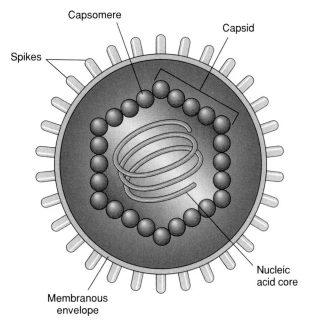

figure 13.2
Structure of an Enveloped Animal Virus
This drawing shows the major components that may be part of an enveloped animal virus. The nucleic acid core surrounded by a capsid, composed of capsomeres, constitutes the nucleocapsid. Some animal viruses may also have a membranous envelope surrounding the capsid, and glycoprotein spikes with either hemagglutinin or neuraminidase activity.

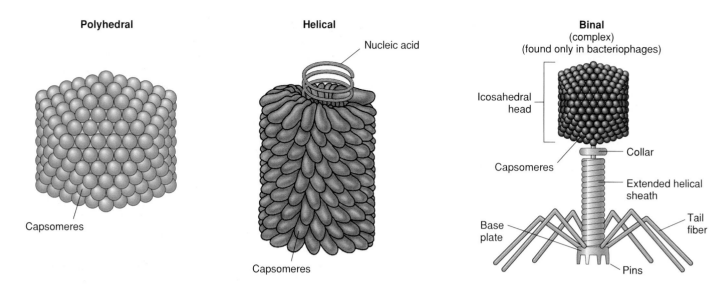

figure 13.3

Basic Nucleocapsid Forms

There are three basic categories of capsid architecture: polyhedral, helical, and binal.

For example, the **adenovirus,** which causes acute respiratory and ocular infections, has a double-stranded DNA genome with about 35 kilobases. These bases code for only a few genes. Even the largest and most complex viruses have genomes coding for fewer than 200 to 300 different proteins. In comparison, the genome of *Escherichia coli* is estimated to contain anywhere from 2,000 to 4,000 genes.

Naked nucleic acids are highly susceptible to degradation by the nucleases that often are found in the environment. To protect their nucleic acids from these enzymes, viruses are constructed with a protein coat surrounding their genome. This protective coat, called a **capsid,** is made up of repeating protein subunits called **capsomeres.** Because some viruses do not devote much of their limited genes for the coding of capsomere proteins, capsomeres in the smaller, more elementary viruses generally are composed of only one type of protein. In the more complex viruses, capsomeres can consist of several different types of proteins. The capsid, in combination with the nucleic acid, is commonly referred to as the **nucleocapsid.**

Some animal viruses (viruses with animal hosts) have a membranous envelope surrounding the capsid. **Matrix proteins** are located in the space between the envelope and the nucleocapsid. They strengthen the envelope and connect it to the nucleocapsid. The envelope is composed of **glycoproteins,** embedded within a **lipid bilayer** that is derived from the host cell's plasma membrane as the virus is released by an extrusion process following virus replication. Glycoproteins are complexes of carbohydrates and proteins. They are located on the outer envelope surface and often take the form of spikes called **peplomers. Influenza viruses** (orthomyxoviruses), for example, have glycoprotein spikes with either hemagglutinin or neuraminidase activity. The hemagglu-tinin spikes on these viruses can bridge red blood cells in clusters to cause **hemagglutination;** as will be discussed later in this chapter, this characteristic can be used to assay for the presence and quantity of virus in a solution. Neuraminidase is an enzyme that is not unique to viruses—it is synthesized by many types of bacteria, including *Vibrio cholerae* and many streptococci. The enzyme cleaves *N*-acetylneuraminic acid residues from glycoproteins. Because glycoproteins containing *N*-acetylneuraminic acid are found in mucus, it has been hypothesized that the neuraminidases of the influenza virus aid in penetration of the mucus of the respiratory tract.

The proteins in the viral envelope are coded by viral genes. The lipids and carbohydrates of the envelope, in contrast, are derived from the host cell. The kinds of lipids and carbohydrates in the envelope thus depend on the type of host cell infected by the virus. The presence of lipids makes enveloped viruses sensitive to disinfection and damage by lipid solvents such as ether.

Viruses Have Three Basic Forms: Polyhedral, Helical, and Binal

Viruses have small genomes and are able to code for only a few structural proteins. This limitation results in viral coats that frequently consist of simple, repeating protein subunits. With this repeating structure and the energy restrictions associated with its assembly, viruses can assume only a few possible symmetrical forms that fall into three basic categories: **polyhedral, helical,** and **binal** (previously known as **complex**) (Figure 13.3). These three terms describe the architectural arrangement of capsids that make up the outer protein coat of the virion.

The electron microscope is invaluable for defining viral morphology. Even before the refinement of techniques for electron microscopy, Francis Crick and James D. Watson in 1956 predicted the shape and structure of plant viruses (viruses with plant hosts). They reasoned that because of the small amount of genetic information available in viruses, only a limited number of different molecules could be used to form the outer covering of a virion. The type of shape that could efficiently utilize such repeating building blocks was one with cubic symmetry. The underlying structure of viruses with cubic symmetrical forms is the icosahedron.

The icosahedron, the most common polyhedral form in viruses, has 20 faces, 30 edges, and 12 vertices. Each face on this geometrical structure is an equilateral triangle and results in a symmetrical framework. The animal viruses adenovirus, papovavirus, and herpesvirus are examples of viral nucleocapsids with an icosahedral shape.

Viral Structure and Function
Viral Structure: Structure and Shape • pp. 3–7

Many plant viruses and some bacterial and animal viruses have helical nucleocapsids. Examples of such viruses are the plant virus **tobacco mosaic virus (TMV)**, the animal virus **rabies virus**, and the bacteriophage **M13.** TMV is rod-shaped and is composed of 2,130 protein subunits stacked in a helical fashion around a central core. The viral genome of single-stranded RNA is wound within the core and among rows of the subunits. The placement of the viral RNA within the protein coat protects the nucleic acid from harmful nucleases in the surrounding environment.

Viruses that have neither polyhedral (icosahedral) nor helical symmetry or have combinations of these forms are termed binal. Examples of binal viruses are the poxviruses and the T-even bacteriophages (**T2, T4,** and **T6**). Poxviruses have an indistinguishable capsid surrounding a nucleic acid core of double-stranded DNA. The poxviruses cause such diseases as smallpox (variola) and cowpox, characterized by pustular skin lesions called **pocks.**

The T-even bacteriophages have a complex structure that consists of three parts: head, sheath, and tail. The phage head has a hexagonal shape and is composed of protein subunits surrounding the viral genome. This phage head is attached to the tail portion via a narrow contractile sheath. The tail actually consists of six thin tail fibers that extend like folding legs from the sheath. These fibers are important for attachment to the bacterial surface during infection. Many bacteriophages differ from this structure and may lack a sheath (bacteriophages **T1** and **T5**), have short tails (bacteriophages **T3** and **T7**), or have no tail (bacteriophage φ**X174**). Other bacteriophages (bacteriophage **fd**) are filamentous and resemble long pieces of insulated wire. The wire represents their nucleic acid surrounded by a tubular protein coat.

Classification of Viruses

Many attempts have been made to develop systems for classifying viruses. Because all viruses must infect a host organism as a prerequisite for reproduction, early systems of viral classification were based on the type of host or the organ system infected: **tobacco mosaic virus, turnip yellow mosaic virus, cowpox virus, adenovirus,** and so on. The problem with this classification system was that many viruses were eventually found to infect multiple organs or have a wide host range.

Viruses sometimes are placed into three separate classes on the basis of their host: animal, plant, or bacteria. Some classification schemes include dividing viruses by the types of diseases they cause. However, such schemes have problems because viruses can produce different clinical signs and symptoms in the same host or in different hosts. Other classification schemes include dividing viruses by their nucleic acid relatedness, viral morphology, and chemical composition. While none of these schemes is completely without problems, virologists generally agree that most viruses can be separated on the basis of certain physical, chemical, and morphological characteristics. These characteristics—type of nucleic acid (DNA or RNA), symmetry of the nucleocapsid (helical, cubical, or binary), presence or absence of an envelope, and capsid size—were initially used by André Lwoff, Robert Horne, and Paul Tournier in 1962 to group viruses into several basic classes (the LHT system). The LHT system of classification does not attempt to show genetic relatedness among viral groups; instead it compares viruses on the basis of common similarities that can be readily observed or chemically measured.

Viral Structure and Function
Viral Structure: Taxonomy • pp. 8–9

In 1966 the International Committee on Nomenclature of Viruses (ICNV) was established to develop a systematic method for viral classification. Subcommittees of the ICNV were formed to specifically classify viruses infecting vertebrates, invertebrates, plants, and bacteria. The name ICNV was changed in 1973 to the International Committee on Taxonomy of Viruses (ICTV), and a fifth subcommittee to consider the classification of fungal viruses was established in 1975.

Bacteriophages Are Classified by Their Morphology and Nucleic Acid Content

Bacterial viruses (bacteriophages) are extensively studied particles that have been instrumental in studies of genetics and molecular biology (see transfer of genetic material, page 255, and recombinant DNA technology, page 281). There have been no specific guidelines for bacterial virus classification; most bacteriophages have been haphazardly named by scientists as new forms were

Viroids and Prions— Agents Smaller Than Viruses

Viruses were long considered to be the simplest infectious agents known. In 1961, however, William Raymer, a plant pathologist with the U.S. Department of Agriculture, discovered an infectious agent simpler than viruses in potatoes with the disease potato spindle tuber. After several years of intensive work, this agent was finally characterized and named a **viroid.** Viroids are several thousand times smaller than viruses and consist of only nucleic acid (Figure 13.4). The viroid originally found in potato spindle tuber disease had a molecular weight of 130,000 daltons in comparison with the 40 million daltons molecular weight of the TMV. The viroid nucleic acid apparently is protected from environmental nucleases by its tightly folded configuration. Viroids are unusually resistant to heat and ultraviolet radiation.

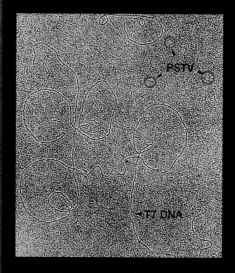

figure 13.4

Electron Micrograph of the Potato Spindle Tuber Viroid (PSTV), Comparing Its Size to the DNA of Bacteriophage T7 of *Escherichia coli* (×20,680).

Viroids are responsible for a number of plant diseases, including hop stunt, avocado sun blotch, cucumber pale fruit, citrus exocortis, potato spindle tuber, and tomato bunchy top. Because their nucleic acids are so small, viroids rely entirely on plant cell enzymes for replication. No viroid has yet been associated with organisms higher than plants.

Scrapie, a neurological disease of sheep, goats, and other animals characterized by twitching, excitability, intense itching, excessive thirst, weakness, and eventually paralysis, is caused by small infectious proteins, with no detectable nucleic acid. These **scrapie-associated fibrils (SAFs),** or **prions** (proteinaceous infectious particle) as they are now called, share the viroids' unusual resistance to heat and ultraviolet radiation. Prions appear to be an exception to the rule that all infectious agents contain nucleic acid as their genetic material (Figure 13.5). One hypothesis for their infectivity is that these particles may be encoded by a latent host gene that is activated upon infection of the host. Other slow, degenerative diseases such as kuru and Creutzfeldt-Jakob disease (CJD) in humans and transmissible encephalopathy in mink are caused by these proteinaceous agents.

CJD, a rare disease that normally is associated with old age, produces sponge-like holes in the brain and leads to depression, memory loss, and rapidly developing, devastating dementia. Although CJD may take up to 50 years to develop, once symptoms appear, death can occur within seven to nine months. Cattle are afflicted with a similar disease, bovine spongiform encephalopathy (BSE), or "mad cow disease," causing twitching, confusion, lethargy, and death. Scientific evidence suggests that humans may contract the disease through eating contaminated meat. BSE is especially a problem in Britain, where the disease first appeared in 1985 in a previously healthy Holstein dairy cow that ate protein meal prepared from sheep tissue, which may have been infected with scrapie. The cow became edgy, uncoordinated, and aggressive. By the end of 1995, researchers had identified 154,592 cattle with BSE on British farms. A public outcry for the British government to slaughter the country's 11 million to 13 million cattle arose in 1996 when several teenagers contracted CJD, which is rarely seen in young people. The teenagers were believed to have contracted the disease from eating contaminated beef. In response to this uproar, the British government agreed to destroy nearly five million cattle over a six-year period in an effort to wipe out mad cow disease.

a.

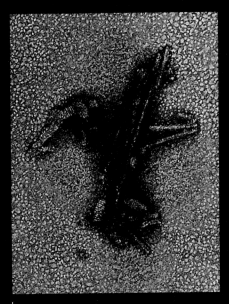

b.

figure 13.5
Electron Micrographs of Prions
a. Prions purified from clinically ill hamster brains infected with prion strain 263K; negatively stained with uranyl acetate (×115,760). b. Prions isolated from human Creutzfeldt-Jakob disease; negatively stained with phosphotungstic acid (×89,670).

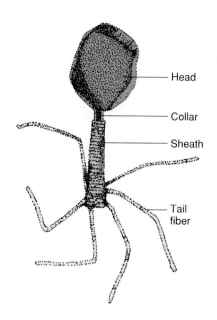

Head

Collar

Sheath

Tail
fiber

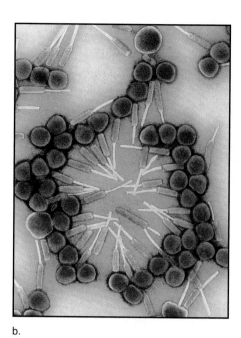

a.

b.

figure 13.6

Electron Micrographs of Bacteriophages

a. Bacteriophage T2 (family: *Myoviridae*) (colorized, ×270,000). b. Bacteriophage P1 (colorized, ×52,857). Both T2 and P1 are double-stranded DNA bacteriophages that infect *E. coli*.

discovered. As a general rule, phage morphology and the content of the phage genome and capsid have been used in such classification systems (Figure 13.6). For example, **SPO1** and **SPO2** are double-stranded bacteriophages that infect *Bacillus subtilis*. *Escherichia coli* is a host for many types of bacteriophages (frequently called **coliphages,** because they infect *E. coli*). The T-even coliphages (**T2, T4, T6**) are double-stranded DNA phages with binal symmetry consisting of a hexagonal head connected by a narrow sheath to long tail fibers. Coliphage φ**X174** contains a single-stranded circular DNA genome within an icosahedral capsid consisting of 12 capsomeres. Coliphage **fd** also contains single-stranded circular DNA, but it is filamentous with a helical symmetry. The DNA of bacteriophage λ is double stranded and linear in the virion but forms a circular structure within the infected host.

Recently the Bacterial Virus Subcommittee of the ICTV began to establish a system for the classification of bacterial viruses. Classification is based on the morphology and nucleic acid content of the virus (Table 13.2). Family names have been established for the more than 2,000 bacteriophages that have been discovered and characterized. The process is slow and tedious but provides us with an orderly system of nomenclature for bacteriophages.

Viruses Infect Other Microorganisms Besides Bacteria

Bacteriophages represent only one category of viruses that infect microorganisms. Other viruses infect cyanobacteria (cyanophages), fungi (mycoviruses), algae (phycoviruses), and protozoa. Cyano-

phages are similar to bacteriophages in structure and mode of infection. It has been suggested that such viruses may be useful in the biological control of algal blooms. Viruses that infect eucaryotic microorganisms (fungi, algae, and protozoa) frequently use their hosts as vectors for transmission to other organisms. The most extensively studied viruses of eucaryotic microorganisms have been the mycoviruses. All that have thus far been discovered contain double-stranded RNA. Viruses that infect fungi generally exist in a latent (present but not active) state and appear to be transmitted through hyphal connections and fungal spores. One mycovirus, **hypovirus,** causes reduced virulence (hypovirulence) when it is transmitted to the chestnut blight fungus *Cryphonectria parasitica*. Virus infection is associated with alterations of signal transduction pathways involved in the expression of fungal genes linked to virulence. This property makes hypovirus potentially useful as a biological agent for control of chestnut blight and other fungal diseases.

Plant Viruses Are Classified into 23 Virus Groups and Two Families

There are over 300 different plant viruses that have been categorized into 23 virus groups and two families by the Plant Virus Subcommittee of the ICTV (Table 13.3). An additional 200 viruses remain unclassified at this time.

Plant viruses do not fit easily into a basic taxonomic scheme and are grouped primarily according to the types of diseases they cause. All but two recognized groups of plant viruses contain RNA

table 13.2

Some Families of Bacteriophages

Family	Morphology	Type of Nucleic Acid[a]	Examples
DNA bacteriophages			
Inoviridae		Circular ss DNA	M13, fd
Microviridae		Circular ss DNA	φX174, G4, M12
Corticoviridae		Circular ds DNA	PM2
Myoviridae		Linear ds DNA	T2, T4, T6, P2
Pedoviridae		Linear ds DNA	T3, T7, P22
Plasmaviridae		Circular ds DNA	MVL2
Styloviridae		Linear ds DNA	λ, T1, T5
RNA bacteriophages			
Cystoviridae		Linear ds RNA	φ6
Leviviridae		Linear ss RNA	Qβ, R17, MS2, f2

[a]ss = single-stranded; ds = double-stranded.

Some Groups of Plant Viruses

Group	Morphology	Enveloped (E) or Naked[a] (N)	Approximate Size of Virion (nm)	Type of Nucleic Acid[b]	Representative Virus
DNA viruses					
Geminivirus		N	2–18	Circular ss DNA	Maize streak
Caulimovirus		N	50	Circular ds DNA	Cauliflower mosaic
RNA viruses					
Almovirus		N	18–58 × 18	Linear ss RNA	Alfalfa mosaic
Bromovirus		N	23	Linear ss RNA	Brome mosaic
Carlavirus		N	690 × 12	Linear ss RNA	Carnation latent
Closterovirus		N	600–2,000 × 12	Linear ss RNA	Beet yellows
Comovirus		N	30	Linear ss RNA	Cowpea mosaic
Cucumovirus		N	30	Linear ss RNA	Cucumber mosaic
Hordeivirus		N	110–160 × 23	Linear ss RNA	Barley stripe mosaic
Ilarvirus		N	26–35	Linear ss RNA	Tobacco streak
Luteovirus		N	25	Linear ss RNA	Barley yellow dwarf
Nepovirus		N	30	Linear ss RNA	Tobacco ringspot
Potexvirus		N	480–580 × 13	Linear ss RNA	Potato X
Potyvirus		N	680–900 × 12	Linear ss RNA	Potato Y
Rhabdovirus			130–150 × 45 430–500 × 110	Linear ss RNA	Lettuce necrotic yellow

table 13.3

Some Groups of Plant Viruses (continued)

Group	Morphology	Enveloped (E) or Naked[a] (N)	Approximate Size of Virion (nm)	Type of Nucleic Acid[b]	Representative Virus
Tobamovirus		N	300 × 18	Linear ss RNA	Tobacco mosaic
Tobanecrovirus		N	28	Linear ss RNA	Tobacco necrosis
Tobravirus		N	46–114 and 180–215 × 22	Linear ss RNA	Tobacco rattle
Tombusvirus		N	30	Linear ss RNA	Tomato bushy stunt
Tymovirus		N	30	Linear ss RNA	Turnip yellow mosaic

[a]Nonenveloped.

[b]ss = single-stranded; ds = double-stranded.

a.

b.

figure 13.7

Examples of Lesions from Virus Infection of Plants

a. Necrotric lesions caused by tobacco mosaic virus infection of the tobacco plant *Nicotiana glutinosa*. b. Leaf color changes caused by tobacco mosaic virus infection of an orchid.

as their genetic material. **Caulimovirus,** a virus that infects cauliflower, possesses double-stranded DNA as its genetic material. **Geminivirus** infects maize and has a single-stranded DNA genome.

Many plant viruses are transmitted by insect or arthropod vectors and can replicate in these vectors; other viruses are transmitted directly from plant to plant. Plant viruses appear to lack specific mechanisms for penetration of plant hosts. In most instances they enter the host as a result of physical damage to the plant cell due to insect injury, weather deterioration, or mechanical abuse from cultivation methods. The infection of plants by viruses leads to a number of different visible symptoms, including discoloration, molting, and yellowing of leaves; abnormal root growth; and reduced fruit production (Figure 13.7). Because viral infections cannot be cured, viral diseases cause major economic losses that amount to millions of dollars annually.

Animal Viruses Are Classified into 18 Families

Many of the viruses that infect vertebrates also infect invertebrates. It is not unusual, for example, for a virus carried by an arthropod (mites, ticks, or mosquitoes) to be transmitted to a vertebrate. The arthropod serves as a **vector,** or vehicle of transmission, for the virus. For this reason, a similar classification system exists for viruses infecting vertebrates and invertebrates.

The ICTV separates viruses infecting vertebrates and other similar hosts into 18 families on the basis of common characteristics (Table 13.4). These 18 viral families are divided into two main

table 13.4

Classification of Animal Viruses

Family	Morphology	Enveloped (E) or Naked[a] (N)	Approximate Size of Virion (nm)	Type of Nucleic Acid[b]	Representative Virus
DNA viruses					
Parvoviridae		N	22	Linear ss DNA	Kilham rat
Adenoviridae		N	70–90	Linear ds DNA	Human adeno 2
Iridoviridae		N	130–300	Linear ds DNA	Tipula iridescent
Hepadnaviridae		E	42	Circular ds DNA	Hepatitis B
Papovaviridae		N	45–55	Circular ds DNA	Polyoma
Herpesviridae		E	150–200	Linear ds DNA	Herpes simplex
Poxviridae		E	200–390	Linear ds DNA	Smallpox
RNA viruses					
Caliciviridae		N	40	Linear ss RNA	Norwalk
Picornaviridae		N	22–30	Linear ss RNA	Polio
Reoviridae		N	60–80	Linear ds RNA	Rotavirus
Arenaviridae		E	50–300	Linear ss RNA	Lassa fever
Filoviridae		E	800–900 × 80	Linear ss RNA	Ebola
Bunyaviridae		E	100	Linear ss RNA	California encephalitis
Coronaviridae		E	60–220	Linear ss RNA	Coronavirus OC43

table 13.4

Classification of Animal Viruses (continued)

Family	Morphology	Enveloped (E) or Naked[a] (N)	Approximate Size of Virion (nm)	Type of Nucleic Acid[b]	Representative Virus
Orthomyxoviridae		E	80–120	Linear ss RNA	Influenza
Paramyxoviridae		E	150–300	Linear ss RNA	Measles
Retroviridae		E	100	Linear ss RNA	Human immunodeficiency virus
Rhabdoviridae		E	70–80 × 130–240	Linear ss RNA	Rabies
Togaviridae		E	40–75	Linear ss RNA	Dengue

[a]Nonenveloped.

[b]ss = single-stranded; ds = double-stranded.

groups: DNA viruses and RNA viruses (Figure 13.8). This system provides perhaps the most systematic approach to vertebrate viral taxonomy at this time. The classification is by no means complete and will undoubtedly be revised as additional viral groups are recognized and characterized.

Some of the animal viruses such as **herpesvirus** and **rhabdovirus** have a wide host range, infecting a wide variety of animals. Other viruses have a narrow host range. For example, human beings are the only known natural host for the **poliovirus** and **rhinovirus** (picornavirus). The **Ebola virus** causes a highly fatal severe hemorrhagic disease in humans, but causes only inapparent infections with no detectable symptoms in African green monkeys, which can also be infected with the virus. Ebola viruses are morphologically and antigenically distinct from the rhabdoviruses (family: *Rhabdoviridae*), and are officially classified as members of a new family called *Filoviridae*. Animal viruses are responsible for many infectious diseases, including influenza (**influenza virus**); measles (**rubeola virus**); benign warts (**papovavirus**); rabies (**rhabdovirus**); encephalitis (**togavirus**); Dengue fever (**Dengue virus**); and acquired immune deficiency syndrome, or AIDS (**human immunodeficiency virus, or HIV**). These and other viral diseases are discussed in detail in Chapter 17.

Propagation and Assay of Viruses

Viruses are submicroscopic, intracellular, obligate parasites that lack many of the enzymes and cellular components (for example, ribosomes) to generate energy, synthesize proteins, and replicate their nucleic acid. Consequently, they require a living host that has these enzymes and cell constituents for replication. A suspension of virus particles cannot be observed under a light microscope, nor can these particles be propagated on a nutrient agar plate, as can most bacteria. The cultivation of viruses requires living host cells.

Viruses Require a Host for Propagation

Because viruses cannot replicate without a host, they also cannot be propagated in the laboratory without a suitable host organism (animal, plant, or bacterium). Extracellular virions do remain viable, but eventually they must find a suitable host for replication. This requirement for a host forms the basis for the cultivation of viruses.

Animal viruses may be cultivated in the laboratory in **animal tissue cultures.** These consist of animal cells that are propagated and maintained under laboratory conditions. Animal tissue cultures are established by treating a specific animal organ with an enzyme such as trypsin to disassociate the tissue mass into individual cells. The cell suspension is then placed into a flat-bottomed container (a flask, bottle, or Petri dish) and covered with a rich liquid medium for the maintenance of the animal cells. A typical maintenance medium usually contains amino acids, vitamins, ions, buffering agents, and animal serum. Antibiotics are also frequently added to such media to decrease or prohibit bacterial contamination. The cell suspension is usually incubated in an atmosphere containing 5% carbon dioxide. As the animal cells grow and divide, they attach to the surface of the container and eventually form a single, continuous layer (**monolayer**) of cells. Cultures initiated from the original host tissue are called **primary cultures.** Tissue cultures are maintained by periodically changing the liquid maintenance medium or, when the cells become too crowded (**confluent**), passing a portion of the cells into a new container to establish **secondary cultures.**

Cells from normal tissues do not grow indefinitely and eventually die after a finite number of passages. Such cells, however, can

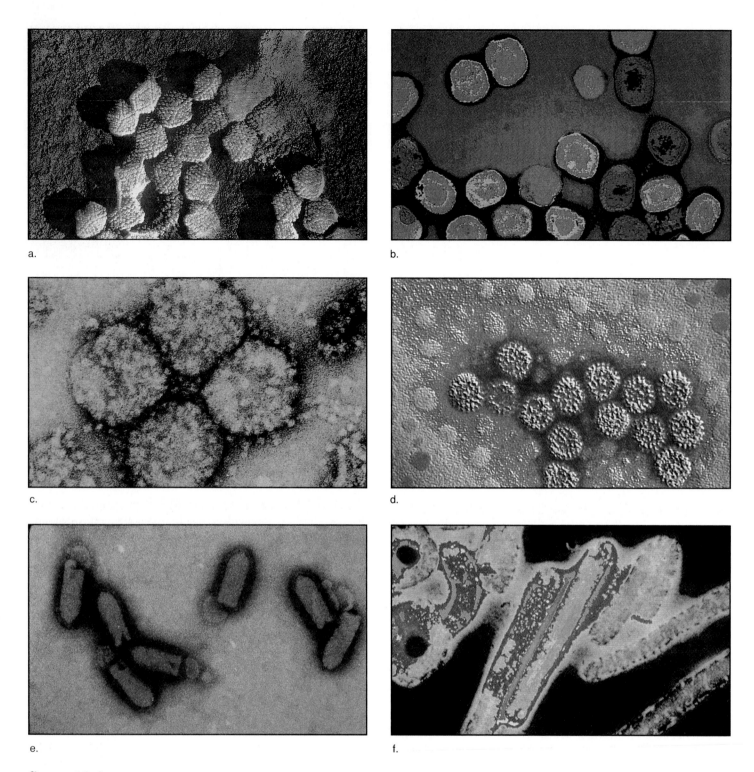

figure 13.8
Electron Micrographs of Some Animal Viruses
a. Adenovirus, a DNA virus (family: *Adenoviridae*) (colorized, ×300,000). b. Vaccinia virus, a DNA virus (family: *Poxviridae*). c. Coronavirus, an RNA virus (family: *Coronaviridae*) (colorized, ×500,000). d. Rotavirus, an RNA virus (family: *Reoviridae*) (colorized, ×48,500). e. Rabies virus, an RNA virus (family: *Rhabdoviridae*) (colorized, ×200,000). f. Ebola virus, an RNA virus (family: *Filoviridae*).

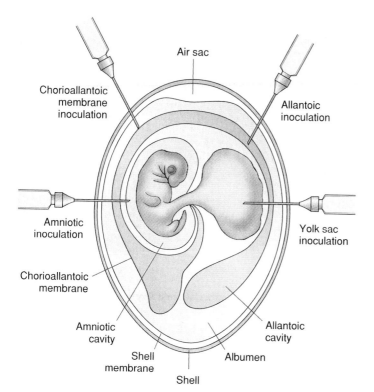

figure 13.9

Embryonated Egg and Sites for Virus Inoculation

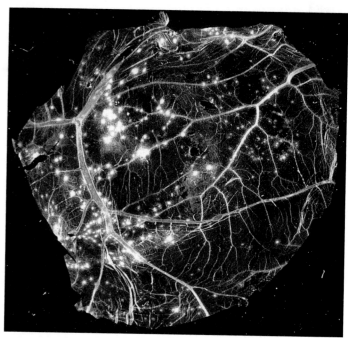

figure 13.10

Chicken Embryo Chorioallantoic Membrane Infected with Smallpox Virus

The opaque lesions on the transparent membrane are pocks.

be preserved by storage at ultralow temperatures in freezers or liquid nitrogen. Alternatively, certain types of cultures on occasion yield exceptional cells that acquire the ability to multiply indefinitely. These cells are used to establish **cell lines** that are immortal and grow continuously under proper maintenance. Examples of cell lines commonly used in research are HeLa cells (originally derived from cancerous cervical tissue of a woman named *Henri-etta La*cks) and BHK cells (isolated from *b*aby Syrian *h*amster *k*idney tissue). These cells are widely used in research because they are readily available and easily propagated.

Viruses inoculated into tissue cultures infect, replicate, and eventually can cause cell lysis. Each virion initially infects only one cell. When this cell lyses, mature viruses are released, and they infect and lyse neighboring cells. Clear areas called **plaques** form where viruses have attacked cell cultures that are overlaid with soft nutrient agar. Because each plaque usually represents a single virion from the original inoculum, the number of plaques can be counted and multiplied by the dilution factor to quantitate the virus concentration in that inoculum.

Viruses also can be propagated in vivo by inoculating them directly onto animal or plant tissue. One animal model commonly used for animal viral studies is the chicken embryo. Embryonated eggs 6 to 12 days old are used for viral inoculations. Sites for inoculation include the chorioallantoic membrane, yolk sac, amniotic fluid, and allantoic cavity (Figure 13.9). The chorioallantoic membrane is inoculated by first disinfecting the shell surface and then drilling a small hole through the shell and the shell membrane.

The chorioallantoic membrane is collapsed by removing air from the air sac of the egg. Virus is deposited onto the collapsed chorioallantoic membrane, and the shell opening is resealed with wax or paraffin to prevent contamination. The embryonated egg is then incubated. As each virus infects cells and replicates, an opaque lesion called a pock appears on the transparent membrane (Figure 13.10). These visible lesions can be enumerated to determine the concentration of viruses in the original inoculum. Pock counting, although used extensively in the past for enumeration of viruses, has now been largely replaced by the plaque technique.

Bacteriophages are propagated in the laboratory on suitable host bacteria. The bacteriophages are first mixed with a suspension of the appropriate host bacteria in melted soft nutrient agar. This phage-bacteria suspension is then poured onto the surface of a Petri dish containing hardened agar. The plate is incubated, and the bacteriophages are allowed to infect and lyse the bacterial cells. Clear plaques form in areas of the bacterial lawn (a layer of confluent bacterial growth) where phages are present (Figure 13.11). Each plaque represents one original phage particle.

Viruses Can Be Assayed by Various Methods

Plaque or pock counting is one method used to quantitate viruses. Viruses can also be assayed by a variety of other methods, including the actual counting of virus particles and detection of virions by hemagglutination.

figure 13.11
Clear Plaques of Bacteriophage λ (left) and Bacteriophage T2 (right) on Lawns of *Escherichia coli.*

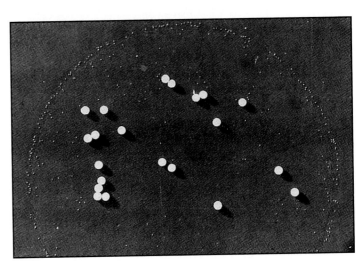

figure 13.12
Assay of Viruses by Electron Microscopy
Particles of poliovirus (small particles) are mixed with polystyrene latex spheres (large particles). The mixture is sprayed in droplets on a supporting membrane grid, dried, and shadowed. The resultant preparation is examined by electron microscopy, and the total number of virus particles is derived from the ratio of latex spheres to virions observed.

One of the first methods used to enumerate viruses was the direct counting of virus particles by electron microscopy (Figure 13.12). Virions are counted by mixing them with a known number of latex spheres and spraying the mixture onto a grid. The coated grid is examined by electron microscopy and the total number of virus particles is derived from the ratio of latex spheres to virions observed. Direct enumeration is rapid but does not distinguish infectious from noninfectious virus particles.

Viruses that possess hemagglutinin spikes (for example, influenza viruses) can be assayed by hemagglutination. The spikes attach to red blood cells and form bridges of virus particles between these cells. The maximum number of red blood cells that can be linked by a single virus particle is two because of the size of the cells. As additional bridges are constructed between virions and red blood cells, large aggregates develop. If the assay is performed in a small, clear tube or well, the aggregates sediment and form a film at the bottom. Red blood cells that do not aggregate simply fall to the bottom of the container, where they roll together and form a visible round red button. Since each virion is assumed to link two red blood cells, the total number of virus particles in a particular suspension can be approximated by diluting the original virus suspension and determining the dilution at which hemagglutination fails to occur. Like electron microscopy, hemagglutination assays do not distinguish between infectious and noninfectious virus particles.

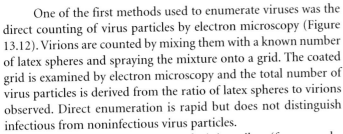

Replication of Viruses

Virus replication is a multistep process that begins with attachment of the virion to the host cell and terminates with the production of progeny viral particles. The steps associated with virus replication are (1) **attachment** of the virion to the host (**adsorption**), (2) **penetra-**tion of the virus or its nucleic acid into the cell (and **uncoating** in animal and plant viruses), (3) **replication** of the viral nucleic acid and **synthesis** of viral proteins, (4) **assembly** and/or **maturation** of new virus particles, and (5) **release** of mature virus particles from the host cell (Figure 13.13). In most viruses, all of the steps except replication of the viral genome are similar. Because viruses may contain single- or double-stranded DNA or single- or double-stranded RNA, replication of the viral genome may occur in a variety of ways.

Virus Replication Begins with Infection of the Host

Virus-host interactions are very specific. The replication cycle for animal and bacterial viruses begins with adsorption, the attachment of the virus to a receptor on the cell surface. In bacteria, a variety of different molecular components can serve as receptors, including flagella, pili, outer membrane proteins, lipopolysaccharides, and teichoic acids. Bacteriophages have specialized structures for adsorption, such as the tail fiber in T-even phages. Phage attachment is immediately followed by injection of the phage nucleic acid into the host cell (Figure 13.14). Only the phage genome enters the cell. The remaining portions of the phage (the capsid, sheath, and tail) remain outside the host.

Most animal viruses lack specialized attachment structures and have attachment sites located throughout the virion's surface. However, enveloped viruses such as orthomyxoviruses and paramyxoviruses attach to the host cell through glycoprotein spikes. Adsorption occurs to specific receptors on the host cell. For example, orthomyxoviruses bind specifically to the sialic acid residues on the plasma membrane. Polioviruses bind only to cells of primates, and preferentially to cells of the central nervous system and cells lining the intestinal tract. Following adsorption, most enveloped viruses enter the cell by **endocytosis,** a process in which the virion is ingested

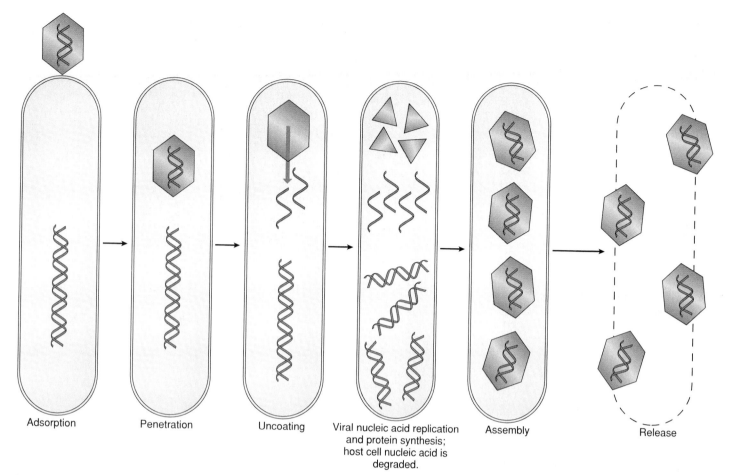

| Adsorption | Penetration | Uncoating | Viral nucleic acid replication and protein synthesis; host cell nucleic acid is degraded. | Assembly | Release |

figure 13.13

Steps Associated with Virus Replication
Virus replication begins with adsorption of the virus to a host cell, followed by penetration of the virus or its nucleic acid into the cell (and uncoating in plant viruses). Viral nucleic acid is replicated and

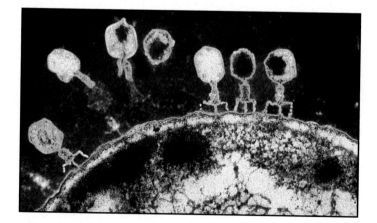

figure 13.14

Electron Micrograph of T4 Bacteriophages Attached to _Escherichia coli_
Bacteriophage attachment to receptors on the host cell surface is followed by injection of the phage nucleic acid into the cell (colorized, ×126,000).

viral proteins are synthesized inside the host cell. New virus particles are assembled and mature viruses are released from the host cell.

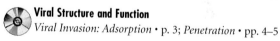 **Viral Structure and Function**
Viral Invasion: Adsorption • p. 3; _Penetration_ • pp. 4–5

into a phagolysosome (a vesicle inside the cytoplasm that fuses with a lysosome) and then later released into the cytoplasm. Some enveloped viruses enter by fusion of the virion envelope with the plasma membrane. Animal viruses that lack an envelope penetrate by endocytosis or by direct passage through the plasma membrane.

Plant viruses enter plant cells through openings in damaged cell walls, often formed by insects feeding on plant tissues. Inside the host cell, the virion is uncoated and its nucleic acid is released into the cytoplasm. The mechanism of uncoating is poorly understood but may involve host cell proteases.

Adsorption, penetration, and uncoating results in a period of the virus replication cycle in which infectivity is lost. During this period, known as the **eclipse,** no mature viral particles are present. In some viruses such as bacteriophage T4, the eclipse is characterized by a degradation of host DNA and a shift of the host metabolic machinery to synthesis of viral proteins. In the case of bacteriophage T4, viral genes are expressed in two stages: early proteins and late proteins. The early proteins are synthesized during the initial stages of viral infection and provide those viral-specific enzymes

Evidence That Bacteriophage Nucleic Acid, Not Protein, Enters the Host Cell During Infection

A major milestone in virology was the experiment of Alfred D. Hershey and Martha Chase in 1952 showing that bacteriophage nucleic acid, not the protein coat, enters the host cell during infection (Figure 13.15). Hershey and Chase took bacteriophage T2 particles with radioactive ^{32}P and others with radioactive ^{35}S. Because all of the phosphorus in a bacteriophage is located in the nucleic acid, and sulfur is found only in the amino acids methionine and cysteine in the protein coat, this labeling was specific for these two phage components. The specifically labeled T2 particles were then used to separately infect two different suspensions of *Escherichia coli*. Following infection, Hershey and Chase separated the infected cells from any unattached virions by low-speed centrifugation. They then subjected the infected host cells to violent agitation in a blender, a process that broke the phage tails and sheared off the empty coats, devoid of DNA, from the host cells. The remaining cells and injected phage material were assayed for progeny phage by the plaque technique.

Hershey and Chase's experiment yielded two significant results. First, they discovered that when ^{35}S-labeled phages were used to infect *E. coli*, 75% to 80% of the radioactive label was lost after the blending step. In contrast, when ^{32}P-labeled phages were used for infection, most (65% to 80%) of the radioactive label remained within the blended cells. Second, the blended cells produced normal phage progeny. These results showed that the protein portion of a bacteriophage remains outside the host cell during infection, whereas the nucleic acid enters the cell. Furthermore, because blended infected cells had lost their previously attached phage coats and tails but could still produce mature phage progeny, the experiment clearly showed that the phage nucleic acid, not the protein coat, carried the genetic information required for replication.

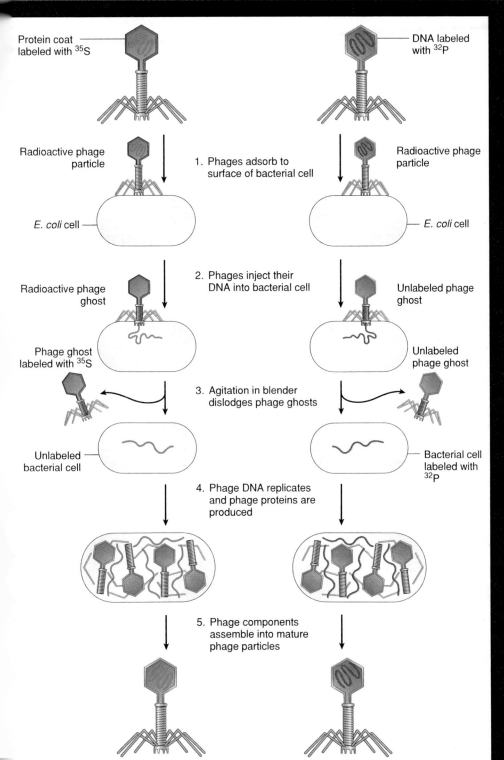

Protein coat labeled with ^{35}S

DNA labeled with ^{32}P

Radioactive phage particle

Radioactive phage particle

1. Phages adsorb to surface of bacterial cell

E. coli cell

E. coli cell

Radioactive phage ghost

2. Phages inject their DNA into bacterial cell

Unlabeled phage ghost

Phage ghost labeled with ^{35}S

Unlabeled phage ghost

3. Agitation in blender dislodges phage ghosts

Unlabeled bacterial cell

Bacterial cell labeled with ^{32}P

4. Phage DNA replicates and phage proteins are produced

5. Phage components assemble into mature phage particles

figure 13.15

Hershey and Chase Experiment
Separate *E. coli* populations are infected with bacteriophage T2 labeled with either ^{35}S-protein or ^{32}P-DNA. After infection, unattached phage are removed by centrifugation, and the infected bacterial cells are agitated in a blender. This breaks the tail by which the phage "ghost" (phage without the nucleic acid) is attached to the cell wall, and the empty phage head, devoid of DNA, is liberated into the medium. Assay of the host cells indicates that 75% to 80% of the attached ^{35}S is stripped off the infected cells by the blender treatment. In contrast, most of the ^{32}P remains in the infected cells. The blender treatment does not affect the host cell's capacity to produce progeny phage.

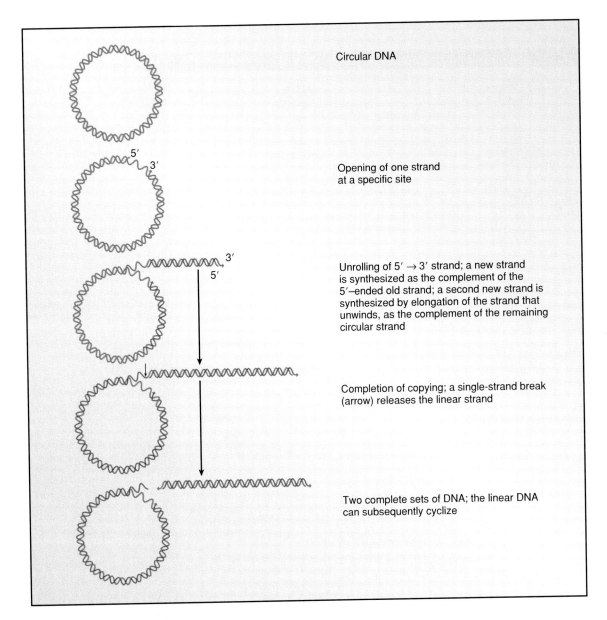

Circular DNA

Opening of one strand
at a specific site

Unrolling of 5′ → 3′ strand; a new strand
is synthesized as the complement of the
5′–ended old strand; a second new strand is
synthesized by elongation of the strand that
unwinds, as the complement of the remaining
circular strand

Completion of copying; a single-strand break
(arrow) releases the linear strand

Two complete sets of DNA; the linear DNA
can subsequently cyclize

figure 13.16
Rolling-Circle Mechanism of DNA Replication

necessary for the replication and continued expression of viral genes. The polymerases and other enzymes required for the transcription are either derived from the host or, in certain instances, are already present in the extracellular virion; these are injected into the cell with the viral nucleic acid. The late proteins are synthesized near the end of the infection process and include viral structural proteins and lytic enzymes.

 **Viral Structure and Function**
Viral Invasion • pp. 1–8

DNA Viruses Have Various Methods for Replication of DNA

The method of replication of DNA viral genomes depends on the type of genome. Most DNA viruses have double-stranded DNA that is either linear (poxvirus, herpesvirus, bacteriophage λ, and T-even phages) or circular (papovavirus and baculovirus).

In most cases, linear DNA is circularized before replication by the joining of short, complementary single-stranded regions on the ends of the DNA. The circularized double-stranded DNA is then replicated by a **rolling-circle mechanism** (Figure 13.16). An endonuclease nicks one strand of DNA, and the 5′ end of this strand is unwound and serves as a template for the synthesis of new DNA. The remaining circular DNA strand also is used as a template to elongate the strand that unwinds. As replication proceeds, a long double-stranded piece of DNA forms, with repetitive copies of the viral genome. This molecule is called a **concatamer** (DNA units linked end-to-end). The concatamer subsequently is cut by an endonuclease to form virion-sized genomes.

Bacteriophage T4 has a linear, double-stranded DNA genome but does not replicate by the rolling-circle mechanism. Instead, the DNA is replicated bidirectionally, as in bacteria. However, unlike

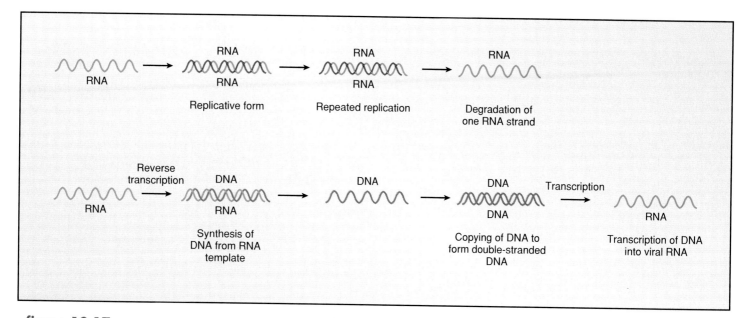

figure 13.17

Replication in Single-Stranded RNA Viruses
Single-stranded RNA viruses replicate their genomes by one of two modes. (a) The RNA is replicated to produce a replicative form, which becomes a template for the synthesis of RNA copies. (b) In retroviruses, the viral RNA genome serves as a template for the synthesis of a DNA molecule by reverse transcription. The DNA is copied to form double-stranded DNA, which then is transcribed into viral RNA.

Viral Structure and Function
Viral Replication: RNA Virus Replication • pp. 14, 16, 19

bacterial DNA replication, bacteriophage T4 DNA replication occurs with repeated initiations so that a concatamer is formed. The concatamer subsequently is cleaved to form virion-sized genomes.

In the case of viruses that have single-stranded DNA, such as bacteriophage φX174, the DNA is converted to a double-stranded form, called the **replicative form,** by host DNA polymerase. The replicative form then replicates by the rolling-circle mechanism, but one of the DNA strands is discarded before new viral particles are assembled.

Viral Structure and Function
Viral Replication: DNA Virus Replication • pp. 1–11

RNA Viruses Have Various Methods for Replication of RNA

Double-stranded RNA viruses, such as reovirus, replicate in a manner similar to double-stranded DNA viruses, except that ribonucleotides instead of deoxyribonucleotides are used as the precursors. Single-stranded RNA viruses replicate in one of two manners. In most cases (togavirus, picornavirus, paramyxovirus, and rhabdovirus), the virion RNA is copied using a viral RNA replicase to form a complementary strand, which serves as a template to make additional copies of the genome. During replication the two complementary strands combine to form an intermediary replicative form. In retroviruses [Latin *retro*, backward] the single-stranded RNA genome (considered to be a plus strand) directly serves as a template for the synthesis of single-stranded DNA (minus strand DNA) using a viral reverse transcriptase (RNA-dependent DNA polymerase) (Figure 13.17). This DNA is then copied to form double-stranded DNA, which may be used as a template for the synthesis of viral RNA or be integrated into the host genome. Many retroviruses are oncogenic, causing sarcomas, leukemias, and lymphomas in animals; it is this integration of viral-coded DNA into the host cell genome that causes cell transformation.

Viral Structure and Function
Viral Replication: RNA Virus Replication • pp. 13–24

Complete Virions Are Formed During Assembly

Viral components synthesized within the host cell must be packaged into complete virions prior to release. In the case of many bacteriophages, this assembly process is spontaneous and begins with the aggregation of capsomere subunits into complete capsids. Viral nucleic acid is then inserted into these empty phage heads, where the nucleic acid is condensed into a tightly packed mass. In complex viruses, the assembly process may involve several sequential steps, which eventually produce complete virions. It is known, for example, that the maturation and assembly of bacteriophage T4 takes place in three separate pathways involving synthesis of the head, tail, and tail fibers.

In enveloped animal viruses, assembly occurs by the wrapping of a portion of the host plasma membrane around the viral nucleoprotein core as it passes through the membrane to the outside of the cell. Prior to this step, viral-specified envelope proteins

One of the procedures often used to study phage growth as a function of time is the **one-step growth curve**, initially described by Max Delbrück and Emory Ellis

One-Step Growth Curve

(Figure 13.18). A heavy suspension of bacteria infected with bacteriophage is incubated for a few minutes to allow phage attachment to the bacteria. The suspension is then diluted several thousandfold and reincubated. Samples are removed from the suspension at different times and assayed for phage particles by the plaque method.

The results of such an experiment show that phage growth can be divided into several distinct phases. The first portion of phage growth is a **latent period** in which the number of plaque-forming units remains constant. During this period, the phages, which have infected the host cells, replicate and synthesize new material. Plating of a sample from the suspension during the latent period produces only one plaque per phage-infected bacterium. Although each infected bacterium may have several phage particles, these are immature at the time of plating. As a consequence, the progeny phages released from each bacterium are confined on the plate within the immediate area of the bacterium, resulting in the formation of a single plaque. In fact, samples taken at the very beginning of the growth curve produce no plaques, because immediately after entry into the host cell, the phage nucleic acid has not had an opportunity to replicate. This initial portion of the latent period, when no phage is produced, is known as the **eclipse period.**

As phage particles mature and are released from lysed bacterial cells, the number of detectable phages in the suspension dramatically increases. This increase, called the **rise period,** continues until all of the infected bacteria have lysed and released their contents of progeny phage. At this time the number of detectable phage particles reaches a plateau. The quantity of phage released from each bacterium can be calculated in a one-step growth curve by determining the ratio of phage particles at the beginning and end of the rise period. This ratio is called the **burst size** and is unique for any given phage-host interaction.

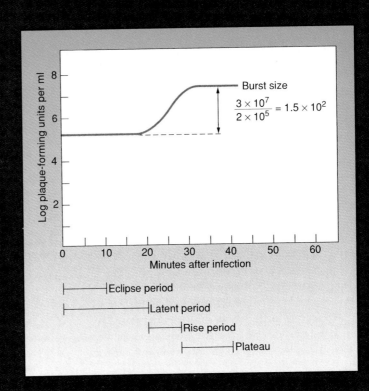

figure 13.18
A One-Step Growth Curve
A one-step growth curve consists of a latent period (which includes an eclipse period), a rise period, and a plateau. The burst size is determined by the ratio of phage particles at the beginning and end of the rise period. In this example, the burst size is 1.5×10^2.

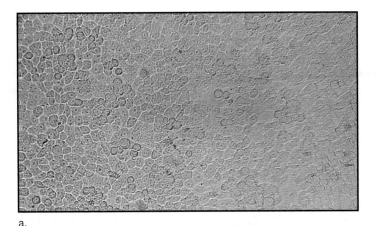

a.

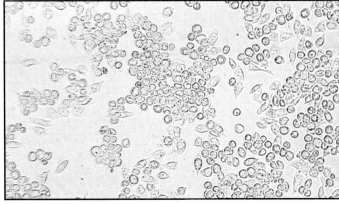

b.

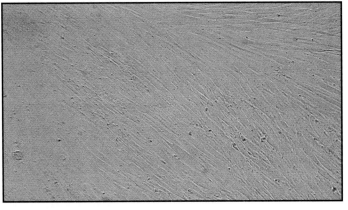

c.

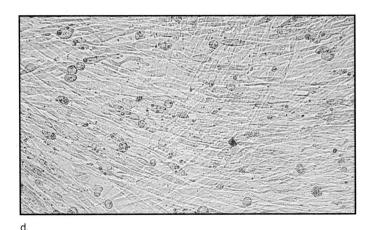

d.

figure 13.19

Examples of Cytopathic Effects

a. Normal human Hep-2 cell line. b. Human Hep-2 cell line infected with herpesvirus. The cytopathic effect is seen as discrete aggregations of cells. c. Normal human diploid cell line. d. Human diploid cell line infected with cytomegalovirus. The cytopathic effect is seen as rounded and enlarged cells with splitting of the cell layer and stranding.

migrate to the plasma membrane and replace the host proteins in the membrane. Thus the viral envelope consists of a combination of viral-specified proteins and host-specified lipids.

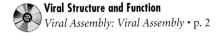

Viral Structure and Function
Viral Assembly: Viral Assembly • p. 2

Intact Virions Are Released from the Host After Assembly

The last step in virus replication is release of the intact virions from the host. The specific mechanism of release depends on the type of virus that has infected the cell. Bacteriophages are released from bacteria with plasma membranes typically disrupted by action of the enzyme lysozyme, one of the late proteins specified by the viral genome (filamentous phages and bacteriophage ϕX174 releases do not involve lysozyme).

Most phage release is an immediate process resulting in rapid cell lysis and death (although some bacteriophages such as M13 have a slow and nondestructive release). In contrast, the liberation of animal viruses is usually a slow process, often continuing for several hours. Animal viruses are either released through ruptures in the plasma membrane or, in the case of enveloped viruses, by a budding process from the host plasma membrane. In this latter instance, the viral envelope is actually derived from the host membrane, which has been modified by the insertion of viral-coded proteins. Because virus particles are gradually released from animal cells, the host cell may continue to metabolize during the process.

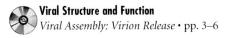

Viral Structure and Function
Viral Assembly: Virion Release • pp. 3–6

Viral Infections in Animal and Plant Cells Can Result in Cytopathic Effects

One of the consequences of viral infections in animal and plant cells is physical damage to the cells. Such morphological changes, which can be seen microscopically and macroscopically, are termed **cytopathic effects (CPEs)**. Examples of CPEs are changes in the structure of the nucleolus; formation of inclusion bodies; and damage to the plasma membrane, with a tendency for cells to become rounded and fuse together to form giant cells, or polykaryocytes (Figure 13.19). CPEs occur early in the infection

One of the problems associated with viral diseases is linked to the nature of their replication. Because viruses are metabolically inert under extracellular the translation of viral mRNA (Figure 13.20). These compounds can also be chemically induced by exposing cells to double-stranded RNA, which suggests

Are Viral Diseases Treatable?

conditions and utilize the host metabolic machinery during replication, antimicrobial agents are generally ineffective in controlling viral infections. Antimicrobial agents could be used to inhibit viral replication within the host cell, but such inhibition would also have to be directed against cellular metabolism and consequently would also be detrimental to the host.

Different substances, however, have been used to control viral infections. These antiviral agents include dyes, interferons, and other chemical compounds. Two of these agents are particularly promising: **acyclovir** and **interferons.** Acyclovir, a nucleoside derivative, has been shown to be of limited effectiveness in controlling genital herpes. Interferons, glycoproteins first described by Alick Issacs and Jean Lindenmann in 1957, also have been used to interfere with viral replication. Interferons are naturally produced by mammalian cells during viral infection and inhibit viral replication, specifically

that viral double-stranded RNAs synthesized during infection may be the actual mediators of interferon induction.

Interferons are cell specific, not virus specific, and are most effective when they are produced by cells of the same species. This specificity of interferons presents problems in the industrial production of these antiviral agents. In the past, interferons were extracted with great expense and difficulty from human white blood cells. Recent advances in genetic engineering, however, have made it possible to genetically manipulate bacteria for the production of this compound. As a result, large quantities of interferon are now available for use in the treatment of viral diseases, including cancer.

Viral Structure and Function
Viral Disease: Viral Therapy •
pp. 19–27

Microbial Pathogenesis
Nonspecific Host Defense:
Cellular Defenses • p. 30

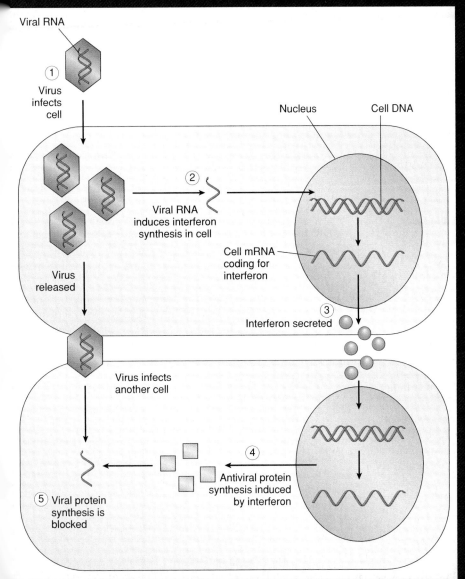

Viral RNA

① Virus infects cell

② Viral RNA induces interferon synthesis in cell

Virus released

Virus infects another cell

⑤ Viral protein synthesis is blocked

Nucleus

Cell DNA

Cell mRNA coding for interferon

③ Interferon secreted

④ Antiviral protein synthesis induced by interferon

figure 13.20
Mode of Action of Interferon
The general sequence of events associated with interferon production and activity follows. (1) Interferon production begins after virus infection of a cell. Double-stranded RNA viruses are particularly good inducers of interferon production. (2) The viral RNA induces interferon synthesis in the host cell. (3) This particular host cell is not protected by the interferon, but the interferon released from the infected cell moves to other uninfected cells and binds to their surfaces. (4–5) These uninfected cells, which now have interferon on their surfaces, are stimulated to produce at least two enzymes: a 2′, 5′-oligoadenylate synthetase that catalyzes the synthesis of an unusual polymer (2′, 5′-oligoadenylate) and a protein kinase. 2′, 5′-oligoadenylate activates an intracellular ribonuclease that degrades viral RNA. The protein kinase is activated only in the presence of double-stranded RNA, which, with the exception of the retroviruses, is formed as an intermediate in the replication of RNA viruses. The activated protein kinase catalyzes the phosphorylation of one of the factors (eIF2a) required for the initiation of protein synthesis. The phosphorylated eIF2a is inactive and, therefore, synthesis of all proteins, including viral proteins, ceases. There are three major classes of human interferons (IFN-α, IFN-β, and IFN-γ), which have as their major cell sources leukocytes, fibroblasts, and lymphocytes, respectively.

process and are believed to result from viral inhibition of host macromolecular synthesis. A visible consequence of CPEs are the plaques or pocks that appear from the viral infection of host cells, resulting from **lytic infections** (infections in which mature progeny virions are released from a lysed cell).

Viral Structure and Function
Viral Pathogenesis: Host Cell Damage • pp. 7–8

Other Consequences of Viral Infections

Lytic infections are only one effect of virus infection. Other possible consequences of virus-host cell interactions are **persistent infections, abortive infections, transforming infections,** and **lysogeny.**

Viruses Can Be Continuously Released from a Host Cell

Persistent infections are those in which viruses are continuously released from a cell that is not immediately killed. Under such conditions the host cell survives and continues to metabolize. An example of a persistent infection is infection of monkey kidney cells by **paramyxovirus SV5** (simian virus 5). Although SV5 causes lytic infection in many cells, it infects monkey kidney cells with practically no cell damage. The infected cells continue to grow and produce virus.

A similar but somewhat different situation occurs in the case of **defective interfering (DI)** viruses. DI viruses lack complete genomes and therefore are unable to multiply without assistance from other virions. However, when DI viruses and normal, infectious viruses simultaneously infect a cell, the result can be a reduction in cell damage and quantity of viruses released—a persistent infection. DI viruses arise from errors in viral replication and are believed to be involved in infections by **measles virus, Newcastle Disease Virus (NDV),** and **Western Equine Encephalitis (WEE) virus.**

Persistent infections also occur when the production of infectious viral particles is dampened by the presence of antibodies or antiviral agents. In some instances a virion may be less virulent in a particular host, resulting in a persistent rather than a lytic infection.

Viral Structure and Function
Viral Pathogenesis: Persistent Infections • pp. 21–23

There Can Be a Significant Reduction or Elimination of Virus Production in the Host

Not all viral infections lead to high-level production of complete, intact virions. Under certain conditions when there is interference in the viral multiplication cycle, virus production can be significantly reduced or completely eliminated. Such abortive infections occur when virions infect nonpermissive (not supporting productive viral infections) or not fully permissive cells, when there is a defect in components required for virus multiplication, or when antiviral agents interfere in the normal infection process.

Virus Infection Can Lead to Cell Transformation

Infections by certain types of DNA or RNA viruses can lead to transforming infections—those causing unregulated cell growth, or **cell transformation.** Transformed cells are distinguished from normal cells by changes in metabolism, the appearance of new antigens on the plasma membranes, altered morphology, and a loss in **contact inhibition.** Contact inhibition is a characteristic of animal cells in tissue culture. Normal cells in tissue culture move in a random amoeboid manner until the ruffled edges on their membranes touch. This contact slows down and inhibits further cell movement and division, resulting in a monolayer of cells. Transformed cells lose this contact inhibition and aggregate into masses of abnormal growth characteristic of **neoplasms,** or **tumors.**

Tumors may be **benign** or **malignant.** Benign (noncancerous) tumors arise when cells lose their ability to stop moving and grow upon contact with similar cell types, but still exhibit contact inhibition with other cell types. Malignant (cancerous) tumors, in contrast, consist of cells that do not respond to contact inhibition from either their own cell types or other cell types. Benign tumors, although rarely fatal, can grow large enough to interfere with normal host function. There are records of benign tumors weighing as much as 20 kg and displacing body organs. Malignant cells spread through the body via other tissues, blood, and the lymphatic system and therefore present a more serious problem to the host. Neoplasms, whether benign or malignant, are further categorized by their location in the organism. **Carcinomas** arise from epithelial tissue; **fibromas** from fibrous connective tissue; **melanomas** from pigment (melanin) cells; and **sarcomas** from connective tissue found in such areas as bone, muscle, and lymph nodes.

Viral Structure and Function
Viral Pathogenesis: Host Cell Damage • pp. 12–13

Oncogenic Viruses Cause Tumors

Both DNA viruses and RNA viruses are known to cause certain types of neoplasms in animals. Examples of **oncogenic** (tumor-causing) DNA viruses are papovaviruses, adenoviruses, and herpesviruses. **Simian virus 40 (SV40),** a papovavirus that was originally isolated from monkey kidney cells, causes tumors in the baby hamster. Adenoviruses of human or animal origin in many instances are oncogenic for newborn rodents. Certain types of herpesviruses are naturally oncogenic for animals and at least two of these viruses, **herpes simplex virus type 2** and **Epstein-Barr virus (EBV),** show possible evidence of involvement in human cancers. Two types of herpes simplex virus infect humans: **herpes simplex virus type 1** causes fever blisters and cold sores, and herpes simplex virus type 2 (or genital herpes, as it is commonly known) is primarily responsible for sexually transmitted genital infections (although herpes simplex virus type 1 can also cause genital infections). Epidemiological evidence has revealed not only that women with herpes simplex type 2 infections have a higher-than-average rate of cervical carcinoma, but also that some women having cervical carcinoma have high herpes simplex type 2 antibody titers. This evidence is only suggestive at this point, not definitive. Epstein-Barr virus, the causative agent of infectious mononucleosis, is another virus associated with human malignant disease. EBV has been found in tumor cells from patients with Burkitt's lymphoma, a neoplasm of lymphoid tissue found especially in children living in central Africa, and nasopharyngeal carcinoma.

 Viral Structure and Function
Viral Disease: Enveloped DNA Virus • pp. 6–8

The retroviruses (RNA viruses) have been found to be oncogenic. One of these oncogenic retroviruses, **Rous sarcoma virus** (an oncornavirus), has been extensively studied. It was first isolated by Francis P. Rous and produces malignant tumors in birds. It also is one of the RNA viruses in which the enzyme reverse transcriptase was first discovered in 1970 by David Baltimore of the Massachusetts Institute of Technology, and Satoshi Mizutani and Howard M. Temin of the University of Wisconsin.

In recent years another retrovirus, the **human T cell lymphotrophic virus (HTLV),** has been strongly implicated in certain forms of human leukemia involving white blood cells called T cells. Robert Gallo of the National Cancer Institute has found HTLV antibodies in individuals with T-cell cancers. These and other findings suggest a clear association between HTLV and T-cell changes. At least three types of HTLVs have thus far been discovered. **HTLV-I** and **HTLV-II** are leukemia-lymphoma viruses. **HTLV-III** (now known as **human immunodeficiency virus,** or **HIV 1**) causes acquired immune deficiency syndrome (AIDS) (see AIDS, page 536). It has been suggested that the AIDS virus may have originated from a species of monkey (the African green) that carries a virus **(STLV-III,** for **simian T lymphotrophic virus type III)** that is remarkably similar to HIV 1. STLV-III does not cause illness in the African green monkey, but other monkeys and humans do get sick from it; AIDS is endemic (continuously present in the population) in central Africa, where the African green monkey is found. A second virus associated with AIDS, **HIV 2,** was discovered in 1986 in prostitutes in Senegal, West Africa. HIV 2 is related to a simian immunodeficiency virus (**SIV**) found in monkeys and appears to produce less severe disease than does HIV 1.

How do these oncogenic viruses transform cells? One theory states that cancer genes (**oncogenes**) on the chromosome cause cell transformation. It is easier to envision how DNA viruses could carry oncogenes in their genome and insert such cancer-causing genes directly into the host genome during infection. RNA viruses, however, present a problem with their RNA genome. For many years scientists questioned how genes in an RNA virus could be inserted into host DNA. It is now known that retroviruses contain reverse transcriptase. This enzyme transcribes viral RNA into DNA, which can subsequently be incorporated into the host DNA. This viral-coded DNA is sometimes known as a **provirus.** Reverse transcriptase, however, is not always associated with transformation; oncogenic RNA viruses possess reverse transcriptase, but not all viruses with reverse transcriptase are oncogenic. The enzyme nonetheless is important in cancer and explains the involvement of retroviruses in this disease.

Cancer Can Be Caused by Nonviral Factors

Viruses are not the only causative agents of cancer. Chemicals, radiation, and environmental factors are known to be intimately associated with different types of human cancers and may, in fact, induce oncogenes. Many of these carcinogens (cancer-causing agents) are commonly found in the environment and in industries. The addition of chemical cancer-causing agents to processed foods for preservation or flavor is particularly alarming and has come under close scrutiny in recent years. Governmental agencies have also monitored chemical pollution of the environment, especially where the chemicals are known tumor-inducing agents.

Cancer is a most-feared disease and is second only to cardiovascular disease as the greatest killer of our population. It is estimated that one out of every four people in the United States will develop cancer during their lifetime, with the majority of these individuals dying from the disease. The mortality rate of cancer in this country has nearly doubled in the last decade.

Despite these alarming statistics, there have actually been significant advances in the detection and treatment of human neoplasia (abnormal tissue growth) in the past few years. Acute lymphatic leukemia, at one time considered to be incurable, can now be treated with chemotherapy, with 50% of treated children living five years or more. Although no definitive vaccine or cure for cancer is available at this time, surgery, radiation therapy, and combination immunotherapy and chemotherapy regimens are often successful in arresting malignancies. As additional information is generated on the epidemiology of cancer—particularly on

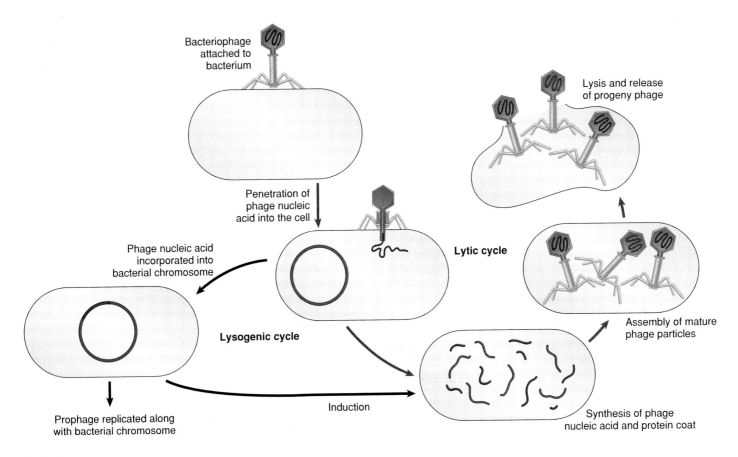

figure 13.21

Lysogeny

In a normal lytic cycle of infection, the bacteriophage replicates within the host cell, and progeny phage are released upon cell lysis. In a lysogenic cycle, the phage nucleic acid is incorporated into the bacterial chromosome and is replicated along with the chromosome. The prophage remains a part of the bacterial chromosome until it is induced to leave and enter a normal lytic cycle.

the role of viruses and other carcinogenic agents in the cause of different cancers—it should be possible to control and even eliminate many forms of this deadly disease.

A Bacteriophage Can Be Incorporated into the Host Chromosome in Lysogeny

A counterpart to transforming infections in eucaryotes is lysogeny in bacteriophage-bacteria interactions (Figure 13.21). Unlike normal lytic cycles of infection, in which bacteriophages replicate within the host cell and their progeny are released upon cell lysis, lysogeny involves incorporation of phage DNA into the host cell genome. This viral DNA, known as a **prophage,** is replicated along with the host DNA and vertically passed to daughter cells.

Not all bacteriophages are capable of lysogeny. Those that can enter such a relationship as well as participate in lytic infections are called **temperate,** to distinguish them from **virulent**

phages, which are capable only of lytic infections. It is difficult to determine if a bacterium carries a prophage. Bacteria containing prophages are called **lysogenic,** or **lysogens,** and can usually be identified as prophage carriers only when the prophage is released from the host genome and produces progeny phage particles upon resumption of the lytic cycle.

Lysogeny occurs with numerous types of bacteria and bacteriophages but has been extensively studied with the temperate bacteriophage λ, which infects *E. coli*. Bacteriophage λ has a shape similar to T-even bacteriophages, with a slender tail and a head filled with linear double-stranded DNA, but no tail fibers. Infection of a sensitive *E. coli* cell by this temperate phage results in one of two possible fates: a normal lytic cycle ensues, or the phage DNA is integrated into the bacterial genome during lysogeny. Whether or not lysogeny occurs is under the control of the infecting phage and depends on the activity of two repressors coded by a regulatory region on the phage genome. Should lysogeny be the choice, the phage DNA is circularized and integrated into the bacterial genome. Insertion of the phage nucleic acid always occurs in the same location on the host genome for any given *E. coli* cell. Whereas the phage genes are usually not expressed in this integrated form, the prophage is replicated with the bacterial DNA.

The prophage remains a part of the bacterial genome until it is **induced** to leave and enter into a normal lytic cycle. Induction occurs either spontaneously or artificially. Prophages can be artificially induced to enter a lytic cycle by exposure of the bacterial cells to a number of different agents that are known to alter DNA. Ultraviolet radiation, mitomycin C, and X rays are among these inducing agents.

An alternative to induction of a prophage is the **curing** of a lysogenic bacterium. Lysogenic bacteria are cured if their prophages are excised from the host genome and do not replicate to form progeny phage. Curing is accomplished by heavy irradiation or chemical treatment of the host cell. The cured cell no longer has a prophage and is susceptible to reinfection by the same phage or other phage types.

A bacterium that has entered into a lysogenic relationship with a particular phage cannot be reinfected with the same phage. This resistance is controlled by the same regulatory region on the phage genome that determines if lysogeny will occur. Occasionally a prophage may undergo mutations in its DNA that render it unable to produce progeny phage upon induction. These **defective prophages** can be recognized because curing of the lysogen may result in lysis, but with the release of incomplete, noninfectious phage particles. If, however, lysogenic bacteria containing such defective prophages are **superinfected** (infected with a different phage), the second phage acts as a **helper,** in some instances donating its genes to complement the missing genes in the defective prophage. The result is the release of a mixture of defective and normal progeny phage from the lysed cell during induction.

Lysogeny is a phenomenon that has important consequences for bacteria. Lysogens become resistant to superinfection by phages of the same type. More importantly, however, bacteria that integrate phage genes into their genome may assume new characteristics, a process known as **lysogenic conversion.** The pathogen *Corynebacterium diphtheriae*, for example, synthesizes a toxin that destroys epithelial cells lining the respiratory tract, resulting in a pseudomembrane characteristic of the disease diphtheria (see diphtheria, page 507). The toxin is not coded by the bacterial genome but by a gene (*tox* gene) on a temperate phage (**prophage β).** Cells of *C. diphtheriae* synthesize toxin only when they have undergone lysogenic conversion. If such cells are cured of their prophage, they no longer synthesize the toxin, and are incapable of causing disease. The expression of the *tox* gene is controlled by the metabolism and physiological state of the host bacterium. Toxin production is regulated by the exogenous Fe^{2+} supply; maximum toxin is produced only when the exogenous Fe^{2+} supply is low. Although it is not exactly known how Fe^{2+} regulates toxin production, one popular theory is that high exogenous Fe^{2+} concentrations induce the prophage to enter a lytic cycle, resulting in nonexpression of the *tox* gene.

Lysogenic conversions are common among toxin-producing bacteria. The scarlet-fever-causing erythrogenic toxin produced by *Streptococcus pyogenes* (see scarlet fever, page 498) and toxins produced by some strains of *Clostridium botulinum* (see botulism, page 517) are coded by prophages. In many cases lysogenic conversions are directly responsible for the virulence of pathogenic bacteria.

In 1957 two physicians, D. Carleton Gajdusek and Vincent Zigas, described a progressive degenerative disease confined to a small population in the eastern highlands of Papua New Guinea. This disease, characterized by cerebral ataxia (incoordination), neuronal loss, and a shivering-like tremor, was named **kuru** ("shivering," or "trembling," in the Foré language of the afflicted people).

Kuru was the first human degenerative disease shown to be caused by a prion, a finding that opened a new, exciting frontier in microbiology. Unlike conventional viral diseases, kuru has unique characteristics (Table 13P.1).

The kuru prion resembles viruses in its filterability (filterable in the 25-nm to 50-nm average pore diameter range), reaches high titers (up to 10^8 to 10^{10} infectious units per gram of tissue) in the brains of experimentally infected animals, and restriction of host range (humans are the only apparent natural hosts, although chimpanzees and other nonhuman primates have been experimentally infected). However, the kuru prion

table 13P.1

Properties of Kuru

- Long incubation period.
- Lack of immunogenicity or host immune response.
- Chronic progressive illness without remission or relapse.
- Induction of spongiform change, neuronal loss, and gliosis.
- Absence of specific viral particle by electron microscopy.
- Specific nucleic acid not yet identified, and resistance to procedures that attack nucleic acids (resistant to low pH, nucleases, ultraviolet irradiation at 237 nm, zinc hydrolysis, photochemical inactivation with psoralens, and chemical modification by hydroxylamine).
- Unaffected by immunosuppression, immunopotentiation, or most antiviral drugs; splenectomy or other factors that alter splenic function may vary the course of infection.
- No cytopathic effect in tissue culture.
- Unusual spectrum of resistance to certain physical and chemical treatments.

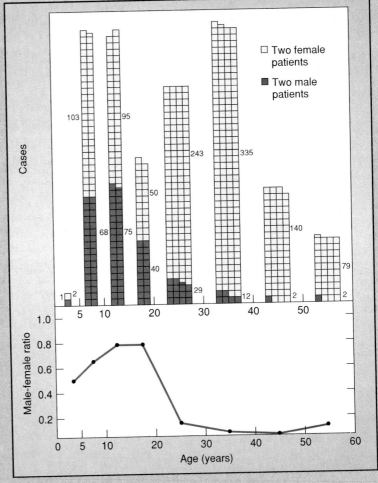

figure 13P.1

Age and Sex Distribution of the First 1,276 Kuru Patients Studied in Early Epidemiological Investigations

These early studies suggested that kuru was confined primarily to women and children. In recent years, age and sex patterns of the disease have completely changed; kuru is now found only among young adults (who were infected as children), not among the children.

has other, unconventional properties that are not associated with viruses. It is unusually resistant to ultrasonication, heat, ionizing radiation, and a variety of chemical agents. Furthermore, the kuru agent is resistant to procedures that attack nucleic acids: it has resistance to low pH, nucleases, ultraviolet irradiation at 237 nm, zinc hydrolysis, and chemical modification by hydroxylamine. This indicates that it does not possess nucleic acid.

Initial epidemiological studies showed that kuru was a disease confined primarily to the Foré women and children (Figure 13P.1). Further studies revealed that the disease was transmitted by ritual cannibalism among the tribes of this region. As a rite of mourning and respect for dead relatives and clansmen, women and children ate the viscera and highly infectious brains of the corpses. With missionary intervention in the late 1950s, cannibalism ceased among the Foré people and the incidence of kuru also decreased (Figure 13P.2). Since 1957 kuru has been disappearing gradually. Today it occurs only among the young adults (who were infected as children 40 years ago), not among the children. No one born in a village since the cessation of cannibalism has ever developed kuru. Because there is no other known reservoir besides humans for the kuru prion, it is predicted that kuru will eventually be eliminated as a disease. However, it will always remain as a model of diseases caused by prions.

Sources

Gajdusek, D.C. ©The Nobel Foundation 1977. Unconventional viruses and the origin and disappearance of kuru. *Science* 197:943–960.

Prusiner, S.B. 1995. The prion diseases. *Scientific American* 72:48–57.

References

Alpers, M.P. 1979. Epidemiology and ecology of kuru. In *Slow transmissible diseases of the nervous system*. Vol. 1, *Clinical, epidemiological, genetic, and pathological aspects of the spongiform encephalopathies*. S.B. Prusiner, and W.J. Hadlow, eds. 67–90. New York: Academic Press.

Gajdusek, D.C., and V. Zigas. 1957. Degenerative disease of the central nervous system in New Guinea: the endemic occurrence of "kuru" in the native population. *New England Journal of Medicine* 257:974–978.

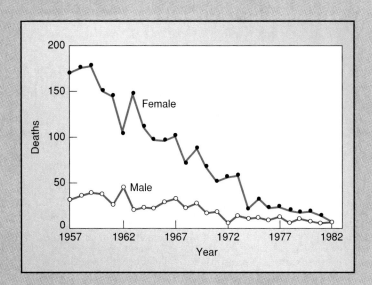

figure 13P.2
The Overall Incidence of Kuru Deaths in Male and Female Patients by Year Since Its Discovery in 1957

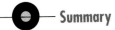

1. Viruses are intracellular parasites that metabolize and reproduce only within a host cell. All viruses have genetic material in the form of DNA or RNA surrounded by a protein coat (capsid). Some animal viruses also have a membranous envelope.

2. Viruses have three basic forms: polyhedral, helical, and binal. Animal and plant viruses display polyhedral or helical symmetry. The T-even bacteriophages have a binal structure that consists of three parts: a head composed of protein subunits surrounding the viral genome, a narrow sheath, and a tail consisting of thin tail fibers.

3. Bacteriophages are classified by their morphology and nucleic acid content. The T-even bacteriophages are binal phages containing double-stranded DNA. Coliphage φX174 contains a single-stranded circular DNA genome within an icosahedral capsid. Coliphage fd is a filamentous bacteriophage with a helical symmetry and single-stranded circular DNA.

4. The plant viruses include Caulimovirus, a double-stranded DNA virus that infects cauliflower, and Tobamovirus, a single-stranded DNA virus that causes tobacco mosaic disease.

5. Animal viruses are responsible for diseases such as severe hemorrhagic fever (Ebola virus), AIDS (human immunodeficiency virus), influenza (influenza virus), and measles (rubeola virus).

6. Animal and plant viruses may be cultivated in the laboratory in tissue cultures that are derived from cell lines. Viruses also can be propagated directly on animal or plant tissue.

7. Bacteriophages are propagated on lawns of suitable host bacteria. Clear plaques form in areas of the bacterial lawn where bacteriophages have infected and lysed the bacterial cells.

8. Virus replication consists of several steps: attachment of the virion to the host (adsorption), penetration of the virus or its nucleic acid into the cell (and uncoating in animal and plant viruses), replication of the viral nucleic acid and synthesis of viral proteins, assembly and/or maturation of new virus particles, and release of mature virus particles from the cell.

9. In addition to lytic infections, virus-host cell interactions can result in persistent infections, abortive infections, transforming infections, and lysogeny.

10. Lysogeny is the incorporation of bacteriophage DNA into the host cell genome. The incorporated bacteriophage DNA, or prophage, remains a part of the cell genome until it is induced to leave and enter into a normal lytic cycle. One consequence of lysogeny is the expression of the bacteriophage genes in the host bacterium, or lysogenic conversion.

EVOLUTION *and* BIODIVERSITY

Ebola virus, Dengue virus, and human immunodeficiency virus are all examples of emerging viruses that threaten the public health and kill thousands of people each year. Forty years ago these viruses either were not known or were not of major concern. Where did these viruses come from and what happened during the past four decades that resulted in this rapid rise in disease cases? Viruses, like procaryotes and eucaryotes, have evolved over the years and, in many cases, have become adapted to modern society. HIV is believed to have arisen from the simian T-lymphotrophic virus type III that infects African green monkeys. Ebola virus, one of the most deadly viruses known, is also carried by the African green monkey and can rapidly spread through a population as it did in Kikwit, Zaire, in the spring of 1995, causing massive hemorrhaging and multiple organ failure, and killing more than 90% of its victims. The Dengue virus, although known for many years, has recently emerged to cause vast epidemics of Dengue fever in Latin America. It seems that new viruses are discovered each year that are among the most dangerous infectious agents known. What appears to be new viruses in many cases are existing viruses that have adapted to changes in environmental conditions and new hosts. Viral genomes, which are significantly smaller than the genomes of procaryotes and eucaryotes, undergo mutations relatively easily. These mutations can lead to increased virulence and new diseases. Viruses may be the smallest infectious agents, but clearly are also among the most dangerous agents known to humankind.

Questions

Short Answer

1. Describe the basic structure of viruses.

2. Explain how viruses differ from other organisms.

3. Why are viruses called "obligate intracellular parasites"?

4. Prove (give evidence) that viruses are living.

5. Prove (give evidence) that viruses are nonliving.

6. Some viruses possess a membranelike envelope. Explain.

7. How do viroids differ from virions?

8. What criteria are used to classify viruses?

9. Why does the host figure so prominently in viral classification?

10. How can these obligate intracellular parasites be studied in the laboratory? What happens to the virus if it is not given a host?

11. How can viruses be enumerated?

12. Identify, in sequence, the steps associated with viral replication.

13. Briefly describe several methods of replication for viral nucleic acids.

14. Which viruses possess reverse transcriptase? What is the role of this enzyme?

15. What would be the fate of the viral progeny if the host cell were lysed prior to their maturation and release?

16. Compare and contrast the lytic and lysogenic life cycles of bacteriophages.

17. How does the life cycle of a plant or animal virus differ from the life cycle of a bacteriophage?

18. What is cell transformation and how is it evidenced?

Multiple Choice

1. T-even bacteriophages are:
   a. binal
   b. helical
   c. polyhedral

2. Which of the following requires host enzymes?
   a. assembly/maturation
   b. attachment (or adsorption)
   c. penetration (and uncoating for some viruses)

   d. replication/synthesis
   e. release

3. Which of the following is an example of a DNA virus?
   a. Ebola virus
   b. herpes simplex virus
   c. influenza virus
   d. human immunodeficiency virus

Critical Thinking

1. Describe early efforts to classify viruses. Discuss the current state of classification for viruses. Describe the system you would like to see in 10 or 20 years.

2. Given the high degree of host specificity, comment on the evolutionary links between viruses and their hosts. Research what we know about restriction endonucleases and transduction. Does this change your answer?

3. Identify several forms of cancer known or suspected to be caused by viral infections. What is necessary to prove the oncogenic role of a virus? What is known about the connection between some viruses and some forms of cancer?

4. Emerging diseases (for example, Ebola, Dengue fever, AIDS) are covered in the newspapers on a daily basis. Why are these diseases "emerging"? From a public health perspective, how would you prevent these diseases? What factors must you overcome to be successful?

 Supplementary Readings

Berns, K.I. 1990. Parvovirus replication. *Microbiological Reviews* 54:316–329. (A discussion of the replication of parvovirus.)

Birge, E.A. 1994. *Bacterial and bacteriophage genetics.* New York:Springer-Verlag. (An in-depth discussion of the genetics of bacteria and bacteriophages, including mechanisms for replication of phage nucleic acids.)

Bishop, J.M. 1989. Viruses, genes, and cancer. *American Zoologist* 29:653–666. (A review of the role of viruses and oncogenes in cancer.)

Diener, T.D. 1980. Viroids. *Scientific American* 244:66–73. (An excellent review article on viroids.)

Fields, B.N., and D.M. Knipe, ed. 1991. *Fundamental virology,* 2d ed. New York: Raven Press. (A comprehensive discussion of animal and human viruses.)

Gallo, R.C. 1987. The AIDS virus. *Scientific American* 256:47–56. (A review of the human immunodeficiency virus.)

Johnson, H.M., F.W. Bazer, B.E. Szente, and M.A. Jarpe. 1994. How interferons fight disease. *Scientific American* 70:68–75. (A review of the mode of action of interferons and their use in the treatment of infectious diseases and some forms of cancer.)

Karam, J.D., ed. *Molecular biology of bacteriophage T4.* Herndon, Va.: ASM Press. (An in-depth discussion of bacteriophage T4 and its value in research and teaching.)

Le Guenno, B. 1995. Emerging viruses. *Scientific American* 73:56–64. (A review of newly discovered hemorrhagic fever viruses, including Ebola virus, Hantavirus, and Marburg virus.)

Levy, J.A., H. Fraenkel-Conrat, and R. Owens. 1994. *Virology,* 3d ed. Englewood Cliffs, N.J.: Prentice-Hall. (A textbook of virology.)

Matthews, R.E.F. 1991. *Plant virology,* 3d ed. New York: Academic Press. (A textbook of plant viruses.)

Porterfield, J.S., ed. 1995. *Exotic viral infections.* London: Chapman & Hall. (A detailed description of diverse groups of viruses that cause a variety of human diseases.)

SYMBIOSIS

chapter fourteen

Mutualism
Nitrogen Fixation
Lichen
Flashlight Fishes
Symbionts of Protozoa
Symbionts of Insects
Symbiosis in Ruminants

Parasitism
Bdellovibrio
Chlamydiae

Commensalism

**Mutualism, Commensalism, and Parasitism—
    Dynamic Relationships**

PERSPECTIVE
Increasing Crop Yield Through Nitrogen Fixation

EVOLUTION AND BIODIVERSITY

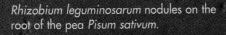

Rhizobium leguminosarum nodules on the
root of the pea *Pisum sativum.*

 Microbes in Motion PREVIEW LINK

Among the key topics in this chapter is the developmental cycle of
Chlamydia. The following section in the *Microbes in Motion* CD-ROM may
be useful as a preview to your reading or as a supplemental study aid:

Miscellaneous Bacteria: Chlamydia (Growth and Metabolism), 9–14.

icroorganisms have evolved over billions of years to adapt themselves not only to the surrounding physical environment, but also to the presence of other organisms living in the environment. This association may be with plants, animals, or other microbes. Microorganism adaptation to living with other forms of life has led to close relationships that often are essential for survival of one or both members.

The term **symbiosis** [Greek *sym*, together, *bios*, life] is used to describe an intimate association between organisms of different species. The German botanist Heinrich Anton de Bary (1831–1888) first used *symbiosis* in 1879 in describing the close relationship he observed between an alga and a fungus. Other biologists later referred to the relationship de Bary studied more specifically as **mutualism**, because in this symbiosis, the alga and the fungus benefited mutually.

Although the term *symbiosis* is usually restricted to mutualism, it may be extended to include **parasitism** and **commensalism**. A parasitic (**antagonistic**) relationship is one in which one organism, the **parasite**, is benefited and the other organism, the **host**, is harmed. The parasite is generally smaller than the host. Parasitic associations are usually of long duration. Some microorganisms are obligate parasites; their existence depends on a successful host-parasite interaction. For example, rickettsiae, chlamydiae, and viruses are obligate intracellular parasites—they are unable to reproduce outside the host. In **commensalism,** one organism is benefited and the other is unaffected. The dependence of anaerobes on aerobes for oxygen removal is an example of commensalism. Growth of the anaerobes is favored by the lowered oxygen tension, but the aerobes are not affected by the relationship, assuming there is no competition for the same available substrates.

Although symbioses can be categorized into these three types of associations, there is often no clear distinction between these categories. The interactions between populations are frequently dependent on environmental conditions; under different environmental conditions, symbiotic relationships can change. For example, when a host is debilitated by an illness or its microflora has changed because of broad-spectrum antibiotic therapy (see broad-spectrum antibiotics, page 125) or immunosuppressive diseases (see immunosuppression, page 478), mutualistic bacteria can become antagonistic (parasitic) and cause serious disease. A nonpathogen in one individual can become a life-threatening pathogen in another individual. This has become an increasingly common phenomenon with the emergence of infectious diseases such as AIDS, enterohemorrhagic *E. coli* food poisoning, and necrotizing fasciitis caused by *Streptococcus pyogenes* (see human microflora, page 440). There is a very fine line between mutualism and parasitism.

Symbioses may be divided into two groups based on the closeness of the association. In **endosymbiosis,** the microorganism grows within the host cell; in **ectosymbiosis,** the microorganism may be attached to the host cell, but it remains outside. The distinction is not always easy to make. For example, the bacterium *Bdellovibrio*, which parasitizes other gram-negative bacteria, penetrates the cell wall but not the plasma membrane of its host.

Microbes exist in symbiosis with many different kinds of organisms, including plants, animals, and even other microbes. In many of these cases, the relationship with the microbe has a direct impact on the life cycle and survival of the partner in the association. Some common types of relationships are described in this chapter.

Mutualism

Mutualism is a common form of symbiosis involving microbes. Mutualistic associations between microbes and plants, animals, and other microbes are widespread in nature and have been extensively studied.

Rhizobium, Frankia, Anabaena, and Other Bacteria Fix Nitrogen Symbiotically

One of the most thoroughly studied symbiotic associations is the one between gram-negative heterotrophic soil bacteria collectively called rhizobia (*Rhizobium, Bradyrhizobium,* and *Azorhizobium*) and plants of the legume family, such as soybeans, alfalfa, clover, and peas. This association results in the formation of **root nodules** and the biological fixation of atmospheric nitrogen into combined nitrogen, a process that only procaryotes perform. Although biological fixation of nitrogen may occur nonsymbiotically in free-living bacteria (see nitrogen fixation, page 576), an estimated five to ten times more nitrogen is fixed symbiotically. The mutualistic association between rhizobia and legumes is highly specific; nitrogen fixation occurs only if a legume is infected by a specific rhizobial species. The relationship is mutualistic because the plant benefits from the bacterial conversion of gaseous nitrogen into a usable combined form. The plant, in turn, provides the bacterium with nutrients for growth and metabolism. Such symbiotic relationships are important in agriculture and provide legumes with a selective advantage that other plants do not have when grown in nitrogen-poor soil (Figure 14.1).

Much of our knowledge of symbiotic nitrogen fixation has been derived from studies of *Rhizobium*. The symbiosis between *Rhizobium* and a legume is initiated by infection of plant root hairs by the bacteria (Figure 14.2). Nearly all soils contain rhizobia, but outside the host plant, the bacteria are incapable of naturally fixing nitrogen. The roots of leguminous plants secrete flavonoid compounds that chemotactically attract the rod-shaped rhizobia to the **rhizosphere** (the region of soil closely surrounding the roots) (see chemotaxis, page 63). The rhizosphere contains a much greater density of bacteria than is found in the adjacent soil; this ability of leguminous plants to attract and stimulate the growth of soil bacteria is known as the **rhizosphere effect.**

a.　　　　　　　　　　　　　　　　　　　b.

figure 14.1

Benefits of Symbiotic Nitrogen Fixation

Inoculating legumes with *Rhizobium* significantly increases available nitrogen for the plant. a. Legumes inoculated with *Rhizobium*. b. Uninoculated legumes.

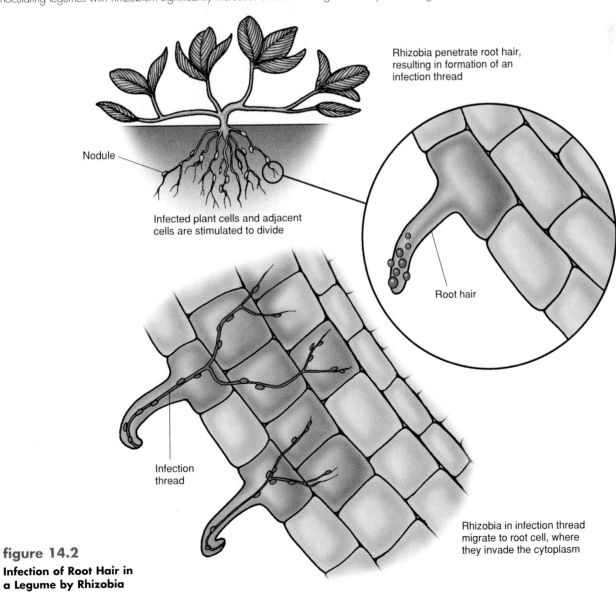

Rhizobia penetrate root hair, resulting in formation of an infection thread

Nodule

Infected plant cells and adjacent cells are stimulated to divide

Root hair

Infection thread

Rhizobia in infection thread migrate to root cell, where they invade the cytoplasm

figure 14.2

Infection of Root Hair in a Legume by Rhizobia

a.

b.

figure 14.3

Rhizobia-Legume Root Nodules

a. Nitrogen-fixing nodules on clover roots. b. Nitrogen-fixing nodules on alfalfa roots.

Rhizobium has a set of nodulation (*nod*) genes, one of which (*nodD*) is constitutively expressed. The *nodD* product (NodD) recognizes specific flavonoids secreted by plant roots. This specific interaction between NodD and particular flavonoids is an important determinant in *Rhizobium*-host plant specificity. In the presence of the flavonoid, NodD binds to a conserved 60-base pair sequence in each *nod* gene promoter and induces transcription of the remaining *nod* genes, which are associated with root hair curling, initiation of nodule development, and other nodulation events.

The *nod* genes are responsible for the synthesis of glycoprotein signal molecules called Nod factors that are secreted by *Rhizobium* and initiate root hair curling. The bacteria infect the plant by penetrating the root hairs and entering via an invagination of the root hair membrane. As this rhizobia-filled area of the membrane extends into and through the root cells, an **infection thread** is established. Only in tetraploid cells (cells containing

twice the normal number of chromosomes) is the invasion successful, causing the formation of red-brown nodules characteristically found in nitrogen-fixing legumes (Figure 14.3). During infection the bacteria enter the cell cytoplasm, where they multiply and are transformed from rods to swollen, misshapen cells called **bacteroids** (Figure 14.4). These are rich in the enzyme **nitrogenase,** which enables them to fix atmospheric nitrogen. The fixation of nitrogen in nodules usually continues until seed formation; it then ceases, and the deteriorating nodules release bacteria into the soil.

Symbiotic nitrogen fixation also occurs with bacteria other than the rhizobia and nonleguminous plants (Table 14.1). *Frankia,* an actinomycete (see *Frankia,* page 343), forms nitrogen-fixing symbioses with the alder (*Alnus*), the Australian pine (*Casuarina*), the autumn olive (*Elaeagnus*), the California lilac (*Ceanothus*), the bayberry (*Myrica*), and over 180 other plants. Plants in symbiotic

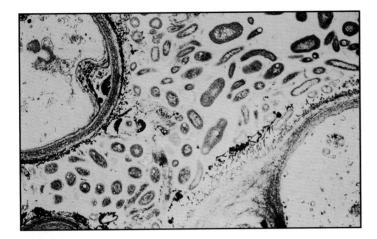

figure 14.4

Bacteroids of *Rhizobium leguminosarum* in an infected pea root

The numerous dark objects are bacteroids (transmission electron micrograph, colorized, ×7,000).

figure 14.5

Frankia-Alder Root Nodules

Nitrogen-fixing nodules on the roots of the alder *Alnus*.

table 14.1

Some Symbiotic Nitrogen-Fixing Procaryotes

Organism	Host
Bacteria	
Azotobacter paspali	Paspalum notatum (sand grass)
Bradyrhizobium japonicum	Glycine max (soybean)
Frankia alni	Alnus (alder)
Frankia casuarinae	Casuarina (Australian pine)
Frankia ceanothi	Ceanothus (California lilac)
Rhizobium leguminosarum	Lens culinaris (lentil), Pisum sativum (garden pea), Vicia faba (broad bean)
Rhizobium lupini	Lupinus (lupine)
Rhizobium meliloti	Medicago sativa (lucerne), Melilotus officinalis (melilot)
Rhizobium phaseoli	Phaseolus (bean)
Rhizobium trifolii	Trifolium (clover)
Azospirillum lipoferum	Digitaria decumbens (tropical/subtropical grass)
Cyanobacteria	
Anabaena	Azollo (water fern)
Nostoc	Gunnera (angiosperm), Macrozamia (gymnosperm)

association with *Frankia* are known as actinorhizal plants (Figure 14.5). The extensive distribution of *Alnus* and *Casuarina* suggests that they play a major role in the nitrogen cycle. *Frankia*-infected alder trees can live in nitrogen-poor regions, while other plants grow poorly in such areas. An alder was the first tree to colonize Mount Saint Helens in southwestern Washington after the 1980 volcanic eruption.

Rhizobia-legume and *Frankia*-plant symbioses are examples of endosymbioses. Although they both form root nodules in the host plant, rhizobia and *Frankia* are quite different and are good examples of convergent evolution. *Frankia* does not form bacteroids, and its appearance inside the root nodule is the same as free-living cells. *Frankia* infects a broad spectrum of plants, whereas rhizobia form symbioses specifically with legumes.

An example of an ectosymbiotic relationship involving nitrogen fixation is the association of *Anabaena azollae* and *Azolla*. *A. azollae* is a nitrogen-fixing cyanobacterium that infects the small floating water fern *Azolla* and provides organic nitrogen to it (Figure 14.6). *Azolla* populations infected with *A. azollae* contain as much as 20 kg of fixed nitrogen per acre. In the Far East, the *Anabaena*-infected *Azolla* are placed in rice paddies and increase rice yields by as much as 150% per year. As the nitrogen-laden ferns die, they release nitrogen for use by the rice plants. More than 3 million acres of rice paddies are planted with *Azolla* in the People's Republic of China, providing 100,000 metric tons of nitrogen fertilizer worth more than $100 million annually.

figure 14.6

Anabaena-Azolla Symbiosis

Scanning electron micrograph showing filaments of the cyanobacterium *Anabaena azollae* in symbiosis with the water fern *Azolla*. The larger, oval cells in the filaments are heterocysts (colorized).

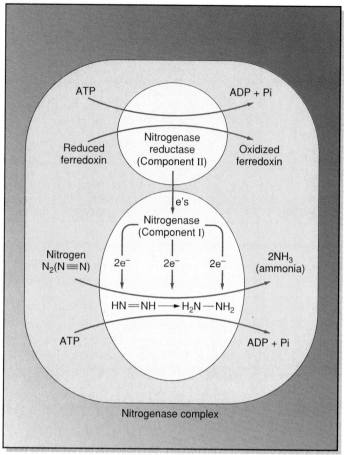

figure 14.7

Scheme for Nitrogen Fixation in Microorganisms

Electrons are transferred from ferredoxin to Component **II** (nitrogenase reductase) of the nitrogenase complex. This transfer is followed by ATP hydrolysis and the transfer of electrons to Component **I** (nitrogenase). The reduced nitrogenase can then reduce nitrogen to ammonia.

Nitrogen Fixation Requires Large Amounts of Energy

The biological fixation of molecular nitrogen requires a large expenditure of energy. As many as 20 moles of ATP in *Clostridium pasteurianum* and 30 moles of ATP in *Klebsiella pneumoniae* are diverted to fixation of 1 mole of nitrogen. This vast amount of energy is needed to reduce the triply bonded nitrogen molecule ($N \equiv N$) to two ammonia (NH_3) molecules. The enzyme associated with this process is **nitrogenase.**

Nitrogenase is actually a binary enzyme (**nitrogenase complex**) consisting of two distinct proteins, Components I (**nitrogenase**) and II (**nitrogenase reductase**). Both components contain iron. The larger of the two proteins (Component I) also has two atoms of the metal molybdenum per molecule. The two proteins are so remarkably similar in different bacteria that Component II from *Azotobacter chroococcum* (a bacterium that is widespread in soil) can be combined with Component I from *Klebsiella pneumoniae* (also a common soil bacterium, but often present in mammalian intestines) and get a fully active hybrid nitrogenase enzyme complex. The complete nucleotide sequence of the *K. pneumoniae* gene region for nitrogen fixation (**nif genes**) has been determined. This region consists of 20 known *nif* genes that encode Component I and Component II of nitrogenase, a cofactor that contributes the iron and molybdenum in Component I, and regulatory proteins (see *K. pneumoniae nif* genes in the Perspective, page 434).

Several essential ingredients are required for the nitrogenase reaction to occur (Figure 14.7). Besides ATP, the reaction requires Mg^{2+} (for ATP hydrolysis reactions) and a reducing agent. Specialized proteins called ferredoxins are the reducing agents for nitrogen fixation in living cells. Electrons are transferred from ferredoxin to nitrogenase reductase. This transfer is followed by ATP hydrolysis and the transfer of electrons to nitrogenase. The reduced nitrogenase can now reduce nitrogen to ammonia. Reduced ferredoxin is thus the source of electrons for the reduction of nitrogen to ammonia.

The nitrogenase enzyme is irreversibly destroyed when exposed to oxygen. How, then, can aerobic microorganisms fix nitrogen? Microbes have evolved various mechanisms to protect the oxygen-sensitive nitrogenase. Facultatively anaerobic bacteria such as *Klebsiella* and *Enterobacter* usually fix nitrogen only anaerobically. Aerobic bacteria and a few facultatively anaerobic nitrogen-fixing bacteria decrease the oxygen level of their environment through respiration. The free-living nitrogen-fixing bacterium *Azotobacter*, for example, has the highest respiration rate of any living organism. High respiration rates rapidly remove oxygen, protecting nitrogenase from damage and also generating the large amounts of ATP required for nitrogen fixation. *Derxia gummosa,* an aerobic, nitrogen-fixing bacterium found in tropical soils in Asia, Africa, and South America, has a normal respiration rate but produces a thick polysaccharide slime layer that restricts the flow of oxygen into the cell. *Anabaena* compartmentalizes the nitrogenase enzyme in specialized cells called heterocysts that lack photosystem II, the oxygen-evolving photosystem. In symbiotic nitrogen fixation by *Rhizobium* species, an oxygen-binding pigment, **leghemoglobin,** removes oxygen from the vicinity of the nitrogenase enzyme while at the same time providing oxygen for energy generation by the nitrogen-fixing bacterium. Leghemoglobin is a red, hemoglobinlike protein produced only in the presence of *Rhizobium*-legume symbiotic interaction. Its production carries across organismal lines: the heme is produced by the bacterium and the globin by the plant.

Nitrogenase is not specific for nitrogen and reduces several other triply-bonded compounds such as cyanide, nitrous oxide, methyl isocyanide, and acetylene. The reduction of acetylene ($HC\equiv CH$) is particularly important because the product, ethylene ($H_2C=CH_2$), is easily and readily detected by gas chromatography. Acetylene reduction and the subsequent appearance of detectable ethylene thus provides a quantitative measure of the nitrogen-fixing capacity of a biological system.

A Lichen Is a Mutualistic Association Between a Fungus and an Alga

The term **lichen** describes a plantlike structure made up of a fungus and an alga or a cyanobacterium living in mutualistic symbiosis. Lichens are common in a variety of mostly hostile environments. More than 20,000 characterized types of lichens are found in such diverse places as deserts and the Arctic and Antarctic, where they frequently are the most noticeable forms of life. Lichens known as reindeer mosses (*Cladonia subtenuis*) cover vast regions of land in the Arctic and are the primary source of food for such animals as reindeer and caribou. These structures grow on bare rocks, tree trunks, soil, and house roofs; they display a variety of colors ranging from red to brown to yellow to black

(Figure 14.8). Algal pigments are primarily responsible for the coloration. In some instances the fungus may also contribute to pigmentation. Lichen pigments are used in dyes, perfumes, and other commercial products.

In the lichen relationship, the fungus, generally an ascomycete, provides a solid support to anchor the lichen to barren surfaces. Fungal hyphae are tightly packed around algal cells to form the body of the lichen, termed the **thallus.** Specialized absorptive hyphae called **haustoria** are often found penetrating the algal cells, serving as channels of nutrient exchange between the fungus and the alga. The fungus additionally is important in the symbiotic relationship because it absorbs water from the environment. Such absorption enables the lichen to survive in environments where water is scarce. The alga, in turn, contributes to the relationship by carrying out photosynthesis and providing organic compounds (as well as combined nitrogen from nitrogen fixation in the case of cyanobacteria) for use by the fungus. The relationship is considered a mutualistic association that is beneficial to both organisms and enables the lichen to survive under conditions that neither the fungus nor alga could tolerate alone.

Because lichens obtain their nutrients virtually only from what is transported via the atmosphere, they are very sensitive to pollution. Lichens were indicators of the 1986 Chernobyl nuclear disaster in the former Soviet Union since they concentrated the radioactivity produced that was swept by air currents around the world, especially to the Arctic. A tragic feature of this nuclear accident was that large portions of the caribou herds of the Laplanders had to be destroyed since they had fed on the contaminated lichens.

A Mycorrhiza Is a Mutualistic Association Between a Fungus and a Plant

Certain types of soil fungi are important in the nutrition of vascular plants. These fungi are closely associated with the roots of these higher plants and significantly increase the absorption area of the roots for minerals and water. This association is known as a **mycorrhiza,** a word that literally means "fungus root."

Two major types of mycorrhizae are found in nature. **Endomycorrhizae** are the more common type and occur in approximately 80% of all vascular plants. The importance of endomycorrhizae was first recognized in orchids, which grew poorly in greenhouses unless they were infected by fungi. In endomycorrhizae, organic compounds secreted into the soil by the plant attract fungi toward the plant roots. The fungal hyphae penetrate the cortical cells of the plant root and extend into the surrounding soil in an endomycorrhiza. Because the hyphae often form intracellular vesicles and branching structures called arbuscules in the cortical cells, endomycorrhizae are also known as vesicular-arbuscular mycorrhizae. In **ectomycorrhizae,** the plant roots are surrounded but not penetrated by fungal hyphae (Figure 14.9). Ectomycorrhizae are typically found in trees and shrubs, particularly in temperate forests. Because mycorrhizae are highly susceptible

a.

b.

c.

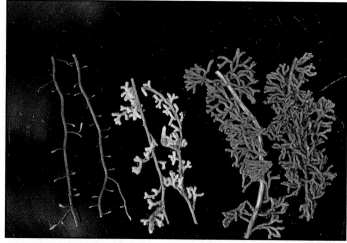

a.

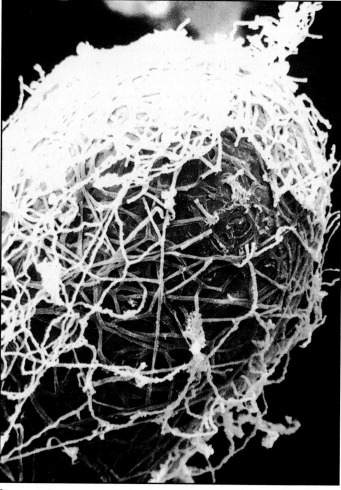

b.

figure 14.8
Lichens in Their Natural Environments
a. Lichens on a cemetery gravestone. b. Fence and tree in Monteverde, Costa Rica, covered with lichens and mosses. c. Limestone walls and stele in an Aztec courtyard at Tikel, Guatemala, covered with fungal growth, algae, and lichens. This growth contributes to the slow breakdown of the limestone.

figure 14.9
Ectomycorrhizae
a. Shown are nonmycorrhizal roots (left), white mycorrhizae (middle), and yellow-brown mycorrhizae (right). b. Scanning electron micrograph of ectomycorrhizae of *Boletus* (a mushroom) on *Populus* (an aspen) (colorized, ×200).

figure 14.10
Effect of Mycorrhiza on Plant Growth
The five-month-old lemon trees on the right were grown in symbiotic association with a mycorrhizal fungus, *Glomus intraradices*. The trees on the left were grown without the fungus.

figure 14.11
Bacterial Luminescence
The flashlight fish *Photoblepharon palpebratus*, showing luminescent bacteria in the light organ.

to damage from acid rain, this susceptibility has raised concerns among environmentalists regarding the burning of fossil fuels and its effect on acid deposition in the atmosphere.

A mycorrhizal relationship is beneficial to both the fungus and the plant. The fungus benefits from the carbohydrates made available to it by the plant, and the plant benefits from the increased absorption area provided by the fungus (Figure 14.10). Most plants grow and easily absorb nutrients from the surrounding environment if such nutrients are readily available. However, it has been experimentally observed that under limited nutrient conditions, certain elements such as nitrogen and phosphorus are not readily transferred from the soil to the plant roots. Plants having a mycorrhizal relationship absorb three to five times as much phosphorus (in the form of phosphates) as uninfected plants. Mycorrhizae are especially important in nutrient-poor and water-limited environments, such as deserts, where plants could not survive without this symbiotic relationship. Fossils of early vascular plants from 400 million years ago show evidence of mycorrhizal symbioses. This has led to the suggestion that the evolution of such associations may have been a critical step in enabling plants to make the transition to a relatively sterile and nutrient-poor soil during colonization of the land.

Flashlight Fishes Have a Symbiotic Relationship with Luminescent Bacteria

One of the most interesting symbiotic relationships involves luminescent bacteria, which include organisms such as *Photobacterium* and *Vibrio*, and fishes of the family *Anomalopidae*. These

fishes are called **flashlight fishes** because of a specialized organ beneath each eye that is packed with luminescent bacteria (Figure 14.11). As many as 10^{10} bacteria per milliliter of fluid are found in these light organs. The light emitted by these bacteria is mediated by the enzyme **luciferase,** which catalyzes the oxidation of reduced flavin mononucleotide (FMNH$_2$) and a long-chained aliphatic aldehyde (RCHO) by the following reaction:

$$FMNH_2 + RCHO + O_2 \xrightarrow{\text{luciferase}}$$
$$FMN + H_2O + RCOOH + h\nu \text{ (light)}$$

Light is emitted continuously by the bacteria, provided oxygen is present, although the light can be turned on and off by a black shutter the fish can raise or lower over each organ. The light emitted by these specialized organs is used to communicate, lure prey, and confuse predators.

Some Bacteria Are Symbionts of Protozoa

One of the most unusual and fascinating mutualistic associations is that of *Paramecium aurelia* and its bacterial endosymbionts (Figure 14.12). These bacteria, designated by letters of the Greek alphabet, are highly host specific (Table 14.2). Infection appears to depend on one or two specific genes in the protozoan.

The first endosymbiont discovered and genetically characterized was the "**killer factor,**" given the designation κ (kappa). Paramecia infected with κ release toxic particles (paramecins) that are not harmful to the killer paramecia, but are lethal for other, **sensitive** strains. Infected cells of *P. aurelia* contain hundreds of κ. Between 5% and 30% of these endosymbionts have refractile (R) bodies, which appear as tightly-rolled ribbons. The R bodies are

a.

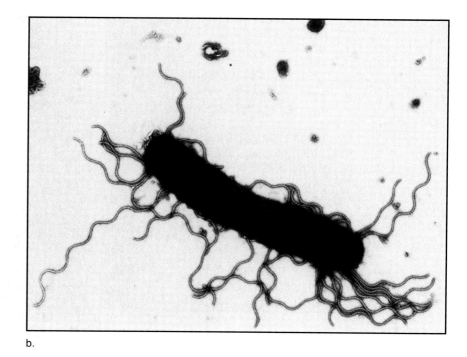

b.

figure 14.12

Endosymbionts of *Paramecium aurelia*

a. *Paramecium aurelia* bearing the endosymbiont *Lyticum flagellatum*. The numerous black rods throughout the cytoplasm are the endosymbionts (×500). b. *Lyticum flagellatum* isolated from *Paramecium aurelia*. Note the peritrichous flagella (×15,000).

table 14.2

Endosymbionts of *Paramecium aurelia*

Endosymbiont	Greek Letter Designation	Description
Caedobacter	κ (kappa)	Rods capable of forming refractile (R) bodies appearing as tightly rolled ribbons inside the host cell; R bodies are associated with toxic particles formed by killer paramecia
Pseudocaedobacter	π (pi)	Slender rods
	ν (nu)	Slender rods
	μ (mu)	Slender rods, often elongated
	γ (gamma)	Diminutive rods, often appearing as doublets
Tectobacter	δ (delta)	Rods with electron-dense material surrounding the outer of their two membranes
Lyticum	λ (lambda)	Motile rods with peritrichous flagella, although motility has not been observed inside the host cell
	σ (sigma)	Largest endosymbionts; curved, flagellated rods

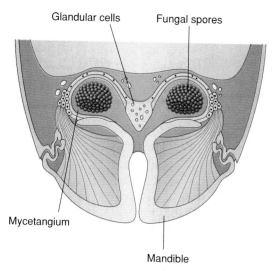

Glandular cells   Fungal spores

Mycetangium

Mandible

figure 14.13

Section Through the Head of the Ambrosia Beetle, *Xyleborus monographus*, Showing the Location of Fungal Spores in Mycetangia

believed to be associated with the activity of toxic particles produced by killer paramecia.

Certain procaryotes are endosymbionts of amoebas. The giant amoeba *Pelomyxa palustris* has two types of endosymbionts: a small procaryote distributed through the cytoplasm and a large procaryote in the vicinity of the nucleus and the glycogen bodies of the amoeba. The large procaryote has internal structural features resembling those found in *Methanospirillum* (domain Archaea) and the cristae of mitochondria, suggesting that these endosymbionts may function as mitochondria, which are absent in *P. palustris*. This role of these endosymbionts supports the hypothesis of the endosymbiotic origin of mitochondria (see endosymbiotic hypothesis, page 206).

Protozoa themselves serve as hosts for symbiotic bacteria. For example, some bacteria produce cellulases and assist protozoa in cellulose digestion. Some termites have intestinal bacteria capable of fixing nitrogen. For example, *Enterobacter agglomerans* fixes nitrogen under conditions of reduced oxygen tension and is frequently isolated from termites. The fixed nitrogen probably serves as a precursor of nitrogen-containing compounds in the termite, and the diet of the termite provides the bacteria with growth-sustaining nutrients.

Some Bacteria Are Symbionts of Insects

Bacteria are endosymbionts to many different types of insects. In some insects such as aphids, the endosymbionts (*Buchnera aphidicola*) supply necessary nutrients (for example, amino acids) that are not synthesized by the insect or not provided in adequate amounts in its diet. Without the endosymbionts, the aphids die. In other insects such as wood-eating cockroaches and termites, the endosymbionts help in digestive functions. Although *Clevelandina, Diplocalyx, Hollandina, Pillotina,* and other similar endosymbionts have never been cultivated from wood-eating insects, it can be easily surmised that they are important in the degradation of wood cellulose (see symbiotic spirochetes, page 317). In both of these examples—the aphids and the wood-eating insects—the insects evolved and readily adapted to their environments by establishing relationships with endosymbionts.

Wood-boring ambrosia beetles maintain a mutualistic type of relationship with fungi (Figure 14.13). The fungi reside in specialized organs called mycetangia, where they are protected from desiccation and other environmental stress. As the beetle bores through dead or decaying wood, the fungi are deposited onto the wood. As the fungi grow and multiply, they are cultivated by the activities of the beetle and then eaten. Ambrosia beetles, incapable of digesting cellulose, rely heavily on the fungal mycelia growing in their tunnels as a food source.

Microbial Symbiosis Assists Ruminants in Digestion

Ruminants are plant-eating mammals that include cows, goats, giraffes, camels, and sheep. Like other mammals, ruminants cannot make cellulases and therefore are incapable of hydrolyzing the β (1, 4) bond in cellulose (see cellulose structure, page 48). However, these animals have developed a mutualistic association with microorganisms that enables them to utilize the cellulose in their diets. This cellulose-decomposing microbial flora is contained within a special organ called the **rumen** in these animals.

The rumen is a large organ (100 liters in a cow and 6 liters in a sheep) that constitutes one portion of the digestive system of ruminants (Figure 14.14). This strongly anaerobic environment contains a microbial population at concentrations of approximately 10^{10} organisms per milliliter. Microorganisms in the rumen include not only cellulose decomposers such as *Bacteroides, Ruminococcus,* and *Butyrivibrio,* but also microbes that digest starch and those that produce methane (Table 14.3). Among the starch decomposers in the rumen are the Bacteria *Bacteroides, Succinomonas,* and *Selenomonas. Methanobrevibacter ruminantium* (domain Archaea) is the principal methane producer. The rumen also contains a large protozoan population, composed almost exclusively of ciliates; these protozoa feed on the bacterial population and perhaps may be active themselves in the decomposition of cellulose and starch in the rumen.

Food digestion by ruminants begins with the grinding and initial mixing of plant material in the mouth. The material enters the rumen, where it is broken down by microbial fermentation. Cellulose and other complex forms of carbohydrates, proteins, and other organic materials are degraded into organic acids and gases. The major organic acids produced as a result of rumen fermentation are acetic, propionic, and butyric acids. These pass through the rumen wall and enter the bloodstream, where they are

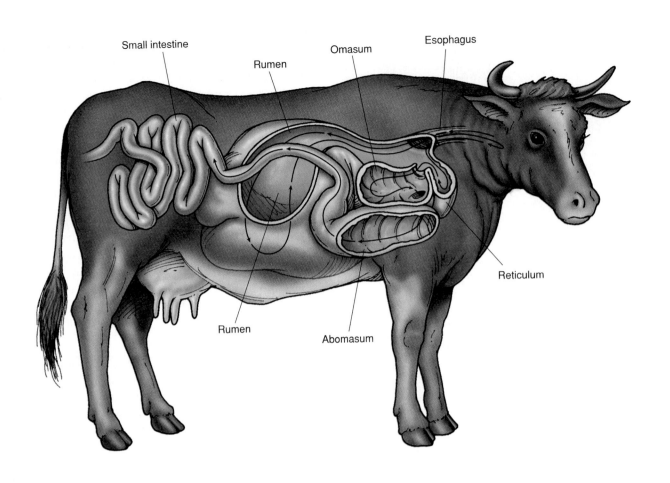

figure 14.14

The Digestive System of a Cow

The different parts of the cow's digestive system are shown, with arrows illustrating the flow of food.

table 14.3

Some Rumen Procaryotes

Substrate	Organism	Fermentation Products
Cellulose	*Butyrivibrio fibrisolvens*	Acetate, butyrate, formate, lactate, CO_2, H_2
	Clostridium lochheadii	Acetate, butyrate, formate, CO_2, H_2
	Fibrobacter succinogenes	Acetate, formate, succinate
	Ruminococcus albus	Acetate, formate, CO_2, H_2
Lactate	*Peptostreptococcus elsdenii*	Acetate, butyrate, propionate, valerate, CO_2, H_2
	Selenomonas lactilytica	Acetate, succinate
Pectin	*Lachnospira multiparus*	Acetate, formate, lactate, CO_2, H_2
Starch	*Bacteroides amylophilus*	Acetate, formate, succinate
	Bacteroides ruminicola	Acetate, formate, succinate
	Selenomonas ruminantium	Acetate, lactate, propionate
	Streptococcus bovis	Lactate
	Succinimonas amylolytica	Acetate, propionate, succinate
CO_2 and H_2, or formate	*Methanobrevibacter ruminantium*	Methane
	Methanomicrobium mobile	Methane

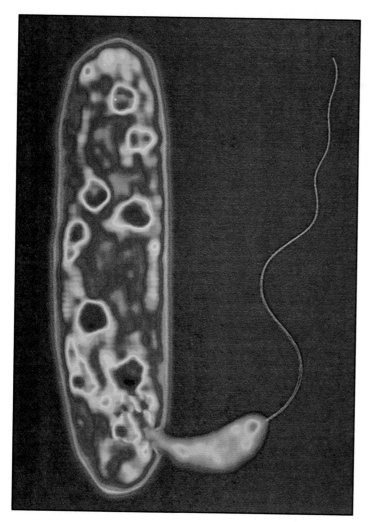

figure 14.15

Electron Micrograph of *Bdellovibrio bacteriovorus* Attacking *Pseudomonas*

The polar-flagellated *Bdellovibrio* is shown boring through the outer membrane and cell wall of its prey (colorized).

from the decomposition of plant materials by the symbionts. Furthermore, the microorganisms, when digested, are sources of vitamins and proteins. There are also advantages for the microorganisms. Although some are eventually degraded, they are provided with a favorable environment for growth, rich in fermentable carbohydrates and maintained at a constant temperature.

Parasitism

Parasitism is a symbiosis that, when microorganisms are involved, is usually associated with diseases in plants and animals. Viruses, bacteria, protozoa, and fungi may all be parasites that cause disease in other organisms. However, many microbes that can act as parasites are also found in soil and other environments and may be significant in the composition of ecological communities.

Whether or not a microorganism becomes a parasite depends on many factors, including the immunological state of the potential host, the presence or absence of other microorganisms in the host, and the environment. An immunocompromised host, whose immune system is suppressed by drugs, disease, or a genetic defect, is more susceptible to microbial infection than one that has a functional immune system (see immunosuppression, page 478). The presence of certain microorganisms can aid growth (for example, by providing necessary growth factors) or inhibit growth (for example, by unfavorably altering the environment or by secreting inhibitory products) potentially of pathogenic microbes (see microbial antagonism and synergism, page 440). Certain microorganisms are pathogenic under one set of environmental conditions and nonpathogenic or less pathogenic under other conditions. For example, virulence of bacterial pathogens such as *Neisseria gonorrhoeae* is apparently affected by the availability of iron (see role of iron in virulence of *N. gonorrhoeae*, page 100). A diet rich in glucose promotes production of dextran (a polymer of glucose) and adhesion by *Streptococcus mutans* to teeth, resulting in dental plaque (see dextran and adhesion by *S. mutans*, page 448).

Bdellovibrio Parasitizes Gram-Negative Bacteria

Bdellovibrio [Greek *bdella*, leech, Latin *vibrare*, to shake, vibrate] is a curved gram-negative bacterium with the unique property of preying upon other bacteria (Figure 14.15). Bdellovibrios are polarly flagellated and highly motile. They move through the environment in a rapid rotary fashion, with speeds as high as 100 cell lengths per second, and attack other gram-negative bacteria. *Bdellovibrio* feeds by boring through the outer membrane and cell wall of the host, disrupting the plasma membrane and causing leakage of the cytoplasmic contents, and divides only after the host is dead and of no further use.

Bdellovibrio bacteriovorus, the most extensively studied of these parasitic microbes, enters a host cell by penetration of the outer membrane and cell wall (Figure 14.16). A hole is formed through the membrane and wall by enzymatic action of the parasite, as well as possibly by mechanical action through a spinning

distributed throughout the body as a source of energy. Gases such as methane and carbon dioxide are produced in large quantities (60 to 80 liters/day in a cow) by anaerobic fermentation and are released from the rumen by frequent belching.

Microbial cells pass out of the rumen along with any undigested coarse plant material, and then pass back into the mouth, where this material is further chewed as cud. The rechewed material is swallowed again, this time bypassing the rumen and entering the **reticulum,** the **omasum,** and finally the **abomasum,** which is the true stomach of the ruminant. Digestive enzymes in these other organs of the digestive system degrade the microbes and undigested plant material before they are passed to the intestine for absorption.

The rumen exhibits a delicately balanced symbiosis that benefits both animal and microorganism. The ruminant benefits

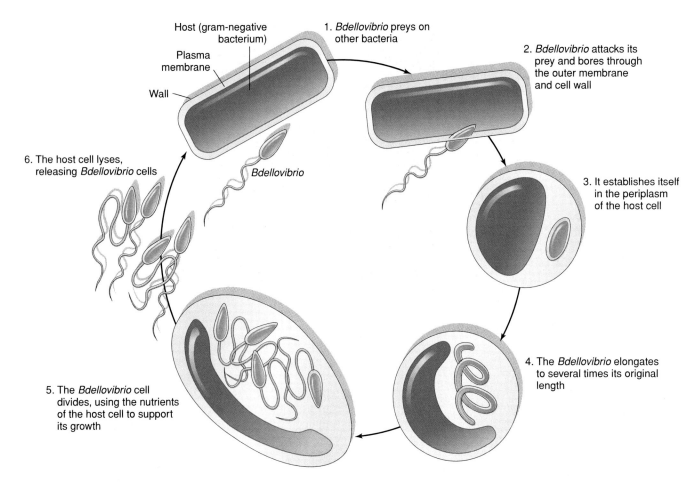

figure 14.16
The Life Cycle of *Bdellovibrio*

motion as the parasite penetrates the surface layer of the host bacterium. The parasitic microbe enters through this hole and lodges itself in the periplasm, between the cell wall and the plasma membrane. Here the *Bdellovibrio* elongates to several times its original length and then divides into several progeny cells, using the metabolic products and cellular components of the host cell to support its growth. The entire process takes approximately four hours and requires an actively metabolizing host cell. After cell division the progeny bdellovibrios are liberated as the host cell lyses.

Bdellovibrios are widespread in nature and are easily isolated from sewage, natural waters, and soil by methods like those used for bacteriophages. Soil or water samples are filtered through a membrane filter with pores small enough to retard larger bacteria, yet of sufficient diameter to allow passage of the minute bdellovibrios. The filtrate is inoculated onto an agar plate seeded with host bacteria. Bdellovibrios present in the inoculum form plaques on the bacterial lawn and can be isolated from these plaques. Most strains of *Bdellovibrio* are obligate parasites, but a few are host independent. These strains grow heterotrophically without the assistance of parasitic relationships with host cells. Because they prey upon other bacteria, bdellovibrios are sometimes called predators rather than parasites.

Chlamydiae Are Animal Parasites with Unusual Developmental Cycles

Chlamydiae are obligate intracellular parasites that cause disease in birds and in humans and other mammals. These small (0.2 to 1.5 µm in diameter) coccoid microorganisms have an affinity for epithelial cells of mucous membranes and are considered energy parasites because they are unable to synthesize high-energy phosphate bonds and are completely dependent on the host for ATP.

The chlamydiae have an unusual developmental cycle. They alternate between two forms: the **elementary body (EB)** and the **reticulate body (RB)** (Figure 14.17). The EB is a round particle 200 to 300 nm in diameter and is specialized for extracellular survival. The RB is a larger (500 to 1,000 nm in diameter) flattened noninfectious particle that is involved in intracellular multiplication. Differences in the structure and composition of the cell envelopes of the EB and RB probably account for their different properties. The EB has a rigid and impermeable cell envelope, whereas the RB cell envelope is fragile, permeable, and easily disrupted.

The first event in *Chlamydia* replication is the attachment of an infectious EB to a protein receptor on a susceptible host cell (Figure 14.18). The EB is phagocytized and enters the cell

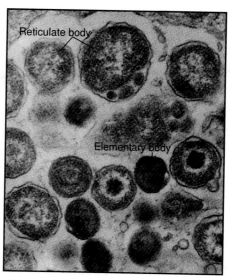

figure 14.17

Chlamydia trachomatis in a Host Cell

An electron micrograph of *C. trachomatis* in the cytoplasm of a host cell.

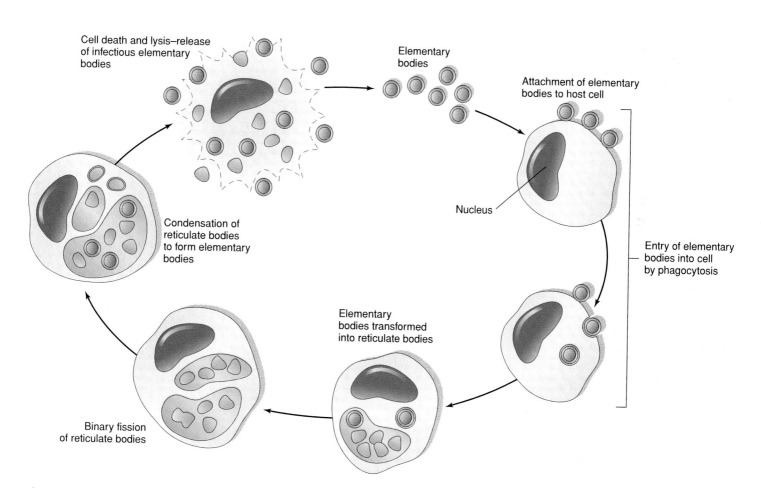

figure 14.18

Chlamydia Developmental Cycle

The infectious form of *Chlamydia*, the elementary body (EB), is taken up into the animal host cell by phagocytosis. Inside the cell, the EB is transformed into the larger, noninfectious reticulate body (RB) inside the phagosome. The RB undergoes continued binary fission to produce EBs, which are released upon cell death and lysis. The entire developmental cycle takes about 40 hours.

Germ-Free Animals

Germ-free animals are laboratory animals that contain no microbial flora. Germ-free rats, mice, guinea pigs, and other animals are obtained by removing the fetus aseptically from the mother by cesarean section. The fetus is placed in a sterile chamber that is aerated with filtered air (Figure 14.19). Sterile food in the form of a liquid nourishment duplicates the components of natural mother's milk and must be hand-fed to the infant for the first few weeks of life. After this period, the infant is able to feed itself on sterile food and water. Germ-free animals can be maintained and propagated by natural reproduction after an initial colony has been established.

Germ-free animals are used in research to study the effects of the normal microbial flora on disease, symbiotic relationships, and contribution of vitamins and other nutrients to the host. Such animals must be fed vitamin K in their diets. This vitamin routinely is supplied by *Escherichia coli*, a normal inhabitant of the intestine (see vitamin K, page 516). Germ-free animals, furthermore, are extremely susceptible to pathogenic microbial infection. Their immune systems are poorly developed because they have not been exposed to a normal flora, which induces protective immune responses. Germ-free animals were instrumental in showing that dental caries are caused by strains of *Streptococcus* and not simply by diet. Germ-free animals fed on high sucrose diets do not develop dental caries. However, if the animals are inoculated with certain strains of streptococci, tooth decay occurs (see *Streptococcus mutans* and dental caries, page 514).

Germ-free animals have been valuable for studies of microbial diseases and microbe interactions with a host. These animals provide scientists with insight into the role of natural microbial populations in the general health and nutrition of a host animal.

figure 14.19
Isolation Unit for Maintenance of Germ-Free Animals
Animals inside the unit are provided with filtered air and sterile food. Germ-free animals can be maintained and propagated by natural reproduction after the initial colony has been established.

cytoplasm in a phagosome, or inclusion body. Within eight hours after its entry into the cell, the EB is transformed into the larger noninfectious RB inside the phagosome. The RB undergoes a series of binary fissions to form smaller, dense-centered particles. At 20 hours after infection, these newly formed EBs are very much in evidence and outnumber the RBs. At about 40 hours, the host cell lyses and releases the EBs.

The interaction of chlamydiae with host cells is purely parasitic. The host cell is harmed by the association, whereas the chlamydiae, which are unable to synthesize their own high-energy compounds, benefit from their parasitism of the host's metabolic energy.

 **Miscellaneous Bacteria**
Chlamydia: Growth and Metabolism • pp. 9–14

Commensalism

Commensalism [Latin *com,* together, *mensa,* table] is a symbiotic relationship in which one organism is benefited and the other organism is unaffected. It is similar to a mutualistic relationship; the two often are difficult to differentiate. The extent of the benefit that one or the other organism receives from the association determines if the relationship is mutualistic or commensalistic.

Commensalism occurs commonly in soil as some organisms degrade complex molecules like cellulose or starch to simpler compounds used by other microbes. For example, many fungi are able to digest cellulose to glucose, which then can be catabolized by other microorganisms.

Commensalism is also involved in the production of growth factors by microorganisms. Some microbes produce and excrete growth factors, such as amino acids and vitamins, that are used by other microorganisms. The presence of these growth factors allows fastidious microorganisms to develop in otherwise inhospitable environments.

Many of the microorganisms found in the normal human flora are commensals. For example, as oxygen is used up by facultative anaerobes in the gastrointestinal tract, obligate anaerobes such as *Bacteroides* are able to grow. *Bacteroides* benefit from this association, but the facultative anaerobes derive no obvious benefit. Many types of streptococci and staphylococci inhabit the nasopharynx. These bacteria live on the mucus and nutrients in that region of the body, but under normal conditions are of no apparent benefit or harm to the host.

Mutualism, Commensalism, Parasitism—Dynamic Relationships

Although symbioses can be categorized into three types of associations, there is often no clear distinction between these categories. The interactions between populations are frequently dependent on environmental conditions, and under different environmental conditions, symbiotic relationships can change. For example, when a host is debilitated by an illness or its microflora has changed because of broad-spectrum antibiotic therapy (see broad-spectrum antibiotics, page 125), mutualistic bacteria can become parasitic and cause serious disease. *Staphylococcus aureus* is frequently ingested with food and can be found in the human intestine, where it is found in small numbers because of competition for nutrients with other microorganisms. However, when a broad-spectrum antibiotic is administered, drug-resistant staphylococci may overgrow other drug-sensitive bacteria and cause a condition known as staphylococcal enterocolitis (see microbial antagonism, page 441). *Streptococcus pneumoniae* is another example of a bacterium that is found in the upper respiratory tract of many people, but usually causes disease only in those individuals with viral infections of the upper respiratory tract or who are immunocompromised (see *S. pneumoniae,* page 504). *Escherichia coli* is normally a harmless bacterium in the intestine, but if it gains entry to the blood, the bladder, or the spinal cord, it can cause serious and sometimes fatal diseases (see urinary tract infections, page 532). The line between mutualism or commensalism and parasitism is very fine. What can be a nonpathogen under certain conditions can become a life-threatening pathogen under other conditions (see opportunistic pathogens, page 559).

Microbes have an indispensable role in agriculture. Symbiotic nitrogen fixation significantly increases soil fertility, particularly when crop rotation is rigidly practiced, but U.S. farmers still spend billions of dollars annually on nitrogenous fertilizers. These chemical fertilizers are expensive, require sophisticated industrial production facilities, and can pollute the environment. In contrast, crop rotation has been practiced for centuries and has been successful in enhancing soil fertility. Cyclic rotation, however, reduces the amount of time farmland can be used to grow productive crops. If the biological yields of fixed nitrogen could be increased severalfold, many of the current economic problems in crop productivity would be resolved.

There are several approaches to increasing the quantity of nitrogen available for crops. One approach might be to transfer the genes for nitrogen fixation from bacteria into the agricultural crops themselves. The genetically altered plants might then be able to fix nitrogen nonsymbiotically. The transfer of genes is not a novel idea but is difficult, particularly since it involves transfer of genetic material between two different groups of organisms, procaryotes and eucaryotes. Furthermore, the oxygen sensitivity of nitrogenase and the high energy requirements for nitrogen fixation make such a concept impractical. Eucaryotic plants have no provisions for anaerobic compartments like procaryotes, and energy would have to be diverted by the plant from crop yield to nitrogen production. Because of these problems, current research on nitrogen fixation is focused on transferring the genes for nitrogen fixation to other microorganisms.

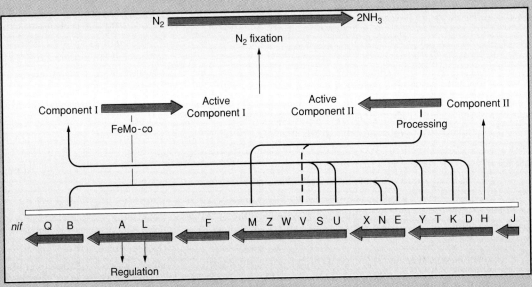

figure 14P.1

Genes of the *nif* Cluster in *Klebsiella pneumoniae*

The 20 identified genes of the *nif* cluster are indicated by capital letters. The seven *nif* operons are indicated below the genes. Component I (nitrogenase) and Component II (nitrogenase reductase) are formed independently to make up the nitrogenase complex. A cofactor known as FeMo-co contributes the iron and molybdenum in Component I.

Summary

1. The term symbiosis refers to an intimate association between organisms of different species. A mutualistic relationship is one in which both organisms benefit from the association. Parasitism occurs when one organism, the parasite, benefits and the other organism, the host, is harmed. In commensalism, one organism is benefited and the other is unaffected.

2. Symbioses may be divided into two groups based on the closeness of the association. The microorganism grows within the host cell in endosymbiosis, whereas the microorganism may be attached to the host cell, but remains outside in ectosymbiosis.

3. *Rhizobium, Bradyrhizobium,* and *Azorhizobium* are gram-negative, heterotrophic, soil bacteria collectively called rhizobia that form mutualistic associations with plants of the legume family, resulting in symbiotic nitrogen fixation. Symbiotic nitrogen fixation also occurs with other nonleguminous mutualistic associations with procaryotes such as *Frankia* and *Anabaena.*

4. Nitrogen fixation requires large expenditures of energy to reduce the triply bonded nitrogen molecule to two ammonia molecules. Nitrogenase, the enzyme that catalyzes nitrogen fixation, is irreversibly destroyed when

The genes for nitrogen fixation, collectively known as the *nif* genes, were first characterized in the 1970s in the bacterium *Klebsiella pneumoniae*. It is now known that the *nif* cluster consists of 20 separate genes organized into seven operons (Figure 14P.1). Plasmids can be constructed to mobilize the *nif* genes and transfer them from *nif*+ strains to *nif*- organisms. This transfer was first accomplished in the 1970s with the transfer of *nif* genes from *K. pneumoniae* into *E. coli*. Similar transfers have since been performed with *Salmonella typhimurium* as a recipient. It has even been possible to introduce *nif* genes into eucaryotic yeasts. Although the modified yeasts possess the genes for nitrogen fixation, they are not able to fix nitrogen. This is likely to be due to difficulties in transcription and translation of bacterial DNA in a eucaryotic cell. These successful transfers of the *nif* genes represent considerable achievements and provide a cornerstone for expanding the range of microorganisms capable of nitrogen fixation.

Another way to increase the availability of nitrogen to crops might be to improve the efficiency of nitrogen fixation in symbiotic relationships between rhizobia and leguminous plants. This could be accomplished by selection among currently available bacterial strains or by the development of new or bioengineered strains that produce larger quantities of ATP to cleave the triple bond in nitrogen gas.

Crop productivity could be increased by extending the host range of nitrogen-fixing bacteria to include cereals and other nonleguminous crops. There is some evidence that rhizobia can enter symbiotic relationships with nonleguminous plants. Some species of *Trema*, a tropical tree, produce nodules when infected with rhizobia. It has recently been discovered that the gram-negative bacterium *Azospirillum* infects the roots of tropical grasses, corn, wheat, barley, and other cereal crops. Inside the root, the bacteria grow *inter*cellularly between root cells, but do not produce nodules. In comparison, rhizobia and *Frankia* grow *intra*cellularly and produce nodules during their symbiotic associations with plants. Grasses and crops infected with *Azospirillum* show increased levels of fixed nitrogen and growth yield. If such symbiotic relationships could be expanded to other plants, increased productivity of economically important crops would be possible without expensive commercial fertilization. It is not known why only certain plants are able to enter into symbiosis with nitrogen-fixing bacteria. It is known that legumes are unique in that their roots secrete large amounts of flavonoids, which chemotactically attract rhizobia and trigger *nod* gene expression in the bacteria. Future studies on plant cell morphogenesis and cell signaling and the induction of *nod* genes in rhizobia may provide insight on extending the host range of symbiotic nitrogen-fixing bacteria.

Sources

Arnold, W., A. Rump, W. Klipp, U.B. Priefer, and A. Pühler. 1988. Nucleotide sequence of a 24,206-base-pair DNA fragment carrying the entire nitrogen fixation gene cluster of *Klebsiella pneumoniae*. *Journal of Molecular Biology* 203:715–738.

MacNeil, D. 1975. Genetics and regulation of nitrogen fixation in *Klebsiella pneumoniae*. In *Microbiology* 1981, D. Schlessinger, ed. 81–84. Washington, D.C.: American Society for Microbiology.

References

Brill, W.J. 1981. Agricultural microbiology. *Scientific American* 245:198–215.

Postgate, J. 1978. *Nitrogen fixation*. London: Edward Arnold.

Roberts, G.P., T. MacNeil, D. MacNeil, and W.J. Brill. 1978. Regulation and characterization of protein products coded by the *nif* (nitrogen fixation) genes of *Klebsiella pneumoniae*. *Journal of Bacteriology* 136:267–279.

exposed to oxygen. In rhizobia-legume symbioses, leghemoglobin binds and removes oxygen from the vicinity of nitrogenase.

5. A lichen is a mutualistic association between a fungus and an alga or a cyanobacterium. The fungus provides a solid support to anchor the lichen to barren surfaces, and the alga or cyanobacterium provides organic compounds through photosynthesis for use by the fungus.

6. A mycorrhiza is a mutualistic association between a fungus and a plant. The fungus benefits from the carbohydrates made available to it by the plant, and the plant benefits from the increased absorption area provided by the fungus.

7. Some luminescent bacteria of the genera *Photobacterium* and *Vibrio* live in symbiotic association with fishes called flashlight fishes. Light emitted by the bacteria, which are located in a specialized organ beneath each eye of the fish, is used by the fish to communicate, lure prey, and confuse predators.

8. Bacteria can be endosymbionts of paramecia, amoebas, and insects. In some insects such as wood-eating cockroaches and termites, the bacterial endosymbionts probably play an important role in the degradation of wood cellulose.

9. Certain microorganisms live in mutualistic relationships in the rumen of plant-digesting mammals such as cows, goats,

giraffes, camels, and sheep. The microbial symbionts decompose cellulose and other plant materials and are also sources of vitamins and proteins for the ruminant. The rumen is a favorable environment for growth of the microbes, rich in fermentable carbohydrates and maintained at a constant temperature.

10. *Bdellovibrio* preys upon gram-negative bacteria by boring through the outer membrane and cell wall of the host bacterium and dividing in the periplasm. After cell division the progeny bdellovibrios are released as the host cell lysis.

11. Chlamydiae are obligate intracellular parasites that alternate between elementary bodies and reticulate bodies during their developmental cycle. These bacteria have an affinity for epithelial cells of mucous membranes and are completely dependent on the host for ATP.

12. Many of the microorganisms in the human microflora are commensalistic. These symbionts live on nutrients in the human body, but under normal conditions are of no apparent benefit or harm to the host.

13. Mutualism, commensalism, and parasitism are dynamic relationships that can change depending on environmental conditions. The line between mutualism or commensalism and parasitism is very fine and can be altered in hosts that are immunocompromised, have other infections, or are administered broad-spectrum antibiotics.

EVOLUTION *and* BIODIVERSITY

There is extensive evidence that mitochondria and chloroplasts evolved from intracellular symbionts (see endosymbiont hypothesis, page 206). Mitochondria and chloroplasts resemble procaryotes in many ways. Both have 70S ribosomes, circular DNA, and similar rRNA sequences. They are approximately the same size and multiply in a similar manner, by simple division. The evolutionary origin of mitochondria furthermore is supported by the absence of any known symbioses between eucaryotic cells lacking mitochondria and aerobic respiring procaryotes. Eucaryotes with mitochondria have apparently outcompeted such associations. Without the evolution of mitochondria and chloroplasts from endosymbionts, it is not inconceivable that contemporary eucaryotes would be nonphotosynthetic and restricted to anaerobic environments. Symbiotic associations not only have important roles in evolution, but also are important in organism diversity. Through symbiotic relationships, metabolic capabilities and ecological habitats are expanded. Organisms that normally are unable to grow in a habitat are able to flourish because of their symbiotic associations with other organisms.

Questions

Short Answer

1. Identify three types of symbiotic relationships.

2. Compare and contrast endosymbiosis and ectosymbiosis.

3. Describe the symbiosis of nitrogen fixation.

4. *Rhizobium* species are aerobic, but nitrogen fixation cannot occur in the presence of oxygen. Explain.

5. Describe the symbiotic relationship that forms lichens.

6. Explain why lichens are sensitive indicators of air pollution.

7. Describe the symbiotic relationship of mycorrhiza.

8. Describe the symbiotic relationship that makes the flashlight fish luminescent.

9. Explain why the bacterial endosymbionts found in paramecia inhabiting cockroaches and termites are of such interest.

10. Ruminants, plant-eating mammals such as cows, do not produce the enzymes necessary for digestion of their food. Explain how this is possible.

11. Describe the symbiotic relationship of *Bdellovibrio* and gram-negative bacteria.

12. Describe the symbiotic relationship of *Chlamydia* and its host.

13. Describe the symbiotic relationship of humans and their normal microflora.

14. Illustrate the dynamic state of symbiotic relationships.

Critical Thinking

1. Discuss the impact of a germ-free world. Is it possible? Is it feasible? Is it advisable?

2. Explain why developing nations have much higher rates of parasitism than the USA. From a public health perspective, how would you change this? What obstacles would you have to overcome?

3. A current trend in agriculture is to use a known parasite to eliminate crop-destroying pests. Discuss the implications of this practice.

Ahmadjian, V. 1993. *The lichen symbiosis*. New York: Wiley. (A comprehensive guide to lichens.)

Allen, M.F. 1991. *The ecology of mycorrhizae*. Cambridge, England: Cambridge University Press. (A detailed discussion of mycorrhizae and their habitats.)

Atlas, R.M., and R. Bartha. 1992. *Microbial ecology: Fundamentals and applications*. 3d ed. Menlo Park, Calif.: Benjamin-Cummings. (A textbook of microbial ecology, with sections on symbiotic relationships.)

Benson, D.R., and W.B. Silvester. 1993. Biology of *Frankia* strains, actinomycete symbionts of actinorhizal plants. *Microbiological Reviews* 57:293–319. (A review on the structure, physiology, metabolism, and genetics of nitrogen fixation by *Frankia*.)

Brewin, N.J. 1991. Development of the legume root nodule. *Annual Review of Cell Biology* 7:191–226. (A comprehensive review article on the chemical and biological processes associated with legume root nodule development.)

Campbell, R. 1985. *Plant microbiology*. London: Edward Arnold. (A review of plant-microbe interactions.)

Cheng, T.C. 1970. *Symbiosis: Organisms living together*. New York: Western Publishing. (A review of major symbiotic relationships and their characteristics.)

Douglas, A.E. 1994. *Symbiotic interactions*. Oxford, England: Oxford University Press. (A concise discussion of microbial symbiotic relationships, including lichens, mycorrhizae, and nitrogen fixation. Metabolic capabilities, nutritional interactions, and ecological consequences of symbiotic relationships are described.)

Hobson, P.N., ed. 1988. *The rumen microbial ecosystem*. London: Elsevier Applied Science. (An extensive review of the rumen and its microbial flora.)

Margulis, L. 1981. *Symbiosis in cell evolution: Life and its environment on the early earth*. San Francisco: W.H. Freeman. (A discussion of how symbiosis may have played a role in the evolution of life.)

Paerl, H.W. 1990. Physiological ecology and regulation of N_2 fixation in natural waters. *Advances in Microbial Ecology* 11:305–344. (A review of nitrogen fixation among organisms in aquatic environments.)

Postgate, J. 1978. *Nitrogen fixation*. London: Edward Arnold. (A detailed description of nitrogen fixation, including a discussion of the chemistry of nitrogenase, the physiology of nitrogen fixation, and the microorganisms associated with nitrogen fixation.)

chapter fifteen

HOST-PARASITE RELATIONSHIPS

Scanning electron micrograph of a macrophage (red) engulfing *Escherichia coli* (green) (colorized).

The Normal Human Microflora
Antagonism
Synergism

General Concepts of Host-Parasite Relationships
Reservoirs
Disease Transmission
Entry to Host
Sequence of Infectious Disease
Parasites

Microbial Factors of Virulence
Mechanisms of Invasion
Toxins

Innate (Nonspecific) Host Resistance
Condition of the Host
Physical Barriers
Protective Factors in Blood
Inflammation Response
Phagocytes

PERSPECTIVE
Dextran-Induced Agglutination of *Streptococcus mutans* in the Formation of Dental Plaque

EVOLUTION AND BIODIVERSITY

Microbes in Motion — PREVIEW LINK

Among the key topics in this chapter are the normal flora of the human body, reservoirs of infection, types of transmission, virulence factors of microbes, and host defenses against microbial diseases. The following sections in the *Microbes in Motion* CD-ROM may be useful as a preview to your reading or as a supplemental study aid:

Anaerobic Bacteria: Normal Flora (Sites), 1–2. *Gram-Negative Organisms:* Normal Flora (Sites), 1–2. *Gram-Positive Organisms:* Normal Flora (Sites), 1–2; *Gram-Positive Cocci*–Streptococcus (Virulence Factors), 15–18. *Epidemiology:* Disease Aquisition, 1–4; Terminology (Carriers), 6–11; Transmission (Horizontal—Contact), 3–12. *Microbial Pathogenesis:* Principles of Pathogenicity (Definitions), 3–5; Adherence (Mechanisms), 2–7; Antiphagocytic Factors, 1–9; Toxins, 1–17; Nonspecific Host Defense (Mechanical/Chemical Control), 4–28.

One of the most dramatic and significant properties of microorganisms is the ability to cause infectious disease. Most microorganisms do not harm humans; in fact, the great majority are beneficial to human beings and higher animals. However, the ability of a minor class of microorganisms to cause disease was a major impetus for the development of the science of microbiology. Great interest still exists today in the study of host-parasite relationships and how they can result in disease. Today scientists are researching newly discovered microorganisms and the diseases they cause, such as HIV and acquired immune deficiency syndrome, Ebola viral hemorrhagic fever, hantavirus pulmonary disease, and *Escherichia coli* O157:H7 foodborne diarrhea. The study of host-parasite relationships and the diseases resulting from them is called **pathogenic microbiology** or **medical microbiology.**

The Normal Human Microflora

Not all relationships between humans and microbes necessarily lead to overt disease. Many species of microorganisms normally inhabit different parts of the human body without causing harm. Microbes that populate the human body in mutualistic or commensalistic relationships are known as the **normal flora,** or more correctly, the **normal microflora** or **normal microbiota** of the body (Table 15.1).

 Anaerobic Bacteria
Normal Flora: Sites • p. 2

 Gram-Negative Organisms
Normal Flora: Sites • pp. 1–2

 Gram-Positive Organisms
Normal Flora: Sites • pp. 1–2

A healthy human fetus is free of microbes before birth; it is only during and after birth that the microflora begin to be established. The first colonization of the infant is derived from the mother's vaginal flora after rupture of the amnion. These microorganisms are transient to the newborn and are replaced soon after birth of the infant with other microorganisms in the surroundings. The infant acquires these microbes by surface contact with the mother and birth attendants, by swallowing microorganisms as part of the diet, and by contacting contaminated objects in the birth (hospital) and home environments.

The microorganisms making up the normal human flora are a large and diverse group. The adult human body is estimated to be made up of approximately 10^{13} human cells, but is also estimated to have about ten times that number of associated microbial cells. Microorganisms typically inhabit all the areas of the body that are exposed or even distantly connected to the external environment: the skin and the mucous membranes of the oral

table 15.1

Normal Human Microflora by Body Region

Body Region	Principal Microflora
Skin	*Corynebacterium xerosis, Propionibacterium acnes, Staphyloccocus aureus, Staphyloccocus epidermidis*
Oral cavity	*Actinomyces* sp., *Bacteroides* sp., *Candida* sp., *Fusobacterium* sp., *Haemophilus* sp., *Lactobacillus* sp., *Neisseria* sp., *Prevotella* sp., *Staphylococcus* sp., *Streptococcus* sp., *Veillonella* sp.
Respiratory tract (primarily the upper part)	*Actinomyces* sp., *Candida* sp., diphtheroids, *Haemophilus* sp., *Neisseria* sp., *Staphylococcus* sp., *Streptococcus* sp.
Digestive system (primarily the large intestine)	*Bacteroides* sp., *Candida* sp., *Citrobacter* sp., *Clostridium* sp., *Enterobacter* sp., *Escherichia coli, Fusobacterium* sp., *Klebsiella* sp., *Lactobacillus* sp., *Peptococcus* sp., *Proteus* sp., *Streptococcus* sp.
Genitourinary system	*Bacteroides* sp., *Candida* sp., *Clostridium* sp., diphtheroids, *Lactobacillus* sp., *Staphylococcus* sp., *Streptococcus* sp., *Trichomonas* sp.

cavity, upper and lower respiratory tract, gastrointestinal tract, and genitourinary tract; each of these body surfaces is colonized by its own unique flora. Other parts of the body, such as the blood and internal organs, normally do not contain microbes; the presence of microbes in these parts is evidence of infection and disease.

The normal flora consists of two groups of microorganisms: (1) **resident flora,** a more or less constant group of microbes that survives and grows in and on the body, and (2) **transient flora,** which may inhabit the body for only short periods of time—a few days or weeks. Although most of these microbes are harmless commensals, a few are opportunistic pathogens. Problems may arise when the delicate balance between microbe and host is disturbed by debilitating disease, antimicrobial drugs, or suppression of the host immune response as in organ transplants, steroid treatment, and AIDS.

Microorganisms Can Compete Antagonistically

Some indigenous microbes provide the host with protection against potentially pathogenic microorganisms through a mechanism known as **antagonism.** Microbial antagonism is the result of competition for nutrients, alteration of the environment (for example, oxygen tension or pH), or the release of products that inhibit other microbes, such as antibiotics and bacteriocins (see antibiotics and bacteriocins, page 125).

For example, the yeast *Candida albicans* is normally a minority member of the human vaginal flora because of competition

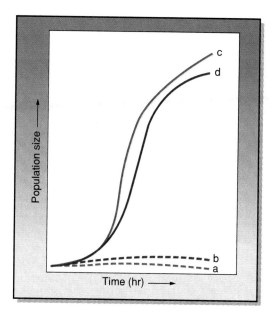

figure 15.1

The Synergistic Relationship Between *Enterococcus faecalis* and *Lactobacillus arabinosus* During Cross-Feeding
In a medium lacking phenylalanine and folic acid, neither *Enterococcus faecalis* (curve a) nor *Lactobacillus arabinosus* (curve b) grows. When inoculated together in the same medium, however, both bacteria grow (curves c and d), because they provide each other with the required growth factors.

with other members of the flora for available nutrients. Lactobacilli are especially prominent among the members of normal microflora in maintaining low populations of yeast. In healthy people, *C. albicans* is unable to multiply rapidly and does not cause problems. However, the prolonged use of antibacterial drugs suppresses growth of bacterial members of the vaginal flora. Without competition, *C. albicans* proliferates and causes vaginitis (inflammation of the vagina).

Clostridium difficile, which received its name because of difficulties early investigators encountered in isolating this slow-growing bacterium, is normally a minor component of the intestinal microflora in 5% to 10% of the human population. When the competing normal microflora is inhibited by antibiotics, the toxin-producing *C. difficile* multiplies and causes a severe inflammation of the colon called pseudomembranous colitis (PMC). PMC is characterized by fever, diarrhea, severe abdominal cramps, and the formation of a pseudomembrane composed of fibrin, leukocytes, and necrotic colonic cells. Although PMC was first described nearly a century ago, it was a rare disease until 1970 when widespread use of antibiotics resulted in a significant increase in the number of disease cases.

Staphylococcus aureus is another member of the normal human flora that usually does not cause problems but has the poten-

tial to do so. This bacterium is frequently ingested with food, and small numbers are found in the human intestine, where they cause no harm. However, during treatment with a broad-spectrum antibiotic (for example, tetracycline), drug-resistant enterotoxin-producing staphylococci may overgrow other drug-sensitive intestinal bacteria and invade the bowel wall to cause acute staphylococcal enterocolitis, a disease characterized by diarrhea, inflammation, and ulcerations of the intestinal mucosa.

Microorganisms Can Grow Synergistically

Some strains or species of microorganisms grow better in combination than alone. This characteristic is known as **synergism.** Synergism and mutualism are similar in that both populations benefit from the relationship. In synergism, organisms accomplish more together than they could independently.

One form of synergism is **cross-feeding,** or **syntrophism,** in which two populations supply each other's nutritional needs. For example, *Enterococcus faecalis* requires folic acid (a vitamin) for growth and *Lactobacillus arabinosus* requires phenylalanine (an amino acid). Neither *E. faecalis* nor *L. arabinosus* can grow alone in a minimal medium lacking folic acid and phenylalanine. However, both organisms can grow together in a minimal medium, because *L. arabinosus* produces folic acid, which is required by *E. faecalis,* and *E. faecalis* produces phenylalanine, which is required by *L. arabinosus* (Figure 15.1). Bacterial synergism is not uncommon, and enables the thriving of two populations that normally might not grow well alone. Frequently, as in the case described, synergistic interactions are important in providing nutrients required for the growth of both populations.

Syntrophism also occurs in the relationship between methanogenic Archaea such as *Methanobacterium, Methanosarcina,* and *Methanococcus,* which produce methane from simple carbon compounds (acetate, carbon dioxide, formate, and methanol), and other microorganisms. Frequently methane formation is accompanied by electron transport phosphorylation via anaerobic respiration in which carbon dioxide is the terminal electron acceptor and hydrogen is the electron donor:

$$4 H_2 + CO_2 \rightarrow CH_4 + 2 H_2O$$

$$\Delta G^{o'} = -32 \text{ kcal/mole}$$

Methanogenic procaryotes are unable to use other, more complex substrates and are therefore almost always found in association with other microorganisms. These other microorganisms metabolize cellulose, starch, proteins, fats, and other complex substances, producing simple waste products (acetate, carbon dioxide, formate, and methanol) that can be used by the methane-producing Archaea. One of the rate-limiting steps in microbial methanogenesis is the breakdown of complex organic material to simpler compounds for use by the methanogens. The methanogens obviously benefit from this association, but so do the other microorganisms, because their waste products are removed.

The Story of Typhoid Mary

A classic example of the role of carriers in disease transmission is the story of Mary Mallon. It began in 1906 when the New York City Department of Health traced an outbreak of typhoid fever in a Long Island home to Mary, the family cook. During its investigation, the health department discovered that Mary had worked from 1896 to 1906 in seven different homes, in which there had been 28 cases of typhoid fever.

Mary was presented with this evidence and asked to submit blood, urine, and fecal specimens for laboratory analysis to determine if she was a carrier of the typhoid bacterium *Salmonella typhi*. Her stool samples were found to contain the bacterium, although she had no visible symptoms of typhoid fever. Because Mary was a carrier of typhoid bacteria and continuously shed the organism, doctors recommended that she have her gallbladder removed to eliminate what they believed to be the source of her infection. Mary became afraid when confronted with this information; it was not made clear to her what the operation was, nor how it would prevent further infections. She refused the operation and was imprisoned to prevent her from spreading the infection.

Mary's case received extensive publicity. An article in the *Journal of the American Medical Association* referred to her as Typhoid Mary, and many newspapers carried stories about her. In 1910 Mary was released from her hospital prison because of public outcry; many people believed that it was wrong to imprison a person who was only a carrier of a disease.

During the next several years, Mary worked as a cook at restaurants, hospitals, and sanitariums and caused typhoid fever outbreaks at each of these places. Public health officials were alarmed by these outbreaks and returned Mary to her prison in 1915, where she remained for the next 23 years until her death at the age of 70. Mary's story illustrates the role of carriers in the spread of infectious diseases. At least 53 cases of typhoid fever and three deaths were traced to Mary Mallon.

General Concepts of Host-Parasite Relationships

Infectious disease begins with colonization of the host by a pathogenic microorganism. The term **infection** refers to the growth of microbes in the tissues of a host. Infection does not necessarily lead to injury of the host. **Infectious disease** occurs when the infecting microbe causes damage to the host organism. The damage may be apparent through visible signs or symptoms (that is, clinical disease) or it may be subclinical or inapparent.

Not all diseases are caused by microorganisms. Diseases caused by microbial agents are designated *infectious* to distinguish them from *noninfectious* diseases (for example, alcoholism, diabetes, and emphysema). In many infectious diseases such as measles, pneumonia, and gonorrhea, the pathogen is transmitted from one individual to another. These diseases are said to be **communicable.** In other, **noncommunicable** infectious diseases, the responsible agent is not transmitted from person to person. For example, *Clostridium tetani*, a soil inhabitant, produces tetanus when it is introduced into a wound or an abrasion. Tetanus is thus an infectious disease, but it is not communicable.

Epidemiology
Disease Acquisition • pp. 1–4

Infectious Microorganisms Persist in Reservoirs

Infectious microorganisms generally persist in nature in **reservoirs of infection.** These reservoirs, which are necessary to perpetuate the microorganism, may be animate or inanimate and are as diverse as humans, animals, arthropods, soil, and water. Microbes generally live and multiply without damaging their reservoirs. The control or elimination of reservoirs is an important part of disease control.

Soil, Water, and Food Are Inanimate Reservoirs

Soil, water, and food are the primary inanimate reservoirs of infection. Soil and water often contain microorganisms that are shed by animals and plants. Sometimes spores of these microbes remain viable for years. Anthrax spores, for example, are found in soil (the reservoir) and can be ingested by grazing animals. The spores germinate in the animal host and may produce disease.

In some cases, environmental conditions influence the spread of disease through inanimate reservoirs. Mastitis, a mammary gland infection of cattle caused by *Streptococcus agalactiae* and other bacteria, is a serious problem in Florida. Heavy rains there, particularly during the summer, turn pasturelands into muddy fields. As infected cattle wade through the mud, *S. agalactiae* on their mammary glands is inoculated into the mud, which becomes a reservoir for infection of healthy cattle.

Waterborne human diseases are usually associated with water that has been contaminated with human or animal feces. Sewage is the most common source of water contamination. Fecal coliform counts (see fecal coliform counts, page 593) of drinking and recreational water provide health departments with indices of fecal contamination. From this information, recommendations regarding the safety of the water can be made. Many pathogenic species of microorganisms survive in water, especially fresh water, for long periods and can cause many different types of waterborne gastrointestinal and other diseases.

Foods are important reservoirs for microorganisms. Foods derived from animals or plants usually have a natural microflora, which is supplemented by other microorganisms acquired by improper handling during or after processing. In some instances, processing can add significantly to the microbial content of food. For example, ground beef has a large surface area that can be extensively contaminated with microorganisms during the grinding process. An ounce of freshly ground beef may harbor several million microorganisms and have a short shelf life because of this contamination.

Human Beings Are an Example of Animate Reservoirs

Human beings are, of course, the most important living reservoirs and sources of human infectious agents. Most human pathogenic microorganisms are adapted to growth and survival on or in the body. Microbes that cause such diseases as gonorrhea, syphilis, and mumps ordinarily infect only humans and do not naturally occur in other reservoirs.

Human beings may harbor an infectious microorganism in their bodies in the absence of disease and spread it to other, susceptible individuals. Such persons are said to be **carriers.** There are three types of carriers: (1) **healthy carriers,** who have no visible signs or symptoms of disease at any time; (2) **incubatory carriers,** who are in the initial stages of disease before clinical symptoms

appear; and (3) **convalescent carriers,** who have recovered from the disease and no longer have symptoms but still carry and shed the infectious agent. The carrier state may be **temporary,** with the microorganism present for a brief period; **intermittent,** with periodic shedding of the microorganism; or **permanent,** with continuous carriage and shedding of the infectious agent. In some diseases, such as typhoid fever and hepatitis, a carrier may remain infectious for many years. The most dangerous carriers are those who shed infectious agents through the respiratory or intestinal tract, particularly people employed in establishments where food can easily become contaminated.

Epidemiology
Terminology: Carriers • pp. 6–7

Animals are also important reservoirs and sources of infectious agents. Animal diseases transmissible to human beings are called **zoonoses** [sing., **zoonosis**] (Table 15.2). Human beings are usually accidental hosts in zoonoses, acquiring them either by contact with an infected animal or through insects that transmit the bacteria from the infected animal. Domestic animals are reservoirs of many diseases, including brucellosis (cattle, sheep, pigs, and goats), anthrax (cattle and sheep), and salmonellosis (chickens, turkeys, rodents, and turtles). These diseases are particularly a problem in developing countries, where milk may not be pasteurized and meat is not screened for potential infectious agents. In the United States, meat routinely is not tested for pathogenic microorganisms, which this has led to outbreaks of food poisoning by bacteria such as enterohemorrhagic *E. coli* (see enterohemorrhagic *E. coli*, page 523). The reservoirs for influenza are pigs and ducks. When pigs and ducks live in close proximity to humans, as in China, the strains of influenza in these two types of animals come in contact and new forms of influenza virus emerge. This explains why almost all new strains of influenza virus originate in China.

Rabies is a zoonosis that can be acquired by the bite of a rabid dog, cat, bat, skunk, or raccoon. Most human cases of rabies in the United States occur via dog bites, although the skunk, raccoon, and fox account for the majority of documented animal cases (about 6,000 a year). The number of cases of human rabies in the United States has decreased to fewer than five cases per year in recent years, although it is estimated that there are as many as 10,000 human cases a year worldwide.

Epidemiology
Terminology: Zoonoses • pp. 10–11

Diseases Can Be Transmitted Directly or Indirectly

To cause disease, an infectious agent must first be transmitted from its reservoir to a susceptible host. Transmission may occur directly from the reservoir to the host or indirectly via an intermediate between the reservoir and the host.

table 15.2

Examples of Zoonoses

Organism	Zoonosis	Reservoir
Bacteria		
Brucella sp.	Brucellosis	Cattle, goats, pigs, sheep
Chlamydia psittaci	Ornithosis, psittacosis	Birds, especially parrots
Coxiella burnetti	Q fever	Cattle, goats, sheep
Francisella tularensis	Tularemia	Wild rabbits
Leptospira interrogans	Leptospirosis	Wild and domestic animals (for example, dogs and cats)
Rickettsia rickettsii	Rocky Mountain spotted fever	Rodents
Rickettsia typhi	Endemic typhus	Rodents
Salmonella sp.	Salmonellosis	Chickens, turkeys, rodents, turtles
Yersinia pestis	Bubonic plague, pneumonic plague, septicemic plague	Rodents
Fungi		
Epidermophyton sp.,	Ringworm	Domestic mammals (for example, dogs)
Microsporum sp.,		
Trichophyton sp.		
Protozoa		
Plasmodium sp.	Malaria	Monkeys
Toxoplasma gondii	Toxoplasmosis	Cats and other mammals
Trypanosoma cruzi	Chagas' disease	Wild mammals
Viruses		
Arbovirus	Western equine encephalitis	Horses, birds
	Yellow fever	Monkeys
Influenza	Influenza	Pigs (swine influenza), ducks
Rhabdovirus	Rabies	Bats, foxes, skunks

Direct Transmission Involves Contact with the Reservoir or Another Host

Some microbial pathogens are transmitted directly to the host. The infectious agents of several sexually transmitted diseases such as gonorrhea (*Neisseria gonorrhoeae*), nongonococcal urethritis (*Chlamydia trachomatis* and *Ureaplasma urealyticum*), and syphilis (*Treponema pallidum*) are sensitive to drying and heat and do not survive long outside the host. These bacteria therefore are transmitted from person to person by sexual intercourse and other intimate contact, entering the body through the genitalia, mouth, or anus. Viral diseases (for example, common colds, influenza, and mumps) are especially likely to be spread by aerosols or by direct contact. It is not unusual for such viral diseases to spread quickly and afflict several members of a family or a closed community.

Other infectious diseases may be acquired through direct contact with infected animals. For example, anthrax (*Bacillus anthracis*) is an occupational hazard for those who work closely with animals or their products (for example, veterinarians, ranchers, and textile workers). In the United States, human pulmonary anthrax may also be contracted by inhalation of spores from contaminated animal products such as skin, wool, hides, and bones. Recent cases of human anthrax in this country have been traced to spores on the leather of imported West Indian bongo drums.

Tularemia (*Francisella tularensis*) is a zoonosis that occurs in rabbits but may be transmitted to humans. Hunters are especially at risk for tularemia by aerosols during skinning, by infection through cuts and abrasions, and by ingestion of improperly cooked rabbit meat or contaminated water. Tularemia is a seasonal disease, with most human cases occurring during the winter because of the hunting season. Tularemia may also be transmitted by indirect contact through bites from ticks and flies.

Indirect Transmission Involves Contact with a Fomite or Vector

Infectious microbes may be transmitted indirectly by inanimate or animate means. An inanimate object that is contaminated with an infectious organism and is associated with its transmission is called a **fomite**. Examples are dishes, clothing, towels, needles, and dust.

Examples of Arthropod Vectors of Human Diseases

Organism	Disease	Vector
Bacteria		
Bartonella bacilliformis	Bartonellosis	*Phlebotomus* (sand fly)
Borrelia burgdorferi	Lyme disease	*Ixodes scapularis* (deer tick)
Borrelia recurrentis	Relapsing fever	*Pediculus* (louse),
		Ornithodorus (tick)
Francisella tularensis	Tularemia	*Dermacentor* (tick)
Rickettsia prowazekii	Epidemic typhus	*Pediculus* (louse)
Rickettsia rickettsii	Rocky Mountain spotted fever	*Dermacentor* (tick)
Rickettsia tsutsugamushi	Scrub typhus	*Trombicula* (mite)
Rickettsia typhi	Endemic typhus	*Xenopsylla* (flea)
Yersinia pestis	Bubonic plague	*Xenopsylla* (flea)
Protozoa		
Entamoeba histolytica	Amebic dysentery	*Musca* (housefly)
Leishmania donovani	Leishmaniasis	*Phlebotomus* (sand fly)
Plasmodium sp.	Malaria	*Anopheles* (mosquito)
Trypanosoma cruzi	Chagas' disease	*Triatoma* (triatomid bug)
Viruses		
Alphavirus	Encephalitis	*Culex* (mosquito)
Flavivirus	Yellow fever	*Aedes* (mosquito)

Conjunctivitis (inflammation of the conjunctiva) is frequently caused by communal usage of towels contaminated with *Chlamydia trachomatis*. Fungal spores are often transmitted on dust that arises from soil impregnated with the spores. Windstorms in the San Joaquin valley in California generate extensive dust clouds that carry spores of *Coccidioides immitis* to susceptible human beings. Populations exposed to these dust clouds have a high incidence of coccidioidomycosis, a disease of the respiratory tract and lungs resembling pneumonia or pulmonary tuberculosis.

Living organisms that are intermediaries in the transfer of infectious agents are called **vectors** (Table 15.3). The most common vectors are arthropods (for example, lice, fleas, or ticks), although other animals (for example, dogs, cats, or rodents) may also serve as vectors. In some cases the microorganism proliferates within the arthropod vector. For example, the malarial parasite *Plasmodium* actually completes part of its life cycle in its mosquito vector (*Anopheles*) after the mosquito is infected by taking a blood meal from a person with malaria. The life cycle is continued when the mosquito takes another blood meal and inoculates the parasite into a susceptible human host.

Organisms that are themselves infected and are also reservoirs of infectious agents for transmission are called **biological vectors.** In other instances a vector may serve as a carrier of a microorganism without being infected. Such carriers are called **mechanical vectors.** Flies frequently carry microbes on their body parts from feces to foods or to human beings. The infectious agent does not enter the fly, but is merely carried on it from one location to another.

Epidemiology
Transmission: Horizontal; Contact • pp. 3–12

Microorganisms Gain Entry to a Host Through a Portal

Infectious agents that are transmitted to a human host enter the host through several possible portals of entry. These entrance routes include the nose, mouth, eyes, ears, genitourinary tract, and open cuts or abrasions on the skin. The portal through which a microbe enters the host and establishes an infection depends to a large extent on the type of infectious agent and its mode of transmission. For example, waterborne or food-borne microbes that cause gastrointestinal diseases typically enter the body through the nose or mouth. Pathogens that are transmitted by biological vectors frequently gain access to the body through punctures in the skin caused by insect bites. Microorganisms that cause sexually transmitted diseases generally enter the body through the genitourinary

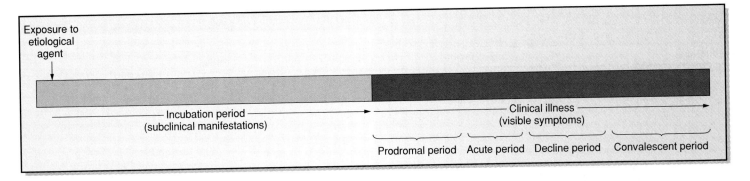

figure 15.2
The Natural History of Infectious Disease
Subclinical pathological changes frequently occur during the incubation period. Most diseases are not diagnosed until after disease signs and symptoms have appeared.

tract, although other orifices may be used, depending on sexual practices. Regardless of the portal of entry used by an infecting microorganism, once the pathogen gains a foothold in the host, disease may occur.

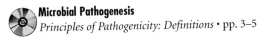 **Microbial Pathogenesis**
Principles of Pathogenicity: Definitions • pp. 3–5

Infectious Disease Follows a Sequence from Infection to Disease Resolution

The **natural history of infectious disease** is the sequence of events that occurs from the infection of a host to disease resolution, either through healing or death of the host (Figure 15.2). The natural history may be divided into two parts with respect to signs and symptoms.

The Parasite Grows and Multiplies During the Incubation Period

Initially a subclinical situation exists, a period frequently called the **incubation period** of the disease. Throughout the incubation period, the parasite can grow and multiply, as well as establish a focus of infection in the host. Often there are no visible signs of disease during this period, but there may be subclinical effects of infection such as toxic damage to tissues and systemic distribution of the pathogen to target organs of the body.

The incubation period varies and depends on the nature and dose of the infecting pathogen, portal of entry, mechanisms for tissue damage, and capability of the host to resist infection. The incubation period is usually short in localized diseases such as typical respiratory or gastrointestinal diseases. Influenza, streptococcal pharyngitis, salmonellosis, and cholera all have incubation periods of only a few days. Some exceptions exist, such as pertussis and tuberculosis, which have longer incubation periods. Systemic infections almost always have an extended incubation period ranging from one week to several months or years. This longer incuba-

tion period, in diseases such as syphilis, typhus, and leprosy, results from the pathogen's mode of growth and reproduction and the host's response.

Symptoms Appear During Clinical Illness

The second part of the natural history of infectious disease begins when clinical symptoms are first apparent. During this time the clinical and laboratory diagnoses are usually performed. The clinical illness may vary from a mild, self-limiting disease to a severe, debilitating disease that can lead to permanent damage or death.

Several distinct stages are associated with the clinical segment. After the incubation period, there generally is a **prodromal period** when symptoms such as headache and malaise (a feeling of ill-being) first appear. Symptoms and signs reach their peak during the **acute period** of disease. As the disease resolves, there is the **decline period,** followed by a **convalescent period** during which the patient regains strength.

Although there are no symptoms of disease during the incubation and convalescent periods, the host can be contagious and spread the infectious agent. Diseases with long incubation periods such as AIDS (see AIDS, page 536) or long convalescent periods such as typhoid fever (see typhoid fever, page 519) can be spread by a contagious host.

Disease Is Not Evident in Latent Infections

Some infections do not follow the normal natural history of disease. **Latent infections** do not immediately produce detectable or overt signs or symptoms. In some cases, as with the herpes simplex virus, which causes cold sores and genital herpes, the pathogen stops reproducing and remains dormant in the host cell until it is activated by such conditions as excessive sunlight, fever, or emotional stress (see herpes simplex infections, page 540). In other cases the pathogen is controlled, but not eliminated, by defensive mechanisms of the body. Infection becomes apparent only when the body's resistance is reduced and the pathogen is able to express itself. In tuberculosis and leprosy, the microbial agent persists in

Virulence of a microbe can be measured in several ways. Laboratory assays for toxin production, observations of cytopathic effects on tissue cultures, or

Measurement of Virulence

studies of lethality to hosts injected with the agent are useful methods of measuring virulence. For example, a pathogen may be inoculated into a group of susceptible animal hosts, and the number or amount of pathogen required to kill 50% of the animals in a given time period is determined. This amount is known as the **lethal dose, the 50% endpoint (LD_{50})** of the micro-

organism for the particular animal host (Figure 15.3).

The LD_{50} value depends on such factors as the culture medium and the growth stage of the inoculated pathogen, the route of inoculation (oral, intravenous, or intraperitoneal), and the characteristics of the animal host (sex, weight, age, species, and strain). When these parameters are specified, and when enough animals and enough different inocula have been used to provide a statistically valid response, the LD_{50} value is a measure of the relative virulence of a microorganism for a particular

animal host. The reason for choosing the 50% endpoint instead of other values for the lethal dose endpoint is that the rate of change in mortality is usually greatest around the 50% part of the dose-response curve.

A different parameter that is sometimes used to measure the transmissibility of a pathogen is the **ID_{50}** (infectious dose in 50% of the animals). This is the quantity of pathogen required to successfully infect 50% of the inoculated animals. Infection is usually detected by subculture of the agent from infected animal hosts. Because most pathogenic microbes have different degrees of virulence in different hosts, the LD_{50} and ID_{50} values for a specific pathogen must be taken in the context of the particular animal host used. Nonetheless, these measurements provide scientists with relative indices of microbial virulence.

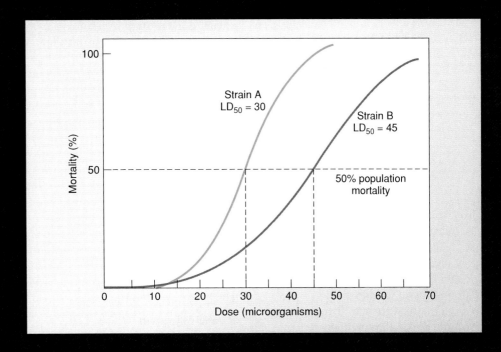

figure 15.3

Determination of the LD_{50}

The LD_{50} of a pathogen is determined by injecting various doses of it into laboratory animals. A dose-response curve is generated, and the LD_{50} is the dose required to kill 50% of the

animals. In the example, the LD_{50} for strain A is 30 microorganisms and for strain B, 45 microorganisms.

the host tissues for several years before signs or symptoms of disease appear (see tuberculosis, page 510, and leprosy, page 527). Latent infections are important because although the host may not undergo any obvious pathological changes, the infectious agent may cause some damage to the host or still be transmissible to other susceptible hosts.

Parasites Must Overcome Host Defenses to Initiate Infection and Disease

Whether infection does or does not develop into a disease depends on a combination both of host and microbial factors. One important factor is the **pathogenicity** of the invading microorganism—the ability of a microbe to establish infection and cause disease in a host. **Virulence** is a quantitative measure of pathogenicity and refers to the capacity of a microorganism to overcome the host's body defenses. Virulence differs among species and even strains of the same species of microbe, and depends in part on the type of host the pathogen infects. Highly virulent pathogens have a high probability of causing disease in a host. Hosts are protected against infection by physical, chemical, and cellular mechanisms. The invading microbe must overcome these mechanisms if it is to successfully establish itself in the host.

Microbial Factors of Virulence

There are two main ways by which microorganisms are able to damage a host. Some pathogens are **invasive;** that is, they invade the host, survive, multiply, and directly harm the host tissues by depriving them of nutrients or physically damaging them. Other pathogens produce **toxins,** biological molecules that are harmful to the host. A few pathogens are both invasive and toxigenic.

Microorganisms Have Many Mechanisms to Assist in Invasion

Microorganisms that are invasive first must survive, and then grow and multiply in host tissues. To survive, invasive microbes must overcome a number of host defensive barriers, including physical obstacles such as the skin and mucous membranes, chemical defenses such as interferon and lysozyme, and phagocytes (host scavenger cells that attack and ingest foreign particles). After the parasite gains access to the host, it may then become established and begin to reproduce. Not all potentially invasive infections develop into diseases. In some cases, the host defenses prevent the parasite from becoming established.

Microorganisms May Have a Preference for Particular Host Tissues or Organs

Many pathogens are **organotropic** (tissue or organ specific) in their infection of a host. For example, *Vibrio cholerae, Shigella dysenteriae, Salmonella typhimurium, Entamoeba histolytica, Giardia lamblia,* and rotaviruses colonize the gastrointestinal tract,

whereas *Streptococcus pneumoniae, Haemophilus influenzae, Ajellomyces capsulatus* (*Histoplasma capsulatum*), and rhinoviruses tend to invade the respiratory tract.

Organotropism may be determined by the nutritional requirements of a parasite, as in disease caused by the gram-negative rod *Brucella.* This bacterium causes infectious abortion in cows, sheep, pigs, goats, and dogs, but not in human beings. This difference in disease manifestation can be explained by the action of erythritol, a four-carbon tetrahydroxy alcohol ($HOCH_2$—$CHOH$—$CHOH$—CH_2OH). The presence of this alcohol in large quantities in the chorion, placenta, and fetal fluids of these other animals, but not in other animal tissues or in human beings, promotes rapid growth of *Brucella.* Humans infected with *Brucella* develop other symptoms of disease (remittent fever, focal necrosis and suppuration, and gastrointestinal symptoms).

Some Microorganisms Adhere to Specific Tissues or Organs

The ability of some microorganisms to adhere to specific tissues or organs is an important determinant for organotropism. Some bacteria such as *N. gonorrhoeae* possess pili (fimbriae) that attach to urogenital mucosal epithelial cells. Apiliated strains are less virulent.

Many bacteria have surface macromolecules that promote adherence to solid surfaces. An adhesive polysaccharide matrix, the **glycocalyx,** extends from the surface of many bacteria. It binds cells together in an aggregate mass, protects the cell from phagocytosis, and is involved in the attachment of cells to solid surfaces.

One of the best-studied examples of adhesion is the attachment of *Streptococcus mutans* to teeth. *S. mutans* produces dextran (a polymer of glucose) from dietary sucrose. This sticky, water-insoluble polysaccharide is thought to be important in the adherence of *S. mutans* to tooth surfaces and in the development of dental plaque (Figure 15.4). Evidence for this comes from observations that mutants of *S. mutans* defective in dextran synthesis do not form plaque on the tooth surfaces of laboratory rats, and from studies with germ-free animals. It furthermore has been found that enzymes that hydrolyze dextran (dextranases) are effective in reducing the quantity of dental plaque in experimental animals. Although *S. mutans* is important in the formation of plaque and the initiation of dental caries, other bacteria such as lactobacilli and actinomycetes are also commonly found on teeth and may contribute to carious enamel lesions. The metabolic acids produced by these bacteria are an important determinant of tooth decay (see dental caries, page 514).

Microbial Pathogenesis
Adherence: Mechanisms • pp. 2–7

Microbial Enzymes Aid in Invasion by Damaging Tissue or Dissolving Materials

Many different types of microbial enzymes are produced that assist microorganisms in invasion (Table 15.4). These enzymes may enhance virulence by causing tissue damage or by dissolving materials that normally would prevent spread of the pathogen through the tissue. **Hemolysins,** enzymes that destroy erythrocytes (red

figure 15.4

Adherence of *Streptococcus mutans* to Tooth Enamel

Dextran produced by *S. mutans* is important in its adherence to tooth surfaces and in the development of dental plaque (colorized, ×31,000).

table 15.4

Microbial Enzymes Associated with Virulence

Enzyme	Mode of Action	Examples of Bacteria Producing the Enzyme
Coagulase	Clots plasma	*Staphylococcus aureus*
Collagenase	Degrades collagen in muscle tissue	*Clostridium perfringens, Pseudomonas aeruginosa*
Deoxyribonuclease (DNase)	Degrades DNA in pus	*Staphylococcus aureus, Streptococcus pyogenes*
Hemolysin	Lyses red blood cells and other cells	*Staphylococcus aureus, Streptococcus pyogenes*
Hyaluronidase	Degrades the ground substance of connective tissue	*Clostridium perfringens, Staphylococcus aureus, Streptococcus pyogenes*
Lecithinase	Splits lecithin in plasma membranes	*Clostridium perfringens*
Leukocidin	Destroys white blood cells	*Staphylococcus aureus*
Staphylokinase, streptokinase	Converts plasminogen to plasmin, dissolving blood clots	*Staphylococcus aureus, Streptococcus pyogenes*

blood cells), are produced by many bacteria, including streptococci and staphylococci (see hemolysis, page 502). Some hemolysins lyse not only erythrocytes, but also other types of cells. For example, *Streptococcus pyogenes* produces two hemolysins, **streptolysin O** and **streptolysin S.** Streptolysin O, so named because it is inactivated by oxygen, causes the release of cytoplasmic granules in leukocytes (white blood cells), resulting in lysis. Streptolysin S, an oxygen-stable hemolysin (the S refers to the fact that the enzyme is produced only when the bacteria are grown in serum-containing media), has a similar though less dramatic effect. Staphylococcal α **hemolysin** disrupts lysosomes and is toxic for a variety of tissue cells, including those found in muscle and the kidneys.

Streptokinase and **staphylokinase** are proteases that convert plasminogen in human serum to plasmin. The fibrinolytic plasmin dissolves blood clots and therefore may promote spread of the bacteria to surrounding tissue. Streptokinase has been successfully used to dissolve blood clots in heart attack victims. In contrast, **coagulase,** an enzyme produced by *Staphylococcus aureus*, converts fibrinogen to fibrin, resulting in plasma clotting. The virulence of staphylococci is correlated with coagulase production. As coagulase is produced, it forms a fibrin wall around skin lesions caused by staphylococci to provide a restricted focus of infection (see staphylococcal pyodermas, page 526). Coagulase may also contribute to pathogenicity by inhibiting phagocytosis through deposition of fibrin on the bacterial cell surface, although this has not been proven.

Some streptococci and staphylococci produce **deoxyribonucleases (DNases),** which degrade DNA. The thick pus that accumulates in areas of streptococcal infections often contains viscous DNA released from dead phagocytes and other cells. The DNases degrade this DNA, liquefy exudates, and promote the movement of the bacteria to other parts of the body.

Leukocidins, produced by *Staphylococcus aureus*, destroy polymorphonuclear leukocytes (PMNs). One leukocidin produced by staphylococci binds to the membrane of the PMN and increases permeability. The leukocidin also causes fusion between the membranes of cytoplasmic granules inside the PMN and the plasma membrane of the cell, resulting in cell disruption and release of the granules away from the cell. Diseases caused by staphylococci are often associated with pus comprised, in part, of these destroyed PMNs. The PMNs protect the body against infection, and their destruction is one reason that staphylococci often successfully invade tissues (see polymorphonuclear leukocytes, page 456, and phagocytosis, page 460).

Streptococci, staphylococci, and clostridia produce **hyaluronidase.** This enzyme is called the spreading factor because it breaks down the ground substance of connective tissue and may be instrumental in the spread of these bacteria through body tissues.

Collagenase, an enzyme that breaks down the collagen framework of muscle tissue, is produced by *Clostridium perfringens*, a causative agent of gas gangrene. The spread of clostridia through tissue in gas gangrene is facilitated by the action of this enzyme.

 Gram-Positive Organisms
Gram-Positive Cocci—Streptococci: Virulence Factors • pp. 15–18

Capsules Protect Microorganisms from Phagocytosis

Some pathogenic bacteria produce thick outer capsules that protect them from phagocytosis and other antimicrobial activities of the host (Figure 15.5). These capsules are composed of polysaccharides, proteins, or both. Capsules are particularly important in the virulence of such bacteria as *Streptococcus pneumoniae, Klebsiella pneumoniae, Neisseria meningitidis, Haemophilus influenzae*, and *Bacillus anthracis*. In these encapsulated bacteria, removal of the capsule by physical or genetic manipulation significantly decreases the virulence of the microbe.

 **Microbial Pathogenesis**
Antiphagocytic Factors • pp. 2–5

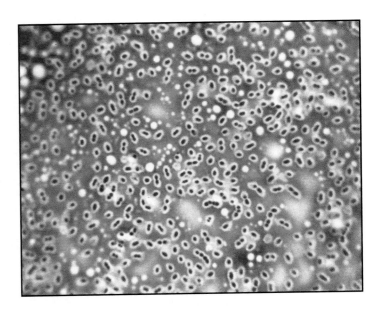

figure 15.5
Micrograph Showing the Thick Capsule of *Streptococcus pneumoniae*

Microbes Produce Two Types of Toxins: Exotoxins and Endotoxins

Toxins are poisons synthesized by living organisms. Microbes produce two kinds of toxins (Table 15.5). An **exotoxin** is a soluble protein released into the surrounding environment by a microorganism during growth and metabolism. An **endotoxin** is the lipopolysaccharide portion of the outer membrane of a gram-negative bacterium. The Lipid A component of lipopolysaccharide is responsible for the toxicity of endotoxin. Unlike exotoxins, which are usually excreted into the environment by viable cells, endotoxins are usually released only upon death and lysis of the bacterial cell. Some gram-negative bacteria such as *Shigella dysenteriae, Vibrio cholerae*, and some strains of *Escherichia coli* produce both an exotoxin and an endotoxin, whereas gram-positive bacteria, which do not have an outer membrane, produce only exotoxins. Endotoxins are a major cause of death in surgical infections, afflicting debilitated hospitalized patients.

Bacterial Exotoxins Are Among the Most Potent Toxins Known

Exotoxins are produced predominantly by gram-positive bacteria and are more specific in their mode of action and generally more damaging than are endotoxins (Table 15.6). Exotoxins may be divided into several categories on the basis of the site affected: **enterotoxins** (small intestine), **neurotoxins** (nerve tissue), and **cytotoxins** (general tissue). *Staphylococcus aureus, Clostridium perfringens, Escherichia coli*, and *Vibrio cholerae* are examples of

table 15.5

Comparison of Endotoxins and Exotoxins

Characteristic	Endotoxin	Exotoxin
Bacterial source	Gram-negative bacteria	Primarily gram-positive bacteria, some gram-negative bacteria
Location in bacterium	Lipopolysaccharide	Product of bacterial cell, released extracellularly
Chemical structure	Toxic activity resides in lipid portion of lipopolysaccharide	Protein
Heat stability	Stable; can withstand 121°C for 1 hr	Unstable; usually destroyed by heating at 60°C to 80°C (except enterotoxins)
Toxicity	Low	High
Toxoid production	No	Yes (except enterotoxins)
Representative symptoms	Fever, inflammation, increased phagocytosis, rash, septic shock	Neurological complications, cell necrosis, loss of fluid and electrolytes from intestines
Representative diseases	Septic abortion	Diphtheria, botulism, tetanus, cholera

table 15.6

Bacterial Exotoxins

Bacterium	Disease	Toxin	Mode of Action
Bordetella pertussis	Whooping cough	Pertussis toxin	Necrosis
Clostridium botulinum	Botulism	Neurotoxin	Blocking the release of acetylcholine, resulting in flaccid paralysis
Clostridium perfringens	Gas gangrene, food poisoning	α-toxin	Lecithinase
		β-toxin	Necrosis
		ε-toxin	Necrosis
		ι-toxin	Necrosis
		θ-toxin	Hemolysis
Clostridium tetani	Tetanus	Neurotoxin	Blocking the release of inhibitory transmitter, resulting in spastic paralysis
Corynebacterium diphtheriae	Diphtheria	Diphtheria toxin	Inhibition of the elongation step in eucaryotic protein synthesis
Escherichia coli (enterotoxigenic strains)	Gastroenteritis	Heat-stable toxin (ST)	Activation of guanylate cyclase, resulting in fluid and electrolyte loss from intestinal cells
		Heat-labile toxin (LT)	Activation of adenylate cyclase, resulting in fluid and electrolyte loss from intestinal cells
Shigella dysenteriae	Bacillary dysentery	Neurotoxin	Inhibition of protein synthesis
Staphylococcus aureus	Pyogenic infections (for example, impetigo)	α-toxin	Hemolysis
		β-toxin	Hemolysis
		γ-toxin	Hemolysis
		δ-toxin	Hemolysis
		Enterotoxin	Unknown
		Leukocidin	Degranulation of leukocytes
Streptococcus pyogenes	Pyogenic infections (for example, impetigo), scarlet fever	Streptolysin O	Hemolysis
		Streptolysin S	Hemolysis
		Erythrogenic toxin	Localized erythematous reactions (abnormal redness of the skin)
Vibrio cholerae	Cholera	Choleragen	Activation of adenylate cyclase, resulting in fluid and electrolyte loss from intestinal cells
Yersinia pestis	Bubonic plague, pneumonic plague, septicemic plague	Plague toxin	Necrosis (possible)

Detection of Endotoxin

Bacterial endotoxin can be detected even in minute concentrations by a laboratory test known as the **Limulus amoebocyte lysate (LAL) assay.** Although the mechanism of the assay is not known, endotoxin causes gelation of the lysate, an aqueous extract of amoebocytes from the blood of the horseshoe crab, *Limulus polyphemus*. An assay is performed by mixing dilutions of the specimen suspected to contain endotoxin with equal volumes of *Limulus* amoebocyte lysate in small test tubes. The mixture is incubated for 60 minutes at 37°C, and the degree of gelation is measured either visually or with a spectrophotometer. The LAL assay detects endotoxin in picogram (10^{-12} gram) quantities.

Clinical evaluations of the assay on blood samples of patients with gram-negative bacteremia (bacteria in the blood) have been somewhat disappointing. The LAL test is positive in only about 50% of these patients. The test also has given false-positive results of tests on blood samples from patients who do not have bacteremia. Despite these shortcomings, the LAL procedure is valuable for the assay and testing of sterile parenteral (non-oral) hospital solutions for the presence of endotoxin, as well as for the detection of gram-negative bacterial meningitis and other diseases.

---

bacteria that may produce enterotoxins. *Clostridium botulinum* and *Clostridium tetani* are neurotoxin-producing bacteria. An example of a cytotoxin is diphtheria toxin, which inhibits protein synthesis in human mucosal cells and in kidney and heart tissue.

Bacterial exotoxins are among the most potent toxins known. One nanogram (10^{-9} g) of botulinum toxin is a lethal dose for a guinea pig. One milligram of botulinum toxin could kill 1,000 human beings.

Exotoxins are soluble protein molecules that may be rendered harmless by digestive enzymes (acids) as well as by boiling. However, some exotoxins, such as botulinum toxin, are made more potent by partial proteolysis. Since exotoxins are small, soluble proteins, they are very immunogenic and elicit a strong immune response. Exotoxins can be converted to **toxoids** by treatment with formalin (formaldehyde) or other chemicals (for example, alum). Toxoids are toxin preparations that have lost their toxic activity but still are able to elicit the host immune response. Toxoids are used as vaccines for diseases such as diphtheria and tetanus (see toxoid, page 478).

Many exotoxins are coded for by plasmid genes (*E. coli* heat-stable and heat-labile enterotoxins, *S. aureus* enterotoxin, and tetanus toxin), and exotoxin production is lost if the plasmid is lost. Other exotoxins are coded for by genes in the bacterial chromosome (choleragen) or in a prophage in the chromosome (botulinum toxin and diphtheria toxin). The conferring of toxigenicity on a bacterium by a bacteriophage is called **lysogenic conversion** (see lysogenic conversion, page 258). The lysogenic bacteriophage contains the gene coding for the toxin and expresses it during the prophage stage. Lysogenic conversion is an exception to the general principle that phage genes are not expressed in the lysogenic state. *Streptococcus pyogenes, Clostridium botulinum,* and *Corynebacterium diphtheriae* are examples of bacteria that undergo lysogenic conversion.

Microbial Pathogenesis
Toxins: Exotoxins • pp. 10–17

Bacterial Endotoxins Are Part of the Gram-Negative Outer Membrane

Endotoxins, unlike the heat-labile protein exotoxins, are composed of heat-stable lipopolysaccharides that are part of the outer membrane of gram-negative bacteria (see outer membrane, page 57). Their toxicity resides in the lipid portion of the lipopolysaccharide molecule. Because they are structural components of the cell envelope, endotoxins generally are not released into the environment until the cell lyses, although sometimes endotoxin is released by growing bacteria without cell lysis. Endotoxins cannot be converted to toxoids, are less toxic than typical exotoxins, and generally have the same mode of action regardless of the bacterium that produces them.

Host Resistance and Tuberculosis

Tuberculosis, caused by *Mycobacterium tuberculosis* (see tuberculosis, page 510), is an example of an infectious disease in which a person's innate and acquired resistance influences susceptibility to infection. Tuberculosis infects 1 billion people worldwide, and 3 million people annually die of the disease, but not everyone is equally susceptible to infection.

In the United States, tuberculosis is more prevalent among people of black African descent, American Indians, and Eskimos than people of European descent. For example, a survey of men entering the U.S. Navy between 1958 and 1969 showed that 12.4% of recruits of black African descent tested positive for tuberculosis compared with 3.8% of recruits of European descent. This variation in resistance is likely associated with the historical exposure of different populations to *M. tuberculosis*. Tuberculosis was a disease common in Europe during the seventeenth century and earlier (the seventeenth century English author John Bunyan wrote of tuberculosis as "the agent of all these men of death"). Consequently, descendants of the European population may have developed a higher immunity to the tubercle bacillus than descendants of other populations that were not exposed to tuberculosis until the last two or three centuries, when they came in contact with Europeans.

Epidemiological evidence suggests that tuberculosis mortality and case rates are also affected by socioeconomic factors, age, and occupation. Tuberculosis has long been associated with malnutrition, overcrowding, and poverty. A study of New York City school personnel in 1973 to 1974 showed that the frequency of tuberculosis was four times higher (22.4% versus 5.5%) in the poorest areas than in areas with the highest socioeconomic standards, after adjustments for differences in population composition by age, sex, and race. Frequently in areas of poverty there is decreased innate host resistance to disease because of malnutrition and inadequate medical care. The risk of exposure to disease is increased from overcrowding and lack of sanitary facilities.

Other factors, including age and occupation, may also play a role in the development of tuberculosis. The incidence of tuberculosis infections increases with age, leveling off at about 40 to 50 years. The disease is especially severe in infants, presumably because of their immature immune mechanisms. Pulmonary tuberculosis rates are high among miners and foundry workers, whose lungs are constantly exposed to irritants and particles from the work environment. Although some of these variations in resistance may be associated with differences in populations, there is no doubt that resistance in humans varies more strikingly in tuberculosis than in most other infectious diseases.

The effects of bacterial endotoxins on a host are more difficult to assess than the effects of exotoxins, since less is known about the mechanism of endotoxin activity. Endotoxins stimulate a large number of host metabolic and immunologic events (Table 15.7). Hosts exposed to endotoxins generally show symptoms such as fever, inflammation, increased phagocytosis, rash, and septic shock (shock associated with the multiplication of bacteria in the blood).

Microbial Pathogenesis
Toxins: Endotoxins • p. 9

Endotoxins induce fever by causing the release of **endogenous pyrogens** (temperature-elevating substances) from polymorphonuclear leukocytes. The pyrogens act at the hypothalamus, which regulates body temperature, causing conservation of body heat and a rise in temperature. Inflammation, increased phagocytosis, and rashes are related to the effects of endotoxins on the host immune system. Septic shock is caused by an increase in capillary permeability resulting in reduced blood pressure. Pooling of the blood (blood stasis) occurs in various organ systems. Because of this blood stasis, wastes and toxic metabolic products accumulate, and nutrients and oxygen do not reach the tissues. Prolonged inflammatory and pyrogenic effects of endotoxins can lead to severe and irreversible damage to body tissues and organs.

Microbial Pathogenesis
Toxins: Endotoxins • pp. 2–9

Pathogenic microorganisms can damage a host by direct invasion, by the production of toxins, or by both means. Some bacterial products directly assist in the invasion and establishment of the pathogen in the human body. Exotoxins and endotoxins produced by microbes can have serious, sometimes fatal consequences

table 15.7

Consequences of Endotoxin Action in Humans

Consequence	Mechanism
Fever	Release of endogenous pyrogens (from polymorphonuclear leukocytes) that affect the hypothalamus
Inflammation	Activation of polymorphonuclear leukocytes by the presence of toxin
Increased phagocytosis	Activation of alternative complement pathway (see Chapter 16)
Rash	Hemorrhage and coagulation in capillaries of the skin
Septic shock	Increased capillary permeability with disseminated intravascular coagulation and blood stasis

in microbial disease. No single factor determines the eventual outcome of microbial infection, but these microbial characteristics play a significant role in determining the course of disease.

Innate (Nonspecific) Host Resistance

The many different virulence factors of microorganisms appear to give them an advantage in host-parasite relationships. Most host organisms, however, have defense mechanisms to resist invasion by microbial parasites. Were it not for these host mechanisms of resistance, illness and disease would be rampant in the world, and many more hosts would perish.

Host defenses can be divided into two basic categories: (1) **innate (nonspecific) resistance** and (2) **acquired (specific) resistance,** or **immunity**. Defenses that provide general host protection against any type of pathogen are called nonspecific. Innate resistance does not depend on prior exposure of the host to the invading parasite for activation. Acquired resistance, in contrast, occurs only after the host has been exposed to a pathogen or its products. Acquired resistance involves specific proteins (antibodies) and white blood cells (T lymphocytes) as well as monocytes. Nonspecific defenses are characteristic to all animals. In comparison, the specific immune response is an evolutionary characteristic of vertebrates, and partly responsible for the adaptive success of vertebrates.

Factors associated with nonspecific host resistance to invasion by a pathogen include the general health and physiological condition of the host; physical barriers to invasion, such as the skin and mucous membranes; antimicrobial substances produced by the host; components of the blood; inflammation; and phagocytosis. Nonspecific resistance typically involves general defense mechanisms designed to protect the host against many types of pathogens rather than a specific pathogen. Such nonspecific resistance provides the host with immediate protection

against a broad spectrum of invading parasites until immune mechanisms are able to mount a defense directed specifically against the entering pathogen.

The General Health and Physiological Condition of the Host Can Influence the Course of a Disease

A person's overall health and physiological condition can significantly influence the course of a disease. Factors such as age, sex, race, nutrition, stress, and occupation are important determinants in disease development.

The incidence of disease is generally higher in the young, whose immune systems are not fully developed, and in the very old, whose immune systems are less effective, especially to chronic diseases. Many infectious diseases, such as the childhood diseases measles, mumps, and chickenpox, confer lifelong immunity after exposure. Other diseases, such as those involving the urinary and respiratory tracts, do not permit lasting immunity and tend to be chronic in older persons.

Incidence of certain diseases among people are modified by factors such as hormonal composition, ethnicity, and behaviors. Breast cancer has a markedly higher incidence in females, whereas cancer of the larynx and esophagus occurs more frequently in males. A possible explanation for the higher incidence of larynx and esophageal cancer in males is that historically more men than women tended to smoke. People of black African descent have substantially higher incidence rates of tuberculosis than people of European descent. The difference may be due to the history of exposure of the two populations. Tuberculosis has been epidemic in Europe for centuries, whereas the disease became epidemic in sub-Saharan Africa only after 1900 (see host resistance and tuberculosis, page 453).

Malnutrition and occupation often are contributing factors to disease incidence and severity. For example, measles and tuberculosis are common and more likely to be severe in countries where there is inadequate diet and limited medical care services. In contrast, these diseases are rarely fatal in the areas of the world where adequate medical services are available, although in recent years the number of cases and deaths from tuberculosis has increased in the United States because of drug-resistant strains of *Mycobacterium tuberculosis* and increased susceptibility in immunocompromised individuals (see tuberculosis, page 510). Many specific diseases are directly related to occupational exposures. Among these diseases are hepatitis in nurses, medical technologists, and dental practitioners and brucellosis in veterinarians, livestock raisers, and meat cutters. If occupational risks of exposure to infectious agents are identifiable, corrective measures can be taken to minimize them. For example, today dental practitioners routinely use gloves and face masks to reduce the risk of infection by oral pathogens present in patients.

Age, sex, race, nutrition, and occupation are only a few of the host factors that may influence the incidence of infectious diseases. The role of these and other epidemiological factors in disease development is discussed in Chapter 18.

figure 15.6

Scanning Electron Micrograph of Ciliated Cells in the Mucous Membrane of the Trachea (colorized, ×4,170)

Physical Barriers Protect the Human Body Against Infectious Microorganisms

The human body harbors many different types of microorganisms, including some that are potential pathogens. Despite the constant contact of these microbes, few cause problems because of the protection afforded by the skin and mucous membranes. These tissues provide a first line of defense. Problems arise when microbes penetrate these external surface layers.

Skin consists of tightly packed epithelial cells stacked in continuous strata, an effective barrier if unbroken. However, microbes that gain entry through breaks in the skin can travel to other parts of the body via the blood or lymphatic systems and cause serious diseases. Microbial infection due to damaged skin can be minimized by the application of iodine or other antiseptics to the lesion (see antiseptics, page 122).

The orifices of our bodies are convenient portals of entry for invading microorganisms. The mucous membranes lining these openings and body cavities are easier to penetrate than unbroken skin. However, infection of the mucous membranes is kept at a minimum by nonspecific anatomic defenses. Cells in the mucous membrane of the respiratory tract, for example, secrete thick, viscous mucus that entraps foreign particles, dust, and microorganisms. Cilia (minute, hairlike projections on the respiratory epithelium) beat synchronously and carry this particle-laden mucus to the nose and throat, away from the lungs (Figure 15.6). Mucus is swallowed, and microorganisms trapped in it are destroyed primarily by the acidic pH of the stomach. Most microbial pathogens are unable to survive the harsh environment of the stomach, but neutralization of gastric acid by antacids can increase susceptibility to certain diseases. Certain strains of *Vibrio cholerae*, for example, must be ingested at high doses of 10^8 cells or greater before infection occurs. The infecting dose is lowered to 10^4 cells if the acid of the stomach is neutralized. Gastric acidity thus plays an important role in host defense against cholera.

Other body fluids also serve to cleanse the body of foreign materials. The periodic flow of urine through the urethra, tears washing over the surface of the eyeball, secretion of vaginal fluids, vomiting, and peristalsis are examples of such cleansing actions.

Antimicrobial Substances Produced by the Body Protect Against Invasive Microorganisms

Our bodies produce a number of antimicrobial substances that combat invasive microorganisms. One such inhibitor is the enzyme lysozyme. Lysozyme, commonly found in tears, saliva, and nasal secretions, hydrolyzes cell wall peptidoglycan in bacteria, resulting in cell lysis (see lysozyme, page 59). The antiviral agent interferon, naturally produced by the body in response to a viral infection, is an important line of defense and can prevent the replication of viral pathogens within cells (see interferon, page 406).

Sebaceous glands in the skin secrete an oily substance, sebum, that forms a thin, protective film on the skin's surface. Sebum is composed of fats and oils, which are inhibitory for most bacteria and fungi. The salts in sweat also are antimicrobic.

Microbial Pathogenesis
Nonspecific Host Defense: Mechanical/Chemical Control • pp. 4–20

Human Blood Contains Protective Factors

The average adult human has approximately 5 to 6 liters of blood circulating through the arteries, capillaries, and veins. Blood consists of a fluid matrix, or plasma, and a cellular portion. Plasma, the portion of blood that remains after the cells are removed, contains fibrinogen, a protein that is instrumental in clotting. If plasma is allowed to clot, the fluid portion that remains after removal of the clotted material is called **serum.** The serum portion of plasma contains several different kinds of proteins, among them **antibodies** and **complement.** Antibodies are molecules that react specifically with substances called **antigens** and are associated with immunity (see antibodies, antigens, and complement, page 468).

Formed elements (cells and their derivatives) constitute approximately 45% of the volume of blood, and plasma makes up the remaining volume. The formed elements consist of three major types of cells: **erythrocytes, platelets,** and **leukocytes** (Table 15.8). Erythrocytes, or red blood cells, contain hemoglobin and transport oxygen and carbon dioxide in the circulatory system. Platelets, also known as thrombocytes, are small, fragile, cytoplasmic cell fragments involved in blood coagulation and transportation of the vasoconstrictor serotonin.

Leukocytes, commonly known as white blood cells because they lack pigment, make up only a small fraction (less than 0.2%) of the total number of cells in blood. Despite their small numbers of only 5,000 to 9,000 cells per cubic millimeter of blood, leukocytes are vital in combating infection. Leukocytes stained with Wright's stain (a methyl alcoholic mixture of an acid and a basic dye) are differentiated into five basic cell types: **basophils, eosinophils, neutrophils, lymphocytes,** and **monocytes.**

Granulocytes Contain Stainable Cytoplasmic Granules

The first three cell types—basophils, eosinophils, and neutrophils—are collectively known as **granulocytes** because they possess stainable cytoplasmic granules. Granulocytes have a short life span (seldom more than two weeks) and are named by their staining characteristics. Basophils, which make up 0.5% to 1% of leukocytes, retain the basic dye in Wright's stain, and their granules stain a deep purple. Basophil granules contain the vasodilator histamine and other substances, the release of which during antibody-mediated hypersensitive reactions causes increased capillary permeability and other symptoms of inflammation.

Eosinophil granules attract acid stains and stain a bright red with eosin-containing dyes. Eosinophils account for 2% to 4% of the total white cell count in blood and have an affinity for antigen-antibody complexes, which they phagocytize. Eosinophils gather at tissue sites of antibody-mediated allergic reactions and release enzymes that degrade histamine, thereby limiting the pathological effects of these allergic reactions. Eosinophils also are associated with immunity to some parasites, particularly helminths (parasitic worms), as well as having some detoxification functions.

The most numerous type of white cell in the blood is the neutrophil. Neutrophils make up approximately 50% to 70% of all the leukocytes, with 3,000 to 7,000 neutrophils in each cubic millimeter of normal blood. Neutrophils are commonly called **polymorphonuclear leukocytes (PMNs)** because of their multilobed nucleus. Neutrophil granules attract both acidic and basic dyes, giving them a neutral lavender tint.

Neutrophilic granulocytes play a prominent role in phagocytosis and the destruction of foreign particles in the body. They normally circulate in relatively low numbers in the bloodstream and protect against microbial invasion. However, during acute infections, the number of white blood cells (particularly neutrophils) dramatically increases, a condition known as **leukocytosis.** Excess neutrophils are drawn from reserves in the bone marrow (where they are produced) and are chemotactically attracted to the sites of injury or infection. Neutrophils squeeze through the capillary walls into the tissue (a process called *diapedesis*) and phagocytize foreign particles or microbes. Hydrolytic enzymes released from neutrophil granules aid in the digestion of ingested materials.

Leukocytosis is frequently used as an aid in the clinical diagnosis of many diseases, including tuberculosis, pneumonia, mononucleosis, leukemia, and meningitis. In some severe infections such as malaria, typhoid fever, measles, and influenza, total leukocyte counts actually decrease to below 5,000 per cubic millimeter of blood. This condition is known as **leukopenia.** As soon as the bone marrow produces sufficient leukocytes, leukopenia is replaced by leukocytosis.

Formed Elements of Blood

Cell Type		Number in Human Blood (Per Cubic mm)	Function
Erythrocytes (red blood cells)		4.5 million to 6.0 million	Transport of oxygen and carbon dioxide in the circulatory system
Platelets (thrombocytes)		140,000 to 340,000	Blood coagulation, carriers of serotonin
Leukocytes (white blood cells) 1. Granulocytes a. Basophils (0.5%–1% of all leukocytes)		5,000 to 9,000	Involvement in hypersensitive reactions, inflammation
b. Eosinophils (2%–4% of all leukocytes)			Phagocytosis, involvement in hypersensitive reactions, inflammation
c. Neutrophils (polymorphonuclear leukocytes; 50%–70% of all leukocytes)			Phagocytosis, inflammation
2. Agranulocytes a. Lymphocytes (B and T lymphocytes; 20%–25% of all leukocytes)			Antibody synthesis, cell-mediated immunity
b. Monocytes (macrophages; 3%–8% of all leukocytes)			Phagocytosis

Agranulocytes Do Not Contain Cytoplasmic Granules

Lymphocytes and monocytes are white blood cells called **agranulocytes** because they do not possess cytoplasmic granules. Certain agranulocytes, unlike granulocytes, are long-lived cells.

A lymphocyte is a small mononuclear cell with a nucleus that occupies most of the cell volume. Lymphocytes account for 20% to 25% of all leukocytes. They mature either in the bone marrow (B lymphocytes) or in the thymus (T lymphocytes). Lymphocytes are nonphagocytic and are involved in antibody formation and cell-mediated immunity (see B lymphocytes and T lymphocytes, page 468).

Monocytes, which make up 3% to 8% of all leukocytes, have a single, kidney-shaped nucleus and are active in phagocytosis. They are produced in the bone marrow and released into the blood. Circulating monocytes migrate into the tissues, where they differentiate into **macrophages,** which have a primary function of phagocytosis and destruction of invading microorganisms. Macrophages are also important participants in immune responses.

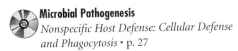

Microbial Pathogenesis
Nonspecific Host Defense: Cellular Defense and Phagocytosis • p. 27

The Reticuloendothelial and Lymphatic Systems Are Primary Defenses of the Body

The **reticuloendothelial (RE) system,** also called the **mononuclear phagocytic system,** is one of the primary defenses of the body (Figure 15.7). The RE system consists of two types of phagocytic cells: **wandering macrophages** and **stationary,** or **fixed, macrophages.** Wandering macrophages are mobile and migrate through the tissues to infected areas. There are two types of fixed macrophages. **Reticulum cells** form a loose, underlying structural network in the connective tissue that extends throughout the body. **Endothelial cells** are found in the spleen, liver sinusoids (Kupffer cells), lymph nodes, adrenal gland, and pituitary gland, where they are arranged as lining cells.

The **lymphatic system** consists of an interconnecting network of vessels that collects excess fluid, called the **lymph,** from the body's tissues (Figure 15.7). The excess fluid originates in the blood, but escapes into interstitial spaces in the tissue through capillary walls and is returned to the blood circulatory system through lymphatic vessels. One-way valves in the vessels keep the lymph moving away from the tissue. The returning lymph, which carries bacteria, viruses, and other potential pathogens from the tissues, flows through small, bean-shaped structures called **lymph nodes,** the spleen, appendix, adenoids, tonsils, and other tissues. Lymph nodes contain lymphocytes and macrophages that help clear the lymph of its pathogens. Occasionally the lymph nodes become infected and swollen because of large numbers of microorganisms entering the nodes.

Inflammation Is the Body's Response to Injury, Irritation, or Infection

Inflammation is the body's nonspecific response to injury, irritation, or infection caused by a physical, chemical, or biological agent (Figure 15.8). The hallmark of inflammation is dilation and increased permeability of the blood vessels in the affected area, producing the four cardinal symptoms of edema, erythema, pain, and heat. The resultant increased blood flow contributes to erythema (reddening), whereas the loss of plasma and leukocytes into the tissue by diapedesis yields edema (swelling) and pyogenesis (pus formation). Leukocytes are attracted to the damaged site by chemotactic factors that may include bacterial products, chemicals released from inflamed tissues, and complement-derived peptides. These leukocytes phagocytize and destroy the foreign particles associated with the inflammation. Living, dying, and dead leukocytes (primarily neutrophils) combine with fluid, tissue debris, and living or dead microorganisms (in microbial infections) to produce a **purulent** (pus-filled) **exudate.** Basophils, mast cells (nonmotile connective tissue cells found next to capillaries), and other cells release large amounts of chemical substances called **mediators** at the inflammatory site. These include histamine (the main chemical mediator), serotonin, prostaglandins, cytokines, and bradykinins which, among other effects, are vasodilators that increase blood vessel permeability and enhance the inflammatory response.

The elevation of body temperature (fever) is another important aspect of inflammation, although not always occurring. Fever is a response to chemical substances called endogenous pyrogens that are released by leukocytes interacting with bacterial endotoxins. Endotoxins also can act as pyrogens in gram-negative infections. Pyrogens act upon the hypothalamus region of the brain to cause a rise in body temperature. Elevated body temperatures enhance phagocytic activity and may also reduce the growth rate of pathogenic bacteria.

In the early stages of inflammation, neutrophils predominate in the exudate. This type of inflammation, which has a rapid onset and persists for a few days or weeks, is called **acute. Chronic inflammation** remains for several months or years and occurs when the injuring agent (or its products) persists in the tissue and the host is unable to effectively respond to the injury. Lymphocytes and macrophages are the predominant cell types in chronic inflammation. Connective tissue proliferation is prominent in this type of inflammation and is an attempt by the host to isolate the damaged area. A **subacute inflammation** is an intermediate form of inflammation and usually lasts for a few weeks to a few months.

All of these events combine to give characteristic inflammatory symptoms of erythema, edema, pyogenesis, fever, and pain. The inflammatory response is a nonspecific mechanism by which

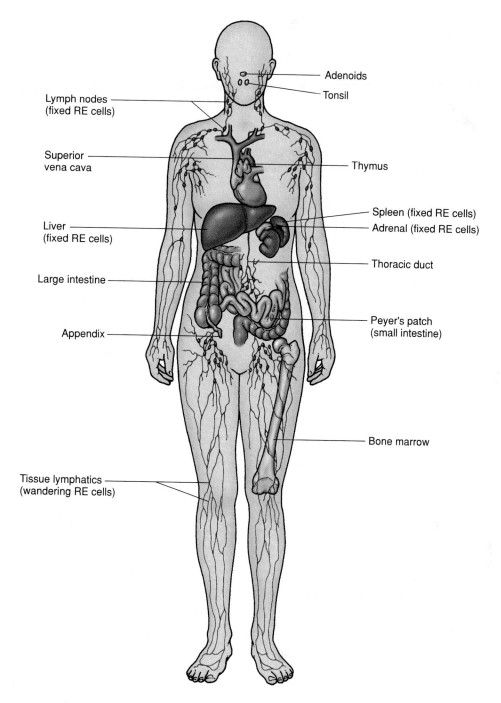

Adenoids

Tonsil

Lymph nodes
(fixed RE cells)

Superior
vena cava

Thymus

Spleen (fixed RE cells)
Adrenal (fixed RE cells)

Liver
(fixed RE cells)

Thoracic duct

Large intestine

Peyer's patch
(small intestine)

Appendix

Bone marrow

Tissue lymphatics
(wandering RE cells)

figure 15.7

The Reticuloendothelial and Lymphatic Systems

The reticuloendothelial (RE) system is composed of fixed and wandering macrophages that are located at various body sites. The lymphatic system consists of a network of lymphatic vessels that drain fluid from the tissue into the lymph nodes. The system also consists of the spleen, blood lymphocytes, adenoids, tonsils, appendix, and Peyer's patches.

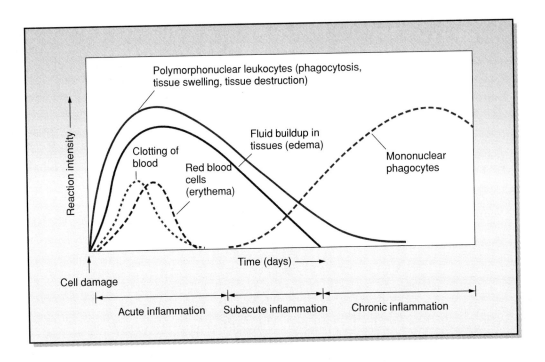

figure 15.8

Events Accompanying Inflammation

Inflammation is characterized by dilation and increased permeability of the blood vessels in the affected area of the body. The sequence of events accompanying inflammation is shown.

the host contains and eliminates bacterial pathogens and other injurious materials. However, prolonged inflammation can be harmful to the host, causing persistent pain, progressive tissue destruction, and potential organ damage and loss of function. Anti-inflammatory drugs are used to minimize some of these effects of inflammation. Antihistamines block the effects of histamines. Aspirin and corticosteroid hormones are anti-inflammatory agents that can reduce the pain and discomfort associated with inflammation.

Phagocytes Digest Bacteria, Viruses, and Other Foreign Materials

Phagocytosis [Greek *phagein*, to eat, *cytos*, cell] is the process of ingestion and digestion by cells of a substance such as a bacterium, virus, or other foreign material. As part of the nonspecific defense of a host, phagocytosis is an important mechanism that removes undesirable and harmful particles from the body. Different types of blood cells may be involved in phagocytosis, but the most common phagocytes are polymorphonuclear leukocytes and macro-phages. Phagocytes are attracted to microbial pathogens or injured tissue by the same chemotactic substances that are associated with inflammation. The process of phagocytosis can be divided into three stages: (1) adherence, (2) ingestion, and (3) digestion.

Phagocytosis begins with the phagocyte's recognition of, and adherence to, the infectious agent (Figure 15.9). Chemotactic factors released by infectious microbes are believed to attract phagocytic cells. Phagocytes have surface receptors that enable them to attach nonspecifically to bacteria and other foreign particles. Phagocytosis is reduced by the presence of capsules or other similar coverings on the bacterial surface, but phagocyte attachment is greatly enhanced in the presence of specific proteins called **opsonins.** Opsonins such as antibodies and the complement fragment C3b bind to bacterial surfaces to form a bridge between the surfaces and the phagocytes, thereby promoting phagocytosis (see complement, page 481). The importance of C3b opsonization can be seen in those rare individuals who are genetically deficient in complement component C3: they are more susceptible to recurrent bacterial infections.

After attachment, the phagocyte proceeds to ingest the particle by extending pseudopods around it. The phagocyte's plasma membrane invaginates and surrounds the particle, which enters the cytoplasm in a phagocytic vesicle (**phagosome**). Inside the cytoplasm, the vesicle fuses with lysosomes to form a larger structure called a **phagolysosome.**

Within the phagolysosome, bacteria and other microorganisms are killed (digested) by lysosomal enzymes such as lysozyme, proteases, nucleases, and lipases, as well as by highly toxic oxidative substances. During phagocytosis, there is a burst of metabolic

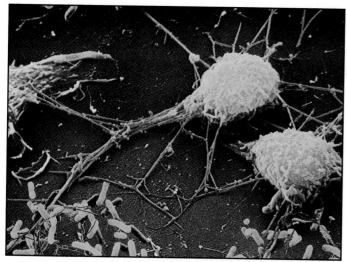

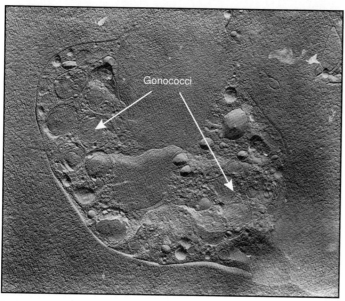

Gonococci

figure 15.9

Phagocytosis

a. Scanning electron micrograph of macrophages attacking *Escherichia coli* cells (blue) (colorized, ×2,150). b. A freeze-etched preparation showing a phagocyte with several *Neisseria gonorrhoeae* cells (gonococci) inside phagosomes (×13,940).

Microbial Pathogenesis
Nonspecific Host Defense: Cellular Defenses and Phagocytosis • p. 28

activity inside the phagocyte, resulting in increased oxygen uptake. As a consequence, oxygen is converted to superoxide free radical, hydrogen peroxide, and hydroxyl radical, which contribute to potent microbicidal activity in the phagolysosome. An NADP oxidase system converts oxygen to superoxide free radical. However, the superoxide free radical is unstable at the acid pH of the phagocyte and is converted to hydrogen peroxide. Another enzyme system, the myeloperoxidase system, uses the hydrogen peroxide and halide ion (chloride and iodide) to generate hypochlorous acid and cause lethal halogenation of bacteria and viruses. The production of lactic acid during the metabolic burst lowers the pH within the phagolysosome to 4.0. This lowered pH not only is directly harmful to some pathogens, but also is the pH at which many lysosomal enzymes are most active.

Microbial Pathogenesis
Nonspecific Host Defense: Cellular and Phagocytosis • p. 28

Most microorganisms ingested by phagocytes are immediately destroyed by this activity. However, some microbes survive and multiply within phagocytes. *Salmonella typhi* (cause of typhoid fever), *Mycobacterium tuberculosis* (cause of tuberculosis), and *Brucella abortus* (cause of brucellosis) are examples of bacteria that survive and multiply intracellularly. These bacteria are protected within the phagocyte from antibodies and antibiotics and may even use the mobile phagocyte to migrate to other tissues. The mechanism for intracellular survival is unknown, although one possibility is that bacterial metabolic products may inactivate harmful lysosomal enzymes.

Other microbes produce proteins called leukocidins that destroy phagocytes. For example, *Staphylococcus aureus* (cause of toxic shock syndrome, skin infections, and food poisoning) produces a leukocidin called the Panton-Valentine leukocidin that lyses phagocytes, resulting in decreased host resistance.

Microbial Pathogenesis
Antiphagocytic Factors: Avoiding Phagocytosis • pp. 6–9

Phagocytosis is a highly effective host defense mechanism against invasion by foreign particles. It is an important part of innate resistance and enables the host to counter infection. Innate host resistance mechanisms provide the body with an array of defenses against invasion by pathogenic microorganisms. Such resistance includes the general health and physiological condition of the host, human body barriers such as the skin and body fluids, which aid in preventing infection, and antimicrobial substances such as lysozyme, interferon, and sebum, which prevent growth of microorganisms. Components of the blood are a major line of host defense against microbial invasion and include antibodies formed specifically against antigens, and leukocytes that are associated with phagocytosis. All of these mechanisms (except antibody production and T-lymphocyte activation) are designed to protect nonspecifically against many types of pathogens until the host is able to develop an acquired resistance directed against the specific invading pathogen. The immune response associated with acquired resistance is discussed in the next chapter.

Dextran-Induced Agglutination of *Streptococcus mutans* in the Formation of Dental Plaque

One of the most significant discoveries in oral microbiology was the correlation between *Streptococcus mutans* agglutination by dextrans and the formation of dental plaque. For many years *S. mutans* had been thought to be associated with microbial dental plaque. However, it was not until 1969 that a series of experiments by Ronald J. Gibbons and Robert J. Fitzgerald provided evidence that *S. mutans* could agglutinate in the presence of dextran and adhere to teeth.

Strains of *S. mutans* grown in glucose broth for 18 hours were washed twice with saline and resuspended in a phosphate buffer to a cell density of 2×10^9 organisms per milliliter. When dextran was added to the suspension, agglutination occurred within a few seconds. This dextran-induced agglutination was observed only with *S. mutans* strains and not with strains of other bacteria including *Streptococcus bovis*, *Streptococcus sanguis*, or *Leuconostoc mesenteroides* (Table 15P.1).

Dextran-induced agglutination of *S. mutans* occurred over a pH range of 5 to 10 (Table 15P.2) and appeared to be specific for dextran but not other polysaccharides (Table 15P.3). Agglutination was particularly intense between pH 8 and 9, with the addition of only 6 ng of dextran causing agglutination at pH 8.5. Agglutination was specific for dextran of high molecular weight (2×10^5 or 2×10^6 daltons). Agglutination of *S. mutans* also occurred after the addition of sucrose but not of glucose or fructose. The addition of dextranase to cells agglutinated by dextran resulted in dispersal of the organisms.

To determine whether dextran-induced adherence of *S. mutans* could be a factor in dental plaque buildup, Gibbons and Fitzgerald prepared fluorescein-labeled *S. mutans* and incubated these cells at 35°C for 20 to 40 minutes with dextran-containing plaques formed in vitro on wires by cultures of *S. mutans*. The plaques fluoresced intensely when they were incubated with fluorescein-labeled suspensions of *S. mutans*, but not with heat-inactivated cells or with labeled *Lactobacillus casei*. Furthermore, fluorescein-labeled cells of *S. mutans* adhered to tooth surfaces treated with dextran of high molecular weight but did not adhere to untreated tooth surfaces.

These experiments by Gibbons and Fitzgerald were instrumental in showing that dextran-induced agglutination of *S. mutans* could be one of the factors associated with the formation of microbial dental plaques. Because of landmark studies such as these, we now are able to develop vaccines against *S. mutans* that may one day eliminate dental plaque.

Source

Gibbons, R.J., and R.J. Fitzgerald. 1969. Dextran-induced agglutination of *Streptococcus mutans*, and its potential role in the formation of microbial dental plaques. *Journal of Bacteriology* 98:341–346.

References

Gibbons, R.J., and J. van Houte. 1975. Dental caries. *Annual Review of Medicine* 26:121–136.

Hamada, S., and H.D. Slade. 1980. Biology, immunology, and cariogenicity of *Streptococcus mutans*. *Microbiological Reviews* 44:331–384.

table 15P.1

Induction of Agglutination by Dextran of High Molecular Weight

Organism	Agglutination[a]	
	2 Hrs	18 Hrs
Streptococcus mutans 6715	4+	4+
S. mutans K1-R	±	4+
S. mutans AHT	3+	4+
S. mutans BHT	±	2+
S. mutans 6927	3+	4+
S. mutans HS6	1+	4+
S. mutans SL	±	3+
S. mutans LM7	±	2+
S. mutans GS5	——	±
S. mutans GS5-MR	3+	4+
S. mutans E49	4+	4+
S. mutans OMZ-61	4+	4+
Streptococcus sanguis 10558	——	——
Streptococcus bovis ATCC 9809	——	——
Leuconostoc mesenteroides	——	——
Enterococcus faecium	——	——
Enterococcus faecalis	——	——
Streptococcus 2M2	——	——
Streptococcus 4M4	——	——
Streptococcus SS2	——	——
Serratia marcescens	——	——
Escherichia coli	——	——
Lactobacillus casei	——	——

[a]Agglutination determined after the addition of 100 µg of dextran with a molecular weight of 2×10^6 to buffered (pH 7.0) cell suspensions. Agglutination graded as ±, 1+, 2+, 3+, or 4+.

Effect of pH on Dextran-Induced Agglutination of *Streptococcus mutans* Strain 6715

Dextran (mol wt 2 × 10⁶) Added per Milliliter	pH[a]									
	5.0	6.0	6.5	7.0	7.5	8.0	8.5	9.0	9.5	10.0
None	—	—	—	—	—	—	—	—	—	—
100 µg	3+	3+	3+	3+	4+	4+	4+	4+	3+	2+
20 µg	2+	3+	3+	3+	4+	4+	4+	4+	3+	1+
4 µg	2+	2+	2+	3+	4+	4+	4+	4+	3+	1+
0.8 µg	1+	1+	1+	2+	3+	4+	4+	4+	3+	—
160 ng	—	—	—	1+	2+	3+	3+	3+	2+	—
30 ng	—	—	—	—	1+	2+	3+	2+	±	—
6 ng	—	—	—	—	—	±	1+	—	—	—
1.2 ng	—	—	—	—	—	—	±	—	—	—
0.24 ng	—	—	—	—	—	—	—	—	—	—

[a]Agglutination determined after 2 hrs of incubation and graded as ±, 1+, 2+, 3+, or 4+.

Agglutination of *Streptococcus mutans* Strain 6715 upon Addition of Various Carbohydrates

Additive	Concentration (µg/ml)	Agglutination after 2 hrs[a]	
		pH 7.0	pH 8.5
None	—	—	—
Dextran, mol wt 2 × 10⁶	1,100	4+	4+
Dextran, mol wt 2 × 10⁵	1,100	2+	3+
Dextran, mol wt 2 × 10⁴	1,100	—	±
Levan	1,100	—	—
Starch	1,100	—	—
Dextrin	1,100	—	—
Inulin	1,100	—	—
Agar	1,100	—	—
Agarose	1,100	—	—
Sucrose	1,000	4+	3+
Sucrose	1,100	±	—
Glucose	1,000	—	—
Fructose	1,000	—	—
Maltose	1,000	—	—
α-methyl glucoside	1,000	—	—

[a]Agglutination graded as ±, 1+, 2+, 3+, or 4+.

Summary

1. Microorganisms that populate the human body in mutualistic or commensalistic relationships are known as the normal microflora. This microflora is established in a human being during and after birth. Prior to birth, a healthy human fetus is free of microbes.

2. Some microorganisms compete antagonistically with other microorganisms. Other microbes grow better in combination than alone, a characteristic known as synergism. One form of synergism is cross-feeding, or syntrophism, in which two populations supply each other's nutritional needs.

3. The term infection refers to the growth of microbes in the tissues of a host. Infectious disease occurs when the infecting microbe causes damage to the host organism.

4. Infectious microorganisms persist in nature in either inanimate (for example, soil, water, and food) or animate (for example, animals and human beings) reservoirs of infection. Animal diseases transmissible to human beings are called zoonoses.

5. Some diseases are transmitted by direct contact with the reservoir or another host. Other diseases are transmitted indirectly by an inanimate object, or fomite, or by a living organism, or vector.

6. The natural history of infectious disease is the sequence of events that occurs from the infection of a host to disease resolution, either through healing or death of the host. This sequence consists of the incubation period, the prodromal period, the acute period, the decline period, and the convalescent period. Some infections are latent and do not immediately produce detectable or overt symptoms.

7. The pathogenicity of a microorganism is its ability to establish infection and cause disease in a host. Virulence is a quantitative measure of pathogenicity and refers to the capacity of a microorganism to overcome the host's body defenses.

8. Microorganisms harm a host by invasion and/or toxin production. Factors that are associated with invasion include a preference by the microbe for certain host tissues or organs, ability to adhere to tissues or organs, production of enzymes that damage tissue or dissolve materials, and production of capsules to protect the microbe from phagocytosis.

9. Two types of toxins are produced by microorganisms: exotoxins, which are soluble proteins released into the surrounding environment, and endotoxins, which are the lipopolysaccharide portions of the outer membranes of gram-negative bacteria.

10. Host defenses against microbial invasions consist of innate (nonspecific) resistance, which does not depend on prior exposure to the pathogen and provides general protection against any type of pathogen; and acquired (specific) resistance, or immunity, which involves antibodies, T lymphocytes, and monocytes, and occurs only after the host has been exposed to a pathogen or its products.

11. Factors associated with nonspecific host resistance include the general health and physiological condition of the host, physical barriers to invasion, antimicrobial substances produced by the host, components of the blood, inflammation, and phagocytosis.

EVOLUTION *and* BIODIVERSITY

The association between host and parasite is a constantly changing, dynamic relationship. As humans age, their bodies are altered and their environments change. Microorganisms are constantly evolving, showing remarkable ability to adapt to these changing conditions. The emergence of antibiotic-resistant bacteria and newly discovered pathogens in recent years is evidence that microbes can easily adapt to many different environments and hosts. For example, antibiotic-resistant bacteria evolved as a result of the widespread and sometimes inappropriate use of antimicrobial agents. Pathogens can also acquire antibiotic resistance genes through genetic transfer from other microbes in the environment, resulting in the selection of drug-resistant organisms (see antibiotic resistance, page 130). With international travel and commerce, infectious agents such as HIV, Ebola virus, *Plasmodium*, and *Vibrio cholerae* can be disseminated to new geographical areas and susceptible hosts. Occasionally the evolution of a virus or a microbe may result in a new variant capable of causing disease. HIV 1 and HIV 2 may have evolved from similar, nonpathogenic viruses in monkeys (see HIV, page 409); the strain of group A *Streptococcus* that causes necrotizing fasciitis may also have developed in this manner (see necrotizing fasciitis, page 499). In the presence of such a wide diversity of microorganisms, humans have also evolved cellular and immunological mechanisms to prevent infection and thus have been successful in maintaining a delicate balance between health and disease.

Questions

Short Answer

1. Compare and contrast normal flora and transient flora.

2. Compare and contrast normal flora and opportunistic pathogens.

3. Explain how antagonism between bacteria can benefit us.

4. In most cases, microorganisms alter their microenvironment to prevent other microorganisms from growing in their niche; however, this is not always the case. Provide examples of synergism.

5. Compare and contrast infection and infestation.

6. Compare and contrast communicable and contagious.

7. Identify several reservoirs of infection.

8. Compare and contrast carriers and fomites.

9. Give several examples of modes of transmission.

10. Identify several examples of portals for infectious agents.

11. Briefly describe the natural history of an infectious disease.

12. Discuss the time course for the appearance and duration of signs and symptoms for infectious diseases.

13. Compare and contrast pathogenicity and virulence.

14. Identify several virulence factors.

15. Compare and contrast endotoxins and exotoxins.

16. Compare and contrast innate resistance and acquired resistance.

17. Identify several physical barriers to disease.

18. Identify several nonspecific antimicrobial substances produced by the body to resist infection.

19. Identify the specific defenses against infection.

20. Briefly describe the process of inflammation and explain its importance.

Multiple Choice

1. Which of the following exhibited signs and symptoms of a disease, but now appears healthy?
   a. convalescent carrier
   b. healthy carrier
   c. incubatory carrier
   d. None of the above.

2. Cockroaches and household flies are _____ for the transmission of disease.
   a. carriers
   b. fomites

   c. biological vectors
   d. mechanical vectors

3. During which of the following stages will signs and symptoms of disease begin to appear?
   a. acute
   b. convalescent
   c. decline
   d. incubation
   e. prodromal

4. Which of the following produces a neurotoxin?
   a. *Clostridium tetani*
   b. *Corynebacterium diphtheriae*
   c. *Streptococcus pyogenes*
   d. *Yersinia pestis*

Critical Thinking

1. Why is it important to know the Gram reaction of a bacterium before selecting antimicrobial therapy? If the Gram reaction is unknown, how could you safely proceed with antimicrobial therapy?

2. What traits would you expect to find in a super-pathogen designed for germ warfare? How would you defend against such an organism?

3. There are approximately 10^{13} cells in your body, and 10^{14} microorganism cells living on or in your body. Explain how this is possible. Identify the areas of your body that are sterile (devoid of normal flora).

4. You have just signed up with the Peace Corps to aid the people in a remote village. Your team has been instructed to prepare a health and hygiene plan before you arrive. Identify the topics of concern and prepare an outline for this plan. (Keep in mind that there will be language barriers and cultural differences.)

 Supplementary Readings

Brubaker, R.R. 1985. Mechanisms of bacterial virulence. *Annual Review of Microbiology* 39:21–50. (A concise summary of nonspecific host defense mechanisms and bacterial virulence factors.)

Miller, V.L., J.B. Kaper, D.A. Portnoy, and R.R. Isberg, ed. 1994. *Molecular genetics of bacterial pathogenesis*. Washington, D.C.: American Society for Microbiology. (A comprehensive overview of the progress made in recent years in understanding genetic mechanisms of microbial pathogenesis.)

Rietschel, E.T., and H. Brade. 1992. Bacterial endotoxins. *Scientific American* 267:54–61. (A review of bacterial endotoxins and their modes of action.)

Roth, A.J., C.A. Bolin, K.A. Brogden, F.C. Minion, and M.J. Wannemuehler, ed. 1995. *Virulence mechanisms of bacterial pathogens*, 2d ed. Washington, D.C.: American Society for Microbiology. (A discussion of the mechanisms of host-pathogen interactions and strategies to overcome bacterial virulence mechanisms.)

Salyers, A.A., and D.D. Whitt. 1994. *Bacterial pathogenesis: A molecular approach*. Washington, D.C.: American Society for Microbiology. (A textbook on how bacteria cause disease.)

Stephen, J., and R.A. Pietrowski. 1986. *Bacterial toxins*, 2d ed. Washington, D.C.: American Society for Microbiology. (A detailed review of the mode of action and the role in disease of bacterial toxins.)

chapter sixteen
16

IMMUNOLOGY

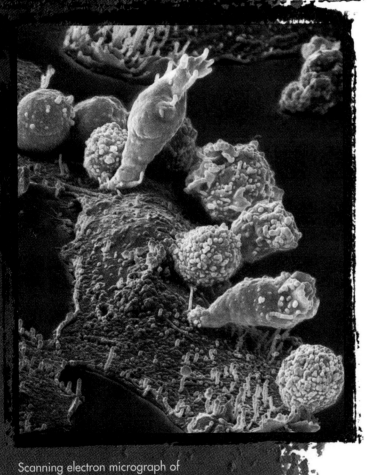

Scanning electron micrograph of cytotoxic T cells (green) attacking a cancer cell (yellow) (colorized).

Humoral Immunity
Lymphoid Tissue
Antigens
Antibodies
Antibody Diversity
T-Cell Subpopulations
The Major Histocompatibility Complex
Activated B Lymphocytes
Humoral Immunity and the Host Defense
Complement

Cell-Mediated Immunity
Cytokines
Delayed Hypersensitive Responses
Other Manifestations
Assays to Measure Cell-Mediated Immunity

In Vitro Antibody-Antigen Reactions
Precipitation
Immunoelectrophoresis
Agglutination
Neutralization
Complement to Detect Antigen-Antibody Reactions
Immunofluorescence
Radioimmunoassay
Enzyme-Linked Immunosorbent Assay

PERSPECTIVE
The First Identified Cases of AIDS in the United States

EVOLUTION AND BIODIVERSITY

 Microbes in Motion — PREVIEW LINK

Among the key topics covered in this chapter are antigen-antibody interactions, the development and use of vaccines, natural killer cells, and the use of antigen-antibody interaction in viral detection. The following sections in the *Microbes in Motion* CD-ROM may be useful as a preview to your reading or as a supplemental study aid:

Viral Structure and Function: Viral Pathogenesis (Host Cell Response), 14–18; Viral Detection (Detection Antibody), 22–23, (Detection Antigen), 5–9. *Vaccines:* Vaccine Development (Terminology), 7–11; Vaccine Targets (Bacterial Targets), 2–12.

he human body is exposed to an almost limitless diversity of foreign particles during a lifetime, including pathogenic microorganisms and viruses, yet remarkably it is able to protect itself against this myriad of invaders. In many cases, infectious microbes are part of our microflora but are kept in check by our immune system. This immune system acts as a molecular sentinel, guarding our bodies against invaders. How does it accomplish this difficult task? This chapter will explore the basic elements of the immune system and its responses to foreign substances.

The study of immune mechanisms is called **immunology.** Unlike innate (nonspecific) resistance, which encompasses a wide variety of mechanisms directed against all invading organisms rather than any specific invader, immunity involves acquired (specific) resistance. It is a characteristic unique to vertebrates.

There are two major components of the human immune system. **Humoral** [Latin *humor,* a liquid] **immunity** is associated with proteins called **antibodies.** Antibodies are directed against specific foreign particles called **antigens** (**anti**body **gen**erator). Antibodies are found predominantly in blood and body fluids. The human immune system also has **cellular** aspects, referred to as **cell-mediated (cellular) immunity.** Cellular immunity depends on the interaction of antigens with host cells known as T lymphocytes.

Humoral Immunity

Humoral immunity develops from the formation of specific antibodies against antigens. This type of immunity effectively protects a person from many types of microorganisms that would normally cause disease.

Lymphoid Tissue Is Important in the Immune Response

When the body is invaded by a foreign substance, it may induce humoral and cell-mediated immune responses. The body's lymphoid tissue plays an important role in these types of immune responses. Lymphoid tissue represents about 2% of the total body weight, mostly located in the lymph nodes and thymus. Additional tissue is in the spleen, bone marrow, gastrointestinal tract, and other mucosae.

T Lymphocytes Are Associated with Cell-Mediated Immunity; B Lymphocytes Are Associated with Antibody Production

The lymphoid tissue produces two types of cell populations derived from bone marrow stem cells: **T lymphocytes (T cells)** and **B lymphocytes (B cells).** T (thymus-derived) lymphocytes mature in the thymus and are responsible for cell-mediated immunity and the regulation of all immune responses. Thus the removal of the thymus in a newborn animal markedly reduces its cell-mediated immune response. B lymphocytes mature in the bone marrow and are associated with antibody production. The

term B (bursa of Fabricius) lymphocyte came from the discovery that in birds, these cells are produced in the bursa of Fabricius, a small piece of lymphoid tissue attached to the posterior region of the intestine. Mammals do not have a bursa for B cell production; instead, these cells are produced in the fetal liver and, after birth, in the bone marrow.

Antigens Induce and React with Antibodies

Humoral immunity is activated in response to the presence in the body of an antigen. An antigen is defined as any molecule that (1) induces the production of an antibody, (2) is able to react with that antibody, (3) is a macromolecule with a high molecular weight (>10,000 daltons), and (4) is usually foreign to the host. Thus an antigen must have **immunogenicity,** or the ability to elicit antibody formation. Antigens may be bacteria, fungi, protozoa, or other cells, as well as viruses and molecules such as bacterial toxins, proteins, or carbohydrates.

Although an antigen molecule is large, an antibody reacts with only a small part of it, called the **epitope,** or **antigenic determinant** (Figure 16.1). Antigenic determinants are portions of proteins, carbohydrates of low molecular weight, or other simple constituents of an antigen. Slight changes in the chemical composition or physical configuration of determinant sites can alter the immunologic properties of an antigen. A single antigen may have one or multiple determinant sites. The term **valence** refers to the number of determinant sites on an antigen. Large antigenic molecules are multivalent and able to elicit formation of several different kinds of antibody after introduction into a host. Most determinant sites are located on the surface of antigens, but some are hidden within them and exposed only if the antigenic molecule is hydrolyzed. Hydrolysis of this type may occur via enzymes within phagocytic cells, and results in increased efficacy of antibody action.

 Viral Structure and Function
Viral Pathogenesis: Host Cell Response • pp. 14–18

Haptens are antigenic determinants that by themselves are incapable of stimulating an antibody response. A hapten can be made immunogenic by covalently linking it to a large, immunogenic carrier molecule. For example, the hapten 2,4-dinitrophenyl (molecular weight, 184 daltons) is not immunogenic alone, but is able to stimulate antibody production when it is combined with the carrier bovine serum albumin (molecular weight, 66,000 daltons). Although haptens by themselves are not immunogenic, they can combine with antibodies already produced against the hapten-carrier molecule.

Antibodies Are Glycoproteins That Bind Specifically with the Antigen That Stimulated Their Production

Antibodies are glycoproteins produced in response to antigens and directed specifically against antigenic determinant sites. Antibodies are generally found in the body fluids and the fluid portion of blood. Normal human serum contains a number of different types of proteins. When subjected to electrophoresis, these serum proteins are separated into four major fractions: albumin, α-globulin, β-globulin, and γ-globulin. Most antibodies are found within the

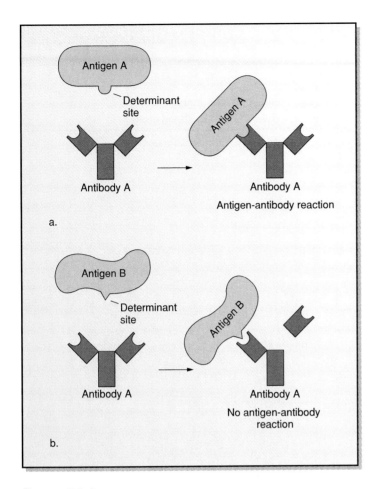

figure 16.1

Antigen-Antibody Binding
a. Antibodies react with a determinant site on the antigen molecule.
b. This reaction is specific and does not occur if the antigenic
determinant site is not recognized by the antibody.

γ-globulin fraction of the serum. The terms **immunoglobulins
(Ig)** and **gamma globulins** frequently are used to denote the antibody fraction of serum proteins. Serum containing antibodies that
react with a specific antigen is referred to as **antiserum; antitoxin**
is often used to describe serum with antibodies against specific
toxin molecules.

Immunoglobulins Are Divided into Five Major Classes

Immunoglobulins are divided into five major classes on the basis of
their physical, chemical, and immunologic properties: **IgG, IgM,
IgA, IgE,** and **IgD** (Table 16.1). **IgG** is the predominant circulating
immunoglobulin and constitutes 80% of all the antibodies normally found in the human body. It binds to microorganisms to enhance their phagocytosis and lysis, reacts with surface antigens on
bacteria to activate complement, and is the major humoral line of
defense in the body. IgG is produced late in the immune response
and can contribute to persistent immunity. It is the only immunoglobulin able to cross the placenta, and it provides maternally
acquired passive immunity in utero and to the neonate at birth.

IgM is generally the first antibody to be produced during an
antibody response. It is able to bind to bacterial surface antigens to
activate complement. IgM is the largest of the immunoglobulins;
it has a molecular weight of 900,000 daltons and ten antigen-
binding sites. The five monomeric units of IgM are held together
by disulfide bonds and a polypeptide J chain.

IgA is found not only in serum (serum IgA), but also in body
secretions (secretory IgA). Serum IgA is a monomer, whereas
secretory IgA is a dimer consisting of two monomers joined by a
polypeptide J chain and a secretory component polypeptide that is
probably wrapped around the dimer. IgA is the principal immunoglobulin in colostrum (mother's milk), saliva, tears, and gastrointestinal and respiratory secretions. It is often the first existing
antibody to contact invading microorganisms, in contrast to IgM,
which is the first antibody to be specifically produced in response
to invaders. IgA in colostrum provides temporary passive immunity to nursing newborn infants. IgA binds to surface antigens of
microorganisms and prevents the adherence of these microbes to
the mucosal membranes of the respiratory, gastrointestinal, and
genitourinary tracts.

IgE is found in low concentrations throughout the body and
is involved in Type I hypersensitive reactions, where IgE molecules
primarily attach to tissue mast cells and basophilic leukocytes.
When IgE on the surface of these cells binds antigen, the antibody-
antigen complexes stimulate the cells to release vasoactive amines
such as histamine and serotonin. These chemicals are responsible
for the symptoms of hypersensitive reactions. Serum IgE levels
may increase severalfold in allergic individuals. IgE levels also
increase in individuals with intestinal parasitic infections, which
suggests a role in immunity against parasitic diseases.

IgD constitutes approximately 1% of the total serum immunoglobulins. Its role in immunity is unclear, although its presence
on the surface of certain types of lymphocytes (B lymphocytes)
suggests that it may serve as a receptor for antigen and may regulate the synthesis of the other immunoglobulins.

Immunoglobulins Have Light and Heavy Chains Composed of Constant and Variable Regions

Immunoglobulins have a basic structure made up of four polypeptide chains—two **light (L) chains** and two **heavy (H) chains.** Each
of the smaller (light) chains has a molecular weight of 25,000 daltons and is identical in immunoglobulins of all five classes. The
larger (heavy) chains have a molecular weight of 50,000 to 77,000
daltons each and are structurally distinct for each class. IgG, IgD,
and IgE occur as monomers of the four-polypeptide-chain structure, whereas IgM occurs as a pentamer, and IgA occurs as a
monomer, dimer, and even higher multimer.

Most of the information on antibody structure has been obtained from IgG, the most prevalent immunoglobulin in the body
(Figure 16.2). The heavy and light chains of IgG are symmetrically
arranged to form a T- or Y-shaped molecule with a flexible "hinge"
near the middle of the heavy chains. The four polypeptides of IgG
contain intrachain and interchain disulfide bridges. Digestion of
IgG by the proteolytic enzyme papain and mild reduction of certain
disulfide bonds cleave the immunoglobulin into three fragments.

Classes	Structure	Molecular Weight (Daltons)	% of Total Antibody	Serum Level (mg/ml)	Number of Antigen-Binding Sites	Heavy Chains	Light Chains	Major Characteristics
IgG		150,000	80	3	2	γ	κ or λ	Major circulating antibody
IgM	J chain	900,000 (pentameter)	10	1.5	10	μ	κ or λ	First antibody to be specifically produced during immune response
IgA	J chain	160,000 (monometer) 385,000 (dimer)	5–15	1.5–4	2 4	α	κ or λ	Often the first antibody to contact invading microorganisms; major secretory antibody; exists as monometer in serum and as dimer in secretions
IgE		190,000	0.002–0.05	0.0001–0.0003	2	ε	κ or λ	Involved in Type I hypersensitive reactions
IgD		185,000	1	0.03	2	δ	κ or λ	Present on surfaces of lymphocytes

Two of these fragments are identical and contain the amino (NH_2) ends of both heavy and light chains. These identical fragments are called **Fab (Fragment, antigen-binding).** The Fab portions of the antibody molecule bind to antigens. **Bivalent** immunoglobulins such as IgG have two identical Fab sites, each of which is able to bind specifically to an antigenic determinant.

The third fragment released upon papain digestion is the **Fc (Fragment, crystallizable).** The Fc region does not combine with antigen, but it does bind to complement in complement activation and to Fc receptors on macrophages, neutrophils, and other cells. The latter role contributes to enhanced phagocytosis.

Studies of the amino acid sequences of immunoglobulins have shown that the heavy and light chains have both variable and constant regions. The variable regions of the light and heavy chains are located in the amino-terminal half of the Fab portion of each immunoglobulin chain. The amino acid sequences in the variable regions—particularly in three "hypervariable" regions—are responsible for the specificity of the antibody's combining sites and the diversity of antibody molecules. **Idiotype** refers to the variable region of an antibody molecule.

The amino acid sequences in the constant regions determine other properties of an antibody. The constant region of the heavy chain specifies the five major classes of immunoglobulins. There are five types of heavy chains—gamma (γ), mu (μ), alpha (α), delta (δ), and epsilon (ε)—corresponding to the five immunoglobulin classes IgG, IgM, IgA, IgD, and IgE, respectively. Although constant within a given class, the amino acid sequence of the constant region varies considerably from one class to another. Within certain immunoglobulin classes, variations in the heavy chain structure give rise to subclasses. For example, human IgG can be grouped into four subclasses: IgG_1, IgG_2, IgG_3, and IgG_4. These subclasses make up, respectively, 70%, 19%, 8%, and 3% of circulating human IgG. There also are two subclasses of human IgA (IgA_1 and IgA_2). Subclasses have not yet been detected for the other immunoglobu-

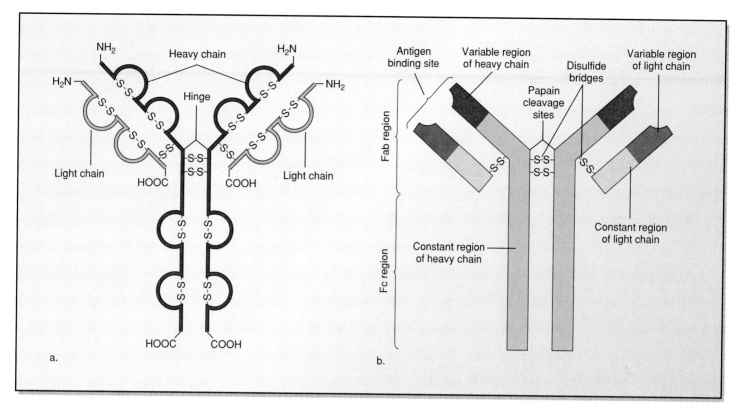

figure 16.2

Structure of Immunoglobulin G (IgG)

IgG consists of two heavy chains and two light chains containing intrachain and interchain disulfide bridges. a. IgG structure showing intrachain and interchain disulfide bridges. b. Alternative simplified IgG structure showing only the disulfide linkages between chains. The

Fab and Fc regions of the antibody molecule are illustrated. Variable regions of the chains are responsible for the specificity of the antibody's combining sites and the diversity of antibody molecules.

lin classes. **Isotype** refers to the constant-region determinants for an immunoglobulin class or subclass.

The light chains of an immunoglobulin exist in two distinct forms: **kappa (κ) chains** and **lambda (λ) chains,** which differ considerably in their amino acid sequences. An antibody contains either two κ or two λ light chains (or multiples of these), but not one of each type. In humans, approximately 60% of the total immunoglobulins contain κ chains and 40% have λ chains. In addition to the light chains, a polypeptide J (joining) chain occurs in the structure of IgA and IgM. The J chain links both of the monomeric units of the IgA dimer molecule and two of the five monomeric units of the IgM pentamer molecule.

Antibody Diversity Is Made Possible by Gene Rearrangement

An average human being has only about 100,000 genes, but can make specific antibodies against literally millions of different antigens over a lifetime. How is this possible?

This question was answered in part by Susumu Tonegawa of the Basel Institute for Immunology in 1976 when he discovered that the constant and variable regions of immunoglobulin molecules are encoded in different parts of the chromosome and are rearranged in B cells during their development to produce an infi-

nite diversity of antibodies. Tonegawa was awarded the 1987 Nobel Prize in physiology or medicine for his discovery.

During B cell development, the DNA sequences for each heavy chain are randomly assembled from a V (variable) gene segment, a J (joining) gene segment, a D (diversity) gene segment for the variable region, and a single C (constant) gene segment for the constant region in the germ-line B-cell chromosome (Figure 16.3). The V, J, D, and C gene regions in the germ-line B-cell chromosome are separated by introns that are typical of eucaryotic DNA (see introns and exons, page 223). During gene rearrangement, the randomly selected V, J, and D gene segments are cut and joined onto the C gene segment, a process called combinatorial joining because different combinations of the gene segments can be joined. The active gene is then transcribed to form a primary RNA transcript that is subsequently spliced to yield the final mRNA. The DNA sequences for each light chain are randomly selected in a similar manner, but from a V gene segment and a J gene segment for the variable region and a single C gene segment for the constant region.

This chance rearrangement of gene segments provides all possible gene combinations. For example, 300 different V gene segments, 12 different D gene segments, 4 different J gene segments, and 1 C gene segment in mouse immunoglobulin generate 14,400 different varieties ($300 \times 12 \times 4 \times 1$) of heavy chains and 1,200 different varieties ($300 \times 4 \times 1$) of light chains. Together, the varieties

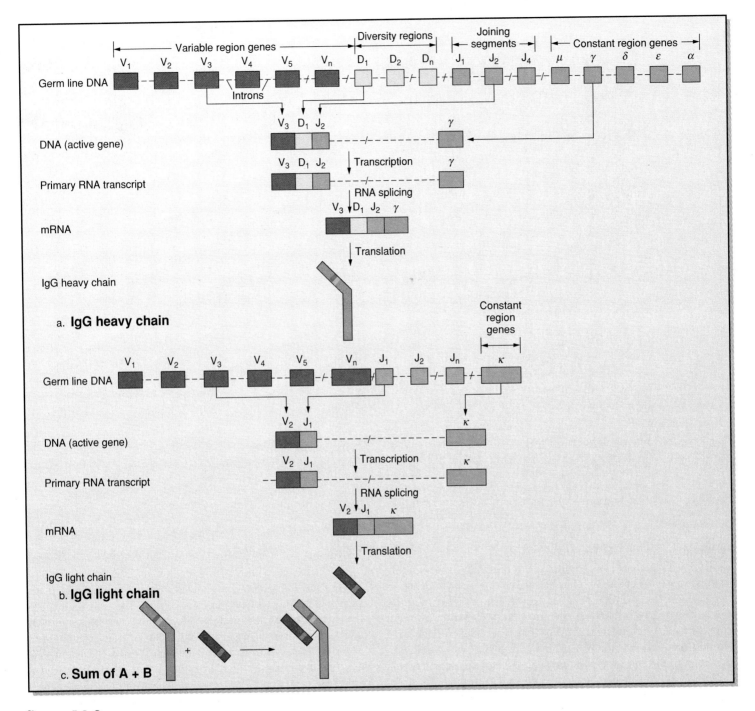

figure 16.3

Generation of Antibody Diversity by Combinatorial Joining
a. Randomly selected V, J, and D gene segments are cut and joined onto the C gene segment to form an active DNA, which is transcribed to form a primary RNA transcript. The RNA transcript is spliced to form mRNA, which is then translated to the heavy chain. b. Randomly selected V and J gene segments are cut and joined onto a C gene segment to eventually produce the light chain. c. Heavy and light chains are joined to form the antibody molecule.

Superantigens

Generally very few T lymphocytes react when the immune system is exposed to a microbial antigen. However, some bacterial exotoxins function as **superantigens** and stimulate the immune system to produce large numbers of T cells that are ineffective in the immune response. Toxic shock syndrome toxin, staphylococcal exfoliatin, and staphylococcal enterotoxin A (all produced by *Staphylococcus aureus*), and streptococcal exotoxin A (produced by *Streptococcus pyogenes*) are examples of superantigens. These antigens bind simultaneously to the class II MHC molecule and to the β chain of the T cell, resulting in the secretion of high levels of cytokines and the production of many T cells. This excess T lymphocyte stimulation and cytokine production results in an overwhelming cell-mediated response characterized by symptoms such as fever, nausea, vomiting, and shock. With so many T cells activated, many of them die, leaving the body susceptible to growth of the invading pathogen and infection by other microbes.

of heavy chains and light chains allow for the existence of 17,280,000 (14,400 × 1,200) possible antibody gene combinations. Additional antibody diversity is generated by the random addition of nucleotides by a terminal deoxynucleotidyl transferase during D-J and V-D-J joinings in heavy chain genes, and by a high rate of somatic mutation arising during B cell development that alters individual nucleotides in variable gene regions of germ-line DNA.

Different T-Cell Subpopulations Perform Different Functions

T cells do not function directly in the production of antibody but are essential in the humoral immune response to a large number of antigens (Table 16.2). The importance of T cells in antibody production was discovered in the 1960s through experiments in which mice were x-irradiated to kill most of their lymphocytes. When these irradiated mice were injected with antigen and either thymus cells or bone marrow cells, they were unable to make antibody. However, mice given mixtures of thymus cells and bone marrow cells produced large quantities of antibodies. These experiments showed that T cells provided by the thymus are important in the humoral response to most antigens.

T lymphocytes have a surface receptor consisting of a dimer, a β and an α polypeptide chain held in place and embedded in the plasma membrane of the lymphocyte. These receptors, called T-cell receptors (TCR), show antigen specificity. T cells activated through their surface receptors form clones of T cells from a common ancestor. T-cell–specific surface proteins, called CD4 (CD for cluster designation) and CD8, distinguish two populations of T lymphocytes. CD8-containing T lymphocytes consist of **cytotoxic T (T_C) cells,** which interact with and destroy cells containing antigen on their surfaces. CD4-containing T lymphocytes consist of **T helper (T_H) cells,** which produce various cytokines, stimulate B lymphocytes to produce antibodies, and are involved in cell-mediated immunity. **Cytokines** are low-molecular-weight proteins that regulate important biological processes such as cell growth, cell activation, tissue repair, immunity, and inflammation (Table 16.3). The principal sets of cytokines are **interferons (IFNs), inter-**leukins (ILs), colony stimulating factors (CSFs), and **tumor necrosis factors (TNFs).** Cytokines produced by lymphocytes are called **lymphokines.** There are several subsets of T_H cells. **T_H1 cells** are associated with cytotoxicity and local inflammatory reactions, delayed hypersensitive responses, and production of IFN-γ and IL-2. **T_H2 cells** activate B lymphocytes, produce IL-4, IL-5, IL-6, and IL-10, and function primarily in humoral immunity.

The Major Histocompatibility Complex Plays a Role in the Immune Response

The ability of cytotoxic T cells to recognize and react with antigens on target cell surfaces depends on a set of cell surface **histocompatibility antigens,** or **HLAs** (**h**uman **l**eukocyte **a**ntigens). Cells in the human body have HLAs on their surfaces. Prior to birth, the body recognizes and removes or inactivates T lymphocytes specific for these antigens. This makes it possible for the immune system to distinguish between "self" and "nonself" antigens. This immunologic tolerance is important; otherwise, the immune system would attack "self" antigens and could cause severe damage. These antigens are coded by a group of genes called the **major histocompatibility complex (MHC)** genes. Cells that are tumorous or infected with viruses display tumor or viral antigens associated with self MHC molecules. These changes are significant enough to be detected by cytotoxic T cells, leading to the destruction of the target cell.

In humans the MHC gene cluster is located on chromosome 6. At least 50 different genes encode HLAs, resulting in a wide variety of MHC protein molecules that exist in the human population. With the exception of identical twins, the chance of two individuals having identical MHC genes and, consequently, identical MHC molecules is extremely small.

The MHC genes encode three types of proteins: class I, class II, and class III. Class I MHC proteins are found on the surfaces of all nucleated cells in the body and serve to identify these cells as "self." Class I proteins also elicit an immune response when injected into a host with antigenically different class I proteins. If tissue grafts are made between two organisms and there are differences in

Comparison of B Lymphocytes and T Lymphocytes

Type	Site of Maturation	Type of Immunity	Half-life	Mobility	Function
B lymphocytes	Bone marrow or lymphoid tissue	Humoral	Short (days to weeks)	Relatively localized·	Differentiate into antibody-secreting plasma cells
T lymphocytes	Thymus	Cell-mediated	Long (months to years)	Widely distributed	Are involved in delayed hypersensitivity
					T helper (T_H) cells: activate B lymphocytes, produce cytokines, and are involved in cell-mediated immunity
					Cytotoxic T (T_C) cells: destroy cells with antigen on their surfaces

Examples of Major Cytokines

Cytokine	Cell Source	Effects
Interleukin-1	Macrophages, B cells	Activation of lymphocytes, stimulation of macrophages
Interleukin-2	T cells	T-cell growth factor
Interleukin-3	T cells	Colony-stimulating factor
Interleukin-4	T cells	B-cell growth factor
Interleukin-5	T cells	B-cell growth factor
Interleukin-6	T cells, B cells, macrophages	B-cell growth factor
Interleukin-7	Bone-marrow stromal cells	Stimulates growth and differentiation of B cells
Interleukin-8	Macrophages, skin cells	Chemotaxis of neutrophils
Interleukin-9	T cells	Induces proliferation of some T helper cells in the absence of antigen, promotes growth of mast cells
Interleukin-10	T cells	Inhibition of cytokine synthesis
Interleukin-11	Bone-marrow cells	Stimulates B cell development
Interleukin-12	Monocytes	Induction of T_H1 cells
Interleukin-13	T cells	Blocks IL-12 production; regulator of inflammatory response
Interleukin-14	T cells	Induces proliferation of activated B cells
Tumor necrosis factors α and β	Macrophages, lymphocytes, T cells	Activation of macrophages, granulocytes, and cytotoxic cells
Interferon-α/β	Macrophages, lymphocytes	MHC class I induction, antiviral effect
Interferon-γ	T cells, natural killer cells	MHC induction, macrophage activation, endothelial cell adhesion
Monocyte colony stimulating factor	Monocytes	Stimulates division and differentiation of monocytes
Granulocyte colony stimulating factor	Macrophages	Stimulates division and differentiation of macrophages
Migration inhibition factor	T cells	Inhibits migration of cells

Monoclonal Antibodies

In 1975 scientists of Britain's Medical Research Council under the direction of Cesar Milstein and Georges Kohler discovered a new technique for producing clones of antibody-producing cells that synthesize antibodies of a single specificity. These **monoclonal antibodies** are produced by fusing a cancer cell from a mouse to an antibody-producing cell.

A mouse is injected with a specific antigen. Several days after inoculation, the spleen is removed, and B lymphocytes, which normally migrate to the spleen, are fused with rapidly dividing mouse cancer cells called **myelomas.** Myeloma cells are used because they can produce huge quantities of immunoglobulin and can live indefinitely in cell culture. This fusion results in hybrid malignant cells (**hybridomas**) that not only can proliferate indefinitely, but also are able to produce large amounts of homogenous antibodies. Clones of the same hybridoma all manufacture antibodies of a single specificity, called monoclonal antibodies because they are derived from a single hybridoma cell.

Monoclonal antibodies provide scientists with modern-day magic bullets that recognize antigens from various types of diseases. Monoclonal antibodies, furthermore, are used in diagnosis of such diseases as gonorrhea and rabies, and in the preparation of vaccines for microbial diseases. Because the monoclonal antibody is specific for a single antigenic determinant, it is more discriminatory than polyclonal antibodies in recognizing specific antigens. Perhaps the greatest hope is that monoclonal antibodies will provide a treatment for cancer. For example, a monoclonal antibody specific for cancerous cells can be tagged with a radioisotope to detect a tumor in a patient. The antibody can then be armed with a powerful toxin or chemical to specifically seek out and kill the cancer cell, leaving other cells untouched. The discovery of the principle of monoclonal antibody production is a significant milestone in microbiology; for this achievement Milstein and Kohler were awarded the Nobel prize in medicine in 1984.

---

histocompatibility antigens, these differences can lead to a response by cytotoxic T cells and graft rejection. This is the basis for HLA typing prior to organ transplants.

Class II proteins are found on the surface of cells called **antigen-presenting cells (APCs).** APCs, which include B lymphocytes, dendritic cells, and macrophages, are involved in presenting antigen to T helper cells. T helper cells activated in this manner secrete IL-2, which in turn activates B cells. Class III MHC genes encode the C4 protein of the complement pathway as well as factor B of the alternative complement pathway.

Activated B Lymphocytes Develop into Antibody-Secreting Plasma Cells

The **clonal selection theory** of Sir Macfarlane Burnet explains how genetically diverse B lymphocytes develop into antibody-secreting plasma cells (Figure 16.4). Burnet, an Australian virologist and physician, proposed that early in its differentiation, each lymphocyte developed an ability to recognize a specific antigen, using a specific receptor protein on the cell surface. Both T cells and B cells have surface receptors that recognize and bind specific antigens. Each cell can only recognize a single antigenic determinant. In the case of B lymphocytes, the receptor is an immunoglobulin (monomer IgM and IgD). There is a separate B cell for each antigenic determinant. As a specific antigen binds to these receptors, the B lymphocyte is activated.

B cells are also activated by T-dependent antigens that are associated with T helper cells. These antigens are first taken up by an antigen-presenting macrophage, which processes and presents the antigen and its MHC to a T helper cell. The T helper cell recognizes the antigen associated with the MHC molecule on the surface of the antigen-presenting cell. The macrophage secretes interleukin-1, which stimulates the T helper cell to divide and secrete interleukin-2. Interleukin-2, along with other cytokines secreted by the T helper cell and APC, activates the B cells.

According to the clonal selection theory, some of the activated B cells multiply and mature into clones of antibody-secreting **plasma cells.** Each plasma cell is derived from the same ancestral cell, so all cells in a clone have the same surface receptors and antigenic specificity. Plasma cells derived from the same clone are committed to making antibody of a particular specificity. A mature plasma cell survives for only a few weeks, but during this life span synthesizes antibodies at a rate of about 2,000 molecules per cell per second.

Some B lymphocytes formed during clonal selection do not differentiate into plasma cells; they differentiate into special lymphocytes called **memory cells.** These longer-living memory cells are activated so that if the host is reexposed to the same antigenic determinant, the cells react to that antigenic grouping and mount a secondary antibody response. This secondary response is more rapid and intense than the primary immune response.

Humoral Immunity Is an Important Part of Host Defense Against Microbial Invasion

Immunoglobulins bind to antigenic determinant sites with a close and specific fit by van der Waals forces, electrostatic forces, hydrophobic interactions, and hydrogen bonds. This combination is comparable to a key fitting a lock.

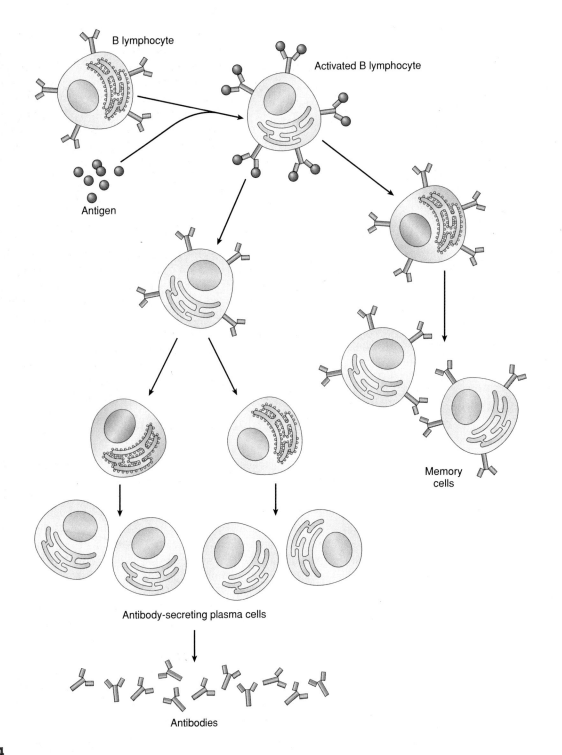

B lymphocyte

Activated B lymphocyte

Antigen

Memory cells

Antibody-secreting plasma cells

Antibodies

figure 16.4

Clonal Selection Theory for Antibody Synthesis

B lymphocytes exposed to specific antigens are activated to produce clones of antibody-secreting plasma cells. Some of these B lymphocytes differentiate into memory cells for an anamnestic (memory) secondary response to antigenic stimulation.

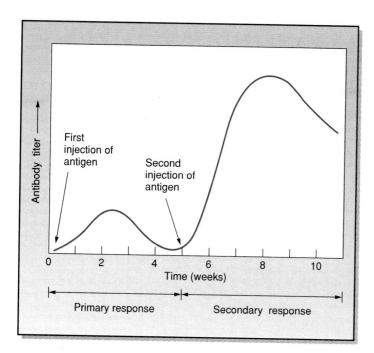

figure 16.5

Immunologic Response to Antigenic Stimulation
The primary immunologic response to an antigen develops over a period of one to three weeks. The secondary response develops much faster and to a higher antibody titer.

The prime purpose of the immunologic response is host defense. In humoral immunity, immunoglobulins bind to microbial cells and promote their phagocytosis, or they combine with microbial toxins or viruses and neutralize them. Immunoglobulins are multivalent and combine with antigens to form large aggregates that are easily phagocytized. Such complexes also immobilize microbes, minimizing their spread through the body. Toxins bound by specific antibodies are frequently neutralized. Antibodies that attach to viruses reduce their ability to attach to or penetrate host cells. Humoral immunity is thus an important part of host defense against microbial invasion.

Immunoglobulins Are Formed in Response to Antigenic Stimulation

The immunologic response to antigenic stimulation can be measured by the antibody level, or **antibody titer**, in the serum. On initial exposure to an antigen, the B cells take a few days to differentiate into antibody-secreting plasma cells. Then there is a slow, gradual rise in antibody titer, followed by a gradual decline (Figure 16.5). This is the **primary response** of the B lymphocytes to antigenic stimulation; the contact with that particular antigen makes the host reactive to it. If the host is exposed a second time to the same antigen, there is generally a more intense, rapid **anamnestic response** [Greek *anamnestikos*, easily recalled]. This is also called the **secondary response** of antibodies; it is more intense and rapid due to the response of the memory cells that remain from the prior response to the specific antigen.

The anamnestic response of a host to a particular antigen is the basis for **immunization,** in which the body receives an initial small dose of antigen that sensitizes the immune system for antibody production. If we are subsequently exposed to the same antigen, the body calls forth the rapid, intense, secondary antibody response. A similar result is obtained with the administration of a **booster dose** of antigen (such as tetanus toxoid antigen). These are usually given to individuals who might require elevated antibody titers for protection against a serious or highly infectious disease.

Vaccines
Vaccine Development: Terminology • pp. 7–8

Immunity Can Be Naturally or Artificially Acquired

Immunity can be naturally or artificially acquired. **Naturally acquired active immunity** is produced in a person through natural infection. A natural active immune state develops in a person following contact with the antigen. The host actively produces antibody, or a cell-mediated immunity to the antigen is developed, or both. The immunity is long-lasting.

If acquired immunity to an antigen is temporarily produced in a person by the transfer of antibodies from one person to another person by natural means, the immunity is known as **naturally acquired passive immunity.** Such an immune state only lasts as long as the donor antibodies are present in the host. The recipient of these antibodies is passively protected by the donor's antibodies. Naturally acquired passive immunity occurs when maternal antibodies are transferred through the placenta to the fetus. These maternal antibodies provide the fetus and immunologically immature neonate with humoral protection during the first few months of life.

Artificially acquired passive immunity occurs when antibodies formed in other hosts are introduced into a new host. An example of artificially acquired passive immunity is the injection of pooled human γ-globulin into a person who has been exposed to hepatitis A virus (HAV). Infection with HAV is so widespread that the sera of many individuals contain anti-HAV antibodies, and a relatively high titer usually can be obtained through concentrated pooled sera. Because the half-life of most human antibodies in the circulation is only a few weeks, passive immunity provides only temporary protection against an antigen (while an individual is producing his or her own antibodies through natural immunity). Naturally acquired active immunity is long term because of the anamnestic response of B memory cells.

Vaccines Protect a Host Through Artificially Acquired Active Immunity

A **vaccine** is a material administered to a subject to produce **artificially acquired active immunity.** Three types of vaccines are commonly used in immunization: live, attenuated vaccine; killed vaccine; and toxoid, or subunit, vaccine (Table 16.4). A **live, attenuated vaccine** consists of live microorganisms that have been rendered avirulent or with reduced virulence. Avirulent strains are obtained by mutation or passage on laboratory media for genetic selection. Since the injected microbe is still alive, it multiplies in the host and provides a source of

table 16.4

Examples of Vaccines

Organism	Disease	Vaccine Type
Bacteria		
Bordetella pertussis	Whooping cough	Killed cells of *B. pertussis*
Clostridium tetani	Tetanus	Toxoid, prepared by treatment of *C. tetani* exotoxin with formalin
Corynebacterium diphtheriae	Diphtheria	Toxoid, prepared by treatment of *C. diphtheriae* exotoxin with formalin
Salmonella typhi	Typhoid fever	Killed cells of *S. typhi*
Vibrio cholerae	Cholera	Killed cells of *V. cholerae*
Viruses		
Aphthovirus	Foot-and-mouth disease	Subunit
Arbovirus	Yellow fever	Live, attenuated
Hepatitis B virus	Hepatitis B	Subunit
Poliovirus	Polio	Live, attenuated
Rubeola virus	Measles	Live, attenuated

Vaccines
Vaccine Targets: Bacterial Targets • pp. 2–12

antigen for stimulation of protective antibody production, and a source of memory cells for subsequent infection by the virulent strain. Live, attenuated vaccines are extremely effective in preventing disease, but concern always exists that back mutation of the injected microbe to a virulent form may occur or that some hosts may be unable to cope with even an attenuated organism. Live vaccines have been used for immunization against poliomyelitis, rubeola, and mumps.

A **killed vaccine** contains the virulent infectious agent killed by chemicals or radiation. Heat is generally not used for inactivation since it may damage antigenic determinants on the microorganism. Killed vaccines are safer to administer than live vaccines, but do not normally provide long-term immunity. Repeated immunization by booster doses is often required to maintain immunity. Killed vaccines are usually injected for protection against typhoid, cholera, and influenza.

A third class of vaccines includes **toxoids** of exotoxins and **cellular components** of microorganisms. Toxoids are exotoxins that have been chemically modified to nontoxic forms (usually by treatment with formalin or similar compounds) but still retain their immunologic characteristics. Toxoids are used as vaccines for diphtheria and tetanus. Cellular components of microorganisms are sometimes used to elicit antibody responses. When only certain portions of the microbe are used, such vaccines are extremely specific in the types of antibodies produced in the host. Recombinant DNA procedures are now used to manufacture microbe-produced viral proteins for vaccines against such diseases as hepatitis B and foot-and-mouth disease. Such a **subunit vaccine,** which contains only portions of the virus that are harmless, is currently in large-scale efficacy trials for AIDS (see AIDS vaccine, page 540).

Vaccines
Vaccine Development: Composition • pp. 9–10

The Goal of Immunization Is Herd Immunity

The goal of public health or veterinary immunization programs is to immunize a large enough proportion of the susceptible population that the spread of an infectious agent is greatly reduced or eliminated. If the number of individuals susceptible to a disease is decreased through immunization, the chances of contact between infected and susceptible persons also are reduced. The population then provides **herd immunity.**

The proportion of individuals in a population that must be immunized to establish effective herd immunity depends on the pathogen, the size and type of the population, and the characteristics of the disease, as well as its transmission pattern. For most infectious diseases, herd immunity can be attained by immunizing 70% of the susceptible population. However, highly infectious diseases such as measles require a 95% to near 100% immunization level of the susceptible population to establish herd immunity.

Vaccines
Vaccine Development: Herd Immunity • p. 11

The Host Immune Response Is Sometimes Naturally or Artificially Suppressed

In certain situations, an immune response cannot be mounted to a given antigen. **Agammaglobulinemia** is a genetic disease characterized by a total or partial inability to synthesize γ-globulins. It was the first immunodeficiency disease described and is believed to involve an arrest of B-cell maturation. Patients with agammaglobulinemia are highly susceptible to microbial infections unless given prophylactic antibiotics or γ-globulin. Prognosis is good when these therapeutic regimens are followed.

Types of Hypersensitive Reactions

Type of Reaction	Type of Antibody	Characteristics	Examples
Immediate hypersensitivity (antibody mediated)			
Anaphylactic reactions (Type I)	IgE	Antigen combines with IgE associated with mast cells and basophils, resulting in the release of histamines, prostaglandins, leukotrienes, and other chemical substances	Insect stings, bronchial asthma, hay fever, food allergies, drug reactions
Cytotoxic reactions (Type II)	IgG, IgM	Circulating antibodies react with tissue cells or particles, resulting in complement-mediated cytolysis or increased phagocytosis	Transfusion reactions, agranulocytosis
Immune-complex reactions (Type III)	IgG, IgM	Antibody combines with large quantities of antigen to produce antibody-antigen complexes that trigger the release of histamines and other chemical substances, resulting in tissue necrosis and inflammation	Arthus reaction, serum sickness
Delayed hypersensitivity (cell mediated)			
Delayed hypersensitive response	No antibody	Antigen sensitizes T lymphocytes, resulting in the release of cytokines and other chemical substances from the sensitized T lymphocytes	Tuberculin reaction
Transplant rejection	No antibody	Foreign antigen in donor tissue causes recipient's T lymphocytes to release cytokines and other chemical substances, resulting in the rejection of the donor tissue	Graft rejection

Sometimes it is necessary to intervene clinically and suppress the immune system of a patient, particularly in tissue or organ transplants between genetically different individuals. Recipients of such transplants normally would reject the foreign tissue or organ. The immunologic response causing such rejection results primarily from the action of sensitized cytotoxic T cells. Upon contact with specific cell graft antigen, cytotoxic T cells cause damage to the membranes of the graft cells by releasing cytokines and other substances. The ideal donor for transplanted material is an identical twin, because no foreign graft antigens would be transplanted to the recipient. Ideal tissue donors are rare, however, and in most cases the recipient's immune system must be suppressed to permit the transplant to "take." Agents used for immunosuppression include corticosteroids, ionizing radiation, and antilymphocytic serum (animal serum that contains antibodies to human lymphocytes). Unfortunately, immunosuppressive agents may also suppress the patient's entire immune response. The patient then becomes highly susceptible to microbial infection and is referred to as a **compromised host.** Cyclosporin A, a recently discovered fungal peptide, is revolutionizing the field of organ transplantation. It is highly specific in its mode of action—suppressing the activity of cytotoxic T lymphocytes without affecting humoral immunity—and has been successfully used to maintain liver and kidney transplants.

Antigenic Stimulation Sometimes Results in Antibody-Mediated Hypersensitive Responses

A body's immune response to antigenic stimulation sometimes results in injury to the host. This type of damage is classified as **hypersensitivity** because it results from an abnormally sensitive host response to an antigen. In hypersensitivity the host must first be sensitized to an antigen; the hypersensitive reaction occurs after a second exposure. There are several types of hypersensitivity reactions (Table 16.5). **Antibody-mediated hypersensitivities** are sometimes called **immediate hypersensitivities** because symptoms occur within a few minutes to hours after host contact with the antigen. Other hypersensitive reactions are **cell-mediated immune responses** and are called **delayed hypersensitivities** because symptoms appear in hours to days after antigen contact. Cell-mediated immune responses are discussed later in this chapter.

Anaphylactic reactions [Greek *ana*, away from, *phylaxis*, protection], also known as **Type I hypersensitive reactions,** are caused by the interaction of antigen with antibodies (IgE) associated with mast cells (nonmotile connective tissue cells found next to capillaries) and basophils. Antigens that function in this manner are called **allergens.**

Anaphylactic reactions occur only after the body is initially sensitized with antigen, prompting the formation of IgE molecules that attach to IgE-Fc receptors on the surfaces of basophils and mast cells. Mast cells and basophils both contain cytoplasmic granules filled with histamine and other chemical mediators (Table 16.6). IgE-cell complexes are found throughout the body and cause no harm unless there is a second exposure of the body to the antigen. Within minutes of the second exposure, the antigen combines with these complexes, causing the release of histamine, prostaglandins, leukotrienes, and other chemical substances.

Anaphylactic reactions can be either systemic or localized. Systemic anaphylaxis is a generalized reaction that occurs when a person sensitized to an allergen such as an antibiotic (for example, penicillin) and insect venom (for example, bees, wasps, or hornets) is reexposed to it. The release of histamine and other chemical mediators from

table 16.6

Chemical Mediators

Mediator	Location	Physiological Effects
Eosinophil chemotactic factor of anaphylaxis (ECF-A)	Mast cells and basophils	Attraction of eosinophils and enhancement of complement activity
Heparin	Mast cells and basophils	Anticoagulant effects
Histamine	Mast cells	Vasodilation, increased capillary permeability, chemokinesis, bronchoconstriction
Leukotriene	Mast cells	Increased capillary permeability; bronchoconstriction
Prostaglandin	Mast cells	Vasodilation, bronchoconstriction, chemokinesis
Serotonin	Blood platelets	Increased capillary permeability

mast cells causes blood vessels to dilate and become more permeable, and increases sensory nerve sensitivity. These effects cause increased localized blood flow and the movement of protein and water through the distended blood vessels to the tissue, resulting in edema (an accumulation of excessive amounts of fluid in the tissue), inflammation, and shock. Shock results from a dramatic drop of arterial blood pressure and insufficient blood flow to vital organs, and can be fatal.

Localized, or atopic [Greek *atopos*, out of place], anaphylaxis is associated with limited production of IgE and localized mast cell involvement. Some examples of atopic reactions are hay fever, food allergies, and brochial asthma. Hay fever, which is characterized by coughing, sneezing, congestion, and itchy and tearing eyes, is caused by initial exposure to airborne allergens such as ragweed pollen, dust, fungal spores, and animal dander that sensitize mast cells in the eyes, nose, and upper respiratory tract. In food allergies, a person allergic to seafood, cow's milk, chocolate, or other foods develops symptoms of diarrhea, vomiting, and urticaria (hives) after eating the food. Antihistamines help relieve the symptoms of hay fever and food allergies by blocking the vasoactive action of histamines released from mast cells. Bronchial asthma occurs when allergens similar to those causing hay fever cause release of chemical mediators from IgE-sensitized mast cells in the lower respiratory tract, resulting in constriction of the bronchial tubes and inflammation of the airways. Symptoms include shortness of breath, bronchiospasms, and increased mucus production. Since histamine is not one of the chemical mediators released, antihistamines are not of value in alleviating the symptoms of asthma. Bronchodilators, which relax the bronchial muscles; expectorants, which promote the dissolution and discharge of mucus that accumulates in the alveoli; and anti-inflammatory agents such as steroids, which reduce inflammation of the airways, are helpful in relieving the symptoms associated with asthma.

Skin tests, in which small amounts of suspected allergens are inoculated into the skin, are used to identify the responsible allergen (Figure 16.6). Sensitivity to the antigen is indicated by the rapid appearance of a red raised area, or wheal and flare reaction, at the site of inoculation. In addition to avoiding contact with the responsible allergen, attempts can also be made to desensitize the individual. Desensitization involves injecting repeated doses of the allergen to stimulate the production of IgG antibodies instead of IgE antibodies. The IgG antibodies intercept and neutralize the allergens before they have a chance to interact with IgE-sensitized mast cells.

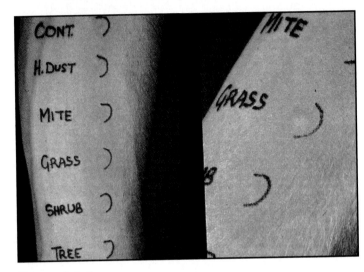

figure 16.6
Skin Tests Used to Diagnose Allergies
Suspected allergens are inoculated into the skin. Sensitivity to a specific allergen is indicated by the rapid appearance of a red raised area.

Type II allergic reactions, or **cytotoxic reactions,** occur when circulating antibodies (IgG or IgM) react with tissue cells or particles, resulting in (1) complement activation and cytolysis or (2) opsonization and increased phagocytosis of the particle. An example of a Type II reaction is the lysis of incompatible red blood cells that occurs with mismatched blood transfusions. Another example is the Rh incompatibility that occurs between mother and fetus. Such incompatibility may lead to serious reactions. For example, in Rh incompatibility, antibodies against Rh factor are present in the maternal Rh-negative blood and pass through the placenta to attack the red blood cells of the fetus. The infant may be born with **hemolytic disease of the newborn,** which is manifested by anemia, jaundice, edema, and enlargement of the spleen and liver.

Type III antibody-mediated hypersensitive reactions are characterized by **immune complexes.** IgG and IgM are associated with type III reactions. Two types of reactions are representative of immune complexes: the **Arthus reaction** and **serum sickness.** The Arthus reaction is named after the French bacteriologist Nicholas M. Arthus (1862–1945), who first observed it in animals. This localized inflammation occurs when immune complexes of antigen, antibody, and complement form. Polymorphonuclear leuko-

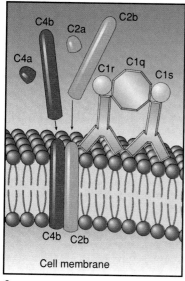
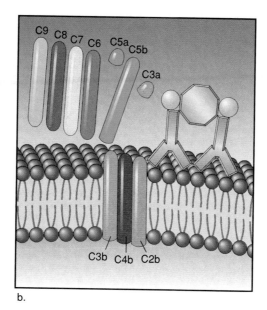
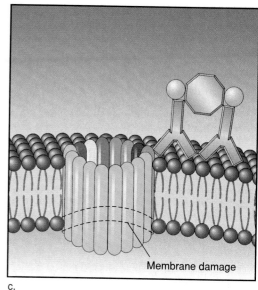

a.

Cell membrane

b.

C3b C4b C2b

c.

Membrane damage

figure 16.7

The Classical Complement Pathway

The mechanism of the classical complement pathway is shown, using sheep erythrocyte as the antigen. a. Complement proteins C1q, C1r, and C1s bind to the cell membrane to form a C1 complex. C4b binds to the cell membrane, followed by binding of C2b to C4b.

b. Next, C3b binds to C4bC2b. C5b then binds to the cell membrane. The fixation of the C5b is followed by the addition of C6, C7, C8, and C9. c. The fully assembled C5b–C9 membrane-attack complex results in membrane damage.

cytes are attracted to the affected area, which leads to necrosis and inflammation. Serum sickness occurs when large amounts of foreign antigen are intravenously injected and result in the formation of immune complexes. These immune complexes frequently lodge in the kidneys and joints and cause such symptoms as nephritis and painful joints. Serum sickness historically was caused by injection of antitoxins against diphtheria, tetanus, or other bacterial diseases into presensitized individuals. From 1920 to 1940, these antitoxins were prepared in horses, and the injected horse serum proteins elicited the hypersensitive response. Today similar allergic reactions may be encountered to penicillin and other drugs.

Complement Is a Group of Proteins That Augment the Action of Antibodies

Complement is the term used to describe a group of about 20 serum proteins that complement, or mediate, the action of specific antibodies in their destruction of bacteria and other particles. These proteins, unlike antibodies, are present in serum independent of host immunologic response. Furthermore, certain components of complement are heat labile and are inactivated after 30 minutes at 56°C (antibodies remain active at this temperature).

The complement system is important in host defense and participates in several types of nonspecific reactions, including: (1) enhancement of phagocytosis, (2) lysis of some bacteria, (3) neutralization of viruses, and (4) tissue inflammation. Nearly all of the complement proteins exist in serum as inactive molecules that are activated through a series of cascading reactions that may follow one of two pathways: the **classical pathway** and the **alternative pathway** of complement activation.

Complement May Be Activated by Interaction with Antigen-Bound Immunoglobulins

The classical pathway of complement activation is triggered by the binding of an immunoglobulin molecule to an antigen (Figure 16.7). Only antibodies of the IgM and IgG classes participate in complement activation. The complement components associated with activation are designated C1 through C9 (C stands for complement). The mechanism of the classical pathway has been defined in vitro using sheep erythrocytes as antigen, and consists of three distinct stages: (1) recognition (C1), (2) enzymatic activation (C4, C2, and C3), and (3) membrane attack (C5 through C9).

The recognition unit of the classical pathway consists of three proteins (C1q, C1r, C1s) that make up the C1 complex. The binding of antibody to antigen causes a conformational change in the antibody structure that exposes a receptor on the antibody molecule for C1. C1q attaches to the antibody via this receptor. With the binding of C1q to the antibody-antigen complex, C1r is split and C1s undergoes enzymatic cleavage.

The cleavage of C1s activates this component of the C1 complex to subsequently cleave a small peptide (C4a) from the N-terminal of C4. The resultant C4b fragment fixes to the erythrocyte cell membrane. The C4a fragment is released to the surrounding fluid. Activated C1s next cleaves C2 to produce C2a and C2b; C2a is lost to the fluid phase. C2b binds to the C4b fragment already attached to the cell membrane to form a C4bC2b enzymatically active complex that cleaves C3 into two fragments, C3a and C3b. C3a (anaphylatoxin) is a potent peptide that causes the release of histamine from mast cells to produce vasodilation, increased capillary permeability, chemotaxis of phagocytes, and inflammation.

C3b binds to C4bC2b, producing the complex C4bC2bC3b—termed C5 convertase because it splits C5 to C5a and C5b.

The cleavage of C5 by the C4bC2bC3b complex is the first step in the membrane attack stage of the classical pathway. The C5b fragment formed by this cleavage fixes to the membrane and subsequently binds, in succession, C6, C7, C8, and as many as 10 to 16 molecules of C9. The fully assembled C5b-C9 structure, called the **membrane-attack complex (MAC),** disrupts the membrane lipid bilayer and forms a lytic transmembrane pore (Figure 16.8). Water, ions, and small molecules diffuse freely through the membrane, resulting in cell lysis.

Activation of the Alternative Complement Pathway Is Independent of Immunoglobulins

The alternative complement pathway (also called the **properdin** pathway after a serum protein complex that is involved in the reactions of the pathway) occurs in the absence of antibody-antigen complexes. It is activated by bacterial, fungal, and plant polysaccharides, lipopolysaccharides, and other molecules, and by certain antibody aggregates that do not bind C1q and therefore are unable to initiate the classical pathway.

The alternative pathway bypasses the initial sequence of events from C1 to C2 in the classical pathway and involves properdin and two other serum proteins designated Factors B and D (Figure 16.9). The complement protein C3 has an unstable thioester bond that can be spontaneously hydrolyzed to continuously generate low levels of an activated form of C3 called C3i in serum. This spontaneous activation of C3 is known as **tick-over activation.** In the alternative pathway, Factor B binds with C3i to form a complex C3iB, from which a small fragment Ba is cleaved by Factor D to leave C3iBb. C3iBb is a C3 convertase that splits C3 to C3a and C3b. Most of the C3b molecules in the fluid phase are hydrolyzed and inactivated, but those that come in contact with a target cell surface such as a bacterial plasma membrane will bind to it. Although C3b may bind to mammalian plasma membranes, high levels of sialic acid in these membranes inactivate the bound C3b molecules. C3b molecules bound to the target cell surface combine with Factor B to form C3bB, which in turn is cleaved by Factor D to produce C3bBb. Properdin combines with C3bBb to stabilize the complex. C3bBb is a C3 convertase that splits C3 to more C3b, which combines with Factor B in a continuous positive feedback loop. C3bBb may also bind C3b to produce C3bBbC3b, a C5 convertase that splits C5 to C5a and C5b. At this point the alternative pathway follows the sequence of membrane attack that occurs in the classical pathway. Because the alternative pathway does not require antibody, it is an example of nonspecific resistance.

Microbial Pathogens
Nonspecific Host Defense: Cellular and Phagocytosis • p. 29

Cell-Mediated Immunity

Whereas humoral immunity is concerned primarily with the elimination of pathogenic bacteria, bacterial toxins, and viruses in body fluids such as the blood or mucous secretions, **cellular**

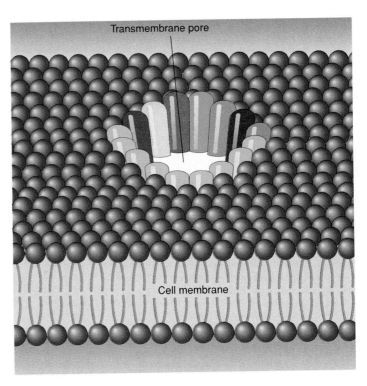

figure 16.8
Transmembrane Pore Formed by Membrane-Attack Complex
The membrane-attack complex disrupts the membrane lipid bilayer and forms a lytic transmembrane pore.

immunity, or **cell-mediated immunity (CMI),** is effective against intracellular pathogens and against fungi and protozoa. Inside the host cell, the pathogens are sheltered from the immunoglobulins of humoral immunity. CMI is designed to respond to this invasion, using T lymphocytes. The cell-mediated immune system is also associated with the immune response during tissue transplantation and is the body's defense against foreign tissue.

Cell-mediated immune reactions involve interactions between antigen and cells of the immune system. Like humoral immunity, CMI is an important part of the host defense against invasion. Unlike humoral immunity, which can be passively transferred in serum containing antibodies, CMI cannot be passed in serum. However, it can be transferred using lymphocytes from a sensitized host. A number of different types of cells, including T lymphocytes, macrophages, killer cells (K cells), and natural killer cells (NK cells), are responsible for CMI. CMI responses are associated with recovery from many types of infections, particularly those in which microorganisms multiply and survive intracellularly in the body.

Cellular immunity is important in host defense against many different types of microbial diseases, including tuberculosis, leprosy, candidiasis, coccidioidomycosis, mumps, measles, and toxoplasmosis. In addition, certain drugs, dyes, and chemicals, such as those found in poison ivy, poison oak, insecticides, jewelry, and coins, are able to trigger CMI. The term **contact dermatitis** is used to describe delayed skin reactions caused by contact with these substances.

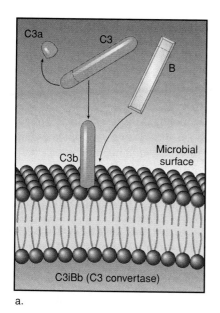
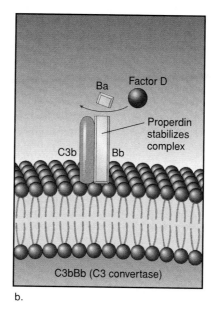
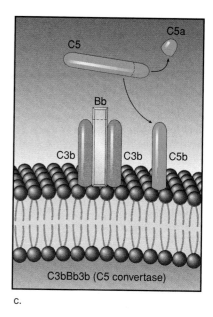

figure 16.9

The Alternative Complement Pathway

a. Following tick-over activation, C3b molecules attach to the target cell surface. b. Factor B attaches to C3b. Fragment Ba is cleaved by Factor D to leave C3bBb. C3bBb, a C3 convertase, splits C3 to form more C3b, which in turn combines with Factor B in a feedback cycle. The binding of properdin to the C3bBb stabilizes the complex.

c. C3bBb binds to C3b to produce C3bBb3b, a C5 convertase that splits C5 to C5a and C5b. C5b binds to the cell membrane. The events then follow the classical complement pathway (see steps b and c in Figure 16.7).

Cytokines Activate Macrophages and Increase Their Phagocytic Activity in Cell-Mediated Immunity

Cell-mediated immunity involves cytotoxic and other T lymphocytes that respond specifically to antigen (Figure 16.10). Like the B lymphocytes, T lymphocytes have surface receptors that bind to antigen. When stimulated by an antigen, T lymphocytes respond by dividing and differentiating into immunologically active T helper cells that release cytokines. Among the cytokines involved in CMI are: (1) **macrophage chemotactic factor (MCF),** which attracts macrophages to the infection site; (2) **macrophage migration inhibitory factor (MIF),** which prevents macrophages from leaving the infection site; (3) **macrophage activating factor (MAF),** which alters macrophages immunologically and increases their phagocytic activity; and (4) **interleukin-2,** which stimulates cytotoxic T cells to divide. Recent evidence indicates that interleukin-2-activated cytotoxic cells kill cancer cells in human tumors and may be an effective weapon against cancer.

In CMI, cytokines activate macrophages and increase their phagocytic activity. These activated macrophages then kill or inhibit growth of the invading microbe. In addition, cell-mediated immune responses result in the generation of cytotoxic T cells, which direct their activity against cancerous cells and virus-transformed cells. The cytotoxic T cells can recognize viral antigens on the surface of target cells and destroy the cells by releasing lymphotoxins, which associate with and damage the target cell's plasma membrane.

Delayed Hypersensitive Responses Are Caused by the Release of Cytokines from Sensitized T Cells

One manifestation of CMI is the **delayed hypersensitive** response. Delayed hypersensitivity was first observed by Edward Jenner (1749–1823) in 1798 during his immunization studies with cowpox virus. Jenner, a British physician, discovered that inoculation with cowpox virus, which is closely related to smallpox virus but does not cause illness in humans, was effective in immunizing a person against smallpox. This procedure came to be known as vaccination [Latin *vacca,* cow]. Jenner further noted that previously immunized individuals who were revaccinated developed inflammation lesions 24 to 48 hours later at the site of the second vaccination. Similar results were observed nearly 100 years later in 1890 by Robert Koch (1843–1910) when he inoculated *Mycobacterium tuberculosis* into the skin of guinea pigs with tuberculosis. Localized red lesions appeared within 24 to 48 hours at the site of inoculation. This delayed hypersensitive response became the basis for the tuberculin skin test used to screen for tuberculosis (Figure 16.11).

Both of these reactions are characteristic of delayed hypersensitive responses; the cells responsible for delayed hypersensitivity are T lymphocytes. Animals that have a nonfunctional thymus or have had their thymus surgically removed during fetal or neonatal life no longer exhibit delayed hypersensitive reactions.

It is postulated that the mechanism of delayed hypersensitivity involves recognition and attachment of T lymphocytes to antigens that enter the body. The antigens bind to receptor sites on

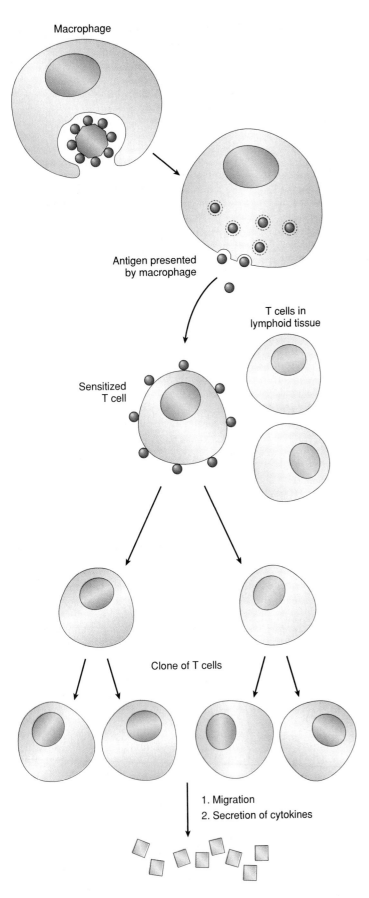

Macrophage

Antigen presented
by macrophage

T cells in
lymphoid tissue

Sensitized
T cell

Clone of T cells

1. Migration
2. Secretion of cytokines

the surfaces of the T lymphocytes, causing the lymphocytes to become activated, or **sensitized.** The sensitized lymphocytes increase in size and number and attract other lymphocytes to the site of antigen entry, where the lymphocytes release cytokines associated with inflammation.

Not all CMI reactions are beneficial to the host. Release of lysosomal enzymes and toxic substances by macrophages and T lymphocytes aggregated at antigen sites can lead to extensive tissue damage. This damage is evident in necrotic lesions that sometimes form at these inflamed sites.

Cell-Mediated Immunity May Be Manifested in Other Ways

A number of responses other than delayed hypersensitivity are evoked by CMI. Killer cells (K cells) are cells that look very much like lymphocytes, but lack the surface characteristics of either B or T lymphocytes. For this reason, they often are called **null cells.** K cells are important in cellular protection against viral infections. As antibody binds to viral antigen, the K cell binds to the Fc portion of the antibody. The virus lyses as a result of this binding and the release of chemical substances from the K cell. This type of toxicity is known as **antibody-dependent cell-mediated cytotoxicity (ADCC).**

Another type of null cell, the natural killer cell (NK cell), is normally found in humans. Even in the absence of sensitizing antigen, these cells are capable of destroying tumor cells and cells infected by enveloped viruses such as mumps or herpes simplex. The activity and specificity of NK cells is significantly enhanced by the antiviral substance interferon. Little is known about the mechanism of NK cell cytotoxicity, but these cells are important in natural defense against certain types of viruses and cancers.

Viral Structure and Function
Viral Pathogenesis: Host Cell Response • pp. 17–18

CMI also plays a principal role in graft rejection. Host rejection of grafted tissue is a major problem in tissue transplantation. Individuals, even within the same species, develop different antigens in their tissues. As a result of these differences, the transplantation of tissue often results in recipient rejection of the grafted tissue. This rejection is mediated by a number of different mechanisms, including one similar to delayed hypersensitivity. Here T lymphocytes migrate to the graft and release cytokines. In some instances, if immunocompetent tissue is transplanted to a host that is not immunocompetent, the graft sees the recipient as foreign and produces T lymphocytes that attack the host tissues. Regardless of the source of the lymphocytes, the result of this cell-mediated activity can be rejection. For this reason, immunosuppression treatment is frequently necessary for a successful transplant. Such treatment normally involves the use of agents to reduce the number and activity of T lymphocytes. Antilymphocyte

figure 16.10
Mechanism of Cell-Mediated Immunity

In cell-mediated immunity, T lymphocytes bind to antigen and become sensitized. The stimulated T cells then divide and differentiate into immunologically active cells that secrete cytokines.

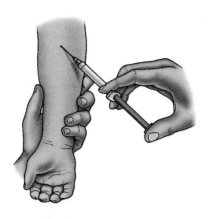

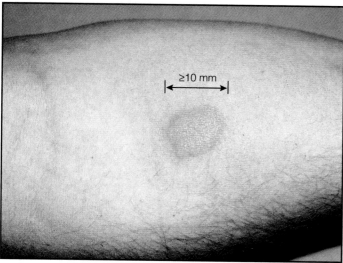

Positive reaction

figure 16.11

Tuberculin Skin Test

In the tuberculin skin test, antigen (purified protein derivative, or PPD) is injected into the superficial layers of the skin of the forearm. After 48 hours, the area of induration (thickening of the skin) is measured. A positive reaction is one in which the diameter of induration (not the redness of the skin alone) is 10 mm or more.

Miscellaneous Bacteria
Mycobacteria: Antigenic Structures • p. 17

globulin, for example, is antiserum prepared from horses that is directed against lymphocytes. Prednisone, a steroid, suppresses T lymphocyte and antibody formation.

Assays to Measure Cell-Mediated Immunity Depend upon Delayed Hypersensitive Skin Reactions or Detection of Sensitized Lymphocytes

Compared with tests for the measurement of antibodies in humoral immunity, there are relatively few clinical tests to measure CMI. Skin tests are the most common. A battery of antigens is injected intradermally, and a delayed hypersensitive reaction (reddening and swelling) is observed 24 hours later for those antigens to which the individual has developed CMI. The antigens used commonly include those for tetanus, mumps, tuberculosis (purified protein derivative), and *Candida* infections. Skin tests, however, require purified antigen and depend on patients' responding to the antigen without harmful side effects.

Laboratory tests have been developed that detect the presence of macrophage migration inhibitory factor (MIF). Lymphocytes from a patient are mixed with sensitizing antigen and then incubated with guinea pig macrophages. The extent of macrophage migration inhibition is then measured against control suspensions containing nonsensitized lymphocytes.

Sensitized lymphocytes can also be detected by changes in their morphology and metabolism. Such changes are determined by incubating patient lymphocytes in the presence of sensitizing antigen. Increased rates of RNA and DNA synthesis in sensitized lymphocytes are measured by radiolabeling experiments using labeled RNA and DNA precursors.

In Vitro Antibody-Antigen Reactions

The antigen-antibody reactions we have discussed in humoral immunity occur in vivo (in the host). Antigen-antibody reactions can also be demonstrated in vitro (in a test tube or artificial environment). The area of microbiology concerned with the observation of in vitro antigen-antibody reactions is called **serology.** Serological methods for the detection of microbial antigens can be used in the diagnosis of microbial diseases. In recent years such techniques have become increasingly important for the rapid diagnosis of diseases. A few of the more commonly used techniques will be discussed.

Soluble Antigens Mixed with Multivalent Antibodies Form Large, Precipitable Aggregates

In **precipitation,** multivalent soluble antigens mixed with multivalent antibodies form large aggregates that precipitate from solution. These precipitates are easily seen with the naked eye and are indicative of the antigen-antibody reactions. The antibodies that participate in these reactions are called **precipitins.** As precipitins combine with a number of determinants on the antigens, a visible precipitate forms.

The principle of precipitation is used in several types of laboratory tests. In the **precipitin ring test,** or **precipitin interface test,** antiserum containing specific antibody is allowed to enter a thin glass tube by capillary action. Soluble antigen is then layered onto the surface of the antiserum in the tube. The two solutions are permitted to react for approximately 30 minutes. During this time, antibody and antigen diffuse toward each other until they meet. At this point, if the antigen and antibody are specific for each other (homologous), a visible precipitate develops at the interface of the antigen and antibody solutions (Figure 16.12). In the past the capillary precipitin test has been used for the grouping of streptococci using group-specific antibodies prepared in rabbits; today streptococcal grouping is routinely performed using coagglutination or other more rapid procedures.

The precipitation method is also performed in agarose or other similar gels. The **Oudin single diffusion test,** described by Jacques Oudin in 1946, is performed by adding antiserum to liquid agarose in a narrow-bore test tube. The antiserum-agarose

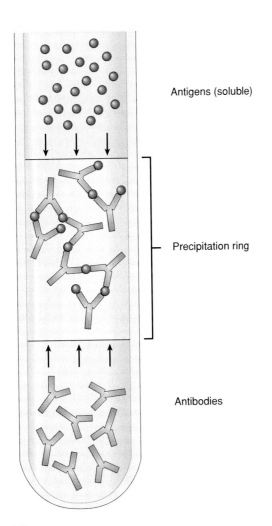

Antigens (soluble)

Precipitation ring

Antibodies

figure 16.12

Precipitin Ring Test

In the precipitin ring test, antibody and antigen are placed next to each other in a capillary tube. The antibody and antigen diffuse toward each other, and if they are specific for each other, a visible precipitate develops at the interface of the antibody and antigen solutions.

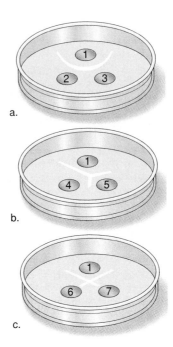

figure 16.13

The Ouchterlony Test

Well 1 contains antibodies to antigens in wells 2 through 7. a. Line of identity showing a single band of antigen-antibody precipitate. The antigens in wells 2 and 3 are serologically identical. b. The antigens are cross-reacting, causing the lines of precipitate to partially fuse, forming a spur. The antigens in wells 4 and 5 are serologically partially identical. c. Two separate lines of precipitate form and cross in a common area. The antigens in wells 6 and 7 are serologically unrelated.

mixture is allowed to gel. Antigen is then overlaid on the agarose gel, and the tube is allowed to stand for several hours or days. Precipitation occurs as the antigen diffuses through the gel toward the antibody. The Oudin test was modified by Oakley and Fulthorpe, who placed antiserum in agarose at the bottom of a test tube, overlaid it with a second layer of agarose, and added antigen on top of the entire gel. Antigen and antiserum move toward each other through the upper agarose layer in this **double diffusion test.** Orjan Ouchterlony developed a variation of this double diffusion technique in 1948. The **Ouchterlony test** is performed in a Petri dish filled with agarose. Small wells are cut in the agar, and antigen and antiserum are placed in different wells. As antigen and corresponding antiserum migrate toward each other, thin lines of precipitation form in the gel (Figure 16.13). The Ouchterlony technique is particularly useful in comparing the specificity of different types of antibodies and antigens. The different types of diffusion tests are collectively called **immunodiffusion tests.**

Antigens Can Be Separated by Immunoelectrophoresis

Immunoelectrophoresis is a procedure used to identify antigens in complex mixtures. The antigen mixture is placed into a well cut into an agarose layer on a glass slide. An electric current is applied to the agarose layer, causing the antigens to migrate and separate in the gel on the basis of their electrophoretic mobility. The current is discontinued, and antiserum specific for the antigens under investigation is added to a trough. The antibodies and antigens diffuse through the gel and form precipitin bands at their interface (Figure 16.14). Immunoelectrophoresis is especially useful to detect a large number of proteins in serum.

Insoluble Antigens Mixed with Multivalent Antibodies Form Aggregates Detectable by Agglutination

Antigens that remain on the surface of cells or are insoluble are detectable by **agglutination** reactions. These were among the first laboratory tests used for the diagnosis of microbial diseases and are still used to diagnose diseases caused by *Salmonella, Vibrio, Rickettsia,* and other microbes. In the case of *Salmonella,* agglutination is used to separate the bacterium into more than 2,200 different serotypes. There are several versions of the agglutination test.

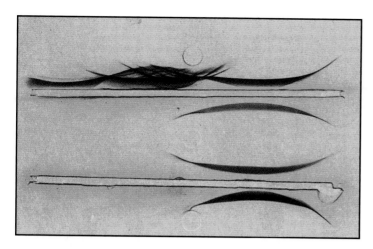

figure 16.14

Immunoelectrophoresis of Human Serum and Human Albumin
Human serum, purified human albumin, and highly purified human albumin (antigens) were placed in the three wells, top to bottom. The antigens were separated according to their charge by electrophoresis. Antihuman serum from rabbit (antibody against human serum proteins, prepared in a rabbit) was then added to each of the two troughs between the wells. The antigens and antibodies form precipitin bands, with a full range of reaction shown between human serum (with all of its proteins including albumin) in the top well and the antihuman serum in the top trough. A single precipitin band is seen between purified or highly purified human albumin (middle and bottom wells) and the antihuman serum (top and bottom troughs). Immunoelectrophoresis is useful in the comparison of mixtures of antigens as found in serum.

The original agglutination technique, used for the laboratory diagnosis of typhoid fever, was called the **Widal test** after its developer, the French physician Fernand Widal (1862–1929). In this test, the patient's serum (or a dilution of it) is mixed in a test tube with a standard suspension of *Salmonella typhi*. The antibody-bacteria mixture is allowed to react for a few minutes and is then read for agglutination.

Slide agglutination tests today use the same principle as the Widal test. In a common agglutination test for *Salmonella*, bacteria from a colony on an agar plate are mixed on a glass slide with a drop of *Salmonella*-specific antisera. The mixture is rocked back and forth for one minute and then examined for agglutination.

A variation of the agglutination test is **coagglutination.** Antigen is mixed with antibodies linked to heat-killed *Staphylococcus aureus* cells. The antibodies are attached by their Fc portions to a surface component of the *S. aureus* cell called Protein A. The antigen-antibody-*S. aureus* complex forms a visible clump that appears within one to two minutes (Figure 16.15). A similar procedure, **latex agglutination,** uses the same principle, with antibodies attached to latex spheres instead of bacterial cells. Coagglutination and latex agglutination are both rapid serological procedures that are commonly used for the identification of *Streptococcus*, *Neisseria gonorrhoeae*, *Haemophilus influenzae*, and other bacteria.

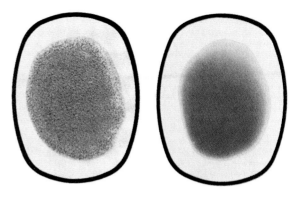

figure 16.15

Coagglutination Test
A positive coagglutination reaction, indicated by the visible lattice formed from the combination of antibody and antigen, is shown on the left, and a negative reaction is shown on the right.

The agglutination of red blood cells is known as **hemagglutination.** Certain types of viruses, such as measles, mumps, and influenza viruses, have the ability to agglutinate red blood cells. This characteristic is used to detect specific viral antibody in serum in a procedure called the **hemagglutination inhibition test.** Serum suspected of containing viral antibodies is mixed with the virus particles, and red blood cells are then added. If the serum contains antibodies, the virus particles are neutralized and hemagglutination is inhibited. Red blood cells can be treated with tannic acid or chromic chloride to noncovalently adsorb antigens such as proteins. When such antigen-coated red blood cells are mixed with the corresponding antibodies for the antigen, aggregates form that are visible to the naked eye. This type of agglutination is known as **indirect hemagglutination** or **passive hemagglutination.**

Neutralization Tests Are Used to Detect Toxins and Viruses or Their Antibodies

Neutralization tests are used for the detection of toxins and viruses, or antibodies to them. As an example, serum from a patient is mixed with known virus and the mixture is inoculated into laboratory mice. If the patient's serum contains antibodies against the virus, the virus is neutralized and the mice are protected.

Complement Can Be Used to Detect Antigen-Antibody Reactions

Most of the antigen-antibody reactions that have been discussed are easily visible with the naked eye. In many cases, however, and particularly with viruses, such reactions are not readily seen. The **complement fixation assay** can be used to detect these reactions (Figure 16.16).

The complement fixation test is performed in two stages. In the first stage, patient serum suspected of containing particular

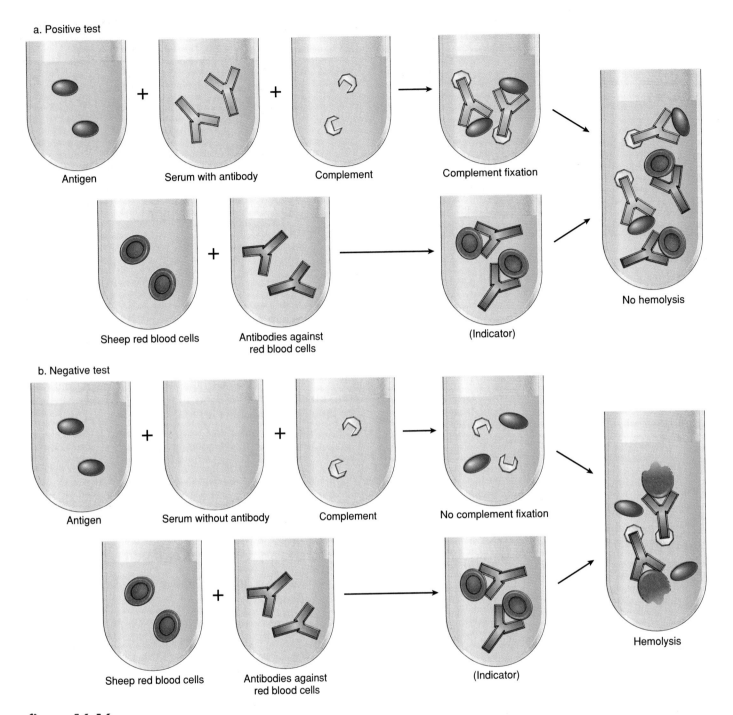

a. Positive test

Antigen + Serum with antibody + Complement → Complement fixation

Sheep red blood cells + Antibodies against red blood cells → (Indicator)

No hemolysis

b. Negative test

Antigen + Serum without antibody + Complement → No complement fixation

Sheep red blood cells + Antibodies against red blood cells → (Indicator)

Hemolysis

figure 16.16

Complement Fixation Assay

The assay is based on the fixing of complement to antigen-antibody complexes. a. If complement is fixed (used up), there is no hemolysis of red blood cells, and the assay is positive for the presence of antibody. b. If complement is not fixed, there is hemolysis, and the assay is negative for the presence of antibody.

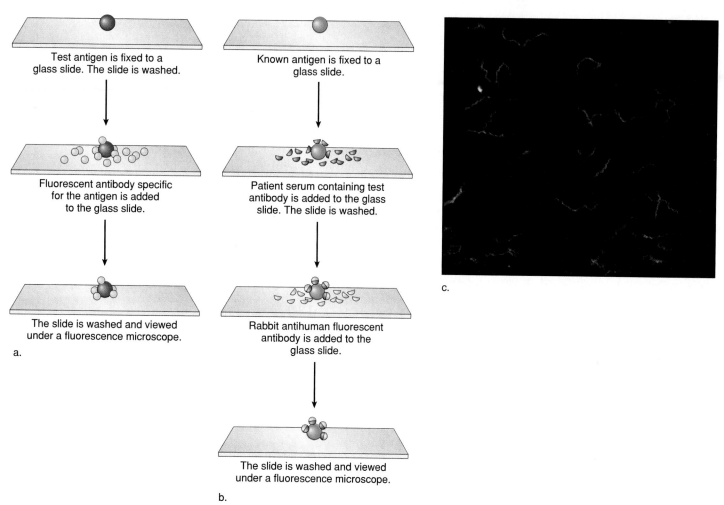

Test antigen is fixed to a glass slide. The slide is washed.

Fluorescent antibody specific for the antigen is added to the glass slide.

The slide is washed and viewed under a fluorescence microscope.

a.

Known antigen is fixed to a glass slide.

Patient serum containing test antibody is added to the glass slide. The slide is washed.

Rabbit antihuman fluorescent antibody is added to the glass slide.

The slide is washed and viewed under a fluorescence microscope.

b.

c.

figure 16.17

Immunofluorescence

a. In direct immunofluorescence, fluorescent antibody specific for the test antigen combines with the antigen, and fluorescence is observed after the last washing step. b. In indirect immunofluorescence, test serum, containing antibody specific for the known antigen, combines with the antigen. Rabbit or goat antihuman antibody labeled with fluorescein isothiocyanate then binds to the test antibody-antigen complex, and fluorescence is seen after the last washing step. c. Immunofluorescent stained cells of *Treponema pallidum*.

antibodies is heated at 56°C for 30 minutes to inactivate native complement; the heating does not affect antibodies in the serum. Next, the serum is diluted and mixed in a test tube or microtiter plate with known amounts of complement (usually from a guinea pig) and antigen. The mixture is incubated, and the presence of uncomplexed complement is determined in the second stage.

The second stage involves the addition of a complement-detecting indicator system to the mixture. The indicator system consists of sheep red blood cells and antibody against them. Complement fixed in the first stage of the procedure is not available to attach to the sheep red blood cell–antibody complexes; thus hemolysis does not occur (a red button of intact blood cells is seen at the bottom of the test tube or microtiter plate well). If hemolysis is observed (no red button), complement was not fixed in the first stage, and the patient serum does not contain the antibody in

question. The available complement attaches to the sheep red blood cell–antibody complex, resulting in lysis of the red blood cell (no red button).

Viral Structure and Function
Viral Detection: Detection Antibody • pp. 22–23

Fluorescent Dyes Are Used in Immunofluorescence to Detect Antigens or Antibodies

The development of the fluorescence microscope provided serologists with a tool for the rapid detection of antigens or antibodies from host specimens. The **fluorescent antibody technique,** or **immunofluorescence method,** uses a known preparation of antibodies labeled with a fluorescent dye such as fluorescein isothiocyanate (FITC) to detect antigen or antibody (Figure 16.17). In the direct fluorescent antibody test, fluorescent antibodies are used to detect specific antigens in cultures or smears. In the indirect fluorescent antibody test, specific antibody from patient serum is bound to antigen on a slide. The antibody-antigen complex is then detected with a fluorescent antibody.

In direct immunofluorescence, antigen from a clinical specimen or a colony is alcohol fixed or air dried on a glass slide. Known fluorescein-labeled antibody is added to the slide and incubated with the fixed antigen. The slide is then washed and examined under a fluorescence microscope. Antigen that specifically reacted with the labeled antibody appears as bright, glowing, yellow-green fluorescence under the microscope; there is no fluorescence if no antigen was recognized by the antibody. In the latter case, the antibody was removed from the slide during the washing step of the protocol. The direct fluorescence staining procedure can be used for the identification and detection of many different antigens using different antibodies.

The indirect fluorescence procedure can be used for the detection of *Treponema pallidum* antibodies from the sera of patients with syphilis. *T. pallidum* organisms are fixed onto a glass slide. The smear is covered with patient serum and incubated for 30 minutes. Next, the slide is washed, and rabbit or goat antihuman immunoglobulin labeled with fluorescein isothiocyanate is added to the slide. The mixture is incubated again, washed, and examined under a fluorescence microscope. Slides containing patient serum positive for treponemal antibodies will fluoresce. These antibodies remain on the slide after reacting with the fixed treponemal antigen and are recognized as antigen by the fluorescein-labeled, antihuman immunoglobulin.

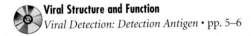
Viral Structure and Function
Viral Detection: Detection Antigen • pp. 5–6

Radioactivity Is Used in Radioimmunoassay to Detect Small Amounts of Antigen

Radioimmunoassay (RIA), a technique introduced by Rosalyn Yalow and Solomon Berson in 1959, detects very small amounts of antigen by the measurement of radioactivity. The technique was originally developed to detect minute amounts of insulin in the blood of diabetics. Known amounts of radioactively labeled insulin and unlabeled insulin antibodies are mixed with a sample of patient serum. The labeled insulin and the patient's insulin in the serum compete with each other for the insulin antibodies. The more insulin that is present in the serum, the fewer labeled insulin that will attach to the antibodies. The quantity of insulin in the serum can be determined by separating antigen-antibody complexes from the mixture and measuring the amount of antibody-bound labeled insulin. The RIA technique is sensitive enough to measure picogram levels of antigen.

Enzymes Are Used in Enzyme-Linked Immunosorbent Assays to Detect Antigens or Antibodies

A technique that is similar in principle to the RIA is the **enzyme-linked immunosorbent assay (ELISA).** One of the most common forms of ELISA is the **sandwich technique,** which "sandwiches" antigen between antibodies (Figure 16.18). Known antibody is immobilized on the surface of a well in a polystyrene plate, and test antigen is added. If the antigen is homologous, it attaches to the immobilized antibody. Alkaline phosphatase-labeled specific antibody is added, resulting in an enzyme-antibody-antigen-antibody sandwich. Finally, *p*-nitrophenyl phosphate (the substrate for alkaline phosphatase) is added. The *p*-nitrophenyl phosphate is enzymatically split by alkaline phosphatase, producing *p*-nitrophenol, a yellow compound that can be measured spectrophotometrically.

The ELISA is frequently used to detect viral antibodies and antigens. The procedure is easily learned and does not require the expensive equipment, materials, or safety considerations often associated with immunofluorescence and the use of radioisotopes for radioimmunoassay.

An example of a commonly used ELISA is the indirect ELISA for the diagnosis of AIDS (see AIDS, page 536). The AIDS ELISA uses HIV lysates from infected lymphocytes immobilized in microtiter plate wells to detect HIV antibodies in patient sera. The test is rapid, sensitive, and easily performed, and can be used as a screening test for blood. The AIDS ELISA occasionally gives false positive results and in advanced AIDS cases, in which the patient produces few antibodies to HIV or to any other antigen, the test may not detect the low levels of HIV antibodies. In such cases, an alternative, more accurate procedure, the **Western blot,** is used to diagnose AIDS. In the Western blot procedure, purified HIV proteins are subjected to electrophoresis on a polyacrylamide gel. Electrophoresis separates the proteins into several distinct bands. The proteins are transferred by blotting to nitrocellulose paper. Patient serum is added to the antigen-containing nitrocellulose paper, followed by the addition of enzyme-linked antihuman IgG and the enzyme substrate. HIV antibodies in the patient serum are detected by a color reaction as in the indirect ELISA. The Western blot procedure is more accurate than the indirect ELISA because it detects antibodies to specific HIV antigens that have been electrophoretically separated. For this reason, a positive AIDS Western blot result is generally considered unequivocal.

Viral Structure and Function
Viral Detection: Detection Antigen • pp. 7–9

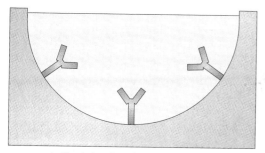

Antibody is adsorbed to well surface
(solid phase)

↓ Wash

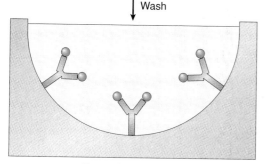

Test antigen is added

↓ Wash

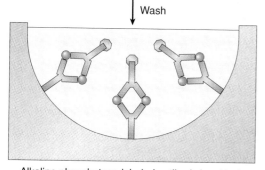

Alkaline phosphatase-labeled antibody is added,
resulting in a double antibody sandwich

↓ Wash

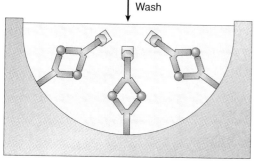

p-nitrophenyl phosphate substrate is added,
and any yellow color is measured
spectrophotometrically

Key:

 Test antigen

Antibody

Alkaline
phosphatase-labeled
antibody

p-nitrophenyl
phosphate substrate

a.

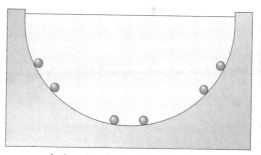

Antigen is adsorbed to well surface
(solid phase)

↓ Wash

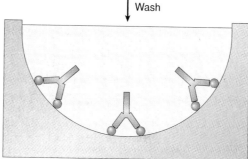

Test antibody is added

↓ Wash

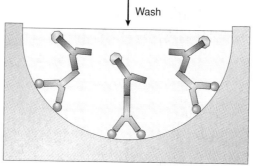

Alkaline phosphatase-labeled
antihuman antibody is added

↓ Wash

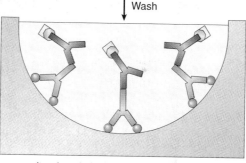

p-nitrophenyl phosphate substrate is added,
and any yellow color is measured
spectrophotometrically

Key:

 Antigen

Test antibody

Alkaline
phosphatase-labeled
antihuman antibody

p-nitrophenyl
phosphate substrate

b.

c.

figure 16.18

Enzyme-Linked Immunosorbent Assay (ELISA)

a. In the direct ELISA, the test antigen is sandwiched between two layers of antibodies (double antibody sandwich). The presence of the antigen is indicated by the alkaline phosphatase–catalyzed conversion of *p*-nitrophenyl phosphate to *p*-nitrophenol. b. In the indirect ELISA, antibody in the test serum is detected by its binding to antigen immobilized in a microtiter plate well. The test antibody is then detected by the addition of rabbit or goat antihuman antibody conjugated with alkaline phosphatase, which converts *p*-nitrophenyl phosphate to *p*-nitrophenol. c. An automated ELISA reader used to spectrophotometrically detect the formation of *p*-nitrophenol in microtiter plate wells.

The First Identified Cases of AIDS in the United States

During the winter of 1981, Michael Gottlieb, an immunologist at the University of California, Los Angeles (UCLA), noticed something strange occurring among four of his patients. These four previously healthy homosexual men had contracted *Pneumocystis carinii* pneumonia, extensive mucosal candidiasis, and multiple viral infections (Figure 16P.1). Although these were not unusual diseases, all four patients showed a similar severe acquired T-cell defect.

The four patients consisted of: (1) a 33-year-old homosexual man with recurrent fever, mucocutaneous candidiasis, *Herpes simplex* infection, and *P. carinii* infection; (2) a 30-year-old homosexual man with daily fever, malaise, and suspected infection with *P. carinii*; (3) a 30-year-old homosexual man with oral thrush, leukopenia, *P. carinii* infection, adenovirus infection, Kaposi's sarcoma, and a history of weight loss; and (4) a 29-year-old homosexual man with high fever, lymphadenopathy, fatigue, oral thrush, *P. carinii* infection, and a history of weight loss. All four patients had marked leukopenia (reduced white blood cell counts), with white cell counts ranging from 1,600 to 4,800 cells/mm^3 (compared with normal white cell count of 4,800 to 10,800 cells/mm^3). The percentage of B lymphocytes was normal in all patients except patient 1, who had 2% to 3% B lymphocytes (compared with the normal 5% to 10%). Cytomegalovirus was recovered from tissue specimens or urine of all four patients.

Of particular interest was the discovery that all four patients had reduced percentages and total numbers of T lymphocytes in the peripheral blood (Table 16P.1). T helper

figure 16P.1

Transmission Electron Micrograph of *P. carinii* in Alveoli
Three alveoli are seen. Two of the alveoli have numerous *P. carinii* trophozoites (T) and a collapsed cyst (C) form. The third alveolus contains a macrophage (M) (×3,000).

cells (CD4-containing T lymphocytes) were practically absent in all of the patients. The percentage of a cytotoxic T cell subset (CD8-containing T lymphocytes) was increased, although the absolute numbers of these lymphocytes were below normal. All four patients had a low ratio (0 : 0.18) of T helper to cytotoxic T cells; a ratio of 1 : 6 is normal.

Because this illness was observed in homosexual men, it was assumed that it involved a sexually transmitted infectious agent or exposure to a common environmental factor. Cytomegalovirus initially was believed to be associated with the illness, since it had been isolated from all four patients. However, there had been no previous reports of depressed T-cell numbers in cytomegalovirus-infected patients, so the illness did not appear to be a traditional cytomegalovirus-induced disease. The virtual absence of T helper cells was con-

cluded to be the major contributing factor to the severe immune deficiency observed in the patients.

It was not until several months later, as additional cases of similar illnesses were reported in medical journals, that a pattern of infection began to develop. We now know that the illness seen by Gottlieb was acquired immune deficiency syndrome (AIDS). His four patients were among the first reported cases of AIDS in the United States.

Source

Gottlieb, M.S., R. Schroff, H.M. Schanker, J.D. Weisman, P.T. Fan, R.A. Wolf, and A. Saxon. 1981. *Pneumocystis carinii* pneumonia and mucosal candidiasis in previously healthy homosexual men: Evidence of a new acquired cellular immunodeficiency. *The New England Journal of Medicine* 305:1425–1431.

table 16P.1

Characterization of T Lymphocyte Subsets

	% Lymphocytes Reactive with Monoclonal Antibodies by Lymphocyte Subset			Ratio of T Helper Cells to Cytotoxic T Cells
Group	T Cells	Cytotoxic T Cells	T Helper Cells	
Patient				
1	45	57	0	0
2	47	52	0	0
3	49	57	10	0.18
4	67	47	2	0.04
Mean ± SD	52[a] ± 10.1	53.3[b] ± 4.7	3.0[b] ± 4.76	0.05[a] ± 0.08
Normal subjects (n = 16 [mean ± SD])	71.0 ± 10.0	28.0 ± 8.0	46.0 ± 12.0	1.6 ± 0.74

[a]Significantly different from the value in normal subjects (p < .003).
[b]Significantly different from the value in normal subjects (p < .0001).

Summary

1. The study of immune mechanisms is called immunology. Humoral immunity is associated with proteins called antibodies that are directed against specific foreign particles called antigens.

2. T lymphocytes mature in the thymus and are responsible for cell-mediated immunity and the regulation of all immune responses. B lymphocytes mature in the bone marrow and are associated with antibody production.

3. Immunoglobulins are divided into five major classes: IgG, IgM, IgA, IgE, and IgD. IgG is the predominant circulating antibody and constitutes 80% of all antibodies found in the human body.

4. Immunoglobulins have a basic structure consisting of two light chains and two heavy chains. The Fab portions of the antibody molecule bind to antigens, whereas the Fc regions bind to complement and to Fc receptors on macrophages, neutrophils, and other cells.

5. Antibody diversity is made possible by gene rearrangement during B cell maturation. DNA sequences for the heavy chain of an antibody are randomly assembled from a V (variable) gene segment, a J (joining) gene segment, a D (diversity) gene segment for the variable region, and a single C (constant) gene segment for the constant region. DNA sequences for the light chain are randomly assembled from V, J, and C gene segments.

6. The major histocompatibility complex (MHC) is a cluster of genes coding for three types of cell surface proteins: Class I proteins, which serve to identify one's own cells as "self"; Class II proteins, which are important in antigen presentation to T lymphocytes; and Class III proteins, which are associated with complement.

7. The clonal selection theory states that B cells activated by antigens multiply and mature into clones of antibody-secreting plasma cells. Some B cells formed during clonal selection differentiate into memory cells that provide an anamnestic response to future exposure to the same antigenic determinant.

8. CD4-containing T lymphocytes consist of T helper (T_H) cells, which produce various cytokines, stimulate B lymphocytes to produce antibodies, and are involved in cell-mediated immunity. CD8-containing T lymphocytes consist of cytotoxic T (T_C) cells, which interact with and destroy cells containing antigen on their surfaces.

9. A vaccine is a material administered to a subject to produce artificially acquired active immunity. Three types of vaccines are commonly used in immunization: live, attenuated vaccine; killed vaccine; and toxoid, or subunit, vaccine.

10. Antigenic stimulation sometimes results in antibody-mediated hypersensitive responses. In antibody-mediated hypersensitivity, the host must first be sensitized to an antigen. Subsequent exposure to the antigen evokes a host response within a few minutes to hours.

11. Complement is a group of about 20 serum proteins that complement, or mediate, the action of specific antibodies. Complement proteins are activated by one of two pathways: the classical pathway and the alternative pathway.

12. Cell-mediated immune reactions involve interactions between antigen and cells of the immune system. A number of different cells, including T lymphocytes, macrophages, killer cells, and natural killer cells, are responsible for cell-mediated immunity. One form of cell-mediated immunity is the release of cytokines by antigen-stimulated T lymphocytes. These cytokines activate macrophages and increase their phagocytic activity.

13. Delayed hypersensitive response is a manifestation of cell-mediated immunity in which T lymphocytes that have been activated, or sensitized, by antigens release cytokines, which cause inflammation.

14. Serology is the area of microbiology concerned with the observation of in vitro antibody-antigen reactions. Serological methods for the detection of microbial antigens are useful in the clinical laboratory for diagnosis of microbial diseases.

EVOLUTION *and* BIODIVERSITY

Immunity, or acquired resistance, occurs only in vertebrates. However, an examination of all living organisms suggests that there has been an evolutionary progression toward the sophisticated immune systems of vertebrates, in general, and specifically of mammals. This evolution of the immune system begins with procaryotes, which are capable of recognizing and responding to stimuli in chemotaxis and phototaxis. Invertebrates lack antibodies and lymphocytes, but have phagocytic cells that form the first line of defense against microorganisms. Invertebrates furthermore produce a variety of chemical compounds, including lysozyme, bactericidins, lysosomal enzymes, agglutinins, and cytokinelike molecules that stimulate phagocytosis. Lower vertebrates such as fish and amphibians have functional complement pathways, major histocompatibility complexes, and cytokines that function similar to those in mammals. IgM, but not other antibody classes, also is found as the major serum antibody of fish. Sharks have a cluster arrangement of immunoglobulin gene subunits, suggesting the evolutionary development of immunoglobulin diversity at this level of vertebrates. As mammals evolved from these lower vertebrates and were exposed to a wide variety of foreign particles, including microorganisms, there must have been tremendous evolutionary pressures to have led to the development of fully functional immune systems in these mammals.

Short Answer

1. Compare and contrast antibodies and antigens.

2. Compare and contrast humoral immunity and cellular (or cell-mediated) immunity.

3. Identify the location(s) and explain the role(s) of lymphoid tissue.

4. Why are antibodies referred to as immunoglobulins?

5. Compare and contrast antiserum and antitoxin.

6. Identify five classes of antibodies.

7. Describe the structure of an antibody.

8. Explain how antigenic specificity of the antibodies is possible.

9. Explain the role of T cells in immunity.

10. Identify several types of T cells.

11. Explain the role(s) of the major histocompatibility complex.

12. Explain what cytokines are and why they are important.

13. Briefly explain the clonal selection theory.

14. Compare and contrast the primary and secondary immune responses.

15. Identify the four types of acquired immunity and their stimuli.

16. Identify three types of vaccines commonly used in immunization. Explain why unaltered, live organisms are not used.

17. If you needed to boost someone's immune system, what options are available? When would it be appropriate to suppress someone's immune system?

18. Why is hypersensitivity a bad thing?

19. Describe what complement is and explain why it is important.

20. Identify several types of in vitro immunological assays.

Multiple Choice

1. Which of the following is able to cross the placenta, thus giving protection to newborns?
   a. IgA
   b. IgD
   c. IgE
   d. IgG
   e. IgM

2. Which of the following is responsible for allergies?
   a. IgA
   b. IgD
   c. IgE
   d. IgG
   e. IgM

3. Which of the following stimulates antibody production?
   a. antigens
   b. B cells
   c. T_H cells
   d. T_C cells

4. Which of the following provides long-term immunity?
   a. artificially acquired active immunity
   b. artificially acquired passive immunity
   c. naturally acquired active immunity
   d. naturally acquired passive immunity

5. Vaccination for_____ depends on the use of a toxoid.
   a. measles
   b. pertussis
   c. polio
   d. tetanus

6. Which of the following occurs in response to a transfusion with the wrong type of blood?
   a. anaphylactic reaction
   b. cytotoxic reaction
   c. delayed hypersensitive reaction
   d. antibody-mediated hypersensitivity reaction
   e. All of the above.

Critical Thinking

1. There is a growing movement in the United States to forgo childhood vaccinations. Because entry into public schools requires documentation of immunization, many parents are opting to homeschool or send their children to church schools. Discuss the impact this will have on the public health of these children and the population at large.

2. Can you develop a vaccine without identifying the etiological agent? If so, is this prudent?

3. Do you think researchers today could get away with a public demonstration of vaccination, such as that performed by Edward Jenner? Explain why or why not.

4. Research several autoimmune diseases. For each, state the immunological disorder and explain how you might treat the affected individual.

Benjamini, E., and S. Leskowitz. 1988. *Immunology, a short course,* 2d ed. New York: Wiley-Liss. (A concise textbook of immunology.)

Coleman, R.M., M.F. Lombard, and R.E. Sicard. 1992. *Fundamental immunology,* 2d ed. Dubuque, Iowa: Wm. C. Brown. (A textbook of immunology.)

Janeway, Jr., C.A., and P. Travers. 1994. *Immunobiology.* Hamden, Conn.: Garland Publishing. (A textbook of immunology.)

Kuby, J. 1994. *Immunology,* 2d ed. New York: W.H. Freeman. (A textbook of immunology.)

Milstein, C. 1980. Monoclonal antibodies. *Scientific American* 243:66–74. (A review of the production and significance of monoclonal antibodies.)

Roitt, I.M. 1991. *Essential immunology.* 7th ed. Oxford, England: Blackwell Scientific. (An immunology textbook, with emphasis on the clinical aspects of immune response, immunity to infection, transplantation, and autoimmunity.)

Roitt, I., J. Brostoff, and D. Male. 1996. *Immunology,* 4th ed. St. Louis: Mosby. (An in-depth textbook of immunology, with very good illustrations.)

Rose, N.R., B.C. DeMacario, J.L. Fahey, H. Friedman, and G.M. Penn, ed. 1992. *Manual of clinical laboratory immunology.* 4th ed. Washington, D.C.: American Society for Microbiology. (A detailed laboratory guide to immunological procedures.)

Smith, K.A. 1990. Interleukin-2. *Scientific American* 262:50–57. (A review of interleukin-2.)

chapter seventeen

INFECTIOUS DISEASES

The Respiratory Tract
Respiratory Tract Microflora
Respiratory Tract Pathogens
Diseases Caused by Other Microorganisms

The Oral Cavity and Digestive System
Oral Cavity Microflora
Dental Diseases
Oral Diseases Caused by Other Microorganisms
Digestive System Microflora
Food Infection/Intoxication
Escherichia coli
Gastrointestinal Diseases Caused by Other
    Microorganisms

Diseases of the Skin and the Genitourinary System
Skin Microflora
Microbial Antagonism and the Physical Barrier
The Male and Female Genitourinary Systems
Acquired Immune Deficiency Syndrome (AIDS)

PERSPECTIVE
Plasmid Mediated β-Lactamase Production in *Neisseria
    gonorrhoeae*

EVOLUTION AND BIODIVERSITY

Scanning electron micrograph of the
protozoan *Trichomonas vaginalis* that
causes the sexually transmitted disease
trichomoniasis. Note the characteristic
flagella and undulating membranes
(×12,000).

Microbes in Motion — **PREVIEW LINK**

Among the key topics covered in this chapter are the bacterial, fungal,
viral, and protozoal diseases of the respiratory, gastrointestinal, skin, and
genitourinary systems. Zoonoses are also covered. The following sections
in the *Microbes in Motion* CD-ROM may be useful as a preview to your
reading or as a supplementary study aid:

Gram-Positive Organisms: Gram-Positive Cocci—Streptococci, 2–10,
(Virulence Factors, 17–23; Gram-Positive Cocci—Staphylococci (Diseases),
13–18; Gram-Positive Bacilli (*Corynebacterium*), 2–12. *Gram-Negative
Organisms:* Gram-Negative Bacilli—Aerobes (*Pseudomonas*), 2–8,
(*Legionella*), 10–19, (*Bordetella*), 20–28; Bacilli—Facultative Anaerobes
(*Enterobacteriaceae*), 3–23, (*Vibrionaceae*), 29–38, (*Pasteurellaceae*),
40–47, (*Sprillaceae*), 60–64; Gram-Negative Cocci (*Neisseria*), 4–11.
Anaerobic Bacteria: Gram-Positive Bacilli (Sporeforming), 9, 16–19.
Miscellaneous Bacteria: Mycobacteria (all topics); Zoonoses (*Borrelia*, 7–19,
(*Rickettsia*), 37–52; Treponemes (Taxonomy), 3–9. *Fungal Structure and
Function:* Diseases (Inhalation), 22–36, (Occupation), 16–21. *Viral Structure
and Function:* Viral Diseases (Enveloped DNA virus), 6–9, (Enveloped
ssRNA virus), 12–16; Viral Replication (RNA Virus Replication), 22–24.
Parasitic Structure and Function: Intestinal Protozoa (Epidemiology), 16–27.

On September 16, 1995, a 54-year-old woman living in Hidalgo County, Texas, was bitten by a mosquito. Several days later she developed high fever, intense headaches, sharp muscle and joint pain, and diarrhea. The woman, who recovered, was the first confirmed case of dengue fever contracted inside the United States since 1986. Dengue fever, a painful mosquito-borne disease often called breakbone fever because patients describe the joint pain as so severe that it feels like their bones are breaking, strikes more than 100 million people annually in 100 countries. Occasionally, in children, dengue virus can cause a severe and potentially fatal hemorrhagic fever characterized by bleeding, convulsions, and shock. Although it is epidemic throughout Latin America, dengue fever has not been a problem in the United States until recently. This tropical disease, along with yellow fever, was nearly eradicated at the turn of the century by attacks on the vector, the *Aedes aegypti* mosquito, which transmits both diseases from person to person. In 1985 portions of the United States became infested with the Asian tiger mosquito, *Aedes albopictus*, which can also transmit dengue fever. *A. albopictus* was introduced into the United States that year in a shipload of waterlogged tires from Asia and has since spread from Texas to as far north as Illinois. Because of diseases like dengue fever, the World Health Organization has created an Office of Emerging Diseases to track, control, and prevent infectious diseases that plague the human population.

Infectious diseases can be acquired by different routes, including (1) the respiratory tract, (2) the oral cavity and digestive system, and (3) the skin and genitourinary system. The preceding chapters have described how the human body is able to protect itself against microbial infection via innate and acquired responses. When these defenses are not effective in curtailing infectious disease, damage to the host can result. This chapter will discuss some representative examples of pathogenic microbes that are capable of causing disease.

The Respiratory Tract

The respiratory tract consists of upper and lower sections. The nose, throat, middle ear, and eustachian tubes connecting the middle ear to the pharynx are the principal structures in the upper respiratory tract (URT). Coarse hairs and cilia filter air moving through these passages and constantly remove foreign particles.

The lower respiratory tract (LRT) is composed of the larynx, trachea, bronchial tubes, and lungs. These structures serve the vital function of gas exchange, so microbial infections here present a serious threat to general health. Ciliated epithelium lining the larynx, trachea, bronchi, and bronchioles helps prevent microorganisms from reaching the lungs. Alveolar macrophages in the lungs cleanse them of dust particles and microorganisms that escape the ciliary action of the respiratory tract.

Many Different Types of Microorganisms Are Normally Found in the Upper Respiratory Tract

The lower respiratory tract is normally free of microbial contamination because of the constant removal of foreign particles by the cilia. In contrast, the upper respiratory tract is inhabited by many different types of microorganisms. These microbes, which include streptococci, staphylococci, neisseriae, diphtheroids, and yeasts, are established early in life. Some of these organisms are potentially pathogenic but do not cause problems because their numbers are effectively controlled by the presence of other microbes, host antibodies that limit them to the respiratory tract, or both. Occasionally—and particularly in immunologically compromised hosts—members of the resident flora overgrow and cause disease.

Respiratory Tract Pathogens Are Highly Contagious and Transmissible

Many bacterial diseases occur in the respiratory tract. The pathogens responsible for these diseases are highly contagious and easily transmitted in aerosols (Tables 17.1 and 17.2). Respiratory tract infections often are difficult to control in a population and may lead to epidemics.

Streptococcal Pharyngitis Is a Common Bacterial Upper Respiratory Tract Infection

Upper respiratory tract infections can be caused by viruses (for example, adenovirus, coxsackievirus, rhinovirus, and influenza virus), fungi (for example, *Candida albicans*), protozoa (for example, *Trichomonas tenax*), and bacteria (for example, *Haemophilus influenzae, Corynebacterium diphtheriae, Mycoplasma pneumoniae, Streptococcus pneumoniae,* and *Streptococcus pyogenes*). More than 90% of upper respiratory tract infections are caused by viruses. **Streptococcal pharyngitis,** or **strep throat,** is the most common form of bacterial upper respiratory tract infection. This acute inflammatory disease is caused by *Streptococcus pyogenes,* also known as group A *Streptococcus.* Streptococci, along with staphylococci and neisseriae, are called **pyogenic cocci** [Greek *pyon,* pus] because many of their diseases are associated with pus formation.

S. pyogenes is typically transmitted by infected droplets in the air. Symptoms of streptococcal pharyngitis usually appear within one to three days after infection and are characterized by fever, inflammation of the mucous membrane of the throat, and glandular swelling. Most cases of strep throat are self-limiting and do not require treatment with antimicrobial agents, although if the three symptoms described are seen, treatment is recommended. Occasionally streptococcal pharyngitis can lead to more serious diseases, most commonly scarlet fever, rheumatic fever, and acute glomerulonephritis, affecting other parts of the body.

Scarlet fever is caused by lysogenic strains of *S. pyogenes* that synthesize an exotoxin designated erythrogenic toxin [Greek *erythros,* red, *genic,* producing] because of the red cutaneous rash pro-

Microbial Pathogens of the Upper Respiratory Tract

Organism	Disease
Bacteria	
Corynebacterium diphtheriae	Diphtheria
Haemophilus influenzae	Epiglottitis, sinusitis (also meningitis)
Mycoplasma pneumoniae	Primary atypical pneumonia
Neisseria gonorrhoeae	Pharyngitis
Staphylococcus aureus	Sinusitis
Streptococcus agalactiae	Neonatal respiratory distress (also neonatal meningitis and septicemia)
Streptococcus pneumoniae	Sinusitis
Streptococcus pyogenes	Pharyngitis (also rheumatic fever, scarlet fever, and acute glomerulonephritis)
Streptococcus sp.	Pharyngitis, sinusitis
Fungi	
Candida albicans	Thrush (oral candidiasis)
Protozoa	
Trichomonas tenax	Trichomoniasis
Viruses	
Adenovirus	Pharyngitis
Coronavirus	Common cold
Coxsackievirus	Pharyngitis
Herpes simplex virus	Fever blisters, pharyngitis, gingivostomatitis
Influenza virus	Influenza
Rhinovirus	Common cold

duced throughout the body. There are at least three immunologically distinct types of toxin (A, B, and C) produced by different strains of streptococci. Erythrogenic toxin production is under the genetic control of a temperate bacteriophage carried by the bacterium; when the virus is removed, erythrogenic toxin is not produced. Phage conversion of nontoxigenic strains to lysogenic, toxigenic strains is not unusual; it occurs in *C. diphtheriae* and other bacterial pathogens. The erythrogenic toxin produced by *S. pyogenes* often enters the bloodstream, spreads throughout the body, and causes the generalized skin rash and strawberry-colored tongue characteristic of scarlet fever (Figure 17.1). In the past, scarlet fever was a severe and often fatal childhood disease. Now, because of the effectiveness of penicillin therapy, scarlet fever is rarely fatal in the United States.

Rheumatic fever and **acute glomerulonephritis** are nonsuppurative (non-pus-forming) sequelae of streptococcal pharyngitis. Rheumatic fever is characterized by lesions that appear in the heart, joints, and skin two to three weeks after the appearance of acute pharyngitis. Not all aspects of the pathogenesis of rheumatic fever are

known, but streptococcal-induced antibodies that cross-react with normal body tissues are important in the disease. People infected with *S. pyogenes* who develop rheumatic fever usually have high levels of antibodies against streptolysin O, a hemolysin produced by these bacteria. The measurement of **anti-streptolysin O (ASO) levels** in serum is widely used as a diagnostic aid for streptococcal infections. A rise in ASO titer or a persistently elevated ASO titer in serum over a period of several weeks is considered diagnostic for a recent *S. pyogenes* infection. Acute glomerulonephritis is most often seen in children and may be associated with symptoms of edema, hypertension, and blood in the urine (hematuria). Although in most cases the disease follows pharyngitis, it may also occur after a cutaneous streptococcal infection.

Some highly virulent strains of *S. pyogenes,* known as invasive group A streptococci, are capable of causing severe and life-threatening infections such as necrotizing fasciitis (rapid destruction of fibrous tissue that encloses and separates muscles). These bacteria, commonly called "flesh-eating bacteria," affect between 500 and 1,500 people annually in the United States. Symptoms of

table 17.2

Microbial Pathogens of the Lower Respiratory Tract

Organism	Disease
Bacteria	
Bordetella pertussis	Whooping cough (pertussis)
Chlamydia psittaci	Ornithosis, psittacosis
Haemophilus influenzae	Pneumonia (also meningitis)
Klebsiella pneumoniae	Pneumonia
Legionella pneumophila	Legionellosis
Mycobacterium tuberculosis	Tuberculosis
Mycobacterium sp. (atypical mycobacteria)	Chronic pulmonary disease resembling tuberculosis
Mycoplasma pneumoniae	Primary atypical pneumonia
Neisseria meningitidis	Pneumonia (also meningitis)
Streptococcus pneumoniae	Pneumonia
Fungi	
Ajellomyces capsulatus (*Histoplasma capsulatum*)	Histoplasmosis
Ajellomyces dermatitidis (*Blastomyces dermatitidis*)	Blastomycosis
Coccidioides immitis	Coccidioidomycosis
Pneumocystis carinii	Pneumonia
Viruses	
Adenovirus	Pneumonia
Influenza virus	Influenza, pneumonia
Respiratory syncytial virus	Pneumonia

necrotizing fasciitis include fever, severe pain and swelling, and rapid invasive infection of tissue, especially muscle tissue. Streptococcal toxic shock-like syndrome, similar in symptoms to staphylococcal toxic shock syndrome (see staphylococcal toxic shock syndrome, page 536) but caused by *S. pyogenes*, killed Muppets creator Jim Henson in 1990.

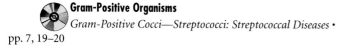

Gram-Positive Organisms

Gram-Positive Cocci—Streptococci: Streptococcal Diseases • pp. 7, 19–20

S. pyogenes is easily identified in the laboratory by its Gram stain morphology, hemolytic pattern on blood agar, and other properties. Streptococci are gram-positive, generally grow in chains, and do not produce catalase, an enzyme that catalyzes the reaction: $2 H_2O_2 \rightarrow 2 H_2O + O_2$ (Figure 17.2). These characteristics are useful in distinguishing streptococci from other upper respiratory tract bacteria such as neisseriae (gram-negative diplococci) and staphylococci (gram-positive, catalase-positive cocci growing in clusters). On blood agar *S. pyogenes* produces β-hemolytic colonies that are smaller in diameter (0.5 mm after 24 hours of growth) than staphylococcal colonies (1 to 3 mm after 24 hours of growth).

The **bacitracin disk test** permits presumptive identification of *S. pyogenes* (Figure 17.3). Most strains of group A streptococci are more susceptible than other streptococcal groups to the antimicrobial agent bacitracin. This susceptibility can be deter-

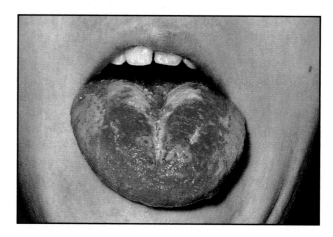

figure 17.1

The Strawberry-Colored Tongue Associated with Scarlet Fever

mined by streaking a pure culture of streptococci on a blood agar plate and placing a disk containing 0.04 unit of bacitracin on the inoculated plate. The agar plate is incubated for 24 hours at 37°C and examined. No growth of group A streptococci occurs around the bacitracin disk. The bacitracin disk test, however, is only a **presumptive test** for the identification of group A streptococci

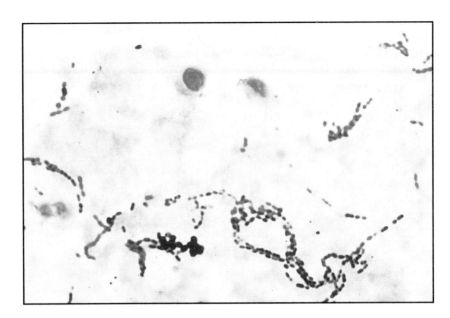

figure 17.2
Gram Stain of *Streptococcus pyogenes* from Sputum Showing a Chain of Gram-Positive Cocci (×1,000)

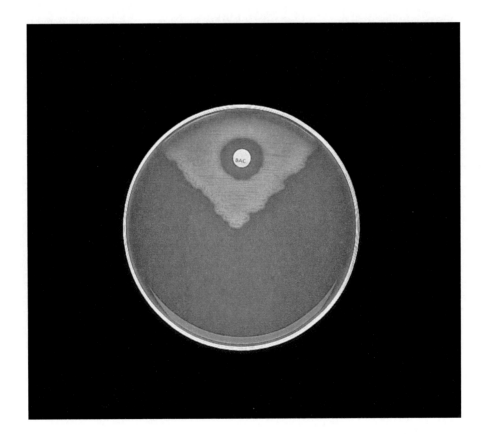

figure 17.3
Bacitracin Disk Test
Streptococcus pyogenes is streaked on a blood agar plate and a bacitracin disk is placed on the inoculated plate. After incubation, no growth of *S. pyogenes* is observed around the bacitracin disk, indicating a positive test result.

The genus *Streptococcus* consists of gram-positive cocci that grow in chains, particularly in liquid culture. A useful criterion for the differentiation of streptococci is their

17.4). This system of hemolysis, proposed by James H. Brown in 1919, provides a convenient means for the preliminary differentiation of streptococci.

incomplete lysis, sometimes accompanied by a green discoloration caused by the reduction of hemoglobin in the red blood cells to methemoglobin. Because of this green discoloration, the α-hemolytic streptococci are often referred to collectively as the **viridans streptococci** [Latin *viridis*, green]. A variation of α hemolysis is **α' hemolysis,** or **wide-zone hemolysis,** in which a small zone of α hemolysis surrounds the colony, with a zone of complete hemolysis extended further out into the medium. When examined only macroscopically, α' hemolysis can be confused with β hemolysis. **γ-hemolytic** bacteria do not synthesize hemolysins and thus have no zones of hemolysis around their colonies.

Classification of Streptococci

hemolytic activity on blood agar that contains 5% (volume/volume) defibrinated sheep blood. These bacteria can be divided into three types on the basis of their hemolytic patterns on blood agar (Figure

β-hemolytic streptococci synthesize hemolysins (streptolysins), which completely lyse erythrocytes to produce clear zones around colonies on blood agar. **α-hemolytic** organisms produce a zone of

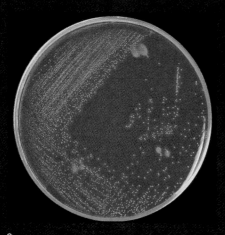

a.

b.

c.

figure 17.4
Hemolytic Patterns of Streptococci

a. β-hemolytic streptococci on a blood agar plate. β-hemolytic *Streptococcus* colonies are surrounded by wide, clear zones of hemolysis in which no red blood cells are visible. b. α-hemolytic streptococci on a blood agar plate. α-hemolytic *Streptococcus* colonies are surrounded by narrow, partial zones of hemolysis often accompanied by green discoloration. c. γ-hemolytic enterococci on a blood agar plate. No hemolysis is seen around colonies of *Enterococcus* (previously called *Streptococcus*).

hemolytic patterns of streptococci are easily differentiated and provide a rapid method for initial identification of streptococci. Most, although not all, pathogenic streptococci are β hemolytic and produce streptolysins that are toxic for red and white blood cells and other cell types.

Streptococci can be further divided into distinct immunologic groups (Table 7.3). In the early 1930s, Rebecca Lancefield separated the streptococci initially into five groups designated A through E on the basis of carbohydrate antigens in the bacterial cell wall. Later, this classification was expanded to include other groups. Strains belonging to groups A, B, C, D, F, and G are most commonly encountered in human streptococcal diseases. In addition to these major groups, group A *Streptococcus (S. pyogenes)* is divided into more than 55 immunologic types based on the M antigen, a protein antigen in the fimbriae on the cell surface. Group B *Streptococcus (S. agalactiae)* is divided into at least eight immunologic types based on polysaccharide capsular antigens. *Typing* of streptococci is helpful in epidemiologic tracing of disease outbreaks, but is not used as frequently as immunologic *grouping* to identify these bacteria. Streptococci can be easily grouped using commercial latex agglutination, coagglutination, or ELISA systems.

Although immunologic grouping and hemolytic patterns do not replace the conventional binomial system of nomenclature used in *Bergey's Manual of Systematic Bacteriology*, they do provide the microbiologist with a relatively easy and descriptive method to identify streptococci.

Gram-Positive Organisms
Gram-Positive Cocci—Streptococci: Taxonomy • pp. 2–10

table 17.3

Immunologic Groups of Streptococci Associated with Human Disease

Group	Species	Hemolysis	Disease
A	*S. pyogenes*	β	Pharyngitis, rheumatic fever, scarlet fever, acute glomerulonephritis
B	*S. agalactiae*	β	Neonatal respiratory distress, meningitis, and septicemia
C	*S. dysgalactiae* *S. equi* *S. equisimilis* *S. zooepidemicus*	β	Pharyngitis, tonsillitis, pneumonia, endocarditis, bacteremia, postoperative wound infection (primarily *S. equisimilis*)
D	*S. bovis*	α, β, or γ	Bacteremia
F	*S. anginosus*	α, β, or γ	Abscess, sinusitis, meningitis
G	No species designation	β	Pharyngitis, sinusitis, septicemia, pneumonia, meningitis

table 17.4

Causes of Bacterial Pneumonia

Bacterium	% of All Bacterial Pneumonia Cases	Susceptible Population
Streptococcus pneumoniae	60%–70%	Adults, especially the elderly and those with cardiac or pulmonary disease
Mycoplasma pneumoniae (primary atypical pneumonia)	20%	Older children and young adults
Legionella pneumophila	4%–10%	All ages
Haemophilus influenzae	1%	Children
Klebsiella pneumoniae	0.5%–4%	Children

because other streptococci may also be bacitracin sensitive. **Confirmatory tests** for the identification of group A streptococci rely on serological determinations and are more accurate. Coagglutination, latex agglutination, immunofluorescence, and counterimmunoelectrophoresis are examples of confirmatory serological tests presently used for the identification of *S. pyogenes* and other streptococci. These tests are more accurate than the presumptive bacitracin disk test, but generally are more expensive to perform.

Bacterial Pneumonia Frequently Follows Viral Infections of the Upper Respiratory Tract

Pneumonia is an inflammation of the lung that can be caused by a bacterium or virus but is most frequently caused by *Streptococcus pneumoniae*, or pneumococcus (Table 17.4). Bacterial pneumonia is often a secondary infection that follows viral infections of the upper respiratory tract. Mucous secretions laden with bacteria are carried to the lungs, where the bacteria establish an active infection. With disease progression, serous fluid accumulates in the lungs to produce edema. Complications of bacterial pneumonia include meningitis (inflammation of the membranes of the brain or spinal cord), pleurisy (inflammation of the pleura), abscess formation, and septicemia (infection of the blood).

 S. pneumoniae organisms are encapsulated, gram-positive, lancet-shaped diplococci that usually are seen in sputum samples from patients with pneumococcal pneumonia (Figure 17.5). Since many other gram-positive bacteria, including other species of streptococci, are commonly found in the upper respiratory tract, *S. pneumoniae* must be further identified by laboratory tests. On the surface of blood agar, virulent, encapsulated strains of *S. pneumoniae* produce round, glistening α-hemolytic colonies. Furthermore, pneumococci are optochin sensitive and bile soluble. **Optochin** is a chemical, ethylhydrocupreine hydrochloride, that inhibits growth of *S. pneumoniae*. The optochin

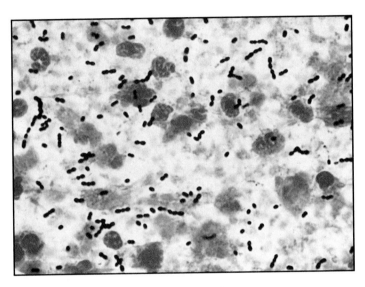

figure 17.5

Gram Stain of *S. pneumoniae* from Sputum Showing Encapsulated Gram-Positive Cocci in Pairs and Short Chains

disk test is performed in a manner similar to the bacitracin disk test. A paper disk containing 5 μg of optochin is placed on a blood agar plate streaked with a suspected culture of *S. pneumoniae*. The plate is incubated for 24 hours in 5% to 10% CO_2. *S. pneumoniae* is susceptible to optochin, resulting in growth inhibitory zones of 18 mm or more in diameter around the disk. *S. pneumoniae* also produces an autolytic enzyme, *N*-acetylmuramyl-L-alanine amidase, that solubilizes the peptidoglycan of the cell wall, resulting in cell lysis. The enzyme is activated by surface-active agents such as bile and bile salts. This property is the basis for the **bile solubility test,** in which suspected pneu-

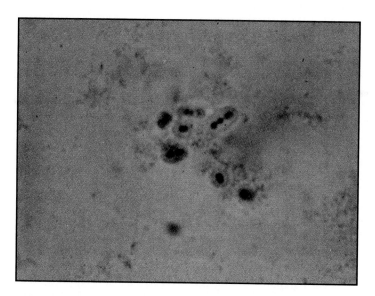

figure 17.6
Light Micrograph Showing the Quellung Reaction
Cells of *Streptococcus pneumoniae* are surrounded by apparently swollen capsules from the quellung reaction.

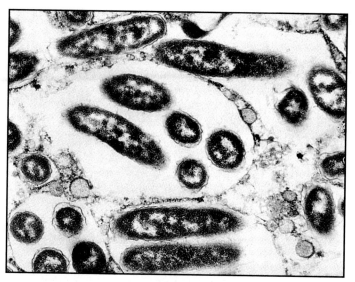

figure 17.7
Electron Micrograph of *Legionella pneumophila*
Note the rod-shaped morphology of the bacterium (×25,000).

mococcal cells are incubated in the presence of sodium deoxycholate, a bile salt, and the extent of cell lysis is determined visually.

S. pneumoniae is heavily encapsulated and can be microscopically identified by marked swelling of its capsule in the presence of antisera directed against capsular antigens (Figure 17.6). This capsular swelling is called the **quellung reaction** [German *quellen,* to swell] and was first described by Neufeld in 1902. The reaction is performed by mixing young cultures of encapsulated bacteria with antiserum in methylene blue (to enhance the visibility of the reaction) and observing the extent of capsular swelling in a wet mount under an oil-immersion lens.

In the United States, the incidence of pneumococcal pneumonia is estimated to be 1 per 1,000, with a case-fatality rate of 5% to 7%. The pneumococcus is the most common pathogen cultured from the respiratory tract in infants and causes serious infections, including pneumonia and bacteremia (the presence of viable bacteria in the blood). Pneumococci are present in the nose, throat, or both of about one-fourth of all healthy people and are more prevalent in the winter months. Outbreaks of pneumococcal infections often occur in families or in institutions where the carriage rate is high. A vaccine consisting of capsular polysaccharides from the 14 most common pneumococcal serotypes is commercially available and is effective in reducing the incidence of pneumococcal pneumonia and bacteremia in high-risk individuals such as the elderly and immunocompromised people.

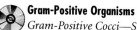 **Gram-Positive Organisms**
Gram-Positive Cocci—Streptococci: Virulence Factors • p. 17; *Diseases* • pp. 22–23

Legionellosis Was First Characterized During an Outbreak in Philadelphia

Legionellosis, or **Legionnaires' disease,** is a form of pneumonia caused by *Legionella pneumophila,* a strictly aerobic, gram-negative rod (Figure 17.7). *L. pneumophila* was first isolated from the lungs of victims by the Centers for Disease Control and Prevention in Atlanta in 1976, approximately six months after an outbreak in Philadelphia (see legionellosis, page 556). It is now classified as a member of the family *Legionellaceae.*

Legionellosis has an incubation period of two to ten days, followed by headache, rapidly rising fever (104°F to 105°F), chills, muscular aches, and a nonproductive (dry) cough. Chest pain, abdominal pain, diarrhea, or a combination of these symptoms may also be present. The disease is best described as a bronchopneumonia that can become confluent and resemble lobar pneumonia.

L. pneumophila is a fastidious bacterium, but it can be isolated and maintained on media containing L-cysteine and iron salts, such as charcoal yeast extract (CYE) agar and Mueller-Hinton agar base supplemented with L-cysteine and ferric pyrophosphate. Colonies become visible after four to five days of growth at 35°C under increased carbon dioxide tension. Laboratory diagnosis is usually by one of several methods: (1) direct immunofluorescence of sputum, respiratory secretions, pleural fluid, or lung tissue using a fluorescent antibody specific for *L. pneumophila;* (2) serodiagnosis by an indirect fluorescence test, with a fourfold rise in titer of paired serum samples (one from the acute phase and one from the convalescent phase) indicative of legionellosis; (3) gas-liquid chromatography, with identification by

Group B *Streptococcus*, or *Streptococcus agalactiae*, historically has been known as one of several types of bacteria that cause **mastitis** in cattle (another causative agent is *Staphylococcus aureus*). In recent years, however, physicians have seen an increased rate of serious human neonatal disease caused by this bacterium. The disease, **group B streptococcal disease**, exists as two distinct clinical forms: **early-onset disease** and **late-onset disease** (Table 17.5).

Early-onset disease is transmitted from an infected mother to her fetus, resulting in respiratory distress or septicemia in the newborn. Symptoms of early-onset disease generally appear within a few hours of birth. Late-onset disease is characterized by a purulent type of meningitis. Because late-onset disease symptoms appear several weeks after birth, it is believed that the disease may be caused by nosocomial (hospital-acquired) transmission of the bacteria, possibly among infants in a newborn nursery.

An estimated 7,500 infants develop symptoms of group B streptococcal disease annually in the United States; most have the early-onset form of the disease. Although about 20% of women have vaginal group B streptococci, few develop any symptoms of disease. Most colonized infants are also asymptomatic; only a small proportion of them (0.5% to 2.0%) develop group B streptococcal disease, yet it is a leading bacterial cause of death among newborns. The mortality rate is estimated as 5% to 20% among newborns. There is little information regarding the virulence of group B streptococci. Recent evidence, however, suggests that the bacteria are invasive and that infants who are symptomatically infected may be heavily colonized with the bacteria and lack adequate levels of protective antibody.

Group B Streptococcal Neonatal Disease

table 17.5

Early-Onset and Late-Onset Group B Streptococcal Disease

Characteristic	Early-Onset Disease	Late-Onset Disease
Onset	Usually within 24 to 48 hrs after birth	7 or more days after birth
Risk factors	Maternal complications during delivery such as prolonged rupture of membranes, fever, and premature onset of labor; preterm deliveries	None identified
Mode of acquisition	Vertical transmission from mother to fetus	Infections in the hospital or at home
Symptoms	Respiratory distress; septicemia	Meningitis

cellular fatty acid composition (*L. pneumophila* does not have the fatty acids characteristic of Lipid A of gram-negative endotoxin); (4) an ELISA to detect *L. pneumophila* antigens in urine and other specimens; or (5) a DNA probe to directly detect *L. pneumophila* in patient specimens.

Although most cases of legionellosis are mild and self-limited, the mortality rate may be as high as 15% to 25% in untreated cases. Mortality in *Legionella* infections is particularly high in immunocompromised or elderly people. Erythromycin, rifampin, and tetracycline are effective in therapy. *L. pneumophila* produces a β-lactamase that degrades penicillin, so this antibiotic is not appropriate for therapy.

There are approximately 700 diagnosed cases of legionellosis annually in the United States. Because the disease is difficult to distinguish clinically from pneumonia caused by other microorganisms and because mild cases may not be reported, the actual number of cases is believed to be higher. A milder form of Legionnaires' disease, called **Pontiac fever** because it was first discovered in Pontiac, Michigan, also exists. Pontiac fever is characterized by a short incubation period (6 to 48 hours) and all the symptoms of Legionnaires' disease, but there is no clinical evidence of pneumonia. Epidemiologic evidence strongly indicates that *L. pneumophila* and other species of the genus occur commonly in moist soil and bodies of water, and that transmission occurs through aerosols, partic-

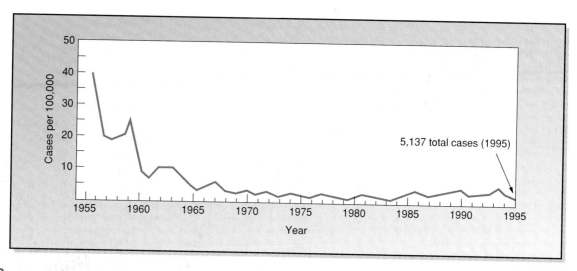

figure 17.8

Pertussis Cases in the United States, 1955 to 1995

The decline in the rate of pertussis is due primarily to compulsory vaccination, which began in the 1940s.

ularly of contaminated water from air-conditioning cooling towers, whirlpool baths, wastewater, shower heads, and nebulizers. There is no evidence of person-to-person transmission.

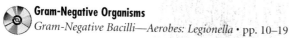

Gram-Negative Organisms
Gram-Negative Bacilli—Aerobes: Legionella • pp. 10–19

Pertussis Is Characterized by Spasmodic Coughing and a Distinctive Whoop

Pertussis, or **whooping cough,** is an acute respiratory disease caused by the gram-negative coccobacillus *Bordetella pertussis.* The bacterium is transmitted primarily by infected droplets in the air and is highly communicable. Whooping cough begins as a mild cough and develops into severe coughing accompanied by vomiting. The coughs are spasmodic, with gasps for air between coughs resulting in a whooping sound characteristic of the disease. A heat-labile exotoxin as well as an extracytoplasmic adenylate cyclase that increases cyclic AMP levels are produced by *B. pertussis* and are thought to contribute to the pathogenesis of whooping cough by causing tissue damage in the respiratory tract. *B. pertussis* is organotropic; its preferential attachment to bronchial epithelial cells is an important virulence determinant. Mortality from pertussis is high in infants (over 70% of deaths occur in children less than one year of age) and decreases dramatically with age.

The spasmodic coughing and whoop of pertussis are clinically diagnostic for the disease. The bacterium can be cultured on specialized media containing blood, potato extract, and glycerol (Bordet-Gengou agar). Specimens for cultures are taken by inserting a swab through the nostril into the nasopharynx while the patient coughs. The swab is removed and passed through a drop of penicillin solution on the Bordet-Gengou agar plate during the streaking for isolation. The penicillin inhibits contaminant bacteria found in the nasopharynx. *B. pertussis* appears as small, glis-

tening, pearl-white colonies after about three days of incubation. Final identification of the bacterium is by immunofluorescence or agglutination tests.

Mild cases of pertussis require no treatment. Serious cases of the disease, particularly in infants, are treated with erythromycin. Children are regularly immunized by the injection of heat-inactivated whole bacteria in a diphtheria-tetanus-pertussis (DTP) vaccine, which also contains diphtheria and tetanus toxoids. Immunization has contributed to the dramatic decline in pertussis cases and deaths in the United States (Figure 17.8). In recent years rare neurological complications resulting from the pertussis component of the vaccine have been observed, although general pertussis vaccination of children still is recommended.

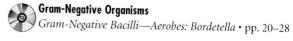

Gram-Negative Organisms
Gram-Negative Bacilli—Aerobes: Bordetella • pp. 20–28

A Pseudomembrane Is Formed in Diphtheria

Diphtheria is a disease that is characterized by a localized throat infection, resulting in necrosis, inflammation, and the formation of a spreading, grayish, pseudomembrane on mucous membranes of the throat and nasal passages. This pseudomembrane is composed of fibrin, necrotic epithelial cells, polymorphonuclear leukocytes, macrophages, and diphtheria bacteria. As the disease progresses, the membrane may become dislodged, resulting in suffocation.

The causative agent of diphtheria is *Corynebacterium diphtheriae,* a gram-positive, pleomorphic, rod-shaped bacterium. *C. diphtheriae* is an obligate aerobe that produces intracellular polyphosphate granules called **Babès-Ernst bodies.** These negatively charged phosphate granules stain preferentially with positively charged (basic) dyes such as methylene blue and toluidine blue and are a diagnostic aid in identification (Figure 17.9). An

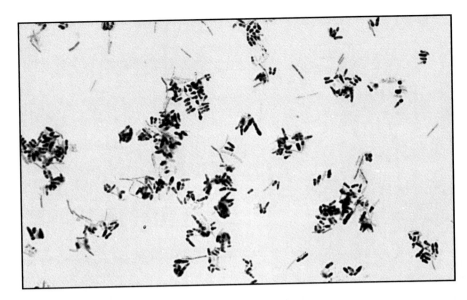

figure 17.9
Granule Stain of *Corynebacterium diphtheriae*
Dark blue granules are shown inside *Corynebacterium diphtheriae* cells.

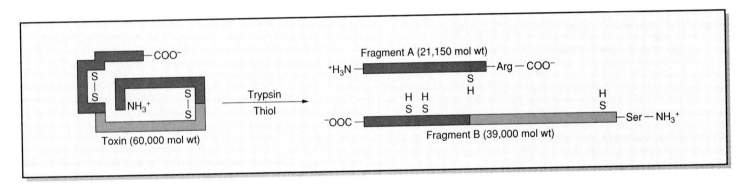

figure 17.10
Structure of Diphtheria Exotoxin
The intact toxin is cleaved to two fragments by trypsin or trypsinlike proteases in the presence of thiol (which reduces disulfide bonds). Fragment A contains the ADP-ribosylating activity, which is masked in the intact toxin. Fragment B interacts with specific receptors on sensitive eucaryotic plasma membranes and enables fragment A to enter the cell cytoplasm.

exotoxin produced by *C. diphtheriae* is the major determinant in the pathogenesis of diphtheria.

Only strains of *C. diphtheriae* lysogenic for a toxin-gene-bearing prophage produce exotoxin (see lysogenic conversion, pages 258 and 411). The toxin is a protein with a molecular weight of 60,000 daltons that is cleaved by trypsinlike proteases into two fragments, fragment A and fragment B (Figure 17.10). Fragment B has two important functions in toxin activity. It recognizes and binds to specific receptors on sensitive eucaryotic plasma membranes, and it facilitates the transport of fragment A across the plasma membrane into the cytoplasm. The membrane receptors have been identified, but it is not known precisely how fragment A crosses the membrane. There are theories that the toxin may be phagocytized and then released from a vesicle in the cell cytoplasm, or that fragment B may interact with membrane components to form a channel across the membrane for the entry of fragment A. Inside the cytoplasm, fragment A catalyzes a reaction transferring the ADP-ribosyl group from NAD to elongation factor 2 (EF2), which is involved in extension of the polypeptide chain during protein synthesis in eucaryotes:

$$NAD^+ + EF2 \rightarrow ADP\text{-ribosyl-}EF2 + nicotinamide + H^+$$

The modified EF2 is unable to participate in the translocation step of eucaryotic protein synthesis, resulting in inhibition of protein synthesis and cell death. The dead epithelial cells in the

Discovery of the Diphtheria Exotoxin

The discovery of the diphtheria toxin is a fascinating account in history. Diphtheria was described by Hippocrates in the fourth century B.C. However, it was not until 1826 that the French physician Pierre Bretonneau (1778–1862) identified diphtheria as a specific disease and distinguished it from other infections of the throat. Bretonneau observed the distinctive gray membrane in the throat of patients with the disease and named the disease diphtheritis [Greek *diphthera*, skin, membrane].

Edwin Klebs (1834–1913), a German-American pathologist, first described the diphtheria bacterium in stained smears from diphtheritic membranes in 1883. A year later the German bacteriologist Friedrich Loeffler (1852–1915) isolated the bacterium in pure culture. In his experimental investigations with guinea pigs, Loeffler noticed that although the bacterium multiplied only at local sites of inoculation, symptoms of disease were seen at other locations (heart, liver, kidney, and other organs). He surmised that these symptoms were caused by a toxin secreted by the bacterium and transported by the bloodstream.

The exotoxin of *C. diphtheriae* was discovered in culture filtrates in 1888 by the French bacteriologists Pierre Paul Émile Roux (1853–1933) and Alexandre Émile John Yersin (1863–1943). Roux and Yersin demonstrated that injection of the purified toxin into animals led to the systemic manifestations of diphtheria. This discovery led to the development of diphtheria antitoxin by Emil von Behring (1854–1917), a German physician, which revolutionized the treatment of diphtheria. For this pioneering work, von Behring was awarded the first Nobel Prize in Psychology or Medicine in 1901.

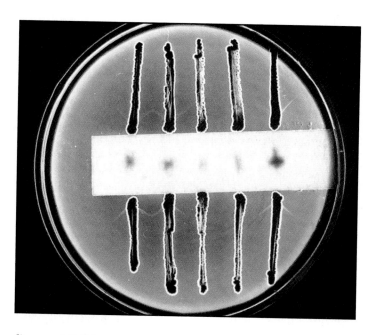

figure 17.11

Elek Gel Diffusion Test for Diphtheria Exotoxin

Five strains of *Corynebacterium diphtheriae* have been streaked perpendicular to an antitoxin-impregnated strip of filter paper. White lines of precipitate form where exotoxin meets with the antitoxin in the agar. Such lines are seen angling off streaks 1, 2, and 5 (from left to right), but not streaks 3 and 4. Strains corresponding to streaks 3 and 4 therefore are nontoxigenic.

throat form part of the pseudomembrane that is characteristic of diphtheria.

C. diphtheriae, unlike *S. pyogenes*, remains localized in the upper respiratory tract and seldom infects other parts of the body. The diphtheria exotoxin, however, may disseminate via the bloodstream to other parts of the body to cause degenerative lesions. The most serious complications of diphtheria are those affecting the heart and the peripheral nerves. Diphtheria toxin causes myocardial degeneration that can result in heart failure. Neurological complications can lead to paralysis of the soft palate, difficulty in swallowing, and nasal regurgitation of fluids.

Diphtheria is diagnosed in the laboratory by culture of the organism and identification of gram-positive pleomorphic bacteria containing Babès-Ernst bodies. Most corynebacteria, as well as other bacteria including certain staphylococci and mycobacteria, reduce the tellurite ion to metallic tellurium, producing gray-black colonies on potassium tellurite-containing agar. Because nontoxigenic *C. diphtheriae* and other corynebacteria may be isolated from the upper respiratory tract, definitive identification of lysogenized *C. diphtheriae* is made by assaying for the production of exotoxin. Exotoxin is assayed by inoculating guinea pigs or by the **Elek gel diffusion test** (Figure 17.11). This test is performed by placing a sterile strip of paper impregnated with diphtheria antitoxin on agar medium with peptone maltose and antitoxin-free calf serum. The suspected culture of *C. diphtheriae* is streaked at

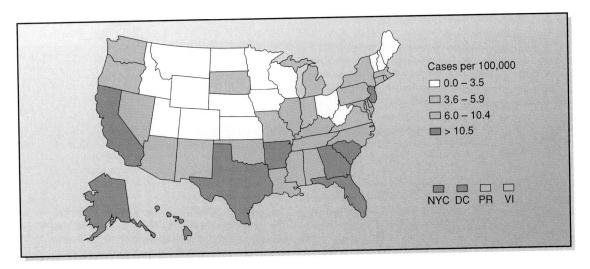

figure 17.12
Tuberculosis Case Rates, by State, in the Continental United States, 1994

right angles across the paper strip and the plate is incubated for 24 hours. Visible lines of antibody-antigen precipitate appear on plates inoculated with toxin-producing bacteria.

At one time diphtheria was a leading cause of death among young children in the United States. The disease is now effectively controlled by the DTP vaccine, which includes diphtheria toxoid. The annual incidence of diphtheria in the United States has significantly decreased from 50,462 cases in 1933 (before general use of the DTP vaccine) to two reported cases in 1994 and no reported cases in 1995.

Gram-Positive Organisms
Gram-Positive Bacilli: Corynebacterium • pp. 2–12

Tuberculosis Is Characterized by the Formation of a Tubercle in the Lung

Tuberculosis is a widespread disease that infects 1 billion people (20% of the world's human population) worldwide. Nearly 20,000 new active cases are diagnosed each year in the United States, where 2,000 deaths occur annually from tuberculosis (Figure 17.12). Recently there has been a resurgence of tuberculosis in the United States and worldwide due to the emergence of drug-resistant strains and other factors. Tuberculosis is also now the leading cause of death among people with AIDS.

Human tuberculosis is caused by *Mycobacterium tuberculosis* and several other species of *Mycobacterium*, including *M. bovis* (a problem in countries where raw milk is consumed) and *M. africanum* (isolated in certain parts of Africa). Tuberculosis is primarily an infection of the lungs (pulmonary tuberculosis), although other portions of the body may be infected through transport of bacteria by the blood or lymphatic system (disseminated or miliary tuberculosis).

M. tuberculosis is usually transmitted by the release of contaminated sputum or droplets from an infected person's sneezes or coughs. The bacterium enters its host through the respiratory tract and sets up foci of infection in the lungs. Alveolar macrophages enter these infected areas and phagocytize the mycobacteria, but many ingested mycobacteria survive and grow within macrophages. A lesion characteristic of tuberculosis, the **tubercle,** forms where the mycobacteria grow. The tubercle is a small nodular lesion surrounded by a fibrous layer of cells. Often the tubercle remains small and inactive, and there are no further signs of disease. In other cases the tubercle becomes calcified and may show up on chest X-ray films. This diagnostic tool is useful only if lung tubercles are large enough and show sufficient amounts of calcification to be visible. Dormant tubercles in adults may become active again in reactivation tuberculosis, which can lead to additional tissue damage.

M. tuberculosis is an obligate aerobic bacterium that typically forms slightly bent or curved, slender bacillus cells when grown in a living host. On enriched artificial media, the bacterium frequently grows as rods or long, slender, cordlike (serpentine) strands.

One of the most striking characteristics of mycobacteria is their high lipid content. Lipids, particularly long-chain saturated fatty acids called mycolic acids, account for up to 60% of the dry weight of the mycobacterial cell envelope. In comparison, other

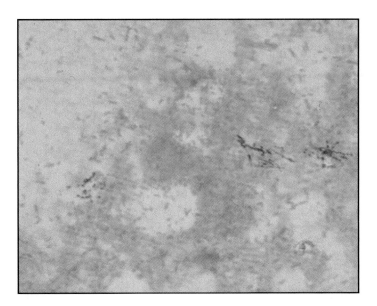

figure 17.13
Acid-Fast Stain of *Mycobacterium tuberculosis*
Acid-fast *Mycobacterium tuberculosis* cells appear as red with the acid-fast stain.

bacteria have from 1% to 20% dry weight lipid content in their cell envelopes. The high lipid content of the mycobacterial cell envelope makes the bacterium resistant to adverse environmental conditions. The lipids also contribute to the hydrophobic nature of the cell surface, which makes the cell strongly impermeable to conventional stains and nutrients and accounts for its slow growth rate (12-hour generation time in rich media).

M. tuberculosis can be identified in the laboratory by examining a sputum sample for **acid-fast bacteria** (Figure 17.13). *M. tuberculosis,* because of the high lipid content of its cell envelope, retains the dye carbolfuchsin even after a destaining step with acid alcohol (3% HCl, 95% EtOH) (see cell envelope of *Mycobacterium,* page 331). Mycolic acids in the envelope are unique to mycobacteria and bind to the carbolfuchsin during the acid-fast stain. Most other bacteria, with lower contents of lipid in their cell envelopes, destain rapidly. However, the acid-fast property is only a presumptive test for mycobacteria since other bacteria, including *Nocardia,* are also acid-fast.

Mycobacteria are further identified by inoculating the sputum sample onto a specialized medium, such as Löwenstein-Jensen medium, which contains egg yolk as a source of lipids to promote the growth of the mycobacteria. The culture is incubated for two to six weeks in 5% to 10% carbon dioxide. Mycobacteria are highly aerobic organisms that grow better when incubated in the presence of elevated carbon dioxide. Even with increased carbon dioxide, colonies of *M. tuberculosis* appear on laboratory media only after several weeks of incubation. This slow growth is related to the reduced diffusion of nutrients through the lipid-containing cell envelope. This unique envelope also endows the bacterium with unusual resistance to strong acids and alkalis, a property often used to separate mycobacteria selectively from other microorganisms in sputum samples. Sputum may be treated with 0.1 N NaOH before inoculation onto growth media to kill other alkali-sensitive bacteria. Colonies of mycobacteria that grow on selective media are further identified with biochemical tests.

The **tuberculin skin test** is a useful screening procedure to determine if a person has been exposed to *M. tuberculosis* (see tuberculin skin test, page 483). A small amount of purified mycobacterial antigen, tuberculin (also called purified protein derivative, or PPD), is injected into the superficial skin layers of the forearm. The injected site is inspected after 48 hours for an area of induration at least 10 millimeters in diameter. Persons previously exposed to *M. tuberculosis* become sensitized to tuberculin and develop a delayed-type hypersensitive reaction. The tuberculin skin test has certain limitations. It does not differentiate past from recent infections, and cross-reactions occur in people who have been exposed to other species of mycobacteria. A positive skin test is usually confirmed by other diagnostic tests, such as culture and chest X-ray examination.

The treatment of tuberculosis is possible with streptomycin, rifampin, isoniazid, ethambutol hydrochloride, and other chemotherapeutic agents. Therapy is usually extended for a period of many months to a few years because of the slow metabolism of the mycobacteria. In recent years *M. tuberculosis* has become more resistant to drugs that historically have been successful in treatment. This increased resistance has been of great concern to physicians, and alternative drugs are being considered for treatment of tuberculosis.

Protective immunization prior to infection is possible with a vaccine called **bacille Calmette-Guérin (BCG).** The vaccine contains live, attenuated *M. bovis* and induces active immunity to *M. tuberculosis* in most people. It is widely used in Europe and other parts of the world, but is not used in the United States; because tuberculosis is less common in the United States, there is some question of its value. Vaccination with BCG vaccine eliminates the usefulness of the tuberculin skin test, because all vaccinated individuals develop a positive tuberculin reaction.

 **Miscellaneous Bacteria**
Mycobacteria (all) • pp. 1–27

Bacterial Causes of Meningitis[a]

Bacterium	Frequency of Isolation by Age Group (%)			
	Premature and Neonate	2 to 60 Months	5 to 40 Years	Greater than 40 Years
Haemophilus influenzae	—	60	5	2
Neisseria meningitidis	—	20	45	10
Staphylococcus sp.	5	—	10	13
Streptococcus pneumoniae	5	10	20	50
Streptococcus sp. (primarily S. agalactiae)	23	2	5	5

[a]Does not include viruses, which cause 30% to 40% of all meningitis.

Inflammation of the Meninges Is Known as Meningitis

Meningitis, or inflammation of the meninges (the membranes that envelop the brain and spinal cord), can be caused by many types of microorganisms (Table 17.6). These microbes are usually transmitted through contaminated air droplets and enter the body via the respiratory tract. They then may enter the bloodstream and spread to the meninges to cause meningitis.

Meningococcal meningitis, or **cerebrospinal fever,** is a form of meningitis frequently associated with closed populations such as prisons, military camps, and schools. The causative agent is *Neisseria meningitidis,* a gram-negative diplococcus. *N. meningitidis* usually inhabits the nasopharynx without causing any symptoms because of host immunity. However, asymptomatic carriers of this bacterium may transmit it to individuals without adequate levels of immunity, resulting in disease. Meningococcal disease is characterized initially by nasopharyngeal infection, followed by multiplication of the bacteria in the blood (**meningococcemia**) and then an acute purulent meningitis. In severe cases, neurological complications may also occur. Most of the symptoms of meningococcal meningitis are due to an endotoxin that causes extensive vascular damage. The bacterium is encapsulated and is thus protected from phagocytosis.

Gram-Negative Organisms
Gram-Negative Cocci: Neisseria • pp. 10–11

Meningitis also may be caused by the gram-negative rod *Haemophilus influenzae.* This bacterium is an agent of meningitis in children under the age of five, and the majority of cases are caused by serotype b. *Haemophilus* also causes septic arthritis, otitis media (infection of the inner ear), epiglottitis, and pneumonia in children.

Haemophilus influenzae requires two growth factors: a heat-stable X factor (hemin) and a heat-labile V factor (NAD). This characteristic may be used to differentiate *H. influenzae* from other bacteria. Coagglutination and agglutination with specific antisera are also used to identify *H. influenzae* and *N. meningitidis.*

A vaccine, the Hib (*H. influenzae* serotype b) vaccine, is available to immunize children and has significantly reduced the incidence of disease. Prior to the introduction of the vaccine, *H. influenzae* caused approximately 20,000 infections annually in the United States.

Gram-Negative Organisms
Bacilli—Facultative Anaerobes: Pasteurellaceae • pp. 40–47

Respiratory Tract Diseases Can Be Caused by Other Microorganisms

Histoplasmosis is a mycotic disease caused by the dimorphic fungus *Ajellomyces capsulatus* (also called *Histoplasma capsulatum*). This organism is present in soil (particularly soil contaminated with bird droppings) and is transmitted by inhalation of conidia produced by the organism during its mycelial phase. Histoplasmosis often begins with a mild respiratory infection with few or no symptoms. Lung lesions resembling the lesions of tuberculosis develop and can become calcified and fibrous. Histoplasmosis is generally a self-limited lung disease, but in a few infected individuals it can progress into a disseminated form involving the liver, lymph nodes, and adrenal glands. Histoplasmosis is a widespread disease in the United States, particularly in the Ohio and Mississippi river valleys, with as many as 30 million people infected (Figure 17.14). Histoplasmosis is also called "spelunker's disease," as the fungus can be inhaled by cave explorers who come into aerosol contact with infected bat droppings.

Coccidioidomycosis, caused by the dimorphic fungus *Coccidioides immitis,* is a pulmonary disease that occurs commonly in the southwestern United States and northern Mexico. Airborne arthrospores of the fungus are inhaled and carried to the lungs, where lesions similar to those of histoplasmosis develop. Most infections of coccidioidomycosis are self-limiting, with such symptoms as fever, cough, and weight loss. Complications, involving dissemination of the disease to other parts of the body, occur rarely.

Fungal Structure and Function
Diseases: Environmental Diseases—Inhalation • pp. 22–30, 35–36

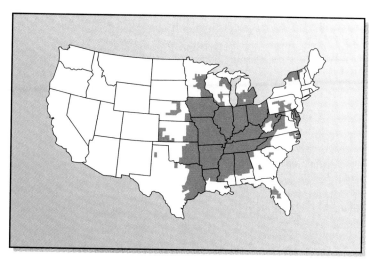

figure 17.14
Histoplasmosis Cases in the United States
The colored areas show the geographic distribution of histoplasmosis.

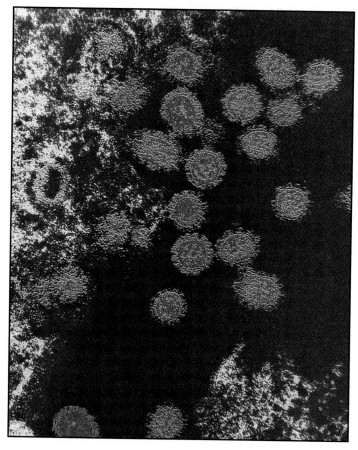

figure 17.15
Electron Micrograph of Influenza Viruses
Glycoprotein spikes containing hemagglutinin or neuraminidase activity are visible on the surface of the virions (colorized, ×36,300).

Influenza is an acute viral respiratory disease characterized by fever, chills, headache, generalized muscular aches, and a frequent cough. Influenza affects people of all ages, but is particularly severe for the very young, the very old, and people with other, complicating diseases. Outbreaks of influenza, which usually occur from early fall to late spring, may affect millions of people. For example, it is estimated that there were 200 million cases and 20 million deaths worldwide from the devastating influenza pandemic of 1918 to 1919.

Influenza is caused by an enveloped single-stranded RNA-containing virus that exists as three major antigenic types: A, B, and C. These types in turn are divided into several subtypes on the basis of antigenic differences in two types of glycoproteins, hemagglutinin and neuraminidase, which occur as spikes on the viral surface (Figure 17.15). Influenza viruses undergo cyclic antigenic shifts, primarily through mutations in the hemagglutinin and neuraminidase genes. These antigenic shifts must be considered in the preparation of vaccines for immunization against influenza.

One complication of influenza, and to some extent, chickenpox, is **Reye's syndrome,** a degenerative disease of the central nervous system. Reye's syndrome occurs primarily in children 14 years of age or younger and can lead to swelling of the brain and a fatty degeneration of the liver. The incidence of Reye's syndrome in the United States is one to two cases per 100,000 children per year. For reasons yet unknown, there is an increased incidence of Reye's syndrome among children with influenza or chickenpox who have been given salicylate-containing products such as aspirin. Consequently, physicians warn parents to be cautious in the use of aspirin for children who have influenza or chickenpox.

Viral Structure and Function
Disease: Enveloped ssRNA Virus • pp. 12–16

The respiratory tract is a major entry point for many microbial pathogens. Although the anatomy and defenses of the respiratory system are designed to prevent infections, some microorganisms are able to successfully evade these defenses. Respiratory tract diseases frequently are highly contagious and can spread rapidly within a population. Some can have serious consequences for the host, resulting in tissue damage, inflammation, and sometimes death. In some cases, such as in pertussis, the microbial pathogen has a preference for tissues of the respiratory system, and tissue damage is localized. In other instances, such as in diphtheria and streptococcal infections, the pathogen or its toxin can disseminate and cause serious disease in other parts of the body.

Immunization has significantly reduced the incidence of diseases caused by respiratory tract pathogens. Vaccines are now available for pneumococcal pneumonia, pertussis, diphtheria, influenza, and in some countries, tuberculosis. Because of immunization, diseases such as pertussis and diphtheria, which at one time affected large numbers of children in the United States, are not as common today.

The Oral Cavity and Digestive System

The human oral cavity and digestive system are in constant contact with microorganisms in ingested food and water. Most of the microorganisms in food and water are harmless to human beings. However, a few are harmful and can cause some of the most serious diseases known. Diseases of the digestive system are the second most frequent cause of illness in the United States, exceeded only by respiratory diseases.

Many Different Types of Microorganisms Are Normally Found in the Oral Cavity

The oral cavity consists of the mouth, teeth, and upper portion of the pharynx. Many different microbes compose the normal oral flora. Most are acquired through the ingestion of liquids, food, and other particles.

The oral cavity is colonized during and after birth by different microorganisms. Conceivably, a wide variety of microorganisms may find their way to the oral cavity, depending on the environment and diet of the infant. Most of these microbes are washed away by saliva, food, and liquids, and are not retained. Others, particularly aerobic organisms, adhere to the soft and smooth surfaces of the mouth by means of pili. It is not until teeth appear and anaerobic pockets are formed via the accumulation of plaque on tooth surfaces that other, anaerobic microorganisms begin to form part of the oral flora.

Streptococcus sanguis and *Streptococcus mutans* are two of the first bacteria to colonize the early teeth. These and other species of streptococci ferment carbohydrates to acids, which contribute significantly to dental caries. Later in life other types of bacteria, including *Actinomyces, Bacteroides, Fusobacterium, Lactobacillus,* and *Prevotella* join the streptococci as part of the resident oral flora. Anaerobic bacteria such as *Actinomyces, Bacteroides, Fusobacterium,* and *Prevotella* grow in the deeper layers of surface plaque and in abscesses, protected from the harmful effects of oxygen. The normal oral flora is as diverse as the diet of the host and is subject to constant change over a lifetime.

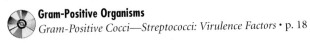

Gram-Positive Organisms
Gram-Positive Cocci—Streptococci: Virulence Factors • p. 18

Dental Diseases Are Caused by an Accumulation of Microorganisms on the Surface of Teeth

Most diseases of the oral cavity are caused not by exogenous pathogens, but by endogenous members of the resident flora (Table 17.7). Cuts or nicks in the gums and underlying blood vessels can carry pathogens in the oral cavity to other parts of the body, causing disseminated or systemic infections.

table 17.7

Diseases Caused by Flora of the Oral Cavity

Organism	Disease
Actinomyces israelii	Gingivitis, periodontitis
Actinomyces naeslundii	Gingivitis, periodontitis
Actinomyces viscosus	Gingivitis, periodontitis
Prevotella melaninogenica	Gingivitis, periodontitis
Candida albicans	Thrush (oral candidiasis)
Fusobacterium nucleatum	Gingivitis
Lactobacillus sp.	Dental caries
Streptococcus mutans	Dental caries
Streptococcus sanguis	Dental caries
Veillonella sp.	Dental abscesses, bite wounds

Dental plaque is a sticky film composed of microorganisms embedded in a polysaccharide/glycoprotein matrix on the surface of teeth. Plaque forms when teeth and gingiva (gums) are not properly and frequently cleaned. Approximately 60% to 70% of the volume of plaque is made up of bacterial cells, mostly gram-positive cocci and filamentous forms that may reach a thickness of 300 to 500 cells. Dental plaque begins with the colonization of the tooth's enamel surface by gram-positive *S. sanguis* and *S. mutans,* as well as by gram-positive rods such as *Actinomyces.* Plaque accumulations can extend into the gingival crevice, where they may provoke inflammation and bleeding. The flora of dental plaque changes. As plaque matures and the redox potential is lowered, the flora shifts from the facultatively anaerobic streptococci and *Actinomyces* toward primarily gram-negative, obligately anaerobic rods and spirochetes. With time, the relatively soft plaque material begins to harden into **calculus,** deposits containing calcium and phosphate. Calculus is difficult to remove by brushing and usually requires mechanical removal, or debridement.

The term **biofilm** is used to describe encased microorganisms attached to a surface by adhesive polysaccharides produced by the cells. Dental plaque is an example of a biofilm that is found in the human body. Biofilms also occur in natural environments on pipes, rocks, boats, and other surfaces to which microorganisms can adhere. In some instances, biofilms can impair the flow of water, oil, or other liquids through encrusted pipes or accelerate deterioration of metal or wooden surfaces through action of microbial metabolic by-products.

Dental plaque that is not removed from tooth surfaces can eventually lead to **dental caries,** the result of destruction of the enamel, dentin, and/or cementum of teeth by acid-forming bacteria. Experiments with germ-free animals show that *S. mutans* is the principal bacterium responsible for caries. As this bacterium adheres to teeth, it produces acids from carbohydrates, and these acids attack and erode tooth enamel (Figure 17.16). Table sugar (sucrose, a disaccharide) is a contributing factor to dental caries.

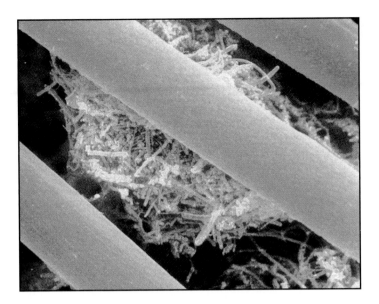

figure 17.16

Dental Floss Showing Particulate Matter Residing Between a Person's Teeth

Bacteria such as *Streptococcus mutans* are part of the particulate matter seen in this scanning electron micrograph. Routine flossing as part of proper dental hygiene loosens and removes the particulate matter between teeth (colorized, ×2,000).

S. mutans splits sucrose, by means of the enzyme glycosyl transferase, into the monosaccharides glucose and fructose and further metabolizes these sugars to lactic acid, which contributes to enamel demineralization. Sucrose also is converted into extracellular dextran, a predominantly $\alpha(1, 6)$ polymer of glucose, by dextransucrase, an enzyme produced by *S. mutans*:

$$\text{n-sucrose} \xrightarrow{\text{dextransucrase}} \text{dextran} + \text{n-fructose}$$

Dextrans are strongly adhesive polysaccharides and promote the attachment of *S. mutans* to tooth surfaces. A reduction of sucrose in the diet reduces acid and dextran production by *S. mutans* and results in a decrease in dental caries.

The term **periodontal disease** is used for diseases that affect the supporting structures of teeth: gingiva, cementum, and supporting bone. Periodontal disease is common among the elderly but can affect people of all ages. Pyorrhea, characterized by abscesses of the gingiva and underlying bone, is the major cause of tooth loss after the age of 30. **Gingivitis** is the most common of the periodontal diseases. This inflammation of the gingiva results from the accumulation of dental plaque in the gingival crevice. **Acute necrotizing ulcerative gingivitis (ANUG),** or **trench mouth,** is an infection that develops abruptly and involves necrosis of the interdental epithelium of the gums. Ulceration and inflammation of the gums are other signs associated with this disease. Spirochetes are thought to be responsible for ANUG, although they have not yet been cultivated from persons with the disease. Stress, anxiety, and poor diet are conditions that appear to predispose people to contracting the disease.

Dental caries and periodontal disease can be controlled by proper dental hygiene. Fluoridation is frequently used to reduce the incidence of caries. Fluoride replaces the hydroxyl groups in the tooth, forming the compound fluoroapatite in place of hydroxyapatite. Fluoroapatite is less soluble at low pHs and thus reduces the extent of tooth erosion by acid. Fluoride also is somewhat antibacterial.

Vaccines to immunize people against *S. mutans* are under study. A goal of such vaccines would be to increase levels of secretory IgA specific for *S. mutans* in the saliva and reduce caries.

Oral Diseases Can Be Caused by Other Microorganisms

Candida albicans and other species of *Candida* are frequently present on the mucous membranes of the mouth, vagina, and intestinal tract but do not normally cause problems. These organisms become invasive when the host has a debilitating disease (for example, diabetes), is immunologically suppressed, or is treated with broad-spectrum antibiotics. Under these conditions, candidiasis can develop. Oral candidiasis, commonly known as **thrush,** is characterized by the inflammation of the oral mucosa and the appearance of a grayish-white exudate. Thrush occurs most frequently in the very young or the very old. In infants, thrush is usually contracted from a *Candida*-contaminated birth canal or contaminated hospital equipment. Elderly people with severe, wasting diseases are also highly susceptible to the disease. Thrush can develop into esophagitis, a common condition in AIDS and other immunosuppressed patients.

Mumps is an acute contagious viral disease of the salivary glands. The disease is caused by an enveloped RNA-containing paramyxovirus that is spread through direct contact with saliva and respiratory secretions. Following entry via the respiratory tract, the mumps virus multiplies in the respiratory tract epithelium and cervical (neck) lymph nodes and establishes an infection of the salivary glands via the bloodstream. Mumps manifests itself primarily with pain and swelling of the salivary glands accompanied by fever and pain when chewing or swallowing. The mumps virus may infect other glandular tissues; complications include orchitis (inflammation of the testes), meningitis (inflammation of the membranes of the brain or spinal cord), encephalitis (inflammation of the brain), or pancreatitis (inflammation of the pancreas). Mumps most often affects children between the ages of five and 15, but the disease is most severe in adults.

The Digestive System Is Complex and Contains a Wide Variety of Microorganisms

The digestive system of vertebrates is complex and consists of the esophagus, stomach, small intestine, and large intestine. Several other structures, including the mouth, pharynx, liver, gallbladder, and pancreas, are associated with digestion and assist in the breakdown of food.

The large intestine of the human digestive system functions like a large, natural chemostat. Food particles entering it fuel the

table 17.8

Bacterial Food Infections

Bacterium	Disease	Incubation Period	General Features of the Disease
Campylobacter jejuni	Gastroenteritis	2 to 10 days	Severe, cramping abdominal pain; bloody diarrhea; chills; fever
Escherichia coli (enterotoxigenic strains)	Traveler's diarrhea	1 to 2 days	Abdominal pain, diarrhea, chills, fever, headache, vomiting
Escherichia coli (enterohemorrhagic strains)	Hemolytic-uremic syndrome; diarrhea	1 to 6 days	Lysis of red blood cells and kidney failure; hemorrhagic colitis without invading cells of the colon; bloody diarrhea
Salmonella typhi	Typhoid fever	1 to 2 weeks	Abdominal distension and tenderness, fever, headache, malaise, appearance of a rash (rose-colored spots) on the abdomen
Shigella boydii, Shigella dysenteriae, Shigella flexneri, Shigella sonnei	Shigellosis (bacillary dysentery)	1 to 14 days	Abdominal cramps, fever, diarrhea, stools containing blood and mucus, dehydration
Vibrio cholerae	Cholera	2 to 5 days	Abdominal cramps, diarrhea, nausea, vomiting, copious fluid and electrolyte loss with stools containing sloughed intestinal epithelial cells (rice water stools), dehydration

growth and metabolism of the resident flora. Very few bacteria reside in the stomach, although acid-tolerant lactobacilli and streptococci are occasionally found. *Helicobacter pylori* also occurs in the stomach and is the leading cause of peptic ulcers (see *Helicobacter pylori*, page 522). The pH becomes less acid in the lower small intestine, and the numbers and diversity of the normal flora increase. The large intestine may have as many as 10^{12} microorganisms per gram of fecal matter. Here anaerobes predominate, outnumbering other bacteria by a ratio of 1,000 to 1 or more. Facultative anaerobes such as *Escherichia, Enterobacter, Citrobacter, Proteus, Lactobacillus,* and *Streptococcus* consume available oxygen and reduce the atmosphere to one more suitable for anaerobic growth. The major anaerobes inhabiting the large intestine are *Bacteroides, Fusobacterium, Peptococcus,* and *Clostridium*.

The intestinal flora is established early in life from bacteria present in food particles. Microorganisms are protected from stomach acids by entrapment in large food particles or by the number of organisms ingested. For example, cholera occurs only when large numbers of *Vibrio cholerae* (10^8 organisms) are ingested. The infecting dose is lowered to 10^4 organisms when consumed in a mild alkali solution to neutralize gastric acidity.

The intestinal flora has a significant influence on the health and physiological condition of the human host. Vitamins K and B$_{12}$ are provided almost exclusively by the normal bacterial flora. Intestinal microbes also retard growth of potential pathogens via various mechanisms of bacterial antagonism.

Bacterial Diseases of the Digestive System Arise from Food Infection or Food Intoxication

Microorganisms that enter the gastrointestinal tract via contaminated food and water can cause food-borne illness that is commonly referred to as **food poisoning**. There are two categories of food poisoning: **food infection** and **food intoxication (food poisoning)**. Food infection occurs when the infectious agent establishes an active infection in the small intestine; disease occurs only after the pathogen has multiplied. *Salmonella typhi, Campylobacter jejuni, Shigella dysenteriae, Vibrio cholerae,* and *Escherichia coli* are some bacteria that cause food infections (Table 17.8).

Food intoxication is a different kind of disease; symptoms are caused by the consumption of food or water contaminated with toxic products of organisms. Food intoxication is caused by such bacteria as *Clostridium botulinum, Clostridium perfringens,* and *Staphylococcus aureus* (Table 17.9). The symptoms and organisms of food infection and food intoxication may be different and may require different approaches for control and treatment.

table 17.9

Bacterial Food Intoxications

Bacterium	Disease	Incubation Period (hr)	Toxin	General Features of the Disease
Clostridium botulinum	Botulism	18 to 36	Neurotoxin	Weakness, dizziness, severe dryness of the mouth and pharynx, blurred vision, peripheral muscle weakness
Clostridium perfringens	Food poisoning	5 to 15	Enterotoxin	Abdominal pain, diarrhea
Staphylococcus aureus	Food poisoning	1 to 6	Enterotoxins A, B, C_1, C_2, D, and E	Abdominal cramps, nausea, vomiting, diarrhea

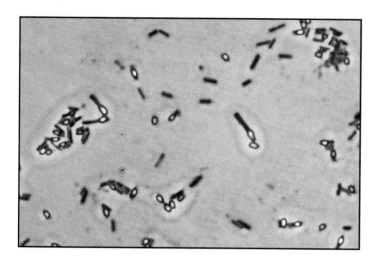

figure 17.17

Gram Stain of *Clostridium botulinum*
Note the prominent oval, subterminal endospores.

Food-Borne Botulism Is Caused by a Potent Neurotoxin Produced by Clostridium botulinum

The most severe form of food intoxication is **botulism** [Latin *botulus*, sausage] caused by the anaerobic, gram-positive sporeformer *Clostridium botulinum* (Figure 17.17). Botulism was first noted in the eighteenth and nineteenth centuries as a disease associated with the ingestion of blood sausage and other meats. Blood sausage was prepared by filling animal casings, or intestines, with blood and ground meats, boiling the sausage for a short period of time, and then smoking it over a wood fire. Spores of *C. botulinum* are heat resistant and able to survive all but pressure-sterilization temperatures. The preparation procedures for blood sausage thus were ideal for the survival and germination of botulinum spores.

Botulism as a food-borne disease is uncommon now in the United States; fewer than 50 cases are seen each year. Botulism usually is caused by the consumption of fish, which may contain *C. botulinum* in the intestine, and improperly preserved home-canned fruits and vegetables. Eight immunologically different types of toxin—types A, B, C_1, C_2, D, E, F, and G—are produced by different strains of *C. botulinum*. In the United States, types A, B, and E are responsible for most human cases of botulism. Clinical symptoms of botulism include weakness, dizziness, severe dryness of the mouth and pharynx, blurring of vision, and peripheral muscle weakness, and begin 18 to 36 hours after ingestion of contaminated food. Intoxication is caused by a potent neurotoxin produced by the bacterium. The neurotoxin is heat labile and can be inactivated by boiling suspect foods prior to ingestion. However, the toxin is stable to gastric acid and proteases and, if present in the active form in ingested food, can pass intact through the stomach.

Infant botulism, first described in 1976, is another form of botulism. The disease strikes infants, usually within the first few months of life, and is probably acquired through the ingestion of foods (commonly honey), soil, or other materials contaminated with endospores of *C. botulinum*. The spores reach the intestinal tract, germinate, multiply, and produce toxin. Symptoms associated with infant botulism are constipation, generalized weakness, weakened neck muscles, flaccid paralysis, respiratory problems, and impaired sucking ability. Only infants are affected, probably because their competing intestinal flora is less well-established.

Botulinum toxin is one of the most potent naturally produced toxins known. Less than 1 µg of toxin is enough to kill a human being. The toxin binds rapidly and irreversibly to neuromuscular junctions and blocks the release of the neurotransmitter acetylcholine. The result is paralysis. Immediate diagnosis and treatment are necessary to prevent death. The earlier the diagnosis and initiation of treatment, the better the prognosis. Diagnosis is generally by clinical symptoms. Botulinum toxin may be detected by its effect on laboratory mice. They are inoculated with suspect food or patient serum containing the toxin and with one of the three antisera (anti-A, anti-B, or anti-E). If toxin is present and not neutralized, the mice develop paralysis and die within a few days.

Because of the severity of botulism, patients are treated with antitoxin immediately upon clinical diagnosis and before laboratory identification of the bacterium or toxin. Respiratory assistance may be required for victims with neurological complications. Antibiotics are not used in therapy, because the disease is an intoxication, not an infection.

Anaerobic Bacteria
Gram-Positive Bacilli: Sporeforming • pp. 9, 16–19

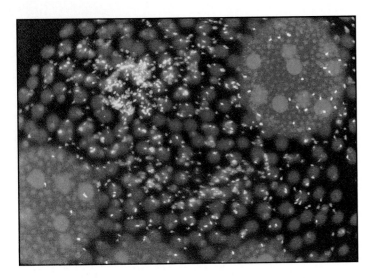

figure 17.18

Fluorescent Antibody Stain of *Vibrio cholerae* (Green Cells) Attached to *Volvox* Species Isolated from River Water in Matlabmatlab, Bangladesh

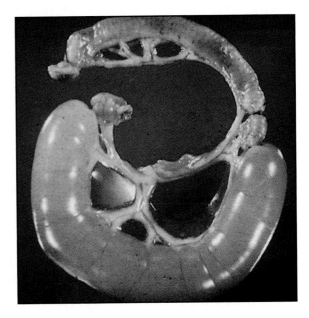

figure 17.19

Fluid Accumulation Caused by Choleragen

Rabbit ileal loop inoculated with choleragen, along with a control. The swelling is due to fluid accumulation caused by the cholera enterotoxin.

Staphylococcal Food Poisoning Is Typically Caused by Consumption of Contaminated Food That Has Been Improperly Refrigerated

Food intoxication is also caused by enterotoxin-producing strains of *Staphylococcus aureus*. This gram-positive bacterium is usually seen as clusters of cocci. *S. aureus* is distinguished from the less pathogenic *S. epidermidis* by its production of coagulase, an enzyme capable of clotting plasma. Six immunologically distinct protein enterotoxins (A, B, C_1, C_2, D, and E) produced by *S. aureus* are implicated in outbreaks of staphylococcal food intoxication.

Typically staphylococcal food intoxication follows the consumption of improperly refrigerated foods such as custards, pastries, salad dressings, sliced meats, and meat products contaminated by food handlers who are carriers of the bacteria. The bacteria grow in the food and produce toxin. Nausea, abdominal cramps, vomiting, and diarrhea usually appear within one to six hours after ingestion of the contaminated food. Most cases of staphylococcal food intoxication, although painful, are not life threatening. The disease resolves without treatment within a few days. Staphylococcal enterotoxins are heat stable at 100°C for 30 minutes. Food must be properly refrigerated both before and after cooking to minimize the chances of food intoxication or kept above 65.6°C if it is to be served hot.

Cholera Is Caused by Consumption of Water Contaminated by Vibrio cholerae

Cholera is a disease that occurs most often in developing countries, where fecal matter frequently contaminates water supplies.

Although few cases occur in the United States, several deaths from cholera and choleralike disease along the Gulf coast occur each year that originate from contaminated shellfish.

Cholera is caused by the gram-negative curved rod *Vibrio cholerae* (Figure 17.18). The species is divided into several serogroups on the basis of the O carbohydrate antigen in the outer membrane lipopolysaccharide (see lipopolysaccharide structure, page 60). Classic cholera is caused by cholera vibrios belonging to O antigen group 1 (*V. cholerae* O1). Symptoms include nausea, diarrhea, vomiting, abdominal cramps and, in severe cases, copious fluid loss from the intestine. These symptoms are the result of the action of cholera enterotoxin (choleragen), which activates adenylate cyclase in the plasma membranes of intestinal cells. Adenylate cyclase converts ATP to cyclic AMP, which is normally produced in small amounts via stimulation by hormones. However, in the presence of choleragen, cyclic AMP is produced in abnormally large amounts and causes uncontrolled loss of fluids and electrolytes into the intestinal lumen (Figure 17.19). Cholera patients may lose as much as 10 to 15 liters of fluid each day (an average of 2.4 liters of water is normally lost from the human body daily), which causes severe dehydration. The liquid feces contain large numbers of vibrios—as many as 10^6 per milliliter or more. The stools are called "rice-water" stools because epithelial cells slough from the intestinal mucosa and resemble rice particles.

V. cholerae is easily recognized and identified in the stools of cholera patients. The bacterium is comma shaped, has a darting motility, and can be stained with fluorescein-labeled specific antisera. Biochemical tests and agglutination assays are used for confirmatory identification.

The treatment of cholera requires the immediate replacement of lost fluids and electrolytes. Intravenous or oral administration of solutions containing glucose, sodium chloride, potassium chloride, and sodium bicarbonate is used for therapy. Antimicrobial agents may be given to eliminate vibrios more quickly from the intestinal tract. Antimicrobial chemotherapy, however, is not a substitute for fluid replacement therapy, which is the most critical and life-saving.

Gram-Negative Organisms
Bacilli—Facultative Anaerobes: Vibrionaceae • pp. 29–38

Typhoid Fever Is Transmitted via Contaminated Water and Food

Typhoid fever is a systemic infection caused by the gram-negative rod *Salmonella typhi*. The disease begins insidiously, usually with fever, abdominal pain, and headache. In untreated patients, the bacteria multiply in the gastrointestinal tract and then may enter the intestinal lymphatics, from which they disseminate via the blood to other areas of the body. Complications include intestinal hemorrhaging or perforation of the intestinal wall. Rose-colored spots, a result of hemorrhaging, may appear on the abdomen. Approximately 37% of those who recover from the disease become asymptomatic carriers and shed large numbers of organisms in their feces because the bacterium takes up residence in the gallbladder (see Typhoid Mary, page 442).

S. typhi is a particularly hardy bacterium, and its ability to survive in the environment for many months explains its transmission via contaminated water and food. In developing countries without adequate sewage treatment, typhoid fever remains a major health problem affecting thousands of people. Salmonellae produce several virulence factors, including a capsule that protects the bacterium after phagocytosis and a pyrogen-inducing endotoxin, which presumably is responsible for the fever in typhoid fever.

Typhoid fever is diagnosed by isolating *S. typhi* from blood, urine, or feces. Salmonellae found in feces only may be indicative of a carrier rather than an active disease state, so blood cultures are particularly important. To isolate the organism from feces, it is selectively grown in media that inhibit other gram-negative enteric bacteria and is identified by biochemical and agglutination tests.

An example of a selective medium is Salmonella-Shigella (SS) agar, which contains sodium citrate, bile salts, and brilliant green dye to inhibit gram-positive bacteria and most other (primarily lactose-fermenting) gram-negative enteric bacteria. On SS agar, *Salmonella* (and also *Proteus*) produces hydrogen sulfide, which reacts with the iron (ferric citrate) in the medium to produce a dark-centered colony. The periphery of the colony is clear, because *Salmonella* does not ferment lactose. Lactose-fermenting bacteria such as *Escherichia coli* and *Klebsiella* form an acid that reacts with neutral red (a pH indicator) to yield pink or red colonies. *Shigella* does not produce either hydrogen sulfide or acid and forms a white colony.

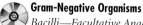

Gram-Negative Organisms
Bacilli—Facultative Anaerobes: Enterobacteriaceae • pp. 13, 15–17, 19–20

Salmonellosis Is Typically Caused by Consumption of Contaminated Meat Products, Poultry, and Eggs

A number of other *Salmonella* (primarily *Salmonella enteritidis* and its numerous serotypes) cause more common and milder forms of gastrointestinal disease, commonly known as **salmonellosis**, or ***Salmonella* gastroenteritis.** Approximately 40,000 cases of salmonellosis are reported in the United States each year. However, as most cases are not reported, the actual number of infections may be as high as 3 to 4 million annually. One of the largest reported outbreaks of salmonellosis occurred in early August, 1995, in West Palm Beach, Florida. Laboratory tests confirmed 189 cases of salmonellosis in diners who ate at a Mexican restaurant from August 3 to 8. Another 2,400 people who ate at the restaurant during this period developed symptoms of gastroenteritis. The infection was traced to improper food handling, including failure to store food at proper refrigeration temperatures and absence of a salad bar sneeze guard to prevent contamination of food. The restaurant was closed for nine days, the problems were corrected, and the employees received food-safety training.

The symptoms of salmonellosis are milder than those of typhoid fever and include moderate fever, nausea, abdominal cramps, and diarrhea. An enterotoxin recently isolated from *S. enteritidis* is believed to be responsible for the diarrhea. Salmonellosis is generally self-limiting and runs its course in one to four days. Because the disease is brief and limited to the gastrointestinal tract, antibiotic therapy is usually not necessary.

Contaminated meat products, poultry, and eggs are normally the culprits in salmonellosis. Although salmonellae are destroyed by cooking food at 68°C, the food can become contaminated after cooking. Contaminated food left at room temperature is particularly susceptible to multiplication of the salmonellae. For this reason, it is recommended that prepared foods be properly refrigerated to prevent increases in bacterial numbers.

Recent evidence from the Centers for Disease Control and Prevention (CDC) has indicated that raw, intact eggs harbor *S. enteritidis*. It is estimated that 0.01% (one in 10,000) of all eggs contain *S. enteritidis*. In a 1994 report, the CDC noted that salmonellosis cases had increased substantially since 1985 and that much of this increase could be attributed to consumption of raw or undercooked eggs. During 1985 to 1993, a total of 504 *S. enteritidis* outbreaks affecting 18,195 people and causing 62 deaths were reported to the CDC. Of 233 outbreaks that could be traced to a specific food source, 193 (83%) were associated with eggs. The CDC now recommends that eggs be refrigerated to prevent proliferation of *Salmonella* and be cooked thoroughly to kill *Salmonella*.

Shigellosis Is Caused by Food or Water Contaminated by Shigella

The gram-negative, nonmotile bacterium *Shigella* is responsible for the disease **shigellosis (bacillary dysentery)**, a severe diarrhea. Poor sanitary conditions are a major contributing factor in shigellosis. Bacteria are usually ingested in fecally contaminated food or water.

Poultry and *Salmonella* Infections

Salmonella is a leading cause of gastroenteritis. In the United States, the most common source of *Salmonella* infection is turkey, followed by beef and chicken.

People in the United States eat nearly 5 billion chickens a year, and the U.S. Department of Agriculture (USDA) estimates that up to 25% of all chickens leaving processing

during processing, but because of the high volume (each inspector looks at 75 chickens per minute), they are limited to looking for bruises or other injuries that may indicate disease. Consequently, many chickens contaminated with *Salmonella* pass inspection and reach the consumer.

Most outbreaks occur because of improperly cooked or processed meat. Salmonellae grow in turkey that is not thoroughly cooked and then is allowed to sit at room temperature. Beef, particularly ground beef, that is extensively handled and subject to contamination during grinding and processing can be a problem when eaten rare. Most food safety experts recommend that ground beef be cooked to 77°C (a temperature high enough to kill *Salmonella*) or until it is well done.

One area of public concern regarding *Salmonella* is contamination of chicken.

plants are infected with *Salmonella*. Chickens become exposed to the bacteria in many ways. Salmonellae, which are commonly found in feces, water, soil, and feed, accumulate on the feet or feathers of the birds and can contaminate the chickens when they are killed, plucked, and prepared for sale. Contamination of the meat can also occur during processing. High-speed machines that can process up to 80 chickens a minute sometimes press the birds so hard that feces spurt from the cloacae and intestines and contaminate the carcasses. USDA inspectors examine each bird visually

Fortunately, although the rates of contamination in poultry are higher than in other foods, the incidence of salmonellosis from eating poultry is not any higher than the incidence from eating other products, because it takes approximately 1,000,000 food-borne salmonellae to cause human infection. Nonetheless, the USDA is committed to reducing *Salmonella* contamination of poultry and is taking steps to culture samples of random birds during processing. If culture samples yield conclusive results of *Salmonella* contamination, the processor would be required to take measures to eliminate contamination.

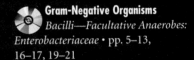 **Gram-Negative Organisms**
Bacilli—Facultative Anaerobes: Enterobacteriaceae • pp. 5–13, 16–17, 19–21

In the United States, *Shigella sonnei* and *Shigella flexneri* are the most frequent causes of shigellosis (73% and 25% of all cases, respectively), followed by *Shigella boydii* and *Shigella dysenteriae*. Most disease outbreaks in this country occur in closed groups such as day-care centers, hospitals, mental institutions, and cruise ships. Infected children and food handlers are responsible for disease transmission in these settings. In developing countries, where sanitary conditions are poor, *S. dysenteriae* and *S. boydii* cause most cases of shigellosis.

Shigellosis is characterized by fever, abdominal cramps, diarrhea, and the appearance of blood and mucus in the feces. There can also be fluid and electrolyte loss, leading to dehydration. Unlike the salmonellae, which penetrate the intestinal mucosa and enter the lamina propria, the shigellae do not penetrate beyond the submucosa. Epithelial cell invasiveness is a key factor in the development of shigellosis, but a potent exotoxin (Shiga toxin) produced by the bacterium may also be involved.

Although *S. dysenteriae* is excreted in large numbers in the feces of diseased individuals, the bacterium remains viable for only a short time there. Thus immediate or rapid culture from stool specimens is important to ensure identification, which is by bio-

chemical and agglutination tests. Shigellae do not ferment lactose and do not produce hydrogen sulfide. Thus on SS agar, shigellae produce colorless colonies, and can be differentiated from dark-centered *Salmonella* colonies and red or pink colonies of lactose-fermenting bacteria.

Many cases of shigellosis are mild and require no treatment. In severe cases, treatment involves fluid and electrolyte replacement and antibiotic chemotherapy.

 Gram-Negative Organisms
Bacilli—Facultative Anaerobes: Enterobacteriaceae • pp. 5–6, 14–17, 22–23

Campylobacter *Is an Increasingly Recognized Cause of Intestinal Infections*

Campylobacter is a gram-negative bacterium that causes abortion and infertility in cattle, sheep, and other animals. In young cultures, the bacteria are shaped like commas, the letter S, or flying sea gulls. As the cultures age, filamentous and spherical forms may appear (Figure 17.20).

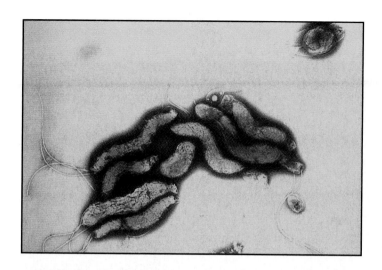

figure 17.20

Scanning Electron Micrograph of *Campylobacter jejuni*
Note the spiral morphology of the cell (colorized, ×10,260).

t a b l e 17.10

Comparisons of Enterotoxigenic *E. coli* (ETEC), Enteropathogenic *E. coli* (EPEC), Enteroinvasive *E. coli* (EIEC), and Enterohemorrhagic *E. coli* (EHEC)

Characteristic	ETEC	EPEC	EIEC	EHEC
Disease	Traveler's diarrhea	Infantile diarrhea	Dysenterylike diarrhea	Hemolytic-uremic syndrome; diarrhea
Age group affected	All ages	Infants	All ages	All ages, particularly infants under one year of age
Toxin	Heat-labile enterotoxin (LT), heat-stable enterotoxin (ST)	Possibly LT and ST	None isolated	Enterotoxin similar to Shiga toxin
Pathogenic mechanism	Enterotoxins	Invasion, possibly enterotoxins	Invasion	Damage to intestinal cells, leading to their necrosis and sloughing
Source of infection	Contaminated food and water	Hospital acquired	Contaminated food	Improperly cooked ground beef
Disease symptoms	Abdominal pain, diarrhea, chills, fever, headache, vomiting	Diarrhea, infrequent vomiting and fever	Abdominal cramps, bloody diarrhea, fever	Lysis of red blood cells and kidney failure (hemolytic-uremic syndrome) Hemorrhagic colitis without invading cells of the colon, bloody diarrhea

Two *Campylobacter* species cause most cases of human disease: *C. fetus* and *C. jejuni.* Although both species may be transmitted to humans from animal reservoirs, *C. jejuni* has a broader animal reservoir, which includes cattle, sheep, poultry, and dogs.

Human *C. jejuni* infections are characterized by intestinal mucosal invasion accompanied by hemorrhagic inflammation and crypt abscess formation. The major clinical symptoms are fever; severe, cramping abdominal pain; and bloody stools. Because these symptoms are also seen in other enteric diseases, laboratory diagnosis of *C. jejuni* is important to differentiate it from *Salmonella, Shigella,* and other enteric organisms.

C. jejuni is usually isolated by stool culture and plated onto a selective isolation medium containing antibiotics to inhibit growth of normal stool flora. *C. jejuni* is incubated at 42°C to differentiate it from *C. fetus,* which does not grow at this temperature. Incubation is under reduced-oxygen conditions (5% to 8% oxygen), because the bacteria are microaerophilic. Colonies are mucoid, spreading, and grayish.

Infected animals are important reservoirs of *Campylobacter;* contaminated milk, food, and water are major sources of infection. Direct contact with infected animals (cattle, sheep, dogs, and poultry) can also lead to disease transmission. Although the importance of human *C. jejuni* gastroenteritis has only recently been recognized, epidemiological surveys indicate that *Campylobacter*-induced infections may be as common as gastrointestinal diseases caused by *Salmonella.*

 **Gram-Negative Organisms**
Bacilli—Facultative Anaerobes: Spirillaceae • pp. 52–59

Escherichia coli Is Responsible for Many Different Gastrointestinal Diseases

Four groups of *Escherichia coli* strains are associated with human intestinal diseases (Table 17.10). **Enterotoxigenic *E. coli* (ETEC)** produces a heat-labile and a heat-stable enterotoxin and is

An Infectious Cause of Ulcers

An estimated 20 million people in the United States either have or are expected to develop peptic ulcers. Peptic ulcers, which previously were thought to be caused primarily by stress and anxiety, are now known to be primarily caused by *Helicobacter pylori*, a gram-negative, micro-aerophilic spiral bacterium (Figure 17.21). In 1982 two Australian physicians, Barry J. Marshall and J. Robin Warren, repeatedly attempted to culture a spiral-shaped bacterium that they had seen in gastric biopsy tissue from 100 patients with peptic ulcers, but were unsuccessful. Then the hospital microbiology laboratory was swamped by an outbreak of antibiotic-resistant *Staphylococcus aureus* and their agar plates with the mysterious bacterium were inadvertently left in the incubator for several days. When they later examined the plates, Marshall and Warren discovered that *H. pylori* had grown up. Following years of skepticism about these findings, an extensive scientific study was published in 1993 in the *New England Journal of Medicine* that showed that 48 of 52 peptic ulcer patients treated with antibiotics were cured of their ulcers. In comparison, 52 peptic ulcer patients receiving a placebo, or non-antibiotic-containing control pill, were not cured.

These findings were instrumental in reconsideration of the treatment of ulcers. *H. pylori* is thought to be the cause of 70% to 90% of peptic ulcers (ulcers can also be caused by autoimmune disease and nonsteroidal anti-inflammatory drugs such as aspirin) and may be associated with stomach cancer (a National Institutes of Health panel has found a relationship between *H. pylori* and stomach cancer, and the World Health Organization considers *H. pylori* to be a dangerous carcinogen). Physicians now include *H. pylori* among the possible causes of peptic ulcers and include antibiotics among the treatment regimens. With antibiotics, the need for ulcer surgery has been reduced, along with medical costs for expensive treatments.

Gram-Negative Organisms
Bacilli—Facultative Anaerobes:
Spirillaceae • pp. 60–64

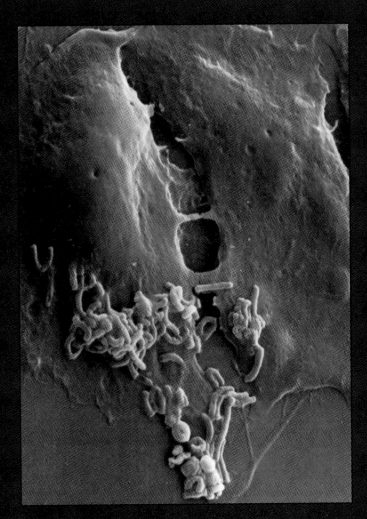

figure 17.21
Scanning Electron Micrograph of *Helicobacter pylori* Adhering to Gastric Cells (×3,441)

During the 1940s and 1950s, certain serotypes of *E. coli* were discovered to cause outbreaks of diarrhea in hospital newborn nurseries. These bacteria, now known as **enteropathogenic *E. coli* (EPEC),** are frequently transmitted nosocomially by direct contact with infected hospital personnel or via fomites. EPEC illness is a serious problem among newborn infants, especially in underdeveloped countries where sanitary conditions are poor and hospital-acquired infections are common. Invasiveness and adherence of EPEC to the small intestinal wall are important determinants of virulence. There is also some evidence that EPEC may have gained plasmids from ETEC strains and thus are able to produce heat-labile and heat-stable enterotoxins.

The third group of *E. coli* associated with diarrhea is **enteroinvasive *E. coli* (EIEC).** EIEC are invasive like the shigellae and produce dysenterylike disease, but they do not cause disease as often as the EPEC and ETEC.

Enterohemorrhagic *E. coli* (EHEC), discovered in 1982, almost all belong to the serological type O157:H7 (lipopolysaccharide O antigen 157, flagellar H antigen 7). These bacteria produce a toxin closely related to Shiga toxin. Some people infected with EHEC develop hemolytic-uremic syndrome, which is characterized by lysis of erythrocytes and kidney failure. Infants and the elderly are especially susceptible to EHEC infections. A 1993 outbreak of EHEC in Washington affected more than 400 people and was traced to improperly cooked hamburger served by a fast-food restaurant chain. A 1996 outbreak in Japan affected more than 8,000 people, including schoolchildren, and appeared to be linked to contaminated school lunches. The best preventive measure for EHEC infections is to thoroughly cook meat. DNA diagnostic probes or immunologic tests for specific antigens or toxins are often used to identify EHEC and other *E. coli* strains associated with intestinal diseases.

Gram-Negative Organisms
Bacilli—Facultative Anaerobes: Enterobacteriaceae • pp. 3–4, 10–11, 15–18

Gastrointestinal Diseases Can Be Caused by Other Microorganisms

Giardiasis is a diarrheal disease caused by the flagellated protozoan *Giardia lamblia* (Figure 17.22). *G. lamblia* has a distinctive appearance and is found in two forms. The vegetative form (trophozoite) causes infection; the resting form (cyst) is used for survival and is passed from host to host. Most cases of giardiasis are asymptomatic. In symptomatic cases the infected person usually experiences chronic recurring nausea, mild abdominal cramps, and diarrhea. Giardiasis is particularly prevalent in countries with poor sanitary measures and can be a significant hazard to travelers in foreign countries. In recent years the incidence of giardiasis in the United States has increased. An important reservoir of *Giardia* in the United States is wild animals such as raccoons, bears, beavers, and muskrats, which shed the *Giardia* cysts in feces, thus contaminating rivers and other bodies of

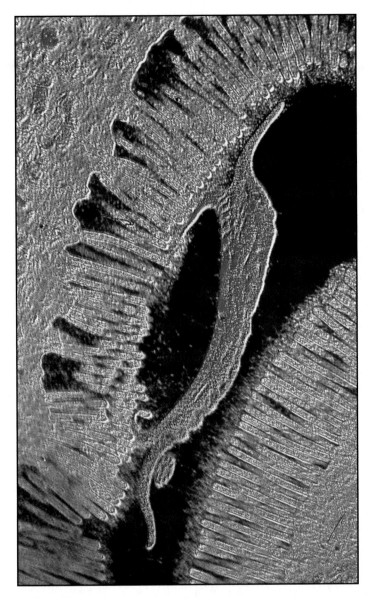

figure 17.22

Electron Micrograph of Intestine Infected with *Giardia lamblia*
The protozoan (orange) is shown adhering to the intestinal wall by two sucking disks. Fingerlike villi (pink) lining the inner surface of the intestine are visible (colorized).

responsible for traveler's diarrhea and infant diarrhea in developing countries. Outbreaks of traveler's diarrhea usually involve contaminated water supplies or food in developing nations. The heat-labile enterotoxin (LT) resembles cholera toxin in its mechanism of action. It activates adenylate cyclase to catalyze the conversion of ATP to cyclic AMP. The heat-stable enterotoxin (ST) stimulates intestinal guanylate cyclase, an enzyme that converts GTP to cyclic GMP. Increased intestinal cyclic GMP and cyclic AMP levels cause a hypersecretion of fluids and electrolytes. The genetic information for these enterotoxins resides on plasmids rather than on the *E. coli* chromosome.

table 17.11

Comparison of Hepatitis A; B; and C

Characteristic	Hepatitis A	Hepatitis B	Hepatitis C
Disease	Infectious hepatitis	Serum hepatitis	Posttransfusion hepatitis
Incubation period	15 to 50 days	50 to 160 days	15 to 140 days
Route of infection	Fecal-oral	Injection, blood transfusion, fecal-oral	Blood transfusion, fecal-oral
Onset	Sudden	Insidious (gradual)	Insidious (gradual)
Number of reported cases in the United States, 1995	31,582	10,805	4,576

water. People in the outdoors (for example, backpackers) using unpurified water become infected. Person-to-person transmission of *Giardia* occurs in day-care centers, mental institutions, and other custodial institutions. One can examine stools for trophozites and cysts or use an ELISA to detect *Giardia* antigen from infected hosts.

Entamoeba histolytica is a protozoan responsible for the food-borne infection **amebic dysentery,** or **amebiasis.** This widespread disease is associated with poor sanitary conditions, particularly in developing tropical countries. Contaminated fresh vegetables and water are common vehicles of transmission. Infections are often asymptomatic, but people with symptomatic disease have diarrhea, abdominal cramps, and feces containing blood and mucus and may progress to hepatic, pneumonic, and neural amebiasis.

Parasitic Structure and Function
Intestinal Protozoa: Epidemiology • pp. 16–27

Viral hepatitis is a disease that can be caused by at least five different hepatitis viruses, types A; B; C; D; and E (Table 17.11). Hepatitis A virus causes **infectious hepatitis,** which is usually acquired orally. It is an RNA enterovirus. The virus infects the intestine and spreads by the blood to other organs, including the liver, kidney, and spleen. Symptoms of infectious hepatitis include loss of appetite, nausea, diarrhea, fever, general discomfort, and jaundice. Outbreaks of infectious hepatitis occur through fecal contamination of food and water.

Hepatitis B virus, which is responsible for the disease **serum hepatitis,** is a DNA virus. It is acquired by sexual contact with an infected person, through injection by a contaminated syringe needle, or by perinatal transmission through the blood of an infected mother to her fetus. Serum hepatitis is a more serious disease than infectious hepatitis; it has a mortality rate as high as 10% in comparison to a mortality rate of less than 1% in infectious hepatitis. Chronic hepatitis B infections can lead to cirrhosis and cytopathic changes in liver cells, which may be associated with liver cancer. Serum hepatitis is an occupational disease of health professionals because of their constant contact with blood and blood products. The disease also occurs frequently among intravenous drug users.

A subunit vaccine, consisting of hepatitis B envelope glycoprotein antigens (HbsAg) produced by genetically engineered yeast cells, is available for immunization (see subunit vaccines, page 478). The vaccine, available commercially under the tradenames Recombivax HB and Engerix-B, contains no blood fragments and is inherently safer than the previous hepatitis B vaccine that was made from viral antigens present in the blood of chronic virus carriers.

Hepatitis C (previously called non-A, non-B hepatitis), is a recently discovered RNA virus that is a major cause of posttransfusion hepatitis. Current evidence indicates that this infection is common and can cause chronic liver disease.

The oral cavity and digestive system are common routes of human infection by microorganisms. Throughout life, these regions of the human body are in contact with many different types of microorganisms. The normal flora of the oral cavity and digestive system, established soon after birth, is acquired through the environment and diet of the infant. Many of these microbes are of benefit to the host, providing protection through antagonism and natural immunity. Occasionally, however, microbial pathogens are ingested.

Although our bodies develop protective immunity against most of these microbes, some pathogens occasionally are able to evade this immune response and cause disease. In the oral cavity, these diseases include dental caries and periodontal disease, which develop from the adherence of microorganisms to the teeth. In the digestive system, diseases are caused by microorganisms ingested through contaminated food or water. Most forms of gastrointestinal infections are mild and self-limiting, although some, such as cholera and typhoid fever, are more serious and even life threatening. Unlike many other types of infections, oral and gastrointestinal infections can usually be prevented by careful hygiene and diet.

Diseases of the Skin and the Genitourinary System

The skin is a flexible, membranous covering of the human body that normally acts as an effective barrier to microbial infection. If, however, the skin is damaged through cuts, wounds, or burns, or is

Microbial Pathogens of the Skin

Organism	Disease
Bacteria	
Mycobacterium leprae	Leprosy (enters through skin)
Propionibacterium acnes	Acne (possible contributing factor)
Pseudomonas aeruginosa	Opportunistic infections, especially in burn victims
Rickettsia akari	Rickettsialpox
Rickettsia prowazekii	Epidemic (louse-borne) typhus
Rickettsia rickettsii	Rocky Mountain spotted fever
Rickettsia tsutsugamushi	Scrub typhus
Rickettsia typhi	Endemic (murine) typhus
Staphylococcus aureus	Furunculosis, impetigo, scalded skin syndrome
Streptococcus pyogenes	Erysipelas, impetigo
Fungi	
Candida albicans	Candidiasis
Epidermophyton sp.	Ringworm (tinea)
Microsporum sp.	Ringworm (tinea)
Pityrosporum orbiculare	Pityriasis versicolor
Trichophyton sp.	Ringworm (tinea)
Viruses	
Herpes simplex virus	Herpetic cold sores (herpes simplex)
Varicella-zoster virus	Chickenpox (varicella), shingles (zoster)
Papovavirus (papillomavirus)	Warts (papilloma)
Paramyxovirus (measles virus)	Measles (rubeola)
Togavirus (rubella virus)	German measles (rubella)

penetrated by insect or animal bites, a portal of entry is established for infection by microorganisms.

The skin, the largest organ in our body, consists of two main cell layers: the thin outer epidermis and a thicker underlying dermis. Most pathogens cannot penetrate unbroken skin, although hookworm larvae (*Ancylostoma duodenale* and *Necator americanus*) and the bacterial agent of tularemia (*Francisella tularensis*) are purported to do so. Other organisms such as staphylococci and some fungi can reside deep within hair follicles, causing them to serve as reservoirs of infection.

Few Species of Microorganisms Normally Inhabit the Skin

The skin is an inhospitable environment for most microbes because of its acidity and low moisture content. The few species of microorganisms that are found regularly on the skin are adapted

for survival in this relatively harsh habitat and are present in strikingly large numbers. Their numbers can range from a few thousand per square centimeter on the back and arms to as many as several million per square centimeter in such humid areas as the scalp, groin, and axillae (armpits). Vigorously scrubbing the skin with detergents or disinfectants, as in surgical procedures, temporarily reduces the microbial population. However, the resident flora is rapidly reestablished by microorganisms hidden in the hair follicles, sweat glands, and other areas. The normal skin flora consists primarily of diphtheroids, staphylococci, and various types of yeasts. Most of these microorganisms are found in those areas of the skin that have some moisture—the hair follicles, scalp, axillae, genital and urinary areas, and face.

Diphtheroids (coryneforms)—gram-positive, pleomorphic bacteria that resemble *Corynebacterium diphtheriae* but are less virulent—are the principal skin inhabitants. The diphtheroid *Propionibacterium acnes* (previously called *Corynebacterium acnes*) is found in largest numbers on the skin, residing in the relatively anaerobic environment of the hair follicles. It is reported to be a contributing factor to acne, because it most commonly exists in those areas of the skin prior to this condition. Other diphtheroids, such as *Corynebacterium xerosis*, are aerobic and colonize moist areas of the skin, where they actively metabolize lipids. The odiferous fatty acid by-products of this lipid metabolism are responsible for human body odor. Deodorants are formulated with ingredients that selectively inhibit gram-positive bacteria and reduce body odor.

The second largest group of bacteria on the skin are the staphylococci. *Staphylococcus epidermidis* is a salt-tolerant bacterium commonly found on the surface of the skin. Sweat, which serves an excretory and cooling function, contains salts. As sweat evaporates, the concentrated salts present an inhospitable environment for most bacteria, but not staphylococci. *S. epidermidis* is a frequent contaminant of nose, throat, urogenital, and rectal cultures. Although *S. epidermidis* generally is nonpathogenic, it can cause minor abscesses and serious systemic diseases such as postoperative endocarditis.

A number of different types of yeasts are found on the skin. One example is *Pityrosporum ovale*, a small budding yeast commonly found on the scalp and oily skin areas. It is not pathogenic and is considered part of the normal flora of skin.

Microbial Skin Pathogens Must Overcome Microbial Antagonism and the Physical Barrier of the Skin to Cause Infection

The normal flora of the skin is instrumental in limiting colonization by other, more pathogenic microorganisms. Metabolic end products, such as acidic compounds and fatty acids synthesized by resident microbes, suppress the growth of other organisms. Despite this microbial antagonism and the physical barrier of the skin to infection, many types of microorganisms can cause skin diseases (Table 17.12).

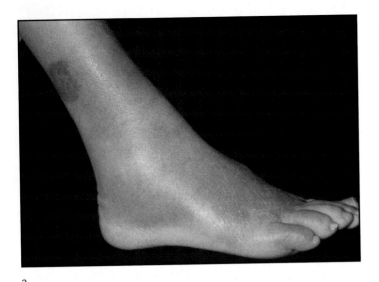

a.

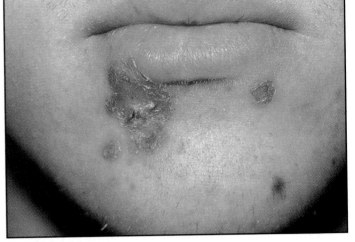

b.

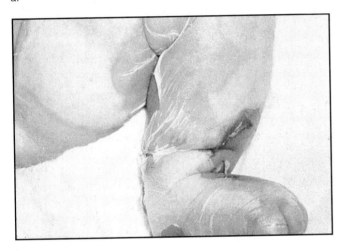

c.

figure 17.23

Pyodermas Caused by *Staphylococcus aureus* and *Streptococcus pyogenes*

a. Erysipelas on the foot. Note the large edematous patches.
b. Impetigo on the face. Note the characteristic encrusted pustules.
c. Scalded skin syndrome of an infant. Note peeling of the superficial skin layer.

Pyodermas Are Typically Caused by Staphylococcus aureus *or* Streptococcus pyogenes

Pyodermas [Greek *pyon*, pus, *derma*, skin] are inflammatory skin infections caused by pus-forming bacteria such as *Staphylococcus aureus* or *Streptococcus pyogenes* (Figure 17.23). These infections can occur in any part of the body and may range from a small pimple to a very large lesion with rapid tissue destruction and abscess formation. The major forms of pyoderma are furunculosis, erysipelas, impetigo, and scalded skin syndrome.

Furunculosis is almost exclusively a staphylococcal disease that originates by infection of a hair follicle. The infection spreads into the subcutaneous tissue and results in the formation of a suppurative lesion, a boil or furuncle. Sometimes there is deep-seated infection of several contiguous hair follicles, interconnected by abscesses to form carbuncles. Carbuncles usually occur in the upper back and neck, where the tissue is thick and elastic. Most small furuncles recede without therapy, but those containing large quantities of pus must be surgically drained.

Erysipelas is a skin disease caused by *S. pyogenes* and other streptococci. Erysipelas is sometimes called St. Anthony's fire because of the characteristic bright red edematous patches that appear on the skin. Sharply defined lesions occur on the face and extremities. The disease is usually a secondary infection that results from complications of surgical or traumatic wounds.

Impetigo is another secondary skin disease that is caused by *S. pyogenes* and *S. aureus*. The disease begins as a vesicle containing clear fluid that changes to a small, elevated portion of skin containing pus (a pustule). Pustules eventually rupture and become encrusted. The vesicular and pustular lesions contain large numbers of bacteria and are highly contagious. As the lesions rupture, the bacteria spread to other areas of the body and establish new foci of infection. Impetigo is most frequently seen in children under the age of ten and is associated with close contact and poor sanitary conditions.

S. aureus is responsible for a series of epidermal diseases collectively known as **scalded skin syndrome,** so called because the skin lesions resemble tissue that has been scalded with boiling water. The scalded skin syndrome is considered to be a form of impetigo in which layers of epidermis are peeled off as a consequence of necrosis. These diseases occur primarily in infants and children under four years of age.

S. aureus and *S. pyogenes* produce a variety of toxins and enzymes that may be responsible for the symptoms of these skin diseases. These products include hemolysins, leukocidins, kinases,

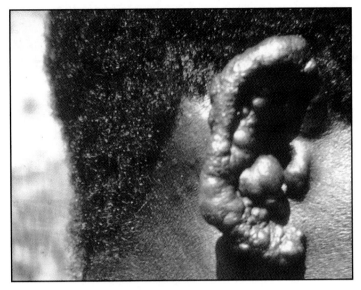

a.

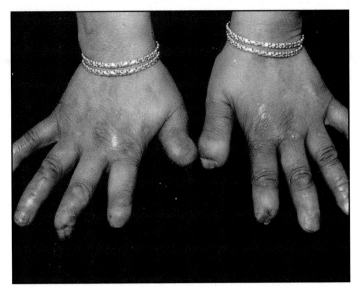

b.

figure 17.24

Lepromatous Leprosy

a. Nodular skin lesions characteristic of lepromatous leprosy. Each nodule contains large numbers of *Mycobacterium leprae*.

b. Deformed hands, a result of lepromatous leprosy.

and nucleases (see microbial enzymes, page 448). A staphylococcal exotoxin, **exfoliatin,** is responsible for the pathogenesis of scalded skin syndrome. Exfoliatin is produced only by certain strains of staphylococci (identified by their susceptibility to lysis by specific bacteriophages) and cleaves adjacent cells in the granular layer of the epidermis. In addition to these enzymes and toxins, **M proteins** distributed on the cell surface and encapsulation of *S. pyogenes* are major virulence factors that protect this bacterium from phagocytosis. Capsule formation also may be important for the virulence of staphylococci. Although only a few strains of *S. aureus* are encapsulated, these are more virulent for mice.

Because β-hemolytic streptococci are very susceptible to antimicrobial agents, most streptococcal infections are treated by penicillin, although in recent years some penicillin-resistant strains have been isolated from people. Treatment of staphylococcal diseases is complicated by the fact that most strains of staphylococci, unlike streptococci, are resistant to many antimicrobial agents commonly used in chemotherapy. Many staphylococci produce penicillinase (β-lactamase) and are thus resistant to β-lactam antibiotics. The clinical management of staphylococcal diseases requires that the drug susceptibilities of these bacteria be determined before chemotherapy.

 Gram-Positive Organisms
Gram-Positive Cocci—Staphylococci: Diseases • pp. 13–18

Leprosy Is a Skin Disease That Can Develop into One of Several Phases

Leprosy, also called **Hansen's disease** after the Norwegian physician G. Armauer Hansen who first described the etiologic agent in 1874, is a cutaneous disease caused by the rod-shaped, acid-fast bacterium *Mycobacterium leprae*. In biblical times, leprosy was considered to be a highly contagious and chronic disease that required isolation of patients.

Today people with noncontagious phases of leprosy are no longer isolated. Although the mechanism by which leprosy is transmitted is unknown, one theory is that the bacteria penetrate the skin through a cut or abrasion following prolonged, close skin-to-skin contact. Viable *M. leprae* can be found in nasal discharges of those who are actively infectious, and this may be another mechanism of transmission. Inside the body, the bacteria slowly multiply (the average generation time is 12.5 days in the mouse footpad). The incubation period of leprosy is generally three to five years, but it may be as long as 20 to 40 years.

The earliest lesion is referred to as **indeterminate leprosy** because it usually consists of only hypopigmented macules with small numbers of bacteria. In up to 75% of the cases, indeterminate leprosy is self-healing, and the disease progresses no further. In other cases, including those that are not treated, the disease may subsequently develop into one of several phases, varying in severity and contagiousness. These phases include (1) the **lepromatous phase,** characterized by severe, often disfiguring lesions and large numbers of bacteria inside **lepra cells** (macrophages with large amounts of mycobacteria and a characteristically foamy cytoplasm); (2) the less serious **tuberculoid phase,** in which there are few lepra cells and bacteria and the skin lesions are large, granulomatous, and blanched; and (3) the **borderline phase,** an intermediate phase between lepromatous and tuberculoid phases (Figure 17.24). There also is peripheral nerve involvement leading to loss of sensation in affected areas of the skin in tuberculoid leprosy. The two extreme phases of leprosy

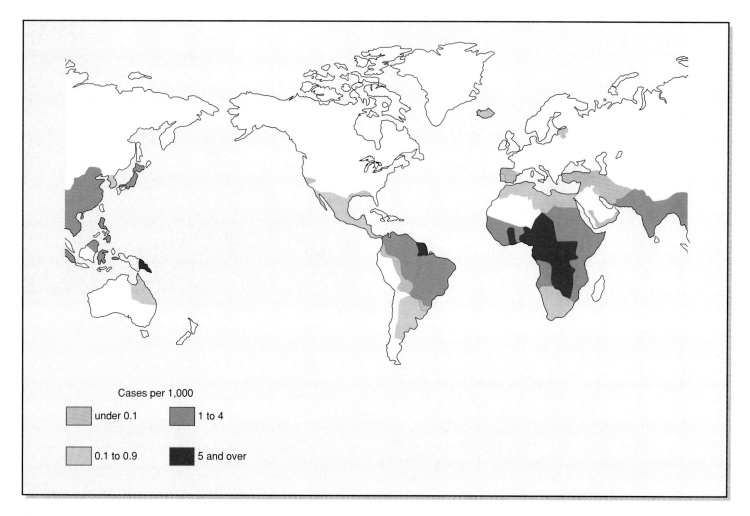

figure 17.25

Distribution of Leprosy in the World

There are an estimated 15 to 20 million cases of leprosy worldwide, primarily in tropical and subtropical countries.

reflect the host response to *M. leprae*. Patients with tuberculoid leprosy, often called the healing phase of leprosy, have stronger and apparently more effective cell-mediated immune responses to *M. leprae* than patients with lepromatous leprosy.

M. leprae cannot be grown on artificial laboratory media (although it can be grown in cell culture and in animals such as the mouse and the armadillo), so a diagnosis of leprosy is most often made by observing acid-fast bacilli in skin or nasal scrapings. Therapy is usually with rifampin, a bactericidal antibiotic, or dapsone, a bacteriostatic sulfone. Prolonged treatment is required to ensure complete recovery and prevent relapses.

Leprosy is most prevalent in the tropics and subtropics, especially Africa, south and southeast Asia, and parts of South America. Although there are fewer than 3,000 leprosy patients in the United States (all of whom have emigrated from other countries), there are

an estimated 15 to 20 million cases throughout the world today (Figure 17.25). The World Health Organization has set as a goal the prevention of all new cases of leprosy by the end of the century.

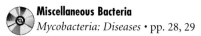

Miscellaneous Bacteria
Mycobacteria: Diseases • pp. 28, 29

Pseudomonas aeruginosa Is an Opportunistic Pathogen That Frequently Causes Infection and Death in Burn Victims

Pseudomonas aeruginosa is a common bacterium found in soil, water, plants, hospital environments, and in the intestines of about 10% of healthy people. It is ordinarily harmless, but can act as an opportunistic pathogen that causes disease in compromised and debilitated hosts whose immunity to the organism is low. This bacterium uses a wide variety of organic compounds and can

table 17.13

Characteristics of Some Rickettsial Diseases

Disease	Agent	Insect Vector	Reservoir	Geographic Distribution
Typhus				
Epidemic (louse-borne) typhus	*Rickettsia prowazekii*	*Pediculus humanus* var. *corporis* (human body louse)	Humans	Worldwide
Endemic (murine) typhus	*Rickettsia typhi*	*Xenopsylla cheopis* (rat flea)	Rats	Worldwide
Spotted fever				
Rocky Mountain spotted fever	*Rickettsia rickettsii*	*Dermacentor andersoni* (wood tick), *Dermacentor variabilis* (dog tick)	Dogs and other small mammals, birds	Western hemisphere
Rickettsialpox	*Rickettsia akari*	*Allodermanyssus sanguineus* (mite)	Mice	Northeastern United States, Russia
Scrub typhus	*Rickettsia tsutsugamushi*	*Leptotrombidium akamushi* (mite)	Small wild rodents, birds	Asia, Australia

multiply in any type of moist environment, including swimming pools, whirlpool baths, humidifiers, respiratory support equipment, and even weak antiseptic solutions. *P. aeruginosa* causes opportunistic infections in areas of the body normally devoid of a normal flora. For example, burn victims, particularly those with third-degree burns, are susceptible to serious and often fatal *P. aeruginosa* infections. Persons with various types of cancer, cystic fibrosis, and those on chemotherapy are also at risk. The pathogenicity of *P. aeruginosa* is due to many different extracellular products. At least two different proteases are produced by the bacterium and may be responsible for the hemorrhagic skin lesions seen in some infections. Exotoxin A, which has a structure and mode of action similar to that of diphtheria toxin, is thought to be a major virulence factor. The exotoxin inhibits protein synthesis by the same mechanism as diphtheria toxin (modification of elongation factor 2).

P. aeruginosa is identified by its Gram stain morphology (gram-negative motile rods) and by its ability to utilize certain types of carbohydrates by oxidative pathways. Many strains produce a water-soluble blue pigment, pyocyanin, and a water-soluble fluorescent pigment, pyoverdin. *P. aeruginosa* isolates also have a characteristic strong, grapelike odor. The bacterium is morphologically similar to members of the family *Enterobacteriaceae*, but can be differentiated from the enteric bacteria by aerobic (oxidative, nonfermentative) metabolism, polar monotrichous flagellation, and the production of cytochrome c oxidase. Enteric bacteria ferment carbohydrates and are facultative anaerobes with peritrichous flagella when motile.

The enzyme cytochrome c oxidase, synthesized by *Pseudomonas* and some other bacteria, is part of the cytochrome system in aerobic organisms. This enzyme activates the oxidation of reduced cytochrome c by molecular oxygen. The absence or presence of cytochrome c oxidase can be determined by the oxidase test (see oxidase test, page 166).

Pseudomonads often carry antibiotic resistance plasmids (see R plasmids, page 261) and are among the most resistant of all microorganisms to antimicrobial agents. Gentamicin, amikacin sulfate, and carbenicillin are some agents that have been successfully used to treat pseudomonad infections.

Gram-Negative Organisms
Gram-Negative Bacilli—Aerobes: Pseudomonas • pp. 2–8

Rickettsiae Are Obligate Intracellular Parasites That Can Cause a Variety of Skin Rashes

The genus *Rickettsia* consists of a group of bacteria that are primarily obligate intracellular parasites. Most rickettsiae are transmitted by arthropods (lice, fleas, ticks, or mites). Following direct inoculation into the skin, the rickettsiae proliferate in endothelial cells of the vascular system and eventually cause widespread small lesions. Rickettsiae are thought to produce a toxin (although none has yet been isolated) that causes an increase in capillary permeability and delayed hypersensitivity.

These microorganisms are responsible for a number of arthropod-borne diseases, including **typhus** and **Rocky Mountain spotted fever** (Table 17.13). Typhus exists in two forms. **Epidemic typhus,** also called **louse-borne typhus,** is caused by *Rickettsia prowazekii* and is transmitted by the human body louse. It is characterized by a body rash that appears on the trunk and spreads to the extremities. Severe cases of louse-borne typhus can lead to myocardial and renal failure.

Endemic typhus, or **murine typhus,** is a natural infection of rats transmitted to humans by the rat flea. The disease is caused by *Rickettsia typhi* and has symptoms similar to those of louse-borne typhus.

Rocky Mountain spotted fever is caused by *Rickettsia rickettsii* and transmitted by ticks. Although the disease was first reported at the turn of the century in the Rocky Mountain states

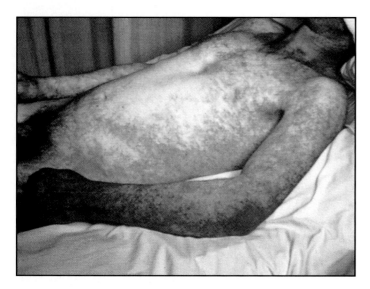

figure 17.26
Rash Associated with Rocky Mountain Spotted Fever
The characteristic Rocky Mountain spotted fever rash first appears on the wrists and ankles and spreads within a few hours from the extremities to the trunk.

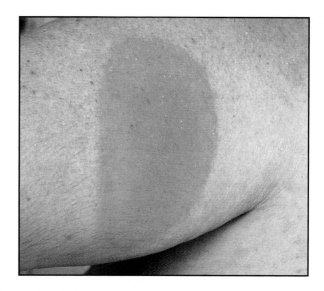

figure 17.27
Skin Rash Associated with Lyme Disease
The rash develops as concentric red rings resembling a "bull's eye."

of Idaho and Montana, it has since been found in almost every state of the United States and in other countries. Today most cases occur in mid-Atlantic and southeastern states. Rocky Mountain spotted fever is manifested by a fever, headache, and rash. The rash, unlike that of typhus, begins on the ankles and wrists and spreads from these areas to the trunk (Figure 17.26).

Control of arthropod-borne diseases is best achieved by eradication of the arthropod vector. This generally involves the use of insecticides and, in the case of louse-borne typhus, massive delousing of a population. Vaccines that are now available for certain types of rickettsial diseases permit the development of antibodies and active immunity in high-risk populations.

Miscellaneous Bacteria
Zoonoses: Rickettsia • pp. 37–52

Lyme Disease Is a Tick-Borne Disease Caused by Borrelia burgdorferi

Lyme disease is another tick-borne disease caused by the spirochete *Borrelia burgdorferi* (see *Borrelia burgdorferi,* page 316). The disease, which was first discovered in 1975 during an outbreak in the town of Old Lyme, Connecticut, is generally transmitted to humans by the deer tick, *Ixodes scapularis.* Lyme disease begins as a skin rash with concentric red rings resembling a "bull's eye" at the site of the tick bite, followed by fatigue, fever, headache, and arthritis (Figure 17.27).

The white-footed mouse, *Peromyscus leucopus,* is the primary reservoir for *B. burgdorferi,* although deer and other wild mammals, as well as birds, can serve as reservoirs. Humans are accidental hosts for the deer tick, which acquires the bacterium from these reservoirs. In some areas of the eastern United States, up to 80% of ticks may be infected with *B. burgdorferi.* Like rickettsial diseases, Lyme disease is prevented through vector control and personal protection measures such as wearing light-colored clothing, tucking pants into socks, and using insect repellents on clothing and exposed skin.

Miscellaneous Bacteria
Zoonoses: Borrelia • pp. 7–19

Skin Diseases Can Be Caused by Other Microorganisms

Skin, hair, and nail diseases caused by fungi are collectively known as **dermatomycoses.** Superficial dermatomycoses, which occur in the surface layers of the skin, usually involve infections of the hair shaft. *Pityrosporum orbiculare,* part of the normal flora of the skin and scalp of healthy humans, also causes a disease called **pityriasis versicolor.** This disease is primarily cosmetic and is characterized by brown scaly patches of skin on the neck, trunk, and arms. **Ringworm,** or **tinea,** appears initially as a small localized lesion, followed by a ring of visible inflammation with a central region of healing (Figure 17.28). The skin around the inflamed area is scaly and may be covered with pustular patches. Causative agents of the disease include members of the genera *Microsporum, Trichophyton,* and *Epidermophyton. Candida albicans* is responsible for the disease **candidiasis.** Cutaneous candidiasis generally occurs in moist areas of the skin, particularly in hosts who are diseased, immunologically compromised, or on prolonged antibacterial therapy. The symptoms of candidiasis range from simple, localized

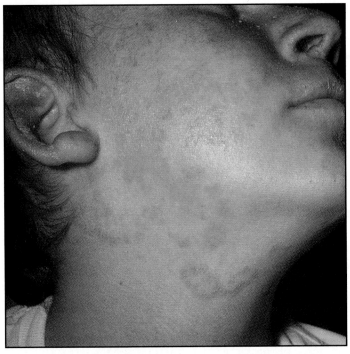

a.

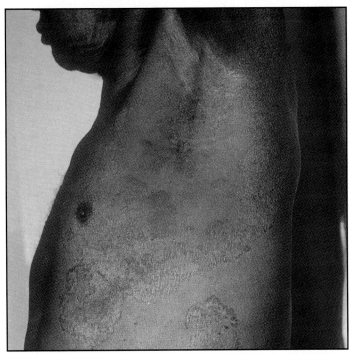

b.

figure 17.28
Ringworm (tinea)
a. Ringworm on the face and neck of a child (tinea corporis).
b. Ringworm scattered over the body (tinea corporis).

lesions to serious systemic infections. Diabetics and pregnant women particularly are susceptible to mucocutaneous and urinary tract infections with *Candida*.

 Fungal Structure and Function
Diseases: Environmental Diseases—Occupation • pp. 16–21

A number of viruses cause lesions of the skin. The more common viral skin diseases are **chickenpox** (varicella), **measles** (rubeola), **German measles** (rubella), **herpetic cold sores** (herpes simplex), and **warts** (papilloma). Chickenpox, or varicella, is a mild, self-limited childhood disease that arises from an infection of the respiratory system. The varicella zoster virus (VZV), which is responsible for the vesicular lesions in chickenpox, remains viable and in a latent form in ganglionic nerve cells following the childhood disease. The VZV virus may become activated in adults who undergo some form of trauma, such as serious illness or stress, resulting in the painful disease **shingles.** Symptoms of shingles include tenderness, pain, and sometimes paralysis along the superficial nerves in the area fed by the infected ganglionic nerves.

Measles, or rubeola, is a highly contagious disease that is spread by the respiratory route. Initial symptoms include a low-grade fever, dry cough, sore throat, and Koplik's spots (small red patches with white centers on the oral mucosa). As the virus moves into the blood, it is disseminated to the lymphoid tissue and skin. The typical red, maculopapular rash appears, first on the face and neck and then throughout the body. Measles is caused by a paramyxovirus called the measles virus. The disease begins with multiplication of the virus in the respiratory tract, followed by its dissemination via the bloodstream (**viremia**) to the skin. The typical viremia associated with measles probably contributes to the long-lasting immunity conferred by the disease. Because humans are the only reservoirs for measles, the persistence of the measles virus in a population depends on a large group of susceptible people. Vaccination, particularly of children, significantly reduces the incidence of measles. The MMR (measles, mumps, rubella) vaccine is recommended for all healthy children at 15 months.

German measles (rubella) resembles measles, but it is milder and for many years was believed to be harmless. However, in 1941, Australian ophthalmologist Norman McAllister Gregg noted that women contracting rubella during the first trimester of pregnancy often gave birth to infants with congenital defects. It is now known that the rubella virus, classified as a single-stranded RNA-containing togavirus, causes birth defects involving the eyes, ears, heart, and brain of infants (collectively known as **congenital rubella syndrome**). The virus is transmitted to the fetus from maternal blood across the placenta during pregnancy. Otherwise the virus is transmitted by direct contact, through nasal secretions, and by aerosols. It enters through the respiratory tract and infects the mucous membranes of the upper respiratory tract. It then appears in the blood (viremia) and can spread to other organs in the body. Although the virus multiplies in many organs, few clinical signs are manifested except for the characteristic rash that appears 14 to 25 days after infection. In the past, outbreaks of rubella were not uncommon. An epidemic in the United States during 1964 resulted in congenital defects in approximately 20,000 infants. A live attenuated viral vaccine (MMR vaccine) is now available to prevent rubella. With widespread use of the vaccine in the United States since 1969, the number of cases of rubella has decreased from more than 57,000 cases (1969) to fewer than 300 cases per year. In addition, congenital rubella today is a rare disease, with ten or fewer cases reported each year in the United States.

Fever blisters, or cold sores, may be found on oral tissues and on the skin. These blisters are caused by herpes simplex virus type 1, an enveloped double-stranded DNA virus. A person's first encounter with this virus is usually during childhood. After this initial infection, the virus hides in a latent form and can reappear throughout one's lifetime as a consequence of such precipitating causes as diet, trauma, sunlight, cold, or emotional disturbances. Although most herpes lesions are mild, some can be severe and dissemination to visceral organs and other portions of the body may occur. Herpes infections are extremely contagious if active lesions are present; the virus can be transmitted by oral and respiratory secretions.

Warts are benign tumors that are usually spread through direct contact. They are caused by the papilloma group of small, circular double-stranded DNA viruses and often regress spontaneously. Warts can also be removed surgically or by treatment with liquid nitrogen or chemicals such as salicylic acid.

Viral Structure and Function
Viral Diseases: Enveloped DNA Virus • pp. 6–9

The Male and Female Genitourinary Systems Differ in Their Normal Microbial Flora

The genitourinary system in humans serves two primary functions: (1) collecting, concentrating, and excreting wastes while regulating the concentrations of body fluids and electrolytes and (2) producing gametes for propagation and, in the female, supporting and nourishing the developing embryo. The genitourinary system is open to the external environment and thus provides a portal of entry for potential disease-causing microorganisms.

The male urethra (18 to 20 cm in length) is normally free of bacteria, although its external opening may be contaminated with normal skin inhabitants such as *Staphylococcus epidermidis*. The female urethra, which has a length of only 2.5 to 3.0 cm and is close to the anal and vaginal openings, has an extensive normal flora consisting of enterics, neisseriae, mycobacteria, gram-negative anaerobes, *Candida* (a yeast), and many other microbes. The shorter length of the urethra is a factor in the higher frequency of female urinary tract infections.

The male reproductive system is generally not contaminated with microorganisms. In contrast, the vagina, which is located next to the anus, is exposed to many different types of microbes. In adults, the vagina has a varied flora consisting of lactobacilli,

table 17.14

Six Most Common Agents of Urinary Tract Infections in Humans

Causative Agent	Frequency of Isolation from Urinary Tract Infections (%)
Escherichia coli	90
Proteus mirabilis	3
Klebsiella pneumoniae	2.5
Enterococcus faecalis	2
Enterobacter aerogenes	1
Pseudomonas aeruginosa	0.5

staphylococci, streptococci, diphtheroids, *Bacteroides, Clostridium,* and yeasts. The vaginal flora is influenced by hormonal and pH changes and varies throughout life. A few weeks after birth, maternally derived estrogens cause glycogen to be deposited in the epithelial cells of the neonate's vagina. Lactobacilli predominate at this time of life, converting the glycogen to lactic acid. The resulting acidic pH of the vagina restricts the flora to lactobacilli until the estrogen levels begin to decrease a few weeks later. As estrogen levels decrease and the vaginal pH rises to neutrality, other microorganisms, including diphtheroids, streptococci, and staphylococci, become established.

During puberty in females, when estrogen levels increase again, lactobacilli predominate, and the pH of the vagina is maintained at acidic levels except at ovulation. After menopause, the pH of the vagina increases to neutrality, and the vaginal flora becomes mixed and resembles the flora of childhood.

Escherichia coli *Is the Most Common Cause of Urinary Tract Infections*

Contamination of the urethra and obstructions of the urinary tract are common underlying causes of urinary tract infections. *Escherichia coli* is the most frequent cause of acute urinary tract infections, but many other bacteria also contribute to disease, including *Proteus mirabilis, Klebsiella pneumoniae, Enterococcus faecalis, Enterobacter aerogenes,* and *Pseudomonas aeruginosa* (Table 17.14). Whenever urine is retained in the urinary tract, bacteria may multiply and eventually spread to other parts of the urogenital system, including the bladder and kidneys. Urinary tract infections fall into several categories: the most common, **cystitis** (bladder infection), and the less common, **pyelonephritis** (kidney infection). **Bacteriuria** is the presence of bacteria in the urine and may reflect cystitis, pyelonephritis, or both.

Urinary tract infections are generally diagnosed by quantifying and identifying the significant bacteria present in normally sterile urine. The urine must be properly collected to avoid contamination by bacteria normally found on the external portions of the urethra. The outer portion of the urethra (and in females, the surrounding structures) should be thoroughly cleansed and the ure-

table 17.15

Microorganisms Causing Sexually Transmitted Diseases

Organism	Disease
Bacteria	
Calymmatobacterium granulomatis	Granuloma inguinale
Chlamydia trachomatis	Lymphogranuloma venereum, nongonococcal urethritis (NGU)
Gardnerella vaginalis	NGU
Haemophilus ducreyi	Chancroid
Neisseria gonorrhoeae	Gonorrhea
Staphylococcus aureus	Toxic shock syndrome
Treponema pallidum	Syphilis
Ureaplasma urealyticum	NGU
Fungi	
Candida albicans	Genital candidiasis
Protozoa	
Trichomonas vaginalis	Trichomoniasis
Viruses	
Herpes simplex virus	Genital herpes
Human immunodeficiency virus (HIV)	Acquired immune deficiency syndrome (AIDS)
Papovavirus (papillomavirus)	Genital warts

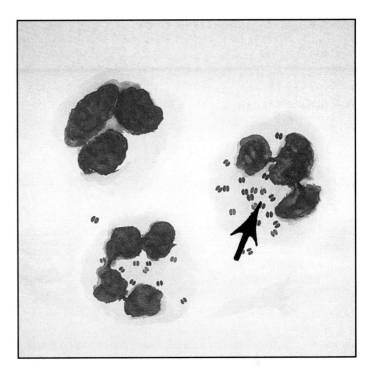

figure 17.29

Gram Stain of *Neisseria gonorrhoeae* from Urethral Exudate
Gram-negative kidney-bean-shaped diplococci (indicated by arrow) are visible inside polymorphonuclear leukocytes.

thra flushed with an initial voiding of urine before the urine is collected into a sterile container. Urines containing 100,000 bacteria or more per milliliter are considered to be indicative of bacteriuria.

Bacteria that cause urinary tract infections are usually susceptible to antimicrobial agents and can be easily treated. Sometimes it is necessary to correct conditions that predispose us to these infections, such as anatomical abnormalities or obstructions of the urethra.

Infectious Diseases of the Genital Tract Are Called Sexually Transmitted Diseases

Infectious diseases of the genital tract are routinely sexually transmitted and are collectively classified as **sexually transmitted diseases (STDs)** (Table 17.15). The terminology is somewhat misleading, since it has been shown that STDs can also be transmitted by genital contact with oral orifices, open skin lesions, and even, in some cases, inanimate objects. STDs are not confined to the genital tract—they also can appear in other areas of the body, including the oral cavity, anus, conjunctiva, and skin. Regardless of the site of disease, STDs constitute one of the most prevalent types of infectious diseases of those reported and recorded in the United States and other countries today. These diseases are of considerable concern to public health agencies.

Neisseria gonorrhoeae *Causes Gonorrhea, a Disease That Is Characterized by Purulent Discharge and Dysuria in Males and That Can Lead to Salpingitis in Females*

Gonorrhea is a disease that has existed throughout human history. Historical records indicate that such famous people as Napoléon, Charles IX and Henry IV of France, and Richelieu had gonorrhea. The causative agent, *Neisseria gonorrhoeae,* was described in 1879 by Albert Neisser. However, the disease was clinically described as early as A.D. 130. Galen, a Greek doctor known as the "Prince of Physicians" who treated the Roman gladiators, first coined the word *gonorrhea* [Greek *gonorrhoia,* flow of seed] to describe the purulent exudate characteristic of gonorrhea. The disease initially was thought to be an early manifestation of the more serious disease syphilis. It was not until the eighteenth century that the two were distinguished as separate and distinct.

N. gonorrhoeae is an aerobic, gram-negative diplococcus (Figure 17.29). Like many other sexually transmissible agents, it is extremely fastidious. It is susceptible to desiccation and is killed at temperatures of 40°C to 43°C. Transmission of the bacterium is usually through sexual contact. Attachment of *N. gonorrhoeae* by pili to the mucosal epithelium, where it causes inflammation, is considered to be an important part of its pathogenesis. The bacterium produces an IgA protease that specifically cleaves and inactivates this immunoglobulin. An endotoxin is also produced by *N. gonorrhoeae*, but this has not yet been shown to be a virulence determinant.

Gonorrhea in males is typically a urethritis characterized by a purulent urethral discharge and dysuria (difficulty or pain in urination). Males with untreated gonorrhea can develop inflammatory diseases such as prostatitis and epididymitis.

Unlike males, females with gonorrhea are usually asymptomatic because of the shorter female urethra. Nevertheless, the disease is more serious in women. If untreated, it can frequently lead to salpingitis, or pelvic inflammatory disease, a condition in which the fallopian tubes are inflamed. Although salpingitis may be caused by other bacteria such as *Chlamydia trachomatis* and *Mycoplasma hominis*, many cases are caused by *N. gonorrhoeae*. Salpingitis often leads to sterility and is a major concern of public health officials.

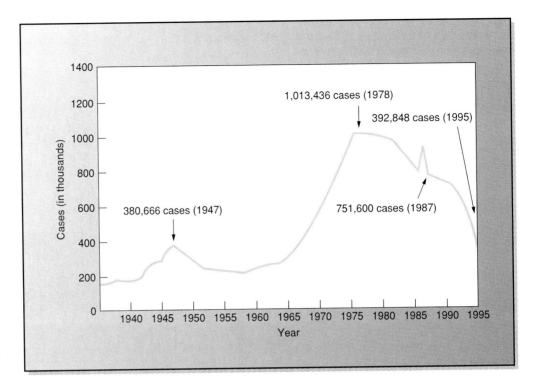

figure 17.30

Reported Cases of Gonorrhea in the United States

The number of cases of gonorrhea in the United States steadily increased over 25 years and has only recently begun to decrease in the past 20 years.

Besides genital gonorrhea, *N. gonorrhoeae* also causes other forms of disease, including gonococcal pharyngitis; gonococcal conjunctivitis; and disseminated gonococcal diseases such as septicemia, meningitis, and arthritis. Newborn infants may contract ophthalmia neonatorum, an inflammation of the conjunctivae, as a result of passage through an infected birth canal. The incidence of ophthalmia neonatorum is low in the United States because immediately after birth an erythromycin ointment or a 1% silver nitrate solution is applied to the eyes of newborn infants.

Gonorrhea is diagnosed by the examination of smears of patient exudate for the presence of gram-negative diplococci inside polymorphonuclear leukocytes. Such a diagnosis is only presumptive, however, since other species of neisseriae and other genera of bacteria have a similar morphology. An oxidase test is of use in separating the oxidase-positive neisseriae from other types of diplococci. Confirmatory identification, however, can only be achieved through other biochemical or coagglutination tests.

Gonorrhea is usually treated with penicillin. **Penicillinase-producing strains of *N. gonorrhoeae* (PPNGs)** are generally treated with ceftriaxone or other agents that are not affected by the plasmid-mediated β-lactamase (see β-lactamase, page 125). PPNGs were first seen in the United States in 1976, probably introduced by soldiers returning from southeast Asia. Nearly 100% of *N. gonorrhoeae* isolates in southeast Asia are penicillin-resistant because of the broad nonprescription use of penicillin. As penicillin is indiscriminately used without diagnosis, resistant strains of gonococci

are selected in the population and eventually predominate. The number of cases of PPNGs has increased in the United States from fewer than 100 reported cases in 1976 to as high as 11% of all isolates in some geographical areas. Although the number of gonorrhea cases has decreased in recent years in the United States, public health officials are concerned about the continued increase in percentage of resistant strains (Figure 17.30).

Gram-Negative Organisms
Gram-Negative Cocci—Neisseria • pp. 4, 6–9

Syphilis Generally Follows Three Clinical Stages

Syphilis occurs less frequently than gonorrhea but is a more serious and potentially fatal disease. The disease is caused by the spirochete *Treponema pallidum*, an invasive organism that proliferates and causes damage in infected tissue.

Syphilis was epidemic in Europe during the latter part of the fifteenth century, with the first recorded occurrence in 1497. Some historians believe that syphilis was introduced into Europe by members of Columbus' crew who acquired the disease while in the West Indies. Alternatively, *T. pallidum* may have been a mutation of a less virulent strain of the bacterium transported from Africa to Europe by migrating armies and civilians.

The clinical course of untreated syphilis can vary considerably in different patients, but generally follows three stages (Figure 17.31).

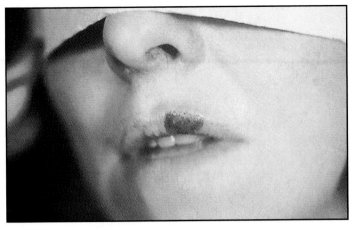

a.

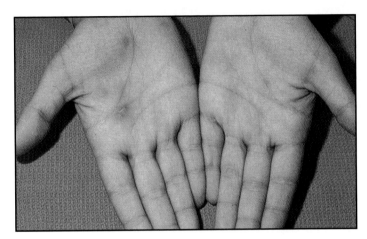

b.

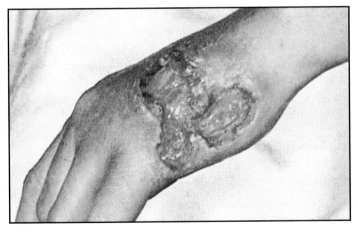

c.

figure 17.31

Characteristic Lesions Associated with Various Stages of Syphilis

a. Chancre of the primary stage on the lip. b. Skin rash of the secondary stage on the hands. c. Gumma of the tertiary stage on the hand.

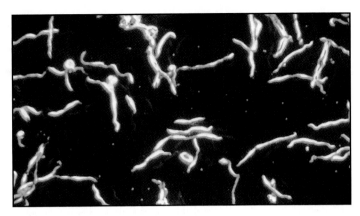

figure 17.32

Darkfield Smear of *Treponema pallidum*
Note the characteristic slender spiral cells.

The first stage, **primary syphilis,** is characterized by the appearance of a painless, small, primary lesion (**chancre**) at the site of inoculation after an incubation period of two weeks to three months. Treponemes grow and multiply in the lesion during the primary stage, and although the chancre heals spontaneously, they will escape to the blood and disseminate throughout the body if the person remains untreated. Several weeks or months after the appearance of the chancre, the **secondary stage** of syphilis begins. A generalized skin rash develops and may be accompanied by a low-grade fever and lesions in body organs. The primary and secondary stages are considered highly contagious; large numbers of treponemes are found in the lesions. Patients who remain untreated after the secondary stage can progress to the **tertiary stage.** This third stage, which may occur several years after the initial appearance of the disease, frequently results in serious dermal lesions called gummas, neurological syphilis (tabes dorsalis) through invasion of the central nervous system, and often-fatal cardiovascular syphilis.

T. pallidum readily passes the placental barrier and can infect a developing fetus, causing **congenital syphilis.** Although some neonates with congenital syphilis are stillborn or have massive lesions, others may not develop such lesions or other symptoms until later in life. The severity of symptoms in congenital syphilis depends on the extent of infection in the mother. Congenital syphilis can lead to scarring of the cornea (interstitial keratitis) and congenital malformation such as notched incisors (Hutchinson's teeth) and bowing of the leg (saber shin).

The diagnosis of syphilis depends on the stage of the disease. Primary and secondary syphilis frequently can be diagnosed by dark-field or immunofluorescent examination of exudate from lesions (Figure 17.32). Serological tests for syphilis generally are useful only after antibodies appear—usually one to three weeks after the initial lesion develops. Among the more common serological tests are the Venereal Disease Research Laboratory (VDRL) test and the rapid plasma reagin (RPR) test, both of which are designed to detect **reagin,** an antibody that appears in infections by treponemes and other infectious agents, including some bacteria, viruses, and protozoa. Reagin, although not an antibody made specifically against *T. pallidum,* is easy to detect serologically. However, false indications of syphilis may be obtained with other diseases, including yaws, malaria, systemic lupus erythematosus, and leprosy.

Both syphilis and gonorrhea caused by non-PPNG are easily treated with penicillin. Because one antibiotic successfully eliminates the two types of STDs, penicillin treatment of a patient with either syphilis or gonorrhea normally cures him or her automatically of the other disease, if present.

Miscellaneous Bacteria
Treponemes: Taxonomy • pp. 3–9

Nongonococcal Urethritis May Be Caused by Different Types of Bacteria

The term **nongonococcal urethritis (NGU),** or **nonspecific urethritis (NSU),** is used to describe any inflammation of the urethra not caused by *N. gonorrhoeae.* These terms are used when no gonococci can be seen in, or recovered from, urethral exudate. NGU mimics gonococcal urethritis except that the discharge is slight and mucoid rather than purulent as in gonorrhea. NGU may be caused by different types of bacteria, including *Chlamydia trachomatis, Ureaplasma urealyticum* (a type of mycoplasma), and streptococci. *C. trachomatis* is presently isolated in 50% or more of NGU cases and is clearly a major cause of the disease.

In the past, diagnosis of NGU was difficult because chlamydiae cannot be easily cultivated in the laboratory. Today they can be identified by ELISAs and immunofluorescence; thus cultivation in tissue culture is no longer necessary. NGU is of major concern to public health agencies, as it is estimated that as many as 10 million people in the United States annually may have the disease. Women especially are at high risk because they can develop salpingitis from their infections.

Toxic Shock Syndrome Is a Disease Caused by Staphylococcus aureus

Toxic shock syndrome (TSS) is a disease caused by toxigenic strains of *Staphylococcus aureus.* The disease was first described and named in 1978, although it is believed that cases occurred as far back as 1920. TSS gained notoriety in 1980 when an epidemic occurred in the United States, affecting 941 people (primarily menstruating women) and resulting in 73 deaths.

The disease has a short incubation period (12 to 24 hours), and its symptoms include a sudden onset of high fever, vomiting, diarrhea, muscular pains, and rapid progression to hypotension (low blood pressure) and shock. A skin rash

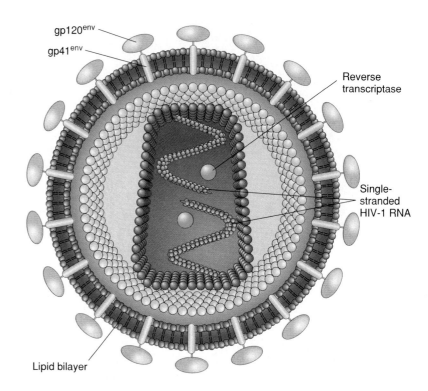

figure 17.33

Human Immunodeficiency Virus-1 (HIV-1)

HIV-1 is an enveloped retrovirus that has two copies of a single-stranded RNA genome. The icosahedral structure contains 72 external spikes, including two major envelope proteins, gp120 and gp41.

appears that resembles the rash of scarlet fever. An exotoxin, toxic shock syndrome toxin-1 (TSST-1), has been isolated from *S. aureus* strains causing toxic shock syndrome and appears to be responsible for the disease symptoms. It is believed that superabsorbent tampons, which may concentrate nutrients, may provide a rich environment in the vagina to promote the growth of toxigenic staphylococci and toxin production. Although TSS affects only a small percentage of women (and men), it is still considered to be a serious disease warranting monitoring by public health agencies.

Acquired Immune Deficiency Syndrome (AIDS) Results in a Deficiency of CD4+ Cells

Acquired immune deficiency syndrome (AIDS) is a slowly progressive disease that is characterized by severe immunodeficiency. AIDS was first recognized as a distinct disease in 1981 (see first identified cases of AIDS, page 492), but the virus causing AIDS was not isolated until later. At least two viruses, human immunodeficiency virus-1 (HIV-1) and HIV-2, are associated with AIDS (see HIV-1 and HIV-2, page 409). AIDS is caused primarily by HIV-1. HIV-2, which was discovered in 1986 in prostitutes in Senegal, West Africa, appears to produce less severe disease than does HIV-1.

HIV-1 is an enveloped retrovirus that has two copies of a single-stranded RNA genome, each containing 9,749 nucleotides (Figure 17.33). There are 72 spikes on the surface of the virus,

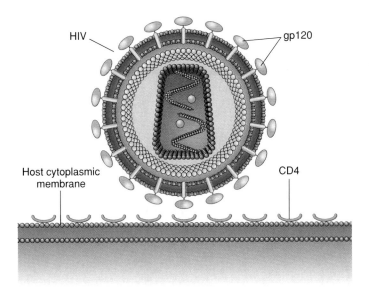

HIV

gp120

Host cytoplasmic membrane

CD4

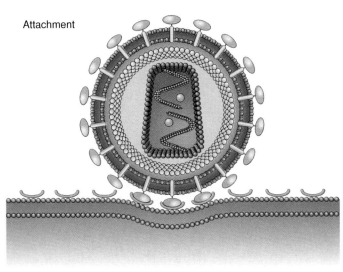

Attachment

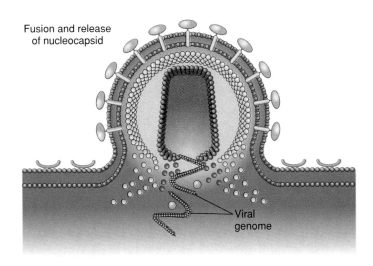

Fusion and release of nucleocapsid

Viral genome

including two major envelope proteins, gp120 (a glycoprotein with a molecular weight of 120,000 daltons) and gp41 (a glycoprotein with a molecular weight of 41,000 daltons).

Human immunodeficiency viruses have CD4-containing cells, which include T helper cells, as their major targets (see CD4-containing T lymphocytes, page 473). HIV-1 has a specificity for the CD4 molecule on CD4$^+$ cells, which serves as a cell surface receptor for the virus. The gp120 envelope glycoprotein of HIV interacts with and binds to the CD4 molecule, resulting in fusion of the membrane of the virus and the cytoplasmic membrane of the CD4$^+$ cell (Figure 17.34). The viral nucleic acid and reverse transcriptase are released into the cell.

Inside the infected host cell, the viral RNA is copied into a single strand of DNA by the viral reverse transcriptase (Figure 17.35). The viral RNA is degraded by a ribonuclease component of the reverse transcriptase, and the DNA strand is copied to form double-stranded DNA (see replication of RNA viruses, page 403). The double-stranded DNA may be used as a template for synthesis of viral RNA (unintegrated provirus) or may be integrated into the host genome (integrated provirus). An integrated provirus can remain in a latent stage for long periods of time and the host cell may show no sign of infection. In an actively infected host cell, the unintegrated provirus' DNA is transcribed into RNA and the RNA is translated into viral enzymes and structural proteins. Molecules of gp120 are incorporated into and expressed on the surface of the host cell. Infectious virions are assembled and released by budding from the host cell (Figure 17.36).

Viral Structure and Function
Viral Replication: RNA Virus Replication • pp. 22–24

The result of active infection of a CD4$^+$ cell is destruction of the cell. The precise mechanism for cell destruction is not known, but one theory is that gp120 envelope glycoproteins embedded in the surfaces of infected CD4$^+$ cells may signal immune system killer cells to seek out and destroy the infected cells (see killer cell, page 484). Infected cells also fuse with other uninfected CD4$^+$ cells to form multinucleated giant cells called syncytia that eventually die.

AIDS patients have extremely low levels of CD4$^+$ cells, a factor considered when diagnosing AIDS. The criterion for severe immunosuppression in HIV-infected persons is a CD4$^+$ T-lymphocyte count below 200 cells per cubic millimeter of blood (a healthy person has between 600 and 1,200 CD4$^+$ cells per cubic millimeter of blood). AIDS patients progress through several stages of decreasing immune function. As the CD4$^+$ cell counts fall below normal levels, the first symptoms may be

figure 17.34
Attachment and Entry of HIV into a CD4$^+$ Cell

Viral Structure and Function
Viral Invasion: Adsorption • p. 3; *Penetration* • p. 6

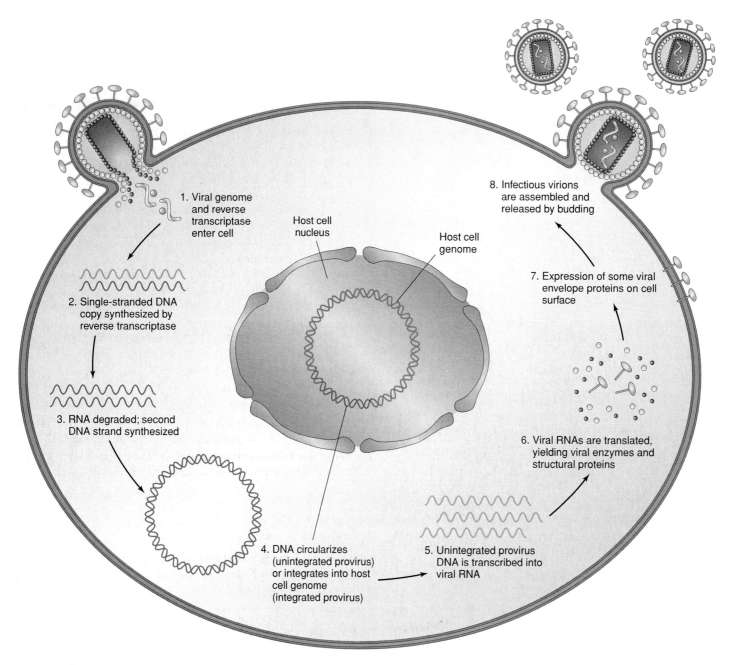

1. Viral genome and reverse transcriptase enter cell

2. Single-stranded DNA copy synthesized by reverse transcriptase

3. RNA degraded; second DNA strand synthesized

4. DNA circularizes (unintegrated provirus) or integrates into host cell genome (integrated provirus)

5. Unintegrated provirus DNA is transcribed into viral RNA

6. Viral RNAs are translated, yielding viral enzymes and structural proteins

7. Expression of some viral envelope proteins on cell surface

8. Infectious virions are assembled and released by budding

Host cell nucleus

Host cell genome

figure 17.35
Replication of HIV

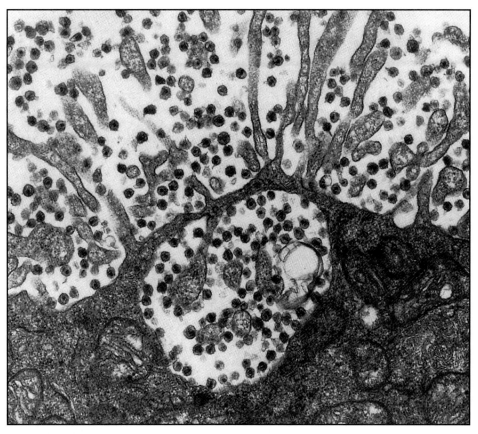

a.

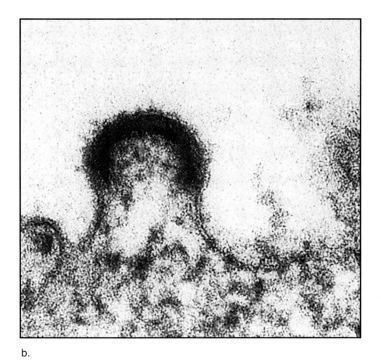

b.

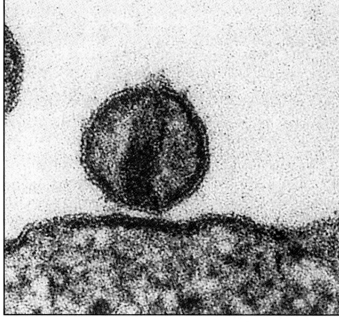

c.

figure 17.36

Electron Micrographs of HIV Infecting CD4⁺ Cells

a. A CD4⁺ cell from human umbilical cord blood producing large numbers of HIV particles. The fingerlike processes are projections from the plasma membrane. One enclosure is formed by these fingerlike projections (×10,650). b. A budding virus particle (×45,600). c. A mature HIV particle with characteristic rod-shaped core (×45,600).

lymphoadenopathy, or swollen lymph nodes. The first opportunistic infections in AIDS patients usually appear at CD4$^+$ cell counts of between 200 and 400 cells per cubic millimeter. These infections become more severe and the body goes into systemic immune deficiency as CD4$^+$ cell counts fall below 200 cells per cubic millimeter. Because CD4$^+$ cells are central to the immune response, AIDS patients are more susceptible than healthy persons to Kaposi's sarcoma (a cancer of the blood vessels of the skin or internal organs), invasive cervical cancer, cryptococcosis (fungal infections), pulmonary tuberculosis, and opportunistic diseases such as candidiasis and *Pneumocystis carinii* pneumonia.

Although people with AIDS make antibodies to HIV, these antibodies do not appear to stop the progression of the disease. There are several possible explanations: (1) the AIDS virus is in very small quantities in the blood and does not elicit high antibody titers; (2) the virus avoids detection by hiding inside the target cell's DNA; (3) the virus thrives in T_H cells and macrophages, the very cells that should destroy it; and (4) the virus changes its antigenic structure frequently, thereby thwarting any effort by antibodies to recognize a particular antigen.

HIV infection is diagnosed by an indirect ELISA and by the Western blot (see ELISA and Western blot, page 490). The ELISA, which uses HIV lysates from infected T lymphocytes to detect HIV antibodies in patient sera, is used as a screening test. Although the ELISA is rapid, sensitive, and easily performed, it occasionally gives false positive results and, in advanced AIDS cases in which the patient produces few antibodies to HIV or any other antigen, may not detect the low levels of HIV antibodies. In these instances the Western blot, which detects antibodies to specific HIV antigens that have been electrophoretically separated, is used to confirm HIV infection.

As of June, 1996, 548,102 Americans had been diagnosed as having AIDS, 343,000 (63%) of whom had died (Figure 17.37). The World Health Organization estimates that there have been 4.5 million AIDS cases worldwide and that 20 million adults and 1.5 million children are infected with HIV. It is estimated that by the year 2000 there may be 40 million people infected worldwide with HIV, including 10 million children.

The compound azidothymidine (AZT) has been used to treat patients with HIV infection. AZT, an analog of the base thymine, blocks reverse transcriptase, interrupting nucleotide assembly during reverse transcription from viral RNA to DNA. AZT is not a miracle cure; serious side effects, including bone marrow suppression, have been reported, and the long-range effects from use of the compound are unknown. Nonetheless, initial clinical results indicate that AZT restores some cell-mediated immunity in AIDS patients and may be useful in suppressing the HIV in them until a more effective cure can be found.

In 1995 the first new type of therapy to treat HIV infection since AZT was approved by the U.S. Food and Drug Administration. The drug 3TC blocks a protein vital to the synthesis of HIV. Clinical studies have shown that when 3TC is used in combination with AZT, HIV reverse transcriptase that had developed resistance to AZT became resensitized to AZT, patient CD4$^+$ levels increased, and HIV levels in patient blood dropped by as much as 90%. Based on these results, 3TC is now recommended in combination with AZT for initial AIDS therapy.

Currently a subunit AIDS vaccine is undergoing large-scale efficacy trials in Thailand. The vaccine is composed of the gp120 envelope glycoprotein antigen that remains constant despite mutational changes by the virus. Such a subunit vaccine is attractive because it may protect against different variants of HIV.

Genital Herpes Is a Recurrent Disease

Herpes simplex virus type 2, and sometimes type 1, causes an STD of the genitalia called **genital herpes.** The disease begins with the appearance of small, localized lesions on the genitalia that resemble blisters or cold sores. After four to five days, the lesions rupture into painful ulcers that release the virus in very large numbers.

Approximately 50% of people with active genital herpes complain of some type of systemic symptom, which might include headache, fever, aches and pains, and malaise. There may be itching and a burning sensation upon urination. The lesions normally last for one to three weeks and then disappear as the virus moves into a latent form, sequestered in the nerves of the lumbar spinal cord. Genital herpes is particularly a problem in pregnancy, because the virus may be transmitted to the newborn infant via the birth canal. This infection in newborns is severe and often fatal. Symptoms include loss of weight, vomiting, and diarrhea. Death is frequently due to dissemination of the virus to the brain, resulting in encephalitis. Mortality may be as high as 70% in untreated infants.

Genital herpes is a recurrent disease that may reappear throughout life, usually under stressful conditions. Like other viral diseases, genital herpes is difficult to treat. Idoxuridine creams and phototherapy using proflavin dyes with exposure to light may offer some relief from active forms of the disease. More recently, pharmaceutical companies have focused their efforts on the use of interferon and the drug acyclovir as possible treatments for genital herpes. Acyclovir inhibits replication of the virus. Topical administration of acyclovir reduces the healing time in primary herpes but is ineffective in recurrent disease. Recent evidence suggests that oral or intravenous administration of acyclovir may be more effective in the treatment of recurrent episodes.

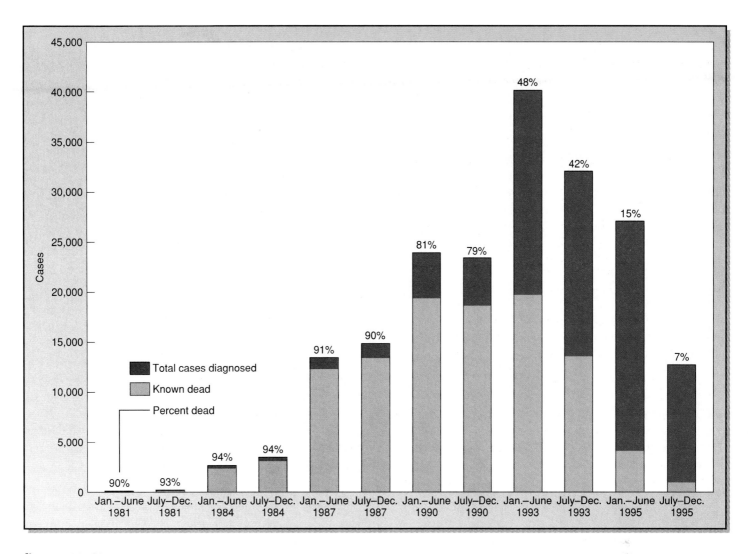

figure 17.37
Reported AIDS Cases in the United States by Half-Year of Diagnosis, 1981 to 1995

The Centers for Disease Control and Prevention estimate that genital herpes infects 20 million U.S. citizens, or nearly one person in ten. This is a matter for concern, since most people are infected for life and may transmit the infection to others. Genital herpes also seems to be associated with a higher-than-normal incidence of cervical cancer, although it has not been proven that herpes simplex virus causes the cancer. Many companies are working on a herpes vaccine. If perfected, such a vaccine may be useful in controlling the spread of genital herpes in the population.

Skin and genitourinary diseases are among the most prevalent human diseases in the world. Diseases such as leprosy, chickenpox, cystitis, gonorrhea, nongonococcal urethritis, and genital herpes affect millions of people annually. Although in most in-

stances these diseases are not life threatening, they nonetheless are an economic and sociological burden. Other diseases, such as syphilis and acquired immune deficiency syndrome (AIDS), are regarded as more serious. Syphilis, caused by *Treponema pallidum,* affects fewer people than does gonorrhea or other STDs, but it can produce serious consequences, including slow neurological degeneration and potentially fatal heart disease. AIDS, caused by human immunodeficiency virus (HIV), is one of the most deadly diseases ever discovered. HIV affects the host immune system by reducing the T helper cell population in the body. As a result, people with AIDS are more susceptible than normal, healthy persons to opportunistic diseases such as Kaposi's sarcoma, candidiasis, and *Pneumocystis carinii* pneumonia.

Emerging Infectious Diseases

Infectious diseases have plagued humans for centuries, yet it seems that scientific advances have successfully prevented the uncontrollable spread of disease agents.

Despite the improvements in medical care and surveillance in recent years, new and dangerous diseases continue to emerge at an alarming rate. As the world's population continues to increase and international travel becomes more common, diseases that at one time were considered to be restricted to remote geographical areas are now capable of rapid dissemination.

This was dramatically demonstrated by the sudden outbreak of Ebola virus in Kikwit, Zaire, in the spring of 1995, which caused massive hemorrhaging and multiple organ failure, and killed more than 90% of its victims (see Ebola virus, pages 395 and 550), and the outbreak of the initially mysterious "Four Corners Disease" in the United States.

Named for the remote southwestern region where Arizona, Colorado, New Mexico, and Utah meet, "Four Corners Disease" is caused by a hantavirus—named after the Hantaan River in Korea where the virus caused hemorrhagic fever in American soldiers—a single-stranded RNA virus belonging to the family *Bunyaviridae*. In May of 1993, forty cases of hemorrhagic fever caused by hantavirus and characterized by a flulike illness that progressed to respiratory failure were reported. This disease, which can be spread by rodents, had been reported in other countries but had not previously been seen in the United States. Hantavirus hemorrhagic fever is one example of a growing list of emerging infectious diseases that are now beginning to be recognized (Figure 17.38 and Table 17.16).

The Centers for Disease Control and Prevention (see Centers for Disease Control and Prevention, page 550) and other health agencies have developed a strategic plan involving surveillance, research, and prevention to meet the challenge of these potentially epidemic disease threats. Implementation of this plan will help ensure immediate and successful responses to emerging infectious diseases.

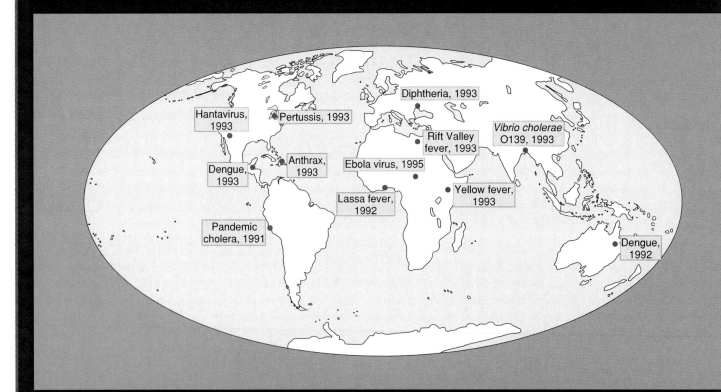

figure 17.38

Examples of Emerging and Resurgent Infectious Diseases in the 1990s

Recent Examples of Emerging Infections and Probable Factors in Their Emergence

Infection or Agent	Factor(s) Contributing to Emergence
Viral	
Argentine, Bolivian hemorrhagic fever	Changes in agriculture favoring rodent host
Bovine spongiform encephalopathy (cattle)	Changes in rendering processes
Dengue, dengue hemorrhagic fever	Transportation, travel, and migration; urbanization
Ebola, Marburg	Unknown (in Europe and the United States, importation of monkeys)
Hantaviruses	Ecological or environmental changes increasing contact with rodent hosts
Hepatitis B, C	Transfusions, organ transplants, contaminated hypodermic apparatus, sexual transmission, vertical spread from infected mother to child
HIV	Migration to cities and travel; after introduction, sexual transmission, vertical spread from infected mother to child, contaminated hypodermic apparatus (including during intravenous drug use), transfusions, organ transplants
HTLV	Contaminated hypodermic apparatus, other
Influenza (pandemic)	Possibly pig-duck agriculture, facilitating reassortment of avian and mammalian influenza viruses[a]
Lassa fever	Urbanization favoring rodent host, increasing exposure (usually in homes)
Rift Valley fever	Dam building, agriculture, irrigation; possibly change in virulence or pathogenicity of virus
Yellow fever (in "new" areas)	Conditions favoring mosquito vector
Bacterial	
Brazilian purpuric fever (*Haemophilus influenzae*, biotype *aegyptius*)	Probably new strain
Cholera	In recent epidemic in South America, probably introduced from Asia by ship, with spread facilitated by reduced water chlorination; a new strain (type O139) from Asia recently disseminated by travel (similarly to past introductions of classic cholera)
Helicobacter pylori	Probably long widespread, now recognized (associated with gastric ulcers, possibly other gastrointestinal disease)
Hemolytic-uremic syndrome	Mass food processing technology allowing contamination of meat (*Escherichia coli* O157:H7)
Legionella (Legionnaires' disease)	Cooling and plumbing systems (organism grows in biofilms that form on water storage tanks and in stagnant plumbing)
Lyme borreliosis *(Borrelia burgdorferi)*	Reforestation around homes and other conditions favoring tick vector and deer (a secondary reservoir host)
Streptoccoccus, group A (invasive; necrotizing)	Uncertain
Toxic shock syndrome (*Staphylococcus aureus*)	Ultra-absorbency tampons
Parasitic	
Cryptosporidium, other waterborne pathogens	Contaminated surface water, faulty water purification
Malaria (in "new" areas)	Travel or migration
Schistosomiasis	Dam building

[a]Reappearances of influenza are due to two distinct mechanisms: Annual or biennial epidemics involving new variants due to antigenic drift (point mutations, primarily in the gene for the surface protein, hemagglutinin) and pandemic strains, arising from antigenic shift (genetic reassortment, generally between avian and mammalian influenza strains).

Source: Data from Stephen S. Morse, Factors in the emergence of infectious diseases, *Emerging Infectious Diseases* 1(1): 7–15 (January–March 1995).

Gonorrhea is one of the most prevalent diseases in the United States, with 1 million cases (400,000 reported and an estimated 600,000 unreported cases) occurring annually. In 1976 a new strain of *Neisseria gonorrhoeae* that is resistant to penicillin appeared in the United States. These bacteria were named penicillinase-producing *Neisseria gonorrhoeae* (PPNG) because they produce a penicillinase (β-lactamase) that inactivates penicillin by cleaving the β-lactam ring of the antibiotic. In 1977 Lynn Elwell, Marilyn Roberts, Leonard Mayer, and Stanley Falkow at the University of Washington School of Medicine reported the existence of plasmids carrying the structural genes for β-lactamase in PPNG strains.

The *N. gonorrhoeae* strains examined by Elwell and colleagues included two antibiotic-susceptible strains (F62 and KH45), a penicillin-resistant strain (IPL) originally isolated in London from the vagina of a woman with pelvic inflammatory disease, and four penicillin-resistant isolates (CDC01, CDC66, CDC67, and CDC36N) supplied by the Centers for Disease Control and Prevention (CDC) in Atlanta. The CDC strains were isolates from military men returning from southeast Asia, and an isolate from a sexual contact.

Plasmid DNA was extracted with ethanol from lysates of *N. gonorrhoeae* strains and characterized by agarose gel electrophoresis (Figure 17P.1). In addition to the bacterial chromosome, four distinct covalently closed circular plasmids were detected by gel electrophoresis: a 24.5-megadalton plasmid, a 4.4-megadalton plasmid, a 3.2-megadalton plasmid, and a 2.6-megadalton plasmid. The 24.5-megadalton and 2.6-megadalton plasmids were found in both penicillin-susceptible and penicillin-resistant strains of *N. gonorrhoeae* and therefore were considered typical of gonococci. The 4.4-megadalton plasmid was found only in penicillin-resistant gonococci and therefore

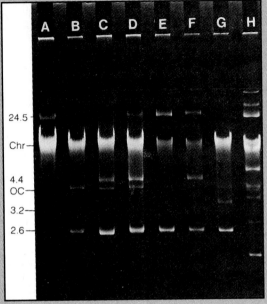

figure 17P.1

Agarose Gel Electrophoresis of Ethanol-Precipitated DNA from Lysates of Penicillin-Susceptible (Penˢ) and Penicillin-Resistant (Penʳ) Strains of *Neisseria gonorrhoeae*

(A) Strain KH45 (Penˢ) containing the 24.5-megadalton plasmid. (B) Strain F62 (Penˢ) containing the 2.6-megadalton plasmid. (C) Strain CDC66 (Penʳ), Southeast Asia isolate. (D) Strain CDC67 (Penʳ), Southeast Asia isolate. (E) Spontaneous Penˢ derivative of strain CDC01. (F) Strain CDC01 (Penʳ), from case contact of patient infected in Southeast Asia. (G) Strain IPL (Penʳ), London isolate. (H) Standard plasmid DNAs, ranging in size from 62 megadaltons (uppermost band) to 1.9 megadaltons (lowest band). Molecular weights (megadaltons) are listed on the left. Chr is the chromosome and OC is the open circular form of the 2.6-megadalton plasmid.

Summary

1. Infectious diseases can be acquired by different routes, including the respiratory tract, oral cavity, digestive system, skin, and genitourinary system.

2. Examples of microorganisms that cause diseases of the respiratory tract are:
   Streptococcus pyogenes (streptococcal pharyngitis, scarlet fever, rheumatic fever, and necrotizing fasciitis)
   Streptococcus pneumoniae (bacterial pneumonia)
   Legionella pneumophila (legionellosis)
   Bordetella pertussis (pertussis, or whooping cough)
   Corynebacterium diphtheriae (diphtheria)
   Mycobacterium tuberculosis (tuberculosis)
   Neisseria meningitidis (meningococcal meningitis)

   Ajellomyces capsulatus, also called *Histoplasma capsulatum* (histoplasmosis)
   Coccidioides immitis (coccidioidomycosis)
   Influenza virus (influenza)

3. Examples of microorganisms that cause diseases of the oral cavity and the digestive system are:
   Streptococcus mutans (dental caries)
   Candida albicans (oral candidiasis, or thrush)
   Clostridium botulinum (botulism)
   Staphylococcus aureus (staphylococcal food intoxication)
   Vibrio cholerae (cholera)
   Salmonella typhi (typhoid fever)
   Salmonella enteritidis (salmonellosis)

was associated with the penicillin-resistant phenotype. This association was strengthened by the observation that a spontaneous penicillin-susceptible derivative of strain CDC01 lacked the 4.4-megadalton plasmid. The IPL strain isolated in London did not carry the 4.4-megadalton plasmid but harbored a 3.2-megadalton plasmid.

Penicillin-resistant bacteria of the family *Enterobacteriaceae* have a transposable 3.2-megadalton DNA sequence (TnA) for the β-lactamase gene on R plasmids. It has been suggested that the β-lactamase genes on gonococcal plasmids may have been acquired from an enteric species either by conjugation or through the translocation of the TnA DNA sequence into a res-

table 17P.1

Hybridization Between ³H-labeled RSF1050 Plasmid DNA and Whole-Cell DNA

Source of Unlabeled DNA	Relative DNA Sequence Homology with ³H-Labeled RSF1050 Plasmid DNA (%)[a]
Escherichia coli W1485 (RSF1050)	100
Haemophilus influenzae (RSF007)	51
Neisseria gonorrhoeae F62 (Pen[s])[b]	1
Neisseria gonorrhoeae 45 (Pen[s])	1
Neisseria gonorrhoeae 67 (Pen[r])	22
Neisseria gonorrhoeae 66 (Pen[r])	20
Neisseria gonorrhoeae IPL (Pen[r])	20

[a]The degree of DNA-DNA hybridization was assayed using ³H-labeled plasmid DNA from *E. coli* W1485 and unlabeled plasmids from *N. gonorrhoeae* and *H. influenzae*.

[b]Pen[s] = penicillin susceptible; Pen[r] = penicillin resistant.

ident *N. gonorrhoeae* plasmid. To determine if any or all of the TnA sequence was present in PPNG strains, Elwell and colleagues hybridized ³H-labeled plasmid DNA from *Escherichia coli* (plasmid RSF1050 containing the TnA sequence as 50% of its DNA) with unlabeled plasmids from *N. gonorrhoeae* and an unlabeled R plasmid (plasmid RSF007 containing the entire TnA sequence) from *Haemophilus influenzae* (Table 17P.1). Labeled RSF1050 plasmid DNA shared no DNA sequences in common with the penicillin-susceptible *N. gonorrhoeae* strains carrying either the 24.5- or 2.6-megadalton indigenous plasmids. In contrast, RSF1050 DNA hybridized 20% to 22% with DNA from the PPNG strains CDC66, CDC67, and IPL. DNA homology of 51% was obtained between RSF1050 DNA and *H. influenzae* RSF007 DNA. On the basis of this 51% homology, it was estimated that the 4.4- and 3.2-megadalton gonococcal R plasmids contained 39% to 43% of the TnA sequence and therefore may have acquired this β-lactamase-specifying sequence from an enteric species or *H. influenzae*.

These studies by Elwell and colleagues provided important evidence of plasmid-mediated β-lactamase production in *N. gonorrhoeae*. Furthermore, the hybridization experiments supplied a clue concerning the possible origin of the gonococcal R plasmids.

Sources

Elwell, L.P., M. Roberts, L.W. Mayer, and S. Falkow. 1977. Plasmid mediated β-lactamase production in *Neisseria gonorrhoeae*. *Antimicrobial Agents and Chemotherapy* 11:528–533.

Roberts, M., L.P. Elwell, and S. Falkow. 1977. Molecular characterization of two β-lactamase-specifying plasmids isolated from *Neisseria gonorrhoeae*. *Journal of Bacteriology* 131:557–563.

Shigella dysenteriae and other species (shigellosis, or bacillary dysentery)
Campylobacter jejuni (gastroenteritis)
Escherichia coli (ETEC, EPEC, EIEC, and EHEC) (various gastrointestinal diseases)
Giardia lamblia (giardiasis)
Entamoeba histolytica (amebiasis, or amebic dysentery)
Hepatitis viruses (hepatitis)

4. Examples of microorganisms that cause diseases of the skin and the genitourinary system are:
Staphylococcus aureus (furunculosis, impetigo, scalded skin syndrome)
Streptococcus pyogenes (erysipelas, impetigo)
Mycobacterium leprae (leprosy)

Pseudomonas aeruginosa (opportunistic skin infections)
Rickettsia (Rocky Mountain spotted fever, typhus)
Borrelia burgdorferi (Lyme disease)
Varicella zoster virus (chickenpox, shingles)
Herpes simplex virus (herpetic cold sores, genital herpes)
Papillomavirus (warts)
Escherichia coli (urinary tract infections)
Neisseria gonorrhoeae (gonorrhea)
Treponema pallidum (syphilis)
Chlamydia trachomatis and *Ureaplasma urealyticum* (nongonococcal urethritis)
Staphylococcus aureus (toxic shock syndrome)
HIV-1 and HIV-2 (acquired immune deficiency syndrome)

EVOLUTION and BIODIVERSITY

The diverse diseases caused by pathogenic microorganisms and viruses suggest that these pathogens have readily adapted to changing host conditions and environments. In recent years, it has become evident that a single species of a microbe does not define a singular disease. For example, *Streptococcus pyogenes* causes diseases ranging from streptococcal pharyngitis to scarlet fever to rapidly fatal necrotizing fasciitis. Different strains of *Escherichia coli* are responsible for different types of human intestinal diseases. HIV-1 can exist as an integrated provirus in a latent stage for long periods of time without any sign of host infection and then emerge as an active, deadly virus. Ebola virus infection is rapidly fatal for some people, but not for other people. In some cases, these disease differences can be attributed to specific proteins or toxins produced by the pathogen or to variations in host immunity. In other cases, the reasons for differences in disease manifestations are not known. It is known that microorganisms are dynamic forms of life, capable of evolution and adaptation. It is apparent that as the spectrum of infectious diseases continues to change, humans remain vulnerable to a wide variety of new, emerging diseases.

Questions

Short Answer

1. Strep throat is caused by _____.
2. How are scarlet fever and rheumatic fever related to strep throat?
3. What is the most common etiological agent for pneumonia?
4. What is the etiological agent of legionellosis?
5. What is the etiological agent of whooping cough?
6. What is the etiological agent of diphtheria?
7. What is the etiological agent of tuberculosis?
8. What is the etiological agent of meningococcal meningitis?
9. Identify two fungi responsible for respiratory disease.
10. What is the etiological agent of influenza?
11. What is the principal etiological agent of dental caries?
12. What is the etiological agent of thrush?
13. What is the etiological agent of mumps?
14. Identify several etiological agents that may cause food infections.
15. Identify at least three etiological agents of food poisoning.

16. What differentiates these four strains of *Escherichia coli*: ETEC, EPEC, EIEC, and EHEC?
17. Identify two protozoa responsible for gastrointestinal disease.
18. Which virus causes foodborne infectious hepatitis?
19. Identify three microorganisms which can cross unharmed skin.
20. What is the etiological agent of acne?
21. Identify the etiological agents generally responsible for erysipelas and impetigo.
22. What is the etiological agent of Hansen's disease?
23. Identify the etiological agent most likely to affect burn patients.
24. What are the etiological agents of typhus?
25. What is the etiological agent of Lyme disease?
26. Identify the most common etiological agents of ringworm.
27. What is the etiological agent of common warts?
28. What is the etiological agent of shingles?
29. What is the etiological agent of gonorrhea?
30. What is the etiological agent of syphilis?
31. What is the etiological agent of AIDS?

Critical Thinking

1. Explain why many of the pathogens affecting the respiratory tract are gram positive. Explain why many of the pathogens affecting the digestive system are gram negative.
2. Why is tuberculosis so much more damaging than other respiratory infections? Why is tuberculosis so difficult to treat? (Give biological, as well as sociological, reasons.)
3. Explain why an influenza vaccine is feasible while one for the common cold is not. Why must the influenza vaccine be changed every year?
4. Explain why gonorrhea, though easy to diagnose, remains prevalent today. (Give biological, as well as sociological, reasons.) Explain why gonorrhea was epidemic during the 1960s, though syphilis was not. Do you expect the incidence of gonorrhea and syphilis to decline throughout the 1990s? Explain why or why not.
5. Compare and contrast the designations "HIV-positive" and "AIDS." Does the distinction make a difference to (a) the patient, (b) the course of treatment, (c) the friends and family of the patient, or (d) people unaware of the patient's disease? Explain.

Supplementary Readings

Baron, E.J., L.R. Peterson, and S.M. Finegold. 1994. *Bailey & Scott's diagnostic microbiology,* 9th ed. St. Louis: Mosby. (A textbook of diagnostic microbiology, describing techniques for isolation and identification of clinically significant microorganisms.)

Bloom, B.R., ed. 1994. *Tuberculosis, pathogenesis, protection, and control.* Washington, D.C.: American Society for Microbiology. (An in-depth discussion of *Mycobacterium tuberculosis,* including its genetics, physiology, immunology, pathogenesis, and reemergence as a major pathogen.)

Davis, B.D., R. Dulbecco, H.N. Eisen, and H.S. Ginsberg. 1990. *Microbiology,* 4th ed. Philadelphia: J.B. Lippincott. (A comprehensive textbook of medical microbiology.)

Genco, R., S. Hamada, T. Lehner, J.R. McGhee, and S. Mergenhagen, ed. 1994. *Molecular pathogenesis of periodontal disease.* Washington, D.C.: American Society for Microbiology. (A textbook on microbial virulence factors associated with periodontal disease.)

Hamann, B. 1994. *Disease: Identification, prevention, and control.* St. Louis: Mosby. (A concise review of infectious and noninfectious diseases.)

Howard, B.J., J.F. Keiser, T.F. Smith, A.S. Weissfeld, and R.C. Tilton, ed. 1994. *Clinical and pathogenic microbiology,* 2d ed. St. Louis: Mosby. (A textbook of clinical microbiology.)

Isenberg, H.D., ed. 1992. *Clinical microbiology procedures handbook.* Washington, D.C.: American Society for Microbiology. (An extensive handbook on procedures used in clinical microbiology laboratories.)

Koneman, E.W., S.D. Allen, V.R. Dowell, Jr., W.M. Janda, H.M. Sommers, and W.C. Winn, Jr. 1992. *Color atlas and textbook of diagnostic microbiology,* 4th ed. Philadelphia: J.B. Lippincott. (A textbook of diagnostic microbiology, containing color photographs of diagnostic tests to identify clinically significant microorganisms.)

Larone, D. 1995. *Medically important fungi, a guide to identification,* 3d ed. Washington, D.C.: American Society for Microbiology. (A laboratory manual describing procedures for the identification of clinically significant fungi.)

Miller, V.L., J.B. Kaper, D.A. Portnoy, and R.R. Isberg, ed. 1994. *Molecular genetics of bacterial pathogenesis.* Washington, D.C.: American Society for Microbiology. (A comprehensive overview of the progress made in recent years in understanding genetic mechanisms of microbial pathogenesis.)

Murray, P.R., E.J. Baron, M.A. Pfaller, F.C. Tenover, and R.H. Yolken, ed. 1995. *Manual of clinical microbiology,* 6th ed. Washington, D.C.: American Society for Microbiology. (An extensive clinical laboratory manual describing standard procedures used for the identification of microorganisms.)

Pershing, D.H., T.F. Smith, F.C. Tenover, and T.J. White, ed. 1993. *Diagnostic molecular microbiology, principles and applications.* Washington, D.C.: American Society for Microbiology. (A comprehensive manual of nucleic acid techniques used to identify microorganisms.)

Rietschel, E.T., and H. Brade. 1992. Bacterial endotoxins. *Scientific American* 267:54–61. (A review of bacterial endotoxins and their modes of action.)

Roth, A.J., C.A. Bolin, K.A. Brogden, F.C. Minion, and M.J. Wannemuehler, ed. 1995. *Virulence mechanisms of bacterial pathogens,* 2d ed. Washington, D.C.: American Society for Microbiology. (A discussion of the mechanisms of host-pathogen interactions and strategies to overcome bacterial virulence mechanisms.)

Salyers, A.A., and D.D. Whitt. 1994. *Bacterial pathogenesis: A molecular approach.* Washington, D.C.: American Society for Microbiology. (A textbook on how bacteria cause disease.)

Sherris, J.C., ed. 1994. *Medical microbiology, an introduction to infectious diseases,* 3d ed. Norwalk, CT: Appleton and Lange. (A comprehensive textbook of medical microbiology.)

Wachsmuth, I.K., P.A. Blake, and O. Olsvik, ed. 1994. Vibrio cholerae *and cholera, molecular to global perspectives.* Washington, D.C.: American Society for Microbiology. (A comprehensive review of *Vibrio cholerae,* including its genetics, pathogenesis, and role in disease.)

chapter eighteen

PUBLIC HEALTH AND EPIDEMIOLOGY

Some General Concepts of Epidemiology
Epidemics
Morbidity and Mortality Rates

Epidemiological Analysis of Diseases
Disease Rates and Time
Disease Rates and Place
Disease Rates and People

Collection of Epidemiological Data
Biases
Screening and Diagnostic Tests vs. Interviews

Observational and Experimental Studies
Retrospective and Prospective Studies
Experimental Studies

PERSPECTIVE
Risk Factors Associated with Toxic Shock Syndrome in Menstruating Women

EVOLUTION AND BIODIVERSITY

Maximum Containment Laboratory at the Centers for Disease Control and Prevention. Some of the most dangerous pathogens in the world are investigated in this laboratory.

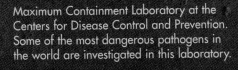

Microbes in Motion ———————— **PREVIEW LINK**

Among the key topics covered in this chapter is the terminology used in the study of disease acquisition. The following section in the *Microbes in Motion* CD-ROM may be useful as a preview to your reading or as a supplementary study aid:

Epidemiology: Disease Acquisition (Communicable), 3.

On the outskirts of Atlanta, next to Emory University and nestled in Georgia's wooded landscape, is a series of governmental buildings housing numerous scientific laboratories and offices. Here, thousands of microbiologists, physicians, statisticians, and social workers study all human diseases, ranging from mild food poisoning to virulently lethal AIDS. This cadre of workers constitutes the federal **Centers for Disease Control and Prevention (CDC),** which is responsible for detecting, monitoring, and controlling diseases in the United States.

Epidemiologists (or disease detectives, as they often are called) at the CDC monitor the nation's health from their headquarters in Atlanta, but they are not confined to these buildings. If there is an outbreak of penicillinase-producing *Neisseria gonorrhoeae* or even a single case of food-borne botulism, the epidemiologists are sent out into the field to analyze the problem and bring information and suspect samples back to the CDC (Figure 18.1).

Such was the case in May, 1995, when several people mysteriously died in the small agricultural city of Kikwit in Zaire, Africa. Victims developed headache, fever, diarrhea, and massive hemorrhaging from every body orifice. Death occurred within days. Physicians and nurses at the small, ill-equipped medical clinic where the patients had been taken were puzzled by the rapidly developing symptoms and, in fact, several medical workers who had come in contact with some of the first victims also succumbed to the same disease and died. Within hours the CDC was notified and a specially equipped international response team was immediately flown to Kikwit. At the same time, blood samples from ill patients were rushed in special containers to the CDC laboratories in Atlanta for analysis.

On May 10 antigen and antibody analysis by enzyme-linked immunosorbent assay (ELISA) and the polymerase chain reaction (PCR) confirmed that the Kikwit outbreak was caused by Ebola virus, the same virus that was responsible for a similar hemorrhagic fever outbreak in Zaire 19 years earlier resulting in 233 deaths among 296 reported cases. Ebola virus is a filovirus (see Ebola virus, page 395) with a natural reservoir that is not known (Figure 18.2). Infection by this virus leads to lesions that appear to eat through flesh, causing death by massive bleeding from the eyes, ears, and other body orifices. Transmission of Ebola virus occurs through close personal contact with infectious blood or other body fluids. The high fatality rate and the lack of specific treatment or a vaccine necessitates working with the virus in the highest level (biosafety level 4) biocontainment facility.

Meantime in Kikwit, CDC scientists dressed in protective body suits isolated suspected Ebola virus victims and quarantined the entire city. Apparently transmission of the virus had been aided by the reuse of unsterilized syringes and surgical instruments in the Kikwit clinic. With the influx of medical supplies, including gowns, masks, boots, and gloves, the Ebola epidemic soon ebbed, but not before more than 100 people (nearly 90% of the cases) had died.

The CDC's involvement in identifying the cause of this disease outbreak and limiting its potential spread is only one example of the agency's work. It not only investigates novel diseases such as Legionnaires' disease and AIDS; it also constantly monitors more common diseases such as influenza, measles, tuberculosis, and gonorrhea. Through the efforts of the CDC and other health agencies, childhood diseases such as mumps and rubella have been effectively controlled in the United States. The CDC is also involved in overseas projects, assisting Indonesia in a poliomyelitis epidemic, for example, and Zanzibar in malaria control. The CDC works closely with state health agencies, assisting them in the identification of disease-causing microorganisms and educating health professionals in diagnostic and epidemiological procedures.

Data gathered by the CDC are disseminated to health agencies and the medical community through many different publications. A key publication is the ***Morbidity and Mortality Weekly Report* (MMWR),** which is published weekly and contains up-to-date morbidity and mortality data on communicable diseases and informative articles on unusual or interesting diseases (Figure 18.3).

The CDC is only one example of the many public and private agencies that have contributed to the improvement of health worldwide. The science of protecting and improving community health through organized preventive medicine efforts is known as **public health.** In countries where public health measures are stringently implemented, death rates from many infectious diseases have decreased significantly. Public health measures include the control of the spread of disease by immunization, identification of the origin of specific diseases, and education of the public on the control of disease vectors. The study of the source, occurrence, distribution, and control of disease in a population is known as **epidemiology.**

Some General Concepts of Epidemiology

Although the science of epidemiology is a modern one, the origins and early interest in the occurrence and control of diseases developed many years ago. As early as the fourth century B.C., Hippocrates recognized an association between the environment and disease. In his *Airs, Waters, and Places,* Hippocrates emphasized the importance of considering changes in seasons, temperatures, water, soil, and other environmental factors in human disease development.

Humans have long been curious about the origin and control of diseases. This curiosity led to the establishment of epidemiology, a field in which health professionals trace diseases and recommend measures of control. Diseases studied by epidemiologists include infectious diseases and those not communicable, such as most cancers, vitamin deficiencies, alcoholism, and cardiovascular conditions.

Epidemiologists may be interested in an outbreak of food intoxication or an increase in the rates of tuberculosis in a small

figure 18.1

Centers for Disease Control and Prevention (CDC) in Atlanta
Several buildings housing scientific laboratories and offices make up the CDC.

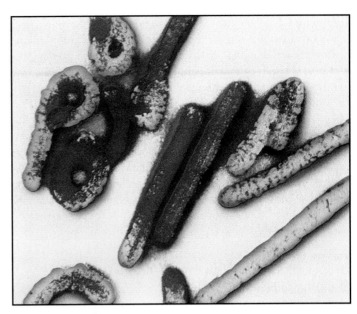

figure 18.2
Electron Micrograph of Ebola Virus (colorized)

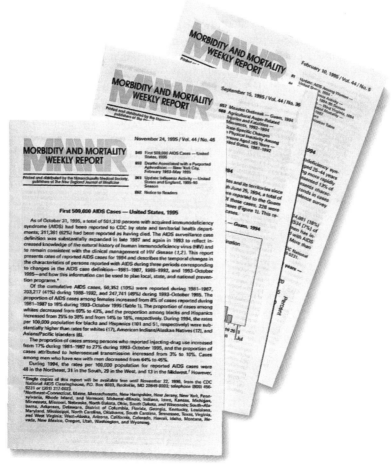

figure 18.3

Morbidity and Mortality Weekly Report, **a Publication of the Centers for Disease Control and Prevention**

community. They might investigate the demographics of a population or subpopulation that has higher-than-normal rates of lung cancer or cardiovascular disease. Unlike physicians, who diagnose and treat diseases of individuals, and unlike microbiologists, who identify the causes of diseases, epidemiologists are concerned with the sources and prevention of diseases in populations. Three types of health professionals—physicians, microbiologists, and epidemiologists—work closely together to eliminate disease. Physicians and microbiologists provide epidemiologists with data on disease cases. The epidemiologists use these data to determine disease sources and trends and to make recommendations on measures for disease control. Epidemiologists approach diseases from an inductive point of view—they collect and analyze data to derive conclusions about the characteristics of a particular disease.

An Epidemic Is a Markedly Increased Occurrence of a Disease in a Population

Several terms are used in epidemiology to describe the outbreak of a disease. An **epidemic** is a markedly increased occurrence of a disease in a particular population during a specified period. An epidemic may encompass any time period, from several hours to several years. There are two principal types of epidemics.

Common-source epidemics are those in which a group of people is exposed at one time to a particular disease agent from a common source of contamination. The exposure generally is brief, and all individuals within the group develop the disease at approximately the same time. Outbreaks of food intoxication are examples of common-source epidemics. Figure 18.4 depicts a common-source outbreak of botulism that occurred between March 31 and April 6, 1977. On March 31, in Oakland County, Michigan, two persons were hospitalized with signs and symptoms of botulism, and 13 others were identified with similar symptoms. All 15 persons had eaten at the same Mexican restaurant. Over the next seven days, an additional 44 cases of botulism were confirmed. It was determined that all had consumed hot sauce with jalapeño peppers at the implicated restaurant. Investigators from the local county health department inspected the restaurant, identified and seized 147 jars of home-canned jalapeño peppers, and ordered the restaurant to cease operation. Laboratory analysis confirmed the presence of *Clostridium botulinum* in the home-canned jalapeño peppers and in the stool specimens of some patients. This was the largest outbreak of botulism that had occurred in the United States since reporting of the disease began in 1899. The abrupt rise and decline of the epidemic curve in this example, generally having a single peak, is characteristic of an epidemic having one common source.

Propagated epidemics occur when an infectious agent is transmitted from one host to another either through direct contact

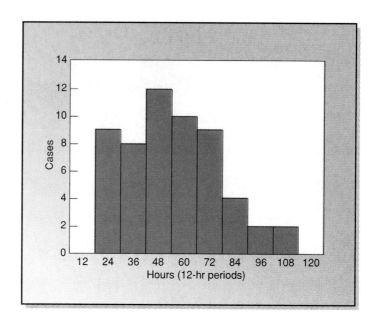

figure 18.4

Cases of Botulism During Food-Borne Outbreak, by Period Between Eating at Restaurant and Onset of First Neurological Symptoms, Oakland County, Michigan, March 31 to April 6, 1977

or via a vector (see vector, page 445). New cases of disease in propagated epidemics occur sporadically and depend on the communicability and virulence of the infectious agent. Consequently, epidemic curves of a propagated epidemic have multiple peaks that have gradual declines due to the dilution of susceptible hosts through acquired immunity. The patterns of propagated epidemics are often complex and difficult to determine. In comparison, the patterns are less complex in common-source epidemics and can usually be traced to a common origin. Figure 18.5 shows the pattern of reported measles cases in Dade County, Florida, from June 19 to December 6, 1986. During this period, 183 cases of measles were reported to the Dade County Department of Health. The index (initial) case of measles occurred in a four-year-old unvaccinated child who had acquired measles while on vacation in Honduras and developed a rash on June 19, one day after her return to Florida. The disease was transmitted to other children via contact in day-care centers, homes, schools, shelters for children, and hospitals. The characteristic, propagated epidemic curve shows the sporadic nature of the disease cases during the outbreak and a gradual decline in cases following a peak in early October.

Epidemics that affect several countries or major portions of the world are termed **pandemics.** During the fourteenth century,

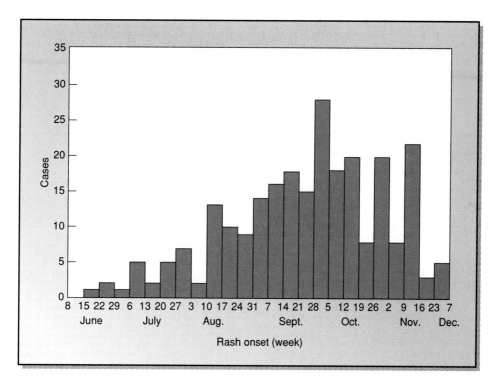

figure 18.5
Reported Measles Cases by Date of Rash Onset, Dade County, Florida, June 19 to December 6, 1986

bubonic plague spread rapidly through Europe, killing 25 million people and seriously debilitating millions more. Because this outbreak affected populations in many European countries, it is an example of a pandemic. The term **endemic** is used to describe diseases that are constantly and often insidiously present in a population. Endemic diseases generally involve fewer people than an epidemic and do not manifest a dramatic rise and decline in case numbers. Tuberculosis, measles, and gonorrhea are examples of diseases that are endemic in the United States.

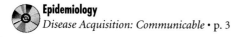
Epidemiology
Disease Acquisition: Communicable • p. 3

Morbidity and Mortality Rates Are Used by Epidemiologists to Determine the Trend of Disease

Two terms that reflect the trend of diseases are **mortality** and **morbidity** (Table 18.1). Mortality refers to the number of deaths in a population, whereas morbidity describes the number of cases of disease in a population. Epidemiologists use rates, or proportions (for example, 10 cases of syphilis per 100,000 population), instead of simple counts (for example, 10 cases of syphilis) to express mortality and morbidity so that comparisons of groups and diseases may be more meaningful and useful.

Mortality Data Reflect the Number of Deaths in a Population

The **mortality rate** is defined by the following formula:

$$\text{Mortality rate (per 1,000)} = \frac{\begin{array}{c}\text{number of deaths in the population}\\\text{during a specified period of time}\end{array}}{\begin{array}{c}\text{number of persons in the population}\\\text{during that period}\end{array}} \times 1,000$$

The numerator of the rate consists of the number of deaths that occur in a specified population. The denominator is determined from a census or other estimate of the population. In a large population, which is constantly changing because of births, deaths, emigrations, and immigrations, the denominator is obtained from the estimated population at the middle of the year (July 1). Mortality rates in many less-developed countries are underestimated frequently because of inaccurate diagnosis or underreporting of diseases.

Mortality data are usually obtained through death certificates or other documents and can be made specific for different parameters. For example, the mortality rate from tuberculosis in the United States was 0.7 per 100,000 population in 1990. The mortality rate for tuberculosis in that year was 0.6 per 100,000 white males, 3.1 per 100,000 black males, 0.4 per 100,000 white

table 18.1

Morbidity and Mortality Rates of Selected Notifiable Diseases in the United States

Disease	Morbidity Rate (Prevalence Rate) per 100,000 Population		Mortality Rate per 100,000 Population	
	1994	1940	1994	1940
Acquired immune deficiency syndrome (AIDS)	30.07	NAª	3.140	NA
Brucellosis	0.05	2.51	0.000	0.086
Diphtheria	0.00	11.77	0.000	1.104
Gonorrhea	168.40	133.26	0.000	0.318
Malaria	0.47	59.21	0.000	1.093
Measles (rubeola)	0.37	220.66	0.000	0.535
Pertussis	1.77	139.35	0.000	2.217
Poliomyelitis	0.00	7.43	0.001	0.778
Salmonellosis	16.64	NA	0.000	NA
Shigellosis	11.44	13.26	0.000	1.864
Smallpox	0.00	2.12	0.000	0.011
Syphilis	32.00	358.39	0.000	14.404
Tuberculosis	9.36	78.05	0.734	45.795
Typhoid fever	0.17	7.43	0.000	1.044

ªData not available.

females, and 1.4 per 100,000 black females. Such data specificity provides epidemiologists with comparative information on disease death rates in different population groups.

Morbidity Data Reflect the Occurrence of Disease in a Population

Two types of rates are used to express the morbidity of a disease: **prevalence rate** and **incidence rate.** The prevalence rate describes the occurrence of existing cases of disease in a population during a specified period of time:

$$\text{Prevalence rate (per 1,000)} = \frac{\text{number of cases of disease present in the population at a specified time}}{\text{number of persons in the population at that specified time}} \times 1,000$$

The incidence rate is the number of *new cases* of disease in a population from among previously nondiseased persons in a given period of time:

$$\text{Incidence rate (per 1,000)} = \frac{\text{number of new cases of a disease occurring in the population during a specified period of time}}{\text{number of persons in the population during that period of time}} \times 1,000$$

Incidence is used to determine the rate at which new illness develops, and is a direct measure of the probability, or **risk,** of developing a disease over a period of time. Prevalence (P) is related to incidence (I) and average duration (D) of disease and equals the incidence rate times the average duration of the disease ($P = I \times D$). For example, a disease with an incidence rate of 5 per 1,000 and an average duration of two years has a prevalence rate of 10 per 1,000.

Like mortality rates, morbidity rates (prevalence and incidence rates) may be specified by age, sex, race, geographical area, and other parameters. Morbidity data are obtained from hospital records, insurance records, screening surveys, and reports submitted by private physicians and health departments to public health agencies. Epidemiologists use morbidity data to monitor the health status of a population.

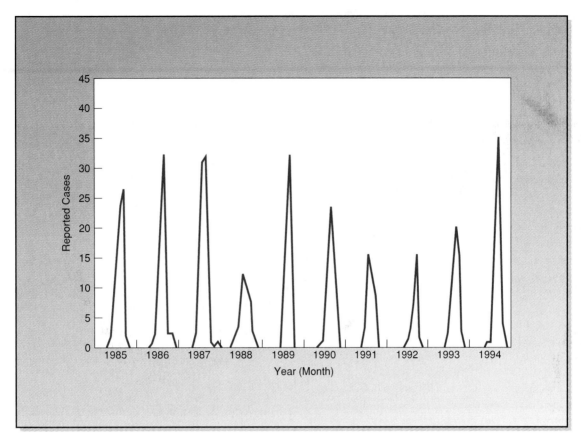

figure 18.6

Reported Cases of Arthropod-Borne Encephalitides in the United States by Month of Onset, 1988 to 1994

The seasonality associated with encephalitis cases reflects increased arthropod activity during warm-weather months.

Epidemiological Analysis of Diseases

An epidemiologist considers many different factors when investigating an outbreak of disease. The most significant are the time and place of the outbreak and the people affected.

Disease Rates May Change with Time

An epidemiologist is interested in increases or decreases in disease rates over a period of **time.** Changes in morbidity or mortality rates with time suggest that the contributing factors for a disease have also changed. Some diseases, such as viral influenza, occur in cycles. Influenza appears in cyclic intervals because of

regular shifts in the antigenic composition of influenza viruses. These patterns of antigenic shifts are predictable and can be used to plan immunization schedules for future potential influenza epidemics.

Time is also important in diseases that have seasonal fluctuations. Diseases transmitted by arthropod vectors generally occur during periods of the year when temperature and humidity are favorable to these carriers. For example, arthropod-borne encephalitides occur most frequently during the summer because the arthropod vectors of these diseases, mosquitoes, are active at that time of the year (Figure 18.6). The incidence of arthropod-borne diseases can be significantly reduced through the control or eradication of the transmitting vector.

Disease Rates May Be Influenced by Geography

Place is an important epidemiological factor; some diseases are more common in certain geographical areas. In the United States, for example, STDs occur more frequently in the Southeast than elsewhere (Figure 18.7). High rates of gonorrhea, syphilis, and other STDs may reflect the transient and mobile population in the Southeast, insufficient education about STDs, greater sexual activity, or combinations of these and other factors.

Clustering of disease cases in a particular location can provide epidemiological evidence of a disease-related problem in that region. For example, the outbreak of Legionnaires' disease in 1976 was confined primarily to members of the American Legion attending a convention in Philadelphia. Most of those who later showed symptoms stayed at or near the headquarters hotel, the Bellevue-Stratford. This clustering of cases led epidemiologists to investigate the hotel and vicinity. Although a definitive cause was never identified, the common assumption from the epidemiological data was that the responsible bacterium, *Legionella pneumophila,* originated from the air-conditioning cooling towers of the hotel. The experience gained in the Philadelphia outbreak was useful in tracing similar outbreaks of legionellosis or related diseases in Memphis and Atlanta in 1978 and, remarkably, in understanding an occurrence in Pontiac, Michigan, in 1968. In the Pontiac outbreak, patient sera were collected and stored at the CDC. These samples were tested after 1976 and were found to contain antibodies to *Legionella.* Because Pontiac fever was a milder disease than Legionnaires' disease and had no known fatalities, it may have been caused by a species related to *L. pneumophila.* Today, Legionnaires' disease and Pontiac fever are considered to be separate diseases.

Disease Rates May Be Affected by the Characteristics of a Population

The third epidemiological factor is perhaps the most important, yet most varied—**people.** Although people can be characterized in many ways, epidemiologists generally are interested in three main characteristics of an exposed population: age, sex, and race.

Morbidity and Mortality Rates Are Generally Higher Among Infants and the Elderly

Age is a personal variable that directly affects morbidity and mortality rates in a population. In most populations the mortality rate is high in infancy, declines markedly between the ages of 5 and 14, and then increases again later in life (Figure 18.8). Infants generally have higher mortality rates than children and young adults, because infants have immature immune systems and must rely on maternal antibodies received in utero for immunologic protection. As infants are exposed to a childhood disease such as mumps dur-

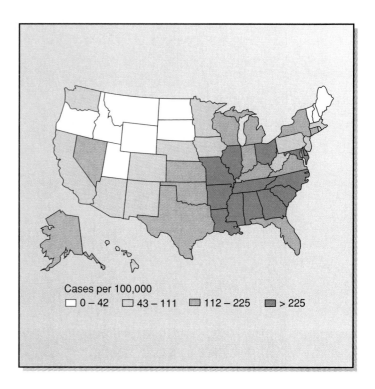

figure 18.7

Distribution of Gonorrhea in the Continental United States
Reported cases per 100,000 population by state are shown for 1995. Note the high case rate in the southeastern states.

ing the first few years of life, they develop lifelong immunity against mumps. Increases in morbidity and mortality rates with increasing age can be attributed to factors such as smoking, diet, and occupational risks, decreases in immunologic defenses, and changes in the nutritional and physiological status.

Age can also reflect occupational exposures. Human brucellosis and anthrax occur primarily in adult males who have direct contact with animal reservoirs of these diseases. Air traffic controllers, who are subject to high levels of occupational stress, have higher rates of hypertension than do people in most other occupations.

Morbidity and Mortality Rates May Differ Among Males and Females

The sex of individuals often has a significant effect on morbidity and mortality rates (Table 18.2). One disease associated with gender is breast cancer, a disease primarily found in females. In contrast, respiratory forms of cancer affecting the larynx, esophagus, tongue, and lung cause, on the average, three to four times more

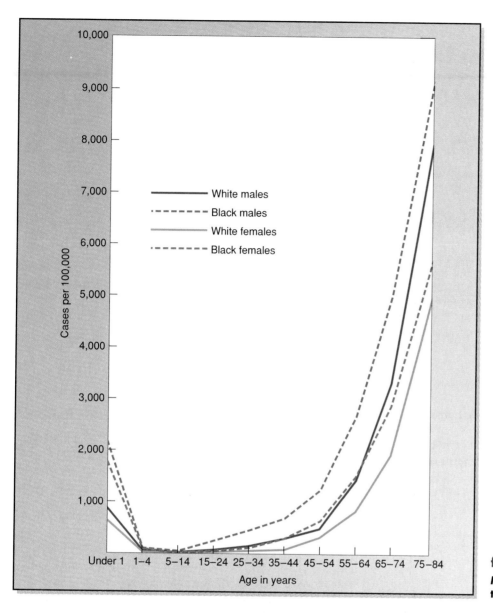

figure 18.8

Mortality Rates by Age, Race, and Sex in the United States, 1990

table 18.2

Mortality Rates for Selected Causes for Males and Females, 1990

Cause	Mortality Rate per 100,000			
	White Male	Black Male	White Female	Black Female
Breast cancer	0.2	0.3	35.9	29.0
Diabetes mellitus	16.5	21.1	20.5	31.5
Influenza	0.7	0.3	1.2	0.3
Leukemia	9.0	5.9	7.1	4.6
Meningitis	0.4	1.1	0.3	0.8
Pneumonia	30.7	28.5	35.2	20.9
Rheumatic fever and rheumatic heart disease	1.6	1.0	3.6	1.8
Septicemia	6.6	11.3	8.1	11.5
Tuberculosis	0.6	3.1	0.4	1.4
Viral hepatitis	0.7	0.7	0.5	0.7
All causes	930.9	1008.0	846.9	717.9

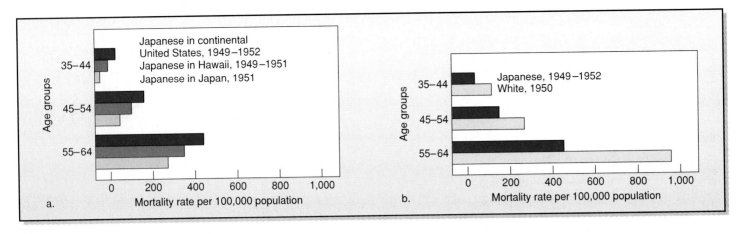

figure 18.9

Mortality Rates for Heart Disease Among Japanese and Whites in the United States

a. Mortality rates for heart disease among adult Japanese males in the continental United States, Hawaii, and Japan by age group, 1949 to 1952. b. Mortality rates for heart disease among adult Japanese and white males in the continental United States by age group, 1949 to 1952.

deaths in males than in females. In the past, respiratory forms of cancer were more common in males because more males smoked than females and also because of occupational exposure to asbestos, industrial chemicals, and coal mine dust, although equal opportunity laws are changing this pattern.

Morbidity and Mortality Rates May Differ Among Races

Many diseases differ markedly in their frequency and severity between different ethnic groups. Blacks develop hypertension, cervical cancer, and syphilis more frequently than whites, and also have substantially higher mortality rates from these diseases. Whites, in comparison, have higher death rates for arteriosclerotic heart disease and leukemia.

In some cases, differences in morbidity and mortality rates among races can be attributed to genetic differences. Sickle-cell anemia, a disease confined mainly to blacks, particularly those of West African descent, is an inherited disorder of the blood that is genetically transferred from one generation to the next. In other cases, immunologic differences between populations affect susceptibilities to particular diseases. American Indians and Eskimos are more susceptible to tuberculosis, presumably because members of these population groups have been exposed to the tubercle bacillus only in the last several hundred years and therefore have neither as long a history of association with the disease as the other populations nor the natural immunity developed over time (see box on host resistance and tuberculosis, page 453).

The relationship between certain diseases and racial groups is often difficult to define. Such epidemiological variables as occupation, environment, socioeconomic status, family size, and education can be influenced by race. The ethnic stock of an individual must furthermore be considered in any study of racial relationships. For example, a comparative study of the frequency of coronary heart diseases and stroke in Japanese men living in the continental United States, Hawaii, and Japan was conducted from 1949 to 1952. This study showed that Japanese men in the United States and Hawaii had higher heart disease mortality rates than males in Japan. Japanese males in the United States, however, had lower mortality rates from heart disease than white males in the U.S. population (Figure 18.9). Since the Japanese in the continental United States and Hawaii are migrants, these data suggest that environmental factors, living habits, or both may be etiologically associated with heart disease mortalities.

Morbidity and Mortality Rates May Be Affected by Other Variables in the Population

Other variables are often considered by epidemiologists in comparing people. Among these are marital status, personal habits, diet, and maternal age at the time of birth. An epidemiologist must be able to interpret morbidity and mortality data within the limitations of these variables.

Collection of Epidemiological Data

Epidemiological data are generally collected in one of two ways. People can be interviewed and the results of the interview considered in data analyses. Populations can also be screened with diagnostic tests to determine the incidence of disease.

Nosocomial, or hospital-acquired, infections are a serious problem in the hospital community. It is estimated that 5% to 10% of persons admitted to hospitals

Nosocomial Infections

contract infections while there. Each year approximately 2 million patients in the United States develop nosocomial infections that add an average of three days to their hospital stays and cost insurance companies, the government, and private individuals an additional $1 billion in hospital expenses. Of these 2 million nosocomial infections, more than 20,000 (1 in 100) result in death and another 60,000 (3 in 100) contribute to the death of the patient.

Hospitals are an ideal environment for the transmission of infectious agents. Patients in a hospital are exposed to a wide variety of potential pathogens. Hospital infections are transmitted to patients by medical personnel (physicians, nurses, and other hospital staff), contaminated objects (respirators, urinary catheters, surgical instruments, linens, and dressing materials), and, occasionally, contaminated foods and medication. Frequently a nosocomial infection develops because the patient's own flora is introduced into a sterile environment in the body. The microorganisms responsible for most nosocomial infections are hardy, resistant to disinfection, and able to survive for long periods of time in the hospital environment. These **opportunistic pathogens** (organisms that normally are nonpathogenic but are capable of causing disease in a compromised host) frequently infect the respiratory, genitourinary, and gastrointestinal tracts (Table 18.3).

Control measures for nosocomial infections include surveillance of hospital personnel and equipment, use of aseptic techniques in handling medical equipment and supplies, and isolation of infected patients. The recognition of nosocomial problems is important, and the education of employees is a central part of hospital infection control programs.

Today most hospitals have established programs for the surveillance and control of nosocomial infections. Hospitals accredited by the Joint Commission on the Accreditation of Hospitals (JCAH) must meet minimum standards for an infection control program. These minimum standards include establishment of an infection control committee with representatives from various hospital departments, including nurses, physicians, laboratory technologists, hospital administrators, housekeepers, and engineers. In addition, many hospitals have an infection control practitioner on their staff. The practitioner oversees infection prevention and control, and coordinates epidemiologic investigations and surveillance programs. In hospitals where stringently enforced infection control programs have been implemented, the incidence of nosocomial infections has dramatically decreased.

table 18.3

Examples of Opportunistic Microorganisms Responsible for Nosocomial Infections

Nosocomial Infection	Opportunistic Microorganism
Upper respiratory infections	Haemophilus influenzae
	Streptococcus pneumoniae
	Streptococcus pyogenes
Lower respiratory infections	Klebsiella pneumoniae
	Myxoviruses
	Paramyxoviruses
	Pseudomonas aeruginosa
	Streptococcus pneumoniae
Gastroenteritis	Campylobacter jejuni
	Escherichia coli
	Salmonella
	Shigella
Septicemia	Candida albicans
	Escherichia coli
	Pseudomonas aeruginosa
	Staphylococcus aureus
Burns	Escherichia coli
	Pseudomonas aeruginosa
	Staphylococcus aureus
Wounds	Escherichia coli
	Pseudomonas aeruginosa
	Staphylococcus aureus
	Streptococcus pyogenes
Urinary tract infections	Enterobacter aerogenes
	Enterococcus
	Escherichia coli
	Proteus
	Pseudomonas aeruginosa

table 18.4

Sensitivity and Specificity for Screening Tests
The four possible test results are shown.

Result of Screening Test	Disease State	
	Disease	**No Disease**
Positive	True positive (TP)	False positive (FP)
Negative	False negative (FN)	True negative (TN)

$$\text{Sensitivity} = \frac{TP}{TP + FN}$$

$$\text{Specificity} = \frac{TN}{TN + FP}$$

There Are Potential Biases in Interview Surveys

Interview surveys are easy to administer and are relatively inexpensive, but their validity and accuracy depend on a number of assumptions. It must be assumed that the interviewee is truthful in answering survey questions. In some instances, as in follow-up reports on STDs, the person interviewed may be reluctant to provide information or may give inaccurate information intentionally. In other instances, the interviewee may be unable to recall events of epidemiological importance. This often occurs in surveys of food poisonings, in which some people cannot remember the foods they have consumed.

Another problem associated with interviews is the scale of the survey. Enough people must be interviewed to make the survey statistically valid, but the number should not be so large as to make the survey too expensive or burdensome. The participants in the survey should represent a cross section of the population.

Screening and Diagnostic Tests Are More Objective Than Interview Surveys

The uncertainty of information obtained by interview surveys has led to the use of **screening procedures** or **diagnostic tests** to obtain data for epidemiological analysis. Such procedures are designed to detect unrecognized disease in a population. Although these protocols are often expensive and difficult to administer, they are more objective than interview surveys and provide results that can be easily confirmed.

Three criteria are important in screening or diagnostic tests: (1) accuracy or validity; (2) reliability or reproducibility; and (3) yield, or amount of disease that is detected in a population. Tests that do not meet these criteria are limited in the quality of information they can generate.

A Desirable Test Is Accurate (Sensitive and Specific)

The accuracy of a screening or diagnostic test is determined by its **sensitivity** and **specificity.** The sensitivity of a test is defined as the ability to correctly identify those people who have the disease. Specificity is the ability of a test to correctly identify those who do *not* have the disease.

Both sensitivity and specificity for a screening or diagnostic test can be determined by running the test on a population group and comparing the results with the actual condition of individuals in the population, as determined by some other definitive testing procedure. There are four possible results from such a comparison (Table 18.4). Those who are known to have the disease and test positive by the diagnostic test are said to be **true positives;** those who do not have the disease and test positive are **false positives.** People who do not have the disease and are negative by the diagnostic test are called **true negatives,** whereas those who have the disease and test negative are **false negatives.**

These values can then be used to determine sensitivity and specificity by the following formulas:

$$\text{Sensitivity} = \frac{\text{true positives}}{\text{true positives} + \text{false negatives}} \times 100$$

$$\text{Specificity} = \frac{\text{true negatives}}{\text{true negatives} + \text{false positives}} \times 100$$

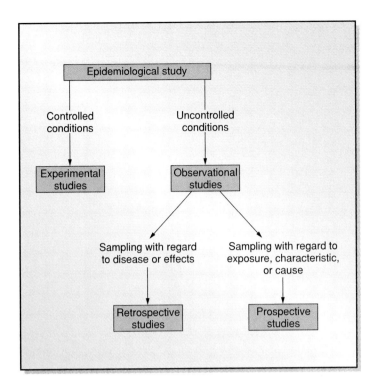

figure 18.10
Observational and Experimental Studies

Laboratory tests that have high sensitivity and specificity values are considered to be valid and accurate. Such tests can accurately identify individuals with a particular disease and only those individuals.

Screening tests are often combined to increase their sensitivity or specificity. There are two forms in which tests can be combined: **tests in series** and **tests in parallel.** A person is considered positive when examined by tests in series only if all the tests are positive. If any test result is negative, the individual is said to be negative for the disease. Tests for tuberculosis are usually performed in series. Because the tuberculin skin test may give false-positive results from infections with other species of mycobacteria, a tuberculin-positive test is followed up with more specific tests, such as chest X-ray examinations and bacteriological culture and identification of *Mycobacterium tuberculosis.*

With tests in parallel, a person is considered positive if any of the test results is positive. An example of tests in parallel is the use of screening tests for breast cancer. Breast cancer can be detected by physical examination, mammography, and thermography. Each of these procedures is useful in the diagnosis of breast cancer, and a positive result with any of the three procedures is an indication for further studies by biopsy.

Tests in series are used to increase the specificity of a screening procedure—only those people who are positive by all test results are called positive. The sensitivity of screening is generally increased through tests in parallel—all people identified as having a disease by one or more test results are considered positive.

A Test Should Be Reliable, Reproducible, or Both

The **reliability,** or **reproducibility,** of a test increases when the same test is performed several times on a person. Multiple testing is particularly valuable in minimizing observer variability when recording or analyzing test results. This variability is most noticeable in interpretations of X-ray films. Studies have repeatedly shown that radiologists frequently have different interpretations of films, particularly when there is little or no previous knowledge of patient history or clinical symptoms. Such variability can be reduced through independent interpretations of films, as well as by confirmation of interpretations by other clinical or diagnostic test results.

A Screening Test Should Be Performed on a Population with a High Frequency of Disease

For screening tests to be useful, they must be performed on populations with a high enough frequency (prevalence) of a particular disease to yield statistically significant data (the **yield** of a test). The yield of a screening test is defined as the amount of previously unrecognized disease that is detected and treated as a result of the screening. Diseases such as rabies or leprosy, which are rarely encountered in the United States, are not present in a large enough segment of the population to make screening of the disease economically feasible. Screening programs for some rare diseases nonetheless may be warranted if the consequences of the disease are severe and can be averted through detection.

Observational and Experimental Studies

The ultimate goal in gathering disease data is to determine the causal relations between population characteristics and the occurrence of a specific disease. There are basically two types of epidemiological studies to establish such associations: **observational** and **experimental** (Figure 18.10). The major difference between these two is that the epidemiologist can specify or manipulate the variable conditions in experimental settings, but is unable to control these variable conditions in observational studies.

Observational Studies Are of Two Types: Retrospective and Prospective

Observational studies may be **retrospective (case-control studies)** or **prospective (incidence studies).** A retrospective study is one in which comparisons are made between a group

table 18.5

Fourfold Table for a Retrospective Study

	Number of Individuals	
Characteristics	With Disease (Cases)	Without Disease (Controls)
With	a	b
Without	c	d

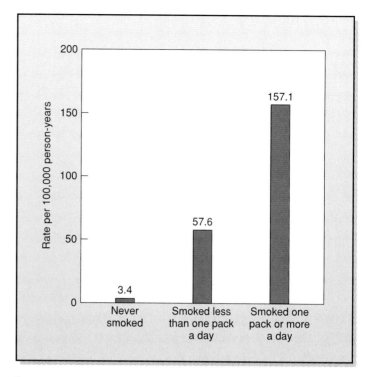

figure 18.11

Mortality Rates from Malignant Neoplasm of the Lung Among Smokers and Nonsmokers

of persons who have the disease (**cases**) and people who do not have the disease (**controls**). The presence or absence of the characteristic of interest is compared in cases and controls, usually with the assistance of a fourfold contingency table (Table 18.5). If the frequency of the characteristic in people with the disease (a/a + c) is statistically significantly different from the frequency of the characteristic in people without the disease (b/b + d), a statistical association can be said to exist between the disease and the characteristic. Any such comparison must consider whether the characteristic interacted before or after disease occurred in the cases. Furthermore, care must be taken in the selection of cases and controls for retrospective studies to ensure that subjects in the two groups are randomly chosen to avoid biased data. However, even with random selection, the case and control groups may differ in age, race, or sex. These differences can be avoided by matching cases and controls for these factors so that each case has a paired control. An example of a retrospective study is the association that was discovered in the United States between homosexuality and AIDS. This association still exists, but AIDS is now spreading through other population groups, including drug addicts and heterosexuals.

Inferences derived from retrospective studies are often confirmed by prospective studies. A prospective study starts with a group of people (a **cohort**) who are free of a given disease, but who vary in exposure to the suspected disease-associated characteristic. The cohort is followed over time to determine the rate at which disease develops (the incidence) in relation to the characteristic. Unlike retrospective studies, in which cases and controls are examined for past exposure to a factor, prospective studies are concerned with following a cohort to determine the frequency of disease with exposure to a factor. Thus the time of exposure to the factor before disease onset is never in doubt.

One of the most extensive prospective studies ever made in the United States was conducted by the American Cancer Society. The society recruited 18,000 volunteers in 1952 and asked each volunteer to survey ten white men, 50 to 60 years of age, and annually record their smoking histories and deaths from lung cancer (malignant neoplasm of the lung). After 44 months of data collection, the findings of this study indicated a correlation between the number of cigarettes smoked and the risk of death from lung cancer (Figure 18.11).

A major advantage of prospective studies is that the cohort is defined before disease develops (Table 18.6). Therefore these studies are less likely than retrospective studies to be influenced by knowledge that disease exists in examinations of causal relationships. Prospective studies also permit the observation of other possible outcomes. For example, the American Cancer Society study of smokers and nonsmokers also revealed an association between smoking and other diseases, such as coronary heart disease and emphysema. The main disadvantages of prospective studies are that they are more expensive to perform than retrospective studies, they are subject to attrition (loss of people during the study be-

table 18.6

Summary of Advantages and Disadvantages of Prospective and Retrospective Studies

Type of Study	Advantages	Disadvantages
Retrospective study	Relatively inexpensive	Incomplete information
	Smaller number of subjects	Biased recall
	Relatively quick results	Problems of selecting control group and matching variables
	Suitable for rare diseases	Yield of only relative risk
Prospective study	No bias in association of disease factor and disease	Possible bias in ascertainment of disease
	Yield of incidence rates, as well as relative risk	Requirement for large numbers of subjects
	Possible yield of associations with additional diseases as a by-product	Long follow-up period
		Problem of attrition
		Changes over time in criteria and methods
		Relatively expensive

table 18.7

Incidence of Dental Caries in Newburgh and Kingston, New York, 1954 to 1955

Age	Number of Children with Permanent Teeth		Number of DMF Teeth[a]		DMF Teeth per100 Children with Permanent Teeth	
	Newburgh[b]	Kingston	Newburgh	Kingston	Newburgh	Kingston
6–9	708	913	672	2,134	98.4	233.7
10–12	521	640	1,711	4,471	328.1	698.6
13–14	263	441	1,579	5,161	610.1	1,170.3
15–16	109	119	1,063	1,962	975.2	1,648.7

[a]DMF includes permanent teeth decayed, missing (lost subsequent to eruption), or filled
[b]Sodium fluoride was added to Newburgh's water supply beginning May 2, 1945.
From David B. Ast and Edward R. Schlesinger, The conclusion of a ten-year study of water fluoridation, *American Journal of Public Health* 46(3):265 (March 1956). Copyright © 1956 American Public Health Association. Reprinted by permission.

cause of lack of interest, migration, or death), and they often require long periods of observation. Prospective studies also are not suitable for diseases of low incidence, since a large number of people must be included in the cohort to obtain a statistically sufficient number of disease cases.

The Investigator Specifies or Manipulates Conditions in Experimental Studies

Sometimes an epidemiologist does not obtain sufficient data from either a prospective study or a retrospective study and must conduct an **experimental study.** In planned experiments, the investigator controls the influence of etiologic factors or preventive measures on the study population.

An example of an experimental study is the study conducted in Kingston and Newburgh, New York, in the 1940s and 1950s, to determine the effect on dental caries of adding fluoride to the municipal water supply. In this classic experiment, sodium fluoride was added to Newburgh's water supply in 1945. Ten years later, the incidence of dental caries in children was found to be lower in Newburgh than in Kingston (Table 18.7). Experimental studies such as the Kingston-Newburgh study provide epidemiologists with scientific evidence for determining preventive measures and public health policies.

PERSPECTIVE

Toxic shock syndrome (TSS) gained notoriety in 1980 when 941 people contracted the disease, including 73 who later died. Most (85%) of these people were women who were menstruating at the time they became ill. Of these menstruating women, 98% eventually were found to have used highly absorbent tampons. This association of tampon use and TSS came from a retrospective study conducted by Kathryn Shands and colleagues in the Bureau of Epidemiology at the CDC.

Shands and colleagues studied 52 cases, aged 12 to 49 years, with TSS and 52 age- and sex-matched controls. The cases were chosen from a group of 100 women suspected of having TSS as reported to the CDC from May 23 to June 28, 1980. The medical records of these suspected cases were reviewed or their clinical histories were discussed in detail with the attending physician, and 52 cases that met the TSS criteria established by Shands and other physicians were selected for further study (Table 18P.1). A telephone questionnaire was administered to cases and controls; it included questions about their marital status, fertility, contraceptive methods, frequency of sexual intercourse, sexual intercourse during menstruation, use of tampons or sanitary napkins during menstruation, brand and absorbency of tampon or napkin used, and use of deodorant tampons. In a related study, vaginal cultures were collected before the administration of antibiotics from 64 suspected cases of TSS. Vaginal cultures also were collected from 71 controls composed of menstruating women visiting family planning clinics or private physicians' offices in major U.S. cities. Cultures were transported to the CDC, where they were processed by standard bacteriological procedures. *Staphylococcus aureus* isolated from these cultures was identified by hemolysis on blood agar plates, fermentation of mannitol, catalase production, and coagulase production.

table 18P.1

Case Definition of Toxic Shock Syndrome

Fever: temperature ≥38.9°C

Rash: diffuse macular erythroderma

Desquamation of the palms and soles one to two weeks after the onset of illness

Hypotension: systolic blood pressure ≤90 mm Hg for adults or below the fifth percentile by age for children below 16 years of age, orthostatic drop in diastolic blood pressure ≥15 mm Hg from lying to sitting, or orthostatic syncope

Multisystem involvement—three or more of the following:

  Gastrointestinal: vomiting or diarrhea at onset of illness

  Muscular: severe myalgia or creatine phosphokinase level at least twice the upper limit of normal for laboratory tests

  Mucous membrane: vaginal, oropharyngeal, or conjunctival hyperemia

  Renal: blood urea nitrogen or creatinine level at least twice the upper limit of normal for laboratory tests for urinary sediment with pyuria (≥5 white cells per high-power field) in the absence of urinary tract infection

  Hepatic: total bilirubin, serum aspartate aminotransferase (SGOT), or serum alanine aminotransferase (SGPT) at least twice the upper limit of normal for laboratory tests

  Hematologic: platelets ≤100,000 per cubic millimeter

  Central nervous system: disorientation or alterations in consciousness without focal neurological signs when fever and hypotension are absent

Negative results on the following tests, if obtained:

  Blood, throat, or cerebrospinal-fluid cultures

  Rise in titer to Rocky Mountain spotted fever, leptospirosis, or rubeola

The mean age of the cases and controls in the first study was 27 years. The mean duration of menstruation (plus or minus the standard deviation) was 5.7 ± 1.0 days for cases and 5.2 ± 1.4 days for controls. The mean duration from the onset of menstruation to the onset of illness (TSS) was 3.8 days in the cases (Figure 18P.1). All 52 cases (100%) used tampons during the menstrual period coincident to the time of disease onset, as compared to 44 (85%) of 52 controls. The eight controls who did not use tampons used minipads or napkins. Seventeen cases (33%) and 17 controls (33%) used minipads, napkins, or both in addition to tampons. Furthermore, among case-control pairs (44 pairs) in which both the case and the control used tampons, more cases than controls (42 cases versus 34 controls) used tampons every night and day during the menstrual period. In the second study, *S. aureus* was isolated from 62 of 64 women (97%) with TSS and from 7 of 71 women (10%) in the control group.

These data provided the first epidemiological evidence of the association of tampon use and *S. aureus* with TSS. It is an example of how the CDC, through its network of cooperating clinics and physicians, can trace the source and cause of a disease such as toxic shock syndrome.

Source

Shands, K.N., G.P. Schmid, B.B. Dan, D. Blum, R.J. Guidotti, N.T. Hargrett, R.L. Anderson, D.L. Hill, C.V. Broome, J.D. Band, and D.W. Fraser. 1980. Toxic-shock syndrome in menstruating women: Association with tampon use and *Staphylococcus aureus* and clinical features in 52 cases. *New England Journal of Medicine* 303:1436–1441.

Reference

Todd, J., M. Fishaut, F. Kapral, and T. Welch. 1978. Toxic-shock syndrome associated with phage-group-1 staphylococci. *Lancet* 2:1116–1118.

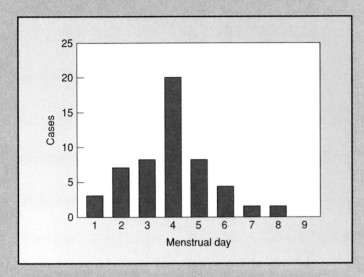

figure 18P.1

Day of Menstrual Period on Which Clinical Illness Began

Summary

1. Public health is the science of protecting and improving community health through organized preventive medicine efforts. Epidemiology is the study of the source, occurrence, distribution, and control of disease in a population.

2. An epidemic is a markedly increased occurrence of a disease in a particular population during a specified period. Common-source epidemics are those in which a group of people is exposed at one time to a particular disease agent from a common source of contamination. Propagated epidemics are those in which an infectious agent is transmitted from one host to another either through direct contact or via a vector.

3. The mortality rate is determined from the ratio of the number of deaths in a population during a specified time period to the number of persons in the population during that period. The morbidity prevalence rate is determined from the ratio of the number of cases of disease present in a population at a specified time to the number of persons in the population at that time. The morbidity incidence rate describes the number of new cases of disease occurring in a population during a specified period of time.

4. When investigating an outbreak of disease, an epidemiologist is concerned with the time and place of the outbreak and the people affected.

5. Epidemiological data can be collected by interview surveys and by screening procedures or diagnostic tests. Three criteria are important in screening or diagnostic tests: (1) accuracy or validity; (2) reliability or reproducibility; and (3) yield, or amount of disease that is detected in a population.

6. The accuracy of a screening or diagnostic test is determined by its sensitivity and specificity. A sensitive test correctly identifies those people who have the disease. A specific test correctly identifies those people who do not have the disease.

7. Observational and experimental studies are used by the epidemiologist to determine the causal relations between population characteristics and the occurrence of a specific disease. The epidemiologist can specify or manipulate conditions variable in experimental settings, but is unable to control these variable conditions in observational studies.

EVOLUTION *and* BIODIVERSITY

Public health is a relatively new science, with beginnings traced to the epidemiological investigations by British physician John Snow during an outbreak of cholera in London in 1854. Snow, a founding member of the London Epidemiological Society, which was organized in 1850 as the first known epidemiologic society, showed an association between cholera deaths and water provided by a specific water company. As a result of Snow's investigations and the increasing worldwide interest in infectious diseases and their causes, public health departments were established in Great Britain and other countries to monitor the health of communities. During the past 150 years, public health has evolved in the United States to local, state, and federal health agencies that oversee the nation's health. These organizations range from county health departments, which perform valuable functions such as immunizations, prenatal clinics, and sexually transmitted disease screenings, to the Centers for Disease Control and Prevention, which tracks, identifies, and controls diseases such as influenza, viral hemorrhagic disease, AIDS, cholera, and tuberculosis. These public agencies, aided by private foundations such as the March of Dimes and the American Lung Association and by international organizations such as the World Health Organization, play key roles in educating the public and controlling the spread of diseases. As new infectious diseases are identified, these diverse groups will continue to determine the etiology of diseases as well as recommend measures for their control and prevention.

Questions

Short Answer

1. How does epidemiology differ from microbiology or medicine in general?

2. What distinguishes an epidemic from an endemic disease?

3. Compare and contrast common-source epidemics and propagated epidemics.

4. What distinguishes a pandemic from an epidemic?

5. Compare and contrast morbidity and mortality.

6. Compare and contrast prevalence and incidence.

7. What is meant by the term "notifiable disease"?

8. In addition to the number of cases of a disease, what major factors would interest an epidemiologist?

9. Identify several characteristics of individuals that may be of interest to an epidemiologist.

10. How do epidemiologists gather data?

11. Compare and contrast sensitivity and specificity as they relate to epidemiological tests.

12. Compare and contrast true positives and false negatives.

13. Compare and contrast true positives and false positives.

14. How do experimental epidemiological investigations differ from observational ones?

15. Compare and contrast retrospective and prospective studies.

Critical Thinking

1. Just one hundred years ago, infectious diseases were the leading cause of death. Explain why. How has this changed during the past one hundred years?

2. Infectious diseases are still a major cause of death in developing countries. Explain why. How should this issue be addressed?

3. Why are people in developed countries more likely to acquire exotic or tropical diseases? How should this issue be addressed?

4. Explain how advances in medicine have contributed to the rise in nosocomial infections. Prepare a plan for your local hospital or nursing home to prevent nosocomial infections.

5. When is it necessary to quarantine an individual? Give several examples of diseases that warrant quarantine.

 Supplementary Readings

Anderson, R.M., and R.M. May. 1992. Understanding the AIDS pandemic. *Scientific American* 266:58–67. (A review of the spread of AIDS.)

Brachman, P.S. 1991. *Bacterial infections of humans: Epidemiology and control.* New York: Plenum Medical Books. (A discussion of the epidemiology of human bacterial diseases.)

Giesecke, J. 1994. *Modern infectious disease epidemiology.* Boston: Little, Brown, and Company. (A textbook on the fundamental concepts of epidemiology.)

Lilienfeld, A.M., and D.E. Lilienfeld. 1980. *Foundations of epidemiology.* 2d ed. New York: Oxford University Press. (A well-written and extensive textbook describing the basic concepts of epidemiology.)

Mauser, J.S., and S. Kramer. 1985. *Epidemiology: An introductory text.* Philadelphia: Saunders. (A textbook presenting the basic concepts of epidemiology, with study questions and problems.)

Streiner, D.L., G.R. Norman, and H. Munroe-Blum. 1989. *Epidemiology.* Toronto: B.C. Decker. (A concise textbook of epidemiology.)

chapter nineteen

19

MICROBIAL ECOLOGY

Water pollution control and sewage treatment facility, Canada.

Biogeochemical Cycles

The Carbon Cycle
Primary Producers
Herbivores and Carnivores
Microorganisms

The Nitrogen Cycle
Nitrogen Fixers
Ammonia

The Phosphate Cycle

The Sulfur Cycle

Microbes and Soil
Composition and Physical Properties of Soil
Microorganisms in Soil
Bacteria and Leaching of Metals from Low-Grade Ores
Hydrocarbon-Degrading Microorganisms
Microorganisms as Biological Pesticides

Microbes and Water
Microorganisms in Marine Habitats
Microorganisms in Freshwater Habitats
Chemical and Biological Contaminants of Water
Sewage Treatment

Microbes and the Air

PERSPECTIVE
A Fluorogenic Assay to Detect Coliforms

EVOLUTION AND BIODIVERSITY

 Microbes in Motion ──────────── PREVIEW LINK

Among the key topics covered in this chapter is the outbreak of cryptosporidiosis in Milwaukee in 1993. The following section in the *Microbes in Motion* CD-ROM may be useful as a preview to your reading or as a supplementary study aid:

Parasitic Structure and Function: Intestinal Protozoa (Epidemiology), 16–23.

Without microorganisms, chemical elements would not be recycled and nutrients would not be available to sustain life. Due to the grave problem of progressive loss of natural resources, there has been renewed interest in the study of **microbial ecology,** which is concerned with the interactions of microorganisms with the biotic (living) and abiotic (nonliving) components of the environment. Diverse microbial populations interacting within a habitat constitute a **community,** and this community, together with the abiotic factors in the environment that supply it with the raw materials for life, forms an **ecosystem.** Such associations not only affect the populations within a biological community, but also can alter the chemistry of the surrounding environment. This chapter will consider the importance of these interactions and their effects on the environment.

Biogeochemical Cycles

Biogeochemical cycles describe the movement of *chemical* elements through the *biological* and *geological* components of the world. The atmosphere (the gaseous mass surrounding the earth), lithosphere (the earth's crust), and hydrosphere (water) comprise the geologic portion of these cycles. The biological part consists of living organisms, which are classified as producers, consumers, or decomposers.

Although all living organisms play some role in biogeochemical cycles, microorganisms are indispensable in the recycling process. They serve not only as producers and consumers, but also have unique roles as biological decomposers. Bacteria and fungi in particular break down chemical substances in dead, decaying material into simpler compounds or elements that other organisms within the ecosystem use for metabolism. The conversion of organic matter to minerals and other inorganic materials is called **mineralization.** It occurs principally because of the microbial decomposition of dead plants and animals and provides continuity to the recycling of elements.

Chemical elements are not all recycled at the same rate. Differences in recycling are necessary to accommodate various needs of organisms for the chemical elements. For example, the vegetation of the earth consumes approximately 1.3×10^{14} kg of CO_2 annually. In comparison, most minor or trace elements are required in significantly smaller quantities by living organisms. A simple synthetic growth medium for *Escherichia coli* may contain 10,000 µg/ml of glucose as a carbon and energy source, but only 1/1,000 (10 µg/ml) of that concentration of iron in the form of $FeSO_4$.

Six major elements (carbon, nitrogen, sulfur, phosphorus, oxygen, and hydrogen) make up 97% of the dry weight of a bacterium and are required in large amounts for growth (Table 19.1).

table 19.1

Chemical Elements Occurring in a Procaryotic Cell

Element	% Dry Weight[a]
Carbon	50.0
Oxygen	20.0
Nitrogen	12.0
Hydrogen	10.0
Phosphorus	4.0
Sulfur	1.0
Potassium	<1.0
Sodium	<1.0
Calcium	<0.5
Magnesium	<0.5
Other elements	<0.5

[a]Percentage of dry weight of cell mass of *Escherichia coli.*

These elements are recycled at higher rates than other elements. Minor elements (for example, magnesium, potassium, and sodium) and trace elements (for example, aluminum, copper, selenium, and zinc) are chemicals required in smaller quantities by living organisms and therefore are recycled more slowly over time. Some exceptions exist to this recycling pattern. Iron and manganese are minor/trace elements but are constantly converted from one form to another through oxidation-reduction reactions (see oxidation-reduction, page 145). Calcium, a minor element that normally is not required in large quantities by microorganisms, nevertheless has an important role in endospore heat resistance (see endospore heat resistance, page 68). Molybdenum, iron, and cobalt are associated with nitrogen fixation. Magnesium and various other minor elements play important roles as cofactors that bind to enzyme surfaces and are required for enzymatic activity (see cofactors, page 145). This chapter focuses on the biogeochemical cycles of four important elements: carbon, nitrogen, phosphorus, and sulfur.

The Carbon Cycle

The most important single element in the biosphere is carbon. It is the backbone of organic compounds and constitutes approximately 40% to 50% of the dry weight of living tissues. There are more compounds made of carbon than of all other elements combined. Most of the carbon on earth is stored in the form of fossil fuels—coal, peat, oil, and natural gas. The remaining carbon is found as part of living or decaying organisms, atmospheric CO_2, or dissolved forms of CO_2 (bicarbonates and carbonates). The cycling of carbon in the environment is closely linked with the flow of energy, so carbon cycling is critically important to all forms of life.

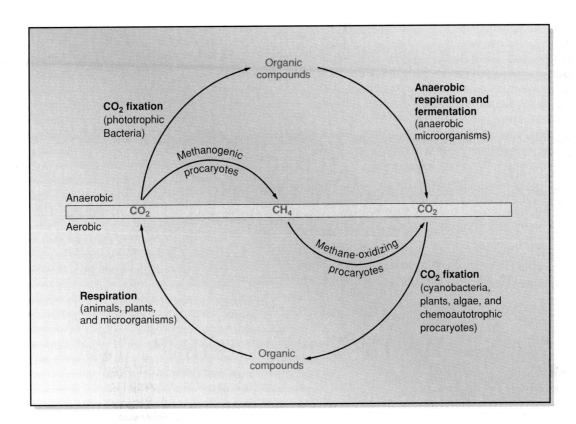

figure 19.1
The Carbon Cycle

The carbon cycle involves three groups of organisms: producers, consumers, and decomposers. Because organic compounds also serve as food and energy sources, this cycle parallels food webs and patterns of energy flow in an ecosystem and often involves the same or similar metabolic pathways and organisms.

The ultimate source of all carbon is CO_2. Not only is it a raw material for photosynthesis, but it is also a major waste product of respiration. It is formed from the decay of organic matter and from the combustion of carbonaceous fuels. Yet this inert, colorless, odorless gas comprises only 0.032% of the earth's atmosphere. With the finite quantity of atmospheric CO_2 and its rapid rate of consumption by autotrophic bacteria, algae, and plants, clearly the amount of available CO_2 would soon become exhausted without extensive transformation of the element carbon.

Carbon transformation occurs via a **carbon cycle,** which revolves around CO_2 (Figure 19.1). Organisms known as **primary producers** fix atmospheric CO_2 into organic forms of carbon. The primary producers include not only photosynthetic organisms—plants, algae, cyanobacteria, and phototrophic bacteria (in order of magnitude of CO_2 fixed annually)—but also chemolithotrophs

and chemoorganotrophs that fix CO_2 during metabolism. Carbon dioxide is also anaerobically reduced to methane by the methanogens (domain Archaea) (see methane production from wastes, page 598).

The carbon that is fixed into organic matter is eventually returned to the atmosphere as gaseous CO_2 in one of three ways: (1) animals, plants, and microorganisms evolve CO_2 during respiration; (2) plants are consumed, and a significant portion of their carbon is oxidized as food by herbivores and eventually carnivores (**consumers**); or (3) upon the death of animals and plants, the dead tissues undergo decomposition by degrading microorganisms (**decomposers**), which recycle the carbon into the atmosphere in the form of CO_2. This series of events completes the carbon cycle.

Primary Producers Fix Atmospheric Carbon Dioxide into Organic Compounds

Autotrophic bacteria, especially the cyanobacteria, fix vast amounts of CO_2. The role of photoautotrophic and chemoautotrophic bacteria in CO_2 fixation has previously been discussed

Most of the CO_2 in the atmosphere is derived from either the decomposition of organic matter by microorganisms or the release of CO_2 as a by-product of cellular

The Greenhouse Effect

metabolism. The levels of atmospheric CO_2 would remain relatively constant if these were the only sources of the gas.

During the past 100 years, however, the level of atmospheric CO_2 has increased from 0.029% to the current 0.032%. It is estimated that by the year 2000, the level of CO_2 in the atmosphere will be nearly 0.04%. The source of this additional CO_2 is the burning of fossil fuels, which releases large amounts of CO_2 into the atmosphere (up to 6 billion metric tons of carbon per year). Also contributing to the rising CO_2 composition of the air is deforestation. A reduction in vegetation means fewer plants that remove CO_2 from the atmosphere through photosynthesis.

Although the level of increased atmospheric CO_2 may seem small, it is constantly increasing at a significant rate and contributes to a phenomenon known as the **greenhouse effect** (Figure 19.2). As the sun's rays enter the atmosphere, they pass through the CO_2 layer. The solar rays striking the ground are reflected back as longer-wavelength infrared radiation (heat). The atmospheric CO_2 layer, although transparent to visible light, is not transparent to infrared rays and reflects much of these rays down toward the ground. As a consequence, the atmosphere acts like a greenhouse; the CO_2 insulation blanket traps heat, resulting in increased warming on the earth. Carbon dioxide, although the most significant, is not the only gas that contributes to the greenhouse effect. Chlorofluorocarbons used in refrigeration and cooling units, as well as nitrous oxide, also trap heat.

One benefit of elevated temperatures would be extended growing seasons and increased crop productivity in many areas of the world. However, scientists are concerned that even a small increase in temperatures may eventually cause the polar ice caps to melt. The result would be a dramatic increase in sea levels and possible flooding of coastal cities around the world. Warmer temperatures could also extend the life spans and habitats of insects that destroy crops and spread diseases normally restricted to tropical and subtropical areas.

In contrast, the burning of fossil fuels increases the particulate matter in the atmosphere. Such debris reflects light from the sun back into space and limits the amount of light and heat reaching the earth. This can lower the atmospheric temperature. An example of this effect was the eruption of the Mexican volcano El Chichonal in 1982. The vast quantities of dust released into the atmosphere by this eruption blocked a portion of the sun's rays, resulting in a localized temperature drop.

The net effect of these two opposing processes still is not known. Possibly, plants may increase their rate of CO_2 fixation in response to the higher levels of atmospheric CO_2. The oceans' ability to absorb vast quantities of CO_2 and remove it from circulation may minimize any increases in CO_2 levels. Many unresolved questions remain regarding the consequences of the greenhouse effect. What is certain is that human resourcefulness and the Industrial Revolution have significantly modified our environment and the ecosystems within it.

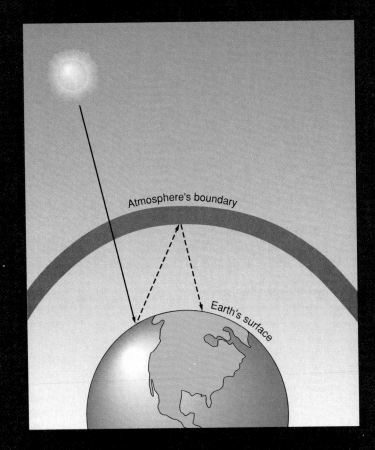

Atmosphere's boundary

Earth's surface

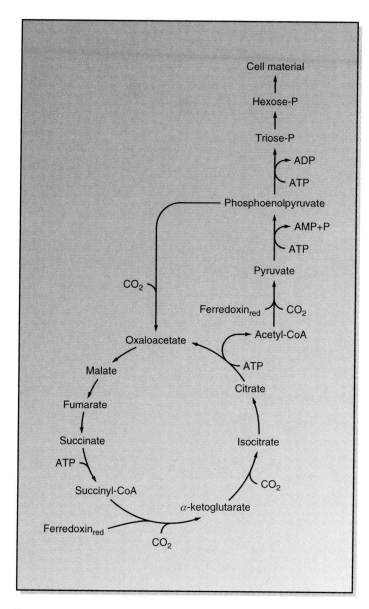

figure 19.3

The Reductive Carboxylic Acid Cycle

Carbon dioxide is fixed by a reversal of the TCA cycle in photoautotrophic bacteria.

thiosulfatophilum, a green sulfur bacterium that photosynthesizes anoxygenically in the anaerobic portions of aquatic environments and in muds, and requires ATP, NADH + H$^+$, reduced flavin, and reduced ferredoxin. Ferredoxin is reduced in a light-dependent reaction coupled with the oxidation of H$_2$S. The reduced ferrodoxin is approximately equal to H$_2$ in reducing power and thus is able to serve as an electron donor for the reduction of CO$_2$. Carboxylation of substrates occurs at four locations in the reductive carboxylic acid cycle. Reduced ferrodoxin participates in two of these carboxylation reactions (reductive carboxylation of acetyl-CoA to pyruvate and of succinyl-CoA to α-ketoglutarate) that otherwise would be thermodynamically infeasible. Both reactions in the corresponding tricarboxylic acid cycle involve essentially nonreversible oxidative decarboxylations. The other two carboxylations in the reductive carboxylic acid cycle are the carboxylation of phosphoenolpyruvate to oxaloacetate and of α-ketoglutarate to isocitrate. The reductive carboxylic acid cycle probably occurs in other photosynthetic bacteria either as a sole pathway for CO$_2$ fixation or in association with the Calvin cycle. This cycle has not yet been found to exist in photoautotrophic organisms other than bacteria.

The purple and green photoautotrophic bacteria are confined to restricted habitats because of their requirements for reduced compounds and anaerobic environments for photosynthesis. These microorganisms are, for all practical purposes, exclusively found in aquatic habitats at depths where there is a proper combination of light penetration and anaerobiosis to permit anoxygenic photosynthesis.

Some photosynthetic bacteria, called photoorganotrophs or photoheterotrophs, use light as an energy source and organic compounds as carbon sources. The purple nonsulfur bacteria *Rhodospirillaceae* are predominantly photoorganotrophs, capable of using simple organic compounds such as methanol, acetate, or formate as the sole carbon sources for phototrophic growth. These bacteria grow anaerobically in the light, using organic compounds either as carbon sources or as electron donors for CO$_2$ assimilation. Many of these bacteria are also able to grow microaerophilically to aerobically in the dark as chemoorganotrophs, using organic compounds as both energy and carbon sources.

Chemoautotrophs Use Chemical Compounds as a Source of Energy for Carbon Dioxide Fixation

Chemoautotrophic bacteria, in comparison to photoautotrophs, are widely distributed in the natural environment and occur in freshwater ponds and springs, acid drainage water discharge from mines, and soil. Nitrifying bacteria, for example, are found in soil and aerobic waters, although usually in small numbers. Sulfur-oxidizing bacteria have a more restricted habitat and reside in waters containing oxygen and H$_2$S. *Thiobacillus thiooxidans* and *Thiobacillus ferrooxidans,* which are used in mining operations, are frequently found in acid mine drainage streams (see leaching of metals from low-grade ores, page 582).

Unlike photoautotrophs, which derive their energy from sunlight, chemoautotrophs obtain their energy from the oxidation of chemical compounds (NH$_3$, NO$_2^-$, CH$_4$, H$_2$S, H$_2$). These

(see carbon dioxide fixation, page 188). Photoautotrophy is a characteristic that exists among plants, algae, and certain types of bacteria. Chemoautotrophy, in comparison, is unique to a relatively small group of bacteria. All of these primary producers are important in CO$_2$ fixation.

Photoautotrophs Use Light as a Source of Energy for Carbon Dioxide Fixation

Besides the Calvin cycle, which is present in autotrophs, certain photosynthetic bacteria fix CO$_2$ by a pathway that is basically a reversal of the tricarboxylic acid cycle (Figure 19.3). This reductive carboxylic acid cycle was first discovered in 1966 in *Chlorobium*

chemical compounds generally arise from transformations of organic matter or from mineralization of organic substances. Chemoautotrophy accounts for only a small percentage of the CO_2 fixed; nevertheless, chemoautotrophs have a wide distribution and contribute heavily to the global fixation of CO_2.

Heterotrophic Microbes Are Also Capable of Carbon Dioxide Fixation

Heterotrophic CO_2 fixation is an important way for microorganisms to synthesize intermediates of the TCA cycle from other chemical compounds. An example of metabolic CO_2 fixation is the synthesis of oxaloacetate in heterotrophic organisms. There are two primary mechanisms by which CO_2 is metabolically fixed. The first type of mechanism is driven by the substrate phosphoenolpyruvate (PEP) in the reaction:

$$\text{Phosphoenolpyruvate} + CO_2 \xrightarrow[\text{acetyl-CoA}]{\text{PEP carboxylase}} \text{oxaloacetate} + P_i$$

A variation of this reaction is catalyzed by the enzyme PEP carboxykinase:

$$\text{Phosphoenolpyruvate} + ADP + CO_2 \xrightleftharpoons[Mg^{2+}]{\text{PEP carboxykinase}} \text{oxaloacetate} + ATP$$

In the second type of mechanism, oxaloacetate is formed from the fixation of CO_2 with pyruvate. This mechanism requires ATP to produce a biotin-enzyme-CO_2 complex that carboxylates pyruvate:

$$ATP + \text{pyruvate} + CO_2 \xrightleftharpoons[\text{biotin, } Mg^{2+}, \text{ acetyl-CoA}]{\text{pyruvate carboxylase}} \text{oxaloacetate} + ADP + P_i$$

The oxaloacetate formed by either type of mechanism is used to keep the TCA cycle functioning. Oxaloacetate can also be transformed to other intermediates in the TCA cycle or other metabolic pathways.

Methanogens Reduce Carbon Dioxide to Methane

The methanogens (for example, *Methanobacterium*, *Methanococcus*, *Methanospirillum*, and *Methanosarcina*), which comprise one of the main branches of the domain Archaea, anaerobically reduce CO_2 to CH_4. The electrons for this reduction come from H_2. Although methanogenic CO_2 reduction is a form of anaerobic respiration, methanogens do not possess a typical electron transport chain with cytochromes, quinones, and flavoproteins. The reduction of CO_2 to CH_4 nonetheless appears to involve some type of electron transport system, since ATP is synthesized in the process. This electron transport is associated with at least two cofactors

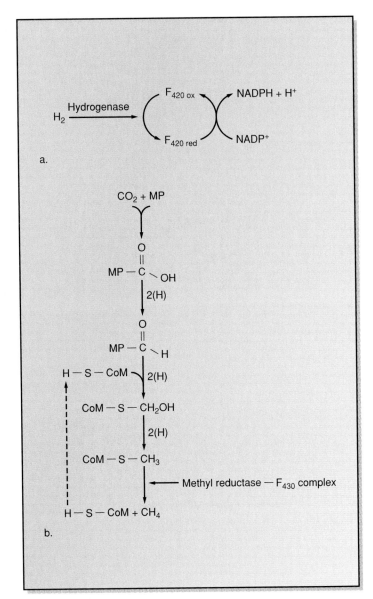

figure 19.4

Biogenesis of Methane

a. H_2 uptake is coupled with the reduction of $NADP^+$ to $NADPH + H^+$ via the electron carrier F_{420}. b. Bound CO_2 is transferred by carriers and is eventually reduced to CH_4. MP = methanopterin; H-S-CoM = reduced coenzyme M.

that are found in methanogens: Factor 420 (F_{420}) and coenzyme M (2-mercaptoethanolsulfonic acid). Two other cofactors, F_{430} and F_{342}, have been identified in methanogens and are also believed to be associated with methanogenesis.

The reduction of CO_2 to CH_4 is thought to occur in several steps, beginning with the coupling of H_2 uptake to the reduction of $NADP^+$ to $NADPH + H^+$ via the electron carrier F_{420} (Figure 19.4). Carbon dioxide is bound by F_{342}. The bound CO_2 is transferred by

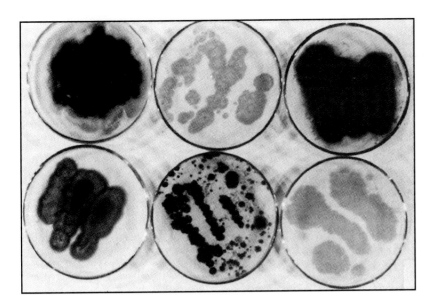

figure 19.5

Different Strains of *Polyangium cellulosum* on Filter Paper
Note the dissolution of the filter paper in areas of bacterial growth.

several carriers, including coenzyme M and F_{430}, and is eventually reduced to CH_4, with electrons for the reduction coming from either reduced F_{420} or NADPH + H^+. The methane produced by methanogens can be chemically oxidized in the atmosphere to carbon monoxide and is indirectly an air pollutant.

Methanogenic procaryotes are considered to have been among the earliest living organisms on earth, because they grow anaerobically and synthesize ATP from the reduction of CO_2 with H_2 as an electron donor—conditions of CO_2 and H_2 that prevailed on the primitive earth. Methanogens are found in anaerobic habitats rich in organic matter that nonmethanogenic procaryotes ferment to CO_2 and H_2. These habitats include swamps, marshes, marine sediments, the intestinal tract and rumens of animals, and the anaerobic sludge of sewage treatment plants.

The amount of CO_2 fixed by heterotrophs and methanogens is quite small. Autotrophs, particularly photoautotrophs, are responsible for most of the conversion of CO_2 to organic carbon. Approximately 75 billion metric tons of carbon are fixed annually by this process. The carbon fixed by autotrophic organisms is ultimately returned to the atmosphere by action of consumers and decomposers.

Herbivores and Carnivores Contribute to the Carbon Cycle by Consuming Organic Carbon and Returning It to the Atmosphere as Carbon Dioxide

Herbivores and other organisms, including microorganisms, consume organic matter formed directly or indirectly by CO_2 fixation. Herbivores, in turn, serve as food sources for carnivores. Part of the organic carbon consumed by these animals is returned to the atmosphere as CO_2 through respiration. The remaining carbon is either excreted as waste products or remains fixed in the tissues until death. Decomposers degrade the carbon in fecal waste and dead or decaying organic matter and recycle it into the atmosphere in the form of CO_2. This completes the carbon cycle.

Microorganisms Contribute to the Carbon Cycle as Decomposers

Microorganisms are important at all stages of the carbon cycle, but they are particularly important as decomposers. The biological degradation of organic materials is not a simple process; many different types of bacteria and fungi must participate. The recycling of organic carbon to CO_2 occurs either aerobically (respiration) or anaerobically (fermentation, anaerobic respiration). The type of degradation depends on the material degraded, the environment, and the decomposers. Most organic matter is readily degraded by heterotrophic microorganisms, but some requires more specialized degradation.

Cellulose Is Hydrolyzed to Glucose

Cellulose, the major constituent of plant cell walls, is degraded specifically by such bacteria as *Clostridium*, *Cytophaga*, *Polyangium*, *Ruminococcus*, *Bacteroides*, *Nocardia*, and *Streptomyces* and by the fungi *Aspergillus* and *Trichoderma* (Figure 19.5). Cellulose is a polymer of glucose containing as many as 15,000 glucose units bound together by β(1, 4) linkages. Microorganisms decompose cellulose by first hydrolyzing the insoluble polysaccharide to the water-soluble dimer cellobiose with extracellular cellulases. The resultant cellobiose is further hydrolyzed by the enzyme cellobiase to glucose, which then is absorbed by the decomposer or remains in the external carbon pool.

Many fungi degrade cellulose for its carbon and energy. This is in sharp contrast to the bacteria, a group in which cellulose decomposition is relatively rare. Cellulose-decomposing bacteria include aerobic, mesophilic soil organisms (for example, *Cytophaga*, *Polyangium*, *Nocardia*, and *Streptomyces*), anaerobic spore-formers (for example, *Clostridium*), and bacteria residing in the rumen (for example, *Clostridium*, *Ruminococcus*, and *Bacteroides*). Some cellulolytic soil bacteria and fungi are plant pathogens. The cellulases they excrete break down the cell walls of plants and facilitate the penetration of the microorganism into the host plant. Cellulolytic rumen bacteria degrade cellulose into products that frequently are used as substrates by other rumen microorganisms (see rumen microorganisms, page 427). Although these bacteria constitute only a small proportion (5% or less) of the total microbial population in the rumen, they are critical in controlling not only the supply of substrates for other rumen microorganisms, but also the rate of breakdown of food in the host animal's diet and thus the whole digestion process.

Chitin Is Hydrolyzed to N-Acetylglucosamines

Chitin (poly *N*-acetylglucosamine) is a polysaccharide found in the exoskeletons of insects and crustacea and in the cell walls of many fungi. This polymer is similar in structure to cellulose, with an acetylamino ($-NHCOCH_3$) unit replacing one of the hydroxyl groups of each glucose in cellulose. The breakdown of chitin involves extracellular chitinases that hydrolyze this tough, rigid, insoluble polysaccharide into water-soluble *N*-acetylglucosamines and glucosamines, which are absorbed by chitin-decomposing organisms and are metabolized via glycolysis.

Chitin decomposition is brought about mostly by the actinomycetes (for example, *Streptomyces* and *Nocardia*) and other soil and marine bacteria (for example, *Cytophaga*, *Pseudomonas*, *Vibrio*, *Clostridium*, and *Bacillus*) and fungi (for example, *Mortierella*, *Trichoderma*, and *Verticillium*). Arable soils may contain up to 10^6 chitinoclastic microorganisms per gram of soil. The actinomycetes generally are the predominant chitin digesters in such soils, provided that the soil is amply aerated to support the growth of these aerobic bacteria. The chitin-containing cell walls of chitinoclastic fungi would appear to be susceptible to the action of chitinases excreted by these organisms. However, such digestion is minimized by the presence of a second major polysaccharide, glucan (composed of D-glucose units). Only cells containing both chitinase and glucanase are lysed by the action of these enzymes.

Some Synthetic Chemicals Are Resistant to Microbial Degradation

The metabolic diversity of microorganisms makes them excellent decomposers. Certain kinds of natural compounds are decomposed slowly (for example, the decomposition of lignin to humus), whereas some are nondegradable by microorganisms. Many types of synthetic chemicals, particularly pesticides such as 1,1,1-trichloro-*bis*-(*p*-chlorophenyl)-ethane (DDT), are highly resistant to microbial degradation. This resistance to degradation, or **recalcitrance,** can have many possible causes. Recalcitrance may be due to the complexing of certain chemical groups (for example, aromatic rings or halogens) to a compound to render it resistant to enzymic attack, the inability of the compound to be assimiliated by the microbial cell, or the failure of the compound to elicit a degradative enzyme response.

Recalcitrant compounds accumulate and can contaminate portions of the food web and persist in the tissues of animals. This is a problem in ecosystems. Fortunately, the chemical industry has introduced compounds with improved biodegradability. For example, the substitution of methoxy groups for the two chloro groups in the basic structure of DDT resulted in formation of 1,1,1-trichloro-2,2-*bis*-(*p*-methoxyphenyl)-ethane, a synthetic chemical substance that is less resistant to microbial degradation.

The Nitrogen Cycle

Molecular nitrogen (N_2) is the most abundant gas on earth and constitutes nearly 80% of the gases in the earth's atmosphere. Despite the abundance of elemental atmospheric nitrogen, the combined forms of nitrogen (those found in inorganic and organic compounds) are relatively scarce in the soil and water. Nitrogen is an important constituent of proteins, nucleic acids, and many coenzymes. Few living organisms, however, can directly use elemental atmospheric nitrogen. Therefore the cyclic transformation of molecular nitrogen to combined nitrogen must occur in what is called the **nitrogen cycle** (Figure 19.6). Each year approximately 10^8 to 10^9 tons of nitrogen are transformed by the nitrogen cycle. The biogeochemical recycling of nitrogen requires microbial activities and consists of four stages: nitrogen fixation, ammonification, nitrification, and denitrification.

Only Certain Procaryotes Can Fix Nitrogen

Atmospheric nitrogen is inert—its triple bonds ($N\equiv N$) are extremely difficult to break and require much energy. Because of its high energy requirements, the fixation of atmospheric nitrogen into combined nitrogen by living organisms is a process restricted to only certain procaryotes. Nitrogen is also fixed abiotically by lightning discharges, ultraviolet radiation, electrical sparking, artificial combustion, and industrial production of fertilizers. However, these nonbiological processes account for only a very small portion of all nitrogen fixation. Over 150 million metric tons of nitrogen are fixed each year, and of this amount, more than 90% are biologically fixed.

The bacterial fixation of nitrogen occurs by symbiosis (mutually beneficial associations between bacteria and the roots of specific host plants) or nonsymbiotically with free-living bacteria. Given that it takes 12 to 24 ATP molecules to produce a single molecule of ammonia during biological nitrogen fixation, the process is not widespread among procaryotes but does occur among different Bacteria as well as Archaea (for example, the methanogens). The basic mechanism for molecular nitrogen fixation is similar in both symbiotic and nonsymbiotic nitrogen fixation (see nitrogen fixation, page 418).

The most extensively studied symbiotic nitrogen fixers are *Rhizobium* and *Bradyrhizobium*, which form associations with leguminous plants. In terrestrial habitats, these symbiotic bacteria fix nitrogen at rates of two to three orders of magnitude higher than rates shown by free-living, nitrogen-fixing soil bacteria. In aquatic environments, cyanobacteria (for example, *Anabaena* and *Nostoc*) fix nitrogen at ten times the rate exhibited by free-living, nitrogen-fixing soil bacteria.

Bacteria that fix nitrogen nonsymbiotically contribute only a small amount to the total nitrogen fixed annually. *Clostridium pasteurianum*, the first nitrogen-fixing microbe to be discovered, fixes

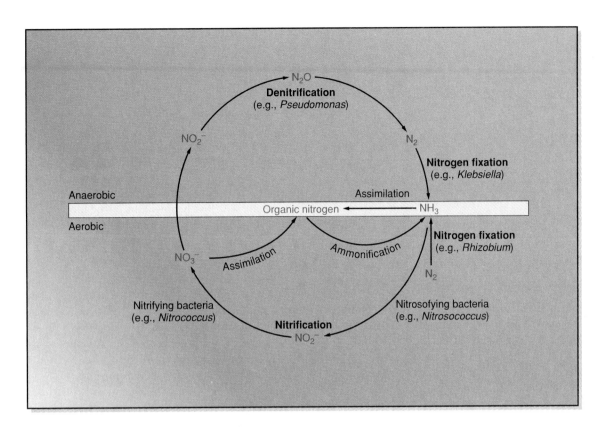

figure 19.6
The Nitrogen Cycle

figure 19.7
A Bloom of Cyanobacteria on Onondaga Lake, New York

less than 1/1000 the amount of nitrogen per year than is produced in a symbiotic leguminous association. Despite this modest amount of fixation, considerable attention has been given to non-symbiotic nitrogen fixation in soil because of the importance of nitrogen to crop production. Nonsymbiotic nitrogen fixers are a diverse group and include obligate anaerobes (*Clostridium* and *Desulfovibrio*), facultative anaerobes (*Bacillus, Klebsiella,* and *Enterobacter*), aerobes (*Azotobacter, Azospirillum, Azomonas,* and *Beijerinckia*), and phototrophs (*Chromatium, Chlorobium,* and *Rhodospirillum*). In most cases, inefficient ATP production, utilization, or both limits the amount of nitrogen fixed by these bacteria.

The cyanobacteria are an exception. Because they use radiant energy to produce ATP, they are not under the same energy restrictions as other, nonphototrophic bacteria when light is available. The cyanobacteria are therefore significant contributors to soil nitrogen as fixers of CO_2 and molecular nitrogen, and are often the first organisms to colonize isolated or infertile areas (Figure 19.7).

Ammonia Is Formed During Decomposition of Organic Nitrogenous Compounds

Combined nitrogen taken up by organisms eventually finds its way into proteins, nucleic acids, and other organic matter. These organic nitrogenous compounds ultimately are decomposed. The first step in decomposition is the formation of ammonia (NH_3) or ammonium ions (NH_4^+). This process, known as **ammonification,** occurs in bacteria, plants, and animals. An example of ammonification is the decomposition of the common excretory product urea to ammonia and carbon dioxide:

$$O=C \begin{matrix} NH_2 \\ \\ NH_2 \end{matrix} + H_2O \xrightarrow{urease} 2\,NH_3 + CO_2$$

$$2\,NH_3 + 2\,H_2O \longrightarrow 2\,NH_4^+ + 2\,OH^-$$

Ammonification is common in well-aerated soils rich in nitrogenous compounds. Soil bacteria and fungi are important in ammonification. Most of the released ammonium ions are immediately recycled into organic matter and assimilated by living organisms, thereby continuing the transfer of nitrogen within the nitrogen cycle.

Ammonia Is Oxidized to Nitrate in a Process Known as Nitrification

Ammonia that is not incorporated into organic compounds or lost in the atmosphere is oxidized to nitrate, a process known as **nitrification.** Nitrification actually occurs in two steps. The first step is the oxidation of ammonia or ammonium ions to nitrite:

$$NH_3 \text{ or } NH_4^+ \xrightarrow{\text{nitrosofying bacteria}} NO_2^-$$

The second step completes nitrification when nitrite is oxidized to nitrate:

$$NO_2^- \xrightarrow{\text{nitrifying bacteria}} NO_3^-$$

Nitrification is limited primarily to bacteria that are autotrophic. These organisms use the energy derived from the oxidation of ammonia or ammonium ions to fix carbon dioxide into organic matter. Different groups of bacteria carry out the two separate steps of nitrification. Bacteria known as **nitrosofying bacteria** oxidize ammonia or ammonium ions to nitrite. Examples are bacteria of the genera *Nitrosomonas, Nitrosococcus, Nitrosolobus,* and *Nitrosospira.* The second part of nitrification, the oxidation of nitrite to nitrate, is performed by **nitrifying bacteria,** which include the genera *Nitrobacter, Nitrococcus,* and *Nitrospina.* A few heterotrophs such as *Arthrobacter* and the fungus *Aspergillus* are also capable of nitrification. Heterotrophic nitrification, however, has only a minor role in the oxidation of nitrogenous compounds.

Nitrification is strictly an aerobic process that occurs optimally under conditions of neutral pH. The biological formation of nitrites and nitrates in soil can be both beneficial and detrimental. Although plants can utilize ammonium as a nitrogen source, nitrate is more effectively assimilated. Nitrifying bacteria thus enrich the soil with nitrates and reduce the need for expensive commercial fertilization.

The nitrates that are produced by nitrification, however, are not readily retained in the soil. Unlike the positively charged ammonium ions, which readily bind to negatively charged clay particles in soil, the negatively charged nitrates rapidly wash away in rainwater during storms. These nitrates, as well as nitrites, can infiltrate the soil and reach groundwater, where they can constitute a health hazard if they accumulate to high levels. Nitrites in such groundwaters can be converted to carcinogenic nitrosamines. Nitrates in drinking water can also be a problem if they are reduced to nitrites in the stomach. The normal human adult stomach has an acid pH and thereby minimizes the chances of such reduction. The infant stomach, however, has a more neutral pH, which can result in the reduction of nitrates to nitrites. These nitrites can combine with the hemoglobin in the blood of these infants and result in decreased oxygen flow and respiratory distress.

Denitrification Completes the Recycling of Nitrogen in the Nitrogen Cycle

The nitrates that are formed as a result of nitrification are later reduced either to NH_3 or to N_2. If NH_3 is produced, the process is known as nitrate ammonification. This process is widespread among bacteria, but because nitrate ammonification does not result in the formation of nitrogen gas, it is not as important as **denitrification** in the recycling of nitrogen.

Denitrification is the formation of gaseous nitrogen (N_2), as well as possibly the intermediates nitric oxide (NO) and nitrous oxide (N_2O), from nitrite or nitrate. The various products formed during denitrification are released to the atmosphere or water. Nitrogen gas, the major product of denitrification, is made available once again for nitrogen fixation. Nitric oxide and nitrous oxide, although not produced as frequently as nitrogen gas, react with ozone in the atmosphere and reduce the ozone layer. The ozone protects the earth from excessive solar ultraviolet radiation—a reduction in its size could have detrimental effects. Therefore discharges of nitric oxide and nitrous oxide, predominantly from industrial and automobile emissions, are closely monitored by environmental agencies.

Oxygen inhibits the synthesis of nitrate reductase, the first enzyme in denitrification, and therefore the process of denitrification is strictly an anaerobic process. Under anaerobic conditions, nitrate serves as an alternative electron acceptor to oxygen, and the nitrate is reduced to nitrogen gas. This process occurs only in bacteria and is termed **dissimilatory nitrate reduction.** This is an example of anaerobic respiration. In **assimilatory nitrate reduction,** nitrate is reduced to ammonia, which is then assimilated into organic compounds (see assimilatory nitrate reduction, page 82). The same kind of enzyme, nitrate reductase, is responsible for the first step in both dissimilatory and assimilatory nitrate reduction. The dissimilatory enzyme, however, is membrane bound and oxygen sensitive, whereas the assimilatory enzyme is soluble and sensitive to high levels of ammonia.

Pseudomonas is the most common denitrifying organism. Other bacteria, including *Achromobacter, Alcaligenes, Bacillus, Flavobacterium,* and *Thiobacillus,* also denitrify. Because the nitrate reductase in denitrification is oxygen sensitive, denitrification generally occurs in environments with low levels of oxygen, such as waterlogged soils.

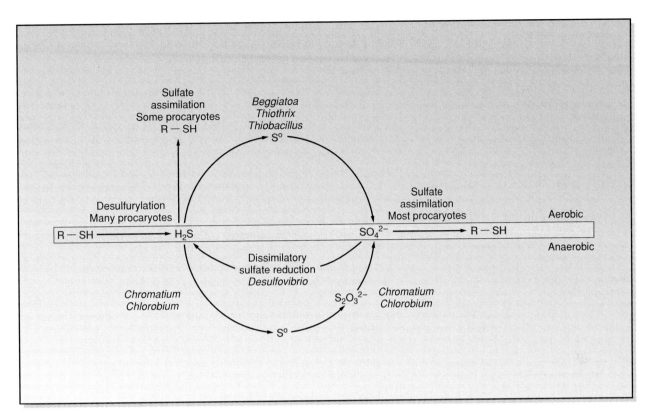

figure 19.8
The Sulfur Cycle

Denitrification is detrimental to soil fertility, since it removes the important plant nutrient nitrate. Nonetheless, denitrification is an important component of the nitrogen cycle, providing a continuous flow of nitrogen in the ecosystem.

Microbial soil activities can benefit farming. For example, plowing fields creates an aerobic soil environment that favors ammonification and nitrification and discourages denitrification, thereby maximizing the amount of ammonium ions and nitrate in the soil.

The Phosphate Cycle

Phosphorus is an important element in living cells. It is found in cells not as the volatile, free, uncombined element, but rather as organic phosphate complexes or phosphate ions. It forms the backbone of nucleic acids and is an integral part of the high-energy bonds in ATP, phosphoenolpyruvate, acetyl phosphate, and other phosphorylated, high-energy compounds. Plasma membranes contain phosphorus in the form of phospholipids, and many coenzymes, including NAD and FAD, have phosphorus as part of their structures.

Phosphorus is a reactive element and does not occur in the uncombined form in nature. The element is generally associated with rocks or minerals as insoluble salts of calcium, iron, magnesium, and aluminum. Rocks common in Florida, Tennessee, Montana, and Idaho, for example, typically contain apatite, a calcium phosphate salt. The human body is estimated to

contain approximately 500 g of phosphorus, mostly in the form of calcium phosphate in bones and teeth.

Although phosphorus is abundant in the ecosystem, it is a limiting factor for the growth of many procaryotic and eucaryotic organisms, since much of this phosphorus is bound in insoluble salts. This is particularly evident in aquatic environments, where the insoluble phosphate salts precipitate into the sediment, resulting in a reduction of algal biomass.

Microorganisms solubilize phosphate salts as part of the **phosphorus cycle** and make the phosphorus available for themselves and other organisms. The solubilization of phosphate salts usually occurs by the action of organic acids produced during microbial metabolism. Solubilized phosphates can then be assimilated by microorganisms and other organisms in the vicinity. Upon death and decay of the organism, the phosphates once more enter the environment and are recycled.

The Sulfur Cycle

Sulfur, unlike phosphorus, is abundant and widely distributed in nature. Reservoirs of sulfur include fossil fuels, elemental sulfur deposits, rocks, and minerals. Sulfur naturally occurs in three different and common oxidation states: 0 (S^0, or elemental sulfur), -2 (S^{2-}, or inorganic sulfides), and $+6$ (SO_4^{2-}, or sulfates). Transformations among these states that occur aerobically and anaerobically make up the **sulfur cycle** (Figure 19.8).

Sulfur is an essential component of living cells because it occurs in amino acids (cysteine, cystine, and methionine), coenzymes (thiamine, biotin, and coenzyme A), and other cellular compounds. Plants and other macroorganisms assimilate inorganic sulfur-containing compounds, but only microorganisms are able to use elemental sulfur. Microorganisms (particularly phototrophs and chemolithotrophs) are capable of converting sulfur from one form to another and thus are indispensable in sulfur recycling.

Elemental sulfur and sulfides are used as energy sources by some chemolithotrophs. Bacteria such as *Beggiatoa* and *Thiothrix,* usually found in sulfide-rich environments, use hydrogen sulfide as a source of electrons in their metabolism. The hydrogen sulfide is oxidized to elemental sulfur, which either accumulates in the cell or is further oxidized to sulfate. Although *Beggiatoa* was the bacterium used by Sergei Winogradsky to formulate his concept of chemolithotrophy (see chemolithotrophy, page 80), many strains of *Beggiatoa* cannot be grown under strictly autotrophic conditions. These strains obtain their energy from the oxidation of hydrogen sulfide but derive their carbon from organic compounds and therefore are examples of lithotrophic heterotrophs, or **mixotrophs.** Mixotrophs such as *Beggiatoa* are unable to grow on a completely inorganic medium with reduced sulfur compounds and CO_2, but can grow if acetate or some other assimilable organic compound is added to the medium as a carbon source. *Beggiatoa* will also grow in the absence of reduced sulfur compounds, using acetate as the sole carbon and energy source.

Thiobacillus thiooxidans is an acidophilic bacterium that often grows on sulfur contaminants of coal deposits. It aerobically oxidizes elemental sulfur and sulfides (for example, FeS_2, iron pyrite, or fool's gold) to sulfates by the following reaction:

$$FeS_2 + H_2SO_4 + 1/2\, O_2 \rightarrow FeSO_4 + 2\, S^0 + H_2O$$

$$2\, S^0 + 2\, H_2O + 3\, O_2 \rightarrow 2\, H_2SO_4$$

As a consequence of this reaction, sulfuric acid is leached from coal deposits into mine water, resulting in the acidification of water and the surrounding soil. This thick acid leachate leaves a yellow stain, called "yellow boy" by U.S. miners. Acid induction by microbes seriously damages the environment and is a major concern of environmentalists. Microbial sulfur oxidation can also contribute to a lowering of soil pH, and elemental sulfur is sometimes added to alkaline soil to lower its pH. Sulfur-oxidizing bacteria are quite active in soils and aquatic habitats, and the sulfates produced from their activity can be assimilated by other organisms. Within soils, sulfur oxidation produces large amounts of sulfuric acid, which solubilizes and mobilizes phosphates and other minerals that are beneficial to microorganisms and plants.

Sulfates also form during the anaerobic oxidation of sulfides and thiosulfates by purple and green phototrophic bacteria. These photosynthetic bacteria use H_2S or S^0 instead of H_2O as reducing agents for CO_2 fixation. *Chromatium* and *Chlorobium* form elemental sulfur as an intermediate product during such oxidation and deposit it either intracellularly, in the case of *Chromatium,* or extracellularly, in the case of *Chlorobium.*

The sulfates produced by aerobic or anaerobic oxidations are either assimilated into cellular proteins and other chemical compounds or reduced to hydrogen sulfide (H_2S). **Dissimilatory sulfate reduction** to H_2S commonly occurs during the degradation of amino acids. Furthermore, some bacteria use sulfate as a terminal electron acceptor in anaerobic respiration. *Desulfovibrio* is an obligate anaerobe common in aquatic environments that uses sulfate as a terminal electron acceptor. Recently *Desulfovibrio* has been used in bioremediation programs to treat acid mine wastes. Sulfate reduction by *Desulfovibrio* neutralizes the acid water and precipitates out metals in addition to removing the sulfates.

Sulfides that form after the reduction of sulfates, the degradation of amino acids, or other processes can react with metals to form insoluble metal sulfides. Sulfides also are recycled by chemolithotrophs and phototrophs into elemental sulfur and sulfates in the sulfur cycle.

A large portion of the sulfur in the biosphere is found in fossil fuels. When fossil fuels are burned, the sulfur is oxidized to sulfur dioxide (SO_2), a toxic substance. In the atmosphere, SO_2 combines with H_2O to form sulfurous acid (H_2SO_3). The resulting acid rain causes accelerated corrosion and is a health hazard, particularly for people with respiratory problems. Because of this, pollution standards in recent years have included restrictions on sulfur dioxide emissions.

Microbes and Soil

Most people think of microorganisms as agents of disease. However, microbes play other important roles in the **biosphere**—that portion of the earth (consisting of soil, water, and air) on which life exists. Microorganisms are intimately associated with each of the components of the biosphere and significantly influence the composition and quality of this environment. These microbial activities can be either beneficial or detrimental. Microorganisms help in the decomposition of toxic wastes and environmental pollutants, in the efficient utilization of limited natural resources, and in transformations of chemical substances that can be used by other organisms. Microbes, however, can also excrete toxic products that are harmful to the biosphere.

The renowned soil microbiologist Jacob Lipman (1874–1939) once remarked that "a soil devoid of microorganisms is a dead soil." Lipman was acutely aware of the important role of microorganisms in the environment, and his statement accurately describes the close association of microbes with soil. Fertile soil contains a wider variety of microorganisms than is found in other types of environments, and it is this diversity that makes soil an ideal habitat for other organisms.

Distribution of Microorganisms in Soil

| Depth (cm) | Organisms per Gram of Soil | | | | |
	Aerobic Bacteria	Anaerobic Bacteria	Actinomycetes	Fungi	Algae
3–8	7,800,000	1,950,000	2,080,000	119,000	25,000
20–25	1,800,000	379,000	245,000	50,000	5,000
35–40	472,000	98,000	49,000	14,000	500
65–75	10,000	1,000	5,000	6,000	100
135–145	100	400	—	3,000	—

From Martin Alexander, *Introduction to Soil Microbiology,* 2nd ed. Copyright © 1977 John Wiley & Sons, Inc., New York. Reprinted by permission of John Wiley & Sons, Inc.

Soil Varies in Its Composition and Physical Properties

Soil is that portion of the earth's crust consisting of organic and mineral matter and capable of supporting life. The soil layer of the earth ranges in thickness from a few inches to several feet. The composition and physical properties of soil vary, and depend on geographic location, geologic age, and the biological inhabitants.

Climate and availability of water are two of the geographic factors that influence the characteristics of soil. Fertile soils generally are located in areas of moderate climates and abundant water supplies. Here, there are concentrations of organic matter from decomposing plants, animals, and microbes. Microorganisms that decompose are important in the recycling of chemical elements and soil rejuvenation.

Soil characteristics also depend on the geologic age of the land. Geologic eras are delineated by periods of erosion, deposition of continents and mountains, and changes in the mineral and chemical composition of the earth. Such changes significantly influence the composition of the soil—particularly its mineral content, which is derived from the weathering of rocks. As soil ages, it is altered by climate, vegetation, and physical forces. Geologically young soil has a greater tendency than older soil to reflect its parent rock materials.

As a consequence of these geographic and geologic factors, soils differ in their physical properties and chemical content. Soils vary in temperature, pH, humidity, oxygen level, and organic concentrations. The extent of these variations depends upon the soil's location, age, and environment. Soil differences affect the numbers and types of macroorganisms and microorganisms that inhabit the soil. Microorganisms, because of their diversity and resistance, adapt more easily to changes in soil conditions than macroorganisms, so microbes typically are found in high numbers in soils.

There Are Considerable Numbers and Varieties of Microorganisms in Soil

Many kinds of bacteria, fungi, algae, and protozoa are found in soil (Table 19.2). Because soils differ in their chemical and physical properties, the types and quantities of microorganisms in them also differ. Soils that are located in temperate environments and are rich in organic matter generally provide more favorable habitats for microbes. A fertile soil will display a vertical variation of microorganisms, with the greatest number and greatest diversity of organisms in the upper, aerobic, organically rich surface layer rather than in the lower, relatively anaerobic, less-fertile layers.

Soil Bacteria Are Diverse in Their Metabolism

Bacteria are numerically the dominant microorganisms in soil and are responsible for many of the biochemical changes in soil. The most common soil bacteria are *Arthrobacter, Bacillus, Pseudomonas, Agrobacterium, Alcaligenes, Flavobacterium, Streptomyces,* and *Nocardia.* Although the majority of these bacteria are aerobes or facultative anaerobes, obligate anaerobes such as *Clostridium* and *Desulfovibrio* are also found in soil. Anaerobes generally inhabit soils inundated with water, which provides anaerobic environments. Air in the soil is displaced by water, through which oxygen diffuses slowly.

Soil bacteria are especially noted for their diverse metabolisms, a necessary diversity because the organic nutrients in soil vary. *Pseudomonas cepacia,* for example, can utilize well over 100 different types of carbohydrates as energy and carbon sources, and *Bacillus* can utilize such substrates as starch, cellulose, and gelatin. *Arthrobacter* also is nutritionally diverse and can degrade various herbicides and pesticides, caffeine, nicotine, and phenol.

Soil generally provides a favorable habitat for the growth and proliferation of heterotrophic microorganisms that can use organic material found near the soil surface. The numbers of microorganisms in soil vary, depending on the type and condition of the soil. Higher numbers of microbes usually occur in soil permeated with plant roots than in soil devoid of roots. This characteristic, known as the **rhizosphere effect,** is a consequence of the excretion of organic matter by plant roots (see rhizosphere, page 418). The chemicals include compounds such as carbohydrates, amino acids, and vitamins that in many cases enhance the growth of the surrounding microbial population. This results in a higher concentration of microorganisms in the vicinity of the plant roots. The degree of the rhizosphere effect depends on the type and physiological state of the plant. Microorganisms around the roots benefit from this interaction and can provide the plant with essential nutrients through their metabolism (for example, leguminous associations).

Fungi, Algae, and Protozoa Also Occur in Soil

Fungi account for a large part of the microbial population in well-aerated, cultivated soil. Although not as numerous as the bacteria in most soils, fungi make up a significant part of the total biomass because of their large size and extensive network of filaments. Plate counts of soil fungi are misleading, since a colony appearing on agar may arise from a single spore, a long filament, a fragment of a vegetative mycelium, or other fungal structures. An estimate of fungal biomass in soil is usually determined by measuring the lengths of their mycelia.

Because fungi are heterotrophic, they depend upon oxidizable organic matter for energy and therefore are found in large numbers in rich, organic soil. Fungi also predominate in acidic soil (below pH 5.5), which bacteria generally do not tolerate. The bacteria, however, proliferate in neutral and moderately alkaline soils (between pH 6 and 8), not because fungi are unable to grow at these pHs, but because they cannot effectively compete against the rapidly growing, numerous bacteria. *Penicillium* and *Aspergillus* are among the most common fungi isolated from soil.

Eucaryotic algae and cyanobacteria are found in large numbers in the upper layers of soil, where water is abundant and light is accessible. Under such circumstances these photosynthetic microbes contribute to the organic matter of the soil. Photoautotrophic algae usually are the first organisms to appear in barren or denuded areas (for example, after a massive burning or volcanic eruption). As the algae die, they provide a source of organic carbon for plants, and vegetation begins to appear again. Environmental factors have only a minor influence on the distribution of algae, since, unlike other organisms, algae do not require a source of organic carbon because they are autotrophic. Light accessibility is a limiting factor in the distribution of algae, which is reflected in the decline in algal growth and numbers below the soil surface. Algae are found at soil depths of 1 m or more, but light does not penetrate at these depths, so they grow heterotrophically or are metabolically inactive.

Protozoa are found in greatest abundance near the surface of soil (within the top 15 cm), where bacteria and other microorganisms occur in large numbers and provide an adequate food supply on which protozoa prey. The water availability and amount of organic matter from decaying vegetation are also greatest at the soil surface and undoubtedly contribute to the density of protozoa. Protozoa consume large numbers of bacteria and appear to limit the density of bacteria in soil. Flagellated protozoa (for example, *Allantion, Bodo, Cercobodo,* and *Tetramitus*) dominate the flora of terrestrial habitats. Soil can also be a reservoir for pathogenic protozoa such as *Entamoeba histolytica* (the cause of amebic dysentery) and *Naegleria fowleri* (the cause of meningoencephalitis). These pathogens may contaminate vegetables fertilized with human or animal excrement and be transmitted to humans.

Different types of viruses persist in soil. Many of these are bacteriophages that infect soil bacteria; bacteriophages specific for *Agrobacterium, Azotobacter, Clostridium, Bacillus, Nocardia, Streptomyces,* and other soil bacteria have been isolated. Current evidence indicates that these bacteriophages are not numerous. Viruses that cause human, animal, and plant diseases are also found in soil and are of agricultural and public health importance. For example, hepatitis virus discharged from septic tank effluents may migrate to groundwater and present a hazard to people drinking the water. Tobacco mosaic virus and other plant viruses remain infective in soil for months or years and can cause significant economic loss to farmers; thus the detection and monitoring of such viruses in soil is important.

Bacteria Are Important in the Leaching of Metals from Low-Grade Ores

The recovery of copper from drainage waters of mines is a practice that dates back nearly 3,000 years. The central role of bacteria in extracting copper and other metals from low-grade ores, however, was not realized until only 35 years ago. The metal content of ores varies and in some cases is too low to justify normal commercial mining. Because of limited metal resources, the mining of low-grade ores is more common today than in the past.

Leaching is commercially used for the extraction of copper, lead, zinc, and uranium from sulfide-containing ores. The process utilizes the ability of a bacterium such as *Thiobacillus ferrooxidans* or *Thiobacillus thiooxidans* to oxidize these ores. *T. ferrooxidans* obtains its carbon from carbon dioxide and its energy for growth from the oxidation of either iron or sulfur. Iron in the form of ferrous ion (Fe^{2+}) is converted into the ferric form (Fe^{3+}). Different forms of sulfur, including elemental sulfur (S^0), sulfides, and thiosulfate, are oxidized in steps, with sulfate ion (SO_4^{2-}) as the final product of oxidation. *T. ferrooxidans* is acidophilic (optimum pH, 2.5 to 5.8) and generally is found in low-acid environments such as hot springs and sulfide ore deposits, which contain high concentrations of sulfuric acid. *T. thiooxidans* is similar to *T. ferrooxidans* in its characteristics and oxidation of sulfur compounds. Although metal sulfides oxidize spontaneously in the presence of oxygen, the rate of oxidation significantly increases by the presence of these bacteria. The bacteria involved in the leaching of metals from low-grade ores are among the most remarkable organisms known. They are chemolithotrophic bacteria that derive their energy from the oxidation of inorganic substances. Many of these bacteria are also autotrophic and obtain their carbon from carbon dioxide.

One of the more general applications of microbial leaching is in copper mining. Low-grade copper ores contain less than 0.5% copper in the form of either chalcocite (Cu_2S) or covellite (CuS). Copper can be extracted from these sulfide-containing ores by several methods. The most common is oxidation of the copper ore by ferric iron (Fe^{3+}) generated through the oxidative action of *T. ferrooxidans* on ferrous iron (Fe^{2+}). The source of the ferrous iron in this process is pyrite (FeS_2), a frequent contaminant of ores. The bacteria oxidize pyrite (also known as fool's gold because of its pale brass-yellow color and brilliant metallic luster) to Fe^{3+} by the general reaction:

$$8\ Fe^{2+} + 2\ O_2 + 8\ H^+ \rightarrow 8\ Fe^{3+} + 4\ H_2O$$

The Fe^{3+} is then used to oxidize covellite, for example, in a separate chemical reaction:

$$CuS + 8\ Fe^{3+} + 4\ H_2O \rightarrow Cu^{2+} + 8\ Fe^{2+} + SO_4^{2-} + 8\ H^+$$

The Fe^{2+} formed in this reaction is reoxidized to Fe^{3+} by *T. ferrooxidans,* thereby continuing the leaching process.

In commercial-scale operations for the microbial leaching of copper, crushed low-grade copper ore containing *T. ferrooxidans* is placed in a large pile. Dilute sulfuric acid is sprayed over the pile and allowed to trickle down through the ore. The spraying of the acid is an important part of the leaching process, since it introduces oxygen (required for the oxidation reaction) into the solution. The acid conditions established by the sulfuric acid promote growth of the bacteria and result indirectly in the oxidation of copper sulfide in the ore. The final copper-rich leachate is collected and further processed to purify the copper. Additional processing involves reprecipitation of the metal and recycling of the liquid. The liquid contains H_2SO_4 and Fe^{2+} and can be held in an oxidation pond, where the Fe^{2+} is reoxidized to Fe^{3+} by *T. ferrooxidans.* Rejuvenated liquid is then recycled through the ore pile.

Microbial leaching of low-grade copper ores is important in the mining industry; an estimated 25% of all copper is recovered by this process. The recovery of other metals by leaching involves similar designs and will undoubtedly become more universal as metal resources are depleted and low-grade ores are commercially mined.

Microbial leaching, however, presents problems such as the production and discharge of sulfuric acid. These discharges lower the pH of the surrounding soil and, when released into rivers and streams, injure or kill the biological inhabitants there.

A similar environmental problem is acid mine drainage. *T. ferrooxidans,* the same bacterium that is used in commercial microbial leaching, attacks sulfide mineral contaminants in coal, including pyrite and other metal sulfides. When the coal is unmined and below the ground under anaerobic conditions, this obligate aerobe is inactive. When coal is exposed to oxygen in mining, however, *T. ferrooxidans* begins to grow and oxidize the sulfide minerals. The sulfuric acid, produced by oxidation, drains from the mines and pollutes rivers and streams in the immediate vicinity. This is a serious problem in some regions of the world. The high acidity of these rivers and streams kills animal life and vegetation. There is also considerable damage from acid mine drainage to pipes, mining rails, and other equipment (Figure 19.9).

Different methods are used to control acid mine drainage. Mines can be sealed to minimize drainage and limit the oxygen needed by *T. ferrooxidans.* Alternatively, acid mine drainage can be neutralized by the addition of lime to the discharge. This is time-consuming and expensive because lime must be constantly added to maintain neutralizing power. Regardless of the control methods, acid mine drainage continues to be a major environmental problem, particularly in the northeastern United States, where commercial mining is extensively practiced.

Hydrocarbon-Degrading Microorganisms Are Potentially Useful in Oil Spill Cleanups

The oil crisis of the 1970s made the world realize that oil reserves are a limited natural resource and must be carefully conserved. The hydrocarbons in petroleum can serve as carbon and energy sources for certain microorganisms, including *Pseudomonas, Nocardia,* and *Streptomyces.* Fortunately these oil-degrading microorganisms are obligate aerobes and are unable to attack petroleum deposits buried deep in the earth. Once these deposits are recovered, they become contaminated with bacteria and fungi and are subject to gradual microbial degradation. Microorganisms proliferate at fuel-water interfaces. For this reason the petroleum industry takes special efforts to prevent seepage of water into fuel storage tanks. Under some circumstances, however, water finds its way into storage tanks through seepage, carryout from the washing of fuels, or condensation of atmospheric moisture in the headspace of storage tanks. Microorganisms then grow and cause deterioration of fuel tank linings, corrosion of metals, and formation of thick, gelatinous sludge. These problems can be minimized by using antimicrobial inhibitors in wells and storage tanks, filtrating fuel to remove any contaminating microorganisms, and carefully monitoring fuel storage tanks to prevent undue contamination by water.

Not all microorganisms are harmful to the petroleum industry. For example, they can help clean up oil spills, a serious environmental problem. Over 10 million metric tons of petroleum pollute the marine environment each year. According to the National Academy of Sciences, accidental oil spills account for less than 7% of the petroleum that contaminates the environment. The remaining oil comes from natural oil seepages, sewage discharges, storm water runoff, the flushing of ships' bilges, and normal discharges during routine petroleum industry operations. Some of the oil that is spilled in an aqueous environment evaporates by weathering processes or is degraded by photochemical oxidation. A large portion of the remaining oil is removed by

figure 19.9
Acid Mine Drainage from a Strip Mine

microbial oxidation. Natural oxidation of oil by microorganisms is slow. No single microbe possesses all the enzymes required to degrade the different types of hydrocarbons that might be present in a petroleum mixture. Furthermore, seawater contains suboptimal levels of nitrogen and phosphorus to support rapid growth of microorganisms. These problems are overcome by inoculating oil spills with different types of oil-degrading microbes and adding nutrients to promote microbial growth.

Recently, strains of bacteria have been used to specifically degrade hydrocarbons in accidental oil spills. One such bacterium, *Pseudomonas putida*, carries genes on plasmids that code for enzymes that rapidly and effectively degrade oil components. Although bacterial degradation of oil is still of limited use in oil cleanup, use of this bacterium under controlled conditions provides scientists with an important biological tool to assist in cleansing an oil-contaminated environment. *P. putida* is the first living organism to be patented by the U.S. Patent and Trademark Office. The recognition of a microorganism as a patentable product of biotechnology has far-reaching economic consequences.

Microorganisms Are Useful as Biological Pesticides

The use of pesticides to control insects and other pests that endanger food crops has dramatically improved productivity of our farmlands. However, although chemicals are generally effective in pest control, there are many problems in their use. Most chemical pesticides have broad lethal effects and harm beneficial insects as well as pests. An increased use of pesticides leads to the development of resistance in insects and the potential for immediate or long-term harm to other organisms, including humans. Some pesticides are resistant to biological degradation (recalcitrance) and persist in the soil for many years. DDT, once a commonly used insecticide, remains in the soil for ten years. Chlordane, another insecticide, persists for over 12 years. The accumulation of these toxic chemicals presents serious hazards to the inhabitants of environments in which these chemicals are found. Toxic chemicals also may seep into the groundwater and eventually to water bodies.

For these reasons microorganisms are seen as the potential pesticides of the future. Some microorganisms have narrow host

table 19.3

Representative Biological Pesticides

Pesticide	Pest Organism/Disease
Bacteria	
Bacillus lentimorbus	Japanese beetle/milky white disease
Bacillus popilliae	Japanese beetle/milky white disease
Bacillus thuringiensis	Alfalfa caterpillar
	Blackfly/vector for river blindness
	Cabbage worm
	California oakworm
	Corn earworm
	Cotton bollworm
	Gypsy moth
	Mosquito/vector for malaria
	Tobacco budworm
Enterobacter aerogenes	Locust and grasshopper
Fungi	
Arthrobotrys dactyloides	Pineapple root knot nematode (Hawaii)
Beauveria bassiana	Codling moth
	Colorado beetle (Russia)
	Corn earworm
Hirsutella	Citrus rust mite (Russia)
Metarrhizium	Leafhopper (Brazil)
Protozoa	
Microsporidia	Silkworm/pébrine
Viruses	
Baculovirus	Corn earworm
	Cotton bollworm
	Soybean pod-feeding caterpillar
	Spruce and pine sawflies
Myxoma virus	European rabbit

ranges and can be used to selectively kill or injure specific pests. Insects rarely develop resistance to these microbes. Few microbial pesticides are toxic or pathogenic to humans, animals, or plants. More than 1,500 microorganisms or microbial by-products are potential pesticides. Microorganisms that could be used as pesticides include bacteria, fungi, and viruses, although most have not been developed for commercial use (Table 19.3).

One bacterium, *Bacillus thuringiensis*, is marketed successfully as an insecticide for use on crops, trees, ornamental plants, and stored grain products. This gram-positive, spore-forming bacillus is effective as an insecticide because of a parasporal glycoprotein crystal that is synthesized within the organism during its sporulation cycle. The glycoprotein crystal is actually a protoxin (a toxin precursor) that is activated after ingestion of the bacterium by a susceptible insect. *B. thuringiensis* is particularly effective in the control of such pests as the cabbage worm, bollworm, alfalfa caterpillar, tobacco budworm, and gypsy moth. Humans and other animals are unaffected by the protoxin, probably because gastric acids or proteases degrade the protoxin to a nontoxic form.

Other *Bacillus* species cause milky white disease in Japanese beetles, which feed voraciously on plants and can cause extensive economic losses. *Bacillus popilliae* and *Bacillus lentimorbus* infect the larvae of Japanese beetles, causing them to turn a milky white; thus these bacteria are effective in controlling this pest. Although the use of biological pesticides is not currently widespread, this method will be important in the future for the natural control of harmful pests.

Microorganisms not only serve as natural pesticides, but many heterotrophs also break down **xenobiotics,** or synthetic chemical compounds not formed by natural biosynthetic processes. Consortiums of microbes are usually needed to break down xenobiotics. For example, a consortium of soil bacteria consisting primarily of *Pseudomonas* species has been reported to rapidly mineralize the herbicide 2,4-dichloro-phenoxyacetic acid

(2,4-D). Other bacteria are capable of either cometabolizing chemicals or using them as nutrients. **Cometabolism** is a process in which a chemical is transformed but does not directly serve as a nutrient source for a microorganism. This phenomenon was first described in cultures of *Pseudomonas methanica* that grew on methane as the sole carbon and energy source but could concomitantly oxidize other compounds, including ethane to acetic acid and propane to propionic acid. Neither ethane nor propane alone supports the growth of *P. methanica,* but the two hydrocarbons can be metabolized in the presence of methane. More recently, cometabolism has been found to be a mechanism by which some bacteria are able to degrade such pesticides as DDT, heptachlor, and endrin.

Pesticides are also used as nutrients by microorganisms to support growth. The yeast *Lipomyces starkeyi* degrades paraquat (1,1′-dimethyl-4,4′-bipyridylium-2A) and uses it as a nitrogen source. Monuron (3-[*p*-chlorophenyl]-1,1-dimethylurea), another herbicide, is used as a carbon source by soil bacteria and fungi.

In both cometabolism and metabolism, pesticides are metabolically converted to nontoxic products by one of several different types of reactions: (1) degradation of a complex substrate into simple products, (2) detoxication of a chemical substance to a nontoxic compound, and (3) conjugation of the pesticide with other compounds or cell metabolites. Some pesticides are poorly biodegradable and may require degradation by more than one type of microorganism. Nonetheless, the biodegradation of pesticides and other toxic synthetic chemicals is important for the removal of these harmful substances from polluted environments.

Microbes and Water

Water covers 70% of the earth's surface and is also present in varying amounts in the atmosphere. It is an essential component of all cells and is a requirement for life. The water composition of a cell varies from 45% to 95%.

The water on the earth is constantly recycled in the **hydrologic cycle.** The cycle begins with the evaporation of water from oceans, lakes, and other water bodies into the atmosphere. Atmospheric water returns to the earth by precipitation as rain, snow, or hail. The water percolates through the soil and back into the bodies of water.

Water is of two major types: salt water and fresh water. Saltwater and freshwater organisms are quite different, and saltwater organisms differ from one another based upon osmotic properties.

A Wide Variety of Microorganisms Are Found in Marine Habitats

Salt water contains a significant level of dissolved salts. The major bodies of salt water are oceans, seas, estuaries, and certain saltwater lakes. Oceans are the largest and most varied of the salt waters; they cover an area of 361 million sq km and have a total volume of 1,347,000,000 cu km, with an average depth of 3,730 m.

The greatest recorded ocean depth is 11,000 m at the Mariana Trench, southwest of Guam. Ocean salinity ranges from 3.3 to 3.8 grams of dissolved salts per 100 grams of water, with an average of 3.5 grams per 100 grams of water. Ninety percent of the oceans' salts are salts of chlorine, sodium, magnesium, sulfur, calcium, or potassium. Other minerals such as phosphorus, nitrogen, and iron are found in low concentrations, usually at trace levels of 1 ppm or lower. The pH of salt water remains relatively constant, at a range of 8.3 to 8.5. The temperature of seawater fluctuates, depending on location, season, and depth.

Estuaries are partially enclosed coastal bodies of water that separate marine waters from inland sources of fresh waters. As a consequence of this geographical location, estuaries have salinity levels that range from less than 0.5 to 2.5 grams of dissolved salts per 100 grams of water at their mouths, where the water flows into the marine environment. Estuaries are ecologically sensitive environments, and serve as habitats for fish, fowl, and a wide variety of marine life. In recent years environmentalists have become increasingly concerned about the quality of life in estuaries seriously damaged by human impact, including overdevelopment and pollution from industrial and wastewater discharges.

Many kinds of microbes reside in marine habitats. Algae are common in these environments and provide significant organic carbon for marine life. Algae and small animals often are associated as plankton. Because of their dependence on light as a source of energy, algae occur primarily in the upper strata of the oceans and seas. Although they constitute a vital part of the food chain in marine environments, they also can be a nuisance and threat to other forms of life. Nutrient levels in such environments can significantly increase as a result of sewage plant discharges or urban/suburban runoff. Under such conditions, algal blooms are common. As these increased quantities of algae grow and eventually die, available oxygen is absorbed and lower quantities of oxygen are available for other inhabitants of the marine environment.

The bacterial population in estuaries consists of *Pseudomonas, Flavobacterium,* and *Vibrio,* as well as enteric organisms. The quantities and types of organisms vary, and depend on the tide, rainfall, salinity, depth, and temperature. Bacterial quantities are higher at low tide, probably a result of bacteria flowing from freshwater sources. Storms significantly increase the quantities of bacteria present in estuaries. Most of the bacteria found in water runoff come from animal and fowl fecal matter deposited on the ground. Sometimes overflow from sewage systems contributes to these higher numbers of bacteria in the water. The numbers and types of bacteria in water depend to a large extent on the physical parameters of the water—salinity, temperature, dissolved oxygen, and pH.

Recently there has been concern about certain pathogens discovered in marine environments, particularly *Vibrio cholerae, V. parahaemolyticus,* and hepatitis viruses. Surveys of estuaries along the Gulf coast of the United States indicate that *V. cholerae* is abundant, and there have been several cases of human cholera reported in Florida, Louisiana, and Texas. Many of these cases have been linked to the ingestion of raw or improperly cooked shellfish con-

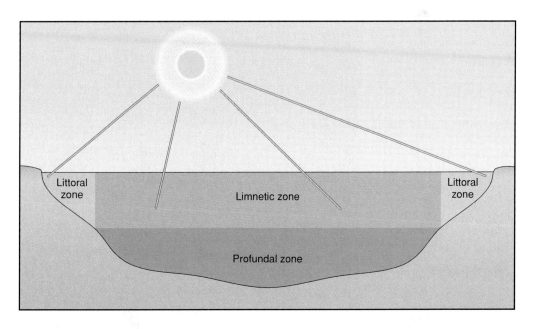

figure 19.10

Light Penetration Zones of a Freshwater Lake

Light penetrates the littoral and limnetic zones, but not the profundal zone of a lake.

taining toxigenic strains of *V. cholerae* (see *V. cholerae*, page 518). Most of the *V. cholerae* isolated from the natural environment are nontoxigenic and do not produce enterotoxin. Presently the potential threat to the public health of large numbers of nontoxigenic *V. cholerae* in estuarine environments is unknown.

Many Different Types of Microorganisms Are Also Found in Freshwater Habitats

Freshwater habitats (for example, rivers, streams, swamps, marshes, and lakes) contain a wide variety of microorganisms. Most rivers and many lakes are fed from springs. Groundwater and spring water usually have low levels of nutrients and microorganisms because of the filtration effect of soil. Organisms found in these waters include those of the genera *Micrococcus, Pseudomonas, Serratia, Flavobacterium, Chromobacterium,* and *Achromobacter.*

As these waters reach the surface or inland bodies of water, they become contaminated with different types of microorganisms. Rivers—flowing water in close contact with the soil—may contain large numbers of soil bacteria (*Bacillus, Actinomyces,* and *Streptomyces*), fungi (*Polyphagus, Penicillium,* and *Aspergillus*), and algae (*Microcystis* and *Nostoc*). These microorganisms frequently impart an earthy odor and taste to river water. Rivers also receive high concentrations of bacteria and agricultural chemicals through surface runoff water from adjoining soil during heavy rains and irrigation. In many countries, rivers are heavily polluted with sewage bacteria, especially *Escherichia coli, Enterococcus fae-*

calis, Proteus vulgaris, Clostridium species, and other intestinal bacteria. Although many pollutants are removed from rivers during their natural flow, the presence of these intestinal bacteria can be a public health hazard.

Freshwater lakes are relatively stagnant bodies of water that can be divided into zones of light penetration and temperature (Figure 19.10). Most lakes are surrounded by rooted vegetation in a large **littoral zone** along the shore. Light penetrates the shallow littoral zone and the open-water **limnetic zone** but is unable to reach the **profundal zone** in the deep portions of many lakes. Some lakes may also be stratified by temperature differences. In the summer the upper layer of a lake (the epilimnion) is warmed by sunlight, whereas the lower layer (the hypolimnion) is characterized by lower temperatures.

As lake vegetation and animal life decompose, their organic matter provides a source of nutrients. Lakes that have very high concentrations of nutrients (particularly nitrogen and phosphorus) are termed **eutrophic** and have large and active biological communities. These lakes generally have low oxygen concentrations because of extensive microbial decomposition of organic matter. In comparison, lakes that receive small amounts of nutrients and are nutrient sparse are **oligotrophic.** Some species of bacteria are especially adapted to the low levels of nutrient concentration in oligotrophic habitats but may be inhibited by higher nutrient levels. For example, the stalked bacterium *Caulobacter* grows well in oligotrophic lakes, but growth is inhibited in environments containing 1% (weight/volume) or more of organic material. *Caulobacter* has an extension of the cytoplasm called a prostheca that provides it with a greater surface area for the absorption of nutrients from the environment.

The microflora of a lake is determined to a large extent by the lake's nutrient content, thermal stratification, and light compensation level (the depth of effective light penetration). Cyanobacteria and algae are abundant in the littoral and limnetic zones but are found in fewer numbers in the profundal zone. Photoautotrophic bacteria (*Chlorobium, Rhodopseudomonas,* and *Chromatium*) are found at lower depths where light is still available but oxygen tensions are reduced. At these depths the bacteria use reduced organic and inorganic substances as electron donors. Large numbers of heterotrophs occur just below the zone of maximum photosynthetic activity, and colorless sulfur bacteria

(*Thiospira, Thiothrix,* and *Thioploca*) and sulfate-reducing organisms (*Desulfovibrio*) are found in the deeper, low-oxygen layers of lakes. Chemolithotrophic bacteria (*Nitrosomonas, Nitrobacter,* and *Thiobacillus*) are also found in freshwater bodies and contribute significantly to the cycling of nitrogen, sulfur, and other inorganic substrates.

The microorganisms in rivers, lakes, streams, swamps, and other freshwater environments are an important source of food for other aquatic organisms. Cyanobacteria and algae can be the dominant phytoplankton in freshwater habitats. These and other microorganisms frequently are the beginning of the food chain in an aquatic environment. This is particularly true in eutrophic lakes that have high nutrient concentrations and large numbers of bacteria, fungi, and algae (Figure 19.11).

Chemical and Biological Contaminants Affect the Quality of Water

Fresh water is a precious resource that must be conserved and closely monitored for chemical pollutants and microbial contamination. Less than 2% of the world's water is **potable** (suitable for drinking), and this water must serve the needs of nearly 6 billion people.

Each day in the United States, over 25 billion gallons of water are used. Only a small fraction of this water is actually consumed; the rest is used for other domestic purposes, including bathing, laundering, and irrigation. In most communities, used water that is not consumed is collected by an elaborate system of pipes and other conduits called sewers and discharged at a central point. This spent water in sewers is often joined by rainwater and groundwater, as well as industrial liquid wastes. This **sewage** is unsuitable for human consumption and use, and generally contains hazardous chemicals and pathogenic microorganisms that render it dangerous. Sewage must be treated to remove these chemicals and microbes before the water can be reused.

Two major kinds of contaminants affect the quality of water: **chemical** and **biological.** Chemical contaminants are divided into two categories: inorganic and organic. Examples of inorganic contaminants are metals such as iron, lead, manganese, cadmium, mercury, copper, and zinc. Many of these inorganic chemicals are byproducts of industrial processing or mining whereas others such as lead, copper, and zinc are often found as materials in sewer pipes and are released into water by corrosion. Pipe corrosion is frequently caused by the metabolic action of bacteria, particularly sulfate-reducing and sulfur-oxidizing bacteria. Anaerobic sulfate-reducing bacteria oxidize iron and convert sulfates to hydrogen sulfide, which in turn converts iron in sewer pipes to iron sulfide, resulting in pipe deterioration. Sulfur-oxidizing bacteria produce sulfuric acid from the oxidation of hydrogen sulfide. The sulfuric acid corrodes these pipes made of copper, zinc, or other substances.

Organic contaminants in water consist of pesticides, petroleum wastes, detergents, and industrial chemicals. Many of these organics are not easily decomposed by microorganisms and

figure 19.11
Algal Bloom in a Polluted Eutrophic Lake

therefore constitute a major problem in sewage treatment. For this reason, legislation has been introduced in recent years to reduce the quantity of nonbiodegradable substances discharged into sewage.

The most prevalent biological contaminants in sewage are microbes, particularly bacteria and viruses. The primary source is feces—both animal and human. Most of the bacteria carried by storm runoff originate from animal fecal matter. Studies have shown that during storms, the water that drains off the land and into sewage systems also carries a large quantity of bacteria and chemicals. The chemicals include pesticides applied to lawns, chemical wastes near industrial plants, and organic matter deposited on the ground by different sources.

In addition to chemical and biological contaminants, physical properties also affect the quality of biological life in water. Among these are pH, temperature, dissolved oxygen concentration, and salinity.

Safe Drinking-Water Standards

| | Guidelines | |
Constituent	Environmental Protection Agency	World Health Organization
Microorganisms		
Most probable number (MPN) procedure	<2.2 MPN/100 ml in 90% of samples per month	<2.2 MPN/100 ml in 95% of samples per year
Membrane filter (MF) procedure	1 coliform/100 ml per month	
Inorganic chemicals		
Arsenic	0.050 mg/l	0.050 mg/l
Barium	1.000 mg/l	
Cadmium	0.010 mg/l	0.010 mg/l
Chromium	0.050 mg/l	
Cyanide		0.050 mg/l
Lead	0.050 mg/l	0.100 mg/l
Mercury	0.002 mg/l	0.001 mg/l
Nitrate	10.000 mg/l	
Selenium	0.010 mg/l	0.010 mg/l
Silver	0.050 mg/l	
Organic chemicals		
Chlorinated hydrocarbons		
Endrin	0.0002 mg/l	
Lindane	0.004 mg/l	
Methoxychlor	0.100 mg/l	
Toxophene	0.005 mg/l	
Chlorophenoxys		
2,4-D	0.100 mg/l	
2,4,5-TP silvex	0.010 mg/l	

Biochemical Oxygen Demand Is One Method to Monitor Water Quality

Several different methods are used to determine the effectiveness of sewage treatment in the removal of chemicals and microorganisms from water. These are not designed to detect all chemicals and microbes, but primarily those that might constitute a danger to the public health (Table 19.4).

One common parameter used to determine the purity of water is the **biochemical oxygen demand (BOD).** Also known as the biological oxygen demand, the BOD is defined as the quantity of oxygen required to meet the metabolic demands of microorganisms oxidizing organic matter in the water. Effective sewage treatment significantly reduces the amount of organic matter, and hence the BOD, in sewage. The BOD of a water sample is determined by aerating it, measuring the amount of oxygen in the sample before incubation, placing the sample in a sealed container, and incubating the container for five days at 20°C. During this five-day period, microorganisms in the water grow and oxidize any organic materials in it. After the incubation period, the BOD of the water can be determined by measuring the quantity of residual oxygen in the container. Water that is effectively treated should have an 80%

Chesapeake Bay: Example of a Revitalized Estuary

One of the largest and economically important estuaries in the United States is Chesapeake Bay (Figure 19.12). Located off the Atlantic Ocean in eastern Maryland

protozoan parasite, *Haplosporidium nelsoni*. The parasite enters the oyster through the gills and infects the gills and digestive tract.

estuaries to their original pristine conditions, it certainly is possible to reverse many of the problems currently associated with these environments.

and Virginia, the bay is approximately 320 km long and from 5 to 50 km wide, with a total area of 8,386 sq km. Many rivers, including the Potomac, Rappahannock, York, and James, feed into the bay, which eventually opens into the Atlantic Ocean through a 19-km-wide mouth. Chesapeake Bay is not only an important habitat for marine life, but also is heavily used for commercial fishing and recreation. Oyster harvesting is an important industry of the bay; millions of dollars of oysters are harvested each year in Maryland and Virginia.

Chesapeake Bay, however, has problems. In recent years its oyster harvest has been sharply reduced from 3.5 million bushels of oysters each year in 1980 to less than half that amount today. Most of the oysters currently harvested are only empty, greenish-gray shells. The oysters have been killed by a microscopic

Many different factors have contributed to the increased population of *H. nelsoni* in Chesapeake Bay. Droughts and reduced entry of fresh water from the surrounding watershed increased salinity levels in the bay, which has promoted growth of the parasite. Elevated nutrients brought about by runoff from farmlands and sewage plants promote more extensive growth of algae and other marine plant life in the bay. As these algae and plants grow and die, they use up the oxygen that otherwise would be used by other forms of marine life, including oysters.

The net effect of these different conditions is the gradual deterioration of Chesapeake Bay. The problem is not unique to Chesapeake Bay; similar conditions, although not as severe, are occurring in other estuaries and are of concern to environmentalists. Although it may not be possible to restore

Such reversal has taken place with Chesapeake Bay. In 1983 the governors of Maryland, Pennsylvania, and Virginia joined with the mayor of the District of Columbia and the administrator of the Environmental Protection Agency (EPA) in signing the Chesapeake Bay Agreement, a pact pledging to restore the health of the bay. Since then, millions of dollars have been spent to clean up the bay by controlling waterfront development, curbing industrial dumping into the bay, and changing farming practices. Already there are signs of improvement in the bay's condition. The amount of underwater seagrasses has increased and certain types of marine life, such as the blue crab, remain abundant. Scientists and environmentalists realize that restoration of Chesapeake Bay will take many years, but these first encouraging signs of the bay's revitalization indicate that such efforts are paying off.

to 90% reduction in BOD, with a final BOD of 100 to 300 mg/l. Such reductions in the BOD usually involve several stages of sewage treatment to remove organic materials.

Concentrations of certain inorganic and organic chemicals in water can be used as indices of water quality. Common inorganic chemicals analyzed in water are lead, cadmium, mercury, and nitrate. These inorganics are determined by colorimetric methods or by instrumental analysis. Organic chemicals generally are extracted and concentrated prior to identification by chromatography or other methods. The detection of chemical pollu... ...n be expensive, particularly when tests are done ...variety of chemicals.

Indicator Organisms Are Frequently Used to Monitor Bacterial Contamination of Water

It is economically prohibitive and theoretically impractical to monitor water for every conceivable type of microorganism. For this reason, certain **indicator organisms** are monitored in water. Indicator organisms, although usually not pathogenic, nonetheless provide a representative index of the extent of water contamination by pathogenic microbes. Pathogens also are generally at a much lower concentration than the indicator organisms in mammalian intestines, although the pathogens can be of sufficient virulence that only a relative few are needed to initiate disease. The

figure 19.12
Sampling Chesapeake Bay
a. A multiple water sampler is deployed in Chesapeake Bay. b. Water samples are retrieved and filtered to collect bacteria and viruses.

indicator organisms generally used in water quality monitoring are those that are associated with the gastrointestinal tract and fecal matter. Since many waterborne pathogens are also found in the gastrointestinal tract and cause gastrointestinal diseases, high numbers of these indicator organisms in water suggest that the water has been fecally contaminated.

The most common group of indicator organisms used in water quality monitoring are the **coliforms,** bacteria that are gram-negative, aerobic or facultatively anaerobic, non-spore-forming rods that ferment lactose with gas production within 48 hours at 35°C. This broad classification includes primarily those bacteria found in the family *Enterobacteriaceae,* including *Es-*

cherichia coli, Enterobacter aerogenes, and *Klebsiella pneumoniae.* These organisms are representative of bacteria normally present in the intestinal tracts of birds and mammals, so they provide a general, albeit adequate, index of fecal contamination of water.

Two procedures are used to test water for the presence and quantity of coliforms: the most probable number (MPN) procedure and the membrane filter (MF) procedure (Figure 19.13). The MPN procedure involves the inoculation of different volumes of the water sample—for example, 10 ml, 1 ml, and 0.1 ml—into a series of lauryl tryptose broth tubes (see most probable number procedure, page 111). The broth cultures are incubated at 35° ± 0.5°C for 48 ± 3 hours and then examined for gas production

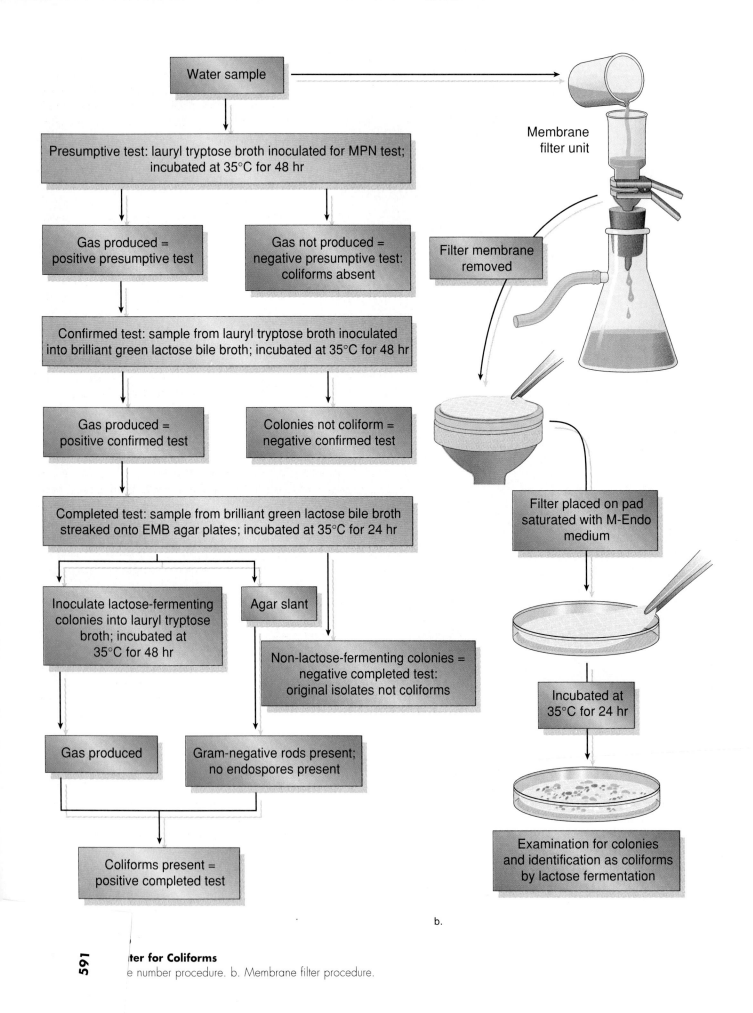

Water sample

Presumptive test: lauryl tryptose broth inoculated for MPN test; incubated at 35°C for 48 hr

Gas produced = positive presumptive test

Gas not produced = negative presumptive test: coliforms absent

Confirmed test: sample from lauryl tryptose broth inoculated into brilliant green lactose bile broth; incubated at 35°C for 48 hr

Gas produced = positive confirmed test

Colonies not coliform = negative confirmed test

Completed test: sample from brilliant green lactose bile broth streaked onto EMB agar plates; incubated at 35°C for 24 hr

Inoculate lactose-fermenting colonies into lauryl tryptose broth; incubated at 35°C for 48 hr

Agar slant

Non-lactose-fermenting colonies = negative completed test: original isolates not coliforms

Gas produced

Gram-negative rods present; no endospores present

Coliforms present = positive completed test

Membrane filter unit

Filter membrane removed

Filter placed on pad saturated with M-Endo medium

Incubated at 35°C for 24 hr

Examination for colonies and identification as coliforms by lactose fermentation

b.

ter for Coliforms
e number procedure. b. Membrane filter procedure.

figure 19.14

Analysis of Water by the Membrane Filter Procedure

Water samples are filtered, and the filters are then placed into small Petri dishes on sterile absorbent pads saturated with a selective growth medium such as M-Endo medium.

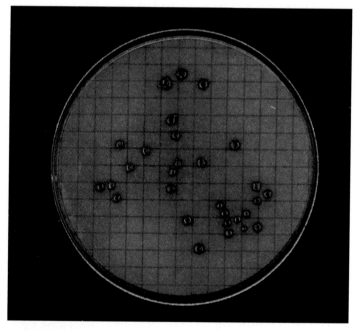

figure 19.15

Membrane Filter Showing Coliform Colonies Growing on M-Endo Medium

Coliforms often produce colonies with a metallic sheen from reaction with the basic fuchsin in the medium.

as a result of lactose fermentation. Broth cultures showing gas production are presumed (**presumptive test**) to contain coliforms. The patterns of growth in tubes containing different dilutions of the water sample provide an estimate of coliform concentration in the original sample. The results are confirmed (**confirmed test**) by inoculating lauryl tryptose broth cultures showing any amount of gas into tubes of brilliant green lactose bile broth. These broth cultures are incubated at $35° \pm 0.5°C$ for 48 ± 3 hours, and the formation of gas in any amount in these cultures constitutes a positive confirmed test. A **completed test** is performed by streaking positive brilliant green lactose broth cultures for isolation onto plates of selective media, usually Levine's eosin-methylene blue (EMB) agar. EMB agar contains the dyes eosin and methylene blue, which inhibit most gram-positive bacteria. Bacteria that strongly ferment lactose produce colonies with a characteristic green metallic sheen, caused by accumulation of eosin and methylene blue in the colonies at acidic pHs. Weak lactose fermenters produce purple colonies. Lactose-fermenting colonies on the Levine's EMB plates are picked and inoculated into tubes of lauryl tryptose broth. Cultures that produce gas from the fermentation of lactose in lauryl tryptose broth within 48 ± 3 hours and are found to be gram-negative, non-spore-forming rods in the EMB agar culture are considered positive for coliforms in the completed test.

The MF technique is less time-consuming than the MPN technique and is used to test relatively large volumes of water. In the MF technique, a known volume (usually 100 ml) of the water sample is filtered through a membrane filter (Figure 19.14). The filter is then placed into a small Petri dish on a sterile absorbent pad saturated with a medium selective for growth of coliforms, such as M-Endo medium. M-Endo medium contains sodium sulfite, sodium lauryl sulfate, sodium desoxycholate, and basic fuchsin to inhibit the growth of gram-positive bacteria. Lactose-positive coliforms form dark red colonies as a result of acid and aldehyde production from lactose in the medium. Strong acid producers, such as *Escherichia coli,* produce a metallic sheen from reaction with the basic fuchsin (Figure 19.15). The colonies are counted to provide an estimate of the concentration of coliforms in the original water sample. These results are usually considered definitive, although in almost all instances additional confirmed and completed tests are suggested. The MF procedure is rapid and involves fewer steps than the MPN technique, but is not as accurate when water samples contain insoluble materials that may interfere with filtration.

Fecal Coliform Counts Are Used to Indicate Fecal Contamination in Water

Coliform counts provide a general indication of fecal contamination in water. The coliform group, however, constitutes a large class of bacteria, some of which may not be derived from intestinal sources. Simply monitoring water samples for total coliform counts thus may not be an accurate index of water contamination by fecal matter. For this reason, water quality laboratories frequently will examine water samples for **fecal coliforms**—those coliforms that grow at a temperature of 44.5°C. Since intesti

Cryptosporidiosis

The importance of identifying specific microbial pathogens in potable water was accentuated in Milwaukee during the spring of 1993. That year an outbreak of cryptosporidiosis in Milwaukee's municipal water supply reached epidemic proportions and forced public officials to close down the city's two water purification plants to avoid massive infection of the city's population. The culprit, *Cryptosporidium*, is a protozoan that causes abdominal cramps, profuse diarrhea, vomiting, and fever, especially in infants and the elderly.

People with normal immune systems generally are mildly ill for only a few days. However, in immunocompromised individuals such as those with AIDS, the diarrhea can be protracted and life threatening. Over 400,000 people in the Milwaukee outbreak developed prolonged diarrheal disease—4,400 required hospitalization.

Cryptosporidium is particularly a problem in municipal water systems because the parasite resists the chlorine used to eliminate bacteria in the water and is not detectable by the coliform test. *Cryptosporidium* is found in the intestines of animals, including humans. In the Milwaukee outbreak, it is believed that the source of the protozoan was water that had drained from farm pastures in to the Milwaukee River during the heavy spring rains. This farm runoff probably carried protozoa from the feces of infected animals.

The Milwaukee epidemic was only one of several similar infectious disease outbreaks of potable water that has occurred in the United States in recent years. Incidents such as this stress the continuing vigilance that must be maintained to ensure the safety of our public water supplies.

 Parasitic Structure and Function
Intestinal Protozoa: Epidemiology • pp. 16–23

---

tract inhabitants generally will grow at this elevated temperature, whereas environmental coliforms do not, testing water for fecal coliforms is one way to determine the extent of fecal contamination in the water. Fecal coliforms are easily detected in water samples by further testing coliforms for growth at an elevated temperature of 44.5 ± 0.2°C in EC broth (a lactose broth containing bile salts to inhibit gram-positive bacteria). Lactose fermenters are not inhibited in EC broth, and produce gas during the fermentation of lactose. Fecal coliform counts are especially important in tests of natural waters, wastewater effluent, bathing waters, and seawaters. Potable waters are generally monitored for both total and fecal coliforms, since such waters should have no coliforms of any kind.

The total and fecal coliform tests are universal methods for monitoring bacterial contamination of water. There has been evidence in recent years, however, that coliform counts—total or fecal—may not always be indicative of the presence of potentially pathogenic microorganisms in water. Other inhabitants of the intestinal tract (*Clostridium perfringens* and *Enterococcus*) may also be used as indicators of fecal contamination. Alternatively, water may be monitored for the presence of specific bacterial pathogens (for example, *Vibrio cholerae, Salmonella enteritidis, Pseudomonas aeruginosa,* and *Chromobacterium violaceum*), pathogenic protozoa (for example, *Giardia lamblia, Entamoeba histolytica,* and *Cryp-ium*) and pathogenic viruses (for example, hepatitis virus,

poliovirus, and other enteroviruses). In certain instances, the presence of these pathogens may not parallel total or fecal coliform counts. Identification of these specific pathogens is time-consuming and more expensive than coliform tests, so most laboratories do not monitor water for them. This is particularly true of viruses, which generally are in such low concentrations in natural waters that they must first be concentrated from large volumes—often exceeding several hundred liters—prior to detection and enumeration. However, the isolation of such specific pathogens is probably a more accurate index of the public health risks associated with contaminated water.

Sewage Treatment Removes Microorganisms and Chemicals from Sewage

The effective treatment of sewage to remove pathogenic microorganisms and toxic chemicals is an important consideration in any municipal water supply system (Figure 19.16). In the past, sewage could be discharged into large bodies of water with only minimal treatment. The dilution of the sewage by immense amounts of water was generally sufficient to render harmless any chemical or biological agents in the discharge. Today, however, the treatment of domestic sewage has become a major ecological problem in the United States, particularly in urban areas, where commercial development and population densities often exceed the capacities

594

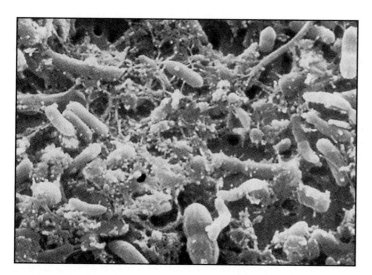

figure 19.16

Raw Sewage Filtered onto a Membrane Filter

Note the large number of filamentous and rod-shaped bacteria (×7,750).

suspended solids are allowed to settle as primary sludge. Materials such as oil or grease, which float on the surface, are removed with a skimmer. The settled sludge is collected and may be buried in a sanitary landfill or placed in an anaerobic sludge digester for further treatment. Anaerobic bacteria in the digester decompose the insoluble organic materials, producing methane and carbon dioxide in the process. The liquid wastewater remaining in the settling tank or basin is then ready for secondary treatment. The fluid from the primary treatment, called the **effluent,** has had a 30% to 40% reduction in its BOD.

In some treatment systems, the primary effluent is chlorinated to kill potentially pathogenic microorganisms and then discharged into receiving waters. Although this effluent is relatively noninfectious, it is still high in organic matter and may significantly affect the ecology of the water body. Effluents containing large amounts of phosphorus may cause a closed water body such as a lake to change from an oligotrophic state to a eutrophic one. This results in the massive development of planktonic algae (algal blooms), which may spoil bathing and recreational areas because of the discoloration and stench from rotting algal debris. In some lakes and in open systems such as flowing waters (rivers and streams), the organic materials in effluents with a high BOD are quickly metabolized by heterotrophic organisms, which take up large quantities of oxygen. The water soon develops an oxygen deficit, and as a consequence, there may be fish kills.

Organic Matter in Sewage Is Oxidized by Microorganisms in Secondary Treatment

In secondary treatment, organic matter in the wastewater from primary treatment is further oxidized by microorganisms. Municipalities use a number of different types of aerobic processes for the treatment of sewage. The two most common are the **trickling filter process** and the **activated sludge process** (Figure 19.18).

A trickling filter consists of a large tank or basin filled with a bed of crushed stone, gravel, slag, or other porous material. Sewage is sprayed in a fine mist over the rocks. As the sewage trickles down through the bed, organic matter clings to the rocks, where it is digested by heterotrophic microorganisms. The microorganisms are contained within a biofilm produced by slime-forming bacteria such as *Zoogloea,* and the organic matter is oxidized to gases and inorganic products.

In the activated sludge process, sewage is mixed with slime-forming bacteria (*Zoogloea*) in a large aeration tank. As the mixture is aerated, large flocs, or clumps, form. These clumps contain not only the original slime-forming bacteria, but also a large population of heterotrophs, which oxidize the organic matter within these clumps. The activated sludge system resembles a continuous culture system, since additional wastewater is continuously pumped into the tank and the treated water is removed into a holding tank, where the flocs are allowed to settle. A portion of the

of sewage treatment plants. In such municipalities, excess waste is often diverted into rivers, streams, and lakes, where toxic chemicals and pathogenic microorganisms in the waste can present problems to the general public health. Although modern, efficient waste treatment facilities minimize this problem, considerable work still remains to be done on improving sewage treatment.

Sewage treatment is classified as primary, secondary, or tertiary, depending on the extent of waste treatment and removal (Figure 19.17). **Primary sewage treatment** involves the removal of suspended solid and floating material. In **secondary sewage treatment,** biological agents are used to further purify the sewage. Additional purification, either through filtration or chlorination by **tertiary sewage treatment,** results in water that is pure and safe enough to return to municipal water supplies. Tertiary treatment processes, although expensive, have nonetheless become popular in many communities as one method to regenerate dwindling water resources.

Suspended Solids Are Removed from Raw Sewage in Primary Treatment

Primary treatment involves the removal of suspended solids from raw sewage. As raw sewage enters the sewage treatment plant, it passes through screens and grit chambers that remove large objects such as rocks, bottles, paper, and pieces of metal or wood. The screened water is then sent to settling tanks or basins, where the

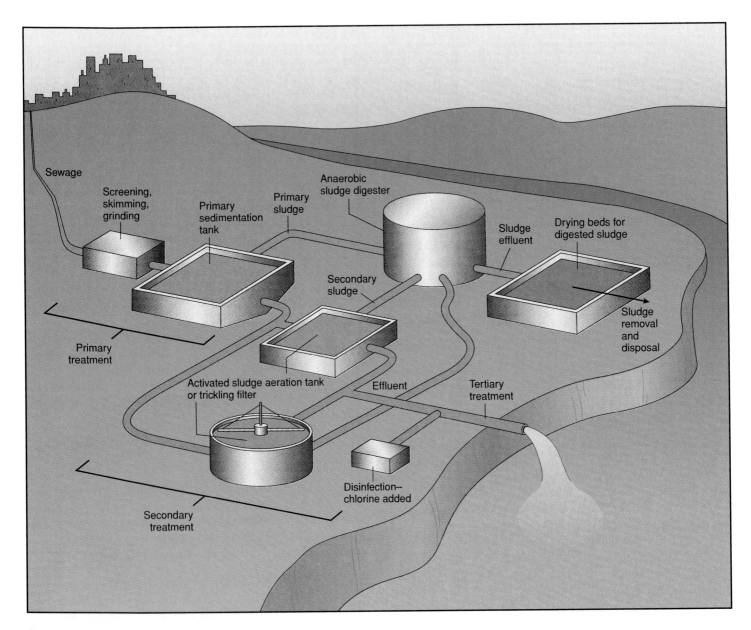

figure 19.17
Stages Involved in Sewage Treatment

a.

c.

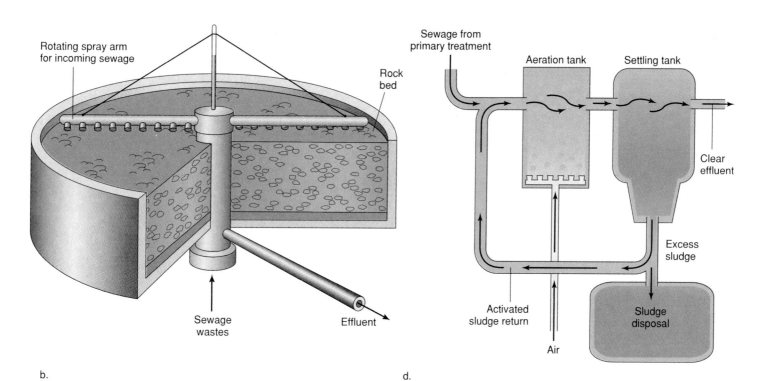

Rotating spray arm
for incoming sewage

Rock
bed

Sewage
wastes

Effluent

b.

Sewage from
primary treatment

Aeration tank

Settling tank

Clear
effluent

Excess
sludge

Activated
sludge return

Air

Sludge
disposal

d.

figure 19.18

**Trickling Filter and Activated Sludge Facilities for Secondary
Sewage Treatment**

a. Trickling filter. b. Diagram of a trickling filter. c. Activated sludge
aeration tank. d. Diagram of an activated sludge aeration system.

Methane Production from Wastes

The microbial conversion of complex organic wastes to methane and carbon dioxide involves three groups of microorganisms: (1) fermentative procaryotes, which produce mainly short-chain organic acids, CO_2, and H_2 from the organic waste materials; (2) acetogenic procaryotes, which further oxidize these organic acids and other compounds such as lactate and ethanol to acetate, CO_2, and H_2; and (3) methanogenic procaryotes, which utilize acetate and a few other simple compounds (H_2, CO_2, formate, methanol, methylamines, and carbon monoxide) to produce methane. The methanogens (domain Archaea) are obligate anaerobes that are slow growers (one species has a doubling time of nine days) and therefore represent the rate-limiting step in the anaerobic digestion of organic wastes.

The process of anaerobic digestion of waste has been used in sewage treatment for over 100 years. The conventional anaerobic sludge digester typically consists of a large holding tank in which sludge is heated to 35°C to 60°C, mixed, and held for 15 days (Figure 19.19). During this holding period, microbes oxidize the organic wastes to produce methane and CO_2. The methane produced is used to heat the biological reactor (bioreactor). The primary advantages of a thermophilic (50°C to 60°C) bioreactor are the rapid growth of thermophilic acid-consuming organisms and the decreased sludge retention times.

A modification of the conventional bioreactor is the anaerobic activated sludge process, which pumps effluent from the bioreactor to a settling tank. Settled sludge in the tank is returned to the bioreactor to maintain a high concentration of active mass, which is required for efficient methane production and system economy.

In recent years, anaerobic methane fermentation processes have been used in many communities to convert animal and plant waste into energy. These systems involve mixing manure or plants with wastewater containing organic residues in a bioreactor. As the anaerobic microorganisms break down the volatile solids, gases (60% methane and 40% CO_2) are released through the top of the reactor. The methane is used to fuel steam-producing boilers and generators. Digested solids removed from the reactor can be used as a feed supplement for livestock. A typical bioreactor can process 25 tons of manure a day, generating $158,000 worth of methane and $380,000 worth of nutrient-rich by-products annually. Currently there are about a dozen major anaerobic digesters producing methane on a commercial basis. These bioreactors are helping turn waste into energy and other useful products.

---

settled floc material is recycled into the tank as an inoculum to continue the process. The remaining floc is further treated or removed for burial or incineration.

The activated sludge process is efficient—able to reduce BOD by up to 90%—and is a very popular method for handling large volumes of sewage rapidly. The effluent from the activated sludge process contains low levels of organic matter, but like the trickling filter process, may contain high levels of inorganic materials.

One problem that sometimes occurs in the activated sludge process is **bulking,** which occurs when such filamentous bacteria as *Beggiatoa, Sphaerotilus,* and *Thiothrix* form loose flocs that do not rapidly settle. Bulking can thus result in reduced clarification of wastewater. The exact causes of bulking are not known, but are believed to be associated with wastewater conditions that promote the growth of these filamentous bacteria.

Inorganic Salts, Organic and Inorganic Suspended Solids, and Poorly Biodegradable Organic Materials Are Removed from Sewage in Tertiary Treatment

Although effluents from secondary treatment have a low BOD, they may contain eutrophication-inducing inorganic salts (phosphorus and nitrogen compounds), organic and inorganic suspended solids, and poorly biodegradable organic materials. Advanced or tertiary treatment processes are designed to reduce or eliminate these materials and depend more on physical and chemical processes than biological processes.

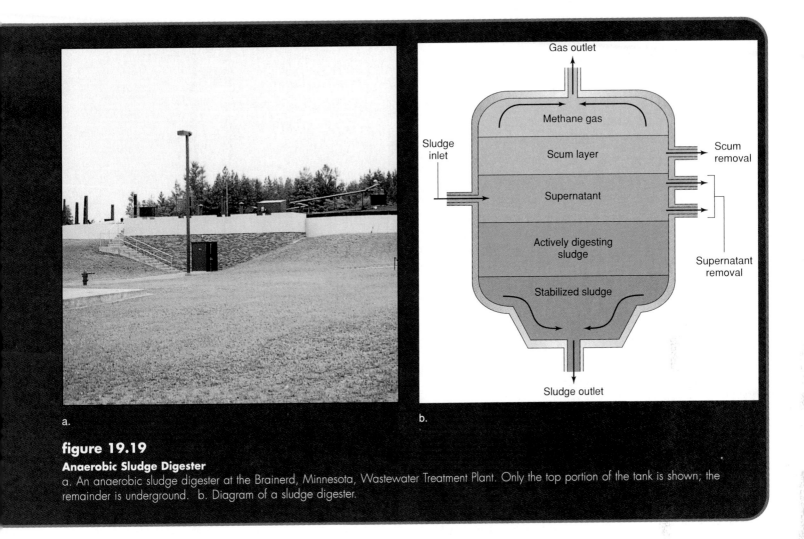

figure 19.19

Anaerobic Sludge Digester

a. An anaerobic sludge digester at the Brainerd, Minnesota, Wastewater Treatment Plant. Only the top portion of the tank is shown; the remainder is underground. b. Diagram of a sludge digester.

For phosphorus elimination, the phosphates are converted to poorly soluble aluminum, calcium, or iron compounds and removed by precipitation. Nitrogen in sewage effluent is removed primarily through nitrification by microorganisms. Because nitrosofying and nitrifying bacteria are slow-growing organisms, the extent of nitrification during tertiary treatment depends on adequate treatment plant design and the proper removal of sludge so that these bacteria are grown under optimal conditions. Otherwise, large amounts of nitrogen compounds may escape tertiary treatment and be released in the effluent. Suspended solids are eliminated in sewage through filtration or sedimentation. Poorly biodegradable substances can be removed by the use of specialized microorganisms capable of using them as substrates. Chlorine is frequently added to tertiary-treated effluent to kill any remaining microorganisms.

The Septic Tank Is Frequently Used in Rural Areas for Sewage Treatment

The **septic tank** is a small-scale anaerobic treatment process that commonly is used in rural areas (Figure 19.20). It is simple, inexpensive, and satisfactory if properly operated. The septic tank consists of an underground sedimentation container into which sewage from a home enters and is retained for a short time. The organic matter in the sewage settles to the bottom of the tank, where it is covered by a thin organic film that excludes oxygen. Anaerobic bacteria in the sediment digest the organic matter into simpler chemical compounds and gases. The gases are then discharged through a vent in the tank. Liquids in the tank rise and overflow through an outlet pipe and are distributed in the surrounding soil. As the water trickles through the

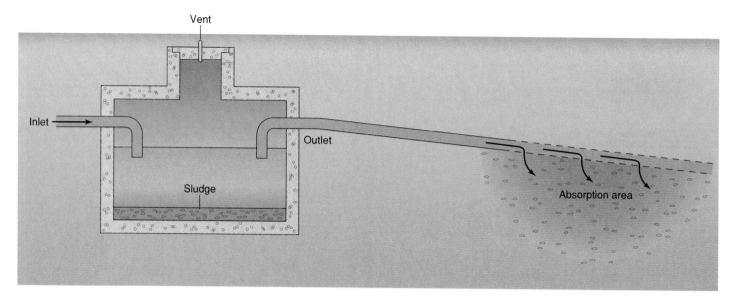

figure 19.20

Diagram of a Septic Tank

Organic matter in sewage settles to the bottom of the tank. There it is digested by anaerobic bacteria into simpler compounds and gases, which are discharged through a vent. Liquids in the tank overflow and are discharged to the surrounding soil. Undigested sludge must be routinely removed from the tank.

soil, any remaining organic matter is decomposed by aerobic procaryotes.

The septic tank theoretically should be adequate for the efficient treatment of small amounts of sewage. Problems are often encountered, however, with the types of soil through which the overflow water flows. Soil that has a high clay content, for example, generally retains water and forms an anaerobic environment that retards the growth of aerobic digesters. Not all bacterial pathogens are removed by such treatment. Septic tanks therefore should not be located near water supplies. Undigested solids in the bottom of a septic tank must be periodically removed. Despite these problems, the septic tank is still widely used in rural sewage disposal.

Oxidation Ponds Are Used in Some Communities to Handle Small Loads of Sewage

Small communities and isolated areas frequently use oxidation ponds for treatment of wastewater. Sewage is channeled into an initial pond, where the sludge settles out. The liquid portion of the sewage is then pumped into adjacent series of ponds, where aeration allows bacterial growth and degradation of organic matter. These secondary ponds often are seeded with algae, which provide oxygen for the growth of aerobic, heterotrophic bacteria. Such oxidation ponds, although not as elaborate as municipal treatment plants, nonetheless are generally adequate and relatively efficient in handling small loads of sewage.

Examples of Microorganisms Found in the Troposphere

1,000 m over Maine/ Greenland	150–5,500 m over Washington, D.C.	300–3,000 m over Nashville, Tenn.	1,500 m over Northwest Territories, Canada
Cladosporium	Acremoniella	Actinomyces	Achromobacter
Leptosphaeria	Alternaria	Aspergillus	Corynebacterium
Macrosporium	Penicillium	Bacillus	Flavobacterium
Mycosphaerella	Pestalozzia	Cladosporium	Sarcina
		Fusarium	
		Penicillium	

Source: P. H. Gregory, *Microbiology of the Atmosphere,* 2nd ed., Leonard Hill Books, Great Britain, 1973.

Microbes and the Air

The atmosphere has several characteristic layers, each defined by its temperature and height above the earth's surface. The **troposphere,** the layer closest to the earth, extends to an altitude of 8 to 12 km. The air in the troposphere is constantly in motion, and the temperature decreases with increasing altitude, reaching a low of –57°C at the apex of this region. Beyond the troposphere is the **stratosphere,** which extends up to approximately 50 km and has a temperature range of –80°C to –10°C. Above the stratosphere is the **ionosphere,** where temperatures can reach a maximum of several thousand degrees.

Microorganisms are not found in the upper regions of the atmosphere because of the temperature extremes, scarcity of available oxygen, absence of nutrients and moisture, and low atmospheric pressures. Microorganisms are frequently found in the lower portion of the troposphere, where they are dispersed by the air currents. Most of the viable microbes present in the troposphere are either spore formers or microbes that are easily dispersed in the air (Table 19.5). The relatively low humidity in the atmosphere (especially during daylight hours) and ultraviolet rays from the sun limit the types and number of microorganisms that are able to survive in this part of the biosphere. The atmosphere, nonetheless, serves as an important medium for dispersing many types of microbes to new environments.

Many microbial diseases are transmitted through the air. Microorganisms are expelled in aerosals during sneezing, coughing, or even normal breathing. The incidence of diseases caused by airborne transmission can be reduced by covering one's nose and mouth during coughing or sneezing and by the use of face masks.

Coliforms have long been used as indicator bacteria to monitor sanitation and the microbiological quality of food and water. The most probable number (MPN) method and the membrane filter (MF) technique are two accepted methods used in the United States for the detection of total coliforms and fecal coliforms. The MPN procedure is laborious and time-consuming, requiring 120 hours or longer to complete. The MF technique is more accurate and less time-consuming than the MPN method, but is not suitable for food samples and for water samples containing large amounts of suspended solids, silt, algae, and bacteria, since these materials interfere with filtration and colony development. Recently a fluorogenic assay for the detection of coliforms, specifically *Escherichia coli*, was reported. This assay is based on the hydrolysis of 4-methylumbelliferyl-β-D-glucuronide (MUG) by β-glucuronidase to yield 4-methylumbelliferone, which is visible by long-wave ultraviolet light. The assay was developed by Peter Feng and Paul Hartman at Iowa State University.

Feng and Hartman compared the accuracy of the MUG assay with conventional coliform detection methods, including the MF and the MPN procedures. With the MF technique, 100 μg of MUG per milliliter was incorporated into M-Endo medium before the broth was dispensed onto a sterile absorbent pad. A filter containing a mixed inoculum of *Escherichia coli* and *Enterobacter aerogenes* was incubated on top of the medium-saturated pad. With the MPN procedure, MUG was incorporated at a final concentration of 100 μg per ml into lauryl tryptose broth (LTB). The LTB-MUG medium was dispensed into test tubes containing Durham vials to measure gas production. Overnight broth cultures (0.1 ml) were inoculated into LTB-MUG medium and conventional LTB medium.

With the membrane filter technique, the presence of MUG distinguished *E. coli* (a fecal coliform) from *E. aerogenes* (a nonfecal coliform). *E. coli* colonies on M-Endo medium fluoresced when exposed to ultraviolet light. *E. aerogenes* colonies were nonfluorogenic. With the MPN procedure, *E. coli* produced equivalent amounts of gas in LTB-MUG medium and conventional LTB medium. In LTB-MUG medium, the presence of *E. coli* could be detected by the appearance of fluorescence throughout the broth when examined under ultraviolet light (Figure 19P.1).

An anaerogenic strain of *E. coli* did not produce gas in LTB-MUG medium or LTB medium, but did produce fluorescence in the LTB-MUG medium.

Ten water samples collected from rivers, streams, and the Ames, Iowa, wastewater treatment plant were inoculated into LTB-MUG media using a five-tube MPN series, and incubated at 35°C for 48 hours. Tubes that were gas positive, fluorescence positive, or both were subcultured into EC broth and incubated at 44.5°C to confirm the presence of fecal coliforms. Food samples (ground beef, chicken giblets, raw milk, and frozen broc-

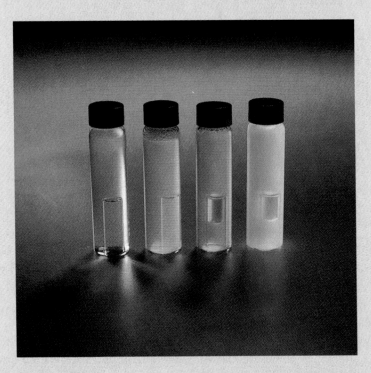

figure 19P.1

Incubated MUG Tubes Viewed Under Ultraviolet Light

From left to right: (1) sterile, uninoculated broth; (2) growth, no gas, no fluorescence (a noncoliform); (3) growth, gas, no fluorescence (*Enterobacter aerogenes*, a nonfecal coliform); (4) growth, gas, fluorescence (*Escherichia coli*, a fecal coliform). Tubes were incubated for 24 hours at 35°C.

coli) were homogenized in a blender (25-g samples mixed with 225 ml of 0.1% peptone water), diluted, and inoculated (0.1 ml) into LTB-MUG tubes as described for the water samples. For both the water samples and the food samples, approximately 90% (range: 34% to 100%) of the LTB-MUG tubes showing both gas production and fluorescence contained fecal coliforms as determined by gas production in EC broth (Tables 19P.1 and 19P.2). Fluorescence caused by β-glucuronidase activity typically occurred after only 24 hours of incubation in LTB-MUG and before gas production from the fermentation of lactose.

These data indicated that the fluorogenic assay is an accurate, rapid alternative to conventional MF and MPN procedures. Furthermore, the fluorogenic assay is less time-consuming than the MPN procedure and can be used on samples that normally could not be processed by the MF procedure. The MUG-based assay is now available commercially and offers an attractive alternative to other coliform detection methods.

Source

Feng, P.C.S., and P.A. Hartman. 1982. Fluorogenic assays for immediate confirmation of *Escherichia coli. Applied and Environmental Microbiology* 43:1320–1329.

References

Manafi, M., W. Kneifel, and S. Bascomb. 1991. Fluorogenic and chromogenic substrates used in bacterial diagnostics. *Microbiological Reviews* 55:335–348.

Moberg, L.J. 1985. Fluorogenic assay for rapid detection of *Escherichia coli* in food. *Applied and Environmental Microbiology* 50:1383–1387.

Robison, B.J. 1984. Evaluation of a fluorogenic assay for detection of *Escherichia coli* in foods. *Applied and Environmental Microbiology* 48:285–288.

table 19P.1

Results of EC Broth Fecal Coliform Confirmatory Tests on LTB-MUG Broth Tubes Obtained from the MPN Analysis of Water Samples

LTB-MUG Reaction	Number Tested	Positive EC Broth Test
Gas (+), fluorescence (+)	72	61 (85%)
Gas (+), fluorescence (−)	72	7 (9%)
Gas (−), fluorescence (+)	17	0 (0%)

table 19P.2

Results of EC Broth Fecal Coliform Confirmatory Tests on LTB-MUG Broth Tubes Obtained from the MPN Analysis of Food and Milk Samples

Sample	LTB-MUG Reaction[a]	Number Tested	Positive EC Broth Test
Ground beef	Gas (+), fluorescence (+)	63	63 (100%)
	Gas (+), fluorescence (−)	22	0 (0%)
Chicken giblets	Gas (+), fluorescence (+)	97	86 (89%)
	Gas (+), fluorescence (−)	4	0 (0%)
Raw milk	Gas (+), fluorescence (+)	33	27 (82%)
	Gas (+), fluorescence (−)	55	3 (5%)
Frozen broccoli	Gas (+), fluorescence (+)	3	1 (34%)
	Gas (+), fluorescence (−)	83	0 (0%)

[a]Of 196 gas (+), fluorescence (+) samples, 177 (91%) had positive EC broth tests. Of 164 gas (+), fluorescence (−) samples, 3 (2%) had positive EC broth tests.

Summary

1. Microbial ecology is the study of interactions of microorganisms with the biotic (living) and abiotic (nonliving) components of the environment. Biogeochemical cycles describe the movement of chemical elements through the biological and geological components of the world.

2. Microorganisms play important roles in the carbon cycle, nitrogen cycle, phosphate cycle, sulfur cycle, and other biogeochemical cycles.

3. The biosphere is that portion of the earth on which life exists, and consists of soil, water, and air. Bacteria are numerically the dominant microorganisms in soil, and their diverse metabolisms are responsible for many of the biochemical changes in soil.

4. Bacteria are used in the leaching of metals from low-grade ores, the degradation of hydrocarbons in oil spills, and the biological control of insects.

5. Potable water is water that is suitable for drinking. Used water that is not consumed becomes sewage. Sewage must be treated to remove chemicals and microbes before the water can be reused.

6. One of the parameters to determine the purity of water is the biochemical oxygen demand (BOD). The BOD is the quantity of oxygen required to meet the metabolic demands of microorganisms oxidizing organic matter in the water. Effective sewage treatment significantly reduces the amount of organic matter, and hence the BOD, in sewage.

7. Bacterial contamination of water is monitored by testing for indicator organisms called coliforms. Coliforms are gram-negative, aerobic or facultatively anaerobic, non-spore-forming rods that ferment lactose with gas production within 48 hours at 35°C. Fecal coliforms have the same characteristics, but grow at 44.5°C. Coliform and fecal coliform counts can be determined by the most probable number (MPN) procedure and the membrane filter (MF) procedure.

8. Sewage treatment consists of three stages. Primary sewage treatment removes suspended solids and floating material. Secondary treatment further purifies the sewage by use of microorganisms to oxidize organic matter. Tertiary treatment removes inorganic salts, organic and inorganic suspended solids, and poorly biodegradable organic materials.

9. Microorganisms are frequently found in the lower portion of the troposphere, the layer of the atmosphere that is closest to the earth and extends to an altitude of 8 to 12 km. Airborne microorganisms are capable of transmitting diseases through the air.

EVOLUTION *and* BIODIVERSITY

Microorganisms are ubiquitous; they are found on land, in water, and in the air. As inhabitants of the biosphere, microbes contribute significantly to the composition and quality of our environment. In soil, microbes metabolize a wide variety of chemical compounds. Some of these soil microorganisms, such as *Thiobacillus ferrooxidans* and *Thiobacillus thiooxidans*, obtain their energy from the oxidation of inorganic substances and are commercially valuable in the leaching of metals from low-grade ores. Although microbial leaching is of economic benefit, it can lead to damage of the environment from acid by-products. Other microorganisms such as *Bacillus thuringiensis* are potentially important pesticides. Biological pesticides are appealing because they do not have the broad lethality of chemical pesticides and therefore are safe for humans and animals. In water, phototrophic microorganisms such as algae and the cyanobacteria serve as primary producers, fixing carbon dioxide into organic forms of carbon for use by heterotrophs. Other microorganisms are important in the breakdown of organic compounds during sewage treatment. Although air contains fewer microorganisms than soil or water, the atmosphere nonetheless is a medium by which infectious agents can be dispersed and cause disease. This diversity of life in soil, water, and air is possible only with microorganisms.

Questions

Short Answer

1. How does a community differ from a population?

2. What is the ultimate source of all carbon?

3. What is the role of primary producers in the carbon cycle? What is the most common means of primary productivity?

4. How do herbivores and carnivores contribute to the carbon cycle?

5. How do microorganisms contribute to the carbon cycle?

6. What is meant by the term "recalcitrant"?

7. Identify the four stages of the nitrogen cycle.

8. _____ and _____ are primarily responsible for nitrogen fixation.

9. _____ is primarily responsible for denitrification.

10. Explain why an abundance of nitrifying bacteria is both good and bad for the farmer.

11. Why is phosphorus a "limiting nutrient"? How is it made available?

12. Why are microorganisms essential to the sulfur cycle?

13. When is soil more likely to be anaerobic?

14. What is meant by the rhizosphere effect?

15. At what pH are fungi likely to outnumber bacteria in soils?

16. Are algae and cyanobacteria likely to exist in soil?

17. Are protozoa likely to exist in soil?

18. How are microorganisms harmful to the petroleum industry? How are they helpful?

19. In addition to reducing the need for fertilizers, how can microorganisms help agriculture?

20. Identify some of the bacteria commonly found in estuaries.

21. Explain why groundwater and well water generally have low levels of microorganisms.

22. Identify several types of bacteria which indicate the presence of sewage in waterways.

23. Identify several common biological and chemical contaminants of water.

24. Identify the most common means of monitoring water quality.

Multiple Choice

1. The most prevalent element in the biosphere is:

a. carbon
b. nitrogen
c. phosphorus
d. sulfur

2. Which of the following involves the formation of nitrate from ammonia?

a. ammonification
b. denitrification
c. nitrification
d. nitrogen fixation

3. Which of the following soil microorganisms is anaerobic?

a. *Bacillus*
b. *Clostridium*
c. *Pseudomonas*
d. *Streptomyces*

4. Microbial leaching is economically important for the mining of:

a. copper
b. gold
c. mercury
d. silver

5. Environments low in nutrients are referred to as:

a. eutrophic
b. littoral
c. limnetic
d. oligotrophic

6. Which of the following refers to chlorination or filtration?

a. primary sewage treatment
b. secondary sewage treatment
c. tertiary sewage treatment

Critical Thinking

1. Explain why soil is generally free of pathogenic bacteria, even though the victims of infectious diseases are buried. Discuss alternatives to burial.

2. Given what you have just learned, explain why clays are less fertile than other soils. Can you correct this problem? Explain.

3. Discuss the use of microorganisms to clean up oil spills. Why are many people resistant to such strategies? What evidence supports their concerns? How can you counter their concerns in a practical manner?

4. You suspect a local stream as the source of a recent outbreak of disease, yet you cannot find a source of human sewage. Do you discount the stream? Why or why not?

5. Investigate your local water/wastewater utility. How is your sewage treated? Why? How is your drinking water treated? Why? What would you change? Why? What would you change if your community grew by 10,000 people? Why? What would you change if your community population shrank to less than 1,000 people? Why?

Supplementary Readings

American Public Health Association. 1992. *Standard methods for the examination of water and wastewater,* 18th ed. Washington, D.C.: American Public Health Association. (A procedural manual for chemical and bacteriological analyses of water and wastewater.)

Atlas, R.M., and R. Bartha. 1993. *Microbial ecology: Fundamentals and applications,* 3d ed. Menlo Park, Calif.: Benjamin-Cummings. (A basic textbook of microbial ecology.)

Brierley, C.L. 1983. Microbial mining. *Scientific American* 247:44–53. (A review of the roles of bacteria in the leaching of metals from low-grade ores.)

Ford, T.E., ed. 1993. *Aquatic microbiology: An ecological approach.* Cambridge, Mass.: Blackwell Scientific Publications. (A discussion of microorganisms in aquatic environments.)

Lynch, J.M., and N.J. Poole. 1979. *Microbial ecology: A conceptual approach.* New York: Wiley. (A discussion of microorganisms in their natural environments: in soil, water, air, animals, plants, and extreme habitats.)

Miller, L.K., A.J. Lingg, and L.A. Bulla, Jr. 1983. Bacterial, viral, and fungal insecticides. *Science* 219:715–721. (A description of the use of microorganisms as insecticides, with emphasis on their molecular and biochemical mode of action.)

Mudrack, K., and S. Kunst. 1986. *Biology of sewage treatment and water pollution control.* Chichester, England: Ellis Horwood. (A review of microbiological processes of sewage treatment, with comparison of processes for the treatment of sewage and sewage sludge.)

Paul, E.A., and F.E. Clark. 1989. *Soil microbiology and biochemistry.* New York: Academic Press. (An in-depth textbook on the biochemistry of soil microbes.)

Rheinheimer, G. 1991. *Aquatic microbiology,* 4th ed. New York: Wiley. (A discussion of the ecology of aquatic microorganisms.)

Walker, N., ed. 1975. *Soil microbiology.* New York: Wiley. (A review of soil microorganisms.)

FOOD AND INDUSTRIAL MICROBIOLOGY

Milk and Dairy Products
Raw Milk
Fermented Milks and Milk Products
Cheese Production

Meat
Spoilage
Preservation and Reduction of Spoilage

Poultry and Seafood

Canned Foods

Fermented Foods

Alcoholic Beverages
Beer Production
Wine Production
Distilled Beverage Production

Industrial Processes
Culture Medium
Microorganisms for Industrial Fermentation
Culture Conditions
Culture Methods

Enzymes
Microbial Amylases
Microbial Proteases
Other Microbial Enzymes of Industrial Importance

Metabolic Products
Primary Metabolites
Secondary Metabolites

Single-Cell Protein

PERSPECTIVE
Single-Cell Protein: Food of the Future

Laboratory technician preparing fermentation equipment used to produce recombinant hepatitis B vaccine.

EVOLUTION AND BIODIVERSITY

 Microbes in Motion ⸺ `PREVIEW LINK`

Among the key topics covered in this chapter are an outbreak of salmonellosis traced to contaminated milk and an outbreak of listerosis. The following sections in the *Microbes in Motion* CD-ROM may be useful as a preview to your reading or as a supplementary study aid:

Gram-Negative Organisms: Bacilli—Facultative Anaerobes (*Enterobacteriaceae*), 2–7, 16–23.

Gram-Positive Organisms: Gram-Positive Bacilli (*Listeria*), 13–21.

Nearly 6,000 years ago, the Egyptians stumbled upon a flavorful brew that was made by allowing grain to stand for a period of time. This beverage not only had an enticing taste, but also seemed to lift the spirits. Little did these ancient revelers realize that they were drinking a fermented concoction that was a predecessor of today's $50 billion beer industry—an industry that produces over 500 million barrels of beer annually and heavily relies upon microorganisms for its existence.

Beer brewing is only one small part of an industry that uses microbes for the production of various foods and drinks. These minute, invisible organisms are used extensively for the manufacture of dairy products, cheeses, bread, sauerkraut, pickles, beer, and wine. The uncontrolled growth of microorganisms in foods can also cause spoilage. Today the food microbiologist is concerned not only with the production of foods by microorganisms, but also with the extension of the shelf life of these foods.

Milk and Dairy Products

Milk was among the first agricultural products. Its nutritional properties make it an excellent source of carbohydrates, proteins, minerals, and vitamins. These same properties also make milk an excellent medium for the growth and proliferation of microorganisms.

Raw Milk Contains Microorganisms Introduced from the Udder and During Human Handling

Milk is normally sterile before it is secreted by the cow. At the time it is drawn, milk becomes contaminated by microorganisms introduced from the udder and during handling. The numbers of bacteria that are found in drawn milk vary from cow to cow, but should not exceed recommended standards for plate counts when drawn from a healthy udder. Milk is graded by the U.S. Public Health Service according to bacterial numbers by standard plate count (Table 20.1). Grade A pasteurized milk and milk products should have no more than 20,000 bacteria per milliliter, whereas Grade A dry milk products are required to have no more than 30,000 bacteria per gram. Generally milk of high quality should have only a few hundred bacteria per milliliter, whereas milk that is of poor quality and heavily contaminated may have millions of bacteria per milliliter.

Raw milk from a healthy cow contains *Streptococcus lactis* and other lactic-acid–producing streptococci and lactobacilli, as well as coliform bacteria and psychrophilic gram-negative rods (*Pseudomonas*, *Alcaligenes*, and *Flavobacterium*). These bacteria are usually introduced into milk through contaminated milking utensils, dust, and manure. Unless milking equipment is sterilized, the surfaces easily become coated with large numbers of bacteria.

Milk sours and spoils because of the growth of lactic acid and bacteria. The first evidence of milk spoilage is a sour flavor caused by acid formation, followed by coagulation of milk to give a solid curd. In raw milk, the souring is caused primarily by homofermentative *S. lactis*, which produces mainly lactic acid. However, many bacteria other than lactic acid bacteria can cause acid fermentation in milk, especially if the temperature and other conditions are unfavorable for growth of the lactic acid bacteria. Coliforms produce some lactic acid and considerable amounts of other products such as acetic acid, formic acid, and butanediol.

An additional source of contamination is dairy workers who handle the milk. Carriers of *Salmonella*, *Corynebacterium diphtheriae*, *Shigella dysenteriae*, or other disease-causing microorganisms may transmit these bacteria to milk. Other pathogens are introduced into milk by infected cows. Bovine tuberculosis, undulant fever (caused by *Brucella abortus*), and Q fever (caused by *Coxiella burnetii*) may be transmitted to humans by the milk of an infected animal. Processed milk may also be contaminated by malfunctions in processing equipment or by processing system uncleanliness. State health inspectors routinely inspect milk processing plants, but sometimes contamination is undetected. When this happens there may be epidemics of milk-borne infections.

All non-spore-forming pathogenic bacteria and many nonpathogenic bacteria in raw milk are destroyed by pasteurization (see pasteurization, page 119). The two pasteurization methods used by the dairy industry are (1) the low-temperature holding (LTH) method, in which milk is heated to 62.8°C (145°F) for 30 minutes, and (2) the high-temperature short-time (HTST), or flash pasteurization method, which exposes milk to a temperature of 71.7°C (161°F) for 15 seconds. The guidelines for pasteurization temperature and time were originally established with *Mycobacterium tuberculosis* and later modified with *Coxiella burnetii*, since these were considered to be the most heat-resistant pathogens likely to be found in milk. Pasteurization eliminates these and other pathogens and also extends the shelf life by reducing the numbers of spoilage bacteria without altering the flavor or nutritional value of milk.

A form of milk sold in Europe and recently in the United States is **ultra high-temperature (UHT) milk.** Raw milk is heated to 137.8°C (280°F) for a few seconds (lower temperatures are used for standard pasteurization) and then packed in a sterile airtight container. Because the ultrahigh temperature kills bacteria that sour milk, the UHT product can be stored without refrigeration for several months. Once opened, the milk is exposed to bacteria on the carton exterior and must be refrigerated. The opened milk has a shelf life of 10 to 21 days, depending on such factors as the number and types of bacterial contaminants and the temperature of the refrigerator.

The Phosphatase Test Measures Efficiency of Milk Pasteurization

Phosphatase, an enzyme present in raw milk, is destroyed by pasteurization; tests for its absence in milk ensure that pasteurization has occurred properly. The **phosphatase test** depends on the ability of phosphatase to degrade disodium phenylphosphate to phenol and sodium phosphate. The phenol liberated by this reaction can be estimated and provides a measure of phosphatase activity. The phosphatase test is simple, yet it provides valuable information on the efficiency of milk pasteurization.

Microbiological Standards for Milk and Dairy Products

Product	Maximum Permissible Number of Microorganisms per Milliliter	
	Total Bacteria	Total Coliforms
Grade A raw milk (before mixing with other batches of milk)	100,000	
Grade A raw milk (after mixing with other batches of milk)	300,000	
Grade A pasteurized milk	20,000	10
Grade B raw milk	600,000	
Grade B pasteurized milk	40,000	
Certified raw milk	10,000	
Condensed milk	30,000	10
Nonfat dry milk	30,000[a]	
Ice cream	50,000	
Frozen yogurt	50,000	

[a]Per gram

Examples of Fermented Milks and Their Starter Cultures

Fermented Milk	Substrate	Starter Cultures
Acidophilus milk	Skim milk	Lactobacillus acidophilus
Bulgarian milk	Skim milk	Lactobacillus bulgaricus
Cultured buttermilk	Skim milk	Leuconostoc cremoris or Leuconostoc dextranicum
		Streptococcus cremoris or Streptococcus lactis
Kefir	Cow, sheep, or goat milk	L. bulgaricus
		S. lactis
		Lactose-fermenting yeasts
Koumiss	Mare milk	L. bulgaricus
		S. lactis
		Lactose-fermenting yeasts
Sour cream	Cream	L. cremoris or L. dextranicum
		S. cremoris or S. lactis
Yogurt	Concentrated skim milk	L. bulgaricus
		Streptococcus thermophilus

Bacteria Are Used to Produce Fermented Milks and Milk Products

Although some microorganisms may make milk unfit for human consumption, others are important in the manufacture of milk products. Fermented milks, butter, and cheese are some of the milk products made by microbial activities.

Lactic acid-producing bacteria are responsible for the fermentation of milk. For thousands of years, the souring or fermenting of milk has been one way to preserve it and to produce beverages with distinctive flavors. The fermentation of the milk sugar lactose results in the accumulation of lactic acid, a substance that acts as a natural preservative of milk. Other products formed from this fermentation provide the characteristic aromas and flavors of fermented dairy products (Table 20.2).

Fermented milks are produced through the addition of **starter cultures** of lactic acid bacteria. Starter cultures acidify the milk and produce flavoring compounds during the fermentation. Cultured buttermilk and sour cream both use *Streptococcus lactis* or *Streptococcus cremoris* for production of lactic acid and

Salmonellosis Outbreak from a Dairy Plant

The largest salmonellosis outbreak in U.S. history occurred in Illinois during the spring of 1985. More than 175,000 people in the Midwest were stricken with diarrhea, nausea, vomiting, and fever as a result of milk contamination at the Jewel Food Stores' Hillfarm dairy plant near Chicago.

The Hillfarm Dairy, located in the suburb Melrose Park, is a state-of-the-art milk-processing plant that is computerized and capable of producing 175,000 gallons of milk a day. Prior to the outbreak, the plant had safely produced billions of gallons of milk for 18 years. However, starting in late March of 1985, cases of *Salmonella typhimurium* infection were reported to the Illinois State Health Department. These cases increased in number, and by April 9, more than 1,500 people were confirmed as ill from salmonellosis.

The outbreak was traced to the Hillfarm Dairy, and immediately scientists from the state health department, CDC, Food and Drug Administration (FDA), and Jewel probed the plant section by section, searching for the cause of the contamination. Over 400 miles of stainless steel pipes, hundreds of air valves, several giant holding tanks, and more than 600 large metal plates that heat and cool milk during pasteurization were carefully scrutinized for traces of *Salmonella*. Because the Hillfarm Dairy is such an efficient and well-maintained plant and has a closed processing system, there initially were speculations of sabotage. However, after thousands of cultures were taken, investigators concluded that the contamination was not sabotage. It was not a superbug or heat-resistant strain of *Salmonella*, and it was not a failure of the pasteurization process.

The contamination was a unique microbiological-engineering phenomenon, traced to two air-pressure valves in a small section of the plant. These two valves separate raw from pasteurized milk. The valves are normally closed to keep the two milks separate, but are opened when the processing plant is cleaned. The valves were repeatedly subjected to pressure testing with a bright red dye and initially were found to function properly. But upon further testing, investigators observed leaking between the two milk lines and concluded that this was the primary source of the milk contamination. The investigation also revealed that milk often was held unrefrigerated in this piece of pipe for considerable lengths of time, providing ideal conditions for bacterial growth.

Several interesting facts were uncovered during the course of this investigation. The enzyme phosphatase was not detected in batches of milk contaminated with *Salmonella*. This absence of phosphatase, an enzyme that is present in raw milk and is used for pasteurization quality control, puzzled scientists. It was concluded that phosphatase was present in the contaminated milk, but at a level far below the level of detection. Plasmid analyses of the *S. typhimurium* responsible for the Hillfarm outbreak matched those of a *Salmonella* strain isolated from another outbreak in August 1984 in the Chicago area. Although Hillfarm milk was implicated in the 1984 outbreak, there was little evidence to support the implication.

The Hillfarm salmonellosis outbreak was an expensive epidemic. It cost the dairy, which shut down April 9 after the outbreak began, $3.5 million and affected thousands of people who drank the contaminated milk. *Salmonella* was responsible for at least two deaths and was a contributing factor in several other deaths. This incident shows the serious problems that result from food contamination.

 Gram-Negative Organisms
*Bacilli—Facultative Anaerobes:
Enterobacteriaceae* • pp. 2–7,
16–20

Leuconostoc cremoris or *Leuconostoc dextranicum* for production of flavoring compounds (acetic acid and acetylmethylcarbinol). The bacteria are added either to skim milk (cultured buttermilk) or to cream (cultured sour cream).

Yogurt, a fermented, slightly acid, semisolid food, is prepared by first evaporating some of the water from skim milk. *Streptococcus thermophilus* and *Lactobacillus bulgaricus* are then added to the concentrated skim milk solids, and the mixture is incubated at 45°C for several hours. The characteristic flavor of yogurt comes from the accumulation of lactic acid and acetaldehyde produced by *L. bulgaricus*. Fruit is frequently added to yogurt to mask the sharp taste of acetaldehyde.

Other fermented milks that are produced through bacterial activity are acidophilus milk, Bulgarian milk, kefir, and koumiss. Acidophilus milk is essentially milk that has been sterilized and inoculated with *Lactobacillus acidophilus*. Sterilization of the milk is necessary prior to inoculation, because *L. acidophilus* is easily overgrown by other microorganisms. The resultant product is of therapeutic use in restoring the normal intestinal flora following antibiotic therapy. Bulgarian milk, prepared by inoculating milk

Examples of Cheeses and Microorganisms Associated with Their Production

Type	Name	Microorganism
Soft, unripened	Cottage	Leuconostoc citrovorum Streptococcus lactis
	Cream	Streptococcus cremoris
	Neufchâtel	Streptococcus diacetilactis
Soft, ripened	Brie	Brevibacterium linens Penicillium camemberti Penicillium candidum S. cremoris S. lactis
	Camembert	P. camemberti P. candidum S. cremoris S. lactis
	Limburger	B. linens S. cremoris S. lactis
Semisoft, ripened	Blue	Penicillium roqueforti S. cremoris S. lactis
	Brick	B. linens S. cremoris S. lactis
	Monterey Jack	S. cremoris S. lactis
	Muenster	B. linens S. cremoris S. lactis
	Roquefort	P. roqueforti S. cremoris S. lactis
Hard, ripened	Cheddar	Lactobacillus casei S. cremoris Streptococcus durans S. lactis
	Colby	L. casei S. cremoris S. durans S. lactis
	Edam	S. cremoris S. lactis
	Gouda	S. cremoris S. lactis
	Swiss	Lactobacillus helveticus Propionibacterium shermanii or Lactobacillus bulgaricus and Propionibacterium freudenreichii S. lactis
Very hard, ripened	Parmesan	L. bulgaricus S. cremoris S. lactis
	Romano	L. bugaricus Streptococcus thermophilus

with *L. bulgaricus,* is similar to commercial buttermilk but differs in its higher acidity and lack of aroma. Kefir and koumiss are beverages popular in parts of Europe. Kefir is prepared from cow, goat, or sheep milk by mixed lactic acid and alcoholic fermentation. Bacteria (*S. lactis* and *L. bulgaricus*) produce the acid, and yeasts produce the alcohol. In the milk, the organisms form small granules, or kefir grains, which can be removed, stored, and reused as starter cultures. Koumiss is prepared in a similar manner, but from mare milk. Both kefir and koumiss are slightly effervescent beverages of low alcoholic (1% to 2%) content.

The Manufacture of Butter Involves Bacteria

Butter is an inefficient but popular way to store milk fat. It is made by churning cream until butterfat globules are separated from the liquid buttermilk portion. Butter may be prepared from sweet cream or from cream seeded with starter cultures. *Streptococcus diacetilactis* and *S. lactis* are examples of bacteria used as starter cultures. Diacetyls produced by these bacteria give butter its characteristic flavor and aroma. The high fat and low water contents of butter discourage microbial growth and give butter a longer shelf life than fresh milk.

Microorganisms Are Important in Cheese Production

Cheese is thought to have originated in Asia 8,000 years ago. It was used by Roman soldiers as a high-protein storage form of milk during the invasion of Europe. Today cheese remains a popular food, with nearly 2,000 varieties manufactured around the world.

Most cheeses are made from cow milk (either whole or skimmed), although some are produced from cream or the milk of other animals. In Europe and southwestern Asia, sheep and goats are the major source of milk for cheese production. Reindeer, water buffalo, llamas, and camels provide milk for cheese in other parts of the world. The type of milk used influences the flavor, aroma, and pigmentation of cheese. Cheese made from goat milk or sheep milk is spicier and sharper in flavor than cheese prepared from cow milk because of the higher content of caproic, caprylic, and capric acids in the former. These fatty acids give a pungent aroma to the milk. Cow milk contains varying amounts of β-carotene, which gives cheese a yellowish color, while milk from other animals (goats, sheep, and water buffalo) has little or no β-carotene and yields cheeses that are basically white.

Cheeses can be broadly divided into two groups—**unripened cheeses** and **ripened cheeses** (Table 20.3)—which differ in the length of the aging process. Ripened cheeses are aged for extended periods of time (1 to 16 months), whereas unripened cheeses are obtained after milk curdles, without any further curing.

Cheese is made by adding a starter culture of lactic acid bacteria to fresh milk to curdle it; the milk is warmed to a temperature conducive to growth of the bacteria (Figure 20.1). *S. lactis* or *S. cremoris* are usually used for curdling at temperatures of 20°C to 37°C. Some cheeses (for example, Swiss cheese) are prepared by cooking

figure 20.1

Steps in Cheese Making

a. Starter culture is added to the milk to assist in curdling it. b. A milk-clotting enzyme such as rennin is added to coagulate the milk casein, forming a custardlike mass. c. Cutting begins the process of separating the liquid (whey) from the milk solids (curd). d. The curd and whey are cooked and stirred until the desired temperature and firmness of the curd is reached. e. Whey remains after the curd is tightly formed. This by-product is drained and may be used to make ricotta or other whey cheeses or as a starting material for the manufacture of industrial chemicals such as alcohol and lactic acid. f. The curds are salted to inhibit growth of undesirable micro-organisms, control the moisture content, and contribute to the flavor of the cheese. g. Pressing determines the characteristic shape of the cheese and helps complete the curd formation. h. The cheese is then moved to a humidity- and temperature-controlled room for ripening.

An Outbreak of Listeriosis Caused by Contaminated Milk

Listeria monocytogenes is a gram-positive bacterium that usually causes septicemia, meningitis, or abortion in humans. Because the bacterium is widely distributed in nature and can be recovered from such diverse sources as human and animal feces, sewage, insects, and plants, its mode of transmission is not completely understood. In recent years human listeriosis has increased in incidence, including some food-related outbreaks. One such outbreak of milk-borne listeriosis occurred in Massachusetts.

During the summer of 1983 (June 30 to August 30), 49 cases of human listeriosis were reported to the Massachusetts Department of Public Health. These 49 cases constituted an attack rate of 1 per 118,000 and was four to five times higher than rates reported in previous summers. Forty-two of the cases occurred in adults; the remaining seven cases involved mother-infant pairs. All of the adults had pre-existing illnesses or conditions causing immunosuppression. Disease symptoms included meningitis, septicemia, and abortion.

An epidemiological survey showed no geographical clustering of cases. Case-control studies ruled out person-to-person, airborne, or waterborne transmission of the disease. It was not acquired through consumption of fresh vegetables, meats, salads, or unpasteurized dairy products. A trend was detected, however, with the purchase of whole or 2% pasteurized milk by 14 of 19 cases (as compared with 11 of 38 neighborhood-matched controls) at a particular foodstore chain. From this information, it was determined that the milk was the source of the *Listeria.*

The milk implicated in this outbreak was traced to a single plant—a modern, clean, well-run facility with records indicating proper product pasteurization. During September, milk samples were taken from farm suppliers, the milk cooperative collecting the milk from farms, and the pasteurizing plant. *L. monocytogenes* was isolated from 15 of 124 samples obtained before pasteurization, but was not isolated from any pasteurized samples. In retrospect, however, the investigators recognized that the outbreak had ended by September and that no further cases of listeriosis were associated with consumption of the milk sampled.

Although the source of *Listeria* responsible for this outbreak was never determined, the investigation revealed that raw milk containing *L. monocytogenes* was a potential source of the disease. It was concluded that large amounts of *Listeria,* excreted in milk by cattle, may survive pasteurization, and that the postpasteurization storage of contaminated milk at refrigeration temperatures may result in the growth of bacteria. A small infectious dose may be all that is necessary to cause disease in some individuals, such as the immunocompromised adults in the Massachusetts outbreak. Since this report, outbreaks of listeriosis caused by milk and milk products such as cheese have occurred in other states. It is now recognized that *L. monocytogenes* may be transmitted by milk and milk products and that pasteurization may not effectively eliminate all listerias from these products.

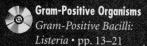

Gram-Positive Organisms
Gram-Positive Bacilli:
Listeria • pp. 13–21

---

curds at higher temperatures (50°C to 54°C). In these cases, starter cultures consist of bacteria that tolerate and grow at these higher temperatures (*S. thermophilus, L. bulgaricus,* or *Lactobacillus helveticus*). The enzyme rennin is also added to coagulate the milk casein and form a smooth, solid curd. Rennin previously was obtained from the stomachs of calves but now can be extracted from fungi.

During the incubation period, a watery fluid called **whey** is separated from the solid curd. The final moisture of the cheese can be changed by varying the temperature and stirring speed of the curd-and-whey mixture. After the moisture has been reduced to the desired level, the whey is removed, and the hot, soft curd is salted by immersing the immature cheese in saturated brine.

Salting inhibits the growth of undesirable microorganisms, controls the moisture content, and contributes to the flavor of cheese.

Following salting, the cheese is ripened. Microbes used in ripening may be introduced at almost any time of the preparation, depending on the type of cheese desired. Some microorganisms (for example, propionibacteria, lactobacilli, and streptococci for Swiss cheese) are added to the milk before the curdling process, whereas others (for example, *Penicillium roqueforti* for Roquefort and other blue cheeses) are inoculated into the blocks of curd.

Several things happen during ripening. Many of the lactic acid bacteria present in the cheese from the starter culture die and release intracellular enzymes that transform fats, carbohydrates, and proteins in the curd. Ripening microorganisms added to the cheese ferment carbohydrates and lactic acid remaining in the curd and produce chemical compounds that give the cheese its distinct aroma, flavor, and physical characteristics. For example, the holes or eyes of Swiss cheese are formed from carbon dioxide gas bubbles generated by *Propionibacterium*. The blue color and distinct flavor of Roquefort cheese and other blue cheeses result from growth of the blue-colored mold *P. roqueforti*. Roquefort cheese is made and ripened in limestone caves in the region around Roquefort, France, and the cheese gets its name from this village. The length of the ripening period determines the hardness of the cheese. Hard cheeses (cheddar, Swiss, Parmesan, and Edam) undergo less protein breakdown and are ripened for periods of 2 to 16 months. Semisoft cheeses (Roquefort, blue, Muenster, and brick) are ripened for shorter periods of time (usually two to three months) and soft cheeses (Limburger, Brie, and Camembert) have an even shorter ripening period. Unripened cheeses such as cottage cheese, ricotta, and cream cheese require no aging and may be used immediately after the curdling process.

Microorganisms in Meat and Meat Products

Meat and meat products are constantly exposed to microorganisms and consequently can spoil rapidly unless properly stored and refrigerated. The inner flesh of healthy animals is usually free of microbes and is contaminated only after it has come in contact with butchering implements (knives and saws) during bleeding, skinning, and cutting. As meat is cut, there is repeated inoculation of the exposed surfaces. Ground beef and sausages are especially susceptible to microbial contamination, because grinding increases surface area for microbial growth. Contamination can be minimized by regularly cleansing grinding utensils, knives, and saws or immersing them in solutions of hot water, but this is rarely done. Meat is an ideal culture medium for bacteria and other microorganisms because it is rich in protein and minerals, high in moisture, and is at a favorable pH for microbial growth.

Microorganisms Are Responsible for Meat Spoilage

The spoilage of meats can occur aerobically or anaerobically (Table 20.4). Under aerobic conditions, bacteria and yeasts may cause the production of surface slime or molds, changes in meat pigments, decomposition of meat fats, and the appearance of undesirable odors and tastes. Surface slime may be caused by species of *Pseudomonas, Achromobacter, Streptococcus, Leuconostoc, Bacillus,* and *Micrococcus.* Fungi growing aerobically on the surface of meats can cause spotting. These spots may be black (*Cladosporium herbarum*), white (*Sporotrichum carnis*), or green (*Penicillium* species). Although localized spots of surface spoilage can be trimmed, surface spoilage proceeds at a rapid rate and causes appreciable deterioration of meat.

Most raw meats have a red color, called the **bloom,** when they are fresh. This color is due to a pigment, myoglobin, that is

table 20.4

Microorganisms Associated with Meat Spoilage

Conditions	Type of Spoilage	Some Responsible Microorganisms
Aerobic	Surface slime	Achromobacter sp. Alcaligenes sp. Bacillus sp. Leuconostoc sp. Micrococcus sp. Pseudomonas sp. Streptococcus sp.
	Spotting	Cladosporium herbarum Penicillium sp. Sporotrichum carnis
	Changes in color	Lactobacillus sp. Leuconostoc sp.
	Rancidity	Achromobacter sp. Pseudomonas sp.
	Souring	Micrococcus sp. Pseudomonas sp.
Anaerobic	Souring	Clostridium sp.
	Putrefaction	Clostridium sp. Proteus sp. Pseudomonas sp.

present in the meat. Myoglobin is red in its oxidized form and brown in its reduced form. Bacteria (*Lactobacillus* and *Leuconostoc*) cause greening, browning, or graying of meat by the production of hydrogen sulfide and peroxides that reduce myoglobin. Many spoilage bacteria and yeasts have lipases that hydrolyze unsaturated fatty acids off of fats in meats. These fatty acids become more susceptible to oxidative rancidity. *Pseudomonas* and *Achromobacter* are examples of bacteria that cause rancidity of fats. Besides rancidity, bacteria and yeasts also produce undesirable odors and tastes in meat as they grow. **Souring,** the term used to describe these odors and tastes, results from volatile acids (for example, acetic, formic, butyric, and propionic acids) that are formed during microbial metabolism of meat proteins and carbohydrates.

Putrefaction is the anaerobic decomposition of meat proteins with the production of hydrogen sulfide, indole, mercaptans, ammonia, amines, and other foul-smelling compounds. Putrefaction is usually caused by *Clostridium,* although some facultative anaerobes are putrefactive. Gas formation (CO_2 and H_2) frequently accompanies putrefaction by clostridia.

Many Methods Can Be Used to Preserve Meats and Reduce Spoilage

The spoilage of most meats can be reduced by prompt chilling of the meat at freezing or near-freezing temperatures. The shelf life of fresh meats normally is only three to ten days at 5°C. Depending

figure 20.2
Salting Ham

Type of Food	Type of Spoilage	Some Responsible Microorganisms
		table 20.5
		Microorganisms Associated with Spoilage of Poultry and Fish
Poultry	Slime, odor	*Achromobacter* sp. *Pseudomonas* sp. *Salmonella* sp.
Fish	Discoloration	*Micrococcus* sp. *Pseudomonas* sp. *Serratia* sp.
	Rotting	*Flavobacterium* sp. *Micrococcus* sp. *Pseudomonas* sp. *Serratia* sp.

on the number of microorganisms present, meat can be stored for approximately 30 days at a temperature of 0°C. Refrigerated storage of meat will not kill pyschrophilic bacteria (*Pseudomonas, Achromobacter,* and *Flavobacterium*) that cause spoilage, but it does effectively reduce their growth (see psychrophiles, page 90).

Frozen storage of meat at –20°C further extends the shelf life to several months. Freezing kills many of the spoilage bacteria, but sufficient numbers remain so that appreciable growth occurs if thawing is done slowly. Meat deteriorates in quality during freezing not so much because of microbial damage, but because of gradual decreases in flavor and odor due to dehydration (freezer burn) and oxidation. Such losses can be minimized by careful wrapping of meat prior to freezing. Microbial growth is reduced in aged beef (beef that is aged at refrigeration temperature for several weeks) by maintaining the relative humidity at 80% to 90% in the aging chambers.

The dehydration of meats has been practiced for centuries. Native Americans hung strips of beef in the sun to produce jerky. Some types of sausage are preserved by drying. Today freeze-drying of meats and other foods is a practical and popular method to preserve them without changing their flavor or texture. Dehydrated foods can be stored for long periods of time without refrigeration. Dehydration reduces the moisture content of the food and prevents growth of microorganisms (see water activity, page 92).

Dehydration is frequently used with the curing of meats (Figure 20.2). In the past, beef and pork were cured with salt (sodium chloride) to preserve these meats without refrigeration. Today other curing agents (sugar, sodium nitrate, sodium nitrite, and sometimes vinegar) are used for flavoring as well as preserving meats. Sodium chloride and sugar serve as flavoring agents and as preservatives by lowering the water activity (a_w). For example, the salts, sugar, and meat protein lower the a_w in cured meats such as hams. Sugar is also an energy source for nitrate-reducing bacteria in the curing solution. Sodium nitrite is an inhibitor of some anaerobic bacteria, including *Clostridium,* and also is a source of

nitric oxide, a color fixative that gives cured meats their color. Sodium nitrate is a reservoir for the formation of nitrite by bacterial reduction. Nitrites and nitrates are effective preservatives. However, their use in curing has been restricted in recent years because these compounds can react with secondary and tertiary amines to form nitrosamines, which are highly carcinogenic.

Most meats are smoked after curing to further preserve them and to give them a characteristic flavor. The heat of the smoking process, together with chemicals released from wood smoke and the drying of the meat, prevent microbial growth during storage. One of the chemicals identified in wood smoke is formaldehyde. Although present in small quantities, formaldehyde is the primary bactericidal agent in wood smoke.

Chemicals have been widely used in the past to preserve meat, but strict regulations in recent years by the U.S. Food and Drug Administration (FDA) have curtailed their use. Sodium benzoate, calcium propionate, boric acid, and sorbic acid are some of the chemicals still used in food preservation. These chemicals are organic compounds that inhibit microbial growth by interfering with cell metabolism or the integrity of the plasma membrane. Ionizing radiation has also been used experimentally with meats but has not been practical because it is expensive and causes undesirable changes in color and flavor.

Poultry and Seafood

Poultry and fish are major sources of animal protein in the human diet. However, these foods can be easily contaminated during processing. As poultry is processed, microorganisms from the skin, feathers, feet, and intestinal tract are potential sources of contaminants. Microorganisms commonly found on processed poultry include bacteria of the genera *Pseudomonas, Achromobacter, Flavobacterium,* and *Micrococcus* (Table 20.5). Spoilage of poultry is usually caused by *Pseudomonas,* although *Achromobacter* and yeasts may also be involved. Prompt chilling of the bird after processing can retard this spoiling.

Antibiotics such as chlortetracycline and oxytetracycline at levels not exceeding 7 ppm may also be used to retard bacterial growth. The antibiotics can be applied to the surface of pieces of meat or be fed to the animal before slaughter. Currently an estimated 2.5 million kilograms of antibiotics are fed to cattle, poultry, and other animals every year in the United States. It is thought that 75% to 80% of all cattle and poultry are given antibiotics sometime during their lifetimes. Although antibiotics promote the general health of an animal and reduce the numbers of potential spoilage bacteria in its intestinal tract, the FDA is concerned that such indiscriminate use of antibiotics in animal feed may lead to the appearance of resistant strains of bacteria. Problems with antibiotics were first reported in 1954, when fatalities among calves in a California dairy herd were traced to salmonellae that had become resistant to the antibiotics. More recently, in 1983, at least 18 people in the Midwest became ill by a tetracycline-resistant *Salmonella* strain traced to ground beef. The CDC later discovered that most of the victims ate ground beef made from cattle that had been fed tetracycline.

The microbial flora of seafood depends upon microorganisms present in the water. Fish generally are covered with bacteria of the genera *Pseudomonas, Achromobacter, Vibrio, Flavobacterium, Sarcina, Micrococcus,* and *Serratia.* The numbers of bacteria on the skin of newly caught fish range from a few hundred to several million per square centimeter. Most of these bacteria are removed by washing, but the remaining ones rapidly spread to the flesh after the fish is opened. Fish is highly susceptible to microbial spoilage, so special precautions must be taken to minimize microbial growth. Freshly caught fish should be rapidly chilled in ice. Additional means of preservation include salting, drying, heating, and freezing.

Spoilage, when it does occur, consists of autolysis, oxidation, or bacterial activity. As unsaturated fats in fish oils are oxidized, trimethylamine is released to give a stale fish odor. The bacteria most often involved in spoilage are those that are part of the natural flora of the external slime of fish. During spoilage, fish enzymes cause autolytic changes and release nitrogenous compounds (for example, amino acids and amines) and carbohydrates, which then are used by the bacteria. Bacteria metabolize these substrates and produce odorous products such as amines, ammonia, aldehydes, sulfides, and indole. Oysters generally remain in good condition as long as they are in the shell and kept alive at cool temperatures. However, oysters decompose rapidly when they are dead, after shucking.

Canned Foods

Nicholas Appert (c. 1750–1841) was awarded a prize of 12,000 francs in 1810 by Napoléon Bonaparte for the development of a method for preserving foods by heating them in cork-stoppered, widemouthed glass bottles. Ever since then, canning has been a popular method for food preservation. Today commercial canning involves heating canned goods at a high temperature long enough to destroy any spores of *Clostridium botulinum* (see botulism, page 517). Process heating temperatures vary from 100°C for high-acid foods (pH below 4.0) to 121°C for low-acid foods (pH above 4.5). The National Canners Association recommends a thermal process time for low-acid foods that is sufficient to reduce a population of 10^{12} *C. botulinum* spores to one surviving spore. Because 10^{12} spores is an improbably large population, this criterion ensures the safety of canned foods with respect to *C. botulinum* contamination. The canning industry seeds test cans during processing with the relatively heat-resistant indicator species *Bacillus stearothermophilus* as a quality control measure. If spores of *B. stearothermophilus* have survived the processing, the entire batch must be discarded.

Modern commercial canning procedures are highly reliable, and very few cases of botulism are traced to commercial canning. Most cases are caused by improper home canning, in which insufficient heat was applied, particularly for low-acid foods such as meats, fish, vegetables, and the new varieties of low-acid tomatoes.

Growth of most microorganisms is prevented in canned foods not only by the heating process, but also by the low intrinsic pH of high-acid foods and by the removal of oxygen during canning. When spoilage of canned foods occurs, it usually is caused by a hydrogen swell, resulting from the pressure of hydrogen gas released by the action of acid from the food on the metal of the can. Microbial spoilage of canned foods is rare, but it can occur if microbes either survive the commercial canning process or enter the container through a leak after heat treatment (Table 20.6). Endospore-forming bacteria such as *Bacillus* and *Clostridium* are usually the culprits in underprocessed canned foods, whereas non-spore-forming bacteria are associated with entry through leaks in canned foods. Acids (lactic acid) and gases (H_2, CO_2) produced by microbial metabolism are responsible for the sour odor and gaseous swell that are characteristic of spoiled canned foods.

Fermented Foods

Many important foods are made by microbial fermentations. These include meats, vegetables, fruits, and breads (Table 20.7).

Several different types of sausage are prepared by allowing meat to be fermented by microorganisms. *Pediococcus cerevisiae* is generally used to produce salami, Lebanon bologna, and summer sausage. The acids formed from heterolactic acid fermentation preserve the meat and also give it a distinct, tangy flavor (see pH and bacterial growth, page 91).

When cabbage is shredded, salted, and packed tightly, the first organisms to multiply are species of *Leuconostoc.* The salt draws out the plant juices containing the sugars and helps control the growth of other bacteria. Under these conditions *Leuconostoc* proliferates, producing acetic acid, lactic acid, ethanol, mannitol, carbon dioxide and other compounds that contribute to the flavor

Microbiology of Canned Food Spoilage

Type of Canned Product	Type of Spoilage	Some Responsible Microorganisms
Low-acid products, pH above 4.5 (for example, asparagus, corn, and peas)	Flat sour Swelling	Bacillus stearothermophilus Clostridium nigrificans Clostridium sporogenes Clostridium thermosaccharolyticum
High-acid products, pH below 4.0 (for example, fruit and tomato juices)	Flat sour Swelling	Bacillus thermoacidurans Clostridium butyricum

t a b l e 20.7

Examples of Fermented Foods Prepared Using Microorganisms

Fermented Food	Starting Material	Microorganisms Involved in Fermentation
Beer	Grains	Saccharomyces sp.
Bread		
rye	Wheat flour	Lactobacillus brevis Leuconostoc mesenteroides
sourdough	Wheat flour	Lactobacillus sanfrancisco Saccharomyces exiguus
white	Wheat flour	Saccharomyces cerevisiae
Cured ham	Pork	Aspergillus sp. Penicillium sp.
Olives	Fresh olives	Lactobacillus plantarum L. mesenteroides
Pickles	Cucumbers	L. plantarum and other species Pediococcus sp.
Poi	Taro root	Lactobacillus sp.
Sauerkraut	Cabbage	L. plantarum and other species Leuconostoc sp.
Sausage	Pork and beef	Pediococcus cerevisiae
Soy sauce	Rice, soybeans	Aspergillus oryzae Lactobacillus delbrueckii Saccharomyces rouxii
Wine	Grape juice	S. cerevisiae Saccharomyces champagnii

of sauerkraut. The pH of the cabbage is lowered to about 3.5 by the acids. At this pH, species of *Lactobacillus* (primarily *Lactobacillus plantarum*) begin to multiply and produce additional amounts of lactic acid, which further lowers the pH to about 2.0. The lactobacilli also utilize the mannitol and thus remove the bitter flavor of the sauerkraut. The fermentation is finally stopped by canning or refrigerating the sauerkraut.

Pickles are prepared by placing cucumbers in barrels or tanks of brine, where *L. plantarum*, as part of the cucumber's normal flora, begins to ferment the sugars. As the cucumbers are fermented, they become soft from the acid. After fermentation the pickles are soaked to remove excess salt before they are made into sour, sweet, or other types of pickles.

Microorganisms are useful in two ways in bread making: (1) they may produce gas to raise or leaven the dough and (2) they may produce metabolic by-products that contribute to the flavor of bread. Leavening of bread is usually accomplished through the action of *Saccharomyces cerevisiae* and other yeasts, which ferment the sugars in the dough and produce carbon dioxide and ethanol. The carbon dioxide causes the dough to rise and gives the product the desired loose, porous texture. The ethanol evaporates and the yeasts are killed during baking. Although yeasts may contribute to the flavor of bread through products released during the fermentation of sugars, bacteria growing in the dough contribute the most to flavor. For example, the tangy or sour flavor of rye bread comes from the growth of heterofermentative lactic acid bacteria such as *Lactobacillus brevis* and *Leuconostoc mesenteroides*.

Alcoholic Beverages

The production of alcoholic beverages is an economically important industry that relies heavily on the use of microorganisms. The most significant alcoholic beverages are beer, wine, and distilled beverages.

Yeasts Are Used in Beer Production to Ferment Grains

Beer is made from the fermentation of barley grains by yeasts. Grains contain starch, which cannot be used directly by yeasts; thus the starch must first be broken down to fermentable sugars by a process called **malting.** In malting, the barley grains are germinated and crushed to release amylases and proteases. The malt then is mashed and gradually heated with water to allow the amylases to convert the starch to dextrins and maltose and the proteases to convert the proteins to more-soluble molecules. The liquid portion, called the **wort,** is then separated from the insoluble grain solids, mainly the barley husks. The wort is cooked with **hops,** the dried petals from the vine *Humulus lupulus* (Figure 20.3). Hops are added as a flavor enhancer and antimicrobial agent. The antimicrobial action of hops is due to their α resins.

After several hours of cooking, the wort is transferred to a fermentation vat, and yeast is added in large quantities (a process called **pitching**). Two types of yeast are generally used in beer brewing. Top-fermenting yeasts (*Saccharomyces cerevisiae*) are vigorous fermenters grown at relatively high temperatures (14°C to 23°C). These yeasts are uniformly distributed throughout the fermenting wort and are carried to the surface by the carbon dioxide produced during fermentation. The product formed from these yeasts has a uniform turbidity—heavy beers of high alcoholic content, such as ale. A higher concentration of hops is used in the production of ale and contributes to its tart taste. Bottom-fermenting yeasts (*Saccharomyces carlsbergensis*) ferment more slowly and at lower temperatures (6°C to 12°C). These yeasts aggregate during fermentation and settle to the bottom of the fermentation vat. Bottom-fermenting yeasts produce beer of lighter alcoholic content, such as lager.

Following fermentation, the green beer, as it is now called, has a harsh taste that can be removed by aging for a few weeks to several months in refrigerated storage tanks. During the aging process, proteins, yeasts, and resins precipitate from the beer, resulting in a mellowing of the flavor. The mature beer is separated from the precipitate and filtered. The finished product is carbonated and then pasteurized at 60°C to 61°C to remove viable yeast.

Low-carbohydrate (low calorie) "light" beers currently represent a significant share of the beer market and are made with yeasts that ferment carbohydrates in the wort completely to alcohol and carbon dioxide, reducing the carbohydrate concentration in the beer to a minimum. Although brewing yeast strains are capable of hydrolyzing many carbohydrates in the wort, they are unable to ferment wort dextrins. One technique for producing low-carbohydrate beers is to add fungal glucoamylases to the wort during fermentation to hydrolyze wort dextrins to fermentable sugars. However, glucoamylase is not completely inactivated by normal pasteurization temperatures, and its presence in the finished product can result in the hydrolysis of residual dextrins to glucose, producing an undesirable sweetness. One potential procedure to overcome this problem is to use a brewing yeast strain that possesses amylolytic activity and thus is able to hydrolyze the dextrins during the brewing process. The species *Saccharomyces diastaticus* possesses the extracellular enzyme glucoamylase and has been investigated for its potential use as a brewing strain.

figure 20.3
Brewing Beer
Wort is vigorously boiled with hops in large brew kettles. The hops impart fragrance and flavor to the final product.

figure 20.4
A Bank of 10,000- to 200,000-Liter Fermentors Used for the Commercial Growth of Microorganisms

Unfortunately, *S. diasticus* strains produce beers with a characteristic phenolic off-flavor. However, the gene responsible for this off-flavor (*POF*, or phenolic off-flavor, gene) has been identified, and in the future it may be possible to genetically construct a *S. diasticus* strain containing the genes for glucoamylase production but not for the phenolic off-flavor.

Yeasts Are Used in Wine Production to Ferment Fruit Sugars

Wine is made by yeasts that ferment simple sugars (for example, fructose and glucose) in grapes and other fruits. Wine making is a simpler process than beer making, because the inoculum, *Saccharomyces cerevisiae* var. *ellipsoideus,* is either already present on the skins of the grapes or is added. The grapes are crushed to release their juice, called **must.** For white wine, the must is separated from the **pomace** (the skin, pits, and stems) by pressing before fermentation is started. For red wine, the must is fermented with the pomace, which produces the red color; after fermentation, the pomace is removed by pressing. The fermented wine is then placed in vats to clarify and age. During aging, suspended solids are sedimented, and the wine slowly clears. The wine is periodically filtered from the sediment and added back to the top of the storage vat, a procedure called **racking.** Fine wines are usually aged for several months or years to improve their flavor and odor (**bouquet**).

Yeasts Are Used in the Production of Distilled Beverages

Yeast fermentation is limited by the quantity of alcohol present in the medium, and at 18%, alcohol fermentation stops. For this reason, distillation is required for the production of hard liquors.

Distilled beverages are made by heating a fermented liquid at temperatures high enough to volatilize the alcohol, which is then condensed and concentrated by a distillation apparatus. The final product depends on the type of liquid distilled. Whiskey is made from fermented malt. Brandy is a product of fruit or fruit juice, rum is produced from molasses, and vodka can be made from a variety of grains. Because the fermented liquid is concentrated during distillation, distilled beverages have much higher alcoholic contents than beer or wines. The yeasts used in the production of distilled liquors are usually special distiller strains of *Saccharomyces cerevisiae,* which give high yields of alcohol.

Industrial Processes

Industrial microbiological processes are large-scale operations carried out in immense fermentors that hold 10,000 to 200,000 liters (approximately 2,500 to 50,000 gallons) of liquid (Figure 20.4). Raw materials, microorganisms, and environmental conditions must be combined in the right proportions in fermentation to efficiently obtain high yields of the desired product. This is the job of the industrial microbiologist, who selects culture media, microbial strains, and proper growth conditions for fermentation.

The industrial use of microorganisms is not new. Microbes have been used for centuries to make food and alcoholic beverages, produce chemicals, and manufacture commercially important products. Microorganisms are ideal for such purposes because their small size and high metabolic rates make them exceptionally efficient living factories. Recent advances in genetic engineering have generated new industrial applications of microorganisms and have been responsible for the development of a new discipline,

table 20.8

Composition of Beet Molasses and Milk Whey

Composition of Beet Molasses		Composition of Milk Whey	
Constituent	% of Dry Weight	Constituent	% of Dry Weight
Sucrose	48–50	Lactose	64–79
Raffinose	1	Protein	10–14
Glucose + fructose	1	Soluble nitrogenous compounds	2.3
Nitrogen compounds	12–13	Lipids	1.4–3.2
Glutamate	3.5	Mineral salts	8.6–12.7
Other amino acids (aspartate, asparagine, alanine, glycine)	5.5		
Betaine	3.25–4.25		
Ash	11–12		
Other	10–15		

biotechnology, which is concerned with the industrial applications of microorganisms. Today microorganisms can be programmed to perform specific functions in industrial microbiology. As a consequence, many commercial products of economic value are now made wholly or in part by microbes. These products fall into three major categories: (1) enzymes synthesized by the microbe, (2) metabolic products, and (3) the microbial cells themselves.

A Desirable Culture Medium Is Inexpensive, yet Contains All Essential Ingredients

Growing microorganisms in large quantities is not an easy or inexpensive task. The chemical composition of the culture medium used in industrial fermentation is very important from the standpoint of economy and utility. The culture medium should be inexpensive, yet contain all the essential ingredients for microorganisms to produce large quantities of the desired product. The brewery industry has successfully grown large quantities of yeast for years; now that knowledge is applied to the growth of other microorganisms for commercial use.

The carbon source is the largest component of the culture medium. If carbon is limiting, the total biomass of the culture is correspondingly reduced (see carbon source for microbial growth, page 81). Pure sugars such as glucose and sucrose are usually too expensive to use industrially, so a cheaper source of carbon is employed. Molasses (a by-product of the cane and beet sugar industry), milk whey, and starch (cereals and potatoes) are examples of alternative carbon sources frequently used in industrial fermentations (Table 20.8). Beet molasses is composed of approximately 50% sucrose and also has large quantities of nitrogenous compounds. Cane molasses contains 62% sugar (sucrose, glucose, and fructose) and is rich in vitamins. Milk whey, a by-product of cheese manufacture, is an inexpensive source of lactose and proteins. Starch is useful as a carbon source only after it has been pretreated with plant enzymes or microbial amylases to hydrolyze this polysaccharide into simple utilizable sugars.

An organic or inorganic nitrogen source is added to the culture medium (see nitrogen source for microbial growth, page 81). In the antibiotic industry, corn steep liquor is often used. Corn steep liquor (a by-product of maize starch production) contains 7% to 8% nitrogen and is rich in amino acids, nucleosides, and vitamins. Ammonia is a common inorganic nitrogen source that is easily assimilated by microorganisms.

Minerals and microbial growth factors may be included in the culture medium (see growth factors for microbial growth, page 84). Requirements for phosphorus and sulfur are satisfied by the addition of phosphates and sulfates. Some microorganisms require vitamins and enzyme cofactors for maximal growth. Growth factors are usually present in the materials used as sources of carbon and nitrogen, although they may also be added in the form of yeast or liver extract as a supplement to the medium.

The prepared culture medium is sterilized to prevent contamination. Different methods of sterilization are used for the large volumes of liquids used in industrial fermentations. One method is the direct injection of steam into the medium. A second method is the circulation of steam through pipes immersed in or surrounding the culture medium. Both of these methods take several hours, and the extended heating can cause denaturation of proteins and destruction of growth factors. An alternative method that takes less time and does not decrease medium quality is continuous sterilization: the medium is passed through a heat exchanger, in which it is heated at 120°C for a short time (ten minutes) and then rapidly cooled (Figure 20.5).

Microorganisms for Industrial Fermentation Are Carefully Selected

An inoculum for a 50,000-liter fermentor should be approximately 5% to 10% of the volume of the fermentor. It is prepared in several stages. Microbial growth from an agar slant or plate is inoculated into a liquid medium in a small Erlenmeyer flask. After several hours, the exponentially growing culture is used to inoculate a small fermentor

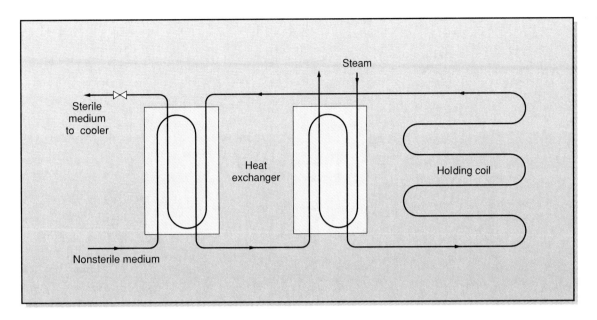

figure 20.5

Continuous Sterilization of Culture Media

In continuous sterilization, the culture medium is passed through a heat exchanger in which it is heated at 120°C and then rapidly cooled.

(300 to 500 l). Growth from this fermentor is then used to inoculate a larger fermentor (3,000 to 5,000 l) and the resultant culture is used as an inoculum for the production vessel. Care must be taken at all stages of inoculum preparation to avoid contamination.

The microorganisms used in industrial fermentations are carefully selected to ensure that they have the appropriate characteristics for scaleup growth. To be effective, the microorganism should have a high degree of genetic stability and be able to produce large quantities of the desired end product. The microbe should also be easy and inexpensive to cultivate and have a high rate of reproduction. Yeasts are often more useful than bacteria for industrial fermentations, because they can be grown to high densities while continuously excreting the desired product, are stable with respect to mutations, and do not produce endotoxins that can complicate product purification. Suitable strains of microorganisms may be isolated from natural environments or may be custom-made in the laboratory. For example, strains of *Bacillus subtilis* and *Bacillus licheniformis* used for the commercial production of alkaline proteases were isolated from hippopotamus dung from a Copenhagen zoo. The original culture of *Penicillium chrysogenum* used in penicillin production was obtained from a moldy melon purchased in Peoria, Illinois. This fungus was mutated in the laboratory by radiation and chemical agents more than 20 times to eventually develop a strain capable of producing several thousand times more penicillin than the original melon isolate.

Historical methods to improve strains by random mutations and selection are rapidly being replaced by newer techniques involving genetic recombination. Significant advances in biotechnology now make it possible to create desirable genetic alterations in microorganisms using such techniques as transformation, phage-mediated transduction, plasmid-mediated conjugation, and other recombinational methods. Recombinant DNA technology, unlike relatively hit-or-miss methods of mutation and selection, can be used to custom-make microorganisms that can produce large quantities of specific proteins.

Culture Conditions Are Carefully Controlled in Industrial Fermentors

Industrial fermentors are usually tall, stainless steel, cylindrical vessels with a height-to-diameter ratio of at least 5 to 2. The culture medium in these large fermentors must be aerated to provide the oxygen required for growth of aerobic organisms. Pure oxygen is too expensive to use for aeration, so sterile air is usually injected under pressure into the culture medium. Aeration is accompanied by vigorous agitation to obtain maximum growth rates. Foaming may develop from proteins and microbial products in the culture medium and is reduced by the use of antifoam agents (vegetable or animal oils, long chain alcohols, or detergents).

Most microorganisms have an optimum pH and temperature for growth. Because microbial metabolism often produces heat and leads to a change in pH of the medium, the temperature and pH must be monitored and adjusted. Hydrochloric acid, sodium hydroxide, or ammonia are used to adjust the pH of the medium. Optimum growth temperature is maintained by circulating warm or cold water through coils or jackets on the fermentation vessel.

Different Methods Are Used to Culture Microorganisms Industrially

Industrial fermentations may be carried out in several different ways, depending on the microbe and type of product produced. Solid media or semi-solid media are infrequently used for

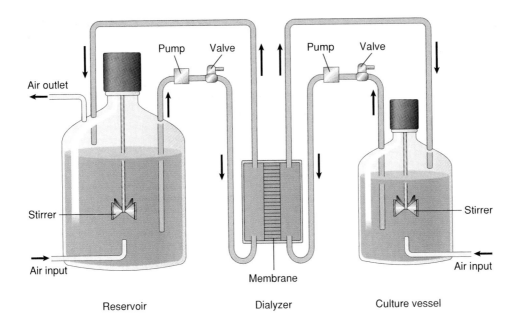

figure 20.6

Dialysis Culture

In dialysis culture, nutrients and metabolites are exchanged through a dialyzing membrane. Nutrients pass through the membrane from the reservoir to the culture vessel, and metabolites pass through the membrane in the opposite direction.

fermentation processes in Western countries because of these media's space requirements and harvesting difficulties. The Japanese, however, obtain fungal amylases from *Aspergillus oryzae* grown on a mixture of bran and starch. Fungal mycelia that develop on this culture medium are dried and ground to yield a crude amylase preparation at low cost.

Most industrial fermentations are carried out in **batch cultures** or in **continuous cultures**. In batch fermentations, the inoculated culture is grown, and the desired product is extracted by centrifugation, filtration, or other methods (see batch culture, page 115). Continuous cultures are useful in situations where either a culture condition (for example, pH or temperature) or the growth substrate permits growth of only the target organism (see continuous culture, page 118). For example, continuous culture is used in the production of yeast, beer, and lactic acid and in the production of single-cell protein from methanol. Continuous culture techniques are more susceptible to contamination and to the selection of mutants with an increased growth rate, but they are desirable because they yield many times more product per unit time than do batch cultures.

Dialysis culture and the use of **immobilized enzymes** and **immobilized microbial cells** are two innovative concepts used in industrial processes. In the dialysis culture technique, microorganisms are suspended in a vessel of liquid medium that is separated by a membrane from a second, larger vessel containing fresh medium (Figure 20.6). Nutrients and metabolites are exchanged through the membrane. Dialysis culture is used in situations where buildup of toxic metabolites can inhibit microbial growth. For example, *Lactobacillus delbrueckii* produces lactic acid from glucose metabolism, but the acid inhibits growth of the bacterium. The concentration of lactic acid can be reduced below the inhibitory concentration of 10 g/l by dialysis culture, thereby permitting continued growth of the bacterium and increased recovery of lactic acid from glucose metabolism. Threonine biosynthesis in *Escherichia coli* can be increased severalfold by removal of threonine through dialysis culture. This removal prevents feedback inhibition of the enzyme homoserine kinase in the threonine biosynthetic pathway. Dialysis culture is also used in the recovery of products such as enzymes or toxins; 80 times more botulinum toxin is recovered by dialysis culture than by conventional methods. The toxin is nondiffusible and therefore is concentrated in the growth chamber during culture.

The industrial application of immobilized enzymes and immobilized microbial cells to prepare and recover chemical products is relatively new. This technology involves adsorbing microbial enzymes or cells onto a solid support such as cellulose particles or polyacrylamide beads. A substrate solution is then passed over the support. The substrate attaches to the immobilized enzymes or cells and is converted to the desired end product.

table 20.9

Applications of Immobilized Microbial Cells

Enzyme Reaction	Application	Microorganism	Immobilization Method
Oxidoreductase	L-sorbose $\xrightarrow{\text{L-sorbose dehydrogenase}}$ L-sorbosone	*Gluconobacter melanotenus*	Polyacrylamide
Transferase	$NAD^+ + ATP \xrightarrow{\text{NAD}^+ \text{ kinase}} NADP^+ + ADP$	*Achromobacter aceris*	Polyacrylamide
Hydrolase	Lactose + H_2O $\xrightarrow{\beta\text{-galactosidase}}$ D-glucose + D-galactose	*Lactobacillus bulgaricus* *Saccharomyces lactis*	Polyacrylamide Cellulose triacetate
	Penicillin G + H_2O $\xrightarrow{\text{penicillin amidase}}$ 6-aminopenicilloic acid + phenylacetic acid	*Escherichia coli*	Polyacrylamide
Lyase	L-aspartate $\xrightarrow{\text{aspartate 4-decarboxylase}}$ L-alanine + CO_2	*Pseudomonas dacunhae*	Polyacrylamide
	Indole + L-serine $\xrightarrow{\text{tryptophan synthase}}$ L-tryptophan	*Escherichia coli*	Polyacrylamide
Isomerase	D-glucose $\xrightarrow{\text{glucose isomerase}}$ D-fructose	*Actinomyces* sp.	Anion exchange resin

Immobilized industrial systems are economical because product is continuously produced in a minimal amount of space. Immobilized enzymes and microbial cells were first used in the late 1960s and early 1970s to prepare amino acids, and are now used to produce other compounds such as sugars, organic acids, and alcohols (Table 20.9).

Enzymes

Enzymes—the protein molecules that catalyze reactions and transform chemicals—are widely used in industry in food production, textile manufacture, and the production of pharmaceuticals. The present world market for industrial microbial enzymes generates sales of nearly $200 million.

There are several reasons for the industrial production of microbial enzymes. Microorganisms grow rapidly and can be genetically programmed to synthesize large quantities of a particular enzyme. Furthermore, the enzymes that are formed are usually excreted into the medium, where they can easily be recovered. Patents have been granted for more than a thousand different microbial enzymes, but the enzymes of greatest industrial importance today are α-amylase, protease, pectinase, invertase, and glucose oxidase (Table 20.10).

Microbial Amylases Are Used in Bread Making, Beer Production, Manufacture of Sugar Syrups, and Textile Manufacture

α-amylase is an enzyme that hydrolyzes the $\alpha(1,4)$ linkage in starch to produce dextrin, maltose, and oligosaccharides. *Aspergillus niger*, *Aspergillus oryzae*, and *Bacillus subtilis* are the principal microorganisms used for the commercial production of amylases.

Amylases are important in bread making, beer production, the manufacture of sugar syrups, and textile manufacture. They are added to bread dough to promote the breakdown of starch to sugars, which are then fermented by yeast to produce CO_2. The brewing industry uses amylases to produce beers without dextrins (the low-calorie beers). These same enzymes are used by soft drink manufacturers for the production of sugar syrups from starch. Several steps are involved in this process, starting with the hydrolysis of starch by α-amylase to maltose and oligosaccharides. These products are reduced to glucose by the enzyme glucoamylase, and the glucose is then isomerized to fructose, the sweet sugar that is found in many soft drinks. The confectionery industry uses amylases in a similar manner to produce chocolate syrup from cocoa. Bacterial amylases, which are more stable at high temperatures, are used by the textile industry in a manufacturing step known as desizing. Threads used in weaving are prestrengthened with starch. After weaving has been completed, the cloth is soaked in a dilute solution of amylases to remove the starch.

Microbial Proteases Are Used in the Clothing Industry, in Photography, and in Stain Removal

Proteases are enzymes that cleave peptide bonds in proteins. Bacterial proteases (from *Bacillus subtilis* and *Bacillus licheniformis*) and fungal proteases (from *Aspergillus niger* and *Aspergillus oryzae*) are widely used in industry. Protease treatment of animal hides during tanning gives the leather a soft texture and is an attractive alternative to the traditional tanning process of soaking hides in dog feces and urine. In the clothing industry, proteases are used to dissolve a gummy substance (sericin) that surrounds silk fibers. Proteases are employed in photography to hydrolyze the gelatin base of photographic films to recover their silver content. These enzymes are also popular for removing proteinaceous stains such as blood, mucus, and chocolate from clothing.

table 20.10

Industrially Useful Microbial Enzymes

Industry	Application	Enzyme	Source
Analytical	Sugar determination	Glucose oxidase	Fungi
		Galactose oxidase	Fungi
Baking and milling	Bread baking	Amylase	Fungi, malt
		Protease	Fungi
Brewing	Mashing	Amylase	Bacteria, malt
		Glucoamylase	Fungi
	Oxygen removal	Glucose oxidase	Fungi
Carbonated beverage	Oxygen removal	Glucose oxidase	Fungi
Cereal	Breakfast food manufacture	Amylase	Fungi, malt
Chocolate, cocoa	Syrup manufacture	Amylase	Bacteria, fungi
Coffee	Coffee bean fermentation	Pectinase	Fungi
Confectionery, candy	Soft-center candy manufacture	Invertase	Fungi
	Sugar recovery from scrap candy	Amylase	Bacteria, fungi
Dairy	Cheese production	Rennin	Animals, fungi
	Modifying milk fats for flavor	Lipase	Animals, fungi
Dry cleaning	Spot removal	Amylase	Bacteria, fungi, pancreatin
		Protease	Bacteria, fungi, pancreatin
Fruit and fruit juice	Clarification, filtration, concentration	Pectinase	Fungi
Pharmaceutical and clinical	Manufacture of digestive aids	Amylase	Bacteria, fungi, pancreatin
		Protease	Bacteria, fungi, bromelain, papain, pancreatin, pepsin
Photographic	Recovery of silver from spent film	Protease	Bacteria
Textile	Desizing of fabrics	Amylase	Bacteria, malt, pancreatin
		Protease	Bacteria, fungi, pancreatin

Other Microbial Enzymes Are of Industrial Importance

Pectinases break down pectins found in plant tissues. These enzymes, which are industrially prepared from *Aspergillus* and *Penicillium*, are used commercially to clarify wines and fruit juices. Invertase is produced by *Saccharomyces cerevisiae* and *Saccharomyces carlsbergensis*. This enzyme hydrolyzes sucrose to the more soluble fructose and glucose that are used as liquid centers in chocolates. Fungi (*Aspergillus niger* and *Penicillium notatum*) produce glucose oxidase, an enzyme that oxidizes glucose to gluconic acid with the simultaneous production of hydrogen peroxide. Glucose oxidase is used in the food industry to remove glucose from dried egg white to prevent the discoloration that normally would occur from the combination of glucose with amino acids. Glucose oxidase is also used in combination with catalase to eliminate oxygen from mayonnaise, fruit juices, and other foods; the oxygen in these foods causes discoloration and deterioration.

Many microbial enzymes are also used in medicine. Amylases, proteases, and lipases sometimes are used as digestive aids in individuals with deficiencies in the production of digestive enzymes. For example, some people have a deficiency in β-galactosidase and are intolerant to milk sugar (lactose). As a result of this enzyme deficiency, the lactose in the milk is not hydrolyzed and not absorbed by the body. Instead, intestinal bacteria ferment the lactose, producing considerable amounts of gas in the large intestine. Lactose intolerance can be overcome by the addition of fungal lactase to milk. The fungal lactase hydrolyzes lactose, thereby permitting its products to be absorbed in the intestine. Streptokinases (produced by streptococci) injected intravenously transform plasminogen to plasmin (fibrinolysin). The plasmin causes digestion of fibrin, the substance responsible for blood clots. L-asparaginase, an enzyme that hydrolyzes asparagine to aspartic acid and ammonia, appears to be

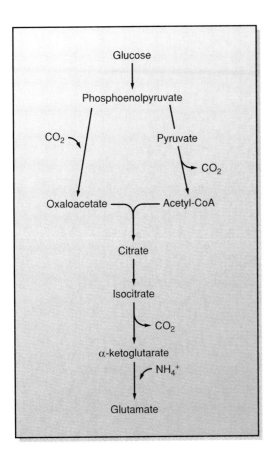

figure 20.7
Pathway for Biosynthesis of Glutamate in *Corynebacterium glutamicum*

effective in the treatment of leukemia. Neoplastic leukemia cells are unable to synthesize L-asparagine and require exogenous sources of this amino acid. Asparaginase reduces the quantity of exogenous asparagine available to these cancerous cells.

Metabolic Products

Many different compounds produced by microorganisms are useful commercially. These products may be divided into primary metabolites and secondary metabolites.

Primary Metabolites Are Metabolic Compounds Involved in Cell Growth or Function

Primary metabolites are compounds involved in cell growth or function and are found in the biochemical pathways of actively growing organisms. Amino acids, organic acids and alcohols, polysaccharides, and vitamins are among the more industrially useful primary metabolites produced by microorganisms.

Microorganisms Produce Industrially Important Amino Acids

Microorganisms produce all of the common amino acids, but glutamate and lysine are the two used most often in the food industry. World production of glutamate exceeds 100,000 tons per year. In the form of monosodium glutamate, this amino acid is a flavor enhancer in foods. Glutamate is commercially produced by using *Corynebacterium glutamicum*, a bacterium that yields up to 60 grams of amino acid per liter of culture. This high yield of glutamate is possible because *C. glutamicum* has an incomplete TCA cycle and lacks the enzyme α-ketoglutarate dehydrogenase (Figure 20.7). This causes glutamate to accumulate. Glutamate, however, is not excreted by the cell unless the growth medium contains suboptimal amounts of the vitamin biotin. Without an adequate supply of biotin, the cells make membranes deficient in phospholipids. These phospholipid-deficient membranes are leaky and permeable to glutamate. The addition of penicillin or detergents has the same effect, causing the cell to become leaky (penicillin inhibits wall synthesis, and detergents disrupt the membranes) and release glutamate, but this requires additional purification steps to remove the penicillin or detergent.

A mutant of *C. glutamicum* is used in the industrial production of another amino acid, lysine. Lysine is used to fortify bread and breakfast cereals and also as an animal food supplement. Lysine is one of the products of a branched pathway in *C. glutamicum* (Figure 20.8). The other products of the pathway are methionine, threonine, and isoleucine. By preventing synthesis of homoserine through mutation, scientists are able to make *C. glutamicum* synthesize lysine without the synthesis of the other three amino acids.

Microorganisms Are a Source of Industrially Important Organic Acids and Alcohols

Microorganisms frequently synthesize organic acids and alcohols as by-products of carbohydrate metabolism. At the turn of the century, organic alcohols and solvents such as ethanol, butanol, acetone, and 2,3-butanediol were produced by microbial fermentations. Today these compounds are chemically synthesized because such production is less expensive than microbial fermentations.

Many organic acids, however, continue to be industrially made by microorganisms. One of the most widely used organic acids is citric acid. More than 100,000 tons of citric acid are manufactured annually throughout the world, and all but a fraction of this amount is produced by fermentation processes. Citric acid sequesters metal ions and is extensively used for this purpose in the food, pharmaceutical, cosmetic, and other industries. The food industry adds citric acid to soft drinks, jams, jellies, candies, and preserved fruits. It is also used in medicines, cosmetics, shampoos, electroplating of metals, leather tanning, and even removal of iron deposits in pipes in the oil industry. Citric acid historically has been obtained from citrus fruits. Its commercial preparation uses *Aspergillus niger* grown in a well-aerated fermentor (*A. niger* is a highly aerobic fungus). Molasses or corn starch is used as a sugar source, and the pH of the culture is maintained at 3.5 or less to prevent the formation of oxalic and gluconic acids at the expense of citric acid. Citric acid is isolated by precipitation from the culture broth using lime.

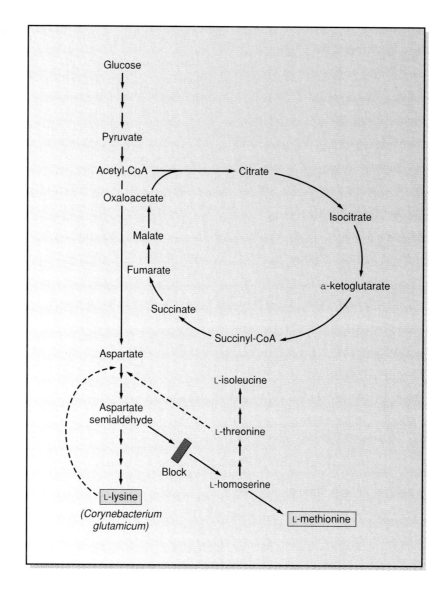

figure 20.8

Branched Pathway for L-Lysine Synthesis in *Corynebacterium glutamicum*

L-lysine is commercially made by using a mutated *Corynebacterium glutamicum* strain that is unable to synthesize L-homoserine, which is a precursor for L-methionine, L-threonine, and L-isoleucine. This mutation prevents L-threonine from reaching levels that could lead to feedback repression of one of the enzymes involved in the biosynthesis of L-lysine.

Itaconic acid (methylene succinic acid), which accumulates in cultures of *Aspergillus terreus,* is used in the manufacture of plastics, artificial dentures, and certain synthetic fibers. Lactic acid is another organic acid used in plastics. It is also used in leather tanning and in the food industry as a preservative. *Lactobacillus delbrueckii* is used for the industrial production of lactic acid.

Industrially Important Polysaccharides Are Synthesized by Microorganisms

Polysaccharides derived from plants and seaweeds have been used for thousands of years. The ancient Egyptians used gum arabic as a component of embalming fluid. A sweetened gel made from red sea-weeds was an important staple in the diet of the Chinese during the time of Confucius. Today polysaccharides are used in industry as food stabilizers, marine drill lubricants, and thickening and suspending agents in foods, cosmetics, and paints. Microbial polysaccharides frequently can be produced from cheap raw materials and therefore are attractive substitutes for water-soluble vegetable gums.

Polysaccharides are synthesized in different forms by microorganisms as protection against adverse environmental conditions. Pathogenic bacteria frequently produce thick polysaccharide capsules to insulate them from phagocytosis. Microbes in natural environments such as soil, lakes, and ponds often secrete polysaccharide slime layers. Some microorganisms are able to metabolize their own polysaccharides as a reserve source of energy.

table 20.11

Vitamins Made by Microbial Fermentation

Vitamin	Culture	Medium	Fermentation Conditions	Yield
Carotene (precursor of vitamin A)	*Blakeslea trispora* or *Mycobacterium smegmatis*	Molasses, soybean oil, β-Ionone, thiamin	Aerobic, 72 hrs at 30°C	1 g/l 0.007 g/l
Cyanocobalamin (vitamin B$_{12}$)	*Propionibacterium freudenreichii*	Glucose, corn steep liquor, betaine, cobalt	Anaerobic, 3 days at 30°C; aerobic, 2 days at 30°C	20 mg/l
	Propionibacterium shermanii	Glucose, corn steep liquor, ammonia, cobalt	Anaerobic, 3 days at 30°C; aerobic, 4 days at 30°C	23 mg/l
	Streptomyces olivaceus	Glucose, soybean flour, cobalt, mineral salts	Aerobic, 4 days at 28°C	5.7 mg/l
5-keto gluconic acid (in vitamin C synthesis)	*Gluconobacter oxidans* subsp. *suboxidans*	Glucose, CaCO$_3$, corn steep liquor	Aerobic, 33 hrs at 30°C	100%, based on substrate used
L-sorbose (in vitamin C synthesis)	*Gluconobacter oxidans* subsp. *suboxidans*	D-sorbitol, 30% corn steep liquor	Aerobic, 45 hrs at 30°C	70% based on substrate used
Riboflavin	*Ashbya gossypii* or *Eremothecium ashbyii*	Glucose, corn steep liquor	Aerobic, 7 days at 36°C	4.25 g/l

Microbial polysaccharides can be divided into two major groups: **homopolysaccharides,** consisting of one type of sugar in a repeating structure, and **heteropolysaccharides,** containing two or more kinds of sugars. Dextrans (α-linked polymers of glucose) were the first microbial homopolysaccharides to be commercially produced. Although many different types of bacteria are capable of synthesizing dextrans, only two (*Leuconostoc mesenteroides* and *Leuconostoc dextranicum*) are used in the commercial production of dextran. Dextrans were developed as blood plasma extenders in the 1950s but are not as widely used today.

Xanthan gum, a complex heteropolysaccharide made by *Xanthomonas campestris,* is one of the microbial polysaccharides widely used by industry today. *X. campestris* is a bacterium originally isolated from the rutabaga plant. The gumlike polysaccharide that it produces increases solution viscosity and so is added to oil well drilling fluids, textile printing pastes, ceramic glazes, toothpastes, sauces and gravies, pourable salad dressings, and dairy products.

Microorganisms Are Sources of Vitamins

Vitamins are essential animal nutrients. Prototrophic microorganisms synthesize small amounts of vitamins for their own growth and metabolism. These same microorganisms can be engineered to produce large amounts of a particular vitamin commercially. Although most vitamins are chemically synthesized, a few can be made less expensively by microbial fermentation (Table 20.11).

Vitamin B$_{12}$, or cyanocobalamin, is required by humans and other animals for erythrocyte maturation, proper functioning of the nervous system, and protein and fat metabolism. Although vitamin B$_{12}$ is found abundantly in animal flesh and animal products and normally is not a dietary problem, strict vegetarians and people unable to absorb this vitamin must have supplements of it. Pernicious anemia is a disease that is associated with a deficiency of vitamin B$_{12}$.

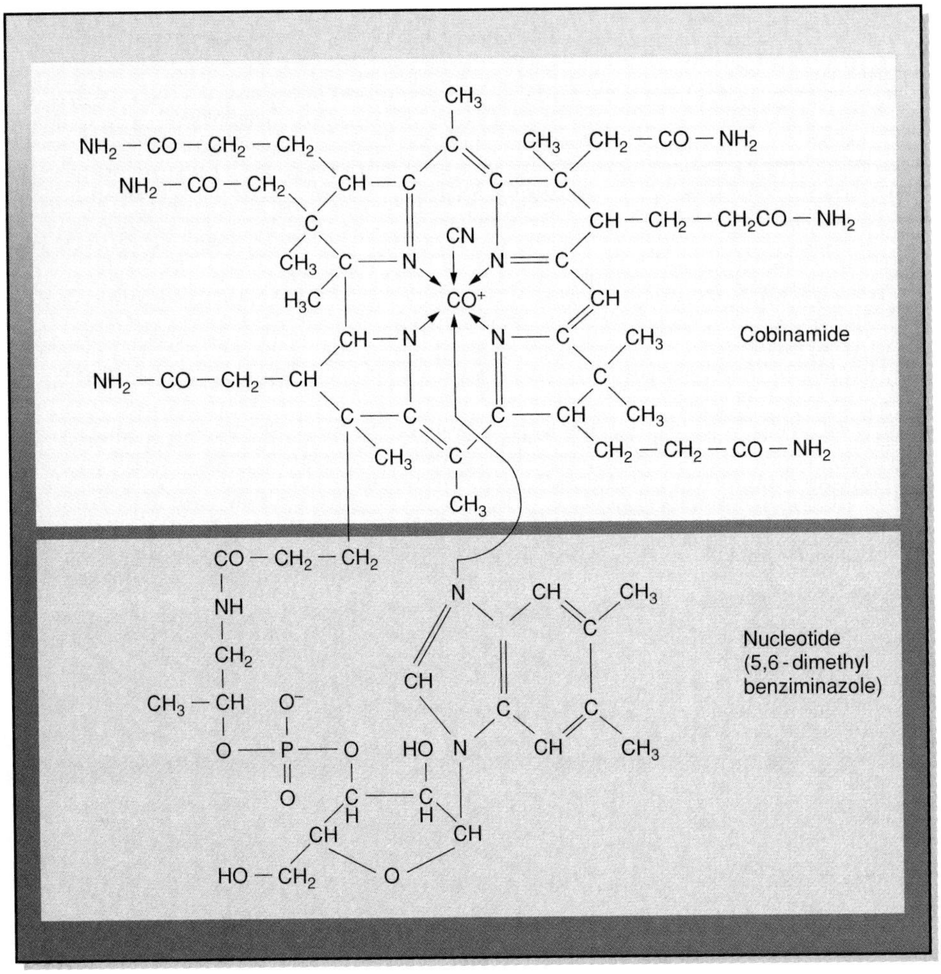

figure 20.9
Structure of Cyanocobalamin (Vitamin B$_{12}$)

The vitamin B$_{12}$ molecule consists of cobinamide linked to a nucleotide (Figure 20.9). The cobinamide has a central atom of cobalt attached to a cyanide group and surrounded by four reduced pyrrole groups.

Propionibacterium freundenreichii, P. shermanii, Pseudomonas denitrificans, and *Streptomyces olivaceus* are four microbial sources of vitamin B$_{12}$. *P. freundenreichii* is grown in two stages (anaerobically for three days and aerobically for two days) in the commercial production of vitamin B$_{12}$. Cobinamide formation is inhibited by oxygen, so the bacterium is grown first under anaerobic conditions in a culture medium containing glucose, cobalt, and betaine (a quaternary ammonium salt that furnishes methyl groups to the vitamin struc-

ture). After cobinamide is synthesized, the bacterium is shifted to aerobic conditions. The nucleotide is formed and links to cobinamide to form cobalamin. The endocellular cobalamin is released from the cells by acidifying and heating the culture. The cobalamin is then separated from the cell debris and combined with potassium cyanide to produce cyanocobalamin.

Riboflavin is an essential component of flavoprotein coenzymes. Riboflavin is commercially produced using fungi such as *Ashbya gossypii* and *Eremothecium ashbyii*. The fungus is grown aerobically in a medium containing glucose and corn steep liquor. After a seven-day incubation period, the cells are concentrated and used as a feed supplement for animals.

table 20.12

Some Antibiotics Produced by Microorganisms

Antibiotic	Cell Target	Microorganism Source
Amphotericin B	Sterols in fungal plasma membrane	*Streptomyces nodosus*
Bacitracin	Cell wall	*Bacillus subtilis*
Cephalosporin	Cell wall	*Cephalosporium* sp.
Chloramphenicol	Ribosome	*Streptomyces venezuelae* or chemical synthesis
Chlortetracycline	Ribosome	*Streptomyces aureofaciens*
Erythromycin	Ribosome	*Streptomyces erythraeus*
Griseofulvin	Microtubules in fungi (possible)	*Penicillium griseofulvin* *Penicillium nigricans* *Penicillium urticae*
Kanamycin	Ribosome	*Streptomyces kanamyceticus*
Neomycin	Ribosome	*Streptomyces fradiae*
Novobiocin	DNA synthesis	*Streptomyces niveus* *Streptomyces spheroides*
Nystatin	Sterols in fungal plasma membrane	*Streptomyces noursei*
Oxytetracycline	Ribosome	*Streptomyces rimosus*
Penicillin	Cell wall	*Penicillium chrysogenum*
Polymyxin B	Plasma membrane	*Bacillus polymyxa*
Streptomycin	Ribosome	*Streptomyces griseus*

Secondary Metabolites Are Products of Pathways That Apparently Are Not Associated with Primary Cellular Processes

Secondary metabolites are produced during the stationary phase of growth and are products of pathways that seemingly are not associated with primary cellular processes (see secondary metabolites, page 204). Antibiotics are one type of secondary metabolite produced by microorganisms. Since the discovery of penicillin by Alexander Fleming in 1928, over 7,000 naturally occurring and 30,000 semisynthetically derived antibiotics have been discovered (see antibiotics, page 125). Less than 150 of these antibiotics are produced on a commercial scale today, yet this handful of antibiotics has led to a billion-dollar business annually in the United States alone. Antibiotics are the most important class of pharmaceuticals made by microorganisms. A few of these antibiotics are listed in Table 20.12.

Penicillin Is Produced by Penicillium

Penicillin is produced by growing *Penicillium chrysogenum* in glucose and corn steep liquor supplemented with phenylacetic acid, the precursor of the benzyl side chain of penicillin G. The fermentation process takes approximately 200 hours. Glucose is rapidly fermented during the early stages of growth; it is only later in growth that penicillin, a secondary metabolite, is produced. When the fermentation is complete, the liquid medium containing the penicillin is separated from the fungal cells on a rotary vacuum filter drum. The penicillin is harvested from the filtrate by chemical extraction and represents a product that is 99.5% pure.

Cephalosporins Are Produced by Cephalosporium

Strains of the mold *Cephalosporium* produce a group of antibiotics referred to as the cephalosporins. These are similar in structure to penicillin but are generally less susceptible than penicillin to

Recombinant DNA Techniques for the Production of Antibiotics

In the past, most antibiotics have been produced by microorganisms chosen for high productivity by random mutation and selection methods. Although such procedures may have generated strains that produced large amounts of commercially important antibiotics, they were inefficient and time-consuming.

Many of these classical methods have been replaced in recent years by genetic recombination techniques that are used to produce microorganisms with commercially desirable genetic traits. Among the potentially more promising recombinational procedures are (1) shotgun cloning, which uses randomly fragmented DNA from a donor organism to transform the recipient organism;

(2) direct gene cloning, which involves the insertion of specific genes into organisms to alter their genotypes; and (3) protoplast fusion, which combines protoplasts and DNA of two species to produce a hybrid organism.

Protoplast fusion has potential widespread applications in biotechnology and has been used very successfully with *Streptomyces* species. In protoplast fusion, protoplasts of cells from two different species—obtained by the use of specific lytic enzymes or inhibitors of cell wall synthesis—are suspended in a high osmotic solution to maintain cell integrity. A mixture of the protoplasts is treated with polyethylene glycol to enhance aggregation. Within these aggregates, some adjacent proto-

plasts (and their DNA) fuse together. The protoplasts are transferred to an osmotically stabilized growth medium without polyethylene glycol, and the fused proto-

plasts regenerate their cell walls and form a hybrid organism with recombinant DNA.

Unlike other methods of genetic exchange, which consist of the transfer of only a small part of the donor chromosome, protoplast fusion results in the temporary formation of a quasi-diploid organism containing the complete chromosomes of both cells. Although this diploid state is only temporary, the possibilities for recombination are greater than with other procedures. It is now possible through protoplast fusion to create new organisms capable of synthesizing a new generation of hybrid antibiotics.

---

hydrolysis by β-lactamase (Figure 20.10). Microbiological processes for the production of cephalosporins resemble those used for penicillin production.

Tetracyclines Are Produced by Streptomyces

The tetracyclines are a group of antibiotics that have a common hydronaphthacene skeleton and are effective against numerous gram-negative and gram-positive organisms (Figure 20.11). Chlortetracycline (Aureomycin), one member of the tetracyclines, is prepared from *Streptomyces aureofaciens* grown in a culture medium containing starch, ground nut meal, corn steep liquor, beet molasses, calcium carbonate, and potassium chloride. Potassium chloride provides the chloride ions in the structure of chlortetracycline. The calcium carbonate not only maintains the pH of the medium, but also serves another important role. Tetracycline in concentrations of greater than 100 mg/l inhibits the growth of *S. aureofaciens*. The polyenolketone structure in the tetracycline molecule makes the antibiotic a powerful chelating agent that

probably removes important trace ions from the culture medium. The calcium in the medium precipitates tetracycline as an insoluble calcium complex and enables the organism to grow continuously and produce tetracycline in quantities as high as 3 g/l.

Streptomycin Is Produced by Streptomyces

Streptomycin and similar antibiotics are commercially produced using *Streptomyces griseus,* a common soil bacterium (see soil bacteria, page 581). Although streptomycin is produced near the end of the exponential phase of growth and during the stationary phase, the breakdown of the filamentous form of the organism during the stationary phase makes filtration and recovery of the antibiotic more difficult. It is therefore desirable that high yields of streptomycin be obtained early in growth. Production of streptomycin, however, is subject to catabolite repression, so the concentration of glucose in the culture medium must be closely monitored during growth to ensure optimal production of streptomycin.

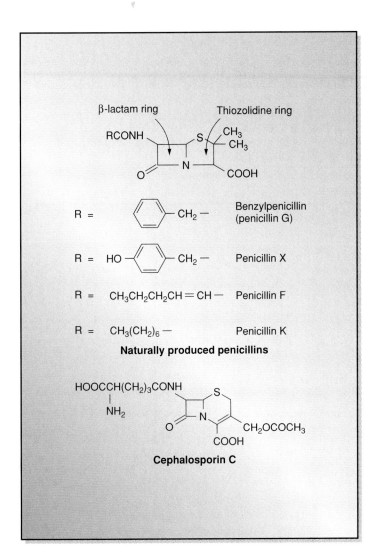

β-lactam ring   Thiozolidine ring

RCONH

CH_3
CH_3
COOH

R = [benzene] CH_2—   Benzylpenicillin (penicillin G)

R = HO—[benzene]—CH_2—   Penicillin X

R = $CH_3CH_2CH_2CH=CH$—   Penicillin F

R = $CH_3(CH_2)_6$—   Penicillin K

Naturally produced penicillins

$HOOCCH(CH_2)_3CONH$
NH_2
CH_2OCOCH_3
COOH

Cephalosporin C

figure 20.10
Structures of Some Naturally Produced Penicillins and Cephalosporin C

H_3C OH
R^1 R^2 $N(CH_3)_2$
OH
OH
$CONH_2$
OH   O   OH   O

R^1	R^2	
H	H	Tetracycline
Cl	H	Chlortetracycline (aureomycin)
H	OH	Oxytetracyline (terramycin)

figure 20.11
Structures of the Tetracyclines

table 20.13

Comparison of Protein Production Rates in Animals, Plants, and Microorganisms

Organism (1,000 kg)	Protein Produced (kg/day)
Beef	1
Soybean	10
Yeast	10^5
Bacteria	10^{11}

Single-Cell Protein

Microbial cells have a high protein content (as much as 50% to 70% protein) and synthesize their proteins much more rapidly than do plants or animals (Table 20.13). Microorganisms therefore represent a rich source of protein for human or animal consumption. This use of microorganisms as a potential food source is a relatively recent development. The term **single-cell protein (SCP)** was coined by C.L. Wilson at the Massachusetts Institute of Technology in 1966 to describe dried microbial cells that can be used for food or feed.

The consumption of microorganisms is not new. They are consumed in bread (yeast), milks and cheeses (*Streptococcus* and *Lactobacillus*), and mushrooms (basidiomycetes). However, it was not until World War II, when the Germans began to purposely cultivate microorganisms (particularly *Candida utilis*), that they were seriously considered for direct consumption. During the 1960s, when petroleum was less expensive, there was considerable interest in producing SCP from bacteria grown on hydrocarbons and chemicals derived from them (for example, methanol and ethanol). The most significant advances in SCP production have been made by Britain's Imperial Chemical Industries Ltd., which has used methanol to grow *Methylophilus methylotrophus*. The bacterium, after being grown in huge fermentation tanks, is dried and sold as the protein-rich product Pruteen.

Yeasts represent an excellent source of SCP because they use a wide variety of substrates and also have a high vitamin content. The Boise-Cascade Corporation in Oregon has constructed a yeast plant with a yearly capacity of 4.5 million kilograms that produces *Candida utilis* from sulfite liquor. The sulfite liquor is a by-product from an ammonia-base pulp process and normally would be burned as fuel. Instead, the liquor is now used as a substrate for SCP production.

The main disadvantage in commercial production of SCP for food today is cost; production processes are expensive and require large investments of capital. Nonetheless, microorganisms represent a potential solution to the world food crisis.

Single-Cell Protein: Food of the Future

According to estimates by the U.S. Bureau of the Census, the world's population will exceed 6 billion by the year 2000, and an average 4.5 people will be born every second. At the present rate of growth, this projected population will require an extensive amount of food for sustenance. Traditional sources of food products, such as agriculture, husbandry, and fishery, may be inadequate to meet this need. One potential source of human food is microorganisms in the form of single-cell protein (SCP). This is not a new idea; in fact, it has been advocated as an alternative food source for many years. Only recently, however, with increasing concern about limited world food production, have microorganisms been seriously considered as a major source of food.

For microbial biomass to serve as a foodstuff, several major factors must be considered. First, a suitable carbon source must be used in SCP production. Carbon sources may be divided into several categories: high-energy sources (ethanol, methanol, natural gas, acetic acid, and oil), wastes (molasses, sulfite waste liquor, whey, and fruit wastes), and renewable plant resources (sugar and starch). Of these various potential substrates, sugars are readily utilized by most microorganisms. However, the use of sugars for the production of SCP is economically infeasible. Ethanol is attractive as a possible SCP substrate because it is relatively inexpensive; it is used primarily as a substrate by microorganisms, thus reducing the probability of culture contamination; and residual amounts can easily be washed from cells. Furthermore, the use of ethanol raises few toxicological problems, a major consideration in our regulatory-conscious society. Another potential SCP substrate is oil and natural gas hydrocarbons. It has been estimated that only 2% (2 billion tons) of the oil extracted annually in the world would be sufficient to generate 25 to 30 million tons of yeast protein, which could feed 2 billion people per year.

A second consideration in SCP production is the selection of the microorganism (Table 20P.1). Three groups of microorganisms might be used as a source of food protein: yeast, bacteria, and algae. Of these three groups, yeasts—particularly those of the genera *Saccharomyces*, *Torulopsis*, and *Candida*—have been most extensively studied and considered as SCP sources. Yeasts have a high protein content (dry yeast contains approximately 50% protein) and have well-balanced amino acid compositions. Bacteria also have high protein contents (72% to 78% protein) and thus are another appealing source of SCP.

A third factor to consider in using microorganisms as food sources is the effect of their chemical constituents on human nutrition. Experiments at the University of California,

table 20P.1

Nitrogen and Protein Contents of Microbial Cells Compared with Selected Foods of Animal and Plant Origin

Source	Nitrogen (%)	Crude Protein (%)
Filamentous fungi	5–8	31–50
Algae	7.5–10	47–63
Yeast	7.5–8.5	47–53
Bacteria	11.5–12.5	72–78
Milk	3.5–4.0	22–25
Beef	13–14.4	81–90
Egg	5.6	35
Rice	1.2–1.4	7.5–9.0
Wheat flour	1.6–2.2	9.8–13.5
Cornmeal	1.1–1.5	7.0–9.4

Berkeley, and at the Massachusetts Institute of Technology have shown that the uptake of excessive amounts of nucleic acid can result in increased levels of uronic acids in the blood and urine. In other studies, human volunteers fed *Chlorella* developed serious gastrointestinal disorders. The alga *Spirulina* may be best suited for human consumption because it produces no detectable effects when included in the diet at levels of 40 to 60 g/day. This observation is not surprising, since *Spirulina* is routinely consumed by people in Chad and was also eaten by the ancient Aztecs in Mexico.

Despite these potential problems, SCP is a promising alternative to the increasingly limited conventional supply of food. As the world population increases and outpaces traditional agricultural foodstuffs, microorganisms are certain to become a major source of protein for human consumption.

Source

Kharatyan, S.G. 1978. Microbes as food for humans. *Annual Review of Microbiology* 32:301–327.

References

Ciferri, O. 1983. *Spirulina*, the edible microorganism. *Microbiological Reviews* 47:551–578.

Khilberg, R. 1972. The microbe as a source of food. *Annual Review of Microbiology* 26:427–466.

Summary

1. Microorganisms are used to produce fermented milks and milk products such as acidophilus milk, yogurt, butter, and cheese.

2. Meat and meat products are constantly exposed to microorganisms and consequently spoil rapidly unless properly stored and refrigerated. Drying and curing are two methods to preserve meats without refrigeration.

3. Microorganisms are useful in the preparation of fermented foods such as sausages, cabbage, and pickles. Yeasts used in bread making produce gas to raise or leaven the dough, and produce metabolic by-products that contribute to the flavor of bread. Yeasts are also used in beer production to ferment grains and in wine production to ferment fruit sugars.

4. Microorganisms are used in industry to produce materials of economic value such as enzymes, amino acids, organic acids and alcohols, vitamins, and antibiotics. Industrial-scale growth of microorganisms requires the use of the appropriate culture medium, carefully selected strains that are genetically stable and capable of producing large quantities of the desired end product, and optimum growth conditions.

5. Microbial cells have a high protein content and therefore represent a rich source of protein for human or animal consumption. The term single-cell protein is used to describe dried microbial cells that can be used for food or feed.

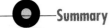

EVOLUTION and BIODIVERSITY

The origins of food and industrial microbiology date back thousands of years ago to the early Egyptians, who used yeast as a leavening agent for bread and in the manufacture of beer and wine. With time, the large-scale, controlled growth of microorganisms was found to be useful in producing economically valuable products such as antibiotics, chemicals, enzymes, alcoholic beverages, and food supplements. In recent years recombinant DNA technology has created genetically improved microorganisms that are highly efficient in their metabolism and capable of synthesizing commercially desirable products, including some that had never before been possible to produce by microbes. Scientists are now able to transfer genes for insulin or interferon production into harmless bacteria. Bacteria and yeasts can be genetically manipulated to produce large amounts of a specific vitamin or antibiotic. Genetic engineering has also produced microbial vaccines against human diseases such as hepatitis B, influenza, and cytomegalovirus, as well as against animal foot-and-mouth disease. Industrial microbiology has ancient roots, but through biotechnology, the field has evolved into one of the most dynamic areas of science. As more is learned about the genetics and physiology of vastly diverse microorganisms, this knowledge undoubtedly will be used to further benefit humankind and the world.

Questions

Short Answer

1. Identify microorganisms that may be present in raw milk.

2. Describe the conditions for pasteurization.

3. Why is milk pasteurized?

4. Why isn't raw milk used for fermentation?

5. How do ripened and unripened cheeses differ? Give a few examples of each.

6. Identify microorganisms that may be present in meats.

7. How can you prevent the spoilage of meats?

8. Why are poultry and other animals routinely fed antibiotics?

9. What criteria are used in canning?

10. Why have fermented foods been popular throughout the ages? What are their limitations?

11. Identify several fermented foods.

12. What organism is generally involved in bread making? What does its name tell you about this organism? What other fermentation product is attributed to this organism?

13. Identify several products of interest by industrial fermentors.

14. Describe some of the problems encountered by industrial fermentors.

15. Identify several industrially important enzymes.

16. Why are enzymes important industrially?

17. Why are microorganisms used as a source of vitamins?

18. What microorganisms are commonly used to produce antibiotics?

19. Discuss the use of microorganisms as a food source.

Multiple Choice

1. Which of the following will sterilize milk?

    a. HTST
    b. LTH
    c. UHT
    d. All of the above.
    e. None of the above.

2. *Lactobacillus bulgaricus* and *Streptococcus thermophilus* are likely to produce:

    a. acidophilus milk
    b. butter
    c. kefir and koumiss
    d. yogurt

3. You can produce your own _____ if you add *Lactobacillus cremoris* and *Streptococcus lactis* to cream.

    a. cultured buttermilk

    b. kefir

    c. koumiss

    d. sour cream

4. Brie, Camembert, and Limburger cheeses are:

    a. soft, unripened

    b. soft, ripened

    c. hard, ripened

    d. very hard, ripened

Critical Thinking

1. You learned earlier that most tissues in your body are sterile. Explain, then, why we are so worried about microorganisms on or in meat. What is the source of these organisms?

2. Earlier in this course, you learned about the metabolic process called "fermentation." How does "industrial fermentation" differ? Is "industrial fermentation" technically accurate?

3. Plan a picnic or party where all foods and beverages will be fermented. Be sure to plan for entrees, side dishes, and desserts!

4. Other than foods, what examples of industrial fermentation can you find in your grocery store? Visit your local grocer and make a list!

5. Prior to the 1800s, there was little need to pasteurize milk before consumption. Explain why. Why are fermented milk products less likely to contain pathogens than raw milk?

 Supplementary Readings

Banwart, G.J. 1989. *Basic food microbiology,* 2d ed. New York: Chapman & Hill. (A textbook of food microbiology.)

Frazer, W.C., and D.C. Westhoff. 1988. *Food microbiology,* 4th ed. New York: McGraw-Hill. (A textbook of food microbiology.)

Glazer, A., and H. Nikaido. 1995. *Fundamentals of applied microbiology.* New York: W.H. Freeman and Company. (A textbook that explores the many areas that comprise industrial microbiology, including the use of microorganisms in the production of food, pesticides, vaccines, and antibiotics.)

Glick, B.R., and J.J. Pasternak. 1994. *Molecular biotechnology, principles and applications of recombinant DNA.* Washington, D.C.: American Society for Microbiology. (A comprehensive textbook on molecular processes used in industrial microbiology.)

Hugo, W.B., and A.D. Russell. 1992. *Pharmaceutical microbiology,* 5th ed. Oxford, England: Blackwell Scientific Publications. (An extensive discussion of industrial production of antimicrobial agents.)

International Commission on Microbiological Specifications for Foods. 1995. *Microorganisms in foods 5.* New York: Chapman & Hill. (The fifth volume in a series, focusing on the major microorganisms associated with food-borne diseases.)

Jay, J.M. 1992. *Modern food microbiology,* 4th ed. New York: Chapman & Hill. (A comprehensive textbook of food microbiology.)

Lieve, L., ed. 1985. *Microbiology 1985.* Washington, D.C.: American Society for Microbiology. (A selection of papers on the genetics and molecular biology of industrial microorganisms.)

Mountney, G.J., and W. Gould. 1988. *Practical food microbiology and technology,* 3d ed. New York: Van Nostrand Reinhold. (A compendium of the microbiology and technology of different food products, including discussions of the composition of foods and microorganisms associated with foods.)

Nachamkin, I., M.J. Blaser, and L.S. Tompkins, eds. 1992. *Campylobacter jejuni, current status and future trends.* Washington, D.C.: American Society for Microbiology. (A review of the role of *Campylobacter jejuni* in foods.)

Patel, P. 1995. *Rapid analysis techniques in food microbiology.* New York: Chapman & Hill. (A discussion of techniques used for rapid identification of microorganisms in foods.)

Peppler, H.J., and D. Perlman, eds. 1979. *Microbial technology: Microbial processes.* New York: Academic Press. (A review of microorganisms used for the production of insecticides, organic acids, single-cell proteins, amino acids, antibiotics, microbial enzymes, and other commercial products.)

Peppler, H.J., and D. Perlman, eds. 1979. *Microbial technology: Fermentation technology.* New York: Academic Press. (A summary of microorganisms used in the food and drink industry and the technology involved in microbial fermentation processes.)

Priest, F.G., and I. Campbell, eds. 1995. *Brewing microbiology,* 2d ed. New York: Chapman & Hill. (An in-depth discussion of the microbiology of brewing.)

Speck, M.L., ed. 1984. *Compendium of methods for the microbiological examination of foods,* 2d ed. Washington, D.C.: American Public Health Association. (A comprehensive laboratory manual on the isolation and identification of microorganisms from foods.)

Wiseman, A., ed. 1983. *Principles of biotechnology.* London: Surrey University Press. (A review of the principles and applications of industrial microbiology.)

Classification of Procaryotes According to *Bergey's Manual of Systematic Bacteriology*

Bergey's Manual of Systematic Bacteriology classifies procaryotes into 33 sections. The *Manual* is divided into four volumes: Volume 1 (*Gram-Negative Bacteria of General, Medical, or Industrial Importance*), Volume 2 (*Gram-Positive Bacteria Other Than the Actinomycetes*), Volume 3 (*The Archaea, Cyanobacteria, and Remaining Gram-Negative Bacteria*), and Volume 4 (*The Actinomycetes*). There have been several changes in procaryotic taxonomy since *Bergey's Manual of Systematic Bacteriology* was published, which was between 1984 and 1989. Some of these changes can be found in the ninth edition of *Bergey's Manual of Determinative Bacteriology*, published in 1994. *Bergey's Manual of Determinative Bacteriology* was written to serve as an aid for identifying procaryotes, whereas *Bergey's Manual of Systematic Bacteriology* was written to provide information on the classification of procaryotes.

Volume 1. Gram-Negative Bacteria of General, Medical, or Industrial Importance

Section 1. The Spirochetes

Order 1. *Spirochaetales*
   Family I. *Spirochaetaceae*
      Genus *Spirochaeta*
      Genus *Cristispira*
      Genus *Treponema*
      Genus *Borrelia*
   Family II. *Leptospiraceae*
      Genus *Leptospira*
Other Organisms
   Hindgut Spirochetes of Termites and *Cryptocercus punctulatus*

Section 2. Aerobic/Microaerophilic, Motile, Helical/Vibrioid Gram-Negative Bacteria

   Genus *Aquaspirillum*
   Genus *Spirillum*
   Genus *Azospirillum*
   Genus *Oceanospirillum*
   Genus *Campylobacter*
   Genus *Bdellovibrio*
   Genus *Vampirovibrio*

Section 3. Nonmotile (or Rarely Motile), Gram-Negative Curved Bacteria

   Family I. *Spirosomaceae*
      Genus *Spirosoma*
      Genus *Runella*
      Genus *Flectobacillus*
   Other Genera
      Genus *Microcyclus*
      Genus *Meniscus*

      Genus *Brachyarcus*
      Genus *Pelosigma*

Section 4. Gram-Negative Aerobic Rods and Cocci

Family I. *Pseudomonadaceae*
   Genus *Pseudomonas*
   Genus *Xanthomonas*
   Genus *Frateuria*
   Genus *Zoogloea*
Family II. *Azotobacteraceae*
   Genus *Azotobacter*
   Genus *Azomonas*
Family III. *Rhizobiaceae*
   Genus *Rhizobium*
   Genus *Bradyrhizobium*
   Genus *Agrobacterium*
   Genus *Phyllobacterium*
Family IV. *Methylococcaceae*
   Genus *Methylococcus*
   Genus *Methylomonas*
Family V. *Halobacteriaceae*
   Genus *Halobacterium*
   Genus *Halococcus*
Family VI. *Acetobacteraceae*
   Genus *Acetobacter*
   Genus *Gluconobacter*
Family VII. *Legionellaceae*
   Genus *Legionella*
Family VIII. *Neisseriaceae*
   Genus *Neisseria*
   Genus *Moraxella*
   Genus *Acinetobacter*
   Genus *Kingella*
Other Genera
   Genus *Beijerinckia*
   Genus *Derxia*
   Genus *Xanthobacter*
   Genus *Thermus*
   Genus *Thermomicrobium*
   Genus *Halomonas*
   Genus *Alteromonas*
   Genus *Flavobacterium*
   Genus *Alcaligenes*
   Genus *Serpens*

Genus *Janthinobacterium*
Genus *Brucella*
Genus *Bordetella*
Genus *Francisella*
Genus *Paracoccus*
Genus *Lampropedia*

Section 5. Facultatively Anaerobic Gram-Negative Rods

Family I. *Enterobacteriaceae*
  Genus *Escherichia*
  Genus *Shigella*
  Genus *Salmonella*
  Genus *Citrobacter*
  Genus *Klebsiella*
  Genus *Enterobacter*
  Genus *Erwinia*
  Genus *Serratia*
  Genus *Hafnia*
  Genus *Edwardsiella*
  Genus *Proteus*
  Genus *Providencia*
  Genus *Morganella*
  Genus *Yersinia*
Other Genera of the Family *Enterobacteriaceae*
  Genus *Obesumbacterium*
  Genus *Xenorhabdus*
  Genus *Kluyvera*
  Genus *Rahnella*
  Genus *Cedecea*
  Genus *Tatumella*
Family II. *Vibrionaceae*
  Genus *Vibrio*
  Genus *Photobacterium*
  Genus *Aeromonas*
  Genus *Plesiomonas*
Family III. *Pasteurellaceae*
  Genus *Pasteurella*
  Genus *Haemophilus*
  Genus *Actinobacillus*
Other *Genera*
  Genus *Zymomonas*
  Genus *Chromobacterium*
  Genus *Cardiobacterium*
  Genus *Calymmatobacterium*
  Genus *Gardnerella*
  Genus *Eikenella*
  Genus *Streptobacillus*

Section 6. Anaerobic Gram-Negative Straight, Curved, and Helical Rods

Family I. *Bacteroidaceae*
  Genus *Bacteroides*

Genus *Fusobacterium*
Genus *Leptotrichia*
Genus *Butyrivibrio*
Genus *Succinimonas*
Genus *Succinivibrio*
Genus *Anaerobiospirillum*
Genus *Wolinella*
Genus *Selenomonas*
Genus *Anaerovibrio*
Genus *Pectinatus*
Genus *Acetivibrio*
Genus *Lachnospira*

Section 7. Dissimilatory Sufate- or Sulfur-Reducing Bacteria

Genus *Desulfuromonas*
Genus *Desulfovibrio*
Genus *Desulfomonas*
Genus *Desulfococcus*
Genus *Desulfobacter*
Genus *Desulfobulbus*
Genus *Desulfosarcina*

Section 8. Anaerobic Gram-Negative Cocci

Family I. *Veillonellaceae*
  Genus *Veillonella*
  Genus *Acidaminococcus*
  Genus *Megasphaera*

Section 9. The Rickettsias and Chlamydias

Order I. *Rickettsiales*
Family I. *Rickettsiaceae*
Tribe I. *Rickettsieae*
  Genus *Rickettsia*
  Genus *Rochalimaea*
  Genus *Coxiella*
Tribe II. *Ehrlichieae*
  Genus *Ehrlichia*
  Genus *Cowdria*
  Genus *Neorickettsia*
Tribe III. *Wolbachieae*
  Genus *Wolbachia*
  Genus *Rickettsiella*
Family II. *Bartonellaceae*
  Genus *Bartonella*
  Genus *Grahamella*
Family III. *Anaplasmataceae*
  Genus *Anaplasma*
  Genus *Aegyptianella*
  Genus *Haemobartonella*
  Genus *Eperythrozoon*
Order II. *Chlamydiales*

Family I. *Chlamydiaceae*
  Genus *Chlamydia*

Section 10. The Mycoplasmas

Division *Tenericutes*
  Class I. *Mollicutes*
    Order I. *Mycoplasmatales*
      Family I. *Mycoplasmataceae*
        Genus *Mycoplasma*
        Genus *Ureaplasma*
      Family II. *Acholeplasmataceae*
        Genus *Acholeplasma*
      Family III. *Spiroplasmataceae*
        Genus *Spiroplasma*
      Other Genera
        Genus *Anaeroplasma*
        Genus *Thermoplasma*
        Mycoplasma-like Organisms of Plants and Invertebrates

Section 11. Endosymbionts

A. Endosymbionts of Protozoa
  Endosymbionts of ciliates
  Endosymbionts of flagellates
  Endosymbionts of amoebas
  Taxa of endosymbionts:
    Genus *Holospora*
    Genus *Caedibacter*
    Genus *Pseudocaedibacter*
    Genus *Lyticum*
    Genus *Tectibacter*
B. Endosymbionts of Insects
  Blood-sucking insects
  Plant sap-sucking insects
  Cellulose and stored grain feeders
  Insects feeding on complex diets
  Taxon of endosymbionts:
    Genus *Blattabacterium*
C. Endosymbionts of Fungi and Invertebrates Other Than Arthropods
  Fungi
  Sponges
  Coelenterates
  Helminthes
  Annelids
  Marine worms and mollusks

Volume 2. Gram-Positive Bacteria Other Than the Actinomycetes

Section 12. Gram-Positive Cocci

Family I. *Micrococcaceae*
  Genus *Micrococcus*
  Genus *Stomatococcus*

Genus *Planococcus*
  Genus *Staphylococcus*
Family II. *Deinococcaceae*
  Genus *Deinococcus*
Other Genera
  Genus *Streptococcus*
    Pyogenic Hemolytic Streptococci
    Oral Streptococci
    Enterococci
    Lactic Acid Streptococci
    Anaerobic Streptococci
    Other Streptococci
  Genus *Leuconostoc*
  Genus *Pediococcus*
  Genus *Aerococcus*
  Genus *Gemella*
  Genus *Peptococcus*
  Genus *Peptostreptococcus*
  Genus *Ruminococcus*
  Genus *Coprococcus*
  Genus *Sarcina*

Section 13. Endospore-Forming Gram-Positive Rods and Cocci

Genus *Bacillus*
Genus *Sporolactobacillus*
Genus *Clostridium*
Genus *Desulfotomaculum*
Genus *Sporosarcina*
Genus *Oscillospira*

Section 14. Regular, Nonsporing, Gram-Positive Rods

Genus *Lactobacillus*
Genus *Listeria*
Genus *Erysipelothrix*
Genus *Brochothrix*
Genus *Renibacterium*
Genus *Kurthia*
Genus *Caryophanon*

Section 15. Irregular, Nonsporing, Gram-Positive Rods

Genus *Corynebacterium*
  Plant Pathogenic Species of *Corynebacterium*
Genus *Gardnerella*
Genus *Arcanobacterium*
Genus *Arthrobacter*
Genus *Brevibacterium*
Genus *Curtobacterium*
Genus *Caseobacter*
Genus *Microbacterium*
Genus *Aureobacterium*
Genus *Cellulomonas*

Genus *Agromyces*

Genus *Arachnia*

Genus *Rothia*

Genus *Propionibacterium*

Genus *Eubacterium*

Genus *Acetobacterium*

Genus *Lachnospira*

Genus *Butyrivibrio*

Genus *Thermoanaerobacter*

Genus *Actinomyces*

Genus *Bifidobacterium*

Section 16. The Mycobacteria

Family *Mycobacteriaceae*

Genus *Mycobacterium*

Section 17. Nocardioforms

Genus *Nocardia*

Genus *Rhodococcus*

Genus *Nocardioides*

Genus *Pseudonocardia*

Genus *Oerskovia*

Genus *Saccharopolyspora*

Genus *Micropolyspora*

Genus *Promicromonospora*

Genus *Intrasporangium*

Volume 3. The Archaeobacteria*, Cyanobacteria, and Remaining Gram-Negative Bacteria

Section 18. Anoxygenic Phototrophic Bacteria

Purple Bacteria

Family I. *Chromatiaceae*

Genus *Chromatium*

Genus *Thiocystis*

Genus *Thiospirillum*

Genus *Thiocapsa*

Genus *Lamprobacter*

Genus *Lamprocystis*

Genus *Thiodictyon*

Genus *Amoebobacter*

Genus *Thiopedia*

Family II. *Ectothiorhodospiraceae*

Genus *Ectothiorhodospira*

Purple Nonsulfur Bacteria

Genus *Rhodospirillum*

Genus *Rhodopila*

Genus *Rhodobacter*

Genus *Rhodopseudomonas*

Genus *Rhodomicrobium*

Genus *Rhodocyclus*

*An older term for the Archaea

Green Bacteria

Green Sulfur Bacteria

Genus *Chlorobium*

Genus *Prosthecochloris*

Genus *Pelodictyon*

Genus *Ancalochloris*

Genus *Chloroherpeton*

Multicellular, Filamentous, Green Bacteria

Genus *Chloroflexus*

Genus *Heliothrix*

Genus *"Oscillochloris"*

Genus *Chloronema*

Genera Incertae Sedis

Genus *Heliobacterium*

Genus *Erythrobacter*

Section 19. Oxygenic Photosynthetic Bacteria

Group I. Cyanobacteria

Subsection I. Order *Chroococcales*

Genus *Chamaesiphon*

Genus *Gloeobacter*

Synechococcus-group

Genus *Gloeothece*

Cyanothece-group

Gloeocapsa-group

Synechocystis-group

Subsection II. Order *Pleurocapsales*

Genus *Dermocarpa*

Genus *Xenococcus*

Genus *Dermocarpella*

Genus *Myxosarcina*

Genus *Chroococcidiopsis*

Pleurocapsa-group

Subsection III. Order *Oscillatoriales*

Genus *Spirulina*

Genus *Arthrospira*

Genus *Oscillatoria*

Genus *Lyngbya*

Genus *Pseudanabaena*

Genus *Starria*

Genus *Crinalium*

Genus *Microcoleus*

Subsection IV. Order *Nostocales*

Family *Nostocaceae*

Genus *Anabaena*

Genus *Aphanazomenon*

Genus *Nodularia*

Genus *Cylindrospermum*

Genus *Nostoc*

Family *Scytonemataceae*

Genus *Scytonema*

Family *Rivulariaceae*
    Genus *Calothrix*
  Subsection V. Order *Stigonematales*
    Genus *Chlorogloeopsis*
    Genus *Fischerella*
    Genus *Stigonema*
    Genus *Geitleria*
Group II. Order *Prochlorales*
  Family *Prochloraceae*
    Genus *Prochloron*
    Genus *"Prochlorothrix"*

Section 20. Aerobic Chemolithotrophic Bacteria and Associated Organisms

A. Nitrifying Bacteria
  Family *Nitrobacteraceae*
  Nitrite-oxidizing Bacteria
    Genus *Nitrobacter*
    Genus *Nitrospina*
    Genus *Nitrococcus*
    Genus *Nitrospira*
  Ammonia-oxidizing Bacteria
    Genus *Nitrosomonas*
    Genus *Nitrosococcus*
    Genus *Nitrosospira*
    Genus *Nitrosolobus*
    Genus *"Nitrosovibrio"*
B. Colorless Sulfur Bacteria
    Genus *Thiobacterium*
    Genus *Macromonas*
    Genus *Thiospira*
    Genus *Thiovulum*
    Genus *Thiobacillus*
    Genus *Thiomicrospira*
    Genus *Thiosphaera*
    Genus *Acidiphilium*
    Genus *Thermothrix*
C. Obligately Chemolithotrophic Hydrogen Bacteria
    Genus *Hydrogenobacter*
D. Iron- and Manganese-Oxidizing and/or Depositing Bacteria
  Family *"Siderocapsaceae"*
    Genus *"Siderocapsa"*
    Genus *"Naumanniella"*
    Genus *"Siderococcus"*
    Genus *"Ochrobium"*
E. Magnetotactic Bacteria
    Genus *Aquaspirillum* (*A. magnetotacticum*)
    Genus *"Bilophococcus"*

Section 21. Budding and/or Appendaged Bacteria

I. Prosthecate Bacteria
  A. Budding Bacteria
    1. Buds produced at tip of prostheca
      Genus *Hyphomicrobium*

      Genus *Hyphomonas*
      Genus *Pedomicrobium*
    2. Buds produced on cell surface
      Genus *Ancalomicrobium*
      Genus *Prosthecomicrobium*
      Genus *Labrys*
      Genus *Stella*
  B. Bacteria that Divide by Binary Transverse Fission
    Genus *Caulobacter*
    Genus *Asticcacaulis*
    Genus *Prosthecobacter*
II. Nonprosthecate Bacteria
  A. Budding Bacteria
    1. Lack peptidoglycan
      Genus *Planctomyces*
      Genus *"Isosphaera"*
    2. Contain peptidoglycan
      Genus *Ensifer*
      Genus *Blastobacter*
      Genus *Angulomicrobium*
      Genus *Gemmiger*
  B. Nonbudding, Stalked Bacteria
    Genus *Gallionella*
    Genus *Nevskia*
  C. Other Bacteria
    1. Nonspinate bacteria
      Genus *Seliberia*
      Genus *"Metallogenium"*
      Genus *"Thiodendron"*
    2. Spinate bacteria

Section 22. Sheathed Bacteria

    Genus *Sphaerotilus*
    Genus *Leptothrix*
    Genus *Haliscomenobacter*
    Genus *"Lieskeella"*
    Genus *"Phragmidiothrix"*
    Genus *Crenothrix*
    Genus *"Clonothrix"*

Section 23. Nonphotosynthetic, Nonfruiting, Gliding Bacteria

Order I. *Cytophagales*
  Family *Cytophagaceae*
    Genus *Cytophaga*
    Genus *Capnocytophaga*
    Genus *Flexithrix*
    Genus *Sporocytophaga*
  Other genera
    Genus *Flexibacter*
    Genus *Microscilla*
    Genus *Chitinophaga*
    Genus *Saprospira*

Order II. *Lysobacterales*
  Family *Lysobacteraceae*
    Genus *Lysobacter*
Order III. Beggiatoales
  Family Beggiatoaceae
    Genus *Beggiatoa*
    Genus *Thioploca*
    Genus *Thiothrix*
    Genus *"Thiospirillopsis"*
Other families and genera
  Family *Simonsiellaceae*
    Genus *Simonsiella*
    Genus *Alysiella*
  Family *"Pelonemataceae"*
    Genus *"Pelonema"*
    Genus *"Peloploca"*
    Genus *"Achroonema"*
    Genus *"Desmanthos"*
Other genera
  Genus *Toxothrix*
  Genus *Leucothrix*
  Genus *Vitreoscilla*
  Genus *Desulfonema*
  Genus *Achromatium*
  Genus *Agitococcus*
  Genus *Herpetosiphon*

Section 24. Fruiting, Gliding Bacteria: The Myxobacteria

Order *Myxococcales*
  Family I. *Myxococcaceae*
    Genus *Myxococcus*
  Family II. *Archangiaceae*
    Genus *Archangium*
  Family III. *Cystobacteraceae*
    Genus *Cystobacter*
    Genus *Melittangium*
    Genus *Stigmatella*
  Family IV. *Polyangiaceae*
    Genus *Polyangium*
    Genus *Nannocystis*
    Genus *Chondromyces*

Section 25. Archaeobacteria (the Archaea)

Group I. Methanogenic Archaeobacteria
  Order I. Methanobacteriales
    Family I. *Methanobacteriaceae*
      Genus *Methanobacterium*
      Genus *Methanobrevibacter*
    Family II. *Methanothermaceae*
      Genus *Methanothermus*
  Order II. *Methanococcales*
    Family *Methanococcaceae*
      Genus *Methanococcus*

Order III. *Methanomicrobiales*
  Family I. *Methanomicrobiaceae*
    Genus *Methanomicrobium*
    Genus *Methanospirillum*
    Genus *Methanogenium*
  Family II. *Methanosarcinaceae*
    Genus *Methanosarcina*
    Genus *Methanolobus*
    Genus *Methanothrix*
    Genus *Methanococcoides*
Other Taxa
  Family *Methanoplanaceae*
    Genus *Methanoplanus*
  Other Genus *Methanosphaera*
Group II. Archaeobacterial Sulfate Reducers
  Order *"Archaeoglobales"*
    Family *"Archaeoglobaceae"*
      Genus *Archaeoglobus*
Group III. Extremely Halophilic Archaeobacteria
  Order *Halobacteriales*
    Family *Halobacteriaceae*
      Genus *Halobacterium*
      Genus *Haloarcula*
      Genus *Haloferax*
      Genus *Halococcus*
      Genus *Natronobacterium*
      Genus *Natronococcus*
Group IV. Cell Wall-less Archaeobacteria
    Genus *Thermoplasma*
Group V. Extremely Thermophilic S^0-Metabolizers
  Order I. *Thermococcales*
    Family *Thermococcaceae*
      Genus *Thermococcus*
      Genus *Pyrococcus*
  Order II. *Thermoproteales*
    Family I. *Thermoproteaceae*
      Genus *Thermoproteus*
      Genus *Thermofilum*
    Family II. *Desulfurococcaceae*
      Genus *Desulfurococcus*
    Other bacteria
      Genus *Staphylothermus*
      Genus *Pyrodictium*
  Order III. *Sulfolobales*
    Family *Sulfolobaceae*
      Genus *Sulfolobus*
      Genus *Acidianus*

Volume 4. The Actinomycetes

Section 26. Nocardioform Actinomycetes

    Genus *Nocardia*
    Genus *Rhodococcus*

Genus *Nocardioides*
Genus *Pseudonocardia*
Genus *Oerskovia*
Genus *Saccharopolyspora*
Genus *Faenia*
Genus *Promicromonospora*
Genus *Intrasporangium*
Genus *Actinopolyspora*
Genus *Saccharomonospora*

Section 27. Actinomycetes with Multilocular Sporangia

Genus *Geodermatophilus*
Genus *Dermatophilus*
Genus *Frankia*

Section 28. Antinoplanetes

Genus *Actinoplanes*
Genus *Ampullariella*
Genus *Pilimelia*
Genus *Dactylosporangium*
Genus *Micromonospora*

Section 29. Streptomyces and Related Genera

Genus *Streptomyces*
Genus *Streptoverticillium*
Genus *Kineosporia*
Genus *Sporichthya*

Section 30. Maduromycetes

Genus *Actinomadura*
Genus *Microbispora*
Genus *Microtetraspora*
Genus *Planobispora*
Genus *Planomonospora*
Genus *Spirillospora*
Genus *Streptosporangium*

Section 31. Thermomonospora and Related Genera

Genus *Thermomonospora*
Genus *Actinosynnema*
Genus *Nocardiopsis*
Genus *Streptoalloteichus*

Section 32. Thermoactinomycetes

Genus *Thermoactinomyces*

Section 33. Other Genera

Genus *Glycomyces*
Genus *Kibdelosporangium*
Genus *Kitasatosporia*
Genus *Saccharothrix*
Genus *Pasteuria*

The Mathematics of Bacterial Growth

Exponential Growth

Bacteria grow and reproduce by binary fission, resulting in population increases by a factor of 2 with each generation. The total number of cells in a bacterial population in exponential growth starting with one cell may be expressed by the mathematical equation

$$N = 1 \times 2^n \qquad (1)$$

where N is the total number of cells in the population and the exponent (n) represents the number of generations. Because bacterial populations generally are inoculated with not one cell, but hundreds or thousands of cells, Equation 1 may be rewritten as

$$N = N_0 \times 2^n \qquad (2)$$

where N_0 is the initial population number at time t_0, N is the number of cells in the population at time t, and n is the number of generations between time t_0 and time t.

Equation 2 can be used to solve for the number of generations (n). First, the logarithm of Equation 2 is taken to derive the mathematical expression

$$\log_{10} N = \log_{10} N_0 + n \log_{10} 2 \qquad (3)$$

The equation is expressed logarithmically to accommodate the enormous increases in population during exponential growth. Equation 3 is rearranged to solve for n:

$$n = \frac{\log_{10} N - \log_{10} N_0}{\log_{10} 2} \qquad (4)$$

The equation can be simplified by substituting the value of $\log_{10} 2$, which is 0.301:

$$n = \frac{\log_{10} N - \log_{10} N_0}{0.301} \qquad (5)$$

The number of generations per unit time, or the exponential growth rate (R), can be derived from Equation 5:

$$R = \frac{n}{t - t_0} = \frac{\log_{10} N - \log_{10} N_0}{0.301(t - t_0)} \qquad (6)$$

Equation 6 can be used to determine the number of generations that have occurred in a population between time t_0 and time t, if N_0 and N are known. For example, a bacterial culture containing an initial population of 100 cells and a population of 100,000,000 cells ten hours later would have an exponential growth rate (R) of 2.0 generations/hour:

$$R = \frac{\log_{10}(100,000,000) - \log_{10}(100)}{0.301(10 \text{ hours})}$$

$$R = \frac{8 - 2}{3.01 \text{ hour}}$$

$$R = 2.0 \text{ generations/hour}$$

The generation time (g) of the population can be determined by the reciprocal of the exponential growth rate:

$$g = \frac{1}{R}$$

$$g = \frac{1}{2.0 \text{ generations/hour}}$$

$$g = 0.50 \text{ hour/generation, or 30 minutes/generation}$$

During exponential growth in a batch culture, the rate of increase of cells in the population is proportional to the number of cells present at any particular time and can be expressed in mathematical terms:

$$\frac{dN}{dt} = kN \qquad (7)$$

where N is cell number or some other measurable parameter, t is time, and k is the constant of proportionality, or *specific growth-rate constant*. Rearrangement of Equation 7 produces

$$\frac{dN}{N} = kdt \qquad (8)$$

Integration of Equation 8 between the limits of t_0 and t and N_0 and N yields

$$\ln N - \ln N_0 = k(t - t_0) \qquad (9)$$

Conversion of Equation 9 from natural logarithms to logarithms to the base 10 gives the following expression:

$$\log_{10} N - \log_{10} N_0 = \frac{k}{2.303}(t - t_0) \qquad (10)$$

Equation 10 can be used to determine the growth-rate constant (k) if the initial number of cells in the population (N_0) and the population at time t are known. The growth-rate constant of a culture containing an initial population of 100 cells and a population of 100,000,000 cells ten hours later is 1.38 hour^{-1}:

$$k = \frac{\log_{10} N - \log_{10} N_0(2.303)}{t - t_0}$$

$$k = \frac{(8-2)(2.303)}{10 \text{ hours}}$$

$$k = 1.38 \text{ hour}^{-1}$$

The specific growth-rate constant is a reproducible description of the growth rate of a culture in a particular environment. The relationship between k and the generation time (g), or the time required for the number of the cells to increase by a factor of 2, can be derived from Equation 9. When N_0 has increased to $2N_0$, $t - t_0$ becomes g:

$$\ln N - \ln N_0 = k(t - t_0) \qquad (9)$$

$$\ln 2N_0 - \ln N_0 = kg$$

or

$$kg = \ln \frac{2N_0}{N_0}$$

$$kg = \ln 2$$

$$g = \frac{\ln 2}{k}$$

$$g = \frac{0.693}{k}$$

In the example given, where $k = 1.38 \text{ hour}^{-1}$, the generation time would be 0.50 hour:

$$g = \frac{0.693}{1.38 \text{ hour}^{-1}}$$

$$g = 0.50 \text{ hour, or 30 minutes}$$

Continuous Culture

Growth rate remains constant during exponential growth and then abruptly drops to zero when nutrients are depleted in batch culture. In continuous culture of bacteria, fresh medium is continuously fed to a culture, and the volume is kept constant by removing spent medium. The culture is in steady state when the system reaches an equilibrium point wherein cell numbers and nutrient levels in the culture vessel become constant and the growth rate is just sufficient to replace cells lost through the overflow.

In a chemostat, growth rate is controlled by the rate at which fresh medium is added to the growth vessel, and population density is controlled by a growth-limiting concentration of a specific required nutrient (for example, a carbon source such as glucose). This relationship can be expressed mathematically:

$$D = \frac{F}{V} \qquad (11)$$

where F is the flow rate (measured in units of culture volumes per time), V is the volume of culture in the vessel, and D is the dilution rate (expressed in units of time^{-1}), or the number of volumes of medium that pass through the culture vessel per unit of time.

The dilution rate (D) of a chemostat can be expressed as

$$\frac{dN}{dt} = DN \qquad (12)$$

Previously growth rate was expressed as

$$\frac{dN}{dt} = kN \qquad (7)$$

Any changes in bacterial concentration in a chemostat would be defined as

changes in cell concentration = growth – dilution

or

$$\frac{dN}{dt} = kN - DN$$

If the dilution rate (D) is kept constant in a chemostat, there is no change in bacterial numbers and a steady state is reached in which

$$\frac{dN}{dt} = 0$$

and

$$kN = DN$$

or

$$k = D \qquad (13)$$

Equation 13 states that the growth rate of a culture in a chemostat is determined by the dilution rate.

The chemostat is a constant, stable system because it is self-correcting. In a batch culture, nutrients are depleted as the organisms grow, resulting in parallel decreases in nutrient concentration and growth rate. In a chemostat, however, there is an excess of all nutrients except a single required nutrient, which becomes the growth-limiting factor. The limited availability of this nutrient controls bacterial growth. By controlling the dilution rate (D), the rate of bacterial growth (k) may also be controlled.

For example, if the rate of addition of nutrient to the growth vessel in a chemostat is decreased, the rate of cell loss from the vessel would also decrease, resulting in an increase in the cell density in the culture. The more dense culture would utilize the limiting nutrient at a faster rate, resulting in decreased concentration of the nutrient in the vessel. This would cause a decrease in the growth rate of the culture until it matched the rate of cell loss through the overflow. An opposite series of events would occur if the rate of addition of medium to the culture vessel were increased. Thus the growth rate of the culture in a chemostat adjusts to changes in the rate of cell loss to maintain a constant culture density.

This relationship between growth rate and nutrient concentration is illustrated by the equation

$$k = k_{max} \frac{C}{K_s + C} \qquad (14)$$

where k is the specific growth rate at limiting nutrient concentration (C), k_{max} is the growth rate at saturating concentration of the nutrient, and K_s is a constant numerically equal to the substrate concentration at which $k = \frac{1}{2}k_{max}$. Equation 14 is an adaptation of the Michaelis-Menten equation used in enzyme kinetic analyses.

Substituting D for k in Equation 14 gives

$$D = k_{max} \frac{C}{K_s + C} \qquad (15)$$

and solving for C gives

$$C = K_s \frac{D}{k_{max} - D} \qquad (16)$$

Equation 16 states the fundamental relationship between nutrient concentration (C) in the growth vessel and dilution rate (D). As can be seen from Equation 16, at steady state, nutrient concentration (C) depends on the dilution rate (D) in a chemostat.

Microbiology Internet Resources

The Internet is rapidly becoming a rich resource for the study of microbiology. If you have an Internet connection and a web browser, an incredible amount of information is available to you. For this text-book, the publisher maintains a specific Lim *Microbiology* Web site that provides a number of resources for both student and professor alike. Take a look at it (http://www.wcbp.com/cellm/lim) and use it to amplify and expand the content of the book, to test your knowledge of the material, and to explore the full depth and breadth of the exciting science of microbiology.

Additionally, listed below are some general or entry sites related to microbiology. This printed version appears on the Lim Web site as well, so that you can check into the site and link directly to these pages, as well as to new or updated listings.

General Sites:

1. **Microbiology and Virology Virtual Library.** This is probably the most extensive list of World Wide Web sites related to microbiology. If you click on the word *BioScience* in the title, you will be sent to a search engine for the page. This site is very reliable with quick access. [*http://golgi.harvard.edu/biopages/micro.html*]

2. **Microbial Underground.** Contains lists of sites related to medical, microbiological, and molecular biology. This also contains an on-line course in medical bacteriology. Depending on your location, you can access the United Kingdom or the U.S. mirror site for optimum browsing performance.[*http://www.qmw.ac.uk/~rhbm001/index.html*]

3. **BIOSCI.** This is the BIOSCI home page which includes all of the bionet usenet newsgroups. You can access current newsgroups, view the newsgroup archives, post messages, etc. The BIOSCI user address database is also available through this site. [*http://www.bio.net*]

4. **Bugs in the News.** A collection of microbiology articles for lay audiences. A great way to introduce the non-scientist to microbiology! [*http://falcon.cc.ukans.edu/~jbrown/bugs.html*]

5. **American Society for Microbiology.** This is the home page of the American Society for Microbiology. The page contains a good search engine and additional links of interest to practicing microbiologists. [*http://www.asmusa.org*]

Microscopy:

1. **Microscopes and Microscopy.** Links to microscope-related web sites, newsgroups, and FTP sites. A listing of current meetings and courses related to microscopy is also included. [*http://www.ou.edu/research/electron/mirror*]

2. **Yahoo: Microscopy.** From the creators of web indices, the Yahoo list of links related to microscopy. Includes a number of good microscopic image (both light and electron) archive sites. [*http://www.yahoo.com/text/Science/Chemistry/Microscopy*]

Virology:

1. **WWW Server for Virology.** This is a web server devoted to virology, located at the Institute of Molecular Virology at the University of Wisconsin. A complete collection of research and educational links related to virology. A very reliable, popular site. [*http://www.bocklabs.wisc.edu/Welcome.html*]

2. **WWW Virology Servers.** Located at the Garry Lab at the Tulane University School of Medicine. A well-organized list of world-wide virology servers. This site also has a United Kingdom mirror for readers in Europe. [*http://www.tulane.edu/~dmsander/garryfavweb.html*] [*http://www-micro.msb.le.ac.uk/335/garryfavweb/garryfavweb.html*]

Immunology:

1. **British Society for Immunology.** Links to immunology related sites on the web. Unfortunately, there just isn't a lot out there on immunology. [*http://194.128.227.252/uk/society/bsi/bsilinks.htm*]

Diseases/Epidemiology:

1. **World Health Organization.** Information and updates about WHO's activities. Everything from disease prevention to

outbreak control is posted at this site. This site was invaluable during the recent Ebola outbreak in Zaire. [*http://www.who.ch*]

2. **Centers for Disease Control.** Information from the CDC on disease monitoring. Links include Morbidity and Mortality Weekly. [*http://www.cdc.gov*]

3. **Epidemiology Virtual Library.** Epidemiology links according to country, university, or disease. Access to job listings and Usenet newsgroups included. [*http://chanane.ucsf.edu/epidem/epidem.html*]

4. **TB/HIV Research Laboratory.** This site currently includes information on TB only. Results of research at Brown University and links to related sites. [*http://www.brown.edu/Research/TB-HIV_Lab*]

Specific Organisms:

1. **Actinomycetes/Streptomycetes.** Research related links such as researcher directories, meetings, and abstracts. An online library and methods section is included. Links to actinomycete related WWW sites available. [*http://molbio.cbs.umn.edu/asirc*]

2. **Yeast Virtual Library.** WWW links relating to budding, fission, and *Candida*. Budding yeast entries include data on the genome sequencing effort and protein database. Links to other yeast support sites like ATCC, bionet.molbio.yeast, and researcher directory included. [*http://genome-www.stanford.edu/VL-yeast.html*]

3. **Mycology Virtual Library.** A thorough collection of mycology related links on the web. An index is available to make your searching easier. [*http://www.keil.ukans.%7Efungi/*]

Taxonomy:

1. **Tree of Life.** Eventually this site will contain an on-line listing of the genetic and evolutionary relationship of every living organism. This site is under construction and not all-inclusive, but keep checking. [*http://phylogeny.arizona.edu/tree/phylogeny.html*]

Chemistry for the Microbiologist

All living organisms, including microorganisms, are composed of basic chemical building blocks that form the structures, macromolecules, enzymes, and other constituents of the cell. An understanding of the chemical principles of these molecules is necessary to understand microbiology. This appendix provides a brief summary of the essential concepts of chemical molecules, including a discussion of atoms, elements, molecules, and atomic bonding. The chemistry of macromolecules is discussed in chapter 3.

Atoms and Elements

Matter is composed of fundamental substances called **elements.** An element is a chemical substance that cannot be broken down further into simpler substances by ordinary chemical or physical means. Although there are 92 naturally occurring elements, six elements (carbon, nitrogen, sulfur, phosphorus, oxygen, and hydrogen) make up 97% of the dry weight of a bacterium. Each element is given a one- or two-letter abbreviation of its English or Latin name (Table A.1). For example, C is the chemical symbol for carbon, N for nitrogen, S for sulfur, Mg for magnesium, and so forth.

An **atom** is the smallest unit of an element that can combine with another element. An atom is composed of three types of subatomic particles: **protons,** which are positively charged; **electrons,** which are negatively charged; and **neutrons,** which have no charge (Figure A.1). Protons and neutrons, which have about 1,840 times the mass of an electron, are found in the central core, or **nucleus,** of the atom and impart a net positive charge to the nucleus. The negatively-charged electrons orbit the nucleus. The number of protons equals the number of electrons in all atoms, making the atom neutral in charge.

The **atomic number** of an atom is determined by the number of protons in the nucleus of the atom and ranges from one (in a hydrogen atom) to more than one hundred (in the largest atoms). The **atomic weight** of an atom is determined by the total number of protons and neutrons. Each atom has its own atomic number and atomic weight. For example, the atomic number of carbon, which has 6 protons, is 6 and its atomic weight is 12. The atomic number of oxygen, which has 8 protons, is 8 and its atomic weight is 16. Although the number of protons in an element and its atomic number are constant, the number of neutrons may vary, resulting in **isotopes** of an element with different atomic weights.

table A.1

The Major Elements and Their Characteristics

Element	Atomic Symbol	Atomic Number	Atomic Weight	Ionized Form
Calcium	Ca	20	40.1	Ca^{++}
Carbon	C	6	12.0	–
Chlorine	Cl	17	35.5	Cl^-
Copper	Cu	29	63.5	Cu^+, Cu^{++}
Hydrogen	H	1	1.0	H^+
Iodine	I	53	126.9	I^-
Iron	Fe	26	55.8	Fe^{++}, Fe^{+++}
Magnesium	Mg	12	24.3	Mg^{++}
Nitrogen	N	7	14.0	–
Oxygen	O	8	16.0	–
Phosphorus	P	15	31.0	–
Potassium	K	19	39.1	K^+
Sodium	Na	11	23.0	Na^+
Sulfur	S	16	32.1	–

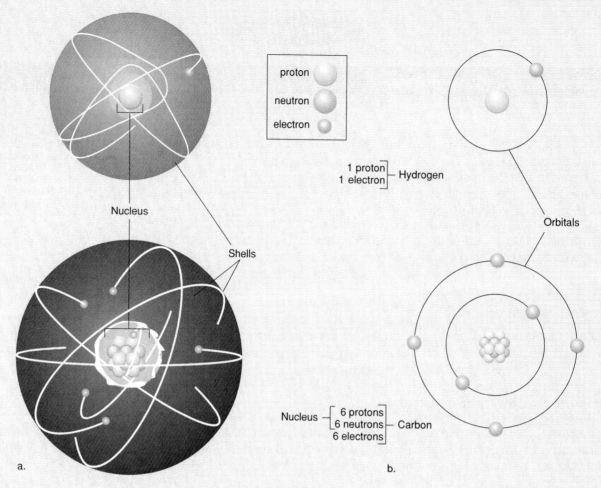

figure A.1
Structures of Atoms

a. Models of a hydrogen atom and a carbon atom, indicating the positions of protons and neutrons in the nucleus and of electrons in orbitals and shells. b. Models of a hydrogen atom and a carbon atom, indicating the numbers and arrangements of electrons in orbitals.

Carbon-14 (^{14}C) is a **radioactive isotope** of carbon that has 6 protons and 8 neutrons, resulting in an atomic weight of 14. Because of these additional neutrons, the atomic nucleus of ^{14}C is highly unstable and can disintegrate, resulting in the release of energy and an electron (beta particle). Radioactive isotopes such as ^{14}C are useful for labeling biological materials because the release of beta particles can be easily detected. Radioactive emissions are also harmful to living systems because they can damage or kill biological materials.

Electron Shells

Electrons orbit the nucleus of an atom at different energy levels, or **shells** (Figure A.2). The shells closest to the nucleus have the lowest energy, whereas the furthest shells have the greatest energy. Each shell holds a characteristic maximum number of electrons. The innermost shell can contain up to two electrons; the second, eight electrons; the third, eighteen electrons. The outer shells are filled only after the innermost shells have been filled. When its outermost shell is filled, an atom is in its most chemically stable configuration. The number of electrons in the outermost shell of an atom, or its **valence,** determines the binding capacity of the atom. For example, an oxygen atom (valence of 2) has six electrons in its outermost shell and can accept two additional electrons in this shell. Carbon (valence of 4) has four electrons in its outermost shell and can accept four additional electrons. The **electron configuration** of an atom is important because it determines the chemical properties of the atom and its ability to form bonds with other atoms. The principle types of chemical bonds are ionic bonds, covalent bonds, and hydrogen bonds.

Ionic Bonds

When atoms lose or gain electrons, they become charged and are called **ions.** An examination of the sodium atom and the chlorine atom illustrates the differences between an atom and an ion (Figure A.3). When the sodium atom (Na) loses an electron to another atom, the sodium atom becomes a positively charged ion, or

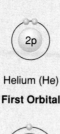

Helium (He)
First Orbital

Carbon (C)   Nitrogen (N)   Oxygen (O)
First and Second Orbitals

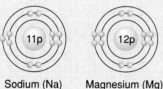

Sodium (Na)   Magnesium (Mg)   Phosphorus (P)   Sulfur (S)   Chlorine (Cl)
Second and Third Orbitals

figure A.2

Models of Elements and Shells

The shells are filled by electrons as the atomic numbers increase. Electrons appear in pairs, except for elements with incompletely filled outer shells.

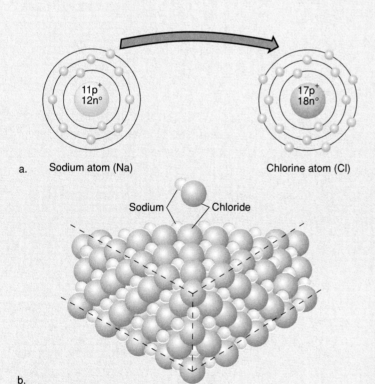

a.   Sodium atom (Na)   Chlorine atom (Cl)

b.

figure A.3

Ionic Bonding Between Sodium and Chlorine.

a. When an ionic bond forms between sodium and chlorine, sodium loses its single outer orbital electron to chlorine. b. This reaction results in a solid crystal complex that interlinks sodium and chlorine atoms.

cation (Na$^+$). When the chlorine atom (Cl) accepts an electron from another atom, the chlorine atom becomes a negatively charged ion, or **anion** (Cl$^-$). Because they are of opposite charge, cations and anions are attracted to each other. The binding of a cation (for example, Na$^+$) with an anion (Cl$^-$) constitutes an **ionic bond** and the molecule formed is called an **ionic compound** (NaCl). Ionic bonds form between atoms with valences that complement each other. Atoms in ionic compounds represent a more stable configuration than the separate, unbonded atoms. However, ionic bonds readily dissociate in water because of the tendency of water to move between the ions.

Covalent Bonds

Covalent bonds form when two atoms share electrons in their outer shells. A simple example of a covalent bond is the joining of two hydrogen atoms to form a molecule of hydrogen gas (H$_2$) (Figure A.4). Each hydrogen atom (valence of 1) has one electron in its shell and can accept an additional electron to fill the shell. The two hydrogen atoms share their single electrons in a covalent bond. Another example of covalent bonding is the sharing of electrons between carbon (valence of 4) and hydrogen (valence of 1). A molecule of methane (CH$_4$) is formed when one carbon atom and four hydrogen atoms share electrons.

A **single covalent bond,** expressed as a single line between atoms (H–H), is formed when one pair of electrons is shared between two atoms. When two pairs of electrons are shared, the bond is called a **double covalent bond** and is represented by a double line (C=C). Three pairs of electrons are shared in a **triple covalent bond** (N≡N). Triple covalent bonds are very stable and difficult to break.

When electrons are shared unequally between atoms, **polar covalent bonds** are formed and the molecule containing these bonds is called a **polar molecule.** One part of a polar molecule has a positive charge and another part has a negative charge. A **nonpolar molecule** is one that is electrically neutral and electrons are shared equally between the atoms. Covalent bonds are strong and do not break unless they are exposed to high temperatures, strong chemicals, or enzymes. For this reason, covalent bonds are commonly found in DNA, RNA, proteins, and other macromolecules that form the basis of biological systems.

Hydrogen Bonds

Hydrogen bonds develop when a positively charged hydrogen atom covalently bonded to oxygen, nitrogen, or fluorine is attracted by the negatively charged portions of other polar molecules such as the oxygen portion of a water molecule. When this occurs, a weak bond forms between the hydrogen atom and the other negatively charged molecule. An example of hydrogen bonding occurs between water molecules (Figure A.5). When molecules of water are close to each other, the hydrogen atom of one water molecule is attracted to the negative charged oxygen atoms of the other. Although a single hydrogen bond is weak, many such bonds can hold molecules together. Extensive hydrogen bonding holds together the two strands of DNA and the structures of many proteins.

Water

Water, which makes up 45% to 95% of a cell, is the most common liquid on earth and is the principal constituent of living systems. Although it is a relatively small and simple molecule, water is an excellent solvent and plays an important role in facilitating chemical reactions. Many of the unique physical and chemical properties of water are explained by its molecular structure.

A water molecule is composed of one atom of oxygen and two atoms of hydrogen. Each of the hydrogen atoms is linked to the oxygen atom by a covalent bond. The water molecule has an equal number of electrons and protons and, therefore, is neutral in charge. Because the oxygen atom attracts electrons more strongly than the hydrogen atom, the shared electrons of the covalent bonds spend more time around the oxygen nucleus than they do around the hydrogen nucleus. As a consequence, water is a highly polar molecule.

The polar nature of water enables each water molecule to form hydrogen bonds with four other nearby water molecules. This explains the relatively high boiling point (100° C) of water and the large amount of heat energy (540 calories of heat to evaporate 1 gram of water) required to separate water molecules from each other to form water vapor. Because of this strong attraction between water molecules, water is an excellent temperature regulator and helps the cell maintain a constant temperature that is important for biological activities and chemical reactions.

The polarity of water molecules is responsible for its capacity as a universal solvent. As ionic substances, such as sodium chloride (NaCl), are dissolved in water, the polar water molecules cluster around and separate the ions (Figure A.6). The positive sodium ions (Na^+) are attracted to the negatively charged oxygen atom of water and the negative chloride ions (Cl^-) are attracted to the positively charged hydrogen atoms. As a result, the Na^+ and Cl^- of NaCl are totally surrounded by water and are dissolved. Polar substances, such as carbohydrates and simple alcohols, tend to be attracted to water molecules and are called **hydrophilic**

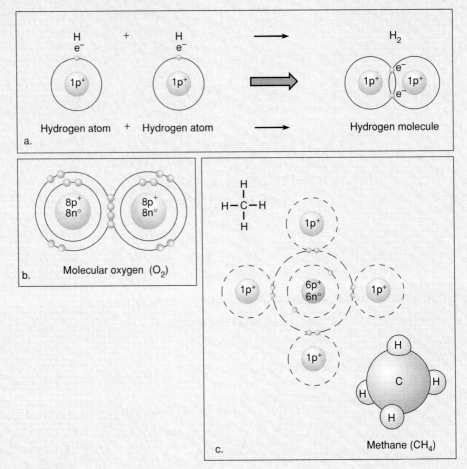

figure A.4
Examples of Covalent Bonding
a. A single covalent bond is formed when two hydrogen atoms share their electrons.
b. A double covalent bond is formed between two oxygen atoms that share four electrons.
c. Simple, working, and three-dimensional models of methane. The carbon atom has four electrons to share and the hydrogen atoms each have one electron to share.

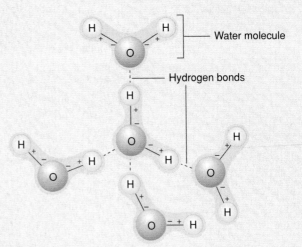

figure A.5
Hydrogen Bonding in Water
The positively charged hydrogen atom of one water molecule is attracted to the negatively charged oxygen atom of another water molecule. Each water molecule forms hydrogen bonds with four other adjacent water molecules.

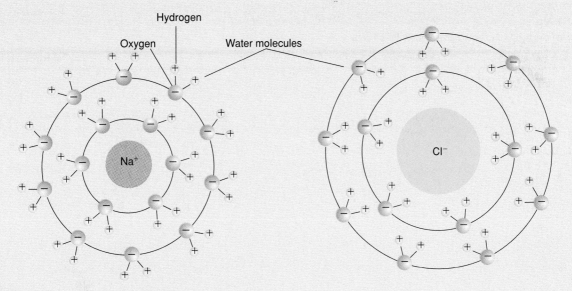

figure A.6

Dissolution of Sodium Chloride in Water

As sodium chloride (NaCl) is dissolved in water, the polar water molecules cluster around and separate the sodium ion (Na⁺) and the chloride ion (Cl⁻).

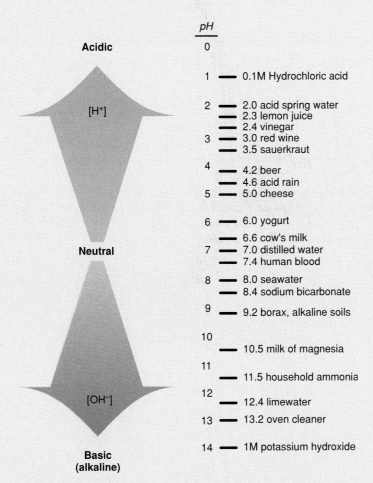

pH	
Acidic	0
	1 — 0.1M Hydrochloric acid
[H⁺]	2 — 2.0 acid spring water
	— 2.3 lemon juice
	— 2.4 vinegar
	3 — 3.0 red wine
	— 3.5 sauerkraut
	4 — 4.2 beer
	— 4.6 acid rain
	5 — 5.0 cheese
	6 — 6.0 yogurt
	— 6.6 cow's milk
Neutral	7 — 7.0 distilled water
	— 7.4 human blood
	8 — 8.0 seawater
	— 8.4 sodium bicarbonate
	9 — 9.2 borax, alkaline soils
	10
	— 10.5 milk of magnesia
	11 — 11.5 household ammonia
	12 — 12.4 limewater
	13 — 13.2 oven cleaner
[OH⁻]	14 — 1M potassium hydroxide
Basic (alkaline)	

figure A.7

The pH Scale

The pH scale, showing approximate pH readings for various substances.

[Greek *hydros*, water, *philos*, to love], whereas substances that are entirely or largely nonpolar, such as fats and oils, tend to be repelled by water and are called **hydrophobic** [Greek *phobos*, fear]. **Amphipathic** compounds contain both hydrophilic and hydrophobic groups. An example is the phospholipids in the plasma membrane.

Another characteristic of water molecules is their slight tendency to to **ionize**, or dissociate into ions: hydrogen ions (H⁺) and hydroxide ions (OH⁻). Only a small but constant number of water molecules ionize in pure water. This number is constant because the tendency of water to dissociate into ions is balanced by the tendency of the ions to reunite to form water. As a consequence, pure water is said to be neutral, neither acid nor alkaline.

Acids, Bases, and pH

When an ionic substance is dissolved in water, it may change the proportion of H⁺ and OH⁻. For example, when hydrochloric acid (HCl) dissolves in water, it dissociates into H⁺ and Cl⁻. As a consequence, an HCl solution contains more H⁺ than OH⁻ and is considered **acidic**. However, when potassium hydroxide (KOH) dissolves in water, it is separated into K⁺ and OH⁻, resulting in a **basic** solution that has more OH⁻ than H⁺.

The acid and base concentrations of solutions are measured by the **pH scale,** which ranges from 0 (the most acidic) to 14 (the most basic) (Figure A.7). The midpoint of this range (pH 7) represents a neutral solution in which the concentrations of H⁺ and OH⁻ are equal. The symbol **pH** stands for the negative logarithm of the hydrogen ion concentration in moles per liter. As Table A.2 indicates, the greater the hydrogen ion concentration, the lower the pH value. Because the pH scale is logarithmic, a change in one

Hydrogen Ion and Hydroxide Ion Concentrations on the pH Scale

g/l of Hydrogen Ions	Logarithm	pH	g/l of Hydroxide Ions
1.0	10^0	0	0.00000000000001
0.1	10^{-1}	1	0.0000000000001
0.01	10^{-2}	2	0.000000000001
0.001	10^{-3}	3	0.00000000001
0.0001	10^{-4}	4	0.0000000001
0.00001	10^{-5}	5	0.000000001
0.000001	10^{-6}	6	0.00000001
0.0000001	10^{-7}	7	0.0000001
0.00000001	10^{-8}	8	0.000001
0.000000001	10^{-9}	9	0.00001
0.0000000001	10^{-10}	10	0.0001
0.00000000001	10^{-11}	11	0.001
0.000000000001	10^{-12}	12	0.01
0.0000000000001	10^{-13}	13	0.1
0.00000000000001	10^{-14}	14	1.0

whole pH unit represents a tenfold change in hydrogen ion concentration. A solution with a pH of 5 has a hydrogen ion concentration 10 times greater than a solution with a pH of 6.

Living systems are sensitive to changes in pH and most biological activities take place at a pH between 6 and 8. The notable exceptions are chemical reactions that occur in the acidic environment of the stomach in humans and other animals. Although some microorganisms live in acidic or alkaline environments, they maintain a near neutral pH within the cell by controlling the movement of H^+ across the plasma membrane. Other microorganisms produce acidic end products during metabolism, which lower the environmental pH. Extreme pH changes during growth of these microorganisms can be minimized by the addition of buffering agents such as phosphates or carbonates to the growth medium. **Buffers** are salts of weak acids or bases that combine with or release hydrogen ions in response to changes in pH, thereby slowing change in the hydrogen ion concentration.

Answers to End-of-Chapter Questions

Chapter 1

Short-Answer Questions

1. bacteriology, virology, mycology, phycology, protozoology, genetic engineering, medical microbiology, immunology, microbial ecology, food microbiology, industrial microbiology, etc.

2. Robert Hooke in 1665.

3. Anton von Leeuwenhoek in 1674.

4. The theory of spontaneous generation states that life can arise from nonliving matter. This theory dates back hundreds of years before the 1600s.

5. Redi's experiment involved three jars of meat: the first was covered with paper, the second was covered with gauze, and the third was left uncovered. He concluded that macroorganisms (e. g., flies) could not spontaneously generate.

6. Critics discounted Spallanzi's experiment because he excluded the "vital force" (now known as oxygen).

7. Pasteur used a "swan neck flask" to allow air to freely enter.

8. The germ theory states that microorganisms might cause disease. This theory developed, due to the efforts of scientists such as Pasteur and Koch, in the late 1800s.

9. Koch's postulates state that (1) the same microbe must be found in all cases of the disease, (2) that microbe must be isolated from the diseased host and grown in pure culture, (3) that microbe must be inoculated into a healthy host and shown to cause the same disease, and (4) that microbe must be reisolated from the inoculated host.

10. The golden age of microbiology was 1876 to 1906. During this time, many diseases were proven to be caused by microorganisms.

11. Elie Metchnikoff discovered phagocytes in 1884.

12. Edward Jenner developed vaccination in 1798.

13. Viruses were found to be particles of protein and nucleic acids.

14. Restriction endonucleases are enzymes that cleave DNA at specific sites. They are important because they can be used to isolate and rearrange genes.

15. Infectious diseases mentioned in this chapter include anthrax, trachoma, syphilis, gonorrhea, smallpox, polio, and Ebola.

16. Noninfectious diseases mentioned in this chapter include cardiovascular diseases, diabetes, black lung disease, and certain types of cancer.

17. Infectious diseases are the leading cause of death today.

Chapter 2

Short-Answer Questions

1. Macroorganisms include all plants and animals. Microorganisms include bacteria, algae, fungi, protozoa, and viruses.

2. Size, shape, and arrangement are some of the basic traits used to identify bacteria.

3. The basic shapes are coccus and bacillus; however, there are many variations on these fundamental shapes: coccobacillus, spirilla, spirochete, pleomorphic, etc.

4. Six types of microscopes are bright-field, dark-field, fluorescence, phase-contrast, transmission electron, and scanning electron microscopes.

5. Leeuwenhoek's microscope was little more than a magnifying glass. Modern bright-field microscopes have two or more lenses with a light source.

6. Major features of a bright-field microscope include the light source, iris diaphragm, condenser lens, stage, objective lenses, body tube, and ocular (or eyepiece) lenses.

7. The total magnification is 950×.

8. Immersion oil is used with the *oil-immersion lens* (the lens with high magnification and high numerical aperture). The oil is used to prevent defraction of the light and, thereby, to increase resolution.

9. Light microscopes can see objects of 0.2 μm, while electron microscopes can see objects as small as 0.001 μm.

10. Simple stains are important for determining morphology; differential stains are important for distinguishing between bacteria based on their staining characteristics; and special stains enhance a special feature of the cell.

11. The Gram stain remains the most commonly used staining procedure.

12. Dark-field microscopy offers the advantage of high contrast, enabling the observation of small or thin structures (for example, flagella).

13. Phase-contrast is advantageous for viewing living specimens and for amplifying differences in contrast within the cells.

14. Fluorescence microscopy is especially useful for diagnosis by immunofluorescence.

15. Clear imaging of surface features is possible with SEM, while TEM enables clear imaging of internal features. Unfortunately, electron microscopes are too large and expensive for routine use.

16. The arrangement of cells depends on several factors, including their division pattern and growth conditions.

Multiple-Choice Questions

1. d

2. d

3. b

4. b

Chapter 3

Short-Answer Questions

1. Bacteria and Archaea are procaryotic cells. Algae, fungi, and protozoa are eucaryotic cells.

2. Viruses are acellular particles; therefore, they exhibit neither cell type.

3. Eucaryotic cells have membrane-enclosed organelles; procaryotic cells do not.

4. Proteins (composed of amino acids), nucleic acids (composed of nucleotides), carbohydrates (composed of monosaccharides and disaccharides), and lipids (composed of glycerol and fatty acids) are the four major groups of macromolecules.

5. Water, vitamins, and inorganic compounds (for example, salts) are also major cell constituents.

6. All procaryotic cells have a plasma membrane, cytoplasm, and nucleic acids (DNA and RNA). Organelles (for example, chloroplast, mitochondria, nucleus, etc.) will not be found in any procaryotic cells.

7. Examples of organelles include nucleus, mitochondria, chloroplast, microbodies, vacuoles, lysosomes, Golgi complex, and cisternae.

8. Archaea, although they are procaryotic, differ from Bacteria by their chemical composition and preference for extreme environments.

9. Bacterial membrane lipids have ester linkages between glycerol and the fatty acids. Archaeal membrane lipids have ether linkages between glycerol and repeating branched aliphatic chains of isoprene. Eucaryotes have similar membranes with the addition of sterols.

10. Procaryotic cell walls, generally present, are composed mainly of peptidoglycan. Eucaryotic cell walls, when present, are composed of cellulose or chitin. Peptidoglycan and lipopolysaccharide are unique to bacteria.

11. The gram-positive cell wall is composed of a thick layer of peptidoglycan, along with teichoic acid. The gram-negative cell wall includes a thin layer of peptidoglycan, without the teichoic acid, surrounded by an outer membrane of lipopolysaccharide.

12. With the Gram staining procedure, gram-positive bacteria retain the primary dye (crystal violet) and appear purple, while gram-negative bacteria are decolorized (by alcohol) then counterstained pink (by safranin).

13. A capsule is a polysaccharide and/or protein layer found outside the cell wall of some bacteria. It may be used for protection against phagocytes, adhesion to surfaces, prevention of dehydration, or storage of nutrients.

14. A pilus is a proteinaceous appendage used to attach to another cell.

15. Bacterial flagella are single protein filaments, attached by a hook mechanism, for the purpose of locomotion. Eucaryotic cells may have flagella or cilia that exhibit a complex 9 + 2 arrangement of filaments.

16. The nucleoid is the "nuclear region" of a bacterial cell (the area with the cell's chromosome). The nucleus, found only in eucaryotic cells, is a membrane-enclosed compartment containing the cell's chromosomes.

17. A plasmid is "extrachromosomal" DNA. While it is also circular and double-stranded, the plasmid contains extra, nonessential genes to benefit the cell.

18. Ribosomes are RNA-rich particles in which amino acids are assembled to make proteins. Procaryotic ribosomes are smaller (70S) than eucaryotic ribosomes (80S).

Multiple-Choice Questions

1. d

2. c

3. b

4. c

5. b

Chapter 4

Short-Answer Questions

1. energy, carbon, nitrogen, phosphorus, sulfur, some organic growth factors, and minerals

2. chemoheterotrophs

3. chemolithotrophic

4. assimilatory nitrate reduction, ammonia assimilation, nitrogen fixation

5. Amino acids, purines, pyrimidines, and vitamins must be provided if they cannot be produced by the bacterium.

6. passive diffusion, facilitated diffusion, active transport, and group translocation

7. passive diffusion

8. active transport (especially the Na^+/K^+ pump)

9. Both require carrier proteins; however, facilitated diffusion occurs with the concentration gradient and requires no energy while active transport uses energy to bring substances against the concentration gradient.

10. Group translocation chemically alters the substrate as it is brought across the membrane, whereas active transport expends energy to bring an unaltered compound across the membrane.

11. temperature, pH, available water, osmotic concentration, and oxygen tension

12. Psychrophiles have more unsaturated fatty acids in their membrane to maintain fluidity at low temperatures, while thermophiles have enzymes and proteins, which are more heat stable.

13. Bacteria may release acids to lower the pH or raise the pH by releasing ammonia and/or carbon dioxide.

14. Water activity is the water available to an organism and it depends on the concentration of solutes (for example, salts, sugars).

15. Many forms of oxygen can be lethal to cells (for example, O_2^-, O_2^{2-}, OH^-). Most cells have enzymes to convert these toxic radicals to harmless compounds, but obligately anaerobic bacteria lack the enzymes that would make life in an oxygenated environment safe.

16. Complex media are made from undefined sources (for example, beef or yeast extract) and contain a variety of ingredients. Synthetic media are made by adding exact amounts of known ingredients.

17. Enrichment media facilitate the growth of slower, more fastidious bacteria. Selective media inhibit the growth of unwanted organisms, enabling the isolation of the desired bacteria. Differential media neither enhance nor inhibit the growth of bacteria, but allow one to distinguish between types of bacteria based on some observable trait.

18. Agar is routinely used as a solidifying agent. It melts at 97° C to 100° C, but solidifies at 45° C to 47° C.

19. Aseptic technique is critical to prevent infection and contamination of materials, in addition to achieving pure culture and identifying microorganisms.

Multiple-Choice Questions

1. c

2. b

3. d

4. d

5. c

Chapter 5

Short-Answer Questions

1. Bacterial cells divide by binary fission.

2. Bacterial cells grow very little in size. "Growth" generally refers to population size.

3. 5 hours

4. direct microscopic count, viable count, MPN, turbidity, cell mass, metabolic activity

5. Pour plates are best for determining the viable count of anaerobic cultures. Streak plates are best for heat-sensitive cultures.

6. (a) measure metabolic activity

   (b) membrane filtration

   (c) direct microscopic count (Note: Turbidimetric measurement is only quick if you already have a standard curve for that culture.)

7. The phases are: lag, exponential, stationary, and death phase. The only phase lacking cell division is the lag phase.

8. Although bacteria do not reproduce during the lag phase, they have been observed using nutrients and building cellular components (for example, proteins and DNA).

9. Exponential growth occurs during ideal conditions. In a batch culture, as in nature, the accumulation of wastes and depletion of nutrients prohibits indefinite exponential growth.

10. Exponential growth can be maintained in a chemostat or turbidostat, both of which are designed to add fresh media while removing wastes and cells.

11. (a) dry heat

    (b) UV radiation

    (c) autoclaving

12. Betadine, mouthwash, hydrogen peroxide, and alcohol are examples of antiseptics. Alcohol, iodine, bleach, and copper sulfate are examples of disinfectants.

13. inhibition of cell wall synthesis, damage to the plasma membrane, inhibition of protein and nucleic acid synthesis, and inhibition of cell metabolism

14. Drug resistance has become widespread due to the overuse of antimicrobial agents, especially broad-spectrum drugs, as well as the inappropriate prescription of antibiotics for viral infections, widespread use of antibiotics in animal feed, and failure to complete the prescribed course of therapy.

15. Chemicals used to control microbial growth are "antimicrobial agents." Antibiotics are those antimicrobial agents which are produced by other microorganisms and administered to treat disease.

16. The Kirby-Bauer test is a disk diffusion test to determine the susceptibilities of microorganisms to antimicrobial agents. This test is a rapid, convenient method to select the best treatment for an infection.

Multiple-Choice Questions

1. c
2. e
3. d
4. b

Chapter 6

Short-Answer Questions

1. The first law of thermodynamics states that energy can neither be created nor destroyed. The second law of thermodynamics states that entropy (the measure of disorder) is increasing.

2. Enzymes are protein (or, in the case of ribozymes, RNA) catalysts that lower the activation energy to facilitate chemical reactions.

3. The rate of an enzymatic reaction will decrease if the temperature rises or falls, the pH rises or falls, the substrate concentration falls, the product concentration rises, or the concentration of inhibitors rises.

4. Both bind to the enzyme; however, competitive inhibitors compete for the substrate's binding site whereas noncompetitive inhibitors bind to a different site on the enzyme.

5. Feedback inhibition involves allosteric enzymes that are inhibited by end products of the biosynthetic pathway. This mechanism is an important method of conserving energy—turning off a pathway once it is no longer needed.

6. Both reactions are coupled and involve the transfer of electrons. Oxidation involves the loss of electrons whereas reduction involves the gain of electrons.

7. NAD, NADP, FAD, and FMN are common electron carriers that transfer electrons from their initial donor to a final acceptor. Their association with certain pathways results in energy conservation and ATP formation.

8. High-energy compounds, having one or more phosphate bonds, include ATP, phosphoenolpyruvate, acetyl phosphate, and 1,3-diphosphoglycerate.

9. Anabolism and catabolism are chemical reactions carried out by the cell's enzymes. Anabolic reactions are those that require energy to build molecules whereas catabolic reactions are those that break down molecules and release energy.

10. ATP generation occurs by substrate-level phosphorylation, oxidative phosphorylation, and photophosphorylation.

11. Both are methods of carbohydrate catabolism. Fermentation involves organic molecules as both electron donors and electron acceptors. Respiration, however, involves organic electron donors and external inorganic electron acceptors (for example, O_2).

12. Bacteria may use the Embden-Meyerhof, Entner-Doudoroff, pentose phosphate, and/or phosphoketolase pathways. The Embden-Meyerhoff pathway is the major pathway, but others are important for their ability to recycle cofactors (for example, $NADPH+H^+$) or convert hexoses to pentoses.

13. Bacterial fermentation may result in nonacidic end products. In addition to the most common example, alcohol, some bacteria produce substances like butanediol and acetoin.

14. Glucose is oxidized to 2 pyruvate molecules, 2 ATP, and 2 $NADH+H^+$ via the Embden-Meyerhof pathway. Each pyruvate molecule is converted to acetyl-CoA, along with 1 CO_2 and 1 $NADH+H^+$. Each acetyl-CoA is then oxidized by the TCA cycle to 2 CO_2, 1 ATP (GTP), 1 $FADH_2$, and 3 $NADH+H^+$. The reduced electron carriers enter the electron transport chain, forming additional ATPs. A total of 38 ATPs is produced from the complete catabolism of glucose.

15. Both processes involve the oxidation of an organic substrate (for example, glucose) with an inorganic final electron acceptor. The final electron acceptor for aerobic respiration is oxygen while anaerobic respiration uses nitrate, carbonate, or sulfate as the final electron acceptor. Anaerobic respiration generally produces less energy than aerobic respiration.

16. Chemolithotrophs obtain their energy from the oxidation of inorganic compounds (for example, H_2, H_2S, NH_3, NO_2^-, and Fe^{2+}) while chemoorganotrophs obtain their energy by oxidizing organic compounds.

Multiple-Choice Questions

1. c
2. b
3. d
4. b

Chapter 7

Short-Answer Questions

1. Plants, algae, and cyanobacteria carry out oxygenic photosynthesis using two photosystems (photosystems I and II) in the Hill-Bendall scheme. Oxygen is produced from photolysis of water. ATP is produced during cyclic and noncyclic photophosphorylation. $NADP^+$ is reduced to $NADPH + H^+$ during noncyclic photophosphorylation. Phototrophic bacteria carry out anoxygenic photosynthesis in cyclic photophosphorylation involving only photosystem I. ATP is synthesized, but $NADP^+$ is not reduced to $NADPH + H^+$ except when there is an ATP-requiring reversal of the electron transport chain.

2. Carotenoids and biliproteins are found in cyanobacteria. Only carotenoids occur in phototrophic bacteria.

3. H_2O is the electron donor; CO_2 is the electron acceptor.

4. The light and dark reactions of photosynthesis are not always coupled. The light reaction generates ATP and $NADPH + H^+$. Although CO_2 fixation, or the dark reaction, requires ATP and $NADPH + H^+$, these requirements may be fulfilled by other metabolic pathways. Photoheterotrophic bacteria obtain their energy through the light reaction of photosynthesis, but obtain their carbon from chemical compounds. Chemoautotrophic bacteria fix CO_2, but do not participate in the light reaction of photosynthesis.

5. Photolysis is the light-dependent splitting of H_2O that replenishes electrons lost from the reaction center chlorophyll P680 in photosystem II during the light reaction of photosynthesis in the Hill-Bendall scheme. Oxygen is released during photolysis.

6. Magnesium is the central atom in chlorophyll (and bacteriochlorophyll) and, therefore, is important for the light reaction of photosynthesis.

7. Yes. Reduced compound "X" has a more negative redox potential (−0.41 V) than $NADP^+$ (−0.32 V) and, therefore, can donate its electrons to $NADP^+$.

8. The reaction: 1,3-diphosphoglycerate → glyceraldehyde-3 phosphate is the reduction step in the Calvin cycle. Electrons are supplied by $NADPH + H^+$.

9. Cellulose-digesting bacteria are found in soil and in the intestinal tracts of animals that feed on grass and plant materials (ruminants).

10. Glycogen storage granules serve as a source of carbon and energy for bacteria. Bacteria utilize endogenous glycogen by phosphorylysis of glycogen to glucose-6-phosphate.

11. Gluconeogenesis is the biosynthesis of glucose from noncarbohydrate compounds through reversal of the Embden-Meyerhof pathway. Not all steps of the Embden-Meyerhof pathway are reversible. Three steps are thermodynamically irreversible but are bypassed in gluconeogenesis by using alternative enzymes and pathways. See Figure 7.16.

12. Microbial proteases are used to soften and prepare animal hides for tanning; as an addition to washing detergents to dissolve mucus, blood, and other proteinaceous stains; and as meat tenderizers.

13. The Stickland reaction is a coupled system occurring in some species of *Clostridium* for the dual fermentation of alanine and glycine. In the process of donating electrons to glycine, alanine is deaminated to pyruvate. In turn, glycine accepts the electrons and is deaminated to acetate. See Figure 7.18.

14. Amino acid decarboxylase tests are performed by inoculating bacteria into a base medium containing low amounts of glucose, a 1% solution of an amino acid (arginine, lysine, or ornithine), and the pH indicator bromcresol purple. After incubation, the medium will be either yellow or purple. Bacteria that do not decarboxylate the amino acid in the medium will catabolize glucose, forming acidic end products that lower the pH of the medium, resulting in a yellow color. Bacteria that do decarboxylate the amino acid will initially produce acid from glucose, but as the amino acid is decarboxylated to putrescine or cadaverine and CO_2, the pH of the medium rises, resulting in a purple color. Diamines such as putrescine and cadaverine are unstable in the presence of oxygen and, therefore, the test is performed under anaerobic conditions by overlaying the surface of the medium with paraffin or mineral oil after inoculation.

15. Vertebrates lack the enzymes isocitrate lyase and malate synthase necessary for the glyoxylate cycle.

16. CO_2 cannot serve as an energy source because it is too oxidized.

Multiple-Choice Questions

1. a
2. c
3. d
4. c
5. c

Chapter 8

Short-Answer Questions

1. nucleotide, codon, gene, chromosome, genome

2. Both processes are involved in protein synthesis. Transcription is the formation of an RNA sequence complementary to the DNA, whereas translation is the formation of an amino acid sequence based on the RNA.

3. Both refer to the traits encoded by an organism's DNA; however, the genotype indicates the actual genetic composition of an organism, whereas the phenotype represents those traits that are expressed or observable.

4. Hershey and Chase proved the role of DNA in heredity by studying replication of the T2 bacteriophage which is made of only DNA and protein. By alternately isotope-labeling DNA and protein, they found that only DNA was passed on to progeny.

5. Covalent bonds between the sugar of one nucleotide and phosphate of the next nucleotide form the backbone of each DNA strand. Two strands of DNA, running in opposite directions, are held together by hydrogen bonds between adjacent bases (guanine to cytosine, thymine to adenine). Stacking forces between the bases cause the DNA to make a complete turn every ten bases.

6. Histone proteins, found in eucaryotes, neutralize the negative charges of the phosphodiester bonds between nucleotides to facilitate coiling of DNA into compact chromosomes.

7. Meselson and Stahl isotope-labeled *E. coli* DNA. After one generation of growth in fresh media, they extracted and denatured the DNA. They found that half of the strands were labeled and half were unlabeled, thereby proving that each DNA strand is used as a template for synthesis.

8. Initiator proteins recognize a specific DNA sequence as the origin of replication. Unwinding at this site is facilitated by DNA helicase and topoisomerases so that primase may begin synthesis. Synthesis proceeds in both directions from the origin; however, each strand is synthesized in the 5' to 3' direction.

9. Okazaki fragments refer to the short segments of DNA synthesized and annealed by ligase. Since the enzymes which synthesize DNA can only do so in the 5' to 3' direction, synthesis of the lagging strand must be achieved in short segments as the replication fork opens.

10. DNA synthesis actually begins with an RNA primer synthesized by primase, an RNA-polymerizing enzyme. This primer (approximately 10 nucleotides long) is later excised, probably by DNA polymerase I.

11. Eucaryotic DNA consists of exons and introns, both of which are transcribed into mRNA. The noncoding regions are excised from mRNA before translation. Procaryotic DNA and RNA are not divided into exons and introns.

12. Each mRNA is a complementary copy of a gene used as the template for synthesis of a protein. tRNAs carry amino acids to the site of protein synthesis. rRNAs form ribosomes, the site of protein synthesis, which catalyze the polymerization of amino acids.

13. RNA polymerase recognizes promoter sequences (Pribnow box and −35 sequence) on the DNA strand to be transcribed and synthesizes mRNA in a 5' to 3' direction.

14. Once mRNA synthesis begins in procaryotes, ribosomes attach to the mRNA and begin protein synthesis.

15. The Shine-Dalgarno sequence, a short purine-rich region preceding the initiation codon, specifies the ribosome binding site. Anticodons of aminoacyl-tRNAs bind to codons on the mRNA and peptide bonds are formed between adjacent amino acids, thereby causing the release of tRNAs. The polypeptide or protein is released when the ribosome reaches a termination codon.

16. Expression of some traits (for example, lactose fermentation) requires several genes, known as an operon. One of the genes encodes for a repressor protein. While expression of this trait is not required, the repressor protein binds to the operator site, thereby preventing transcription of the other genes. If expression is required, a metabolite binds to the repressor, thereby altering its shape to prevent its binding and allow transcription of the other genes.

17. In some operons (for example, biosynthesis of amino acids), there is also a leader sequence which codes for a polypeptide repeating the same amino acid. When that amino acid is abundant, the leader sequence is synthesized and it results in termination of transcription of other genes in the operon.

Multiple-Choice Questions

1. e

2. d

3. d

4. d (Note: Examples *a*, *b*, and *c* do *not* alter the specified amino acid. Example *d* results in termination.)

5. b

Chapter 9

Short-Answer Questions

1. Procaryotes are not capable of sexual reproduction; therefore, they rely on mutation and genetic exchange for evolution.

2. DNA replication is highly efficient; however, errors do occur about once in every 10^9 bases copied. The rate will increase in the presence of mutagens (for example, UV light, base analogs, intercalating agents).

3. Because of redundancy in the genetic code, base substitution at the third position of a codon may not alter the amino acid sequence.

4. The insertion or deletion of a single nucleotide (or any number other than 3) will alter the reading frame, thereby making it likely that nonsense codons will arise prematurely.

5. Changing a single nucleotide can be a "lethal mutation." Mutations can be helpful, especially if they reverse a previously harmful error.

6. UV light causes covalent bonds between adjacent pyrimidines, thereby preventing the activity of DNA and RNA polymerases. Base analogs replace the appropriate nucleotide and result in base pairing errors during replication. Intercalating agents bind between base pairs, eventually displacing single bases and leading to frame shift mutations.

7. Many mutations are caught and repaired by the cell's repair mechanisms (for example, excision repair, photoreactivation, glycosylases, SOS) before they can become permanent.

8. Replica plating (inoculating and incubating cultures on a complex, enriched medium and a defined medium simultaneously) is a common strategy for identifying auxotrophic mutants.

9. Some mutations result in changes in cellular or colonial morphology (for example, motility, arrangement, pigmentation, encapsulation).

10. conjugation, transformation, and transduction

11. Genetic material brought into a cell is usually degraded by exo- and endonucleases. In order to effect a permanent change, the new DNA must be integrated into the chromosome by recombination. Although extrachromosomal DNA may be stabilized and remain for many generations, it may also be degraded, expelled, or lost over time.

12. Transfection is a special form of transformation, involving naked phage DNA and *Escherichia coli*, which can be induced in the laboratory. In the case of transduction, progeny phage have accidentally packaged bacterial DNA along with their own. This is then transferred to the next host cell of the phage.

13. Generalized transduction involves the random packaging of bacterial DNA fragments by a phage. Specialized transduction requires a temperate phage (one able to integrate its genome into the host's genome) and involves the removal of specific bacterial genes (that is, those genes next to the site of integration).

14. Plasmids are, literally, extrachromosomal pieces of DNA. Except for being much smaller, they are structurally identical to chromosomes.

15. They are important to the cell because they contain extra genes that may provide a survival advantage (antibiotic resistance, heavy metal resistance, synthesis of bacteriocins or toxins, and so forth).

16. Many plasmids (for example, the F plasmid) carry genes for products which facilitate cell attachment and the transfer of plasmids between cells. Some plasmids (referred to as

episomes) can integrate into the chromosome, thereby facilitating the transfer of chromosomal genes and increasing the frequency of recombination among recipient cells.

Multiple-Choice Questions

1. b
2. c
3. b
4. d
5. b

Chapter 10

Short-Answer Questions

1. Genetic engineering was made possible by science's discovery and exploitation of the mechanisms of recombination used by cells (especially the discovery of restriction endonucleases).

2. Restriction endonucleases recognize and cleave specific sequences of double-stranded DNA, called palindromes, allowing researchers to map and manipulate DNA in vitro. For the bacterial cell, restriction endonucleases may be a defense against the invasion of bacteriophages.

3. DNA from different sources can be cleaved with the same restriction endonucleases. When the DNA fragments are mixed together, the "sticky ends" overlap due to complementary base pairing. Researchers can bind these fragments together with the enzyme ligase, thereby forming a new (recombinant) DNA molecule.

4. Vectors are self-replicating DNA molecules (for example, a virus or plasmid) used to transport DNA.

5. The target gene may be taken directly from genomic DNA, synthesized from an mRNA template, or synthesized in vitro if the gene sequence is known.

6. Discovery of the viral enzyme reverse transcriptase enabled researchers to synthesize genes, properly called complementary DNA, from mRNA.

7. Electrophoresis is a technique for separating DNA molecules, based on their size and electrical charge, through a gel. Since DNA is negatively charged, it is loaded into the gel near the negative electrode and it will migrate toward the positive electrode, with the smallest pieces traveling farthest.

8. Both procedures employ electrophoresis followed by hybridization with an isotope-labeled gene or "probe"; however, Southern blotting is done for DNA pieces while Northern blotting is done for RNA pieces.

9. Vectors, carrying the target gene, may be introduced into the host by transfection or transformation.

10. Incorporation of the target gene may be observed by monitoring the cells for production of the target gene product. If the gene has previously been isolated or sequenced, a probe may be used to confirm incorporation.

11. A selective marker, such as antibiotic resistance, is the gene that will be cleaved for insertion of the target gene. Recombinant DNA is identified by "insertional inactivation" of the marker trait.

12. Polymerase chain reaction is a technique for copying (or amplifying) DNA that involves: (1) denaturation of the original double-stranded DNA, (2) annealing oligonucleotide primers to the single strands of DNA, and (3) allowing DNA polymerase to extend the primers. This process will repeat itself, copying (or amplifying) the target DNA molecule billions of times in only an hour.

13. DNA fingerprinting relies on the uniqueness of each person's DNA. The technique involves: (1) digesting a DNA sample with restriction endonucleases, (2) separating the fragments (RFLPs) by electrophoresis, and (3) detecting by Southern blotting with a probe for repeat sequences (known as VNTRs).

14. Genetic engineering for transgenic organisms allows the rapid production of large numbers of organisms with the desired trait while traditional selective breeding programs require several generations of breeding to develop a population in which the preferred trait is predominant.

Multiple-Choice Questions

1. e

2. e

Chapter 11

Short-Answer Questions

1. Both refer to the classification of organisms; however, nomenclature is the method of assigning genus and species names to each organism while taxonomy is the method of grouping organisms based on their similarities and differences.

2. Whittaker proposed 5 kingdoms based on phenotypes (for example, cell structure and nutritional mode): Monera, Fungi, Protista, Plantae, and Animalia. Woese proposed 3 kingdoms based on genotypes (specifically, 16S rRNA sequences): Archaea, Bacteria, and Eucarya.

3. Woese's system is phylogenetic; it analyzes evolutionary relationships between organisms.

4. The classical approach emphasizes cell structure and morphology, as well as metabolic and other physiological characteristics. Numerical taxonomy also relies on phenotypes, but determines a percent similarity between two organisms by comparing equally weighted traits. Molecular approaches rely on DNA or RNA sequencing or hybridization studies to determine the percent similarity between two organisms.

5. Flowcharts, or dichotomous keys, are routinely used for bacterial identification in diagnostic microbiology. These road maps, which consider the presence or absence of various traits, are advantageous for relatively quick identification of previously described species. However, because they weigh some traits (for example, morphology and Gram reaction) more heavily than others, an error in these observations can lead to drastically erroneous conclusions.

6. Both are procaryotic, with a single circular chromosome, but lacking a nucleus or other organelles. Bacteria typically have cell walls of peptidoglycan and muramic acid, membranes with ester-linked straight chained hydrocarbons, and sensitivity to antibiotics. Archaea, however, have cell walls lacking peptidoglycan or muramic acid, membranes with ether-linked aliphatic chained hydrocarbons, and resistance to antibiotics.

7. Gram-negative Bacteria, gram-positive Bacteria other than Actinomycetes, Archaea, and Cyanobacteria.

8. (See Table 11.6.)

9. (See Table 11.12.)

10. (See Table 11.14.)

11. (See Table 11.18.)

12. Both are procaryotic; however, mycoplasmas have sterols in their membranes and lack cell walls entirely, while other bacteria lack sterols and possess a cell wall.

13. Both are procaryotic parasites; however, *Rickettsia* are transmitted by arthropod vectors, while *Chlamydia* are transmitted by direct person-to-person contact. *Rickettsia* and *Chlamydia* are obligate intracellular parasites.

14. Both exhibit filamentous growth; however, actinomycetes are procaryotic and fungi are eucaryotic.

Multiple-Choice Questions

1. c

2. a

3. c

4. b

5. a

Chapter 12

Short-Answer Questions

1. Bacteria, unlike other microorganisms, are procaryotic.

2. Algae, unlike other microorganisms, are photosynthetic.

3. Cyanobacteria were once considered to be algae because of their photosynthetic ability; however, it was later discovered that they lack organelles (even chloroplasts), so they were reclassified as procaryotes.

4. Euglenophyta; cellulose

5. Phaeophyta; Pyrrophyta

6. Fungi resemble plants in many ways (nonmotile eucaryotic organisms having cell walls); however, they are not photo-synthetic and their cell walls are primarily composed of chitin.

7. While both are forms of fungi, molds are the filamentous hyphae and yeasts are unicellular forms.

8. Fungi give us beer, wine, bread, many cheeses, and antibiotics.

9. Among the few bacteria that produce spores, these haploid structures are made internally (endospores) for survival. Most fungi produce spores externally (exospores) for repro-duction, either sexually or asexually (haploid or diploid, respectively).

10. Slime molds lack the characteristic cell wall; they are motile and phagocytic. They are still considered to be fungi because they produce fruiting bodies and spores typical of fungi.

11. Protozoa are motile, nonphotosynthetic, eucaryotic unicellular organisms lacking cell walls.

12. cysts; trophozoite

13. Most protozoa are free-living and of no medical significance; however, *Cryptosporidium, Toxoplasma, Giardia, Trypanosoma, Plasmodium,* and *Entamoeba* are well-known pathogens.

Multiple-Choice Questions

1. b

2. b

3. d

4. c

Chapter 13

Short-Answer Questions

1. Viruses are composed of nucleic acid (either DNA or RNA, but not both) surrounded by a protein coat (called a capsid).

2. The most unique feature of viruses is that they are acellular (that is, they have neither a plasma membrane nor any constituents of a typical cell).

3. Viruses are unable to perform any of the functions of life (biosynthesis, replication, etc.). In order to propagate, they must enter a host cell and utilize the host's enzymes and metabolites.

4. Viruses possess genetic information that, with the help of a host cell, is used to propagate.

5. Outside of a host cell, viruses do not replicate DNA, use energy, carry out biosynthesis, or perform any other function typical of living organisms.

6. If present, an envelope is a fragment of the host cell's membrane. It covers each new virus particle as the particle exits the host. The viral genome cannot encode for the synthesis of the envelope.

7. Viroids are infectious agents composed only of RNA, unlike virions, which are complete virus particles.

8. Viruses are classified by their host (bacteria, plant, or animal), morphology (binal, helical, or polyhedral), nucleic acid type (DNA or RNA), and nucleic acid structure (single vs. double stranded, linear vs. circular).

9. Most viruses are extremely host specific; some are even tissue specific for a given host.

10. Viruses can only be cultivated inside their host: the whole organism, an embryonated egg, or an appropriate tissue culture. Most viruses can survive outside their host for long periods of time; however, they are dormant.

11. Viral plaques or pocks can be enumerated by the naked eye or virus particles can be enumerated with an electron microscope. Some viruses can also be enumerated by a hemagglutination assay.

12. attachment (or adsorption), penetration (and uncoating for some viruses), replication/synthesis, assembly/maturation, and release

13. Viral DNA may be replicated by the rolling circle mechanism or bidirectionally. Viral RNA is replicated in a manner similar to the rolling circle mechanism, though single-stranded (+ strand) RNA viruses must first synthesize a complementary strand (–strand). Retroviruses have a –strand which is used as a template for DNA synthesis.

14. Reverse transcriptase is a viral enzyme found in retroviruses. It is necessary for the synthesis of DNA from RNA.

15. If the host is lysed prior to maturation and release, the viral progeny would not be viable. Only complete viral particles can survive outside the host.

16. Both life cycles involve attachment (or adsorption), penetration (and uncoating for some viruses), replication/synthesis, assembly/maturation, and release; however, the lytic phage proceeds through this cycle uninterrupted whereas the lysogenic (or temperate) phage may integrate its nucleic acid into the host's chromosome (after penetration, before replication/ synthesis) for an indefinite period of time.

17. The ultimate fate of the phage's host cell is death. This is not usually the case with plant and animal viruses, which tend to trigger a variety of cytopathic effects (for example, cell transformation, formation of inclusion bodies, damage to the membrane, or other changes in structure or function).

18. Cell transformation (not to be confused with genetic transformation) is a cytopathic effect of some viruses on their host and is evidenced by changes in the host's metabolism, appearance of new antigens on the host's membrane, changes in morphology, and loss of contact inhibition (leading to tumor formation).

Multiple-Choice Answers

1. a

2. d

3. b

Chapter 14

Short-Answer Questions

1. commensalism, mutualism, parasitism

2. Both refer to close relationships between different species; however, endosymbiosis refers to one organism living within another while ectosymbiosis refers to one organism living outside the other.

3. Although nitrogen fixation does occur among free-living cells, it is favored by symbiosis. Legumes are plants that produce specialized root hairs that are infected by *Rhizobium* species. The bacterium *Rhizobium* is provided with nutrients for its growth and a suitable environment by the plant and, in turn, the plant fixes atmospheric nitrogen (N_2) into organic nitrogen compounds usable by the plant.

4. Leghemoglobin, an oxygen-binding pigment, enables the oxygen-sensitive nitrogenase to fix nitrogen. Leghemoglobin is a complex molecule requiring both bacterial and plant components to function (specialized heme from *Rhizobium* and globin from the legume).

5. Lichens are the aggregate of certain fungi and algae or cyanobacteria. The algae or cyanobacteria photosynthesize to produce food for themselves and their host while the fungi provide a place for algal growth. This relationship is an example of mutualism.

6. Dependent on air transport for nearly all of their nutrients, lichens generally die in the presence of air pollution.

7. Mycorrhizae are the result of a mutualistic association between a fungus and plant (as seen with orchids). The fungus receives carbohydrates from the plant and the plant benefits from an increased surface area for absorption.

8. Flashlight fish of the family *Anomalopidae* possess a specialized organ, filled with luminescent bacteria, below their eyes. The light is emitted continually by the bacteria, which benefit from the protection of their home, but the fish raises or lowers a shutter to control the signal light for communication, to lure prey, and to confuse predators.

9. The relationships between *Paramecium aurelia* and its bacterial endosymbionts are of interest because of their high degree of specificity and apparent genetic determination, as well as the role these bacteria appear to play as mitochondria for their host.

10. Ruminants do not produce cellulase, the enzyme necessary for hydrolyzing plant matter. The rumen of these animals is inhabited by a variety of microorganisms (including cellulose and starch digesters) that break down plant material into nutrients that can be absorbed through the intestinal wall and utilized by the animal.

11. By boring into the periplasm and disrupting cytoplasm for their own use, *Bdellovibrio* parasitizes gram-negative bacteria.

12. *Chlamydia*, unable to synthesize its own ATP, is an obligate intracellular parasite of humans, other mammals, and birds.

13. Under normal conditions, most of the human normal flora are commensals, benefiting from their habitat in the body without helping or harming the body.

14. Symbiotic relationships are held in balance by environmental factors and are subject to change; for example, many commensals become parasites in an immunocompromised host.

Chapter 15

Short-Answer Questions

1. Both inhabit our bodies without harm; however, normal flora are permanent residents whereas transient flora, picked up from our surroundings, are removed or outcompeted within a few days or weeks.

2. Both are microorganisms which live on or in us, usually without causing harm. However, conditions may change (we may be immunocompromised) so that otherwise harmless "normal flora" take advantage of the opportunity and become pathogens.

3. In most cases, our normal flora are antagonistic to potential pathogens; that is, by using certain nutrients, releasing metabolic products, and altering their environment, normal flora help prevent the growth of potential pathogens.

4. *Enterococcus faecalis,* which requires folic acid for growth, and *Lactobacillus arabinosus,* which requires phenylalanine, support the growth of each other by sharing these nutrients. Another example involves methanogenic bacteria which produce methane from simple carbon compounds and live off of the wastes of other microorganisms that require methane.

5. Both terms refer to the growth of other organisms associated with our bodies; however, infection refers to the growth of microorganisms (especially bacteria) in our tissues and infestation refers to the growth of insects on our tissues.

6. Both terms refer to diseases that can be transmitted from host to host; however, contagious diseases are communicable diseases that are easily transmitted.

7. Reservoirs of infection, places where infectious agents may be found, may be animate (for example, humans or other animals) or inanimate (for example, soil, water, and food).

8. Both are a means of transmission for some pathogen; however, carriers are infected people and fomites are inanimate objects (for example, door knobs, utensils) that harbor the pathogen.

9. directly (for example, sexual contact, hand-to-hand contact) or indirectly (for example, fomites, vectors)

10. nose, mouth, eyes, ears, genitourinary tract, open cuts or abrasions on the skin

11. incubation, prodromal, acute, decline, convalescence

12. For acute diseases, signs and symptoms appear and subside quickly. For chronic diseases, symptoms appear and subside slowly. For latent diseases, symptoms are very slow to appear (months or years).

13. Both terms refer to the microorganism's ability to cause disease. Pathogenicity is the ability to establish infection and cause disease, whereas virulence is a measure of the microorganism's pathogenicity (that is, the more virulent the microorganism is, the easier it is to overcome the host's defenses).

14. Virulence factors generally refer to invasiveness (via flagella, glycocalyx, etc.) and toxin production (for example, hemolysin, kinase, coagulase).

15. Both are substances produced by the microorganism that are poisonous to the host. Endotoxin is the Lipid A component of the gram-negative cell wall, while exotoxins are products excreted from the cell (enterotoxins, neurotoxins, cytotoxins, etc.). Furthermore, endotoxins are heat stable, while exotoxins (usually made of protein) are heat labile. While endotoxins are only associated with gram-negative bacteria, exotoxins may be produced by any bacterium.

16. Both refer to the host's ability to prevent and/or control an infection. Innate resistance is those traits, present at birth, that are nonspecific, whereas acquired resistance is those traits, developed in response to exposure, that are specific.

17. skin, mucous membranes, cilia, pH, flow of fluids

18. lysozyme, interferon, sebum, salts, bile

19. antibodies and T cells

20. Inflammation is the body's nonspecific response to injury, irritation, or infection caused by a physical, chemical or biological agent. The cardinal signs (edema, erythema, pain, and heat) are due to the release of chemical factors and, subsequent, migration of phagocytes to the area. It is the phagocytes that carry out the work of repair.

Multiple-Choice Questions

1. a

2. d

3. e

4. a

Chapter 16

Short-Answer Questions

1. Both play a role in our immune response; however, antigens are substances (protein, carbohydrate, or lipid) that elicit an immune response, and antibodies are the substances (protein) that specifically fight antigens.

2. Both are specific against antigens; however, humoral immunity refers to the antibodies that are produced, while cellular immunity refers to the T cells that respond.

3. lymph nodes, thymus, spleen, bone marrow, GI tract, and other mucosae; Lymphoid tissue is the site of T cell maturation (bone marrow for B cells and thymus for T cells).

4. Antibodies are, literally, immune proteins.

5. Both refer to sera containing antibodies; however, antisera generally contains antibodies against a specific antigen, while antitoxin is sera containing antibodies against a toxin.

6. IgA, IgD, IgE, IgG, IgM

7. Shaped like a Y or T, antibodies are composed of two light chains and two heavy chains of polypeptides.

8. Antigenic specificity is possible due to the rearrangement of genes during B cell maturation.

9. T cells help regulate the activity of B cells, as well as attack foreign or altered cells directly.

10. T_H cells stimulate B cells to produce antibodies and are associated with delayed hypersensitivity, cytotoxicity and local inflammatory reactions, and production of certain cytokines; T_C cells attack foreign or altered cells directly.

11. MHC allows our immune system to distinguish self from non-self.

12. Cytokines are a variety of proteins which, generally, stimulate B cells.

13. The clonal selection theory states that B cells activated by antigens multiply and mature into clones of the antibody-secreting plasma cells. Some B cells become memory cells to enable a more rapid response upon future encounters with the same antigen.

14. Both refer to the production of antibodies as stimulated by antigens; however, the primary response is slower and depends more on IgM, while the secondary response is faster and depends more on IgG.

15. Artificially acquired active immunity is stimulated by a vaccine. Artificially acquired passive immunity is administered as serum. Naturally acquired active immunity is stimulated by an infection. Naturally acquired passive immunity is transferred from the mother (transplacentally and via breastfeeding).

16. Vaccines may use live, attenuated organisms, killed organisms, toxoids, or subcellular components. Since vaccines are made to prevent serious, life-threatening diseases, it would not be prudent to administer a live, unaltered organism.

17. You might boost someone's immune system by administering serum or, if appropriate, a booster vaccine. It is appropriate to suppress someone's immune system after a tissue or organ transplant to prevent rejection.

18. Hypersensitive reactions may involve pain, swelling, and even death.

19. Complement is actually a class of about 20 proteins and plays a role in mediating antibody activity.

20. In vitro immunological assays include the precipitin ring test, diffusion tests, immunoelectrophoresis, agglutination, hemagglutination, neutralization, complement fixation assays, fluorescent antibody techniques, radioimmunoassays, and ELISAs (enzyme-linked immunosorbent assays).

Multiple-Choice Questions

1. a

2. c

3. c

4. a and c

5. d

6. b

Chapter 17

Short-Answer Questions

1. Group A streptococci, also known as *Streptococcus pyogenes*.

2. Scarlet fever is caused by a strain of *Streptococcus pyogenes* carrying a lysogenic phage with the gene for erythrotoxin. Rheumatic fever is a sequelae of strep throat; that is, it is an infection of the heart that occurs when strep throat is not properly treated.

3. *Streptococcus pneumoniae*

4. *Legionella pneumophila*

5. *Bordetella pertussis*

6. *Corynebacterium diphtheriae*

7. *Mycobacterium tuberculosis*

8. *Neisseria meningitidis*

9. *Histoplasma capsulatum* and *Coccidioides immitis*

10. Influenza viruses

11. *Streptococcus mutans*

12. *Candida albicans*

13. paramyxovirus

14. *Salmonella typhi, Campylobacter jejuni, Shigella dysenteriae, Vibrio cholerae, Escherichia coli*

15. *Clostridium botulinum, Clostridium perfringens, Staphylococcus aureus*

16. These four strains can be differentiated on the basis of their virulence factors (toxins and invasiveness).

17. *Giardia lamblia, Entamoeba histolytica*

18. hepatitis A virus

19. *Ancylostoma duodenale, Necator americanus, Francisella tularensis*

20. *Proprionibacterium acnes*

21. *Streptococcus pyogenes*

22. *Mycobacterium leprae*

23. *Pseudomonas aeruginosa*

24. Epidemic typhus is caused by *Rickettsia prowazekii*, while endemic typhus is caused by *Rickettsia typhi.*

25. *Borrelia burgdorferi*

26. Ringworm, or tinea, is caused by the fungi *Microsporum, Trichophyton,* and *Epidermophyton.*

27. papilloma virus

28. chickenpox, or varicella-zoster, virus

29. *Neisseria gonorrhoeae*

30. *Treponema pallidum*

31. human immunodeficiency virus

Chapter 18

Short-Answer Questions

1. Doctors diagnose and treat diseases, microbiologists identify the causes of diseases, and epidemiologists study the source, transmission, and prevention of diseases.

2. Endemic diseases are those always present in a given population, whereas epidemic diseases are those with a rapid rise in incidence.

3. Common-source epidemics are those in which a group of people is exposed to the same reservoir, generally at the same time (for example, food poisoning). Propagated epidemics are those in which the disease is transmitted through a group of people, generally by direct contact or a vector (for example, malaria).

4. An epidemic across a large geographical area is a pandemic.

5. Both are terms reflecting the rate of a disease (for example, number of cases per 1,000 population). Morbidity refers to the rate by which people become infected with a specific disease, whereas mortality refers to the rate at which people die from that disease.

6. Both terms refer to the rate of morbidity; however, prevalence is a measure of the cases present during a given time period, while incidence is a measure of the new cases occurring during that period.

7. Notifiable diseases are the diseases (approximately 20) that the CDC is tracking. Doctors across the country are required to report cases of these diseases so that the CDC may accurately assess trends.

8. Time (comparing year to year or season to season), place (geographic location), and people (characteristics of the population) are major factors for study by epidemiologists.

9. age, race, gender, socioeconomic standing, smoking, diet, occupation, travel history, stress, and prior infection.

10. interview studies, screening procedures, and diagnostic tests, as well as historical research

11. Both refer to screening or diagnostic testing; however, a sensitive test is one that correctly identifies people who have the disease and a specific test correctly identifies people who do not.

12. Both people have the disease; however, true positives exhibit a positive result with the diagnostic test and false negatives exhibit a negative result.

13. Both people exhibit a positive result to the diagnostic test; however, true positives actually have the disease and false positives do not.

14. The epidemiologist can manipulate variables in an experimental investigation, but not in an observational one.

15. Both are examples of observational epidemiological investigations; however, retrospective studies match cases (who have the disease) with controls (who do not have the disease), while prospective studies follow the incidence of the disease in cohorts (people who do not have the disease, but may vary in other factors) over a period of time. Retrospective studies examine past exposure, whereas prospective studies look at future exposure.

Chapter 19

Short-Answer Questions

1. A community is a collection of various microorganisms, whereas a population is a collection of the same species.

2. CO_2

3. carbon fixation; photosynthesis

4. Also known as consumers, herbivores and carnivores utilize organic carbon and return carbon dioxide to the atmosphere.

5. Microorganisms are important to all phases of the carbon cycle, especially as decomposers.

6. Recalcitrant compounds are those that are resistant to degradation.

7. nitrogen fixation, ammonification, nitrification, denitrification

8. *Rhizobium, Bradyrhizobium*

9. *Pseudomonas*

10. While nitrifying bacteria produce ammonia and nitrates, thus decreasing the need for fertilizers, these nitrates are rapidly washed away by rain and may contaminate groundwater.

11. Although all organisms require phosphorus (for membranes, nucleic acids, etc.), available phosphorus is in short supply and only a few microorganisms can solubilize it from inorganic forms (for example, rocks and bones).

12. Although many organisms can assimilate inorganic sulfur, only certain microorganisms are able to use elemental sulfur.

13. Soil is likely to be anaerobic when it is wet.

14. The rhizosphere effect refers to the abundance of microorganisms found surrounding the roots of plants. This is primarily due to the secretion of nutrients by the plant.

15. Soils with a pH below 5.5 are more likely to host fungi.

16. Algae and cyanobacteria are likely to exist in the upper layers of soil where sunlight is available.

17. Protozoa are likely to exist in the upper layers of soil, where moisture and nutrients abound.

18. Microorganisms are abundant at the oil/water interface and active degraders of the oil. This very trait is being exploited, however, to clean up oil spills.

19. Microorganisms are being studied for use as biological pesticides.

20. *Pseudomonas, Flavobacterium, Vibrio*

21. filtration effect

22. *Escherichia coli, Enterococcus faecalis, Clostridium* species

23. Biological contaminants are primarily sewage bacteria. Chemical contaminants include organic (detergents, pesticides, petroleum, etc.) and inorganic (iron, lead, mercury, etc.) substances.

24. BOD and MPN

Multiple-Choice Questions

1. a

2. c

3. b

4. a

5. d

6. c

Chapter 20

Short-Answer Questions

1. *Streptococcus lactis* and other lactic-acid producing streptococci and lactobacilli, as well as coliforms, *Pseudomonas, Alcaligenes, Flavobacterium, Salmonella, Corynebacterium diphtheriae, Shigella dysenteriae, Brucella abortus,* and *Coxiella burnetii*

2. In LTH pasteurization, milk is heated to 62.8°C for 30 minutes, in HTST pasteurization, milk is heated to 71.7°C for 15 seconds, and in UHT pasteurization, milk is heated to 137.8°C for just a few seconds.

3. The temperatures and holding times represent the conditions necessary to eliminate the most heat-resistant pathogens (for example, *Coxiella burnetii*). This prevents the likelihood of disease transmission.

4. Although raw milk does contain lactobacilli, and lactobacilli are responsible for most fermented milk products, it also contains other bacteria which may be harmful. Furthermore, it is nearly impossible to control fermentation to the desired product without selectively introducing specific strains of microorganisms (generally as starter cultures).

5. Unripened cheese (for example, cottage cheese, cream cheese, ricotta) is the product right after curdling has occurred; whereas ripened cheese (for example, Colby, Swiss, cheddar) has aged for 1 to 16 months.

6. *Pseudomonas, Achromobacter, Streptococcus, Leuconostoc, Bacillus, Micrococcus, Cladosporium herbarum, Sporotrichum carnis, Penicillium* species

7. Although these methods do not eliminate spoilage microorganisms, refrigeration or freezing will extend the shelf life of meats by slowing growth. Dehydration, or freeze drying, effectively prevents the growth of microorganisms. Curing, with salt or sugar, prevents the growth of most microorganisms. Smoking or the addition of preservatives will also extend the shelf life of meats.

8. Livestock are fed antibiotics not because they are sick but because the antibiotics help reduce the chance that consumers will acquire infections.

9. Canning involves heating foods (100°C for acidic foods, 121°C for nonacidic foods) to ensure the elimination of *Clostridium botulinum* spores.

10. Without the advantages of refrigeration, fermentation was a major method of food preservation. Although generally successful, fermentation was not guaranteed. Conditions or indigenous microorganisms may not lead to proper fermentation, and spoilage (particularly by fungi) is still common once the container is opened.

11. beer, bread, olives, pickles, sauerkraut, poi, soy sauce, wine

12. *Saccharomyces cerevisiae;* it ferments sugars; beer and wine

13. enzymes, amino acids, organic acids and alcohols, vitamins, and antibiotics

14. Industrial fermentation is generally a large-scale operation in order to be economically feasible. Many fermentation reactions are prohibited by this scale; therefore, it is important to select an organism that has a high degree of genetic stability and that is able to produce large quantities of the desired produce, even in large containers. In addition, optimum conditions of temperature, oxygen, and pH, must be met and maintained.

15. amylase, glucose oxidase, invertase, pectinase, protease, rennin

16. These enzymes play important roles in food production, textile manufacturing, and the production of pharmaceuticals.

17. In addition to their ease of production, microorganisms are the logical source of many vitamins. Even within our own bodies we can find microorganisms producing vitamins that we require but are unable to produce.

18. The fungus *Penicillium* and the bacterium *Streptomyces* are the most common sources of antibiotics.

19. Microorganisms make an excellent food source because of their high protein content; however, SCP production is still expensive.

Multiple-Choice Questions

1. e

2. d

3. d

4. b

GLOSSARY

A

abortive infection An infection in which there is interference in the viral multiplication cycle, leading to a significant reduction or complete elimination of virus production.

abortive transduction Transduction in which the transducing phage DNA is not incorporated into the host chromosome, resulting in unilinear transmission of the phage DNA to only one of the two daughter cells formed at cell division.

accessory pigment A pigment other than chlorophyll that harvests light energy and transfers it to chlorophyll.

acidophile An organism that thrives at pHs as low as 1.

acquired immune deficiency syndrome (AIDS) An infectious disease syndrome caused by human immunodeficiency virus that results in a significant deficiency of $CD4^+$ cells and increased susceptibility to opportunistic infections and some forms of cancer.

acquired resistance Specific resistance of the body to a pathogen occurring only after host exposure to the pathogen or its products; also called specific resistance.

activated sludge process An aerobic sewage treatment process in which sewage is mixed with slime-forming bacteria in a large aeration tank to break down organic matter in the sewage.

activation energy (E_a) Energy required to initiate a chemical reaction.

active site The site on an enzyme that recognizes and binds to substrate.

active transport The energy-requiring, carrier-mediated movement of molecules across a membrane against a concentration gradient.

acute period The period of disease when signs and symptoms reach their peak.

Adansonian taxonomy *See* **numerical taxonomy.**

aerobe An organism that requires molecular oxygen for growth.

aerobic respiration Catabolic reactions producing ATP in which either organic or inorganic compounds are the primary electron donors and molecular oxygen is the terminal acceptor in electron transport.

aerotaxis The movement of organisms in response to oxygen.

agammaglobulinemia An immunodeficiency disease characterized by an inability or a decreased ability to synthesize γ-globulins.

agar A polysaccharide extract of red algae that is used as a solidifying agent in microbiological media.

agglutination The formation of visible clumps by the cross-linking of antigens and their corresponding antibodies.

agranulocyte A leukocyte that does not possess cytoplasmic granules; lymphocytes and monocytes are agranulocytes.

alga(pl. algae) A photosynthetic eucaryotic microorganism.

algicidal Having the property of killing algae.

algistatic Inhibiting algae.

alkalophile An organism that lives and grows between pH 8.5 and 11.5.

allosteric effector A substance that binds to the allosteric site of an allosteric enzyme, causing a change in the conformation of the enzyme active site and inhibition or activation of enzyme activity.

allosteric enzyme An enzyme with a binding site (active site) for the substrate and a different site (allosteric site) for binding an allosteric effector.

alpha (α) hemolysis Lysis of red blood cells characterized by incomplete zones of clearing, sometimes accompanied by a green discoloration, surrounding microbial colonies growing on blood agar.

alternative complement pathway An antibody-independent pathway for complement activation that bypasses the sequence of events from C1 to C2 in the classical complement pathway and involves properdin and two other serum proteins designated Factors B and D; also called the properdin pathway.

Ames test A test to screen mutagens by the incidence of back mutations in histidine-requiring auxotrophs of *Salmonella typhimurium.*

amino acid An organic compound of low molecular weight containing a carboxyl group and an amino group.

aminoglycoside One of a group of antimicrobial agents that reversibly bind to the procaryotic ribosome to inhibit protein synthesis. Streptomycin, gentamicin, and kanamycin are examples of aminoglycosides.

ammonification The formation of ammonia or ammonium ions from nitrogenous compounds by microbial action.

amphibolic pathway A metabolic pathway that functions in both anabolism and catabolism.

anabolism The synthesis of new substances from precursors, usually requiring energy.

anaerobe An organism that grows in the absence of molecular oxygen.

anaerobic respiration Catabolic reactions producing ATP in which either organic or inorganic compounds are the primary electron donors and an inorganic compound other than molecular oxygen is the terminal acceptor in electron transport.

anamnestic response The recall by the immune system of a prior response to a specific antigen; also called secondary response.

anaphylactic hypersensitive reaction A violent immune response to an antigen-IgE antibody reaction, characterized by increased permeability of capillaries, edema, and inflammation; also called type I hypersensitive reaction.

anaplerotic reaction A reaction that replenishes intermediates removed from metabolic pathways.

anoxygenic photosynthesis Photosynthesis that occurs in the absence of molecular oxygen and during which molecular oxygen is not produced.

antagonism The killing, inhibition, or injury of one microorganism by another.

antibiotic A chemical, produced by a microorganism, that inhibits or kills other microorganisms.

antibody A protein produced by the body in response to an antigen and directed specifically against antigenic determinant sites; also called immunoglobulin.

antibody-mediated hypersensitive reaction *See* **immediate hypersensitive reaction.**

antigen Any molecule that induces an antibody to be produced specifically against it and that is able to react with that particular antibody.

antigenic determinant *See* **epitope.**

antigen-presenting cell (APC) A cell that processes and presents antigen to T lymphocytes.

antimicrobial agent A chemical that inhibits or kills microorganisms.

antisense DNA or RNA A single-stranded DNA or RNA with a base sequence complementary to a targeted gene's RNA message that can bind the target RNA and inhibit it.

antiseptic An antimicrobial agent that is used on external body surfaces.

antiserum (pl. **antisera**) Serum that contains antibodies.

antitoxin An antibody against a toxin molecule, capable of reacting with and neutralizing the toxin.

aplanospore A nonmotile spore formed by algae during reproduction.

Archaea An evolutionarily distinct domain of procaryotes distinguished from the domain Bacteria by characteristics such as rRNA sequences, lack of muramic acid in the cell wall, and ether instead of ester bonds in membrane lipids.

Arrhenius plot A plot of the logarithm of the growth rate versus the inverse of the growth temperature.

arthrospore A fungal spore formed by fragmentation of vegetative hyphae and not resistant to heat or drying.

artificially acquired passive immunity Immunity that is acquired when antibodies formed in other hosts are introduced into a new host.

ascus (pl. **asci**) A saclike, ascospore-containing structure in fungi of the phylum *Ascomycota*.

aseptic technique The procedure for handling cultures in such a manner as to eliminate contamination by undesired microorganisms.

assimilatory nitrate reduction The reduction of nitrate to ammonia with the incorporation of the nitrogen into cellular materials.

assimilatory sulfate reduction The reduction of sulfate to hydrogen sulfide with the incorporation of the sulfur into cellular materials.

atmosphere The gaseous mass surrounding the earth.

atomic force microscope A type of microscope in which magnifications of several millionfold are achieved by using a minute probe to trace the outline of atoms on a specimen's surface.

attenuated vaccine A vaccine consisting of live microorganisms that have been rendered avirulent or with reduced virulence, but that still are antigenic; also called live, attenuated vaccine.

attenuation A regulatory mechanism involving the control of gene expression by termination of transcription.

autoclave An instrument for sterilizing materials by high temperature, pressure, and flowing steam.

autogamy A type of reproduction in which the micronucleus divides into two separate nuclei, which join to form a zygote nucleus. The cell then divides into two daughter cells.

autotroph An organism that uses carbon dioxide as a sole source of carbon.

auxotroph A mutant that has one or more growth factor requirements.

B

Babès-Ernst body *See* **metachromatic granule.**

bacillus (pl. **bacilli**) A cylindrical or cigar-shaped bacterial cell.

back mutation A subsequent mutation that reverses the effects of the original mutation.

Bacteria Name for domain of procaryotes other than those of the domain Archaea.

bactericidal Having the property of killing bacteria.

bacteriochlorophyll A photosynthetic pigment found in bacteria.

bacteriocin A protein produced by bacteria that kills other, closely related bacteria.

bacteriocinogenic plasmid A plasmid that carries genes for the synthesis of bacteriocins.

bacteriolytic Having the property of killing bacteria by lysis, or dissolution, of the cell.

bacteriophage A virus capable of infecting a bacterium; also called phage.

bacteriorhodopsin A light-harvesting purple pigment in *Halobacterium* that picks up protons and transports them across the bacterial plasma membrane.

bacteriostatic Having the property of inhibiting bacteria.

bacterium (pl. **bacteria**) A procaryotic microorganism.

bacteroid An irregular-shaped *Rhizobium* cell found in root nodules.

balanced growth Microbial growth in which there is an orderly increase in the DNA, RNA, and protein of the cell population.

basidiospore A sexual fungal spore produced by fungi of the phylum *Basidiomycota*.

basophil A granulocyte that takes up basic dyes.

batch culture Growth in a closed system, affected by nutrient limitation and waste product accumulation.

beta (β) hemolysis The complete lysis of red blood cells, characterized by clear zones surrounding microbial colonies growing on blood agar.

beta (β) oxidation The stepwise oxidation of fatty acid into 2-carbon fragments.

biliprotein A water-soluble accessory pigment for photosynthesis; also called phycobilin.

binary fission An asexual process of cell division in which a single cell divides into two separate and equal daughter cells.

binding protein   A soluble protein in the periplasm of gram-negative bacteria that binds to molecules and functions in their active transport across the membrane.

biochemical oxygen demand (BOD)   The quantity of oxygen required by microorganisms to oxidize organic matter in water; also known as biological oxygen demand.

bioenergetics   The study of energy transformations in living systems.

biofilm   Microorganisms attached to a surface by adhesive polysaccharides produced by the microbial cells.

biogeochemical cycle   The recycling of chemical elements through the biological and geologic components of the world.

biotechnology   The discipline dealing with the manufacture of commercially valuable products by the use of microorganisms.

B lymphocyte   A lymphocyte derived from the bone marrow and that can mature into antibody-secreting plasma cells; also called B cell.

bottom-fermenting yeast   Yeast that aggregates and settles during fermentation to produce a clarified beer low in alcohol content.

bright-field microscope   A type of microscope in which the specimen is viewed against a light background.

broad-spectrum antimicrobial agent   An antimicrobial agent that is effective against many different types of microorganisms.

broth   A liquid growth medium.

bulking   An activated sludge processing condition in which filamentous bacteria form loose flocs that do not rapidly settle, resulting in reduced clarification of wastewater.

butanediol fermentation   A fermentation in which the products are 2,3-butanediol, ethanol, carbon dioxide, and small quantities of lactate, succinate, acetate, and H_2.

C

calorie   The quantity of heat energy required to raise the temperature of 1 g of water 1°C.

calorimeter   A device to measure the quantity of energy within a chemical substance by the complete combustion of the substance.

Calvin (C_3) cycle   A sequence of chemical reactions used by autotrophs to fix carbon dioxide into organic compounds.

capping   In eucaryotes, the addition of 7-methylguanine at the 5' terminus of mRNA shortly after initiation of RNA synthesis, apparently to protect the RNA from degradation by nucleases.

capsid   The protein coat that surrounds the nucleic acid in viruses.

capsomere   An individual protein subunit of the capsid in viruses.

capsule   The layer of polysaccharide, protein, or glycoprotein external to the cell wall; also called slime layer.

carbohydrate   An organic compound composed of carbon, hydrogen, and oxygen in the approximate ratio of 1:2:1.

cardinal temperatures for growth   Minimum, maximum, and optimum temperatures for growth.

caries   A tooth disease characterized by destruction of the enamel, dentin, and/or cementum of the tooth by acid.

carotenoid   A lipid-soluble accessory pigment for photosynthesis.

carrier   A host that is infected with an infectious microorganism in the absence of disease, but that can spread the infectious agent to other susceptible hosts.

catabolism   The biochemical reactions associated with the breakdown of chemical compounds.

catabolite repression   The suppression of gene activity in the presence of glucose or some other readily utilizable energy source.

catalyst   A chemical agent that increases reaction rates without itself being changed.

cDNA library   A DNA library that contains cDNA constructed from mRNA.

cell envelope   The outer covering of a bacterium, consisting of the plasma membrane; the cell wall; and, in gram-negative bacteria, the periplasm and an outer membrane.

cell line   Culture of cells that, under proper maintenance, is immortal and grows continuously.

cell transformation   Unregulated cell growth in which there is a loss in contact inhibition and possibly also changes in cell morphology, the appearance of new antigens on the plasma membrane, and changes in cell metabolism.

cellular immunity   Immunity that is mediated by T lymphocytes and macrophages; also called cell-mediated immunity.

cell wall   The structure outside the plasma membrane that protects the cell and is responsible for shape and rigidity.

Centers for Disease Control and Prevention (CDC)   The federal agency responsible for the detection, monitoring, and control of diseases.

central dogma   The scheme of events describing the flow of genetic information from DNA to RNA to proteins.

chemical energy   Energy residing in the bonds of chemical compounds and released through the dissociation of these bonds.

chemiosmosis   The establishment of a proton gradient across a membrane and the synthesis of ATP as this gradient is dissipated from the transport of protons back across the membrane; also called proton motive force.

chemoautotroph   An organism that obtains its energy from the oxidation of chemical compounds and that uses carbon dioxide as a sole source of carbon.

chemoheterotroph   An organism that obtains its energy from the oxidation of chemical compounds and that requires organic forms of carbon.

chemolithotroph   An organism that obtains its energy from the oxidation of inorganic compounds; also called lithotroph.

chemoorganotroph   An organism that obtains its energy from the oxidation of organic compounds; also called organotroph.

chemostat   A continuous culture device with growth controlled by the flow rate of the system and the concentration of a limiting nutrient.

chemotaxis   The movement of organisms in response to a chemical stimulus.

chemotherapeutic agent   An antimicrobial agent that is selectively toxic for microorganisms but does not harm the host.

chemotroph   An organism that obtains its energy from the oxidation of chemical compounds.

chimera   A plasmid containing nucleic acid from two different organisms.

chlorobium vesicle   A specialized structure associated with photosynthesis in phototrophic bacteria of the families *Chlorobiaceae* and *Chloroflexaceae;* also called chlorosome.

chlorophyll A photosynthetic, light-harvesting pigment having a porphyrin ring structure with magnesium as a central atom and phytol as a side chain.

chloroplast The pigmented, membrane-enclosed organelle in algae and plants that is the site of photosynthesis.

chlorosome *See* **chlorobium vesicle.**

chromatin The readily stainable DNA-protein complex of the chromosomes in the nucleus of eucaryotic cells.

chromoplast A pigmented plastid in algae and plants, responsible for the bright colors seen in flowers, fruits, and autumn leaves.

cilium (pl. **cilia**) A short, hairlike structure found in large numbers on the cell surface of eucaryotic cells, and associated with locomotion.

citric acid cycle *See* **TCA cycle.**

classical complement pathway The antibody-dependent pathway for complement activation.

clonal selection theory A theory that clones of B and T cells arise from a single cell that has been stimulated by an antigen.

clone A population of cells derived from a single cell.

clustering of cases A larger-than-normal number of disease cases in a particular geographic area.

co-agglutination The formation of visible clumps by the cross-linking of antigens and corresponding antibodies that are linked by their Fc portions to the Protein A surface components of heat-killed *Staphylococcus aureus* cells.

coagulase An enzyme produced by pathogenic staphylococci that causes clotting of plasma.

coccobacillus (pl. **coccobacilli**) A plump, cigar-shaped bacterial cell.

coccus (pl. **cocci**) A spherical bacterial cell.

codon A sequence of three adjacent bases on mRNA that codes for an amino acid or the initiation or termination of protein synthesis.

coenocytic Multinucleated (organism or cell).

coenzyme An organic molecule of low molecular weight that binds to an enzyme and participates in the enzymatic reaction, often by accepting and donating electrons or functional groups.

cofactor An inorganic or organic molecule of low molecular weight that binds to an enzyme and participates in the enzymatic reaction, often by accepting and donating electrons or functional groups.

cohort The defined population group that is followed in prospective studies.

colicin A bacteriocin produced by *Escherichia coli.*

coliform A gram-negative aerobic or facultatively anaerobic non–spore-forming, rod-shaped bacterium that ferments lactose with gas production within 48 hours at 35°C.

colony (pl. **colonies**) The visible population of cells arising from a single cell and growing on a solid culture medium.

colony stimulating factor (CSF) A cytokine that stimulates division and differentiation of certain cells such as monocytes and macrophages.

col plasmid A plasmid carrying genes for the synthesis of colicins.

co-metabolism The metabolic transformation of a substance that does not serve as a source of nutrients to the microorganism.

commensalism A symbiotic relationship in which one organism benefits and the other organism is unaffected.

common-source epidemic disease A disease outbreak in which a group of people is exposed at one time to a specific disease agent.

communicable disease A disease that can be transmitted by an infectious agent from one individual to another.

competence The physiological state in which a bacterium is able to take up and incorporate donor DNA.

complement A group of thermolabile proteins found in serum that interacts with antibodies to promote phagocytosis and lysis of bacteria.

complementary DNA (cDNA) A DNA copy of mRNA.

complementary DNA library (cDNA library) A DNA library that contains cDNA constructed from mRNA.

complex (undefined) medium (pl. **complex media**) A culture medium, the exact composition of which is not known.

compound microscope A microscope with two or more sets of lenses.

condenser The part of the microscope that focuses light from the light source onto the specimen.

conidiophore A fungal hypha with condiospores at the tip.

conidiospore, conidium (pl. **conidia**) An asexual fungal spore, resistant to heat and drying.

conjugation The joining of two cells with transfer of genetic material.

conjugative plasmid A plasmid that carries a sequence of genes called the *tra* genes that are associated with conjugative transfer.

constitutive Having continuous gene expression.

contact inhibition The cessation of animal cell movement and division as a result of cell-to-cell contact.

continuous culture A microbial culture with the volume and phase of growth maintained at a constant level by the addition of fresh medium and the removal of an equal volume of spent medium and old cells.

convalescent period The period of disease when recovery occurs.

corepressor A substance that acts with a repressor to prohibit gene expression.

coryneform *See* **diphtheroid.**

cosmid A plasmid vector that contains λ phage *cos* (cohesive end) sites incorporated into the plasmid DNA.

crista (pl. **cristae**) The convoluted internal mitochondrial membrane.

cross-feeding The phenomenon in which the growth of an organism is dependent on the provision of one or more nutrients or growth factors supplied by another organism; also called syntrophism.

cyclic photophosphorylation The synthesis of ATP through cycling of electrons in light-dependent reactions of photosynthesis.

cyst A dormant stage formed by some bacteria and protozoa that is resistant to dessication.

cytochrome An electron transport molecule consisting of a porphyrin ring with a central iron atom conjugated to a protein.

cytogamy The fusion of two cells without the exchange of nuclei.

cytokine A low-molecular-weight protein that regulates important biological processes such as cell growth, cell activation, tissue repair, immunity, and inflammation.

cytokinesis The separation of the cytoplasm of a dividing cell into two equal daughter cells following nuclear division.

cytopathic effect (CPE) The visible morphological change to cells growing in tissue culture.

cytoplasmic membrane *See* **plasma membrane.**

cytotoxic hypersensitive reaction An exaggerated immune response caused by the interaction of antibodies with tissue cells and complement; also called type II allergic reaction.

D

dark-field microscope A type of microscope in which the specimen appears light against a dark background.

death phase (of population growth curve) The portion of the population growth curve where there is a net decrease in viable cell numbers; also called decline phase (of population growth curve).

decarboxylase An enzyme catalyzing the removal of the carboxyl group from an amino acid, resulting in the formation of an amine or a diamine.

decimal reduction time (DRT) The time required to kill 90% of the microorganisms in a suspension at a specific temperature; also called D value.

decline period The period of disease when symptoms subside.

decline phase (of population growth curve) *See* **death phase (of population growth curve).**

delayed hypersensitive reaction An allergic reaction mediated by T lymphocytes.

deletion The removal of DNA segments from the chromosome.

dendrogram A diagrammatic matrix used to illustrate similarity levels in numerical taxonomy.

denitrification The anaerobic conversion of nitrate into nitrogen gases.

deoxyribonucleic acid (DNA) A type of nucleic acid that carries genetic information. DNA contains deoxyribose sugars (in contrast to RNA, which contains ribose sugars).

dermatomycosis A fungal disease of the dermis.

diatom A member of the algal division *Chrysophyta*, having two thin, overlapping shells of silica.

diauxic growth A growth pattern in which there are two exponential phases of the population growth curve as the cells use one substrate and then the other.

differential medium A culture medium that distinguishes different types of microorganisms by their metabolism of media components and subsequent colony appearance.

dikaryon A fungus with paired but not fused nuclei, derived from different parent hyphae.

dimorphic A fungus exhibiting both the mold and the yeast growth types under different environmental conditions.

dinoflagellate An alga of the division *Pyrrophyta*, having a spinning type of motility as the result of its flagellation.

diphtheroid A gram-positive pleomorphic bacterium that resembles *Corynebacterium diphtheriae* but is not virulent; also called coryneform.

diploid Having two sets of chromosomes.

disease A condition of ill health in a living organism that may be caused by an infectious agent (an infectious disease).

disinfectant An antimicrobial agent that is used on nonliving surfaces.

disk diffusion method A method to determine the antimicrobial susceptibility of a microorganism, using disks impregnated with known concentrations of antimicrobial agents.

dissimilatory nitrate reduction The reduction of nitrate to nitrite, with nitrate used as an electron acceptor.

DNA fingerprinting A technique for the identification of repetitive DNA sequences that occur in the genomes of humans and other eucaryotes.

DNA library A collection of cloned DNA fragments from the genome of an organism.

domain The highest level of classification, consisting of the Bacteria, the Archaea, and the Eucarya.

doubling time *See* **generation time.**

dry weight The weight of cells minus any moisture.

D value *See* **decimal reduction time.**

E

ectomycorrhiza (pl. ectomycorrhizae) A symbiotic association between a fungus and plant roots in which the plant roots are surrounded but not penetrated by fungal hyphae.

ectosymbiosis A symbiosis in which the microorganism may be attached but remains external to the host cell.

Einstein *See* **photon.**

electromagnetic energy Light energy that travels in photons.

electron microscope A type of microscope in which magnifications of 100,000× or greater are achieved by replacing glass lenses with electromagnetic lenses and light with electrons.

electron transport chain A series of electron carriers through which electrons are transported from a substrate to a final electron acceptor, typically resulting in ATP synthesis.

electroporation The creation of small holes in the plasma membrane of a cell by electric shock to permit the entry of DNA.

Elek gel diffusion test A test in which an antitoxin-toxin reaction for *Corynebacterium diphtheriae* is detected by precipitation on an agar plate.

elementary body (EB) A small infectious body that is specialized for extracellular survival in the *Chlamydia* developmental cycle.

Embden-Meyerhof pathway A sequence of chemical reactions in which one molecule of glucose is oxidized to two molecules of pyruvate; a glycolytic pathway.

empty magnification An increase in magnification without an accompanying increase in resolution.

endemic disease A disease that is continuously present in a population.

endergonic A chemical reaction having a positive change in free energy and requiring input of energy to proceed.

endogenote A recipient cell's genome in which the donor DNA can be integrated during the transfer of genetic material from one cell to another.

endomycorrhiza (pl. endomycorrhizae) A symbiotic association between a fungus and plant roots in which fungal hyphae penetrate the plant roots.

endospore A spore formed within a cell.

endosymbiont hypothesis A theory that procaryotes living as endosymbionts inside eucaryotes evolved into chloroplasts and mitochondria.

endosymbiosis   A symbiosis in which the microorganism grows within the host cell.

endothermic   A chemical reaction proceeding with absorption of heat.

endotoxin   A toxin associated with the lipopolysaccharide portion of the outer membrane of gram-negative bacteria; the lipid A component of lipopolysaccharide is responsible for the toxicity of endotoxin.

energy   The capacity to do work.

enrichment culture   A culture medium that enhances the growth of specific types of microorganisms while often inhibiting the growth of other organisms.

enterohemorrhagic *Escherichia coli* (EHEC)   Strains of *E. coli* that produce a toxin closely related to Shiga toxin and can cause hemolytic-uremic syndrome, which is characterized by lysis of erythrocytes and kidney failure.

enteroinvasive *Escherichia coli* (EIEC)   Strains of *E. coli* that are invasive and cause hemorrhagic enterocolitis.

enteropathogenic *Escherichia coli* (EPEC)   Strains of *E. coli* that commonly cause diarrhea in newborn infants.

enterotoxigenic *Escherichia coli* (ETEC)   Strains of *E. coli* that synthesize heat-stable (stable toxin, ST) and heat-labile (labile toxin, LT) enterotoxins and cause traveler's diarrhea.

enterotoxin   An exotoxin that affects the small intestine.

enthalpy (H)   The total amount of energy released during a reaction.

Entner-Doudoroff pathway   A sequence of chemical reactions in which glucose is oxidized to 6-phosphogluconate and subsequently to pyruvate.

entropy (S)   Randomness or disorder in a system.

enzyme   A protein (or, in the case of ribozymes, RNA) catalyst found in living systems.

enzyme-linked immunosorbent assay (ELISA)   A laboratory technique used to detect antibodies or antigens in a sample through an enzyme that causes a color change in its substrate.

eosinophil   A granulocyte that takes up acid dyes, especially eosin.

epidemic disease   A disease that affects a large segment of the population within a region at one time.

epidemiology   The science dealing with the incidence, distribution, and control of disease in a population.

episome   Plasmid DNA that has been incorporated into the bacterial chromosome.

epitope   The portion of an antigen that determines immunologic specificity; also called antigenic determinant.

erysipelas   A skin disease characterized by invasion of the subcutaneous blood vessels, causing the appearance of red patches.

erythrocyte   A red blood cell.

estuary   A partially enclosed coastal body of water that separates marine waters from inland sources of fresh water.

Eucarya   The phylogenetic domain containing eucaryotes.

eucaryote   A cell or organism having a membrane-enclosed nucleus, specialized membrane-enclosed organelles, and a high level of internal structural organization not found in a procaryote.

eutrophic   Containing a high concentration of nutrients.

exergonic   A chemical reaction having a negative change in free energy and proceeding with liberation of energy.

exfoliatin   Staphylococcal exotoxin responsible for the pathogenesis of scalded skin syndrome.

exogenote   The donor DNA introduced into a recipient cell during the transfer of genetic material from one cell to another.

exon   Translatable portions of mRNA.

exosporium   The loose outer envelope that surrounds an endospore.

exothermic   A chemical reaction proceeding with the release of heat.

exotoxin   A toxic soluble protein produced by a microorganism and released into the environment.

exponential phase (of population growth curve)   The period of population growth in which the population is actively growing at a constant rate; also called logarithmic phase (of the population growth curve).

expression vector   A vector that contains not only the target gene but also regulatory sequences that can be used to control expression of the gene.

extrinsic membrane protein   A protein found in the lipid bilayer of the plasma membrane and that can be removed by changes in pH or ionic strength; also called peripheral membrane protein.

F

facilitated diffusion   The carrier-mediated movement of molecules across a membrane along a concentration gradient.

facultative anaerobe   An organism that grows in the presence or absence of oxygen, respiring in the presence of oxygen and fermenting in its absence.

fat   A triglyceride containing an ester of glycerol and three fatty acid molecules.

fatty acid   A straight-chain lipid having a carboxyl group at one end.

F^+ cell   A cell that contains the F factor on a plasmid.

F^- cell   A cell that lacks the F factor.

F' cell   A cell in which the F factor and host genes are on a plasmid.

fecal coliform   A coliform that grows at a temperature of 44.5°C.

feedback inhibition   Inhibition by an end product of the activity of an enzyme in a metabolic pathway.

fermentation   The anaerobic oxidation-reduction of carbohydrates with organic compounds as electron donors and electron acceptors.

F factor   The fertility factor that is found in the donor bacterium during conjugation; also called sex factor.

fimbria (pl. fimbriae)   *See* **pilus.**

first law of thermodynamics   The law stating that energy can neither be created nor destroyed in any transformation.

flagellin   The protein subunit of a bacterial flagellum.

flagellum (pl. flagella)   A long, thin appendage used for cellular motility.

fluorescence microscope   A type of microscope in which objects absorb ultraviolet or near-ultraviolet light and emit visible light.

fomite   An inanimate object contaminated with an infectious microorganism.

food infection   A disease occurring when an infectious agent ingested with food or water establishes an active infection in the small intestine.

food intoxication A disease occurring when food or water containing toxic products of microorganisms is consumed.

fractional sterilization *See* **tyndallization.**

free energy (G) The portion of energy released during a reaction that can be used for work; expressed as G° if determined under standard conditions of 1 atmosphere pressure and 1 molar concentration, and as G°, if determined under standard conditions and at pH 7.0.

fruiting body A large, specialized structure containing spores and produced by some procaryotes (myxobacteria) and some fungi.

fungicidal Having the property of killing fungi.

fungistatic Inhibiting fungi.

fungus (pl. fungi) A nonmotile eucaryotic microorganism that has a cell wall but does not contain chlorophyll.

furunculosis An infection of the hair follicle, causing the formation of a boil or furuncle in the underlying subcutaneous tissue.

G

gametangium (pl. gametangia) A specialized fungal reproductive structure located at the tip of the hypha and containing the gamete.

gamete A haploid reproductive cell that unites with another gamete to form a diploid zygote.

gamma globulin (γ-globulin) A protein fraction of blood rich in antibodies.

gamma (γ) hemolysis No lysis of erythrocytes.

gas vacuole An internal vacuole that retains gas and gives buoyancy to the cell.

gene The segment of a chromosome that is transcribed into mRNA coding for a single polypeptide.

gene amplification The process in which a cell produces large quantities of a specific gene product.

gene cloning The insertion of a target gene into a vector for introduction into a cell, where the gene is amplified and expressed.

gene gun A device that shoots DNA-coated microprojectiles into a cell without killing the cell.

generalized transduction The transfer of random fragments of host DNA by transduction.

general recombination Recombination involving a reciprocal exchange between two homologous DNA molecules; also called homologous recombination.

generation time The time required for a population to double in number; also called doubling time.

gene therapy The treatment of genetic diseases by replacement of dysfunctional genes.

genetic engineering The deliberate modification of the genetic makeup of a cell or organism.

genetic recombination The process through which genetic material from two separate cells is brought together.

genetics The field of biology that deals with mechanisms responsible for the transfer of traits from one organism to another.

genome One complete set of genes.

genomic library A DNA library that contains DNA fragments from a digested genome.

genotype The genetic composition of an organism.

genus (pl. genera) The taxonomic category of related organisms, below family level and above species level.

germ-free animal An animal that contains no microbial flora.

germicide The general term used to describe antimicrobial agents that kill microorganisms.

germination The outgrowth of spores to vegetative cells.

germ theory of disease A theory that microorganisms cause disease.

gingivitis Inflammation of the gum.

gluconeogenesis The biosynthesis of glucose from noncarbohydrate compounds through reversal of the Embden-Meyerhof pathway.

glycocalyx The matrix of polysaccharides on the surface of bacteria, instrumental in binding cells together in an aggregate mass to protect them from phagocytosis and assist them in attachment to a solid surface.

glycolysis The oxidation of glucose to pyruvate; *see* Embden-Meyerhof pathway for an example.

glycoprotein An organic compound consisting of carbohydrate and protein.

glyoxylate cycle An anaplerotic pathway that channels 2-carbon compounds such as acetate and acetyl-CoA into the TCA cycle.

Golden Age of Microbiology An era from 1876 to 1906 during which the causes of most bacterial diseases were discovered.

Golgi complex (Golgi apparatus) The membrane-enclosed organelle associated with the packaging and secretion of substances from the cell interior through the membrane to the exterior.

Gram stain A fundamental bacteriological staining technique to differentiate bacteria that retain the primary stain of crystal violet (gram-positive bacteria) from bacteria that are decolorized and take up the secondary stain of safranin (gram-negative bacteria).

granulocyte A leukocyte possessing cytoplasmic granules; basophils, eosinophils, and neutrophils are granulocytes.

granum (pl. grana) The chloroplast structure consisting of stacked layers of thylakoids.

greenhouse effect An increase in land temperatures on the earth, resulting from increased levels of atmospheric carbon dioxide.

group translocation The transport mechanism in which the substrate is chemically modified during movement across the membrane and the substrate can accumulate against a concentration gradient within the cell.

growth The orderly increase in the cell mass of an organism.

growth factor An organic nutrient required by microorganisms for growth, but not necessarily synthesized by them.

H

halophile An organism that is osmophilic and has a specific requirement for sodium chloride.

haploid Having one set of chromosomes.

hapten A substance that combines with a specific antibody but cannot by itself induce the formation of antibodies.

haustorium (pl. haustoria) The special absorptive hypha of a fungus.

hemagglutination The agglutination of red blood cells.

hemolysin An enzyme that breaks down red blood cells.

herd immunity The immunization stage at which a large enough percentage of a susceptible population is immunized to significantly reduce or eliminate the spread of an infectious agent.

heterofermentation Fermentation of glucose or other carbohydrates to a mixture of products.

heteropolysaccharide A polysaccharide consisting of two or more types of sugars.

heterotroph An organism that obtains its carbon from organic compounds.

hexose monophosphate shunt *See* **pentose phosphate pathway.**

high-efficiency particulate air (HEPA) filter A filter that removes 99.97% of particles 0.3 μm and larger from air.

high-frequency recombination (Hfr) cell A cell in which the F factor is incorporated into the host chromosome, permitting the transfer of chromosomal genes to another cell.

high-frequency transduction An increased frequency of specialized transduction through the production of lysate containing a high percentage of transducing bacteriophage.

Hill-Bendall scheme The series of steps hypothesized to be responsible for electron transfer, ATP synthesis, and NADP reduction in the light reaction of photosynthesis.

histocompatibility antigen *See* **human leukocyte antigen.**

histone A positively charged protein associated with eucaryotic DNA.

homofermentation Fermentation of glucose or other carbohydrates to a single product, lactate (lactic acid).

homologous recombination *See* **general recombination.**

homopolysaccharide A polysaccharide consisting of one type of sugar in a repeating structure.

hops The dried petals from the vine *Humulus lupulus,* added to wort as a flavor enhancer and antimicrobial agent in beer brewing.

Human Genome Project A concerted international effort to clone, map, and sequence the entire human genome.

human leukocyte antigen (HLA) An antigen on the surface of human cells that is recognized by immune system cells and therefore is important in regulation of the immune response; also called histocompatibility antigen.

humoral immunity Immunity associated with antibodies produced against antigens.

hyaluronidase An enzyme that degrades hyaluronic acid.

hybridoma A hybrid cell formed by the fusion of an antibody-producing lymphocyte and a rapidly dividing cancer cell and used continuously to manufacture identical antibodies.

hydrologic cycle The cycling of water in the biosphere, beginning with the evaporation of water from oceans, lakes, and other surface bodies, followed by precipitation of atmospheric water and percolation of the water through the ground and back into the bodies of water.

hydrosphere The aqueous envelope of the earth, including oceans, lakes, streams, and underground water.

hydrothermal vent A natural vent on the sea floor that releases superheated water rich in hydrogen sulfide and other reduced inorganic compounds.

hypersensitivity A state of abnormal susceptibility to an antigen, leading to an exaggerated immune response.

hypha (pl. **hyphae**) A fungal filament.

I

idiotype The variable region of an antibody molecule that determines its specificity.

immediate hypersensitive reaction An allergic reaction involving IgE antibodies; also called antibody-mediated hypersensitive reaction.

immobilized enzymes (immobilized microbial cells) Enzymes or microbial cells that are adsorbed onto a solid support (for example, cellulose particles or polyacrylamide beads). A substrate in solution is passed across the support and is changed into product by the immobilized enzymes or cells.

immunity The specific resistance of a host to a pathogen.

immunization The process or procedure by which a subject is rendered immune.

immunogenicity The ability to stimulate antibody production.

immunoglobulin (Ig) *See* **antibody.**

immunological tolerance The inability to recognize antigens for antibody production.

immunology The science that deals with the study of resistance of a host to infection.

impetigo A bacterial skin disease characterized by pus-filled lesions that rupture and become encrusted.

incidence rate The number of new cases of disease in a population during a specified time period; a type of morbidity rate.

inclusion body An internal cell component that accumulates lipids, polysaccharides, or inorganic compounds; also called storage body.

incubation period The time interval between infection and the first appearance of disease symptoms.

induced mutation A mutation occurring as a result of chemical or radiation effects on the chromosome.

inducer A substance that causes structural genes to be expressed.

induction The process by which an enzyme is synthesized in response to the presence of a substance (the inducer).

industrial microbiology The technology of using microorganisms to produce a product of commercial value.

infection The presence of viable microorganisms in a host.

infectious disease A disease that can be transmitted from one person, animal, or plant to another.

infectious dose, 50% end point (ID_{50}) The quantity of pathogen required to infect successfully 50% of the inoculated animals within a given period of time.

inflammation The host's nonspecific response to injury, irritation, or infection, characterized by dilation and increased permeability of the blood vessels in the affected area.

inhibitor A chemical substance that prevents or slows an enzyme reaction by binding to the enzyme and preventing substrate binding.

innate resistance Host resistance not directed against any particular pathogen; also called nonspecific resistance.

insertion   The addition of DNA segments into the chromosome.

insertion sequence   A genetic element inserted into portions of a procaryotic chromosome, and that can control gene expression.

integral membrane protein   *See* **intrinsic membrane protein.**

interference contrast microscope   A type of microscope in which contrast is achieved by destructive and/or additive interference of light waves.

interferon   A glycoprotein produced by cells during viral infection and able to inhibit viral replication.

interleukin   A cytokine that regulates growth and proliferation of lymphocytes.

interrupted mating   The process in which conjugation between an Hfr cell and an F⁻ cell is interrupted at different times to permit mapping of the Hfr chromosome.

intrinsic membrane protein   A protein embedded in the interior of the lipid bilayer of the plasma membrane, and that can only be removed by the use of detergents or nonpolar solvents; also called integral membrane protein.

intron   A nontranslatable portion of mRNA.

ionosphere   The uppermost layer of the atmosphere, above the stratosphere.

iris diaphragm   The part of the microscope that controls the diameter of light leaving the condenser and striking the specimen.

isotype   The constant-region determinant for an immunoglobulin class or subclass.

K

kappa (κ) chain   One of two types of light chains occurring in an antibody molecule.

karyokinesis   A process involving the division and separation of chromosomes in eucaryotes.

killed vaccine   A vaccine that consists of dead microorganisms.

killer strain (of *Paramecium*)   A paramecium that liberates toxic particles that are harmful to other, sensitive paramecia.

kilocalorie (kcal)   The quantity of heat energy required to raise the temperature of 1 kg of water 1°C.

kinetic energy   Energy of work or motion.

Kirby-Bauer test   A disk diffusion test in which zones of inhibition are measured around disks to determine the susceptibilities of microorganisms to antimicrobial agents.

Koch's postulates   A set of criteria for proving that a microorganism causes a particular disease.

Kornberg enzyme   A synonym for DNA polymerase I, an enzyme that fills in the gaps of Okazaki fragments during DNA replication and also repairs ultraviolet light-inflicted DNA damage (named after Arthur Kornberg).

Krebs cycle   *See* **TCA cycle.**

L

lagging strand   The designation for the discontinuous daughter strand in DNA replication.

lag phase (of population growth curve)   The period of population growth in which there is no increase in cell numbers.

lambda (γ) chain   One of two types of light chains occurring in an antibody molecule.

laser-scanning confocal scanning microscope   A type of microscope in which magnifications of several millionfold are achieved by scanning a cone of laser light across a specimen. The light is focused through a pinhole aperture to a specific point within the specimen and the imaging light from the specimen is focused through a second (confocal) pinhole to a detector.

latent infection   An infection that does not immediately produce detectable or overt symptoms.

latex agglutination   The formation of visible clumps by the cross-linking of antigens and corresponding antibodies that are attached to latex spheres.

leaching   The extraction of metals from low-grade ores.

leader peptide   In attenuation, a short peptide rich in tryptophan that is synthesized when tryptophan is abundant.

leading strand   The designation for the continuous daughter strand in DNA replication.

leghemoglobin   An oxygen-binding protein produced during symbiotic nitrogen fixation

by *Rhizobium* and a legume; the heme portion of the protein is produced by the bacterium and the globin by the plant.

lethal dose, 50% end point (LD$_{50}$)   The quantity of a pathogen required to kill 50% of the inoculated animals within a given period of time.

leucoplast   A nonpigmented plastid found in plants, associated with the synthesis and storage of starch, proteins, and oils.

leukocidin   A bacterial product that kills leukocytes.

leukocyte   A white blood cell, such as a neutrophil, basophil, eosinophil, lymphocyte, or monocyte.

leukocytosis   An abnormally high number of leukocytes.

leukopenia   An abnormally low number of leukocytes.

L-form   A bacterium that normally has a cell wall, but has lost the ability to synthesize the wall.

lichen   A plantlike structure consisting of a fungus and an alga living in mutualistic symbiosis.

light microscope   A type of microscope that uses light as its source of illumination. The four types of light microscopes are: bright-field, dark-field, fluorescence, and phase-contrast.

limnetic zone   The zone of a lake to which light penetrates.

***Limulus* amoebocyte lysate (LAL) assay**   An assay to detect endotoxin using extracts of amoebocytes from the horseshoe crab *Limulus polyphemus*.

lipid   A fat or fatlike molecule that is insoluble in water but soluble in organic solvents.

lipopolysaccharide (LPS)   A lipid-carbohydrate molecule consisting of three basic parts (O-specific side chain, core poly-saccharide, and Lipid A), found as part of the outer membrane of gram-negative bacteria.

lithosphere   The solid portion of the earth.

lithotroph   *See* **chemolithotroph.**

littoral zone   The zone of a lake near to the shore.

live, attenuated vaccine   *See* **attenuated vaccine.**

logarithmic phase (of population growth curve)   *See* **exponential phase (of population growth curve).**

lophotrichous A tuft of flagella on the cell.

L-phase variant *See* **L-form.**

luciferase The enzyme that catalyzes light formation in luminescent bacteria.

lymphocyte An agranulocyte involved in antibody production and cell-mediated immunity.

lymphokine A chemical mediator released by T lymphocytes.

lyophilization A freeze-drying technique for long-term preservation of microorganisms.

lysogenic conversion The expression of phage genes while the phage exists in a prophage state in the host chromosome.

lysogeny The state in which a bacterium carries a prophage.

lysosome A eucaryotic organelle containing hydrolytic enzymes.

lysozyme An enzyme, commonly found in body secretions (tears and saliva) and egg white, that hydrolyzes the ß-1,4 linkages between *N*-acetylmuramic acid and *N*-acetylglucosamine in cell wall peptidoglycan.

lytic cycle A virus life cycle that results in the release of virus upon lysis of the host cell.

M

macroorganism An organism that can be seen with the unaided eye.

macrophage A phagocytic cell.

macrophage activating factor (MAF) A cytokine that alters macrophages immunologically and increases their phagocytic activity in cell-mediated immunity.

macrophage chemotactic factor (MCF) A cytokine that attracts macrophages to the infection site in cell-mediated immunity.

macrophage migration inhibitory factor (MIF) A cytokine that prevents macrophages from leaving the infection site in cell-mediated immunity.

magnetosomes Intracellular crystal particles of the iron oxide magnetite (Fe_3O_4) that act as magnets and help bacteria orient their movement along magnetic fields.

magnetotaxis The movement of organisms in response to a magnetic field.

major histocompatibility complex (MHC) A cluster of genes coding for cell surface molecules that serves as a unique marker of the individual.

malting A step in beer brewing involving the germination and crushing of barley grains to release amylases and proteases.

mast cell A nonmotile connective tissue cell found next to capillaries.

matching coefficient (S_S) The comparison of positive and negative characteristics among organisms in numerical taxonomy.

meiosis Reduction division by which the chromosome number is halved in eucaryotic cells.

melting temperature (T_m) The temperature at which there is a sharp increase in absorption of double-stranded DNA at 260 nm as the strands separate.

membrane attack complex (MAC) The complement complex components (C5b–C9) that disrupt the plasma membrane and create a lytic transmembrane pore.

memory B lymphocyte A B lymphocyte that mounts a rapid and enhanced secondary antibody response when challenged a second time with the same antigen.

mesophile An organism with an optimum growth temperature of 20°C to 40°C.

messenger RNA (mRNA) RNA that carries the genetic message from DNA and serves as a template for translation of the message into proteins.

metabolism The catabolic and anabolic reactions in a cell.

metachromatic granule An inclusion body containing polyphosphate granules and found in cells of *Corynebacterium* and other bacteria: also called Babès-Ernst body.

methyl-accepting chemotaxis protein (MCP) Protein in the plasma membrane or periplasm that is involved in transmitting signals for movement of the flagellum; this protein is alternatively methylated or demethylated in response to the binding of chemoreceptors to specific chemicals; also called transducer.

methylotroph An organism that uses 1-carbon compounds as its sole carbon and energy sources.

methyl red test A laboratory test to differentiate between mixed acid fermenters and butanediol fermenters by use of a pH indicator, methyl red.

microaerophile An organism that grows best at low oxygen concentrations.

microbial ecology The field of science concerned with the interactions of microorganisms with the living and nonliving components of the environment.

microbiology The study of organisms too small to be seen with the unaided eye, specifically viruses, bacteria, fungi, algae, and protozoa.

microorganism An organism that normally cannot be seen without the aid of a microscope.

microscope An instrument that uses lenses to magnify an object too small to be seen with the unaided eye.

microtubule A hollow protein cylinder in a eucaryotic cell associated with the movement of chromosomes, flagella, and cilia.

mineralization The conversion of organic matter to minerals and other inorganic materials.

minimal bactericidal concentration (MBC) The lowest concentration of an antimicrobial agent that kills the test microorganism.

minimal inhibitory concentration (MIC) The lowest concentration of an antimicrobial agent that inhibits growth of the test microorganism.

missense mutation A mutation in which there is a change in the amino acid coded by the affected region of the chromosome.

mitochondrion (pl. mitochondria) The membrane-enclosed organelle in a eucaryotic cell that is the site of respiration and ATP synthesis.

mitosis The process in eucaryotic cells involving separation of replicated chromosomes and subsequent cytoplasmic division, resulting in two identical daughter cells.

mixed acid fermentation The fermentation of glucose, resulting in the formation of lactate, succinate, and acetate, as well as ethanol, carbon dioxide, and H_2.

mixotroph An organism that uses inorganic compounds as its energy source and organic compounds as its carbon source.

mold A fungal growth type having a vegetative structure consisting of hyphae.

molecular approach The method of procaryotic taxonomy in which genetic similarity is used to classify organisms.

monoclonal antibody An antibody produced from a single hybridoma and its clones.

monocyte An agranulocyte that has a single nucleus and is active in phagocytosis.

monolayer A single layer of cells.

mononuclear phagocytic system *See* **reticuloendothelial (RE) system.**

monotrichous   Having one flagellum on a cell.

Morbidity and Mortality Weekly Report (MMWR)   A weekly publication of the Centers for Disease Control and Prevention that contains up-to-date morbidity and mortality data on communicable diseases.

morbidity rate   *See* **prevalence rate, incidence rate.**

mortality rate   The number of deaths that result from a disease per unit population during a specified time period.

most probable number (MPN)   A statistical expression of cell density in a suspension.

multitrichous   Having more than one flagellum on a cell.

mutant   An organism that differs from its parent because of a change in the DNA.

mutation   An inheritable change in a chromosome.

mutualism   A symbiotic relationship in which organisms living together benefit one another.

mycelium (pl. **mycelia**)   A mass of hyphae.

mycorrhiza (pl. **mycorrhizae**)   A symbiotic association between a fungus and plant roots.

myeloma   A plasma tumor cell composed of cells derived from tissues of the bone marrow.

myxospore   A resistant cell formed by myxobacteria.

N

narrow-spectrum antimicrobial agent   An antimicrobial agent that is effective against specific microorganisms or groups of microorganisms.

natural history of disease   The sequence of events that occurs from the time a parasite infects a host to the time that the disease is resolved, either through recovery or death of the host.

naturally acquired active immunity   The production of antibodies in an individual in response to antigenic stimulation through natural infection.

naturally acquired passive immunity   Immunity that is acquired by the transfer of antibodies from one person to another person by natural means.

negative control   A regulatory system in which there is enzyme synthesis in the absence of the controlling factor.

negative stain   A stain in which the background rather than the specimen is stained; in electron microscopy, a stain in which the specimen is electron transparent (light) against an electron-dense (dark) background.

neurotoxin   A toxin that affects nerve tissues.

neutrophil   A granulocyte with multilobed nuclei that is the main phagocyte in the blood; also called polymorphonuclear leukocyte (PMN).

nitrification   The oxidation of ammonia to nitrite and of nitrite to nitrate.

nitrogenase   An oxygen-labile enzyme responsible for microbial nitrogen fixation.

nitrogen fixation   The reduction of nitrogen gas to ammonia.

nomenclature   The systematic naming of organisms.

noncommunicable disease   A disease that is not transmitted by an infectious agent (for example, alcoholism, mental disease, and diabetes).

nonconjugative plasmid   A plasmid that does not have *tra* genes, but can have *mob* (mobilization) genes that allow them to take advantage of the *tra* function of a conjugative plasmid for transfer between bacteria.

noncyclic photophosphorylation   The synthesis of ATP through the transfer of electrons eventually to NADP in light-dependent reactions of photosynthesis.

nongonococcal urethritis (NGU)   Inflammation of the urethra caused by a microorganism other than *Neisseria gonorrhoeae;* also called nonspecific urethritis (NSU).

nonhistone protein   A protein associated with eucaryotic DNA, differentiated from histone protein.

noninfectious disease   A disease that is not transmitted from one person, animal, or plant to another.

nonsense codon   *See* **termination codon.**

nonsense mutation   A mutation that results in the formation of a nonsense codon and premature termination of protein synthesis.

nonspecific resistance   *See* **innate resistance.**

nonspecific urethritis (NSU)   *See* **nongonococcal urethritis.**

normal flora   Organisms that normally colonize a host without causing disease.

Northern blot   A technique by which RNA fragments, separated by electrophoresis, are immobilized on a paper sheet and detected with a labeled nucleic acid probe.

nosocomial   Originating in a hospital; also called hospital-acquired.

nucleic acid   A class of molecules consisting of nucleotides joined by phosphodiester bonds (for example, DNA and RNA).

nucleic acid probe   A single-stranded nucleic acid segment with a base sequence that is complementary to a specific DNA or RNA sequence.

nucleocapsid   Viral nucleic acid surrounded by a capsid.

nucleoid   The region in a procaryotic cell that contains its genetic material.

nucleoside   A purine or pyrimidine base joined to a pentose.

nucleosome   The basic structural subunit of chromatin, containing approximately 200 base pairs and histone proteins.

nucleotide   A nucleoside containing one or more phosphate.

null cell   A cell that resembles a lymphocyte but lacks the surface characteristics of either B or T lymphocytes.

numerical aperture   The lens property, equivalent to n sin θ, where n is the refractive index of the medium between the lens and the specimen, and θ is one-half the angle of light entering the objective lens.

numerical taxonomy   The method of bacterial taxonomy in which observable characteristics carry equal weights in the comparison and classification of organisms; also called Adansonian taxonomy.

nutrient   A chemical substance used by an organism for cellular growth and activity.

O

objective lens   The lens that is closest to the specimen in a compound light microscope.

ocular lens   The lens that is nearest the eye in a compound light microscope.

Okazaki fragment   A short DNA fragment (1,000 to 2,000 nucleotides long in procaryotes) synthesized discontinuously on one strand of double-stranded DNA (named after Reiji Okazaki).

oligotrophic   Containing a low concentration of nutrients.

oncogene   A cancer gene in the chromosome.

one-step growth curve   A graphic curve characterizing bacteriophage growth.

operator   The region on the chromosome that controls expression of structural genes in an operon.

operon   A unit on the chromosome under the control of an operator.

opportunistic pathogen   An organism that normally is nonpathogenic but is capable of causing disease in a compromised host.

opsonin   A substance that renders antigens more susceptible to phagocytosis.

opsonization   The process by which an antigen is altered in such a manner that it is more readily engulfed by phagocytes.

optochin   Chemical (ethylhydrocupreine hydrochloride) used for the identification of *Streptococcus pneumoniae.*

organotroph   *See* **chemoorganotroph.**

organotropic   Tissue or organ-specific.

osmophile   An organism that requires an environment of high osmolarity for growth.

osmotolerant   Tolerating high osmotic pressures.

outer membrane   The exterior portion of the gram-negative envelope that consists of a phospholipid bilayer interspersed with proteins and lipoproteins.

oxidase test   A test used to detect the presence of cytochrome c oxidase by the oxidation of *N,N*-dimethyl-*p*-phenylenediamine to a purple-blue product, indophenol.

oxidation   A reaction involving loss of electrons.

oxidation-reduction   A reaction involving the transfer of electrons from one molecule to another; also called redox reaction.

oxidation-reduction potential   A measure of the tendency of a substance to donate or accept electrons, as determined relative to the standard hydrogen electrode; also called redox potential.

oxidative phosphorylation   The synthesis of ATP from ADP and inorganic phosphate, coupled with a membrane-associated electron transport chain and the generation of a proton motive force across the membrane.

oxygenic photosynthesis   Photosynthesis with the production of oxygen.

P

palindrome   A double-stranded nucleic acid sequence that exhibits twofold symmetry around a central axis.

pandemic disease   An epidemic disease that affects several countries or major portions of the world.

parasitism   A symbiotic relationship in which one organism benefits and the other is harmed.

passive diffusion   The movement of small molecules across a membrane down a concentration gradient.

Pasteur effect   The inhibiting effect of oxygen upon glucose fermentation.

pasteurization   A process of treating liquids with heat below the boiling point to kill most pathogenic and spoilage microorganisms.

pathogenicity   The ability of an organism to infect and establish disease in a host.

penicillinase-producing *Neisseria gonorrhoeae* (PPNG)   Strains of *N. gonorrhoeae* that produce ß-lactamase (penicillinase), an enzyme that breaks down the ß-lactam ring of penicillin.

pentose phosphate pathway   A sequence of chemical reactions in which glucose-6-phosphate is oxidized to 6-phosphogluconate, which subsequently is decarboxylated to ribulose-5-phosphate; also called hexose monophosphate shunt, phosphogluconate pathway, and Warburg-Dickens pathway.

peptidoglycan   The rigid layer of bacterial cell walls, consisting of *N*-acetylglucosamine, *N*-acetylmuramic acid, and a few amino acids.

periodontal disease   A disease that affects the supporting structure of the tooth (gingiva, cementum, and supporting bone).

peripheral membrane protein   *See* **extrinsic membrane protein.**

periplasm   The region between the plasma membrane and the outer membrane in the cell envelope of gram-negative bacteria.

peritrichous   Having flagella around the entire cell.

permease   A membrane-bound protein carrier involved in the transport of molecules.

pertussis   Whooping cough.

phage   *See* **bacteriophage.**

phagemid   A hybrid vector containing DNA from a filamentous bacteriophage and a plasmid.

phagocyte   A cell that ingests bacteria, foreign particles, and other cells.

phagocytosis   The process of ingestion and digestion by phagocytes of solid substances such as bacteria, foreign particles, and other cells.

phagolysosome   The vacuole that forms from the fusion of a phagosome with a lysosome inside a phagocyte.

phagosome   A phagocytic vesicle formed by the invagination of the plasma membrane of a phagocyte around a bacterium, virus, or other foreign material.

phase-contrast microscope   A type of microscope that amplifies small differences in refractive indices to increase the contrast between the specimen and the surrounding medium.

phenol coefficient   The ratio of the effectiveness of a chemical agent to that of phenol for a test organism.

phenom   A category used to group similar organisms in numerical taxonomy.

phenotype   Observable characteristics of an organism.

phosphogluconate pathway   *See* **pentose phosphate pathway.**

phosphoketolase pathway   A glycolytic pathway in which glucose is converted to pyruvate and ethanol, with xylulose-5-phosphate formed as an intermediate. Xylulose-5-phosphate is cleaved by phosphoketolase to acetyl phosphate and glyceraldehyde-3-phosphate, which then are further broken down to ethanol and pyruvate, respectively.

photoautotroph   An organism that obtains its energy from light and uses carbon dioxide as its sole carbon source.

photoheterotroph   An organism that obtains its energy from light and its carbon from organic compounds.

photolysis   The light-dependent splitting of water.

photon   A discrete packet of electromagnetic energy; also called Einstein.

photooxidation   The light-dependent destructive oxidation of cellular structures or chemical compounds.

photophosphorylation   ATP synthesis using light energy and electron transport.

photoreactivation   The process by which a light-activated enzyme, photolyase, repairs pyrimidine dimers in DNA by cleaving the covalent linkages between the damaged pyrimidines.

photosynthesis The conversion of light energy into chemical energy either with (oxygenic) or without (anoxygenic) the production of oxygen. The chemical energy can be used to fix carbon dioxide into organic compounds.

photosystem An arrangement of chlorophylls and accessory pigments into groups of 250 to 400 molecules that gather electromagnetic energy and transfer excited electrons to a reaction center chlorophyll.

phototaxis The movement of organisms in response to light.

phototroph An organism that obtains its energy from light.

phycobilin *See* **biliprotein.**

phylogeny The study of comparative evolutionary relationships among organisms.

pilus (pl. **pili**) An appendage found on the surface of some procaryotes; pili are involved in conjugation and in attachment; also called fimbria.

plankton A free-floating aquatic community, composed of microscopic cyanobacteria, algae, plants, and/or animals.

plaque A clear area in a confluent lawn of bacteria or tissue culture cells caused by viral infection and cell lysis; a sticky film composed of microorganisms embedded in a polysaccharide/protein matrix on the surface of teeth.

plasma The fluid matrix of blood; plasma contains fibrinogen, antibodies, complement, and other components.

plasma cell An antibody-secreting cell formed from B lymphocytes.

plasma membrane The thin layer surrounding the cell cytoplasm, consisting primarily of phospholipids and proteins, that acts as a semipermeable barrier for the passage of molecules; also called cytoplasmic membrane or cell membrane.

plasmid Extrachromosomal DNA found in the cytoplasm of some procaryotes.

plasmodium (pl. **plasmodia**) Streaming protoplasmic mass, part of the life cycle of slime molds.

plastid A membrane-enclosed organelle found in algae and plants. There are three types of plastids: chloroplasts, chromoplasts, and leucoplasts.

platelet A small blood cell involved in blood coagulation and the transportation of serotonin; also called thrombocyte.

pleomorphic Having more than one distinct form.

pock An opaque lesion caused by a viral infection of the chorioallantoic membranes of embryonated eggs; a pustule on the surface of the skin.

point mutation A mutation involving a single base change in the chromosome.

polymerase chain reaction A laboratory technique to synthesize within a short period of time many copies of a DNA molecule. DNA is amplified by a cyclic repetition of three steps: (1) the original double-stranded DNA molecule is denatured at high temperature, (2) oligonucleotide primers are annealed to the DNA at low temperature, and (3) the primers are extended on the DNA template by a DNA polymerase.

polymorphonuclear leukocyte (PMN) *See* **neutrophil.**

polypeptide An organic compound composed of amino acids linked by peptide bonds, but of lower molecular weight than a protein.

polyribosome Two or more ribosomes joined together in a chain with mRNA.

pomace The skin, pits, and stems of grapes.

positive control A regulatory system in which the presence of a controlling factor turns on protein synthesis.

positive stain A stain in which the specimen is stained and appears dark against a light background.

potable Suitable to drink.

potential energy Stored energy or energy of position.

pour plate An agar plate that has been prepared by adding a dilute microbial suspension and melted agar into a sterile Petri dish.

prevalence rate The occurrence of disease in a population during a specified time period; a type of morbidity rate.

Pribnow box A consensus sequence (TATAAT) in DNA that is a binding site for RNA polymerase and located ten bases before the point of initiation of transcription.

primary metabolite A chemical compound associated with cell growth or function.

primary producer An organism capable of converting carbon dioxide to organic carbon.

prion An infectious particle containing protein but no nucleic acid; also called scrapie-associated fibril (SAF).

procaryote A cell or organism with no nuclear membrane, specialized membrane-enclosed organelles, or the extensive internal structural organization typically found in a eucaryotic cell.

prodromal period The period in the natural history of disease when the first signs and symptoms of disease appear.

profundal zone The zone of deep water in a lake beyond the depth of effective light penetration.

promoter A region on the chromosome, adjacent to the operator, to which RNA polymerase binds.

propagated epidemic disease A disease in which the disease agent is transmitted from one host to another either through direct contact or via a vector.

properdin pathway *See* **alternative complement pathway.**

prophage A temperate bacteriophage that is integrated into the host chromosome.

prospective study An observational epidemiological study that follows a group of people (a cohort) over a period of time to determine the rate at which a particular disease develops in relation to a specified characteristic.

protein An organic compound composed of amino acids linked by peptide bonds.

proton motive force *See* **chemiosmosis.**

protoplast A cell that has lost its cell wall but remains intact in an isotonic environment.

protoplast fusion A recombinant DNA technique in which protoplasts of two species are combined to produce a hybrid organism containing the chromosomes of both cells.

prototroph An organism with no additional requirements for growth.

protozoan (pl. **protozoa**) A motile eucaryotic microorganism that does not have a cell wall or chlorophyll.

pseudopodium (pl. **pseudopodia**) A cytoplasmic extension produced by amoebae during locomotion.

psychrophile An organism with an optimum growth temperature of 10°C to 20°C.

public health   The science of protecting and improving community health through education and organized preventive medicine efforts.

pure culture   A population of cells that arises from one cell.

purine   A cyclic nitrogenous compound (for example, adenine and guanine) that is found in nucleic acids, many coenzymes, and certain antibiotics.

putrefaction   The anaerobic microbial decomposition of meat proteins with the production of hydrogen sulfide, indole, mercaptans, ammonia, amines, and other foul-smelling compounds.

pyoderma   An inflammatory skin infection caused by pus-forming bacteria.

pyogenic   Pus-forming.

pyrimidine   A cyclic nitrogenous compound (for example, cytosine, uracil, or thymine) that is found in nucleic acids, many coenzymes, and certain antibiotics.

pyrogen   A substance that can cause a rise in body temperature.

Q

quellung reaction   The apparent swelling of the bacterial capsule by the interaction of anti-capsular antibody and capsular antigen.

quinone   An electron carrier that is a lipid-soluble substance of low molecular weight.

R

racking   The procedure in wine making in which the wine is drawn from the sediment and added back to the top of the storage vat.

radioimmunoassay (RIA)   A laboratory technique that uses a radioactively labeled antigen or antibody to compete with and measure the concentration of unlabled antigen or antibody in a sample.

reaction center   A chlorophyll in the photosystem that receives electrons from other chlorophylls and accessory pigments and passes these excited electrons on to the transport chain.

reading frame   The arrangement of nucleotides in mRNA so that a particular pattern of codons is established for translation.

reading frame shift   A change in the reading of codons during translation so that an entirely new protein is formed.

reagin   A tissue antibody produced in response to damaged tissue components resulting from infection by treponemes and some other infectious agents.

recalcitrance   The total resistance of a chemical to microbial degradation.

recombinant DNA technology   The techniques used to manipulate genes in living organisms to carry out genetic engineering.

redox potential   *See* **oxidation-reduction potential.**

redox reaction   *See* **oxidation-reduction reaction.**

reduction   A reaction involving the uptake of electrons.

refractile (R) body   A paramecium endosymbiont form that appears as tightly rolled ribbons and is believed to be associated with the toxic particles (paramecin) released by killer paramecia.

replica plating   The technique in which replicate plates are inoculated from a master plate, using a pad of sterile velvet to transfer colonies of microorganisms.

replicative form   A double-stranded nucleic acid that serves as an intermediate form in the replication of single-stranded viral nucleic acids.

replicon   A stretch of DNA that is synthesized from a replication origin.

reporter gene   A gene on a cloning vector that codes for an easily detectable trait.

repression   The process by which enzyme synthesis is inhibited by the presence of a substance, the repressor.

repressor   The regulator gene product that interacts with the operator to control gene expression.

reservoir of infection   An animate or inanimate source of infectious agents.

resident flora   A more or less constant group of organisms that is a part of the normal flora.

resistance plasmid (R plasmid)   A plasmid that carries genes for antibiotic resistance and *tra* genes for conjugative transfer.

resolving power   The ability to distinguish two objects as separate and distinct entities; also called resolution.

respiration   Catabolic reactions producing ATP in which either organic or inorganic compounds are the primary electron donors and either molecular oxygen (aerobic) or another inorganic compound such as nitrate or sulfate (anaerobic) is the terminal electron acceptor.

restricted transduction   *See* **specialized transduction.**

restriction endonuclease   A nuclease that recognizes and, internally in the DNA, cleaves specific sequences known as palindromes.

reticulate body (RB)   A large, noninfectious particle that is involved in intracellular multiplication in the chlamydia developmental cycle.

reticuloendothelial (RE) system   The system of macrophages found in the spleen, thymus, lungs, lymph nodes, bone marrow, and liver; also called mononuclear phagocytic system.

retrospective study   An observational epidemiological study in which comparisons are made between people with a particular disease (cases) and people without the disease (controls) with respect to a specified characteristic.

reverse transcriptase   RNA-dependent DNA polymerase that uses RNA as a template to synthesize DNA.

rhizoid   A rootlike fungal extension that anchors mycelia to the surface.

rhizosphere   The region of soil closely surrounding the roots of a plant.

rhizosphere effect   The presence of increased concentrations of microorganisms around plant roots as a result of organic matter excreted by the roots into the surrounding soil.

rho factor (ρ)   The protein that unwinds the RNA-DNA hybrid in rho-dependent termination of transcription.

ribonucleic acid (RNA)   A type of nucleic acid involved in protein synthesis. Three types of RNA exist in a cell: mRNA, rRNA, and tRNA. RNA contains ribose sugars (in contrast to DNA, which contains deoxyribose sugars).

ribosomal RNA (rRNA)   RNA that is associated with ribosomes.

ribosome   A cell component consisting of two subunits and containing proteins and rRNA. Ribosomes are the sites of protein synthesis.

ribozyme   An RNA molecule that can catalyze a chemical reaction.

RNA primer   A short segment (approximately 10 nucleotides long) of RNA that precedes or primes newly synthesized DNA.

RNA processing   The modification of the RNA transcript before translation.

RNA splicing The excising of introns and rejoining of exons in mRNA after transcription and before translation.

rod A cylindrical bacterial cell.

rolling circle model A mode of DNA replication that begins with the cutting of one parental strand of the double helix; replication begins as the cut strand is rolled off the double helix.

rumen A special organ in ruminants in which cellulose is broken down by microorganisms.

ruminant A plant-eating mammal such as a cow, goat, or sheep that has a special organ called the rumen.

S

salmonellosis A gastrointestinal disease caused by *Salmonella;* also called *Salmonella* gastroenteritis.

saprophyte An organism living on dead or decaying organic mater.

scalded skin syndrome An epidermal disease of infants caused by *Staphylococcus aureus* and characterized by lesions resembling tissue that has been scalded with boiling water.

scanning tunneling microscope A type of microscope in which magnifications of several millionfold are achieved by shuttling electrons between a minute probe and a specimen's surface.

scrapie-associated fibril (SAF) *See* **prion.**

secondary metabolite A product (for example, an antibiotic) of a metabolic pathway that is not associated with primary cellular processes.

secondary response *See* **anamnestic response.**

second law of thermodynamics The law stating that all processes occur in such a manner that there is a total increase in entropy.

selective medium A culture medium that favors the growth of specific microorganisms while inhibiting the growth of undesired microorganisms.

semiconservative replication DNA replication, with each strand serving as a template for the synthesis of a new daughter strand.

sensitive strain (of *Paramecium*) A paramecium that is harmed by toxic particles released by killer strains of *Paramecium*.

sensitivity of a test The ability of a test to correctly identify those people who have a particular disease.

septicemia A systemic disease in which microorganisms actively multiply in circulating blood.

septic tank A type of anaerobic sewage treatment process commonly found in rural areas.

septum (pl. septa) A cross-wall.

serum (pl. sera) The fluid portion of blood remaining after plasma has clotted.

sewage The liquid waste material carried by a system of pipes and other conduits called sewers.

sex factor *See* **F factor.**

sexually transmitted disease (STD) A disease that is transmitted by sexual contact.

shigellosis Gastrointestinal disease caused by *Shigella* species; also called bacillary dysentery.

Shine-Dalgarno sequence A short, purine-rich region preceding the initiation codon on procaryotic mRNA that binds to the 16S rRNA of the 30S ribosomal subunit and serves to thread the mRNA into the ribosome.

shotgun cloning The enzymatic cleavage of an entire donor genome into small DNA fragments and the insertion of these fragments into cloning vectors.

shuttle vector A vector that is used to move DNA between two unrelated organisms.

siderochrome A siderophore that is a hydroxamate of low molecular weight.

siderophore An iron-binding compound that transports iron into the cell.

signature sequence A short, unique nucleotide sequence that is found in the 16S rRNA of certain groups of procaryotes.

silent mutation A mutation that has no effect on the amino acid coded by the affected area of the chromosome.

similarity coefficient (S$_J$) The comparison of positive characteristics among organisms in numerical taxonomy.

single-cell protein (SCP) Dried microbial cells that are used for food or feed.

slime layer *See* **capsule.**

SOS response An inducible repair system that excises and repairs DNA damage.

souring The formation of undesirable odors and tastes in meat resulting from volatile acids produced during microbial metabolism of meat proteins and carbohydrates.

Southern blot A technique by which DNA fragments, separated by electrophoresis, are immobilized on a paper sheet and detected with a labeled nucleic acid probe; named after Edward M. Southern.

specialized transduction The transfer of a specific portion of host DNA by transduction; also called restricted transduction.

species A taxonomic category just below genus level and describing organisms with a similar phenotype.

specificity of a test The ability of a test to correctly identify those people who do not have a particular disease.

specific resistance *See* **acquired resistance.**

spheroplast A cell that has lost a portion of its cell wall but remains intact in an isotonic environment.

spirillum (pl. spirilla) A spiral-shaped bacterial cell.

spontaneous generation A theory that living organisms can arise from nonliving matter.

spontaneous mutation A mutation that occurs naturally, usually as a result of errors in DNA replication or from mistakes during genetic recombination.

sporangiophore A fungal hypha with sporangia at the tip.

sporangiospore An asexual fungal spore formed within a sac called the sporangium.

sporangium (pl. sporangia) A specialized fungal reproductive structure containing sporangiospores; the outer envelope-like covering of an endospore.

spore The resistant resting body formed by some cells in response to unfavorable environmental conditions.

sporogenesis The process of spore formation.

spread plate An agar plate on which a dilute microbial suspension has been spread evenly across the surface to obtain isolated colonies.

staphylokinase A *Staphylococcus* protease that converts plasminogen in human serum to plasmin, thereby dissolving blood clots.

stationary phase (of population growth curve) The period of population growth in which there is no change in cell numbers.

steady state The state of growth in a continuous culture when cell numbers and nutrient levels in the culture vessel become constant.

sterilization   The killing of all viable organisms.

sterol   A type of lipid lacking fatty acids, routinely found in eucaryotic plasma membranes but rarely in procaryotic plasma membranes.

Stickland reaction   The sequence of chemical reactions in clostridia in which pairs of amino acids are fermented, with one amino acid serving as the electron donor and the other serving as the electron acceptor.

storage body   *See* **inclusion body.**

strain   A population of cells that are all derived from a common ancestor and retain the characteristics of the ancestor.

stratosphere   The layer of the atmosphere above the troposphere and extending up to approximately 50 km.

streak plate technique   A method used to obtain isolated colonies by streaking a mixed culture across the surface of an agar plate with an inoculating loop.

streptokinase   A *Streptococcus* protease that converts plasminogen in human serum to plasmin, thereby dissolving blood clots.

streptolysin O   An oxygen-labile hemolysin produced by *Streptococcus pyogenes*.

streptolysin S   An oxygen-stable hemolysin produced by *Streptococcus pyogenes*.

structure analog   A chemical compound that structurally resembles a cellular metabolite and competes with this metabolite in cellular enzymatic reactions.

substrate   The molecule on which an enzyme acts.

substrate-level phosphorylation   The addition of phosphate onto an organic compound.

subunit vaccine   A vaccine consisting of parts of a microorganism.

sulfonamide   An antimicrobial agent that is a structure analog of *p*-aminobenzoic acid.

superantigen   A bacterial protein that stimulates the immune system to produce large numbers of T lymphocytes.

suppressor mutation   A change in DNA that overcomes the effect of an original change in DNA, ultimately causing no alteration in the original gene.

surfactant   A surface-active compound.

swarmer cell   A flagellated daughter cell released from the parent cell.

symbiosis   A relationship between two or more organisms.

synchronous culture   A culture in which all the cells of the population are in the same stage of growth.

synergism   A symbiotic relationship in which different species of microorganisms living together benefit one another and grow better together than separately.

syngamy   Sexual reproduction involving union of two gametes to form a zygote.

synthetic drug   A man-made chemical used internally to inhibit or kill microorganisms.

synthetic (chemically defined) medium   A growth medium, the exact chemical composition of which is known.

syntrophism   *See* **cross-feeding.**

T

taxonomy   The systematic categorization of organisms into a coherent scheme.

TCA cycle   The sequence of chemical reactions in which pyruvate is oxidized to carbon dioxide and reduced coenzymes are produced; also called Krebs cycle and citric acid cycle.

T cytotoxic (T$_C$) cell   A CD8-containing T lymphocyte that interacts with and destroys cells containing antigens on their surfaces.

teichoic acid   An acidic polysaccharide of repeating subunits of glycerol or ribitol, joined by phosphodiester linkages, found in the cell wall and plasma membrane.

temperate phage   A bacteriophage that can become integrated into the host chromosome instead of immediately lysing the host.

termination codon   One of three codons (UAA, UAG, and UGA) that does not code for a specific amino acid and therefore signals the termination of protein synthesis; also called nonsense codon.

tests in parallel   Tests performed in which any test yielding a positive result is sufficient evidence for a disease state.

tests in series   Tests performed in which all tests must yield positive results to confirm a disease state.

thallus (pl. thalli)   The plantlike structure of a fungus or alga.

T helper (T$_H$) cell   A CD4-containing T lymphocyte that stimulates B lymphocytes to produce antibodies and is involved in cell-mediated immunity.

thermal death time (TDT)   The shortest time required to kill all of the microorganisms in a suspension at a specific temperature.

thermodynamics   The study of energy transformations.

thermophile   An organism with an optimum growth temperature of 40°C or higher.

thrombocyte   *See* **platelet.**

thrush   A *Candida*-induced inflammation of the oral mucosa.

thylakoid   Membranous vesicles containing photosynthetic pigments.

T lymphocyte   A lymphocyte that is differentiated in the thymus and is important in humoral and cell-mediated immunity; also called T cell.

top-fermenting yeast   Yeast that rises to the surface of the beer during fermentation to produce beer that is uniform in turbidity and high in alcohol content.

toxoid   A toxin that has been treated to destroy its toxic activity but not its antigenic properties.

transcription   The synthesis of RNA from DNA.

transducer   *See* **methyl-accepting chemotaxis protein (MCP).**

transduction   The exchange of DNA by use of a bacteriophage.

transfection   The uptake of naked bacteriophage DNA through modified cell envelopes of competent bacteria; the insertion of cloning vectors into mammalian cells by endocytosis.

transferrin   An iron-binding protein found in the blood of vertebrates.

transfer RNA (tRNA)   RNA that carries amino acids to ribosomes during protein synthesis.

transformation   The insertion of naked extracellular DNA from a donor cell into a competent recipient cell.

transgenic animal or plant   An animal or plant that has obtained new genetic information from the insertion of foreign DNA.

transient flora   Organisms that inhabit a host for only a short period of time as part of the normal flora.

translation   The synthesis of protein from mRNA.

transmission of disease   The mechanism by which infectious agents are transferred from one host or object to a susceptible host.

transpeptidation   The penicillin-sensitive step in cell wall synthesis involving cross-linking of adjacent tetrapeptides.

transposon   A gene-containing element that can be moved among nucleic acids.

trickling filter   An aerobic sewage treatment process in which sewage is sprayed onto a bed of crushed rocks.

troposphere   The layer of the atmosphere closest to the earth and extending to an altitude of 8 to 12 km above the surface.

tube dilution method   A method in which tubes containing serial dilutions of an antimicrobial agent are inoculated with a test microorganism to determine antimicrobial susceptibility.

tubercle   A small nodular lung lesion formed in tuberculosis.

tuberculin   The *Mycobacterium tuberculosis* antigen used in the tuberculosis skin test.

tumor necrosis factor (TNF)   A cytokine that activates macrophages, granulocytes, and cytotoxic cells.

turbidostat   A continuous culture device in which the flow rate of fresh medium is automatically adjusted to maintain turbidity levels as measured by a light-sensing device.

tyndallization   A sterilization process in which material is heated to 100°C for 30 minutes on three consecutive days; also called fractional sterilization.

type I hypersensitive reaction   *See* **anaphylactic hypersensitive reaction.**

type II allergic reaction   *See* **cytotoxic hypersensitive reaction.**

U

uncoupler   A chemical agent that causes the membrane to be leaky to protons, thereby uncoupling ATP synthesis from proton movement across the membrane.

V

vaccine   A material administered to a subject to induce artificially acquired active immunity.

vacuole   A membrane-bound cell component.

valence   The number of antibody-binding sites on an antigen.

vector   A living organism that is an intermediary in the transfer of infectious agents; a vehicle for the introduction of a target gene into a cell during gene cloning.

vegetative cell   A cell that is engaged in growth, metabolism, and reproduction.

vehicle   An inanimate object capable of transmitting infectious agents.

viable count   A count of visible colonies growing on agar medium in a Petri dish.

viroid   Infectious RNA, devoid of a protein coat and several thousand times smaller than a virus.

virulence   A quantitative measure of pathogenicity; the capacity of a microorganism to overcome the body defenses of the host.

virulent phage   A bacteriophage that is capable of lytic infections.

virus   A submicroscopic filterable agent consisting of either RNA or DNA surrounded by a protein coat.

W

Warburg-Dickens pathway   *See* **pentose phosphate pathway.**

water activity (a_w)   The available water in an organism's surroundings; water activity is determined by measuring the relative humidity in the air space in a substance's environment.

western blot   A technique by which proteins are separated and immobilized on a paper sheet and detected by reaction with a labeled antibody.

white blood cell   *See* **leukocyte.**

wild type   The unaltered original genotype.

wobble   The ability of the third base on the 5' end of tRNA anticodon to pair with two or more bases at the 3' end of the mRNA codon.

wort   The liquid portion of malt, consisting of a dilute solution of sugars that is fermented to form beer.

Y

yeast   A unicellular eucaryotic fungal cell.

yeast artificial chromosome (YAC)   A vector that consists of a linear plasmid that is comprised partially of a yeast chromosome.

Z

zoonosis (pl. **zoonoses**)   An animal disease transmissible to humans.

zoospore   A flagellated spore formed by algae and some fungi during reproduction.

zygospore   A resistant fungal spore arising from a zygote.

zygote   A diploid cell resulting from the union of two haploid gametes.

CREDITS

Photos

Chapter 1 **Opener:** © K.G. Murti/Visuals Unlimited; **1.1a:** © David M. Phillips/Visuals Unlimited; **1.1b:** © Science VU/Visuals Unlimited; **1.1c:** © David M. Phillips/Visuals Unlimited; **1.1d:** © K.G. Murti/Visuals Unlimited; **1.1e:** © Dr. Anne Smith/SPL/Photo Researchers, Inc.; **1.2:** Courtesy of Bausch & Lomb; **1.3:** © Volker Steger/Peter Arnold, Inc.; **1.5, 1.6:** The Bettmann Archive; **1.8:** © Historical Pictures/Stock Montage; **1.9:** © Science-VU/Visuals Unlimited; **1.10:** © Omikron/Science Source/Photo Researchers, Inc.; **1.11:** © Oliver Meckes/Photo Researchers, Inc.; **1.12:** NASA; **1.13:** © NIH/ Custom Medical Stock Photos; **1.14a:** © Vanessa Vick/Photo Researchers, Inc.; **1.14b:** © Ken Graham/Tony Stone Images

Chapter 2 **Opener:** Courtesy of Dr. Gene Michaels, University of Georgia; **2.1:** Esther R. Angert & Norman R. Pace; **2.2:** Courtesy of Lilly Research Laboratories, Indianapolis, Indiana. Photo by Dr. Daniel C. Williams; **2.4a–c:** © David M. Phillips/Visuals Unlimited; **2.4d:** © Paul W. Johnson/Biological Photo Service; **2.4e, 2.5:** © David M. Phillips/Visuals Unlimited; **2.6:** Courtesy of Nikon; **2.11:** © CDC/Biological Photo Service; **2.12:** © ASM/Science Source/Photo Researchers, Inc.; **Table 2.1(both):** © G.W. Willis/Biological Photo Service; **2.14a–d:** © David M. Phillips/Visuals Unlimited; **2.16a, b:** Courtesy of Carl Zeiss, Inc.; **2.17a:** © J.J. Cardamore/ Biological Photo Service; **2.17b:** Courtesy of H. Farzadegan and I.L. Roth, University of Georgia; **2.17c:** Courtesy of Robley C. Williams, Virus Laboratory and Department of Molecular Biology, University of California, Berkeley; **2.17d:** Courtesy of H. Farzadegan and I.L. Roth, University of

Georgia; **2.17e:** Courtesy of Henry C. Aldrich, University of Florida; **2.18:** © Driscoll, Youngquist, and Baldeschwieler, Caltech/SPL/Photo Researchers, Inc.; **2.19a, b:** From A. Ciancio, "Microscopy and Analysis," *Atomic Force Microscopy of Microorganisms,* 42:25, 1994. © Microscopy and Analysis; **2.19c:** © Michael Abramsan/Time Magazine; **2P.2:** Roger M. Cole, USPHS

Chapter 3 **Opener:** © CNRI/SPL/Photo Researchers, Inc.; **3.7a, b:** Courtesy of David L. Balkwill, Florida State University; **3.10a:** From John G. Holt, *Shorter Bergey's Manual of Determinitive Bacteriology,* 8/e, Plate 1.3, fig. 1 © Williams & Wilkins; **3.10b:** From John G. Holt, *Shorter Bergey's Manual of Determinitive Bacteriology,* 8/e, Plate 12.2, fig. 9 © Williams & Wilkins; **3.15b:** © S.C. Holt/Biological Photo Service; **3.15f:** © John J. Cardamone, Jr./Biological Photo Service; **3.16:** © David M. Phillips/Visuals Unlimited; **3.18a:** © S.C. Holt/Biological Photo Service; **3.19:** © Raymond B. Otero/Visuals Unlimited; **3.20:** © Fred Hassler/Visuals Unlimited; **3.22a–c:** CDC; **3.25a:** D. Balkwill and D. Maratea; **3.25b:** Courtesy of Y. Gorby; **3.26a:** © S.C. Holt/Biological Photo Service; **3.27:** CDC; **3.28:** Courtesy of Daniel D. Jones, The University of Alabama at Birmingham. Reprinted with Permission of the *Canadian Journal of Microbiology;* **3.29a:** © T.J. Beveridge/Biological Photo Service; **3.29b:** Courtesy of H. Farzadegan and I.L. Roth, University of Georgia; **3.32a:** From Moberly, Shafa and Gerhardt, *Journal of Bacteriology* 92:223, 1966. © American Society for Microbiology; **3.33a:** © Jack Bostrack/Visuals Unlimited; **3.33b:** © John Cunningham/Visuals Unlimited; **3.33c:** © George Wilder/Visuals Unlimited; **3.37b:** © William Dentler/Biological Photo Service

Chapter 4 **Opener:** Courtesy of Remel, Lenexa, KS; **4.1:** © WHOI/Carl Wirsen/Visuals Unlimited; **4.10:** © Carroll Weiss/Camera MD Studios; **4.11a:** Courtesy of Remel, Lenexa, KS; **4.11b:** © Christine L. Chase/Visuals Unlimited; **4.13:** Courtesy of Remel, Lenexa, KS

Chapter 5 **Opener:** © CNRI/SPL/Photo Researchers, Inc.; **5.1:** © John J. Cardamone, Jr./Biological Photo Service; **5.6:** © Jack Bostrack/CBR Images; **5.12a:** Courtesy of Yellow Springs Instrument Co.; **5.28:** © Carroll Weiss/Camera MD Studios; **5.29:** Courtesy of bioMérieux Vitek, Inc.; **5.30:** Courtesy of Becton Dickinson Company

Chapter 6 **Opener:** © Leonard Lessin/Peter Arnold, Inc.; **6.26:** © T.W. French/Visuals Unlimited

Chapter 7 **Opener:** © M.I. Walker/Photo Researchers, Inc.; **7.5:** © S.C. Holt/Biological Photo Service; **7.6a:** From John G. Holt, *Shorter Bergey's Manual of Determinitive Bacteriology,* 8/e, Plate 1.4 © Williams & Wilkins; **7.6b:** From John G. Holt, *Shorter Bergey's Manual of Determinitive Bacteriology,* 8/e, Plate 1.3, fig. 1 © Williams & Wilkins; **7.6c:** From John G. Holt, *Shorter Bergey's Manual of Determinitive Bacteriology,* 8/e, Plate 1.2, fig. 1 © Williams & Wilkins; **7.11a:** From John G. Holt, *Shorter Bergey's Manual of Determinitive Bacteriology,* 8/e, Plate 1.1, fig. 4 © Williams & Wilkins; **7.11b:** From John G. Holt, *Shorter Bergey's Manual of Determinitive Bacteriology,* 8/e, Plate 1.1, fig. 5 © Williams & Wilkins; **7.11c:** From John G. Holt, *Shorter Bergey's Manual of Determinitive Bacteriology,* 8/e, Plate 1.1, fig. 7 © Williams & Wilkins; **7.22:** Courtesy of Dr. Elmer W.

Illustrations

INDEX

Page numbers followed by a *t* indicate tables; page numbers followed by an *f* refer to figures; page numbers in boldface refer to major discussions.

Abbe, Ernst, 23
Abbe condenser, 24
Aberration
    chromatic, 22–23
    spherical, 22, 23f
Abomasum, 428f, 429
Abortive transduction, **258**, 259f
Abortive viral infection, **408**
Abscess, 325, 503t, 504
Absorbance, 113
Absorption spectra
    of accessory pigments, 181f
    of bacteriochlorophylls, 180f
    of chlorophylls, 180f
Acanthopeltis, 355
Accessory pigment, **179–80,** 181f, 184
Acellular slime mold, 371–72, 371f
Acetaldehyde, 156f, 159f
Acetate, **157**, 161f, 192f
    from rumen fermentation, 427–29, 428f
    Stickland reaction, 196, 196f
Acetivibrio, 325t
Acetoacetate, 160f
Acetoacetyl-ACP, 202
Acetoacetyl-CoA, 160, 160f
Acetobacteraceae, 320t
Acetogen, 598
Acetoin, 161
Acetone, **157–60**, 160f
Acetyl-ACP, 202
Acetyl-CoA, 160–63f, 162, 192f, 200–201f, 201–2, 573, 573f
    energy from, 148t
    Stickland reaction, 196, 196f
    structure of, 148t
Acetyl-CoA carboxylase, 202
N-Acetylglucosamine (NAG), 46, 52, 54f, 339, 576

N-Acetylmuramic acid (NAM), 46, 52, 54f, 56
N-Acetylmuramyl-L-alanine amidase, 117, 504
N-Acetylneuraminic acid, 386
Acetyl phosphate, 148, 154, 156f, 158, 159–60f, 192f, 196, 196f
O-Acetylserine, 83
N-Acetyltalosaminuronic acid, 339
Acholeplasma blastoclostricum, 90f
Achromatic lens, 22
Achromobacter, 578, 587, 601t, 614–15t, 615–16
    A. aceris, 623t
Acid, 651–52
Acidaminococcus, 326
Acid drainage water, 573
Acid-fast microbes, 331, 331f, 342, 511
Acidic dye, 25
Acid mine drainage, 171–72, 172f, 326, 583, 584f
Acid mine waste, 580
Acidophile, 91
Acidophilus milk, 154, 331t, 609t, 610
Acne, 525, 525t
ACP. *See* Acyl carrier protein
Acquired immune deficiency syndrome. *See* AIDS
Acquired immunity, 454
Acrasiomycota, 366t, **369–73**
Acremoniella, 601t
Acridine, 248, 248t
Acrosis, 366t
Actinomadura, 342t
Actinomyces, 440t, 514, 587, 601t, 623t
    A. israelii, 514t
    A. naeslundii, 514t
    A. viscosus, 514t
Actinomycetes, **341–43**, 342t, 343f
    classification of, 640–41
    with multilocular sporangia, 342t, 641
    nocardioform, 640–41
Actinomycin D, 125t, 129

Actinoplanes, 342t, 343
Actinoplanetes, 342t, 641
Actinopolyspora, 342t
Actinorhizal plant, 421
Actinosynnema, 342t
Activated sludge process, 338, 595–98, 597f
Activation energy, 142–43, 142f
Active immunity
    artificially acquired, **477–78**
    naturally acquired, **477**
Active site, 143
Active transport, 85, **86–88**, 87f
Acute glomerulonephritis, 499, 499t, 503t
Acute inflammation, 458, 460f
Acute lymphatic leukemia, 409
Acute necrotizing ulcerative gingivitis, 515
Acute period of disease, 446, 446f
Acyclovir (Zovirax), 132, 132t, 406, 540
Acyl carrier protein (ACP), 202
Adansonian taxonomy. *See* Numerical taxonomy
ADCC. *See* Antibody-dependent cell-mediated cytotoxicity
Adenine, 44, 46f
Adenoids, 459f
Adenosine deaminase deficiency, 292
Adenosine-3'-phosphate-5'-phosphosulfate, 83, 83f
Adenosine-5'-phosphosulfate (APS), 83, 83f
S-Adenosylmethionine, 63
Adenoviridae, 394t
Adenovirus, 385f, 386–87, 396f, 408, 500t
Adenylate cyclase, 238, 518, 523
Adhesion of bacteria, 448, 449f
ADP, 148t
ADP-glucose, 194
ADP-glucose pyrophosphorylase, 194
Aerial mycelium, 343, 343f, 361
Aerobe, 93

Aerobic, gram-negative rods and cocci, 316t, **319–21,** 320t, 321f
Aerobic chemolithotrophic bacteria, 332t, **336–37,** 336f, 639
Aerobic/microaerophilic, motile, helical/vibrioid gram-negative bacteria, 316t, **318–19**, 319f, 635
Aerobic respiration, 150, **168**
Aeromonas, 322t
Aerosol transmission, 444
    microbes in space, 10
Aerotaxis, 65
Aerotolerance, 93
Aflatoxin, **368**
African sleeping sickness, 132, 376, 377t
Agammaglobulinemia, **478**
Agar, 94–95, **98,** 190t, 191, 355, 359
Agarase, 190t
Agarose gel electrophoresis, 274, 275f
Age of patient, morbidity and mortality rates and, 556, 557f
Agglutination, of *S. mutans* by dextran, 462, 462–63t
Agglutination reaction, **486–87**
Agmatine, 198, 199f
Agmatine ureohydrolase, 198, 198f
Agmenellum, 50f
Agranulocytes, 457t, **458**
Agriculture
    applications of recombinant DNA technology in, **297**
    microorganisms in, 11–12
Agrobacterium, 581
    A. tumefaciens, 297, 299f
AIDS, 409, 446, 533t, **536–40,** 536–39f. *See also* Human immunodeficiency virus
    cases in United States, 540, 541f

diagnosis of, 537–40
ELISA for, 490, 540
eucaryotic microbe infections in, 376
first cases identified in United States, **492–93,** 492f, 493t
morbidity and mortality rate of, 554t
opportunistic infections in, 540
thrush in, 515
treatment of, 540
tuberculosis and, 510, 540
western blot for, 490, 540
AIDS vaccine, 478, 540
Air, microbes and, **601,** 601t, 604
Airborne transmission, 444, 601
Ajellomyces, 361
   A. capsulatus, 360t, 361, 448, 500t, 512
   A. dermatitidis, 360t, 500t
Akinete, 335, 335f
Alanine
   in prodigiosin synthesis, 204, 205f
   in Stickland reaction, 196, 196f
   structure of, 43t
   synthesis of, 197t
D-Alanine, 52, 54, 54f, 56
D-Alanine carboxypeptidase, 56
Alanine dehydrogenase, 205f
Alcaligenes, 171t, 172, 578, 581, 608, 614t
   A. eutrophus, 154t
Alcohol(s), as disinfectants and antiseptics, 122t, **123**
Alcoholic beverages, 368, **617–19**
Alcoholic fermentation, 157, 157f
Aldolase, 152
Alfalfa crown wart, 360t, 365
Alfalfa mosaic virus, 392t
Algae, 178, **350–59**
   cell walls of, 352, 354t
   chloroplasts of, 352
   classification of, **352–59,** 353t
   economic importance of, **355**
   enumeration of, 113
   flagella of, 354t
   in freshwater habitats, 587–88
   in lichens, **423,** 424f
   in marine habitats, 586
   morphology and structure of, **350–52,** 351f
   multicellular, **357–59**
   psychrophilic, 90
   reproduction in, **352**
   in soil, 582

as source of single-cell protein, 632, 632t
unicellular, **352–59**
Algal bloom, 390, 586, 588f, 595
Algicide, 122
Algin, 354t, 355
Alkaline protease, commercial production of, 621
Alkalophile, 92
   extreme, 93
Alkylating agents
   as disinfectants and antiseptics, 122t, **124**
   as mutagens, 247–48
Allantion, 582
Allergen, 479
Allergy, 469, 480
   diagnosis of, 480, 480f
Allolactose, 233–34, 234f
Allosteric enzyme, 143, 144f, 145
Almavirus, 392t
Almond leaf curl, 360t
α-helix, 44, 45f
Alphavirus, 445t
Alternaria, 601t
Alternation of generations, 352, 353f
*Alu*I, 273t
Alveolar macrophages, 498
Amanita, 366t
   A. muscaria, 369, 369f
   A. virosa, 369
α-Amanitin, 228
Amantadine, 132, 132t
Amber, 330
Ambrosia beetle, 427, 427f
Amebic dysentery, 132, 376, 377t, 445t, 524
Ames, Bruce, 251
Ames test, **251,** 251f
Amikacin, 125t, 128, 128t
Amino acid, 42
   codons for, 214t
   commercial production of, **625–26**
   content of cells, 81t
   degradation of, 194–97
   required, 84
   structure of, 43t
   synthesis of, **197,** 197t
   transport of, 88
Aminoacyl-tRNA, 232
Aminoacyl-tRNA synthetase, 226
p-Aminobenzoic acid, 85t, 129, 129f
Aminoglycoside, 128, 128f, 128t
2-Aminopurine, 246–47, 248t
Ammonia
   as energy source, 80
   as nitrogen source, 82, **82,** 82f

from organic compounds, **577–79**
   oxidation to nitrate, **578**
Ammonia-oxidizing bacteria, 53f, 171t, **172,** 336, 573–74
Ammonification, 578–79
Amobarbital (Amytal), 168t
Amoeba, 375t, 376
   endosymbionts of, 427
Amoebidium, 366t
Amoebobacter, 333t
Amphibolic pathway, 149, 194
Amphipathic compound, 651
Amphotericin B, 125t, 129–32
   commercial production of, 629t
   mechanism of action of, 126, 629t
   structure of, 127f
Ampicillin, 126f
Ampullariella, 342t
Amylase, 47, 190, 190t
   industrial uses of, **623–24,** 624t
Amytal. *See* Amobarbital
Anabaena
   nitrogen fixation by, **418–23,** 419–22f, 421t, 576
   A. azollae, 421, 422f
Anabolism, 149
Anaerobe, 93
   facultative, 93
   obligate, 93
Anaerobic, gram-negative cocci, 316t, **326,** 636
Anaerobic, gram-negative, straight, curved, and helical rods, 316t, **325,** 325t, 326f, 636
Anaerobic respiration, 150, **168–70,** 168t, 171, 326, 570f
Anaerobic sludge, 575
Anaerobic sludge digester, 595, **598,** 599f
Anaerobiospirillum, 325t
Anaerovibrio, 325t
Anamnestic response, 477
Anaphylactic reaction, 479, 479t
   localized, 479–80
   systemic, 479–80
Anaphylactic shock, 480
Anaplerotic reaction, 194
Ancalochloris, 333t
Animalcules, 2, 13
Animal feed, antibiotics in, 131, 136, 616
Animalia, 306–7, 307t
Animal reservoir, 443
Animal tissue culture
   primary culture, 397

secondary culture, 397
   virus propagation in, 395–97
Animal virus, 386, 396f
   adsorption and penetration of host cells, 400
   classification of, **393–95,** 394–95t
   enveloped, 385f
   propagation of, 395–97
   release from host cell, 405
Anion, 649
Annular diaphragm, 28, 29f, 30
Anoxygenic photosynthesis, 184, 573
Anoxygenic phototrophic bacteria, **332–33,** 332–33t, 333f, 638
Antagonism, microbial, **440–41**
Antenna pigment, 184
Antheridia, 367f, 369, 370f
Anthrax, 5, 7t, 328, 443–44, 542f, 556
Antibiotic(s), 122, 125–32, 125t
   commercial production of, **629–30,** 629t
   discovery of, 14
   indiscriminate use of, 131, 136, 534, 616
   production of, 262, 343
   recombinant DNA techniques, **630**
Antibiotic resistance, **130–31,** 130–31f, **261–62,** 268, 464
Antibody, 8, 456, **468–71.** *See also* Immunoglobulin
   diversity of, **471–73,** 472f
   induction of, **468**
   monoclonal, **475**
   polyclonal, 475
   production of, **468**
Antibody-antigen reaction, **468,** 469f
   in vitro, **485–90**
Antibody-dependent cell-mediated cytotoxicity (ADCC), 484
Antibody-mediated hypersensitivity, 479, 479t
Antibody titer, 477, 477f
Anticodon, 225f, 226
Antifungal agent, **129–32**
Antigen, 8, 456, 468
   definition of, 468
   immunogenicity of, 468
   induction of antibodies, **468**
   reaction with antibodies, **468,** 469f
   valence of, 468

Antigenic determinant, 468, 469f
Antihistamine, 480
Antilymphocyte globulin, 484–85
Antimicrobial agents, **122–33**
Antimicrobial substance,
        produced by body, **456**
Antimicrobial susceptibility
        testing, **132–33,** 133f
    automated methods of,
        134, 134f
    disk diffusion method of, 132,
        133, 133f
    tube dilution method of, 132,
        133, 133f
Antimycin A, 168, 168t
Antiprotozoan agent, **129–32**
Antisense DNA, 293–94
Antisense RNA, **293–94,** 294f
Antiseptic, **122–25,** 122t, 455
Antiserum, 469
Anti-streptolysin O titer, 499
Antitoxin, 469
Antiviral agent, **132,** 132t,
        406, 407f
Aphid, endosymbionts of, 427
Apicomplexa, 375t, **376–77,** 377t
Aplanospore, 352
Apochromatic objective, 23
Appendaged bacteria. *See*
        Budding and/or
        appendaged bacteria
Appendix, 459f
Appert, Nicholas, 616
AP-phosphosulfate reductase, 83f
Apple rust, 360t
APS. *See* Adenosine-5'-
        phosphosulfate
AP site, 249
Aquaspirillum, 319, 337
    A. magnetotacticum, 65, 66f
Ara-A. *See* Vidarabine
Arabinose, 192f
Arber, Werner, 9, 272
Arbovirus, 444t
Archaea, 18, 42, 102, 307, 307t,
        332t, **339–41**
    cell walls of, 54, 102, 339–40
    classification of, 640
    distinguishing characteristics
        of, 313t
    groups in, **340–41**
    habitats of, 339, 339f
    membrane lipids of, 51, 51f
    morphology and physiology
        of, **339–40**
    phylogenetic tree, 313f
    plasma membrane of, 340
    RNA polymerase of, 224,
        312, 314f
    universal phylogenetic tree,
        311–12, 312f

Archaeoglobus, 313f, 340, 340t
Arenaviridae, 394t
Argentine hemorrhagic fever, 543t
Arginine
    decarboxylation of, 198, 199f
    structure of, 43t
    synthesis of, 197t
Arginine decarboxylase, 199f
Arginine dihydrolase, 198, 199f
Arrhenius, Svante, 89
Arrhenius plot, 89, 90f
Arsenic, 143
Arthrobacter, 50f, 154t, 331,
        578, 581
    A. luteus, 273t
Arthrobotys dactyloides, 585t
Arthroderma, 360t
Arthropod-borne disease, 445,
        445t, 529–30, 529t,
        555, 555f
Arthrospore, 364
Arthus, Nicholas M., 480
Arthus reaction, 480–81
Artificially acquired immunity,
        477–78
    active, **477–78**
    passive, **477**
Ascocarp, 366
Ascogonium, 367f
Ascomycete, 366, 367f
Ascomycota, 360t, **365–69,** 366t
Ascospore, 366, 367f
Ascus, 366, 367f
Aseptic technique, **98,** 99f
Ashbya gossypii, 627t, 628
L-Asparaginase, 624
Asparagine, 43t, 197t
Aspartate
    in purine synthesis, 202, 203f
    in pyrimidine synthesis,
        202, 203f
    structure of, 43t
    synthesis of, 197t
Aspartate-4-decarboxylase, 623t
Aspergillus, 364f, 366t, 575,
        578, 587
    aflatoxin and, **368**
    preparation of fermented
        foods, 617t
    in soil, 582
    in troposphere, 601t
    water requirement for
        growth, 92t
    A. flavus, 3f, 368
    A. niger, 361f, 623–25
    A. oryzae, 617t, 622–23
    A. terreus, 626
Assimilatory nitrate reduction,
        82, 170, 578
Assimilatory sulfate reduction,
        83, 83f, 170

Asthma, 480
Athlete's foot, 360, 360t
ATM. *See* Atomic force
        microscope
Atom, 647–48, 648f
Atomic force microscope (ATM),
        34, 35f
Atomic nucleus, 647
Atomic number, 647, 647t
Atomic weight, 647, 647t
ATP, 148
    energy from, 148–49,
        148t, 149f
    inhibition of
        phosphofructokinase
        by, 152
    structure of, 139f, 148–49,
        148t, 149f
    synthesis of
        in aerobic respiration, **168**
        in anaerobic respiration,
            168–70
        in β-oxidation, 200f, 202
        in electron transport
            chain, 165–68, 167f,
            175f
        in Embden-Meyerhof
            pathway, 150–52, 151f,
            175f
        in Entner-Doudoroff
            pathway, 153, 153f
        in fermentation, **168**
        in homofermentative vs.
            heterofermentative
            microorganisms,
            158–59
    mechanisms of, **149**
    by oxidative
        phosphorylation, 149,
        161–71
    in phosphoketolase
        pathway, 154
    by photophosphorylation.
        See
        Photophosphorylation
    in photosynthesis, 182–88
    in Stickland reaction,
        196, 196f
    by substrate-level
        phosphorylation, 149,
        150–61, 150f
    in tricarboxylic acid cycle,
        162, 163f, 175f
    use in carbon dioxide fixation,
        189, 189f
    use in nitrogen fixation,
        422–23, 422f
    yield from glucose oxidation,
        variations in, 170f, **171**
ATPase, 166, 167f
Atraclyloside, 168t

Attenuation, **236,** 237f
Attenuator, 236, 237f
Attractant, 63–65, 64f
Autoclaving, **119,** 120f, 122
Autogamy, 373
AutoMicrobic System, 134, 134f
Autoradiography, 220, 276,
        276–77f
Autotroph, **81,** 81t, 174, 188
Auxotrophic mutant, 249–50
Avery, Oswald, 216
Avocado sun blotch, 389
Axial filament. *See* Flagella,
        periplasmic
Azidothymidine (AZT), 132,
        132t, 540
Azithromycin, 129
Azomonas, 577
Azorhizobium, 418
Azospirillum, 435, 577
    A. lipoferum, 421t
Azotobacter, 83, 320, 577
    A. chroococcum, 154t, 422
    A. paspali, 421t
Azotobacteraceae, 320t
AZT. *See* Azidothymidine

Babès-Ernst body, 66, 507,
        508f, 509
Bacillary dysentery, 322, 322t,
        451t, 516t, **519–20**
Bacille Calmette-Guérin
        (BCG), 511
Bacillus
    anaerobic respiration in,
        168–70
    chitin degradation by, 576
    denitrification by, 578
    endospores of, 68, 68f
    endospores in fossil bee,
        330, 330f
    in freshwater habitats, 587
    glucose metabolism
        in, 154t
    in meat spoilage, 614t
    nitrogen fixation by, 577
    peptidoglycan of, 56
    poly-β-hydroxybutyrates
        of, 66
    in soil, 581
    in troposphere, 601t
    water requirement for
        growth, 92t
    B. alcalophilus, 92
    B. amyloliquifaciens H, 273t
    B. anthracis, 7t, 328, 444, 450
        endospores of, 68f, 70f
        Gram stain of, 25f
    B. brevis, 14
    B. cereus, 68f
    B. coagulans, 90f

B. fastidiosus, 58f
B. globisporus, 90f
B. lentimorbus, 585, 585t
B. licheniformis, 621, 623
B. megaterium, 63f, 75, 75f, 75t
B. polymyxa, 126, 629t
B. popilliae, 585, 585t
B. sphaericus, 330
B. stearothermophilus, 90f, 617t
B. subtilis, 19f, 21f, 105f, 106t, 621–23, 629t
   endospores of, 71f, 75, 75f, 75t, 329f
   as host for cloning, 285
   sporulation in, 226
   transformation in, 253
B. thermoacidurans, 617t
B. thuringiensis, 297, 585, 585t, 604
Bacitracin, 125, 125t, 629t
Bacitracin disk test, 500, 501f
Backbone fever. *See* Dengue fever
Back mutation, 246
Bacteremia, 505
Bacteria
   in freshwater habitats, 587–88
   in marine habitats, 586
   in microbial leaching, **582–83**
   in soil, 581–82
Bacteria (domain), 18, 307, 307t
   distinguishing characteristics of, 313t
   membrane lipids of, 51, 51f
   phylogenetic tree of, 314, 315f
   RNA polymerase of, 224
   universal phylogenetic tree, 311–12, 312f
Bactericide, 122
Bacteriochlorophyll, **179,** 179–80f, 332
Bacteriochlorophyll a, 179, 179–80f, 332
Bacteriochlorophyll b, 179, 332
Bacteriochlorophyll c, 179, 332
Bacteriochlorophyll d, 179, 332
Bacteriochlorophyll e, 179, 332
Bacteriochlorophyll P870, 186, 186–87f
Bacteriocin, 262
Bacteriocinogenic plasmid, **262**
Bacteriophaeophytin, 186f
Bacteriophage, 384
   adsorption to host cell, 399, 399f
   assembly of, 403
   classification of, **387–90**
   as cloning vectors, 274, **281–85,** 283–84f, 287
   defective, 255–56

discovery of, 9
electron micrograph of, 9f
families of, 391t
helper, 258, 411
lysogenic cycle of, 255, **410–11,** 410f
lytic cycle of, 255, 256f, 410–11, 410f
nucleic acid entry into host cell, **400,** 401f
one-step growth curve, 404, 404f
propagation of, 398, 398f
release from host cell, 405
in soil, 582
temperate, 255–58, 257f
transducing, 255. *See also* Transduction
Bacteriophage β, 258
Bacteriophage ε, 258
Bacteriophage f2, 391t
Bacteriophage fd, 387, 390, 391t
Bacteriophage G4, 391t
Bacteriophage lambda, 391t
   Charon phages, 283
   as cloning vector, 281–85, 283–84f
   λ*dg,* 258
   λgt11, 283, 287
   lysogenic cycle of, 410–11
   lytic cycle of, 410–11
   transducing, 258
Bacteriophage M12, 391t
Bacteriophage M13, 279, 283, 385f, 387, 391t
Bacteriophage MS2, 391t
Bacteriophage MVL2, 391t
Bacteriophage P1, 390f
Bacteriophage P2, 391t
Bacteriophage P22, 254, 254f
Bacteriophage φ6, 391t
Bacteriophage φX174, 387, 390, 391t, 402–3
Bacteriophage PM2, 391t
Bacteriophage Qβ, 391t
Bacteriophage R17, 391t
Bacteriophage SPO1, 390
Bacteriophage SPO2, 390
Bacteriophage T1, 387, 391t
Bacteriophage T2, 385f, 387, 390, 390f, 391t, 400, 401f
Bacteriophage T3, 387
Bacteriophage T4, 245f, 387, 390, 391t, 399f, 401–3
Bacteriophage T5, 387, 391t
Bacteriophage T6, 387, 390, 391t
Bacteriophage T7, 387
Bacteriorhodopsin, 166–67
Bacteriuria, 532
Bacteroid, 19, 420, 421f

Bacteroides, 315f, 325, 325t, 326f, 433, 514, 516, 575
   microflora, 440t
   in rumen, 427
   B. amylophilus, 428t
   B. fragilis, 262f
   B. ruminicola, 428t
Bactoprenol, 57f
Baculovirus, 585t
Bakteriologische Diagnostik, 314
Balanced growth, 107
Balantidiasis, 377t
Balantidium, 375t
   B. coli, 377t
Baltimore, David, 409
*Bam*HI, 273t
Barley stripe mosaic virus, 392t
Barley yellow dwarf virus, 392t
Bartnicki-Garcia, S., 378
Bartonella bacilliformis, 445t
Bartonellosis, 445t
Basal body, 62, 62f, 74f
Base, 44, 46f, 651–52
Base analog, 246
Base pairing, 46, 47f, 216f, 217–18, 224, 246–47
   wobble hypothesis, 230, 230t
Base substitution, **246**
Basic dye, 25
Basidia, 368
Basidiocarp, 362–63f
Basidiomycete, 362–63f
Basidiomycota, 360t, **365–69,** 366t
Basidiospore, 368
Basophils, 456, 457t, 479
Batch culture, 115, 117
   industrial, 622
Bauer, Alfred, 133
B cells. *See* B lymphocytes
BCG. *See* Bacille Calmette-Guérin
Bdellovibrio, 318–19, 430
   parasitism on gram-negative bacteria, **429–30,** 429–30f
   B. bacteriovorus, 429–30, 429f
Beauveria bassiana, 585t
Bee, fossil, 330
Beer, 608, 617t
   green, 618
   light, 618
   production of, **617–19,** 619f, **623–24**
Beet molasses, 620
Beet yellows virus, 392t
Beggiatoa, 80, 580, 598
Behring, Emil von, 8
Beijerinck, Martinus Willem, 8–9
Beijerinckia, 577
Bendall, Fay, 184
Benign tumor, 408
Benzalkonium chloride, 124f

Benzethonium chloride, 124f
Benzylpenicillin. *See* Penicillin G
Berg, Paul, 9–11, 274
Bergey, David H., 314
Bergey's Manual of Determinative Bacteriology, 314–15, 635
Bergey's Manual of Systematic Bacteriology, 315, 635
Berson, Solomon, 490
Betadine. *See* Povidone-iodine
β-oxidation, **200–202,** 200f
β-sheet, 44, 45f
BHK cells, 397
Bicarbonate, 570
Bifidobacterium bifidus, 154
Bile solubility test, 117, 504–5
Biliprotein, 179–80
Binal virus, **386–87,** 386f
Binary fission
   in bacteria, 105f, **106–7,** 107f
   in protozoa, 373
   in yeasts, 365
Binding protein, 88
Binnig, Gard, 34
Binomial system, 306
Biochemical oxygen demand (BOD), **589–90**
Biochemical unity, 174
Biocontainment, 297, 549f, 550
Bioenergetics, 140, 141f
Biofilm, 514
Biogeochemical cycles, **570**
Biological pesticide, 297, **584–85,** 585t, 604
Biological safety cabinet, 120
Biological vector, 445
Bioreactor, 598
Bioremediation, 11–12, 12f, 14
Biosphere, 580
Biotechnology, 620
Biotin, 85t, 145t, 574, 580
Bite wound, 514t
Bivalent immunoglobulin, 470
Black Death. *See* Bubonic plague
Blade (algae), 359, 359f
Blakeslea trispora, 627t
Blastomyces dermatitidis. See Ajellomyces dermatitidis
Blastomycosis, 360t, 500t
Blood, formed elements in, 456, 457t
Blood agar, 95, 96f, 502, 502f
Blood stasis, 453
Blood transfusion, mismatched, 480
Bloom
   algal. *See* Algal bloom
   on raw meat, 614
Blue cheese, 611t, 613–14

B lymphocytes, 457t, 458, **468**
  comparison to T
    lymphocytes, 474t
  development into plasma
    cells, **475**, 476f
  development of, 471, 472f
BOD. *See* Biochemical oxygen
    demand
Bodo, 582
Body fluids, as physical barrier to
    disease, 455
Body odor, 525
Boil, 526
Boletus, 424f
Bolivian hemorrhagic fever, 543t
Bone marrow, 459f, 468
Booster dose, 477–78
Borderline leprosy, 527
Bordet, J., 7t
Bordetella, 321
  B. pertussis, 7t, 451t, 500t,
    507, 507f
Bordet-Gengou agar, 507
Borrelia, 316–17, 318t
  B. burgdorferi, 295, 316, 445t,
    530, 530f
  B. recurrentis, 445t
Borucki, Monica, 330
Botulinum toxin, 266, 517
Botulism, 7t, 328, 411, **517,** 517t,
    552, 552t, 616
Bouquet (wine), 619
Bovine growth hormone, 297
Bovine spongiform
    encephalopathy,
    389, 543t
Boyer, Herbert, 11, 272, 274
Bradyrhizobium, 418, 576
  B. japonicum, 421t
Brandy, 619
Brazilian purpuric fever, 543t
Bread making, 617, 617t, **623–24**
Brefonneau, Pierre, 509
Brevibacterium linens, 611t
Brewer's yeast. *See Saccharomyces
    cerevisiae*
Brewing, 608, **617–19,** 619f
Brick cheese, 611t, 614
Brie cheese, 611t, 614
Bright-field microscope, 20,
    21–25, 22f, 29–30f, 31t
  illumination of specimen,
    24–25, 25f
  magnification by lenses, **22–23**
  resolution of, **23–24,** 24f
Broad-spectrum antibiotic, 125,
    433, 441
Brome mosaic virus, 392t
5-Bromouracil, 246, 248t
Bromovirus, 392t
Bronchial asthma, 480

Bronchodilator, 480
Broth culture, 94–95
Brown, James H., 502
Brown algae. *See Phaeophyta*
Brown rot of peaches, 360t
Brucella, 321, 444t
  B. abortus, 461, 608
Brucellosis, 443, 444t, 454, 461,
    554t, 556
Bt-toxin, 297
Bubonic plague, 322t, 323, 323f,
    444–45t, 451t
Buchnera aphidicola, 427
Budding
  in bacteria, 337
  in eucaryotes, 73–74
  in protozoa, 373
  in yeast, 362f, 365, 365f
Budding and/or appendaged
    bacteria, 332t, **337,**
    337f, 639
Buffer, 92, 652
Bulgarian milk, 154, 331t,
    609t, 610
Bulking, 338, 598
Bunyaviridae, 394t
Buoyancy, **66–68**
Burkitt's lymphoma, 409
Burnet, Sir Macfarlane, 475
Burn victim, 319
  nosocomial infections, 559t
  P. aeruginosa infections in,
    528–29
Burrill, 7t
Burst size, 404, 404f
Butanediol fermentation, 157f,
    160–61, 161f
Butanol, 160f
Butter, **611**
Buttermilk, cultured,
    609–10, 609t
Butyraldehyde, 160f
Butyrate, **157–60,** 160f
  from rumen fermentation,
    427–29, 428f
Butyrate fermentation, **157–60,**
    157f, 160f
Butyrivibrio, 160, 325t, 427
  B. fibrisolvens, 428t
Butyryl-CoA, 160f

C1 complex, 481
C3b opsonization, 460
Cadaverine, 198, 199f
Caedobacter, 426t
Cairns, John, 220
Calcium
  cellular content of, 84t,
    570, 570t
  characteristics of
    element, 647t

function in cells, 84t
  requirement for, 84
Calcium carbonate, in cell
    envelope, 373
Calcium dipicolinate, 68–70, **75,**
    75f, 75t, 343
Calculus (dental), 514
Caliciviridae, 394t
California encephalitis virus, 394t
Calvin, Melvin, 188
Calvin cycle, 189, 189f
*Calymmatobacterium
    granulomatis,* 533t
Camembert cheese, 361f,
    611t, 614
cAMP. *See* Cyclic AMP
Camplydiscus hibernicus, 356f
Campylobacter, 319, **520–21**
  C. fetus, 19f, 319, 521
  C. jejuni, 319, 516, 516t, 521,
    521f, 559t
Cancer
  nonviral factors in, **409–10**
  virus-related, **408–9**
Candida, 361, 366t, 440t
  C. albicans, 360t, 365f, 440–41,
    499t, 514t, 515, 525t,
    530, 533t, 559t
  C. utilis, 631
Candidiasis, 525t, 540
  cutaneous, 530–31
  genital, 533t
Candle jar, 93, 94f
Cane molasses, 620
Canned food, 121, 552, **616,** 617t
Cannibalism, 413
Cano, Raúl, 330
CAP. *See* Catabolite activator
    protein
CAP-cAMP complex, 238, 239f
Capsid, 385f, 386
Capsomere, 385f, 386, 386f
Capsule, 50f, 58f, **61–62,** 61f,
    71f, 76t
  of pathogenic bacteria,
    450, 450f
Capsule stain, 28, 28f, 61
Carbamoyl phosphate, 202, 203f
Carbenicillin, 126f
Carbohydrate
  content of cells, 81t
  degradation of, **190–92,** 190t
  functions in cells, **46–48**
  metabolism of, **189–94**
  structure of, **46–48,** 48f
  synthesis of, **192–94**
Carbolfuchsin, 511
Carbolic acid, 122t
Carbon
  cellular content of, 84t,
    570, 570t

characteristics of
    element, 647t
  function in cells, 84t
Carbonate, 570
Carbon cycle, **570–76,** 571f
Carbon dioxide
  atmospheric, 571–72
  from butyrate, 157–60, 160f
  in carbon cycle, 570–76
  as carbon source, 81, 81t
  as electron acceptor, 170
  greenhouse effect and, 572
  from lactate, 157
  from organic matter, **575**
  from pyruvate, **157**
  reduction to methane,
    574–75, 574f
  regulation of fungal
    dimorphism, **378,**
    378–79f, 379t
  requirement of *N.
    meningitidis,* **135,** 135t
  from tricarboxylic acid cycle,
    162, 163f
Carbon dioxide fixation, 81, 178,
    178f, 183, **188–89,**
    188–89f
  in carbon cycle, **571–75,** 571f
  energy for
    chemical compounds,
    573–74
    light, **573,** 573f
    heterotrophic, **574**
Carbon source, 46–47, **80–81,** 620
Carbuncle, 526
Carcinogen, 246, **409–10**
  Ames test for, 251, 251f
Carcinoma, 408
Cardinal temperatures, 89
Carlavirus, 392t
Carnation latent virus, 392t
Carnivore, **575**
Carotene, 180, 181f, 352, 357
  commercial production
    of, 627t
Carotenoid, 72, 179–80, 352,
    354t, 357
Carotenol, 180
Carrageen, 355
Carrier-mediated transport,
    86–88
Carrier of disease, **442–43,** 519
  convalescent, 443
  healthy, 443
  incubatory, 443
  intermittent, 443
  permanent, 443
  temporary, 443
Case, 562
Case-control study, **561–63**
Catabolism, 149

Catabolite activator protein (CAP), 236–38, 239f
Catabolite repression, **236–38,** 238–39f
Catalase, 93
Catalyst, 142
Cation, 649
Cauliflower mosaic virus, 392t
Caulimovirus, 392t, 393
Caulobacter, 92t, 337, 587
Cavanaugh, Collen, 344
C₃ cycle, 189
C$_3$ cycle, 189
CD4 marker, 473, 492–93, 536–40
CD8 marker, 473, 492–93
CDC. *See* Centers for Disease Control and Prevention
cDNA. *See* Complementary DNA
Cech, Thomas R., 229
Cell
  definition of, 42
  first, 42
Cell arrangements, **19–20,** 20f
Cell death, 117
Cell envelope, **48–49**
  of gram-negative bacteria, 58f, 60, 60f
  of gram-positive bacteria, 58f, 60
  of mycobacteria, 511
  of protozoa, 373
  structures located outside of, **61–63**
Cell lines, 397
Cell lysis, 59
Cell mass, **107**
  determination of, **114**
Cell-mediated immunity, 468, **482–85**
  measurement of, **485**
  mechanism of, 484f
Cell membrane. *See* Plasma membrane
Cell numbers, **107**
  counting with microscope, **108**
  increases in, **108–12**
Cellobiase, 190, 575
Cellobiose, 190, 575
Cell sap, 73
Cell shape, **52–56**
Cellular component vaccines, 478
Cellular slime mold, 350, 372–73, 372f
Cellulase, 47, 190t, 575
Cellulose, 47, 48f, 70, 72, 352, 354t, 357, 361
  degradation of, 190, 190t, 427, 428t, 433, **575,** 575f
  structure of, 190, 190t
Cell wall
  algal, 352, 354t

archaeal, 54, 102, 339–40
bacterial, **48–49,** 50f
bacteria lacking, **59**
in cell shape and conformation, **52–56**
comparison of Bacteria, Archaea, and Eucarya, 313t
eucaryotic, 52, **70–72,** 71f, 76t
growth in streptococci, **36–37,** 36–37f
inhibition of synthesis of, 125, 125t
procaryotic, 52, 71f, 76t
structure of, 46–47
synthesis of, **56,** 57f
Centers for Disease Control and Prevention (CDC), 550, 551f, 566
Central dogma, **212,** 212f
Central spore, 70
Centriole, 71f, 72
Centrosome, 71f, 72
Cephalosporin
  commercial production of, **629–30,** 629t
  mechanism of action of, 125, 125t, 629t
  structure of, 631f
Cephalosporium, 629–30, 629t
Cercoboda, 582
Cerebrospinal fever, 512
Cervical cancer, 409
Cetylpyridinium chloride, 124f
cfu. *See* Colony-forming unit
C gene, 471, 472f
Chagas' disease, 444–45t
Chaining, of streptococci, 20, 20–21f, 36–37, 36–37f
Chancre, 535
Chancroid, 533t
Charon phage, 283, 287
Cheddar cheese, 331t, 611t, 614
Cheese. *See also specific types of cheese*
  production of, 157, 331t, 368, **611–14,** 611t, 612f
  ripened, 611, 611t, 613–14
  unripened, 611, 611t
Chemical bonds. *See specific types*
Chemical energy, 140
Chemical mediators, 480, 480t
Chemical messenger, 42t
Chemical preservatives, 615
Chemical reaction
  energy requirements of, 142f
  rate of, 142
Chemical work, 140
Chemiosmosis, ATP synthesis via in electron transport, **165–68,** 167f

in photophosphorylation, **187–88**
Chemistry basics, 647–52
Chemoautotroph, 81, 81t, 571–72, **573–74**
Chemoheterotroph, 81, **150**
Chemolithotroph, 80–81, 81t, **171–72,** 171t
  aerobic, 639
  facultative, 172
Chemoorganotroph, 80, 81t
Chemoreceptor, 63, 65
Chemostat, 118, 118f, 643–44
Chemotaxis, **63–65,** 64f
Chemotherapeutic agent, 125–32, 125t
Chemotroph, 80, 81t, 141f
Chesapeake Bay, **590,** 591f
Chester, Frederick D., 314
Chester's Manual, 314
Chestnut blight, 360t, 390
Chicken, contaminated with *Salmonella,* 520
Chickenpox, 132t, 513, 525t, 531
Chimera, 281, 282f
Chitin, 72, 361
  degradation of, 190t, 191, **576**
  structure of, 190t
Chitinase, 190t, 576
Chlamydia, 315f, 326
  developmental cycles of, **430–33,** 431f
  shapes of, 19
  C. psittaci, 327t, 444t, 500t
  C. trachomatis, 295, 327t, 431f, 444–45, 533t, 536
Chlamydiae, 316t, **326,** 327t, 636–37
Chlamydomonas, 92t, 351f, 352, 354f
  C. nivalis, 358f
Chloramphenicol, 125
  commercial production of, 629t
  mechanism of action of, 125t, 128t, 129, 629t
  production of, 343
Chlordane, 584
Chlorella, 188, 632
Chlorination, 595, 599
Chlorine
  characteristics of element, 647t
  as disinfectant, 122t, 124
Chlorobiaceae, 182
Chlorobium, 333t, 577, 580, 587
  C. limicola, 188f
  C. thiosulfatophilium, 573
Chlorobium vesicle, **182,** 183f
Chloroflexaceae, 182
Chloroflexus, 333, 333t

Chloroherpeton, 333t
Chloromonas
  C. brevispina, 358f
  C. granulosa, 358f
Chloronema, 333t
Chlorophene, 123f
Chlorophyll, 72, **178,** 184, 334
  absorption spectra of, 180f
  structure of, 178, 179f
Chlorophyll a, 178, 179–80f, 184, 352, 354t, 357
Chlorophyll b, 178, 180f, 335, 352, 354t, 357
Chlorophyll c, 178, 352, 354t
Chlorophyll d, 354t
Chlorophyll P680, 184–85, 185f
Chlorophyll P700, 184–86, 185f
Chlorophyta, 352, 354t, **357–59**
Chloroplast, 71f, 72, 77, 178, **180,** 182f, 352
  evolution of, 206, 436
  structure of, 73f
Chloroquine, 132
Chlorosome, 182, 332
Chlortetracycline, 616, 629t, 630
Cholera, 7t, 322t, 323, 446, 451t, 516, 516t, **518–19,** 542f, 543t
  Snow's investigation of, 566
Choleragen, 451t, 518, 518f
Cholera toxin. *See* Choleragen
Cholera vaccine, 478, 478t
Cholesterol, 48, 49f
Chondrus, 355
Chorioallantoic membrane, 397, 397f
Christispira, 316–17
Chromatiaceae, 182, 333, 333t
Chromatic aberration, 22–23
Chromatic lens, 23f
Chromatin, 223
Chromatium, 333t, 577, 580, 587
  C. vinosum, 188f
  C. violaceum, 594
Chromobacterium, 587
Chromoplast, 72
Chromosome, 212, 313t
  of *E. coli,* 218–19, 218f
  eucaryotic, 72
Chronic inflammation, 458, 460f
Chroococcaceae, 335, 335t
Chroococcus turgidus, 334f
Chrysophyta, **352–57,** 354t
Chrysosporium, 92t
Chytridiomycota, 360t, **365–69,** 366t
Cidal agent, 122
Cilia, eucaryotic, **73,** 74f, **377**
Ciliated epithelium, of respiratory tract, 498
Ciliophora, 375t, **377,** 377t

Ciprofloxacin, 125t, 129
Circular DNA, 220, **223**
Cisternae, 73
Citrate, 162, 163f
  commercial production of, 361f, 626
Citrate synthase, 162
Citrate utilization test, 324f
Citric acid cycle. *See* Tricarboxylic acid cycle
Citrobacter, 263t, 324f, 440t, 516
Citrulline, 198, 199f
Citrulline ureidase, 199f
Citrus bacterial leaf spot, 320, 321f
Citrus canker, 320
Cladonia subtenuis, 423
Cladosporium, 601t
  C. herbarum, 614t
Clarithromycin, 129
Class (taxonomy), 306t
Classical approach, to procaryotic taxonomy, **308,** 308t
Classification, 307t, 347
  of algae, **352–59,** 353t
  of animal viruses, 393–95, 394–95t
  classical approach to, **308,** 308t
  of fungi, **365–73,** 366t
  identifying and placing microorganisms in classification systems, 307–11
  of microorganisms, **306–11**
  using nucleic acid comparisons, **309–11,** 309–10f
  using numerical taxonomy, **308–9,** 308f, 309f
  of plant viruses, 392–93t
  of procaryotes, **314–15,** 635–41
  of protozoa, **375–77,** 375t
  of viruses, **387–95**
Clevelandina, 317, 427
Clinical illness, **446,** 446f
Clonal selection theory, 475, 476f
Cloning. *See* Gene cloning
Cloning vector, 274
  bacteriophage, 274, **281–85,** 283–84f, 287
  cosmid, 285
  insertion into host cell, **285–87**
  phagemid, 285
  plasmid, 274, **279–81,** 281f, 287
  selective marker on, 279
  Ti plasmid, 297, 299f
  yeast artificial chromosome, 285

Clonothrix, 20
Closterovirus, 392t
Clostridium, 93, 328, 516
  butyrate formation in, 160f
  cellulose degradation by, 575
  chitin degradation by, 576
  endospores of, 68
  fermentation pathways in, 157f
  in freshwater habitats, 587
  in meat spoilage, 614t
  microflora, 440t
  nitrogen fixation by, 83, 577
  in soil, 581
  Stickland reaction in, 196
  C. botulinum, 7t, 106t, 121, 411, 516, **517,** 517f, 517t
    endospores of, 71f, 517, 517f, 616
    lysogenic conversion in, 452
    toxins of, 266, 451t, 452
  C. butyricum, 617t
  C. difficile, 441
  C. lochheadii, 428t
  C. nigrificans, 617t
  C. pasteurianum, 576–77
  C. perfringens, 7t, 449t, 450, 516, 517t, 594
    India ink stain of, 28f
    toxins of, 450–52, 451t
  C. sporogenes, 92t, 617t
  C. tetani, 7t, 442
    electron microscopy of, 33f
    endospores of, 71f
    toxins of, 451t, 452
  C. thermosaccharolyticum, 617t
Clothing industry, **624**
Club root of cabbage, 360t
Coagglutination test, 487, 487f
Coagulase, 449t, 450, 518
Coal, sulfur-containing, 580
Coal refuse pile, 102
Cobalt, 570
Coccidioides immitis, 360t, 445, 500t, 512
Coccidioidomycosis, 360t, 445, 500t, 512
Coccobacillus, 19, 20f
Coccus, 19, 19–20f
Cockroach, endosymbionts of, 317, 427
Codon, 213, 214f
  initiation, 230
  nonsense, 213, 231f, 232–33
  termination. *See* Codon, nonsense
  wobble hypothesis, 230, 230t

Coenocytic hyphae, 361, 362f
Coenocytic organism, 74
Coenzyme, 84–85, 145, 145t
Coenzyme A, 85, 85t, 145t, 162, 580
Coenzyme F$_{420}$, 340
Coenzyme M, 575
Coenzyme Q, 164–65, 164f, 167f, 170f, 171
Cofactor, 145
Cohen, Stanley, 11, 274
Cohn, Ferdinand, 314
Cohort, 562
Colby cheese, 611t
Cold sore, 446, 525t, 531–32
Cole, Roger M., 36–37
Coleochaete, 358f
Colicin, 262
Coliforms, 96, 113, 322, 443, **590–93**
  determination of
    fluorogenic assay, **602–3,** 602f, 603t
    membrane filter procedure, 591–93, 592–93f, 602–3
    most probable number method, 591–93, 592f, 602–3
  fecal, **593–94**
  total, 591–94
Coliphage, 390
Collagenase, 449t, 450
Colony-forming unit (cfu), 109f, 110
Colony hybridization, 288–89, 289f
Colony morphology, 98, 111f
Colony stimulating factor (CSF), 473
Colostrum, 469
Col plasmid, 261–62, 263t
ColE1, 293
Combinatorial joining, 471, 472f
Cometabolism, 586
Commensalism, 418, **433**
Common ancestor, 174, 311–12, 312f
Common cold, 444, 499t
Common-source epidemic, 552, 552f
Communicable disease, 442
Community, 570
Comovirus, 392t
Competence factor, 252–53
Competent cells, **252–53,** 253f
Competitive inhibitor, 143, 144f
Complement, 456, **481–82,** 494
  alternative pathway of, 481, **482,** 483f

classical pathway of, 481–82, 481f
Complementary DNA (cDNA), 274–78, 278f
Complementary DNA (cDNA) clone, 274
Complementary DNA (cDNA) library, 278, 285, 286f
Complement fixation test, **487–89,** 488f
Completed test, 593
Complex media, **94–95,** 95t
Compromised host, 479
Concatamer, 402
Condenser lens, 21, 22f, 23–24, 24f, 29f
Confirmatory test, 504
Confirmed test, 593
Confluent growth, 397
Confocal microscope, **34**
Congenital rubella syndrome, 531–32
Congenital syphilis, 535
Conidia, 343, 343f, 361f, 364, 364–65f, 366, 367f
Conidiophore, 364, 364–65f
Conidiospore. *See* Conidia
Conjugation, 62
  in bacteria, 252, **263–65,** 264–65f
    mapping genes by, **265**
  in protozoa, 373, 374f
Conjugative plasmid, **260–61**
Conjunctivitis, 445, 534
Consensus sequence, 226
Conservative replication, 220
Constitutive enzyme, 234
Constructive interference, 28
Consumer, 570–71
Contact dermatitis, 482
Contact inhibition, 408
Continuous culture, **118,** 118f
  industrial, 622
  mathematics of, 643–44
Control, 562
Control system
  negative, 236
  positive, 236
Convalescent carrier, 443
Convalescent period of disease, 446, 446f
Copper
  characteristics of element, 647t
  microbial leaching of ores, 582–83
Copper sulfate, 122t, 125
Coprinus, 366t
Corallinaceae, 359
Coralline algae, 359
Coral reef, 350, 359

Core polysaccharide, 60, 60f
Corepressor, **235–36**, 235f
Corn brown spot, 360t, 365
Corn smut, 360, 360f
Corn steep liquor, 620
Coronaviridae, 394t
Coronavirus, 396f, 499t
   coronavirus OC43, 394t
Corrosion, 170, 326, 588
Corticoviridae, 391t
Corynebacterium, 19, 601t
   cell arrangements in, 20
   inclusion bodies of, 66
   C. diphtheriae, 7t, 67f, 331,
      499t, 507–10
      identification of, 507,
        508–9f, 509
      lysogenic conversion in,
        258, **266**, 266–67f,
        411, 508
      toxins of, 451t
   C. glutamicum, 12, 625–26,
      625–26f
   C. xerosis, 440t, 525
Cosmid, **285**
cos site, 285
Cottage cheese, 611t, 614
Coupled reactions, **145–48**
Covalent bond, 649, 650f
   double, 649, 650f
   polar, 649
   single, 649, 650f
   triple, 649
Cowpea mosaic virus, 392t
Cowpox, 8, 387
Coxiella burnetii, 327t, 444t, 608
Coxsackievirus, 499t
Cream cheese, 611t, 614
Creatine phosphate, 148t
Cresol, 122t
Creutzfeld-Jakob disease, 389
Crick, Francis H.C., 216, 230
Criminal, DNA evidence
      against, 292
Cristae, of mitochondria, 72, 73f
Cristispira, 318t
   C. pectinis, 21f
Critical illumination, 24–25, 25f
Critical-point drying, 37
Cronartium ribicola, 360t
Crop, transgenic, 297, 299f
Crop yield, nitrogen fixation and,
    434–35, 434f
Cross-feeding, 441, 441f
Cross-linking bridges, in cell
      walls, 54, 55f
Crotonyl-CoA, 160f
Cryphonectria parasitica, 390
Cryptococcosis, 360t, 540
Cryptococcus neoformans, 360t
Cryptosporidiosis, 376, 377t, **594**

Cryptosporidium, 375t, 376,
    377t, 543t
   in water supply, **594**
CSF. *See* Colony stimulating
      factor
Cucumber mosaic virus, 392t
Cucumovirus, 392t
Cud, 429
Culture
   batch, 115, 117
   continuous, **118**, 118f
   special techniques for, **117–19**
   storage of, **91**
   synchronous, **118–19**, 119f
Culture media. *See* Media
Curds, 611–13, 612f
Curing
   of meat, 615, 615f
   of plasmid, 263
   of prophage, 411
Cyanide, 143, 168, 168t
Cyanobacteria, 180, 332t, **333–36**,
    334f, 638–40
   cell structure in, 50f
   cellular organization of, 335
   cell wall of, 334
   classification of, 335, 335t
   colors of, 334
   endosymbiotic, 206, 206f
   evolution of, 174
   in freshwater habitats, 587–88
   Gram staining of, 60
   habitats of, 334
   nitrogen fixation by, 334–35,
      576–77, 577f
   photosynthesis in, **178**, 182
   procaryotic phylogenetic
      tree, 315f
   in soil, 582
   universal phylogenetic
      tree, 312f
Cyanocobalamin. *See* Vitamin B_{12}
Cyanophage, 390
Cyanophora paradoxa, 206f
Cyclic AMP (cAMP), 518, 523
   in catabolite repression,
      236–38, 239f
Cyclic photophosphorylation,
    185–86f, **186–87**
Cyclosporin A, 479
Cyclotella meneghiniana, 3f
Cyst
   of *G. lamblia*, 523–24
   protozoan, 373
Cysteine, 43t, 83, 197t, 580
Cystic fibrosis gene, 291–92
Cystine, 580
Cystitis, 532
Cystoviridae, 391t
Cytochrome(s), 162, 164, 165f,
    166, 167f

Cytochrome a, 170f
Cytochrome a, a_3, 170f
Cytochrome b, 165, 170f,
    185–86, 185f
Cytochrome b_{556}, 170f
Cytochrome b_{558}, 170f
Cytochrome b_{562}, 170f
Cytochrome b_{565}, 170f
Cytochrome bc_1, 186, 186–87f
Cytochrome c, 164f, 165, 170f
Cytochrome c_2, 186, 186–87f
Cytochrome c, c_1, 170f
Cytochrome c oxidase, 165, 529
Cytochrome c reductase, 165
Cytochrome d, 170f
Cytochrome f, 185, 185f
Cytochrome o, 170f
Cytogamy, 373, 375
Cytokines, 473, 474t, 484f, 494
   activation of macrophages
      by, **483**
   in delayed hypersensitive
      reaction, **483–84**
Cytokinesis, 74
Cytomegalovirus, 405f
Cytomembrane, 336f
Cytopathic effect, **405–8**, 405f
Cytophaga, 338, 575–76
Cytoplasmic streaming, **73**
Cytosine, 44, 46f
Cytotoxic reaction, 479t,
    480–81
Cytotoxin, 450

2,4-D, 586
Dactylosporangium, 342t
Dairy products, **608–14**
   fermented, **609–11**, 609t
   microbiological standards for,
      608, 609t
Dairy workers, 608
Dalgarno, Larry, 232
Dark-field microscope, 20, **28**,
    29–30f, 31t
Dark-field stop, 28
Dark reactions, of
      photosynthesis, 178
Davis, Bernard D., 254
DCMU, 186
ddC. *See* Dideoxycytosine
ddI. *See* Didanosine
DDT, 576, 584, 586
Deamination, 197
Death phase, 115, 116f, **117**
de Bary, Heinrich Anton, 418
Debaryomyces, 92t
Decarboxylase, **198**, 199f
Decimal reduction time. *See* D
    value
Decline period of disease,
    446, 446f

Decline phase of culture. *See*
    Death phase
Decomposer, 570–71, **575–76**
   fungi, 360
   recalcitrant
      compounds, 576
Deep (solid media), 95
Defective interfering (DI)
    virus, 408
Defective phage, 255–56
Defective prophage, 411
Deforestation, 572
Deformylase, 232
Dehydrated food, 93, 615
Dehydrogenation reaction, 145
Deinococci, 315f
Delayed hypersensitive reaction,
    479, 479t, **483–84**
Deletion, 246
Denaturation, of proteins, 44
Dendrogram, 308f, 309
Dengue fever, 395, 498, 542f, 543t
Dengue virus, 395, 395t, 414
Denitrification, 170, 577f, **578–79**
Density-gradient
    centrifugation, 220
Dental abscess, 514t
Dental caries, 62, 432, 448,
    514–15, 514t, 563, 563t
Dental hygiene, 515
Dental plaque, 61, 325, 429, 448,
    449f, **514**
   formation of, 462, 462–63t
Deoxyribonucleic acid. *See* DNA
Deoxyribose, 44
Dermatitis
   contact, 482
   fish, 360t
Dermatomycosis, **530–31**
Dermatophilus, 342t, 343
Derxia gummosa, 423
Desensitization, to allergen, 480
Desiccation resistance, 61, 68
Destructive interference, 28
Desulfovibrio, 170, 326, 577,
    580–81, 588
Desulfurococcus, 313f
Desulfurolobus, 340t
Desulfurylation, 579f
Detergent, 122t, 123
Dextran, 61, 429, 448, 515
   agglutination of *S. mutans* by,
      462, 462–64t
   commercial production
      of, 627
Dextranase, 448
Dextransucrase, 515
Dextrin, 618
D gene, 471, 472f
d'Hérelle, Felix, 9
Diabetes mellitus, 300–301

Diagnostic test
  data for epidemiological analysis, **560–61,** 560t
  reliability of, 561
  sensitivity and specificity of, 560–61
Dialysis culture, 622–23, 622f
Diaminopimelate, 52, 54, 54f, 343
Diapedesis, 456
Diarrhea, 322t
Diatom, 17f, 352–56, 354t, 355–56f
Diatomaceous earth, 356
Diauxic growth, 236–38, 238f
Dickens, Frank, 154
Dictydium, 366t
Dictyosteliomycota, 366t, **369–73**
Dictyostelium, 65, 366t
Dictyuchus, 366t
Dicumarol, 168, 168t
Didanosine (ddI), 132, 132t
Dideoxycytosine (ddC), 132, 132t
Differential media, **95**
Differential stain, 25
Diffusion
  facilitated, 85, **86,** 87f
  passive, 85, **86,** 86f
Digestive aids, 625
Digestive system, **514–24**
  bacterial pathogens of, **516–21**
  disease caused by *E. coli,* **521–23**
  microflora of, 440t, 441, **515–16**
  protozoan pathogens of, **523–24**
  viral pathogens of, **523–24**
Diglyceride, 47
Dihydrouridine, 224, 225f
Dihydroxyacetone phosphate, 151f, 152, 158, 159f, 169, 169f, 191f, 195f
Dihydroxyacetone phosphate/glycerol-3-phosphate shuttle, 169, 169f
Dikaryon, 364
Dimethylguanosine, 224
Dimorphic fungi, 361, **378,** 378–79f, 379t
Dinitrogen, 82
2,4-Dinitrophenol, 168, 168t
Dinoflagellate, 350, 356–57, 356–57f. *See also* Pyrrophyta
Dipeptide, 44, 44f
1,3-Diphosphoglycerate, 148–49, 148t, 150–51f, 152, 153f, 155f, 159f, 189, 189f, 193f
Diphtheria, 7t, 331, 451t, 452, 499t, **507–10,** 542f, 554t

Diphtheria toxin, 8, 266, 411, 451t, 508, 508–9f, **509,** 510
Diphtheria vaccine, 452, 478, 478t. *See also* DTP vaccine
Diphtheroid, 331, 440t, 525
Dipicolinate, 68–70, 68f, **75,** 75f, 75t
Diplocalyx, 317
Diplococci, 20
Diploid, 72–73, 365
Diplomonad, 312f
Direct fluorescent antibody test, 489f, 490
Direct microscopic count, **108,** 109f, 116t
Disaccharide, 46, 48f, 190t, 191
Disease cluster, 556
Disease rate
  changes with time, **555,** 555f
  characteristics of population and, **556–58**
  geography and, **556,** 556f
Disease transmission
  direct, **443–45**
  indirect, **443–45**
Disinfectant, **122–25,** 122t
Disk diffusion method, of antimicrobial susceptibility testing, 132, **133,** 133f
Dissimilatory nitrate reduction, 170, 578
Dissimilatory sulfate- and sulfur-reducing bacteria, 316t, **326,** 636
Dissimilatory sulfate reduction, 170, 579f, 580
Distilled beverages, **619**
Disulfide bond, 45f
DI virus. *See* Defective interfering virus
Division (taxonomy), 306t
DNA, 212, **215–24.** *See also* Transcription
  antiparallel chains in, 218, 221
  antisense, **293–94**
  bacterial, 65
  circular, 220, **223**
  double helix, 34f, 46, 47f, 211f, 216, 216f, 218
  of *E. coli,* 1f
  eucaryotic, 223–24
  G + C content of, **217,** 217f, 309–10, 309f
  gel electrophoresis of, 11f
  inverted repeat, 260
  melting curve of, **217,** 217f, 309–10, 309f
  mitochondrial, 239

proof that it is genetic material, **215–16,** 215f
  recombinant. *See* Recombinant DNA technology
  recombination of. *See* Recombination
  repair of, 222–23, 223t, **248–49,** 248–49f
  replication of, 106–7, 212f
    bidirectional, 220, 220f
    concatamer formation, 402
    direction of, 221, 222f
    discontinuous, **221–22,** 222f, **240,** 241f
    errors in, 246, 268
    in eucaryotes, **223–24**
    initiation of, **220–21**
    lagging strand, 221–22f, 222
    leading strand, 221–22f, 222
    in procaryotes, **223–24**
    RNA primer, 221f, **223,** 224
    rolling-circle mechanism, **223,** 264, 264f, 402, 402f
    semiconservative, **219–23**
    theta replication, 220, 220f
    unwinding, 220–21, 221f
  similarities and differences among organisms, 242
  structure of, 44, 46, 47f, 216–19, 216f
  sucrose gradient sedimentation of, 241f
  supercoiled, 218–19
  transformation with. *See* Transformation, bacterial
  unwinding of, 226
  variable number of tandem repeats, 292, 293f
DNA bacteriophage, 391t
DNA fingerprinting, **292,** 293f
DNA fragments
  separation by gel electrophoresis, 274, 275f
  Southern blot, 276, 276–77f
  with sticky ends, 272, 274, 279, 280f
DNA gyrase, 129
DNA helicase, 220–21, 221f, 223
DNA library, **285,** 286f
DNA ligase, 222, 222f, 248–49, 248f, 274, 278
DNA polymerase I, **222–23,** 233t, 248–49, 248f
DNA polymerase II, **222–23,** 223t

DNA polymerase III, 221f, **222–23,** 223t
DNA probe, 276, 276–77f, 311
DNase, 296t, 449t, 450
  test for production of, 202, 203f
DNA synthesizer, 278
DNA topoisomerase, 221
DNA virus, 384–86
  animal viruses, 394t
  plant viruses, 392t
  replication of DNA, **402–3,** 402f
Domain, 306t
Donor cell, 250
Double covalent bond, 649, 650f
Double diffusion test, 486
Double helix, 34f, 46, 47f, 211f, 216, 216f, 218
Doubling time. *See* Generation time
Doudoroff, Michael, 152
Downy mildew of grapes, 360t, 369
Drinking water, 588
Drinking-water standards, 589t
Drug-resistant microbes, 125, 616. *See also* Antibiotic resistance
  evolution of, 136
Dry heat sterilization, **122**
Dry weight, 42, 114, 116t
DTP vaccine, 478, 507, 510
Dubos, René, 14
Dunaliella, 92t
Dust clouds, 445
D value, 121, 121f
Dye, 25
Dysuria, 533–34

Eberth, 7t
Ebola virus, 394t, 395, 396f, 414, 542, 542f, 543t, 546, 550, 551f
EBV. *See* Epstein-Barr virus
Eclipse period, 401, 404, 404f
*Eco*RI, 272, 273t
Ecosystem, 570
Ectomycorrhiza, 423, 424f
Ectosymbiosis, 418, 421
Ectothiorhodospira, 333t
  E. mobilis, 53f, 183f
Ectothiorhodospiraceae, 333, 333t
Edam cheese, 611t, 614
Edema, 458, 460f, 480
Edwardsiella tarda, 324f
EF. *See* Elongation factor
Effector, 143, 144f
Effluent, 595
Eggs, raw, 519

EHEC. *See Escherichia coli,* enterohemorrhagic

Ehrlich, Paul, 8, 8f

EIEC. *See Escherichia coli,* enteroinvasive

Einstein (unit), 182

Elderly, disease rates among, 556

Electrical work, 140

Electron, 647

Electron acceptor, 145–46
  in anaerobic respiration, 168t
  terminal, 162, 165, 171

Electron carrier, **146–48,** 147f, 162, 165–66, 167f

Electron configuration, of atom, 648

Electron donor, 145–46

Electron microscope, 20–21, **31–37,** 31t, 33f
  counting viruses with, 398, 398f
  scanning, 21, 32, 32f, **36–37**
  transmission, 21, **32–36,** 32f

Electron shell, 648, 649f

Electron transport, 146–48, 162
  in luminescent bacteria, 173, 173f
  reversed, 187, 187f

Electron transport chain, 149, **162–65,** 164–65f, 170f, 171, 175f
  ATP synthesis in, 165–68, 167f
  in photosynthesis, 182, 184, 185f
  uncouplers and, **167–68,** 168t

Electroporation, 287

Elek gel diffusion test, 509–10, 509f

Element, 647–48, 648f

Elementary body, 19, 430–44, 431f

ELISA. *See* Enzyme-linked immunosorbent assay

Elongation factor 2 (EF2), 508–9

Elongation factor eIF2a, 407f

Elongation factor G (EF-G), 232

Elongation factor Ts (EF-Ts), 232

Elongation factor Tu (EF-Tu), 232

Elwell, Lynn, 544

EMB agar. *See* Eosin methylene blue agar

Embden-Meyerhof pathway, 150–54, 151f, 154t, 155f, 158, 175f, 191f, 193f, 194

Embryonated egg, virus propagation in, 397, 397f

Emerging infectious disease, 414, **542–43,** 542f, 543t, 546

Empty magnification, 23

Encephalitis, viral, 395, 445t

Encephalitozoon, 375t

Endemic disease, 553

Endemic typhus, 327t, 444–45t, 525t, 529, 529t

Endergonic reaction, 140–42, 145

Endocarditis, 328, 503t

Endocytosis, 400

Endogenote, 250

Endogenous pyrogen, 453, 458

Endomycorrhiza, 423

Endoplasmic reticulum, 71f, 72

Endospore, **68–70,** 71f
  activation of, 70
  of *B. subtilis,* 329f
  of *C. botulinum,* 517, 517f, 616
  central, 70, 329f
  development of, 68–69f, 69
  in fossil bee, **330,** 330f
  germination of, 69f, 70
  heat resistance of, 68, **75,** 91, 121
  structure of, 70f
  subterminal, 70, 71f, 517f
  terminal, 70, 71f

Endospore-forming gram-positive rods and cocci, **328,** 328t, 329f, 637

Endosymbiont, 316t, **328,** 637

Endosymbiont hypothesis, **206,** 328, 333, 381, 427, 436

Endosymbiosis, 418, 421

Endothelial cells, 458

Endothermic reaction, 141

Endothia parasitica, 360t

Endotoxin, **450–54,** 451t, 454t
  detection of, **452**
  as outer membrane constituents, **452–54**

Endrin, 586

Energy
  chemical, 140
  concepts of, **140–45**
  conservation of, 141
  definition of, 140
  kinetic, 140
  potential, 140
  requirements of chemical reaction, 142f
  transformations of, **140,** 141

Energy source, 46–47, **80–81,** 140, 174

Engelmann, Theodor Wilhelm, 178

Enolase, 151f, 152, 153f

Enoyl hydrase, 200f

Enrichment culture, **95**

Entamoeba, 312f, 375t
  E. histolytica, 132, 376, 377t, 445t, 448, **524,** 594

Enterobacter, 324f, 516, 577
  fermentation pathways in, 157f
  identification of, 161
  microflora, 440t
  pyruvate degradation in, 160
  E. aerogenes, 92t, 532, 532t, 559t, 585t, 591
  E. agglomerans, 324f, 427
  E. cloacae, 263t

Enterobacteriaceae, 322, 322t
  flowchart for identification of, **324,** 324f

Enterobactin, 88

Enterococcus, 130, 559t, 594
  E. faecalis, 262, 263t, 441, 441f, 532, 532t, 587

Enterotoxin, 450, 451t
  of *S. aureus,* 518

Enterotoxin A, 473

Enthalpy, 141–42

Entner, Nathan, 152

Entner-Doudoroff pathway, **152–53,** 153f, 154, 154t

Entropic doom, 141

Entropy, **141–42**

Enveloped animal virus, 385f

Enzyme(s), 42, 42t, **142–45**
  active site of, 143
  allosteric, 143, 144f, 145
  commercial production of, 12
  constitutive, 234
  extracellular, 52, 190
  immobilized, 622–23
  industrial uses of, **623–25**
  inhibitors of, 143
  lock-and-key mechanism of, 143f
  pH effect on, 143
  specificity of, 143
  temperature effect on, 143
  virulence-associated, **448–50,** 449t

Enzyme I, 89

Enzyme IIA, 89

Enzyme IIB, 89

Enzyme IIC, 89

Enzyme-linked immunosorbent assay (ELISA), **490,** 491f
  AIDS, 490, 540
  indirect, 490, 491f
  sandwich technique of, 490, 491f
  western blot technique of, 490

Enzyme-substrate complex, 143, 143f

Eosin methylene blue (EMB) agar, 96, 97f, 593

Eosinophil(s), 456, 457t

Eosinophil chemotactic factor of anaphylaxis, 480t

EPEC. *See Escherichia coli,* enteropathogenic

Epidemic, **552–53**
  common-source, 552, 552f
  propagated, 552, 553f

Epidemic typhus, 445t, 525t, 529, 529t

Epidemiologist, 550

Epidemiology
  data collection in, **558–61**
  definition of, 550
  epidemiological analysis of diseases, **555–58**
  general concepts of, **550–54**
  observational and experimental studies in, **561–63,** 561f
  risk for toxic shock syndrome, 564–65, 564t, 565f
  Snow's investigation of cholera, 566

Epidermophyton, 444t, 525t, 530

Epifluorescence microscope, 29f, 31

Epiglottitis, 499t

Epilimnion, 587

Episome, 262

Epitope. *See* Antigenic determinant

Epstein-Barr virus (EBV), 409

Epulopiscium fishelsoni, 19, 19f

Equilibrium constant (K_{eq}), 142

Equilibrium point, of chemical reaction, **142**

Eremothecium ashbyii, 627t, 628

Erwinia amylovora, 7t

Erysipelas, 7t, 525t, 526, 526f

Erythema, 458, 460f

Erythritol, 448

Erythrobacter, 333t

Erythrocytes, 456, 457t

Erythrogenic toxin, 266
  of *S. pyogenes,* 411, 498–99

Erythromycin
  commercial production of, 629t
  mechanism of action of, 125t, 128t, 129, 629t
  production of, 343

Erythrose-4-phosphate, 155f, 194, 195f, 197t

Escherich, Theodor, 7t

Escherichia, 322, 322t, 324f, 516
  fermentation pathways in, 157f
  lipopolysaccharide of, 60
  pyruvate degradation in, 160
  temperature range of, 91

E. coli, 7t, 322, 516, 546, 591
  attack by macrophages,
    439f, 461f
  bacteriophages of, 245f, 390
  cell wall of, 56
  chemoreceptors of, 65
  chromosome of,
    218–19, 218f
  DNA of, 1f, 65
  DNA melting curve
    for, 217f
  DNA polymerases of,
    222–23, 223t
  DNA replication in, 220
  electron microscopy of,
    3f, 41f
  electron transport chain
    of, 170f, 171
  enterohemorrhagic
    (EHEC), 322, 443, 516t,
    521t, 523
  enteroinvasive (EIEC), 322,
    521t, 523
  enteropathogenic (EPEC),
    262, 521t, 523
  enterotoxigenic (ETEC),
    322, 516t, 521–23, 521t
  flagella of, 63
  in freshwater habitats, 587
  gastrointestinal disease
    caused by, 521–23
  gene mapping in, 265
  generation time of,
    106, 106t
  genes of, 212
  genome sequencing, 291
  glucose metabolism
    in, 154t
  Gram stain of, 27f
  heat-labile toxin of, 451t,
    452, 521t, 523
  heat-stable toxin of, 451t,
    452, 521t, 523
  as host for cloning, 285
  industrial uses of, 623t
  lactose-fermenting, 97f
  lactose uptake in, 88–89
  with luciferase gene,
    287, 288f
  media for, 95t
  microflora, 440t
  nosocomial infections
    with, 559t
  pathogenic, 433
  peptidoglycan of, 54, 54f
  phosphotransferase system
    of, 89
  pH tolerance of, 92t
  plasmids of, 263t, 279–81,
    281–82f
  promoters of, 226

  proteins of, 194
  restriction endonucleases
    of, 273t
  SOS response in, 249, 249f
  temperature range for,
    89, 90f
  toxins of, 450–52, 451t
  tryptophan operon in,
    235–36, 235f, 237f
  in urinary tract infections,
    532–33, 532t
  water content of, 42
Estuary, 586, 590, 591f
ETEC. *See Escherichia coli,*
    enterotoxigenic
Ethanol, 156f, 157, 159f, 161f
  as disinfectant, 122t, 123
Ethidium bromide, 248, 248t, 274
Ethylene oxide, 122t, 124
Ethyl methanesulfonate,
    247–48, 248t
Eubacterium, 160
Eucarya, 18, 307t
  distinguishing characteristics
    of, 313t
  universal phylogenetic tree,
    311–12, 312f
Eucaryotic cells, 18, 350
  cell wall of, 52, 70–72, 71f
  classification of
    eucaryotes, 307
  composition and structure of,
    70–74
  control of gene expression in,
    238–39
  DNA replication in, 223–24
  evolution of, 77, 380
  incorporation of foreign DNA
    into, 287
  locomotion in, 73
  mRNA of, 212, 213f, 229, 233
  organelles of, 72–73
  oxidative phosphorylation in,
    169, 169f
  photosynthesis in, 178, 180
  plasma membrane of,
    70–72, 71f
  procaryotic cells vs., 71f, 76t
  reproduction in, 73–74
  ribosomes of, 65, 224
  RNA polymerases of, 228–29
  transcription in, 213f
  translation in, 213f, 233
Eucaryotic phylogenetic tree, 380
Euglena, 351f, 356f, 357
  E. gracilis, 182f
Euglenophyta, 352–57, 354t
Eutrophic lake, 587–88, 588f, 595
Evolution
  of autotrophy, 174
  of cyanobacteria, 174

  of drug resistance, 136
  endosymbiont hypothesis, 206
  of eucaryotes, 380
  of eucaryotic and procaryotic
    cells, 77
  of immune system, 494
  of mitochondria and
    chloroplasts, 436
  of multicellular eucaryotes, 38
  of viruses, 414
Evolutionary distance, 302, 311
Excision repair, 248, 248f
Exergonic reaction, 140–42, 145
Exfoliatin, 527
Exogenote, 250
Exon, 212, 213f, 224, 229
Exosporium, 69, 69–70f
Exothermic reaction, 141
Exotoxin, 450–54, 451t
Exotoxin A, 473, 529
Expectorant, 480
Experimental study, 561–63, 561f
Exponential growth, 107, 642–44
Exponential phase, 115, 116, 116f
Expression vector, 285
Extracellular enzyme, 52, 190
Extreme alkalophile, 93
Extreme halophile, 93, 312,
    312–13f, 339f,
    340–41, 340t
Extreme thermophile, 90, 90f, 91,
    312, 313f, 314,
    339–41, 340t
Extrinsic protein, 52
Exxon Valdez oil spill, 11–12, 12f
Eyepiece, 22

Fab fragment, 470, 471f
Facilitated diffusion, 85, 86, 87f
Factor 342, 574
Factor 420, 574–75, 574f
Factor 430, 575
Factor B, 482, 483f
Factor D, 482, 483f
Factor VIII, 296t, 296
Facultative anaerobe, 93
Facultative anaerobic gram-
    negative rods, 316t,
    322–23, 322t, 323f, 636
Facultative chemolithotroph, 172
FAD, 85, 85t, 145t, 148, 164
  structure of, 147f
  in tricarboxylic acid cycle,
    162, 163f
Faenia, 342t
Falkow, Stanley, 544
False negative result, 560
False positive result, 560
Family (taxonomy), 306, 306t
Fatty acid, 47, 49f
  β-oxidation of, 200–202, 200f

  saturated, 47
  synthesis of, 202
  unsaturated, 47, 90
Fatty acyl CoA, 200
F⁻ cells, 263–65
F⁺ cells, 263–65
F' cells, 265
Fc fragment, 470, 471f
Fecal coliforms, 593–94
Fecal contamination, 443, 588,
    593–94
Feedback inhibition, 143–45, 144f
Fehleisen, 7t
Feng, Peter, 602
Fermentation, 150, 154
  ATP formation in, 168
  in carbon cycle, 570f
Fermented food, 616–17, 617t
Fermented milks, 609–11, 609t
Ferredoxin, 160f, 164, 185, 185f,
    422–23, 422f, 573, 573f
Fertile soil, 581
Fertilizer, 434
Fever, 454t, 458
Fever blister, 499t, 532
Fibrobacter succinogenes, 428t
Fibroma, 408
Filament, of flagella, 62, 62f
Filamentous bacteria, 19
Fildes, P., 135
Filoviridae, 394t, 395
Filtration, sterilization by, 120
Fimbriae. *See* Pilus
Finkelstein, Richard A., 100
First cell, 42
First law of thermodynamics, 141
Fish
  flashlight, 425, 425f
  pathogens of, 338, 360t
  spoilage of, 615–16, 615t
Fission, 73–74
FITC. *See* Fluorescein
    isothiocyanate
Fitzgerald, Robert J., 461
Five-kingdom classification,
    307, 307t
Flagellation
  lophotrichous, 63
  monotrichous, 62
  multitrichous, 63
  peritrichous, 63, 63f
  polar, 63f
Flagellin, 62, 260, 261f
Flagellum, 62–63, 62f, 305f
  algal, 354t, 357f
  eucaryotic, 73, 74f
  periplasmic, 316, 317f
  rotation of, 63
  structure of, 62, 62f
  whiplash, 365
Flashlight fish, 425, 425f

Flavin adenine dinucleotide. *See* FAD

Flavin mononucleotide. *See* FMN

Flavivirus, 445t

Flavobacteria, 312f

Flavobacterium, 315f, 578, 581, 586–87, 601t, 608, 615–16, 615t

Flavonoid, 420, 435

Flavoprotein, 162, 164, 166, 167f, 170f, 185

Flea vector, 445t

Fleming, Alexander, 14

Flesh-eating bacteria. *See* Necrotizing fasciitis

Floc material, 597–98

Flowchart, for bacterial identification, **324,** 324f

Flucytosine, 132

Fluid mosaic model, 51

Fluorescein isothiocyanate (FITC), 31, 489

Fluorescence, 31

Fluorescence microscope, 20, **31,** 31t

epifluorescence, 29f, 31

transmitted-light, 29f, 31

Fluorescent antibody test, **489–90,** 489f

Fluoridation, 515, 563, 563t

Fluorochrome, 31

Fluorogenic assay, for coliforms, **602–3,** 602f, 603t

FMN, 85, 85t, 145t, 147f, 148, 164

Folic acid, 85t, 129, 145t

Fomite, **444–45**

Food. *See also specific types of food*

irradiated, 122

Food allergy, 480

Foodborne disease, **443**

Food industry, 12

economic importance of algae, 355

Lactobacillus in, 331, 331t

Food infection, **516–21,** 516t

Food intoxication, **516–21,** 517t, 552

Food poisoning, 328, 451t, **516–21,** 517t

Food preservation, 93

Food spoilage, 90, 368

Foot-and-mouth disease vaccine, 478, 478t

Foot-and-mouth disease virus, 9

Forameniferida, 376

Foré people, kuru among, **412–13,** 412–13f, 412t

Forespore, 69, 69f

Formaldehyde, 615

as disinfectant, 122t, 124

Formalin, 124

Formate, 161, 161f

Formic hydrogenlyase, 161, 161f

N-Formylmethionine, 229, 229f, 231f, 232, 312

Formyl-tetrahydrofolate, 202, 203f

Fossil bee, 330

Fossil fuel, 570, 579

burning of, 572, 580

Four Corners disease, 542, 542f

Fourfold contingency table, 562, 562t

F pilus. *See* Sex pilus

F plasmid, 261, 263–65, 263t

F' plasmid, **265**

Fragmentation, 73–74, 352

Frame shift mutation, 246, 247f

Francisella, 321

F. tularensis, 444, 444–45t, 525

Frankia, 342t, 343

nitrogen fixation by, **418–23,** 419–22f, 421t

F. alni, 421t

F. casuarinae, 421t

F. ceanothi, 421t

Franklin, Rosalind, 216

Free energy, 141–42

Free energy changes, 141–42, 146

Free spore, 70

Freeze etching, 36

Freezer burn, 615

Freshwater habitat, **587–88,** 587f

Fried egg colony, 327, 327f

Frosch, Paul, 9

Frost-proof fruits/vegetables, 618

Frozen food, 615

Fructose-1,6-diphosphatase, 193f

Fructose-1,6-diphosphate, 150, 151f, 152, 155f, 158, 159f, 191f, 193f

Fructose-1,6-diphosphate aldolase, 151f, 152, 158, 159f

Fructose-6-phosphate, 150, 151f, 155f, 189, 191f, 193f, 194, 195f

Fruit, frost-proof, **618**

Fruiting body

of bacteria, 338, 339f

of mushroom, 368

of slime mold, 371–73, 371–72f

Fruit juice, 624–25

Frustule, 352–56

Fucoxanthin, 352

Fuel storage tank, 583

Fumarate, 162, 163f, 168t, 170

Fungi, **359–73**

classification of, **365–73,** 366t

dimorphic, 361, **378,** 378–79f, 379t

in freshwater habitats, 587

industrial importance of, 360

in lichens, **423,** 424f

morphological forms of, **361–62**

pathogenic, 132, 360, 360t

reproduction in, **362–65,** 362–63f

in soil, 582

Fungi (kingdom), 307, 307t, **365–69,** 366t

Fungicide, 122

Furuncle, 526

Furunculosis, 525t, 526

Fusarium, 92t, 601t

Fusobacterium, 160, 325, 325t, 326f, 440t, 514, 516

F. nucleatum, 514t

F. polymorphum, 90f

Gaertner, 7t

Gaffky, 7t

Gajdusek, D. Carleton, 412

Galactose, 191f

Galactose oxidase, 624t

Galactose-1-phosphate, 191f

β-Galactosidase, 190t, 191, 233–34, 234f, 283, 283f, 288, 623t

β-Galactoside permease, 233–34, 234f

Galileo, 2

Gallionella, 171t, 172

Gallo, Robert, 409

Gametangia, 353f, 362–63f, 364

Gamete, 72–73, 352, 354f, 364

Gametophyte, 352, 353f

Gamma globulin, 469, 477

Gardnerella vaginalis, 533t

Gaseous requirements, **89–94**

Gas gangrene, 7t, 328, 450, 451t

GasPak anaerobic system, 93, 94f, 142

Gastroenteritis, 322t, 323, 377t, 451t, 516t

Campylobacter, 319

nosocomial, 559t

Salmonella. See Salmonellosis

Gastrointestinal tract. *See* Digestive system

Gas vacuole, **66–68,** 67f

G + C content, of DNA, **217,** 217f, 309–10, 309f

Gel electrophoresis, of DNA, 11f, 274, 275f

Geminivirus, 392t, 393

Gender, disease rates and, 556–58, 557f, 557t

Gene, 212

Gene amplification, 238, 279, 287. *See also* Polymerase chain reaction

Gene cloning, 11, **274–89.** *See also* Cloning vector; Host cells (cloning); Target gene

detecting host cells with cloned gene, **287–89**

obtaining target gene, **274–79**

steps in, 274

target gene incorporation into cloning vector, **279–85,** 280f

vector insertion into host cell, **285–87**

Gene gun, 287, 287f

Gene loss, 238

Gene machine, 278

Gene mapping, by conjugation, **265**

Generalized transduction, **255,** 256f

General recombination. *See* Homologous recombination

Generation time, 106–7, 106t, 116

determination of, 107

Gene rearrangement, 238

immunoglobulin genes, **471–73,** 472f

Gene therapy, **292–93**

Genetic code, **213–15,** 214f, 214t, 230

Genetic engineering, 272. *See also* Recombinant DNA technology

Genetic material, transfer of, **250–65**

Genetics, 212

Gengou, Octave, 7t

Genital herpes, 132t, 406, 446, 533t, **540–41**

Genital warts, 533t

Genitourinary system, **524–51**

microflora of, 440t, **532**

Genome, 212

Genomic library, 285, 286f

Genotype, 212, 306

Gentamicin, 125

mechanism of action of, 125t, 128, 128t

structure of, 128f

Genus, 306, 306t, 308

Genus name, 306

Geodermatophilus, 342t

Geography, disease rates and, **556,** 556f

German measles, 525t, 531

Germ-free animal, **432,** 432f

Germination, of endospores, 69f, 70

Germ theory of disease, 5–7f, 7t

Geukensia demissa, 344, 344–45f, 345t

Geyser, 339f

Giardia lamblia, 132, 375t, 376, 377t, 380–81, 448, 523–24, 523f, 594

Giardiasis, **523–24**

Gibbons, Ronald J., 462

Gingivitis, 514t, 515

Gingivostomatitis, 499t

Gladstone, G.P., 135

Gliding bacteria
    fruiting, 332t, **338**, 339f, 640
    nonphotosynthetic, nonfruiting, 332t, **338,** 338f, 639–40

Gliding motility, 338

Gloeotrichia echinulata, 177f

Glomerulonephritis, acute, 499, 499t, 503t

Glomus intraradices, 425f

Glucaldehyde, 155f

Glucan, 72, 576

Glucoamylase, 624t

Gluconeogenesis, **192–94**, 193f

Gluconobacter
    G. melanotenus, 623t
    G. oxidans, 627t

Glucose, 46
    catabolite repression, 236
    control of fungal dimorphism, **378**, 378–79f, 379t
    in gluconeogenesis, 193f
    metabolism of, 154t
        Embden-Meyerhof pathway, 150–52, 151f
        Entner-Doudoroff pathway, 152–53, 153f
    production in photosynthesis, 188–89, 188–89f
    structure of, 48f

Glucose isomerase, 623t

Glucose oxidase, 624t, 624

Glucose-6-phosphatase, 193f

Glucose-1-phosphate, 190, 191f, 194

Glucose-6-phosphate, 150, 151f, 152, 153f, 155–56f, 158, 159f, 190, 191f, 193f, 194, 195f

Glucose-6-phosphate dehydrogenase, 153f, 156f

Glutamate
    commercial production of, 625–26
    structure of, 43t
    synthesis of, 197t, 625f

D-Glutamate, 52, 54f

Glutamate dehydrogenase, 82

Glutamine
    in purine synthesis, 202, 203f
    structure of, 43t
    synthesis of, 197t

Glutamine synthetase, 82

Glutaraldehyde, as disinfectant, 122t, 124

Glyceraldehyde-3-phosphate, 149, 150–51f, 152, 153f, 154, 155–56f, 158, 159f, 189, 189f, 191–93f, 194, 195f

Glyceraldehyde phosphate dehydrogenase, 151f, 152, 153f

Glycerol, 86, 191f, 200

Glycerol kinase, 200

Glycerol phosphate, 202

Glycerol-3-phosphate, 169, 169f, 191f, 200

Glycerol-3-phosphate dehydrogenase, 169f

Glycine
    in purine synthesis, 202, 203f
    in Stickland reaction, 196, 196f
    structure of, 43t
    synthesis of, 197t

Glycocalyx, 50f, 61, 448

Glycogen, 47, 48f
    degradation of, 190, 190t, 191f
    structure of, 190, 190t
    synthesis of, 194

Glycogen storage granule, 190

Glycolytic pathway, 150

Glycomyces, 342t

Glycoprotein, 46, 386

Glycosylase, 249

Glycosyl transferase, 515

Glyoxylate, 201, 201f

Glyoxylate cycle, **201**, 201f

Goeddel, David V., 300

Golden Age of Microbiology, **7**, 7t

Golden algae. *See Chrysophyta*

Golgi, Camillo, 73

Golgi body, 71f, 73

Gonococcal conjunctivitis, 534

Gonococcal pharyngitis, 534

Gonorrhea, 7t, 130, 320t, 321, 321f, 444, **533–34,** 533–34f, 544–45, 554t
    geographic distribution in United States, 556, 556f

Gouda cheese, 611t

Graft rejection. *See* Transplant rejection

Gram, Hans Christian, 26

Gramicidin, 14

Gram-negative bacteria, 25–27, 27f, **316–28,** 316t, 332t, 638–40
    aerobic/microaerophilic, motile, helical/vibrioid, 316t, **318–19,** 319f, 635
    aerobic rods and cocci, 316t, **319–21,** 320t, 321f, 635–36
    anaerobic cocci, 316t, **326,** 636
    anaerobic straight, curved, and helical rods, 316t, **325,** 325t, 326f, 636
    cell envelope of, 58f, 60, 60f
    cell wall of, 54, 54f
    endotoxins of, 452–54
    facultative anaerobic rods, 316t, **322–23,** 322t, 323f, 636
    nonmotile, curved, 316t, **319,** 319f, 635
    outer membrane of, 57–60
    parasitism by *Bdellovibrio*, 429–30, 429–30f
    peptidoglycan of, 56
    plasmids of, 263t

Gram-positive bacteria, 25–27, 27f, 315f, **328–31,** 637–38
    cell envelope of, 58f, 60
    cell wall of, 54, 54–55f
    cocci, **328,** 328t, 329f, 637
    endospore-forming, **328,** 328t, 329f, 637
    irregular, nonsporing rods, 328t, **331,** 637–38
    regular, nonsporing rods, 328t, **331,** 331t, 637
    peptidoglycan of, 56

Gram stain, **25–27**, 25f, 27f, 60

Gram-variable bacteria, 26

Grana, 72, 73f, 180

Granule stain, 66

Granulocyte(s), **456,** 457t

Granulocyte colony stimulating factor, 474t

Granuloma inguinale, 533t

Grapevine closterovirus, 35f

Green algae. *See Chlorophyta*

Green bacteria, 179, 332, 333t, 573, 580

Green beer, 618

Greenhouse effect, **572,** 572f

Green nonsulfur bacteria, 312f, 315f

Green olives, 331t

Green sulfur bacteria, 187, 188f, 315f, 333, 333t, 573

Gregg, Norman McAllister, 531

Griffith, Fred, 215, 252

Griseofulvin, 132, 629t

Groundwater, 578, 587

Group translocation, 85, 87f, **88–89**

Growth
    bacterial, **106–7**
    balanced, 107
    continuous, 643–44
    control of, **119–22**
    definition of, 106
    diauxic, 236–38, 238f
    exponential, 107, 642–44
    mathematics of, 642–44
    measurement of, **107–14**, 116t
        by changes in cell activity, **114,** 115f, 116t
        by increases in cell numbers, **108–12**
        by turbidimetric methods, **113–14,** 114f
        by weighing cells, **114**
    optimal, 106
    pH effect on, **91–92,** 92t
    quantitative relationships of, **107**
    temperature effect on, **89–91**

Growth curve, **115–17,** 116f
    one-step, **404,** 404f
    synchronous, 118–19, 119f

Growth factors, **84–85,** 433

Growth media. *See* Media

GTP, 162, 232

Guanine, 44, 46f

Guanylate cyclase, 523

Gum arabic, 626

Gumma, 535, 535f

Guttulina, 366t

Gymnodinium, 356f
    G. breve, 356–57

Gymnosporangium juniper-virginianae, 360t

Habitat, 570

Haeckel, Ernst H., 306, 307t

*Hae*III, 273t

Haemophilus, 322t, 440t
    H. aegyptius, 273t
    H. ducreyi, 131f, 533t
    H. influenzae, 273t, 448, 450, 500t, 504t, 512, 512t
        antibiotic resistance in, 131f
        genome sequencing, 291
        growth factors for, 512
        identification of, 487, 512
        nosocomial infections with, 559t
        temperature range for, 90f
        transformation in, 253

Hafnia alvei, 324f

Hahn, Jerome J., 36–37

Hair follicle, 526
Halobacteriaceae, 320t
Halobacterium, 93, 102, 339f,
    340–41, 340t
  purple membrane of, 166–67
  water requirement for
    growth, 92t
Halococcus, 92t, 93, 340–41, 340t
Haloferax, 340t
Halogens, as disinfectants and
    antiseptics, 122t, **124**
Halophile, extreme, 93, 312,
    312–13f, 339f,
    340–41, 340t
Hansen, G. Armauer, 527
Hansen's disease. *See* Leprosy
Hantavirus, 542, 542f, 543t
Haploid, 72
Hapten, 468
Hartman, Paul, 602
Hastings, J.W., 173
Haustoria, 423
Hay fever, 480
Healthy carrier, 443
Heat-killing efficiency,
    measurement of,
    121, 121f
Heat resistance, of endospores,
    68, **75,** 91, 121
Heat shock regulation, 226
Heat sterilization, 121–22, 121f
Heavy chain, 469–70, 470t, 471f
  constant region of, 470, 471f
  variable region of, 470, 471f
Heavy metals
  as disinfectants and
    antiseptics, 122t,
    124–25
  resistance to, 262
HeLa cells, 397
Helical virus, **386–87,** 386f
Helicobacter pylori, 92, 291, 516,
    522, 522f, 543t
Heliobacterium, 333t
Heliothrix, 333t
Helmont, J.B. van, 2
Helper phage, 258, 411
Hemagglutination, 386, 487
Hemagglutination assay, of
    virus, 398
Hemagglutination inhibition
    test, 487
Hemagglutinin, 386, 513, 513f
Hemocytometer chamber, 108
Hemolysin, 448–49, 449t
α Hemolysis, 95, 96f, 502–3,
    502f, 503t
β Hemolysis, 95, 96f, 502–3, 502f,
    503t, 527
γ Hemolysis, 95, 96f, 502–3,
    502f, 503t

Hemolytic disease of
    newborn, 480
Hemolytic pattern, of
    streptococci, **502–3,**
    502f, 503t
Hemolytic-uremic syndrome,
    322, 516t, 521t,
    523, 543t
Hemorrhagic fever, 542, 550
Hepadnaviridae, 394t
HEPA filter. *See* High-efficiency
    particulate air filter
Heparin, 480t
Hepatitis, 443, 454
  infectious, 524, 524t
  serum, 524, 524t
  viral, **524,** 524t, 557t
Hepatitis A virus, 477, 524, 524t
Hepatitis B vaccine, 295, 295f,
    478, 478t, 524, 607f
Hepatitis B virus, 394t, 524,
    524t, 543t
Hepatitis C virus, 524, 524t, 543t
Hepatitis virus, 295, 582, 586, 594
Heptachlor, 586
Herbicide-tolerant plant, 297
Herbivore, **575**
Herd immunity, **478**
Herpes simplex virus, 499t,
    525t, 533t
  herpes simplex virus type 1,
    409, 531
  herpes simplex virus type 2,
    409, **540–41**
Herpes vaccine, 541
Herpesviridae, 394t
Herpesvirus, 387, 395, 405f, 408–9
  antiviral drugs, 132, 132t
Hershey and Chase experiment,
    216, **400,** 401f
Hesse, Frau, 98
Heterocyst, 177f, 334–35,
    335f, 423
Heteroduplex region, 253
Heterofermentative
    microorganism, **158–59**
Heterogamous algae, 352
Heteropolysaccharide, 627
Heterotroph, **81,** 81t, 174, 192
Heterotrophic carbon dioxide
    fixation, **574**
Hexacapsula, 375t
Hexachlorophene, 122t, 123, 123f
Hexokinase, 150, 151f, 153f,
    156f, 193f
Hexose
  control of yeast-mycelium
    dimorphism, **378,**
    378–79f
  degradation of, 191–92, 191f
  synthesis of, **192–94,** 193f

Hexose monophosphate shunt.
    See Pentose phosphate
    pathway
Hexuronic acid, 70
Hfr cells, 265
Hib vaccine, 512
High-efficiency particulate air
    (HEPA) filter, 120
High-energy compounds,
    148–49, 148t
High-frequency recombination,
    264–65
High-frequency
    transduction, 258
High-stringency conditions,
    287–88
Hill, Robin, 184
Hill-Bendall scheme, 184–85, 185f
*Hind*III, 272, 273t
Hippocrates, 550
Hirsutella, 585t
Histamine, 479, 480t
Histidase, 205f
Histidine
  in prodigiosin synthesis,
    204, 205f
  structure of, 43t
  synthesis of, 197t
Histocompatibility antigen
    (HLA), 473
Histone, 72, 72f, 218
Histonelike protein, 65
Histoplasma capsulatum. See
    Ajellomyces capsulatus
Histoplasmosis, 360, 360t, 500t,
    512, 513f
HIV. *See* Human
    immunodeficiency
    virus
Hives, 480
HLA. *See* Histocompatibility
    antigen
Hoffmann, Erich, 7t
Holdfast, 337, 359
Hollandina, 317, 427
Home canning, 616
Homofermentative
    microorganism, **158–59**
Homologous recombination,
    252, 252f
Homopolysaccharide, 627
Honey, 517
Hook, flagellar, 62, 62f
Hooke, Robert, 2
Hookworm, 525
Hop stunt, 389
Hordeivirus, 392t
Horne, Robert, 387
Host cell (cloning)
  cloning vector insertion into,
    285–87

detecting cells with cloned
    gene, **287–89**
Host defense, **448.** *See also*
    Immunity
  against invasive pathogens,
    448–50
  against microbial invasion,
    475–81
  antimicrobial
    substances, 456
  general health and
    physiological
    condition, **454**
  inflammation, 458–60, 460f
  innate, **454–61**
  phagocytosis, **460–61,** 461f
  physical barriers to disease,
    455–56, 455f
  protective factors in blood,
    456–58
  reticuloendothelial and
    lymphatic systems,
    458, 459f
  tuberculosis and, **453**
Host-parasite relationship, 418,
    442–48
Hot springs, 91, 582
Hot sulfur springs, 339,
    339f, 341
HPr protein, 89
HTLV. *See* Human T cell
    lymphotrophic virus
Human adenovirus 2, 394t
Human Genome Project, 291
Human growth hormone,
    296f, 296
Human immunodeficiency virus
    (HIV), 395, 395t, 414,
    533t, 543t
  antiviral drugs, 132, 132t
  attachment and entry into
    CD4 cell, 537f, 539f
  evolution of, 464
  HIV-1, 536–37, 546
  HIV-2, 536
  identification with nucleic
    acid probe, 295
  replication of, 537, 538–39f
  structure of, 536f
Human T cell lymphotrophic
    virus (HTLV), 409, 543t
  HTLV-I, 409
  HTLV-II, 409
  HTLV-III. *See* Human
    immunodeficiency
    virus
Humoral immunity, **468–82**
Hutchinson's teeth, 535
Hyaluronidase, 449t, 450
Hybridization, 288
  colony, 288–89, 289f

molecular taxonomy, 310–11, 310f

subtractive, 278

Hybridoma, 475

Hydrocarbon-degrading microorganisms, **583–84**

Hydrogen
cellular content of, 84t, 570, 570t
characteristics of element, 647t
as electron donor, 80–81, 574–75
in photosynthesis, 186–87, 187f
function in cells, 84t
oxidation of, **172,** 573–74

Hydrogenase, 172, 574f

Hydrogenation reaction, 145

Hydrogen bacteria, 171t

Hydrogen bonds, 649–50, 650f
in base pairing, 216f, 218

Hydrogenobacter, 337

Hydrogenomonas, 172

Hydrogen-oxidizing bacteria, 336–37

Hydrogen peroxide, 93, 461

Hydrogen sulfide, 80
as electron donor, 333, 580
in photosynthesis, 187, 187f
as energy source, 80–81
formation from sulfate, 83, 83f
oxidation of, **171–72,** 573–74

Hydrogen sulfide production test, 324f

Hydrogen swell, 616

Hydrolase, in industrial processes, 623t

Hydrologic cycle, 586

Hydrophilic substance, 650–51

Hydrophobic substance, 651

Hydrothermal vent, 80, 80f, 340–41, 344

Hydroxamate, 88

β-Hydroxybutyryl-CoA, 160f

Hydroxylamine, 247, 248t

Hydroxyl free radical, 93, 461

Hyperchromicity, of denatured DNA, 217, 217f

Hypersensitive reaction, **479–81,** 479t
antibody-mediated, 479, 479t
delayed, 479, 479t
immediate, 479, 479t
type I, 469

Hyphae, 342, 361, 362f, 423

Hyphomicrobium, 19, 337, 337f

Hypochytriomycota, 366t, **369**

Hypolimnion, 587

Hypovirus, 390

Ice cream, 609t

Ich (fish disease), 369

Ichthyophthirius, 375t

ID$_{50}$, 447

Idiotype, 470

Idoxuridine (IUdR), 132, 132t, 540

Ig. *See* Immunoglobulin

i gene, 233–34

IL. *See* Interleukin

Ilarvirus, 392t

Illumination
critical, 24–25, 25f
Köhler, 24–25, 25f
in microscopy, **24–25,** 25f

α-Imino acid, 42

Immediate hypersensitive reaction, 479, 479t
type I, 479, 479t
type II, 479t, 480–81
type III, 479t, 480–81

Immobilized enzymes, 622–23

Immobilized microbial cells, 622–23, 623t

Immune complex, 481

Immune system, evolution of, 494

Immunity
cell-mediated, 468, **482–85**
humoral, **468–82**

Immunocompromised host, 429

Immunodiffusion test, 486

Immunoelectrophoresis, **486,** 487f

Immunofluorescence, 31
to detect antigens or antibodies, **489–90,** 489f

Immunogenicity, 468

Immunoglobulin (Ig), **469,** 470t.
See also Antibody
bivalent, 470
evolution of, 494
formation in response to antigens, **477,** 477f
specificity of, 470
structure of, **469–71,** 470t, 471f

Immunoglobulin (Ig) genes, rearrangement of, **471–73,** 472f

Immunoglobulin A (IgA), **469,** 470t
secretory, 469
subclasses of, 470

Immunoglobulin D (IgD), **469,** 470t, 475

Immunoglobulin E (IgE), **469,** 470t

Immunoglobulin G (IgG), **469,** 470t, 480
subclasses of, 470

Immunoglobulin M (IgM), **469,** 470t, 475, 480, 494

Immunologic response, **475–81**
anamnestic, 477
primary, 477, 477f
secondary, 477, 477f

Immunologic tolerance, 473

Immunology, **7–8,** 468

Immunosuppression, 429, 479, 484–85

Impetigo, 328, 451t, 525t, 526, 526f

Incidence rate, 554

Incidence study, **561–63**

Incident-light excitation fluorescence microscope. *See* Epifluorescence microscope

Incineration, for sterilization, 121

Inclusion body, **65–66,** 71f

Incubation period, **446,** 446f

Incubatory carrier, 443

Indeterminate leprosy, 527

Indicator organism, for bacteria in water, **590–93**

Indirect ELISA, 490, 491f

Indirect fluorescent antibody test, 489f, 490

Indirect hemagglutination, 487

Induced mutation, 246

Inducer, 233–34, 234f

Induction
of prophage, 410f, 411
of protein, **233–34,** 234f, 236

Industrial processes, **619–23,** 619f, 633
conditions in industrial fermentors, **621**
culture methods, **621–23**
media for, **620,** 620t
microorganisms for, 11–12, **620–21**

Infant, disease rates among, 556

Infant botulism, 517

Infecting dose, 455, 516

Infection, 442
latent, **446–48**

Infection control program, 559

Infection thread, 419f, 420

Infectious abortion in cattle, 321

Infectious disease, 442
natural history of, **446–48,** 446f

Infectious hepatitis, 524, 524t

Inflammation, 453, 454t, **458–60,** 460f, 480
acute, 458, 460f

chronic, 458, 460f
subacute, 458, 460f

Inflammatory mediator, 458

Influenza, 443–44, 444t, 446, 500t, 513, 555, 557t

Influenza vaccine, 478, 513

Influenza virus, 386, 395, 395t, 500t, 513, 513f
antigenic shifts in, 513, 555
antiviral drugs, 132, 132t

Infusions, 13

Inhibitor, of enzyme, 143

Initiation codon, 230

Initiation complex, **229–32,** 231f, 233

Initiation factors, 232

Innate host resistance, **454–61**

Inoculating loop, 96, 97f, 121

Inorganic compounds, storage depots for, 65–66

Inorganic ions, content of cells, 81t

Inoviridae, 391t

Insect, bacterial symbionts of, **427,** 427f

Insertion, 246

Insertional inactivation, 281

Insertion sequence (IS), **260,** 261f

Insulin, genetically-engineered, 296–97, 296t, 298–300f, **300–301,** 301t

Interference
constructive, 28
destructive, 28

Interference contrast microscope, 30

Interferon, 132, 406, 407f, 456, 473, 474t, 484, 540
genetically-engineered, 296t, 296

Interleukin (IL), 473

Interleukin-1 (IL-1), 474t, 475

Interleukin-2 (IL-2), 474t, 475, 483

Interleukin-3 (IL-3), 474t

Interleukin-4 (IL-4), 474t

Interleukin-5 (IL-5), 474t

Interleukin-6 (IL-6), 474t

Interleukin-7 (IL-7), 474t

Interleukin-8 (IL-8), 474t

Interleukin-9 (IL-9), 474t

Interleukin-10 (IL-10), 474t

Interleukin-11 (IL-11), 474t

Interleukin-12 (IL-12), 474t

Interleukin-13 (IL-13), 474t

Interleukin-14 (IL-14), 474t

Intermittent carrier, 443

Internal cell structures, **65–68**

Internal membrane system, 52, 53f

Internet resources, 645–46

Interrupted mating
    experiment, 265
Interview survey, **560**
Intracellular parasite, 7, 326, 430
Intrasporangium, 342t
Intrinsic protein, 52
Intron, 212, 213f, 224, 229, 233
Invasiveness of pathogen, **448–50**
Invertase. *See* Sucrase
Inverted repeat, 260
In vitro antibody-antigen
    reactions, **485–90**
Iodine
    characteristics of
        element, 647t
    tincture of, 122t, 124
Ion, 648
Ionic bond, 648–49, 649f
Ionic compound, 649
Ionizing radiation, for
    sterilization, **122**
Ionosphere, 601
Iridoviridae, 394t
Iris diaphragm, 22f, 24–25
Iron, 570
    characteristics of
        element, 647t
    content of cells, 84t
    as energy source, 80
    function in cells, 84t
    oxidation of ferrous iron to
        ferric iron, **172**
    requirement for, 84
    toxin production and, 411
    transport of, **88**
    in virulence of *N. gonorrhoeae,*
        100, 101t
Iron bacteria, 171t, 172
Iron-oxidizing bacteria, 172f, 336–37
Iron-sulfur protein, 162, 164–66,
    167f, 185–86
Irradiated food, 122
Irregular, nonsporing gram-positive
    rods, 328t, **331,** 637–38
Irreversible inhibition, 145
IS. *See* Insertion sequence
Isocitrase, 201
Isocitrate, 162, 163f, 201, 201f
Isocitrate dehydrogenase, 162
Isogamous algae, 352
Isolation unit, for germ-free
    animals, 432f
Isoleucine, 43t, 197t
Isomerase, in industrial
    processes, 623t
Isopropanol, **157–60,** 160f
    as disinfectant, 122t, 123
Isotope, 647–48
Isotype, 471
Issacs, Alick, 406
Itaconic acid, 626

IUdR. *See* Idoxuridine
Ivanowsky, Dmitri, 8

Jacob, François, 233
Janssen, Zacharias, 2
J chain, 471
Jeffreys, Alex, 292
Jenner, Edward, 8, 8f, 483
Jerky, 615
J gene, 471, 472f
Juniper rust, 360t

K_{eq}. *See* Equilibrium constant
Kanamycin
    commercial production
        of, 629t
    mechanism of action of, 125t,
        128, 128t, 629t
Kaposi's sarcoma, 540
Karyokinesis, 74
KDO. *See* Ketodeoxyoctonate
Kefir, 154, 331t, 609t, 610–11
Keilin, D., 164
Kelp, 350, 355
Kepler, Johann, 2
Keratoconjunctivitis,
    herpesvirus, 132t
Ketodeoxyoctonate (KDO), 60
2-Keto-3-deoxy-
        6-phosphogluconate,
        152, 153f
2-Keto-3-deoxy-6-
        phosphogluconate
        aldolase, **152–53,** 153f
5-Keto gluconic acid, 627t
α-Ketoglutarate, 162, 163f, 197t,
        573, 573f
α-Ketoglutarate dehydrogenase,
        162, 625, 625f
Ketone body, 201
Kibdelosporangium, 342t
Kilham rat virus, 394t
Killed vaccine, 478, 478t
Killer cells, 482, 484
Killer factor, 425
Kilocalorie (kcal), 140
Kineosporia, 342t
Kinetic energy, 140
Kingdom, 306–7
Kirby, William, 133
Kirby-Bauer test, 133, 133f
Kitasato, S., 7t
Kitasatosporia, 342t
Klebs, Edwin, 7t, 509
Klebsiella, 83, 322t, 324f
    microflora, 440t
    nitrogen fixation by, 577
    plasmids of, 263t
    K. pneumoniae, 61f, 450, 500t,
        504t, 532, 532t, 559t, 591
    drug resistance in, 262

nitrogen fixation by, 422,
        434–35, 434f
Koch, Robert, 5, 5f, 7t, 98,
        323, 483
Koch's postulates, **5–7,** 6f
Kohler, Georges, 475
Köhler illumination, 24–25, 25f
Koplik's spots, 531
Kornberg, Arthur, 222
Koumiss, 609t, 610–11
Krebs, Sir Hans, 162
Krebs cycle. *See* Tricarboxylic
        acid cycle
Kuru, **412–13,** 412–13f, 412t

Labyrinthomorpha, 375t
Labyrinthula, 366t, 375t
Labyrinthulomycota, 366t, **369**
Lachnospira, 325t
    L. multiparus, 428t
lac-insulin plasmid, 300f
lac promoter, 287
β-Lactamase, 125, 130–31, 527,
        544–45
Lactase, 624
Lactate, 159f, 161, 161f, 192f
    commercial production
        of, 622
    degradation of, 157
    production from
        pyruvate, **154**
    in rumen fermentation, 428t
Lactate dehydrogenase, 154
Lactic acid bacteria, **154,** 331
    growth factors for, 85
    homofermentative and
        heterofermentative,
        158–59
    production of fermented dairy
        products, 609–10, 609t
Lactic acid fermentation, 157f
Lactobacillus, 154, 158, 190, 331
    in digestive tract, 516
    fermentation pathways
        in, 157f
    in food industry,
        331, 331t
    in meat spoilage, 614t
    microflora, 440t
    nucleoside requirement
        of, 84
    in oral cavity, 514, 514t
    pH tolerance of, 92
    preparation of fermented
        foods, 617, 617t
    water requirement for
        growth, 92t
    L. acidophilus, 92t, 106t, 331t,
        609t, 610
    L. arabinosus, 441, 441f
    L. brevis, 331t, 617t

L. bulgaricus, 331t, 609t,
        610–11, 611t,
        613, 623t
    L. casei, 331t, 611t
    L. delbrueckii, 617t, 622
    L. helveticus, 331t, 611t, 613
    L. lactis, 90f, 331t
    L. plantarum, 331t, 617, 617t
    L. sanfrancisco, 617t
Lactococcus, 154, 158
Lactose
    degradation of, 190t, 191
    structure of, 190t
    transport of, 88
    uptake of, 89
Lactose intolerance, 624
Lactose operon, 233–34, 234f,
        236, 238–39f
lacZ gene, 288
Lag phase, **115–16,** 116–17f, 135
Lake, 587–88
    zonation of, 587, 587f
Laminaria digitata, 359f
Lamprobacter, 333t
Lamprocystis, 333t
Lancefield, Rebecca, 503
Laser-scanning confocal scanning
        microscope, **34**
Lassa fever, 394t, 542f, 543t
Latent infection, **446–48**
Latent period, in one-step growth
        curve, 404, 404f
Latex agglutination test, 487
Laverans, 7t
LD$_{50}$, 447, 447f
Leaching. *See* Microbial leaching
Lead, microbial leaching of
        ores, 582
Leader peptide, 236, 237f
Leavening, 157, 368, 617, 633
Lecithinase, 449t
Lederberg, Joshua, 254–55
Leeuwenhoek, Anton van, 2, 4,
        4f, 19
Leghemoglobin, 320, 423
Legionellaceae, 320t
Legionella pneumophila, 500t,
        504t, 505–7, 505f
Legionellosis, 500t, **505–7,** 505f,
        543t, 556
Legionnaires' disease. *See*
        Legionellosis
Legume, 12, 83, 297, 320, 417f,
        418–23, 419–22f,
        435, 576
Lehmann, K.B., 314
Leishmania, 375t
    L. donovani, 376, 377t, 445t
Leishmaniasis, 445t
Lepra cells, 527
Lepromatous leprosy, 527, 527f

Leprosy, 331, 446, 525t, **527–28,** 527–28f
  borderline, 527
  indeterminate, 527
  lepromatous, 527, 527f
  tuberculoid, 527–28
Leptosphaeria, 601t
Leptospira, 316–17
  L. interrogans, 305f, 316–17, 318f, 444t
Leptospirosis, 316–17, 444t
Leptothrix, 20
Leptotrichia, 325t
Lettuce necrotic yellow virus, 392t
Leucine, 43t, 197t
Leuconostoc, 154, 614t, 616, 617t
  L. citrovorum, 611t
  L. cremoris, 609t, 610
  L. dextranicum, 609t, 610, 627
  L. mesenteroides, 154, 617, 617t, 627
Leucoplast, 72
Leukemia, 409, 625
  acute lymphatic, 409
Leukemia-lymphoma virus, 409
Leukocidin, 449t, 450, 451t, 461
Leukocytes, 456, 457t
Leukopenia, 456, 492
Leukotriene, 479, 480t
Levine's EMB agar, 593
Leviviridae, 391t
lexA gene, 249, 249f
L-form. *See* L-phase variant
LHT system, 387
Lichen, **423,** 424f
  as indicators of air pollution, 423
Licmophora, 356f
Light, interactions with matter, 182, 184f
Light beer, 618
Light chain, 469–70, 470t, 471f
  constant region of, 470, 471f
  kappa, 470t, 471
  lambda, 470t, 471
  variable region of, 470, 471f
Light energy, 80–81
Light microscope, 20
  compound, **21–25**
  illumination of specimen, **24–25,** 25f
  magnification by lenses, **22–23**
  resolution of, **23–24,** 24f
Lignin, 72, 190t, 191
Ligninase, 190t
Limburger cheese, 611t, 614
Limiting nutrient, 117, 643–44
Limnetic zone, 587, 587f
Limulus amoebocyte lysate assay, **452**

Lindenmann, Jean, 406
Lindow, Steven, 618
Linear measurement, 18
Linnaeus, Carolus, 306, 307t
Lipase, 200, 460
  industrial uses of, 624t
Lipid
  comparison of Bacteria, Archaea, and Eucarya, 313t
  content of cells, 81t
  with ester linkages, 51
  with ether linkages, 51, 91
  membrane, 340
  metabolism of, **197–202**
  of mycobacteria, 510–11
  storage depots for, 65–66
  structure of, **46–48,** 49f
Lipid A, 60, 60f
Lipid bilayer, 49–51, 51f, 386
Lipid body, 50f
Lipid-degrading bacteria, 95
Lipman, Fritz A., 162
Lipman, Jacob, 580
Lipomyces starkeyi, 586
Lipopolysaccharide (LPS), 46, 57–60, 450–54
Liquid culture, estimating number of viable cells in, **111–13,** 112f, 113t
Lister, Joseph, 123
Listeria monocytogenes, 613
Listeriosis, milk-borne, **613**
Littoral zone, 587, 587f
Live, attenuated vaccine, 477–78, 478t
Loeffler, Friedrich, 7t, 9, 509
Logarithmic phase. *See* Exponential phase
Loose smut of wheat, 360t
Lophotrichous flagellation, 63
Louse-borne relapsing fever, 318t
Louse-borne typhus. *See* Epidemic typhus
Louse vector, 445t
Löwenstein-Jensen medium, 511
L-phase variant, 59
LPS. *See* Lipopolysaccharide
L ring, 62, 62f
Lucibacterium, 173
Luciferase, 173, 173f, 173t, 425
Luciferase gene, 287, 288f, 299f
Luminescent bacteria, **173,** 173f
  in flashlight fish, **425,** 425f
Lung cancer, 562, 562f
Luteovirus, 392t
Lwoff, André, 387
Lyase, in industrial processes, 623t
Lyme disease, 316–17, 445t, **530,** 530f, 543t

Lymph, 458
Lymphatic system, **458,** 459f
Lymph node, 458, 459f
Lymphocytes, 456, 457t, **458**
Lymphogranuloma venereum, 326, 327t, 533t
Lymphoid tissue, **468**
Lymphokine, 473
Lymphoma, Burkitt's, 409
Lyophilized culture, **91**
Lysine
  commercial production of, 625–26
  decarboxylation of, 198, 199f
  structure of, 43t
  synthesis of, 197t, 626f
Lysine decarboxylase, 199f
Lysine decarboxylase test, 324f
Lysis. *See* Cell lysis
Lysogen, 410–11
Lysogenic conversion, **258, 266,** 266–67f, 411, 452, 499, 508
Lysogenic cycle, of bacteriophage, 255, **410–11,** 410f
Lysosome, 71f, 73
Lysozyme, 405, 456, 460
Lytic cycle, of bacteriophage, 255, 256f, 408, 410–11, 410f
Lyticum, 426t
  L. flagellatum, 328, 426f

MAC. *See* Macrophage activating factor
MacLeod, Colin M., 216
Macrolide, 129
Macromolecule, **42–48**
Macronucleus, 373, 374f, 377
Macroorganism, 18
Macrophage(s), 439f, 457t, 458, 459f, 461f, 475, 482
  activation of, **483**
  alveolar, 498
  phagocytic activity of, **483**
  stationary, 458
  wandering, 458
Macrophage activating factor (MAC), 483
Macrophage chemotactic factor (MCF), 483
Macrophage migration inhibitory factor (MIF), 483, 485
Macrosporium, 601t
Mad cow disease, 389, 543t
Maduromycetes, 342t, 343, 641
Madurose, 343
Magnesium
  cellular content of, 84t, 570, 570t
  characteristics of element, 647t

function in cells, 84t
requirement for, 84
Magnetosome, 65, 66f
Magnetotactic bacteria, 336–37
Magnetotaxis, 65, 66f, 337
Magnification, 2, 22–23
  comparison of various microscopes, 31t
  empty, 23
Maize streak virus, 392t
Major histocompatibility complex (MHC), **473–75,** 494
  class I proteins, 473–75
  class II proteins, 475
  class III proteins, 475
Major histocompatibility complex (MHC) genes, 473
Malaria, 7t, 132, 377, 377t, 444–45t, 445, 543t, 554t
Malate, 163f, 201, 201f
Malate synthetase, 201, 201f
Malignant tumor, 408
Mallon, Mary, **442**
Malnutrition, 454
Malonate, 143, 144f
Malonyl-CoA, 202
Maltase, 190t, 191
Malting, 617
Maltose, 190t, 191
Manganese-oxidizing bacteria, 336–37
Mannose, 191f
Mannose-6-phosphate, 191f
M antigen, 503
Marburg virus, 543t
Marine habitat, 171, 575, **586–87**
Marine saltern, 102, 339
Marmur, Julius, 272
Marsh, 575, 587
Marshall, Barry J., 522
Marteilia, 375t
Massa, J., 173
Mast cells, 479
Mastigocladus laminosus, 90f
Mastigophora, 375–76
Mastitis, 443, 506
Matching coefficient, 309, 309t
Maternal antibody, 477
Mating type, 362–63f, 365–66
Matrix protein, 386
Maximum temperature, 89
Mayer, Leonard, 544
MBC. *See* Minimal bactericidal concentration
McCarty, Maclyn, 216
McClintock, Barbara, 260
MCF. *See* Macrophage chemotactic factor
McMunn, C.A., 164

MCP. *See* Methyl-accepting chemotaxis protein
Measles, 454, 525t, 531, 552, 553f, 554t
Measles virus, 395, 395t, 408, 531
Meat, **614–15**
  preservation of, **614–15**, 615f
  spoilage of, **614–15**, 614t
Meat products, **614–15**
Mechanical vector, 445
Mechanical work, 140
Media, 79f, **94–96**
  buffers in, 92
  complex, **94–95**, 95t
  differential, **95**
  for industrial processes, **620**, 620t
  isolation and enrichment of specific microorganisms, **95–96**
  selective, **96**
  sterilization of, 620, 621f
  synthetic, **94–95**, 95t
Medical microbiology, 11, 440
Medicine
  applications of recombinant DNA technology in, **291–97**
  roles of microorganisms in, 11–12
Mefloquine, 132
Meiosis, 74, 76t, 352
Meiospore, 353–54f
Melanoma, 408
Melting curve, of DNA, **217**, 217f, 309–10, 309f
Membrane-attack complex, 481–82f, 482
Membrane filter procedure
  for coliform determination, 591–93, 592–93f, 602–3
  for determination of viable count, **111**, 112f
Memory cells, 475, 476f, 477
Menaquinone, 170f, 171
M-Endo medium, 96, 97f, 593, 593f
Meningitis, 320t, 321, 322t, 557t
  bacterial, 500t, 503t, 504, **512**, 512t
  meningococcal, 512
Meningococcemia, 512
Meningoencephalitis, 376, 377t
Menstruation, toxic shock syndrome and, 564–65, 564t, 565f
Merbromin (Mercurochrome), 122t, 124
Mercurochrome. *See* Merbromin
Mercury, 143
Merozygote, 250

Merthiolate. *See* Thimerosal
Meselson and Stahl experiment, 219–20, 219f
Mesophile, 89, 90f, **91**
Mesosome, 52
Messenger RNA (mRNA), 46, 212, **224**
  capped, 229, 233
  eucaryotic, 212, 213f, 229, 233
  genetic code, 213–15
  leader, 236
  monocistronic, 212
  polyA tail of, 229
  polycistronic, 212, 213f, 233
  procaryotic, 212, 213f
  processing of, 229
  reading frame of, 230
  ribosome binding site on, 232
  splicing of, 229
  synthesis of, 226–28
  template for synthesis of target gene, **274–78**, 278f
Messenger RNA (mRNA) precursor, 229
Metabolic burst, 461
Metabolic diversity, 174
Metabolic plasmid, 263t
Metabolism, 149
  antibiotics that inhibit, 125, 125t, **129**
  concepts of, **149**
  overview of, 207f
Metabolite
  commercial production and usefulness of, **625–30**
  primary, 204, **625–28**
  secondary, **204, 629–30**
Metachromatic granule, 66, 67f
Metal layering, 37
Metaphen. *See* Nitromersol
Metarrhizium, 585t
Metchnikoff, Élie, 7–8
Metchnikovella, 375t
Methane, 573–74
  commercial production of, 598, 599f
  as energy source, 81
  production from wastes, **598**, 599f
Methane-oxidizing bacteria, 81
Methanobacterium, 93, 312–13f, 339–40, 340t, 441, 574
  M. thermoautotrophicum, 91, 341f
Methanobrevibacter, 339, 340t
  M. ruminantium, 427, 428t
Methanococcus, 312–13f, 340, 340t, 441, 574
  M. jannaschii, genome sequence of, 291, 311–12

Methanogen, 312, 313f, 339–40, 340t, **574–75**, 574f, 598
  in carbon cycle, 571
  in rumen, 427
  syntrophism in, 441
Methanolobus, 340, 340t
Methanomicrobium, 340, 340t
  M. mobile, 428t
Methanopterin, 340, 574f
Methanosarcina, 313f, 340, 340t, 441, 574
  M. barkeri, 341f
Methanospirillum, 313f, 340t, 427, 574
Methanothermus, 339, 340t
Methicillin, 126f, 130
Methionine
  as initial amino acid, 229–30
  structure of, 43t, 229f, 580
  synthesis of, 197t
Methyl-accepting chemotaxis protein (MCP), 63, 64f
Methylesterase, 63, 65
7-Methylguanosine, 224, 229
ε-*N*-Methyllysine, 62
Methyl methanesulfonate, 247–48, 248t
N-Methyl-*N*-nitro-*N*-nitrosoguanidine, 248, 248t
Methylococcaceae, 320t
Methylococcus, 321
Methylomonadaceae, 321
Methylomonas, 321
Methylophilus methylotrophus, 631
Methyl red test, 161
Methyl reductase, 574f
Methyltetrahydrofolate, 202, 203f
Methyltransferase, 63
Methyltroph, 81, 81t
Metronidazole, 132
MHC. *See* Major histocompatibility complex
MIC. *See* Minimal inhibitory concentration
Microaerophile, 93
Microbial ecology, 11, 570
Microbial leaching, **582–83**, 604
Microbial mat, 80f, 345f
Microbial plastics, 66
Microbiological assay, 85
Microbiologist, 9
Microbiology
  definition of, 2
  historical perspectives on, **2–9**, 14
  medical, 11, 440
  modern, **9–12**

Microbispora, 342t
Microbody, 71f, 73
Micrococcus, 328, 587, 614–15t, 615–16
  M. cryophilus, 90f
  M. luteus, 170f, 171
Microcyclus, 319
  M. aquaticus, 319f
Microcystis, 587
  M. aeruginosa, 67f, 92
Microflora, **440–41**
  commensalism in, 433
  of digestive system, 441, **515–16**
  of genitourinary system, **532**
  germ-free animal, 432, 432f
  of oral cavity, **514**, 514t
  resident, 440
  of respiratory tract, **498**
  of skin, **525**
  transient, 440
Microgravity conditions, microbes in space, 10
Micromonospora, 342t
Micronucleus, 373, 374f, 377
Microorganism, 18
  multicellular, 18
Microscope, **20–37**. *See also specific types*
  invention of, 2, 4
Microspora, 375t
Microsporidia, 381, 585t
Microsporum, 360t, 444t, 525t, 530
Microtetraspora, 342t
Microviridae, 391t
Middle lamella, 72
MIF. *See* Macrophage migration inhibitory factor
Migration inhibition factor, 474t
Miliary tuberculosis, 510
Milk, **608–14**
  fermented products, 154, **609–11**, 609t
  listeriosis outbreak from, **613**
  microbiological standards for, 608, 609t
  pasteurization of, 119–20, 608, 613
    high-temperature short-time (flash) method, 120, 608
    low-temperature holding method, 119–20, 608
  raw, **608**
  salmonellosis from dairy plant, **610**
  souring of, 191
  spoilage of, 608
  ultra high-temperature, 608
Milstein, Cesar, 475

Mineralization, 570
Minimal bactericidal
    concentration
    (MBC), 133
Minimal inhibitory concentration
    (MIC), 133
Minimum temperature, 89
–35 sequence, 226
Missense mutation, 246, 247f
Mitchell, Peter, 165
Mitochondria, 71f, 72, 77, 162
    DNA of, 239
    evolution of, 206, 436
    oxidative phosphorylation in,
        169, 169f
    in petite mutants, 238–39
    structure of, 73f
Mitochondrial matrix, 72, 73f
Mitosis, 74, 76t
Mixed acid fermentation, 157f,
    160–61, 161f
Mizutani, Satoshi, 409
MMR vaccine, 531–32
MMWR. *See Morbidity and*
    Mortality Weekly
    Report
mob genes, 261
Molasses, 620, 620t
Mold, **361–62,** 361f
Molecular biology, 14
Molecular chronometer, 268, 311
Molecular taxonomy, **309–11,**
    309–10f
    G + C content, 309–10, 309f
    nucleic acid hybridization,
        310–11, 310f
Molybdenum, 570
Monera, 307, 307t
Monilinia fructicola, 360t
Monocistronic mRNA, 212, 213f
Monoclonal antibody, **475**
Monocyte(s), 456, 457t, 458
Monocyte colony stimulating
        factor, 474t
Monod, Jacques, 233
Monoglyceride, 47
Monolayer, 395
Mononuclear phagocytic tissue.
        See Reticuloendothelial
        system
Monosaccharide, 46, 48f
Monosodium glutamate,
        commercial production
        of, 12
Monotrichous flagellation, 62
Monterey Jack cheese, 611t
Monuron, 586
Morbidity, **553–54,** 554t
    age of patient and, 556, 557f
    gender and, 556–58, 557f, 557t
    race and, 557–58f, 557t, 558

Morbidity and Mortality Weekly
    Report (MMWR),
    550, 551f
Morganella morganii, 324f
Mortality, 553
Mortality rate, **553–54,** 554t
    age of patient and, 556, 557f
    gender and, 556–58, 557f, 557t
    race and, 557–58f, 557t, 558
Mortierella, 576
Mosquito-borne disease, 445,
    445t, 498
Most probable number (MPN)
        method, **111–13,** 112f,
        113t, 116t
    for coliform determination,
        591–93, 592f, 602–3
Mozzarella cheese, 331t
MPN method. *See* Most probable
        number method
M protein, 527
M ring, 62, 62f
mRNA. *See* Messenger RNA
MS-2 System, 134
Mucoid colony, 61
Mucor, 92t, 366t
    M. racemosus, 379t
    M. rouxii, 378, 378f, 379t
    M. subtilissmis, 379t
Mucous membrane, 455, 455f
Mucus, 455
Muenster cheese, 611t, 614
MUG assay, 602–3, 602f, 603t
Mullis, Kary, 11, 288
Multicellular eucaryote, evolution
        of, 38
Multicellular, filamentous, green
        bacteria, 333, 333t
Multicellular microorganism, 18
Multiple fission, 373
Multiple resistance plasmid, 262
Multitrichous flagellation, 63
Mumps, 444, 515
Mumps virus, 515
Murine typhus. *See* Endemic
        typhus
Mushroom, 368, 369f
    poisonous, 369
Must, 619
Mutagen, 268
    Ames test for, 251, 251f
Mutagenesis, site-directed,
        278–79, 279f
Mutant, separation from wild
        type cells,
        249–50, 250f
Mutation, **246–50,** 268. *See also*
        specific types of
        mutations
    definition of, 246
    reversible, **246**

Mutation rate, **246–48,** 248t,
        249, 268
Mutualism, 418, **418–29,** 433
Mycelium, 341–43, 343f, 361,
        361–63f, 368
    aerial, 361
    vegetative, 361
Mycetangia, 427, 427f
Mycobacteria, 328t, **331,** 331f,
        510–11, 638
Mycobacterium, 93, 111, 331
    M. africanum, 510
    M. bovis, 510–11
    M. leprae, 331f, 525t, **527–28**
    M. phlei, 217f
    M. smegmatis, 627t
    M. tuberculosis, 7t, 106t, 453,
        461, 500t, 510–11,
        510–11f
    drug-resistant, 131f,
        454, 511
    identification of, 295,
        511, 511f
Mycolic acid, 510–11
Mycoplasma, 316t, 327, **327,**
        327f, 637
    lack of cell wall in,
        59, 59f
    shape of, 19
    size of, 18
    sterols in, 48, 59, 77, 327
    M. genitalium, 291
    M. pneumoniae, 21f, 59f, 327,
        327f, 500t, 504t
Mycorrhiza, **423–25,** 424–25f
Mycosphaerella, 601t
Mycotoxin, 368–69
Mycotypha africana, 364f
Mycovirus, 390
Myeloma, 475
Myeloperoxidase system, 461
Myoviridae, 391t
Myxamoebae, 372–73, 372f
Myxidium, 375t
Myxobacteriales, 338
Myxoma virus, 585t
Myxomycota, 366t, **369–73**
Myxospora, 375t
Myxospore, 338
Myxovirus, 559t

NAD, 85, 85t, 148
    in photosynthesis, 178, 178f,
        182–88
    reduction of
        in Embden-Meyerhof
            pathway, 152
        in Entner-Doudoroff
            pathway, 152–53, 153f
        in pentose phosphate
            pathway, 154

    in phosphoketolase
        pathway, 154
    in tricarboxylic acid cycle,
        162, 163f
    regeneration from NADH,
        154–61
    structure of, 147f
NADH + H$^+$, donation of
        electrons to electron
        transport chain, 166
NADH + H$^+$ dehydrogenase, 165
NAD+ kinase, 623t
NADP, 85, 85t, 145t, 148
    in carbon dioxide fixation,
        189, 189f
    in photosynthesis,
        186–87, 187f
NADP oxidase, 461
Naegleria, 375t
    N. fowleri, 376, 377t
Nafcillin, 125, 125t
NAG. *See* N-Acetylglucosamine
Nalidixic acid, 125t, 129
NAM. *See* N-Acetylmuramic acid
Narrow-spectrum drug, 125
Nathans, Daniel, 9, 272
Natronobacterium, 93
Natronococcus, 93, 340t
Natural killer cells (NK cells),
        482, 484
Naturally acquired immunity, **477**
    active, **477**
    passive, **477**
Naumanniella, 337
Necrotizing fasciitis, 328, 464,
        499–500
Needham, John, 4
Negative control system, 236
Negative stain, 28, 28f, 32
Neisser, Albert, 7t, 533
Neisseria
    culture of, 93
    microflora, 440t
    oxidase test, 166
    temperature range of, 91
    N. gonorrhoeae, 7t, 124, 320t,
        321, 321f, 429, 444, 448,
        499t, **533–34,** 533–34f,
        533t
    antibiotic resistance
        in, 131f
    electron microscopy of, 33f
    identification of, 487
    penicillinase-producing,
        125, 130, 130f, 534,
        544–45, 544f, 545t
    in phagosomes, 461f
    pili of, 62
    temperature range for, 90f
    tetracycline-resistant,
        130, 130f

virulence of, iron and, **100,** 101t

N. meningitidis, 320t, 321, 500t, 512, 512t

carbon dioxide requirement, **135,** 135t

Neisseriaceae, 320t, 321

Nemalion, 355

Neomycin

commercial production of, 629t

mechanism of action of, 125t, 128, 128t, 629t

Neonatal respiratory disease, 499t, 503t, 506, 506t

Neoplasm. *See* Tumor

Nepovirus, 392t

Nested amplification, 291

Neufchâfatel cheese, 611t

Neumann, R.E., 314

Neuraminidase, 386, 513, 513f

Neurological syphilis, 535

Neurospora, 366t, 368

N. crassa, 14

Neurotoxin, 450, 451t, 517

Neutral fat, 47, 49f

Neutralization test, **487**

Neutron, 647

Neutrophils, 456, 457t

Newcastle disease virus, 408

Niacin, 85

Nicotinamide adenine dinucleotide. *See* NAD

Nicotinamide adenine dinucleotide phosphate. *See* NADP

nif genes, 297, 422, 434–35, 434f

NIH guidelines, for recombinant DNA research, 297

Nine-plus-two arrangement, of microtubules, 73, 74f

Nitrate

as electron acceptor, 150, 168, 168t

as nitrogen source, 82, **82,** 82f

in water supply, 578

Nitrate ammonification, 578

Nitrate reductase, 82, 82f, 170, 578

Nitrate reduction

assimilatory, 82, 170, 578

dissimilatory, 170, 578

Nitric oxide, 578

Nitrification, 172, 577f, **578,** 579, 599

Nitrifying bacteria, 53f, 171t, 172, 336, 336f, 573, 577f, 578

Nitrite, 573–74

as electron acceptor, 168t

as energy source, 80

in groundwater, 578

oxidation to nitrate, 172

Nitrite-oxidizing bacteria, 171t, **172,** 336

Nitrite reductase, 82, 82f

Nitrobacter, 92t, 171t, 172, 336, 578, 588

N. winogradskyi, 336f

Nitrococcus, 171t, 172, 336, 578

N. mobilis, 336f

N. oceanus, 53f

Nitrogen

cellular content of, 84t, 570, 570t

characteristics of element, 647t

function in cells, 84t

removal from sewage, 599

Nitrogenase, 82f, 320, 335, 420, 422–23, 422f

assay of, 423

oxygen-sensitivity of, 423

Nitrogenase complex, 422, 422f

Nitrogenase reductase, 422–23, 422f

Nitrogen cycle, 172, **576–79,** 577f

Nitrogen fixation, 12, **82–83,** 297, 343, 417f

abiotic, 576

crop yields and, **434–35,** 434f

by cyanobacteria, 334–35

energy required for, **422–23,** 422f

in nitrogen cycle, **576–77,** 577f

in nonleguminous crops, 435

nonsymbiotic, 576–77

scheme for, 422f

symbiotic, 320, **418–23,** 419–22f, 421t, 576

Nitrogen gas, 170, 578

atmospheric, 576

as nitrogen source, 82, **82–83**

Nitrogen source, **81–83,** 620

Nitromersol (Metaphen), 122t, 124

Nitrosamine, 578, 615

Nitrosococcus, 171t, 172, 336, 578

N. oceanus, 336f

Nitrosofying bacteria, 336, 577f, 578

Nitrosolobus, 171t, 172, 336, 578

N. multiformis, 336f

Nitrosomonas, 92t, 171t, 172, 336, 578, 588

N. europaea, 336f

Nitrosospira, 172, 336

Nitrospina, 171t, 172, 336, 578

Nitrospira, 171t, 578

N. gracilis, 336f

Nitrous acid, 247, 248t

Nitrous oxide, 170, 578

NK cells. *See* Natural killer cells

Nocardia, 171t, 172, 342, 342t, 575–76, 581, 583

Nocardioforms, 328t, 342t, 638, 640–41

Nocardioides, 342t

Nocardiopsis, 342t

Nod factor, 420

nod genes, 420

Nomarski differential interference contrast microscope, 17f, 29–30f, 30

Nomenclature, 306, 306t

Noncommunicable disease, 442

Noncompetitive inhibitor, 143, 145

Nonconjugative plasmid, **260–61**

Nonculturable microbes, 347

Noncyclic photophosphorylation, **184–86**

Nongonococcal urethritis, 326, 327t, 444, 533t, **536**

Nonhistone proteins, 72

Noninfectious disease, 442

Nonmotile, gram-negative curved bacteria, 316t, **319,** 319f, 635

Nonphotosynthetic, nonfruiting gliding bacteria, 332t, 639–40

Nonpolar molecule, 649

Nonsense codon, 213, 231f, 232–33

Nonsense mutation, 246, 247f

Nonspecific resistance. *See* Innate host resistance

Nonsymbiotic nitrogen fixation, 576–77

Norfloxacin, 125t, 129

Nori, 355

Normal flora. *See* Microflora

Norwalk virus, 394t

Nosema, 375t

Nosepiece, 22, 22f

Nosocomial infection, **559,** 559t

Nostoc, 335f, 421t, 576, 587

Nostocaceae, 335, 335t

Notifiable disease, 554t

Novobiocin, 629t

Nuclear envelope, 72

Nuclease, **202,** 460

Nucleic acid. *See also* DNA; RNA

content of cells, 81t

degradation of, 202

functions in cells, **44–46**

metabolism of, 579

structure of, **44–46,** 46f

synthesis of, antibiotics that inhibit, 125, 125t, **128–29,** 128t

Nucleic acid hybridization. *See* Hybridization

Nucleic acid probe, **287–88,** 289f

pathogen identification with, **295,** 296f

Nucleocapsid, 386, 386f

Nucleoid, 18, 50f, **65,** 67f, 71f, 212

Nucleolus, 71f, 72, 228

Nucleoside, 44

Nucleoside-requiring bacteria, 84

Nucleosome, 223

Nucleotide, 44, 46f, 81t

Nucleus, 18, 71f, 72, 313t

Null cells. *See* Killer cells

Numerical aperture, 23–24, 24f

Numerical taxonomy, **308–9,** 308f, 309t

Nutrient broth, 95t

Nutrient deprivation, 70

Nutrient loss, 61

Nutrient transport, **85–89**

Nutritional requirements, **80–85**

Nuttall, 7t

Nyctotherus, 375t

Nystatin, 125t, 129–32

commercial production of, 629t

mechanism of action of, 126, 629t

structure of, 127f

O antigen, 58f, 60, 60f

Objective lens, 21–24, 22f, 24f, 29f

oil-immersion vs. dry, 24

Obligate anaerobe, 93

Obligate parasite, 418

Observational study, **561–63,** 561f

prospective, **561–63,** 562f, 563t

retrospective, **561–63,** 563t

Occupation, contribution to disease, 454

Ocean, 586

Oceanospirillum, 319

Ocular lens, 21–22, 22f, 29f

Oerskovia, 342t

Oil-immersion lens, 24

Oil spill, 11–12, 12f, **583–84**

Okazaki, Reiji, 221, 240

Okazaki fragments, 221–24, 222f, **240,** 241f

2',5'-Oligoadenylate, 407f

Oligomycin, 168t

Oligonucleotide, synthetic, 278–79, 279f

Oligonucleotide primer, in polymerase chain reaction, 288–91, 290f
2',5'-Oligonucleotide synthetase, 407f
Oligosaccharide, 46
Oligotrophic lake, 587
Olives, 331t, 617t
Omasum, 428f, 429
Oncogene, 409
Oncogenic virus, **408–9**
One-carbon compound, 81, 81t
One-step growth curve, **404**, 404f
Oocyst, 377, 377f
Oogamy, 352
Oogonia, 369, 370f
Oomycetes, 369, 370f
Oomycota, 360t, 366t, **369**
Oospore, 369, 370f
Operator, 233, 234–35f, 235
Operon, 233, 234f
Ophthalmia neonatorum, 124
Opportunistic infection, 322t
    in AIDS, 540
    nosocomial, 559, 559t
Opsonin, 460
Optical density, 113
Optimum temperature, 89
Optochin, 504
Optochin disk test, 504
Oral cavity, **514–24**
    bacterial pathogens in, 514–15
    fungal pathogens in, 515
    microflora of, 440t, **514**, 514t
    viral pathogens in, 515
Order (taxonomy), 306, 306t
Organelles, 18, **72–73**, 76t, 313t, 373
Organic acids, commercial production of, **625**
Organic alcohols, commercial production of, **625**
Organic compounds, in carbon cycle, 570–76
Organotropism, **448**
Organ transplant, 479
Oriental sores, 377t
Origin of replication, 220, 224, 279
oriT site, 264, 264f
Ornithine, 198, 199f
Ornithine decarboxylase, 198, 199f
Ornithosis, 444t, 500t
Orthomyxoviridae, 395t
Orthomyxovirus, 400
Orthophenylphenol, 122t, 123f
Oscillatoria, 334
Oscillatoriaceae, 335, 335t
Oscillochloris, 333t

Osmophilic microorganism, 92–93
Osmotic pressure, 52, 92–93
Osmotic shock treatment, 88
Osmotolerant microorganism, 92–93
Osteomyelitis, 328
Ouchterlony, Orjan, 486
Ouchterlony test, 486, 496f
Oudin, Jacques, 485
Oudin single diffusion test, 485–86
Outer membrane, **48–49, 57–60,** 58f
    endotoxin from, **452–54**
Outer sheath, of spirochetes, 316
Outgrowth, 70
Oxacillin, 125, 125t, 126f
Oxaloacetate, 157, 162, 163f, 194, 197t, 201, 201f, 573, 573f
Oxalosuccinate, 162
Oxidase test, **166,** 166f, 321, 529, 534
Oxidation, **145**
Oxidation pond, **600**
Oxidation-reduction reaction. *See* Redox reaction
Oxidative phosphorylation, 149, 161–71, **169,** 169f
Oxidoreductase, in industrial processes, 623t
Oxygen
    cellular content of, 84t, 570, 570t
    characteristics of element, 647t
    effect on microbial growth, 119
    as electron acceptor, 150, 170f
    function in cells, 84t
    release in photosynthesis, 178, 183–84, 185f, 186
    requirement for, **93**
    sensitivity to, **93**
    toxic metabolites of, 93
Oxygen electrode, 115f
Oxygenic photosynthesis, 184
Oxygenic photosynthetic bacteria, 332t, **333–36,** 638–39
Oxygen uptake curve, 114, 115f
Oxytetracycline, 616, 629t

Pace, Norman R., 311
Palindrome, 272
Pandemic, 552–53
Panopoulos, Nickolas, 618
Panton-Valentine leukocidin, 461
Pantothenic acid, 85, 85t, 145t
Papilloma virus, 531

Papovaviridae, 394t
Papovavirus, 387, 395, 525t, 533t
Paracoccus denitrificans, 170f, 171
Paramecin, 425
Paramecium, 375t, 377
    chemotaxis in, 65
    light microscopy of, 30f
    P. aurelia, endosymbionts of, 425–27, 426f, 426t
    P. tetraurelia, 328
Paramylon, 357
Paramyxa, 375t
Paramyxoviridae, 395t
Paramyxovirus, 408, 525t, 559t
Paraquat, 586
Parasite, 418
    obligate, 418
Parasitism, 418, **429–33,** 433
Parasporal glycoprotein crystal, 585
Parmesan cheese, 331t, 611t, 614
Parvoviridae, 394t
Passive diffusion, 85, **86,** 86f
Passive hemagglutination, 487
Passive immunity
    artificially acquired, **477**
    naturally acquired, **477**
Pasteur, Louis, 4, 5f, 13, 119, 152
Pasteur effect, 152, 168
Pasteurella, 322t
Pasteurellaceae, 322, 322t
Pasteur flask, 4, 13, 13f
Pasteuria, 342t
Pasteurization, of milk, **119–20,** 608, 613
Pathogen, 440
    entry into host, **445–46**
    identification with nucleic acid probes, **295,** 296f
Pathogenicity, 448
Pathogenic microbiology, 440
Pathovar, 320
Payne, Shelley, 100
PCR. *See* Polymerase chain reaction
Peach leaf curl, 360t
Pear fire blight, 7t
Pectin, 72, 190t, 354t
    degradation of, 191, 428t
Pectinase, 190t
    industrial uses of, 624–25, 624t
Pectinatus, 325t
Pediococcus, 154, 158, 617t
    P. cerevisiae, 616, 617t
Pedoviridae, 391t
Pellicle, 357, 373
Pelodictyon, 333t
    P. clathratiforme, 183f
Pelomyxa palustris, 427
Pelvic inflammatory disease, 534

Penicillin
    commercial production of, **629,** 629t
    discovery of, 14
    mechanism of action of, 56, **125,** 125t, 629t
    production of, 361f
    resistance to, 130, 130f
    semisynthetic, 126f
    structure of, 125, 126f, 631f
Penicillin amidase, 623t
Penicillinase. *See* β-Lactamase
Penicillinase-producing *N. gonorrhoeae*, 534, **544–45,** 544f, 545t
Penicillin F, 126f, 631f
Penicillin G, 125, 126f, 631f
Penicillin K, 631f
Penicillin V, 125, 126f
Penicillin X, 631f
Penicillium, 365f, 366t, 368
    in freshwater habitats, 587
    in meat spoilage, 614t
    preparation of fermented foods, 617t
    in soil, 582
    in troposphere, 601t
    water requirement for growth, 92t
    P. camemberti, 611t
    P. candidum, 611t
    P. caseicolum, 364f
    P. chrysogenum, 361f, 621, 629, 629t
    P. griseofulvin, 629t
    P. nigricans, 629t
    P. notatum, 14, 361f, 624
    P. puberulum, 368
    P. roqueforti, 611t, 614
    P. urticae, 629t
Pentamidine isethionate, 132
Pentose, 44
    degradation of, 192f
    synthesis of, **194,** 195f
Pentose phosphate pathway, **154,** 155f, **194,** 195f
Peplomer, 386
Peptic ulcer, 92, 516, **522,** 522f
Peptide bond, 44, 44f, 232
Peptidoglycan, 52, 56, 58f
    structure of, 52–54, 54–55f
    synthesis of, **56,** 57f, 125
Peptidyl transferase, 224, 232
Peptococcus, 440t, 516
Peptostreptococcus elsdenii, 428t
Periodontal disease, 325, **515**
Periodontitis, 514t
Periplasm, 58f, **60,** 88
Periplasmic flagellation, 63, 63f
Perkinsea, 377
Permanent carrier, 443

Permease, 86
Peroxidase, 93
Persistent viral infection, **408**
Person-to-person
    transmission, 444
Pertussis, 7t, 321, 446, 451t, 500t,
    507, 507f, 542f, 554t
Pertussis toxin, 451t
Pertussis vaccine, 478t. *See also*
    DTP vaccine
Pestalozzia, 601t
Pesticide, 576
    biological, **584–85,** 585t, 604
    degradation of, 585–86
Petite mutant, 238–39
Petri, Richard J., 95
Petri dish, 95
Petroff-Hausser counting
    chamber, 108, 109f
Petroleum industry, 583–84
Phaeophyta, 352, 354t,
    357–59, 359f
Phage. *See* Bacteriophage
Phagemid, **285**
Phagocytes, **460–61,** 461f
Phagocytosis, 61, 456, 460–61,
    461f, 494
    by protozoa, 373
Phagolysosome, 400, 460–61
Phagosome, 460, 461f
Pharmaceutical industry, 14
Pharyngitis, 499t
    gonococcal, 534
    streptococcal. *See*
        Streptococcal
        pharyngitis
Phase-contrast microscope, 20,
    28–30, 29–30f, 31t
Phase ring, 28–30, 29f
pH effect
    on enzymes, 143
    on growth, **91–92,** 92t, 119
Phenol(s), as disinfectants and
        antiseptics, 122t,
        123, 123f
Phenolate, 88
Phenol coefficient, 123
Phenom, 309
Phenotype, 212, 306
Phenoxymethyl penicillin. *See*
    Penicillin V
Phenylalanine, 43t, 197t
Phenylalanine deaminase
    test, 324f
Phenylalanine tRNA, 225f
Phosphatase, 83
Phosphatase test, **608,** 610
Phosphate bond, high-energy,
    148, 148t, 149f
Phosphate cycle, **579**
Phosphate granule, 507, 508f

Phosphate salts, 579
Phosphodiester bond, 46, 47f, 223
Phosphoenolpyruvate, 89, 148,
    151f, 152, 153f, 155f,
    193f, 194, 197t,
    573–74, 573f
    energy from, 148t
    structure of, 148t
Phosphoenolpyruvate
    carboxykinase, 193f,
    194, 574
Phosphoenolpyruvate
    carboxylase, 574
Phosphoenolpyruvate synthase,
    193f, 194
Phosphofructokinase, 150, 151f,
    152, 193f
Phosphoglucoisomerase,
    150, 151f
Phosphoglucomutase, 190
6-Phosphogluconate, 152, 153f,
    155f, 158, 159f,
    194, 195f
6-Phosphogluconate dehydratase,
    152–53, 153f
6-Phosphogluconate
    dehydrogenase, 156f
Phosphogluconate pathway. *See*
    Pentose phosphate
    pathway
6-Phosphogluconolactone,
    155f, 195f
2-Phosphoglycerate, 151f, 152,
    153f, 155f, 193f
3-Phosphoglycerate, 149,
    150–51f, 152, 153f,
    155f, 188–89, 188–89f,
    193f, 197t
Phosphoglycerate kinase, 151f,
    152, 153f
Phosphoglyceromutase, 151f, 152
Phosphoketolase, **154,** 156f, 158,
    159f, 192f
Phosphoketolase pathway, **154,**
    156f, 158, 191, 192f
Phospholipid, 46–48, 49f
    in plasma membrane,
        49–51, 51f
5-Phosphoribosylpyrophosphate,
    197t
Phosphorus
    cellular content of, 84t,
        570, 570t
    characteristics of
        element, 647t
    function in cells, 84t
    removal from sewage, 599
Phosphorus source, **83**
Phosphorylase, 190, 190t
Phosphotransacetylase, 192f
Phosphotransferase system, 89

Photoautotroph, 81, 81t, 571–72,
    573, 573f
Photobacterium, 173, 425
    P. fischeri, 173
Photography, 623
Photoheterotroph, 81, 81t
Photolyase, 249
Photolysis, 184, 185f, 186
Photon, 182
Photooxidation, 180
Photophosphorylation, 149, 178,
    178f, **182–88**
    cyclic, 185–86f, **186–87**
    mechanism of, **184–86**
    noncyclic, **184–86**
Photoreactivation, 249
Photosynthesis, 53f, 72, **178–82**
    anoxygenic, 184, 573
    in carbon cycle, 571
    carbon dioxide fixation, 178,
        178f, 183, **188–89,**
        188–89f
    in eucaryotes and
        cyanobacteria, **178**
    oxygenic, 184
    photophosphorylation, 178,
        178f, **182–88**
Photosynthetic bacteria, 81
    oxygenic, 332t, **333–36,**
        638–39
Photosynthetic membrane, 182,
    183f, 187
Photosynthetic pigment, **178–79**
Photosystem, **184**
Photosystem I, 184, 185f,
    186, 186f
Photosystem II, 185f, 186
Phototaxis, 65
Phototroph, 80, 81t, 141f
Phototrophic bacteria, anoxygenic,
    332–33, 332–33t, 333f, 638
Phragmidium discoflorum, 360t
pH scale, 651–52, 651f, 652t
Phycobilin, 179–80, 354t, 359
Phycobiliprotein, 352
Phycobilisome, 180
Phycocyanin, 180, 181f, 352
Phycoerythrin, 180, 181f, 352
Phycovirus, 390
Phylogenetic tree, **311–14**
    of *Archaea,* 313f
    of Bacteria, 314, 315f
    eucaryotic, 380
    universal, 311, 312f
Phylogeny, 291, 347
    rRNA and, 307, **311–14,**
        312–13f, 380
Physarum, 366t
    P. polycephalum, 33f
Physical requirements, **89–94**
Physoderma, 360t, 365

Phytophthora, 366t
    P. infestans, 360t, 369
Pickles, 331t, 617, 617t
Pickling, 93
Picoplankton, 336
Picornaviridae, 394t
Picornavirus, 384
Pilimelia, 342t
Pilin, 62, 265
Pillotina, 317, 427
Pilus, **62,** 62f
    sex, 62
    type I, 62
Pinnularia, 351f
Pinocytosis, by protozoa, 373
Pinta, 316, 318t
Pitching, 618
Pityriasis versicolor, 525t, 530
Pityrosporum
    P. orbiculare, 525t, 530
    P. ovale, 525
Plague, 7t
Plague toxin, 451t
Planctomyces, 315f
Plankton, 350, 356, 373, 586
Planobispora, 342t
Planomonospora, 342t
Plantae, 306–7, 307t
Plant cells, genetically-
        engineered, 297
Plant pathogen, 320
Plant virus, 387, **390–93,** 392t
    classification of, 392–93t
    penetration of host cells,
        400–401
Plaque
    dental. *See* Dental plaque
    phage on bacterial lawn, 398,
        398f
    virus in tissue culture, 397
Plasma, 456
Plasma cells, **475,** 476f
Plasma extender, 627
Plasma membrane, **48–49,** 50f,
    58f, 71f, 77
    archaeal, 340
    damage by antibiotics, 125,
        125t, **126–28,** 127f
    eucaryotic, **70–72,** 71f, 76t
    fluidity of, 51
    as permeable barrier, **49–52**
    procaryotic, 76t
    sterols in, 72
    structure of, 49–51
Plasmaviridae, 391t
Plasmid, 65, **258–63**
    chimeric, 281, 282f
    as cloning vector, 274, **279–81,**
        281f, 287
    conjugative, **260–61**
    curing of, 263

integration into chromosome, **262–63**

nonconjugative, **260–61**

replication of, 293–94, 294f

toxin genes on, 452

types of, 263t

Plasmid incompatibility, 262

Plasmid pBB101, 300–301

Plasmid pBF4, 262f

Plasmid pBH1, 300f

Plasmid pBR322, 271f, 279–81, 281–82f, 300, 301f

Plasmid pIA1, 301

Plasmid pIB1, 301

Plasmid pSC101, 262f

Plasmodiophora, 366t

  P. brassicae, 360t

Plasmodiophoromycota, 360t, 366t, **369–73**

Plasmodium, 375t, 377, 377t, 445t

  chloroquine-resistant, 132

  zoonoses, 444t

  P. malariae, 7t

Plasmodium (slime mold), 371–72, 371f

Plasmopara, 366t

  P. viticola, 360t, 369

Plastics, microbial, 66

Plastid, 71f, 72

Plastocyanin, 185, 185f

Plastoquinone, 185–86, 185f

Plateau, of one-step growth curve, 404, 404f

Platelets, 456, 457t

Pleomorphism, 18–19

Plesiomonas, 322t

Pleurisy, 504

Pleurocapsaceae, 335, 335t

PMN. *See* Polymorphonuclear leukocytes

Pneumococcal pneumonia, 504–5, 504f

Pneumococcal pneumonia vaccine, 505

Pneumocystis, 366t

  P. carinii, 360t, 500t

Pneumocystis carinii pneumonia, 376, 492, 492f, 540

Pneumonia

  bacterial, 322t, 328, 499t, 503–4t, **504–5**, 504–5f

  fungal, 360t, 500t

  pneumococcal, 504–5, 504f

  Pneumocystis carinii, 376, 492, 492f, 540

  primary atypical, 18, 59, 327, 500t, 504t

  rates for males and females, 557t

  viral, 500t

Pneumonic plague, 444t, 451t

Pock(s), 387

Pock counting, 397–98, 397f

Poi, 617t

Point mutation, 246

Polar covalent bond, 649

Polar flagellation, 63f

Polar molecule, 649

Polio, 554t

Polio vaccine, 478, 478t

Poliovirus, 3f, 385f, 394t, 395, 398f, 400, 594

Polyangium, 575

  P. cellulosum, 575f

PolyA tail, on mRNA, 229

Polycistronic mRNA, 212, 213f, 233

Polyclonal antibodies, 475

Polyenes, 128

Polyhedral virus, **386–87,** 386f

Poly-β-hydroxybutyrate, 66

Polymerase chain reaction (PCR), **288–91,** 290f

  discovery of, **11**

  nested amplification, 291

Polymorphonuclear leukocytes (PMN), 456

Polymyxin, 125t

  commercial production of, 629t

  mechanism of action of, 126, 629t

  structure of, 127f

Polyoma virus, 385f, 394t

Polyphagus, 587

Polyphosphate granule, 331, 341f

Polyporus, 366t

  P. sulphureus, 369f

Polyribosome, 232–33

Polysaccharide, 46–47, 48f. *See also* Carbohydrate

  commercial production of, **626–27**

  degradation of, **190–92,** 190t

  storage depots for, 65–66

Polysphondylium, 366t

Pomace, 619

Pontiac fever, 506–7, 556

Population growth curve. *See* Growth curve

Porphyra, 355

Portals of entry, **445–46**

Positive control system, 236

Positive stain, 25, 25f

Posttranscriptional modification, 238

Posttranslational modification, 233, 238

Potable water, 588

Potassium

  cellular content of, 84t, 570, 570t

characteristics of element, 647t

  function in cells, 84t

  requirement for, 83–84

Potato blight, 360, 360t, 369

Potato spindle tuber, 388–89, 388f

Potato wart, 360t, 365

Potato X virus, 392t

Potato Y virus, 392t

Potential energy, 140

Potexvirus, 392t

Potyvirus, 392t

Poultry

  contaminated with *Salmonella,* 520

  spoilage of, 615–16, 615t

Pour plate, 98, 110–11, 110–11f

Povidone-iodine (Betadine), 122t, 124

Powdery mildew, 364f

Powdery scab of potatoes, 360t

Poxviridae, 394t

Poxvirus, 384

Precipitation, of antigen-antibody aggregates, **485–86,** 486f

Precipitin, 485

Precipitin interface test, 485

Precipitin ring test, 485, 486f

Prednisone, 485

Presumptive test, 500–504, 593

Prevalence rate, 554

Prevotella, 440t, 514

  P. melaninogenica, 514t

Pribnow box, 226

Primaquine, 132

Primary animal tissue culture, 397

Primary atypical pneumonia, 18, 59, 327, 499t, 504t

Primary metabolite, 204, **625–28**

Primary producer, **571–75**

Primary sewage treatment, **595**

Primary syphilis, 535, 535f

Primase, 221f, 223

Primosome, 223

P ring, 62, 62f

Prion, **388–89,** 389f, **412–13,** 412–13f, 412t

Procaryotic cells, 18

  cell arrangements, **19–20,** 20f, 38

  cell wall of, 52

  chemical composition of, 81t, 84t

  classification of procaryotes, 307, **314–15**

  composition and structure of, **48–70**

  DNA replication in, **223–24**

eucaryotic cells vs., 71f, 76t

  evolution of, 77

  mRNA of, 212, 213f

  photosynthesis in, **179,** 182, **186–87**

  ribosomes of, 65, 224

  shape of, **19,** 19–20f, 38

  size and scale of, **18–19,** 38

  transcription in, 213f

  translation in, 213f

Prochloron, 335

Prochlorophyte, **333–36**

Prochlorothrix, 335

Prodigiosin, **204,** 204–5f

Prodromal period, 446, 446f

Producer, 570

  primary, **571–75**

Profundal zone, 587, 587f

Proline, 42

  in prodigiosin synthesis, 204, 205f

  structure of, 43t

  synthesis of, 197t

Proline oxidase, 205f

Promicromonospora, 342t

Promoter, 233–34, 234–35f, 235

  binding of CAP-cAMP complex to, 238

  binding of RNA polymerase to, **226**

  DNA sequence of, 226

  strong, 287

Propagated epidemic, 552, 553f

Properdin, 482, 483f

Prophage, 255, 258, 410–11, 410f

  curing of, 411

  defective, 411

  induction of, 410f, 411

Prophage β, 411

β-Propiolactone, 124

Propionate, **157**

  from rumen fermentation, 427–29, 428f

Propionibacterium, 157, 157f, 331, 614

  P. acnes, 440t, 525, 525t

  P. freundreichii, 611t, 627t, 628

  P. shermanii, 611t, 627t, 628

Propionic acid bacteria, **157**

Propionic acid fermentation, 157f

Prospective study, **561–63,** 562f, 563t

Prostaglandin, 479, 480t

Prosthecae, 337

Prosthecochloris, 333t

Protease, 194, 460

  industrial uses of, 194–95, **623,** 624t

Protein, **42–44**

  as carbon and energy source, **194–97**

content of cells, 81t
degradation of, **194–97**
denaturation of, 44
functions in cell, 42, 42t
genetically-engineered, **295–97**, 296t
metabolism of, **194–97**
periplasmic, 60
in plasma membrane, 51–52, 51f
single-cell. *See* Single-cell protein
structure of, **42–44**
  primary structure, 44, 45f
  quaternary structure, 44, 45f
  secondary structure, 44, 45f
  tertiary structure, 44, 45f
synthesis of, **229–33**. *See also* Translation
  antibiotics that inhibit, 125, 125t, **128–29**, 128f, 128t
  during lag phase, 117f
  during microbial growth, 114, 115f
  regulation of, **233–39**
Protein kinase, 407f
Proteus, 322t, 324f, 440t, 516, 614t
  plasmids of, 263t
  temperature range of, 91
  P. mirabilis, 263t, 532, 532t
  P. vulgaris, 62f, 92t, 587
Protist, 366t
Protista, 306–7, 307t
Proton, 647
Proton gradient, 86, 88, 166, 187
Proton motive force, 165–66
Protoplasmic cylinder, 316
Protoplast, 59
Protoplast fusion, 630
Prototroph, 250
Protozoa, **373–77**
  cellular structure of, **373**
  classification of, **375–77**, 375t
  pathogenic, 377t
  reproduction in, **373–75**
  in soil, 582
  symbionts of, **425–27**, 426f, 426t
Providencia, 263t, 324f
  P. stuartii, 273t
Provirus, 409
Pruteen, 631
Pseudocaedobacter, 426t
Pseudomembrane, 507–10
Pseudomembranous colitis, 441
Pseudomonadaceae, 319, 320t
Pseudomonas, 81, 171t, 320
  anaerobic respiration in, 168–70

atomic force microscopy of, 35f
  chitin degradation by, 576
  denitrification by, 578
  in freshwater habitats, 587
  glucose metabolism in, 154t
  hydrocarbon degradation by, 583
  in marine habitats, 586
  in meat spoilage, 614t
  in milk, 608
  oxidase test, 166
  oxidation of hydrogen gas by, 172
  plasmids of, 263t
  poly-β-hydroxybutyrates of, 66
  in poultry and fish spoilage, 615–16, 615t
  psychrophilic, 90, 90f
  in soil, 581
  P. aeruginosa, 106t, 107f, 319–20, 525t
    antibiotic-sensitivity of, 126–28
    cell envelope of, 58f
    drug resistance in, 262
    electron microscopy of, 33f
    flagella of, 63, 63f
    infections in burn victims, **528–29**
    nosocomial infections with, 559t
    proteases of, 194
    in urinary tract infections, 532, 532t
    virulence-associated enzymes of, 449t
    in water supply, 594
  P. cepacia, 581
  P. dacundae, 623t
  P. denitrificans, 628
  P. methanica, 586
  P. putida, 263t, 584
  P. saccharophila, 152
  P. syringae, ice⁻, 297, 618
Pseudonocardia, 342t
Pseudopeptidoglycan, 339
Pseudoplasmodium, 372–73, 372f
Pseudopodia, 73, 376
Pseudouridine, 224, 225f
Psittacosis, 327t, 444t, 500t
*Pst*I, 273t
Psychrophile, 89, **90**, 90f
Public health, 550, 566
Puccinia, 366t
  P. framinis, 360t
Pulmonary tuberculosis, 510
Puncture wound, 325
Pure culture, **96–99**

Purine, 44, 46f, 84, **202,** 203f
Purine analog, 132
Purple bacteria, 179, 312f, 315f, 332, 333t, 573, 580
Purple nonsulfur bacteria, 333, 333t, 573
Purple plasma membrane, 166–67
Purple sulfur bacteria, 187, 188f, 333, 333f
Purulent exudate, 458
Pus, 450, 458, 498
Pustule, 526
Putrefaction, 614, 614t
Putrescine, 199f
Pyelonephritis, 532
Pyocyanin, 320
Pyoderma, **526–27,** 526f
Pyogenic cocci, 328, 498
Pyorrhea, 515
Pyridine nucleotide, 162, 164
Pyridoxal phosphate, 85t, 197
Pyridoxine, 85t, 145t
Pyrimidine, 44, 46f, 84, **202,** 203f
Pyrimidine analog, 132
Pyrimidine dimer, 246
Pyrococcus, 340t, 341
  P. woesei, 91
Pyrodictium, 312–13f, 340, 340t
  P. occultum, 90f, 91
Pyroverdin, 320
Pyrrophyta, **352–57,** 354t, 357f
Pyruvate, 151f, 153f, 155–56f, 159f, 192–93f, 573, 573f
  in amino acid synthesis, 197t
  degradation of, **154–61,** 157f
  ethanol and carbon dioxide from, **157**
  reduction to lactate, **154**
  Stickland reaction, 196, 196f
Pyruvate carboxylase, 193f, 194, 574
Pyruvate kinase, 151f, 152, 153f, 193f
Pythium, 366t

Qadri, S.M. Hussain, 204
Q fever, 327t, 444t, 608
Quarantine, before space missions, 10, 10f
Quaternary ammonium compounds, 122t, 123
Quellung reaction, 61, 61f, 505
Quinine, 132
Quinolone, 129
Quinone, 85t, 162, 164, 166, 186, 186–87f

Rabies, 443, 444t
Rabies virus, 387, 395, 395t, 396f
Race, disease rates and, 557–58f, 557t, 558

Racking, 619
Radiation, for sterilization, **122**
Radioactive isotope, 648
Radioimmunoassay (RIA), **490**
Rancidity, of meat, 614, 614t
Raymer, William, 388
R determinant, 262
Reaction center, of photosystem, 184
Reactivation tuberculosis, 510
Reading frame, 230
Reading frame shift, 246
Reagin, 536
recA gene, 249, 249f, 252
Recalcitrant compounds, **576,** 588
Receptor, 42t
Recipient cell, 250
Recombinant DNA, 252
Recombinant DNA technology, 302, 633. *See also* Gene cloning
  applications of, **291–97**
    in agriculture, **297**
    for antibiotic production, **630**
    in medicine, **291–97**
    historical perspectives on, **9–11,** 11f, **272–73**
    public and scientific concerns about, **297**
Recombination
  general. *See* Recombination, homologous
  high-frequency, **264–65**
  homologous, 252, 252f
Red algae. *See* Rhodophyta
Redi, Francesco, 4, 13
Redox pair, 145
Redox potential, **145–46,** 146f
Redox reaction, 145–46, 148
Red tide, 356–57, 357f
Reduction, **145**
Reductive carboxylic acid cycle, 573, 573f
Reductive pentose cycle. *See* Calvin cycle
Refractile body, 425–27, 426t
Refractive index, 24, 24f
Refrigeration, 90, 519, 615
Regeneration, in protozoa, 375
Regular, nonsporing gram-positive rods, 328t, **331,** 331t, 637
Regulator gene, 233–34
Reindeer moss, 423
Relapsing fever, 445t
  louse-borne, 318t
  tick-borne, 318t
Release factor 1 (RF1), 232
Release factor 2 (RF2), 232
Release factor 3 (RF3), 232

Reliability of test, 561
Rennin, 612f, 613, 624t
Reoviridae, 394t
Reovirus, 403
Repellent, 63–65, 64f
Replica plating, 249–50, 250f
Replication fork, 222f, 224
Replicative form, 402–3, 403f
Replicon, 224
Reporter gene, **288**
Repression, 233, **235–36,** 235f
Repressor protein, 233, 234f, 235–36
Reproducibility of test, 561
Reservoir of infection, **442–43**
    animals, **443**
    inanimate, **443**
Resident flora, 440
Resistance plasmid. *See* R plasmid
Resolution, of light microscope, **23–24,** 24f
Respiration, **150**
    aerobic, 150, **168**
    anaerobic, 150, **168–70,** 168t, 171, 326, 570f
    in carbon cycle, 571, 571f
Respiratory syncytial virus, 384, 500t
Respiratory tract, **498–514**
    bacterial pathogens of, **498–512,** 500t
    fungal pathogens of, 500t, 512
    lower, 498
    microflora of, 440t, **498**
    nosocomial infections of, 559t
    protozoan pathogens of, 500t
    upper, 498
    viral pathogens of, 500t, 513
Restriction endonuclease, 9, 272, 273t
    microbial sources of, 273t
    recognitions sequences of, 273t
Restriction fragment length polymorphism (RFLP), **292**
Restriction map, 272
Reticulate body, 19, 430–33, 431f
Reticuloendothelial system, **458,** 459f
Reticulum, 428f, 429
Reticulum cells, 458
Retinal, 166
Retrospective study, **561–63,** 563t
Retroviridae, 395t
Retrovirus, 403, 403f, 409
Reversed electron transport, 187, 187f
Reverse transcriptase, 274–78, 278f, 403, 403f, 409, 536, 538f, 540
    inhibitors of, 132

Reverse transcription, 212f
Reversible inhibitor, 143
Reversible reaction, **142**
Reye's syndrome, 513
RF. *See* Release factor
RFLP. *See* Restriction fragment length polymorphism
R group, 42, 43t
Rhabdoviridae, 395t
Rhabdovirus, 392t, 395, 444t
Rheumatic fever, 328, 499, 503t, 557t
Rh incompatibility, 480
Rhinovirus, 395, 448, 499t
Rhizobia, 418, 435
Rhizobiaceae, 320t
Rhizobium, 83, 320
    glucose metabolism in, 154t
    nitrogen fixation by, 297, **418–23,** 419–22f, 421t, 576
    plasmids of, 263t
    shapes of, 19
    R. leguminosarum, 417f, 421f, 421t
    R. lupini, 421t
    R. meliloti, 421t
    R. trifoli, 421t
Rhizoid, 361, 365
Rhizopus, 362–63f, 366t
    R. stolonifer, 360t, 366
Rhizosphere, 418
Rhizosphere effect, 418, 582
Rho-dependent termination, 228, 228f
Rhodobacter, 333t
Rhodococcus, 342t
Rhodocyclus, 333t
Rhodomicrobium, 333t
Rhodophyta, 180, 352, 354t, **357–59**
Rhodopila, 333t
Rhodopseudomonas, 333t, 587
Rhodospirillaceae, 182, 573
Rhodospirillum, 333, 333t, 577
Rho factor, 227–28
Rho-independent termination, 227–28, 228f
RIA. *See* Radioimmunoassay
Riboflavin, 145t
    commercial production of, 627t, 628
Ribonucleic acid. *See* RNA
Ribose, 44, 46f, 192f
Ribose-5-phosphate, 155f, 192f, 194, 195f
Ribosomal proteins, 65
Ribosomal RNA (rRNA), 46, 65, **224**
    conservation of sequence of, 311

    phylogeny and, 291, 307, **311–14,** 312–13f, 380
    5S, 224, 228
    5.8S, 224
    16S, 224, 232, 311
    18S, 224, 380
    23S, 224
    28S, 224
    sequencing of, **311**
    synthesis of, 226–28
Ribosome, 50f, **65,** 71f, 224. *See also* Translation
    antibiotics acting at, 128, 128t
    A site on, 231f, 232
    comparison of Bacteria, Archaea, and Eucarya, 313t
    eucaryotic, 65, 76t, 224, 233
    procaryotic, 65, 76t, 224
    P site on, 231f, 232
    sedimentation constant of, 65
Ribosome binding site, on mRNA, 232
Ribothymidine, 224
Ribozyme, 42, 229
Ribulose, 192f
Ribulose-1,5-diphosphate (RuDP), 188–89, 188–89f
Ribulose-1,5-diphosphate (RuDP) carboxylase, **188–89,** 188–89f, 344
Ribulose-5-phosphate, 155–56f, 159f, 189, 189f, 192f, 194, 195f
Rice-water stools, 518
Richardson, G.M., 135
Rickettsia, 326, 636–37
    skin diseases caused by, **529–30**
    R. akari, 525t, 529t
    R. prowazekii, 327t, 445t, 525t, 529, 529t
    R. rickettsii, 327t, 444–45t, 529, 529t
    R. tsutsugamushi, 327t, 445t, 525t, 529t
    R. typhi, 327t, 444–45t, 525t, 529, 529t
Rickettsiae, 316t, **326,** 327t
Rickettsialpox, 525t, 529t
Ricotta cheese, 612f, 614
Rifampicin, 129
Rifamycin, 125t, 129
Rift Valley fever, 542f, 543t
Riley, W.H., 173
Ringworm, 360, 360t, 444t, 525t, 530, 531f
Ripened cheese, 611, 611t, 613–14

Rise period, 404, 404f
Risk of disease, 554
River, 587
Rhizidiomyces, 366t
RNA, **224–29.** *See also* Transcription; Translation
    antisense, **293–94,** 294f
    classes of, **224–26**
    messenger. *See* Messenger RNA
    ribosomal. *See* Ribosomal RNA
    structure of, 44, 46
    synthesis during microbial growth, 114, 115f
    transfer. *See* Transfer RNA
RNA I, 293–94, 294f
RNA II, 293, 294f
RNA bacteriophage, 391t
RNA polymerase, 224
    antibiotics that inhibit, 129
    archaeal, 224, 312, 314f
    Bacterial, 224
    binding to promoter, **226,** 233–36, 235f
    chain elongation, 226–27
    comparison of Bacteria, Archaea, and Eucarya, 313t
    eucaryotic, **228–29**
    sigma factor, 224, 226–27
RNA polymerase I, 228
RNA polymerase II, 228
RNA polymerase III, 228
RNA primer, 221f, **223,** 224
RNA replicase, 403
RNase H, 223
RNA virus, 384–86
    animal viruses, 394–95t
    plant viruses, 392–93t
    replication of RNA, **403,** 403f
Roberts, Marilyn, 544
Rochalimaea quintana, 327t
Rocks, 579
Rocky Mountain spotted fever, 326, 327t, 444–45t, 525t, 529–30, 529t, 530f
Rod-shaped bacteria, 19, 19–20f
Rohrer, Heinrich, 34
Rolling circle replication, **223,** 264, 264f, 402, 402f
Romano cheese, 611t
Root hair, 418, 419f, 420
Root nodule, 417f, 418, 419–21f, 420–21
Root rot, 360
Roquefort cheese, 361f, 611t, 613–14
Rose rust, 360t
Rotavirus, 394t, 396f, 448

Rous, Francis P., 409
Rous sarcoma virus, 409
Roux, Émile Pierre Paul, 509
R plasmid, 262, 263t, 545
RPR test, 536
rRNA. *See* Ribosomal RNA
Rubella. *See* German measles
Rubella syndrome, congenital,
 531–32
Rubeola. *See* Measles
RuDP. *See* Ribulose-
 1,5-diphosphate
Rumen, 339, 427, 428f, 575
Ruminant, 190, **427–29,**
 428f, 428t
Ruminococcus, 93, 328, 427, 575
 R. albus, 428t
Run (bacterial movement), 63,
 64f, 65
Rust, 368, 368f
Rye bread, 617t

Saber shin, 535
Saccharomonospora, 342t
Saccharomyces, 366t, 617t
 fermentation pathways
  in, 157f
 water requirement for
  growth, 92t
 S. cerevisiae, 362f, 368,
  617–19, 617t, 624
  genome sequencing, 291
  as host for cloning, 285
 S. champagnii, 617t
 S. diastaticus, 618
 S. exiguus, 617t
 S. lactis, 623t
Saccharopolyspora, 342t
Saccharothrix, 342t
Safety, of recombinant DNA
 technology, **297**
*Sal*I, 273t
Salicylanilide, 168, 168t
Salmonella, 322, 322t, 324f
 antibiotic resistance
  in, 131f
 enrichment culture, 95
 identification of, 487
 lipopolysaccharide of, 60
 nosocomial infections
  with, 559t
 phosphotransferase system
  of, 89
 plasmids of, 263t
 in poultry and fish, 520,
  615–16, 615t
 pyruvate degradation
  in, 160
 temperature range of, 91
 zoonoses, 444t
 S. anatum, 258

S. enteritidis, 7t, 121, 519, 594
S. typhi, 7t, 131f, 323, 442,
 461, 516, 516t, **519**
S. typhimurium, 106t, 323,
 448, 610
 in Ames test, 251, 251f
Salmonella gastroenteritis. *See*
 Salmonellosis
Salmonella-Shigella (SS) agar,
 519–20
Salmonellosis, 7t, 443, 444t, 446,
 519, 554t
 from dairy plant, **610**
Salpingitis, 533–34
Salted meat, 615, 615f
Salting, 93
Salt lake, 102, 339, 586
Salt water. *See* Marine habitat
Saprolegnia, 366t, 369, 370f
 S. parasitica, 360t
Saprophyte, 360
Sarcina, 20, 20f, 601t
Sarcodina, 375–76
Sarcoma, 408
Sarcomastigophora, **375–76,**
 375t, 377t
Saturated fatty acid, 47
Sauerkraut, 331t, 616–17, 617t
Sausage, 616, 617t
Scalded skin syndrome, 525t,
 526–27, 526f
Scaleup, 621
Scanning electron microscope
 (SEM), 21, 32, 32f,
 36–37
Scanning tunneling microscope
 (STM), **34,** 34f
Scarlet fever, 328, 411, 451t,
 498–99, 499t,
 500f, 503t
Scenedesmus, 351f
Schaudinn, F., 7t
Schern, Henry W., 135
Schistosomiasis, 543t
Schizogony, 377f
Schwann, Theodor, 13, 13f
SCID. *See* Severe combined
 immunodeficiency
SCP. *See* Single-cell protein
Scrapie, 389
Screened water, 595
Screening test
 data for epidemiological
  analysis, **560–61,** 560t
 frequency of disease in
  population, 561
 reliability of, 561
 sensitivity and specificity of,
  560–61
 tests in parallel, 561
 tests in series, 561

Scrub typhus, 327t, 445t,
 525t, 529t
Seaweed, 355, 359
Sebaceous gland, 456
Sebum, 456
Secondary animal tissue
 culture, 397
Secondary metabolite, **204,**
 629–30
Secondary sewage treatment,
 595–98, 597f
Secondary syphilis, 535, 535f
Second law of thermodynamics,
 141–42
Secretory IgA, 469
Sedoheptulose-7-phosphate,
 155f, 194, 195f
Selection technique, 249–50
Selective marker, 279
Selective media, **96**
Selenite broth, 95
Selenomonas, 325t, 427
 S. lactilytica, 428t
 S. ruminantium, 428t
SEM. *See* Scanning electron
 microscope
Semiconservative replication,
 219–23
Semilogarithmic graph, 107,
 108f, 116
Semi-solid media, 621–22
Sensitivity of test, 560–61
Septicemia, 322t, 503t, 504, 557t
 nosocomial, 559t
Septicemic plague, 444t, 451t
Septic shock, 453, 454t
Septic tank, **599–600,** 600f
Sericin, 623
Serine, 43t, 197t
Serine sulfhydrylase, 83
Serology, 485
Serotonin, 480t
Serratia, 322t, 324f, 587
 S. marcescens, 202, 203f,
  204, 217f
Serum, 456
Serum hepatitis, 524, 524t
Serum proteins, 468–69
Serum sickness, 480–81
Settling tank/basin, 595
Severe combined
 immunodeficiency
 (SCID), 292
Sewage, 443, 588
 raw, 595f
Sewage treatment, **594–600,** 604
 primary, **595**
 secondary, **595–98,** 597f
 tertiary, 595, **598–99**
Sewage treatment plant,
 569f, 596f

Sewer pipes, 588
Sex pilus, 62, 264, **265,** 265f
Sexually transmitted disease
 (STD), 444, **533,** 556
 bacterial, 533t
 fungal, 533t
 protozoal, 533t
 viral, 533t
Sexual reproduction
 in algae, 352
 in eucaryotic cells, **73–74**
 in protozoa, 373–75
Shadowing, 36
Shands, Kathryn, 564
Sheathed bacteria, 332t, **337–38,**
 338f, 639
Shell, of protozoa, 373
Shellfish
 in Chesapeake Bay, 590
 contaminated, 323, 357,
  586–87
Shiga toxin, 520
Shigella, 322, 322t, 324f
 antibiotic resistance
  in, 131f
 intestinal disease caused
  by, 519–20
 nosocomial infections
  with, 559t
 plasmids of, 263t
 S. boydii, 516t, 520
 S. dysenteriae, 7t, 106t, 322,
  448, 516, 516t
  antibiotic resistance
   in, 131f
  toxins of, 450, 451t
 S. flexneri, 516t, 520
 S. sonnei, 516t, 520
Shigellosis, 7t, 516t, **519–20,** 554t
Shine, John, 232
Shine-Dalgarno sequence, 231f,
 232, 380–81
Shingles, 132t, 525t, 531
Shock
 anaphylactic, 480
 septic, 453, 454t
Shotgun cloning, 285
Shuttle vector, 285
Siderococcus, 337
Siderophore, **88**
Sigma factor, 224, 226–27
 sigma-29, 226
 sigma-32, 226
 sigma-37, 226
 sigma-55, 226
 sigma-70, 226
Signature sequence, 311
Silent mutation, 246
Silica, in cell envelope, 373
Silicon, 84, 352
Silverman, Michael, 63

Silver nitrate, 122t, 124, 534
Simian T lymphotrophic virus (STLV), 409, 414
Similarity coefficient, 308f, 309, 309t
Simon, Melvin, 63
Simple stain, 25
Simple sugar, 46
Single-cell protein (SCP), **631–32,** 631–32t
    substrate for, 631–32
Single covalent bond, 649, 650f
Single-strand DNA-binding protein, 221, 221f
Sinusitis, 499t, 503t
Site-directed mutagenesis, 278–79, 279f
Skimmer, 595
Skin, **524–51**
    bacterial pathogens of, **525–32,** 525t
    fungal pathogens of, 525t, **530–31**
    microflora of, 440t, **525**
    as physical barrier to infection, 455–56
    viral pathogens of, 525t, **531–32**
Skin test, 485
    to diagnose allergies, 480, 480f
    tuberculin, 483, 485f, 511, 561
Slant, 95
Slide agglutination test, 486–87
Slime layer, 334, 338, 423
Slime mold, 350f, **369–73**
    acellular, 371–72, 371f
    cellular, 350, 372–73, 372f
Smallpox, 8, 387, 554t
Smallpox virus, 394t, 397f
Smith, Hamilton, 9, 272
Smittium, 366t
Smoked meat, 615
Smut, 368
Snapping cell division, 331
Snow, John, 566
Snow algae, 358f
S1 nuclease, 278f
Soap, 122t, 123
Soda lake, 92
Sodium
    cellular content of, 570, 570t
    characteristics of element, 647t
    requirement for, 84
Soil
    algae in, 582
    bacteria in, 581–82
    composition and physical properties of, **581**

fungi in, 582
microbes and, **580–82,** 581t, 604
protozoa in, 582
as reservoir of infection, **443**
viruses in, 582
Solemya velum, 344, 344–45f, 345t
Solid media, 94–95, 98, 621–22
Sorbose dehydrogenase, 623t
SOS response, **249,** 249f
Sour cream, 609–10, 609t
Sourdough bread, 617t
Souring
    of meat, 614, 614t
    of milk, 191
Southern blot, **276,** 276–77f
Soy sauce, 617t
Space travel, microbes in space, **10,** 10f
Spallanzani, Lazzaro, 4, 13
Spawn (mushroom), 368
Specialized transduction, **255–58,** 257f
Special stain, 25
Species, 306, 306t, 308
Species name, 306
Specific growth-rate constant, 642–43
Specificity of test, 560–61
Spectrophotometer, 113, 114f
Spelunker's disease. *See* Histoplasmosis
Sphaeromyxa, 375t
Sphaerotilus, 171t, 172, 337–38, 598
    S. natans, 338f
S phase, 224
Spherical aberration, 22, 23f
Spheroplast, 59
Spirillospora, 342t
Spirilla, 19, 19f
Spirillum, 92t
    S. volutans, 319f
Spirochaeta, 316–17, 318t
Spirochetes, 19, 20f, 315f, **316–17,** 316t, 317–18f, 318t, 635
Spirogyra, 351f, 358f
Spirulina, 334f, 632
Spleen, 459f
Spongospora subterranea, 360t
Spontaneous generation, **2–4**
    disproving doctrine of, 4, **13,** 13f
Spontaneous mutation, 246
Spontaneous reaction, 141
Sporangiophore, 364
Sporangiospore, 364–65, 367f
Sporangium, 69–70, 69f, 343, 362–63f, 364, 364f, 369, 370f

Spore
    of *Actinoplanes,* 343
    algal, 352
    eucaryotic, 73–74
    fungal, 362–64f, 364
    of slime mold, 371–73, 371–72f
Spore cell wall, 69, 69–70f
Spore coat, 69, 69–70f
Spore core, 69, 69–70f
Spore cortex, 69, 69–70f
Spore septum, 69, 69f
Spore stain, 70, 329f
Sporichthya, 342t
Sporocarp, 372f, 373
Sporocyst, 377f
Sporogenesis, 69–70, 69f
Sporosarcina ureae, 67f
Sporotrichum carnis, 614t
Sporozoea, 377
Sporozoite, 377f
Sporulation, in *B. subtilis,* 226
Spotting, of meat surface, 614, 614t
Spreading factor, 450
Spread plate, 98, 110–11, 110f
S ring, 62, 62f
SS agar. *See* Salmonella-Shigella agar
Stain(s), **25–28**
    differential, 25
    Gram. *See* Gram stain
    negative, 28, 28f, 32
    positive, 25, 25f
    simple, 25
    special, 25
Stain removal, **623**
Stalked bacteria, 337
Standard hydrogen electrode, 145, 146f
Stanley, Wendell, 9
Staphylococcal enterocolitis, 433
Staphylococcal enterotoxin A, 473
Staphylococcal exfoliatin, 473
Staphylococcal food poisoning, **518**
Staphylococcus, 328, 512t
    cell arrangements in, 20, 20f
    microflora, 440t
    peptidoglycan of, 56
    psychrophilic, 90
    water requirement for growth, 92t
    S. aureus, 7t, 19f, 21f, 96f, 106t, 121, 125, 328, 329f, 433, 441, 461, 499t, 525

antibiotic resistance in, 130, 131f, 262
antimicrobial susceptibility testing, 132f
cell wall of, 54, 54–55f
DNase of, 202
enterotoxin of, 518
food intoxication, 516, 517t
gentamicin-resistant, 130
Gram stain of, 27f
lactose uptake by, 89
methicillin-resistant, 130
microflora, 440t
nosocomial infections with, 559t
penicillin-resistant, 130
plasmids of, 263t
pyodermas caused by, **526–27,** 526f
in sexually transmitted disease, 533t
in toxic shock syndrome, **536**
toxins of, 450–52, 451t
virulence-associated enzymes of, 449t, 450
S. epidermidis, 96f, 328, 440t, 525, 532
Staphylokinase, 449–50, 449t
Starch, 47, 48f
    degradation of, 190, 190t, 427, 428t, 433
    structure of, 190, 190t
Starch granule, 50f
Starter culture, 609–11, 609t
Stationary macrophages, 458
Stationary phase, 115, **116,** 116f
STD. *See* Sexually transmitted disease
Steady state, 118
Stem rot, 360
Sterilization, **119**
Steroid, 47–48, 49f
Sterol, 48
    in eucaryotic plasma membranes, 72, 128
    in *Mycoplasma,* 59, 327
Stickland reaction, **196**
Stigmata, 357
Stigmatella aurantiaca, 338f
Stigonemataceae, 335, 335t
Stipe, 359, 359f
STLV. *See* Simian T lymphotrophic virus
STM. *See* Scanning tunneling microscope
Storage body. *See* Inclusion body
Strain, 306, 306t
    industrial, 620–21
Stramenopila, 366t, **369**

Stratosphere, 601
Strawberry leak, 360t
Streak plate technique, **96–98,** 97–98f
Stream, 587
Strep throat. *See* Streptococcal pharyngitis
Streptoalloteichus, 342t
Streptococcal exotoxin A, 473
Streptococcal pharyngitis, 328, 446, **498–504,** 503t
Streptococcal toxic shock-like syndrome, 500
Streptococci
  cell wall growth in, **36–37,** 36–37f
  chaining of, 20, 20–21f, 36–37, 36–37f
  classification by hemolytic pattern, **502–3,** 502f, 503t
  classification into immunologic groups, 503, 503t
Streptococcus, 328, 499t, 516
  cell arrangements in, 20, 20–21f
  dental plaque and, 61
  group A, 498–500, 504, 543t
  group B, **506,** 506t
  identification of, 487
  in meat spoilage, 614t
  microflora, 440t
  pH tolerance of, 92
  temperature range of, 91
  S. agalactiae, 443, 499t, 503t, **506,** 506t, 512t
  S. anginosus, 503t
  S. bovis, 428t, 503t
  S. cremoris, 609–11, 609t, 611t
  S. diacetilactis, 611, 611t
  S. durans, 611t
  S. dysgalactiae, 503t
  S. equi, 503t
  S. equisimilis, 503t
  S. lactis, 106t, 191, 329f, 608–11, 609t, 611t
  S. mutans, 95, 429, 448, 449f, 514–15, 514t, 515f
    dextran-induced agglutination of, **462,** 462–63t
  S. pneumoniae, 95, 96f, 106t, 117, 329f, 448, 500t, 504, 504f, 504t, 512t
    antibiotic resistance in, 131f

bile solubility test, 117
capsule of, 450, 450f
DNA melting curve for, 217f
Griffith transformation experiment with, 215, 215f, 252
identification of, 504–5, 505f
nosocomial infections with, 559t
quellung reaction, 61f
transformation in, 253
S. pyogenes, 7t, 21f, 95, 96f, 418, 498–504, 499t, 503t, 525t, 546
  cell wall growth in, 36–37, 36–37f
  erythrogenic toxin of, 411, 498–99
  identification of, 500–504, 501f
  lysogenic conversion in, 452, 499
  nosocomial infections with, 559t
  pyodermas caused by, **526–27,** 526f
  toxins of, 266, 451t
  virulence-associated enzymes of, 449, 449t
S. sanguis, 514, 514t
S. thermophilus, 609t, 610, 611t, 613
S. zooepidemicus, 503t
Streptokinase, 296t, 449–50, 449t, 624
Streptolysin, 502
Streptolysin O, 449, 451t, 499
Streptolysin S, 449, 451t
Streptomyces, 342t, 343, 343f, 575–76, 641
  antibiotic production in, 262
  in freshwater habitats, 587
  hydrocarbon degradation by, 583
  in soil, 581
S. albus, 273t
S. antibioticus, 129
S. aureofaciens, 629t, 630
S. erythraeus, 129, 629t
S. fradiae, 629t
S. griseus, 629t, 630
S. kanamyceticus, 629t
S. niveus, 629t
S. nodosus, 629t
S. noursei, 629t

S. olivaceus, 627t, 628
S. rimosus, 629t
S. spheroides, 629t
S. venezuelae, 629t
Streptomycin, 125
  commercial production of, 629t, **630**
  discovery of, 125
  mechanism of action of, 125t, 128, 128t, 629t
  production of, 343
  structure of, 128f
Streptosporangium, 342t, 343
Streptoverticillium, 342t
Stroma, 72, 73f, 180
Stromatolite, 334, 335f
Structure analog, 129
Styloviridae, 391t
Subacute inflammation, 458, 460f
Substage condenser, 24–25, 28
Substrate, 143
Substrate-level phosphorylation, 149, **150–61,** 150f, 162
Subterminal endospore, 70, 71f, 517f
Subtractive hybridization, 278
Subunit vaccine, 295, 478, 478t
Succinate, 157, 161–62, 161f, 163f, 201, 428t
Succinate dehydrogenase, 143, 144f, 162
Succinimonas, 325t, 427
  S. amylolytica, 428t
Succinivibrio, 325t
Succinyl-CoA, 163f, 573, 573f
Sucrase, 190t, 191, 624t, 625
Sucrose, 48f, 190t, 191, 514–15
Sucrose fermentation test, 324f
Sucrose gradient sedimentation, of DNA, 241f
Sugar
  content of cells, 81t
  group translocation of, 89
  transport of, 88
Sugar syrup, **623–24**
Sulfadiazine, 129f
Sulfanilamide, 129f
Sulfasuxidine, 129f
Sulfate, 579–80
  assimilatory reduction of, 83, 83f
  as electron acceptor, 150, 168t, 170
  from hydrogen sulfide, 171–72
  as sulfur source, 83
Sulfate-reducing bacteria, 170, 316t, **326,** 340, 340t, 588, 636

Sulfate reduction
  assimilatory, 170
  dissimilatory, 170, 579f, 580
Sulfide, 579–80
Sulfite reductase, 83, 83f
Sulfolobales, 341
Sulfolobus, 171, 171t, 313f, 339f, 340–41, 340t
  S. acidocaldarius, 91, 312, 314f
Sulfonamide, 125t, **129,** 129f
Sulfur
  cellular content of, 84t, 570, 570t
  characteristics of element, 647t
  elemental, 171–72, 579–80
  function in cells, 84t
Sulfur assimilation, 579f
Sulfur cycle, **579–80,** 579f
Sulfur dioxide, 580
Sulfur granule, 81, 172, 187, 188f, 333, 338, 338f
Sulfuric acid, 583, 588
Sulfur-oxidizing bacteria, 171–72, 171t, 337, 344, 573, 580, 588
Sulfur-reducing bacteria, 316t, **326,** 636
Sulfur source, **83**
Sulfur springs, 171
Superantigen, **473**
Supercoiling, of DNA, 218–19
Superinfection, 411
Superoxide dismutase, 93
Superoxide free radical, 93, 461
Suppressor mutation, 246
Surface antigenicity, 60
Surface-to-volume ratio, 106
Surfactants, as disinfectants and antiseptics, 122t, **123,** 124f
Sushi, 355
Suspended solids, 595
SV5, 408
Svedberg unit, 65
Swamp, 340, 575, 587
Swanson, Robert, 272
Swarm cells, 371f, 372
Swarmer cells, 337–38
Sweat, 525
Swiss cheese, 157, 331, 331t, 611–14, 611t
Symbiont, of protozoa, **425–27,** 426f, 426t
Symbiosis, **417–36.** *See also* Commensalism; Mutualism; Parasitism
  chemoautotrophic bacteria with marine bivalve, **344,** 344–45f, 345t
  coral reefs, 350

Symbiotic nitrogen fixation, 320, **418–23,** 419–22f, 421t, 576
Synchronous culture, **118–19,** 119f
Synchytrium, 365, 366t
S. endoboticum, 360t
Synechococcus lividus, 90f
Synergism, **441,** 441f
Syngamy, 353–54f, 373, 377
Synthetic drugs, antimicrobial, 122, 125–32, 125t
Synthetic media, **94–95,** 95t
Synthetic oligonucleotide, 278–79, 279f
Syphilis, 7t, 316, 444, 446, 490, 533t, **534–36,** 535f, 558
  congenital, 535
  diagnosis of, 535–36
  neurological, 535
  primary, 535, 535f
  secondary, 535, 535f
  tertiary, 535, 535f

Tampon, 564–65
Tanning, 623
Taphrina deformans, 360t
Taq polymerase, 91, 290f, 291
Target gene
  detecting host cells carrying, **287–89**
    nucleic acid probes, **288–89**
    protein product identification, **288**
    reporter genes, **289**
  incorporation into cloning vector, **279–85,** 280f
  obtaining copies of, **274–79**
  separation of DNA fragments by gel electrophoresis, 274, 275f
  synthesis of
    from nucleotides in vitro, **278–79**
    from RNA template, **274–78,** 278f
TATAAT sequence, 226
Taxonomy, 306, 306t, **308,** 308t
3TC, 540
TCA cycle. *See* Tricarboxylic acid cycle
T cell(s). *See* T lymphocytes
T-cell cancer, 409
T-cell receptor (TCR), 473
TCR. *See* T-cell receptor
T-DNA, 297, 299f
TDT. *See* Thermal death time

Tectobacter, 426t
Teichoic acid, 56
Tellurite-containing agar, 509
TEM. *See* Transmission electron microscope
Temin, Howard M., 409
Temperate phage, 255–58, 257f
Temperature
  cardinal, 89
  maximum, 89
  minimum, 89
  optimum, 89
Temperature effect
  on enzymes, 143
  on growth, **89–91,** 119
Temporary carrier, 443
Terminal deoxynucleotidyl transferase, 473
Terminal electron acceptor, 162, 165, 171
Terminal spore, 70, 71f
Termination codon. *See* Nonsense codon
Termite, intestinal bacteria in, 317, 427
Tertiary sewage treatment, 595, **598–99**
Tertiary syphilis, 535, 535f
Test (envelope of protozoa), 373
Tests in parallel, 561
Tests in series, 561
Tetanus, 7t, 328, 442, 451t
Tetanus toxin, 452
Tetanus vaccine, 452, 478, 478t. *See also* DTP vaccine
Tetracycline, 125
  commercial production of, 629t, **630**
  mechanism of action of, 125t, 128, 128t
  production of, 343
  resistance to, 130, 130f
  structure of, 631f
Tetrahydrofolic acid, 85t, 145t
Tetrahymena, 74f, 375t
  T. pyriformis, 3f
  T. thermophila, 374f
  ribozymes of, 229
Tetramitus, 582
Tetrapeptide, 52, 54, 56
Textile manufacture, **623–24**
Thallus, of lichen, 423
Thermal cycler, 291
Thermal death time (TDT), 121
Thermoacidophile, 341
Thermoactinomyces, 342t, 343, 641
Thermoactinomycetes, 342t
Thermococcales, 341
Thermococcus, 312–13f, 340t, 341

Thermodynamics, **140,** 141f
  first law of, **141**
  second law of, **141–42**
Thermomonospora, 342t, 343, 641
Thermophile, 89, 90f, **91**
  extreme, 90, 90f, **91,** 312, 313f, 314, 339–41, 340t
Thermoplasma, 102, 313f, 340t, 341
Thermoproteales, 341
Thermoproteus, 312–13f, 340–41, 340t
Thermosipho, 314
Thermotoga, 312f, 314, 315f
Thermus aquaticus, 90f, 91, 291
Theta replication, 220, 220f
Thiamine, 84–85, 85t, 145t, 580
Thiamine pyrophosphate, 85, 85t, 145t
Thimerosal (Merthiolate), 122t, 124
Thin section, 36
Thiobacillus, 154t, 171, 171t, 344, 578, 588
  T. ferrooxidans, 172, 573, 582–83, 604
  T. thiooxidans, 91, 92t, 337, 573, 580, 582, 604
Thiocapsa, 333t
  T. roseopersicina, 188f
Thiocystis, 333t
Thiodictyon, 333t
Thiogalactoside transacetylase, 233–34, 234f
Thiokinase, 200f
Thiolase, 200f
Thiopedia, 333t
  T. rosea, 333f
Thioploca, 588
Thiospira, 588
Thiospirillum, 333t
Thiosulfate, 81, 580
Thiothrix, 338f, 580, 588, 598
Thiovulum, 344
Thraustochytrium, 366t
Three-kingdom classification, 306, 307t
Threonine
  commercial production of, 622–23
  structure of, 43t
  synthesis of, 197t
Threonine deaminase, 144f
Thrush, 360t, 499t, 514t, 515
Thylakoid, 50f, 72, 73f, 180, 182, 182f, 187, 334–35
Thymine, 44, 46f
Thymus, 459f, 468, 473
Tick-borne relapsing fever, 318t

Tick-over activation, 482
Tick vector, 445t
Tinea. *See* Ringworm
Tinea corporis, 531f
Ti plasmid, 297, 299f
Tipula iridescent virus, 385f
Tissue culture. *See* Animal tissue culture
Tissue plasminogen activator, 296t, 296
T lymphocytes, 457t, 458, **468**
  in AIDS, 492–93, 493t
  CD4, 536–40, 537f
  cell-mediated immunity, **482–85**
  comparison to B lymphocytes, 474t
  cytotoxic T cells, 473, 479, 483, 492–93, 493t
  sensitized, 484–85, 484f
  subpopulations of, **473**
  T helper cells, 473, 475, 483, 492–93, 493t
  T_H1 cells, 473
  T_H2 cells, 473
TNF. *See* Tumor necrosis factor
Tobacco bushy stunt virus, 393t
Tobacco mosaic virus, 8–9, 9f, 383f, 385f, 387, 393f, 393t, 582
Tobacco necrosis virus, 393t
Tobacco rattle virus, 393t
Tobacco ringspot virus, 392t
Tobacco streak virus, 392t
Tobamovirus, 393t
Tobanecrovirus, 393t
Tobramycin, 128
Tobravirus, 393t
Togaviridae, 395t
Togavirus, 395, 525t
Tomato bunchy stunt, 389
Tombusvirus, 393t
Tonegawa, Susumu, 471
Tonsil, 459f
Tonsillitis, 503t
Tooth. *See* Dental *entries;* Oral cavity
Total coliforms, 591–94
Tournier, Paul, 387
Toxic shock-like syndrome, streptococcal, 500
Toxic shock syndrome, 328, 533t, **536,** 543t, **564–65,** 564t, 565f
Toxic shock syndrome toxin, 473, 536
Toxin, 42, 42t, 448, **450–54**
  plasmid-coded, 262, 266, 452
Toxoid, 452, 478

Toxoplasma, 375t
   T. gondii, 376–77, 377f, 377t, 444t
Toxoplasmosis, 376–77, 377f, 377t, 444t
Trace elements, 570, 570t
Trachoma, 326, 327t
Tracking microscope, 63, 64f
tra genes, 260–62
Transaldolase, 195f
Transamination, 197
Transcription, 212, 212f, **226–28**
   coupled with translation, 227f
   elongation stage of, **226–27**
   eucaryotic, 213f
   initiation of, **226**
   procaryotic, 213f
   termination of, **227–28**
      by attenuation, **236**, 237f
      rho-dependent, 228, 228f
      rho-independent, 227–28, 228f
Transcriptional control, 233–39
Transducer, 63
Transducing phage, 255
Transduction, 252, **255–58**
   abortive, **258**, 259f
   discovery of, **254**, 254f
   generalized, **255**, 256f
   high-frequency, 258
   specialized, **255–58**, 257f
Transfection, **255**, 285
   bacterial, 287
   of mammalian cells, 287
Transferase, in industrial processes, 623t
Transferrin, 88
Transfer RNA (tRNA), 46, **224–26**. *See also* Translation
   attachment of amino acids to, 226
   in attenuation, 236
   comparison of Bacteria, Archaea, and Eucarya, 313t
   initiator, 230, 231f, 312, 313t
   modified bases in, 224, 225f
   structure of, 224, 225f
   suppressor mutations, 246
   synthesis of, 226–28
Transformation
   bacterial, 215–16, 215f, **252–55**, 253f, 287
   cell by virus, **408**, 408f
Transforming principle, 252
Transgenic animal, 287
Transgenic crops, 297, 299f

Transglycosylation, 56
Transient flora, 440
Transition (mutation), 247, 248t
Transketolase, 195f
Translation, 212, 212f, **229–33**
   coupled with transcription, 227f
   direction of, 232
   elongation stage of, 231f, **232**
   eucaryotic, 213f, **233**
   initiation of, **229–32**, 231f
   procaryotic, 213f
   rate of, 233
   regulation of, 233–39
   termination of, 213, 231f, **232–33**
Translocase, 232
Transmembrane pore, 482, 482f
Transmission of disease, 464
Transmission electron microscope (TEM), 21, **32–36**, 32f
Transmitted-light fluorescence microscope, 29f, 31
Transpeptidation, 56, 125
Transplant rejection, 479, 479t, 484–85
Transport protein, 42, 42t, 86
Transport work, 140
Transposable element, **260**, 261f
Transposon, **260**, 262
Transversion (mutation), 247, 248t
tra region, 264
Traveler's diarrhea, 322, 516t, 521, 523
Trench fever, 327t
Trench mouth, 515
Treponema, 316–17, 318t
   T. carateum, 316, 317f
   T. pallidum, 7t, 316, 317f, 533t, **534–36**, 535f
      dark-field microscopy of, 28
      detection of, 489f, 490
      nomenclature of, 306t
   T. pertenue, 316
Tricarboxylic acid (TCA) cycle, **162**, 163f, 175f, 201, 201f
Trichoderma, 575–76
Trichodina, 375t
Trichome, 335
Trichomonas, 375t, 440t
   T. tenax, 499t
   T. vaginalis, 132, 376, 377t, 381, 497f, 533t
Trichomoniasis, 376, 377t, 497f, 499t, 533t

Trichophyton, 360t, 444t, 525t, 530
   T. gourvilli, 361f
Trickling filter process, 595–98, 597f
Trifluridine, 132, 132t
Triglyceride, 47
Trinacria regina, 356f
Triose-3-phosphate, 155f
Triose phosphate isomerase, 151f, 152
Triple covalent bond, 649
tRNA. *See* Transfer RNA
Trophozoite, 373, 376–77, 377f
Troposphere, 601, 601t
True negative result, 560
True positive result, 560
Trypanosoma, 132, 375t
   T. cruzi, 444–45t
   T. gambiense, 376, 377t
Trypticase soy broth, 95t
Tryptophan, 43t, 197t
Tryptophan operon, 235–36, 235f
Tryptophan synthase, 623t
Tryptophanyl-tRNA, 236, 237f
TTGACA sequence, 226
Tube dilution method, of antimicrobial susceptibility testing, 132, **133**, 133f
Tubercle, 510
Tuberculin skin test, 483, 485f, 511, 561
Tuberculoid leprosy, 527–28
Tuberculosis, 7t, 331, 446, 454, 461, 499t, **510–11**, 510–11f, 553, 554t, 558, 561
   AIDS and, 510, 540
   bovine, 608
   host resistance and, **453**
   miliary, 510
   pulmonary, 510
   rates for males and females, 557t
   reactivation, 510
Tubifera, 366t
   T. ferruginosa, 350f
Tularemia, 321, 444, 444–45t, 525
Tumble (bacterial movement), 63, 64f, 65
Tumor, 408
   benign, 408
   malignant, 408
   oncogenic viruses and, **408–9**
Tumor necrosis factor (TNF), 473, 474t
Turbidimetric measurements, **113–14**, 114f, 116t

Turbidimetric methods, of automated antimicrobial susceptibility testing, 134, 134f
Turbidity, 113
Turbidostat, 118
Turkey, contaminated with *Salmonella*, 520
Turnip yellow mosaic virus, 393t
Tuttle, Dorothy M., 135
Twort, Frederick W., 9
Tyndall, John, 119
Tyndallization, **119**
Tynovirus, 393t
Typhoid fever, 7t, 322t, 323, 442–43, 446, 461, 487, 516t, **519**, 554t
Typhoid fever vaccine, 478, 478t
Typhoid Mary, **442**
Typhus, 326, 327t, 446, 529, 529t. *See also specific types*
Tyrocidine, 14
Tyrosine, 43t, 197t

Ubiquinone. *See* Coenzyme Q
UDP-*N*-acetylglucosamine (UDP-NAG), 56, 57f
UDP-*N*-acetylmuramic acid (UDP-NAM), 56, 57f
UDP-galactose, 191f
UDP-glucose, 191f
Ulcer, peptic. *See* Peptic ulcer
Ulothrix, 358f
Ultra high-temperature milk, 608
Ultramicrotome, 36
Ultraviolet radiation
   as mutagen, 246, 248t
   for sterilization, **122**
Ulva, 352, 353f
Uncoupler, **167–68**, 168t
Undulant fever, 608
Universal phylogenetic tree, 311, 312f
Unripened cheese, 611, 611t
Unsaturated fatty acid, 47, 90
Uracil, 44, 46f
Uranium, microbial leaching of ores, 582
Ureaplasma urealyticum, 444, 533t, 536
Urease, 578
Urease test, 324f
Urethra, 532
Urethritis, 18, 534
   nongonococcal. *See* Nongonococcal urethritis

Urinary tract infection, 322, 322t, 532, 532t
  caused by *E. coli*, **532–33,** 532t
  diagnosis of, 532–33
  nosocomial, 559t
Urophlyctis, 360t, 365
Urosporidium, 375t
Ustilago, 366t
  U. tritici, 360t
  U. zeae, 360f
U-tube experiment, 254, 254f
UvrABC endonuclease, 248, 248f

Vaccination, 477, 483
  Jenner's experiments with, 8, 8f
Vaccine, **477–78,** 478t. *See also specific diseases*
  genetically-engineered, **294–95,** 295f
  killed, 478, 478t
  live, attenuated, 477–78, 478t
  production of, 467f, 607f
  subunit, 295, 478, 478t
Vaccinia virus, 33f, 385f, 396f
Vacuole, 73, 76t
Vaginal flora, 440–41
Vaginitis, 132, 441
Vairimorpha necatrix, 381
Valence
  of antigen, 468
  of atom, 648
Valerate, 428t
Valine, 43t, 197t
Vancomycin, 125, 125t
Van Ermengem, 7t
van Niel, Cornelius B., 184
Variable number of tandem repeats, 292, 293f
Varicella-zoster virus, 525t, 531
VDRL test, 535–36
Vector, cloning. *See* Cloning vector
Vector of disease, **444–45**
  arthropod. *See* Arthropod vector
  biological, 445
  mechanical, 445
Vegetables, frost-proof, **618**
Vegetative cells, 68
Vegetative mycelium, 361
Veillonella, 326, 440t, 514t
Verticillium, 576
  V. albo-atrum, 364f
Vesicular-arbuscular mycorrhiza. *See* Endomycorrhiza
V gene, 471, 472f
Viable count, **108–11,** 109–11f, 116t
  during death phase, 117

  dilution of bacteria for, 109f, 110
  during lag phase, 117f
  membrane filtration technique for, **111,** 112f
  pour plate technique for, 110–11, 110f
  spread plate technique for, 110–11, 110f
Vibrio, 322t, 425, 576, 586
  V. cholerae, 7t, 323, 323f, 448, 455, 516, 516t, **518–19,** 518f, 586–87, 594
  temperature range for, 90f
  toxins of, 450–52, 451t
  V. cholerae O1, 518
  V. cholerae O139, 542f
  V. marinus, 90f
  V. parahaemolyticus, 63f, 323, 586
  V. vulnificus, 323
Vibrioniaceae, 322–23, 322t
Vidarabine (ara-A), 132, 132t
Viral infection
  abortive, 408
  cell transformation in, **408**
  persistent, **408**
  treatment of, **406,** 407f
Viremia, 531
Viridans streptococci, 502
Virion, 384
Viroid, **388–89,** 388f
Virulence
  definition of, 448
  iron and, in *N. gonorrhoeae,* **100,** 101t
  measurement of, **447,** 447f
  microbial factors of, **448–54**
Virulence plasmid, 263f
Virus, 18
  assay of, **399**
  binal, **386–87,** 387f
  classification of, **387–95**
  comparison with other microorganisms, 384, 384t
  crystallization of, 9
  cytopathic effects of, **405–8,** 405f
  discovery of, **8–9**
  electron microscopy for counting, 398, 398f
  evolution of, 414
  filterability of, 8–9
  helical, **386–87,** 387f
  host for propagation of, **395–98**

  lytic cycle of, 408
  microorganisms infected by, **390**
  nucleic acids of, 384–86
  polyhedral, **386–87,** 386f
  propagation and assay of, **395–98**
  properties of, 384–87
  replication of, **398–408**
    adsorption of, 398, **399–401,** 399f
    assembly of new particles, 398, 399f, **403–5**
    penetration of cell, 398, **399–401,** 399f
    release from host cell, 398, 399f, **405,** 408
    uncoating, 398, 399f, 401
  size of, **8–9, 384,** 385f
  in soil, 582
  structure of, 8–9, **384–86,** 385f
Vitamin, 84–85, 85t, 145
  commercial production of, **627–28,** 627t
Vitamin B$_1$. *See* Thiamine
Vitamin B$_6$. *See* Pyridoxine
Vitamin B$_{12}$, 85t, 516
  commercial production of, 627–28, 627t
  structure of, 628, 628f
Vitamin C, commercial production of, 627t
Vitamin K, 85t, 432, 516
Voges-Proskauer test, 161, 324f
Volvox, 349f
von Behring, Emil, 509

Waksman, Selman, 125
Wandering macrophages, 458
Warburg, Otto, 154
Warburg-Dickens pathway. *See* Pentose phosphate pathway
Warren, J. Robin, 522
Warts, 395, 525t, 531–32
Water
  content of cells, 42, 81t
  content of spores, 68
  microbes and, **586–600,** 604
  polarity of, 650
  properties of, 650–51
  refractive index of, 24
  requirement for microbial growth, **92–93,** 92t
  structure of, 650–51
  as universal solvent, 650
Water activity, 92, 92t
  effect on microbial growth, 119
  preservation of meat and, 615

Waterborne disease, **443,** 518
Waterlogged soil, 578
Water mold. *See* Oomycetes
Water pipes, 588
Water quality, **588–94**
  biochemical oxygen demand, **589–90**
  biological contaminants, 588
  chemical contaminants, 588
  indicator organisms, **590–93**
Watson, James D., 216
Welch, 7t
Western blot, 490
  AIDS, 540
Western equine encephalitis, 444t
Western equine encephalitis virus, 408
Wheal and flare reaction, 480
Wheat rust, 360, 360t
Whey, 612f, 613, 620, 620t
Whiplash flagellum, 365
Whiskey, 619
White bread, 617t
White Cliffs of Dover, 376
White pine blister rust, 360t
Whittaker, Robert H., 307, 307t
Whooping cough. *See* Pertussis
Widal, Fernand, 487
Widal test, 487
Wild type, 246, 249–50
Wilkins, Maurice H.F., 216
Williams, Robert P., 204
Wilson, C.L., 631
Wine, 617t, **619,** 624–25
Winogradsky, Sergei, 80, 178, 580
Wobble hypothesis, **230,** 230t
Woese, Carl R., 307, 307t, 311
Wolinella, 325t
Work
  chemical, 140
  electrical, 140
  mechanical, 140
  transport, 140
Wort, 617–18
Wound infection, 7t, 322t
  nosocomial, 559t

Xanthan gum, 627
Xanthomonas, 154t, 320
  X. campestris, 106t, 320, 321f, 627
Xanthophyll, 180, 352
Xenobiotic, 585
X-gal, 283, 283f
Xylan, 354t
Xylose, 192f
Xylulose, 192f

Xylulose-5-phosphate, 154,
        155–56f, 159f, 191–92,
        192f, 194, 195f

YAC. *See* Yeast artificial
        chromosome
Yalow, Rosalyn, 490
Yaws, 318t
Yeast, 360, **361–62**
    in beer production,
        617–19, 619f
    in distilled beverage
        production, **619**

ethanol production by, **157**
lactose-fermenting, 609t
reproduction in, 365
on skin, 525
as source of single-cell
        protein, 631–32, 632t
in wine production, **619**
Yeast artificial chromosome
        (YAC), **285**
Yeast-mycelium dimorphism,
        378, 378–79f
Yellow boy, 580
Yellow fever, 444–45t, 542f, 543t

Yellow fever vaccine, 478t
Yersin, Alexander J.E., 7t, 323, 509
Yersinia, 322, 322t, 324f
    Y. pestis, 7t, 323, 323f,
        444–45t, 451t
Yield of test, 561
Yogurt, 154, 331t, 609t, 610

Zernicke, Fritz, 28
Zidovudine. *See* Azidothymidine
Zigas, Vincent, 412
Zinc, microbial leaching of
        ores, 582

Zinder, Norton, 254–55
Zone of inhibition, 133, 133f
Zoogloea, 595
Zoonosis, **443,** 444, 444f
Zoospore, 343, 352, 358f, 364,
        369, 370f
Zooxanthellae, 350
Zovirax. *See* Acyclovir
Z-pathway, 184–85, 185f
Zygomycota, 360t,
        365–69, 366t
Zygospore, 354f, 366
Zygote, 73–74, 362–63f, 364